A history of the early medieval siege, c.450–1220

Sieges were the predominant form of warfare across the medieval world, and siege methods and technology developed alongside improvements in defence. This book goes back to the original sources to present a comprehensive view of the whole subject, tracing links across continents and analysing the relationship with changes in the design of town and castle defences, and linking contemporary historical accounts with archaeological studies. It considers the most important questions raised by siege warfare: who designed, built and operated siege equipment? How did medieval commanders gain their knowledge? What were the roles of theoretical texts and the developing science of siege warfare? How did nomadic peoples acquire siege skills? Were castles and town walls built purely of a military purpose, or did they play a symbolic role also?

The volume begins with the replacement of the western Roman empire by barbarian successor states, but also examines the development of the Byzantine Empire, the Muslim Caliphate and its successors, and the links with China, through to the early thirteenth century.

A history of the early medieval siege c.450–1220

Peter Purton

THE BOYDELL PRESS

First published 2009
The Boydell Press, Woodbridge

ISBN 978 1 84383 448 9

The Boydell Press is an imprint of Boydell & Brewer Ltd
PO Box 9, Woodbridge, Suffolk IP12 3DF, UK
and of Boydell & Brewer Inc.
668 Mt Hope Avenue, Rochester, N.Y. 14620, USA
website: www.boydellandbrewer.com

A CIP record for this book is available
from the British Library

Designed and typeset in Adobe Jenson with Clairvaux display by
The Stingray Office, Chorlton-cum-Hardy, Manchester

Contents

List of maps

List of plates

Except for those marked *, all photographs are © the author. Extended captions and photo credits accompany the plates themselves.

Preface

ARCHAEOLOGY AND HISTORY confirm that, from prehistoric times, people created fortified places, and others tried to capture them. Ancient civilisations such as Greece, Rome, Persia, India and China reached high levels of strength, sophistication and advanced technology in both defence and attack.

Two of the last great empires of the ancient world, Rome (in its western division) and Persia, were replaced altogether between the fifth and seventh centuries AD, leaving a much transformed eastern Rome (Byzantium) as the sole memory of the classical world in Europe, Asia Minor, the Middle East and North Africa. China underwent dramatic upheavals, with "barbarian" invasions and new dynasties, while maintaining cultural continuity.

The following thousand-year period is known to Western historians as the Middle Ages, and it had sufficient common characteristics to justify it being seen as another historical era. The term "medieval" has been used to suggest an interval of ignorance, superstition, barbarism and "darkness" between the glories of classical civilisations before, and a more enlightened modern age after, born in Europe with the Renaissance, then exported to other corners of the world (whether wanted or not) through imperialist expansion. This book tries to view medieval history in its own terms. But medieval times *were* ignorant, barbaric and, for most of humanity most of the time, miserable. Only for a period between the eleventh and thirteenth centuries (in Europe) was there sustained economic growth and prosperity, and even then economic development was very slow, despite significant technological innovation that made for substantial improvements in productivity. Normally, much of the surplus that was generated from the economy went (or was expropriated) to sustain the religious institutions and the lifestyle of the numerically relatively small class of rulers. Although medieval societies were different from each other and changed substantially over time, sometimes making the upper levels of the peasantry or townsfolk barely distinguishable (in economic terms) from the lower levels of the nobility, the rural knights, the fundamental shape of society remained the same, and status or caste continued to maintain social distinctions. The glories of the age, the magnificent architecture of cathedrals and major castles, stood in extreme contrast to the misery of life, most of the time, for the overwhelming majority. All the evils of underdevelopment in the twenty-first-century world — vulnerability to famine, disease and infant mortality, short life expectancy, lack of education, and the persistence of superstition — were the common fate of medieval peoples for most of the period.

There were other important distinctions. Economic and social structures differed profoundly from modern industrialised societies. Every early medieval state — be it in Europe or Asia — could only function by devolving power to regional and local élites. These élites were the landowners (both lay and ecclesiastical), and land was the only significant source of wealth, power and influence. To the owners of the land also belonged the right to bear arms and the right to build fortifications - not, as was once commonly argued, to contest the royal power, but as a normal part of the social superstructure.[1] Only towards the end of the time covered by this volume were more centralised, less personal and more bureaucratic forms of government beginning to emerge in some areas.

Today, anachronism abounds in the information about the Middle Ages presented by the media. But at least modern cinema sometimes recognises that siege warfare was the predominant form of war. Modern warfare, following the French Revolutionary and Napoleonic wars and the writings of Clausewitz, was based until the current age on mass armies and decisive battles. Medieval warfare was different in almost every respect (with the exception of the disastrous consequences for those caught up in it), with different objectives and methods. Rarely was annihilation of the enemy through decisive victory the strategy — on the contrary, all good advice of the age proposed battle as the last resort because of the uncertainty of the outcome. Even where battles were "decisive", they rarely led on their own to the end of conflicts. Often they were actually the result of or precursor to sieges. War could have many different objectives, just as at any other time, and the notion that medieval rulers were incapable of military planning is another long-standing myth. Whatever the objective, the securing or defending of strong places was routinely an essential element.

This book therefore is a study of how siege war was conducted, how it developed, what the connection was between evolving techniques of attack and defence, and how fortifications developed. There have been many studies of individual sieges, of siege warfare as an element within the events of particular wars, and of siege weapons. Studies of castles (and, lamentably far behind castles, town walls), have advanced our understanding of the role, function and purpose of these structures out of all recognition. New volumes are added to the shelves every year. What *this* book seeks to do is to place all of this within the context of global development across the Middle Ages. Using contemporaneous evidence, I seek to reconstruct what actually happened, and where archaeology has revealed what stood at the time, the link has been made. I have been able to visit quite a number of the sites described. By comparing this information with developments elsewhere at the same time, a clearer picture of the process of development emerges.

Why cover the whole (then known) world? Because if, at the bottom, society was usually literally parochial, higher up the social ladder it was international. The ruling classes were often well travelled, national borders as known today did not exist (the boundaries of lordship being usually more significant), and ideas,

[1] The development of cities followed different patterns in Europe from that in the Far East. Many studies exist. But in neither did the urban élites become an alternative source of power to the landed class, even in places where cities themselves became, locally, independent, such as in Germany and Italy.

knowledge and experience crossed large distances without trouble. Religious institutions, whether Christian or Muslim, Buddhist or Hindu, ignored state boundaries and provided common culture and languages. People with technical skills might serve many different masters across many different lands, and unlike today's increasingly narrow specialisation, many of them could as well design a cathedral or castle as construct a battering ram, according to their paymaster's needs. Some learnt from textbooks, just as some rulers learnt their military skills from books. Some were of ancient origin (for example Vegetius in Europe or Kautilya in India), while others were newly written by contemporaries. As will be shown, they had wider use than might have been suspected. This was also a process, and I will look at the evolution of the science of siege warfare – poliorcetics.[1]

The novel geographical extent of this book is essential to achieve this book's objective, but it entails its own problems. I have relied on translations into Western languages of the medieval texts from Asia or the Muslim world. These translations cover only a small proportion of the histories and records of China, hence the smaller space devoted to these countries. However, the detail of medieval Chinese accounts puts contemporary Western histories to shame. Similarly, while many Arabic sources have been translated into Western European languages, some major histories remain inaccessible in part. Here it has been necessary to rely on secondary accounts that have used this material.

In contrast, if the evidence from early medieval European sources is limited, as the centuries passed the quantity increases dramatically. Voluminous published texts exist covering every European country, often as a result of decisions made in the nineteenth century to consolidate "national" medieval histories. Mile upon mile of chronicles, records, "lives", letters and other material fill the library shelves, and offer an immense mine of information – often incidentally – about medieval siege warfare. Often (and especially for the earlier period) the detail is minimal; for the most part, chronicles and other records were written for other purposes altogether. But the information that can be gleaned from the detail, sometimes accidentally revealed, constitutes the evidence base for much of what follows here. Adding this to the more substantial information provided by direct accounts of many sieges that have attracted the notice of later historians (and have therefore been discussed many times) enables a broad picture to be formed. To obtain a true picture of medieval sieges, it has been important to spread the net wider than most histories that either concentrate on a few relatively well-known events or limit themselves to one geographical area or one particular time. Many other events were of no less importance, and deserve to be noticed. For every remembered siege there were scores more that attract little attention, but study of which is necessary to grasp the whole picture. Siege warfare was paramount and endemic, and through comprehensiveness, hopefully, will come a fuller understanding, as well as awareness of some past events that were actually of great significance in their time or their region even if largely forgotten now.

Some compromises have been made in presentation. Since the majority of

[1] From the Greek word πολιορκητής, "besieger of cities".

material stems from sources themselves using the Christian calendar, this system has been used. Consistency with the names of people and places is impossible. Generally, I have used the modern name where it exists, but not always. Consider the names of historical Welsh or Irish people or places. The sources used earlier versions, often imposed by a conqueror, later studies "modernised" them, but modern usage has reverted to the original language. Consistency would lead to confusion rather than clarity, tolerance therefore is urged! Many places changed their names several names (often into forms completely different from the first) as a result of boundary changes — the Balkans and Asia Minor (Anatolia) give many examples. I have tried to use one form, offering alternatives in brackets. But even the original names often had different forms, no single one of them being "correct". Measurements have been quoted in both imperial and metric forms throughout. But medieval weights and measures (and currencies) offer their own challenge, with units with the same name having different values altogether. Where experts have identified modern equivalents, I have used them.

I have tried to allow contemporary sources to speak for themselves. Modern studies have been called upon where this is the only way to reach the original texts, when dealing with scientific evidence, or reconstructions. Modern archaeological studies naturally play an important part, although problems with this kind of evidence should not be forgotten.[1] I hope that my own prejudices (which will peep through from time to time) will not prevent readers from reaching their own conclusions on the evidence put before them.

[1] They are discussed in Chapter 9.

Acknowledgements

MANY PEOPLE helped me in this study. Richard Eales, Bob Higham, John Kenyon and Denys Pringle were kind enough to read draft chapters and offer much helpful criticism, or pointers to more recent studies or theories than I had used. Most of their suggestions have been taken on, and I alone am responsible for remaining errors of fact, or outdated theories. It would, in truth, be impossible to keep up to date with every development in studies across three continents covering a thousand years of history, but they have saved me from many mistakes. Sarah Speight kindly provided material in advance of publication. Others have assisted with information on many different aspects: I would mention here just Pam Braddock, Lawrence Butler, Steve Freeman, Peter Humphries, Rob Morgan and Peter Vemming. Boris Krivchenko organised the translation of an important article from Russian. The maps were drawn by Phillip Judge. The staff of the British Library coped with my demands over more than a decade, while the Institute of Archaeology (University College London) and the Warburg Institute allowed me ready access to their collections of journals. Other research was done at King's College London, the School of Oriental and African Studies, the Bodleian Library and the School of Slavonic and East European Studies. Their staff were always helpful. Study tours led by Richard Eales, Pamela Marshall, Denys Pringle and Sarah Speight enabled me to visit many of the medieval fortifications discussed here, while friends and, especially, my extraordinarily willing mother, have driven me to or accompanied me around stone fragments and half-buried mounds aplenty with remarkable patience and self-sacrifice, even if they did not always understand why I wished to go there. My editor, Jeffrey Dean, has saved me from many errors and generally improved the presentation of the book enormously. To all of them I express my sincere gratitude.

Abbreviations

The following abbreviations are used for frequently cited titles in the footnotes. Other abbreviations used only in the Bibliography are listed there.

ASC	*Anglo-Saxon chronicle*
ByS	*Brenhinedd y Saesson*
ByT(H)	*Brut y Tywysogyon* (Hergest version)
ByT(P)	*Brut y Tywysogyon* (Peniarth version)
OV	Ordericus Vitalis, *Ecclesiastical history*
RCHM	Royal commission on (ancient and) historical monuments
VCH	*A history of the county of . . .* (Victoria county histories)

Maps

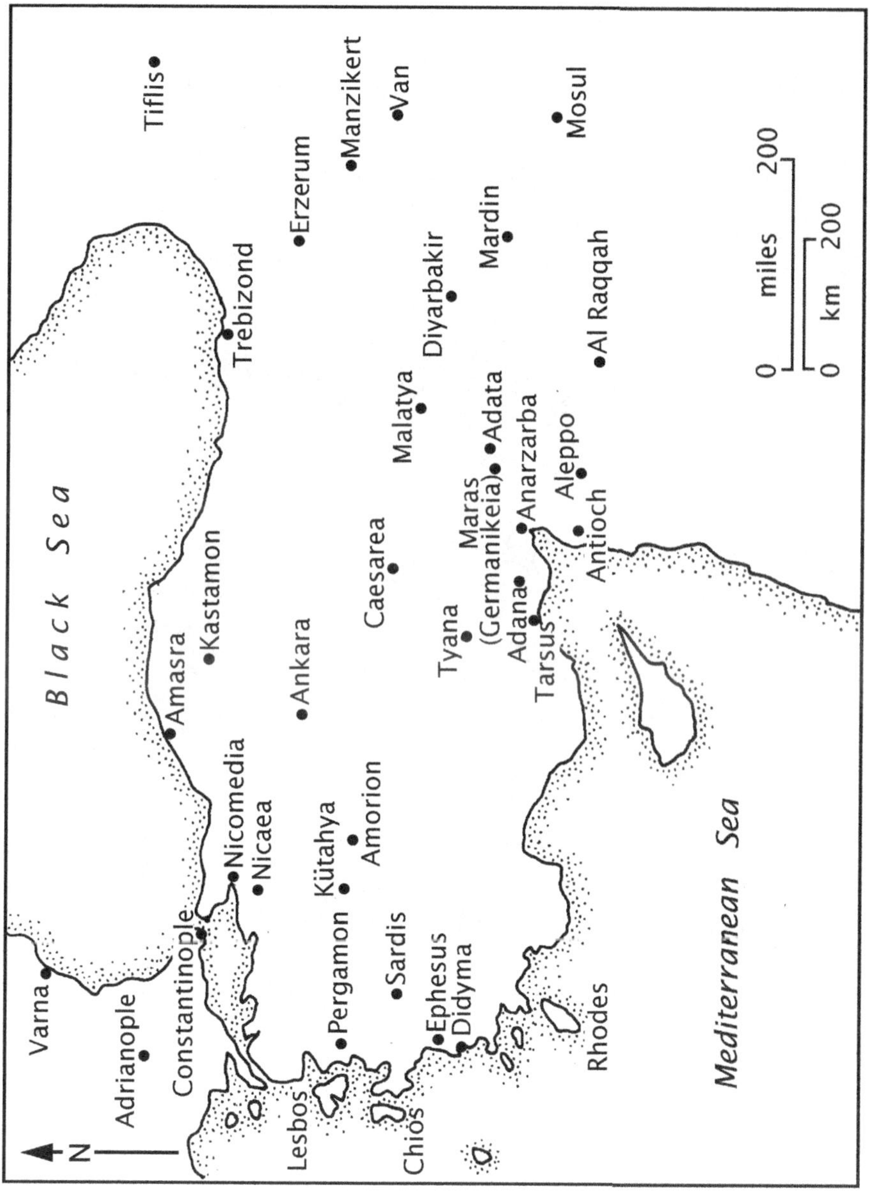

Map 1: Asia Minor

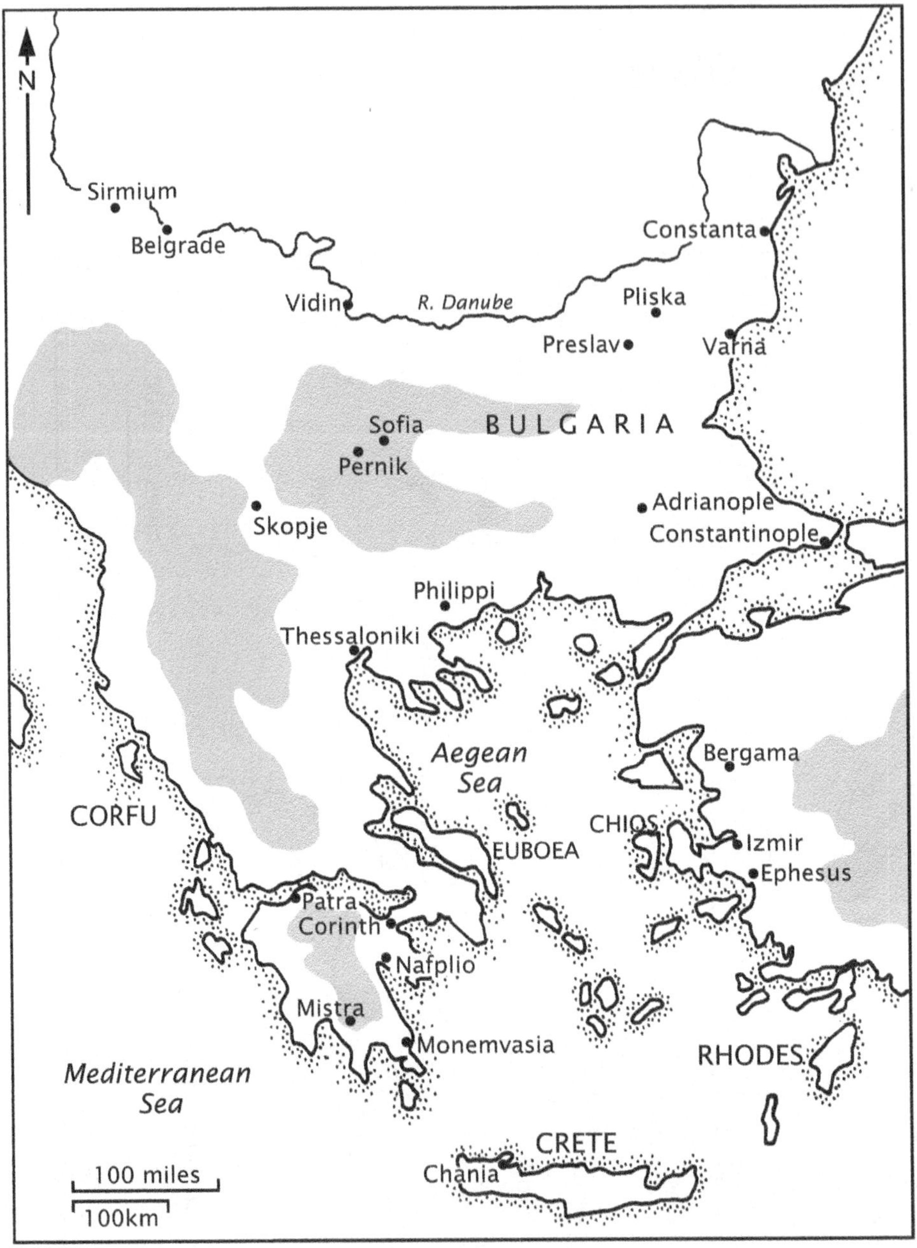

Map 2: The Balkans

Map 3: The Middle East

Trento
Cividale
N
Novara
Bergamo
Treviso
Aquileia
Ivrea
Milan
Verona
Pavia
Cremona
Padua
Mantova
Alessandria
Genoa
Bologna
Ravenna
Zadar
Lucca
Cesena
Rimini
Pisa
Split
Ancona
Adriatic Sea
Narni
Rome
Troia
Capua
Barletta
Aversa
Benevento
Bari
Naples
Melfi
Matera
Salerno
Taranto
Tyrrhenian Sea
Palermo
Messina
Mediterranean Sea
Lentini
Butera
Syracuse
Tunis
Noto
100 miles
100 km

Map 4: Italy

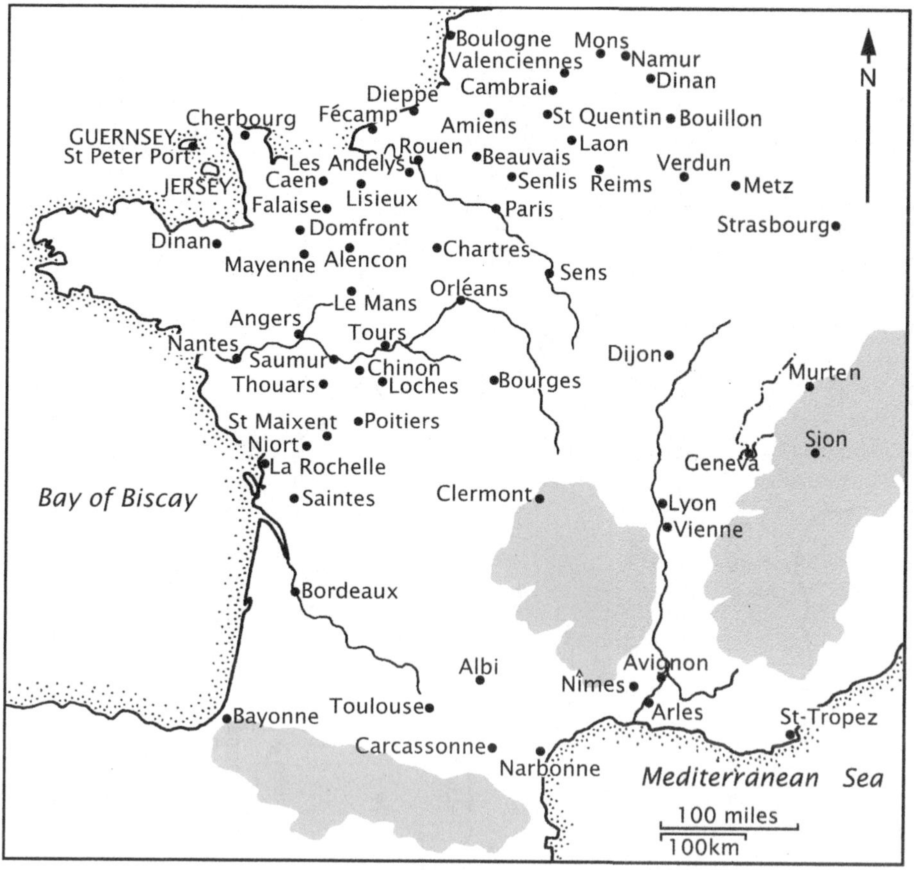
Boulogne
Mons
Valenciennes
Namur
Dinan
Cambrai
Dieppe
Fécamp
Cherbourg
Amiens
St Quentin
Bouillon
GUERNSEY
St Peter Port
Rouen
Laon
Les Andelys
Beauvais
Verdun
JERSEY
Caen
Senlis
Reims
Metz
Lisieux
Falaise
Paris
Domfront
Strasbourg
Dinan
Chartres
Mayenne
Alencon
Sens
Orléans
Le Mans
Angers
Tours
Dijon
Nantes
Saumur
Chinon
Murten
Thouars
Loches
Bourges
St Maixent
Poitiers
Niort
Sion
La Rochelle
Geneva
Bay of Biscay
Saintes
Clermont
Lyon
Vienne
Bordeaux
Avignon
Albi
Nîmes
Arles
Toulouse
Bayonne
St-Tropez
Carcassonne
Narbonne
Mediterranean Sea
100 miles
100km
N

Map 5: France

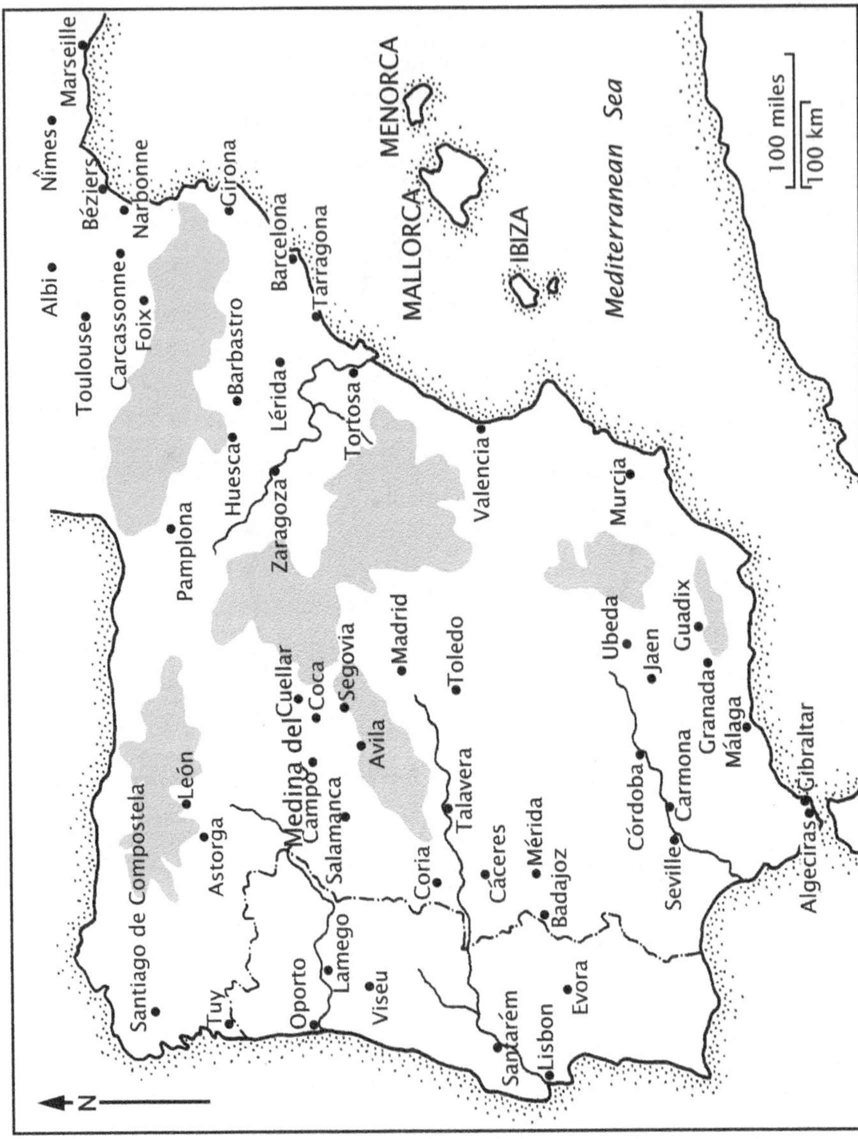
N
Marseille
Nîmes
Béziers
Narbonne
Albi
Carcassonne
Toulouse
Foix
Girona
Barcelona
Tarragona
Barbastro
Lérida
Huesca
Tortosa
Pamplona
Zaragoza
Valencia
MENORCA
MALLORCA
IBIZA
Mediterranean Sea
100 miles
100 km
Santiago de Compostela
León
Astorga
Medina del Campo
Cuellar
Coca
Segovia
Madrid
Avila
Salamanca
Toledo
Talavera
Coria
Cáceres
Mérida
Badajoz
Murcia
Ubeda
Jaen
Guadix
Granada
Málaga
Córdoba
Carmona
Seville
Gibraltar
Algeciras
Tuy
Oporto
Lamego
Viseu
Santarém
Lisbon
Evora

Map 6: Iberia

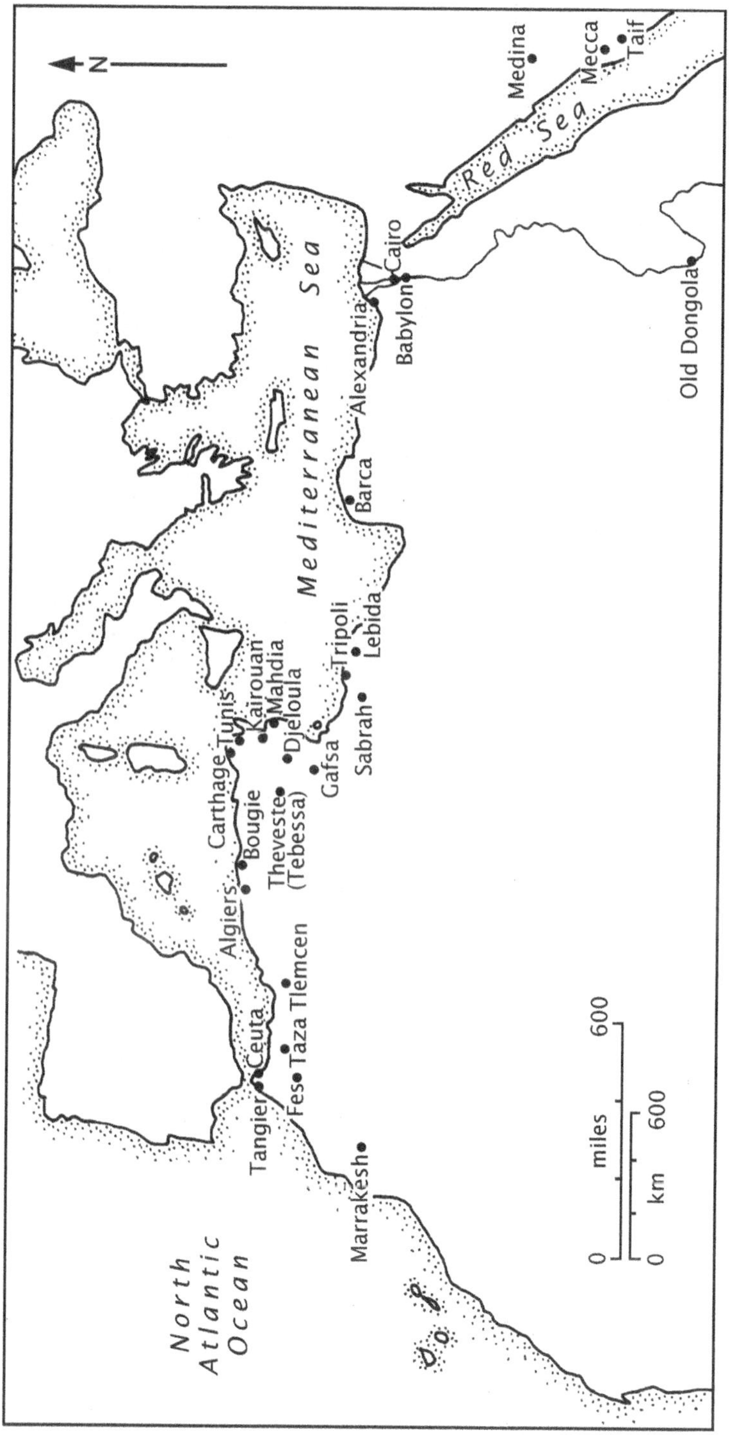

Map 7: North Africa

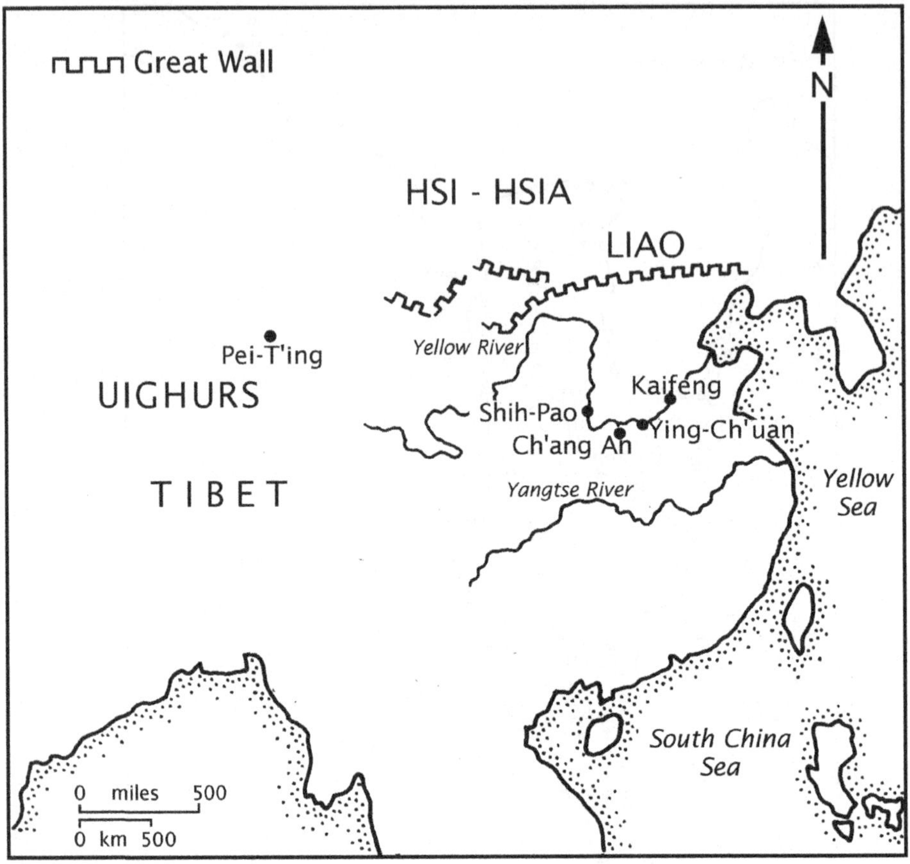
Great Wall
N
HSI - HSIA
LIAO
Pei-T'ing
UIGHURS
Yellow River
Kaifeng
Shih-Pao
Ying-Ch'uan
Ch'ang An
TIBET
Yangtse River
Yellow Sea
South China Sea
0 miles 500
0 km 500

Map 8: China

Map 9: Britain

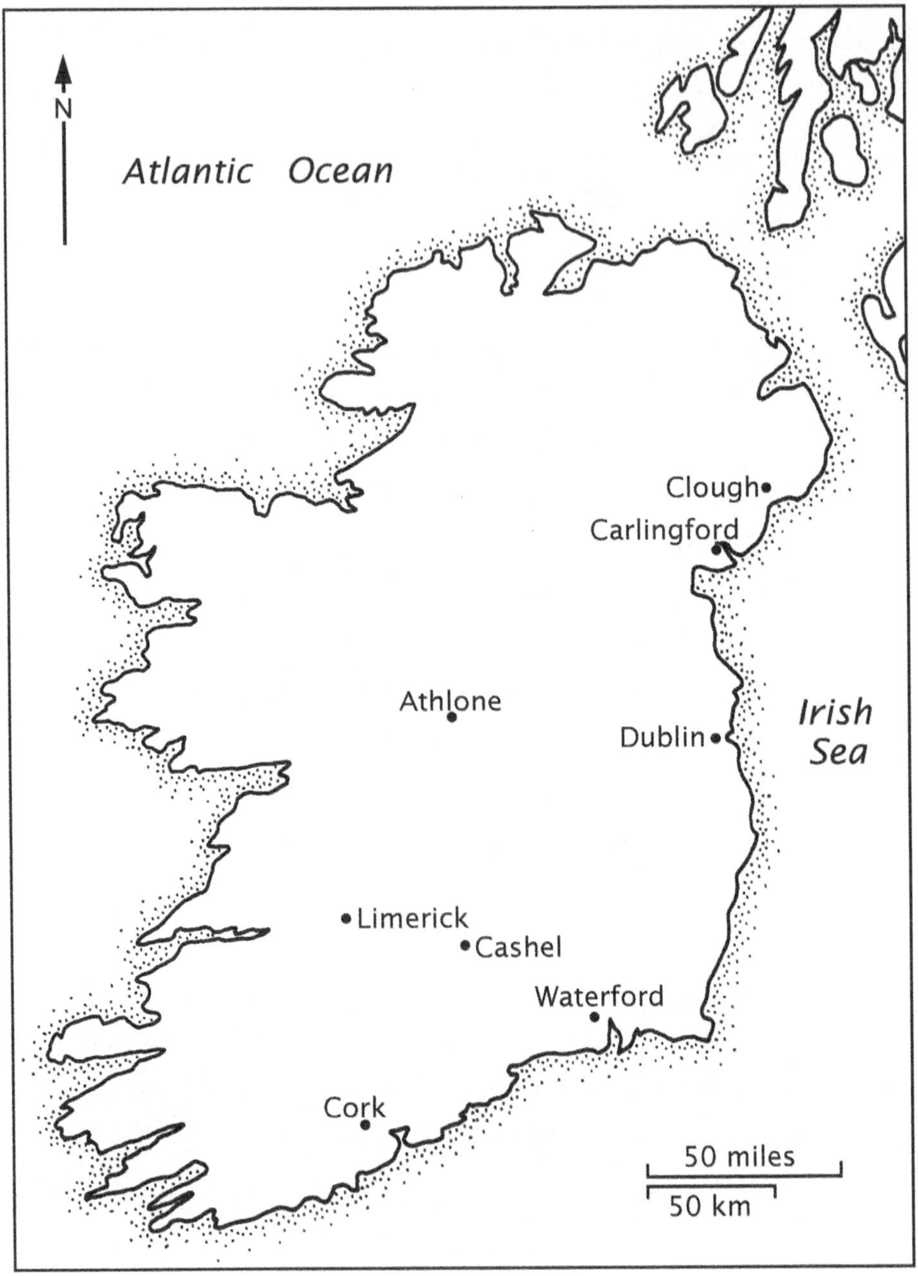

Map 10: Ireland

Map 11: Central Europe

Map 12: North-west Europe

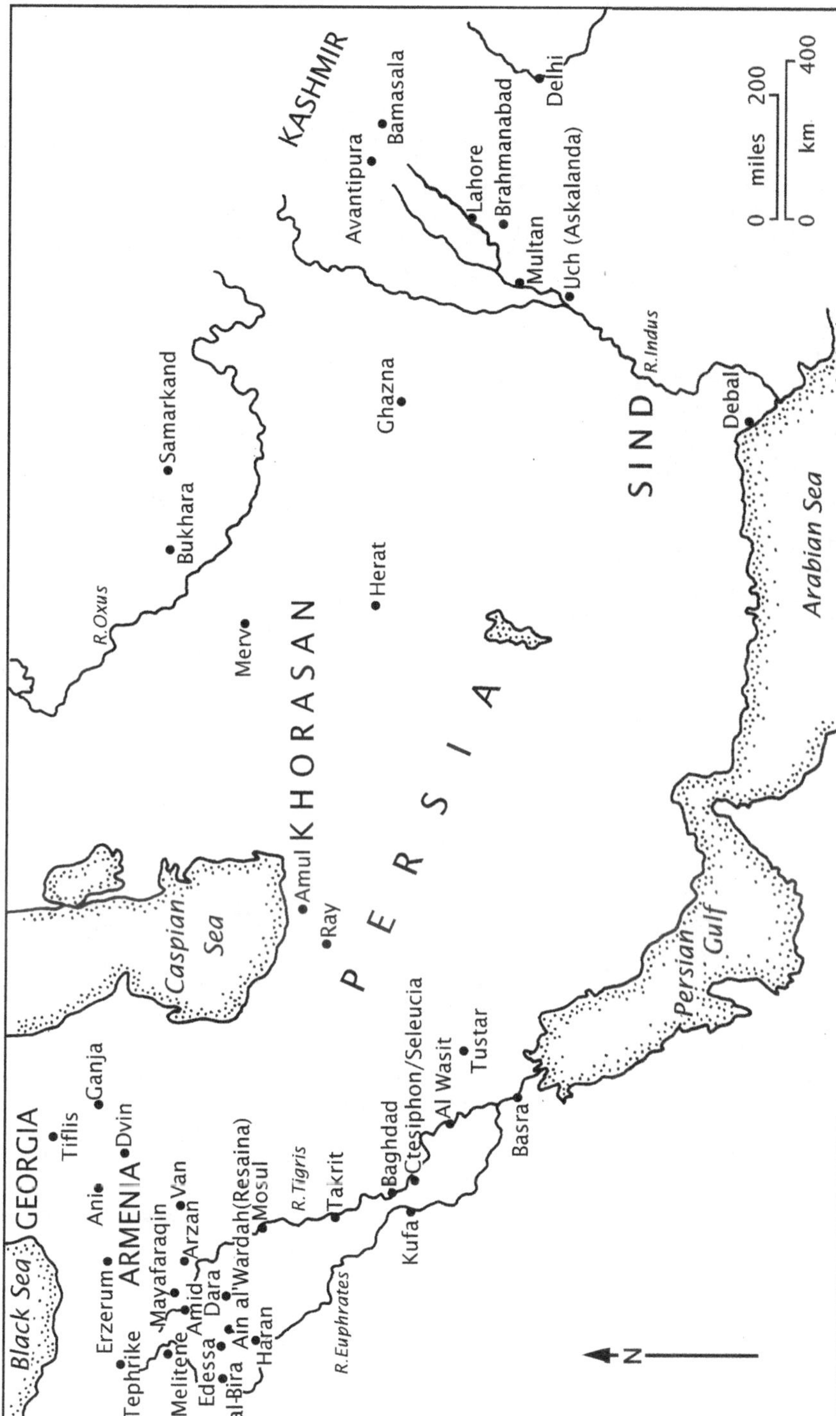

Map 13: Central Asia and Sind

Map 14: Lands of the Rus'

1 After "Rome"

THE WESTERN ROMAN EMPIRE had been disintegrating from a combination of external pressure and internal shrinkage for a long time before the last western emperor was deposed in AD 476. The leaders of the many "barbarian" peoples who had, in reality, exercised power in the provinces for a long time, continued to rule there. Continuity with the greatness identified with Rome was maintained in two institutions: the Christian Church, and the survival of the eastern section of the empire, ruled by emperors based in Constantinople. The subject of this chapter is the extent to which Roman traditions in the attack and defence of fortifications survived into, or were modified, in the time traditionally called the "Dark Ages".

The eastern empire was in no doubt about itself: it *was* Rome, although its language was Greek. Its realm, encompassing the Balkans, Asia Minor, the Middle East, and North Africa, maintained a high level of culture and civilisation. It preserved the military traditions, the architecture and technology, and the professional soldiers and skilled craftsmen. It was, as well, a crossroads between east and west. It would undergo profound social and economic transformation during the centuries to follow, but its existence provided a link with the past that influenced the world around it for another thousand years. In western Europe, Germanic peoples now occupied the land of the old empire. The Franks were in modern (northern) France, the Burgundians west and south of the Alps, the Vandals in North Africa, and the Visigoths in the Iberian peninsula. The Ostrogoths held most of Italy. The Angles and Saxons settled in Britain. Modern scholarship and archaeological investigation have profoundly modified previous conceptions of what these settlements involved. For the most part, it seems that existing aristocratic families were not wiped out, but shared power with the new rulers. Writers drawn from their ranks (lay people or churchmen) often provide the only record of events. The new peoples, through processes still not entirely understood, gradually became merged with indigenous peoples, although sometimes the latter were driven out. While the classic view of the Dark Ages has been substantially revised to recognise much continuity with what went before, it is undoubtedly true that levels of culture and technology declined significantly with the ending of the structures of the previous, centralised regime.

The evidence for the development of fortifications and of the means of capturing them for this first post-Roman age is scantier in all respects than it is for later centuries. Of contemporary written accounts there are a few. It is almost impossible to verify their information from other written sources. With any historical source, judgements have always to be made about reliability, but here there is precious little to base a judgement on. The monastic chronicles that provide so much of the information for later times are almost non-existent. That there are no charters, diplomas, or princely enactments does not mean that none were made, but genuine ones do not survive. The information such sources supply for the policy and practice of government (at all levels) in later ages poses no problem of interpretation here: there is none. Archaeological investigations have provided most of the usable information, but there are still many problems. An earthwork of one age can only be distinguished from one of another if datable finds are excavated. But we are talking of an age where the cultural level had sunk rather low, and rural settlements have left few traces, while in the towns anything from that time has long since been built over, many times.

Assumptions, therefore, have to be made. In doing so, it is important to remember the point that, while, despite the decline in cultural level, it was not the same everywhere, there were outposts of Roman civilisation, and the organisational, technical, and military lessons of Rome did not simply disappear.

Late Roman cities and forts

What did the Romans leave behind for the new barbarian regimes to occupy and fight over? Although the economy of the disappearing empire had been overwhelmingly rural, the culture and the administration had been concentrated in the cities. Cities had also become the natural location for the Church's bishoprics, and with the collapse of the previous regime it was often the bishops who found themselves exercising local authority, based in their cities. No fewer than 112 cities are listed for Gaul alone at the beginning of the fifth century. One important feature about these places as they passed under the rule of new kings was that they were heavily fortified. Although the walls, towers, and gates were in many cases already two centuries old,[1] and had sometimes been constructed hastily, they were built according to long-established methods and had been maintained. At first, the Romans had erected fortifications on the frontiers of their empire as expansion ceased. It is suggested that the majority of serious fort-building on the northern and western frontiers of the empire themselves took place either around AD 260–85, or occurred in a final burst (as it proved to be) under Valentinian I (r. 364–75). During the third century, the invasions of Gaul and other provinces had begun, and the fortification of the interior cities followed. Often, the defences thrown up at this time were themselves strengthened later on, by thickening and raising the walls and adding towers to the circuit. There are numerous surviving remains from which the main features can be established.[2] The height varies considerably, from about 5.5 metres to, according to Gregory

[1] Although some were more recent. Meulet, "Guide archéologique", 276, says that the walls of Chartres were erected in the early fifth century.

[2] For Gaul, see R M Butler, "Late Roman town walls in Gaul", and Blanchet, *Les enceintes romaines de la Gaule*. For Roman fortifications in general, see the excellent study by

of Tours, 9 metres at Dijon.[1] As later rebuilding has often removed the top of Roman walls, it can be difficult to be certain about their original height. Topped by a wall walk and crenellated battlements, they were at least three metres thick, often more. Construction was based on a mortared rubble core faced with ashlar blocks. Often there were tile courses too (which may be seen, for example, at Carcassonne). Towers came in a variety of shapes, square or semicircular, and could be numerous. Sometimes they were hollow, sometimes solid. Another common feature, reflecting both the pressing need for defences and the decline of the civic institutions, or the changing priorities of the now Christian states, was the reuse as building material of stones from the town's buildings. There is less certain evidence of the provision of ditches, although they were sometimes dug, being located at a distance from the wall itself. It is also evident that later defensive works (from the third century onwards) featured wider berms, distancing an enemy from the wall, with examples on the Rhine frontier ranging from 12 to 30 metres wide.[2]

Further east, similar problems faced the empire on its Danube frontier. A study has demonstrated rather more in the way of new building for much longer than in the west, with a variety of solutions according to local conditions and military or other decisions: some city defences were actually expanded, while others were rebuilt to enclose a smaller area, continuously through to the sixth century, although a generally lower standard of work was also common. Some cities remained unwalled. Several cities were certainly military bases (headquarters and barracks), and included arms factories, such as Sirmium (in modern Serbia), a very large and important city whose walls had long since fallen into disrepair but which was reconstructed in the face of the Gothic threat in the late fourth century.[3]

It was not the absence of defences that led Rome itself to suffer sack at the hands of Alaric the Visigoth in 410. At the same time as the construction campaign was underway in Gaul, new walls had been provided for the city by the emperor Aurelian from 272, and these walls are still there, almost nineteen kilometres in length. The original height of these walls was clearly felt to be inadequate, because they were raised twice, first in 309/312 and then again in 402, to reach an impressive fifteen metres. At the same time, more towers were added and the gateways also strengthened. Altogether there are 380 wall towers, thirty

Johnson, *Late Roman fortifications*. There is a detailed study of the development of the cities of Metz, Verdun and Toul in Dollinger-Leonard, "De la cité romaine à la ville médiéval".

[1] Gregory of Tours, *The history of the Franks*, 182 (book III, chap. xix).

[2] Petrikovits, "Fortifications in the north-western Roman Empire", 178 ff., 197–8. The author attributes various of the new features to measures against siege engines and mining, which seems unlikely, given the evidence cited for their opponents' awareness of siege technology at this period.

[3] Mirković, "Sirmium: its history", 45–6. For other examples in the region, see Panaite and Magureanu, "Some observations", 1–4; Poulter, "The use and abuse of urbanism", 101–30.

metres apart. However, the dry ditch outside had to await the attention of the Byzantine general Belisarius in 537.[1]

Similar building activity, although on a rather smaller scale, took place in other parts of the empire. In Britain it is still possible to see the results at a number of cities. The wall of Chester (Deva) was six metres high, and the surviving south-east angle tower rose to ten metres. This tower appears to have been designed to mount catapults.[2] At York (Eboracum), the fortress adjacent to the town was rebuilt in about 300 with a wall less than two metres thick and about six metres high, backed by an earth rampart and with an external ditch five to six metres wide and more than two metres deep. There were four large gatehouses, multiangular corner towers, and projecting polygonal and rectangular wall towers.[3] In London (Londinium), the wall was considerably strengthened in the fourth century, and towers were added to the eastern side of a circuit, which was 3.2 km long. The section that stands near the Tower of London was 4.4 metres to the wall walk, suggesting an original height of about six metres. There was a ditch here too. The towers were thought to have been about eight or nine metres high, and when they were added the original ditch was filled in and a new one dug.

Apart from the cities (*civitates*) with their walls, which were to determine the development and the limits of most of their medieval successors until well into the following millennium, the Romans had also built numerous forts, *castra* or the smaller *castella*. The style and techniques were of course the same, but the function of these garrison forts was different. In Britain we see such forts along Hadrian's Wall, a much older fortification along the northern frontier, and "Saxon shore" forts like those at Pevensey (Plate 6) or Burgh. Here again, study has shown that there was a difference in strength between those erected early in the third century and those constructed later, which had thicker and higher walls and projecting bastions designed to accommodate archers and artillery. The sophistication of the defences has led to speculation as to their function, including the possibility of being designed as fortified ports, or to resist internal enemies in the several civil wars that took place.[4] But of course, in the end, without an army to hold them, such structures did not maintain any continuous existence. However, where they were taken over by the new Germanic settlers, they might become the basis for future medieval settlements.[5]

The presence of such a large number of existing fortifications was to have a profound impact on what happened next, indeed it could not have been otherwise. Several centuries later, the walled towns of Gaul had become the main means of defence of the Franks against the Vikings, while the failure of Anglo-Saxon

[1] Krautheimer, *Rome*, 7–8; Todd, *The walls of Rome*, 20–82. On the ditch, see Llewellyn, *Rome in the Dark Ages*, 61, and Procopius (see below).

[2] Carrington, *English Heritage Book of Chester*, 29–31; Whimperley and Murphy, *The walls of Chester*, 6–7.

[3] For York, see Royal Commission on Historical Monuments, *An inventory of the historical monuments in the City of York*, II: *The defences*, 7.

[4] Pearson, *The Roman shore forts*, 71–2, 136–40.

[5] Knight, *The end of antiquity*, reports Frankish occupation of Roman *castella*.

England to maintain its walled cities (for the most part) compelled the kings of Wessex to adopt alternative strategies. The building style and the form of these fortifications was the pattern followed by medieval stone defences. However, nothing resembling the Roman defence system was to be built in western Europe for a very long time. The resources expended must have been truly prodigious and well beyond the means of the successor states, but perhaps the needs were also different. We will look at the reasons for this.

Walled towns offered no protection for people living in the countryside, which was where the overwhelming majority of people laboured for their living. In a time of insecurity, continued periodic invasions and raiding, and weak or non-existent central authority, people everywhere had to look for ways of finding refuge when needed. One of the most obvious solutions was to reuse what already existed, and there is evidence that iron-age hill-forts, abandoned for centuries, were pressed back into service. In many cases they would form the basis for the hilltop villages that later became such a feature of, for example, medieval Italy. Earthen ramparts could be supplemented with wooden palisades or stone walls, according to the availability of the material locally, and would be of sufficient size to accommodate the inhabitants of local villages. Located on high points, they would also serve to allow for the spotting of trouble at a distance. Although not much researched, there is plenty of evidence of this in Italy.[1]

In Britain, it was recorded (by the contemporary, Gildas[2]) that the inhabitants took to the "mountains hills and valleys" as places of refuge. Many such places have been identified, some of them reused iron-age sites and some new from the fifth century. Varying enormously in size, they are of a common style of earth or stone ramparts (either one or two of these) and ditches. Located in inaccessible places, often by the sea, it is suggested that their purpose was defensive and that the people who ordered their construction were the local potentates, while their lay-out suggests some knowledge of military engineering. South Cadbury was in use during the late fifth and the sixth centuries, and was defended by a timber-laced rampart and a wooden gatehouse. No less than 20 km of timber had been used in the construction. Early annals (although not so early as the fifth century) record the existence of forts and attacks on them. The sixth- or seventh-century *Taliesin* tells of warriors attacking enemies in hill-forts or on ramparts, while similar terms celebrated the defences of Din Eidin (Edinburgh) in the *Gododdin*. These were the means of control of British rulers, as well as a defence against their neighbours, or the Anglo-Saxon invaders. They range in scale from the tiny enclosure at Dinas Powys (Glamorgan) to the mighty rock of Dumbarton, which was the centre of the kingdom of Strathclyde. The conclusion must be that the Romano-

[1] Schmiedt, "Le fortificazione altomedievale in Italia", 859, 893–6. For a good summary, see Hodges and Whitehouse, *Mohammed, Charlemagne & the origins of Europe*, 45–8.

[2] Gildas, *The ruin of Britain*, 28, 98. Gildas, writing around 540, also noted that Britain had been a country ornamented with 28 cities and with "castellis, murorum turrium" and strong buildings (p. 90).

British kingdoms were rather better organised and more effectively defended than Gildas' lament suggested.[1]

In Gaul there were similar developments. The fifth-century writer, wealthy landowner, bishop (and eventually saint) Sidonius Apollinaris, who lived between about 430 and 470, penned a long panegyric to the fortified country estate of Pontus Leontius, described by himself as a *burgus,* which led to the discovery of a Gallo-Roman villa near Blaye overlooking the river Dordogne. The poem described the recently-built luxurious dwelling and estate centre of a powerful Aquitanian nobleman, which had been fortified with battlemented high walls and towers capable of resisting the full range of siege weaponry and techniques of attack then known — *aries, agger, catapulta, testudo, vinea,* and wheeled scaling ladders. Excavations of the headland site found thick masonry walls with square bastions, apparently destroyed after the Franks conquered Aquitaine from the Visigoths in the late fifth century.[2] In Echternach (in modern Luxembourg), alongside an important Roman road, there is a natural hillock under the later church of St Peter and St Paul, where a late Roman *castellum* has been excavated and partially restored. An older fort had been rebuilt in the early fifth century to create a multi-sided stone-walled perimeter of up to 1.5 metres thick enclosing the top of the hill (diameter about 50 metres). There are four rectangular towers of uniform design, large enough to mount small artillery such as *ballistae,* four postern-like narrow gateways, and interior buildings of unknown purpose, with a well. There is evidence of its occupation continuing through to the middle of the fifth century, and of a later Merovingian church inside (Plate 4).[3] However, we depend on literary references (mainly Gregory of Tours) for the existence of rural *castra,* rather than excavated sites.[4]

Against what threat was this array of walls and ramparts intended to protect?

[1] An excellent summary of the evidence for this is Alcock, "The activities of potentates in Celtic Britain", 25–33, which has a map and gazetteer with the dating evidence for 75 such sites. Dunbarton, which is called by Bede a *civitas,* has no surviving remains of this period. Its history is summarised and a reconstruction attempted in McIvor, *Dumbarton Castle,* 14–15. See also, more recently, Dark, *Britain and the end of the Roman Empire,* 146–8, 199–205. For translations of the later accounts, see *The earliest Welsh poetry,* 24, 28 (*Taliesin*); 37, 44, 55 (*Din Eidin*). S Davies, *Welsh military institutions,* is a recent study.

[2] Nicolaï, "La villa gallo-romaine de Pontius Leontius". The poem is in Sidonius, *Poems and letters,* 271–3 (poem XXII, ll. 118, 120–25). "Battlements" is the proposed translation of "propugnacula". Sidonius was fond of demonstrating his knowledge of the classics, and it is entirely possible that the list of engines was there just to show this, rather than really to suggest the strength of the defences.

[3] Metzler, Zimmer and Bakker, *Ausgrabungen in Echternach,* 269–94.

[4] See James, *The Merovingian archaeology of south-west Gaul,* 291–3.

Late Roman siege warfare

Vegetius

Just as the Romans designed and built the *civitates*, *castra*, and *castella*, so also they were the ones who knew how to attack and defend them. Barbarian warlords originated in societies that did not live in walled cities. When they came up against the Roman-built fortifications, there was only one store of knowledge available to instruct them how they, too, should attack and defend them. To what extent did they learn and how far were they able to assemble the skills and resources to conduct siege warfare themselves once they were the masters? To answer these questions, there are a few written accounts that help.

The late Roman army was very different from the legionary army of the early empire. Fortunately, there survives one text that tells us a great deal about how siege warfare was conducted — or rather, how the author thought it should be conducted. Vegetius' *Epitoma rei militaris*[1] is a source that, because of its uniqueness, has been discussed and exploited countless times. It has been dated to the late fourth century or the early fifth, or possibly the reign of Valentinian III (r. 425–55).[2] Its survival also guaranteed its place as a handbook for the whole medieval era. Its value at this point is to show how late Roman siege warfare should have been conducted.

Vegetius devoted chapters to the defence and attack of fortified places. In doing so he gave clear indications of the functions and effectiveness of the standard weaponry available. Advice was given to defenders about laying in supplies, starting to cultivate open spaces, getting hold of plenty of iron and timber. He advised that liquid pitch and oil (*incendiarum oleum*) be obtained, with the purpose of burning the enemies' machines. Round stones should be collected, if possible from rivers "because they are heavier in proportion to their density and more suitable for throwing", both for slingers and for stone-throwing engines. These too were for destroying enemy machines.[3] A chapter was devoted to the importance of collecting supplies of sinews (*nervi*) because these were essential to the operation of the artillery. He pointed out that "horsehair taken from the tail and mane is said to be useful . . . There is no doubt that women's hair has no less virtue in such kinds of torsion-engines."[4] In addition to handheld slings, the main projectile weapons cited by Vegetius were the *ballista* and the *onager* (literally "wild ass"). Both were powered by torsion, that is, twisted skeins of rope or sinews. The *ballista* resembled a frame-mounted crossbow. The arms were powered by two skeins mounted either side of the stock, and it propelled large darts or javelins. The *onager* was a late-Roman weapon, replacing earlier, more complex forms of engines. It had

1 The edition used here is Vegetius, *Epitome of military science*, trans. Milner.

2 See Goffart, *Barbarians and Romans*, 252, and the references cited there for this argument.

3 Vegetius, *Epitome*, 116–18 (book IV, chaps. vii–viii).

4 Ibid., 118 (book IV, ix). Historical examples known to the writer of women's hair being used are found in a number of ancient writers including Caesar.

a single arm mounted in a very strong wooden frame on the ground, which was hurled forward by the torsion power of a skein of the sinews detailed by Vegetius, its forward motion halted by a wooden crossbeam. Its projectile was usually a stone ball. He says that no heavier weapon was known. Its purpose was to "destroy not only the horses and men, but also the machines of the enemy".[1]

Vegetius described the full panoply of weapons and methods. Forms of attack included digging mines or sapping to undermine the wall, and building wooden mobile towers. This last was clearly a very major undertaking, for the finished device has to be able to be higher than the wall, to protect a battering ram on the first level, to have a drawbridge to reach across to the rampart, and a top gallery for soldiers to fire down upon the defenders. The whole machine had to be able to be moved up against the wall. All the time, defenders were able to observe its construction and to defend themselves by a variety of means: trying to burn the tower by projecting incendiary devices or by making a sortie, raising the height of the wall with earth or timber. Rams protected by fireproof coverings were another device, this time for creating a breach in the wall. The methods of defence against such attack included mattresses and hooks lowered from the wall head to obstruct or pull up the ram.

The simplest means of attacking a defended place was either to storm it, or to starve it out. Vegetius discussed how an immediate attack with ladders could succeed if the defence was inexperienced and afraid. He made suggestions as to how to ensure that ladders are constructed to the right height to overtop the wall,[2] something that was not always the case, with disastrous consequences. Starving a place into surrender depended on the besiegers having the time and the logistical resources to outlast the defence. Interestingly, Vegetius discussed this only from the viewpoint of the defence. So it is clear, this is a very practical treatise, one designed to deal with the defensive problems faced at its time (even if some of the descriptions given obviously allude to earlier times). One of the repeated points made by this author in passing was that skilled and experienced people are necessary to operate the equipment. The artillery had to be "tuned very carefully by experts".[3] This was to be a recurring theme throughout the whole middle ages. For the present, when examining the actual siege warfare of the two centuries following the eclipse of the western empire, the availability of such expertise is a basic question always to be borne in mind.

The defences described earlier were designed to accommodate the artillery discussed by Vegetius. It is important to establish what the chief characteristics of this weaponry were, to understand what effect it had. Many scholars have striven to explain this from contemporary sources and from military manuals, all of which are considerably older than Vegetius. The problem of using the sources is to know what each writer meant when using a descriptive term, because the meaning of the words changed over time. It is also not always clear that the writer was talking about something contemporary, or something that he had himself de-

1 Vegetius, *Epitome*, 127 (book iv, xxii).
2 Ibid., 131 (book iv, xxx).
3 Ibid., 130 (book iv, xxix).

rived from still earlier writers. However, by this time the course of developments is generally accepted. The *ballista* has become a weapon designed for use on walls, where its slow operation (requiring a windlass to draw back the cord) mattered less. It shot bolts almost two metres long, which would have had the power to penetrate armour, hides, and possibly wooden stockades[1] — a claim we saw vindicated by Vegetius. The *onager* received its first mention in about 353 by the writer Ammianus Marcellinus. It was certainly described as being in use by the Romans when the victorious Goths tried to storm Adrianople in 378, when Ammianus called it a *scorpio*, "a type of catapult, commonly called an onager".[2] Here, its large stone smashed into the ground harmlessly, but astonished and terrified the Goths. As a single-armed machine it would have needed less maintenance or adjustment than the twin-armed torsion-powered stone throwers it replaced. Ammianus is quoted as saying it needed eight men to wind down the arm, and its immense recoil when fired made it impossible to place on a stone wall because it would dislodge the stones. It would also be very difficult to move around, to change its aim: a model constructed by Frankland-Payne-Gallwey weighed two tons (2,032 kg).[3] There are strong grounds for believing[4] that torsion-powered machines went out of use, at least in the west, during this period. The reasons are evident from the descriptions given. While the solid stone towers added to the walls of Roman defences were quite sufficient to cope with the weight of the *ballista*, it is not at all apparent that they could sustain the "violent shock", to quote Ammianus again, of the *onager*. If this weapon had to be sited on the ground in order to absorb its shock, its use in defending a wall would be limited, unless a satisfactory system of firing "blind" could be devised. No such difficulty of course stood in the way of its use as an offensive weapon. Vegetius, however, does not help us here. His concern is with artillery for the defence, not the attack.

For confirmation of the truth of this major drawback to the *onager*, observe the reconstructed *onager* at Caerphilly Castle (Glamorgan), where it is incorrectly called a mangonel.[5] The machine is considerably smaller than the machines described in the sources, but its recoil is dramatic. It can also be added that its rate of shooting is slow, and the size and range of the projectile, relative to the size and weight of the machine, is poor by comparison with the medieval lever-based weapons by which it now stands. Add to this the necessity for a supply of appropriate sinews, which could break or become distorted because of the weather, and above all the requirement for builders and operators who were experts in their trades, and one can see many practical reasons for doubting that the torsion-powered stone thrower survived long into the "dark ages".

1 Landels, *Engineering in the ancient world*, 105.

2 Ammianus Marcellinus, *Rerum gestarum*, III: 494–5 (book XXXI, 15, xii).

3 Landels, *Engineering in the ancient world*, 132; Frankland-Payne-Gallwey, *The crossbow*, 249–50 and appendix. The discussion is also summarised in Southern and Dixon, *The late Roman army*, 152–9.

4 For example by Chevedden, "Artillery in late antiquity", 132–63, and Purton, "The myth of the mangonel". For further discussion, see Chapter 9.

5 Humphries, *Engines of war*. The "mangonel" shot is 3–4 kg. The *ballista* there is the twin-armed torsion machine and shoots a bolt 1.5 metres long.

The evidence of siege warfare

With so many walled cities awaiting the invader after the solidly fortified frontier had itself been breached, it was inevitable that any serious attacker must address themselves to dealing with defended places, or else find themselves seriously restricted in their movements and vulnerable to attack from all sides. It has always been the case that the willingness of people to defend themselves is the most important factor in deciding whether a town would resist, thus calling for a siege. As Vegetius repeatedly argued, the first step to make resistance possible was to assemble sufficient resources. A second was to have an armed garrison. There are so many brief accounts of cities falling into the hands of rival kings during the fifth and sixth centuries, with no further information provided, that it is impossible to know which factors determined the outcome.

The Goths do not seem to have had much idea of, or appetite for, siege warfare. Of all peoples, they included those whose long association with Rome must have included some opportunity to acquire these skills, unless the Romans had managed deliberately to keep them secret (a practice we shall see attempted later by the Byzantine empire). They gave up the siege of Sens after 30 days, "muttering they were stupid . . . to have contemplated the siege of a city". After their victory at Adrianople in 378, they quickly gave up their attempt to take the city. In 490, Theodoric besieged Ravenna (see below for more on this period), and three years later had still not taken it (admittedly this city had very effective natural protection).[1] Alaric had besieged Rome three times between 408 and 410 and never once stormed it. Ricimer besieged the same city for eight months in 472. Where they were successful, for example when Theodoric captured Belgrade (Singidunum) from the Sarmatians in 472, our source, Jordanes, gives no indication as to how this was done.[2] The Franks, initially at least, seem to have been equally unsuccessful. It took Clovis several years to secure Paris, because he had no means to prevent access by river. But he did manage to capture Verdun (480).

The attackers who gave some evidence of aptitude were the Huns. In their invasion of the Balkans after 441, they are recorded as taking some one hundred towns. In 451, their ruler Attila assailed Orléans with his battering rams, although he did not take it because the arrival of Roman general Aëtius with an army saved it just in time.[3] In 454, Attila's sons captured three towns in Dacia by regular siege, not by direct storm or starvation. In 452, Attila himself had attacked Aquileia (northern Italy). He constructed and brought to bear "all kinds of machines and artillery", where the use of the word *tormentorum* implies, although it does not prove, that torsion artillery is meant. He used "many machines and other devices

[1] For these and other examples, see Elton, *Warfare in Roman Europe*, 82–5.

[2] Jordanes, *De origine actibusque Getarum*, 117 (chap. lv).

[3] Gregory of Tours, *The history of the Franks*, 116 (book II, vi). Gregory assigned the saving of Orléans to the prayers of its saintly bishop, but acknowledges that an urgent plea had been sent to Aëtius at Arles. Sidonius mentioned the siege and the leading role of the bishop, but declined a request to write a history of it: *Poems and letters*, II: 490 (book VIII, letter 15).

suitable for breaking walls". Since we know that even the most powerful artillery of this age was designed to target people and other machines and would be ineffective against a wall, this description must mean battering rams.[1] It has been argued that the historians whose accounts we use were not describing actual events, but reproducing accounts from ancient histories. When there is no corroborative evidence this is always possible. However, this account carries some conviction, at least in broad terms.

There is some debate about how the Huns acquired these skills. They were a nomadic people, although they employed other, subject peoples too. The most likely conclusion is that they got them through prisoners or deserters from the Roman army. This use of skilled engineers from other backgrounds is to be a common feature of medieval siege warfare. In the case of the Huns, it would also explain why having gained access to these skills, they were apparently lost again.

This hypothesis is strengthened by the apparent failure of besieging barbarian armies of any kind to use sapping. Undermining the wall and bringing it down in this way will be a major facet of medieval siege techniques. It does require particular skills to be carried out effectively. If no engineers with the relevant skills were available, it was very unlikely that the method would be tried. The only reference to mining is from the Gothic siege of Rome in 537, and Procopius' account indicates that this was the rather less challenging task of breaking through an outer wall without defences or defenders.[2]

We know that the armies of late Rome still possessed reservoirs of the skills needed. In 438 Aëtius was able to erect a siege tower against a Visigoth-held town. Siege towers were to be a method learnt by the Goths too, although, as will be seen, they were not very successful in using them.

Within a century, it looks as if such skills as had not been entirely forgotten were falling into disuse, except where sieges involved close encounters with Byzantium, which undoubtedly preserved the old methods. Once again, the evidence is capable of different interpretations, and of course the absence of references to particular pieces of equipment or their use is not evidence that they were not used. However, for the most part, the accounts by Gregory of Tours of the deeds (and many misdeeds) of the Merovingian dynasty of kings of the Franks in the sixth century suggest otherwise. Very much a busy man of the world as well as a bishop, Gregory knew of what he wrote. For Italy, we have the account given by Procopius of Justinian's attempt to recreate the old empire, and here the evidence for siege equipment in use is much stronger.

When Clovis (d. 511), the king of the Franks later honoured as the founder of France, was at war with the Burgundians, and besieged the city of Avignon, Gregory tells us that he accepted advice to give up because "the stronghold is too well fortified for you to capture". The next year, the Burgundian king Gundobad

[1] The second quotation is from Jordanes, *De origine actibusque Getarum*, 90 (chap. xii); *The Gothic history*, 113. It reappears in Thuróczy, *Chronica Hungarorum*, drawn from much earlier sources. Much less detailed is the account of Marcellinus comes, *Chronicon*, 75–84. The rest is from Tausend, "Hunnischer Poliorketik", 265–71.

[2] Procopius, *History of the wars*, 223–5 (book v, chap. xxii).

besieged the city of Vienne. The defenders were concerned at the shortage of provisions and expelled the common people: another common feature of sieges. Among those expelled, however, was the engineer in charge of the aqueduct, who in high dudgeon brought the besiegers into the town by showing them a water gate that could be prised open with iron crowbars.[1] So the town had an engineer, and the attackers iron tools. But they relied on this chance event to get in.

Later (582), when Gregory was recording events from his own times, he described how King Chilperic, fearing an alliance of his relations, ordered his "dukes and counts to tell them to repair the walls of their cities, and to shut themselves up in these fortifications".[2] Here again is evidence of the importance of the Roman walls inherited by the Franks, of the need to maintain them, and of the strength their possession offered. It is no wonder that so much of this history revolves around the occupation of the cities. In 586, the Visigoth Recared came out of Spain and "captured the castle [*castrum*] of Cabaret . . . and attacked the castle of Beaucaire near Arles . . . and shut himself up inside the walls of Nîmes", where he obviously felt safe.[3]

The only time Gregory described a siege in detail was at Comminges (in 585). The town, well provided with supplies, was held by a pretender to King Guntram's throne and besieged by Leudegisel, one of the king's counts. After two weeks, "Leudegisel spent his time in preparing new machines with which to destroy the city. He constructed wagons, which he fitted with battering rams and protected with wattle work, old leather saddles and planks of wood, so that the troops could rush forward under their cover to knock down the walls." But the defenders responded with "flaming barrels of pitch and fat, as well as boxes filled with stones", and the attack was repelled. The next day the besiegers "prepared fascines from bundles of sticks and tried to fill in the deep ravine which lies on the eastern side". This also failed. In the end the pretender was tricked into surrender and the townspeople were massacred.[4] But it is clear that the siege techniques of the attackers were not up to breaking through the defences of Comminges, and among all the detail there is no mention of any kind of artillery being used to drive the defenders from the walls and so make the task of breaking them down easier.

Gregory also provided important information about the organisation of Merovingian fighting forces. It is clear, for example, that the cities provided a base for a count and the focal point for the organisation of local levies, who would also presumably provide the garrison when under attack. As with the troops of later Frankish armies, they seem to have been required to arm and supply themselves.[5] What is less evident is where the engineers came from. Perhaps their existence is so obvious for Gregory's contemporaries that they require no comment. On the other hand, the description of the ineffective efforts of the king's army at Com-

[1] Gregory of Tours, *The history of the Franks*, 147 (book II, xxxii–xxxiii).

[2] Ibid., 374 (book VI, xlvii).

[3] Ibid., 462 (book VII, xxx).

[4] Ibid., 420–24 (book VII, xxxvi).

[5] Bachrach, "Military organization in Aquitaine", 4–5, makes brief reference to the situation under the Merovingians.

minges suggests no very high level of expertise. Bachrach argues that Merovingian armies included elements from the existing Roman armies,[1] and if that is the case then it is probable that this is where the siege expertise would have been found. Given the vital importance of the *civitates* and the *castra* to the Frankish kings, as Gregory has illustrated, it seems strange if they took no steps to acquire, retain or pass on these skills. It is suggested that where they could use troops from southern Gaul, which was the most romanised area, they were more likely to have such resources. But professional soldiers and engineers, after all, expect to be paid, and one of the changes of the post-Roman world was the ending of the old imperial tax systems.[2] It would be surprising if there was not a reduction in the numbers of skilled personnel available to the new rulers, but the evidence is simply not there.

Among the Franks' competitors to the south were the Visigoths, who had taken control of Roman Hispania, which included Septimania, a swathe of what is now southern France. They were to rule here until their kingdom was so dramatically and rapidly extinguished by the Islamic invasion of 711. Evidence for their siegecraft comes from a number of surviving texts, mainly from the late sixth and seventh centuries. They confirm what the account of Gregory of Tours suggests, although the absence of detail means conclusions cannot be definite. Certainly, the possession or capture of the cities remained a central objective. The Franks besieged Zaragoza (Caesaraugusta) for forty days in 541, while the Goths attacked and obtained possession of Tarragona. Euric captured Marseilles and "other castles" (*castellis*). Barcelona (Barcinona), a city with late Roman walls, was captured by the Burgundians.[3] King Leovigild also had to deal with many strongly fortified cities, but rarely was he able to take them by siege. In 570, he was repulsed from two such places and resorted to devastating the countryside, in following years he obtained possession by treachery. In 583, fighting a rebellious son and the Suevi (in the north west of the peninsular), he succeeded "now by hunger, now by the sword". When in 589 the Franks set up camp before Carcassonne, he was able to relieve the place and destroy their *castra*.[4] A hundred years on, methods were scarcely more sophisticated. St Julian of Toledo has left vivid accounts of the conflict with a rebel at Narbonne and Nîmes in 673. The defenders relied on archery and hurling stones; the attackers knew to protect their own encampment but relied on storming the walls to get in. King Wamba had to repair the walls of Nîmes after its fall, which implies some kind of siege equipment was employed,

1 Bachrach, *Merovingian military organization*, 4–12, 20–21.

2 See the comments by Pohl, "The barbarian successor states", 44.

3 *Chronicon Caesaraugustanorum reliquiae*, 223; *Chronica gallica*, 665–6. The walled area of Barcelona (Barcinona) at this time was about 825 × 550 metres (9 hectares), erected around AD 260–310 to cover a smaller area than previously. The wall was protected by numerous closely-spaced projecting rectangular towers of small size, suited more to archers than artillery according to the excavator. A Balil, *Las murallas bajo-imperiales de Barcino*, 34–5, 61.

4 Johannes Biclarensis, *Chronica*, 212, 216–18.

but it is not mentioned anywhere, and the account is such as to suggest that it would have been had it been present.[1]

While the Visigoths inherited Roman cities with their existing walls, it seems that they were quite capable of learning from them, too, although surviving remains seem to be few. A *castrum* dated to between 650 and 711 was found at Rosas (in Gerona), which had a standard type of double wall infilled with small stones, gravel, and earth, but without any mortar. The Visigothic city of Recopolis (in Guadalajara) had similar walls, and projecting rectangular towers six metres square. In appearance they must have been very similar to Byzantine constructions, as will be seen.[2]

Their Ostrogothic cousins seem to have turned their attention not only to using their inherited places as means of defence or refuge against other invaders, but also to have undertaken new building work of their own. Comments in Cassiodorus' *Variae* and in the works of the saintly bishop Ennodius of Pavia confirm that Theodoric the Great (ruler of Italy from 489 to 526) put up new defences, and there have been numerous attributions of constructions (all south of the Alps or in the passes) to the period between the arrival of the Goths and their conquest by the Byzantines, although to what extent this consisted of no more than providing garrisons for existing places (which was common, according to the sources) is hard to prove, with place names being another indicator. There were certainly new refuges created: the bishop of Novara built a *castellum* around 490–500, although no remains have been found, while there is a *castrum* of the subdeacon of Milan (no later than 556). Theodoric ordered the restoration of existing walls at Susa, Ivrea, Trent, Cividale, the repair of the walls and "aged towers" of Arles, and the strengthening of the *castellum* at Tortona. Excavation at Verona has identified a number of phases of work later than the third-century circuit of walls, including construction of what must have been a forewall eight to ten metres outside the circuit of Gallienus, described as of poor quality, during the Theodorician period.[3] Procopius recounted how Theodoric marched against his rival Odoacer, besieging him in Ravenna in 489, until a (short-lived) settlement was reached after a blockade lasting into a third year. At the same time, however, the Gothic army had captured other strong towns "in one way or another", with the exception of Ravenna and also Cesena, which could not be approached because of its location protected by the Po and great extents of shoals on all sides. The lake-island fortress of Bolsena was available for use as a prison in 535, having apparently been

[1] Julian, *Historia rebellionis Pauli adversus Wambam*, 776–91. See also E A Thompson, *The Goths in Spain*, 266–7; Reilly, *The medieval Spains*, 21–6, 43.

[2] Olmo Enciso, "Restos defensivos de Recópolis", 67–72.

[3] Settia, "Le fortificazioni dei Goti in Italia"; Cavaliere Manasse, "Le mura teodoriciane di Verona", 634–40. For the original letters of the king to his lieutenants at Tortona, Arles, etc., see Cassiodorus, *The Variae*, 14–15, 65–6 (I.17, III.41, III.44). The walls of Arles resisted attack by the Franks (507–8) and earned the loyal city a remission of taxes by the grateful Theodoric: ibid., 63 (III.32). The anonymous *Chronica theodoriciana* reported that the restoration of Rome's walls was to be paid for with an assignment on the taxes collected on wine, and the new works of the king at Verona and Pavia included new circuits of walls: Anonymus Valesianus II, *Chronica theodoriciana*, 550–51, 552–3 (chaps. lxvii, lxxi).

previously maintained as a place of great strength.[1] Ennodius' panegyric of the great Goth includes references to his work of erecting castles to defend the kingdom, and to his programme of reconstruction of cities lying in ashes, while a life of the bishop mentions the rebuilding of Pavia using "immense engines", and to the strength of the defences of that place, capable of resisting even attacks made with *Balearis fundae* (literally "slings", but perhaps a figure for engines like the *onager*).[2] As always, it is wise to treat such obviously biased statements with doubt, unless verified by archaeology.

Despite a growing number of excavations of *castella* used during this period, little evidence seems to have emerged to enable conclusions to be drawn about exactly when, or by whom, these fortifications — that range from quite large defensible refuges in heavily populated areas, to small garrison forts in strategically important locations — to be drawn. Their excavator concluded that in fact, most if not all of the north Italian locations were late Roman constructions reused by the Goths.[3] Other possible sites date to the period of Gothic rule in Thrace, with the major fortifications at Sadovets (northern Bulgaria) attributed to Bessas, who was recognised as prince by the East Romans around 488, which are part of a network of defended settlements on strong-points above the river Vit. Sadovsko-Kale stood on a terrace 80 by 20 metres, and was strongly protected on the only approachable side. Here stood a wall, now three metres high, but just 1.8 wide, standing on firm foundations and built from mortared, uncut stones. Entrance was gained through a simple gate. Inside was a second wall of similar construction, with at its terminus a five-sided tower corner tower, with walls 2.7 metres thick. The upper levels have all disappeared. Inside were a row of chambers set against the walls. Coins and burning suggest destruction, possibly in the Avar invasion. The work, clearly showing signs of Byzantine influence, but with Gothic attribution relying on circumstantial evidence, was not of great strength or quality, but seems rather strong just to have been a refuge.[4]

In Britain, the absence of records for the fifth and sixth centuries leaves the field open only for speculation. The Anglo-Saxon conquest must have encountered the many walled towns left by the Romans. Whether they were defended by the Britons or abandoned is not possible to determine. Gildas, in his lamentation of the failure of the Britons to defend themselves, notes that all the towns "were laid low by the repeated battering of enemy rams" (*arietes*), but whether this is mere rhetoric or an accurate reflection of the siege-warfare skills of the invaders cannot be determined. From the other evidence of the abilities of barbarian chieftains to capture fortified towns, it is possible.[5]

[1] Procopius, *History of the wars*, III: 7–8 (book V, i), 37 (book V, iv).

[2] Ennodius, *Panegyricus regi Theodorici dictus* and *Vita Epiphani episcopi Ticinensis*, both in his *Opera*, 207, 211, 299.

[3] Brogiolo, "Dwellings and settlements in Gothic Italy", 113–120, 136. The writer had excavated seven *castella*, of which only that at Monte Castello di Gaino (Lake Garda) was possibly built by the Goths. Other recent archaeological reports are listed in Brogiolo's paper.

[4] Welkov, "Eine Gotenfestung bei Sadowetz".

[5] Gildas, *The ruin of Britain*, 27, 98.

Siege warfare in Gothic and Lombard Italy in the sixth and seventh centuries

Procopius, on the other hand, leaves no doubt as to the full panoply of weaponry deployed in the siege operations involved in the struggle between the Ostrogoths and the Byzantines for control of Italy, a struggle to which he was a witness as secretary to the imperial commander Belisarius. It also appears from his account that the Ostrogoths had acquired some of the knowledge as well, which would not be surprising since they had been rulers of the centre of the old empire for half a century. The description of the siege of Rome (537–8) talks of rams, *ballistae*, and *lupi*. Further, the attackers "made wooden towers" equal to the height of the wall, which were wheeled and pushed by oxen. The rams are described as roofed. The Goths had four of them, each propelled by fifty men. They deployed "innumerable fascines" to fill the ditch. In response, Belisarius, commanding the Roman army, "placed machines in the towers which are called *ballistas*". Procopius goes on to describe their construction, which confirms they are of the type we have seen described by Vegetius. Furthermore, on the walls were placed "other machines, hurling suitable stones: they are similar to slings, and are called *onagers*".[1] This appears to contradict what has been said about the problem of mounting the *onager* on top of a wall because of the likelihood of its damaging the wall through the violence of its recoil. However, it was noted earlier that the defences of Rome had been substantially reinforced twice after their original rebuilding, and the walls and towers were a great deal stronger than the walls that surrounded so many Roman cities in the rest of the old empire. It may well be that the strengthening process was carried out precisely in order to enable the walls to mount such weapons. There seems little other reason, given the absence of siege skills among Rome's likely assailants at the time, to have invested so much in raising and thickening walls that were already strong by contemporary standards.

The other main point about the Goths' attempt to recapture Rome from Belisarius is that it failed. Despite substantial numerical superiority, and the apparent ability to construct rams and towers, their attempted storm was a fiasco, being repelled by the missiles and in particular the artillery of the defenders. Their skills did not extend to providing the cover needed to enable them to reach the walls, and if they did have their own artillery, it was not capable of dealing with that on the walls. The defenders shot down the oxen and left the towers stranded. As with the Franks, it looks as if the Goths were still relatively inept at siege warfare despite their long rule in a country where control of the walled cities was the single most element of political power. There is a solitary further reference to the use of *ballistae* by the Goths, where they are described as being mounted in wooden

[1] Procopius, *History of the wars*, III: 205–7 (book V, xxi). He gives a description of the *lupus*, which suggests that this was some kind of wooden portcullis designed to protect the gates. There is a stirring account of this war (which seems to be rather strongly based on Procopius) in the old history of Hodgkin, *Theodoric the Goth*, 317–40.

towers to cover a river crossing, but do not warrant further detail although the ensuing battle is minutely recounted.[1]

Following the epic siege of Rome, the war between Goth and Roman continued until 552. The Gothic ruler Vittigis (r. 536–40), having been beaten at Rome, left substantial garrisons in Chiusi, Orvieto, and Todi and in other places, including 500 men in the fortress of Cesena and at Montefeltro, to hold them against Belisarius while he marched on Ariminium (Rimini) on the coast, seized by an imperial force, to besiege it. Procopius recorded that the Goths had now learnt from their experience at Rome, and proceeded to build a siege tower on four wheels, as high as the wall, but this time (somehow) propelled by men from the inside. It had a wide ladder there as well, to enable large numbers of soldiers to mount to the top, from which they planned to enter the city. The ditch was very shallow. But they were unable to complete the attack before darkness fell, and left the tower close to the ditch overnight. During the night, the garrison commander led out his troops and managed to dig the ditch much deeper, throwing up the spoil on the inside to create another rampart. By the time the Goths realised, it was too late. Although the next day they attacked and filled the ditch with fascines, these proved unable to bear the weight of the tower, which toppled forward. There was then a fierce fight as the Byzantines sallied to prevent the engine being withdrawn, and although they failed to stop this happening, they killed many of the besiegers in the fight, and the latter settled in for a long siege, in which they would eventually have succeeded through hunger had the Byzantine army and fleet not come to the city's relief.[2]

Time and again, the defences of the cities were sufficiently daunting to persuade the Goths to resort to attempts to starve them out, leading to blockades of many months. Sometimes they succeeded; often they were thwarted by the arrival of relief. Frequently, the Goths turned to demolishing the walls of places they secured, in order to prevent their potential reuse. Totila had even begun to pull down the walls of Rome (having finally occupied it) when he was dissuaded, apparently, by an extraordinary appeal from Belisarius, reminding him of how he would appear in history if he was responsible for the destruction of the great city. On this occasion, when Belisarius reoccupied the city, he found a shortage of artisans that prevented him properly restoring the damage.[3] Later, the imperial army recaptured Rome (again) by scaling ladder, when the Gothic garrison found itself too small to protect the whole perimeter at once. In the whole period of the reconquest of Italy by the empire, there is no evidence that the Goths were able to take serious advantage of the existing fortifications to hold onto their rule, or to remove their enemies from them once established.

The year 568 saw the arrival in the Italian peninsula of another Germanic

[1] Procopius, *History of the wars*, v: 409 (book viii, xxxv).

[2] Ibid., iii: 379–93 (book vi, xi–xii) (attack on Ariminium).

[3] Ibid., iv: 205–13 (book vii, vi) (siege of Naples, 543); 229–31, 253–7 (book vii, ix) (siege of Dryus (Otranto)); 245 (book vii, xi) (destruction of the walls of Pesaro and Fano); 361 (book vii, xxiv) (shortage of artisans). See also Marcellinus comes, *Chronicon*, 105–7.

people, the Lombards, whose rule was rapidly to cover most of the region, leaving the Byzantines restricted to southern coastal cities and Ravenna and a precarious protectorate over Rome, which was in practice ruled by the popes. They provide us with evidence for the development of siege warfare in the following century and beyond, until the Franks finally put an end to Lombard rule in the north of Italy in 774. Paul the Deacon has left an invaluable record of their history.

It seems as if the new arrivals were not much better adapted to the methods involved in capturing fortified places than the Goths they supplanted, but at least they record some successes gained in attacking them. In 600/601, the Lombards recaptured rebellious Padua "when fire was thrown into it".[1] How was this achieved? Perhaps it was incendiary devices projected from catapults, but we are not told. A little later (603), King Agilulf captured Cremona after breaking down its walls with rams ("cum arietibus").[2] In about 610, the Avars invaded and devastated north-east Italy and defeated the Lombards, who proceeded to take refuge in Cividale and other fortresses "whose position was impregnable".[3] This was the same response to military weakness as thwarted the Franks a few years earlier.

Paul indicates then that the Lombards at least were able to take advantage of the strong places they had taken over to ward off attacks from other barbarian armies. There is some further evidence to support this written account. The country they had occupied had been at war for a long time, and during this time rival commanders had reverted to destroying existing walls to deny them to their enemies. Furthermore, there were parts of Italy where there were very few fortified cities in the first place.[4] The Lombards, it seems, did not leave their defence to the existence or strength of the already standing Roman walls alone, although they did take advantage of these where they existed.[5] To guard their lands from attack from the north, they built a range of fortifications. Paul is unhelpful in describing them, using terms such as *castrum* and *castellum*. To him it would have been obvious what was meant; to later readers it gives no idea what kind of fortifications they were. A site has been excavated at Ibligo-Invillino in the north-east.[6] It stands on a high hill in a flat plan, overlooking a Roman road and a crossing of the river Tagliamento. The hilltop, rearing 55 metres above the valley, is 650 metres by 100 metres. The *castrum* occupies 200 × 100 metres. Its excavators say that, between both earlier and later occupation evidence (late Roman and Venetian), there is a stone tower that is either late Roman or Lombard. Although this is

[1] Paulus diaconus, *Historia Langobardorum*, 23 (book IV, chap. xxii); *History of the Langobards*, 167.

[2] Paulus diaconus, *Historia Langobardorum*, 28 (book IV, xxviii); *History of the Langobards*, 171.

[3] Paulus diaconus, *Historia Langobardorum*, 37 (book IV, xxxvii); *History of the Langobards*, 180.

[4] Procopius, *History of the wars*, III: 69 (book V, vii), notes the absence of fortifications in the south of the country: "their towns had from old been without walls". See also Pani Ermini, "Città fortificate e fortificazione delle città", 194–204.

[5] D Harrison, *The early state and the towns*, 67–73; Schmiedt, "Le fortificazioni altomedievali in Italia", 906–18.

[6] Fingerlin, Garbsch and Werner, "Gli scavi nel castello di Ibligo-Invillino", 58–130.

a site reported by Paul, this only means that the Lombards used it, not that they built it. There is no evidence of a curtain wall, and the dating seems precarious. Elsewhere, at Castelsegnio (north-west of Milan), a site dating originally to the fourth century has been excavated. It had late-Roman walls and towers, following the line of the hilltop, and a cistern and a church, with a village below. It is said that the church is seventh century, therefore Lombard.[1] All in all, the present level of evidence makes it hard to demonstrate that the new conquerors, although they had to base their control in the existing cities, added very much to what they found, but rather took advantage of what was there, both in the cities and in the *castra* they found in the countryside.

The Lombards were in conflict, both on their northern frontiers and on the south, with the Byzantine emperors.[2] As the seventh century progresses, Paul gives us evidence of a new siege weapon. In 663, the Byzantines threw the head of a Lombard leader into the city they were attacking with a "machine of war called a *petraria*, which commonly we call *mangola*".[3] By the end of the century, the Lombards were themselves using an array of siege weaponry, as in 701 King Aripert attacked Bergamo, "storming it without any difficulty with battering rams and other machines [*manculis*]".[4] This new weapon represented the first significant change in siege warfare and its techniques since the end of the western Roman empire, and for its origins it is necessary to turn our attention to the east and retrace our steps.

The Byzantine empire

The western successor states to Rome added little to what they had taken over, and witnessed significant regression in the level of culture and skills. In the east, there was no break in the continuity of imperial rule. Although the capital was no longer in Rome, the Roman empire continued to exist. It represented the only significant power in the whole of the eastern Mediterranean, the Balkans, Asia Minor, the Middle East, and (after Justinian's reconquest), North Africa. For the same reason, it faced opposing powers in all directions. Justinian directed an attempt to regain the empire's lands in the west. Belisarius restored Roman rule by crushing the Vandal kingdom in Africa with astonishing speed, and came quite close to reinstating imperial rule in the whole of Italy. But there were other enemies too, more powerful than any in the west, of which the Persian empire was the most significant. Meanwhile, control of the Balkans was always prone to disruption by successive waves of new barbarian invasions. Rich although it was by comparison with the barbarian states described already, the

[1] Christie, *The Lombards*, 177–8.

[2] Paulus diaconus, *Historia Langobardorum*, 147 (book v, viii). Compare *Vita Barbati*, 558.

[3] Paulus diaconus, *Historia Langobardorum*, 148 (book v, viii); *History of the Langobards*, 219–20.

[4] Paulus diaconus, *Historia Langobardorum*, 171 (book vi, xx); *History of the Langobards*, 265.

calls on the empire's resources to pay for Belisarius' conquests, and the defences of those conquests, were enormous. To consolidate their hold on what they had gained required a vast building programme. As evidence for this we have, again, the writings of Procopius, and can test his statements against the surviving buildings that have been studied. Between them, it is possible to deduce the specifications for fortifications thought necessary by the Byzantines in the sixth century and beyond. As a general rule, it will be seen that these fortifications were smaller, and built to more modest dimensions, than their predecessors. A number of reasons for this can be advanced. Resources were probably lacking to construct anything more ambitious, and in the case of towns, it looks like the populations they needed to shelter were smaller. At the same time, in some parts of the empire, the threat came from barbarians who lacked sophisticated siege techniques. Against such enemies, the kind of design called for by Vegetius was deemed unnecessary.

The walls of Constantinople

Naturally, there were exceptions. Constantinople stood above all other cities in the Roman world, not least in the strength and magnificence of its triple, concentric land walls with their numerous projecting towers and ditches. These had been built by the emperor Theodosius (405–50). The inner rampart was 12 metres high and five thick, the towers rising a further five metres above that. An open space of 13.5 metres stood between this and the outer wall, four metres high and two thick, of arcaded construction. A further open space, 15 metres wide, led to the ditch, six to eight metres deep and 19 wide, with low walls on each side. The walls were built of regular limestone blocks. The towers had tall, vaulted lower floors supporting platforms from which the defence could be carried out.[1] The immense resources put into their construction were to be put to the test repeatedly, and repaid their builder's care, as will be seen. But the other defences inherited by Justinian were not so strong.[2]

The problem with using Procopius as evidence is that he may well have exaggerated the weaknesses of the pre-existing fortifications in order to enhance the reputation of his emperor. For example, in the east, he says, Justinian fortified the town of Daras (now Dara), which had previously been a village, whose low walls built without mortar would have fallen easy prey to the Persians, who would raise mounds and assault Daras with "all manner of engines". The emperor therefore heightened the walls, had arrow slits inserted into the battlements, and raised the towers to three levels. He dug a moat and built outworks on the weakest side, where the soil was "suitable for mining".[3] However, another source says that

[1] See Krischen, *Die Landmauer von Konstantinopel*, I, and Meyer-Plath and Schneider, *Die Landmauer von Konstantinopel*, II, for a detailed survey of the whole work and of the later additions.

[2] For a summary survey, see Lawrence, "A skeletal history of Byzantine fortification", 180–81; Ravegnani, *Castelli e città fortificate*, 31–2.

[3] Procopius, *Buildings*, 376 (book VI, chap. XXXVII). The term for engines here is *elepolei*, which in other contexts suggests a siege tower.

the town had been founded by the emperor Anastasius in 507, "surrounded with strong fortifications".[1] At Sucidava (Corabia) on the north bank of the Danube, the work of this period was limited to the addition of a circular angle tower built entirely from reused materials to the fortress first constructed by Constantine.[2]

However, there is a consistency to Procopius that must at least contain some truth. On the same Mesopotamian frontier as Daras, Constantina (formerly Antioch-in-Mesopotamia, now Viranşehir) had a wall so low "that it could be scaled with a ladder", and the towers were so far apart that they could not cover the wall in between. The wall was held together with mud and was only one metre thick. The towers ordered by the emperor were again of three storeys and each one represented a *castellum*, an interesting choice of term to describe it. At Martyropolis (Silvan or Miyafarkin between the river Tigris and Lake Van), so weak that it had surrendered to the Persians in 502 without even an attempt at defence, Procopius recorded that Justinian ordered a second stone wall built a metre in front of the existing one, and the gap between the two filled in with rubble to create a defence now three metres thick, and then doubled the three-metre height.[3] These, then, were substantial works, and the reason they needed to be was that the Persians were formidable enemies. It is hard to check the accuracy of Procopius against present remains: carbon dating of a "Justinianic" fort near Thermopylae in Greece gave a much earlier date of around 450.[4] That is not to say that the works against the Persians were not carried out at the emperor's order, but at Palmyra in the Syrian desert, an important city on a junction of trade routes on the frontier, he tells us that the emperor strengthened the defences. Today, what remain are the lower courses of a typical late-Roman wall originally ordered by Diocletian, now under restoration.[5] However, at Caesarea in Palestine, excavation located what seems to have been a fortress significantly strengthened at the time of Justinian, with strong projecting semicircular towers added (and surviving to a height of five metres), with an external diameter of around ten metres, built as part of the conversion of the theatre into a fortress attached to the town wall.[6] Inscriptions confirm that Justinian was also responsible for the work done to reinforce the third-century defences of Miletus in 538, where, speed being of the essence, much of the building material was plundered from the baths and temples of the classical

[1] Evagrius scholasticus, *Ecclesiastical history*, 376 (book III, chap. xxxvii). This version is supported by Zacharias of Mitylene, who says Daras was to serve as a base and arsenal against the Arab tribes, and the emperor sent engineers to lay it out, and stonecutters and masons, and ordered the bishop to see to its construction, which took two or three years: Zacharias, *The Syriac chronicle*, 166 (book VII, chap. vi).

[2] Tudor, "Sucidava", *Dacia* 5–6 (1938), 402; 7–8 (1941), 363.

[3] Daras: Procopius, *Buildings*, 98–109 (book II, i); Constantina: ibid., 133–5 (book II, v); Martyropolis: ibid., 189–91 (book III, ii).

[4] Mentioned ibid., 2–15, 23 (book IV, ii). See Cherf, "A Justinian fortification near Thermopylae"; Cherf, "Carbon 14 and Prokopios' *De aedificiis*".

[5] Procopius, *Buildings*, 177 (book II, xi). See Burns, *Monuments of Syria*, 162–3.

[6] Wiemken and Holum, "The joint expedition to Caesarea Maritima", 12–40; Ringel, *Césarée de Palestine*, 50–51.

city. The gates were converted into citadels, and a castle was constructed directly on the baths, where the walls are three metres thick.[1]

In North Africa, the local neighbouring population (the Moors) was less likely to engage in full-scale siege warfare, and the Byzantine defences accordingly were less strong. Pringle has published a survey of those that remain, while Gsell has studied those within present-day Algeria.[2] Procopius' account is rather sketchy, by comparison with the detail he offers of other building works carried out at Justinian's order. However, he does point out that many walls had been pulled down by the previous regime (the Vandals) and that the emperor's intent was to restore many cities to full defensive capability. At Leptis Magna (Lebida), the city was now apparently largely empty, and the desert had encroached. The emperor ordered that the "the circuit wall (be) built up from the foundations", but covering a much smaller area than before.[3] Carthage was provided with a whole new curtain wall, the old one having "fallen down", and a moat was dug that had not existed previously, the existence of which has been confirmed by excavation. On the north side of the city, where the ditch was shallower, were found remains of two projecting rectangular towers sited close enough to support each other. Within the city, he ordered the construction of a monastery, which was itself walled, making it "an impregnable fortress."[4] This is not the only reference to Byzantine fortifications of monasteries during this period — the famous Sinai Desert monastery of St Catherine's was walled at this time. On the frontiers with the Moors, fortresses (*thouria* in Greek) were built in large numbers, in each of which were stationed garrisons of soldiers. Procopius records a long list of places that received similar defences and garrisons, including Septum (Ceuta), whose Roman walls had apparently been neglected by the Vandals.[5]

A number of interesting questions are raised by this account. How strong were the defences? What was the strategic purpose, if any, of the large number of fortifications? Where did the garrisons come from? Who continued the task of maintenance? An attempt has been made to answer these.[6] Town walls have been located at 23 of the 37 identified cities. The garrisons were made up of regular imperial frontier soldiers, but it was not possible to work out how large these garrisons were. It appears that the walls were put up to protect the wealthiest areas, and to provide a base for the garrisons. They did not represent a defensive barrier, but rather were located where existing settlements were,[7] although the remains confirm Procopius' indication that the sixth-century sites were smaller than their predecessors. It could be, however, that the walled area represented a response to the military needs of the government rather than enclosing the whole popula-

1 Gerkan, *Die Stadtmauern*, 114–17.

2 Pringle, *The defence of Byzantine Africa*; Gsell, *Les monuments antiques de l'Algérie*.

3 Procopius, *Buildings*, 375 (book VI).

4 Procopius, *Buildings*, 381 (book VI). On the excavations, see Coyne, "The fortifications of Byzantine Carthage", 52–3.

5 Procopius, *Buildings*, 389–91 (book VI).

6 Pringle, *The defence of Byzantine Africa*.

7 Ibid., 83, 89, 98.

tion. However, the maintenance of the defences was the responsibility of the city itself. Many other sites, moreover, were simple garrison forts, as little as 40- or 50-metre-sided rectangles with four corner towers.

Since the Byzantine fortifications were built from new, and presumably by engineers brought by Belisarius' army from other parts of the empire, it might be expected that the architectural style would be similar, although obviously influenced by the availability of local building materials. It is also argued that the Moors' lack of siegecraft accounts for the absence of some of the features noticed in Byzantine defences against, for example, the Persians.[1] The walls of the forts are between 1.4 and 2.6 metres, rather less wide than those in the east, but the core was made of concrete, which was stronger than rubble so allowed the wall to be both taller and thinner without being weaker. However, thinner walls would have surely created problems in deploying artillery, which may therefore have been restricted to the towers. The only surviving sixth-century tower is quite small. The common design is a rectangular plan and usually rectangular towers. Walls were up to ten metres high, with towers rising one story higher. No trace has been found of ditches (apart from that specifically mentioned at Carthage). In several sites, there were inner redoubts of small dimensions.[2] Five different designs of gateways have been identified, with portcullises in the larger ones.[3] It has been suggested that an innovation in Byzantine technique here was the use of bent gate passages, adding to the obstacles facing an enemy who had gained entry.[4]

All in all, this survey of Byzantine works in Africa strongly suggests that, leaving aside resource problems on which we have no information, the scale of the work was tailored fairly carefully to cope with the likely threat from the Moors, and in particular in recognition that these foes did not have siege artillery and were not conversant with means of breaking through a wall such as movable battering rams.

Finally in this survey of Byzantine fortresses of the fifth and sixth centuries, the vulnerable Danube frontier calls for some examination. Having been the beneficiary of a great deal of attention, as they faced repeated attacks from the east and north, the Balkans and the provinces south of the great river also received significant new building under Justinian, confirmed not just by Procopius but also by inscriptions and excavations. Numerous town defences were improved, those of several cities such as Serdica (Sofia) receiving an outer wall (*proteichisma*) or thicker or higher walls, confirmed at Serdica by an inscription. At the same time, in line with work being done elsewhere, the walled areas were often reduced, gateways narrowed, and towers had their interiors filled in to turn them into solid bastions.[5] Similarly in Dalmatia, Salona (near Split) had triangular towers added to the Theodosian defences in the sixth century, presenting a dense array of towers

[1] Ibid., 132.

[2] For example, at Tipasa, Kenchir Guessès (now Hennchir Gasses) and 'Aïn el Bordj (Tigiris): Gsell, *Les monuments antiques de l'Algérie*, II: 359–63.

[3] Pringle, *The defence of Byzantine Africa*, 155–61.

[4] Foss and Winfield, *Byzantine fortifications*, 7.

[5] Biernacka-Lubańska, *The Roman and early-Byzantine fortifications*, 218–20.

on the landward side, where the walls were around four metres thick and included much reused spolia.[1] Further south, excavations at the site of the major classical city of Butrint suggested a new circuit of walls around the lower city had been added in the fifth century. They were up to two metres thick, so a very substantial defence. They might also, however, have been of medieval date.[2]

Alongside traditional methods of building fortifications, the Byzantine empire inherited the long Roman tradition of siegecraft. They did not remain tied to old texts, however, and possessed a long line of writers producing manuals taking account of new realities, such as the change in the nature and composition of the armies, and the changing range of enemies. At the end of the sixth century, the emperor Maurice put his name to a new manual, the *Strategikon*[3] composed at some time between 573 and 628. The treatise devoted a whole section (book x) to siege warfare. Many of the precepts were traditional, but there was considerable stress on siege weapons, both for attack and defence. Walls should be protected by breastworks, sacks filled with straw and sand to protect against stones. Rams and other wall-breakers were to be resisted with sacks of stones. Siege towers could be resisted with fire- and stone-throwing catapults. The stone thrower (*petrobolos*) was clearly distinguished from the *ballista*, which could be used as a weapon in the field as well. The machines were operated by specialists.[4] The context does not make it possible to decide whether the artillery was still the *onager* or whether the new, more powerful and simpler weapons which arrived at this time were adopted by the Byzantines straight away. Since the army would have a long tradition of using the former, it might be judged that this was more likely to be in regular use.

Persian wars: Amida to Jerusalem, 502–614

From the third century the Roman empire had its wars with the Persians. Both were sophisticated practitioners of siege warfare and fortification building. "There have been and now are great numbers of engineers in both countries."[5] The Persians' repertoire was on display when they attacked Amida (modern Diyarbakır), a frequent target in the frontier zone. In 502/3, they built a mound against the wall, the defenders responding first by trying to heighten the defence opposite, then by digging a hole in their own wall to get at the base of the mound, into which they tunnelled. When the Persians brought a battering ram up their mound to attack the new wall, which being newly erected crumbled quickly, the defenders collapsed the whole structure from the inside. While besieging the town, raiders were sent against Edessa (Şanlıurfa), where the population hurried to dig ditches and repair the walls, and the old decayed gates had to be removed and blocked with stone; they were unable to replace the gates because they did not have enough

1 Wilkes, *Dalmatia*, 359–62, and fig. 16 (plan).

2 R Andrews et al., "The late antique and medieval fortifications of Butrint", 127–31.

3 Maurice, *Das Strategikon*. See also Aussaresses, *L'armée byzantine à la fin du VIe siècle*.

4 Maurice, *Das Strategikon*, 337 (book x, chap. i), 343–5 (book x, iii).

5 Procopius, *History of the wars*, v: 157 (book VIII, xi).

iron. Cutting down all the foliage outside, and destroying all the buildings that could shelter the Persians, the Edessans successfully deterred a siege.

Back at Amida, the Persians had rebuilt their mound, using stones, timber, and sacks of hair and linen in order to speed up the work. The defence then turned to a giant stone thrower able to hurl blocks weighing 135 kg, which killed many attackers and smashed their battering ram. (The reference in this account must be to the *onager*.) The Persians were considering abandonment when, one night, the guards got drunk and Persian soldiers climbed in by ladder and captured the town. The emperor's forces attacked Amida the following year, using three siege towers. Collecting artisans from the neighbouring towns and villages, the commander set about digging a mine. However, on its first firing, only the outer portion of the wall collapsed. The canny Persian defenders managed to fill their ditch with water, so that they could identify future mining by the trickle away of the water. They held on until, with cannibalism luridly reported in the town, a truce was arranged between the two empires.[1]

The unfortunate city was attacked again in 512 "with fierce assaults of sharp arrows and with battering rams which thrust the wall to overthrow it, and penthouses [*testudines*] which protected those who brought together the materials for the besiegers' mound and raised it up and made it equal to the height of the wall". Their attack lasted three months, but Amida held out. This was because winter weather caused the bow strings to "relax", bundles of rushes hung down from the walls cushioned them against the rams, and the defenders mined under the mound and brought it down. In the end, the Persians got in when treacherously admitted through an aqueduct, and carried out a massacre.[2] In 531, they attacked Martyropolis, which was a base for the Byzantines to threaten Persian territory, and tried to mine it, but were thwarted by the arrival of winter.[3] The Byzantine siege and ultimate capture of Petra in 551 (on the Black Sea frontier near Batumi, Georgia) revealed their skills in wall building and undermining, and in this case the construction by allied barbarians of a portable ram, as the slope did not allow deployment of the traditional kind. The defenders used prodigious quantities of naphtha to try to burn the rams,[4] but the besiegers overcame resistance by breaking through when another portion of the wall was undermined, and a change of wind direction caused the flaming weapons to burn down the Persian structure from which they were being thrown.

As the sixth century turned into the seventh, the two empires found themselves engaged in a massive, long-lasting, and bloody struggle, which was to have disastrous effects in Asia Minor and Byzantium's middle-eastern provinces, was to briefly threaten Constantinople itself, and as a longer term consequence was to leave both empires seriously weakened in the face of a threat of which they were still unaware: Islam.

[1] All from Joshua the Stylite, *The Chronicle*, ed. and trans. Wright, 37–65. The writer was from a monastery near Edessa. See also the new translation by Trombley and Watt.

[2] Zacharias, *The Syriac chronicle*, 153–4 (book VII, iv).

[3] Ibid., 228 (book IX, vi).

[4] Procopius, *History of the wars*, V: 157–71 (book VIII, xi).

In 587, the emperor invaded the Persian domain, and "arrayed his siege engines and machines" against a stronghold on a rock, in Arzanene on the frontier (approximately Batman province, Turkey). The Persians hung robes over the wall to "mitigate the hardness of the discharges" but to no avail, and the place was captured.[1] Two of his commanders tried to take by surprise Matzaras (Maserti) but the defenders responded "now with stones, now with catapults". The wall however was of dry-stone construction, and the Romans were able to mount it by driving in spikes. The defenders pushed the parapet of the wall, "weakened by the Roman bombardment", onto the climbers. After three tries the place was captured.[2] In 573/4, the Persians had attacked Daras, whose rebuilding by the Byzantines Procopius had described, and surrounded the town with "mounds and ramparts". They diverted the water supply, constructed siege towers, and brought up siege engines,[3] including catapults that were mounted on the mound (or two mounds, according to one account), which was higher than the town walls. They finally took it by storm after a siege of five months. The garrison withdrew into the citadel, which was attacked by the same method of preparing a mound and also using stone-throwing catapults.[4] In 604, they captured the place after a very long siege. Clearly, the protection Justinian's engineers had devised against mining more than half a century earlier was not sufficient, as the source, an Armenian bishop, says that the breach was made by mining.[5] It is clear from these accounts that a full range of siege weaponry was deployed by both sides on Byzantium's eastern frontier. The references to a wall being weakened by a bombardment is strange. There is no evidence that the *onager* was capable of damaging a properly constructed wall, and the *ballista* was definitely an anti-personnel weapon. The clues perhaps lie in the statement that the wall was of dry-stone construction. Perhaps, although there is no evidence, it was also quite thin, although it still seems peculiar that the defenders could, or would, actually push over the very parapet designed to protect them from their opponents' missiles. However, there is another reference to a wall crumbling under bombardment in the account of the Byzantine capture of Ocras, a fort opposite Martyropolis, from the Persians in 591.[6]

Later, the Persians themselves were to take the offensive and to drive into the middle east and Asia Minor. In 614 they captured Jerusalem, and the horror produced by their desecration of the Christian holy places ensured that there are a number of contemporary accounts of the event. First the Persian army, having defeated the Byzantines, swept through Syria, easily taking many places (perhaps by offering the kind of safe conduct offered to Caesarea). Arriving before Jerusalem, they tried first of all to take it by direct attack, but were thwarted by

1 Simocatta, *The History*, 68–9 (book I, chap. xviii).

2 Ibid., 69–71 (book I, xviii).

3 Ibid., 1–2 (book I, xi).

4 Evagrius, *Ecclesiastical history*, 435 (book V, x). The further detail comes from the Syrian *Chronicle of* AD *1234*, trans. Chabot, 160–61 (chap. lxvi), who talks of the "balistas quae magnos lapides proiciebant".

5 Sebeos, *Histoire d'Héraclius*, 57 (chap. xxi).

6 Evagrius, *Ecclesiastical history*, 460 (book VI, xv).

a vigorous defence (with "hails of stones") after the inhabitants had overruled the patriarch, who wanted to accept a Persian offer. They then resorted to siege weapons, including *mangana*, and after twenty days the walls were broken and the city was stormed and sacked.[1] The sources agree that the walls were thrown down, and the implication is either that the Persians had acquired unprecedently powerful stone-throwing artillery, or else the artillery was part of a sustained attack involving also other forms of attack. The reference to "mounds" and towers suggests that the besiegers were not relying on a single method of attack, and the very abbreviated and versified form of these accounts, along with the (slight) delay in their composition after the events, must leave unresolved the precise nature of the artillery used by the Persians. Had they already adopted the new artillery first seen in the west at Thessaloniki twenty or thirty years before? Bishop Sebeos, writing in Armenia a generation later but clearly using a different (and now lost) source for this story, reports that the Persians had sapped the foundations and demolished the walls after nineteen days. The detail given is consistent with our other evidence for Persian siegecraft,[2] suggesting a particular expertise in the complex skills of sapping and mining.

Siege of Constantinople, 626

A dozen years later, the war was brought to the walls of Constantinople itself. In 626, the great city was subjected to the first of a number of serious attacks that were to threaten it, always unsuccessfully, until 1204. On this occasion, the Persians had crossed Asia Minor and were established on the southern shores of the Bosphorous. The really serious attack, however, was to be launched from the European side, by their allies the Avars. Again, there is an unusually rich survival of contemporary accounts, which allows for a fairly detailed reconstruction of the events of that momentous year.

The Avars were relatively recent arrivals from the east, who had established themselves in central Europe at the head of a large empire, in which were to be found also many Slavonic tribes. Certainly they had a fairly high technical level: among other things, they have been credited with the introduction of the stirrup.[3] The *Strategikon* had noted them among new opponents whose ways had to be learnt. Until their elimination by Charlemagne, they were a significant force in the

[1] Sophronius, *Vers sur la prise de Jérusalem*, 141–2 (an elegy composed around 620–28, in Greek); Strategios of Mar Saba, *Prise de Jérusalem par les Perses*, 154–5 (French trans. of Arabic version of an originally Greek text), *La prise de Jérusalem par les Perses*, II: 13–14 (Latin trans. of Georgian version of the same text), "Antiochus Strategos' account of the sack of Jerusalem", 506–7 (English trans. of Georgian version). The author of the latter text is identified only as "a holy monk" of Mar Saba in the Arabic version; the Georgian version names the author as "Strategios" in the text and as "Antiochus Strategios" in the superscription, but the identification of Strategios with Antiochus of Mar Saba is uncertain; see *Clavis patrum graecorum*, nos. 7842–3.

[2] Sebeos, *Histoire d'Héraclius*, 68 (chap. xxiv).

[3] Szádeczky-Kardoss, "Der awarisch-türkische Einfluss auf der byzantinische Kriegskunst", 64–70.

region. In 626, a vast Avar host (at least according to the Byzantine contemporary account) advanced on Constantinople. Fortunately for the Roman empire, the Persians played a rather wait-and-see game on the Asian side; had they mobilised ships, the city would have been in trouble. As it was, the emperor Heraclius (r. 610–641) left a strong garrison (12,000 men), provisions, a fleet, and detailed instructions about strengthening the defences. This involved strengthening the bases of the walls (presumably against mining or sapping) and erecting projecting barriers in front, which would have the effect of obstructing rams or siege towers. A large supply of artillery weapons was also called for, both dart and stone throwers.[1] He himself remained outside the city with an army on the Asian side, and would later carry out a victorious campaign to expel the Persians from Byzantine territory.

Two detailed contemporary accounts survive of the Avar attack and its ultimate defeat. Arriving on 29 July after spending a month assembling their forces at Adrianople, two days later they drew up their force along the whole 5.7-kilometer length of the outermost wall, but concentrated in the middle sections. The *Chronicon Pascale* says that initially there were a few siege engines and mantlets, but the next day they "stationed a multitude of siege engines close to each other ... so that those in the city were obliged to station very many siege engines within the wall". The Avars "bound together [the] stone throwers and covered them outside with hides".[2] The whole third day was spent in a duel of artillery and archery. Our other contemporary source then reports that on the fourth day, they arrayed siege towers (*helepolei*) made of wood, which had been prefabricated and brought to the siege on carts. A request to surrender having been turned down, the assault then began in earnest. It lasted until the tenth day, during all which time the attackers used their full array of weapons to press an attack, with a "multitude of projectiles" hurled by the catapults. Crucially, on this day, the Byzantine fleet won a victory over the boats prepared by the Avars, thus preventing any risk of their linking up with the Persians. Finally, the defenders mounted a sortie and succeeded in burning the Avar towers, mobile wooden shelters, and catapults.[3] The *Chronicon Pascale* suggests a slightly different ending: the Avars finally gave up their attempt, and themselves burnt their siege equipment before withdrawing. While on the one hand it is true that if the Avars had brought their equipment very close to the walls, it would have been vulnerable to a sortie, on the other hand if it really did stretch most of the length of the city's land defences, it is difficult to see the garrison being strong enough to risk a sortie on such a wide front. However, what is significant here is that despite the size of the Avar army and its possession of what must have been an immense siege train (despite the probability of the sources exaggerating it in order to greater glorify the defence), organised and prepared

[1] The instructions are given in Georgius Pisida, *Bellum Avaricum*, ll. 266–79; Italian trans., *Panegirici epici*, 270–71.

[2] *Chronicon paschale*, 173–9.

[3] Theodorus syncellus, *L'homélie sur le siège de Constantinople en 626*, pp. xix–xxxv, 20–33. There is an excellent account of the whole siege in Howard-Johnston, "The siege of Constantinople in 626".

even to the extent of prefabricating the towers, they were able to make no dent in Constantinople's defences, never it seems even getting close enough to try to create a breach. Nor was their artillery powerful enough to do the job.

Thessaloniki — a new weapon?

The Avar artillery was not of the old *onager* type with its limited size and impact, suited for anti-personnel use but unlikely to harm any effective fortifications. The evidence is that, instead, they introduced to the west a new design of weapon, although not of their own creation, which would remain the standard design until the twelfth century when it would be superseded by the counterweight trebuchet. The habit has become established of calling this new weapon a traction trebuchet, although this word was unknown to contemporaries. It combined the features of being easier to build and maintain with the potential for propelling much larger projectiles. We have a rare privilege for this period, an actual description of the weapon that leaves no doubt as to what it was. However, nothing can ever be this straightforward, and the source is replete with problems of its own. The collection of homilies known as the *Miracles of Saint Demetrius* was not composed with the aim of being a history, let alone a technical treatise. They were written to demonstrate the miraculous interventions of Thessaloniki's native saint in protecting the city from harm. Their (many) editors suggest that they are a jumble of real historical events (the chronology being checked against other sources) and Demetrius' miraculous doings. Likely to have been penned by Bishop John in around 620, they cover two failed sieges of the city, in 586 or 597 and again in 617/8.[1] The gap between the two events must raise questions about the accuracy of the record of the earlier, and the possibility of anachronisms creeping in. But the bishop claims to be an eyewitness to both, and from the perspective of studying the siege techniques, it is not critical whether descriptions of one siege are jumbled with the other.

Thessaloniki was an attractive target for invaders seeking loot, the second city of the empire but one whose situation was perilously exposed to attackers who had already established themselves to the north. It was therefore also well defended. It is suggested that the walls were erected in 380, but it has also been argued that this happened in response to the Huns in the 440s, when the seating blocks from the hippodrome were cannibalised for the foundations. The walls stretched for eight kilometres and were of strong construction, now varying in thickness from 2.2 to 4.6 metres (as a result of later strengthening), and with some forty square or rectangular towers, although continuous use has obscured exactly what was there when the Avars approached (Plate 1).[2] It would seem also that there was, at

[1] We rely heavily on *Les plus anciens recueils des Miracles de Saint Démétrius*. See also Barišić, "Les Miracles de St. Démétrius comme sources historiques". On the date of the first attack, see Vyronis, "The evolution of Slavic society", 381–2.

[2] Evans, "The walls of Thessalonika"; Tafrali, *Topographie de Thessalonique*, 59–112; Hetherington, *Byzantine and medieval Greece*, 202. There is a plan of the fortifications in Eustathius, *The capture of Thessaloniki*, p. xxiii.

least in part, a lower outer wall about twelve meters in front. Given these defences, no attacker could expect an easy picking, and would need to come prepared. The *Miracles* allow a reconstruction of what happened. The Avars and Slavs arrived without warning, and attempted an assault by escalade, using ladders brought with them. The Saint allegedly materialised to come to the city's rescue, hurling down the ladder, and the enemy withdrew. The Avars then ravaged the country but found it unable to provide the means of sustaining them. They then attacked using "*helepolei*, iron rams, stone throwers and 'tortoises' [movable sheds with protected roofs] covered with cow and goat skins". Under cover of a bombardment of great stones they got close to the outer wall to attack the foundations with picks and levers. They attacked one of the gates with a ram, but took fright and fled, burning their own machine. The defenders then sallied, apparently from the inner toward the outer wall, and, protected by the wall, sapped under the "tortoises".[1]

The next day they appeared with stone throwers. This weapon apparently deserved a detailed description, because unlike the others used on the previous day, the writer must have assumed that his readers would be unfamiliar with it. The catapult, the account of which has been translated from the Greek several times, was quadrangular, with a wide base but narrowing towards the top, using large iron rollers to which were fixed timber beams "similar to the beams of big houses", having at the back a sling, and at the front thick cables, enabling the arm to be raised and lowered, and which threw "enormous blocks into the air with a terrifying noise".[2] Despite the shock no doubt induced by such a new machine, the defenders succeeded in destroying it with a flaming arrow, perhaps from a *ballista*. Undeterred, the attackers returned the next day with more engines, this time protected with fresh skin coverings. The defenders protected their battlements by hanging absorbent material in front of them. The bombardment lasted for hours. However, although the Avars may have known how to construct these machines, their calculations were not so good when it came to operating them: hardly any of the stones actually reached the wall, whereas the defenders' stones (shot from what we are not told) caused casualties among those serving the stone throwers. Only one Avar stone hit, and this apparently demolished the top of the rampart down to the walkway. Unsuccessful therefore despite their new technology, the Avars gave up and withdrew after seven days.[3]

Returning in 617 or 618 for another try, the Avars again brought to bear a wide range of equipment, and refugees from their advance warned the citizens that "a single stone from their engines would ruin the wall". Their stone throwers were higher than the interior wall, they had wheeled rams, rolling ladders, high wooden towers, and engines throwing fire. Once again, Thessaloniki called upon miracles wrought by its saint to ward off the threat, although the account also acknowledges the help of more mundane defences. The siege tower collapsed when moved, and the tortoises were capsized with hooks let down from the wall. Sailors

1 *Les plus anciens recueils des Miracles de Saint Démétrius*, 1: 117, 122, 139–40, 146.

2 Ibid., 143. Also translated in DeVries, *Medieval military technology*, 134. The subject is also treated exhaustively in various publications by Chevedden (see bibliography).

3 *Les plus anciens recueils des Miracles de Saint Démétrius*, 1: 151–9.

who were familiar with the operation of engines arrived and were deployed on the town's own stone throwers. After 33 days the Avars again abandoned their efforts and negotiated a withdrawal.[1]

There are some oddities in these accounts that are hard to explain. It sounds as if the first siege was more of a raid, given that the attackers themselves had little food. If so, would they really have carried a siege train with them? There scarcely seems time in the story for them to have built the machines on the spot. Siege towers in particular cannot be constructed overnight, as we know from many later accounts that tell of many days spent in building them. There might therefore be reason for doubting much of the detail, if it were not for the description of the stone thrower. It is also notable how inept the attackers were despite all this equipment. As with other barbarian peoples, the ineptness is not a surprise. The Avar incursions into the Balkans were accompanied by the Slavs during this century and after. Procopius had already observed during a Slavonic invasion during Justinian's reign that "they captured many fortresses by siege, though they neither had any previous experience in attacking city walls, nor had they dared come down to the open plain".[2] It is widely accepted that the attacks on Thessaloniki were by both Avars and Slavs. If there is any truth in the accounts given in the *Miracles*, the attackers had encountered some significant siege warfare techniques, and knew how to build the machines. Although it was much more powerful than any other artillery known, they did not, however, know how to use it to any great effect.

Rather more effective use of powerful stone-throwing artillery was certainly made by Heraclius' forces in 627/8 when, soon after the successful resistance of Constantinople, the Byzantine forces, having enlisted the help of Turkic allies, were operating in the Caucasus, where the Albanian ruler of Iberia was an ally of the Persians. The invading forces laid siege to Tbilisi, apparently strongly walled, and its resolve reinforced by the arrival of one thousand experienced Persian troops. A chronicler records how the Roman engineers prepared a whole array of machines, by which "they threw stones accurately in order to demolish the circuit wall by means of very large stones". These were clearly skilled and experienced people, since they also organised to use skins full of stones and sand to divert the course of the nearby river to beat against and weaken the curtain. These were obviously effective methods, as the wall began to crumble. However, the defence was ready, and had thrown up a second line of defence behind it. Dispirited by their failure to achieve a breach, the attackers gave up, but the citizens made what proved to be a fatal error by mocking the Turkic king from their battlements. Probably a year or two later, the wrathful Turks returned. The Persian troops had gone, and the besiegers surrounded Tbilisi for two months before storming the walls, now obviously less able to resist, and perpetrated a massacre. Evidently, albeit from only a single account, the Byzantine artillery in the first attack here was

[1] Ibid., II: 198, 202, 213.

[2] Procopius, *History of the wars*, V: 23 (book VII, xxxviii).

of the wall-breaking variety.[1] Certainly not conclusive evidence, therefore, that Heraclius could call upon engineers able to construct and operate the traction-propelled stone thrower, but helpful in consolidating such a case.

Origin of the new artillery

Since the Avars and Slavs did not themselves invent the new form of stone thrower, and since when they first appeared in the Balkans in the mid sixth century they were as ignorant as any other nomadic people of siege-warfare techniques, where did it come from? A case has been made, based on the chronicler Theophylact Simocatta, that they were shown how to make the new stone thrower by a Byzantine soldier named Busas, captured during the Avaro-Slav attack on a fort in 587, in exchange for his life after his former comrades had refused to ransom him. Theophylact says that the weapon was for a "long range" assault but gives no other clue to its identity.[2] Using this new technology, their subsequent attacks were sweepingly successful (for example, capturing "Bonkeis" (unidentified) and forty other forts in Dalmatia in 595, two years before they brought the weaponry to bear at Thessaloniki).[3] It would not be the first time, or the last, that advanced technologies were transmitted in such a way, although there are also suggestions that the Avars had previously tried to obtain access to Byzantine technicians and architects by more peaceful means.[4] But there is a problem with this version, to which there is no obvious answer. For the Byzantine soldier to be acquainted with a new and devastating form of weapon, the imperial army must itself have been in possession of the weapon, and presumably for a few years at least. But there is nothing in the sixth-century texts we have to suggest this. The same, unhelpful terms are used to describe artillery at the end of the century as they are in Procopius, including in Maurice's *Strategikon*. Furthermore, if Byzantine armies were using it from no later than the 580s, how come it was such a surprise to the bishop of Thessaloniki in 597? The source, Theophylact, was not contemporary. He was writing at least a generation later, during the reign of Heraclius, when the new stone-thrower was probably common. The tale makes a good story, but it is not verified anywhere else. It might also appear strange that the Avaro-Slav armies became so proficient (and so quickly) and devastatingly effective in using it in the 580s and 590s, but were so inept at Thessaloniki in 597. If we leave to one side the

[1] Movses Dasxuranc'i, *The history of the Caucasian Albanians*, 85–6, 94. The first part is also printed in *The Roman eastern frontier and the Persian wars*, part 2: 211–12.

[2] Simocatta, *The History*, 66 (book II, xvi). To add to the confusion, the story is an interpolation.

[3] Ibid., 195 (book VII, xii); Vryonis, "The evolution of Slavic society", 385–8. This account has been followed by several later studies, e.g. DeVries, *Medieval military technology*, 134.

[4] Joannes of Ephesus, *Historiae ecclesiasticae pars tertia*, 247 (book VI, chap. xxiv): a request to the emperor Justin to be sent "mechanicos et architectos" to build palaces and baths.

explanation of our author that put it all down to the superhuman intervention of a dead saint, what remains must leave a big question mark over this whole episode.

When we come to examine the wars shortly to take place as the Arabs made their dramatic impact on history, the weight of evidence will suggest that the new weapon became firmly established during the seventh century. Neither the precise date, then, nor the route of its arrival can be determined with certainty. What is certain is that the only place known to have developed this form of artillery was China. It is equally possible that the Avars (with their origins in Central Asia), the Byzantines, or the Persians could have been the first to learn of and make use of the weapon in the western world.

Siege weapons of ancient China

As in the Roman empire, cities were central to the empire of China, providing centres of administration and control as well as residences for the rich. They were walled. Much care was devoted to ensuring their safety, and also to methods required to capture them.[1] Whereas walls had been built of rammed earth in ancient times, baked brick was a common ingredient by the time of the Han dynasty (around 200 BC to AD 200), with various methods of bonding the layers together. The same methods were apparently applied to the defensive walls of cities by the fifth century AD. Between the outer brick or stone facings were rubble cores; towers, sloping towards the top, were set higher than the wall; and gateways were protected by double gate towers. The Great Wall itself had been begun in 350 BC, and there was a major building period around AD 540–60, after which nothing much was done until the age of the Ming, from which most of the present monument dates, so it is hard to determine what the original work looked like, except that it was a labour on a gigantic scale, 3000 km long, and succeeded in its purpose of deterring invasion until the Huns first penetrated it (in the third century).[2] Although most remaining sources apply to the periods of the Tang and then the Song dynasties, there is evidence of a long tradition of siege weapons dating back to before the Han period. It is also likely that China's neighbours would have acquired similar resources through their contact with the empire. Tibet, for example, whose rise to recognition as a great empire on its own account took place during the early seventh century, was in close contact with China, including periods of warfare in which the Tibetans both took Chinese fortresses (and also captured the defences of others of their neighbours), and erected their own, with untranslated Chinese dynastic histories providing much of the surviving contemporary information. It is not possible to prove, but it seems likely that to achieve these results, they would have ensured they had the current technology, as they

[1] Franke, "Siege and defense of towns in medieval China", for an overview (and details of the later medieval period); more recently, Heng, *Cities of aristocrats and bureaucrats*.

[2] Needham, *The shorter Science and civilisation in China*, v: 31–3, 40. For the immense works carried out in 607–8, and their vast costs in human lives, see Bingham, *The founding of the T'ang dynasty*, 19–20.

certainly had the administrative structures in place to run their empire and its powerful army.[1]

As in the west, the Chinese had the crossbow, although theirs was tension- not torsion-powered. For all that, they developed a large variety of crossbows of different sizes and power, the most substantial being major pieces of artillery requiring several operatives and strong frames to mount them on. Some were weapons designed to shoot a large number of missiles simultaneously. Their arrows could reach 300 paces. They were manufactured under the control of the government. These weapons formed a vital part of the armoury of the Chinese army for hundreds of years,[2] and it is easy to understand how, deployed in large numbers, they could either drive an enemy from the wall of a besieged town, or alternatively keep attackers at a safe distance.

Besides the crossbow, the Chinese had developed a stone-throwing catapult operating on the principle of a sling at the end of a long wooden arm pulled down by the action of operators at the other end. Accounts and descriptions of the varieties of machine in use become common from the Tang dynasty onwards, with the first illustrations that survive dating from the eleventh century. The first clear description was given in 759 by Li Ch'uan.[3] Accounts show that these weapons could be deployed in their hundreds, both for attack and defence. Only a state with large resources would be able to organise the production and operation of such arrays of artillery — resources that the Chinese empire, unlike any other on earth at the time, was able to find. References to their use in sieges of earlier times, and details like the name of an engineer from the fifth century, reinforce the conclusion that the shortage of surviving material does not mean that the weaponry was not available at a time when the Roman empire was still using torsion weapons.[4]

From these later accounts it is possible to deduce what this catapult, which had many names in Chinese, most of them deriving from the word *p'ao*, "throw", actually did. Their construction used bamboo for the throwing arm, and the power of the machine depended, in part, on how many of these were bound together, and in part on how many operators were used to pull down the arm. Information given in the *Wu ching tsung yao* of 1044 indicates both the range and the number of men needed to operate the weapons. These varied from the fifty men needed to operate a single-arm weapon (called "Whirlwind"), throwing a light projectile (four pounds or 1.8 kg) around 85 yards, to a seven-arm monster needing 250 men and hurling 125 pounds (56 kg) a similar distance.[5] It is apparent that the range of

[1] C I Beckwith, *The Tibetan empire in Central Asia*, 13–54, and the sources listed there.

[2] See Hong, *Weapons in ancient China*, 199, 234–5, for illustrations of the firing mechanisms of such weapons as have been discovered.

[3] Needham and Yates, *Science and civilisation in China*, v/6: 204; and the earlier study, Needham, "China's trebuchets". The illustrations have been produced many times, including in Peers, *Imperial Chinese armies*, II: 590–1260, 24–5.

[4] Needham and Yates, *Science and civilisation in China*, v/6: 211.

[5] See the table ibid., 216–17. There seems to be some uncertainty about the calculation of the range, the Chinese word *pu* being interpreted by Franke, "Siege and defense of towns in medieval China", 168, as giving a considerable shorter range.

all these weapons was limited, and would leave their crews extremely vulnerable to crossbows shooting from a besieged town. Indeed, the later sources devote considerable attention to the use of the catapults in defence. They argue that they should never be placed on the wall, but inside, using an observer to direct their aim. Their main targets are the machines of the enemy for the larger machines, and his men for the smaller. For the attacker, the main purpose was to hurl incendiary devices into the besieged town, where the prevalence of wooden buildings made this a serious worry for the defence. Detailed instructions are also given about the ammunition. Uniform size was essential for consistent aiming, and round shot best for the longest range. But it was also argued that it was better to make one's shot from baked clay than to use stones, which could be reused by the enemy.[1]

The evidence for Chinese siege techniques of the sixth century suggests that, apart from the difference in the size of the forces available to Chinese generals, the means of waging war were not so different from their western equivalents in Persia and Byzantium. In a civil war, a new fortress built by the eastern Wei ruler Yu-wen T'ai at Yu-Pei was the focus in 546 of blockade, then a siege of 50 or 60 days, by the western Wei emperor Kao Huan. A siege mound was thwarted by the defenders raising the height of their towers with timber, after the attackers had got close by diverting the adjacent river away. Kao Huan turned to mining, having ten mines dug against one stretch of the rampart, but the defenders dug their own tunnel parallel to the line of the wall, intercepted the attackers (possibly identifying them with the method recommended by ancient writers of using pots covered with hide to pick up vibrations) and burnt them out. Subsequent efforts included another mound, an attack with battering rams, thwarted by being absorbed by a screen of large curtains, and finally no fewer than 21 tunnels launched simultaneously at all four sides of the fortress. The defences, however, consisted of rammed earth, so mining only induced localised collapses, and the defenders held on from inside the palisades they had already erected inside the walls. The siege was finally abandoned with immense loss.[2] An account of a rather more successful siege of 548/9 (Ying-ch'uan) tells of the use of mounds, extending ladders, and incendiaries, and a kind of siege tower protected by breastworks.[3] There are accounts of catapults breaching walls (in 573), and in the same year one general's biography tells of how he resisted overwhelming numbers in a siege involving no fewer than 70 mines.[4]

In 617 no fewer than 300 machines were employed in the capture of the western Sui capital.[5] This already ancient city on the Yellow River, Ch'ang-an (now Xi'an), was immense, covering 84 square km. The city walls (nearly six metres high) were of yellow earth, with impressive extra structures and brick facings at

[1] Franke, "Siege and defense of towns in medieval China", 169.

[2] Wallacker, "Studies in medieval Chinese siegecraft: the siege of Yü-pi".

[3] Wallacker, "Studies in medieval Chinese siegecraft: the siege of Ying-ch'uan".

[4] "Lebensbeschreibungen von Heerführen und Würdenträgern des Hauses Sui", 335–6. The successful commander was Teu-lu-tsi.

[5] Needham and Yates, *Science and civilisation in China*, v/6: 211; Bingham, *The founding of the T'ang dynasty*, 104.

the gates, and an outer ditch four metres deep.[1] It fell despite its strength to the power of the besiegers and their arsenal, after a siege of five weeks.

This, then, was the new weapon that appeared suddenly in the west. Before the discovery of the description in the *Miracles of Saint Demetrius*, it was generally thought that the route was via the Turks of central Asia to the Arabs, during the seventh century. For example, large number of catapults accompanied an expedition westwards in 638.[2] But there had been plenty of contact before, often with the Huns before their westward move. An account of an expedition of 38 BC tells how the Chinese used their crossbows to ensure capture of a town in Sogdia.[3] It is therefore very likely that China's western neighbours would have encountered the empire's siege weapons. What we cannot now say is when the technology of the manually-operated catapult was adopted in the eastern Mediterranean, and how it got there. But get there it certainly did, and the story of its adoption in the west is of considerable significance, although the sources continue to pose as many questions as they answer.

But even though the Arabs may not have been the agents of the introduction of the new weapon, their conquests and the resistance of Christian states provide the next part of the history.

1 Heng, *Cities of aristocrats and bureaucrats*, 9–10; Schafer, "The last years of Ch'ang-an", 133–4.

2 Needham and Yates, *Science and civilisation in China*, v/6: 215.

3 Ban Gu, *History of the former Han dynasty*, II: 283.

PLATE 1. Thessaloniki (Greece). The early Byzantine curtain wall (frequently repaired) was similar to that of Constantinople, with a low outer wall 12 metres in front. The walls and towers were of great strength and can be seen here marching uphill toward the citadel. The fifteenth- or sixteenth-century Trigonion Tower is in the distance.

Plate 2. Nicaea (İznik, Turkey), with early Byzantine defences, was taken by the crusaders in 1097 and was capital of the rump Byzantine empire after the Fourth Crusade.

Plate 3. Antioch (Antakya, Turkey). The now much-ruined Byzantine citadel stands in isolation on a steep hill far above the city below.

PLATE 4. Echternach (Luxembourg). A late Roman *castellum* was reused in Merovingian times (sixth century). The church is younger, but was built on a much older structure.

PLATE 5. York (England). The small tower has been dated to the seventh century, an Anglo-Saxon addition to the Roman walls, which lie under the rampart on which the thirteenth-century walls now stand.

Plate 6. Pevensey (England). One of the Roman "Saxon Shore" forts, the powerful walls and towers continued in use as the outer ward of the medieval castle and sheltered William the Conqueror's troops in 1066.

PLATE 7. Attacks on cities during the Carolingian period. The illustration suggests unsophisticated techniques, not always the case according to the evidence. From the Golden Psalter of St. Gallen, Cod. Sang. 22, p. 141. © Stiftsbibliothek St. Gallen.

PLATE 8. St Hilarion Castle (Cyprus). The outer walls may date from the Byzantine reconquest after 965 and are similar to the tenth-century Byzantine walls of Kyrenia (Girne), but may also date to any time up to the twelfth century.

Plate 9. "Danes Castle" at Exeter (England) may have been a siege castle erected by King Stephen's army during the siege of 1138.

Plate 10. Clough Castle, Co. Down (Ireland), is a traditional motte and bailey. Archaeologists found a pit on top of the motte, believed to be for archers.

Plate ii. Château-sur-Epte (Normandy): a motte of the eleventh century and tower of the twelfth. The castles on the river Epte represented the nearest thing to a "defensive line" known to that age.

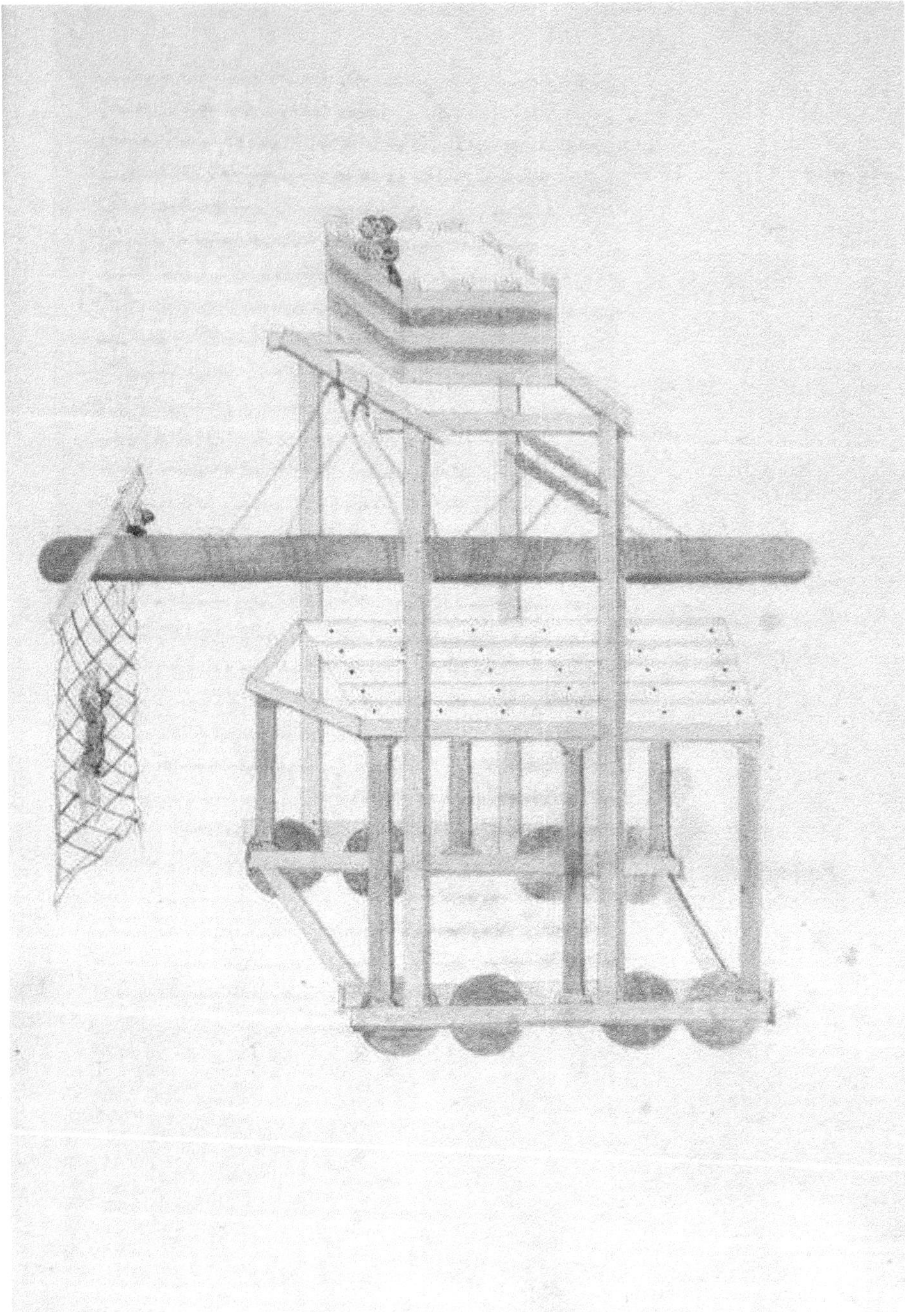

PLATE 12. A multi-function eight-wheeled siege tower and battering ram from an eleventh-century Byzantine version of the text on siege techniques by Hero (as copied in the sixteenth century). The illustration suggests a number of different operations that were, in practice, unlikely to take place within a single structure. BL Add. MS 15276, fol. 12[r].

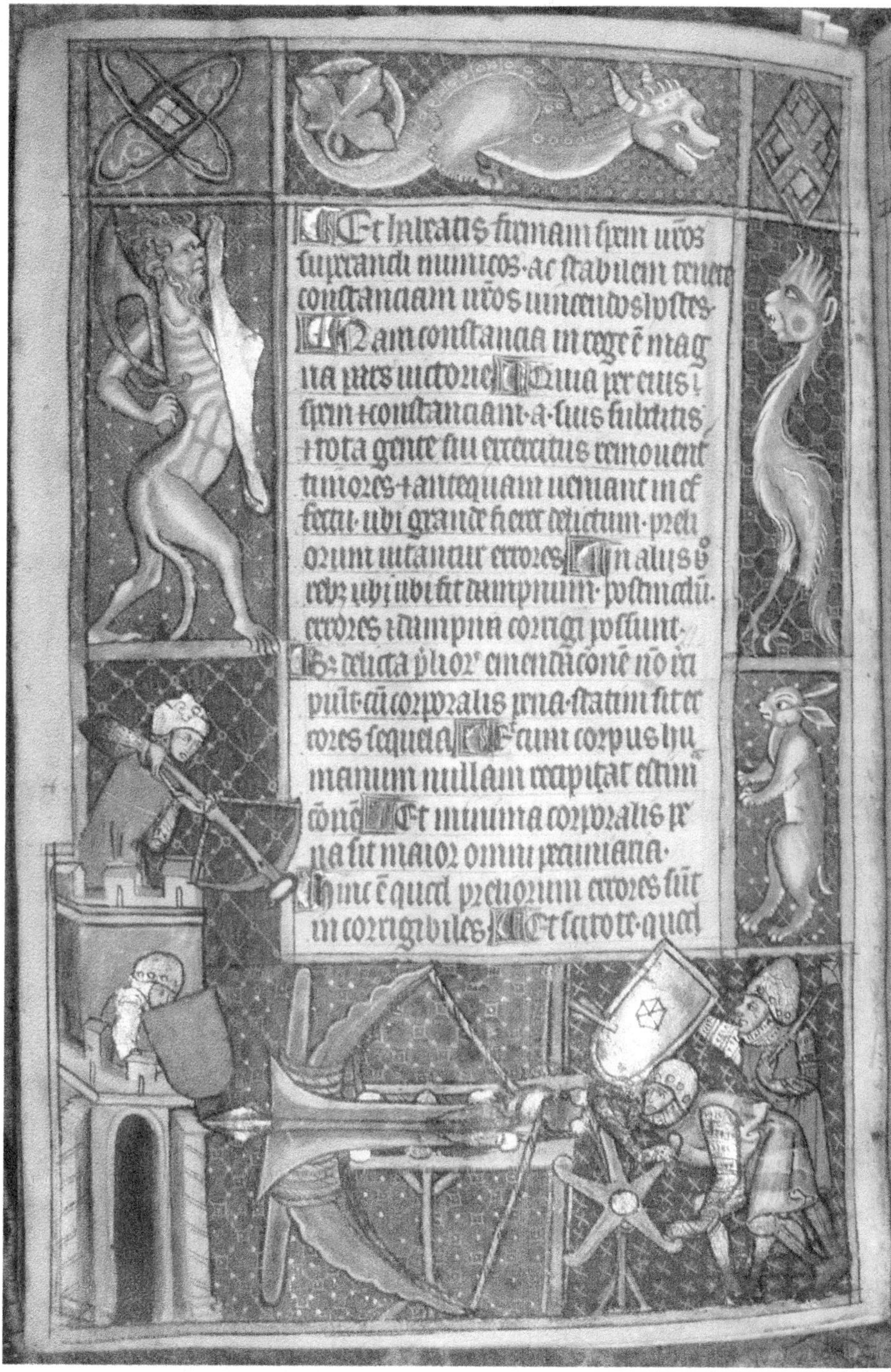

Plate 13. This large ballista is being wound with a winch while the city is defended by a crossbowman. With the permission of The Governing Body of Christ Church, Oxford, Milemete MS 92, fol. 68ᵛ.

PLATE 14. Château-Gaillard (France), Richard the Lionheart's castle built in the 1190s to protect Rouen: here showing from left to right the outer, middle, inner wards and the donjon. It fell to King Philip Augustus.

Plate 15. White Castle (Gwent, Wales), has a number of thirteenth-century arrow loops placed in shooting positions, such as this, that could protect two archers.

Plate 16. Laon (France). The Porte de Soissons was probably built under Philip Augustus in the early thirteenth century. Note the generous provision of arrow loops.

2 The Arab conquests

Mohammed and the early campaigns and conquests

As the great empires of Persia and Rome (Byzantium) fought each other to a standstill in the Middle East, and "barbarian" successor states began to carve out their spheres of influence in Europe, a new force emerged unseen by any of these. Within a few years starting in the fourth decade of the seventh century, the now Muslim Arabs were to create a vast new power that swept the Byzantines from provinces they had held for seven hundred years, and utterly destroyed the Persian kingdom. Pressing on, the Muslim armies were to reach deep into central Asia in one direction, and to eliminate the Visigothic kingdom in Iberia in the other. This phase of conquests ran out of steam after a century of expansion, but the Islamic presence transformed the classical world. At the beginning, although religious inspiration undoubtedly played a significant if intangible part in the conquests, it is necessary to offer some more material judgements to explain how two of the world's most sophisticated, cultured, and developed states collapsed before the impact of the Arabs.

The frontier against which the Arabs first pressed was almost unguarded. The immense drain of the wars with the Persians had led the Byzantines to give very low priority to Palestine and Syria. The Persian conquest of the area and its subsequent recapture had left many places defenceless. Byzantine garrisons were few and far between. There were watchtowers, but their upkeep was the responsibility of the local landowners. What was there was usually small, and as likely had a role in allowing the government to police trade routes to the Mediterranean coast as anything military.[1]

Emperor Heraclius, so recently the saviour of Constantinople and victor over the Persians, was making his decisions on the basis of apparently sound arguments. The Arab tribes beyond the frontier were sometimes a nuisance, but were certainly not a threat. Indeed, defence for much of the time involved hiring them, or buying them off. Their level of technology and culture was low, as Procopius

[1] See Harper, *Upper Zohar*, 115. This was a 26-metre-square quadrangular stone fort with walls of 1–1.5 metres thickness and four small corner towers, apparently destroyed in an earthquake but occupied between the fifth and early seventh centuries; King, Lenzen and Rollefson, "Survey of Byzantine and Islamic sites in Jordan", 414.

had observed before. Expensive fortresses, needing garrisons of professional soldiers, were quite unnecessary. There were, anyway, plenty of walled cities in the area against which backward raiders would batter in vain, having no knowledge of siegecraft.[1] The limited resources of the empire to face had to face east, not south.

Arab sources for the seventh century (of the Christian era) are few, and none of them are contemporary. They rely upon the oral transmission of accounts back to actual witnesses. The earliest compositions based on these traditions make no attempt to decide which of several versions to believe, but recount them all. It has been argued that information passed on this way has been subjected to so many changes as to be highly unreliable. Where there is corroboration, this is vital. Where there is not, it must be remembered that the source may be, in effect, fictitious.

The early history of the rise of the new religion and the conquest by Mohammed of the Hejaz, insofar as it can be reconstructed from such sources, bears out the proposition that there was little enough culture or technology in this area. Mecca, Yathrib (soon to be renamed Medina), and Ta'if were the only towns, and they were small; and most of the population was nomadic. However, to the south was to be found the centuries-old civilisation of the Yemen, and Yemeni Arabs were to play a vital role in the spread of Islam. Yemen was characterised by a high level of prosperity, on the basis of which arose an urban culture, literacy, and the mastery of technology. Towns were provided with mortared stone curtain walls. The area had been conquered by the Abyssinians in 520 and then by the Persians in 570, with significant disruption and destruction, but the gap between its cultural level and that of its northern neighbours remained substantial.[2]

Mohammed's first military efforts point to the difference between the skills and resources available to his followers and those of the great empires shortly to be overturned by the Islamic armies. In AD 627, Mohammed successfully defended Medina against his enemies, the Quraysh. An earthen rampart and a trench were sufficient. Even this limited advantage was gained through the advice of an outsider, on this occasion a Persian.[3] Three years later (630), the Muslims attacked the town of Ta'if. Abu Bakr led the force. Showing their inexperience, the attackers first set up camp too close to the town, and were driven back after taking casualties from arrows. The next stage was an exchange of archery, in which the defenders came off best. We are told that, following his capture of Mecca, the prophet had sent two of his followers to the Yemen "to learn the use of the testudo, the catapult and other instruments", but they had not yet returned. So the Persian Salman, perhaps the same man who had given such effective advice at Medina, helped them to build a stone-throwing catapult, called in one source a *madjanik* (*manjanīq*). However, as they were amateurs, it had no significant effect. They

[1] Kaegi, *Byzantium and the early Islamic conquests*, 47, 51, 56, 60; Nicolle, *Yarmuk, 636 AD*, 32; Mayerson, "The first Muslim attacks on South Palestine", 180–85.

[2] Chelhod et al., *L'Arabie du Sud*, 1: 41–8, 195–215. On the excavations at the city of Ṣan'ā', see Sergeant and Lewcock, "The church (al-Qalīs) of Ṣan'ā'", 44.

[3] The story is told in J J Saunders, *A history of medieval Islam*, 29. See also Gabrieli, *Muhammad and the conquests of Islam*, 114.

then built a cowhide shelter to cover their approach to the gate, but the defenders succeeded in burning it with red-hot scraps of iron. After twenty days of siege, Mohammed gave the order to withdraw.[1]

If the reference to the stone-thrower is not anachronistic, what type was it? The name is unhelpful. *Manjanīq* is a word used for the mangonel, itself a type of *onager*. But it is also the generic term used overwhelmingly by later writers for a stone-thrower. We are told it was built by a Persian, confirming the absence of skills among the Muslims, and this makes it unlikely that it could have been a torsion weapon, and far more likely that it was a manually-operated single-armed lever weapon of the type recently introduced from the east. Its ineffectiveness, even against a place so weakly protected as Ta'if was, is telling. The reference to sending people to the Yemen is intriguing. If it happened, when did they return, and how was their information turned into practical results? It may be that the now rapid movement to the north, bringing conflict with both the Byzantines and Persians, provided the opportunity to demonstrate these skills. Or it may be they were actually learnt through the Islamic conquest itself. The settled Arab tribes that lived around the frontiers of Byzantium were urban. Some lived in towns that were defensible. These areas, as much as the Yemenis from the south, may have provided the technological know-how, and perhaps equipment itself from Byzantine arsenals. Here, as in other spheres, the Muslims proved themselves to be fast learners, and they most successfully absorbed and deployed the various talents of those who fell under their sway.[2]

The first Muslim raids in Palestine had weakened and probed the Byzantine defences in 633/4.[3] In 634/5, Arab armies won a series of victories over Byzantine forces. The conquest of Damascus and Emesa (Hims) followed. A number of accounts of these events survive, from which it is possible to attempt a reconstruction of the siege techniques displayed by the Muslim forces, and examine the reasons for their success.

Damascus was the major city of Syria, and it fell to the Muslims in September 635. The city lies in flat terrain and also does not have significant water defences. It was, however, large (1330 × 850 metres), and well protected with high walls and seven gates. A separate *castrum* for the garrison had been built in the third century. Once thought to lie under the later medieval citadel, this is now doubted.[4] If it was still there, it played no recorded part in the siege. The Byzantine garrison, perhaps of 12,000, should have been sufficient against an Arab force not large enough to enclose the place completely. Dionysius of Tel-Maḥrē's Syriac chronicle says that the gates were closed against the Arabs, who attacked the city vigorously,

[1] Ibn Isḥāq, *The life of Mohammad*, 587–91 at 587; *Mohammed der Prophet*, ed. Weil, 235–6; Akram, *The sword of Allah*, 111–15. Ibn Isḥāq was not a contemporary, but was born half a century after the death of Mohammed.

[2] Nicolle, *Yarmuk, 636 AD*, 39; Jandora, "Developments in Islamic warfare", 101–4.

[3] Mayerson, "The first Muslim attacks on South Palestine", 174–99.

[4] Sauvaget, *Les monuments historiques de Damas*, 39–44; Omran and Dabboura, *The citadel of Damascus*, 1995, 7–13, concluding that there was no citadel in Damascus before the eleventh century; King, "The defences of the citadel of Damascus", 60; Burns, *Monuments of Syria*, 87.

whereupon Damascus, seeing no hope of succour, surrendered on good terms.[1] Balādhurī (not a contemporary, but relying on a now lost earlier history) adds that the Arabs actually gained entry by use of ladders at night, whereupon the garrison fled; then the city agreed to surrender, after some four months of siege.[2] Ṭabarī, the ninth-century compiler, says that although Damascus had been unprepared for siege, Heraclius had ordered in reinforcements and provisions, and following the start of a blockade, there had been weeks of exchanges of archery and the defeat of various sorties by the garrison. This was followed by the night attack with rope ladders thrown up onto the battlements (we are not told how they were thrown up a high wall), and the negotiated surrender of the city to the besiegers on the opposite side of the city. The Arabs then proceeded to the important walled city of Emesa and blockaded it, repelling sorties by the garrison. In one account, the Muslims feigned a withdrawal, then turned on their pursuers and defeated them, an event that precipitated the surrender of the city.[3] In another version, there was an earthquake overnight that brought down part of the wall, whereupon the garrison capitulated. The Arabs then attacked the two remaining Byzantine-held cities of the province, Chalcis and Caesarea. The first surrendered after a siege of ten days; the second apparently fell to assault after the defenders had first given battle outside the walls.[4]

These accounts all demonstrate that the Arabs had not significantly advanced either their knowledge of siegecraft since their earlier encounters with defended towns, nor had they managed to take advantage of any equipment or experts captured from the enemy. It is remarkable how successful they were. Success was to be confirmed. Heraclius ordered a major counterattack, but in 636 the Byzantine army came to grief at the Battle of Yarmūk, and in its aftermath the victorious Arabs rapidly reoccupied Damascus and Emesa (hereafter Hims) and completed the conquest of Palestine and Syria, with the Byzantines retreating dispirited before them and strong cities like Edessa simply negotiating terms of capitulation. There are accounts of strong Byzantine resistance in the east of the region, for example in the late Roman fortress of Tikrīt, that successfully held up the Arabs until the garrison was betrayed by the local Arab population itself going over to the besiegers, whereupon the important city of Mosul also quickly capitulated.[5] There is no report of the Arabs' having been able to do anything other than to blockade the towns.

The Arabs then turned their attention both east and west. In one direction stood Byzantine Egypt and North Africa, populous, rich, and still in touch with Constantinople because of Byzantine sea power. To the east, and directly facing them across a very long frontier, lay the Persian empire of the Sassanid kings. One army marched west and in 640 accomplished, against all the odds, the conquest of

[1] Dionysius of Tel-Maḥrē, *The secular history*, 150–55.

[2] Balādhurī, *Origins of the Islamic state*, 186–7.

[3] Akram, *The sword of Allah*, 398.

[4] Ṭabarī, *Chronique*, IV: 159–61.

[5] Ibid., 170.

Egypt. We shall return to these events. Meanwhile, conflict with the Persians had been taking place simultaneously with the struggle for control of Syria.

The defeat of the Persians

The overthrow of the Persian kingdom involved a series of major campaigns crowned by victory in battles. It also involved the capture of major defended cities. However, the edifice of empire itself was already dangerously weakened before the Arabs gave it a final push. The effectiveness and power of the onslaught was as much a shock to the Persians as to the Byzantines. It seems that the long-standing policy was to expect nomadic raids but little more, and to deter these by establishing a ditch-and-rampart line along the western frontier. It is not clear whether this was still maintained at the time of the Muslim attack, but if it was, it would not have represented an obstacle to an army.[1] Having lost its last war with Byzantium, and suffered a series of internal upheavals, there was now a boy king, who was under the control of a regent who lost his life in the decisive defeat by the Arabs at the Battle of Qādisiyyah in 637. Further defeats left the field open to the invaders to tackle the cities. These stages of the war took place not in Persia proper but in its western provinces of Mesopotamia (modern Iraq), where loyalty to the King of Kings was less powerful a motive for risking one's life.[2] The picture of these operations is no different than elsewhere. In the early attacks, the towns of Iraq did not resist. When they did, their assailants were impotent. The Arabs were held up at the town of Dumat al-Jandal for months, until their commander, Khālid ibn al-Walīd, himself arrived with the main army and took the place by storm.[3]

The Arabs attacked the Persian capital at Ctesiphon on the river Tigris. It may be significant that the reinforcements sent by the Caliph to his army in Mesopotamia included a large force from the Yemen.[4] The challenge posed by the city should have been substantial had it been defended vigorously. The capital was in fact not one town but several, linked together and straddling both sides of the river. Ctesiphon itself covered 58 hectares and was surrounded by a semicircular towered wall. On the west bank stood Seleucia, even larger, with walls made of brick.[5]

The Arab army attacked, "bombarding its people with catapults, closing in on them with armoured siege devices". Ṭabarī, whose account this is, reports another source saying "the Muslims bombarded them with catapults and smaller mangonels"; the commander, Sa'ad, ordered more catapults built, so another twenty were set up. Under cover of the bombardment, the army marched forward, and set scaling ladders against the walls. They found that the Persians had fled. It is recorded that they had relied on Persian deserters to build the siege artillery.

[1] Nyberg, "Die Sasanidische Westgrenze"; Frye, "Sasanian system of walls".

[2] There is an excellent account of the Arab victory in Glubb, *The great Arab conquests*, 189–201.

[3] Ṭabarī, *Chronique*, IV: 101.

[4] Glubb, *The great Arab conquests*, 189.

[5] Christensen, *L'Iran sous les Sassanides*, 379–82.

Ctesiphon itself fell when the Arabs managed to cross the river further up: the garrison did not wait to challenge them. The defeat of Persia's armies and the loss of its capital did not end the war, but thereafter the conquest proceeded with an air of inevitability.

In 638 or 639 the Muslims besieged Shūshtar, fighting the garrison outside, and driving it back into the city. The besiegers gained entry because someone revealed to them a way in by "the outlet of the water".[1] At 'Ain al Wardah (Ra's al 'Ayn near Ceylanpınar, the Roman Resaena), Balādhurī reports a previous account as saying that the defenders closed the gates of the town and set up weapons called *'arrada*, and by their stones and arrows killed many of the besiegers. The town fell, however, by capitulation.[2]

Arab armies still relied to a large extent on direct attack with ladders, or the good fortune of finding a surprise way of entering an enemy fortification. There are references to mantlets or other devices to enable them to get closer to the wall. The references to siege artillery are tantalisingly inexact and non-contemporary. Leaving aside the earlier attempts to familiarise themselves with existing techniques from Yemen, the Arabs would soon have encountered the best technology available, and much would likely have fallen into their hands. It is therefore quite possible that within a few years, artillery could have been deployed even though the references to it are scarce. It is possible that Arabs were themselves learning to use it, perhaps under the guidance of more experienced operators from the newly conquered provinces. But even Ṭabarī's version of one successful siege suggests that wherever there may have been large-scale use of artillery or of any sophisticated siege techniques, it was only because of the assistance of experts from elsewhere. The skills of siegecraft remained beyond the reach of most Muslim commanders in the first years of the conquests in the east.

Egypt and North Africa

Following the victories in Mesopotamia, and with the consolidation of the newly conquered territories in Syria and Palestine, it was inevitable that Arab eyes would turn towards Egypt. Byzantine rule in Egypt may have appeared vulnerable, in this case because of the disaffection of the Coptic Christian population from the orthodox regime imposed by Constantinople. Although the country did not tamely submit to the invaders, a large part of the people did not flock to defend the existing regime. For all that, Egypt appeared a daunting challenge. There was a substantial force of professional soldiers, a number of powerful fortifications, a dense network of towns in the Nile delta that would deny the Arab armies the chance of fast-moving and mobile operations, and the control of the sea by the Byzantines that offered the prospect of reinforcement. Nonetheless, the Caliph agreed to the attempt, and in 640 the Arabs won a battle at Heliopolis that had the important effect of keeping the dispirited defenders behind their walls.

[1] Ṭabarī, *The History*, XIII, trans. G H A Juynboll, 134–6.

[2] Balādhurī, *Origins of the Islamic state*, 276.

The first of these walls to offer serious resistance was the fortress of Babylon on the Nile (close to the later Cairo). Enough remains of the ruins of Babylon to establish what a formidable obstacle this was. Built in the traditional late Roman form, in the shape of an irregular quadrangle, the walls of brick and stone were nearly two metres thick, with four projecting square towers to the south and east, the west wall standing on the Nile itself, and an enormous gateway, with its own port on an arm of the river, on the remaining side. Apparently there was a portcullis slot. Making the whole complex even stronger, there was an island in the Nile opposite the fortress. A wet moat surrounded the whole.[1] The Arab army under 'Amr ibn al-'Ās had come into possession of Byzantine siege artillery through its capture at towns that had already fallen. This time, however, unlike in the east, they do not seem to have had access to people experienced in operating it, because against the fortress it was ineffective, while the defenders' weapons were not. Arab progress therefore was slow, while the Byzantines, who may not have been completely resolute, were none the less under strict orders not to surrender, although no significant effort was made to raise the siege. In due course, the level of the Nile dropped and the besiegers were able to cross the ditch. A storming attempt with scaling ladders gained them access to the wall top, only to find that the garrison had erected blockades to isolate the stretch. However, from this position, the garrison made an offer to surrender that was accepted (April 641). The siege had lasted seven months.[2]

After such an experience, the invaders must have been strongly motivated to continue their attack. Babylon was a minor target by comparison with the chief city of Egypt, Alexandria, to which 'Amr now turned his attention. This was one of the most splendid cities of the world. The circuit of its walls was around sixteen kilometres, it had a large garrison (certainly larger than the Arab army), and was well provided with artillery atop the ramparts. Even if it was substantially reinforced, the attacking army (not more than 20,000 at the start) could not have entertained the idea of enclosing the city. It seems that 'Amr tried first of all a direct assault, relying on the zeal of his troops and the lower morale of the garrison. This, however, was thrown back by stones and arrows from the wall top. The garrison clearly knew how to use its artillery in text book fashion. 'Amr, on the other hand, had neither the siege train nor the numbers to take the city. Instead, he encamped at a safe distance from the wall and maintained some kind of blockade. His troops had the fighting spirit to defeat Byzantine attempts at sallies, crucially ensuring that they remained sitting tight inside. Meanwhile, the Arabs raided at will across the delta, taking towns where they were not defended, but not where they were. Eventually, internal dissension and treason led to an extraordinary surrender by

[1] Descriptions based on a survey of the remains: Butler, *The Arab conquest of Egypt*, 238–43, plan on p. 240.

[2] Ibn 'Abd al-Ḥakam, *La conquête de l'Égypte*. This, though a source of the ninth century, is the earliest surviving account. There are various translations of different sections of his work (see Bibliography); Butler, *The Arab conquest of Egypt*, 249–71.

the Byzantine governor, which included the whole of Egypt, in November 641 after a blockade of many months, with the Byzantine forces withdrawing by ship.[1]

This was not quite the end of the story, because not all the many towns still in Byzantine hands were quite so inclined to give in, but details are not given of their surrender during 642. The following year, 'Amr pressed his advance westwards into the province of Pentapolis. There is no report of any serious opposition at the town of Barca (Al Marj), but Tripolis (Tripoli) was more strongly fortified. Once again, it is apparent that the Arabs had no siege train with them. Instead, they succeeded in capturing the place after several fruitless weeks, when some soldiers found a gap in the defences where the walls came down to the sea, sneaked in, raising their loud battle cry, whereupon the garrison fled to its ships. The next town, Ṣabrātah, was not expecting the enemy to appear so soon, and had its gates open to allow access for its cattle. Instead, the Arabs came sweeping down after a night march before they could be closed.[2] Thus easily was the Arab empire extended along the North African shoreline, even without the benefit of a siege train or engineers able to operate it.

Beyond Pentapolis lay the province of Africa (Ifrīqīya in Arabic) from which the modern continent derives its name. The Roman governor, Gregory, had declared his independence of Constantinople and ruled the whole stretch of the shore of modern Libya and Tunisia. His capital was at Carthage. The first Arab invasion of this area seems to have taken place in 647/8. The various sources (none contemporary) jumble together a number of expeditions in different years. This was the area where Procopius had praised the great works of fortification carried out by Justinian a century before. Initially, having defeated and killed Gregory in battle, the Arabs accepted an offer of tribute and withdrew. But they returned in a series of expeditions over subsequent years. These seem mostly to have been raids seeking to plunder the rich province. It was not therefore in the attackers' plans to become tied down in front of defended towns, although it does seem there were conquests of a number of walled towns.[3] A legendary account of the capture of Tébessa (Theveste — one of the places refortified by Justinian) tells the same overall story. The ramparts being constructed of great stones, it would have been a long task to besiege it. Instead, the Arabs obtained entry by plotting with someone inside who opened the gates.[4] Other towns like Aïn Djeloula were

[1] Ibn 'Abd al-Ḥakam, *La conquête de l'Égypte*, 221–8; John of Nikiû, *Chronicle*, 180–81; Butler, *The Arab conquest of Egypt*, 291–320.

[2] Ibn 'Abd al-Ḥakam, *Conquête de l'Afrique du Nord et de l'Espagne*, 37–9; Ibn al-Adhārī, *Histoire de l'Afrique et de l'Espagne*, 1–2; Butler, *The Arab conquest of Egypt*, 428–30.

[3] Caudel, *Les premières invasions arabes*, using unpublished manuscript accounts along with published ones, gives a detailed reconstruction of the series of attacks. A number of towns in modern Libya were taken in the expedition of 666 (ibid., 96–8) without any details being given as to how this was achieved. The policy of not becoming held up by defended towns is confirmed explicitly by 'Ubayd Allāh, "Un nouveau récit", 36.

[4] Cherbonneau, "Relation de la prise de Tébessa", 229, 236–8. Tébessa was a rectangle some 320 × 280 metres, with two gates and 14 square towers, the walls averaging two metres thickness. An inscription dates its construction to 535: Gsell, *Les monuments antiques*, II: 358.

carried by storm.[1] There was a change of approach in around 670, when the city of Kairouan (in modern Tunisia) was established as a military settlement close to the frontier of the reduced Byzantine domain. This represented a significant change from the previous strategy, and demonstrates also that the Arabs had the skills to plan and construct a major fortified site. It has been noted how Kairouan was well sited for its apparent purpose, with neighbouring pasturage for horses and camels, a location at a vital crossroad, and excellent all-round command of the territory.[2] Briefly driven out by the Moors and Byzantines, the Arabs quickly regained control and completed the work. The advance would be renewed in the last decades of the seventh century.

Southwards to Nubia

South of Egypt lay the Christian kingdoms of Nubia (approximately, modern Sudan), the histories of which are difficult to recover from the written record, and have been explored largely through archaeological excavation of a number of significant sites of towns along the river Nile between Aswan, north of the first cataract of the Nile, and the sixth cataract, north of modern Khartoum, a (straight line) distance of more than 1,000 km. From north to south, the kingdoms were Nobatia, Makuria, and Alwa (or Alodia), and the largest site, probably the capital of Makuria, was (Old) Dongola. There had been continuous interaction between Byzantine Egypt and the lands to their south — apparently including raiding expeditions — and this was maintained under the new regime. The only sources, none earlier than the ninth century, suggest an Arab raid from Aswan in 641/2, as soon as the new rulers had taken over, which was rebuffed by the Nubians, and then a more serious expedition ten years later (652). The source used in Maqrīzī's fourteenth-century compilation stated that 'Abdullāh ibn Sa'ad ibn Abī al-Sarḥ besieged Dongola and bombarded it with *manjaniq*, which were unknown to the Nubians. A church roof was smashed by its stones, whereupon the king arranged peace.[3] The Nubian kingdoms were sustained until eventual conquest by the Mamluk regime in the late thirteenth century, and some have concluded that the detail of the 652 attack has been mixed up with later events.

Archaeological discoveries at Old Dongola have confirmed the presence of a strongly fortified town by the seventh century. Sections of wall and towers have survived up to six metres high, and around four metres wide at the bottom, and were apparently once considerably higher. Well constructed of large mud bricks set in mortar, the walls were additionally faced with stone. Corner towers were of differently rounded shapes, but were all strongly projecting (more than eight

[1] Ibn 'Abd al-Ḥakam, *Conquête de l'Afrique du Nord et de l'Espagne*, 45; Ibn al-Adhārī, *Histoire de l'Afrique et de l'Espagne*, 3–4; John of Nikiû, *Chronicle*, 195; Caudel, *Les premières invasions arabes*, 94.

[2] On the foundation of Kairouan: Ṭabarī, *The History*, XVIII, trans. M G Morony, 102; Caudel, *Les premières invasions arabes*, 103–6.

[3] Reprinted in *Oriental sources concerning Nubia*, 639; Vantini, *Christianity in the Sudan*, 65. For a comprehensive modern account, see Welsby, *The medieval kingdoms of Nubia*.

metres), and of around six metres in width. Along the north wall were two further projecting towers between the corners, spaced irregularly. Reused columns were present in the work. The excavator suggested that the towers were additions, and the dating evidence, although not conclusive, made it realistic to suggest that they were added following the events of 652. This in turn suggests that the Arab attack had been serious enough to persuade the Nubians of the need to undertake substantial additional building.[1] As to the catapults, the discovery of the remains of a church outside the walls destroyed at around this time, gave credence to part of the story at least. If the other evidence from the original conquest of Egypt does not suggest initial familiarity with artillery, it is quite possible for them to have acquired that within ten years and to have used a weapon at Dongola. Perhaps it was transported by boat along the river, or else conveyed in parts by camel.

To the north and east

Meanwhile, scarcely pausing to consolidate their immense conquests in Syria, Iraq, and Iran, the Arabs resumed offensive operations. Their targets now were the rest of the Byzantine empire, in Asia Minor and the Mediterranean, culminating in another siege of Constantinople itself. At the same time, Arab armies continued to march east and north-east from Iran, subduing the remaining provinces of the Persian empire and penetrating deep into central Asia.

A major element in Arab warfare comprised raiding, with armies seeking to amass plunder (or tribute) and thus impoverish their opponents and enrich themselves and the new empire (a share being reserved for the caliph) rather than conquer territory directly. This may have been the motive for an attack on Cyprus, involving the first use of sea power by the Arabs. Fleets came from both Palestine and Egypt, in such numbers that the Byzantines apparently fled rather than resist. Much of the island it seems was unfortified and fell without a blow to the invaders. Since Cyprus had been surrounded on all sides by provinces of the Roman empire since time immemorial, the lack of defences is not surprising. However there was resistance from the "city" of Lapathos (modern Lapta: from the ruins, it had been an ancient city, but evidence for its seventh-century occupation is scarce), on the north coast. So the Arabs began to bombard it "with catapults from all round",[2] leading it to surrender, prudently, seeing as there was no prospect of being relieved. These events took place in 648. It is probable that if this happened — and the attack and use of weapons are corroborated by the source later used by the Byzantine chronicler Theophanes[3] — the invaders had brought the siege engines with them. Having but recently secured the contents of the great Byzantine cities and arsenals of Syria and Egypt, they had by now, perhaps, become

[1] Godlewski, "Old Dongola: the early fortifications". The year-on-year accounts can be found in Godlewski's other publications in the Bibliography. For the similarity of the remains to late Roman defences, see Monneret de Villard, *La Nubia medioevale*, inventory of sites in vol. I. At that time, the walls of Old Dongola had not been uncovered.

[2] Dionysius of Tel-Maḥrē, *The secular history*, 177.

[3] Theophanes confessor, *Chronicle*, 478.

familiar with their use. Alternatively, having enlisted local people as auxiliaries in their armies, they may have brought non-Arab engineers with them to ensure that, having transported large machines across the sea, they could be certain that they would be put to effective use. Since the ships themselves must have been constructed and crewed by people who already had the skill to do so, it would not be surprising if the artillerists also came from this source. Another history reveals Arab willingness to use ships and artillery, this time in combination. In 653, according to the Armenian Sebeos, Arab ships from Alexandria penetrated to the Byzantine city of Chalcedon. These ships had "mangonels installed on them, machines for throwing fire and machines for throwing stones". On this occasion, they were dispersed by a storm.[1]

The most sustained conflict between Roman and Arab took place on the frontiers of Asia Minor. Following the loss of Syria, the Byzantines had withdrawn to the Taurus mountains, abandoning the once rich province of Cilicia, preferring to leave this as a devastated zone to impede further Arab advances. This did not prevent regular Arab incursions into the heartlands of Anatolia. In 664, they managed to occupy Amorium, one of the keys of the defensive network being established, and proceeded from there, according to the Syriac Maronite Chronicle, to attack the town of Synnada (Şuhut), the seat of a bishopric. The chronicler provides a fascinating story. A master carpenter from Paphlagonia approached 'Abd al-Raḥman ibn Khalid, the Arab commander, and said "I will make you a catapult [the Greek word is *manganike*] capable of taking this fortress. Ibn Khalid gave orders for long logs . . . So he made a catapult such as they had never seen before." It was installed before the gates of the fortress, the defenders, trusting in their impregnability, let them get close. "Ibn Khalid's men then drew back their catapult; a rock rose up in the air and hit the gate . . . The second fell short . . . The third fell shorter. The men above jeered and cried out 'pull your weight, you are drawing badly'. They used their own catapult to propel a huge rock down onto Ibn Khalid's catapult from above, wrecking it."[2]

The description is clearly of a manually-operated lever catapult, and the absence of skill of the operators meant that they were unable to achieve a consistent aim, even from relatively close. It was built according to the design of a Paphlagonian, clearly indicating that citizens of the Byzantine empire were accustomed to such machines, just as the successful operation of the defenders' weapon, apparently scoring with its first shot, shows that the garrison were masters of the equipment too. What is odd is the suggestion that the Arabs had never seen such a machine before. This seems highly unlikely. But it could still be the case that there was no one with this Arab force who knew themselves how to build or work the stone thrower. Nonetheless, the evidence grows to suggest that during these decades of the second half of the seventh century, the Arabs began to master siegecraft for themselves.

[1] Sebeos, *Histoire d'Héraclius*, chap. xxxvi, p. 141.

[2] "Maronite chronicle", 34–5.

Byzantine defences

Driven back on all frontiers, the Byzantine empire was in trouble. Unlike the Persian realm, however, the last king of which perished in 651, it was to survive the crisis, to stabilise its frontiers, and ultimately resume offensive operations. The fact that Constantinople lay a long distance from the threatened frontier was vital. The immense defences had already saved the empire in 626. The city contained a large area where agriculture could be maintained, and this had the effect that, with sensible advance provisioning, an attacker was likely to run out of food before the city did. In the 670s, it was to see off a rather hesitant Arab attempt, more of a distant blockade, where the distance of the Arabs from their bases, and Byzantine fleets, were significant factors. Greek fire, the mysterious substance that could burn on water and was therefore such a lethal weapon in sea battles, appeared at this time. The Byzantines were, remarkably, to hold onto the secret of making this weapon for centuries.[1] However, the use of inflammatory missiles based on naphtha was widespread, and was to become a specialism of Muslim siegecraft, and we have already seen references to some kind of incendiary devices in use even this early. Such weapons could be propelled in a number of ways, such as hand-held slings, or from artillery, with greater range and presumably greater safety for those involved.

The empire also gave attention to defences of the frontiers. The army was reorganised in a system of "themes" (*themata*), with forces assigned to specific areas for local defence, and a concentration in the areas most exposed to attack. The changing strategic position after the successful defence of Constantinople in 718, with the restoration of a more even balance between the two sides, determined the steps now needed. Thus troops were assigned in small units to frontier defences, with districts having a main fortified base not always identical with existing cities. Work also had to be done on the fortifications. The defences ordered by Justinian were now old, and many fortifications had been destroyed in the long wars with the Persians. The need to combat Arab raiding parties, now the most common form of warfare, called for new approaches. One part of the new system was the construction of fortifications (the *kastron*) whose walls and garrisons were to provide some protection for local villages. Many defences were now constructed on the tops of hills, as a refuge from raiders. There had been a general trend of smaller settlements for many years, and the combination of town and citadel, sometime with the town itself actually inside the citadel, was the natural response to the collapse of the Anatolian frontier.[2] Although Byzantine sources imply that defensive works were carried out at the command of the emperor, it has been argued that many of these hilltop sites were fortified at local initiative, the positions

[1] There is a large literature on this subject. Haldon and Byrne, "A possible solution to the problem of Greek Fire", argued it was crude oil projected from a bronze pump and ignited as it emerged. The subject is reviewed in B S Hall's new introduction to Partington's classic *History of Greek Fire and gunpowder*, pp. xxii–xxv. See further Chapter 9 below.

[2] Gregory, "Kastro and Diateichisma", 235–7, 252.

themselves having no strategic purpose, and while their inaccessibility contradicts a military function, it increased their value as a simple refuge.[1] The frontier, in a great stretch of territory from Cilicia to Armenia, was reduced to a devastated, depopulated buffer zone, and the territory remained in that condition until the Byzantine resurgence of the tenth century.[2]

There was little leisure for building works. New works were undertaken with a core of strong lime mortar and rubble, using timber bonding courses as recommended by Vitruvius. Such methods were economical and quick, and where defences remain from this period, it is usually of these cores. Towers were increasingly of the open-backed style, simple designs.[3] Direct evidence of the building of the seventh century is limited, in particular the new defences. In those sites that have been studied, the datable remains are usually later.[4] They may replace defences of this time, but it cannot be proven. In many places, given the shortage of time and resources, late Roman walls would no doubt have had to continue to serve. In other places, new building, such as at the citadel at Nicomedia (İzmit), the work was "sloppy".[5]

In general, then, the Byzantines maintained the traditions of the Roman empire while adjusting them to cope with present calamities. There were great cities in Asia Minor, however they might have shrunk during the sixth and seventh centuries, and they required defending. Not all were mere refuges. Some do provide the dating evidence enabling them to be assigned to this time. Sardis, destroyed in 616, acquired a citadel on the hill above the ancient city site around 660, which demonstrated considerable sophistication. Strong walls three metres thick were "carefully and soundly built" into the steep hill. Approachable from only one side, where there was a massive bastion (although the date of this is unclear) and three triangular projections with openings for the use of the *ballista* battery, the construction here was of stone blocks and a vaulted brick superstructure.[6]

Amorium was a vital fortress in the defences, becoming the main military centre of the theme of Anatolikon. Excavations suggest that the seventh century defences were limited to the hill top, rearing above the lower town where the defences are no earlier than the ninth century.[7] So this was a traditional late Roman building, not strengthened against the new aggressors, perhaps indicating the belief that the original defences were adequate to ward off attackers more likely to be set on booty than a regular siege.

At Ephesus, the town retreated into a walled complex on the hill above the ancient city, and a separate walled town by the harbour. New fortifications were

1 Barnes and Whittow, "Yılanlı kalesi", 200.

2 Haldon and Kennedy, "The Arab–Byzantine frontier", 80 ff.

3 Foss and Winfield, *Byzantine fortifications*, 15–27.

4 Foss, *Survey of medieval castles of Anatolia*, I: *Kütahya*, 123.

5 As described in Foss, *Survey of medieval castles of Anatolia*, II: *Nicomedia*, 31–9, 42. The style of illustrations of town walls in this period remained that of late Roman traditions: Ehrensperger-Katz, "Les representations de villes fortifiées", 11.

6 Foss, *Byzantine and Turkish Sardis*, 53–9; see also the good discussion of Byzantine responses by Whittow, *The making of Orthodox Byzantium*, 122–3, 268.

7 Harrison et al., "Amorium excavations 1989[–1993]", esp. Lightfoot, 44 (1994), 111.

erected after ravaging by the Arabs in the 650s. The massive walls of the old theatre were made use of, being incorporated into the defences of the lower town. The area enclosed was severely reduced but still amounted to a square kilometre. The rubble core of the wall of the citadel was four metres thick, and the builders also used spoils from the temple of Artemis. There were rectangular gate towers. The building style was very similar to that of Sardis, and of Pergamon, nearby. Ephesus also became the capital of a theme, but it seems that the builders here had the resources, and perhaps saw the need, to strengthen significantly the defences of the ancient site.[1]

Ankara (Ancyra) had changed hands during the Persian wars, having been captured and burnt in 622. The Arabs repeated the treatment in 654. Presumably the Byzantine defences, still there, followed this event, perhaps being built in the reign of Emperor Constans II (656–61), although an early or mid-eighth century date as also been suggested. The outer wall has square towers every 40 metres, the inner enclosure has pentagonal towers every 20 metres, and there is an inner castle which is rectangular, 350 × 150 metres. The facade is built of great blocks from ancient monuments with brick courses above. The main gate is heavily fortified. Thus massively defended, Ankara resisted every attack until 838.[2]

The city of Amastris (Amasra) on the Black Sea coast may also have benefited from imperial attention. Distant though it was from Arab threats (although a raid is reported in 727), it may have been refortified as a major naval base during this time. Roman building materials were reused on a significant scale in the walls and towers of the inner fortress, which had dimensions of some 400 × 100 metres. The core was mortared rubble recovered from classical buildings and included, for example, seats from the theatre. The wall survives in places to a height of ten metres above the present ground level. Rectangular projecting towers are between six and nine metres wide and project between four and six metres. They are closely spaced (all between eleven and thirty metres). Immense blocks were found making up some of the roofs, suggesting that they were specifically designed as supports for artillery, perhaps still torsion. There are arrow slits. There was a slighter outer wall on the landward side, running parallel, and a ditch in front of that. The original gates were simple passages but involved an angled approach.[3]

So, while there was a significant upgrading of the fortifications of towns which had not previously found themselves in the front line, there was no change in the general approach to the work, walls and towers being comparable in size to their predecessors, and with accommodation for defensive artillery. With some exceptions, the area to be protected had shrunk, but was still of a size to enclose a civilian population, thus confirming that the defences served the purpose of both habitation and the base of a military garrison. Whilst questions of resources, and

1 Foss, *Ephesus after antiquity*, 103–11.

2 Foss, "Late antique and Byzantine Ankara", 30–31, 70, 72, 74–9; Jerphanion, *Mélanges d'archéologie Anatolienne*, 144–222, provides a detailed description of all the remains.

3 Crow and Hill, "The Byzantine fortifications of Amastris"; Crow and Hill, "Amasra", esp. 7–12.

the need for haste, will have decided what was built, there does not seem in the late seventh century to be any feeling that the new enemies required a different approach to fortification from that laid down in late Roman military manuals.

The next phase of conquest

The Muslim advances resumed in the second half of the seventh century, following internal conflicts around the succession to the caliphate, which were resolved through civil war resulting in the establishment of the Umayyads in 661. Equilibrium had been established along the frontiers of Asia Minor, where warfare was dominated by raiding, but Arab armies continued to drive to the east and north east. There is an account however of one serious siege operation against the Byzantines in the compilation of Bar Hebraeus (Gregorius Abū'l-Faraj bin Hārūn al-Malaṭī). The caliph Mu'āwiya (died 680) invested Caesarea in Palestine by land and sea from December until May. "Although 72 engines were hurling stones at its walls, no breach was made." The town fell when the Arabs "dug a hole under the wall" and entered it at the same time as those outside scaled the walls by ladder. They were apparently unable to take advantage of this entry, being pinned down on the wall. The garrison fled by boat.[1] It is a jumbled account written many centuries later, but the suggestion both that the Arabs had siege artillery, and that it was not remarkably effective, is consistent with the other evidence. The suggestion that they had now mastered mining techniques is one of the earliest references to this, and perhaps tied in with evidence for this advance from the caliphate's eastern frontier as well.

In the 680s, using their base at Kairouan, Arab armies began to drive north and west to complete the defeat of the Byzantines in North Africa. The latter played a remarkably passive role. Repeatedly, the Byzantine forces were driven up to the walls of their fortresses, but the Arabs did not have the means, or perhaps the time, to try to capture them. Instead, they drove far to the west, reaching Tlemcen and Tangier in 681. 'Uqba ibn Nāfi''s daring march involved battles and the capture of towns held by the local tribes about which no details survive.[2] Overthrown by a Moorish and Byzantine army that intercepted and killed their commander in 682, the Arabs were back with a reinforced army drawn from Egypt and Syria in 688. Defeating the Moors, this army then captured Carthage itself (695). Stung into action, a new Byzantine army was despatched and Carthage was reoccupied in 697 by the patriarch John. But the following year, and for the last time, a new Arab army besieged Carthage, and forced the defenders to surrender. After fifty years, the strongly fortified province of Africa finally fell. For most of that time, the Byzantine forces had failed to dispute control of the territory with the invader, but had retreated behind their walls, confident from experience that they were

[1] Bar Hebraeus, *Chronography*, 97.

[2] 'Ubayd Allāh, "Un nouveau récit", 38.

safe there. In the end, the approach proved a failure, and the powerful capital city was compelled to negotiate with the Arabs.[1]

The original accounts give little practical detail of the capture of the cities. There is evidence for growing use of siege artillery and growing competence from another event of this time, on this occasion (in 683) in another rebellion, when Mecca itself was the prize. The army of Syria "established siege engines and hurled stones on the Ka'ba which broke the columns of the mosque". They also hurled vases of burning naphtha which destroyed the holy site. Perhaps by way of diminishing the horror of this event for his readers, the machines, the chronicler says, were commanded by an Abyssinian infidel.[2] Although Mecca was hardly a major fortress, this story rings true and suggests that Muslim forces were now able to attack cities with a greater variety of methods than in the early years of the conquest.

The completion of the conquest of the empire of the Sassanids had involved the occupation of Tabaristan, the province immediately south of the Caspian Sea, whose capital, Amol, had been of recent Persian foundation, protected by a wall of baked brick, so wide that three horsemen could ride along it side by side, and a moat a bowshot wide (although one suspects a little exaggeration from the non-contemporary historian). When Tabaristan fell to the Muslims, they established their capital in Gilan and "filled the land with lofty castles".[3] In the 670s, Arab forces having established themselves in the north-eastern provinces of the old Persian realm began to press into central Asia, bringing them into conflict both with the nomadic Turkish tribes of the area, and the settled urban centres of the trade route to China. Some of these sites (such as Merv, a major oasis now in Turkmenistan) were not cities in the established sense, but were scattered settlements built around the fortified dwelling of a local landlord.[4] The conquest of such places was quite a different task from taking a city. However, evidence of the seventh-century defences is rare.

The earliest Arab moves, as elsewhere, seem to have been large scale raids, although we are told they had ballistas in 672. The city of Bukhara, their main objective, was saved on this occasion, but surrendered in 675, only to revolt once the army had left. Thirty years later the Arabs returned in a serious attempt to conquer Transoxiana. The Emir of Khorasan, Qutaiba, crossed the Oxus in 706. The inhabitants of the town of Baikand fortified themselves against him. The defences were strong. The siege lasted 50 days; then the Arabs succeeded in digging under the wall and undermining it, storming the town through the breach thus created.[5] The population was spared. If this tenth-century account is accurate, this (along with the reference mentioned earlier in Bar Hebraeus) would be one

[1] For the sequence of events, here simplified, see the sources cited by Caudel, *Les premières invasions arabes*, 125–74.

[2] Ṭabarī, *The History*, XX, trans. G R Huwting, 117; Ṭabarī, *Chronique*, V/VI: 57.

[3] Ibn Isfandiyār, *History of Tabaristán*, 25–6, 97.

[4] Hiebert, "The oasis and city of Merv", 113–21, on Russian excavations at the site.

[5] Narshakhī, *The history of Bukhara*, 37–8, 44–5. See also Vámbéry, *History of Bokhara*, 29.

of the earliest references to the successful employment of the technique of mining by the Arabs. In later times, the skill of the Khorasanis in this form of attack was well known. It may be that this was a tradition dating back to the Persian empire and enlisted by the Arab conquerors.[1] On this occasion, Bukhara itself adopted the prudent course and surrendered. There are descriptions of the defences of this city, but they all relate to later rebuilding, so it is not possible to consider how strong Bukhara was at this time. Certainly, according to Narshakhī,[2] the citadel and walls seem to have been constructed or reconstructed under Arab rule.

Qutaiba continued his advance. Siege machines were now deployed regularly against the towns that resisted him in 709, and although no details are given of the sieges, the Arab commander was successful.[3] In 711, he penetrated as far as Samarkand with an army of 20,000 men. Blockading the city from all sides, he was alert both to the risk of sorties by the besieged, and the likelihood of attempts at relief from nearby towns. Following his defeat of a relieving army, he turned at once to the city, pressing the attack with great vigour and battered the walls with his machines, soon creating a breach. Upon this, the inhabitants negotiated surrender.[4] There being no supply of stone in the region, the walls were of baked brick, strong but also friable, otherwise it would not have been likely that the Arab stone-throwers would have created a breach; or perhaps bombardment was combined with other forms of attack, such as mining. Whatever the actual methods, it is clear that this commander and his army knew their siegecraft.

Sind

The conquering drive of the Muslims was also directed against the province of Sind, at the mouths of the river Indus, where it collided with the fractured small kingdoms that had succeeded to the great Indian empire of an earlier age. Late in the seventh century, Muḥammad ibn Qāsim marched on Debal (south-east of Karachi), transporting his materials, including siege equipment, by sea, to begin the conquest of the region.

What he confronted is not very clear, as there are scant remains from this period. Certainly, there was official guidance left to the modern generation of Indian rulers by their classical forebears, in the form of the *Arthaśāstra* written by Kauṭilya, which provided advice on all aspects of statecraft, including how and where to build fortifications. The highly literate and cultured Indian ruling caste would have been familiar with these Sanskrit texts, but whether the states of the seventh century had maintained ancient walls was another matter. There is also little evidence of the level of military technology available, although it is likely that the same quality of expertise was available as elsewhere in the world at the time, northern India in particular being a crossroads for trade between Central Asia,

1 Spuler, *Iran in früh-islamischer Zeit*, 498.
2 Narshakhī, *The history of Bukhara*, 24–5, 34–5.
3 Ṭabarī, *Chronique*, v/vi: 153.
4 Ibid., 157–9. On this campaign, see also Gibb, *The Arab conquests in Central Asia*, 38–44.

China, and the Arab world. But exactly what kind of artillery the Indian rulers possessed is not known, although the generic use of the word "sling" might suggest something like the Chinese weapon. The broader term *yantra* probably meant any kind of machine, and whether the classical weapons that certainly included stone throwers and arrow-shooters continued in use remains unclear. The evidence does seem to suggest that most of the main dynasties maintained their own arsenals to support standing armies that sometimes included foreign mercenaries, who may have enabled cultural diffusion. The classical lists of officers included commanders whose title is translated as "commander of catapults and engines".[1] Building techniques were relatively simple. Stone was quarried and shaped by masons according to a well-established traditional style, often highly decorative, but relying on gravity instead of mortar to remain standing, and the many constructional benefits of the arch were unknown to the Hindus. Since the Muslim conquerors frequently demolished the places they captured and then used them as quarries for building anew, the two building styles being completely incompatible on many grounds including those of religious sensibility, we do not know what Muḥammad ibn Qāsim had to attack.[2]

Ibn Qāsim dug a trench around Debal and set up an enormous stone thrower called "the bride", which allegedly needed 500 men to operate it. This great weapon, however, was used merely to knock down the flagpole on the city's main temple, whereupon the Muslims took the place by escalade.[3] The tale's attempt to demonstrate the superiority of Islam must make its historicity doubtful! Attacks on Sind continued, with the regular deployment of *manājanīq* to bombard the Indian defences, often making use of higher ground to fire over the walls. Even so, those places that were either better defended, or stronger walled, were generally taken by close blockade rather than assault. When the town of Kiraj (Chitor, Rājasthān) rebelled against the new rulers, the governor left in control, Junaid, recaptured it by using rams to breach the walls. In 712, naphtha arrows were used to ward off Indian elephants, and miners were used in a combined attack with artillery to breach the walls of Rāwar. The city of Brāhmanābād, however, resisted attack for six months, but the more important city of Alor (Mansūra) had already surrendered after a month's blockade. At Askalanda (Uch), there was fierce resistance, and a rare reference in these Muslim accounts to the Indians deploying their own artillery on the walls (incidentally suggesting that it must have been of relatively small size). There are two differing accounts of the capture of Multān. One says that it fell when the besiegers found a way to cut the city's aqueduct. The

[1] *Dauhsādhyasādhanika*, according to Bhakari, *Indian warfare*, 49, citing U Ghoshal, *Studies in Indian history and culture* (Bombay and Calcutta, 1957), 452.

[2] P Brown, *Indian architecture: Buddhist and Hindu periods*, 65–6; Cousens, *The antiquities of Sind*, 71; Kauṭilya, *Essentials of Indian statecraft*, 78–80. For the probability that the *gophana* or sling in Indian sources included some form of stone-throwing artillery, see Gode, "The history of the sling", 85–8. For the other material in this paragraph, see Bhakari, *Indian warfare*, 34–49, 99–113, where other kinds of equipment are named and described. Much of his evidence dates from much later, however, and we cannot automatically read back their existence in the eighth century.

[3] Balādhurī, *Futūhu-l buldān*, 129.

other says there was a two month long attack with artillery used on both sides, until deserters showed the Muslims where they could mine, which they did and brought about the rapid collapse of the rampart and fall of the city.[1]

New attacks on Byzantium

The turn of the eighth century saw renewed attacks from the caliphate on the east Roman empire. These culminated in another failed attack on the capital city itself, for which there are a number of sources. In 704, the Cilician frontier town of Mopsuestia (Misis/Yakapınar) was captured. The Arabs did not merely destroy the place or loot it. In 706, they rebuilt the walls and installed a garrison. This is more early evidence of the willingness of the conquerors to build to defend what they had taken.[2] We will look more closely at the development of Arab fortress-building in due course. Meanwhile, in 706/7 the attacks continued, and again fortresses were taken.[3] In 707, the target was the important town of Tyana (Kemerhisar), which fell to the Arab commander Maslama ibn 'Abd al-Malik, who followed up his victory by taking others. At Tyana, a Byzantine source tells, the Arabs "threw down part of the walls by means of their siege engines, but were unable to achieve further successes". However, relief attempts failed, and the inhabitants surrendered as their provisions ran out.[4] Tyana was to become an important military base for the Arabs, at which they themselves were to undertake significant building work in the following century.[5]

The growing threat to Constantinople itself was recognised. If the previous Muslim attack in the 670s had been a relatively ineffective blockade, this time the emperor took no risks with the security of the capital. "Expecting the Saracen attack", Anastasius II (r. 713–15) "restored carefully the walls of the city and refurbished the military engines", according to the patriarch Nicephorus. He also laid in great stores of provisions.[6] Quoting a now lost source, Theophanes recounted that the emperor "set up on the towers catapults for darts and stones and other engines".[7] The Byzantine arsenals were well equipped with such weaponry, it seems.

The preparations were not before time. The storm broke in August 717. The caliph 'Abd al-Malik sent his son and commander Maslama across Asia Minor to attack Constantinople. He had been provided with a large army that included a large number of artisans, thousands of camels and asses loaded with provisions, arms, engines of war, and naphtha. Anticipating a long campaign, he had apparently been told to ravage the land and to sow it, and to build accommodation for

1 Ibid., 122–5; *Chach-nāma*, 156–60, 170–72, 193–205.

2 Bar Hebraeus, *Chronography*, 105.

3 The source is al-Wāqidī, as quoted by Ṭabarī and translated by Brooks, "The Arabs in Asia Minor", 191–2.

4 Brooks, "The Arabs in Asia Minor", 192; Nicephorus of Constantinople, *Short history*, chap. 44, pp. 105–7. Nicephorus was born in 758.

5 Ṭabarī, *The History*, XXXII, trans. C E Bosworth, 198–9, details extensive work done in 833 to create a defensive site one mile square.

6 Nicephorus, *Short history*, chap. 49, p. 117 (the year 714).

7 Theophanes, *The Chronicle*, 534.

the troops over the winter. He cut off the city on the landward side with a "wide trench with a breast-high parapet of dry stone" (Theophanes). From this position of security he launched attacks upon the city using his siege weapons, of which he had many kinds (Nicephorus). Once again the Byzantine capital was preserved, though. The Arabs never secured command of the sea, so were unable to threaten the city, and its enormous population, with hunger. On the other hand, the besiegers themselves were unable to maintain adequate supplies for their host, and hunger led to disease. Conditions were worsened by bad weather. They sustained their efforts for 13 months before being ordered to retreat in disarray and with great loss of life, caused rather by disease, starvation, and storms at sea than direct enemy action.[1]

For all the number of accounts of this famous siege, there is remarkably little detail about the actual course of events, in contrast to the Avaro-Slav attack of 626. There are undoubtedly clues in the religious tract attributed to the patriarch Germanus II, who attributed the whole success of the defence to the city's protectress, the Virgin Mary, perhaps deliberately avoiding mention of the emperor, who was then on the other side in the developing theological schism over icons. The patriarch refered to the inertia of the "machines deployed for the assault" and to the failure of the Arabs to "launch a single engine of war against the city".[2] The ineffectiveness of the weapons of the Avars has already been noted. Ninety years on, there had not been much in the way of technological advance, and even if the stone throwers of 718 were more powerful and better manned, they would still have been inadequate against the walls of Constantinople. However, the Arabs were by now familiar with these defences and, as we have seen, had come well equipped. Most likely, they had not been able to get close enough to the walls to attempt to breach them, and were thus compelled to the ultimately unsustainable method of blockade. The Arabs did not try again to capture the capital.

Defeat at Constantinople was not perhaps surprising. The most powerful siege methods known would have been thwarted by the defences of the city, and forcing it to surrender through hunger depended on being able to prevent its receiving supplies, while simultaneously being able to feed, in the field, an army large enough to threaten its massive walls. Less substantial fortifications were however now vulnerable to Arab attacks, as was shown at Nicaea (İznik) a few years later (725/6), although this city's defences were themselves not inconsiderable. Two sources provide accounts of this attack that, while not entirely consistent with one another, enable something of a reconstruction. Nicaea, an important town that may have felt that its distance from the frontier made it fairly safe, was ringed with late Roman walls (Plate 2). The existing defences date from the rebuilding carried out by the Seljuks following the siege of 1097. The older defences were a single

[1] Theophanes, *The Chronicle*, 545–6; Brooks, "The campaign of 716–18", 19–26; Nicephorus, *Short history*, chap. 54, p. 123; Dionysius of Tel-Maḥrē, *The secular history*, 212; Bar Hebraeus, *Chronography*, 167; Ṭabarī, *The History*, XXIV: 39–42.

[2] Germanus of Constantinople, "Homélie sur la déliverance de Constantinople", §§ 11, 18, pp. 201, 203. There is also the suggestion (p. 203) that Greek Fire played a key part in the defence at sea.

wall with a circuit of about five kilometres, 3.6–3.8 metres thick, and nine metres high to the wall walk. There were numerous gates and towers of late Roman style.[1] The Arabs attacked for forty days. During the siege, one of the defenders (an iconoclast — this being the moral and purpose of telling the tale, presumably) was struck by a "stone discharged from a siege engine, and it broke his head and face" (Theophanes). Capture was apparently effected when the Arabs "dug down the wall" and gained entry, prompting the garrison to flee by ship across the lake against which the town stood. These two glimpses suggest that the attackers had artillery (although the missile described suggests this was of a lighter character), and that they knew how to undermine walls, assisted here by the absence of an outer wall and perhaps of a ditch.[2]

The range of expertise increasingly displayed by Arab armies, and also the limitations of eighth century siege techniques, is confirmed by another account. In 750, the Umayyad caliphate was extinguished following a revolution launched from the eastern provinces that brought the Abbasid dynasty to power. In the civil war between the two, the important town of Wāsiṭ on the river Tigris in Mesopotamia (Iraq) was attacked by Abbasid forces in August 749. The place had been an important military establishment for the Arabs and was well protected by water, both the river and a canal. The defenders at least were well supplied with stone throwers, which were used to good effect against the attackers. There were fierce battles beneath the walls, and by boat, and there are accounts of nocturnal fighting illuminated by torches hung from the walls. The defences of Wāsiṭ were clearly too strong for the weapons and techniques available to the Abbasid army, and it can be imagined that the water defences prevented any attempts at undermining. The siege lasted for eleven months and was brought to an end more through unrest and dissension in the garrison than by the effectiveness of the attack.[3]

Into Europe

By the time of the change of dynasty in the eastern heartland of the Arab empire, the area subject to Islam had also undergone a dramatic extension in the west, with the crossing by Ṭāriq ibn Ziyād of the straits later named Gibraltar after him and the spectacular collapse of Visigothic rule in the former Roman province of Hispania.

While the Arabs had been extending their control along the North African coast, dealing along the way with the Berber tribes who controlled the area, the Visigoths' main concerns had been to the north, both with areas of the peninsula that they did not control, and with the Frankish kingdoms of Gaul. Their domain, it will be recalled, crossed the Pyrenees and extended to the river Rhône. In very short order, the Franks were suddenly to find their Visigothic rivals replaced by Muslims.

The Visigothic kingdom was in considerable disarray, with succession disputes

1 Schneider and Karnapp, *Die Stadtmauer von İznik*, 1–8, 10–11.
2 Theophanes, *The Chronicle*, 560; Bar Hebraeus, *Chronography*, 110.
3 Detailed account in Elad, "The siege of Al-Wasit", 59–70.

and civil conflict. The cultural level had developed but little from that previously considered.[1] The once larger Roman cities had shrunken considerably in population, although some new building, copying existing examples, had taken place.[2] The speed of the Muslim conquest (perhaps three or four years) suggests that the warriors available to the Visigoths were relatively few in number. The accounts give stories of battles fought and towns captured, but few details of any sustained attempt to defend the late Roman walls that might, if properly secured and provisioned, have held up the invaders. Some towns were found abandoned. At Córdoba, an Arab force gained entry because the wall was broken and they were able to climb over. The garrison was said to be only 400 strong. Carmona was too strong to assault (its ancient ramparts were "well-built") and was taken by stratagem. Seville surrendered after several months. At Mérida, where late Roman defences "of very strong masonry" stood in the way, the Arabs attacked by building a *dabbaba*, a kind of mobile tower. It sounds like some kind of mobile battering ram, as its purpose was "to open a breach",[3] and because it was held up by poor ground and eventually destroyed by a Visigothic sortie. It looks like the early defeat of the Visigothic army and death of King Roderic, along with the fragmentation of the ruling élite perhaps initially unaware that the Arab attack was more than a raid, had been catastrophic in preventing a sustained resistance that might have taken advantage of the presence of so many fortified sites.[4] It has been argued that the obligation on towns to supply help for royal expeditions that characterised the medieval Spanish kingdoms was rooted in Visigothic times,[5] but the effect of this on the Arab conquest is not evident.

The actual chronology of the conquest is impossible to reconstruct, but it is apparent that the Arab army (which had initially at least been largely composed of Berber tribesmen), operating under the command of Ṭāriq ibn Ziyād and the governor of Ifriqiya, Mūsā ibn Nuṣair, and then after their recall Mūsā's son 'Abd al-'Azīz, spread out rapidly to secure the major cities, many of which did not resist but made their own treaties with the conquerors.[6] Perhaps the dispersal of some of these urban forces in the first phase of conquest critically weakened the defences of the first cities in the way of the advance. Certainly it was not long before the independent Christian territory was limited to the mountainous north-west (As-

1 See Chapter 1.

2 Reilly, *The medieval Spains*, 21–6; Jiménez Garnica, "Settlement of the Visigoths", 98, 119.

3 Ibn al-Athīr, *Annales du Maghreb & de l'Espagne*, 46–7. Ibn al-Athīr was a thirteenth-century chronicler.

4 Ibn 'Abd al-Ḥakam, *Dhikr fatḥ al-Andalus*, tells of the battles but mentions no sieges. The details here are from a much later source, *Ajbar machmuâ*, 23–9. Ibn al-Quṭīyah, *The history of the conquest of al-Andalus*, 5, 19, adds a few incidental details. The quoted descriptions are from the tenth-century chronicler Rāzī: "La 'Description de l'Espagne'", 84 (Mérida), 94 (Carmona).

5 Powers, *A society organized for war*, 127. It has also been argued that the main infantry force of the Visigoths were slaves; see E A Thompson, *The Goths in Spain*, 265–7.

6 There are numerous modern accounts. For one based on Arabic sources and from an Arab viewpoint, see Ṭāha, *The Muslim conquest and settlement of North Africa and Spain*, 93–8.

turias), while the victorious Muslim armies consolidated their rule in their new province of al-Andalus and then pressed on towards and beyond the Pyrenees. Here they encountered the substantial late-Roman cities and their defences, the walls in all probability little changed since the third century, although perhaps rather better maintained or repaired, having been fought over so frequently by Frank, Burgundian, and Visigoth.[1]

In 721, ten years after the Muslim forces set foot in Spain, their armies were fighting for control of Narbonne and Toulouse. Narbonne was taken and the population carted off as slaves, suggesting that the town had not surrendered but had been stormed. The Arab commander Al-Samḥ ibn Malik al-Khawlānī then attacked Toulouse and deployed a range of siege machinery including *fundae*, which in the context are far more likely to have been lever-based stone throwers than handheld slings, although this cannot be proven;[2] and it took a rapidly assembled army of Franks and Aquitanians to drive them off. It is on balance more likely that the Arab attacks on southern Gaul and Septimania were raids seeking booty rather than permanent conquests, just as they were on the caliphate's other frontiers during this time, in which case so formidable an obstacle as Toulouse would have been a target more for the wealth it contained than for any other reason. Nonetheless, siege machinery would not normally accompany a raiding force unless it was light enough to be easily transported. Clearly, even if the Arabs had not had to deploy many of their acquired skills in siegecraft during their first years in Spain, they had not failed to bring these arts with them.

They may also have encountered such skills amongst their opponents. It is likely that the Visigoths had knowledge of artillery, as there are accounts of stone throwers and what sound like *ballistae* in Bishop Julian's account of the rebellion against King Wamba involving siege operations at Nîmes and Narbonne in 673.[3] It is impossible to know whether the late-Roman *onager* was still in use, although for all the reasons previously suggested this seems unlikely, in which case the new artillery must have arrived at this end of the Mediterranean at some point. There is a hint in these accounts of the greater than usual power of the weaponry, which implies we are not talking about the relatively small missiles of the late-Roman torsion weapons, and also an indication of a high trajectory, with stones crashing over the walls and into the city. However, these hints do not sit easily with the general picture we have of Visigothic skills, and it may be that access to siege weapons depended on finding experts from other sources to build and operate it.

[1] Thompson, *The Goths in Spain*, 266–7, and the sources cited there.

[2] "iam dictus dux exercitus tolosam usque pervenit eamque obsidione cingens fundis et diversis generum machinis expugnare concevit"; *Continuatio byzantia arabica*, 358. The identical text appears in the *Continuatio hispana* (ibid.); the two texts are edited in the *Corpus scriptorum muzarabicorum*, I, as *Chronica byzantina-arabica* (p. 14) and *Chronica muzarabica* (p. 38). See also *Chronicon Moissiacense*, 290.

[3] Julian, *Historia rebellionis Pauli*, 776–9: "sed tantos imbres lapidum intra urbem concutiunt, et clamore vocum et stridare petrarum civitas ipsa submergi aestimaretur" (Narbonne); "confluentes undique nostri cum fragore vocum muros urbis petrarum jactibus petunt, missilibus quibusque constitutes per murorum spiculis sagittisque propellant" (Nîmes).

The victory of Charles Martel over the Muslims at Poitiers in 732 was subsequently glorified in western writing as the salvation of Christendom, although in fact the "Saracens" he defeated were another raiding force rather than an army bent on permanent conquest. Certainly the Frankish victory did not end Arab-led incursions into the region. Another account tells of a raid on Provence in 737, leading to an attack on Avignon, a "very strong" fortress that was now taken "by fraud". The Frankish response was prompt, and a powerful force set about recapturing the place, but had to do so by rather more strenuous means than the Arabs had, establishing a full encirclement and deploying machines before succeeding.[1]

A point of temporary equilibrium had now been reached, although it may not have appeared so at the time. During the first century or so of Islam, a nondescript army of desert tribesmen had humbled one great empire, shearing it of Syria, Palestine, Egypt, and North Africa, and even challenged the capital itself. Another once great kingdom had disappeared altogether. Essential to this progress had been the ability to absorb and use effectively the skills, arts and peoples of the newly conquered territories, even if the direct evidence has been limited to casual references.

The fortifications of the Arabs in the first one hundred and fifty years

Nowhere is the reliance of Islam on those it conquered clearer than in the arts and architecture that came rapidly to flourish under its rule. Nor is it a surprise that building styles should be heavily influenced by the culture of the particular province in question, be it Persian, Byzantine, Syrian, Egyptian, or Yemeni. The rapid establishment of stable government by the caliphs, with tax collection and the encouragement of arts and sciences, was naturally reflected in some splendid architecture, of which the Dome of the Rock in Jerusalem (around 690) and the Umayyad mosque in Damascus (around 706) are magnificent examples.[2] There is no question that, in contrast to the non-Byzantine Christian world, there were the resources and the skilled masons available for widespread construction in stone, and the new buildings created in the caliphate were of higher quality than anything contemporary in western Europe. It is rather less clear, however, how far this extended to defensive structures. With the empire constantly expanding, the need for defences may have seemed secondary, although there are sufficient references to such work to suggest that it was part of the scheme of things. In addition, in all those places that were to remain important, substantial rebuilding over the centuries has obscured any work from this first period. On the Byzantine frontier, the nature of the conflict was such that the Muslim frontier was usually based in cities and fortresses recently held by the enemy, in which case there may

[1] From *Ajbar machmuâ*, 170; see Chapter 3.

[2] See Ettinghausen and Grabar, *The art and architecture of Islam*, 17–23, for a summary.

not have been much call for anything other than repair and upkeep work. However, it seems likely that the Umayyads were responsible for some new defences at Amman in Jordan, where the ruined Roman defences of the town and citadel were substantially restored. The curtain walls were from 2.5 to 4 metres thick, with shallow buttresses added to strengthen the new facing, and at least four of the projecting rectangular towers of the upper citadel were of this period, one being closely dated to 724–44.[1] In al-Andalus the presence of "numerous cities and excellent fortresses" was apparently cited by Mūsā as a reason for the caliph not to withdraw from the newly conquered provinces.[2] The cities remained the centres of Muslim rule, government, and military operations.

The many rural complexes, some of them based in former Byzantine fortresses, others newly built, sometimes had defences (for example a complete walled enclosure) and sometimes also a small fortified structure resembling a later medieval castle. But they were primarily agricultural centres, boasting of houses and baths as well, and the residence of a powerful landowner. Some were in effect, desert palaces. They sometimes resembled caravanserais providing shelter on desert roads. Their defensive purpose was minimal.[3] However, in the frontier regions where the Byzantine side was providing itself with a network of local defences, on the Muslim side similar steps were being taken as the military balance became more even. Small *ḥuṣūn* were built, increasingly so under the Abbasids, and these had a clear military character, as they were populated by settlers from other regions, enticed by salaries or sometimes grants of land, but largely held by single men.[4] In al-Andalus, as the displaced new Umayyad caliphate asserted itself, there was also a widespread use of rural *ḥuṣūn*, the function of which seems to have been as centres for tax collection as well as popular refuges, others being used as centres of local revolts against the centralised state and being destroyed as a consequence if not required by the government thereafter.[5]

In the urban centres, little survives from this early period. We have noted the wall at Amman, and the arsenal and defended harbour at Tunis was started in around 732,[6] but most defensive building work known from either archaeology or descriptions dates from later. Islamic rule continued to make use of the cities, but they took different forms and designs, though the defences continued to be a crucial state responsibility.[7] But the Abbasid fortifications of Baghdad, after they moved their capital there, were to astonish everyone who saw them, ensuring a number of descriptions. This was in all respects an exceptional construction, begun from scratch in 758 (or 762) at the order of Caliph al-Manṣūr and ready for occupation seven years later. The earliest detailed description, however, comes

1 Northedge, "The fortifications of Qal'at 'Amman", 437–57.

2 Ibn al-Quṭīyah, *The history of the conquest of al-Andalus*, 29.

3 Lézine, "Sur deux châteaux musulmans d'Ifriqiya"; Sauvaget, "Châteaux umayyades de Syrie", 9–45.

4 Haldon and Kennedy, "The Arab–Byzantine frontier", 109–12.

5 Acién Almansa, "Sobre la función de los *ḥuṣūn*", 263–6.

6 Bakrī, *Description de l'Afrique septentrionale*, 90. This eleventh-century source also describes the walls at Tripoli, Gabès, Aïn Djeloula, Sousse and Ceuta.

7 As with the cities of Syria: Kennedy, "From *polis* to *madina*", 4–26.

from 150 years later. According to Ya'qūbī, al-Manṣūr assembled "engineers, men with a reputation in the art of construction and surveyors skilful in measuring lengths and surfaces". Located where the rivers Tigris and Euphrates almost met, the new city was perfectly round, with four gates each two kilometres apart (but Ya'qūbī's description gives a total diameter for the whole complex of 2,900 metres; the inner citadel had a diameter of 2,286 metres), protected by double doors of iron. The walls, made of local brick, were no less than 30 metres high to the top of the merlons, 45 metres thick at the bottom, narrowing to twelve metres at the top. The bricks were themselves immense, being half-metre cubes, and no less than 162,000 were needed for each course. Bonding was achieved with reeds. As if this was not enough, there was an outer wall some 50 metres in front, with solid round towers, and in front of that a wet moat fed from a canal. According to another description, the dimensions were somewhat less (the main wall "only" 18 metres high). The workforce required must have been immense, although the 100,000 suggested as permanently employed, apart from its suspicious roundness, makes one wonder where they all came from, and how they were fed and housed during the work. The outer defence was removed at the end of the ninth century, by which time the whole vast complex was surrounded by markets and dwellings appropriate to such a metropolis, both government centre and trading emporium.[1] This immense monument seems actually to have been rapidly swallowed up as the city grew, and to have fallen into disuse after its walls were stormed in civil strife in 811. Certainly, nothing survives to confirm or change the picture. However, it can be shown that other architectural features like bent entrances (found at some of the desert castles) and machicolations were known to builders at this time.[2]

With substantial resources in terms of wealth and skilled workers available, the Arabs were both building on existing sites, and erecting new fortifications. In al-Andalus, perhaps spurred on by the presence of a large conquered population, the construction of a citadel (*alcazar*) to secure control in the cities from an early date is hinted at. When Mūsā and Ṭāriq were recalled, the commander "strengthened the fortresses of al-Andalus". In 763, the new Umayyad ruler and founder of a long-lasting dynasty, 'Abd al-Raḥman I, taking refuge from the Abbasids who had now gained the caliphate, was besieged in the citadel of Carmona, which must have been built between 711 and that year. The famous citadel of Toledo was first erected inside the town, in order to control it, in 797.[3] Similar reasons seem to have lain behind the building of the new citadel of Mérida, reconstructed after its initial capture in 713, and strengthened in the next century in the form of a fairly primitive structure designed to ensure control of the water supply, and probably with a strong symbolic function as well.[4] It would be strange if the building

[1] Wiet, *Baghdad*, 8–16; the Arabic sources are analysed in Le Strange, *Baghdad during the Abbasid caliphate*, 15–26, 42–5. There are considerable discrepancies in the overall dimensions that cannot now be resolved.

[2] Creswell, "Fortification in Islam before A.D. 1250", 89–90, 100.

[3] Ibn al-Quṭīyah, *The history of the conquest of al-Andalus*, 21 (Musa), 75 (Carmona), 105 (Toledo).

[4] Valdés Fernández, "El urbanismo islámico de la Extremadura leónesa", 160–63. In

materials used were not from local sources and the builders recruited from local artisans for the construction of the defences of the new regime, even if for quite some years the main enemies were to be found in rebellion and feuding between the different Arab and Berber factions seeking control of the rich new territory. Certainly, a detailed study of techniques of building of foundations (for example, the use of reinforced layers of stones to underpin the walls and towers, and of projecting layers of stone at the foot of walls to provide greater stability, in particular on mountain sites), shows continuity with not just the previous regime, but also with their Roman predecessors.[1]

Important military centres were built such as Kūfa and Wāsiṭ in Mesopotamia, Fuṣṭāt (Cairo) in Egypt, and Kairouan in Africa. Tunis was founded in order to provide a secure naval base for newly-conquered Ifriqiya in about 700, and a thousand Coptic families were settled there from Egypt.[2] Kūfa also supports the contention that Arab conquerors had to employ local skills. They had to "assemble everyone who knew architecture" to establish the enceinte. Similar terms are used to describe the beginning of work at Kairouan.[3] The late-eighth-century wall of Arab Tudela in al-Andalus was two metres thick, making it similar to the late Roman walls of the country.[4] Adequate survivals of fortifications from the time of the first settlements in al-Andalus are rare, as we have seen. Such poor scraps of information as do survive, or have been studied archaeologically, do not allow a comprehensive understanding of the fortifications constructed by the victorious Arabs of the period of the conquests, unlike the far greater store of information (and surviving remains) that date from later times.

Conclusion

In less than a century, the old post-Roman world had been transformed out of recognition. From Asia Minor to the Pyrenees, the new Arab empire held sway, and reached simultaneously to the borders of India and China. Drawing initially on the culture and skills of the new lands now under its control, this empire rapidly became a new centre of civilisation. Among the skills taken on board from their new subjects were the arts of siege warfare, in which the Muslim Arabs had been originally ignorant. Building skills, including defensive works, would follow. Having had to confront and overcome both Iran and Byzantium, the new

south-eastern Spain, the rebellions of the eighth century led to the destruction of many existing urban centres and a hiatus before their replacement in different form during the following centuries: see Gutiérrez Lloret, "Ciudades y conquista", 137–51. For a general picture comparing Spain with Syria, see Kennedy, "From antiquity to Islam", 53–62. (All three articles are published in the same volume.)

1 Martínez Lillo, "Estudio sobre ciertas elementos y estructuras de la arquitectura militar andalusí", 11–32.

2 Ṭāha, *The Muslim conquest and settlement of North Africa and Spain*, 71–2.

3 Ṭabarī, *La chronique*, v/vi: 172 (the year is 640/41); Ṭabarī, *The History*, xviii, trans. M G Morony, 102 (the year 670/71); Ibn al-Athīr, *Annales du Maghreb & de l'Espagne*, 19.

4 Maldonaldo, "La muralla primitiva árabe de Tudela", 30.

power had now posed a threat to both Italy and Gaul. Here, another new empire was about to take shape, the creation of which would set the map of Europe for the medieval period, in which the centres of power would shift from the Mediterranean to the north. The realm of Charlemagne, based on the newly reunified kingdom of the Franks, would face not only south but east and north as well.

3 The age of the Carolingian empire

Background

WESTERN EUROPE in the two centuries after 476 had entered what used to be called the Dark Ages. The darkness of the time was as much to do with the shortage of surviving information as to the reality of the time itself, and certainly archaeology has cast more light than before on the sixth and seventh centuries. For all the revisions, though, the kingdoms of the Franks and Visigoths, the Anglo-Saxons and Lombards, still appear as poor, backward and little developed. Whereas the view that siege warfare had little place in any of these new realms has been challenged,[1] it remains to study developments in this area as the empire of the Franks achieved its apogee at the turn of the ninth century, with economic development and greater political stability creating the conditions for a revival of culture, art and architecture.

The continued use of walled Roman cities as vital centres of power, particularly in the heavily romanised south of Gaul and northern and central Italy, made it inevitable that in the numerous wars following the collapse of imperial rule methods of defending and attacking these cities must have developed. The use by Germanic warlords of the surviving and inherited skills of masons and artisans was to be expected. At the same time, it has been shown that the Goths in Italy and Spain and the Franks in Gaul, although much of their warfare did revolve around the cities, were able to deploy only crude and limited versions of the weaponry and techniques of the Roman empire, while in Britain the Anglo-Saxons, having originated from areas less touched by Rome, leave no evidence even of these vulgarised skills.

If the Merovingian dynasties managed to maintain a relatively peaceful existence (except for wars between wings of the ruling family) in the seventh century,[2] and if the Anglo-Saxons conducted their conflicts largely outside the remains of the Romano-British cities, a decline in the call for skills of building defences, and of defending and attacking them, might be expected. There is no reason, however, to argue for their complete disappearance, and evidence for the continued existence of some of these skills is to be found from the late seventh century onwards. What we do not know is whether skilled craftsmen had to be imported or whether there were native artisans who could be mobilised. We know that,

[1] D R Hill, "Siegecraft from the sixth to the tenth century".
[2] A good summary in Fouracre, "Frankish Gaul to 814".

however limited their numbers, craftsmen continued to find a market for their skills with the ruling families and church, which alone had the resources for undertaking substantial building work. We know that the establishment of the Carolingian empire created the environment for an expansion (called traditionally a renaissance) of art and architecture. We also know that the written evidence for this comes from people whose concern was to glorify the Carolingian dynasty, and this was the more effectively done by downplaying the skills and culture of their predecessors. It is not to diminish the achievements of Charlemagne (r. 768–814) and his forebears Charles Martel (d. 741) and Pippin III (d. 768) to suggest that their works were not entirely created from scratch. In fact, of course, the Carolingian family came to power by a prolonged struggle to achieve pre-eminence in the kingdom ruled by the later Merovingians. It was in 751 that they arranged papal authority to set aside the last king of that enfeebled dynasty and took over a reunited Frankish realm in the north and centre of France. The course of their rise to power, and its establishment in the south of Gaul, was marked by incessant warfare, just as the creation of their new empire, stretching from the Elbe to the Ebro, was the result of continuous and successful wars of conquest.

Siege warfare and the Carolingians

Pippin I was no stranger to attacking fortified towns, although no details are given of his battle in 680 at the *castrum* of Namur, or, in 689, his defended camp at the emporium of Dorestad.[1]

Much of the evidence comes from the Carolingians' warfare in Aquitaine, where the local rulers were unwilling to give up their independence. In 721, Charles Martel had besieged Angers but returned home without success. Late Roman defences had proven their worth again. But in 735 he marched to the river Garonne and occupied Blaye (described as a *castrum*) and Bordeaux (an *urbs*), and other towns in the area. It is not said whether he had to fight for them. The following year he captured Lyons (an "urbs munitissima") and penetrated as far as Marseilles and Arles.[2] In 737, he had to recover fortified towns seized by Arab (or Arab-led) raiders from across the Pyrenees. The recapture of Avignon sounds like a full-scale operation, and the description of the use of siege weaponry implies that this was not anything unusual. Indeed, it looks as if the Frankish army knew well how to deploy to attack such defences (Plate 7). The terms used suggest battering rams and means of scaling the walls.[3] Charles went on from this success to capture Narbonne as well, deploying a similar approach if the two sources are to be believed. These places taken in short order by the Franks were all cities with late Roman walls.[4] We know that these same towns had been the subject of

[1] *Annales Mettenses priores*, 17; Fredegar, *Chronicorum liber quartus*, 84.
[2] *Annales Mettenses priores*, 28, 29; Fredegar, *Chronicorum liber quartus*, 91.
[3] *Annales Mettenses priores*, 29; Fredegar, *Chronicorum liber quartus*, 94.
[4] R M Butler, "Late Roman town walls in Gaul", 25–50.

frequent attack in the sixth century, and at that time had usually resisted successfully except when taken through treachery. We also know that at least their Arab defenders (although we do not know whether they were small in number) were probably familiar with siege warfare. Charles Martel's successes therefore suggest a considerable development in these skills.

Success did not stop there, either. In 741 Charles had to besiege Laon, and then turn his attention south again in the following year, and among other fortresses (*firmitates*) conquered the *castrum* of Loches. The monk notes, with obvious pride, how *virilis* the conqueror was! Following expeditions against the Saxons, the Carolingians came south again in 753, when Pippin III besieged Narbonne (which must therefore have been recovered by the Arabs), but in this occasion failed to take it.[1]

In 761 Pippin's first targets were Bourbon and Clermont. At the first, it is recorded that he destroyed the *castrum*, while at the second he stormed it and burnt it down. The next year, he was back on a trail of mayhem, if the chronicler is not exaggerating, "destructisque castellis et munitionibus", finishing off with the conquest of the "very strong" city of Bourges and the destruction of the *castrum* of Thouars, "of which there was none stronger in Aquitaine". The capture of Bourges apparently involved the surrounding of the place with fortifications and the deployment of "machines and all kinds of weapons". Once again, this implies the use of battering rams to create the breach, as we have not yet met any examples of mining being carried out in the west. After its capture, the king ordered the repair of the walls, perhaps because of the damage inflicted during the siege.

In a telling phrase, the chronicles state that the Aquitanian ruler noted Pippin's use of machines at Clermont and Bourges, and realised he could not stop him. Instead, he ordered the destruction of defences he could no longer defend, an order that Pippin reversed, obviously now determined to use the repaired walls as a bulwark for the extension of Carolingian control.[2]

Many issues are left unclear by these accounts. The Franks must have had the experience and the expertise to capture fortified sites, but we do not know where they acquired it from. It required skill and prior knowledge to build equipment suitable for storming or breaching late Roman walls, and the resources to pay for the builders of the equipment. From what is known of Frankish armies of the eighth century, they consisted mainly of foot soldiers, and would be based on the followings of the major nobility. The main incentive to campaign was the prospect of securing booty. The cities, especially those of the south, were held by local levies, originally all free Frankish settlers, responsible for equipping and feeding themselves, although later on this was not necessarily the case. It was perhaps from the cities already held by the Franks that the skills involved in siege warfare were drawn, or even perhaps from descendants of the original Romanised garrisons, as Bachrach has suggested. In some respects the siegecraft inherited by

[1] *Annales Mettenses priores*, 32, 43; Fredegar, *Chronicorum liber quartus*, 98.

[2] For all the above, *Annales Mettenses priores*, 51–4; Fredegar, *Chronicorum liber quartus*, 111–14.

the Carolingians might be said to be based on that of Rome.[1] However, while the sources give testimony to the knowledge of battering rams and scaling ladders, they make no reference to Frankish use of artillery, even though it seems to have been known to their southern Visigothic neighbours. The absence of a reference does not mean the absence of the weapons, but the nature of the accounts suggests that had Charles and Pippin used *ballistae* or stone-throwers consistently, we should have heard of it. It may therefore be safe to conclude from the absence of any evidence, and given the likely decline in the considerable technical skills needed to build, maintain and use it, that the torsion-operated *onager* went out of use in the Frankish kingdom at some point during the reign of the Merovingians. The new, simpler lever artillery had not yet reached here from the east, either through Byzantine south Italy or via the Arabs of al-Andalus. Or if it had, either it was not needed yet, or else the expertise to construct and operate it was still lacking. Artillery simply appears in the chronicles a little later, without any indication of its first arrival or use.

The chronicle accounts are also unhelpful about the state of the defences against which the Franks had to use their skills and equipment. References to how strong the various fortifications were may reflect the writer's determination to glorify the conqueror rather than actual knowledge of the state of the walls. Indeed, given the unlikelihood of the chroniclers' ever having seen many of the places described, it is far more likely that such descriptions were based on ritualistic formulas. By now, the walls of the former Roman *civitates* were more than 450 years old, and we simply do not know whether they had been maintained. Nor do we know whether the garrisons of the various towns were of a sufficient size to defend the entire length of the ramparts. Some of the cities were of considerable size,[2] and their populations had probably shrunk rather than grown during the intervening periods.

Cities were not the only defences encountered. There were also former Roman *castra*. That at the site of Loudun was an irregular enceinte on a hill top, and coin evidence indicates that a smaller palace within it was used by the Merovingians throughout.[3] Not all the defences were dependent on Roman survivals, and there are suggestions that some rural sites may have been later, although definitive proof is lacking. A study of the other sites mentioned in the chronicles has shown instead that there were a large number, constructed on roughly similar lines. Fournier examined the sites of all the places mentioned in the chronicle accounts of Pippin's campaigns in the Auvergne during the wars with Aquitaine in the mid eighth century.[4] His account makes clear that the south of Gaul was well provided with defended headland sites. What is not so clear is whether they were

[1] Bachrach, *Merovingian military organization*, 14, 110–11, 127; Bachrach, "Military organization in Aquitaine", 1–11.

[2] Bordeaux: 80 acres (32 ha); Carcassonne: 17.5 acres (7 ha) (here the Roman wall was perhaps rebuilt by the Visigoths); Chartres was 800 × 250 metres. R M Butler, "Late Roman town walls in Gaul", 28–30; Blanchet, *Les enceintes romaines de la Gaule*, 203.

[3] Charbonneau-Lassay, "Les châteaux de Loudun", 116–17.

[4] Fournier, "Les campagnes de Pepin le Bref".

of recent or ancient construction, reoccupied at times of danger such as Pippin's invasions, and this may never be known.[1] They were not part of any scheme of regional defence, with all the implications for central control and planning implied by such a phrase, but local defences used either by local landowners or villagers (or perhaps both), their sites dictated by geography, proximity to local settlements, or quite simply the existence of older ramparts there. They would not call for the deployment of particular siege-warfare skills to capture them, the deployment of such skills being required only by the cities with their late Roman stone walls and towers, and only there if the stonework had been maintained in decent repair rather than being allowed to decay, or perhaps being robbed for building materials. Many city walls would be found to be in pressing need of repair when they were later needed against the Vikings. The Carolingians, however, recognised the value of city walls during their conquests and took the steps necessary to render them defensible again after capture. We have suggested that this was because of the damage inflicted during capture. It is also possible that they were in bad repair before capture and needed repair to be make them useful again.

Charlemagne

The empire of Charles the Great and his immediate successors set the shape of western Europe for centuries to come, and later writers looked back to it as a golden age. We may doubt if the Frisians, Saxons, Bavarians, Avars and Lombards, previously independent, saw it that way at the time. Their incorporation into the empire was the result of conquest and in many cases, particularly for the Saxons, of brutal and bloody repression. The story is told by the victors, and their natural bias needs to be borne in mind at all times. For the reign of Charlemagne we have a few chronicles and "biographies", and some fascinating preserved enactments of the great king. For the ninth century, the sources increase in number and so do the documents. They have all been mined repeatedly by historians. There are also a growing number of excavated sites that can be traced back to Charlemagne and his sons and grandsons, providing for the first time since the Romans a clear indication of the type of defences being thrown up.

There was a military expedition in almost every year of Charlemagne's long reign. Facing him across the Rhine, which represented a frontier of romanised Gaul, were the Germanic tribes who did not know walled towns. They were not, however, defenceless, and earthwork ring-forts seem to have been a common feature. Certainly, the Frankish campaigns against the Saxons involved several attacks on and captures of such sites. Eresburg and Syburg were two such fortifications. The Franks were familiar with the nature of these defences, as they had encountered them before. In 748, Pippin had captured a Saxon *castrum* and for forty days "destroyed their castella".[2] Charlemagne's repeated incursions into Saxony in the 770s involved capturing them again. In 776, the Saxons rose in

[1] James, *The Merovingian archaeology of south-west Gaul*, 291–3.
[2] *Annales Mettenses priores*, 41.

revolt and attacked the garrison left by the Franks at Eresburg. The chronicle says that they prepared catapults but that these did more harm to the attackers than to the defenders of the fortress, indicating that they must have been stone throwers.[1] The annals use the word *petraria*, confirming that was a stone thrower, but they are of later compilation, which makes them of questionable authority. The notion of barbarian peoples trying to emulate their "civilised" neighbours and failing is already a common theme among chroniclers. Nonetheless, the reference is there and offers evidence of Charlemagne having a siege train with his army, and of the Saxons trying to copy it.

There is further evidence for this in the Carolingian capitularies. Two dating from the first decade of the ninth century communicate to the king's bishops, counts, abbots and nobles what they must send to his army in the way of soldiers and equipment.[2] The list includes slings (*fundibalis*) "and those men who know well thereof how to throw them" and wagons with a whole range of tools such as hammers, axes, shovels and picks. Whether for having the equipment to fortify an overnight camp or to prepare to attack a walled or palisaded earthwork, such a considered list is evidence of the careful thought given to the need to set out well prepared for the eventuality.

The Franks in Italy

Even if the Saxons were not opponents sophisticated in siege warfare or building techniques, their conquerors had ample experience from elsewhere of these skills. Another highly urbanised area offered yet further opportunity to test out these skills: Italy. In the eighth century, the Italian peninsula was divided between separate Lombard realms north and south of Rome, a tenuous and under-resourced Byzantine province in the far south and at Ravenna, and an area in the centre effectively ruled by the popes. As Arab power spread across the Mediterranean, Italy also became vulnerable to seaborne raids from the Muslims. The chief cause of intervention by the early Carolingians, however, was the threat posed to the papacy by the Lombards, which the Byzantine forces were too feeble to prevent. In 755, Pope Stephen II had turned instead to King Pippin, writing of the devastation wrought by the Lombard king Haistulf, who for 55 days had pressed on Rome "night and day with the greatest fury incessantly with diverse engines".[3] We noted in Chapter I the growing familiarity of the Lombards with siegecraft. At the end of his long record of the Lombards, Paul the Deacon reported King Aripert's capture of Bergamo at the beginning of the eighth century, "storming it without any difficulty with battering rams and various machines of war", including

[1] *Annales regni Francorum*, 44; trans. in *Carolingian Chronicles*, 53.

[2] *Capitularia regum Francorum*, I, no. 75 (p. 168) and no. 77 (p. 171). Discussed by Halphen, *Charlemagne et l'empire carolingien*, 169; France, "The military history of the Carolingian period".

[3] *Epistolae merovingici et karolini aevi*, 499; *Pauli continuatio Casinensis*, 199.

perhaps stone-throwing catapults (*manculis*).[1] There are accounts of the capture (716) of the *castrum* of Cumae, and they had taken Narni from the Pope (718) and besieged Ravenna for several days.[2] The need to maintain or restore old walls was a major secular concern of the Popes. Gregory II (r. 715–31) had ordered work on the walls of Rome, but was prevented from completing it. His successor Gregory III (r. 731–41) resumed the work, paying for the workers' rations and the limestone out of papal revenues. He also had the walls of Civitavecchia, reportedly quite destroyed, rebuilt "stoutly" from their foundations.[3]

The Franks responded more than once to papal requests for help. Pippin's army was powerful enough to bring the Lombards to terms, although the Franks surrounded but did not attempt to capture the powerful capital city of Pavia, staying there only a few days. Although the much later account of this makes an implied contrast between this and Charlemagne's later operation, to the detriment of the former, it may just be that Pavia was too strong to be assaulted and that Pippin was not seriously contemplating conquest.[4] His campaign involved getting among the Lombards' fortified places, as we are told he destroyed a number of them.[5] No further details are offered, but the impression is given that Pippin's army was quite capable of storming, or forcing to surrender, lesser defended sites, but a major walled city in a strong position was out of reach; and with the initial objective secured, a treaty was concluded. Pavia had been the main Lombard capital since the mid seventh century, presumably because of both its strength and its location. The Roman walls have disappeared, but it seems that the town had a rectangular plan, on the banks of the river Ticino. The circuit of less than 3,000 metres made it smaller than some other north Italian cities (for example, Milan[6] and Bologna). There is an unresolved debate as to whether the walls were extended by Theodoric, or whether he simply restored the existing rampart and gates originally built in the third century.[7]

When Charlemagne brought the Frankish army back in 773, it was a different story. An enormous Frankish army was mobilised from all quarters of the growing kingdom to descend on northern Italy, culminating in a siege of Pavia. This time the victory was conclusive, with the city being captured through a long siege. It seems that Pavia was cut off with a strong encircling rampart and that the Lombard king Desiderius, perhaps recognising the hopelessness of his cause, with the Frankish host large and well entrenched, decided to surrender, although only after

1 Paulus diaconus, *History of the Langobards*, 265 (book VI, chap. XX); *Historia Langobardorum*, 171: "cum arietibus et diversis belli machinis."

2 *The lives of the eighth-century popes*, 5, 9.

3 Ibid., 27–8.

4 Einhard, *The life of Charlemagne*, 60.

5 *Annales Mettenses priores*, 48–9.

6 The walls of Milan were rebuilt and extended in the late fourth century and stretched 4,500 metres; they may have been refurbished after 568: Calderini, "Milano archologica", 493–507; Arslan, "L'architettura dal 586 al mille", 502.

7 Hudson, *Archeologia urbana*, 14, 42 (plan, p. 62); Maccabruni, *Pavia*, 14–15; Fagnani, "Il tracciato delle mura romane di Ticinum", argues for extensions of the ramparts by Theodoric. The later sources disagree.

a long resistance ended by hunger and disease.[1] Other Lombard cities, including what the *Lives of the Popes* called "the strongest" at Verona, rushed to make terms. The surrender of Verona must have been a relief to Charlemagne, as it had truly vied with Pavia for impregnability. Here there are remains of the walls to permit substantiation of the chroniclers' claims. The third-century walls themselves represented a strengthening of earlier defences, and the rampart was extended to enclose the famous amphitheatre. The town stood in a loop of the river Adige, with a stream cutting off the whole site. On the far bank, connected with two Roman stone bridges, the walls culminated in an *arx* or citadel. These walls were to serve Verona until the massive rewalling carried out by the city's commune in the twelfth century.[2] This time, however, they were not defended.

To secure his new north Italian domain, Charlemagne left garrisons in Pavia and in cities that had revolted like Cividale and Treviso.[3] The Lombard kingdom in northern Italy was unable, finally, to resist the resources mobilised by Charlemagne's expanded realms. The collapse and disappearance of the kingdom in a single campaign was dramatic testimony to the superior resources of the Franks, rather than to any superiority of technique or skill. Unfortunately, the sources only allow us to guess at the siege skills and techniques available to Charlemagne in his conquest of the Lombards, as they so often do with his other victories. The history of the creation of the Carolingian empire is largely a history of the capture of cities, towns and *castra*. In common with much of the later medieval period, actual battles were few and far between. Charlemagne campaigned almost every year of his long reign, yet fought only a small handful of battles. Most of the later studies of his army have dealt with the arms and equipment of his soldiers, and discussed the question of the developing role of the mounted warrior. But the empire was made, and secured, by the capture and garrisoning of defended settlements and *castra*, operations in which foot soldiers, engineers and artisans and workmen were the vital participants.

The Avars

Against the Avars, occupying the territory of modern Hungary, different approaches were needed. There were no Roman cities here, but instead the Avars had constructed massive fortifications designed to protect their families and their goods, including, according to the accounts of the Franks, the immense wealth they had plundered from their neighbours over the years. Notker reports being given a description by a witness of a great earth and stone ring, six metres high and

1 *Pauli continuatio tertia*, 213.

2 Barbetta, *Le mura e le fortificazione di Verona*, 21–30 (plan, p. 11).

3 *Carolingian chronicles*, 50, 53; Einhard, *The life of Charlemagne*, 61; Notker, *Charlemagne*, 164–5; *The lives of the eighth-century popes*, 136. Notker recounts the story of Desiderius' despair as the immense size and power of the Frankish host unwinds as he watches from a high tower of Pavia, including specific reference to the baggage train. He also confirms that the capture was brought about by a blockade that must have been at a little distance, because he says Pavia was captured "without any shedding of blood, but simply by his [Charlemagne's] military skill."

the same width, topped by an impenetrable hedge.[1] Great logs of oak, beech and fir made the outer frame, filled in with stones and clay. Though their adventures had brought them up against powerful stone defences, and they had adapted their techniques to deal with them, the Avars had not applied the same skills to their own home territory, perhaps because of the absence of suitable material or of suitably skilled builders. The rings represented barricades, but the account suggests that they could not actually be defended. Charlemagne's wars against the Avars lasted over a long period (788–803) but in the end these defensive "rings" did not prevent their defeat and the ending of their power.[2] It seems that here, the Franks decided to leave the territory devastated, plundered of its wealth and depopulated rather than build new forts for garrisons.

Campaigns against the Arabs

More traditional defences were encountered in Charlemagne's and his son Louis's conflicts with al-Andalus. The most famous incident of these wars may have been the ambush of the rearguard of the withdrawing army by the local Basques at the pass of Roncesvalles in 778, but the campaigning that had preceded and followed this episode centred around the attack and defence of strong places. The terrain of the struggle was the land that became, for the Franks, the "Spanish march", and for the Arabs also represented a frontier area in which their control was not consolidated. The effective boundary of al-Andalus became the river Ebro, with a powerful fortress at Tortosa representing a vital military base. Further upriver, Zaragoza was the seat of a powerful local lord. Further east, Lérida and Huesca were also important Muslim settlements. It is necessary to assume that it is safe to push backwards in time the tenth-century account of Rāzī to imagine the position at the end of the eighth.[3] After the loss of Narbonne to the Franks, walled cities at Gerona and Barcelona became the front line. However, we should not imagine that early medieval states had either the conception, or indeed the control, to create strategic lines of defence. Control was exercised locally, and fortified places served either as refuges, or as centres of political power, from which control of local areas was wielded. In al-Andalus, the recently established Umayyad power resided in distant Córdoba. The opponents facing the Franks were local.

Charlemagne's first campaign in 778 used two large armies. He captured and destroyed Pamplona, and the Arab leaders, faced with such overwhelming power, made peace.[4] For the expeditions of Charlemagne's son, the future Louis the Pious, between 800 and 811, there is considerably more detail available, although

[1] Notker, *Charlemagne*, 135–6.

[2] For an account, see Pohl, *Die Awarenkrieg Karls des Grossen*; Bowlus, *Franks, Moravians and Magyars*, 48–9.

[3] Rāzī, "La 'Description de l'Espagne'", 72–3. Rāzī describes an unidentified place called "Barbutania", which he explains caused the Muslims much difficulty in capturing it, but thereafter "they made it a barrage against the Christians" (p. 75); Manzano Moreno, *La frontera de al-Andalus en época de los Omeyas*, 76–98.

[4] *Carolingian chronicles*, 56.

the evidence is sometimes confusing.[1] Louis had launched a major raid in 800 and secured the verbal submission of Zado. In 803 he returned to Barcelona with an army drawn from his own subkingdom of Aquitaine, Burgundy, Gascony and Provence, besieging it for seven months until the food ran out. The defences of Barcelona have been described, occupying part of the heart of the modern city. Although then it was a small provincial town, it was strongly situated and obviously considered important enough to justify a long siege. The walls dated probably from the third century, with a mix of round and square towers and gates guarded by round or octagonal towers. Where the later bishop's palace stood was apparently a citadel added to the circuit.[2]

The so-called Astronomer's *Life of Louis* and the Moissac Chronicle both suggest that the capture was brought about through an effective blockade, with the town closed off on all sides to prevent access or egress. It may be that part of the army was deployed to prevent relief, as the garrison appealed to Córdoba for help. Ermoldus Nigellus' verse biography suggests attempts at attack with missiles and slings but that the walls were too strong for the machines,[3] implying that late Roman walls could resist these Carolingian stone throwers. After the capture, a garrison was left. Obviously, Louis wanted to hold onto Barcelona. The same accounts also indicate several attacks on Tortosa. If the first two failed against the town's defences, the third was pressed with great vigour. Here we have explicit references to rams, mangonels and mobile shelters, as a result of which the walls were breached and the town surrendered after forty days of siege.[4] Oddly, this glorious success is not reported elsewhere. Another Frankish force tried its luck against Huesca, without success. In 811, a treaty was arranged.

So, faced with strong walls and experienced defenders, the Carolingians deployed an array of weaponry and skills, but were still unable to capture a small though strong town except by lengthy blockade. In this, however, they succeeded, despite the heavy demands for resources and provisions that such a long operation must have called for. But their attempts to extend their conquests further on seem to have met varying success, despite a predisposition to support them among many local leading families,[5] indicating the continued problem of warfare in areas with strong (albeit now old) defences.

[1] The sources are: *Chronicon Moissiacense*; Astronomus, *Vita Hludovici imperatoris*; Ermoldus Nigellus, *Carmen elegiacum de rebus gestis Ludovici Pii*. There is a clear analysis of events by Auzias, "Les sièges de Barcelone, de Tortosa et d'Huesca".

[2] Described by Puig i Cadafalch, Falguera and Goday, *L'arquitectura romànica a Catalunya*, 1: 61–2, 74–9; see also Balil, *Las murallas bajo-imperiales de Barcino*.

[3] *Chronicon Moissiacense*, 307; Astronomus, *Vita Hludovici*, 612–13; Ermoldus, *Carmen elegiacum*, 582–6: "machinas nulla valet murorum frangere".

[4] Astronomus, *Vita Hludovici*, 615: "adeo illam arietibus, mangonibus, vineis et ceteris argumentis lacessivit et prostravit muralibus".

[5] Abadal, 'La domination carolingienne en Catalogne', 319–24.

Defensive structures at the time of the early Carolingians

The history so far shows the concern of the Carolingian rulers to secure their control through the capture, and holding, of fortified places. In Italy and Spain, this revolved mainly around the existing stone-walled cities. It is not clear how the garrison was maintained, as the soldiers necessarily returned home after the campaigning season, reverting to their role as peasant farmers. For a very long time, the original walls of Roman cities had to suffice, too. The evidence for additions by post-imperial rulers is scanty at best. This should not be a surprise. If the skills required for stone quarrying and stone masonry survived after the period of late antiquity, the social organisation for major building projects had disappeared. Continuation of stonework depended on the presence of both local sources of stone (the Roman systems of long-distance transport having long since vanished), and on a wealthy ruler to provide resources. What happened to the craftsmen where these two factors were absent is unknown.[1] Ordinary manual labour was perhaps easier to find, as it seems that the custom of having to provide labour services survived in varying degrees. We have already seen the popes mobilising the Roman people to repair the old walls of their city on more than one occasion. Facing the danger of regular raids by the Arabs instead of the threat from the north Italian Lombards, the popes found they had a continuing secular role in providing for the protection of their churches and the city. In the early 840s, Gregory IV had to restore the fortifications of Ostia, which he did by raising the height of the walls, with "crenellations and trap doors, and on top he arranged catapults".[2] Perhaps Ostia and Rome provide the only definite example of a new building exercise in the whole of these centuries. In 847, following a disastrous Arab attack in which St Peter's had been sacked and looted, Pope Leo IV secured the financial help of the emperor Lothair (r. 840–55) to create a circuit of new walls around the Vatican, previously unprotected. Accounts show that a massive mobilisation of the pope's own resources was necessary: large numbers of labourers had to be drawn from the church's farms and those of monasteries, with each group assigned a stretch of wall. The work took until 852. Constructed of reused bricks, soft tufa stone and tiles, the wall, up to eight metres high, perhaps two and a half metres thick and three kilometres long, with 22 towers (*turres castellatae*) and four gates, represented a massive and partially still surviving circuit. It would seem from surveying the remains that the construction was modelled on the existing Aurelian walls of Rome itself.[3] However, the quality of work was substantially

[1] J B Ward-Perkins, "Quarries and stoneworking in the early Middle Ages", 537–42.

[2] *The lives of the ninth-century popes*, 68.

[3] See Gibson and Ward-Perkins, "The surviving remains of the Leonine wall", for a detailed survey; also Krautheimer, *Rome, profile of a city*, 119 (illustrated); Llewellyn, *Rome in the Dark Ages*, 264–5. Krautheimer seems not to have distinguished the original ninth-century wall from later additions. However, no one has actually been able to carry

inferior, and although this is explained by the need for haste, it is also likely to be because the skilled builders were in short supply.

It has been deduced that, if the Aurelian model was followed closely, the towers would have been roofed, and equipped with late Roman-style openings for the operation of ballistas. If that were so, the defences could not have mounted lever-type stone throwers. Nothing remains above the height of the wall-walk of the Leonine walls to confirm or refute the supposition. The admittedly scant other references we have cited suggest that the new type of weapon had been available in Italy for a long time, and it might be thought unlikely that in the desire to slavishly follow the Aurelian model, the pope's builders would have created something that made impossible the use of current siege weaponry as a defence.[1]

References to other more recent defences are too sparse, and remains on the ground too hard to trace, to establish to what extent more local building by Arabs or Lombards was part of the scheme. But Roman walls were reused elsewhere in the expanding empire too. In Bavaria, for example, brought under Charlemagne's control in 788, the former ruler Tassilo had apparently been using Roman *villae publicae* as his palaces. The Franks continued the trend, developing a number of palaces (called *curtes* in the sources), and in particular used the palace at Regensburg as their local headquarters and capital. New work included a circuit of walls along with a new chapel built from the stones of the old work. The hilltop site on the north bank of the Danube was described as impregnable, being built of large squared stones and with numerous towers. Further work would be done by the German king Arnulf at the beginning of the tenth century.[2] As the eastwards drive continued, other former Roman palaces were pressed into service, their provision of solid stone walls being too tempting to ignore, but usually evidence of Carolingian usage is scant, so conclusions can only be tentative: such as with the site of Baden in the new East March (Austria), which was identified as a *palatio* of Carloman's in 869, for which defensive features have been suggested, but not actually found.[3] Another possibly Carolingian fortress was built in the (later) bishopric of Győr at Sabaria (Szombathely), which seems to have comprised an inner ditched enclosure, with poorly constructed walls around two metres thick of a round defensive wall, surviving from much later medieval building work. Sabaria was a royal residence in 791, the head of a margraviate in 828, and a *civitas* by 860, lending strength to the attribution of serious building work to this era.[4]

Beyond the Danube there were no Roman walls, and here the conquests involved capturing the defended sites of other peoples, and reusing these sites, or creating new ones, as centres of the new empire. As an example of the latter there is the account of a campaign in 809 in which the emperor directed a count named

out excavations on the wall. On the theme generally, see B Ward-Perkins, *From classical antiquity to the Middle Ages*, 192–9.

1 Gibson and Ward-Perkins, "The surviving remains of the Leonine wall", 236.

2 Fastlinger, "Karolingische Pfalzen in Altbayern", 233–66. The contemporary descriptions are from Arbeo, *Vitae sanctorum Haimhrammi et Corbiniani*, 32, 35–6.

3 Halmer, *Burgen und Schlösser zwischen Baden, Gutenstein, Wiener Neustadt*, 16–17.

4 Tóth, "Die karolingische Burg von Sabaria-Szombathely", 151–73.

Egbert to build a fortress at a place identified as Eresfrith near the river Elbe, in the frontier area between Saxony and the Slavs to the east, after first exploring the site to determine its suitability. The presence or availability of skilled craftsmen to carry out the work was also indicated by a later reference. In 821, a bishop was accused of encouraging a rebellion in distant Pannonia by helping the leader construct *castella* by providing him with craftsmen and builders.[1]

Identifying the sites is rather more difficult, but in recent years careful excavation and the use of dendrochronology to obtain reliable dating has greatly improved the knowledge of fortifications of this period. Otherwise, left just with the earthworks, telling the age and provenance of a hill-fort is a hazardous exercise. In the areas later called Holstein and Mecklenburg alone, and at this time the land of Slav tribes, there have been identified no fewer than six hundred ring-forts dating before the year 1000.[2] A substantial Slav fort/settlement at Starigard (Oldenburg in Holstein), the main town in Wagria, was burnt and rebuilt on several occasions during this period. It had the form of a lozenge 400 metres long, and 100 metres across the middle, with substantial ditches and ramparts. Another major centre has been dated by dendrochronology to 819 at Lübeck, a strongly fortified trading post and princely dwelling now dangerously exposed to the encroachment of the East Franks.[3] Whether they were built for protection specifically against the Franks, or for other reasons, we cannot know, but they certainly served this role.

The Moravian "empire"

Further south, however, mention must be made of the works of the rulers of the Moravians, another Slavic grouping that established a substantial dominion east of Bavaria and astride the Danube from the 820s (although its actual extent is, remarkably enough, the source of unresolved scholarly dispute), and which kept going into the tenth century under a series of strong princes. A large number of enclosed sites have been excavated from this time and area. Of different sizes and apparently different functions, they have been classified as variously: multiply-fortified sites, acting as royal centres; simpler, provincial settlements; small enclosures serving as forts; and even smaller sites identified as lordly residences. There were also places identified as forming a frontier defence against the Franks to the west. The major settlement at Mikulčice had a complex plan and three separate areas, all surrounded by rivers. The principal fortification had an area of 7.2 hectares and was surrounded by a massive earth and timber rampart fronted by a stone wall, and contained the remains of stone-built palaces and churches dated to the mid ninth century. It has been argued that the abandoned ancient city of Sirmium may have been the chief base of the powerful Moravian rulers Rastislav and Svatopluk (Zwentibold), its ready supply of cut stone being an invaluable

1 *Carolingian chronicles*, 91, 109. The term used for builders was "murarios", implying a specific skill in wall-building.

2 Brather, "Karolingerzeitlicher Befestigungsbau", 115–20.

3 Gabriel et al., *Starigard/Oldenburg*, plan, p. 21; Fehring, "Origins and development of Slavic and German Lübeck", plan, p. 255.

source of building material for their own defences,[1] although there is no clear evidence for this. By contrast, some of the sites identified as having a purely military function were around 100 metres across. But they shared a common form of rampart.[2] These were formidable sites, representing centres of power and settlement for a developing state, and powerfully fortified according to traditional methods.

The "Saxon" frontier

The Saxons resisting Charlemagne's invasions had taken to reusing ancient hilltop earth and timber forts. Generally, these were very large, and served a multitude of purposes. They could be a refuge for the neighbouring population, a centre for administration by the rulers, as well as centres of crafts and trade, playing the part of a town. There were also smaller sites, serving as residences for the aristocracy. Sites used just as refuges would not have been permanently occupied. Frankish sites on the Saxon frontier have been described as ducal or royal hillforts, called in the sources *oppidum, castrum* or *castellum*. Such were the Marienburg at Würzburg, dating to the seventh century, while the Büraburg (in Hesse) has been shown to have been occupied from the late seventh to the ninth centuries. It featured mortared ramparts, ditches, a bastion tower and gates. Resisting Saxon attack in 774, the Büraburg site is 1,100 metres long and occupies eight hectares. It fills two-thirds of the hilltop, with an outer work occupying the rest. The oldest rampart (late seventh century) was 1.5 metres thick. Further strengthening widened it some decades later, and in about 750 it was again widened, to 2.7 metres. There were at least three quadrangular towers. There was a ditch 4.5 metres deep, and inside the foundations of many buildings. Mortared walls had been unknown in Germania before the Carolingians, and the similarity of style of the wall and towers to late Roman forts was notable. Whether the builders came from the once-Romanised areas west of the Rhine, where examples still stood around them, or whether the rulers brought skilled craftsmen from other parts of their empire, is impossible to say, but clearly the late Roman model remained the one to follow.

Some of the old Saxon sites were reoccupied by the Franks and became in their turn the nucleus for new settlement and the development of both new centres of administration (with the power in the hands of the new local lord, often the bishop), and of trade,[3] and sometimes the sites were new. An example is Hamburg, for long a frontier town against the Slavs, where the existence of a *castellum* is affirmed by the charter of Louis the German in 842. Another charter of the

[1] Bowlus, *Franks, Moravians and Magyars*, 174. Bowlus makes a strong case for the centre of the Moravian realm being south of the Danube, not, as the traditional argument goes, in modern Moravia.

[2] Staňa, "Mährische Burgwalle im 9. Jahrhundert", 158–90.

[3] Very good summary in Fehring, *The archaeology of medieval Germany*, 90–108; see also Uslar, *Studien zu frühgeschichtlichen Befestigungen*, 62–4. For the Büraburg, see Wand, "Die Büraburg bei Fritzlar", 175–85. See also Langen, "Die Bedeuting von Befestigungen", 185–93.

same king had given to the church of Regensburg land described as "the place where stood the ancient *castrum* called Herilungoburg".[1]

Some of the fortresses ordered by Charlemagne to secure his conquests can be located and examined. One was built at Hohbeck (on the banks of the river Elbe near Hannover) in 810. The remains are of a small, rectangular site 70 × 170 metres occupying 1.2 hectares. Ditches were dug on two sides. The earthen rampart six metres wide was reinforced with wooden posts and stabilisers. There had been a two storey tower, six metres square, that been burnt and rebuilt. Stonework is rare in the Lower Saxony Frankish fortresses, and dating remains difficult because the type of hilltop ring-wall, often serving the purpose of a refuge, remained popular until the twelfth century.[2]

The arrival of a new threat was to keep the question of fortifications high on the agenda of Carolingian rulers. The Vikings had begun their raiding at the end of the eighth century, when of course the full danger they posed to the lands they were presently just plundering was not apparent. Nonetheless, Charlemagne ordered measures astutely designed to minimise the effectiveness of the seaborne raiders, by having forts and watchtowers built at the mouths of rivers and estuaries. The same order was applied to the coast of Septimania, and was meant to stretch along the whole Mediterranean shoreline as far as Rome.[3] The enemy here were Arab raiders. He also commanded the repair of the old Roman ramparts of towns that one must assume had become neglected, as a result of there being no further apparent use for them, such as at Boulogne.[4]

Carolingian renaissance

Charlemagne and his successors encouraged a substantial revival of culture and the arts through the patronage of their courts, spreading out through the Church and embracing the higher nobility. They assembled scholars from around Europe at the court and supported the collection and copying of classical texts. The subsequent influence of late Roman and Byzantine styles in art and building is apparent.[5] The influence of Roman styles on Carolingian fortifications has been noted, and it is hardly surprising. At a time when encouragement to restore the glories of Rome was forcefully driven from the very top, and artisans had been brought up with the remains of such building work, anything else would have been inconceivable, and we have also seen that the example that may have been set by the more cultured and developed Byzantines, highly visible in Italy, was also largely rooted in Roman tradition. It may also have been the case that this was reinforced by what scientific or technical training there was, as the

1 *Ludowici germanici diplomata*, 248 (charter of 8 June 842), 10 (6 October 832).

2 Heine, "Frühmittelalterliche Burgen in Niedersachsen", 137–43.

3 Einhard, *The life of Charlemagne*, 71–2.

4 Heliot, *Histoire de Boulogne et du Boulonnais*, 60.

5 See of J Beckwith, *Early medieval art*, chap. 1; McKitterick, *The Frankish kingdoms under the Carolingians*, 144–52.

classical texts, such as the *De architectura* of Vitruvius, were among those newly copied. Did this extend to siegecraft? The works of Vegetius were rediscovered in the ninth century. A survey of the surviving manuscripts found that several ninth-century texts had originated at the monasteries of St Denis (Paris), Laon and St Gallen (Reichenau). Archbishop Hrabanus Maurus recommended the text to Lothair II (855/6) and copied him some 14 chapters; Charles the Bald had been the recipient of the work perhaps a decade or more earlier.[1] The part copied for Lothair included the chapters dealing with the organisation and discipline of the Roman army,[2] thus also reflecting the author's original main purpose. Would Carolingian rulers have turned to such an old treatise for help with siege warfare, given that their experience was of frequently dealing with the same walled cities that Vegetius discussed? This was the only military tract around, and given the enthusiasm for rediscovering ancient wisdom, it might have been used for the education of young princes (as opposed to their learning the skills on campaign). Certainly, copies of Vegetius were to become increasingly common as the years passed.[3]

Viking attacks

Vegetius would not have prepared the rulers of the new states developing out of Charlemagne's empire for their new enemy, the Vikings. The first raids by Norsemen were recorded in the late eighth century, and their impact grew and expanded throughout the following century. Starting with expeditions set on plundering rich and exposed targets, the attackers moved on to sending larger armies that sometimes encamped for the winter in their chosen targets and continued their operations for years at a time. Eventually, raiding was to develop into settlement and trading, with dramatic effects across Europe from Dublin to Kiev. Viewed through the eyes of monastic chroniclers, the onslaught was a relentless pagan scourge, and for all the efforts of historians to discount the natural bias of the written sources, there can be little doubt that the consequences for the settled communities of Europe of receiving a Viking attack were often catastrophic.

How was it that kingdoms that so recently had led an irresistible expansion against less civilised peoples proved so powerless against these attacks? The outward drive of the Carolingians had come to an end with the turn of the century, and a new emphasis on defending what was there is apparent in the pronouncements of Charlemagne and his successors.[4] Ironically, the wealth being created in the new kingdoms made them, in turn, a tempting target for aggressive neighbours. At the same time, there appears to have been a significant weakening of the

[1] *Epistolae karolini aevi*, III: 515, 619.

[2] Vegetius, *De procinctu romanae miliciae*.

[3] See also the discussion by Bachrach, "The practical use of Vegetius' *De re militari*", 239–41. For the distribution of manuscripts, see Schrader, "A handlist of extant manuscripts", 280–83.

[4] For an excellent discussion, see Reuter, "The end of Carolingian military expansion".

military strength of the settled areas of western Europe that had for a long time now been insulated from external attack. Certainly, the sensible instruction of Charlemagne to deal with raiders who came by sea and used rivers as their roads seems not to have been carried out. Perhaps the local resources were lacking, or else local counts appointed by the monarch failed to see the need to act in their own locality. It was not simply that Vikings were able to go where they wished, when they wished, without meeting resistance. Chronicles tell many times of successful resistance, usually organised by the local count. But the general picture was one of mobile raiding parties eluding or defeating the forces mobilised against them. With hindsight, it is easy to see that the early raids were precursors of much worse to follow. At the time, the princes of the Carolingian house were more concerned with fighting one another for their share of Charlemagne's empire, and viewed from Aachen, Tours or Rome, a few hostile attacks on a monastery at Lindisfarne or along the exposed coasts of northern France were scarcely cause for serious alarm. But as the ninth century progressed, and large Viking armies replaced the raiding parties, urgent measures became essential.

There has been debate about the prowess of the Vikings when it came to siege warfare.[1] Simple logic suggests two lines of thought. Raiders looking for quick profit would not wish to halt in front of strong stone walls, and would not have the necessary technology for capturing defended strongpoints. However, where the strongholds were no longer defensible, they would be (alongside rural churches and monasteries) the obvious targets for raiders. No one has doubted the adaptability of the Vikings, and it would not take long for a warlike people to work out how to capture places containing concentrations of the wealth they were seeking: silver and gold, other goods, and slaves.

The men from the north had experience of building earthworks themselves, but not perhaps fortifications. The "Danevirke", the 18-kilometre-long earthwork stretching through the frontier area of southern Jutland, has been dated to the eighth century, but does not appear to have had a defensive purpose. It was not used to resist Louis the Pious in 815. The great circular earthworks such as that at Trelleborg have also been discussed many times. They seem to date from the tenth century and to have been royal centres, occupied for a very short period. It seems safest to conclude that the Norse raiders did not come from societies in which fortifications were commonplace.[2]

It was to be expected that those on the receiving end should seek measures to defend their realms against attack. The efforts made in Francia and in Anglo-Saxon England were to become increasingly substantial from the second half of the ninth century. The approach revolved around a new sort of fortification, in efforts that called for a major collective effort within the realms of Charles the Bald and the house of Wessex. But it was a long time before effective measures were

[1] Steenstrup, "Les normands et leurs invasions", 104, argues that the northmen were "masters of siege warfare". But Griffith, *The Viking art of war*, 159–61, gives a different interpretation.

[2] Roesdahl, "The end of Viking-age fortifications in Denmark", 39–41.

conceived. For decades, local authorities had to make do with what was already there. The truth was that for many places what was there was no longer defensible.

Town walls were perhaps viewed as leftovers from antiquity in the long-peaceful west. They might also stand in the way as towns began to grow, and new suburbs became established. But their destruction could also be a matter of policy. As bishops looked to building or rebuilding cathedrals, the value of the walls as a source of already-cut stone must have been an overwhelming temptation. King Louis the German actually ordered the walls of Regensburg pulled down "to complete the fabric of the church".[1] Louis the Pious had given permission for the bishop of Reims to plunder both walls and bridge for the cathedral (817–25). Similar developments had taken place at Tours and Le Mans,[2] Langres, Melun and Beauvais.[3]

It was not a universal picture. We know of other towns whose late Roman walls stood them in good stead, as they enabled cities to see off the Vikings — for example, the cities of the Moselle, Metz, Toul, Verdun and Trier, pretty much in the centre of the empire,[4] of which only Trier was sacked. Sometimes it was not clear what happened. The Vikings attacked the strongly-sited city of Chartres in 857 or 858, having navigated the river Seine, and apparently broke in through the gates without difficulty, causing mayhem and massacre. They may have been intercepted on their way back to their boats and defeated. The local sources for what happened are all much later in date. However, a careful reconstruction of events combined with what archaeology can show suggests massive destruction and a rebuilding of some of the defences led by the bishop after this disaster, but we do not know what state they were in to start with. The new walls represented an abandonment of a major part of the old city. They proved sufficient when the Norsemen returned, this time in the shape of Rollo the future first ruler of Normandy, in 911. Sources that are, again, much later, talk of the Normans having "machines and catapults" and other siege equipment. This availed them nothing, as a relief army was assembled and defeated the invaders in July 911.[5]

The annals tell continually of Viking attacks that sacked and burned villages, towns and monasteries and churches. They also refer to the destruction of *castella* and *castra*. The sources also talk about rural *villas* that must be the residences of the aristocracy. Sometimes it seems that these houses were fortified, although it is not clear in what way, and they were presumably largely made of wood.[6] As we will see, the right of fortification remained (at least theoretically) in the gift of the king.

How the Vikings captured their targets is not always explained. However, their

1 Notker, *Charlemagne*, 152 (book II, chap. xi).

2 Flodoard, quoted by G Schneider, *Erzbischof Fulco von Reims*, 226; *The annals of St-Bertin*, 163–4; McKitterick, *The Frankish kingdoms under the Carolingians*, 233.

3 Vercauteren, "Comment s'est-on défendu contre les invasions normandes?", 119–20.

4 Dollinger-Leonard, "De la cité romaine à la ville médiévale", 195–225.

5 The events are recreated by Lair, *Le siège de Chartres par les Normands*, 5–50; for the archaeology, see Meulet, "Guide archéologique".

6 Charles the Bald burnt the "strongly-fortified house" of one Egfrith on the Loire in 868, *The annals of St-Bertin*, 143.

willingness to remain in front of defences and preparing means for their capture must have depended on their determination to secure the prize, and their confidence in being able to do so. At Sens, for example, they tried for five months in 879, and prepared machines to help. Divine help, or perhaps the strength of the walls, frustrated the attackers.[1]

In 867, in an attack on the Loire, the Vikings found themselves opposed by the Aquitainian duke, and they took refuge in a large stone-built church. One might guess that either this was a very small raiding party, or they must in fact have used the church as part of a hurriedly constructed earthwork defence of the kind mentioned in many accounts. In this case the Aquitainians tried to use machines against the rampart, but were defeated and their leader killed by a Viking sortie.[2] In 873, King Charles the Bald (r. 840–77) cornered a Norse army at Angers. The Vikings had selected the town because of its apparently impregnable location, had strengthened themselves by renewing the ditches and ramparts, and had devastated the surrounding countryside, denying supplies to their attackers. The king surrounded Angers and attacked with "new and exquisite types of machines". The attack failed however because of strong resistance, and the site was very difficult to approach. This story of Regino of Prüm's has been cited many times, focussing on the "exquisite" new machines and what they might have been. Were they a new kind of stone-throwing catapult? Or would the problem of not being able to approach closely perhaps suggest instead that these were either siege towers (for which there is no evidence of previous Frankish use) or a new type of ram? It is possible that the attackers would have built a new catapult without realising it could not reach the target, although a siege tower might be more likely. No firm answer is possible. We shall find a reference to the stone thrower soon enough, at the siege of Paris, and we know it was already in use in the south and Italy and Spain. So if this was it, it was not new! The other interesting aspect of the account is the apparent familiarity of the Norsemen with basic principles of defensive siege warfare.[3] They clearly understood how to make best use of defences, and how to prevent attackers from getting close enough to do serious damage. If they had not themselves read Vegetius, and this was perhaps unlikely, then they had learnt from experience.

Siege of Paris

The Danish siege of Paris in 885, for which we have Abbo's eyewitness account, written some dozen years after the event, allows us to consider the siege skills of the invaders in front of a major defended target. Abbo's verse account has been the source of numerous scholarly attempts to reconstruct what happened in what was plainly a significant military operation.[4] However, there are enough warning signs to suggest caution about the accuracy of the storyteller. To start with, the

[1] Clarius, *Chronicon Sancti Petri Vivi*, 468.

[2] Regino, *Chronik*, 216–17.

[3] Ibid., 240; Regino, *Chronicon*, 585–6.

[4] Abbo, *Le siège de Paris par les Normands*, to which all references apply; see also Abbo,

statement that the Danes came in 700 ships and numbered 40,000 fighters must be open to doubt.[1] The siege lasted many months, over the winter. How did they feed such a host? Although Abbo says they raided and devastated far and wide, the logistics surely fail to add up. There is another problem, too. Abbo talks of the siege being only of the old city on the Île de la Cité. While that is possible, the space on the island is adequate only for a very small population (perhaps a few thousand). Paris itself had long outgrown the original space with its Gallo-Roman walls, and there were settlements on the banks of the Seine, particularly the right bank. What happened to these suburbs? The evidence of the siege itself suggests that the suburbs were not fortified until afterwards, although a capitulary of 877 orders the building of a *castellum* at St Denis. Whatever was there, it did not obstruct the Norsemen. The size and wealth of the city were what attracted the attack in the first place, along with the strategic reality that since the construction or reconstruction of bridges across the Seine at the order of King Charles the Bald, Paris blocked further Viking raids up the river. What confronted the Danes, therefore, was an island defended by stone walls and perhaps towers of late Roman origin, and two defended bridges, one to each bank. The *grand pont* was a major construction with stone foundations and a wide carriageway. There were towers at the far end of the bridges, apparently of wood on top of stone foundations. The bridges were protected along their lengths with palisades, and the towers seem to have had ditches in front of them. All have disappeared, and the reported discovery of the remains of the larger bridge is not uncontested.[2]

Abbo says the Danish host arrived on 24 November 885, and failed to get the city to surrender. They had marched from Rouen, having been reinforced by a contingent from England, and having forced their way past the Frankish defences at Pont-de-l'Arche (see below), had taken Pontoise. Paris was under the leadership of its count, Eudes (the future king of Francia), and its bishop, Gozlin. We do not know the number of armed men available as a garrison, but it is apparent from Abbo's account that they included skilled makers and users of artillery.

Obviously hoping for a quick victory, the Danes first attacked the tower defending the end of the *grand pont*. This was unsuccessful. Abbo tells that the tower was incomplete, with only the foundations "solidly built", but rapid work was done overnight to add another storey to its height. The attack was resumed the next day, and Abbo talks of slings and ballistas in use. The attackers' objective was to collapse the wall by digging it away with iron spikes, protected by shelters. The defenders created a mix of oil, wax and pitch, which, having been liquefied, was hurled down on the attackers. Despite this, the Danes succeeded in setting

De bello parisiaco. Abbo's account is analysed in Hassall and Hill, "Viking warfare", and much used in Favre, *Eudes, comte de Paris et roi de France*.

[1] Gillmor, "War on the rivers", 86–8, argues that the Viking numbers were more likely around 5,000–8,000.

[2] Lombard-Jourdain, *Aux origines de Paris*, 36–9; Boyer, *Medieval French bridges*, 22–6; Vacquer, "La découverte d'une partie du Grand Pont de Paris". It has been asked why a new bridge was built if the Roman bridge was still there. Perhaps the new bridge was a reconstruction, or a rebuilding on the same site.

alight the gate or door (the passage is somewhat confusing). The wind however changed, the smoke driving back the assailants, allowing the Parisians to repair the damage to the tower.[1] These events took place on 26–7 November. Following the failure of the initial assault, the Danes decided to turn to a more formal siege, establishing a camp and ravaging the country while they began to construct siege weapons. Chiefly this comprised enormous battering rams on sixteen wheels, with a roof. The work was setback when a ballista shot from Paris killed the two builders, suggesting the role of specialists in their construction. However, the weapons were completed after a delay of two months, and in preparing for the inevitable assault the count and bishop deployed catapults the length of the bridge. That the Danes had some considerable experience in such matters is suggested by Abbo's account of the attack that followed on 31 January 886. First they needed to fill the ditch around the tower, which had not been mentioned during the previous assault. However, this time a flat approach was essential if they were to bring up their rams, which they moved up the next day. The defenders had prepared their own equipment against a threat that they had seen being built for some time. There follows a description of a weapon "commonly called a *mangana*" that "threw enormous stones", killing "a crowd of Danes". The attempt to move the rams forward failed.[2] Their next attack involved sending ships filled with burning brush against the bridge, but this was thwarted by rocks in the river. Abandoning their rams (presumably too heavy to move) but pulling back their shelters, the Danes withdrew. Luck finally turned against the brave defenders when the river caused the collapse of the other bridge on February 6, leaving the isolated tower at its end to fall. The conclusion was perhaps an anticlimax. In March, a relief army appeared and the Danish army agreed to withdraw after receiving payment of 60 pounds of silver. The next year, they were back, but the defence was ready, and St Germain had been fortified and again catapults were deployed in the defence. A further treaty resulted.[3]

Abbo's account of the *mangana* is the first explicit reference to this weapon in use in northern Europe. His use of classical Roman authors as models for his text has led to some doubt as to whether his descriptions are of real events or are borrowings from the classics. However, as *mangana* is a medieval term for a lever-type stone thrower, not a term known to classical Latin, and from the description of the effect (leaving aside the likely hyperbole of the "enormous" stones) this is what seems to be meant. Certainly, they worked as defensive weapons, preventing the Danes' attacking the tower with their rams. The use of equipment at the siege is attested in other sources too.[4]

[1] Abbo, *Le siège de Paris*, 20–22 (ll. 80–85, 87–8, 99–101)

[2] Ibid., 42 (ll. 302–4, 356–70).

[3] Ibid., 94 (book II, ll. 386–7).

[4] E.g *Annales vedastini*, ed. Simson, 310, 312.

In England and Ireland

The effect of the Viking attacks on the Anglo-Saxon kingdoms of England is well known. Until the second half of the ninth century, after the raiders had started to encamp over the winter and their raiding parties had become armies, no new defensive structures were undertaken in resisting the threat. But there are accounts of warfare involving the attack and defence of existing strongholds to illustrate the level of knowledge and experience of this kind of struggle. The *Anglo-Saxon chronicle*, thought to date from Alfred's reign, and Asser's *Life* of the great king, give brief details. In 867, the Northumbrians pursued a Viking force into York, with its Roman walls. They were able to breach them "because in those days the city did not yet have firm and secure walls".[1] Or perhaps more exactly, the old walls had not been maintained. The following year the Saxons "besieged the host" in Nottingham but failed to take it, instead making a treaty.[2] The *Chronicle* suggests they did not try very hard. In 877, Alfred pursued a raiding force westwards from the south coast of Wessex, but the Danes managed to obtain refuge in Exeter, and its Roman walls were sufficient to prevent the Saxons from breaking in.[3] In 878, when the raiders attacked Devon, the king's thegns took refuge in a stronghold called Countisbury, which was unfortified "except for ramparts thrown up in our fashion", perhaps implying that this was an old hill-fort reused by the Saxons. In 884, they attacked Rochester, which certainly had once had Roman defences, and constructed a defence for themselves against the gate, but the citizens held on until King Alfred arrived to rescue them.[4] In 893, a Viking force occupied the apparently deserted Roman-walled city of Chester. The pursuing levies were unable to prevent them entering the city, but tried for two days to take it, without success. Maybe they had no experience of siege warfare.[5] An Irish account tells of a different event at Chester at about the same time. The Vikings having failed to win a battle with the defenders outside the walls, turned to undermining the wall. "They would make many hurdles and place props under them, and they would make a hole in the wall underneath them." The Saxons "hurled down huge stones so that they crushed the hurdles on their heads." The Vikings then added columns to strengthen their equipment. The Saxons responded with boiling water, which the besiegers countered by spreading hides over the shelters. Finally the defenders abandoned the town.[6] It is a classic account of a sapping operation and the defence against it.

However, the sum total of these and other, even briefer, references provides evidence only that the Vikings had picked up experience of attacking and defending fortifications, while the Anglo-Saxons had no apparent knowledge of the more sophisticated siege techniques that had reached the northern parts

1 Asser, *Life of King Alfred*, 76 (chap. 27); *ASC*, 68.
2 *ASC*, 68–71 (868).
3 *ASC*, 74 (870).
4 Asser, *Life of King Alfred*, 86–7; *ASC*, 78.
5 Asser, *Life of King Alfred*, 117; *ASC*, 88.
6 *Fragmentary annals of Ireland*, 171–3.

of Charlemagne's domains, even though the walled cities in which the invaders sometimes took refuge must have posed the question of how to tackle them. Probably there was no one with the skill to copy, make or operate siege weapons, although there were strong commercial and diplomatic ties with the countries across the Channel.[1] The only contemporary indication that siege weaponry was known in Anglo-Saxon England derives from two of the surviving Anglo-Saxon riddles, of uncertain date. The riddle to which the answer is clearly "battering ram" is not a surprise: cutting down a tree to use as a means of breaking in a door does not require much inventiveness.[2] That which seems to describe a ballista — "I spit out terrors of spears; the greater my fullness, the greater the success. The master beholds how the war darts fly from my womb"[3] — can only have been devised by someone who knew what a ballista looked like, and assumed their audience would do, too. But when they appeared on the scene must remain a mystery, and the relative backwardness of the Anglo-Saxon kingdoms in this department throughout their five centuries of existence continues to stand out.

In Ireland, the Vikings found a country without towns and left their mark on its future by creating the forerunners of the modern cities of the island. The characteristic dwelling of the Irish was the ring-fort, designed to house a group of families, and ranging in size from fifteen to eighty metres. Some were built of earth (the *rath*) and some of stone (the *castel*). There were thousands of such sites, much to the confusion of those trying to identify Norman earthwork castles of the later conquest. There were also promontory forts similar to those existing in Britain.[4] References in the Irish annals to attacks and defence of fortified sites during the Viking attacks in the ninth and tenth centuries are rare, although they multiply for later periods. The most that can be concluded is that both the Irish sites and the new Viking settlements were largely of earth and timber. We hear of the killing of 400 foreigners (Vikings) and the taking and burning of the fort at Dublin in 943, and of a two-month blockade by the Irish in 966.[5]

The Dublin in question seems to have been established around the year 900, marked off with a low earth bank and palisade, replaced some fifty years later with a wooden wall of wattle, which was not to be replaced in stone for another century and a half. At Waterford, evidence has been found of a huge bank and ditch but its date is uncertain.[6] Cork seems to have started life as a monastery, plundered regularly by the Vikings after 821, and the site occupied by them around 845. It was besieged by the king of Cashel in 848, indicating that some form of defence had been created. It was resettled in the tenth century. Its excavators suggest there was

1 Such also is the conclusion of Peddie, *Alfred, Warrior King*, 146.

2 *Anglo-Saxon poetry*, 305. The editor suggests an eighth-century date for the origin of the riddles.

3 Ibid., 294–5.

4 T B Barry, *The archaeology of medieval Ireland*, 37, 54; Dollen, "Die Entwicklung des Burgenbaus in Irland", 50.

5 *Chronicon Scotorum*, 207, 219.

6 Clarke and Ambrosiani, *Towns in the Viking age*, 104–6.

an earthwork hill-fort of unknown date on the north island.[1] In general, though, little has been found by excavation of the earliest Viking settlements in Ireland.

Bridges and burhs

From the viewpoint of the kings of England and Francia in particular, the establishment of ever larger Norse and Danish armies over the winter in defended camps deep inside their realms from the mid ninth century must have represented a qualitatively different threat from the earlier raiding. In England, it threatened the existence of the Anglo-Saxon kingdoms altogether. At this point, there was a drastic change in the response of the invaded kingdoms to the threat, after many decades in which there was little in the way of centralised action. King Charles the Bald made a serious effort to save his kingdom through instituting a programme of defensive works, in which fortified bridges played a key role, the operation depending on the supply of labour services by the population at large. Across the channel, a similar mobilisation of the people was called for to create a network of fortifications called *burhs*, surrounding the kingdom of Wessex and then extended by Alfred's successors to include areas newly brought under their control. It has been suggested that the two initiatives may have inspired one another, and the link has been extended to include the much later *Burgenordnung* of the German king Henry I, over which much ink has been spilt but not a lot of light cast. Nothing can be proven, and there is little point adding further speculation. It would not be surprising if, with the level of royal control at their disposal, and the technical skills available, the rulers of European states at this time were to arrive at similar actions to defend their kingdoms against external threat.

In 862 King Charles, it has been suggested, "stumbled" on a way of hitting back at Norse raiders when he ordered a bridge placed across the river Marne to block the retreat of raiding ships.[2] Near Pont-de-l'Arche, at the confluence of the rivers Seine, Andelle and Eure, was the old royal town of Pîtres. On 1 June 864, he held an assembly at Pîtres where "he ordered fortifications to be constructed there on the Seine to prevent the Northmen from coming up the river".[3] He issued an edict mobilising the manpower of his kingdom to carry out the works required, which involved creating massive fortifications at each end of the bridge. Progress was less than rapid, as Viking attacks continued to be unhindered, and in subsequent years the king reconvened his nobles and reissued the instructions. In 869, his vassals were ordered to perform garrison duty in the *castellum* of wood and stone he had ordered built.[4] For all the enormous effort called for in creating this fortification, its actual performance was remarkably ineffective. Certainly, it failed to stop the Danish host on its way to Paris in 885, as we have seen.

The site at Pont-de-l'Arche has been studied. Archaeomagnetic dating of mate-

[1] Bradley and Halpin, "The topographical development of Scandinavian and Anglo-Norman Cork", 16–23.

[2] Gillmor, "The logistics of fortified bridge building on the Seine", 88.

[3] *The annals of St-Bertin*, 118.

[4] Ibid., 153–4.

rial uncovered on the banks of the Seine gave a likely date for the works of around 860.[1] Subsequent excavation revealed, on the north bank, a square enclosure with sides of 270 metres (a total circuit of 1080 metres). The rampart was constructed originally as a clay bank. Later, the face was cut away, and tree trunks inserted, four and a half metres long, and an outer facing of stone provided, acting as a revetment to protect the end of the tree trunks. There was evidence that the site had been damaged in about 885. The ramparts were actually levelled in new building in the fourteenth century.[2] This was a massive work, although its height cannot now be reconstructed.

The new work here did not have the intended effect. It was a major undertaking, and the first new public defence in the kingdom known from this period. It would seem that it took a long time to complete, and that when finished, it was not properly defended. The ramparts were of earth and timber, with stone for the facing. No attempt was made to create a stone wall. This suggests the limitations of available skills, and of resources. It also seems that the system of providing sufficient numbers of armed men to garrison the resulting works did not function.

Nonetheless, despite all these problems, the orders issued by the king at this time were not limited to the Seine, but mark the beginning of a new period of building fortifications against the Vikings. Part of this process of building work was simply to stop the slow decay of the existing walled cities. There is a long list of former *civitates* ordered to be restored during the second half of the ninth century: Tours and Le Mans (869), Beauvais and Noyon (880), Reims (883), Langres (887), Orléans and Cambrai (888), Tournai (898).[3] But as we have seen, fortification was no longer restricted to places with existing walls, reflecting a change in the type and number of locations felt to be important enough to provide with defences. At the same time as the work was going on at Pont-de-l'Arche, a similar new work was ordered on the Loire at Les Ponts-de-Cé. The monastery of St Vaast (Arras) was provided with defences (883–7),[4] following the example of that of St Denis, while the abbey of St Bertin (Saint-Omer) also provided itself with defences with which it resisted the Vikings in 891.[5] The cathedral at Nantes in the west was reconstructed under Bishop Fulcher from 900, who provided a *castrum* with walls around it, in which "clerics and lay people could flee for safety, protecting themselves from the Normans".[6]

There is some dated evidence of the new fortifications from other areas exposed

[1] Dearden and Clark, "Pont-de-l'Arche or Pîtres?", 569.

[2] Dearden, "Charles the Bald's fortified bridge at Pîtres", plan; Dearden, "Charles the Bald's blockage of the Seine at Pont de l'Arche". See also the old account by Lot, "Le pont de Pîtres", 535–60.

[3] The evidence from various sources assembled in Vercauteren, "Comment s'est-on défendu?", 124–31.

[4] "Iam monasterium muris circumsepto, turribus affirmato, vallorum ambitu circumdato": *Sermo de relatione beati Vedasti*, 402; Vercauteren, "Comment s'est-on défendu?", 126.

[5] Fulco, the future archbishop of Reims, began the work when he became abbot in 878: G Schneider, *Erzbischof Fulco von Reims*, 228. He also began to fortify Reims from 883 (according to the chronicler Flodoard).

[6] *La chronique de Nantes*, 78 (chap. xxvi).

to Viking attack. While it cannot be proven that the building of these places was connected to the invasions, the coincidence renders other explanations — such as local rivalries — less credible. The process was being followed in the German kingdom as well, as the Vikings were also raiding up the Rhine, with work at Cologne and Mainz, where a new ditch was dug outside the walls (882/3).[1] New building accompanied and followed the Danish attacks of 863 and 884/5. In 863, the raiders fortified themselves on an island in the Rhine at Neuss. In 880, the royal palace at Nijmegen was provided with ramparts. In 882, Duke Henry built the *castra* around Duisburg. In 884, the bishops and counts were ordered to undertake defences. At the turn of the century, Elten was fortified. Excavation showed a palisade behind a ditch six metres wide, and internal buildings including a chapel. The site originated in the late ninth century and continued to be used through the tenth, with woodwork being replaced by stone. At Broich (Mülheim an der Ruhr), the original works of the late ninth century were of earth and timber, being replaced by a stone wall, and large internal buildings, in the tenth.[2]

In Flanders, which was established as a county in its own right in the 860s, new defensive works were undertaken. Count Baldwin I was responsible for the first defences at Bruges, the site of which has long disappeared but was probably under the square known as the Burcht, and may have been a rectangular earthwork bounded by the river. A similar development occurred at Ghent, where early stone buildings underlie the much later donjon, and although there is dispute over exactly how many of the later Flemish towns grew up around a Carolingian *castrum*, it is clear that a fair number of them did.[3]

One of the more unusual types of defence dating from this time is to be found in Zeeland-Flanders. Five or six circular earthworks dated to the late ninth century were found on the islands. The diameters vary from 144 to 265 metres. They were made of clay and must have carried a palisade. The ramparts at the base varied from four to twelve metres wide, and all had two roads across the site, meeting in the middle. No occupation remains were excavated, suggesting their role as a refuge for people and livestock. Their defences were quite powerful. Given the frequency of Viking raids up the river Scheldt, with repeated incursions in the 850s, 860s, 879–885 and 891/2, their purpose seems clear. Middelburg went on to become the main settlement on Walcheren, while the others fell into disuse.[4] Further inland, in the land of the Veluwe lying between the great rivers of the Netherlands, is another earthwork, a four-metre-high horseshoe, with 100 metres internal diameter, fronted by a dry ditch. Numerous theories had been advanced about the origins of the Hunnenschans, but redating of the finds suggests a similar late Carolingian date, when the Veluwe was a centre of iron production and

[1] *The annals of Fulda*, 91.

[2] Binding, "Spätkarolingisch-ottonische Pfalzen und Burgen am Niederrhein", 23–5.

[3] Dhondt, "De vroege topografie van Brugge", 5–7; Heliot, "Sur quelques résidences princières", 27–35; Verhulst, "Les origines et l'histoire ancienne de la ville de Bruges"; Noterdaeme, "De vroegste geschiedenis van Brugge", 186–200; Van Werveke and Verhulst, "Castrum en Oudburg te Gent", 57–8.

[4] Heeringen, "The construction of Frankish circular fortresses", 241–7.

the main road crossed the site. A tie-in with the establishment of a local count seems likely.[1]

Overall, despite the failure of the grand project on the Seine, the defensive measures taken seem to have been relatively effective in keeping out the raiders. But it was at a cost to royal authority. Already in the Edicts of Pîtres, Charles the Bald had had to order the destruction of fortifications created without royal authority. There was no doubt that fortification remained a royal prerogative, but the edicts suggest that the royal power was being ignored. Certainly, later in the century and as monarchical authority further diminished, many of the defensive works seem to have been the result of the initiative of the local count, bishop or abbot.

In Wessex, the result of Alfred's works of fortifications was a remarkable tribute to the ability of a small kingdom to mobilise its human resources. The programme of fortified *burhs* thrown up in the second half of the ninth century, and developed much further in the tenth, is an extraordinary effort, and reflects the high degree of royal authority, as well as the building skills, of the Anglo-Saxon kingdom in the face of the same threat of extinction that had already overcome the kingdoms of eastern and northern England. The evidence of the work is in the written sources, the remains on the ground, and in a document known as the *Burghal Hidage*, usually dated to around 911–19 but more recently to the late 870s, which lists the numbers of hides of land associated with the provision of defenders for each of 33 *burhs*, at a ratio of one defender per hide. The total number of men thus assigned amounts to more than 27,000, which is estimated to be one-fifth or one-sixth of the male population. This was a public duty associated with possession of land, and was a continuation of a long-existing obligation.

A detailed study of the remains suggests a close identity between the document and the sites. There is a wide variety in the length of the ramparts, because while the defensive plan was centrally ordained, and the sites are approximately twenty miles apart, sufficiently indicating that they followed a centrally-given instruction, the construction at each site was determined by what was there already, and the need for speed. Some of the burhs were Roman walled sites repaired, as at Bath, Chichester, Exeter and Winchester. Others were promontory sites, perhaps with Iron-Age forts on them. Others were built anew. There are no remains of some sites, and others remain unidentified. At some, such as Wareham and Wallingford, there are significant remains of powerful earthworks. Wareham has a bank 18 metres wide and now almost three metres high behind a ditch, with a length of more than 2,100 metres. Wallingford's rampart, 3,000 metres in length, is eight metres high behind a three-metre-wide ditch. The defences of other sites included turf walls at Christchurch, a turf revetment with later stonework at Cricklade, with a walkway behind and a double ditch, and something similar at Lydford.[2] Most of these sites are large enough to serve as a refuge, and the defences were powerful enough to resist raiding parties without siege equipment. Defenders

[1] Heidinga, "The Hunneschans at Uddel reconsidered", 53–9.

[2] See Brooks, "The Administrative background to the Burghal Hidage", 128–31; D Hill, "Gazetteer of sites". The Kent and London sites are not in Hill's list.

equipped with spears and bows would be able to protect them against direct assault. They would presumably therefore have been the responsibility of the local levies, while the better-armed would form the king's army in the field. The average garrison has been estimated at about 900.

Was there a foreign inspiration for this remarkable network, begun in the 880s? Count Baldwin of Flanders, architect of defences against the Vikings across the channel, has been suggested and there are other contacts with the French kingdom too. The most obvious similarity is between the double burhs built to block rivers, such as those at Buckingham and Hertford and the bridge and fort thrown across the Thames in 895, and Pont-de-l'Arche. It is impossible to confirm. Facing a similar threat, it is not surprising that kings should arrive at similar conclusions, and knowledge of what the other was doing might have been coincidental, or maybe inspirational.[1] There was a major difference, though. The Wessex fortifications remained under central royal control, and a large number of new sites were fortified by Alfred's successors in the continuing and generally successful campaigns they waged to restore control of large areas of the country lost to the invaders.

The role of the new fortresses in these campaigns is clearly recorded: fortress-building by King Edward the Elder (r. 899–924) at Hertford and Witham, and at Tamworth and Stafford by Æthelflæd (913), followed by Eddisbury (Cheshire) and Warwick. In 918, Edward ordered both banks fortified at Buckingham during a stay of four weeks. The programme of works continued at Maldon, Towcester ("reinforced by a stone wall") and the unidentified "Wigangamere". Viking attacks on Towcester and Wigangamere in 921 were defeated. Meanwhile, the Danes set up camp at Tempsford (Bedfordshire), but Edward took it by storm. There is a small earthwork at Tempsford, once identified as this Danish fortification, but it is far too small to have held an army of any size and must be a later, Norman, work. The *Chronicle* records further Saxon works to fortify Colchester, Thelwall, Manchester and Bakewell, and a new bridge across the river at Nottingham.[2] It is an extraordinary level of activity, and it is clear that not only were these new burhs rising everywhere, they were also being garrisoned and defended successfully too.

Highly effective at the time in achieving its objective, this was the first planned system of fortification in the west since the days of the Roman empire.[3] One of the consequences was to delay the introduction of private fortifications (castles) until after the overthrow of the kingdom itself, ironically perhaps, given that the absence of usable fortifications would ease William the Conqueror's path.

[1] For example, Hassall and Hill, "Pont de l'Arche", 189–93; Peddie, *Alfred, Warrior King*, 155; Strickland, "Military technology and conquest", 370–82; Bachrach and Aris, "Military technology and garrison organization", 2–14.

[2] *ASC*, 96, 97, 99, 100, 101–3, 104. On Tempsford, see Dyer, "Danish earthworks". The earthwork is clearly no more than a moated farmstead, and the likely site of the Danish camp has been located on the other side of the river near Sandy.

[3] Strickland, "Military technology and conquest", 371; Schoenfeld, "Anglo-Saxon *burhs* and continental *Burgen*", which also discusses the possible connection with the fortification programme of Henry I in Germany of about 924.

Byzantium facing old and new foes

We left the Byzantines having stabilised a frontier zone with the conquering Arabs in the Middle East, and clinging to their footholds in Italy. However, the Arab conquest of the African coast meant that the Mediterranean itself was now a frontier zone. Although there is evidence that trade and communication continued across the sea despite political and religious divides, the water also provided the routes for conflict in new areas. Italy, as we have seen, was now a target for Muslim raiders as well as the site of rivalry between Lombards, Byzantines and Franks. The rapid absorbing of maritime skills by the Arabs also carried conflict to Mediterranean islands. Faced with this threat, which was permanent although not continuous, and simultaneously with new threats to its Balkan provinces, the east Roman empire had shrunk and substantially changed throughout the seventh century. The city-based civilisation of the late empire had altered dramatically, with a shift to a more rural economy and the shrinking of the old cities. Constantinople remained the great city of the age, but elsewhere the empire would by now have been unrecognisable to the Romans. In one respect, however, Byzantium continued to be fundamentally different from the states emerging in western Europe. It retained a centralised imperial bureaucracy at the heart of its administration, even if in the provinces, powerful aristocratic families wielded great local power. Byzantine armies were still largely paid, professional forces, in contrast to the way that armies were raised in the west, although their numbers, arms and tactics were undoubtedly changing as a result of experience of the many different enemies they faced.[1] A significant consequence of this was the continuation of engineering skills, both for building and for siege-warfare purposes, and the inherited wisdom of past generations continued to be transmitted through a literate ruling elite to its successors.

Accounts of the late eighth century are limited. There are the non-contemporary compilations of the Armenian Ghevond (d. *c.*795), Theophanes, Ṭabarī, Dionysius of Tel-Maḥrē (d. 845), Genesius (tenth century) and the Syrian patriarch Michael (twelfth century). There are also other less detailed Byzantine chronicles, also of later date. The existence of substantial siege-warfare capabilities is attested in these histories. In around 750, Emperor Constantine v captured and ordered destroyed the walls of the town of Melitene (now Malatya), which Ibn Muhammad ordered rebuilt in 760, sending a force of workmen and a protective army.[2] In Armenia, the rebels against the Arabs attacked Dvin with large supplies of "munitions, arms and instruments of war" in the 750s. They blockaded Theodosiopolis (Erzerum) and raised "bastions and batteries around the town, succeeded in making breaches in the outer wall, while they threw, with the aid of machines, enormous blocks of stone into the interior which killed many". However, despite keeping up the attack throughout a winter, the town held out. An

[1] Haldon, "Some aspects of Byzantine military technology", 12–43.
[2] Dionysius of Tel-Maḥrē, *Chronique*, 56, 67.

Arab army arrived in the following spring, well equipped with "engines of war" and provided with "numbers of artisans who make and repair arms and instruments of war". Theodosiopolis was relieved, and the Armenians surrendered their forts.[1] In 766/7, in a new invasion of Byzantine territory, a fortress called Qarnah was besieged. Numerous workers had been summoned from Mesopotamia to build a fort for the attackers, and the Arab commander, 'Abbas, requisitioned carts from the local Armenian population to carry cedar wood with which to construct catapults for the purpose of hurling projectiles over the walls. The garrison responded in kind. Next, the attackers constructed a mobile shelter to protect them while filling the ditch, but the defenders managed to destroy it and kill many of the Muslims. In the end, all the while suffering heavy casualties from the Byzantine artillery, the onset of winter cause the siege to be raised.[2]

The Arabs were not always successful. Amorium was a favourite target of their raids into Asia Minor. In about 775, they besieged it unsuccessfully for three months, because of its "formidable walls" as well as its marshy surrounds. Three or four years later, they tried again. First they had attacked Dorylaeum (Eskeşehir), but withdrew after a fortnight short of supplies, the Byzantines having garrisoned forts in the area to prevent pillaging of the countryside. Then they tried Amorium, but again found it "fortified and well armed", so gave up after a day.[3] In 781/2 they attacked a Byzantine fort at Nakoleia (Seyitgazi) using "stone-throwers which hurled huge stones".[4] In 806, the Caliph Hārūn al-Rashīd (r. 786–809) invaded Byzantine territory and besieged the fort known to the Byzantines as Herakleia (Ereğli), which the Greeks acknowledged to be "very strong". Notwithstanding, it fell after 30 days and the Arabs went on to take other places. Herakleia they destroyed, judging that was the best outcome for them. Apparently the Byzantines shared that view, because they reoccupied the site and rebuilt the defences.[5] The clear message from these short accounts is that when setting out to achieve a conquest, Arab armies were formidable in their ability to capture strongholds, but that armies whose primary aim was raiding could be prevented by well-held fortifications, and that competent defenders could deny them access to supplies and force their withdrawal.

References to Arab skills and practice in the late eighth century are little more frequent than from the first century of their empire, but there are increasing references to conscious building campaigns of fortifications or of sites built with mul-

[1] Ghevond, *Histoire des guerres et des conquêtes des Arabes*, 130–48.

[2] Dionysius of Tel-Maḥrē, *Chronique*, 75–6, 81, 89. The construction work carried out in the frontier regions by Muslim commanders is confirmed by Arabic sources collected in *Palestine under the Moslems* (chiefly Balādhurī, writing in 869, and Estakhri in 951), recording references at Adana in 758 and 782 (p. 382), Anarzaba in 796 (p. 388), Malatya (Melitene) in 756, and Mopsuestia (al Massissah) in 703 and 756 after an earthquake (p. 499).

[3] Ghevond, *Histoire des guerres*, 151 (for 775); Theophanes, *The Chronicle*, 624 (for 778/9).

[4] According to a manuscript account of Pantaleon the deacon, cited in Theophanes, *The Chronicle*, 630.

[5] Theophanes, *The Chronicle*, 662, and Ṭabarī, *The early 'Abbāsī empire*, II: 274, agree on the essentials of what happened.

tiple purposes in mind, that were defensible. In the 770s, the Caliph al-Manṣūr (r. 754–75) ordered the city of Rāfiqah (ar-Raqqah) fortified on the pattern of Baghdad, with walls and a moat. The town, although not fully circular, was 1.5 kilometres across and had mud brick walls 5.85 metres thick and small round towers every 35 metres, and a low outer wall.[1] The same were ordered at Basra and Kufa (although both places had fortifications prior to this). Forts were built on the coasts as bases for expeditions to India in 775. In 786/7, Hārūn al-Rashīd removed the garrisons from al-Jazīra (upper Mesopotamia) in order to construct a line of fortifications to protect Antioch from the Byzantines, and in the 790s we learn that a new governor of Khorasan set about the construction of a system of frontier forts (the *ribāṭ*, a fortified religious house for Muslim volunteers). Apart from suggesting a significantly more defensive approach by the Abbasid caliphs, this range of reported building activity must demonstrate the availability of a substantial number of skilled builders and artisans. The earliest surviving versions of Muslim military handbooks date from the time of al-Manṣūr, and a tenth-century text makes reference to lost texts provided to Muslim rulers on "The conduct of wars, storming of castles and cities", translated from an old Sassanian text book by al-Manṣūr's secretary 'Abd al-Jabbār ibn 'Adī, and "Towers for attack, catapults, stratagems and artifices" "in the handwriting of Ibn Khafīf" (Ibn al-Nadīm). It therefore seems likely that the Arab rulers, generally literate and well educated, had access to quite a range of texts translated from Greek, Latin, Persian and Sanskrit, now lost.[2] It has been suggested that the organisation of specific corps of specialist troops (such as sappers) can be dated to the early ninth century, and the few detailed accounts of siege warfare from this time certainly indicate a more professional approach than hitherto.[3]

In 823, the Caliph al-Ma'mūn (r. 813–33) attacked the rebel-held fortress of Balās, and clearly brought a large train of artillery, because there was such a barrage of stones that the Christians were driven to wail and lament, whereupon the merciful emir (so says Michael the Syrian) stopped the bombardment, and instead sapped and mined the wall and gained entry that way.[4] The next target in suppressing this revolt was Kayṣūm, besieged in October 824 by 'Abdallah ibn Tahir. Again, the Arab engines battered the dwellings of the town with immense stones. The defenders tried to use women and children as a human shield against the bombardment, with some effect, because in the end (after a false submission, new revolt and second siege) the town fell to a blockade bringing it to the brink of starvation.[5]

The attention of the Byzantines had meanwhile been seriously distracted by

[1] Creswell, *Early Muslim architecture*, 38–42. Walls survived to a height of ten metres (1940). The last remains were removed by 1950.

[2] Ṭabarī, *The early 'Abbāsī empire*, I: 223; II: 60, 83, 186, 209. Nicolle, *Armies of the Muslim conquest*, 24. Ibn al-Nadīm, *The Fihrist*, II: 737–8. See also Kennedy, *The armies of the caliphs*, 111–13.

[3] Kennedy, *The armies of the caliphs*, 132 (citing Ṭabarī), 185, indicating the rule of al-Mu'taṣim as the crucial time.

[4] Michael the Syrian, *Chronique*, III: 53 (book XII, chap. xii).

[5] Ibid., 55 (book XII, xii).

the three-year revolt, much closer to home, of the Slav general Thomas, against the new emperor Michael II (r. 820–29). Thomas seems to have enjoyed the support of many of the empire's armed forces, since he was able to attack Constantinople on the land side with sufficient forces to blockade the whole length of the city walls, and to deploy batteries of siege engines including stone throwers to assault the Blachernae palace. Michael, however, had sufficient time to bring in troops from the eastern provinces, and weather conditions (as well as the emperor's apparent rediscovery of "Greek Fire" thanks to the astute Egyptian Kallinikos) meant that the attack by sea was repelled, while that by land failed to make any impact, yet again, on the city's walls. Nonetheless, the attack seems to have lasted for up to one year before Thomas gave it up.[1] Apart from the decisive help of Greek Fire, the account is also revealing in its use of the term *elepolei* to suggest, in this instance, powerful stone throwers, notable for having four legs and projecting immense rocks.[2] It has been argued that this reference may mean a different type of artillery, but the account really only suggests a larger, rather than a new, weapon.

Meanwhile, in 824, the Arabs were able to take advantage of this revolt to invade and conquer the island of Crete. This appears to have been a sustained operation, beginning with the capture of a fortified town, then a step-by-step capture of the rest. Byzantine accounts are disappointingly meagre, and the Arab ones also limited, but it must have been an impressively executed operation, with serious consequences in opening up the Aegean to Arab raiders.[3] A century and a half later, its recapture would be the basis for other histories, and incidentally fascinating information about then Byzantine siege warfare, technology and organisation.

The siege of Amorium, 837–8

The warfare in the frontier zone of Asia Minor was continuing. In 832, al-Ma'mūn ordered the construction of two forts against a Byzantine fort, with garrisons, marking an interesting and early use of the "siege castle" method of attack. The following year the Caliph al-Mu'taṣim (r. 833–42) again directed Arab efforts at Amorium. So important was it that the Byzantines risked a pitched battle to relieve it. The resultant defeat left the defenders no option but surrender.[4] However, it seems that the town fell back into the hands of the empire, because we find a new siege taking place in 837/8. This time, the details of Ṭabarī's and Michael the Syrian's accounts are eloquent testimony to the capabilities and limitations of siege warfare at this time. The caliph equipped himself with a large army and ample supplies including iron tools and naphtha, but his progress through

[1] Georgius monachus, *Chronicon breve*, 1003–6, an account possibly dating from the 840s.

[2] Genesius, *Historia*, 1030–39; *On the reigns of the emperors*, 30–36. The history dates from no earlier than a century after the events.

[3] *Theophanes continuatus*; survey of the events by Christides, *The conquest of Crete by the Arabs*, 6–7, 80–81.

[4] Arab account of Ya'qūbī in Vasil'ev, *Byzance et les Arabes*, I: 273–5. Ṭabarī's version is the same.

the mountains towards Amorium was delayed because of the slowness of his rearguard, bringing the mangonels. If the quoted figures of 50,000 soldiers and 30,000 followers are correct, the slow progress is not surprising.

On arrival, al-Muʿtaṣim assigned a stretch of the wall to each part of his army. One section had been weakened by flooding and roughly rebuilt, so the caliph having identified the spot, set up the stone throwers there. The defenders hung down balks of timber but these were shattered by the stones. The first Arab plan was to use powerful catapults, each "manned by four persons", mounted on platforms on wheeled carts. Stuffed sheep skins were used to fill the ditch, and mobile towers (*dabbābat*), each to hold ten men, were readied. However, they became stuck. "Thus the towers, mangonels and scaling ladders . . . remained ineffective and in the end were broken up . . . and burnt." However, the bombardment had created a breach. This, however, was narrow, so the Arabs deployed large mangonels to concentrate on it, presumably to widen it. At this point, perhaps after about two weeks, the defenders at the breach surrendered it.[1] According to Michael the Syrian, the bombardment was effective in driving the defenders from the ramparts. This enabled the attackers to approach with *tripodia* covered in skins, and when the break in the masonry was spotted, the rams were used to force a breach after two days' work, which the defenders gruesomely blocked with corpses. In the end, a traitor gave the Muslims access, and Amorium fell on 3 July 838.[2]

If the catapults abandoned by the attackers were able to be carried on carts and operated by just four men, they were not very large, and it is hard to credit that they were likely to create a breach in a stone wall. Similarly, if the *dabbābat* were only able to protect ten men, they were not siege towers of the type to enable attackers to oversee the walls, and perhaps they should be regarded rather as mobile shelters designed to enable the attackers to reach the foot of the wall. The version offered by Michael suggests just that course of events. The remains of Amorium have been excavated, although numerous rebuildings have made interpretation rather risky. However, the excavator believed that the fortress of 838 was limited to the hilltop upper town. If so, it was a circuit of a little over 900 metres, with towers every 25 metres and several gates. The towers were hollow, rectangular and projecting in typical late-Roman style, and the walls were just two metres thick, dating from around the fifth century. There were internal buttresses, perhaps to support a wall walk.[3] This was not, therefore, an exceptionally strong defence, but had it not been for the weakened section of wall it might have been expected to sustain a longer resistance against the arsenal deployed against it.

The empire did strike back from time to time. In 852/3, a fleet attacked the Nile towns, causing much damage and deploying both mangonels and ballistas.[4] The first of the emperors of the Macedonian dynasty, Basil I (r. 867–86), proved

[1] Ṭabarī, *The History*, XXXIII, trans. C E Bosworth, 100–112.

[2] Michael the Syrian, *Chronique*, III: 98–100 (book XII, chap. XX).

[3] Harrison, Christie, Lightfoot et al., "Amorium excavations". The finding of a coin of the emperor Theophilus (r. 821–42) in remains of a lower-town tower must raise questions about the original conclusion that only the upper town was protected in 838.

[4] Ṭabarī, *The History*, XXXIV, trans. J L Kraemer, 126–7.

a vigorous and successful warrior on more than one front, as is attested by the biography composed by his descendant, the emperor Constantine VII Porphyrogenitus.[1] If the story as written might be biased, the detail offered is none the less unusual and illustrative of contemporary practice. In 873, for example, Basil sent an army to attack Melitene, which first had to build a bridge across the Euphrates, and then take and destroy an Arab fort at Rapsakion. The troops raised from the local themes were sent out to ravage the countryside and to attack and destroy five named frontier forts, while the emperor marched on the city, where the forces based there first resisted him outside the walls before retreating inside. Basil was accompanied on this expedition by his war engines and other siege equipment, the transport of which certainly slowed down his progress, and in the end he gave up his intention of trying to take Melitene, destroying another four frontier forts on the way back.[2] This campaign illustrated the continuing main feature of the warfare on this front, with the control or destruction of frontier forts often the main achievement of any single operation. 878 saw a similar example. Basil marched into hostile territory in Cilicia to attack the city of Germanikeia (Maraş, now Kahramanmaraş), driving the inhabitants inside the walls, then pressed on to attack Adata (al-Hadath, a site later abandoned), which was a strongly walled place. Here he set up siege engines, hoping to storm the place, but the defence was stubborn. With winter looming, once again he gave up the siege and withdrew.[3] Notable here, then, was that Byzantine emperors went to war with siege trains, combining attacks on major places with their more routine plundering of enemy territory. However, on both these occasions, the towns proved too strong even in the face of siege trains, something that perhaps also suggests that logistics meant that these Byzantine armies could not be sustained long enough in the field to maintain a lengthy operation against any stubborn resistance, although frontier forts fell with apparent ease to any serious attack.

Italy

There was regular conflict between the Carolingian kingdom in northern Italy, the papal state in the centre, the Byzantines in the far south and the Lombard rulers in Benevento, Capua, Naples and Salerno. In addition, Arabs from North Africa interspersed their trading links with violent raids. Stone walls and towers ringed the cities, providing a very necessary protection. The country itself was probably both poor and in many places sparsely inhabited. The inhabitants of the towns had retreated behind the walls of defensible *castra*. Occasionally, they provided a target in themselves, as we learn that Lando, who was both count and bishop of Capua, brought an army to Capua to pull down and destroy one of the city's towers.[4] In 821, in a war with Naples, the Lombard Sico deployed "rams and ma-

1 Constantine, *Vom Bauernhof auf den Kaiserthron*.

2 Ibid., 85–6.

3 Ibid., 95. The text noted here that Basil was warned that he was fated not to take the place, but that it would fall to one called Constantine: as it did to the author in 957.

4 *Chronicon salernitanum*, ed. Westerbergh, 95 (chap. 95).

chines" to break down the walls where they bordered the sea, demonstrating his familiarity with the best siege techniques of the age.[1] Capua itself, a very ancient city, had been burnt by its own Arab mercenaries in 841, but was rebuilt a short distance away in the 850s. At the same time, the count had built a strong *castrum* on an adjacent hill.[2] The Arab presence in southern Italy, however, was to bring imperial intervention. In 842, Arab mercenaries took possession of Bari, by a surprise night attack, and of Taranto. They were to remain in control there for thirty years, and used Bari as the base from which they launched raids across the peninsula. Their fortress was very small, being restricted to the narrow headland on the coast, but because of that location the place was, if stoutly defended, very hard to take. Three times, in 851, 857 and 866, the new emperor Louis II (r. 855–75) tried to expel this presence. Three times he was unsuccessful. Finally, in February 871, Bari surrendered to a combined operation of Franks, Lombards and Byzantines, whose fleet was crucial in the victory. It had taken the resources of a Carolingian emperor, a Byzantine emperor and the local princes to drive out what was, in effect, a settled raiding party.[3]

This was the time when the process of "incastellation" began in the Italian countryside. In order to protect the rural population from raids, the authorities were actively encouraged to create defensible refuges. In 866, Louis's capitulary ordering the expedition to the south called on his vassals to "cause the people to reside in *castella* and with peace". The places were built "with palisades, outworks, ditches and ramparts" "to repel the pagans". The countryside became "covered with *castella* and towers".[4] The primary function of these works, it should be clear, was to provide a refuge for a largely unarmed rural population. As with the Saxon burhs, it was presumably expected that they could be a sufficient deterrent. Unlike the burhs, however, there is no indication that there was any kind of broader plan to their positioning or any kind of regulation made as to their manning.

In 894, in the north, German king Arnulf (r. 877–99) crossed the Alps (perhaps using the Brenner Pass) in winter, and attacked Bergamo, held on behalf of the (from Arnulf's viewpoint) usurper king of Italy, Guido. His troops, it seems, basically pulled down the rampart, trying "to dig through the walls built in antiquity", and succeeding,[5] although there is also reference to the use of ladders. The conquest took just two days, despite fierce resistance. Bergamo stood on a steep hilltop, with late-Roman walls 2.5 kilometres in length enclosing 26 hectares. The speed of the capture (along with the hanging of Guido's count) led other towns of northern Italy rapidly to make terms with the conqueror. However, two years

[1] Erchempertus, *Historia Langobardorum beneventanorum*, 238.

[2] *La cronaca dei conti e dei principi longobardi di Capua*, 22. On the whole period, see Kreutz, *Before the Normans*.

[3] *Chronica Sancti Benedicti Casinensis*, 469–71; Kreutz, *Before the Normans*, 43; Constantine, *Vom Bauernhof auf den Kaiserthron*, 102–3.

[4] *Constitutio de expeditione Beneventana* (866), in *Capitularia regum Francorum*, 95; accurately reported by *Chronica Sancti Benedicti Casinensis*, 470; other quotations collected in Cilento, *Italia meridionale longobarda*, 145 ff.

[5] *The annals of Fulda*, 126–7; *Gesta Berengarii imperatoris*, 387 (book III, lines 94–103). Jarnut, *Bergamo*, 8, 25–6; Jarnut, "La conquista di Bergamo", 26–7.

later, Milan fell to Guido's former supporter Lambert, who took the city (according to a much later source) "with hammers and mangonels".[1] In the same year, Arnulf responded to an appeal from the pope to rescue him from the dominant faction in Rome. His troops protected themselves "with shields and wicker coverings . . . They heaped packs and saddles against the walls and mounted it. One group found a beam fifty feet long and used it to batter down a gate." It was not the massive Aurelian defences they were able to overcome, but the more recent Leonine walls that succumbed to this outright assault.[2] The defences of this time could still be taken by a determined attacker with adequate numbers of fighting men and engineers at his disposal. Walls now more than half a millennium old were not able to resist a resolute attack without substantial reinforcement.

Meanwhile, further south, in 871, an Arab army from Sicily marched across Calabria and attacked Salerno. This was a full-scale attack on what was, for the time, a strong, well-sited city. The Arabs deployed siege artillery. The tenth century anonymous compiler of the local chronicle calls them *petraria*, which is one of the earliest uses of this term, but suggests the stone thrower with which we know all the parties to have been familiar. This one may have been of considerable size, as it appears to have been effective in gradually knocking down a tower. This did not succeed owing to strong counter-action by the defenders, and ultimately the siege was ended when neighbouring Amalfi smuggled in supplies and a hurriedly summoned Frankish and Lombard relief army descended from the north.[3]

The conquest of Sicily

Most of the Muslim attacks in southern Italy were intended as raids. In Sicily, however, they had been involved in a prolonged conquest, lasting more than fifty years. The island was, by now, an isolated Byzantine outpost, a long way from help from the centre of the empire. The conquest was piecemeal, city by city, in periods of concentrated attacks with new forces. Palermo had been secured by 832, Ragusa, Butera and Leontini in 846/53, and the major centre of Castrogiovanni in 858. Very little detail survives of the means of conquest, most of the accounts dating from very much later. Nor is the state of the Byzantine defences clear. It seems that either the Arab attackers were not often supplied with siege equipment, or else the defences had been strong. Syracuse drove off the Arabs in 827, assisted by a Byzantine fleet and an onset of disease among the attackers. Messina initially defeated the attackers in 842 before being taken by surprise, Leontini was captured by a ruse when the defenders were drawn outside the town. Butera, however, held out for five months. Castrogiovanni, allegedly very strong, fell to a surprise escalade, and was then refortified and garrisoned. This does not suggest

[1] Landulfus, *Historia Mediolanensis*, 46.

[2] *The annals of Fulda*, 132–3; Liudprand of Cremona, *Antapodosis*, in his *Works*, 23 (book I, chaps. xxvi–xxvii). The quotation is from Liudprand, who was not contemporary but wrote in the middle of the following century.

[3] *Chronicon salernitanum*, ed. Westerbergh, 127 (chap. 113).

that the existing defences were in fact particularly up to date. Noto fell in 863/4 when a way in was shown to the Arabs.[1]

The capture of Syracuse on 21 May 878, representing in effect the completion of the conquest, may even have been something of an accident. An Arab fleet from Tunis had been assembled to meet a Byzantine fleet that never came, and instead assailed the city. There is an eyewitness account by a monk, Theodosius, written while imprisoned in Palermo, addressed to the archdeacon of the archbishop of Syracuse, who had left the city to seek help from the capital. Theodosius describes a vigorous Arab assault strongly resisted by the garrison, with both sides deploying artillery and the besiegers using a wide array of siege equipment. The Arabs pushed up to the walls, seeking to undermine the city's *propugnacula* – a frustratingly ambiguous term that might signify outworks, or bulwarks, or even (although not in this context) hoarding. Artillery was in use, including stone throwers that caused major damage to the defences. The monk describes the horror of being at the receiving end of the bombardment. The garrison used a weapon called a *brachiolla*, which seems to have been a kind of ballista. Having seriously damaged one of the towers, the Arabs attacked it with scaling ladders. However, attempts to capture Syracuse by assault failed, and it fell eventually, after maybe nine or ten months, to a close land and sea blockade. The Byzantines had attempted relief, but their fleet had been delayed.[2] For all the array of weaponry apparently deployed, a city held by a determined garrison, although cut off on all sides, had had to be starved out.

Saint-Tropez

In 889, a small group of Muslim raiders landed near Saint-Tropez and established themselves in a hilltop base accessible through thick brambles by one narrow path. Joined by others and officially backed by the caliph at Córdoba, and taking advantage of the disarray of the local Christian rulers, Muslim forces based here at Fraxinet (now La Garde-Freinet) raided far and wide, establishing bases throughout Dauphiné and controlling (no doubt very profitably) the Great St Bernard pass. In 945 they even took Grenoble. Only in 975 were they finally defeated by Count William of Arles. As with the emirate of Bari, a small group of what were, in effect, pirates, had defied powerful enemies for a long period, relying on an inaccessible rather than a heavily-defended base to remain to carry out their extortion.[3]

[1] *Cronica di Cambridge*, 278–9; Ibn al-Athīr, *Dal Kamil 'at tawarila*, 367–83.

[2] Theodosius monachus, *Epistula de Syracusanae urbis expugnatione*, 259–61, discussed by Lavagnini, "Siracusa occupata dagli Arabi"; *Cronica di Cambridge*, 279; Ibn al-Athīr, *Dal Kamil 'at tawarila*, 396; Constantine, *Vom Bauernhof auf den Kaiserthron*, 118.

[3] Summary in Lebling, "The pirates of St. Tropez", 26–9.

Thessaloniki, 904

Arab fleets continued to threaten the empire for a long time to come. In 904, what must have been a large fleet carried out an audacious raid culminating in the seizure of the city of Thessaloniki. There is considerable doubt as to the authenticity of the one surviving account, which may have been a work of several centuries later, which seems to make a reference to the fall of the city to the Ottomans, but it is also possible that the fifteenth-century manuscript is a reworking of the story of the 904 events. The Arabs had catapults, both ballistas and stone throwers, and attacked the town both from land and sea. We do not know to what extent the defences last attacked in the Avar siege of 597/618 had been improved, but this account says that part of the walls had been neglected, and that the sea walls were lower, although an iron chain protected the harbour. While the besiegers set up a "crowd" of seven catapults around one of the gates, and hurled enormous rocks from their stone throwers, it was apparently by using their ships as the base for mounting ladders that they gained entry to the city, and carted off 20,000 people to slavery.[1]

Bulgars, Slavs and Rus'

The Byzantine empire also faced new challenges from the north. The Slavs had been settling in the Balkan peninsula for a long time. Unlike their counterparts in central Europe, the majority of their settlements were undefended before the ninth century. When they became common, their form was simple, as elsewhere, taking advantage of geography to create defended promontory sites as refuges with ditch, bank and wooden palisade.[2] Settling in the plains of the Danube and Dniester rivers from the late seventh century were the Bulgars. This group, like the others, did not arrive with a high level of technology, but this would soon change. By and large, providing they were left in peace, the Bulgars were not concerned to engage in warfare with the empire to the south, although the emperors still maintained a theoretical claim to all the territory south of the Danube.[3] The sources for what happened, being Byzantine, naturally give a very one-sided account.

In 809 the Bulgars, having completed the destruction of the Avar kingdom begun by Charlemagne, took Serdica (Sofia), apparently "by deceit", and along with it captured one Eumathios, "an expert in engines", while massacring the large Byzantine garrison. In 811, an imperial expedition led by the emperor Nicephorus I (r. 802–11) himself occupied Pliska, the capital of the Bulgar khan Krum (r. c.800–814), a vast earthwork encampment enclosing 23 square kilometres be-

[1] Cameniata, *De excidio Thessalonicae*, 562ff.; on the state of the defences, cols. 535, 547, 559. The method of capture is confirmed by the *Compilation anonyme sur la défense des places fortes*, 414 (§69). The dispute on the source is in Kazhdan, "Some questions addressed to the scholars", 301–9, and Christides, "Once again on Caminiates' 'Capture of Thessalonika'".

[2] Teodor, *The east Carpathian area of Romania*, 4, 57–60.

[3] Shepard, "Slavs and Bulgars", 229–32.

hind a three-metre-high rampart, burning all the houses and "the enceinte made of assembled pieces of wood".[1] The complacent Byzantine army then suffered a crushing defeat in an ambush in which the emperor himself was killed, and his skull was famously turned into a cup for the use of the khan. The next year the Bulgars marched against the Black Sea fortress of Mesembria (Nesebŭr). This was one of a line of posts along the coast held by the empire that must have been one of the main avenues of contact and trade between the empire and the Bulgars. Standing on a peninsula cut off by defences, it was a naturally strong site. However, Krum arrayed "machines and engines in which he had become an expert" and took the place without resistance. This time, the same chronicler tells us, the siege weaponry had been designed by an Arab "highly skilled in engineering" who had defected from Byzantium after receiving bad treatment. It seems that the Bulgars were finding all the help they needed from the opposition![2] It is not surprising that technological information should not remain secret for very long at a time when there was not only regular peaceful contact between Byzantium and its neighbours, but also a procession of Slav and Bulgar people into imperial service.[3] The only issue was whether the "barbarians" had sufficient craftsmen to create and operate the weapons.

Whereas the archaeological evidence (or in this case its absence) suggests that the Slav settlers for the most part continued a basic rural existence and that the old Roman cities of the Balkans (with some exceptions) fell into disuse, a new process of urban development did eventually take place, alongside the development of fortified sites and religious sites, all calling for building skills. But in the kingdom established by the Bulgars, more substantial urban settlements were created rather sooner, and the conversion to Christianity after 865 brought a spurt of building. The Bulgars quickly developed the skills and resources to create stone walls, projecting towers and gates for their towns alongside more primitive earthwork defences, and numerous sites have been excavated where the stone defences have been dated to the time of this, the first Bulgarian empire.[4]

Krum's army descended on Adrianople (Edirne), the major imperial city of the province. The Byzantine commander is ascribed a speech to his troops in which he states that Krum's army had "the machines and instruments suitable for capturing cities: *tribulos, tetrabulos, testitudines,* high ladders, balls (?), levers, rams, ballistas, flame-throwers, *petrarias, scorpiones,* dart throwers, slings . . .".[5] This formidable list reads like an inventory of every piece of siege equipment known (and of course

[1] *Chronique byzantine de l'an 811*, 435.

[2] Theophanes, *The Chronicle*, 666 (808) and 682 (812). On Mesembria's defences, see Oikonomides, "Mesembria in the 9th century", 269–73; Venedikov et al., *Nessèbre*, 1: 136–58, assigns the original defences to the late Roman period, with Byzantine rebuilding in the fifth and sixth centuries, and largely rebuilt in the later ninth century.

[3] Ditten, "Prominente Slaven und Bulgaren", 95–117.

[4] Georgiev, Ivanov and Dzingov, "Archaeological investigations in Bulgaria"; Bošković, Stričević and Nikolajević-Stojković, "L'architecture de la basse antiquité et du Moyen Âge", 156–62; Michailov, "Fouilles des sites du Moyen Âge", esp. 51 (Tŭrgovishte), 52 (Stŭrmen, Nova Cherna), 55 (Sveti Duh, Dernik).

[5] *Historia de Leone Bardae*, 347. The list includes different kinds of artillery, the meaning

may have been a work of the chronicler's imagination rather than a real inventory). Particularly interesting is the array of terms used to list the artillery, implying as it does a range of different weapons, in which the principle distinction among the stone throwers was probably one of size. Despite this enormous arsenal, it was by hunger that the city was defeated.[1]

Pliska and Preslav

Thus a theme we have encountered before is repeated. A "barbarian" people acquired the most sophisticated technology through its contact with and proximity to the empire. Often, the knowledge would have been of little use without the presence of people able to understand how to use it, and the Byzantine chronicles tell us who was responsible for passing on the state secrets. The extent of sophisticated building work carried out in the new Bulgar realm strongly suggests the rapid appearance of native skills. Sofia (Serdica) had late Roman walls and had been repaired by order of Justinian. The Bulgars carried out repair work to the existing circuit of walls and towers over the next period of their occupation but at this time mainly depended on what they found.[2] At Pliska and Preslav, the establishment of a long period of peace with the empire to the south permitted a period of significant new building in the reign of the new khan Omurtag (r. 814–32), while the conversion to Christianity after 865 gave a further impetus to building work. At Pliska, following the destruction in 811, the inner complex was provided with a solid enceinte. The palaces built for the khan here, at Preslav and at Han Krum were of a similar style and scale. Outer enclosures of some 500 × 400 metres were defended by large earth ramparts backed by stone walls. The palaces in the inner enclosures were constructed of large stone blocks and equipped with the latest refinements including hypocausts. Gateways into the complex were built through a gate tower at all three sites. It is not surprising that the style of this work should follow that of the Byzantine cities in the land the Bulgars now occupied, but the speed with which the Bulgar rulers transformed what had once been basic military encampments into sophisticated palaces and cities was remarkable.[3]

Byzantine interests in the lands north of the Black Sea brought them into contact with another new people, the Rus'. The Vikings had penetrated the great rivers of Russia from the north, and established posts for themselves at key riverside positions during the eighth century. The foundation of the forts which were the sites of the future cities of Novgorod, Smolensk and Rostov was attributed to the time of Rurik in the ninth century. The territory into which they were moving was inhabited by Slav tribes, and it seems that the kind of settlement they encountered was typically small and weakly defended. The oldest Slav fortified settlements date from the eighth century, between the rivers Dniepr and Don.

of which is unclear; probably they were different versions of the lever-operated stone throwers.

1 *Theophanes continuatus*, 68; Georgius monachus, *Chronicon breve*, 979–82.

2 Stančeva, "Sofia au Moyen Âge".

3 Stančev, "L'architecture militaire et civile de Pliska et de Preslav", 345–51.

These early sites were small, and took advantage of geographical features such as hills and islands. Here, a simple wooden fence seems to have been added. Where natural defences were lacking, a ditch and rampart would be dug. It is unlikely that the early Viking settlements were much different, although they would develop substantially with the creation of the principality based on Kiev in the tenth century, the beginnings of the Russian state.[1] Prince Oleg, for example, is recorded as beginning the construction of "stockaded towns" in 880.[2] Denied evidence by the total disappearance of constructions in timber from this era, it is nonetheless thought that watch-towers (*vezhi*), tapering upwards from hexagonal or octagonal log bases, and common in Rus' for centuries, might have dated back to before the ninth century. Certainly, local carpentry skills were clearly highly developed in this region.[3]

The first serious encounter with the new power in the north came as a shock to Byzantium. A fleet of Russian ships descended on Constantinople in 860, while the emperor Michael III was on campaign in the east, and ravaged the suburbs. Told of the risk to his capital, the emperor rushed back and the attackers retreated. It seems likely that this was a traditional Viking raid, drawn by the fabled wealth of the great city, but that despite possibly a long stay of several months, their communications and provisions secured by their sea power, they had no means of breaking into the city.[4] The Patriarch Photius declaimed "where are the armies? Where are the arms, the engines" while "an uncaptured army equipped in servile fashion is sneering at (us)?"[5] But his attribution of salvation to the protection of the Virgin is not borne out by the mute but eloquent testimony of the city's walls and towers.

Developments in east Asia

While the landscape of Europe, North Africa, the Middle East and central Asia was undergoing significant changes, there were important developments too at the other end of that vast continent, contemporary with the rule of the Tang dynasty in China (618–907), long recognised as one of the high points of Chinese civilisation. The nature of the surviving source material makes it difficult to reconstruct even the chronology of events, let alone the detail of the cultures, of neighbouring peoples, even though this period saw the emergence of new regional powers, except through Chinese records. Such was the empire of the Uighurs, established in central Asia in the late eighth century in a vast area north of the Yellow River, with Lake Baikal to the north and Lake Balkhash to the west, and Tibet to the south. Having settled down from a previously nomadic exist-

[1] Rappoport, "Die altrussischen Burgwalle", 62–4; Herrmann, "Neue Arbeiten über die altrussischen Wall- und Wehranlagen", 228–9; Vernadsky, *Kievan Russia*, 23.

[2] *The Russian primary chronicle*, 61.

[3] Opolovnikov and Opolovnikova, *The wooden architecture of Russia*, 15–16.

[4] The story is recounted by Vasiliev, *The Russian attack on Constantinople*, 188–228.

[5] Photius, *The Homilies*, 89, 92.

ence, trade, agriculture and then urban development accompanied the process. During the 750s, the Tang suffered a series of significant defeats on several fronts (including by an Arab army in 751), enabling the Uighurs to secure their position by building a capital city, using skills of metallurgy, stonework and sculpture according to archaeological evidence, and learning also from their acquaintance with the Chinese empire of that state's system of fortresses, some of which had been set up half a century before to protect that very frontier. There is little evidence as to the sophistication of these former nomads at defending or attacking fortresses, but sufficient to suggest that they knew how, and it would be reasonable to assume that their models were Chinese.[1]

The same period saw the greatest power of the Tibetan empire, established during the upheaval in China accompanying the change of dynasty from Sui to Tang. During the seventh century, a great area far larger than Tibet's present size was conquered, and in 668 a line of fortresses was constructed to defend against China. Tibetan conquests to the south and west involved capturing cities. Such victories secured control of the whole of the Silk Road in the 670s, although the Tang recaptured the fortresses in 692. In 749, Tang armies succeeded in storming the Tibetan fortress of Shih-Pao (Shibao) city after just three days, but at very heavy cost. There was conflict around the town of Pei-T'ing (Beidong) in 791. In 794, the Tibetans were engaged in capturing towns from the Uighurs.[2] This empire, too, was quite capable of attacking and defending fortresses. Although information on the organisation of the empire is hard to interpret, senior officials included a "Great Chief of Fortresses" alongside numerous ranks of military officials including those in charge of supplies.[3]

There is a little more detail about the situation in a power south-east of the Tang realm: Annam, although based on a Chinese version of the Annamese imperial annals. This records the exploits of the ruler Cao Biền against his southern ("barbarian") neighbours. In 861, Cao Biền stormed a Nam town (unidentified) killing thousands of people. He then ordered the building of the town of Đại La (modern Hanoi), which was, the annals report based themselves on "ancient annals", 6,344 metres around, with walls more then eight metres high and the same wide with wall walks with parapets 1.8 metres high. There were 55 "pavilions" from which to observe the enemy, six "round" gates and three aqueducts. Apparently, in fact, Cao Biền was responsible for the development rather than the original construction,[4] but in any case this was a powerful and impressive piece of work to compare with any in China (and certainly with any new constructions in Europe).

[1] *The Uighur empire according to the T'ang dynastic histories*, 13, 17, 70–71, 111–13, 124, and other sources, mainly Russian, cited there.

[2] Based on C I Beckwith, *The Tibetan empire in Central Asia*, 13–16, 30–34, 47, 64, 80–81, 134, 155–6, citing Tibetan and Chinese language sources. Pei-T'ing: see Ecsedy, "Uigurs and Tibetans in Pei-T'ing", 83–9; Stein, *Innermost Asia*, II: 579–81 (citing the Tang annals). The possibly limited extent of Tibetan technological innovation was studied by L White jr, "Tibet, India and Malaya", 515–26.

[3] Lalou, 'Revendications des fonctionnaires du Grand Tibet", 171–82; Uray, "Notes on a Tibetan military document", 223–9.

[4] *Les annales impériales de l'Annam*, 204–6, 207–10.

It is argued as well that the Tang model lay behind the design of Japan's first imperial capital at Heijō-kyō (Nara), built from 710 and occupied till 784, which had dimensions of 2.6 by three miles (4.3 by 4.8 kms). It was moved to a new site where Kyoto now stands in the 790s, which was larger still, but relied for protection only on a small earth bank and moat. Defence, it seems, was not a major concern even for a city housing imperial palaces and the centre of the administration in Japan.[1] This approach was long lasting. From the mid seventh century, a civilian militia had been responsible for defence. From 702, when the government had been reorganised according to the Tang model, one of the eight boards established was responsible for weapons, fortifications and beacons, including arsenals and the workers employed there. Warfare itself sometimes compelled the empire to look to its armed forces. In 720, during a war with the Ainu people of the north, a fortress was built at Taga and garrisoned with farmer-soldiers. The design of the fortifications themselves seems to have been learned from Korean refugees. Taga was captured by the Ainu in 726 and the defenders massacred. The twenty-year war needed to defeat the Ainu led the Japanese government to instigate a more professional military system, leading to the establishment and rise to prominence over time of a military class, but fortresses remained rudimentary and temporary rather than central to Japanese society.[2]

Conclusion

By the beginning of the tenth century, the countries of Europe and the neighbouring lands looked very different indeed from their late-antique predecessors. Although Charlemagne's single empire rapidly broke up after his death, Carolingian descendants continued to rule in the successor states throughout the ninth century. The new challenge of the Vikings brought significant change again, with a period of regular disruption, dramatic developments in the islands of Britain and Ireland, and new methods of defence. It may have become fashionable to downplay the negative consequences of the Viking raids, and to discount the inevitable bias of the monastic chroniclers, who were both their primary target and their main witnesses. But it is clear that their impact was significant in terms of both the future trading settlements they founded and the more established places they ransacked. Impelled by the danger, rulers who had the duty to protect their subjects were forced to restore old fortifications, but also to allow the construction of new, rural defences where none had existed before. Whereas the urban defences were to be maintained little altered for centuries, the arrival of fortifications in the countryside, built and maintained by local magnates, was to be part of the process of the privatisation of fortification that would flower in the centuries to come.

In terms of technology and design, there was little development during the Carolingian centuries. Where stone was used, the resulting work was less impressive,

[1] J W Hall, "Kyoto as historical background", 6–12.

[2] All from Murdoch, *A history of Japan*, 1: 215–19, and the Japanese sources cited there.

and less skilfully executed, than late imperial works. Shortage of skilled workers and resources explains this. Elsewhere, newly erected works seem to have been mainly of earth and timber, of simple if sometimes powerful design: there was no reason why a work of earth and timber could not be a powerful defence. This retreat from both the scale and the skill of Rome was also apparent in the works of the still-advancing Arabs and of the Byzantine empire. Constantinople had survived the prolonged threat of Islam and its northern neighbours, but only just, and was still on the defensive. The tenth century would see major developments in all areas.

4 The tenth century

Social and political change

The political shape and social systems of Europe underwent substantial changes during the tenth century. The Carolingian dynasty was replaced in France by the Capetians. But the land over which they reigned had seen power decentralised to such an extent that rulers in the provinces exercised as much real power as the king, and Normandy, Anjou and Flanders in particular were effectively independent powers, following their own domestic and foreign policy agendas. Below the level of these local princes, powerful families fought and feuded and schemed to increase their own powers. In every case, control of fortifications was to be a central element of the political equation. But by and large these fortifications were no longer the walled *civitates* that have been familiar throughout the story so far, but newer, smaller fortresses, combining a number of functions: the castles had made their appearance. Alongside them, the political and social system known as feudalism took root and shaped the continent for hundreds of years.[1]

The kings of East Francia or *Germania* had been more successful in preserving at least the outward trappings of Charlemagne's power, as well as securing his imperial title. The kings who originated in the Saxon duchy made Ottonian Germany a beacon for European civilisation. They established their power through a process of continuous struggle both with rivals at home, close neighbours such as the Slavs, themselves evolving towards settled states, and foreign invaders, in particular the Magyars. These struggles, and the need to impose a new royal house in a land where local rivalries and differences were deep rooted, were also reflected in the defences of this century.

[1] Since this is a term for which there are many different definitions, and starting dates, here it will be used in the sense of a system in which (as before) wealth and power were primarily derived from control of the land, and extracted by means of gathering a surplus from the labour of a peasantry which (unlike before) was increasingly unfree (e.g. became serfs). The surplus provided the resources for lords small and large to live, to build, to compete for power and influence. The political aspect derived from the notion that all land belonged ultimately to the highest power, and the king or prince could demand services (particularly military), or more generally, assistance and loyalty, from lower-ranked lords in return for their lands. Previously, such service had been due to a prince as custom or law dictated, but was not necessarily linked to the holding of particular lands. The complex of customs that would evolve later on, such as those around rendition of castles, were unlikely to have been present for most of the tenth century.

On the fringe of the continent, but certainly not isolated from European culture, the Anglo-Saxon kingdom of England was consolidated under the leadership of the royal house of Wessex and became one of the more prosperous and most centralised monarchies of the age. But at the end of the century, under now weak leadership, it was to fall victim once again to attack from Norsemen, who set in train through the conquest by Cnut the series of events leading ultimately to the Norman conquest. Other Vikings meanwhile continued their settlement and plundering in Ireland.

Italy was still divided between Byzantines, Arabs and Lombards in the south. In the north, there was a struggle between various parties competing for the Carolingian inheritance. The German kings, through their relations with the papacy and their imperial claim to rule in Italy, were to begin a long-term involvement in the peninsula that would bring more than one emperor to grief. In the meantime the process of *incastellamento* that we have already seen begun in the previous century became widespread and radically changed the landscape of the country. As in western Francia, power continued to be devolved to lower and lower levels.

The century saw a definitive halt to Arab expansion. In Spain, al-Andalus came into the hands of a new caliphate, the last of the Umayyad dynasty fleeing from the east and bringing some centralisation to a land where power had, again, become the province of local rulers. The Christian states began to stabilise in the north of the peninsula and took the first tentative steps towards a *Reconquista* that would take five centuries.

The most remarkable development of the century was the restoration of Byzantine power. Under a series of effective rulers, the empire once again took the offensive, driving back the Arabs in Asia Minor and in the Mediterranean, and beginning the process that would see the annihilation of the first Bulgarian empire and the restoration of the Danube as its northern frontier. Meanwhile, the Slav and Bulgar realms that had encroached into imperial territory were drawn firmly into the cultural orbit of the empire, and it was the church of Constantinople that oversaw their conversion to Christianity and exercised an immense influence on the newly developing states. This included at a greater distance the rapidly growing power of the Rus' princes of Kiev. That influence included the arts of building. Whatever the pretensions of the Ottonian dynasty in central Europe, and they were undoubtedly high, it was Byzantium that stood above everyone else at the end of the century as the great power of the age.

The Byzantine resurgence in practice

Against the caliphs

A fascinating element of the Byzantine resurgence was the development in this century of a significant quantity of written material on poliorcetics, the science of siege warfare within a general military science, for the first time since the *Strategikon* attributed to the emperor Maurice around the year 600. The reasons for this new interest in recording military advice may be found in the practice of the emperors as they moved onto the offensive.

Frontier warfare had continued along the long Asia Minor frontier between the empire and the caliphs. Towns and cities were taken and retaken, but for a long time the main preoccupation had been warfare restricted to raiding and counter-raiding. Attempts to make more permanent gains had generally been aborted, such as the emperor Basil I (r. 867–86)'s sieges of both Tephrike (Divriği) and Melitene (Malatya). The emir of Tarsus sent a fleet to attack Euripos (i.e. the island of Euboea), but the Byzantine defenders, forewarned, constructed stone throwers and ballistas, and used Greek Fire, to such effect that the raiders were repelled.[1] But generally, warfare consisted of devastating raids in search of plunder and captives to ransom or enslave. They were disastrous for the people in the areas affected, and had the consequence of encouraging the use of defended hill top sites as refuges. But they were not aimed at achieving the permanent conquest of a region. The storming of Melitene in 931 was unusual,[2] although it was then put to use as a major Byzantine base on the frontiers of Mesopotamia.[3] When the ruler in Syria, Sayf al-Dawlah, captured a Byzantine fortress in 947/8, he destroyed the walls rather than garrison it, and in 949 the Byzantines did the same with Theodosiopolis (Erzerum).[4]

A more sustained offensive got under way in the middle of the century. In 959, the emperor organised an expedition to Crete. Its recapture was vital for the security of large parts of the Byzantine empire, and several failed attempts had already been made on it. Leo the Deacon has left a detailed account of the new campaign. After landing, the Byzantines, led by Nicephorus Phocas (emperor, 963–9), besieged the Arabs in Chandax (Candia, modern Iraklion). The city being judged too strong to take by direct assault, as a result of its seaside location, strong walls and double ditch, a siege camp was set up to blockade it from the landward. When a relief army was defeated, the heads of the dead were propelled into the besieged town by catapult to demoralise the garrison. But it did not fall. So, after spending the winter camped around it, the commander ordered the preparation of siege machines: catapults to fire stones, and rams (*elepolei*)[5] with iron heads. The catapults hurled stones to cover the advance of the rams, which began to break into the wall. Simultaneously, engineers began to undermine the wall. The sandstone gave way easily. The resulting hole was stuffed with combustibles and ignited, bringing down two towers and the wall between. The garrison surrendered, and their lives were spared.[6] Corroboration of the details of the siege equipment is provided in the treatise by the emperor Constantine VII Porphyrogenitus (r. 913–59), who listed the materials provided for the expedition. It included a number of different types of ballista and crossbow, wooden towers, *testudos*, and a variety

[1] Scylitzes, *Byzanz, wieder ein Weltreich*, 171–3, 187–9.

[2] Ibid., 263.

[3] As described by Ibn Ḥawqal, *Configuration de la terre*, 181.

[4] Yaḥyā ibn Saʿīd, *Histoire*, 767–8.

[5] On the use and meaning of this word, see comments at the end of this section.

[6] Leo diaconus, *Historia*, 678–98; *Nikephorus Phokas*, 13–32.

of stone-throwing engines called under the generic terms *manganikon, tetrarea, labdarea* and *eilaktion*, there being four of each.[1] These were probably machines of different sizes and capabilities, all of the lever type, and presumably representing equipment built to a standard design. Leo says the machinery was constructed at the siege; Constantine suggests it was transported there. Perhaps it was taken in parts, and assembled when it was clear that it would be required. We will return to the significance of the terms for particular weapons. What is clear here is that the Byzantine imperial arsenal disposed of a considerable range of differentiated siege equipment.[2] From the account of the actual siege representing the culmination of this expedition, the decisive breakthrough, against an apparently well-built and obviously resolutely defended town was achieved through mining or sapping, after several months of otherwise fruitless attack.

The empire's next moves were in Syria. In 959/60, the important Arab fortress of Amida (Diyarbakır) was besieged, and its walls attacked with "iron rams".[3] Ibn Ḥawqal, the contemporary geographer whose *Kitab Surat al-Ardh* is thorough and comprehensive, recorded that this place, on a hill above the river Tigris, was made of black stones, but despite its size and importance had no garrison when he saw it.[4] Tarsus served as the base and starting-off point for raiding expeditions and was equipped to provide for the large numbers of volunteers (*ghazis*) from all over the Muslim world who came to play their part in the holy war against the infidel by joining such expeditions. Nicephorus encircled it in 964. It was well provisioned, and defended by a double wall and ditch, dating to the time of Hārūn al-Rashīd or his sons in the ninth century, and refurbished and repopulated in the 930s. The gates on the inner wall were constructed entirely of iron, those on the outer being plated with iron. Towers on the outer circuit housed artillery (*manājanīq, 'arradat*) and crossbows. The river Cydnus (Tarsus Çay) flowed through the town, and the defenders could divert the water to fill the ditch.[5] Finding this too strong a target on this occasion, he turned his attention to neighbouring towns, taking Adana and other places. At Mopsuestia (Yakapınar), he had to set up the stone throwers. Carefully studying the defences for weak points, the Byzantines carried out an approach at night and sapped the wall, which collapsed "like thunder", whereupon they were able to storm in and capture the town.[6] To undertake a successful night attack was an operation calling for exceptional skill and planning, but it seems that these ingredients must have been present. In 965, Nicephorus returned to Tarsus in the spring, and took it through starvation. One Arab historian described the town as being this time weak and ill provisioned, suggesting that the previous attack may have led to the consumption of reserves, while the new attack was timed

[1] Constantine, *De ceremoniis aulae byzantinae*, 669–72 (chap. 45).

[2] The *eilakton* was probably a light machine with a throwing arm attached to a single pole, the *labdarea* and *tetrarea* were probably heavier, having respectively L-shaped and four-piece trestle frames. See Chevedden, "The hybrid trebuchet", 200.

[3] Ibn al-Azraq al-Fāriqī, in Vasil'ev, *Byzance et les Arabes*, II: 117.

[4] Ibn Ḥawqal, *Configuration de la terre*, 216–17.

[5] Ibid., 181–2; Bosworth, "The city of Tarsus and the Arab–Byzantine frontiers", 277–82.

[6] Leo diaconus, *Nikephorus Phokas*, 54–5 (book III); Yaḥyā ibn Sa'īd, *Histoire*, 793–4.

to take place before the harvest. After the capture, the emperor ordered Tarsus strengthened further, and left a substantial garrison, intending that it should be a major fortress because of its location.[1]

In 968, the Byzantines directed their operations against the area around Antioch (Antakya), thereby returning to lands they had not held since the Arab conquests three centuries earlier. The city was well provisioned and possessed substantial defences, and Nicephorus having resolved, according to Leo, not to damage the defences with siege engines, had to abandon his attempt. Perhaps the emperor did not want to have to defend a city against counterattack with damaged walls, or perhaps, given the enormous length of the walls, he wanted to protect himself first against attempts at relief, although Ibn Ḥawqal noted that no one had lifted a finger to save the city.[2] Anyway, having pulled back from assaulting Antioch, the imperial forces pressed on, trying their luck at Tripoli in Lebanon. However, again the defences proved too strong, and the Byzantine fleet was held up by contrary winds, so instead the army attacked Arqa. The besiegers threw up three forts of their own to isolate the place, then proceeded to break in using rams (*elepolei*). This took nine days. Having secured a substantial territory by these operations, Nicephorus returned to Antioch, and built a strong camp against it. The stone defences were erected in three days. He left a garrison of cavalry and infantry to keep Antioch penned in. The following year, Antioch fell to a night attack with ladders, carefully planned to be the right height for the walls.[3] Meanwhile, in 968 a Byzantine force took Manzikert in the east by undermining the walls.[4]

The story of successes in sieges of Muslim fortresses continued. Nicephorus' murderer and successor as emperor, John I Tzimiskes (r. 969–76), launched a massive offensive against the Muslims in Syria and Palestine, since 969 under the rule of the Egyptian-based Fatimid dynasty.[5] The opposition took refuge in its fortifications, especially those on the coast whence they could maintain communication with Egypt. The emperor was able to secure the submission of many fortified towns and take others by storm, including Baalbek, Jubail and Beirut in 975, although the latter was "heavily fortified". Little time was spent at each site, probably indicating that the walls had not been raised or strengthened and could be taken either by escalade, or sapping, although Ibn Ḥawqal's descriptions do not suggest that the fault lay with the defences, rather with the lack of fortitude of the Muslims. It was noticeable in this context that once again Tripoli remained unconquerable, and that the emperor did not manage to restore Palestine as a whole (and Jerusalem in particular) to imperial control.[6] The city of Aleppo itself fell to the Byzantines, thanks to the "degradation of the rampart",[7] some of the

1 Ibn Miskawayh, *The concluding portion of The experiences of the nations*, v: 222–6.

2 Ibn Ḥawqal, *Configuration de la terre*, 177.

3 Leo diaconus, *Nikephorus Phokas*, 65–73 (book IV),79 (book V); Yaḥyā ibn Sa'īd, *Histoire*, 816, 823; Matthew of Edessa, *Armenia and the crusades*, 21 (chap. 7).

4 Step'anos Taronec'i, *Armenische Geschichte*, 134.

5 The Fatimids were a Shi'a Muslim dynasty that had originated in North Africa and conquered Egypt; see below.

6 Matthew, *Armenia and the crusades*, 28–32.

7 Ibn Ḥawqal, *Configuration de la terre*, 174. Muqaddasī, whose work on geography was

population taking refuge in the citadel. In 986, Bardas Phocas the younger captured Kilis north of Aleppo, and then marched against Apamea, where he arrayed his machines of war, "which threw down several towers", although he was ordered to withdraw before capturing the town. The emperor Basil II (r. 976–1025) attacked Balanias (Bāniyās). He ordered construction of a ram, which battered the walls with such effect that a tower and the adjacent curtain wall collapsed and the town surrendered.[1] Abū'l Wafa' attacked Miyafarkin (Martyropolis, now Silvan) in 978/9 as part of a new civil war between the various wings of the ruling families of the Muslim world. This particular invasion involved an unusual winter operation to isolate the fortress from its territory, then a direct attack. The defenders arrayed their artillery behind their walls to shoot on the besiegers' camp, who responded with their own artillery hurling fire and stones. The capture was actually brought about by a conspiracy in which a citizen led a revolt against the governor and handed the place over.[2]

Defeat of the Bulgarian Empire

The Bulgarian empire, now Christianised, had extended its rule over much of the Balkans, and had reached the Adriatic Sea at Dyrrhachium (Durazzo, now Durrës). Early in the reign of Constantine VII the Bulgarian ruler Simeon I (r. 893–927) brought a great army to Constantinople. He created an offensive earthwork stretching the length of the land walls, but when he saw the strength of the walls, the number of the garrison and the great array of engines including both stone throwers and ballistas, he realised the hopelessness of the task.[3] The city's defensive arsenal had obviously been maintained in good order. Simeon's war continued, with several successes against Byzantine armies. Adrianople fell to him while Romanus I Lecapenus (r. 920–44) held the imperial throne. His own siegecraft seems to have been rather limited. As at the capital, he did not attempt to break into the place. He surrounded the town with ditch and palisade, and waited for it to surrender through hunger.[4]

In 967, the prince of Kiev, Svyatoslav I (r. *c.*963–72), in response to an invitation from Constantinople to attack Byzantium's enemy from the rear, oiled with an enormous payment, captured "80 towns along the Danube" from the Bulgarians.[5] The Rus' prince, however, planned to stay in Bulgaria as the new heart of his own greatly expanded realm.[6] John Tzimiskes found himself therefore in conflict with

begun around this time, described Aleppo as well fortified with a "spacious castle well provided with water": *Description of Syria*, 12.

1 Yaḥyā ibn Sa'īd, *Histoire*, 415–17.

2 Ibn Miskawayh, *The concluding portion of The experiences of the nations*, V: 426–9.

3 Leo grammaticus, *Chronographia*, 1126; Scylitzes, *Byzanz, wieder ein Weltreich*, 238.

4 Scylitzes, *Byzanz, wieder ein Weltreich*, 256.

5 *The Russian primary chronicle*, 84; *La chronique de Nestor*, I: 89–91; *The Nikonian chronicle*, I: 63, 66.

6 A plausible reconstruction of events is given in Stokes, "The background and chronology", 44–53; "The Balkan campaigns", 466–83.

the Rus', with the Bulgarians playing a secondary role as the struggle unfolded in the eastern part of their country. John was able to bring a strong Byzantine army north from Adrianople in April 972. Our witness, Leo the deacon, noted that the baggage and siege equipment followed slowly behind the main force, while the Byzantine fleet accompanied the advance up the Black Sea coast. Having vanquished a Rus' army, the emperor then besieged its remnants inside the walls of Great Preslav. Facing a bombardment of hand-thrown stones, javelins and arrows from the walls, the emperor ordered a storming by escalade which overran the wall and drove the Rus' back into the old walled imperial palace. This was taken by using Greek Fire to burn it down. Svyatoslav meanwhile held out in the former Byzantine Danube fortress of Dorostolon (Drŭstŭr, now Silistra). John arrived and erected ramparts to blockade the city and protect his army, something that the chronicler noted was "customary." The imperial army included a Master of Artillery called Joannes Curcuas, but the city was too strong to assault, and the place finally fell after a long siege, the Rus' having failed to break out despite several vigorous attempts. The recapture of the Black Sea and Danube fortresses was a critical step in the restoration of Byzantine power in the region, and left the Bulgarian empire seriously weakened, but still a powerful force.[1]

It was John Tzimiskes' successor Basil II, later given the title of the "Bulgar-slayer", who devoted his energies to completing the destruction of this dangerous neighbour. Here it is worth noting that the frightening title the emperor was given, earned by his blinding of 15,000 Bulgar prisoners taken in a decisive victory in 1014, reflected very little on his usual approach, which was skilfully to exercise mercy and toleration, not just to secure surrenders but also to offer his captured enemies high rank in his empire. This paid dividends more than once in securing the yielding of powerful or inaccessible fortresses by their proud Bulgarian lords. In the course of numerous campaigns over nearly two decades, several battles and numerous attacks on fortified places, the Bulgarian empire was ground down. Great and Little Preslav, Pliska, Beroe (Boruy, now Stara Zagora) and Skopje were among the places taken by siege. Although the sources give few details of the siege operations, it is clear that the full array of techniques was available, and used when necessary.[2] In 1004, Vidin was only taken after an eight-month siege, being too strong to storm, and this despite the Bulgarian tsar Samuel (r. 976–1014) having launched a diversionary attack against Adrianople, which he took by the carelessness of the defence, and sacked before withdrawing. The castle of Pernik was frequently a target, but was always too strong to capture, and eventually its brave commander (who had also resisted other attempted blandishments) was rewarded for his loyalty to his ruler by high Byzantine rank when he finally yielded

[1] Leo diaconus, *Historia*, 847–55.

[2] The primary source for this period is Scylitzes, only part of whose text has been translated into Latin or any western language. The details for this paragraph come from Cedrenus, *Compendium historiarum*, and Zonaras, *Militärs und Höflinge*, 38–40, 48–56 (book XVII), whose twelfth-century text is drawn from Scylitzes and Psellus. The clearest English language account remains Runciman, *A history of the first Bulgarian empire* (1930), which is thoroughly based on the surviving records in many languages.

on the collapse of his own realm. In 1016, the fortress of Moglena (north-east of Edessa) was only taken after its walls had been undermined. The Byzantines excavated caverns below the foundations that were filled with timber and other combustibles, and their burning brought down the wall above, whereupon the occupants surrendered and were spared as was Basil's usual approach.[1] The Bulgar resistance to conquest was resolute, but the histories do not allow a proper assessment of their skills in defence of their walls. Ultimately, it was their inability to compete with a fully trained and equipped and very well-led Byzantine army that was fatal. But it took a long time. It was the death of the last Bulgarian ruler at Dyrrhachium on the Adriatic in 1018 that completed the process of conquest.

These events confirm that the Byzantine army was possessed of a high degree of skill in conducting siege operations. They were able to form clear judgements as to the strength of the defences opposing them, and how best to overcome them. There must have been a corps of engineers with the army able to construct or assemble the heavy equipment needed at some speed. The preferred way to attack was clearly a battering ram, and from the way the term to describe the weapon had become generalised from that for a siege tower, it is to be guessed that these weapons involved substantial wooden constructions to protect the crews. It is also possible that the term *(h)elepolis* was generally applied by the Byzantines to the most powerful implement of the time, as has been suggested by G T Dennis.[2] The other remarkable skill seems to have been the ability to carry out assaults by night. From the account of operations in Crete it is apparent that artillery was a regular part of the equipment. But it also seems that its role was secondary to the rams and mines. Presumably its purpose continued to be to attack defenders rather than walls, the shower of stones that might be maintained from an experienced crew providing vital cover for the advance of the ram to the foot of the rampart.

The Byzantine resurgence in theory

Modern siege warfare received its first non-Arab textbooks for three centuries during this time, and a substantial number of tenth-century treatises have survived. The emperor Leo VI (r. 886–912) was responsible for the *Tactica* in around 906. While much of the tone of the work relates to the primarily defensive nature of Byzantine–Arab warfare in his time, a section was devoted to siege warfare. The recommendation that the besiegers fortify themselves was a repetition of good advice dating back to the ancients. The main weapons for attack were listed as scaling ladders, rams and *testudines*. If the buildings in the besieged town were of wood, it was recommended to fire them with burning arrows. One of the means

[1] Cedrenus, *Compendium historiarum*, 186–7.

[2] Dennis, "Byzantine heavy artillery", 99–113. Dennis recounts the whole known development of Byzantine siege artillery, drawing substantially on Chevedden. He is undoubtedly correct about the mid-tenth-century application of *helepolis* to mean a ram (p. 103), and Leo's actual words leave not the slightest doubt that this was what he meant, since he described its features in exact detail (Leo diaconus, *Historia*, 695).

of propelling these was catapults, and there was specific reference to the *eilaktion* as a stone thrower. Mining was one of the main means of attack.[1]

The middle of the century saw works by the pseudonymous author known as Hero of Byzantium (the *Parangelmata poliorcetica* and the *Geodesia*), the anonymous *De obsidione toleranda*, and *On skirmishing*. "Hero"'s is a detailed manual explaining how to construct the equipment described, giving the dimensions, and with illustrations. Much attention was paid to the correct construction of mobile wooden shelters to protect miners and sappers, and the right tools for boring away at the wall. The necessary steps to ensure that such machines could be safely moved up to the foot of the wall in the first place, including filling ditches and avoiding dangers posed by caltrops, were given due attention. Having arrived at the wall, the main threat to such *testudines* was understood to be fire. Battering rams (Plate 12) were also vital, if mining had not brought the wall down. Walls made of stone were easier to overcome, because they shattered under the impact of the ram, which should be directed upwards. Against brick walls, on the other hand, ramming would make holes rather than shatter. The chapters dealing with siege towers described the necessary proportions, and some extraordinarily complex machines were illustrated. This section of the text was stated by Hero to be based on ancient sources (notably Hero of Alexandria), which also explains the detailed description then provided of torsion artillery. Hero also described and illustrated a wide range of scaling ladders of different kinds, including obviously more contemporary constructions from the top of which Greek Fire could be launched at the wall top. The writer also believed that a successful commander needed a sound knowledge of mathematics, including of "dioptric direct viewing" in order to be able to calculate the height of the enemy wall, and offered this knowledge with much other mathematical material in his second text, *Geodesia*.[2]

The author of *De obsidione toleranda* concentrated on the defenders' viewpoint. But his argument fitted the logic of current practice. There was a concern to dig ditches and lay out traps, and to deploy stone throwers to prevent the enemy reaching the foot of the wall. In another text, thought to have been a source for *De obsidione toleranda*, there was wide-ranging advice: along with organising large supplies of stones and inflammables on the ramparts to repel attackers, there was advice to close the town's bars to prevent drunkenness among the soldiers! There was also advice on resisting mining. A line of copper objects should be placed on the ground, and an ear put to them could pick up the vibrations from underground digging, allowing the planning of a countermine.[3] The text begins with the recommendation to secure the services of mechanics, engineers, architects and armourers. *On skirmishing* may have been commissioned by Nicephorus Phocas in the 960s. Its chapter on siege warfare gives valuable insights into the

[1] Leo VI, *Tactica*, 886–99.

[2] *Poliorkētika kai poliorkiai* (in Greek), 210–31; trans. as Hero, *Siegecraft*, 38–43 (*Parangelmata*, chaps. 8–11), 47–9 (chaps. 13–14), 51–3 (chaps. 15–17), 57–61 (chap. 20: rams), 73–85 (chaps. 30–39: towers), 91–3 (chaps. 44–5: torsion); 115 (*Geodesia*). Latin trans. as Hero, *Liber de machinis bellicis; necnon Liber de geodaesia* (1572).

[3] "Mémorandum inédit sur la défense des places", 127–30.

contemporary situation. Reflecting the strength of frontier fortresses, it advised that some places "have no reason to fear a siege". Those likely to be attacked needed to be provided with four months' provisions. Because most of "our towns" were located in mountainous terrain which made close encirclement difficult, defenders should make night time infantry sorties to break up the siege. It was vital to destroy all local sources of food for the besiegers. This would compel them to send out foraging parties far and wide, which would be vulnerable to ambush.[1]

The emperor Constantine Porphyrogenitus expressed the view that an essential part of the general's equipment was books: "military manuals, books on mechanics, including siege machinery (*elepoleis*) and the production of missiles and other information . . . relevant to wars and sieges".[2] The emperor Basil II learnt about war partly by reading books, and partly through his own very substantial experience.[3]

Nicephorus Uranus, whose *Tactica* is dated to around 1000, was a serving officer under Basil II and fought against both Arabs and Bulgars. Chapter 65 deals with siege warfare. He started from the viewpoint of the attacker. He advocated raiding to deny provisions to the targeted place and demanded that all commercial traffic be stopped, a real concern because "our people sell them grain and flocks in their love of gain". Having closed in on the target, the first job was to establish camp where there was a water supply, then dig an exterior trench and lay down caltrops. In keeping with what appears to have been standard practice, the advice was then to offer good terms to the defenders, who could keep their possessions as well as their lives. If they refused to surrender, different terms could be offered to different groups within the town, in order to sow dissension. The next step was for the whole army to prepare *laisai*: screens of vines or willow or mulberry trees, woven together in the shape of a house, with a pitched roof and able to cover up to twenty men. These should be fixed close enough to the wall to bombard the enemy with bows and slings, while others attack with sledgehammers and rams. Nicephorus also gave detailed instructions on mining. Having found a good place, one should erect timber frames and mats made from branches to secure the excavation if the soil was loose. The tunnel needed to be deep in case the enemy had "secretly dug a counter tunnel". He states that previous writers had recommended all sorts of different weaponry, but he had tried them all and found that undermining was the most effective.[4]

Nicephorus might also have been the author of a treatise on *Campaign organization and tactics*, probably written in the 990s, and dealing specifically with conditions in the Balkans. Looking at the problem from the opposite point of view from the writer of *On skirmishing*, this author pointed out that it was very difficult for an army to carry more than 24 days' provisions with it, so if there was to be a siege, it was essential to assure communications through the mountain passes for supplies. Whereas campaigning against the Muslims involved territories that

[1] *On skirmishing*, 213, 225.
[2] Constantine, *Three treatises*, 106–7.
[3] Psellus, *Fourteen Byzantine rulers*, 46.
[4] Nicephorus Uranus, *Taktika*, 157–63.

were fertile and prosperous, fighting against the Bulgars was very hard, because "there is a total lack of necessities". The besiegers' camp needed to be just the right distance from the walls, too far to be hit by the garrisons' throwing machines, close enough to protect one's own siege engines from sorties. The engines needed to be set up at once, opposite the weakest points in the defences, so that the precious provisions were not wasted by delays. The writer refers to his own personal experience of ruses, whereby the defenders were enticed into sorties to destroy the siege engines and were then themselves ambushed. Listing all the engines and devices available to besiegers, he refers readers to other and ancient authorities who have dealt with them thoroughly already.[1]

There has been debate as to whether these treatises actually represented the practice of their age, or whether they only reflected the work of ancient authors. Leo, for example, certainly called on the work of Philo, Onasander and of course Maurice, while references have been found in other texts to ancient works by Biton, Apollodorus, Aineias "the tactician" and Hero of Alexandria. However, it was not surprising that the store of classical knowledge should be called upon, as many of the principles and lessons of warfare had changed very little: the authors of *On skirmishing* and *Campaign organization and tactics* explicitly said so. But the tenth-century texts used descriptive terms, and referred to events, and enemies, of their own century. The illustrations that appear in the surviving manuscripts of the eleventh century were largely of contemporary, not ancient, siege equipment and weaponry. For example, we find drawings of lever-based artillery, not the torsion artillery of the Romans (not illustrated although described by Hero), and of Greek Fire being propelled from siphons. There can be no doubt that these texts were written for the use of contemporaries and that they were useful precisely because they drew basic principles from ancient sources, but updated them in light of experience — and their recommendations were followed by educated and literate generals.[2]

Byzantine fortifications of the tenth century.

The retreat of the empire in the face of its many enemies had led to a decline in the quality and scale of fortifications, with the need for haste dictating building styles, and the reduction in resources, and perhaps of skilled workers, was reflected in sloppier craftsmanship, less substantial and weaker walls, and a withdrawal to less accessible hilltops serving as citadels or refuges for the local population. The working assumption was that such places would not need to be strong enough to resist a sustained siege, and it was unlikely that a raiding force would carry siege

[1] *Campaign organization and tactics*, 303–5, 307, 317.

[2] Dain, "Les stratégistes byzantins", 319–71, traces the tradition from Aineias to Nicephorus Uranus; McGeer, "Byzantine siege warfare", deals with the tenth- and eleventh-century writers; Sullivan, "Tenth century Byzantine offensive siege warfare", closely compares the texts with ancient sources and current practice. The drawings of siege equipment are in Hoffmeyer, "Military equipment", 125–37. There is a translation of Aineias as *How to survive under siege*.

equipment or the wherewithal to construct it. Three types of defence have been identified on the eastern frontier. The *phrouria* were forts proper, built in the frontier zone to accommodate an army garrison. These were the type of place whose mainly anonymous fate of capture and recapture was recorded in the campaigns already mentioned. Further back from the frontier, *aplekta* were fortified camps serving as the concentration point for expeditionary forces, while the *kastra* provided fortified places on main roads. Finally, *kataphygia* were towns fortified to offer protection to the rural population. According to *On skirmishing*, they were meant to hold four months supplies to feed the people until the danger was past.[1] There is a fairly clear example of such a refuge at Saruhanlar in Anatolia, where a headland with two sheer sides enclosing an area of 400 × 300 metres had been cut off with a plain wall strengthened with round and square towers, the whole being invisible from below.[2] With the return to a successful offensive strategy and the revival of imperial fortunes one would expect the tenth century to have seen a programme of more substantial building works. But the picture is not uniform.

The island of Cyprus returned to Byzantine control in 965, and it therefore seems reasonable to suggest that at least some of the surviving Byzantine defences date from this time. It is also possible that these works were built at any time between then and the late twelfth century. Inside the massive later walls of Kyrenia Castle can still be found the thin Byzantine walls of the first castle, built as a 80.5-metre square with small, hollow circular corner towers. The southern (land-facing) side was reinforced with two massive solid pentangular beaks, presumably at a later date. On the nearby high summit of St Hilarion, known to have been occupied from the tenth century, the outer circuit of curtain wall has been ascribed to the Byzantines. It is well provided with projecting towers at regular intervals, mostly open-backed, but the rampart itself is rather than less than two metres thick, and the walls of the towers less than one metre, similar to the towers at Kyrenia (Plate 8).[3]

Where there were already existing walls, it would be likely that the Byzantines would have tried to strengthen or if necessary reconstruct rather than build afresh. At Mesembria (Nesebŭr), the important peninsular site on the Black Sea coast that had fallen to the Bulgars early in the ninth century, it seems that new towers, including two of pentagonal shape at the flanks, were added to the late-Roman walls, creating a much stronger defence facing the landward approach and perhaps providing a base for artillery to keep attackers at a safe distance from the wall. The structure of the gateway (a simple passage guarded by two wall towers) was not improved, however. It is hard to determine whether these works were

[1] Dagron, "Guérilla, places fortes et villages ouverts", provides this analysis, which seems to fit the reality.

[2] Foss, *Kütahya*, 119–21.

[3] Jenkins, "Cyprus between Byzantium and Islam". There is no detailed study of the early defences. Megaw, "The arts in Cyprus: military architecture", offers a brief survey of all the fortifications of the island, and there are local guides produced by the North Cyprus Department of Antiquities. The wall measurements are my own.

done under Basil I or later in the eleventh century.[1] Sardis was reoccupied after a long abandonment.[2] At Didyma, the site of the temple was restored in 988 with the creation of a barrier wall 3.7 metres thick. The resulting rectangular block was a powerful construction but lacked any towers or sophisticated defences.[3]

Păcuiul lui Soare (near Ostrov), constructed as a base for the Danube fleet on an island near Dorostolon after 971, was a rare new building. Abandoned in 1091, and now two-thirds submerged, the remains offer a glimpse of the building style of the late tenth century. The shape was the traditional rectangle, 250 × 200 metres. The wall was strong, 4.2 metres thick and estimated to have been 6–7 metres high, and built of large quarried stone blocks of very high quality masonry. There were towers at regular intervals at the corners and along the wall. Elaborate foundations on timber piles had been necessary because the fort was built on sand. The north gate survives. The gate-tower itself sat across the wall, eight metres in front and two and a half behind. Its careful construction included using L-shaped blocks to attach it to the wall. It was closed by a portcullis at the front and a two-leaf door at the rear of the passage, and a drawbridge behind that. This arrangement was unusual, the portcullis normally being an extra defence immediately in front of the door rather than a separate blockage. The style of gateway at Păcuiul lui Soare seems to have also been copied in gates built by the Bulgars at Pliska and Preslav. It has been suggested that the gate-tower design reflected the need to concentrate defensive capabilities using fewer defenders than the late-Roman design.[4] Clearly, there was no shortage of technical and building skills or of financial resources to pay for the work, so it is also possible that it represented simply a change in style. Nonetheless, the elongation of the gate passage, accommodating a doubling of the closures of the passage, represented an increase in the defensive strength of the gate, and its location close to two other projecting towers would have constituted a fairly formidable obstacle, making it easier to protect against attacks aimed at breaking through the gate or mining or sapping the intervening walls.

The prolonged period of warfare in the Balkans naturally led to significant building work throughout the period, and there are remains of new *kastra* in Macedonia too, the legacy of both Nicephorous Phocas' and Basil II's campaigning.[5] At Philippi, the tenth-century building programme involved, in addition to reconstructing the enceinte, providing an outer wall just in front of the city's rampart, and creating two outer defences for the citadel on the hill far above the town. The rebuild used stonework, bricks and tiles from the old city extensively. In the town, towers were spaced every 35–40 metres, and stood astride the rampart. Each of the three gates was of different design. That giving access to the main route to

[1] Venedikov et al., *Nessèbre*, 136–58; Venedikov, Tschimbuleva and Lasarov, *Nessebar*, 71–5.

[2] Foss, *Byzantine and Turkish Sardis*, 61.

[3] Müller-Wiener, "Mittelalterlicher Befestigungen", 39–40.

[4] Popa, "La porte nord de la forteresse byzantine" (based on the detailed report in Popa, "La porte de nord": see in particular the plans on pp. 272 and 283); updated English summary, Popa, "A tenth-century Byzantine castle".

[5] Hattersley-Smith, *Byzantine public architecture*, 180–88.

the south was protected by two large rectangular bastions, 11.6 × 7.15 metres, and the passage was protected both by a portcullis and a strong door. The citadel was formed of three consecutive redoubts. The outer was protected by an enormous projecting solid bastion, 15 metres long, on either side of which was a postern gate. The next courtyard was provided with square and round towers, rebuilt at this time. The inner redoubt was however entirely new, and the most striking feature was a rectangular donjon-like tower, with two storeys connected by a spiral staircase in the thickness of the wall, and absolutely plain walls, standing in the southern corner of a ward some 45 × 35 metres, dated by inscription.[1]

At Cotyaeum (Kütahya) in Anatolia, a simple rectangle of 300 × 350 metres, there was a massive rebuilding with no fewer than 42 new towers added, U-shaped with banded masonry, and much reused material. The towers offered platforms for artillery, retaining chambers for ballistas on one side of the ramparts. Uniquely, there is a concentric outer wall so close to the inner wall that the towers sit astride both walls, which must have created real problems for defenders circulating unless there were side gates in the towers. But there is uncertainty over its date.[2]

In the lands conquered by John Tzimiskes, visible within the powerful crusader castles at Saône (Qal'at Sahyun / Ṣalāḥ el-Dīn) and Bourzey (Qal'at Barzūyā/Mīrzā), although much ruined and still unexcavated, are Byzantine works that must owe their origin to the period following the 970s. The Byzantines constructed new works on these sites. At Saône, the highest point was given a citadel, protected from the accessible eastern side with a double wall. Other walls cut off the western side. The central fort was an irregular rectangle about 30 × 40 metres, with projecting bastions at the four corners. Chapels in the courtyard below the redoubt indicate that this was a permanently garrisoned fortress. A similar central redoubt exists at the much-ruined castle of Bourzey, standing high on a rock over the plain of the river Orontes ('Āṣī), also rectangular with projecting square corner towers, this time just 40 × 20 metres.[3] The masonry work was of small and roughly cut stones, giving the appearance of being hastily constructed. From their design and location it seems clear that their purpose was to hold a small garrison to secure control of the area. Similarly, at Antioch a small citadel was built at the highest point of the already strongly-walled city (Plate 3). Nothing remains of the Byzantine works put up at Tarsus, Adana and other newly captured fortresses.[4]

A similar pattern was to be found in Byzantium's Italian provinces, secured with the recapture of Bari from the Arabs. Walled cities had always been the centres of rule, and from the tenth century the upkeep of the defences was a local responsibility. New defended towns were built. These towns were not on a common plan, but were founded on suitable high ground. They were protected by a ditch, a complete circuit of walls, with one or two gates. Inscriptions record the

[1] Ducouk and Lemerle, "L'acropole et l'enceinte haute de Philippes"; Roger, "L'enceinte basse de Philippes', 20–30.

[2] Foss, *Kütahya*, 64–85, 123.

[3] Saadé, *Le château de Saladin* (guidebook), 10–12; Kennedy, *Crusader castles*, 15–16, 80–82 (plan of Bourzey); Honigmann, *Die Ostgrenze des byzantinischen Reiches*, 97–100.

[4] On which see R W Edwards, *The fortifications of Armenian Cilicia*, 26–32.

reconstruction of the walls of Taranto in both 880 and 965–9, while Brindisi's walls were rebuilt in the tenth century. The only place with a separate citadel was Bari, the capital of the theme. This seems to have comprised a large tower inside a fortified enclosure constructed of well-made limestone blocks.[1] Many of the other places were modest, resources being strictly limited. The defences included solitary towers in villages, with a clear role as a refuge, and in 959 the monastery of San Vincenzo al Volturno was described as *ecclesia cum castello* ("church with castle"). At Zignago in Liguria, Byzantine rectangular towers, constructed of local stone and entirely without any refinements, were erected as part of a scheme of protecting the coast and inland passes from raiders, providing safe accommodation for up to twenty people. The route from Ravenna to Rome was dotted with hilltop *castella*.[2]

A similar process had taken place in Sicily throughout the ninth century, and here, as in al-Andalus, the Arab conquerors moved in to base their rule on the existing Byzantine walled towns.[3] Everywhere, however, as elsewhere in the empire, the new fortifications remained the initiative of the state, and rural defences were to be found only on the frontiers, unlike the *incastellamento* taking place everywhere else in non-Byzantine Italy.[4]

The Byzantine empire possessed a powerful set of fortifications designed to serve a number of roles, and representing, at their strongest, the most powerful work of the non-Islamic world. Masters of the siege techniques of the time, which combined using artillery to assist (or from defence, to prevent) the process of creating breaches by mining or sapping, it would be expected that their defences would be constructed with this in mind. Taking into account the difficulty of precise dating of the few remains of the period, this brief survey of the evidence supports this suggestion, with a general approach of thickening walls and creating obstacles to a hostile approach to the foot of the wall. At the same time, the paramount need to protect the empire's subjects against devastating raids encouraged the continuation of the policy of building rural refuges. This process, however, remained firmly under the control of the emperors and their officials, even if maintaining the walls in every corner of the far-flung realm remained a duty, and presumably a considerable financial burden, on the local population.

The Islamic world

The caliphate of Baghdad had meanwhile fallen on hard times. The 'Abbasid dynasty had completed the transition from the Arab army of the conquests, largely native and tribal, to mercenary and slave armies, largely recruited among the nomadic Turkic tribes of the steppes. The price paid for

[1] Andreassi and Radina, *Archeologia di una città*, 503–4, 539. Nothing remains above ground of the *kastron*.

[2] Christie, "The archaeology of Byzantine Italy", 274–8.

[3] Maurici, *Castelli medievali in Sicilia*, 196–8.

[4] Martin, *La Pouille*, 258–68, for all the above.

maintaining professional, full-time armies was a loss of power and influence to the military commanders, and by the tenth century the caliphs themselves were little more than ciphers. Elsewhere in the empire, power had become firmly located with local commanders based in the cities, such as Damascus, Aleppo and Mosul in Syria, who themselves sought to establish local dynasties at the expense of their neighbours. In this dissipated condition, the Islamic state(s) had been incapable of mounting a serious or successful resistance to the Byzantine counteroffensive.

Ghazni

In the east, the Turkish commander Sebüktegīn (r. 977–97), then his famous son Maḥmūd (r. 997–1030), founded a new dynasty, and built an empire stretching from central Asia deep into India, with what became a highly cultured capital at Ghazni (Afghanistan) from which the dynasty took its name. Their varied territories gave them access to a variety of resources to call upon: local infantry for siege warfare, Khorasani miners, and Persian expertise at building siege equipment, sometimes pulled by elephants. Elephants were recorded as being fitted with "battering rams" in 1033.[1] In such a region, stone buildings were common and called for the regular use of existing siege technology and there are references to engineers and sappers, the *'arrada* and the catapult. When Maḥmūd of Ghazni besieged a city of Sistan, having raised his army from Khorasan, the height of the fortress and the size of the moat prevented capture, while the defenders used catapults to hurl out pots filled with poisonous snakes and scorpions. The troops raised from Daylam (northern Iran) were noted for their particular skills in siege warfare. The sultan's victories continued as, alongside numerous raids that reached deep into Hindustan and plundered an immense wealth, town after town in modern Afghanistan and Pakistan was captured. Multan was captured from its "heretical" Ismaili holder in 1006. Near Herat, attacking a castle, a battery of both "large and small slinging machines" brought down one side of the fortress. In 1037, Maḥmūd's successor as sultan, Mas'ūd (r. 1030–39), marched from Kabul to invest Hansi, where the previously untaken defences were brought down by mines sprung together in no fewer than five places.[2] Further west, Persia and Iraq fell under the rule of the originally barbarian Būyids in the middle of the tenth century, and although there is little information about their engineering skills, it must be assumed that these continued to be available as needed.[3]

The lands of the Caucasus also slipped out of caliphal control into the hands of local dynasties. The chronicles of the rulers in Shirvan and at the walled town

[1] What this actually meant was that elephants themselves were used as battering rams, with their heads protected by metal plates. On the defence of adding spikes to fortress gates (albeit from a later period), see Nossov, *Indian castles*, 23.

[2] 'Utbī, *The Kitab-i-Yamini*, 51–6, 89, 286–7 (example of use of elephants), 384. 'Utbī was a courtier of the dynasty; the translation is from a Persian version of the original Arabic, Baihaḳī, *Tārīkhu-s Subuktigīn*, 124, 140. Bosworth, "Ghaznevid military organisation", 64–8.

[3] Bosworth, "Military organisation under the Būyids", 143–7.

of Derbent on the Caspian sea, and the Shaddādid dynasty established in 971 at the town of Ganja (formerly Elizavetpol'/Kirovabad), founded in 859 ("his castle stood by the gate"), recount small-scale warfare involving all the local and neighbouring rulers including Georgians, Armenians, and eventually Rus'. These meant occasional attacks on the walled towns and the rulers' defended residences, but little in the way of sustained siege warfare.[1]

Fatimids

In the first ten years of their rise to power in North Africa, the new Shi'a dynasty of the Fatimids, whose leader 'Ubayd Allāh al-Mahdī was declared caliph against the 'Abbasids (909), drew their initial strength from Berber tribes, with whose aid they quickly took power in the country. Tripoli held out for a year against a blockade, but the new dynasty soon acquired the resources and skills it may have been lacking at first. Access to building skills was evidenced by the creation of the powerfully-defended city of Mahdia on the coast, protected by a massive land wall 175 metres long across the narrow neck of land, more than eight metres thick, and protected by round towers at each end, and three more plus the double gate-tower housing an immensely tall and long gate passage in between. At some unknown date, possibly contemporary, a weaker but still strong nine-towered outer wall was also built. Although they took over a corps of Turkish troops previously employed by the fallen regime, the Fatimid record against either Byzantines or against rebel Bedouins or Berbers was not impressive. Nonetheless, they must have become quickly familiar with the use of the *manjanīq*, *'arrada*, *dabbāba* and *burj*.[2]

Al-Andalus

The partial exception to this picture of fragmentation, disunity and disintegration was in the Spanish peninsula, where the power of al-Andalus was to be concentrated again in the hands of single rulers who had been able to establish authority over most of the areas conquered in the eighth century, and had acquired their own title of caliph. There is an eleventh century account by Ibn Hayyān of the campaigns of the caliph 'Abd al-Raḥmān III (r. 912–61), who took the title *al-Naṣīr*, "the victorious". Securing his own power against rebels and local rulers in the southern heartlands of al-Andalus itself involved sustained siege warfare, with walled cities representing the base of power and authority. In 913, he destroyed the walls of Écija, "exchanging stone for earth", but maintained the citadel (*alcázar*). The fortresses of Monteleón and Samontia followed, surrendering despite

[1] *A history of Sharvān and Darband*, 25–33; V Minorsky, "New light on the Shaddādids of Ganja", 9–17.

[2] 'Arīb ibn Sa'd, *An account of the establishment of the Fatemite dynasty*, 54–136; Beshir, "Fatimid military organization", 37; Lev, "The Fatimid army", 166–84. Lev concluded that the Fatimids were unfamiliar with artillery until 991 and naphtha until 975, which seems on the face of it unlikely, even if the Berber component of the army was. On Mahdia, demolished by Charles V in 1550 after it resisted his artillery attack, see Lézine, *Mahdiya*, 14, 18–37.

their strength. Juviles resisted, protected by its remoteness and "the difficulty of attacking it with catapults". The site survives: a high triangular hilltop place, only needing stonework on the long southern face. Here there was a substantial wall some 500 metres long, of large stone blocks strengthened with massive rectangular towers. Little can be deduced about the internal structures but they included several cisterns.[1] Undaunted, the commander ordered the construction of a base on which catapults were set up to attack with stones, and he cut off the water supply, leading to its surrender.[2]

Seville was a harder task, requiring a siege of four months, and the following year Carmona, with its powerful defences, was the target. Meanwhile, the Christian king of León, Ordoño II, took advantage of the conflict to attack Évora (August 913) with what is improbably described as a vast army. The chronicler records that the attacker was able to reach the height of top of the walls because of the immense quantity of rubbish outside, allowing his archers to drive the defenders from the wall-top, so that an escalade was possible, and the wall could be breached at a spot where it had only recently been built. Leaving aside the carelessness of the inhabitants when it came to waste disposal, the same picture emerges here as has been seen elsewhere: the primary means of capture was knocking a breach through the wall, in this case by identifying a spot where the wall had only recently been put up. The fall of Évora was a great shock, with everyone rushing to "repair their walls and fortify bulwarks". The emir at Badajoz, where the citadel walls were of earth dating from the original conquest, hurriedly had it rebuilt in stone ten palms wide.[3] In 915, Ordoño resumed his raiding in the province of Mérida, taking rural fortifications. However, the citadel of Mérida itself was too much, and after "admiring its perfect construction" he withdrew. Rāzī was to describe the "very strong masonry" of this symmetrical square enclosure built from reused Roman stonework around 835.[4]

'Abd al-Raḥmān was still preoccupied with defeating rivals in the south. In September 917 he finally took Carmona with a host and "complete equipment", bombarding it with catapults and storming the place on 25 September. Rāzī records that this was an ancient city, "with ramparts well built", which was then demolished by its conqueror.[5] His base secure, 'Abd al-Raḥmān now turned his attention to the Christians. In 920, it is recorded that he defeated Kings Ordoño and Sancho I Garcés of Pamplona in battle, then stormed the castle of Muez, in which the survivors had sought refuge. He then turned his attention to securing

[1] Plan in Bazzana, Cressier and Guichard, *Les châteaux ruraux d'al-Andalus*, 129.

[2] Ibn Hayyān, *Crónica del califa 'Abdarrahmān III*, 52–9.

[3] Although this is difficult to square with accounts of the foundation of Badajoz after 888, when a small town was provided with all the attributes of an Islamic city, a mosque, baths, and walls funded by 'Abd Allāh, an emir from Córdoba, and built by masons sent from the capital, as the local Berber settlers did not apparently have the skills. Picard, "La fondation de Badajoz", 216–26; Ḥimyarī, *La peninsule ibérique au Moyen-Âge*, 58.

[4] Ibn Hayyān, *Crónica del califa 'Abdarrahmān III*, 72–102; Rāzī, "La 'Description de l'Espagne'", 84; Ḥimyarī, *La peninsule ibérique au Moyen-Âge*, 190. Hernández Jiménez, "The Alcazaba of Mérida".

[5] Ibn Hayyān, *Crónica*, 111–12; Rāzī, "La 'Description de l'Espagne'", 94.

the frontiers by ordering the strengthening of the Muslim fortifications. In 922, the target was Monterrubio, an "important place difficult of access and impregnable" whose garrison harried trade between Elvira (Granada) and Jaén. The Muslims besieged it for 35 days, building contravallations to protect themselves, and submitted it to a "devastating bombardment" leading to its surrender: another rare account of a fortification falling to catapult bombardment alone. In 927/8, al-Naṣīr attacked the Christian fortress at Bobastro. For some reason, no attempt was made to storm or breach the defences. Instead, it was isolated by the capture of the surrounding Christian castles, and the building of six forts to isolate the town and "destroy its means of life", something that succeeded when Bobastro surrendered in January 928. In 932, Toledo fell to the caliph after a long siege. We are told that he marvelled at its impregnability and stout walls.[1] The main focus then switched back to the north, with a prolonged effort to capture Zaragoza, which finally fell in 937. During this time the caliph responded to a request for help from the local Muslim leader, Mūsā, by sending him "experts in fort construction". Under the leadership of an "architect" went 30 masons, ten carpenters and fifteen diggers with materials and supplies to construct a castle. In 940, the caliph secured the frontiers by placing mercenary garrisons in the fortresses well supplied with food and all the "things necessary in a siege."[2]

These accounts confirm that the Arab state in al-Andalus was possessed of the most modern and effective means of siege warfare. On occasion, it seems that bombardment by catapults was itself sufficient to bring about the surrender of a fortified place. A similar outcome was recorded, much later, of the capture of a now unknown place in Málaga in 903.[3] We should assume that these were isolated and weak fortifications without hope of relief, because wherever the defences were formidable, the full range of techniques had to be deployed. Unlike in the east, there are no references to rams being used, although the silence of the sources is not necessarily evidence of the absence of the weapon. Certainly, those places too strong to be assaulted, perhaps in part because of their location, were cut off and starved out, and the references to siege-forts suggests both a sophistication of method and considerable resources, to build and garrison such temporary fortifications for many months at a time.[4] A much later compiler tells of the caliph's vast wealth, derived from the centrally-collected land tax, duties on trade and one-fifth of all the booty from the Christian realms. One third of the riches was spent on the army and officials, and one-third on buildings.[5]

The Conqueror was succeeded by his son in 961, who seemed content to rule from the luxurious and highly cultured capital at Córdoba, and devoted his foreign policy to extending his caliphate to Tunisia, whence the Fatimids had

[1] Ibn Hayyān, *Crónica*, 132–64, 259 (Toledo).

[2] Ibid., 290, 343.

[3] Ibn al-Adhārī, *Histoire de l'Afrique et de l'Espagne*, II: 232.

[4] Lévi-Provençal, *Histoire de l'Espagne musulmane*, III: 102, argues for the normal use of rams and sapping as main means of attack but notes there is very little chronicle evidence to support this.

[5] Maqqarī, *The history of the Mohammedan dynasties in Spain*, II: 146.

transferred to Egypt. This involved him in long and expensive campaigns against the local Berber tribes, during which the caliph's armies recorded a number of captures of strongholds.[1] He in turn was followed by a son who was effectively locked away, with real control being wielded by one of his guardians, Ibn Abī 'Āmir, who took control and adopted the title of *al-Manṣūr*, "the victorious". Al-Manṣūr led numerous and vigorous raids against the Christian states, spectacularly sacking Barcelona in 985 and Santiago de Compostella in 997. Barcelona had Roman walls that had required a substantial siege by Louis the Pious. Its capture in 985 was brought about by a sustained attack, apparently of two months' duration, in which artillery played a key role, along with a sea blockade. Following the capture, by storm, it was burnt. The count, Borrell II, had vainly tried to relieve the town from the outside. The defences seem to have been largely as they had been left from the previous time, except for the fact that there were also two "castles", known as the old and new, belonging to the viscounts of the county, the "old castle" being built adjacent to the gate on the eastern curtain wall and the "new" outside the western.[2] The attack on Santiago de Compostela in 997 was an operation giving evidence of the sophisticated planning and resources available to al-Manṣūr. The army had to cross much hostile territory, in the process getting over numerous rivers in its path. The supplies and heavy equipment (including presumably siege weapons) was conveyed by ship to Oporto then up the river Duero to rendezvous with the army. The accounts say many of the Christians' fortifications were taken, but the city of Santiago itself was vacated by its inhabitants, and it was plundered and its fortifications razed to the ground.[3] The gates of the city were carried off by al-Manṣūr as a symbol of his victory.

There was no attempt at permanent conquest following these spectacular incursions. The raids were designed to generate plunder and to consolidate the overlordship of the caliphate over the petty Christian states. This overlordship itself was vulnerable once the highly centralised power at Córdoba dissipated among the diverse regions and peoples of the peninsula, as was to happen in the following century.

The Christian states

The small Christian kingdoms were not in a strong position to challenge the dominance of Córdoba and were mainly on the defensive. These little kingdoms were, however, becoming consolidated. Ordoño I (r. 850–66) and Alfonso III (r. 866–910), who ruled Asturias and León, had made substantial advances in the frontier

[1] Rāzī, *Anales palatinos del califa de Córdoba al-Hakam II*, 169, referring to the capture of a "little less than impregnable" fort called Ḥiṣn al-Karam, which must have been a place of considerable size since the garrison left in it included 1,000 *jinetes*.

[2] The accounts are reviewed in Zimmermann, "La prise de Barcelone par Al-Mansur"; Rovira i Virgili, *Història nacional de Catalunya*, III: 250–54; for the castles, Monreal and Riquer, *Els castells medievals de Catalunya*, II: 116–17.

[3] Maqqarī, *The history of the Mohammedan dynasties in Spain*, II: 193–5 (book VI, chap. VII); Pelayo of Oviedo, *Chronicon regum Legionensium*, 78.

regions with al-Andalus and considerably extended their territory. The ancient Roman cities "ab antiquis desertas" were reoccupied, and new walls and gates constructed to secure them at León, Astorga, Tui and Amaya. Many other *castra* were built, the city of Talamanca captured.[1] There is no detail in the chronicles as to how these conquests were carried out, but the close relations between the Christian and Muslim states over a long period of time must have led to a sharing of technical skills and knowledge, both in terms of building and siege warfare. The establishment of strong walled towns was to provide an important resource for the rulers of the northern states, previously desperately short of resources by comparison with the wealthy south, that would bear fruit in the eleventh century.

Defensive buildings in Spain and al-Andalus

Detailed studies have been carried out of the fortifications of the tenth century that allow a clear picture to emerge of the design and purpose of a range of fortifications, divided into types with their own specific functions. Overall, the picture is one of a wealthy and centralised state able to consolidate its own power and provide a network of defences designed both to protect its own rule, and to give necessary shelter to its population, ever vulnerable to the raiding warfare so common to all the areas we have observed. The evidence of the chronicles has been compared with studies of the remains on the ground to produce a convincing and coherent picture.

As in the east, so in al-Andalus, the basis of Muslim rule was the cities, and of the 40 or so listed by the Arab geographers, no fewer than 23 were new foundations. Of these, seven were established during the ninth century (Tudela, Murcia, Úbeda, Talamanca, Madrid, Lleida and Badajoz, in order of foundation) and seven during the tenth. Some were very large indeed, Córdoba having a perimeter wall 6,300 metres long, followed closely by Seville with just under 6,000 and Toledo with 4,440. The population of the largest (Córdoba) compared with those of the largest cities of the east, Damascus, Fustat (Cairo) and Baghdad. Estimates for population based on the enclosed areas give Seville 80,000 and Toledo 37,000, followed by Almería, founded in 955 by 'Abd al-Raḥmān, and Granada both with 26,000–27,000. These were immense numbers of people by comparison with western and central Europe.[2]

A similar pattern applied to all of them: a walled town, based around the mosque and market, with a separately defended citadel (*alcázar, alcazaba*) for the emir or governor. The new dynasty came from Syria, and it is not a surprise that eastern patterns were followed. Walls were tall and enclosed rectangular interiors, with rectangular projecting towers at regular intervals rising considerably above the height of the wall. The entire defence was conducted from the wall-walk, served with wide crenellations often with pyramid-shaped tops. The whole would be fronted by a ditch where the terrain allowed. Gates were straight entrances

[1] *Die Chronik Alfons III*, 58–9; *Chronicon Albeldense*, 1138; *Historia Silense*, 30, 35.

[2] Torres Balbás, *Ciudades hispanomusulmanas*, 60–69, 106.

between two towers. Doors were protected against burning by the attachment of iron plates. There do not seem to have been portcullises such as in Byzantine fortresses. Sometimes the terrain prevented a regular pattern, and the walls followed the shape of the crest or hilltop. Outer walls such as those occasionally seen in contemporary Byzantine work were unusual, becoming a common feature in the following century. Whilst some of the building was carried out with large, neatly cut masonry blocks, this was not common, with cheaper and easier constructions of rammed earth constructed around a wooden framework, a style common in North Africa (*ṭabīʿa,* Spanish *tapial*), being particularly popular. The citadel of Almería provides a good example of the use of this type of building material,[1] which was immensely durable, as do the earlier parts of the *alcazaba* of the Alhambra of Granada (Plate 23).

Accurate dating of different phases of building in the Arab fortresses is notoriously difficult. On the basis of construction techniques, it has been suggested that some of the defensive techniques once ascribed to the twelfth century may in fact originate in the tenth century. The *torres albarranas* were towers constructed outside of the line of the wall, linked to it by a bridge to give access, and sited to provide flanking protection for vulnerable parts of the curtain wall. It has been proposed that such towers were erected in the tenth century at Trujillo, Marbella and Mérida (whose walls were rebuilt in 835), although most examples are considerably later and another authority argues that these towers were added in the twelfth century. Similarly, while most of the gates were simple straight passages, a gate of Toledo and gates at Zorita de la Canes, Gormaz and Baños de la Encina (see below) were provided with upper chambers over the passage, giving access to slots in the floor, enabling missiles to be directed at anyone in the passage below. These would be the first *machicolations* or "murder holes". The "bent entrances", gate passages that are made to turn at right angles, were once thought to be no earlier than the eleventh century (the Puerta Nueva at Granada), but another of Toledo's gates (Bab al-Qantara) has been dated to the ninth century, where the turn is imposed by the terrain, while Gormaz Castle may be another, tenth-century, example. The model was to be found in Baghdad, and it became popular in al-Andalus under the Almoravids, but it is possible that in the expanding, centralised authority of the new Caliphate in the west that new and very advanced designs would be undertaken.[2] Such measures can only have had a defensive purpose, bent entrances in particular being particularly inconvenient for normal usage!

The citadels were sited on the highest point of land, and had separate defences and sometimes highly sophisticated and advanced features. They served as a residence for the commanding authority, and if necessary as a refuge. The new fortress built near Córdoba was described by Rāzī as the "royal alcázar".[3] Many still stand

[1] Torres Balbás, *Ciudades hispanomusulmanas*, 466, 531, 551–7; Terrasse, "Les forteresses de l'Espagne musulmane", 462–7.

[2] For all the above, Zozaya, "Islamic fortifications in Spain", 636–44, with plans and illustrations; Lafuente and Zozaya, "Algunas observaciones sobre el castillo de Trujillo"; Maldonaldo, *Ciudades y fortalezas*, 252–4, 405.

[3] Rāzī, "La 'Description de l'Espagne'", 63; Jiménez de Rada, *Historia Arabum*, 269.

today. The same approach was applied across the straits, too, as the caliphs extended their rule into north Africa. A large rectangular tower and curtain wall of neatly cut large stone blocks survives the many rebuilds of Ceuta, refortified along with Tangier and Melilla following their occupation after 931.[1] This must have been built on top of the previous Byzantine defences, described by Procopius as a wall ordered by Justinian to make an "impregnable" fortress,[2] as the implication in the eleventh century geographer's description is that the work was new.[3]

On the frontiers facing the Christian states, new fortifications were built in the tenth century as bases for armies and garrisons. Such were Baños de la Encina (in Jaén), and Gormaz (in Soria). The former stands well-preserved on a hilltop. Made of *ṭabī'a,* it is 400 metres long, has a high wall and fifteen towers, each of them small, but close together. Gormaz, erected over an existing defence no later than 965, on a hill overlooking the Duero river, was large enough to shelter an army. Built of masonry, with a perimeter of a kilometre, it has walls ten metres high and 24 well-spaced towers. It is 366 metres long, and its width varies from just 15 metres up to 52. The defences are strongest facing the only means of approach. The main gate of three is monumental, similar in design to the gates of Córdoba. There is a separate citadel at the eastern end, with much stronger and higher walls, which may date from this rebuilding, although there is also much rebuilding work from later centuries. A cistern was provided in each section of the castle.[4] Gormaz was important enough to have the caliph organise a major and successful relief expedition when a coalition of the Christian princes tried to take it in 975.[5] It fell into the hands of the Christians in 1002.

The study by Bazzana, Cressier and Guichard of sites in south-eastern Spain confirms a common pattern of fortification in the countryside. Here, the common features were simpler structures known to the Arabs as *ḥiṣn* (plural *ḥuṣūn*) or *ma'aqul.* The earliest written evidence of their existence dates from the eighth century, as the Muslims began to consolidate their settlement. The normal form was a hilltop fortification with protected access, such as a tower to shield the entrance, standing over the village and its terraces. Following the reconquest, the Christians might add a donjon or a signal tower. Those that functioned as a sanctuary for the village did not necessarily have permanent occupation. The *ma'aqul* was a very simple construction, a wall and, inside, a cistern, serving only as a refuge, while the *ḥiṣn* might have a more substantial function as a centre of administration and lordship as well as defence. In Christian sources, they are called *castra* and they are granted together with their village and its territory in an integral whole. There is

[1] Terrasse, "Un vestige des fortifications oméiyades de Ceuta", 244–6.

[2] Fita, "Ceuta visigoda y byzantina", reconstructs the occupations and building during the Byzantine capture of the area in the sixth century.

[3] Bakrī, *Description de l'Afrique septentrionale,* 235–6.

[4] Torres Balbás, *Ciudades hispanomusulmanas,* 568, with illustrations of Gormaz, pp. 554–6; both sites are also well illustrated in Ortiz Echagüe, *España: castillos y alcázares,* plates 36 (Baños), 219–22 (Gormaz). Description in Gaya Nuño, "Gormaz, castillo califal", 431–44; re-evaluation in Banks and Zozaya, "Excavations in the caliphal fortress of Gormaz", 675–82.

[5] Rāzī, *Anales palatinos,* 263, 276–7.

a considerable density of such sites in Valencia, although they may of course have been established over a prolonged period of time.[1]

Piracés, 20 kilometres south-east of Huesca, is a hilltop *ḥiṣn*, one of eight listed by Rāzī as dependant on the city. In 1006 it resisted an attack from Zaragoza with siege artillery. The ridge on which it sits is 600 metres long, and the settlement is in three sections with a tower at the highest point and the village, with its own wall, at the lowest. This place must have had a number of functions, including a refuge and a link in an inter-visible chain.[2] Similar fortifications have been studied at Peña de San Miguel north of Huesca,[3] and in what would become Portugal, where remains of the period are very scanty, a fort has been identified at Castelinho in the area of Almodôvar that may be tenth century.[4] Confirmation of the general pattern is also offered by La Luz, near Murcia, with its powerful upper defences dating from the ninth or tenth centuries.[5] Other types of defensive works have also been identified. Some fortifications were placed along key routes. Some were designed just for a garrison, others possibly as signal stations, such as La Iglesieta (Huesca).[6]

For the Christian kingdoms, fortifications were a central feature of their expansion. In addition to the chronicle accounts of cities being reoccupied and fortified, there are also references to the building of castles, often placed high above valleys, providing a defence and a look-out post allowing settlement of the valleys and plains of the river Duero. There is one such site at Alba, at an altitude of 1,280 metres, newly built in 872, which resisted al-Manṣūr in 998. With an area of 4,500 square metres, it had a wall following the contours, eight metres wide, of irregular stone blocks without mortar. A tower protects the entrance and it is divided into three sections, the smallest of which was presumably a redoubt or contained a watch tower.[7] In the tenth century, the capital of the kingdom was moved to León and counties were established, each count being based in a fortified administrative centre. These were also sites of hurried construction and little sophistication, using hill tops with natural protection and good visibility. In Huesca, which fell into Christian hands again after 941, castles began to appear linked to the progress of settlement of the region. In the county of Barcelona, with its more established administration and relationship with the Muslims to the south and a frontier defined from about 900, a similar process of castle-building took place. There were none in 900, but by the end of the century almost 50 had been created.

[1] Bazzana, Cressier and Guichard, *Les châteaux ruraux d'al-Andalus*, 45–6, 73, 107–8, 259–87; Bazzana, "Le *hisn* et les *ma'aqil*", 19–25.

[2] Esco Sampériz and Senac, "Un *hisn* de la Marche Supérieure d'al-Andalus", 126–49.

[3] Esco Sampériz and Senac, "Une forteresse de la Marche Supérieure d'al-Andalus", 17–28.

[4] Guerra and Fabião, "Uma fortificação omíada".

[5] Navarro Suárez and Mateo Saura, "La fortificación islámica de La Luz"; Manzano Martínez, Bernal Pascual and Calabuig Jordán, "El castillo de Sta. Catalina del Monte", 111, for a very clear plan.

[6] Bazzana, Cressier and Guichard, *Les châteaux ruraux d'al-Andalus*, 111–12; P Sénac, "Une fortification musulmane au nord de l'Ebre", 128–45.

[7] Bazzana, "Le début du château dans l'Espagne septentrionale", 33–4.

Overall, a pattern emerges of fortifications being thrown up originally as refuges, then passing through royal command into the hands of counts who took their names from their castles, as the centres of territorial organisation as the state expanded and people were encouraged to settle in lands formerly depopulated and ravaged as frontier zones.[1] The places so established may have served the same function as the Muslim cities with their citadels and the rural refuges and garrisons, but were not of anything like the quality of building. This speaks of a considerable disparity of resources, skilled artisans and the money to pay them, and of a considerably lower level of civilisation at this time.

Central Europe

The tenth century witnessed the consolidation of the kingdom of Germany as inheritor of Charlemagne's dignity as Holy Roman Emperor and the emergence of new states in the east, Bohemia, Hungary and Poland. The emperor, however, had to contend with aristocratic families ruling immense territories of their own, too powerful to be compelled to act against their own dynastic interests. Another feature of the century was incessant warfare, with the primary role of the ruler being, as before, war leader. There were always enemies to face somewhere, whether these were invaders from abroad or dynastic rivals at home, and in the years when there was peace at home, this was the opportunity to lead an expedition into foreign parts. Below the level of the great princes, feuding was endemic, and legitimate. The power of the German king did not extend to the right to intervene to prevent such disorder, the Saxon dynasty that provided the royal house throughout the century being itself only the first amongst equals.[2] The accounts demonstrate a growing place for the castle within this warfare as the century progresses.

The Magyars

At the turn of the ninth and tenth centuries, the threat to the west seemed to come mainly from yet another highly mobile invader, this was one coming by land: the Magyars. Having settled in the Carpathian basin during the later ninth century, this grouping of tribes from the east, who fought on horseback using a powerful bow and were capable of covering immense distances, ransacked their way through Germany and northern Italy, and were recorded as penetrating into Francia and even into al-Andalus. As with nomads everywhere, the problem for understanding them is that all the evidence comes from their enemies or victims, while their archaeological remains are of limited help. Notwithstanding this bias, it does look as if the Magyars' raids had everything to do with plunder or tribute

[1] Ibid., 35–40; Gutiérrez González, "Características de las fortificaciones del reino de León", 64–7.

[2] An excellent short summary is Leyser, "Early medieval warfare".

and had no other objective. The German king Henry I himself bought them off for nine years with a substantial monetary payment.[1]

It would be unusual for a nomadic people to have much experience of fortifications, but the Magyars had been in close contact with Bulgaria and the Byzantine empire. The establishment of their frontiers involved a successful struggle with the Bulgarians, whose rapid adoption of building techniques and styles from Constantinople has been noted. In 894 the Magyars had devastated the country, "occupying their *castra* and destroying their cities (*civitates*)".[2] In 903 the Magyar leader Árpád took a *castrum* after three days' siege, and destroyed the walls. Not long after, the Magyar army attacked the *castrum* at Byhor (Cetatea Biharia), fighting a pitched battle in which many men were killed *per balistas*. On the thirtieth day, they managed to fill the ditch and set up scaling ladders.[3] This is from a much later source, but it suggests that this nomadic people knew both how to take the probably earthwork fortresses of the Danubian frontier, and understood the value of setting up defences of their own.

After destroying the Moravian empire at the end of the ninth century, the Magyars turned their attention westwards. In 924 they captured King Berengar's Pavia, apparently by hurling lighted torches over the walls to set the houses on fire. When they penetrated the German kingdom, the response was not dissimilar to the approach taken by the West Saxon kings. But new defences take time to erect, and the monks of Metz had still not completed the walls surrounding their monastery when the Magyars arrived. They had to flee with their relics to the safety of the city walls.[4] The *Life* of Saint Ulrich, bishop of Augsburg, gives useful glimpses into the course of the conflict with the raiders in the 950s. In 953, the bishop ordered the construction of a *castellum* at Menchingen near Augsburg, on a site then deserted, and had a wooden castle erected and the rampart renovated. Two years later the Magyar columns reached Augsburg, and there was a siege. The city had Roman walls and towers, and the bishop's *milites* defended it so well that the attackers withdrew. The bishop then ordered the defences renovated. The next day the Magyars returned to the attack, many bearing *instrumenta* for breaking down the walls, but withdrew on hearing of the approach of King Otto's army.[5] Clearly, although a mobile force not carrying equipment to prosecute a regular siege, the Magyars had the means to undertake an attack on a defended walled city.

In the subsequent Battle of Lechfeld (955), the German king won a complete victory, and forces based in the by now substantial network of fortifications completed the destruction of the invaders as they attempted to return home.[6] The victory marked the end of the Magyar threat to the kingdom. Their rulers now

[1] Bakay, "Hungary", provides a good summary of the emergence of the Hungarian state; see Engel, *The realm of St Stephen*, for a scholarly modern study.

[2] Thuróczy, *Chronica Hungarorum*, I: 70 (chap. 51). This is a very much later compilation.

[3] "Anonymus", *Gesta Hungarorum*, 53, 55, 62–3, 103–4.

[4] *Miracula S. Gorgonii*, 240.

[5] Gerhardus of Augsburg, *Vita sancti Oudalrici episcopi*, 399–401.

[6] Leyser, "The battle at the Lech, 955", 63.

concentrated on establishing their own state, and for the rest of the century almost all of their conflicts were internal, including a fairly violent process of conversion to Christianity, culminating in the crowning of King Stephen in around 1000.

Hungary was far from immune from external influences, and the rulers were far-sighted enough to encourage the immigration of people who could bring them useful skills. Geographically, the state was extremely well protected by dense forests and by mountains on three sides, and by the rivers Danube, Drava and Sava to the south. This meant that an organised system of defence was possible, concentrated on the limited number of access points. The defences involved fortified sites called *castra*. Early on, a *castrum* was built at Komárom. A suitable site was identified at Szabolcs, and with the advice of the council the prince proceeded to order the building of a "very strong" castle, digging a massive ditch and throwing up a rampart of earth. Excavations of the site by the river Tisza discovered a triangular earthwork with sides of 337, 235 and 387 metres and ramparts up to 17 metres high. It has been dated to two periods, the earlier being the mid tenth century.[1] Few sites from this period have survived, but it seems that the process of state formation during the late tenth century under the grand prince Géza (r. *c*.972–97) and King Stephen (István) I (r. 997–1038) involved an attempt to follow the western model, with counts assigned to particular administrative districts, centred on a *castrum*. Tenth-century examples have been studied at Borsod (Edelény) and Pata (Vágpatta in Hungarian). *Civitates* of Roman origin on the Danube were reoccupied.[2] At Győr (Roman Arrabona), the hill above the town had a castrum erected in Stephen's reign for use by his bailiff, which took the form of a 150-metre-square fortification, with a timber fortress on a mound, and the rest of the site filled with other buildings in which settlement took place.[3]

On the frontiers themselves, the chosen method was to set up earth and timber defences to block the passages, in a style similar to those adopted by other nomads like the Avars, and also adopted by Slav groups that were also beginning to establish settled and centralised states at this time, as in Bohemia and Poland. To the east, an additional defence was established through creating a wilderness that would take many days to cross. The passages were blocked with ditches, fences, earthworks and hedges, described in Latin sources as *indagines, obstacula, munitiones*. They remained the critical elements of frontier defence for a long time to come, and were of considerable strength. In the Hungarian case, the whole defensive arrangement was reinforced by the settling of other peoples as border guards and watch-keepers.[4]

[1] Fügedi, *Castle and society*, 16.

[2] Ibid., 26–8; Bakay, "Archäologische Studien zur Frage der Ungarischen Staatsgründung", 105–61.

[3] Gabler, Szőnyi and Tomka, "The settlement history of Győr", 23–4, 42 (plan).

[4] On all the above, see Göckenjan, *Hilfsvölker und Grenzwächter*, 1–10.

New defences in Germany

The threat posed by the Magyars may have been responsible for a massive programme of building defences as refuges for the people, and also as military bases for organising counterattacks, organised by order of the king. Central to the case for this is the *Burgenordnung* (fortification order) of King Henry I in (probably) 925. In this, the king ordered an assembly of his leading subjects to surround all the "conventicula loca privata munitionibus firmis murisque".[1] But archaeology has not succeeded in producing firm evidence to link any of the many fortified sites of the German kingdom with the royal order.[2] Nor has the original order itself survived. But there is little cause to doubt its reality, which fits well in the context of the measures taken at the time to defend settled rural communities against mobile raiders elsewhere. The chronicler Widukind of Corvey noted the development, and also that measures were taken to assign a proportion of the population to the necessary works,[3] in a way reminiscent of the fortress duty required by the Carolingians. One thing is certain: the tenth century witnessed an accelerated process of fortification building. Whether this was specifically tied to the danger posed by the Magyars, or whether it was a more general process connected with struggles with other peoples (for example, the regular conflicts with the Slavs on the north-eastern frontier), or whether it was part of the process of state formation, cannot be resolved.

The chronicles provide evidence of the routine involvement of the Saxon kings in the attack or defence of fortifications. Otto I besieged Mainz, bringing "many machines" to the wall, but the defenders managed to burn them, eventually surrendering after two months. In 954, the king descended on Regensburg ("urbem regiam que dicitur Rainesburg"), occupying the *castra* then besieging the town, but was unable to bring his "multitude of machines" to bear.[4] Eight years later Otto was in Italy. He besieged Berengar's son in his *castellum* with a daily bombardment of "slings and archery and other instruments of war", although again the place resisted the attacks. The term used here for sling, *fundibalus*, may possibly refer to stone-throwing artillery. In 964, he retook Rome using "siege engines and famine".[5] In 998, Otto III turned to crushing opponents of the imperial authority in Rome, led by Crescentius, and succeeded in storming the rebellious nobleman in the Castel Sant'Angelo, the fortification constructed by the popes on the massive Roman mausoleum of Hadrian. This attack involved scaling ladders and "machinis arcem", possibly ballistas, and a wooden tower made of fir, enabling the besiegers to gain entry both at wall level and through the gates.[6] This is a rare

1 *Ex miraculibus S. Wigberhti*, 224–5. The text was written by a monk at Hersfeld in the early years of Otto I.

2 Leyser, "Henry I and the beginnings of the Saxon empire", 12.

3 Widukind, *Res gestae saxonicae*, 432 (book I, chap. 35).

4 Ibid., 454–6 (book III, chaps. 18, 36).

5 Adalbert, *Continuatio Reginonis*, 216–18; Liudprand, *Liber de rebus gestis Ottonis*, in his *Works*, 72.

6 *Annales Quedlinburgenses*, 74; Glaber, *Historiarum libri quinque*, 25–7 (book I, chap. 12).

reference to the use of some kind of siege tower, apparently deployed with success on this occasion. Otherwise, the evidence suggests that the tenth-century kings of Germany were deploying the same weapons and techniques as had existed for a long time.

The written records refer repeatedly to fortifications being created throughout this time. In 908, the Ennsburg was surrounded by a wall, and the bishop of Eichstätt ordered a wall built to create a defence at the Michelsburg.[1] In 931, the king ordered the building of a fortified *urbs* at Meissen, as a defence against the Slavs.[2] In 941, Otto I granted a nobleman a set of properties previously held by his father, Margrave Gero, "cum novo castello in Osterregulun constructo". In 946, he created the bishopric of Havelberg "in castro Havelberg" including the *castrum* and the *civitas* and a list of other *castra*.[3] In 974, Otto II granted Zwenkau to the bishop of Merseburg with a list of responsibilities including "the work of making the town wall". In 979, he granted to the abbey at Gembloux the duty of "faciendo castellum" and gave Wieselburg to the monastery of St. Emmeram, with the now less considerable duty of protecting from the "infestatione Ungrorum", by constructing a *castellum*.[4] Associated with these grants, increasingly spelt out in the charters, were a set of duties on the local population associated with the building and maintenance of the fortress. These systems were called *bannum*. Alongside them, especially in the fortresses created on the frontier with the Slavs, was created a system of castle-guard (*burgward*) in which the bishops and counts had to ensure that the place was garrisoned for a fixed number of weeks at a time on a roster of some kind by *milites* settled in the surrounding area with that duty as a major part of the condition of their holding.[5] The measures on Germany's eastern frontiers were particularly necessary, as there had been in 983 a massive revolt of the Slav population leading to the destruction of many of the new towns and settlements created east and north-east of Magdeburg on the Elbe during the period of eastward expansion.

A high a proportion of the grantees were bishops and abbots of monasteries. These ecclesiastical personages, unlike most of the lay nobility, were under royal control because they were royal appointees, and formed the critical national underpinning for royal and imperial rule. But there is evidence that not only ecclesiastics were in control of fortifications. At the end of the century, in the Rhineland, the rival noblemen Wichmann and Balderich contended for local power, and the local castle site at Elten featured in their struggle. It is clear from the description that a simple earthwork, located on a piece of high ground in a marshy area and

[1] Uslar, *Studien zu frühgeschichtlichen Befestigungen*, 65–7; Charter of King Louis the Child, *Zwentiboldi et Ludowici Infantis diplomata*, 185.

[2] Thietmar of Merseburg, *Chronicon*, 739.

[3] *Conradi I., Heinrici I. et Ottonis I. diplomata*, 126–7, 155–6.

[4] *Ottonis II. diplomata*, 104, 214, 232.

[5] For example, Otto I's grants in 940, 941 and 965 in *Conradi I., Heinrici I. et Ottonis I. diplomata*, 113, 126, 155, 455; Thietmar, *Chronicon*, 803, concerning the Burgward at Merseburg in 997. Burgward and Burgbann are discussed by Leyser, "Ottonian government", 82–4.

protected by a ditch, was involved.[1] The emperor had intervened to bring a temporary halt to the feud, but was not able to prevent it.

Although one cannot distinguish any fortifications specifically built following Henry I's order, excavation combined with written evidence gives a good picture of the style and form of tenth century fortifications built in the area ruled by the German kings. At the Frankenburg at Rinteln, a fortress is dated to around 900. The almost circular site is on a hill, like most of its Carolingian forebears, and a ringwall protects a palace, chapel and tower. This was clearly a residence of royalty. In contrast, the monastery of St. Gallen had a stone wall with thirteen towers erected before 950.[2] Other monasteries provided with solid outer defences at this time included Hersfeld and Corvey, which were assigned the revenues of surrounding areas to provide the necessary support. One of the constructions definitely associated with Henry I is Werla in the northern Harz, which was a major royal palace. The site was occupied from 926 until about 1180. Aerial surveys showed up its shape as having a *Hauptburg* (principal or inner castle) and several *Vorburgen* (outer castles), with walls and ditches. The upper enclosure, on the highest and narrowest point of the hill, was trapezoidal, with a perimeter of around 550 metres. Excavation revealed that there had been a ninth-century earthwork before Henry started his improvements. The king ordered a new stone wall, built with loam mortar, in front of the old rampart, about 1.5 metres thick. There were two half-round and one square tower, unevenly spaced and providing no flanking. There were two entrances. A square tower of this period stood in front of the ditch, adjacent to the gate. The outer enclosure was some 500 ×120 metres, protected by an earth rampart and a ditch of 3.7 metres depth. Further work carried out in the twelfth century considerably improved and strengthened the defences.[3] Werla was a significant centre, a residence used by the king. The defences were undoubtedly an improvement on those he found there, but the remains do not compare with the strength, size or sophistication of the works we have seen being carried out by the Byzantines or Arabs.

Similarly, Quedlinburg was a palace of which the defences were constructed at royal orders. Excavations discovered a ringwall of ashlar replacing an earlier rampart of unhewn sandstone. Inside were a chapel, hall and other buildings.[4] Henry I had other palaces, at Pöhlde, Nordhausen, Pfalz Grona and Duderstadt.

The work at Elten, referred to by Alpertus, has also been excavated. Located on a low hill beside the Rhine, the place had a simple palisade without towers behind a ditch six metres wide and timber buildings.[5] Considerably more substantial in defences was the Rennenburg, in the north-west Rhineland, with a stone wall 1.8 metres thick, with two projecting towers in the middle of the frontage, all

[1] Alpertus of Metz, *De diversitate temporum*, 702–3.

[2] Piper, *Burgenkunde*, 125, 130.

[3] Seebach, "Der Stand der Werla-Forschung", 165–8; Uslar, *Studien*, 68–72.

[4] Erdmann, "Burg und Kirche zu Quedlinburg", 83–97; Erdmann, "Die Burgenordnung Heinrichs I", 134–48.

[5] Binding, "Spätkarolingisch-ottonische Pfalzen und Burgen", 23–5, plans on pp. 26–8.

dated by finds to the first half of the tenth century.[1] Such places, their precise origin unknown, but their small size indicating a local purpose, were springing up everywhere. They would seem to have been defended residences for the local nobleman, whence he, his family and retainers could establish their lordship, collect and protect their assets, and find protection from neighbours. Their size was limited by the resources available to the builder, but also by their purpose. They were very different from the large, isolated refuges built under the Carolingians and the Saxons before them.

More clearly having a social purpose were the works done at the towns, churches and monasteries. Some towns possessing Roman walls had already been strengthened against the Vikings. Cologne, for example, had been rebuilt after 900. Its suburb was given a wall in 950. Further east, facing the Slav frontier, towns were reinforced, or newly founded as bases for royal power. Passau's defences were repaired at Otto II's orders. Vienna was founded in 1000. The bishop's work at Augsburg has been mentioned. New towns would have been very small, established around the church or cathedral, and sometimes references to defences applied only to the rampart thrown up around this nucleus rather than to the whole town, as for example at Constance. New places on the right bank of the Rhine or the left bank of the Danube were begun as earthworks. Magdeburg (founded 937) was an advanced base for German rule against the Slavs. Meissen would serve the same role, having a permanent garrison. At Bremen, the chronicler says the people "began to fortify Bremen itself with an exceedingly strong wall" in the 990s, a reference said to have meant the cathedral complex only.[2]

The role of the church as a keystone of the power of the German kingdom meant that the church buildings themselves were also a focus of that rule, and that, as strong structures in their own right, they should serve a defensive purpose. As elsewhere the defences were likely to be fairly simple, comprising a ditch, and a rampart topped with a fence or hedge, as in the territory that would become Austria, founded as a frontier march facing the Magyars.[3] It is not entirely clear but it seems likely that at this time, unlike in later times, the church building itself did not serve as a defence, although no doubt there were exceptions; the resemblance of church towers to the later, ubiquitous, donjon, has been noticed, and the possibility that the church might be the only stone-built structure in the settlement may have led to its use as a sanctuary from attack.

Slav tribes continued to live on both sides of the new frontier that the German kingdom was trying to push eastwards, and this created a zone in which conflict frequently happened, with the revolt of 983 the climax. It comes as no surprise to find, therefore, that alongside the towns and fortified churches, smaller earthwork fortifications become numerous in the tenth century in this area. The Slavs had constructed strongly-sited or defended villages for centuries. Protected by

[1] Herrnbrodt, "Die frühmittelalterliche Ringwalle des Rheinlandes", 67–71; Hinz, "Rennenburg", 174–6.

[2] Planitz, *Die deutsche Stadt im Mittelalter*, 236–9; Adam of Bremen, *History of the archbishops of Hamburg-Bremen*, 77 (book II, chap. 33).

[3] Kafka, *Wehrkirchen Niederösterreichs*, 4.

dry stone walls originally, the Slavs seem to have adopted the model of western earthwork structures during the later ninth century, wooden frames filled out with earth and stone, preceded by a berm and ditch. It can be hard to distinguish between fortifications constructed by the Slavs and by the Germans, which were often very close together, and of course places occupied by either may have been originally constructed by the other. There is nonetheless a strong suggestion from detailed studies of local areas, and dating by dendrochronology, that periods of building in the tenth century were related to the political process of settlement, revolt and reimposition of rule. A study of a strip of land ten kilometres square on either side of the Elbe river found 30 fortified sites from the late Carolingian and early Ottonian periods, of which 28 were reliably assigned to the Slavs. In the rebuilding following 983, 23 were Slav and eight German. Functions were varied. Some were refuges, others, much smaller, were apparently lordship castles, built to secure the control of the new rulers.[1] But some of the smaller sites have been found from their pottery to have been Slav, so the identification of scale with function is not always convincing. Some suggest the construction of a German *Burg* in a Slav site, indicating the lordship function. The process whereby the Slav territories were settled by large numbers of German immigrants and incorporated into stable political entities ruled by German families was to take several centuries, during which time the reliance on earth and timber for the buildings changed very little.

Italy: "incastellamento" everywhere

As so many rivals fought one another in Italy, and banditry was rife where authority had collapsed, conditions for the people were grim. It is therefore no surprise that, although the cities continued to function as the base of rulers, in the countryside the process of *incastellamento* begun in the previous century now reached most parts of the peninsula. In noting this dramatic development, radically changing the appearance of the countryside and the form of human settlement, it is important to recall that such developments can have a variety of underlying causes and serve more than one purpose. The fundamental process was one of creating secure villages for peasants, alongside a process whereby noble residences might also be made defensible. Very few of the sites noted in the various studies ever served any military function in the strategic sense, because they were not intended to, had only very weak fortifications, and anyway would not normally have had any trained soldiers to assist in their defence. Their role was to shelter the local rural population and its mobile assets. Some did, however, become the foundation for much stronger fortresses later in their existence.

Rudimentary ditches were the first attempts to construct refuges in the flat plain of the Po in the ninth century. At this time, ditches and palisades were the only protection, whether for villages or for the local lord's villa. Their location

[1] Wachter, "Burgen in hannoverschen Wendland", 155–62, with detailed lists and summaries of finds and dating of all the sites; Henning, "Ringwalburgen und Reiterkrieger"; Gringmuth-Dallmer, "Deutsche und slawische Burgen".

was determined by what they were there to protect, not by any strategic or other concern. Licences issued by King Berengar use terms such as *agger, berlisca, celata, fossatum, merlo, muro, propugnaculum, spizata, turri*. Many of the terms meant a wooden defence. There are also references to hedges. Towers were rare in the tenth-century references, little more than a quarter of documented sites having one, although they became increasingly common in the eleventh. In 906, Berengar licensed what became a formidable castle at Nogara "cum berliscis, merulorum propugnaculis atque fossatis". Very few sites occupied more than 20,000 square metres, with about half having an area of between 5,000 and 15,000 square metres.

In the countryside dependent on the ancient city of Bergamo, there is no evidence of any local defences at the start of the tenth century. The first documentary evidence of a *castrum* in this area is in 911, and in 959 a village previously mentioned without any reference to defences is identified as being "next to the castle" that has now acquired its name. Four documented sites were identified as being protected by ditches and hedges or palisades, while four others had walls, a much higher proportion than elsewhere. It has been suggested that these places were close to the city, and therefore copied the urban model. Most of the castles of this area were established by public officials, others by ecclesiastics, and some were held by a consortium of landowners. In the hilly country of Canossa, there are more references to the term *rocca*, which is distinguished from *castrum*, and when both are cited together the implication is of a ramparted village area adjacent to a fortified hilltop. Again, as in the north, defences often comprised only ditches and hedges. The average area enclosed was between 2,750 and 4,400 square metres.[1]

A similar process has been studied in Latium, where a substantial movement of people *per castra* also occurred, although the underlying cause may have been as much a growing population as the needs of defence. Landowners were seeking to attract people to their own lands and offered well-defended villages in new locations as part of the inducement. Given the nature of the country here, building was in stone from the outset.[2] It is likely that the process involved a contract between a lord and the new occupants, in which the lord would reserve part of the new structure for his own use, including, perhaps, the tower, which became much more common in the eleventh century.[3] Since providing a defensible hilltop village was intended to be an incentive to move, defences must have answered a perceived need for security. Even where the nature of the local power differed, the same processes seem to have occurred. Also in central Italy, the local counts were effectively independent in the Marsica district of the Abruzzi. Here too there is evidence of new rural fortifications developing during the tenth century, sometimes reoccupying ancient sites, often on hilltops or mountain ridges. Some were created to serve the military or strategic purpose of the ruling counts and did

1 Settia, "Castelli e villaggi nelle terre Canossiane", 288–97; Settia, "La struttura materiale del castello", 362, 370–75, 378–96, 399; Settia, "Il primo incastellamento nella Bergamasca", 239–43; Menant, "'Fossata cavare, portas erigere'", 277–81.

2 Toubert, *Les structures du Latium médiéval*, 1: 313–33.

3 D Andrews, "The archaeology of the medieval castrum", esp. 314, 316–32.

not change the settlement pattern of the area. Here again, stone was used from the outset.[1]

By the end of the tenth century, the Italian countryside outside the areas that were still subject to Constantinople had been transformed, and ditched, palisaded or walled villages everywhere crowned the hills, or sprouted among the running waters of the plains. Local lords and counts had also begun to provide their residences with fortifications in a way similar to their counterparts in Germany and France, and if during the first half of the century the higher powers had maintained formal control over the licensing of fortifications as an essential attribute of royal authority dating back through their Carolingian predecessors, their writ already did not run everywhere.

The principalities of the Poles, Bohemians and Rus'

The origins of the dynasty that founded the principalities of the Rus' along the great rivers between the Baltic and the Black Sea may never be known with historical certainty, but during the ninth century the Vikings had established trading posts and settlements along the rivers, and these had been provided with protection using the resource most readily available, timber. Having established overlordship over the Slav inhabitants, whose culture and technology had been at a lower level, the ruling élites were able to begin to play a part in affairs beyond their (rather loosely-defined) borders. The Byzantine empire to the south, with its Danubian frontier, and its control of the city of Cherson (near Sevastopol') in the Crimea, was inevitably the main attraction, for trading relations, for occasional warfare, and in due course for cultural links closely related to the conversion of the principality to Christianity. These links would also, in the course of time, provide an architectural stimulus as well.

Whereas most of the regular conflicts of the Rus' were with the nomads of the steppes, their encounters with the Byzantines could not fail to involve dealing with their sophisticated cities and fortresses. The surviving accounts suggest that knowledge of siege techniques throughout the region was very limited. In 946, the young Svyatoslav had failed to capture a Derevlian town after a whole year's siege and had to resort to a trick to take it. In 968, while the prince was engaged in Bulgaria, his own capital of Kiev was besieged by the nomadic Pechenegs, but they too were unable to make any impression on the defences. In 988, his successor Vladimir took on a major walled city when he besieged the Byzantines at Cherson. Demonstrating some advance in techniques, the Rus' prince established a camp a bow shot from the town (rather closer than the Byzantine experts were advising) and threw up an earthwork to shield his forces. He attempted to fill the ditch, but was thwarted by the defenders who managed to excavate all the earth thrown in by the Rus' and bring it into the town. In the end, the prince learnt of the source of the city's water and, having cut this, induced a rapid surrender.[2]

1 Somma, "Strutture fortificate altomedievali", 85–90.

2 *The Russian primary chronicle*, 80, 84, 112; *La chronique de Nestor*, 1: 89–99, 130–31 (Nestor died in the early twelfth century); *The Nikonian chronicle*, 1: 63–4, 98 (Cherson).

The Rus′ princes are credited with major building works. Kiev was the main centre of the realm and would become, in the eleventh century, an immense metropolis. Important developments towards this status took place during the tenth century. Around 900, according to archaeological investigation, the small settlement on a hill on the right bank of the river Dnieper was extended substantially, and provided with an earthwork rampart. The hilltop was the citadel and princely residence, the low-lying Podil the market area below. Then, under Vladimir, the old fortifications on the hill were levelled and replaced with a new rampart mainly of clay, six metres high, nine metres wide at the base. The gate-towers were made of stone and brick, and the whole was protected by a broad moat. The total area covered by the whole settlement had grown to 125 hectares, with a possible population of 14,000.[1] While the defences were of earth and timber, there were stone buildings in the palace complexes.

The princes' use of defences did not stop with the main centre. As a protection against the nomads for the settled population of their emerging state, the princes ordered the construction of defended towns, and hundreds of miles of earthen ramparts south of Kiev to slow down and perhaps channel the advance of raiders. These ramparts were around four metres high, and had ditches in front 12 metres wide.[2] Such long barriers could hardly be defended except in unusual circumstances, and must have served to compel large raiding armies (especially any already burdened with plunder) to follow particular routes. There is little evidence to show what effect they may have had. Fortified towns were established to secure control of newly-conquered territory, as well: in 985 Vladimir captured what were called *castra* by the much later chronicler and occupied them, including Przemyśl, and in 997 invaded Ruthenia, where "he erected the castrum of Perieslav".[3] Excavations at Przemyśl found evidence of an early hilltop defended site occupying a space of 80 × 70 metres with an earth and timber rampart beneath the later medieval walls.[4]

Meanwhile, in the north there was also expansion and development. There had been earthwork fortifications for the Viking settlements on Lake Ladoga, and later at Novgorod, from an early date but the dating from archaeology remains imprecise. Excavations at Staraya Ladoga found an earthwork of the tenth century, a narrow rampart and evidence of a small tower on the island. A hill-fort at Gorodishche-Georgii dated to the early tenth century had a rampart 12–14 metres wide, and more than three metres high from the outside, while another at Sergov-Gorodok (Sergovo) that blockaded the access to the river Ilmen has an area of 30 by 70 metres, and a horseshoe-shaped rampart 10–12 metres wide, protected by a ditch. The oldest archaeological levels at Novgorod itself date from the early tenth century. It is thought that the name, signifying "new fortress", applied

[1] Callmer, "The archaeology of Kiev", 35–43; Mühle, "Die topographisch-städtebauliche Entwicklung Kievs", 350–54.

[2] *The Russian primary chronicle*, 119; *The Nikonian chronicle*, 1: 114; Franklin and Shepard, *The emergence of Rus*, 170–71; Vernadsky, *Kievan Russia*, 73–4.

[3] Długosz, *Annales seu Chronicae*, 1: 199, 200 (book 11).

[4] Żaki, "Topografia wcześniośredniowiecznejo Przemyśla".

exclusively to the citadel (*kremlin*), a single fortified area amongst a number of settlements. However, it has been concluded that the significant development of the city took place in the middle of the eleventh century.[1] Pskov, similarly, had an early earth rampart, a couple of metres wide, with post holes for a palisade, with an internal area (possibly) of three hectares. There is evidence that the earthen rampart at Polatsk, dating from the ninth century, was destroyed in the tenth.[2]

In design, therefore, the tenth century fortifications of the Rus' shared the same basics as their contemporaries in central Europe. Whether town wall or kremlin, available materials and technology were deployed to produce defences more than adequate to deal with the threats they might face. Trading in the rich natural resources of the country was central to the origins and wealth of the ruling élite, and strong walls to protect the resulting urban centres, alongside the princes' own residences, were conscious developments of this time.

The Slav tribes further west were undergoing various different developments. Those in what is now eastern Germany and northern Poland retained a tribal structure and pagan religions. But in what were to become Poland and Bohemia, a process of tribal unification, the establishment of princely dynasties, conversion to the Christian religion, and state-formation took place. In these, the relations with the German kingdom (especially in Bohemia) were particularly important, with the establishment of the dynasty of the Przemyslids with a major fortified centre at Prague, ultimately a constitutional part of the German empire. There is a suggestion in a twelfth-century source that the new dynasty in Prague had a conscious policy of building *castella* or *castra* to provide protection against the Germans, Poles and Moravians with a reference in 977. There are references to strong fortifications scattered throughout the chronicle, but they generally appear as centres of administration rather than strategic defences.[3] Some evidence of construction or reconstruction of hill-forts in the countryside during this period indicates a continued need for refuges as well.[4]

In Poland, the Piast dynasty emerged on top. There was a move from open settlements to the development of hill forts of various sizes and levels of occupation (precise purposes unknown) during the eighth century,[5] and now there was further development of fortifications. At Płock on the river Vistula, a hilltop site with an inner and outer enclosure closely comparable in design to the fortress-palaces of the Ottonians in Germany has been excavated. Beneath continued building work of later centuries were a tenth-century rampart of earth and timber, 17 metres wide.[6] At Swobnica (German Wildenbruch) in western Pomerania, 20

1 Mühle, *Die städtischen Handelszentren der nordwestlichen Rus'*, 50–52, 105; Nosov, "Ryurik Gorodishche", 16–19; Yanin, "The archaeological study of Novgorod", 86–9.

2 Mühle, *Die städtischen Handelszentren*, 175, 207–8.

3 Cosmas of Prague, *Chronica Boemorum*, 50 (book 1, chap. 27; 977) for the *castra* "sited on Mount Osseca [south of Zbraslav] against the Germans", "below the woods" at Litomyšl against the Moravians, and at Kłodzko (Glatz) against the Poles; 64 (book 1, chap. 36) for the "munitissimum castrum" of Dřevíč in 1002.

4 See the examples given in Gojda, *The ancient Slavs*, 27, 35–7.

5 Kobyliński, "Early medieval hillforts in Polish lands", 147–53.

6 Szafrański, "Die frühmittelalterliche Hauptburg Płock in Masowien", 83–7.

kilometres from the river Oder, a prehistoric site, forming a natural triangle 20 metres high and approximately 120 metres along the longer side, had been given a rampart four metres wide with wooden reinforcements, and bastions at each corner. It was dated to the time of the submission of the area to the Piasts in the late tenth century,[1] after which it apparently fell into disuse. At Poznań (Posen) and Wrocław (Breslau), excavations have identified that the fortifications were of exceptional size, with ramparts up to 20 metres wide at the base, and ten metres high. They were also very strongly constructed, making effective use of the type of soil and of timber frames to provide stability. A considerable number of sites were certainly topped off with well-built palisades, properly secured in the rampart and sometimes using stone reinforcement as well (as at Poznań).[2]

A Jewish writer from al-Andalus visited the region and wrote about it. Ibrāhīm ibn Ya'kub portrayed a country of wealth and considerable armed force available to the princes, wearing complete armour. The fortresses were called *-grad*, which the author's editor translated as "big castrum". He described how the Slavs would lay out a quadrangular site, digging a ditch around it, using the spoil to create the rampart and then erecting a high wall on top. Passing through Bohemia, he noted that Prague was made of stone and lime and was very wealthy.[3] So the new Slav realms in central Europe that emerged in the late tenth century followed their nearest model, the German kingdom, in many respects. While efforts would be made to impress with the construction of expensive stone buildings where possible, most of the fortifications were of earth and timber, and seem from the scanty accounts to have been sufficient to meet the needs of their occupants.

"Western Francia"

The vast territories stretching from the coasts of Flanders in the north to the river Ebro in Spain largely acknowledged as king the western successors of Charlemagne, but did not have a single name to describe the realm. France remained in the future. The political history of the tenth century was characterised by a process in which the direct rule of the crowned kings shrank to a territory between the rivers Seine and Loire, sometimes reinforced by the addition of major territories to east or west as a result of genealogical accident or armed struggle. All around, powerful princes tried to establish their own dynasties based on expanding the counties under their own control, building up networks of supporters among the lesser nobility and churchmen, and each step forward for these princes represented a consequent loss of resource, wealth and power for the monarch. Chief players among these regional potentates were the dynasties in Flanders, the Vermandois, Anjou, Aquitaine and Burgundy, while the descendants of the powerful Robertian rulers of the west and north-west in the ninth century went on to obtain the crown itself in 987 and found the Capetian

1 Szafrański, "Swobnica", 43–6.
2 Hensel, "Types de fortifications Slaves du haut Moyen-Âge", 71–8.
3 Ibrāhīm ibn Ya'kub, *Relatio de itinere slavico*, 145–6.

dynasty. At the beginning of the century, Normandy did not yet exist, awaiting the award by King Charles the Simple of lands by the Seine to the Viking Rollo in 911, but by the end of the century the Norman dukedom was also an established party in the power politics of the kingdom.

Chronicle sources for the history of the century are few and scattered geographically: Flodoard and Richer were monks of Rheims in the middle and at the end of the century respectively, Ademar of Chabannes (born in 988) wrote from Aquitaine, while Rodulfus Glaber (born in 980) lived in Burgundy. They provide some idea of the scale and scope of siege warfare of the time, and of the growing phenomenon of the castle.

A major source of contention was the city of Laon. In 938, King Louis IV (r. 936–54) besieged the new *arx* built there by count Heribert, and took it with "many machines" undermining the walls, but it involved an enormous effort. Having secured the place, the king ordered the construction of a citadel of his own. Heribert's citadel — demolished in the fifteenth century — seems to have been a stone-walled enclosure at the western end of the hill top site, equipped with towers. The king's castle, which was repaired in 1207 by Philip Augustus, so must have been a substantial building from the start, was surrounded by a ditch and ramparts. Central to it was the Tour de Louis d'Outremer, 24 metres high, with a diameter of 17 metres, and walls five metres thick at the base. The single door was originally reached by a drawbridge. A strongly-walled palace completed the complex. Sadly, the tower was pulled down in 1831. In 949, the king had to recapture Laon by a perilous night-time escalade, his men succeeding in overpowering the guard and opening the gates, the garrison retreating to the tower.[1] In between (in 946), the old Roman walls of Senlis, defended by the men of Hugh the Great (Capet), had again served to hold off a siege.[2] Laon continued to play a major role in affairs of state. In 988 it fell to King Charles of Lorraine, who ordered its defences strengthened. Supplies were collected from the whole surrounding province, 500 sentries were appointed to guard the walls, and we learn that crenellations had to be constructed on the tower which "was not yet high", and ditches dug, suggesting that the work ordered by King Louis fifty years before had not been proceeding apace. Fascines were made to reinforce the works and ballistas were constructed for the defenders. Smiths were summoned to make missiles. When King Hugh tried to attack, he surrounded the town with ditches and ramparts but could make no impression. Deciding that Laon was impregnable, he withdrew. On his retreat, Charles ordered further restoration of the town walls, "fallen through age"[3] and further work on Louis' tower.

Hugh the Great's expertise, or else his resources for siege warfare, were apparently not as great as the king's. In 932, it took him two months to secure the surrender of the *castellum* of Saint-Quentin, and he went on besiege Château-Thierry for six weeks. Châtillon-sur-Marne, called a *castrum*, fell more rapidly in

[1] Flodoard of Reims, *Les Annales*, 70, 122; Melleville, *Histoire de la ville de Laon*, I: 68–71.

[2] Flodoard, *Les Annales*, 103.

[3] Richer of Saint-Rémy, *Histoire de France*, II: 172–6.

940, while Count Reginald de Roucy managed to scale the walls in a night attack in 949. The next year, King Louis besieged the men of Arnulf, count of Flanders, in the *turris* of Amiens, suggesting that the tower was becoming a more common type of fortification, and in 951 he took Brienne-le-Château by blockade, destroying the fortifications. The next year, the fortress of Mareuil(-sur-Ay) on the river Marne, being built by Reginald, was stormed by royal forces with machines.[1] Flodoard's accounts suggest a number of trends. Strongly-held walled towns were still centres of power, but citadels were being constructed within them, serving not as final refuges, but as a strongpoint for the ruler. The royal forces could still command the skills and resources to build siege engines, the few examples given suggesting rams or shelter for sapping walls. Alongside, there are increasing references to new fortifications, *castella*, the precise form of which at their foundation can be hard to discover, but probably of relatively rapid construction in earth and timber. Ademar confirms this. In 944, his abbot built several *castella* and restored the walls of the monastery of St Maxentius ("castellum in circuitu perfecit"). In 995, the *castrum* of Gençay was destroyed by Count Adalbert, but rebuilt by William of Aquitaine, suggesting a wooden structure.[2] Similar evidence comes from an account of a siege involving Ademar, son of the viscount of Limoges, at Saint-Benoît-du-Sault, with the use of stones and darts as missiles, the burning of the fortress's wooden doors, and a reference to there being a "wooden tower".[3]

Richer gave detailed information on the struggle in the east for control of Lotharingia (Lorraine) involving a complex interplay of Frankish and German interests. Verdun played a major part. The city sits on the Meuse, connected to a merchant's quarter on the far bank by two bridges, and approachable only from one side. While the new suburb was undefended, the old city still possessed its third-century walls, which had escaped the attentions of the Vikings.[4] In 985 it was captured with an army disposing of many and diverse machines of war. The duke of Upper Lorraine, Theodoric (Thierry) I (r. 978–1027), tried to hold on to the city. The defenders collected timber from the Argonne so that they "could oppose with their own machines" the machines of the besiegers. Fascines topped with osier were constructed. Beams tipped with iron and protected from the risk of burning were prepared. There were 100 engines. The army of the French king Lothair (r. 954–86) approached, creating its own fortified encampment. He ordered the construction of a siege tower, a "belfry". This was to be moved by oxen, pulling it by means of ropes and pulleys using a frame set ten feet in the ground for stability. The enemy having built a similar though smaller machine, there was a bloody conflict, ended when iron hooks were fired into the weaker machine and used to pull it over.[5] Verdun surrendered to the king. This is an early reference to

1 Flodoard, *Les Annales*, 53–4, 56, 75, 124, 127, 131, 133–4.

2 Ademar of Chabannes, *Chronique*, 146–7, 156.

3 *Les miracles de Saint Benoît*, 140.

4 Dollinger-Leonard, "De la cité romaine à la ville médiéval", 209–15; R M Butler, "Late Roman town walls in Gaul", 48–50.

5 Richer, *Histoire de France*, II: 131–8 (book III, chaps. 103–7).

the introduction of such siege equipment in the west, and the account also indicates the use of ballistas and crossbows.

But the variety of equipment available to the great princes was certainly considerable. In 991, for example, Robert (r. 996–1031), successor of Hugh Capet (son of Hugh the Great, r. 987–96), alongside Duke Richard II of Normandy, captured Melun, each taking one side of the river Seine and attacking furiously, day and night. Armed ships provided part of the arsenal, and it was the Normans who broke in through one of the gates. Catapults are listed among the weapons.[1] In 997, the same pair found themselves dealing with a revolt in Burgundy, and besieging Auxerre. Unlike their previous efforts, they were unsuccessful whether they tried "force or guile", and they gave up after many failed assaults. They then tried their luck at Bishop Germanus' strongly fortified abbey (*castrum*) and were equally unsuccessful. It sounds as it they were lacking sophisticated siege equipment on these occasions.[2] A few years later, King Robert captured Sens from its count after three days' attack, but was resisted by the count's brother and several knights who held out in the town's tower ("urbis turrim defenderunt"). This was the same tower recorded as being erected in one corner of Sens by Count Robert at the turn of the century and obviously of some size ("turrim maximam").[3]

New fortifications in the west

The task of understanding the new fortifications of the tenth century is complicated by the small number of written sources and the great difficulty of dating the archaeological remains of sites that either had a very short occupation, or else were repeatedly used and built over. Numbers of regional studies have been made, however, that enable conclusions to be reached. As elsewhere in Europe, the evidence indicates the beginning of a new process of building rural defences getting under way in the second half of the century, before really taking off in a big way in the eleventh century.

A careful study of the terminology used by Ademar[4] shows that the words *castrum* and *castellum* were widely used, but that in the sections of his chronicle borrowed from previous writers he redefined a number of sites from the latter to the former description. However, he also used the words interchangeably. The chief distinction is between both these words and urban fortifications. The other significant point is that of these rural fortifications, a significant proportion (around 40%) were described as newly built in this period. Some were recorded as being built and destroyed in very short times, far too short for any kind of construction other than a ditch and rampart.

[1] Richer, *Histoire de France*, II: 270 (book IV, chap. 76); William of Jumièges, *Gesta Normannorum ducum*, 92–3. William is a rather late source but is here corroborated: Clarius, *Chronique de Saint-Pierre-le-Vif de Sens*, 104–5; *Les grandes chroniques de France*, 632.

[2] Glaber, *Historiarum libri quinque*, 78–81 (book II, chap. 16); Clarius, *Chronique de Saint-Pierre-le-Vif*, 106–7.

[3] Clarius, *Chronique de Saint-Pierre-le-Vif*, 106, 110.

[4] Debord, "*Castrum* et *castellum* chez Adémar de Chabannes", 98–108.

This ties in with the survey carried out much earlier by Aubenas of the process of encastellation in France, more concerned with issues of ownership and control.[1] His study of the written evidence indicated that whereas the right to construct fortifications had been a jealously guarded royal prerogative of the Carolingians, the circumstances of the ninth century had seen this power being delegated to counts and bishops. By the following century the right of fortification had become alienated to the local authorities. They, however, for the most part sought to ensure that no one of lower status infringed *their* rights. In Anjou, the powerful counts (represented from the end of the century by the aggressive Fulk Nerra) used the building of castles as a strategic weapon, and their control was entrusted to loyal followers who would in turn found local dynasties. A similar picture applied in Flanders, where the counts would brook no internal challenge to their authority. In the Dauphiné, it was the bishop who divided the land into fiefs and authorised the construction of castles as centres of authority (976). The chronicle of Sens talks of Count Renard building *castra* in his own right at Joigny and Château-Renard along with the tower in Sens itself: however, the source indicates that the building was in defiance of law and property rights, as Renard used seized monastic land for his constructions.[2]

In Poitou, where tenth-century evidence is rare, Lusignan, Niort and Melle were founded around 950, Saint-Maixent in 944. It is not clear at whose initiative they were built, but some were of private construction, and the count, too weak to prevent them, was compelled to accept them in order to preserve the illusion of his authority. Here, the old administrative districts dating to the Carolingians (*vicaria*) were transformed for the most part into castellanies, centred around new castles.[3]

Further south, a slightly different picture emerges of an earlier descent of power from the counts to lower layers of landowners. The counts had to cede ground to viscounts, and they in turn passed on power to the *milites* who were their feudal inferiors. The number of castles increased twofold between 900 and 975, from a very low starting point. In the following period the increase continued, with 50 sites in the Auvergne, Rouergue and Albi, 43 in eastern Languedoc, 31 in western Languedoc, and 25 in Limousin and Quercy. Their holders owed their loyalty to the local magnate, and some held their castellany jointly with others.[4]

In the Maconnais, there were five or six public fortresses entrusted by the count to local magnates, and one, at Lourdon (in the commune of Lournand), was controlled by the monks of the abbey of Cluny.[5] This castle was occupied by a local magnate before 1000, in an attempt to increase his power and wealth by imposing labour services on the peasantry. Here, the abbey was strong enough to resist the attentions of the lay aristocracy. Elsewhere, one can see how the more powerful

[1] Aubenas, "Les châteaux-forts des xe et xie siècles".
[2] Clarius, *Chronique de Saint-Pierre-le-Vif*, 105.
[3] Garaud, *Les châtelains de Poitou et l'avènement du régime féodal*, 16–24.
[4] A R Lewis, "La féodalité dans le Toulousain et la France méridionale", 247–59.
[5] Bois, *The transformation of the year one thousand*, 63, 141–9.

and aggressive magnates could extend their authority by simple use of superior force.

At some point during this time the first mottes appeared. Which was the first? A case has been made for Vinchy (now Les Rues-des-Vignes), which, after its capture by Otto count of the Vermandois in 979, was demolished, including the *agger*. The case is strengthened by the knowledge that the castle was thrown up in a few months and was then levelled.[1] It may be, therefore, that the form of small castle that was to spread famously across much of western Europe in the following century was first created on the plains of Artois. There are other candidates, too, such as La Ferté-en-Bray in Normandy,[2] although all the early sites are in the same general region (north of the Loire and west of the Rhine).[3] The search for origins is not helped by the non-appearance of the word *mota/motta* itself until the eleventh century. Whether the thinking behind the idea was to create a form of citadel for the lord of the castle, with an enclosure of normal type below, or whether the type of soil leant itself to such designs, or whether in flat areas it combined a visual effect of dominance with a good viewpoint, the relative cheapness and speed of construction of the raised mound with its wooden stockade and/or tower and an enclosed courtyard (the bailey) made this distinctive design a popular model, especially with landholders at the bottom of the scale.

It was not the only model followed. The greater lords would still look for something rather more substantial when they had the leisure and the resources to construct it. The early dukes of Normandy created a palace for themselves from 932 onwards at Fécamp, described as a *castrum*. Probably this was a simple ditch and rampart, changed at some uncertain date to stone walls.[4] Dudo, historian of the early dukes, refers to "city walls" and the "walled residence" of Fécamp.[5] Building in stone was well within the capabilities of the greater magnates, even if the inevitable slowness and cost of the work might limit the number of places so constructed. But the tenth century, having given birth to the new earthwork fortress in the as yet unnamed motte, also witnessed the arrival of the great tower or donjon. For this, the counts of Anjou can claim the credit. The rulers of this county pursued an aggressive policy of expansion, combining astute marriages with more direct use of force against their neighbours. Fulk Nerra, count from 987 to 1040, was a rare example of a magnate who used fortifications as a part of a coherent plan of conquest and tenure, as he proceeded to build castles on his frontiers to isolate slices of territory he coveted, then to secure his conquests with new castles, each entrusted to a reliable castellan. This work began in the tenth century. For the same reasons of speed that called for earthwork and timber castles everywhere else when military and defence concerns were uppermost, most of Fulk's founda-

1 Rouche, "Vinchy: le plus ancient château à motte", 364–8.
2 Boüard, "Quelques données françaises et normandes", 10–26.
3 See Colloque de Caen, "Les fortifications de terre en Europe occidentale", 8–10.
4 Renoux, "Recherches historiques et archéologiques sur le château de Fécamp".
5 Dudo of Saint-Quentin, *History of the Normans*, 94, 164.

tions began life as elevated promontory sites, cut off with a ditch, rampart and palisade. Most would then be given mottes or wooden towers.[1]

The history falls within the scope of the next chapter. Here, though, it deserves to be noticed that the stone donjon at Langeais, long recognised as probably the earliest of the great towers built in stone, has been dated dendrochronologically to the 990s.[2] The remains suggest a rectangular tower 7.35 ×16.1 metres inside, of two storeys, with doors at both levels. The basement appears to have been designed for storage; the upper floor was a single chamber with a chimney. Functionally, the stone tower was a residence, not a defence, for which it was singularly ill suited. It would certainly have taken a number of years to complete. It was, in many design respects, the traditional wooden great hall, but built in two storeys. The defence of the castle lay in the familiar promontory site, 400 metres long and not more than 90 metres wide, isolated by a deep ditch and earthwork rampart. A similar construction was begun nearby, at Loches, only a few years later. They too would be the prototypes for an enormous number of donjons in the following two centuries.

Developments in northern and north-western Europe

Absence of written evidence continues to hinder understanding of the processes of social and political development in the Scandinavian world, even while the "Vikings" laid the foundations of new states from Kiev to Normandy. Royal dynasties were becoming established in Denmark, but the archaeological evidence for their building operations remains hard to decipher. The immense circular earthwork fortresses at Trelleborg, Aggersborg, Fyrkat and Nonnebakken have been mentioned. They have been shown to be contemporary with an extension to the frontier rampart, the Danevirke, and the earthwork ramparts of the towns of Hedeby and Århus and to date from the reign of King Harald Bluetooth (r. 958–86). The circular fortresses are impressive, Trelleborg having an internal diameter of nearly 140 metres, while Aggersborg was 240 metres across, and the others rather smaller. These places were protected by a ditch, had two roadways meeting at the centre, and identical interior buildings in the four quadrants. It is now believed that their purpose was to assist the king to control his realm, although how they were to achieve this is not plain. It also seems that they fell out of use very quickly, to usher in a blank period in fortress-building in Denmark.[3] In guessing at models for these works, we have already noticed the

[1] Deyres, "Les châteaux de Foulque Nerra", 10–15.

[2] Impey and Lorans, "Le donjon de Langeais", 15–20, 31–4. The report suggests the possibility that Langeais may have been built by the counts of Blois, who held the area for some time, but the likely duration of the building work points to Fulk as the more likely builder.

[3] For plans of the geometric fortresses, see Higham and Barker, *Timber castles*, 80–81. T E Christiansen, "Archaeology and history", demolished previous theories. Roesdahl, "The

circular earthworks constructed to provide defences against the Vikings in Flanders and the Low Countries.

Danes and English continued to confront one another across the frontier of the Danelaw established in England. Under Æthelstan and his half-brother and successor Edmund, the kings of Wessex extended their control over Northumbria and Mercia. In the process of the conquest of Northumbria, the *burh* of Tamworth was stormed by the Danes, but Edmund then besieged and captured Leicester (943), although the *Anglo-Saxon Chronicle* gives no detail.[1] Later, in the reign of Æthelred the ill-advised (rather than "unready") (r. 978–1016), the Danes resumed a series of raiding expeditions that the kingdom proved quite incapable of resisting. The defensive system set up by Alfred failed, under lesser leadership, to achieve its purpose. There does not seem to have been much engagement of the *burhs*. In 1001, the Danes sailed up the river Exe and attacked the fortress at Exeter, but they were met by "a solid and courageous resistance" and failed, but were not prevented from ravaging the whole country around.[2] In 1013 the Danish king Sweyn Forkbeard carried through the conquest of the enfeebled kingdom, dying in 1014 and leaving it to his son Cnut to consolidate a new dynasty from 1016.

During the same period, the Vikings of Ireland continued their involvement in the conflicts of the island while they secured their settlements on the coast as bases for trade and local rule. Limerick apparently involved stone building, and in around 950 the Irish of Munster fought their way in close behind the fleeing Vikings so that they "were not able to close the gates", and proceeded to slaughter them "in the houses and towers". The "city" of Cork was occupied, then they marched against Cashel, where their opponents included "the Danes of the fortress". At Waterford, they closed the gates and defended the ramparts, but the Irish under Cellachán, king of Munster, "leapt into the town", suggesting that the ramparts were not very strong or high.[3]

An eleventh-century source tells of the exploits of the High King Brian Boru (r. 1002–14). Around 999, he captured Dublin, plundering the "fortress" before burning it to the ground. Obviously it was rebuilt, because in 1013 the king failed in a renewed attack. This time he was only able to blockade the city, and had to withdraw when his own provisions ran short. Brian, however, is credited with the construction of "bridges and causeways and high roads. By him also were strengthened the duns and fortresses and celebrated royal forts of Munster. He built the fortification of Cashel . . . and the royal forts of Munster in like manner."[4] Another source confirms the building campaign, calling the constructions *daingin*

Danish geometrical Viking fortresses and their context", supports the argument for their being built to assert royal control.

1 *ASC*, 111.

2 *ASC*, 133.

3 *Caithreim Cellachain Caisil*, 62–71. The king ruled from 934 to 954; the MS source, however, dates from the fifteenth century.

4 *Cogadh Gaedhel re Gallaibh*, 113, 141, 151. For Dublin, see also *Chronicon Scotorum*, 255, where Brian is credited with burning the fortress.

and *cathair*.[1] There is no way of knowing what these works consisted of. Most likely they continued to be ring-forts, although a royal centre like Cashel could of course have had stone defences. However, there is no indication that defensive building in Ireland, whether of native or of Viking origin, had attained any new levels of sophistication by the end of the century.

The tenth century witnessed the process of formation of a coherent kingdom in North Britain, in which the Scots of the west coast secured control of the wealthy territories of Fife (ruled by the Picts) and of the lowlands generally, coming into regular contact and conflict first with the Anglian kingdoms of Northumbria, then with the Vikings. Fortified sites are rarely closely datable, and the type of most remained the long-lived hill fort. Smaller places might be the residence of a local leader, larger of a king, albeit one whose power extended over only a small area, and depended on the warriors of his own war band. Building materials would depend on what was available locally, with both dry stone walls and earthen ramparts being thrown up. Over the centuries there are accounts of such places being burnt and rebuilt (for example, the earthwork underlying the medieval tower house at Dunollie near Oban, which may have been an important royal site in the kingdom of Dalriada in the eighth century).[2] Dunollie and Dunadd (near Kilmartin), for example, may have been occupied during the rise to kingship of Kenneth MacAlpin (840 onwards). Dundurn in Strathearn, which is on a craggy hilltop in a commanding position to control routes of communication, gives evidence of repeated building of a citadel, the last phase being a rampart of rubble contained by sandstone slabs as revetment. Such places are cited as evidence of growing royal power.[3] The famous rock of Dumbarton, identified by Bede as a powerful fortified centre of kingship and defended several times according to chronicles, fell to the Vikings after a four-month siege in 870, after which it was abandoned for centuries. The inhabitable space on the rock is very limited, although excavation has recovered a mid-seventh-century rampart.[4] Other places, their identification suggested by use of the word *dun*, have also produced mixed results on examination. Dunbar and Edinburgh (*Dynbaer, Din Eidyn*) were noted in the record, with Edinburgh being the subject of attack by the English in 954, but archaeological evidence is lacking.[5]

Around 1006, the new king of Scotland, Malcolm II, launched an attack on the English that culminated in a siege of Durham. No detail is known of what happened, and even the date is not without challenge, except that it seems that an army from Northumbria and Yorkshire was hurriedly assembled under the earl's son Uhtred and inflicted a defeat on the invader. Durham itself had apparently been established a few years before, with a record of the construction of the site for the cathedral under Uhtred's aegis in 995. It must be assumed that the top of

[1] *The annals of Ulster*, 447.

[2] Alcock and Alcock, "Reconnaissance excavations", 2: "Excavations in Dunollie Castle", 117 (1987), 120–30.

[3] Ibid., 3: "Excavations at Dundurn", 119 (1989), 195–214.

[4] Ibid., 4: "Excavations at Alt Clut", 120 (1990), 96–117.

[5] Alcock, "Early historic fortifications in Scotland", 157–77.

the rock now containing the cathedral and castle complex was the settlement in question, a site "fortified by nature but not easily inhabitable" on account of dense woodland that had first to be cleared by a mobilisation of the surrounding population. Presumably, part of the work had involved simple defences to control the access to the top of the site.[1]

Some conclusions

By the year 1000, powerful states existed in central Europe, and Byzantium was resurgent. However, in terms of the skills of building, defending and attacking the fortresses on which royal power was based everywhere, the most sophisticated examples were to be found either in the empire of Constantinople, or in the western Caliphate in al-Andalus. By the end of the 900s, we have seen significant improvements in defensive features. These took the form both of strengthened walls and towers, and also more sophisticated designs for the protection of the gate. Such steps were not made in a vacuum, but represented a response to the improved techniques of siege warfare, evidenced by the flowering of military manuals during this period. In central Europe, where new states to the east were being established on the model (mainly) of the German kingdom, shortage of resources and skills and local conditions determined that most works were of earth and timber. Whereas in the empires continued central control of fortifications ensured that the state was responsible for the works, whether their function was that of frontier watch tower, military base, centre of administration or rural refuge, elsewhere in Europe the collapsing of central authority had already handed regional and local power down to counts and bishops. During the tenth century, in parts of Francia and Italy, fortification building began to be undertaken even lower down the scale, setting a trend that would expand in the next period. Limited resources, and limited local functions, meant that such works were often small and relatively weak. The full arsenal of weaponry and skills available for siege warfare, and apparently at the disposal of all the rulers of the time, were not within the reach of local magnates or their feudal inferiors, and were also not required to attack or defend local earthworks. Meanwhile, a slower rate of change was taking place on the geographical edges of northern and western Europe where access to resources and to building skills was less. New developments beckoned.

[1] Meehan, "The siege of Durham", 2–18. Simeon of Durham, *De obsessione Dunelmi*, 215–16; Simeon, *History of the Church of Durham*, 673 (chap. xxxviii).

5 Shifting balances: the eleventh century

Overview

IF AT THE END OF THE FIRST MILLENNIUM of the Christian calendar the state of the known world did not appear fundamentally different from its condition a hundred years before, with the greatest threat at Rome, Aachen, León or Constantinople appearing still to lie with Islam, by the year 1100 the balance had changed perceptibly. The disintegration of the splendid caliphate of Córdoba into numerous squabbling petty kingdoms created the opportunity for a significant expansion of the Christian states. By the end of the century the Muslim cities of central Spain had fallen into Christian hands. At the same time, in the east, the pre-eminent position of the Byzantine empire crumbled. The victory of the Seljuk Turks at Manzikert (now Malazgirt) in 1071, unlike previous Byzantine military disasters, entailed the loss of most of Asia Minor. Despite the best efforts of the Comnenian dynasty that took over the shaken empire, the greater part of this immense territory proved to have been lost forever. The political power of Constantinople would never again rise to the level it had known in the year 1000. Instead, it would be the rulers of western Europe who would answer the pope's call for a crusade to rescue the holy places in 1095, and in the ensuing clash with the Muslim rulers of the Middle East would generate important developments in the attack and defence of fortifications.

Elsewhere in Europe, the claim of the German kings to a dominant imperial status would hit a serious obstacle in the refusal of the popes to continue old practices in which the Church was an adjunct of secular rule. The ensuing conflicts provided great opportunities for the enemies of the new Salian dynasty. Among those drawn more or less willingly into the cauldron were the Norman adventurers, who dramatically imposed themselves as rulers of the assorted peoples of southern Italy, overcoming in the process Lombard petty rulers, Byzantine provinces and Arab powers. Expanding in all directions, these Norman conquerors drew on the civilisations they had taken over to produce, in the following century, a splendid new kingdom, and one that did not distinguish between the religions of its neighbours in taking advantage of weaknesses to push out its boundaries. At the same time their cousins in north-west France had also changed another political map by their conquest of Anglo-Saxon England. The great wealth and resources of the new kingdom would eventually be mobilised by the Norman kings and their successors to play a role across the whole of western Europe, while their expansionist urges along with their greater wealth and technical superiority spelt serious trouble for the less developed Celtic areas further to the north and west.[1] Across the channel, the Normans became a major player in the complex politics of western Francia, where the trends already established with the end of the Carolingians would continue. It would be a long time before the Capetian kings were able to do much more than compete among sometimes equal and often stronger local princes, and at times it was all they could manage to keep control over their own central territories. It is salutary to remember that these local rulers' ancestors had been royal officials, the counts of Charlemagne or his predecessors and successors, appointed and recallable at will by the king. Now, they were members of dynasties whose power had become transformed into that of owners of their own dynastic lands, and with networks of their own feudal inferiors providing them with a retinue of armed followers.

Western Francia

At the end of the tenth century, the most successful so far of the counts of Anjou, Fulk Nerra, had secured his position at home and begun a long career of extending his rule over the lands of his neighbours. Encroaching, through every means at his disposal, on Aquitaine featured most immediately in the continuous struggles, but without accompanying his diplomatic manoeuvres with military success, Fulk would have failed. In the whole of his long rule (987–1040), Fulk fought only a handful of battles, but his success was grounded in a thorough appreciation of the requirements of siege warfare and of the role of fortifications in extending his rule, and securing it.[2]

Fulk's siege warfare concentrated most often around castles, although he was also capable of taking on the walled cities. But because his approach involved taking over and securing new territories without previous fortifications, the capture and defence of these recent constructions was an essential part of his conflicts. A variety of techniques was revealed, suggesting that Fulk had a fairly broad training. It is clear from his activities that he must have secured the services of specialists as well as relying on his retinue of followers, and this in turn suggests that alongside masons and carpenters skilled in the erection of castles there must have been a supply of engineers able to create siege equipment. As to the methods of siege warfare, it is likely that precepts were handed down as part of the training of dynastic rulers whose success or failure depended significantly on their ability to hold onto their inheritance, and reward their followers with lands of their own, rather than that the young counts sat down and read, or were trained in, Vegetius. But it may well be that the training they received in siege warfare was based in

[1] This term is not intended to convey the idea of a single ancient origin, for the populations of Wales, Scotland and Ireland were, like everyone else, of various and mixed ethnic origins, but as a shorthand to describe the societies of these countries, which were very different from their neighbours, it must suffice.

[2] The most substantial study in English is Bachrach, *Fulk Nerra*, with references throughout.

Vegetius' ideal concepts. Whatever the truth, it is apparent that this dynasty produced an extraordinary line of counts between the tenth and twelfth centuries, all of them experts in conducting war.

The Angevins were familiar with siege equipment for direct attack, but also astute enough to understand how to capture a place too strong for direct assault. In 1015, Fulk laid siege to Montbazon, but recognising that a direct assault was impractical, he constructed a siege camp some half a kilometre off, and the effect of this blockade was to bring about surrender in due course. In 1026, Odo II (Eudes le Champenois), count of Blois and Champagne and one of the great princes of Francia, attacked the castle built by Fulk ten years before at Montboyau, which was part of his network of fortifications designed to isolate Tours. Odo also had his engineers, who constructed a "wooden tower of great height against the donjon of the castle", which was described as "very strong". Presumably, his forces had also secured their camp, because Fulk chose not to try a direct relief, but instead marched on Saumur, which was short of defenders, having supplied its forces to Odo. Deploying unspecified machines, Fulk took the city by overcoming the guards at night, and sacked it.[1] The garrison of the citadel, few in number, could see no prospect of relief, so they speedily surrendered in order to save their lives. Odo was compelled to redeploy from his siege against the Angevin army, during which time the garrison of Montboyau sallied from their fortress and burnt the siege tower. This must have been a remarkable construction to overtop the donjon, which stood on a motte. Its purpose would have been to enable the besiegers to shoot down on the defence, something that would have rapidly made the place untenable, rather than to allow the attackers to cross to the top of the wall, which would have been impossible unless the tower was mobile and the intervening ditch had been filled and the ramparts flattened. By the same token, it could not be moved away to protect it from the garrison's sortie. It has been suggested that the donjon was of stone, although with only ten years having elapsed since Montboyau was first established, and given later evidence of the time it took to raise stone buildings, along with the serious structural problems created by putting up heavy stone buildings on recently created earthworks without allowing time for them to settle, this seems unlikely.

The thwarted count tried to retake Saumur, again raising a wooden siege tower, which however suffered the same fate as the one at Montboyau.[2] Odo tried his fortune again against Amboise in 1027, but the town and the castle were well defended and he abandoned the attempt. In this world of changing alliances, in 1032 Fulk found himself allied with King Henry I of France (r. 1031–60) in trying to retake Sens from Odo, but three days of assaults on the *castrum* failed. A few years later, Fulk had to recapture Langeais, with its novel stone donjon behind powerful earthwork defences. The strength of the site was such that the Angevin count decided not to try to assault it but to set up a blockade with entrenchments to

[1] *Annales Vindocinenses*, 60–61; same text in *Chronica domni Rainaldi*, 10.

[2] *Historia Sancti Florentii Salmurensis*, 276, 280; *Chronica de gestis consulum Andegavorum*, in *Chroniques des comtes d'Anjou*, 53; *Gesta Ambaziensium dominorum*, ibid., 80–81.

prevent any relief and to starve out the fortress, a goal achieved in 1038.[1] A similar attack by the count of Gascony against the new castle thrown up by Fulk's supporters at Germond(-Rouvre), however, failed in its objective.[2] This can only have been an earthwork, but it was sufficient to hold off a powerful lord. Fulk never did manage to capture Tours, being prevented by the common interest of all his neighbours in thwarting such an extension of Angevin territory. The surrender of the city in fact came about in 1044, in exchange for the release of Odo's successor Count Theobald (Thibaut) III, who had been challenged by Fulk's son, Geoffrey II Martel, to battle, and had lost,[3] thus also illustrating the well-known risk of entrusting the outcome of the struggle to the uncertain fortune of a single battle.

Accounts of castle-building at Amboise give clues as to the nature of these castles and their defensibility. Geoffrey Martel granted "the very strong house in the place that up to today is called Fulk's Motte" to a follower of that name in 1060. A little after, there were no fewer than three fortifications in the town, each held by a different lord. Fulk held his strong house; the lord of the town of Amboise, Sulpicius I, had built "a stone tower" in the place "where the wooden house of his brother [Archembald] was"; Arnulf was guardian of the count's "house" (the château). While shared lordship of towns was not unusual, the presence of three fortresses drew the interest of the chroniclers. In 1066 Geoffrey's nephew and successor, Fulk IV le Réchin, attacked Sulpicius and first occupied his own fortification in the town. From this base, his archers and crossbowmen rained missiles on the defenders of the tower, who responded with javelins, arrows and immense stones thrown from the top. The attackers used mangonels and tried to deploy battering rams. In the fighting, the neighbouring houses and church were destroyed. After five weeks, the tower still held out, but a peace was negotiated, much to the relief, we may be sure, of the good citizens of Amboise.[4] The editor of the chronicle speculated that this vigorous account was borrowed from a classical text but was unable to locate a model, so it might even be that the chronicler was writing about what actually happened. The princes of western Francia had access to the full array of siege weaponry available at the time, and they also had the technicians to assemble and operate it. From this brief survey it is evident that the expertise existed to build and deploy rams and stone-throwers, while the wooden siege tower was becoming more of a feature. Mining however receives little notice.

The castles and their role

The construction of castles for the purpose of extending a ruler's power was part of Fulk Nerra's strategy. No doubt, these castles became centres of lordship, but when they were built their purpose was to consolidate that lordship and help protect it against others. This clearly required that they were defensible. Some 30

[1] *Gesta Ambaziensium dominorum*, 82.

[2] *Chronicon Sancti Maxentii Pictavensis*, 392–3.

[3] *Chronica de gestis consulum Andegavorum*, 56–7.

[4] Ibid., 58 (for the grant); *Gesta Ambaziensium dominorum*, 89–92, for the three castles and the battle.

new castles were attributed to Fulk in the chronicles, most of which can be identified. The chroniclers hint at their military purpose. Montrichard was established "lest (Fulk's) enemies descend on Amboise and Loches",[1] while it was noted that the stone tower built by Sulpicius at Amboise was so tall that from the top one could see the city of Tours.[2] To be militarily useful, rapidity of construction was essential. They must, therefore, have been of earth and timber, and often took advantage of naturally defensible promontories. In 1030, Fulk ordered a castle at Saint-Florent-le-Vieil on the site of the monastery: it was an *agger* (a mound) with wooden buildings in a courtyard. The monk who wrote the account could see the result, literally, on his doorstep.[3] As they were consolidated, and as resources became available, it is reasonable to expect that the original works might be gradually replaced in stone.[4] At what time many of the stone donjons (such as those at Chinon, Amboise, Blois, Langeais, Loches, Montbazon and Montrichard) were raised after the foundation of the castles during Fulk's rule, or that of his immediate successors, and whether they were regarded as part of the defences, or rather as potent evidence of Angevin lordship without any defensive function, are other questions. The chronicle said that Hugh II, lord of Amboise, built stone towers at Montrichard and Chaumont in the early twelfth century,[5] and we have seen a stone tower resisting furious attack in Amboise in 1066. The primary role of these stone towers, sometime called *turris*, which might suggest a defensive connotation, and sometimes *aula*, signifying a hall, remains to be determined: the terms themselves are to be found applying to both purposes.

Fulk's new castles were built in order to play a role in the actual conquests, and in their subsequent consolidation. Entrusted to a loyal supporter of the count as castellan, and held by a small number of retainers, supplemented when needed by knights and trained foot soldiers from other castles, or mercenaries in the pay of the prince, the castle could not be left unattended by a hostile force seeking to take control of the area. Either it had to be attacked and taken, or else blockaded to prevent its occupants interfering with the invader's freedom of action, in the same way that Byzantine forts had been intended to force Arab raiders to keep together rather than scatter over the whole countryside. A pattern of castle building, where the sites were located close enough together to enable one to support another, meant that the invader would have to face a task of great difficulty: unless he was strong enough to assail several castles at the same time, which was unlikely, an attack on one would leave him and his communications vulnerable to attacks from forces sheltering in the others. We do not know if Fulk Nerra sat down with

[1] *Gesta Ambaziensium dominorum*, 82.

[2] *Chronicon Turonense magnum*, 117.

[3] *Historia Sancti Florentii*, 281.

[4] Deyres, "Les châteaux de Foulque Nerra", is disputed by Bachrach, "Fortifications and military tactics". In the end, although the dating of the donjon at Langeais to the 990s (Impey and Lorans, "Le donjon de Langeais"), and that of Loches to the period 1012–35 (Mesqui, "La tour maîtresse"), both cases depend on the assertion that stone buildings were militarily more significant than earthwork and timber defences at this time, something the evidence of siege warfare does not sustain.

[5] *Gesta Ambaziensium dominorum*, 114.

a map and worked out such an approach, but this was certainly the effect. Of course, his success was dependent on having a network of loyal knightly families who could be trusted with devolved authority, and continued loyalty from within their fortification. This was a political luxury not available to all eleventh-century princes.

Normandy before the conquest of England

There were times, certainly, when it was not available to the dynasty of counts inheriting the territory around Rouen granted to Rollo the Viking in 911, who somehow hung onto their lands through many a crisis into the eleventh century, and succeeded in extending their control up to the borders of Brittany. The accession of William, the future Conqueror, as a minor in 1035 induced a prolonged period of civil war and outside involvement.

During the rule of Count Richard II (r. 996–1026), a process of organisation of the territory similar to that already taking place all over northern and western Francia took place, with the grant of counties based around castles, in the first instance to close associates and relatives of the ruler. Key centres such as Mortain, Ivry(-la-Bataille), Évreux and Eu were established as the centres of these administrative districts, which in due course became hereditary in the family, or in the office: bishops too were part of the network. In 1027, Richard III besieged his brother Robert in the *castrum* of Falaise, a fortress-palace now long buried under the later stone castle. The attack involved setting up a siege camp and then attacking with battering rams and ballistas.[1] A settlement was reached to end the fighting. Richard's early demise brought Robert (r. 1027–35) to power.[2] When his bastard son William succeeded at the age of eight, trouble and revolts followed. William of Poitiers, a great supporter of the duke, and William of Jumièges gave the details.

In 1047, it was Count Guy, who held the "very strong" castles of Brionne and Vernon from the duke, his own cousin, who revolted. With the help of the French king Henry I, William won a battle at Val-ès-Dunes. Guy retreated to Brionne, a fortress "which seemed impregnable [both] through the nature of the place, and by works". It possessed a "stone *aula* as a citadel for use in fighting" (although the present tower is from the twelfth century), and was protected by the river Risle. For all that, the rebel was persuaded to surrender. William then set about destroying the "new fortresses" which had been built during the sedition. In 1048 he besieged Domfront. It was recorded that the site of this castle did not permit of assault, with steep cliffs and approach only by two narrow paths, so this time the duke ordered the construction of four *castella* around it to blockade the Angevin garrison. Leaving loyal knights in his siege castles, William then saw off the Angevin count Geoffrey Martel (who in fact was distracted by an invasion of the

[1] William of Jumièges, *Gesta Normannorum ducum*, 98.

[2] For a thorough discussion of the period, see Bates, *Normandy before 1066*.

Touraine by the French king). Alençon then fell to William, and Domfront then capitulated.

In 1053 William besieged the count of Arques(-la-Bataille) in his castle, founded about fifteen years previously on a strong hilltop site. Once again, the duke decided against direct assault, and established siege castles at the foot of the hill, leaving garrisons to coop up the defenders (denying both egress and access, according to William of Jumièges). In 1054 William defeated King Henry (now on the other side) at Mortemer. In the meantime, Arques capitulated through hunger. Facing a hostile outside power in the new alliance between Geoffrey Martel and the Henry I, William threw up a (presumably earthwork) fortress at Ambrières(-les-Vallées), which was several miles over the Norman border with the county of Maine which the Angevins attempted to tear down, the defenders resisting with missiles, stones and slingshot. They apparently then attacked the wall with a battering ram, its beam crashing against the defence (which implies, improbably, that it was constructed of stone — perhaps the account is a flight of the chronicler's fancy!) until William returned with an army to rescue them. In 1063, William was able to effect Norman control of Maine, but soon had to deal with a revolt by Geoffrey II of Mayenne, inheritor of royal power in this county. The fortress was very powerful, with a river on one side, and stone walls. "Neither ram nor catapult (*tormentum*) nor any other instrument of war" could be deployed. So the Normans managed to hurl fire into the castle, driving the defenders from the walls, enabling them to rush in. The garrison withdrew to the citadel but surrendered, and William ordered the fire damage repaired. Work on the remains of the castle of Mayenne confirms that it is likely that a powerful stone donjon and *aula* were part of an earlier defended building reconstructed, possibly by the count, early in the eleventh century.[1] William's next siege action involved conflict with Conan II of Brittany, who laid siege to Dol, as illustrated in the Bayeux Tapestry, and the Normans marched to the relief of the castle, attacking the castle of Dinan (shown as a motte like Dol in the Tapestry) in which Conan had vainly sought refuge.[2]

Several points of interest arise from these accounts, even if they were totally biased. Many of the established castles had stone donjons by the mid eleventh century. The ducal palace at Rouen, for example, had a stone complex of buildings including a great tower by around 1024, with the site being extended by William after around 1060.[3] From William of Poitiers it appears that not all stone donjons, even when described as halls, were purely residential. Their role as a citadel was explicitly indicated in the contemporary history. It also seems that where a strong

[1] Although the dating of the various stages is based on some assumptions rather than firm evidence: Early, "Château de Mayenne".

[2] Guilelmus Pictavensis, *Gesta Guillelmi ducis Normannorum et regis Anglorum*, 16–17, 18–19, 20–22, 36, 58–60, 78, 90, 98, 110 (book I, chaps. 7, 9 (Brionne), 10–11, 17, 18–19 (Domfront), 25–7 (Arques), 32–3 (Ambrières), 40 (Mayenne), 45 (Dol/Dinan)); William of Jumièges, *Gesta Normannorum ducum*, 119–20 (book VII, chap. 4), 125–7 (book VII, 9); *The Bayeux tapestry*, plate 23.

[3] Gauthiez, "Hypothèses sur la fortification de Rouen", 62–76. Nothing remains of the original castle, described as comprising donjons, halls and enclosing walls.

location had been chosen, the sophisticated means of assault at the disposal of the princes of northern France could not be brought to bear. The commonest weapons described were rams and stone-throwing artillery, and once more there is no reference to mining. Nor did the Normans or Angevins use siege towers. What does appear as a consistent weapon for an attacker who finds himself unable to subject a fortress to direct assault, and perhaps has a reasonable fear that another army might attempt relief, was the use of siege castles. Duke William used them twice in this account, on both occasions successfully. Provided with reliable garrisons, and for all their rapid construction strong enough to avoid being taken by storm by either garrison or relief force, they proved a very economical way of blockading a defence and forcing its capitulation. Unable to get out, the garrison could neither secure supplies, nor damage the surrounding countryside. This form of attack would be increasingly popular in Anglo-Norman siege warfare in the decades ahead.[1] Also of interest is the reference to the duke's destruction of the "new castles" created while ducal authority was in abeyance: in Normandy as elsewhere, only the duke or count could authorise fortification, and anyone flouting this power would see it fully exerted once it had been re-established.

The location of sites of new castles built under authority gives some clues as to their combined functions. Since the ninth century, the river Epte separated the land known as the Norman Vexin from that owing direct allegiance to the king of France, but the conflict between the two may have led to the fortification of Neaufles-le-Château (now Neaufles-Saint-Martin) after 1050,[2] where a later donjon now stands on a motte in an irregular enceinte. Most of the massive defences of the valley date, however, from the period after the dukes became kings of England, when conflict with the French king became endemic. Some of the first mottes anywhere may, however, date from this time and from this area: that of Gaillefontaine has been ascribed to around 1050.[3] The four siege castles ordered by Duke William in his siege of Domfront, further south, have been identified as mottes. The control won by William over territories to the south was also consolidated with castles, such as that of Bellême, which was originally a platform on a ridge, without a motte, which was abandoned and then reoccupied around 1030.

There is evidence of a large number of castle-foundations from this time, and there are an immense number of earthwork remains: in the present-day department of Orne alone, there are 140 mottes and scores of earthwork enclosures. The few stone castles largely coincide with those held by the dukes.[4] Many of the others are, of course, later than this period. The function of some of the small earthwork enclosures, even after archaeological investigation, remains a mystery.[5] The same construction methods applied whether the place was a castle, a refuge or

[1] Le Maho, "Fortifications de siège et 'contre-châteaux'", 181–7, also analyses this phenomenon.

[2] Pépin, *Gisors et la vallée de l'Epte*, 7–9.

[3] Boüard, "Quelques données françaises et normandes", 24.

[4] Decaens, "Les châteaux du XIe et XIIe siècle", 25–9; Louise, "La mise en place du réseau castral", 11–19.

[5] As evidenced by Decaens, "Les enceintes d'Urville et de Bretteville-sur-Laise", 311–68.

an agricultural animal pen. From its known and long history, for example, it is certain that Le Plessis-Grimoult was an important local centre, but it now comprises a large, featureless irregular earthwork enclosure on the edge of the village. The rampart apparently created at the beginning was not more than two metres high, although a stone wall was added later, along with a stone gateway, all of uncertain date.[1] Research into sites in the neighbouring county of Maine had produced similar conclusions, with promontory sites cut off with ditches and enceintes, or hilltops given earthwork ramparts and then a motte, while stone donjons appear early in the eleventh century, not surprising given the county's position as a frontier between Anjou and Normandy. In Le Mans itself, the long-disappeared Tour Orbrindelle was built as a donjon by Duke William against the Roman walls, with a ditch separating it from the town over which it was designed to exercise control. In Brittany, the other area of regular Norman intervention, there is also charter evidence of the emergence of castellanies, compact territorial lordships based on a castle, from around 1050. However, of the immense number of mottes (166 in Finisterre alone), there is little evidence of their date of foundation.[2]

England: 1066 and all that

The conquest of the rich and powerful kingdom of England by the duke of Normandy in 1066 had consequences that reach far beyond the scope of this study. Historians continue to argue over the extent to which pre-Conquest England was already following the same path of development as its continental neighbours, or was instead pulled out of isolation and modernised by the Normans. The Anglo-Saxon realm was strongly influenced by a wide range of overseas interactions, particularly Scandinavian, but also with mainland Europe through trade and the church. The Anglo-Saxon kingdom retained the central administrative structure that in its near neighbours had been devolved to regional princes, but under Edward the Confessor the greater families were creating their own regional dynasties. It would be the provincial dynast from Normandy, who had spent his early years securing his control over his own feudal inferiors, themselves establishing hereditary dynasties, who would recreate the strong central government that distinguished England from Francia, Germany and Italy.

Despite knowing of the castle as opposed to the traditional *castrum*, the rulers of England before William did not follow this path, either as an instrument of central power or as the source and symbol of private, local lordship. The system of royal *burhs* that had been so important in resisting the Vikings from late in the ninth century through to the middle of the tenth played no such part later on. Within the kingdom, although there were frequent conflicts, neither king nor earl turned to fortifications either to establish or to deny power. In the border areas with the Scots and Welsh, the picture was a little different, but here walls and

[1] Zadora-Rio, "L'enceinte fortifiée du Plessis-Grimault".

[2] Fleury, "Recherches sur les fortifications du Maine"; Fleury, "Les fortifications du Maine"; more recently, Jones, "The defence of medieval Brittany", 150–64.

ramparts served the purpose of refuges for local people from raiding parties. By mid-century this picture was beginning to change, and there were recognisable castles in England before the Norman Conquest: but it is not certain whether any were being adopted by the English aristocracy. Excavation and chronicle accounts of the role of fortification in pre-1066 England allow some likely conclusions.

Pre-conquest castles in England

The three areas of interest are sites in Herefordshire and Essex, excavations of some inland sites, and Dover. King Edward granted lands to several favourites of Norman origin. In 1052, the Welsh prince Gruffudd ap Llywelyn ravaged in Herefordshire, and defeated the local people "with some Normans from a castle".[1] "At that time the foreigners had built a castle in Herefordshire", said the *Anglo-Saxon chronicle*. In that same dramatic year, Earl Godwin of Wessex returned from exile, and the Normans fled London, and "some went west to Pentecost's Castle, some went north to Robert's Castle."[2] These sites have been long identified with the castles known as Ewyas Harold (Pentecost's) in Herefordshire, and Clavering in Essex, although actual excavation evidence of date is lacking. Alongside these two, and attested by references both pre- and post-1066, were Richard's Castle and Hereford itself. The powerful motte-and-bailey castles that stood there and were occupied for a long time to come were also built upon after the Conquest, while the motte at Hereford disappeared. The attribution is rational, and it is clear that there were castles in Herefordshire in the 1050s, even if it is not certain what form they took. If the mottes were original features, they were certainly precocious, since few Norman mottes can be dated before 1050.[3] That at Richard's Castle was originally some nine metres high with a top a mere six metres across, and was once surrounded on all sides by a ditch twelve metres wide and three metres deep.[4] The site at Clavering is of a different form, being an oblong mound some one hundred by sixty metres, lying close to the river Stort and protected mainly by deep ditches, which may have been flooded by the river. The site is very low, and no evidence of walls and buildings has been recovered, but it resembles the fairly simple enclosures already observed across the Channel, and it belonged to Robert fitz Wimarc from before the Conquest.[5]

In 1055, following the burning of Hereford and its cathedral, Earl Harold called out a large army and set up a protective border force to shield Hereford while "an earthwork (was) built around the town,"[6] of which no trace remains under the scant ruins of the later wall.[7] The earl appreciated the benefit of creating a rampart

1 Florence of Worcester, *The Chronicle*, 152; *ASC*, 176 (*sub anno* 1052).

2 *ASC*, 173, 181 (*sub annis* 1051, 1052).

3 King, *Castellarium anglicanum*, 1: 205, 210.

4 Curnow and Thompson, "Excavations at Richard's Castle", 105–12.

5 VCH *Essex*, 1: 291–3; RCHM *Essex*, 1: 70.

6 *ASC*, 186 (*sub anno* 1055). The event is also recorded in Welsh records: *ByS*, 71; *ByT(H)*, 25.

7 Turner, *Town defences in England and Wales*, 205–6, and archaeological reports listed there and in King, *Castellarium anglicanum*, 1: 214–15.

to defend the town, and if it was an earthwork, then the urgency of the situation did not allow of any other option. It has been intriguingly suggested that Harold might have been responsible also for the first earthwork on top of the Herefordshire Beacon, where there still stands a fortress of the shape of a ringwork with an attached ramparted court. The higher work has dimensions of approximately 55 × 25 metres, with a ditch three metres deep, the rampart rising some seven to eight metres. The lower court may have been of ancient origin, but the ringwork suggests a medieval date, and the absence of any other historical reference, or of any record as being part of anyone's lands, reinforces the possibility that it was constructed under Harold's order around this time. Its purpose could only have been military, as a look-out position: certainly no one would want to live in such a barren, exposed and inhospitable position. However, absence of evidence of its date means this can be no more than an interesting hypothesis.[1]

Strength is added to the possibility by the existence in the same county of the several castles built by Norman lords. The difference in purpose of the new mottes (if they were mottes) and the earl's work at Hereford was fairly clear: the castle sites were small and were designed to protect and assert their occupants' dominion. It was the castle at Ewyas Harold (Pentecost's) from which the Norman's retainers had caused mayhem and roused the ire of the local population.

They might, if time had allowed, have started a trend. There is some evidence that earth-and-timber dwellings of local thegns, such as those at Sulgrave and Goltho, may have had ramparts.[2] In one sense, a ramparted enclosure for the local thegn might amount to a castle, as in size, interior buildings and defensibility it differed little from continental sites of the period. Yet it is known that fortifications were still a royal prerogative at this time, so suspicion must remain strong that the protection had the purpose of thwarting outlaws rather than enemies. The building complex must also have demonstrated the power and wealth of the owner. A rampart, on its own, does not automatically qualify as a defence. It cannot be said that Anglo-Saxon lords had set about constructing private castles for either for themselves, or on behalf of a king or earl.

Dover

What stood at Dover when William landed? William of Poitiers said that the victorious duke of Normandy marched on Dover, which the English abandoned, despite the fact that it was "fortified by nature and by works", while the poetic account of the Conqueror's victory attributed to Guy, bishop of Amiens, refers to the fortress perched on the side of a hill, and to its population handing over the "keys of the castrum". The *castellum* was situated on a rock adjacent to the coast, at a great height. After occupying it, William spent eight days strengthening the

[1] Argued by Remfry, *The Herefordshire Beacon*, 5, 18, 22–7.

[2] Beresford, "The excavation of the deserted medieval village of Goltho". However, the dating and phasing has been challenged.

defences.[1] He had already obtained Harold's promise to hand over the castle of Dover. However, it is impossible to confirm what stood there. It was not the same as the Roman fort, which had been on low ground, and it is known that the church was moved into "the castle of Dover" early in Saxon times, implying (perhaps?) some existing defence. In the tenth century this church was called "St Mary in castro". The *Anglo-Saxon chronicle* (for 1051) in its account of a bloody fracas with Count Eustace II of Boulogne's visiting armed retinue implies that there was an "upper town", and Florence of Worcester specifically says that these men were occupying the "castle on the cliff of Dover".[2] Was this a *burh*? It is probable. Excavations recovered, under the immense bank around the church thrown up in the thirteenth century, a bank and ditch eight metres wide and five deep, dated to the mid eleventh century.[3]

Castles and Conquest

The work at Dover would have been one of the very first castles built by the conquerors in England, as part of the slow progress made by William following the Battle of Hastings, designed to secure the key points of the Wessex heartland before taking the crown. The prefabricated fort built at his landing point at Pevensey (within the late Roman fort) and the castle shown being built at Hastings can only have been rudimentary timber palisades, despite the Bayeux Tapestry's illustration of a motte being thrown up by workmen:[4] there was only a matter of a few days between the landing at Pevensey and the move to Hastings, and a few more between then and the Battle of Hastings. The *Carmen de Hastingae proelio* speaks of earthworks thrown up on the shore to protect the ships.[5] William's march to London met no fortification until the city itself was reached. Bishop Guy recorded that the city was protected by powerful walls and the river, and that the duke set up camp at Westminster, whence he organised to invest the city and to order the construction of siege engines, including iron-headed battering rams and *talpas* (literally moles), under cover of which, presumably, it was planned to sap the walls. The threat to the (late-Roman) defences and the absence of a viable alternative produced capitulation, according to the bishop.[6]

It was the wiping out of the royal family leaving no competent successor, rather than the presence or absence of defences, that led to the acceptance of a new king with no further resistance in the south-east. But the Conqueror followed the

[1] Guilelmus Pictavensis, *Gesta Guillelmi*, 210, 212 (book II, 27). Guy of Amiens, *The "Carmen de Hastingae proelio"*, 38.

[2] Florence of Worcester, *The Chronicle*, 151.

[3] Coad, *Dover Castle*, 15–18, 19–22. The excavation is dealt with in R A Brown, *Dover Castle*.

[4] *The Bayeux tapestry*, plates 49–50.

[5] Guy, *The "Carmen de Hastingae proelio"*, 10.

[6] Ibid., 42, 44. The account also speaks of the threat to the "proud tower", the identification of which eluded the editors and possibly strengthens the argument that the poem is not based on fact, but elaboration. However, none of the accounts agree on the precise circumstances of the yielding of the city to the Normans.

practice he had adopted in his homeland by setting down castles to be held by his followers, to secure his control of a new kingdom in which he had no roots and no aristocracy connected to the new royal house. The traditional methods of taking rebel-owned castles back into ducal control, then either granting them to new lords or securing the renewed loyalty of the old did not apply. Instead, it was done by replacing the entire ruling class of Anglo-Saxon England, and granting its estates to his own followers, Normans, Bretons, Flemings and other French adventurers, the new lords having to build a castle to guarantee their own safety and the rule of the new king. Thus the new regime replaced the old. By 1086 William owned 20 per cent of England, the church (William's loyal bishops and abbots) 50 per cent, and alien barons the rest, less than 5 per cent remaining in the hands of the old aristocracy. Part of the process was the carving out of new, compact territorial divisions explicitly based on castles, as a means of securing conquest and, in border areas, expansion.[1] The Conquest did not end with the death of King Harold and his brothers, it only began there. It was not finally safe, viewed from the perspective of the time, until the crushing of the last domestic rebellions and the seeing off of the last Scandinavian fleet.

The newly-reinforced defences at Dover were put to an early test, having to fight off a raid by Eustace of Boulogne, which the Norman garrison repelled. In 1067, King William suppressed a rebellion in the west. The citizens of Exeter held out in the old fortress, which had last seen action in 1001 against the Danes. The old Roman walls had been restored to resist the Normans. William besieged it for several days before finally succeeding in undermining the wall and bringing about surrender. This is the first reference to the use of mining that we have encountered in eleventh-century sieges in north-western Europe. The implication is that the Normans' skill in siege warfare included knowing which method to apply to different types of defences. He left Baldwin of Meules to construct a new castle on the craggy site,[2] one of the first to be started in stone and with the fine gate that can still be seen there.

The next trouble occurred in the north and in the Fens. After an early foray into Yorkshire and the establishment of an earl in York, who built one of the two mottes in the city, close by the original Roman enceinte, in 1068, there was a massive rising in 1069, supported by a Danish army, during which this castle was stormed and the garrison of some hundreds wiped out.[3] The key element of this defence may have been the ditch, because the Normans had apparently burnt down the neighbouring houses to prevent the attackers from using them as material to fill the ditch.[4] There followed in response the infamous "harrying of the north" by the Norman king. The conquest was secured by the establishment of castleries, and, one castle not having proved sufficient to hold York, a second

1 The process is expertly described in Le Patourel, "The Norman colonization of Britain".

2 OV, II: 204, 212 (book IV).

3 *ASC*, 204.

4 Florence of Worcester, *The Chronicle*, 173.

motte was thrown up.[1] Then, not for the first or the last time, the waterlogged land around Ely became a fortress resisting the crown, led by the outlawed Hereward. The king first tried to build a causeway across to the natural fortress, but it collapsed under the weight of the soldiers, so instead he left a blockading force. A captive soldier saw how strong this place of "Alrehede" (Aldreth) was, and that the monks were "well versed in warfare". The swamps and waters made it "much stronger than any castle surrounded by walls." On his second attack, William ordered the raising of mounds, creating four circular timber erections "on which to place the engines (*instrumenta*)". The defenders built outworks against them and drove them off by firing the reeds. Hereward the Wake was not to be beaten by such traditional siege methods,[2] and Ely fell through treachery.

William established strong presences along the borders with the Welsh. In Hereford William fitz Osbern carried the conquest to the Welsh and founded the castle at Chepstow around 1070. Norman knights pressed forward across Gwent and into the lowlands of Glamorgan, founding castles as they went. The great motte inside the Roman walls of Cardiff was probably built around 1081, and formed the centre of a network of earthwork castles created to secure Norman lordships in the best land of the former Welsh principality.[3] Robert fitz Hamon, lord of Glamorgan from 1093, pushed his frontier only as far as the river Ogmore (Ogwr) by the time of his death (1107), but in that area were ten castles, several of them major works soon to be rebuilt in stone.[4] Others seized land in Brecon, Montgomery and along the whole of the previous frontier zone, as far north as Chester, where the Norman castle was established in the old Roman city also around 1070. Under the energetic Robert of Rhuddlan, subordinate to Hugh of Avranches, earl of Chester, the conquerors drove across north Wales as far as the coast opposite Anglesey, building castles as they went "in the French manner, so that they could more easily rule the land". The earliest were at Rhuddlan (the motte known as Twthill south of the Edwardian castle), and Deganwy.[5] One was thrown up on Anglesey (Castell Aberlleiniog in Llangoed), one at Caernarfon (traceable within the Edwardian castle site), one in Merioneth, and another at

[1] OV, II: 218, 222. There is no agreement over whether the second castle was contemporary with the first, or built after William's pacification.

[2] *De gestis Herwardi Saxonis*, 25–30; original Latin in *Gesta Herwardi*, 388. Florence of Worcester, *The Chronicle*, 177.

[3] The best study of the castles is RCHM Wales, *Early castles in Glamorgan*, 8–12.

[4] Cardiff, Newcastle (Bridgend), Coity, Ogmore, Bonvilston, Brynwell, Dinas Powys, St Nicholas, Rumney, Penmark, St Fagans. Six were held by simple military tenure and required to provide castle-guard at Cardiff (ibid., 13.)

[5] Attested in OV, IV: 130.

Bangor,[1] in which were collected "knights, foot soldiers and archers"[2] who cruelly plundered and oppressed the inhabitants.

In England, a baronial conspiracy collapsed, leaving the isolated rebel, Ralph de Gaël, earl of Norfolk, to flee to his castle at Norwich in 1075. This is the first mention of the castle, which presumably consisted of the immense motte and a ditched bailey. Despite their isolation from any chance of relief, despite the flight of the earl himself (gallantly leaving his wife to hold the fort), and although the attackers made repeated attacks with "various machines", the castle only surrendered after three months.[3] The *Anglo-Saxon chronicle* says that the planned rising was foiled by the garrisons of the castles built by the Normans, as well as by the reluctance of others to join in, in which case William's approach of dotting the country with defensible residences held by his barons and knights had paid an early dividend.

Two years later, King William's son Robert Curthose had embarked on a rebellion against his father with the support of some lords on the Norman–French frontiers, having failed in a surprise attack on the "tower" of Rouen, which had been secured by the king's butler. The king confiscated the rebels' lands and used the proceeds to pay for mercenaries to crush them. The castle at Rémalard was besieged using a method already familiar: four siege forts were built against it and garrisoned, leaving the king free to act elsewhere.[4] In 1079, the still unreconciled Robert gained the support of King Philip I of France, and was sent to the castle of Gerberoy, on the French side of the Vexin frontier. This place was very strong, with walls and outworks, and Robert used it as a base for ravaging Normandy. William gathered his forces in response, and placed them in the castles along the frontier, to prevent the attacks. He then took an army, in midwinter, and besieged Gerberoy for three weeks. Reconciliation followed.[5] Clearly, the castle at Gerberoy, well-provisioned and strongly garrisoned, could not be taken in a winter siege when the attackers' own supplies were probably scarce. These two events confirm the thorough appreciation of the value of fortifications in warfare, that the Vexin castles were at the king-duke's disposal, whoever happened to be in possession, and that they could be deployed for military purposes at will.

[1] The castle at Bangor has never been identified. Later writers have all followed the theory of the greatest of early British castle students, Ella Armitage, *The Early Norman castles*, 261, that it may be the nearby motte at Aber (Abergwyngregyn), adjacent, as we now know, to the palace of the princes of Gwynedd recently discovered there. I believe the sources are all clear enough in indicating that this castle was in Bangor, which then existed as a town and a bishopric, and that it was in the now much deteriorated earthworks of the late Roman fort on the ridge high above the town, facing Anglesey (Ynys Môn). Armitage herself never visited the site. Early searches focussed on the stone ruins on the opposite hill, which are clearly not those of a castle, although identified as such in local legend.

[2] *Vita Griffini filii Conani*, 117; *A mediaeval prince of Wales*, 70.

[3] Florence of Worcester, *The Chronicle*, 178; OV, II: 316–18; *ASC*, 210–12.

[4] OV, II: 358–60.

[5] OV, III: 108–10.

The rebellions of 1088–1102

The death of William in 1087 led to civil war between his sons William Rufus and Robert. The progress of the revolts and their suppression can be plotted through the seizure and recapture of castles. In Normandy, Robert of Bellême immediately drove the dead king's watchmen from the castles of Alençon and Bellême, as did other nobles at Évreux and Conches(-en-Ouche). Brionne, where the powerful castle has already featured in our study, was awarded by Duke Robert to Roger of Beaumont, and Ivry, built two generations earlier, went to William of Breteuil. Everywhere, castles previously held by royal garrisons became the bases for noblemen to pursue their feuds and plundering. In 1088, there was a great rebellion designed to remove William II from the throne and replace him with his brother, involving many key noblemen and the late king's half-brother Odo, bishop of Bayeux and earl of Kent. In the west, Robert of Montgomery, earl of Shrewsbury, and Osbern fitz Richard led the revolt, in which the walled city of Bath was burnt but Worcester held out, defended by its bishop. But the vital area proved to be Kent, where Odo and his supporters planned to provide a safe landing for the triumphant arrival of Duke Robert. Odo himself fortified himself in Rochester, where there was a castle and old town walls. King William II assembled what forces he could and marched to deal with his uncle.

On the way, he found the castle at Tonbridge held against him by Gilbert fitz Richard de Clare. This would have consisted at the time of the still-standing motte and bailey, presumably with wooden palisades. Not waiting to construct siege equipment, the king stormed the castles inside two days, and Gilbert, perhaps having few armed men to resist an army, surrendered. Hearing that Odo had left Rochester and gone to join his other uncle, Robert of Mortain, at Pevensey, William diverted his army and arrived there by forced marches. Here, they would have faced the still powerful walls of the old Roman fort, with the castle originally created in 1066 forming a separate citadel in one corner.[1] The walled area was about ten acres, with a wide ditch in front of a solid stone rampart up to four metres thick and at least eight metres high, additionally protected by around 15 projecting round bastions. William prepared his engines, but in fact the castle was cut off from any relief and surrendered after six weeks.

Odo promised to arrange the surrender of Rochester but instead joined the defenders. Inside were 500 knights including leading rebels. William had built two siege castles to help him press the town, preventing any access or egress, and eventually, with no sign of help from across the channel, the rebels had to surrender.[2] The site of this first Rochester Castle is thought to be that known as Boley Hill, outside the line of the city walls, while it is clear that the defence in 1088 involved the whole defended site. It was after 1088 that bishop Gundulf (who also oversaw the work on the Tower of London) was charged with the construction of a new

[1] It is believed that the circuit was complete except on one side, where it could readily have been repaired: Peers, *Pevensey Castle*, 4. There is an updated guide: Goodall, *Pevensey Castle*.

[2] *ASC*, 224; OV, IV: 126; Florence of Worcester, *The Chronicle*, 187–8.

castle, inside the walls on the same site as the present remains of a high stone curtain wall (the magnificent great tower dates from the next century). However, at this time, Gundulf has been attributed with the construction of a rectangular, windowless tower of plain design, of which the decapitated stump remains as part of Rochester cathedral. The function of this tower, with its thick, plain walls, remains uncertain. It may have been a bell tower, or it might have played a part in the plan for the defences. It is even possible it was a pre-Norman tower, around which the bishop placed his cathedral.[1]

Peace was short-lived. William reversed roles and attempted to secure Normandy. In 1090, Duke Robert tried to recover and besieged the mighty stronghold at Brionne. There were only six knights in the castle, but they (and presumably other retainers) resisted strongly. The place fell when the besiegers fired arrows with heated tips into the roof of the hall, which caught fire in the summer heat. William I had spent years thwarted by Brionne; this time it fell in three hours.[2] The following year, the duke attacked Courcy. The background to this was the growing power of Robert of Bellême, who is recorded as extending his control over neighbouring areas in southern Normandy by creating castles, including that of Château-Gontier (Orne). Other local barons banded together against him. Robert secured the aid of the duke. During the three-week siege at Courcy, Robert built a siege tower called a "belfry", from which he hoped to oversee the defence and render it untenable. According to the exciting account of Orderic Vitalis, there were fierce daily battles just outside the walls, between the gate and the siege tower, for control of a bread oven. Following such a battle, one of the defenders succeeded in setting alight the wooden tower. Not long after, news came of the arrival in Normandy of King William, and the siege was abandoned.[3] The struggle also entailed conflict with Henry, the Conqueror's youngest son, and the king abandoned a long siege of his brother's castle at Mont-Saint-Michel (1091). In 1092, William visited the borders with Scotland, and established a castle and settlement at Carlisle.[4]

In 1095 William faced a new rebellion in England, by Robert de Mowbray, earl of Northumbria. Marching north with an army, the king besieged Tynemouth Castle for two months before storming it. He then headed for the ancient fortified site at Bamburgh, which appeared impregnable. He therefore resorted to the now familiar tactic of a siege fort, nicknamed *malvoisin* (bad neighbour). The rebel earl was shortly afterwards captured.[5]

That same year Robert of Bellême was again trying out his siege skills alongside the duke of Normandy and the king of France in seeking to punish a rebellious vassal holding the castle of Bréval. Robert deployed a "most ingenious device", wheeled against the fortress and hurling great stones, and showed the attackers

[1] Brown, *Rochester Castle*, 5–7; Hope, "Gundulf's Tower", 323 ff. The most recent theory is proposed by McAleer, "The so-called Gundulf's Tower", 111–59.

[2] OV, IV: 208.

[3] OV, IV: 230–34.

[4] McCarthy, Summerson and Annis, *Carlisle Castle*, 11–12.

[5] Florence of Worcester, *The Chronicle*, 200–201; OV, IV: 282; *ASC*, 231.

"how to make assaults by which the . . . wall and palisades" could be destroyed. This sounds like catapults and sapping or mining.[1] The same Robert de Bellême went on to design the royal castle built at Gisors in 1097, and in 1098 inherited the earldom of Shrewsbury, whence he began building castles in his lands at Bridgnorth and Quatford in Shropshire.[2] These were to feature prominently in the fateful rebellion of the powerful Bellême clan against King Henry I in 1102, after the king had found him guilty of various offences, including having no licence for Bridgnorth. The earl proceeded to strengthen all his defences: the town and castle of Shrewsbury, Arundel and Tickhill, "supplying them with provisions, engines and arms". He also had to hasten to finish the construction of the defences at Bridgnorth and Quatford. King Henry was not one to allow him the time to do so. He isolated Arundel with siege forts, having found it too strong to be taken at once, sent the bishop of Lincoln to besiege Tickhill, and himself attacked Bridgnorth. He threw up another siege fort but also began to construct engines for an assault. The rebels gave up. The defeated earl was banished.[3] Robert was to be one of a number of experts in castle-building and siege warfare whom we will encounter emerging into the light of the records during the eleventh century.

Wales

By 1093 the invaders had penetrated as far as south west Wales, conquering Pembroke and Cardigan, establishing castles as they went.[4] The reaction came in 1094. In the north, Gruffudd ap Cynan assembled a great force which he directed at the castles, both the symbols and the direct instruments of the oppressive invaders. He assaulted Aberlleiniog on Anglesey for several days. The "Franks" defended themselves from the walls and tower with javelins and arrows and stones hurled from "balistis", but the Welsh won, killing the leader and seneschal and the garrison of 64 and burning the castle. He followed up with attacks on the other castles in Gwynedd that were similarly successful, burning them all.[5]

Elsewhere, the Welsh succeeded in driving back the newcomers, but could not take the castles (1096, according to the *Brut y Tywysogyon*), although in the west only Pembroke Castle managed to hold out, which had to resist a long blockade that only ended when the commander managed to dupe the Welsh into believing that he had ample supplies when in fact he was desperate. This was likely no more than a ditch and palisade across the promontory on which the later stone castle now stands. At Rhyd-y-gors, the new castle, a motte and bailey in design,

[1] OV, IV: 286–90.

[2] OV, V: 214, 224.

[3] OV, VI: 20; Florence of Worcester, *The Chronicle*, 211; *ASC*, 237. The episode is analysed in Hollister, "The campaign of 1102 against Robert of Bellême".

[4] *ByS*, 83, 85; *ByT(H)*, 33; *ByT(P)*, 19; *Annales Cambriae*, 29.

[5] *Vita Griffini filii Conani*, 119; *A mediaeval prince of Wales*, 70–73. The original Welsh has not ballista, but *magnaleu*, which must translate as mangonel. The 1866 text is a sixteenth-century Latin translation of a Welsh version of a lost original mid-twelfth-century Latin version! The date is also disputed.

was abandoned.[1] This would be the future Carmarthen, the key castle for the conquest of west Wales (although probably not on the same site). The return of the king, with an army greatly superior to anything the Welsh could muster, did little to alter the position. The Welsh, with few fixed possessions, simply melted away, to return to ambush the lumbering Anglo-Norman columns, or wait for their departure. It would not be long before Welsh leaders began to emulate their neighbours in castle building, but it did not happen in this century.

The pattern of castle building from the Conquest to Henry I

The benefit to the student of the absence of all but a handful of castles in England before 1066 is that, unlike in mainland Europe, every site must have been built after this date. The difficulty is that, apart from the castles mentioned in the record as being ordered by the king, of which there are a good many, or mentioned incidentally in the Domesday book (of which there are not many) or in early charters, the great majority of earthworks, whether they are mottes or plain enclosures (known as ringworks) cannot be dated at all. The probability is that very few earthwork castles were built after the mid twelfth century, although the position in Wales was different, and in Scotland different again. That still leaves a period of almost a century in which undated earthworks could be placed. Historical logic has always suggested that the main building periods were immediately after the Conquest, and in the "Anarchy" of Stephen's reign. There are around a thousand such sites, now listed comprehensively,[2] and some of them appear on a map in clusters, although such maps can be very misleading given the possible time span of construction and occupation. Is there a pattern?

Castles could have, simultaneously or in sequence, many purposes. The purpose relevant at the time of founding would determine the site. The resources, available skills and local geography and geology might determine the type of building. It was a tradition of all European countries that fortification was a royal monopoly, and it is hardly conceivable that so powerful a king as William I would have allowed anyone to flout this right. But this need not mean anything more in practice than that anyone wanting to build a castle needed the king's formal permission. In the period immediately following 1066, it was rather more probable that William would encourage his followers to secure their conquered lands with castles than that he would object. In all likelihood, most of the castle sites in southern, and inland England can be assigned to the decades following the Conquest, and the absence of later historical record for many of them further suggests that they may have had only brief occupation, being rapidly abandoned when the settlement was secured.

It is also possible that not every site that we now define as a castle was seen as such at the time. The references to ducal control of castle-building in the *Consuetudines et iusticie* of 1091, along with the compilation of customs and laws called

[1] *Annales Cambriae*, 30; *ByT(P)*, 21; *ByS*, 89. An excellent reconstruction of the whole period is given in Lloyd, *A history of Wales*, II: 403–8.

[2] In King's indispensible *Castellarium anglicanum*.

the *Leges Henrici primi*, both indicate that it was only buildings of a certain size of defence that required royal or ducal licence, although they also noted that "castellatio sine licentia" would place a man at the king's mercy. The first text limited creating a ditch in open country to one spade's depth and one palisade, without battlements or wall walk. The text also prohibited anyone from building a castle on a rock or island, nor were they allowed to deny the ruler the use of their castle at will. The second text referred briefly to the king's permission being required for "castellatio trium scannorum", uncertainly interpreted as meaning "three embankments".[1] The simplest interpretation would be that the ruler retained the right to licence constructions that were defensible, and to require that they be made over to him when needed. The criteria for defensibility seem to have been both size (for example the depth of a ditch) but also whether a palisade was able to offer an active defence, with embrasures and a means of defending the top, or was simply a fence. So it may be that a large number of the small and now decayed earthworks around the country were too small to require permission. It is also necessary to stress, however, that the perception that castles were seen as a threat by the crown so were only reluctantly licenced is a myth of later historians.

Another (related) myth was that William would not permit any fortification within twenty miles of London, where the Tower was being raised as a royal palace and fortress. It has always been known that there were also two privately held castles in the city, Baynard's Castle and Montfichet's Tower, both disappeared but the site of the latter recently recovered. The two were very close to one another at the later Blackfriars, and at least one had a motte, and one a large tower.[2] Another, referred to as "Ravenger's little castle" in the records, may have been near the Tower. Nor were these the only castles within the allegedly prohibited zone. While the small motte at Cranford might not have met the definition of a castle, as the remains are too decayed to be sure, that at Ruislip certainly did, being a standard motte and bailey. Since it was given up to Bec Abbey in 1090, it must have dated from soon after 1066.[3] In the other direction there is a very early stone enceinte at Eynsford, with a broad wet ditch, possibly dating from around 1090. The wall still stands nine metres high and is 1.8 metres thick.[4] There is absolutely no reason why the king should have so limited castle building, nor do the facts sustain the assumption.

Urban castles

There is a pattern in the number of castles founded in the towns. More than half of the recorded foundations (those listed as ordered by the king) were in existing towns, and it is apparent that while no attention was paid at all to the state of

[1] "The Norman *Consuetudines et iusticie*", 282; *Leges Henrici primi*, 108–9. See also Yver, "Les châteaux-forts en Normandie", 60–62.

[2] William fitz Stephen, *Description of London*, 49; Watson, "The Norman fortress on Ludgate Hill", 335–41. The sites are clearly marked in Schofield, *The building of London*, 39.

[3] RCHM *Middlesex*, 107–8.

[4] Rigold, *Eynsford Castle*.

town walls (if they existed), every effort was strained to raise a castle. Some 40 out of 108 towns were subjected to castles. As in London, in the first flush of urgency to provide a protected area for the new rulers, a corner of the town could be cut off by a ditch and palisade. Where the original castle was outside the walls (as at Canterbury or Rochester) it was subsequently relocated inside. In many cases a motte would follow. In others there may have a stone tower from the start (Oxford and Exeter). At the outset, existing property rights were simply ignored and houses in the way were just pulled down, as happened at Lincoln. It is noticeable how the foundations were recorded by the chroniclers as the king's army marched across the country. The castles were either where there has been trouble (like Exeter), or they were on major cross country routes. Their construction was ordered by William even when it was entrusted to other Norman lords to carry out and hold. Rapidly, of course, they acquired other functions, especially as centres of administration, but the military may have dominated as the first sod was turned.[1]

Castleries

One of the features already mentioned was the creation in several places of "castleries": the Sussex Rapes, for example, were home to the earliest. The Conqueror created these compact lordships (distinct from the normal pattern of granting his followers sometimes widely-scattered estates) after 1066 and at Domesday, all of Sussex was divided into five consolidated lordships from the coast to the Weald. Each was based around a single major castle (Hastings, Pevensey, Lewes, Bramber and Arundel, although there was also a motte at Chichester). The Isle of Wight was held by the same Fitz Osbern as held Hereford, and he began the castle at Carisbrooke. The holders were all close followers of the Conqueror, and many were related to him. The pattern was broken up with deaths and rebellions, and the carefully constructed lordships were not preserved. It is as clear as can be that this was a deliberate policy of installing reliable supporters in the equivalent of a military district along the coast where William himself had made his invasion. As elsewhere, it was short-lived.[2] Castleries were directly linked to the process of establishing the new regime. Reliable followers were installed as lords in the Welsh borders, with the formation of marches around Hereford, Shrewsbury and Chester. Compact lordships also accompanied the conquest in northern England, as at Holderness (Skipsea Castle) and Richmond (with its early stone great tower and stone curtain wall).[3] The great lords exercised control from their major castles, but they relied on the service of their own feudal inferiors, who were themselves granted lands to support them.

In the marches, these would also lead to new castles. In the Vale of Montgomery, a study has suggested that most of the 12 motte-and-bailey castles, all

[1] Drage, "Urban castles", 117–24, gives an excellent summary of much research. See also English, "Towns, mottes and ring-works", 45–54.

[2] This much-discussed subject is well summarised in Mason, *William the first and the Sussex Rapes*, 3–19.

[3] Pounds, *The medieval castle in England and Wales*, 33–43, gives a clear description.

of them small and the result of sub-infeudation, arose from the occupation of the valley by the earl of Shrewsbury between 1075 and 1102.[1] A massive castle-building programme took place in the Golden Valley of Herefordshire. The area, devastated in war with the Welsh, was described as waste in 1066, but was significantly settled by the time of Domesday. There were castleries at Ewyas Harold and Clifford and altogether 16 motte-and-baileys.[2] In the context of a frontier regularly fought (or at least raided) over these castles retained a military use for a much longer time than the inland sites. Here, lordships were named after the castles, and marcher lords developed their own customs and rules.[3]

Castle building and castle-guard

The labour involved in throwing up a motte was considerable, and while the invading soldiers must have carried out the work at Pevensey and Hastings, they cannot possibly have thrown up the hundreds of fortifications that were erected in the first few years after the Conquest. The only solution is that the Normans invoked the traditional obligation on the peasantry to work on public fortresses (the *burhs*) and pressed them into service as castle-builders. The *Leges Henrici primi* maintained the traditional right of the crown to this and the other duties on subjects of the Anglo-Saxon crown.[4] Exactly how this translated into a duty to build a knight's fortified residence is not obvious, but there is no other answer. As to the designers, it may be assumed that Norman and French lords knew what a castle should look like and could give instructions to recreate what they already knew. Stonemasons with experience of working on cathedrals could certainly create the stone towers, gates and walls in the few cases where these were started directly after the Conquest.

No one knows how many invaders there were, but even 10,000 settlers would have represented a number entirely inadequate to provide a force able to guard all the new castles in the event of serious attack. Instead, the defence of the significant castles was ensured through a rota system known as castle-guard, in which tenants had to take turns to defend their lord's castle. The castleries were also a mechanism to ensure that the chief castle was permanently guarded, often on a cycle of 40 days. Examples are known from the Sussex Rapes, Richmond, Caerleon, Richard's Castle, Ewyas Harold, Clifford and Montgomery (later on). Richmond, for example, had an assignment of 42 knights in summer, but only 26 in winter. Larger numbers were involved at the great royal castles. Dover had an immense garrison of 170 knights drawn from nine baronies, Windsor 73 from four

1 King and Spurgeon, "The mottes in the Vale of Montgomery".

2 G Marshall, "The Norman occupation of the lands in the Golden Valley".

3 See R R Davies, "Kings, lords and liberties", for a good introduction. The debate as to when some of these lordships became palatinates (that is, their lords exercised powers elsewhere reserved for the crown) need not concern us here.

4 *Leges Henrici primi*, 108.

baronies.[1] What happened to the small motte-and-bailey castles in the marches if their lord was on castle-guard duty elsewhere and there was an attack, is not clear. Certainly it would not be long before this onerous burden (as with other feudal duties) was commuted for cash or negotiated away, although there is evidence for its survival, as a duty of care and maintenance, in the Welsh marches well into the thirteenth century.[2] At the Conquest it was the normal system in both England and Normandy.

The Normans in Italy

Meanwhile, as the Normans established themselves in England, their landless cousins and younger brothers carved themselves a kingdom in southern Italy. The precise details of their first arrival are shrouded in myth, but certainly during the early eleventh century increasing numbers took service as mercenaries of the local princes in their struggle, at this time directed at the resurgent power of Byzantium, and they turned the tables. Basil II reinforced his local commander, Basil Boioannes, and in a bloody battle at Cannae in 1018 crushed the Lombards and their Norman auxiliaries. The re-established Byzantine control was secured by the construction of a new fortress-town at Troia,[3] blocking the pass into Apulia through the Appennines. The Normans found for themselves a secure refuge and, their numbers continuing to grow and their fighting reputation undented by Cannae, they found a ready market for their warlike qualities as paid servants of all the local principalities — including, ironically, as a garrison for Troia when Boioannes' army was recalled by the emperor.

Meanwhile, the fall of southern Italy to Constantinople invoked the fear of the Pope and led to the rare sight of an imperial German army in the south in 1022. The new walls of Troia, on its great spur dominating the surrounding plain, held out against siege for several months until the climate and disease persuaded Henry II to abandon the siege. The emperor had ordered machines built, but these had been burnt by the garrison making sorties, so Henry next ordered them made stronger, and covered in raw hides. The walls, however, had "stubbornly resisted" the machines, suggesting that these included battering rams.[4] The Norman garrison earned the Byzantine commander's gratitude. In 1026, when Boioannes recovered Capua after a long blockade, Normans were to be found on both sides. In 1029, the Norman band secured for itself the town and land of Aversa, with its leader, Rainulf Drengot, married into the local aristocracy. From now on they would not look back. Other bands set themselves up in the territories, living by brigandage, and safe from retribution, the successors of the great Basil II at Constantinople having no interest in their Italian province. In 1035, the sons of

[1] Stenton, *The first century of English feudalism*, 192–5; Hollister, *The military organization of Norman England*, 138–60.

[2] See Suppe, "The persistence of castle-guard", 201–21.

[3] *Chronica monasterii Casinensis*, 261 (book II, chap. 51).

[4] Glaber, *Historiarum libri quinque*, 101–3.

Tancred of Hauteville began to arrive from Normandy. In 1038, when Emperor Conrad II returned to the south to end the depredations of the ruler of Capua in favour of Gaimar IV, prince of Salerno, the Normans managed to be on the winning side, and Rainulf had himself made count of Aversa, invested by the emperor himself.[1]

In 1042, Melfi opened its gates to the Normans, despite its strong hilltop site, although it did not have high walls.[2] After a victorious march through the province, the Lombard prince Argyrus, the forces of Bari and other cities, and the Normans laid siege to the Byzantine garrison of Trani. The attack lasted more than a month, and the Normans constructed a wooden siege tower "of a kind never seen by human eyes in modern times".[3] On this occasion the Byzantines brought the siege to an end by buying off Argyrus. The novelty of the siege tower to witnesses clearly suggests that it was not part of the armoury of any of the local powers, whereas we have seen it described as part of Fulk Nerra's repertoire in the 1020's. There is an implication therefore that the technology was imported along with some of the Norman adventurers. Another siege method already identified with western Francia was also deployed by the Normans when they besieged Capua in 1052/3 and again and again over the coming years. Properly defended, Capua was a very strong city, and in recognition of the difficulty of taking it by existing methods of assault, the Normans set up three siege castles.[4] Shortly after, the fortunes of the Normans took several further decisive turns. After capturing the Pope on the destruction of his army at Civitate (now San Paolo di Civitate) in 1053, the decisive step was taken with the alliance forged with the new pope in 1059, whereby the former mercenary cutthroats were recognised as prince of Capua (Richard I Drengot) and duke of Apulia, Calabria and Sicily (Robert Guiscard), although he had never set foot in the island.

Robert Guiscard was obviously keen to claim his new realm, but before dealing with Sicily had to complete the conquest of Byzantine Calabria. This involved a full scale siege of Reggio in 1059, having first secured the surrender of Taranto and Brindisi. The full range of techniques were called upon, with earthwork *chasteaux* thrown up, and, presumably having found a means of approach to within range, he had made "trebuc et autres engins de guerre".[5] The word "trebuchet", unknown in the eleventh century, is the chronicler's translator's description for a stone thrower, using a word familiar to his own audience. As with the siege tower, the stone thrower is not a weapon encountered regularly in this part of the world, although the Byzantines of course had access to it. Reggio fell, the inhabitants terrified by the preparation of the machines. In this same year 1060, as the Normans carried their first attack in their own right on Muslim Sicily, and found that despite deep divisions between the local emirs the several rulers had no inten-

1 For a lively account, see Norwich, *The Normans in the South*.

2 Amatus of Monte Cassino, *Storia de' Normanni*, 77. This source was preserved only in a medieval French version of the original Latin.

3 *Annales Barenses*, 56.

4 Amatus, *Storia de' Normanni*, 188. It is also possible that "siege tower" was meant.

5 Ibid., 228.

tion of letting the island easily out of their hands, a small Byzantine expedition tried to recover Apulia, and hung on in Bari. In 1061, a Norman force had the great good fortune to enter Messina unopposed, and rapidly set about the strengthening of its walls to secure their foothold on the island.

The conquest of Sicily then unfolded over the coming years, with castles and towns surrendering to Norman siege, and the invaders throwing up castles themselves as they went. In 1071 Guiscard's brother Roger (soon to be count of Sicily) attacked the port of Palermo by land and sea, and blockaded it for five months. Towards the end, scaling ladders were prepared and, a diversionary assault being launched on one side of the town, the escalade succeeded on the other side. The main Muslim city of Palermo proved a very hard obstacle. Attacked over a prolonged period, it was finally carried by escalade in 1072 — we are told that 14 ladders were constructed for the task. The defences of the captured city were then reinforced and castles with "robust walls" erected to provide further protection.[1] New castles were built to restrain Muslim attacks from Catania and other unconquered lands, such as at Paternò and Mazara (del Vallo).[2] In 1079, Taormina was attacked through the construction of no fewer than 22 siege forts to blockade it. When Agrigento fell to the count in July 1086, following a siege of three and a half months, after he had deployed numerous engines against the town, he ordered the construction of a castle with "ditch, towers and outworks (*propugnaculis*)" in order to secure his conquest.[3] In 1088, only Noto and Butera continued to resist, and Roger took the latter by deploying once again his siege machinery.[4]

Byzantine forces had been withdrawn to confront the impending threat posed by the arrival of the Seljuk Turks in Asia Minor. Seizing the moment, Guiscard moved a great force to attack Bari, the main fortified base of the empire for centuries and its location, on a peninsula, not long strengthened. An epic siege of enormous length ensued. The Normans set up all their weaponry against the land walls. They built wooden shelters to cover them, constructed a wooden tower to overlook the wall, and ranks of stone-throwing artillery, and exerted every means to create a breach including battering rams. The strength of the defences was too great for them, and the siege towers were being burnt down regularly by the garrison. The defenders also had artillery and appear to have used it to good effect. The long blockade reached its culmination when the Norman fleet won a naval victory, preventing the resupply of the city by sea. A few weeks later, in April 1071, Bari surrendered, and the rule of the Eastern Roman Empire in Italy came to an end.[5] It was then the turn of Salerno, the last remaining independent city-state under its prince Gisulf II. He faced the combined Norman armies, now composed of Normans, Calabrians, Apulians and Sicilian Muslims as well, early in 1076.

[1] Ibid., 279; Guilelmus Apuliensis, *La geste de Robert Guiscard*, 182–3.

[2] Malaterra, *De rebus gestis Rogerii*, 19 (book I, chap. xxv), 53 (book II, xlv), 66 (book III, xv).

[3] Ibid., 87 (book IV, v).

[4] Ibid., 92 (book IV, xii).

[5] Amatus, *Storia de' Normanni*, 248; Guilelmus Apuliensis, *La geste de Robert Guiscard*, 500–502; Malaterra, *De rebus gestis Rogerii*, 48–53 (book II, xxxix–xlv).

The chronicles tell of a vigorous attack using all kinds of siege equipment, including, again, siege towers. But the weapon that succeeded was hunger. The starving people opened a gate to Robert Guiscard in December, after six months' blockade. Gisulf withdrew into the citadel that towered over the city, and from its walls his soldiers hurled down rocks, which destroyed the duke's stone throwers. Cut off and without anyone from whom relief could be expected, this last resort also gave up in 1077. The Norman installed a garrison in the citadel and also built a new castle in the town, doubling his control of Salerno.[1]

In a few decades, the Norman adventurers had created for themselves a rich and diverse state and had defeated in the process both the eastern and western empires, the papacy and the Muslim rulers of Sicily. It was an achievement every bit as astonishing as that of their cousins who had simultaneously acquired the kingdom of England. In the long wars, they had been required to show their mastery of every form of warfare, and in siege warfare had demonstrated, with their deployment of siege towers and artillery, that they were possessed of the most modern techniques. At the same time, the limitations of those techniques when faced with both strong walls securely sited, and a garrison and population willing to hold out, had again been illustrated. With the consolidation of their control of former Byzantine and Muslim states, they would also inherit the world's most advanced building skills and crafts as would soon be demonstrated in their cathedrals, palaces and castles.

Most of the Normans' sieges had been of cities or towns. But much of their own work of fortification consisted of building castles, including within the towns they captured. Castles would act as a means of controlling the conquered towns and territories as well as providing a defensible residence for the new lords. In Sicily, there would be at least 60 by the twelfth century, many of them, such as Adrano and Paternò, based around stone great towers. The existing building styles (and naturally therefore the traditions of the masons and craftsmen on whom the conquerors would have to rely) quickly influenced the type of buildings created, with a strong Muslim influence discernible in such structures as the palace at Palermo.[2] Some at least of the castles built by the Normans during their conquest, however, when speed would have been important, were mottes, of which a few remain,[3] including one belonging to Guiscard himself in Calabria (at Scribla near Cosenza, the only motte in that province[4]). They also brought with them the donjon or great tower: the chronicles tell of the founding of such towers, for example at Messina and Gerace, and the great speed with which they were

[1] Guilelmus Apuliensis, *La geste de Robert Guiscard*, 188–9; Amatus, *Storia de' Normanni*, 353 ff.; *Chronica monasterii Casinensis*, 422.

[2] Maurici, *Castelli medievali in Sicilia*, 194–8; Houben, *Mezzogiorno normanno-svevo*, 159–63.

[3] A J Taylor, "Three early castle sites in Sicily".

[4] Houben, *Mezzogiorno normanno-svevo*, 159; ascribed to 1046–50 by Malaterra, *De rebus gestis Rogerii*, 14 (book 1, xi).

apparently constructed.[1] They represent the link between the Norman families following Guiscard to conquer Sicily and William the Bastard in England.

Taking the attack to Byzantium

The Normans next turned their attention to the empire they had expelled from Italy. Constantinople was by now dealing with the worst crisis it had known, with Asia Minor overrun by the Turks. Not swayed by concern for their fellow Christians, the Norman dukes seized the chance to strike across the Adriatic. Their target was the city of Durazzo (Dyrrhachium, now Durrës), with its strong brick-built walls. Duke Robert set his camp about a bowshot from the walls. Once more the siege tower was built. This is the first time that it is clearly stated that the tower was designed not only to allow attackers to overlook the walls and direct missiles down to clear the wall tops, but also to allow direct access through the siege tower onto the wall itself. Anna Comnena describes how the garrison built their own wooden tower inside the wall, and constructed a strong wooden beam to push out from the wall-top to block and prevent the Normans from opening their tower's door, and succeeded in thus obstructing the assault. The tower was then set alight with the naphtha and pitch made ready on the wall, along with catapults to project it, and by a simultaneous sortie made from a postern gate to attack the base of the tower with axes. In the end, the city fell not through direct attack, but by surrender early in 1082, after the new emperor Alexius I Comnenus (r. 1081–1118) had failed to bring relief to the garrison, his hastily assembled army of mercenaries and conscripts proving no match in battle for the Normans, even though it included a contingent of Varangian guards who were Anglo-Saxon exiles no doubt thirsting for revenge against these Normans.[2]

Around the Byzantine empire: arrival of the Turks

The decline of the empire after Basil II was not precipitate, but the period of short reigns and weak rulers that ensued was to last half a century. Conflict continued on the empire's southern and eastern frontiers, and the rulers of Armenia, and of the principality based on Edessa (Şanlıurfa), were to be the centre of unwelcome attention from all parties during the century. There was war around the city of Edessa in 1031/2 when a new ruler tried to evict the Muslims. During the fighting, a Muslim catapult destroyed the wall of the cathedral of St Sophia by

[1] Malaterra, *De rebus gestis Rogerii*, 76–7 (book III, xxxi, xxxii). Both works are assigned to around 1080.

[2] Anna Comnena, *The Alexiad*, 142–3; Guilelmus Apuliensis, *La geste de Robert Guiscard*, 250–52; Malaterra, *De rebus gestis Rogerii*, 74 (book III, xxvii). Anna's account is not contemporary, but the detail she provides suggests a direct source. It is also evident that the nature of this particular design of siege tower did not come as a surprise to the garrison, implying that it was not new.

a long bombardment.[1] Shortly after, a stronghold at "Berkol" fell to the Muslims, the garrison being "benumbed by wine", but the Byzantines returned the following year with siege engines and destroyed the walls.[2]

One of the most detailed accounts of warfare involving fortifications stems not from an external invasion, but from an attempted coup in 1047 against the emperor Constantine IX (r. 1042–55) by Leo Tornicius, Boioannes' predecessor as Byzantine commander in Italy 30 years before, who managed to convince the armies of the western provinces to support him in an attack on Constantinople. Constantine, lacking much of an army himself, concentrated on the city's great defences, "building up the parts which had been allowed by negligence to fall into disrepair, and planting his stone-throwing machines thick on the ramparts."[3] The pretender settled down outside, assembling his own artillery, and throwing up his own rampart. Overnight, the emperor collected some armed men and got them to dig a ditch outside the walls, but this was so shallow as to prove no obstacle, and Leo missed his chance to enter a city with open gates and no defenders. The following day, the defenders were hurling "great masses of rocks" from the engines but they fell short. "Then those who were working the machine pulled back the sling farther than usual and shot one of the biggest stones at Tornicius himself". It missed, but he fled in panic and his force broke up.[4] Thus the defences of Constantinople, given an edge with artillery and people who knew how to operate it, had saved the emperor from an army.

Rise of the Seljuk Turks

The conversion of the nomadic Turkish tribes to Islam during the tenth century and their increasing recruitment as soldiers and officials for the 'Abbasid caliphate had far-reaching consequences. As the rule of the caliphs atrophied, dynamic Turkish forces assumed control. The Turkish Ghaznevid dynasty was deeply involved in the east, while another tribal grouping, the Oghuz (Oğuzlar), whose leaders included the descendants of one Selçuk (Seljuk), seized power in Transoxiana, taking Merv in 1036, then seized Iran. Taking the side of the caliph in conflicts with supporters of the Fatimid anti-caliphs of Egypt, their leader Tuğrul (grandson of Selçuk) was entrusted with the new office of sultan by the 'Abbasid caliph (1038). Within a few short years new Turkish dynasties would rule from the Yemen to Armenia, from Syria to Samarkand. While their leaders became cultured and civilised, and took on board the learning and institutions of the lands they now ruled, before developing their own civilisation as Byzantine Asia Minor became Turkish Rum, their armies continued to consist of Turcoman nomads, who were skilled in battle but often happier raiding settled populations than living in towns or living by agriculture. While the Seljuk Turks quickly established

[1] Matthew of Edessa, *Armenia and the crusades*, 53; and see Aristakēs Lastivertc'i, *History*, 36.

[2] Aristakēs, *History*, 41.

[3] Psellus, *Fourteen Byzantine rulers*, 209–10.

[4] Ibid., 215–17.

themselves in, and set about maintaining, the towns and cities of Asia Minor (so that their new capital of Konya (Iconium) became a strongly defended and richly endowed centre), the Turcomans remained largely nomadic.[1] They must have employed skilled siege operators to carry out their attacks on walled towns.

Detailed accounts remain of their attack on Manzikert (Malazgirt) in 1054/5. It began with an attempt at mining, but the garrison dug a countermine and captured the miners. The sultan then sent to Baghdad for artillery, including a "terrible and awe-inspiring machine" weighting 15 *adil* (7,500 pounds) and requiring no fewer than 400 men to operate the ropes. The first shot knocked down three watchmen. Inside, a priest then set up another catapult opposite "and with its first shot struck the enemy machine and smashed its tie beam". The Turks tried again, but a courageous sortie from the town by a rider with three bottles of naphtha succeeded in burning the engine. Another account suggests that the Muslim engine was so powerful that, before its destruction, its missile caused a breach to appear in the wall. Since this is the same account that suggests that the defenders' machine was so accurate that its missile could actually hit the besieger's rock as it was launched, one must suspect a little invention here![2]

The Seljuks' engagement in Asia Minor brought them into contact with Byzantium's Armenian and Georgian neighbours, who were themselves bordered by the Muslim principalities of the Caucasus. Basil II had sought to secure this flank of his domain in various conflicts, alternating with alliances, with his fellow-Christians, while King Bagrat III of Georgia (r. 980–1014) had also to fight the Muslims. This conflict involved using siege weaponry that demolished defences, which is described as a stone thrower using the same terminology as Matthew of Edessa. Stronger towns like Tbilisi were starved into submission over two years from 1036 by Bagrat IV (r. 1027–72). At another siege, of a rebel town, the Georgian used his artillery to hurl arrows over the walls. Despite these skills, David II or IV the Builder (r. 1089–1125) had to submit to the Seljuks who had ravaged the country, such that "there were no people except in the citadels". Later, David, like his ancestor Bagrat, set about restoring their rule by capturing a long list of fortified towns, so clearly siege skills were familiar to them both.[3]

In the Caucasus, the key city of Derbent on the Caspian changed hands repeatedly, with its citadel becoming the base for rival rulers. In 1063, a new wall was added to enclose the suburb growing up around Ganja, provided with "strong gates" and a deep ditch. In the neighbouring principality, in 1066, the Turks invaded, and plundered Shirvan, but this seems to have been a raid rather than an attempt at conquest. In 1068, the garrison of Derbent only now strengthened their

[1] Cahen, *The formation of Turkey*, is a masterful summary of the emergence of Turkish Asia Minor.

[2] Matthew of Edessa, *Armenia and the crusades*, 88–9; Aristakēs, *History*, 102–7; *Histoire de la Géorgie*, I: 328, Ibn al-Athīr, *The annals of the Saljuq Turks*, 93, where the attack is described as a raid. Ibn al-Athīr (who is a much later compiler) also describes Tuğrul's entry into Baghdad in 1055, and how he deployed his mangonels when marching on Tikrit in January 1057, leading to the surrender of the town (p. 106).

[3] *Histoire de la Géorgie*, I: 297–305, 318 (Tbilisi, and new walls), 320, 351–60; Ibn al-Athīr, *The annals*, 25 (Tbilisi).

citadel with a moat, but in 1071 a force sent by the sultan Alp Arslan (r. 1063–72) took over the city and demolished the "middle wall". The remaining walls of the town are substantial, being 18–20 metres high, with numerous towers, all made of large stone blocks that lend some weight to the suggestion that they date back to the Persian emperor Khosrau I Anushirvan (Chosroes; r. 531–78). Inside, remains of a mud-brick wall were traced by excavators, suggesting the middle wall mentioned in the chronicles. A remarkable barrier wall, 40 kilometres long, and of hewn stone, ran from Derbent inland, provided with crenellated square towers.[1]

In 1064 the Armenian city of Ani, strongly sited on a steep hill, fell to the Turks of Alp Arslan: Aristakēs tells that this was after a section of the wall had been demolished with a catapult, whereas Matthew of Edessa attributes its fall to the dissension in the garrison. The Muslim sources support Aristakēs' version. They indicate that the sultan had pressed a close attack, building some kind of wooden siege tower, and driving the defenders from the walls with archers and stone throwers, which allowed his men to mine the wall. A sudden collapse of a stretch of wall permitted the Turks to storm Ani. In 1070/1, Alp Arslan attacked Edessa again. After eight days of futile assault he turned to the engines. Catapults and machines were set up while the gardens and vineyards were pulled down in order to provide the timber to fill the town's ditches. A wooden tower was constructed standing on ten wagons, but it collapsed. He then turned to mining, digging seven tunnels under the moat. The defenders countermined, and after some 50 days the Turks abandoned the siege.[2] Following the Battle of Manzikert, Turkish forces took numerous towns: Erzerum (formerly Theodosiopolis), Diyarbakır (Amida), Mardin (Marida), Malatya (Melitene) and many others seem to have fallen without much resistance. In truth, the Turkish commander does not seem to have regarded his victory as being as significant as later historians have done. His family's own chief interests did not lie in Asia Minor. In 1073 he was succeeded by his son Melikşah I (Malik Shah; r. to 1093) who made his capital in Isfahan, where he ordered the building of many new structures including castles for keeping his treasure. Meanwhile, his brother Tutuş was engaged in occupying Syria: in 1077 he took Aleppo and occupied Damascus, after laying siege to the city. In 1083 Harran was recaptured by Sharaf al-Dawla, who was then himself besieging Tutuş in Damascus, and the former was victorious when his engines demolished part of the curtain by bombardment. In 1086 the sultan took Edessa and Manbij. In 1089, the Seljuks again had to fight to recover control of Transoxiana, where they took Bukhara, and Samarkand after a long blockade.[3]

The accounts, penned by contemporaries who were in some instances eyewit-

[1] Minorksy, "New light on the Shaddādids of Ganja", 19–25; *A history of Sharvān and Darband*, 30–40, 86–9 (notes on the Russian excavation and study of the walls).

[2] Matthew of Edessa, *Armenia and the crusades*, 103 (Ani), 130–32 (Edessa); Aristakēs, *History*, 164 (Ani); Ibn al-Athīr, *The annals*, 154–5 (Ani). Byzantine historians, however, back the story of dissension being the cause of the downfall: Attaleiates, *Historia*, 80–82. Attaleiates also described the great strength of the natural and man-made defences.

[3] Nīshāpūrī and Rashīd, *The history of the Seljuq Turks*, 53–61 (a twelfth-century Persian source); Hasan, "Munejjim Bashi's account", 445–55 (a late source); Ibn al-Athīr, *The annals*, 210 (Harran), 240 (Samarkand).

nesses, provide fascinating detail on the skills available in the second half of the eleventh century. They show both sides familiar with the full range of siege equipment and techniques. However, despite the access of the Turks to this equipment, and to mining, the cities under attack often held out. Clearly, the technological sophistication available to the defenders was enough to cope with the means of attack. There are also suggestions that the artillery had become more powerful than before, with several references to the destruction of stone walls. This evidence forms part of the case made by P E Chevedden for the arrival at this time of a new form of catapult, a "hybrid trebuchet", in which a counterweight was added alongside the manual propulsion to produce a more devastating weapon. However, the best that can be said from these examples is that the contradictions between the accounts mean that this remains an open question. More will be said on this subject in Chapter 9. Also of interest is the rare reference to a siege tower at Edessa, albeit one where the method of construction was not strong enough. Had Alp Arslan heard of the towers in use in the west at this time — or had he drawn on more ancient eastern examples?

The degree of knowledge of techniques is apparent. On the Byzantine side, we saw that this was in part book-based. But it was also founded in the thorough siege warfare practice of the previous century, dominated by the use of rams and mining as the key methods of attack, with artillery playing the vital ancillary role of giving cover for the advance of the other forms of attack. To the library of material written when the empire was at its peak in the previous century was joined the *Strategikon*, possibly written by the contemporary commander Kekaumenos, who had fought in Sicily and Asia Minor, and who came from a military family. The sections of the book dealing with sieges concentrate on how to defend against rams and mines, and the giving of practical advice on ways of preventing defeat. There is a heavy emphasis on using trickery and ruses to take enemy fortresses, exemplified by many historical case studies.[1] When Romanus IV was captured and his army destroyed at the Battle of Manzikert in 1071, he was succeeded by Michael VII (1071–8), who had learnt strategy, tactics and siege craft by reading Aelian and Apollodorus.[2] For a while, such knowledge threatened to be academic, as the Turks occupied the lands and cities of Asia Minor, even briefly establishing a capital at Nicaea (İznik), the Pechenegs became the next nomadic group to penetrate the Balkans, and the ferocious Normans, having secured southern Italy and Sicily, crossed the Adriatic to harry the distressed empire.

Byzantine recovery: Alexius Comnenus

The necessary rescue operation was taken over by Alexius I of the dynasty of the Comneni (r. 1081–1118), whose strategy involved a sensible mixed approach of sieges and blockades and pitched battles where the odds were favourable. Alexius

[1] Kekaumenos, *Vademecum*, 58–9, 62 (ruses), 64–5 (protection against rams and mines), 66–9 (other examples). The latest example given is that of Otranto's capture by the Normans in 1070 (p. 63).

[2] Psellus, *Fourteen Byzantine rulers*, 376.

was said to have been a very knowledgeable and effective general by his biographers, albeit one of these was his doting daughter Anna. One of the dedications to him nonetheless stated that his skills included the invention of such war engines as Archimedes would have been proud of.[1] Unfortunately, there is no information as to exactly what these inventions were, and the accounts of his sieges do not support any argument for a decisive new weapon being suddenly available. Instead, traditional Byzantine military methods of the kind advocated by the familiar instructions seem to have sufficed. He dealt with the Norman enclave by a strategy of distant blockades, cutting off the roads to the fortresses and ambushing supplies, so that the garrisons surrendered through hunger, or fled.[2] At Kastoria, he had to set up *helepolei* and rock-throwing catapults, which between them managed to breach the walls. The Normans resisted strongly, so Alexius ordered a lakeside attack from boats, which brought about surrender (late 1083). In 1085, after Robert Guiscard had died, Alexius used the other traditional Byzantine method, bribery, to recover Dyrrhachium itself. In 1090, the emperor sent his kinsman Constantine Dalassenus to recapture Chios, a task carried out by the same combination of *helepolei* and artillery, which caused the collapse of a whole stretch of wall between two towers.[3]

Central Europe

A new family was elected to the kingship in Germany, picking up at the same time the crown of Italy, and soon to add that of Burgundy. The power and wealth of the monarch depended largely on the lands of his own family, and crucially on being able to control the immense lands granted to bishops and abbots. Control of the appointment of these churchmen, who combined their religious functions with those of feudal lords and, often, of royal ministers, was absolutely essential to the German kings if they were to be able to call upon sufficient resources to equal the authority of the great dukedoms into which the country was traditionally divided. It was the misfortune of the Salian kings that in place of the compliant popes who had once provided an additional prop to the royal authority, a movement of reform was advancing through the church. A clash between two irreconcilable objectives would dominate the latter part of the eleventh century and the first part of the twelfth under the later shorthand title of the "investiture contest", although the issues at stake were rather broader than the question of who had the right to invest a bishop.

This remained in the future as Otto III died young in 1002 and the crown passed to a cousin, Henry II (r. 1002–24), whose own childless death would see the Saxon dynasty give place to the Salians. The traditional concerns of the German kings and their great feudal lords remained the same as before, especially along the frontier with the Slavs to the east. Archbishop and duke came together to improve

[1] Euthymius Zigabenus, *Panoplia dogmatica*, 19–20.
[2] Bryennius, *Histoire*, 194 (book II, chap. 24).
[3] Comnena, *The Alexiad*, 181–2, 233ff.

the defences of Hamburg, destroyed in the uprising of 983 and subjected to further raiding in the new century. The archbishop replaced wooden buildings with a squared-stone church accompanied with a strong house "strongly fortified with towers and battlements", and the duke, not to be outdone, provided lodgings for his own men within the same fortified area (about 1035). Excavation has recovered a free-standing circular tower, 19 metres in diameter, with walls four metres thick. It stands behind the outer rampart of the fortified town, between the enclosure of the *Hammaburg* and the church of St Peter, the earliest such structure in northern Europe.[1] The bishop died, however, before fulfilling his plan to encompass the entire town with walls and towers.[2] The dual purpose of stone-built ecclesiastical buildings was no accident but deliberate policy. It was echoed in the construction of the monastery at Bamberg ordered by Henry II in around 1008, on a rock above the town "provided with walls and outworks (*propugnaculis*)" and a strong tower.[3] In 1015, the Slavs raided Meissen, burning the *suburbia*, but, despite climbing to a position above the *castellum* and trying to set alight the presumably wooden defences, were thwarted in their intentions.[4] Meissen, it will be recalled, had been established as a frontier *urbs* in 931 to secure a newly conquered territory. An irregular triangular hilltop (largest dimensions 250 × 50 metres) formed the castle site, and was protected by a simple wall, inside which were a cathedral and houses. Later in the eleventh century, a rectangular free-standing tower 11 × 12 metres would be erected close to the ecclesiastical building.[5]

Henry II had been called upon already to put down revolts in all parts of his sprawling empire. These included dealing with the margrave of Schweinfurt in around 1003. The rebel possessed a string of castles and these were besieged, captured and destroyed by the new king. That at Schweinfurt itself seems to have been on a spur with steep sides, cut off with a stone wall, tower and ditch. Other named castles were similarly sited on spurs, while there was another large castle at Oberammerthal near Amberg, a hilltop site of two courtyards. This was a ninth-century fortification rebuilt in the tenth century, with stone wall and towers.[6] The picture was clear, every German duke, count and margrave not only had castles of their own, but could call upon the loyalty of others the possession of which had been granted to followers. Trouble in Bohemia continued with a disputed succession, and Henry took a Saxon army to restore order, having to capture fortifications as he went.[7] Back in the west in 1006, Henry joined with the French king Robert II and Richard II of Normandy to try to recover the *castrum* of Valenciennes from Baldwin IV of Flanders. This fortress was still holding out in

1 Bohnsack, "Das Fundament eines steinernen Rundturmes"; Heine, "Burgen der salischen Zeit", 39–41.

2 Adam of Bremen, *History*, 105.

3 Frutolf, *Chronica*, 48; Cosmas of Prague, *Chronica Boemorum*, 66.

4 Thietmar of Merseburg, *Chronicon*, 843.

5 Brachmann, "Zum Burgenbau salischer Zeit", 118, 125 (plan), 140.

6 Ettel, "Die Eroberung der Schweinfurter Burgen"; Hermann of Reichenau, *Chronicon*, 654.

7 Adalbert, *Vita Heinrici II. imperatoris*, 688–91.

1007, when the emperor brought Baldwin to heel by seizing his castle at Ghent.[1] In 1017, he faced an invasion by the Bohemians and the Slav Liutizi in which the castles played another role, that of defending the kingdom. The attackers of the castle of Nimptsch (Niemcza) deployed around it for three weeks while preparing "all kinds of instruments", in the hope of burning it. Nimptsch, whose name betrays its origins as a Germanic settlement in Slav territory, was a walled town with a castle in a very strong location, but nothing remains of the eleventh-century work under later rebuilding.[2] Their attack was frustrated, but it would be interesting to know whether the technological skills undoubtedly possessed by the Bohemians through their close association with the empire also extended to the Slav tribesmen.[3]

Henry's successor Conrad II (r. 1024–39) found his problems mainly in Italy, where he had to lead armies to secure his dominion. In 1026 he ordered the destruction of a number of castles held against him, and in 1027 his soldiers found themselves resisting civil unrest in which the inhabitants took advantage of the high walls and towers of Ravenna. In the west in 1033, Conrad was elected king of Burgundy, but had to try to recapture the castles already taken by Odo II of Blois. That at Murten proved the hardest, and winter weather prevented any progress in this year. In 1034, however, Conrad returned, Murten fell, and its defences were destroyed.[4] The year after, there was another conflict with the Liutizi around a castle at Werben (Elbe), built by the emperor, then captured by the Slavs and its defending count killed along with 40 knights, then recaptured, and entrusted it to another count to hold.[5] In 1037 he was back in Italy to deal with another revolt, and had to besiege Ravenna and Milan, a campaign which lasted over a year, and had to sort out sedition in Parma, where he ordered destruction of part of the walls.[6]

The succession of Henry III in 1039 brought no change in the range of problems confronted by the German ruler, with Henry finding it necessary, in his early years, to intervene in Bohemia and Hungary in the east, and there being unrest and feuding in Lorraine to the west. Again, control and capture or recapture of castles played a central part in the royal actions. In 1053 Count Baldwin V of Flanders burnt the *oppidum* of Huy on the Meuse, and in 1055 besieged Antwerp unsuccessfully.[7] In the east Henry, initially in alliance with King Peter I of Hungary, had gone to war with Břetislav I, duke of Bohemia (r. 1034–55), but was held up by the castles defended against him and the blocking of the passes into the

[1] Sigebert of Gembloux, *Chronica*, 354–5.

[2] Lutsch, *Verzeichnis der Kunstdenkmäler*, 413.

[3] Thietmar, *Chronicon*, 855–6.

[4] Wipo, *Gesta Chuonradi II. imperatoris*, 590–93 (book 30–32); Hermann of Reichenau, *Chronicon*, 666, 668; on the castle: Meyer and Widmer, *Das grosse Burgenbuch der Schweiz*, 142.

[5] Wipo, *Gesta Chuonradi II.*, 597–9 (book 33); Frutolf, *Chronica*, 60; *Annales Hildesheimenses*, 37–9.

[6] Wipo, *Gesta Chuonradi II.*, 566, 604; Hermann of Reichenau, *Chronicon*, 678; *Annales Hildesheimenses*, 41; Frutolf, *Chronica*, 51.

[7] Sigebert, *Chronica*, 359–60.

country with barricades (1041/2).[1] Shortly after however, the king was in conflict with the Hungarians. His successful invasion of the Magyar kingdom along the river Danube involved the capture and destruction of what the chronicle calls two "major cities", Hainburg and Pressburg (now Bratislava). Hainburg was a triangular walled town across the river from Pressburg, with a castle built into the most vulnerable angle of the wall, but nothing remains earlier than the thirteenth century.[2] The bishop of Regensburg, to whom Hainburg had been entrusted as an important frontier march, undertook its reconstruction and provided it with a garrison of knights.[3] The conflict continued, and Henry was back in 1052, once again besieging Pressburg. This time it was recorded that he deployed siege engines against the town, for some eight weeks. The Hungarians resorted to destroying the countryside to deny resources to the besiegers, and the disease-afflicted Germans were forced to give up and withdraw.[4]

Henry IV

Henry IV succeeded as a child in 1056. The conflicts in which he was to become involved made those of his immediate predecessors pale into insignificance. For most of his reign, after the end of his minority in 1066, Henry IV would face not just rebels and frontier skirmishes, but open civil war in which a rival king would compete for the crown, usually with the backing of the pope. Just when Henry seemed to have regained control of his kingdom, each of his sons in turn chose to rekindle the wars. Central to the early problems, or at least a spark and symbol of them, were the castles the crown had ordered built in the Harz region. With the warning that none of the sources was remotely "neutral", a hostile account (Bruno of Magdeburg) reported on the construction of new castles "in deserted places" on high and rocky mountains, provided with walls, towers and gates, royal dwellings and monasteries. The results would have been regarded as "strong and beautiful" if they had been intended to oppose the pagans, but they were instead designed to oppress the Saxons, and they were built with the forced labour of free men.[5] Unusually, we know who the designer was: the king's close associate from childhood, Benno, bishop of Osnabrück between 1068 and 1088. Benno was renowned for both his theoretical knowledge of architecture and his practical skills in building design and masonry, having had a hand in the building work at the royal palace at Goslar. According to his biographer, the king entrusted him with the task of providing "all Saxony with new and strong castles". Later, he was put to work on the Salians' favourite cathedral of Speyer.[6] As with his contemporary at Rochester,

[1] Cosmas, *Chronica Boemorum*, 95; Frutolf, *Chronica*, 62; *Annales Hildesheimenses*, 42.

[2] Frutolf, *Chronica*, 64; *Annales Hildesheimenses*, 46; Büttner, *Burgen und Schlösser an der Donau*, 170–71.

[3] Frutolf, *Chronica*, 66.

[4] Hermann of Reichenau, *Chronicon*, 698–700; Thuróczy, *Chronica Hungarorum*, 90–91.

[5] Bruno of Magdeburg, *Liber de bello Saxonico*, 334–5.

[6] Norbert of Iburg, *Vita Bennonis II. episcopi Osnabrugensis*, 382 (Goslar), 386 (skills), 388–90 (castles), 420 (Speyer).

Gundulf, this bishop was a very practical supporter of royal authority. The intention behind the construction of these powerful expressions of royal power must forever remain doubtful, but the fact of their creation in new sites must weight the argument in support of the Saxon claim that they were an attempt to extend, rather than to recover, the crown's rights. Such a major building project must also have required a great deal of labour, and to have employed pressed local people must have added injury to insult. On top of it all, the king proceeded to bring in 70 Swabian knights to garrison Lüneburg Castle: a sensible precaution from his viewpoint, this castle having been seized from the powerful local Billung family, but the last straw for the Saxons, who burst out in revolt in 1073.

We know the names of seven of these places, helpfully listed by Lampert of Hersfeld.[1] Two were actually not new, but had been strengthened. Their locations did not represent some kind of strategic design, covering as they do a very large total area. But they were certainly of great strength. The Harzburg (at Bad Harzburg, Goslar) occupied a hilltop site some 220 × 55 metres, divided into two unequal courts separated by a deep ditch. It was approachable only from one side. The eastern part had a curtain wall originally ten metres high, 1.45 metres thick. There was a free-standing *Bergfried*, a round tower ten metres in diameter with walls two metres thick. There were also palace buildings and a monastery. The role of the *Bergfried* in German castles may be equivalent to that of the donjon: in other words, open to various interpretations. Much of the remaining stonework dates from later rebuildings, but Benno's castle was a massive stone fortress. Of a different style is the Sachsenstein at Walkenried (Osterode). This was built on a promontory protected by sheer cliffs on two sides. The third side was cut off with a massive stone wall and a deep ditch. The wall boasted a gate and a round tower. The gatehouse comprised an entrance chamber flanked by two six-sided towers. Walls were two metres thick. The round tower had a diameter of 13.5 metres. In front stood an outer court,[2] thought to be of more ancient origin. The castle at Heimburg (Wernigerode) has a more traditional form, being sited on an oval hilltop.[3] The style of these palace-fortresses was novel, and set a trend that was repeated throughout the empire where the resources permitted building on such a scale. Bishop Benno's own fortress-monastery at Iburg (Bad Iburg, Osnabrück) seems, naturally enough, to have followed the same pattern and building styles.[4] But to the Saxon lords, whose own castles were smaller and weaker, and rarely built of stone, the new works unfolding before their eyes over a decade or more must indeed have seemed threatening.

In 1073, the Saxons besieged the Harzburg but found it too strong for them. They resorted to a distant blockade by creating their own siege castles on neighbouring heights. They seized Lüneburg, which was without provisions. Heimburg

[1] Lampert of Hersfeld, *Annalen*, 194.

[2] Heine, "Burgen der salischen Zeit", 49–54 (Harzburg), 56–9 (Sachsenstein); Stolberg, *Befestigungsanlagen im und am Harz*, 137, 352–3.

[3] Stolberg, *Befestigungsanlagen*, 160–62 (Heimburg), 148–9 (Hasenburg); Brachmann, "Zum Burgenbau salischer Zeit", 107, 137.

[4] Heine, "Burgen der salischen Zeit", 60–61.

then fell in a few days, to assault carried out under a hail of arrows and stones from *tormentis*. The king found himself helpless to relieve his garrisons, and both Harzburg and Hasenburg (near Buhla) were starved into surrender.[1] Destruction of the hated sites followed. When the king restored his authority in 1076, he returned to re-establish his castles in Saxony. In 1077, the year in which the king had to make his famous submission to Pope Gregory VII at Canossa, Henry faced an offensive by the king elected by the rebels, Rudolf of Rheinfelden. The antiking attacked the city of Würzburg in August, deploying "diverse types of machine" with a view to breaching the walls, but in vain.[2] On the approach of Henry IV he abandoned the siege and his engines. The autumn months saw a new offensive by the king towards Bohemia, in which he deployed his engines to secure the capture of three (unnamed) castles.[3] In 1079, in a struggle involving Henry's appointed duke of Swabia, Frederick of Büren, rebels led by Welf I or IV (r. 1070–77, 1096–1101), seeking to recover his duchy of Bavaria, and the Zähringen family with its base on the edge of the Alps, there was an exchange of control of the city of Ulm. In the two-week siege of a castle (of uncertain identity), Welf attacked it vigorously with "throwing machines that are called *mangones* in German", leading to its destruction by burning.[4] Meanwhile, the citizens of Cologne revolted against their lord, Archbishop Anno II, in 1074, assaulted the bishop's palace built into a corner of the Roman walls, and used battering rams to try to break through the walls. They do not seem to have been very successful, suggesting that the necessary skills were in short supply in the city.[5]

In Italy, Henry's partisans among the north Italian bishops and nobles were opposed by the powerful Tuscan lordship of Countess Matilda and the supporters of the pope. There were divisions within each camp, too. In 1071, the dominant Milanese faction deployed an army to capture a castle of the king's chosen archbishop, Gotifredo da Castiglione, whom they refused to acknowledge. The besiegers threw up a rampart to protect their camp, then pressed their attack "incessantly", and built "petraries and all types of machine, *fundas* and *ballistas*".[6] The balance changed when Henry IV decided to transfer his operations to the peninsula in 1081, where he stayed for several years. The antiking Rudolf had been killed the year before, and, although Henry could not reconquer Saxony, neither could the rebels make progress. The core of the king's problem remained the support given by the pope to the German rebels, so it was to sorting out Gregory that Henry now turned. There were real strategic problems to face. Gregory had powerful armed support as well as his untouchable papal reach, and for once, the citizens of Rome largely backed him. The climate of the Rome region was well known to

1 Bruno, *Liber de bello Saxonico*, 336–9; *Carmen de bello Saxonico*, 3–4 (book I, ll. 85–112); *Annales Hildesheimenses*, 98; Frutolf, *Chronica*, 81–2; Lampert, *Annalen*, 195–212.

2 Berthold of Reichenau, *Annales*, 279, 299–300; Bruno, *Liber de bello Saxonico*, 366; *Vita Heinrici IV. imperatoris*, 422; Frutolf, *Chronica*, 82.

3 Berthold, *Annales*, 302.

4 Ibid., 319. The hilltop castle of Zähringen and other castles of the lordship, dated to this time, are discussed by Ott, "Die Burg Zähringen und ihre Geschichte", 5–13.

5 Lampert, *Annalen*, 242.

6 Arnulf, *Gesta archiepiscopum Mediolanensium*, 24.

be lethal to susceptible foreign armies, and princes, for vital summer months of each year, forcing would-be besiegers to withdraw to cooler northern lands. For that reason, Henry's siege of Rome would last three years in all. For a long time, the massive Leonine walls kept him out, while his troops ravaged the surrounding countryside, destroying castles and towns. In 1082, he nearly managed to break in by surprise attack, but the Roman soldiers were organised by the pope himself to "defend the *propugnacula*." In 1083, he tried again, organising machines to attack from all directions, pressing forward with battering rams and scaling ladders. The defenders continued to resist strongly, driving the Germans back with stones and missiles, until on 2 June 1083 some of the besiegers chanced to find a stretch of wall undefended, scaled up with a ladder and summoned their comrades to help, thus breaking into the city.[1] Gregory fled to his refuge in the Castel Sant'Angelo, whence he appealed desperately to his Norman allies for help. Meanwhile, the victorious king, having failed to capture the papal refuge, decided to built his own fortification on the Palatine hill. Nothing is known of this, because no sooner had Henry withdrawn to the north again then the population pulled it down, his small garrison having succumbed to disease.

Henry was now able to revert to sorting out the new revolts that had broken out in Germany, where the city of Augsburg had been seized by Welf, apparently using "adulterine keys" (1084). The king blockaded the city for two weeks before recovering it on 6 August. Not having been preparing for such an eventuality, it was not stocked with food.[2] Two years on, another revolt was led by Margrave Ekbert II of Meissen in Saxony, requiring the king to lay siege to the latter's castle at Gleichen. The castle was valiantly defended from mid-August to the end of the year, when Ekbert brought an army to relieve it. Gleichen, very close to two other hilltop castles in Thuringia, was first mentioned in 1034. It seems to have possessed a *Bergfried* nine metres square, with walls 1.5 metres thick, and 15 metres high, in the middle of an inner court on the highest point of the hill, with a lower ward below. Its location and defences were powerful enough to hold out against the royal siege train.[3] Henry was next involved in the endless conflict back in Italy, where he faced once again the redoubtable Countess Matilda and her supporters, including his own rebellious son Conrad.

This struggle revolved entirely around siege warfare, as the emperor sought to capture the powerful cities and castles held by Matilda's supporters across northern and central Italy. In 1091, the city of Mantua fell through treachery, but only after a long siege, Henry not having been able to break down the defences. In 1092, Henry besieged *castrum Montisbelli* (Monteveglio), and apparently pressed it with a close blockade to prevent access or egress, and fierce assault with machines. He was only prevented from taking it by the death of a son, leading him to withdraw

[1] *Vita Heinrici IV.*, 430–32; Benzo of Alba, *Ad Heinricum IV. imperatorem*, 657–9; Bonizo of Sutri, *Liber ad amicum*, 613–15 (Benzo and Bonizo were both bishops); Bernold of Constance, *Chronicon*, 437. Sigebert, *Chronica*, 364; Frutolf, *Chronica*, 94–6; *Chronica monasterii Casinensis*, 434; Ekkehard, *Chronicon universale*, 204–5.

[2] *Annales augustani*, 130–31; Frutolf, *Chronica*, 98.

[3] Tettau, *Beschreibende Darstellung*, 401–4; *Annales Sancti Disibodi*, 9.

in grief.[1] His attempts to take Nogara and the castle of Canossa itself, very much the centre of the realm, also came to nought. Canossa stood on a rock, defended by sheer drops on three sides. Excavations of the site (long since ruined) confirm the description given by Lampert of Hersfeld. At the highest point was a citadel, protected by a thick, curving wall. Below was a towered curtain along the cliff edge, constructed by Adalbert Atto, first prince of the dynasty, in the tenth century. The existence of a third wall, as stated in the written account, was strongly suggested too, making in all a very formidable fortress.[2] Ultimately the task of breaking down the countess's dense network of well-defended fortifications proved beyond the German emperor. While many of the castles were weakly-defended places set up to provide a refuge for the rural population, they were nonetheless thick on the ground — there are no fewer than 85 castles *documented* in the Canossa lordship itself, of which at least 50 can be attributed to before the end of the eleventh century.[3]

Henry's long reign did not end peacefully, either. In 1105 his son Henry placed himself at the head of a further revolt of the discontented, backed as before by the papacy. The Saxons and the south German dukes again rose, and the emperor's loyal cities and castles were their targets. The imperial castle at Nuremberg was subjected to siege. Henry was unable to raise the forces to relieve it, and it surrendered after two months.[4] The Kaiserburg at Nuremberg lies over what was there in Henry's time. It appears to have been founded around 1050, and presumably consisted of a rampart around the top of the rock, with chapel and palace buildings inside and perhaps a *Bergfried*. It was strong enough not to fall to assault.[5]

There were soon two kings in Germany: the son Henry forced his father to abdicate, but Henry IV would not concede so easily to Henry V. The old king retained strong support, and the city of Cologne was the scene of the last siege of the reign, in 1106.[6] It was already strongly defended by late Roman walls. The expansion of the city in the tenth century (to a population of perhaps 20,000) had shifted its centre of gravity to a merchant quarter on the river front and had necessitated the building of new walls. To these defences the townspeople added additional works at the king's orders, ditches and outworks, and engines of war. This further walling converted the old square-planned town into a semicircular circuit.[7] The written accounts all suggest that the walling was done on the king's orders in 1106: however it is impossible for such a length of stone wall to have been erected in just a few months, so if the defences that held off his son really were of

1 Donizo, *Vita Mathildis*, 440 ff., 617, 663.

2 Ibid., 123 ff.; Lampert, *Annalen*, 406: "cum castellum illud triplici muro septum esset". Tondelli, "Scavi archeologici a Canossa".

3 Settia, "Castelli e villagi nelle terre canossiane", 282–301.

4 Ekkehard, *Chronicon universale*, 228; *Chronica imperatorum Heinrico V. dedicata*, 232.

5 Ebhardt, *Der Wehrbau Europas*, I: 526–7; Zeune, "Salierzeitliche Burgen in Bayern", 190–91; Grossmann, "Die Kaiserpfalz (Kaiserburg) zu Nürnberg", 103–11. It seems likely that the Salian castle occupied the western end of the rock, with the surviving defensive elements further east (the five-sided tower and the *Bergfried*) being of later date.

6 Grenier, *Quatre villes romaines de Rhénanie*, 128–35.

7 Stehkamper, "Die Stadt Köln in der Salierzeit", 75–80, 111, 121–2; plans on pp. 76, 81.

stone, they must have been being built over a number of years, something that it is quite possible given the apparently strong relationship between king and city after the events of 1074 (in which it seems that the right of fortification of the city was gained from the archbishop). Henry V began a siege in June, but could not prevent regular supplies coming in by river, and himself suffered supply difficulties. His large army tried to break down the gates, to topple the walls with battering rams, and to "throw down" the walls with catapults, but all to no effect. After a few weeks he abandoned the attack, while the sudden death of his father in August brought an end to the civil war.[1]

In tracing the violent events of Henry IV's reign we have noticed the extent to which fortified sites were the key to controlling land, cities and resources, and we have seen examples of where strongly defended fortresses, suitably provisioned, could resist attack from powerful armies mobilised by the greatest lords of the empire, but would eventually fall when not relieved, or where the attacker could sustain the blockade or assault for long enough. However, the occasions when such blockades could be maintained were limited by many other factors. Control of the fortresses, including the right to build new ones, remained in the hands of the king or of the great magnates, including the bishops and abbots. Ecclesiastical lords could be among the greatest builders, both on behalf of an overlord (as with Bishop Benno) as well as on their own behalf: the chronicle of the abbey of St Benedict at Subiaco (in the hills of Lazio beyond Rome) is revealing on this. Throughout the tenth century, it records the acquisition by the abbots of *castra*, as part of the process of *incastellamento*. But in the 1050s, the castle (*rocca*) of Toccanello was strengthened by the addition of a stone wall to the *ancient tower* for the purpose of "defeating the enemy." Facing local rivals, Toccanello was the subject of armed attack in the next decade. Castles became the material form of the ensuing struggle, being built in the style of Fulk Nerra as means of isolating other sites. In 1082, for example, the abbey's foe Ildemund captured the castle of Genna by surprise attack. The abbot ordered the construction of a *munitio* to constrain Genna and eventually recaptured it. Finally, in 1109, with the aid of men from Rome and Campania as well as the abbey's own knights, there was a major onslaught on Ildemund in his own castle "cum balistis et fundabalartis et machinis", which led the nobleman to cede the disputed land to the abbey.[2] Apparently, the means both of attack and of defence were at the disposal of the greater lords as well as the kings.

Central European castle-building: an overview

The previous account has revealed what is known about a number of the major castles built during the eleventh century, and because so little remains that is datable, it is hard to make clear statements. But among the trends, there was a move in favour of building on hilltops, where the shape of the castle would necessar-

[1] *Vita Heinrici IV.*, 462–4; *Annales Hildesheimenses*, 56–7; Ekkehard, *Chronicon universale*, 236; *Chronica imperatorum*, 242.

[2] *Chronicon Sublacense*, 7, 13–14, 17.

ily be determined by the terrain, and a substantial growth in the number of the *Hohenburgen*, especially in eastern and northern Germany, is clear in the century following 1050. The continuation of the independent tower is also evident, with round towers remaining popular. Towers would have two practical roles: residences in their own right, or the *Bergfried*, generally too small inside to be a comfortable dwelling, with possible ceremonial and symbolic uses. The main defence everywhere remained the wall protecting the inner court, where the vital buildings of lordship were sited. The greater castles would be provided with an outer court as well. Mural towers were built but without any regular layout or placing to suggest that the provision of flanking fire was a major concern. Gates of the strength and sophistication of the massive construction with its flanking towers at the Sachsenstein were still rare. The builders were the king, the bishops and the great nobles, dukes and margraves. Lower down the social scale, counts were also building castles, but the knights (the unfree *ministeriales*), who provided castle-guard for their owners, were not yet themselves builders or owners of castles.[1]

In central and southern Germany the picture is not dissimilar to the north. Thirty-nine castles of the age of the Salians have been identified in Hesse, Rhineland-Palatinate and Saarland. Most of them were very small, clearly designed as residences for the nobility and thus distinct from the refuges and royal castles of the previous age. The general pattern was of a simple enclosure with a stone or wooden wall, and a strong residential tower (present in 80% of the examples). The *Bergfried* began to appear later in this period. The *Hohenburgen* — "high castles" — remained an attribute, literally, of the higher nobility.[2]

West of the Rhine, there were some of the earliest castles built by noblemen in Alsace, with examples from 1018 and 1038. There were some 25 such castles in Lorraine in 1000, but none east of the Rhine until after 1050. The general style was of a walled enclosure with a central residential building and a tower to house the gate. The *Bergfried* and other signs of the classical German noble castle were still some time off in this part of the empire.[3]

Nowhere in this survey has the motte appeared. This style of castle was long delayed in arriving in the heartlands of the Germany, but had already made an appearance in the western areas nearest to developments in the French principalities. A significant proportion of the private castles built in Alsace, for example, were mottes,[4] although their earliest arrival date is not certain. During the eleventh century, the process whereby the former collective defences gave way to smaller private fortresses was also clearly marked in the Netherlands, most of which (except Flanders) owed allegiance to the empire. Mottes were created by local lords either inside or outside the previously existing ramparts across the lowlands of Flanders: Douai, Saint-Omer, Lille, Armentières, Ypres (Ieper), Furnes (Veurne), Gistel, Dendermonde, Bergues, Diksmuide, Ghent, Aalst and Ninove. In other places, mottes appeared where there had not been a Carolingian castrum in the

1 Heine, "Burgen der salischen Zeit", 9–79.
2 Böhme, "Burgen der Salierzeit", 78–80.
3 Biller and Metz, "Anfänge der Adelsburg im Elsass", 278–9.
4 Burnouf, "Les mottes castrales en Alsace", 111 (1985), 4–20.

first place. Most are datable (at the earliest) to the late eleventh century.[1] It has been suggested that at least the initial main purpose of construction was defensive, the living space available in the tower on the motte being minimal. A similar picture obtains for example in the county of Loon (in imperial Brabant).[2] In the Rhineland, mottes also became common forms of building for the local nobility, as they moved beyond the simple earthwork enclosures already witnessed in the feuding recorded by Alpertus at the turn of the tenth and eleventh century. It has been suggested that mottes were imitation *Hohenburgen* transferred to the flatlands of the Rhine, and that the name derives from a translation of *mo(n)t(t)es*,[3] although this theory is far from universally accepted. As they became popular, there is considerable evidence of a pattern of throwing up earth around an existing hall or tower to create a motte — the donjon at Doué-la-Fontaine in western France received this treatment. In the low countries, the process has been studied. At Ghent, in the count's castle, a wooden hall of the tenth century (inside the simple Carolingian ramparted court) was replaced with a two-storey stone hall in the eleventh, and a tower. This would be enmotted in the twelfth. At Petegem-aan-de-Schelde, one of the islets forming part of the old *castrum* was enmotted at the end of the eleventh century. On the opposite bank of the river Scheldt, the castle built during the Ottonian dynasty at Ename with a large hall was given a donjon in (probably) the early eleventh century.[4] Further north, castles remained rare in the counties now making up the Netherlands until the twelfth century, but the areas most integrated into developments in the rest of the empire also witnessed the arrival of the castle during this century. The bishops of Utrecht, loyal supporters of Henry IV, began to build castles from the middle of the eleventh century and by the turn of the century were throwing up mottes, particularly suitable in low-lying areas, but also stone towers and halls, although little survives now earlier than around 1150.[5] As elsewhere, while some of the constructions were by and for the great nobles — counts and bishops — many more were being built by local landowners as private fortresses, albeit with the permission of the count.

Eastern Europe and the Rus'

The states of eastern Europe had become established around the year 1000 with some form of feudal social system. Poland, Hungary and Bohemia were considerably influenced by the more established states to their west. In due course the dukes of Bohemia were to become recognised as kings, while also becoming part of the empire. The major external influence on the principality

[1] De Meulemeester, "Mottes castrales du comté de Flandre", 101–6; De Meulemeester, "La fortification de terre", 18–23; De Meulemeester, "Le début du château", 121–7.

[2] De Meulemeester, "Structures défensives et résidences princières".

[3] Janssen, "Mittelalterlicher Burgenbau am Niederrhein", 127–8.

[4] These examples are described in De Meulemeester, Matthys and Poisson, "Structures enmottées", 139–43.

[5] Besteman, "Mottes in the Netherlands", 218.

of Kiev and its northern cousins was, of course, Byzantine. Alongside the cultural and trading relations that developed throughout, there were also more violent encounters, with the German emperors in particular intervening in Bohemia, Poland and Hungary (often to resolve succession disputes). In such a situation, it would be expected that technological knowledge would become readily absorbed, while the basic principles of siege warfare were pretty much constant anyway.

In 1018, the Polish prince Bolesław I (r. 992–1025) had attacked the Russians in Ruthenia, "taking cities in a hostile manner". Bolesław's military power revolved around large retinues of armoured knights and cities and castles held by his close dependants. These resources were also deployed to erect castles, and secure them with a garrison of knights and sufficient supplies.[1]

The Russian principalities were meanwhile becoming established. The non-contemporary *Eymund's saga* tells that Prince Yaroslav I the Wise gathered "an army of farmers" to fell trees and set them up "as the ramparts of a fortress" and had a great ditch dug in front.[2] The picture is confirmed by archaeology. Novgorod, the "new fortress" incorporating several existing settlements, had a citadel (*kremlin*) whose timber defences first date to 1044.[3] In 1032, the prince ordered the creation of new, defended towns at Tartu and elsewhere,[4] as a means of securing the conquest of the region. In the south, the defeat of the Pechenegs around 1036 gave the occasion for a further and significant expansion of the capital city of Kiev, one which fortunately attracted the attention of many then and since. The earthwork ramparts stretched 3.5 kilometres, and they were twenty metres thick at the base, narrowing to 12 at the top. It has been calculated that the wall used 630,000 cubic metres of earth and 50,000 cubic metres of oak, and would have taken 1,000 workers four years to construct. The chronicles claim it was much faster. There was a ditch in front 15 metres wide and eight deep, fed with water by the river Dnieper. The highlight was the famous Golden Gate with its gilded church on the roof. More than 12 metres high and 6.4 wide, it was built of stone in decorated bands. The resemblance of this, and the dome, to Constantinople's walls and St Sophia are no surprise, it having been convincingly argued that Byzantine masons were behind most of the architecture in Kiev since the 990s, and that it was through them that local masons and builders acquired the necessary skills.[5] The city rapidly developed under Yaroslav the Wise to become one of the greatest metropolises of the age.

In 1071, the Polish prince Bolesław II (r. 1058–79) attacked the Ruthenian fortress at Przemyśl. It was well protected with deep ditches, and the river San flowed on the north side. The Poles first had to get across the river in order to

[1] Długosz, *Annales*, I: 283; Gallus anonymus, *Cronicae et gesta*, 22, 25–6, 35. Neither source is contemporary: "Gallus" wrote in the early twelfth century, Długosz in the fifteenth.

[2] *Eymund's saga*, 77.

[3] Yanin, "The archaeological study of Novgorod", 86–9.

[4] *La chronique de Nestor*, 176.

[5] *The Russian primary chronicle*, 137; *La chronique de Nestor*, 176; *The Nikonian chronicle*, I: 132, 141. Poppe, "The building of the church of St. Sophia in Kiev", 30–31; Mühle, "Die topographisch-städtebauliche Entwicklung Kievs", 354–63; R Morgan, "The great golden gate of Kyiv". The gate has been restored.

attack from all sides, which they did for three days "without intermission". The defenders were driven back by darts and arrows enabling the Poles to enter the town, the Ruthenians withdrawing to the citadel, which was protected both by its site and with a tower. It surrendered through hunger and the prince ordered it destroyed. Bolesław continued his campaigning, taking the principal *castra* of Vladimir and Chełm, which were made only of wooden beams, but finding Wohdin (modern Wohyń?) a tougher target, which took six months to starve out.[1] In 1078, the new Russian ruler Izyaslav attacked Chernigov (now Chernihiv). This was another large walled town, with ramparts 2.5 kilometres around and four metres high, which had also benefited from Byzantine building skills inside. The prince, however, captured one of the gates before being driven off.[2]

The next recorded sieges date from the conflict with the Polovtsians (also known as Kipchaks or Cumans), who besieged Torchesk in 1093. The townsmen resisted bravely, killing many of the nomads by shooting from the ramparts. Demonstrating considerable skills if not any possession of siege equipment, the nomads managed to cut off the town's water supply and secure its surrender after nine weeks, provisions sent by Prince Svyatopolk II failing to get through. The following year they allied with Svyatopolk's cousin Prince Oleg to attack Chernigov again, in which Prince Vladimir II Monomakh was ensconced. The suburbs were burnt, but it seems that the rampart was strong enough to enforce a negotiated settlement. In 1095, the Polovtsians besieged Yuryev (now Bila Tserkva) so effectively for a whole summer that the Russian inhabitants were evacuated to Kiev. Meanwhile Svyatopolk joined with Vladimir to attack Oleg in Starodub, which surrendered through hunger after 33 days. The town of Vladimir fell after a seven-week blockade to Svyatopolk in 1099.[3] All of these sieges were blockades, and there is no evidence that engineers were deployed to organise direct assaults on the earthwork defences. Where in other sieges that took place the chronicle states that the town was taken by storm, as when the men of Novgorod took Minsk in 1067, and two princes took Vsevolodsk in 1097,[4] there is no suggestion anywhere of anything other than a direct assault on the rampart. In 1099, Prince David of Chernigov's attack on Vladimir was firmly resisted, although he tried to storm it "near the towers" under the cover of archery, and in the end the townsfolk jumped down from the rampart to join a relief force in putting the besiegers to flight.[5]

While the Russian princes were falling out with one another, the Poles were fighting in the north against the Pomeranians, with the fortress at Nakło nad Notecią (German: Nakel) being a regular target. Duke Władisław I (r. 1079–1002) attacked Nakło in 1092, but the Slavs sallied out, burning the Polish camp,

[1] Długosz, *Annales*, II: 104–5. The fortress at Przemyśl has been excavated, finding a hilltop citadel some 80 × 70 metres with an earth-and-timber rampart: Żaki, "Topografia wcześniośredniowiecznejo Przemyśla", 95–6 (French summary).

[2] *The Russian primary chronicle*, 166; *The Nikonian chronicle*, I: 198; see also Franklin and Shepard, *The emergence of Rus*, 206, on Chernigov.

[3] *The Russian primary chronicle*, 175–9, 182; *La chronique de Nestor*, 243, 249.

[4] *The Nikonian chronicle*, I: 220.

[5] Ibid., 224.

tents and siege engines forcing the abandonment of the siege. Three years later, the Poles attacked a strong Pomeranian fort at Międzyrzecz, constructed of enormous timbers. The defences were attacked with engines (apparently artillery) and the ramparts destroyed in several places.[1] In the following year (1096) the conflict was with the Bohemians. According to Cosmas of Prague, Duke (nominally King) Vratislav II invaded Polish territory and destroyed a *castrum* on the river Nysa (Kłodzka), then proceeded down river and built his own very strong fortification on a cliff at Kamieniec Ząbkowicki. In 1099, he marched into Moravia and rebuilt the castrum at Podivín.[2] In 1100, the duke besieged Vranov nad Dyjí for six weeks, until hunger induced surrender.

The Poles meanwhile continued their conflict with the Slavs of Pomerania, and in 1107 Bolesław III scored a great triumph by capturing the pagan king's capital, Białogard, by assault without using any "instruments or machines", although he had them. In 1108 he invaded again, attacking the *castellum* of Czarnków, where this time he did array "machines of all kinds", and erected four siege towers against the fortress, which, however, he took without assault when the castellan betrayed it. The castle of Wieleń fell to an assault assisted with "machines and instruments" (Gallus). In 1109, it was the turn of Nakło once again, and this time we do hear of machines being deployed and the town surrendering.[3]

The same year, the emperor Henry V brought an army into Silesia, taking and burning a number of "ignoble fortresses" (Długosz), before besieging Glogau (Głogów) on the river Oder. He had obviously come prepared with siege equipment which was now deployed vigorously. There was a ferocious exchange of arrows and bolts from ballistas, while the Germans "fundas cum lapidibus rotabant", made an attempt to sap the walls, and used battering rams against the towers. The garrison had offered to surrender if not relieved but Bolesław would have none of it and ordered them to hold out, which they did sufficiently strongly to persuade Henry to call off the siege.[4]

The Pomeranians meanwhile had not sat quiet under Polish domination, and in 1112 recaptured Nakło. Bolesław besieged it between 29 September and 24 December 1112. Because of its marshy location, he could not bring siege machines to bear so had to give up. In 1113 he returned and attacked Białogard again, which stood in the bend of a river, for eight days. The Poles prepared machines, and so did the Pomeranians. Among the devices were wooden siege towers erected on earthen platforms, which the defenders managed three times to burn, only to see them rebuilt.[5] These accounts of events from 1108–13 provide the first clear references to such devices in Poland, alongside the more traditional rams and

[1] Długosz, *Annales*, II: 176, 186; Gallus, *Cronicae et gesta*, 67.

[2] Cosmas, *Chronica Boemorum*, 164, 169.

[3] Kadłubek, *Chronica Polonorum*, 103–4; Długosz, *Annales*, II: 206, 234–7; Gallus, *Cronicae et gesta*, 114–18 (book II), 128–9 (book III).

[4] Gallus, *Cronicae et gesta*, 135–7 (book III); Długosz, *Annales*, II: 247–53. The editor of Gallus says a passage in Sallust, *De bello Jugurthino*, may have been a model for his account of the siege. But Gallus includes terms that can only have been contemporary.

[5] Gallus, *Cronicae et gesta*, 160–63 (book III); Długosz, *Annales*, II: 265.

artillery. They also hint that the Pomeranians were themselves finding ways of resisting them.

There is evidence of the siege abilities of the Hungarians in an account of their attack on Belgrade (Alba Bulgarica) in 1071. The royal army approached by river and ordered carpenters to construct eight wooden towers double the height of the walls, so that soldiers could hurl stones and arrows into the town. They also prepared *tormenta* to break down the walls, a task which was achieved, but the attackers were unable to exploit the breach. Finally, after three months, they broke in through one of the places where the wall had been demolished.[1] Once again, then, we find siege towers in the east, alongside an account of the creation of a breach in walls which does not sit with our understanding of the effectiveness of artillery at this time. It is possible, therefore, that the weaponry reported in fact comprised battering rams.

Alongside this evidence that the Hungarian rulers had become acquainted with advanced siege methods, however, is evidence that Hungarian castle-building at the beginning of the century was limited.[2] Those that existed tended to belong to the counts, with a role as centres of government and administration being established by the newly christianised monarchy, and they were predominantly earthworks, although a few were built on the remains of ruined Roman Danube forts. As the century progressed, so did the castle-based organisation, with marcher lordships being created in frontier areas also based around castles,[3] and the Magyar word for castle, *vár*, became attached to numerous place names. A system of social organisation was developed by the crown to support the structure, with an appointed set of officials, soldiers, a group of landholders called "castle-servants" who provided the services of smiths, carpenters, tanners and food suppliers, and the free peasants having to pay taxes to the castle and provide other services. It seems clear that the origin of this system must lie in those developed in Frankish lands to the west.[4] Of excavated sites in the north of the country, many take advantage of the terrain and of previously existing earthworks, and are often large enough to provide space for use as refuges. Construction was overwhelmingly of earth and timber.[5] Meanwhile, the crown was availing itself of more substantial buildings, notably at the royal palace at Esztergom, where King Stephen built a defended palace, no doubt designed to enhance the prestige of the monarchy, but also walled for protection. Excavation of the now destroyed site suggests that the previous Roman structures were not used, that stone buildings (of mortared masonry), including a new circular chapel, date from early in the eleventh century, and that the complex was walled.[6]

Close proximity to both the western and the Byzantine empires would have

[1] Thuróczy, *Chronica Hungarorum*, 1: 100.

[2] Fügedi, *Castle and society in medieval Hungary*, 37–9.

[3] Győrffy, "Die Entstehung der ungarischen Burgorganisation". An Arab visitor in 1120 confirmed the association of castle and county: Abū Ḥāmid, *Tuḥfat al-albāb*, 208.

[4] Győrffy, *King Saint Stephen of Hungary*, 111–19.

[5] Nováki, "Die topographischen Eigentumlichkeiten der ungarischen Burgen", 359–66.

[6] Nagy, "Rapport préliminaire des fouilles d'Esztergom".

permitted the kind of sophisticated siege techniques described here to have come from outside, although it is not clear whether the necessary engineering skills were now to be found at home, or had to be purchased from abroad. For defence, the young state still relied predominantly on walls provided by nature. When Henry v invaded the country in 1108, taking sides in a disputed succession, he attacked Pressburg on the Danube, but was thwarted in his siege of the *castrum* by obstructed river crossings and marshland.[1] The armies raised by the kings of Germany from their own and their vassals' *ministeriales* were often too vast to be challenged in the open but sheer numbers could still be resisted from behind strong defences, and those provided naturally could be as effective as the stone walls built by the still limited resources and skills of local, regional and even national lords.

Al-Andalus in retreat

The civilisation and culture of Muslim Spain in the eleventh century was unmatched anywhere except Baghdad, Constantinople and in China. But the centralised political power of the caliphate of Córdoba evaporated with astonishing speed when the dynasty ended in 1031. With no one leader or family able to claim the quasi-theocratic title, and a diverse population over the whole of which no local power would be able to exercise authority, local rulers took control in all of the major cities of al-Andalus and proclaimed themselves kings, supported by different groups of Muslim and non-Muslim communities. These tiny kingdoms, which called *taifas* in Spanish after the Arabic *ṭā'ifa*, would squabble with each other while the frontier of Muslim rule was pushed decisively southwards throughout the century, until a fresh invasion from Africa restored central control and reversed temporarily the tide of defeat. Confident in their cultural superiority, safe behind the great stone walls of their cities, comfortable in their luxurious palaces defended by even stronger walls in the *alcázares*, the *taifa* kings have left no mark in the history books, as their horizons stretched no further than the territory subject to their city's rule.

One of the larger states was the kingdom of Granada, ruled by the Zirid dynasty, and an account has been left of its history by its last ruler, 'Abd Allāh ibn Buluggīn. In 1057, there was a struggle for Málaga, where the citadel (the lower of the two still standing over the city, where original work has been obscured by much rebuilding and, now, restoration) had been built "in a manner no one else could have done", but the town itself surrendered leaving the garrison isolated. The king's uncle then went on to capture the fortified complex at Guadix. Next, he secured the submission of many other castles in the area around Málaga, most by capitulation but some by assault. The form in which these attacks took place is not recorded, but we know that the Muslim armies in Iberia possessed the most sophisticated weaponry and had the wealth to pay for experts to operate it. Much later, in 1089, the Granadan king besieged a Christian garrison in the castle

[1] Ekkehard, *Chronicon universale*, 242; *Chronica imperatorum*, 250.

of Aledo, said to be strong, well-garrisoned and provisioned. King 'Abd Allāh constructed platforms on which to set up the mangonels and ballistas against the most vulnerable areas, but he was unable to deploy "any of the machines to be used for capturing castles", and a wooden siege tower constructed by the besiegers was destroyed by fire by the defenders. The siege therefore was abandoned.[1]

The kings of Granada were also great builders. Most of the Alhambra complex at Granada dates from later centuries, but the Zirids (who ruled until 1090) created a very strong royal fortress, parts of which are still to be found in the northern and southern walls of the *Alcazaba*, constructed in traditional style and with powerful rectangular towers and a sophisticated bent entrance in the Puerta Nueva in the south-west. The whole of the large interior was occupied by numerous buildings to house the great army of those supporting the royal household. The kings also constructed a line of city walls, pierced by several massive gates and protected by projecting towers, running up the hill on the other side of the stream from the Alhambra, to enclose the original Muslim town on the site. These walls, now obscured by houses and hard to observe, are of the same *ṭabīʿa* building style and are powerful defences.

While the Zirids fought with their neighbours, the Christian Reconquest was finally getting under way. Fernando I of León and Castile (r. 1037–65) launched an attack into Portugal in 1057 in which the important towns and castles of Lamego, Viseu, Coimbra (a hilltop fort) and Pena (Penedono) all fell, along with "many other towns and castles". Coimbra took a long blockade, although the king deployed engines and "wooden castles", presumably siege towers. But Lamego fell quickly to an assault using "[siege] towers opposite the towers and diverse types of engines".[2] While none of these were large or significant centres by Muslim standards, their successive conquests by the Leonese king, and the failure of any of the other Muslim rulers to seek to come to their aid, sets something of a pattern for the next two decades. The Christian attacks came across the length of the long frontier zone. In 1064, an army of Normans and Catalans laid siege to the fortified town of Barbastro (in the kingdom of Zaragoza) and took it after 40 days, following the blocking of the town's aqueduct. The victors proceeded to massacre or enslave the population, but on the withdrawal of the foreign troops Aḥmad al-Muqtadir, king of Zaragoza, recaptured the site. Perhaps the most interesting element to this otherwise obscure event was the presence of Normans and Aquitainians in the Aragonese-Catalan army, presaging the role of international armies in the crusades.[3] Christian forces were able to raid as far south as Granada,

[1] 'Abd Allāh ibn Buluggīn, *El siglo XI en 1a. persona*, 114–15, 124–6, 183–9, 205–7.

[2] *Historia Silense*, 72–4 (a source not later than 1100); Pelayo of Oviedo, *Crónica*, 528; *Chronica Gothorum*, 10; *Chronica Naierensis*, 156–7; *Chronicon lusitanum*, 417; *Primera Crónica general de España*, II: 486–8; Ibn 'Idhārī, *al-Bayān al-mughrib*, in *Christians and Moors in Spain*, III: 67–9.

[3] Although it has been shown that this campaign was not itself a crusade: Ferreiro, "The siege of Barbastro", 129–40. The place is mentioned as a castle in the tenth century by Rāzī: "La 'Description de l'Espagne'", 75. The presence of the duke of Aquitaine is confirmed by *Chronicon Sancti Maxentii Pictavensis*, 403; Ibn al-Kardabūs, *Historia de al-Andalus*, 94; Bakrī, *al-Masālik wa'l-mamālik*, in *Christians and Moors in Spain*, III: 71.

attacking Huesca. Cuenca was submitted to a long siege, eventually buying off the attackers with a great tribute.[1]

In 1085, Alfonso VI of León and Castile (r. 1072–1109; king of León from 1065) captured Toledo. He was familiar with its defences, having spent a period of exile there. Since 1075 there had been civil war between various factions in the city, with Alfonso aiding one of the contenders. The Leonese-Castilian king first began a distant blockade, closing all the passes from the Castilian plateau. In June 1084, the full siege began, with the king establishing his headquarters on the hill where the castle of San Servando now stands. The Leonese-Castilian forces were regularly supplied from the north but the city eventually began to starve despite having capacious magazines. At last, with no one moving to challenge the besiegers, the defenders entered into a treaty in which Alfonso granted lives, property and religion in exchange for entry into Toledo on 6 May 1085. Once its possessor, the king began the construction of a new citadel on the site of the old, standing now as the much reconstructed alcázar. Alfonso then headed off for Córdoba, but decided to return to Toledo (according to Ibn al-Athīr) to "gather his engines".[2] The whole of the central plateau of the peninsula was lost to the Muslims, who certainly understood the significance, witnessed if nothing else by their repeated future attempts to retake the city. The symbolic importance was also great: the taking of Toledo had been one of the key steps in confirming the overthrow of the Visigothic monarchy. The strong defences and location of Toledo, however, made it resistant to direct assault, and Alfonso had decided that only a blockade was going to work. He had the luxury of a political situation in which no one tried to lift the siege or even to re-supply the city.

That changed quickly. In 1086, the Almoravids, a fundamentalist (in modern terms) confederation of Berber tribes led by Yūsuf ibn Tāshfīn, arrived from Morocco, and defeated Alfonso in battle. They spent the following years overturning the *taifa* kingdoms and establishing a centralised government. Having established themselves in 1090 in the fortress at Tarifa to secure their base, they set about seizing and holding castles, then turned to Córdoba. Their attack took place along the river front, where a great breach was created (it is not said how), too big for the defenders to patch up, and the great city fell.[3]

Another great Muslim city, Valencia, would soon fall into the hands of Christian forces, although its conquest would be temporary and its captor, Rodrigo Díaz de Vivar, known as El Cid, was more concerned with his private success than that of any cause. The city was large and strong, standing on a plain against a river, with an encircling ditch provided by a dry river bed on the other sides. The walls were powerful *ṭabīʿa* constructions. The Cid followed Alfonso's approach at Toledo, deciding on a blockade, which began in 1093 and included the capture and fortification of neighbouring castles, but he was also fully aware that Yūsuf was

[1] Ibn al-Kardabūs, *Historia de al-Andalus*, 95, 101.

[2] Lévi-Provençal, "Alphonse VI et la prise de Tolède"; ʿAbd Allāh, *El siglo XI en 1a. persona*, 162; Ibn al-Athīr, *Annales du Maghreb & de l'Espagne*, 481.

[3] ʿAbd al-Wāḥid, *Histoire des Almohades*, 119–21.

unlikely to grant him the same freedom as Alfonso had enjoyed. Obviously having a numerous force, he was able to produce a complete encirclement of Valencia, while compelling all the neighbouring districts to supply him with food. In spring 1094, he tried an assault on one of the gates, which was repelled by stones and arrows. He then tried an engine (presumably a ram), but the defenders demolished that too. Finally, hunger did the work and the gates were opened to him on 15 June. The Almoravid relief attempt was ambushed and defeated.[1] Once again, a walled city of size and strength had resisted assault and compelled a would-be attacker to resort to a long blockade. It was recovered after another long blockade (seven months) in 1100/1101.[2]

Even when the target was less formidable, the same methods applied. In the capture of the castle of Ejea in Zaragoza by King Pedro I of Aragon in 1095 (begun by his father Sancho), the strength of the location and the well-provisioned state of the town forced the king into a long siege, ended — so the recording monk describes — by a miraculous break-in, or perhaps by a surprise attack catching the defenders unaware.[3] In 1096, the important northern city of Huesca fell after a siege lasting from May to October. Huesca possessed a circuit of walls erected in the ninth century, with shallow towers, but this was sufficient to hold up the attackers until relief failed.[4] The victors went on to establish or strengthen the important castles at Loarre and elsewhere as a means of securing their new territories. Loarre is an imposing construction (of much more ancient origin) on top of a height with steep slopes and contains a fortified monastery and a circuit of powerful walls and towers mainly dated to the late eleventh century, and demonstrating that the Aragonese knew the latest techniques of castle building. A great square donjon dominates the inner complex, while the outer ramparts, covering the only possible approach, comprise high walls and closely spaced projecting hollow towers, two of which frame one of the gate passages, the other being a form of barbican.[5] The castle was already immensely strong in the eleventh century. Other strongly-defended castles in the same area that played a part in the Aragonese reconquest during the last quarter of the century included Alquézar and Montearagón.[6] By 1101 (when Barbastro again and finally fell to the Aragonese) a circuit of powerful fortifications bordering the Muslim land of Zaragoza had become Christian, and the foundation had been laid for the overthrow of the whole principality in 1118.

Meanwhile, following the capture of Valencia, in 1097 the Cid besieged Al-

[1] *Historia Roderici*, 951–7; the chronology given in this source is described as confused, and is revised, in Huici Miranda, *Historia musulmana de Valencia*, I: 19–34 (description), II: 62–105 (capture); Ibn al-Kardabūs, *Historia de al-Andalus*, 127.

[2] Ibn al-Kardabūs, *Historia de al-Andalus*, 136.

[3] *Relation du siège de la ville d'Exeja en Aragon*.

[4] *Historia de la corona de Aragón*, 60. Ḥimyarī, *La péninsule iberique au Moyen-Âge*, 236, describes a concentric double wall at Huesca.

[5] Del Arco, *El castillo real de Loarre*. Excellent photographs in Ortiz Echagüe, *España*, plates 58–60.

[6] Del Arco, *Algunas indicaciones sobre antiguos castillos*, 20–21; Ortiz Echagüe, *España*, plates 62, 63.

menara (north of Murviedro),[1] but it took three months before the place was captured. In 1098, his target was the castle of Murviedro (now Sagunto). Here, an assault was made with "arrows, darts and all kinds of arms and machines". Even here, this did not succeed and the garrison was given a period of thirty days in which it would surrender if not relieved. While under attack, the count of Barcelona launched an attack on the Cid's castle at Aurepensa (Oropesa del Mar), but the hero (following his somewhat biased biography!) succeeded in both relieving his castle, and ensuring the surrender of Murviedro.[2] The great warrior died in 1099.

The establishment of the Almoravids in al-Andalus slowed the progress of reconquest appreciably. If at the beginning the Berber tribesmen possessed few of the skills of building or taking fortifications, they were not slow to acquire them. Yūsuf founded the city of Marrakech, where the circuit of walls was part of the original plan. The problem of assigning accurate dates to Muslim fortifications continues to make attribution of particular sites to particular times a hazardous exercise, but the castle of Zagora in Morocco has been assigned to the Almoravids, who occupied the area around 1083. Standing on ground rising from a river bank, the whole occupies 12 hectares and has an irregular perimeter totalling 1,350 metres. The wall is around three metres thick, built of stone blocks with rubble infilling. Solid square bastions are widely spaced, one providing for a bent entrance. There may have been a lower outer wall.[3] It was designed to resist desert-based tribal raids on the nearby commercial centre, so was undoubtedly fit for its purpose. The problem for the Almoravids in al-Andalus was not one of lack of defences, but how to prevent them falling to long sieges based on rigorous blockades, which could only be broken by armies risking battles. Unsurprisingly, this seems to have been the pattern of this period.

In addition, al-Andalus was rarely the Almoravids' main concern: this was their African homeland, with a southward rather than northerly focus. The gold-rich kingdoms of the Sudan (as the sub-Saharan area of Africa was known to the Arabs), in particular the then dominant realm of Ghana, were of continuous importance. In this vast area, walled towns were very rare, and even where the rulers had become Muslims, only their palaces were shielded by (mud-brick) walls. Although, alongside a long-standing trading relationship, the Berbers had waged frequent campaigns of raiding and (sometimes) proselytising, the methods of warfare were different in many ways from those prevailing in Europe, and sieges did not feature at all.[4]

Meanwhile, in North Africa, where the great cities dotted the Mediterranean coast, attack could come from both land and sea. In 1087 there occurred a rare

[1] This siege is assigned to 1093, and the capture to King Sancho, by *Historia de la corona de Aragón*, 54, a more reliable version of events.

[2] All this from *Historia Roderici*, 961–7.

[3] Allain and Meunié, "La forteresse almoravide de Zagora".

[4] See the various Arabic descriptions collected in *Corpus of early Arabic sources for West African history*, especially the contemporary accounts of Bakrī (78–80), Idrīsī (110) and Ibn Abī Zar' (247–8).

example of a seaborne Christian attack on a Muslim port, when a combined force of Pisans, Genoese, Romans and Amalfitans crossed the sea to attack the city of Mahdia, built at the end of the tenth century by Fatimid rulers with immensely strong landward walls. En route, the fleet passed the island of Pantelleria, where there was a *castrum* of "wondrous strength", somewhat improbably described as having nothing to compare with it in the whole world, which nonetheless fell to the invaders through the use of tall wooden siege towers. The attack on Mahdia itself proved easier, with the Italians breaking in through the gates in a single day, then proceeding to destroy the towers, leading to the conclusion that it was not really defended on this occasion.[1]

The conflict in the western Mediterranean between and amongst Muslim and Christian forces revolved around the control of the land, which could only be secured by possession of the strong places. When control of whole territories changed hands, new occupiers had to look to defend what they had taken so that, even though the frontier zones across the Iberian peninsula were amongst the most heavily fortified in the world, new work would still be called for. Now it was the turn of the Christian rulers to look to new walls, and one of the most spectacular surviving examples may date from the late eleventh century: the walls of Ávila. The town was repopulated under Alfonso VI, and the walls have been begun in the 1090s and may have been completed in as little as nine years. The circuit is 2,526 metres, the walls on the most vulnerable side are twelve metres high and three thick, with a strongly projecting semicircular tower (itself 6.5 metres thick) every 20 metres. There are 88 wall towers altogether. The two main gates are protected by twin towers of even more massive design, 20 metres high, projecting 13 metres, and 7.4 metres thick. The narrow entrance passages have arches designed to house portcullises. On the less vulnerable southern side, the walls and towers are smaller and more spaced. Much Roman stonework was reused in the building, which must certainly have involved an enormous workforce and absorbed the skills of a large number of stonemasons.[2]

Ávila was only one of a number of major building works carried out on Alfonso's orders. His was a policy of re-establishing settlement in the regions previously lying between the Christian and Muslim states and largely desolate as a result. The conquest of Toledo allowed these areas to be settled again. So the same period witnessed the refortification of Salamanca, Coca, Olmedo, Medina del Campo, Cuéllar, Sepúlveda and Segovia, all supplied with the strong ashlar walls and semicircular flanking towers that seem to be a feature of King Alfonso VI's new building works.[3]

The mobilisation of people that must have been required to do this work gives a hint of the massive involvement of the entire population in defensive duties.

[1] Cowdrey, "The Mahdia campaign of 1087". This includes the text of the *Carmen in victoriam Pisanorum*: see stanzas 16–17 (Pantelleria), 51–4 (Mahdia).

[2] Gómez-Moreno, *Catálogo monumental de la provincia de Ávila*, I: 59–64.

[3] Gutiérrez Gonzalez, "Características de la fortificaciones del reino de León", 63–6, summarises the process. The standing remains of most of these sites date from considerably later rebuilds.

Townspeople certainly had a range of duties comparable to those borne by the freemen of Wessex under Alfred and his successors, and at this time at least the duties were enforced, and carried out. Municipal militias were an essential component of the king's forces. They could be called on to supply forces for royal expeditions, while defensively their tasks included providing sentries both for outlying castles and towers, and for the town's own gates and towers. The maintenance of the walls in good order was also a municipal duty. Such duties may have existed in many places across Europe, but here in central Spain, they were rigorously followed over a long period of time.[1]

Conclusions

Fortification remained a right of government in most of Europe, although where that government had become devolved, control of fortification was devolved with it, as with the counts of Capetian France, the counts, bishops and abbots of Italy and the dukes of the western empire. Only rarely did it remain the prerogative of the crown (as in Norman England or the Byzantine empire). As the Muslim world split into competing dynasties, control of fortification went with the power. During the eleventh century, the gap between the building skills shown on the one hand in the eastern empire and the Muslim world, and on the other in the rest of Europe, began to narrow. In Iberia, it disappeared. There is nothing to distinguish (in art and skill) the walls of Ávila from those of Granada. No doubt this owed much to there being a common source of supply of masons and carpenters. In the west, the skill of stone building demonstrated in the great towers of the counts of Anjou or the dukes of Normandy was of similar quality, and no doubt also owed much to the crafts developing to meet the call from the church for massive new cathedrals across Christendom. Further down the social scale, where smaller landowners were able only to call on the skills of local craftsmen, and the perhaps forced labour of reluctant peasants, works of earth and timber (however expert the woodwork may have been) were the most that could be managed. The greater mottes of England were still formidable fortresses, nevertheless.

In the western empire, castle-building had yet to reach so far down the ladder, with most of the fortresses being built by the king and his greater vassals, and commonly using where available high rocks. These locations also determined that the building would have to conform to the shape of the rock, and often ensured the absence of the flanking towers which had become recognised as critically important for the defence of level sites elsewhere in Europe and the Muslim world. Across the Christian states, though, great towers or donjons were becoming a common feature everywhere. Their very names imply their close connection with lordship: the tower was the symbol of the rule of the individual. There has been an academic debate on the role of the great tower, with the stress now being

[1] All the above from Powers, *A society organized for war*, 15–16, 127, 142–6.

placed on its effect as a symbol (representing visibly the power of the lord) or being used for ceremonial (with carefully designed routes of access culminating in a space that highlighted the lord's dominion), whereas the traditional view that it served as a redoubt of last resort is discredited.[1]

In this debate, in which it is important not to forget that power was necessarily a combination of real and symbolic authority and actual strength, it is necessary to allow both that one building could have several functions, and also that changing circumstances could give one building different functions at different times. Furthermore, similar buildings could be built for different purposes in different places. The chronicle evidence given here for the use of the castle in the wars of the Angevin counts supports the notion that the great towers had multiple roles including a military one. The *Bergfried* in German castles was clearly usually far too small to be a residence, may have had symbolic or ceremonial purpose, but also, from the accounts of its actual use, was a strongpoint of the defence or a tower of final refuge. But the White Tower of the Tower of London was also demonstrably intended as a royal palace, where the primary defensive role was played by the circuit of walls around it. Even here, it is now sometimes argued that its palatial function was quickly abandoned as rulers opted for more comfortable quarters in Westminster.

While building skills had spread and resources had increased to allow for a spread of major stone fortifications during the century, the only new method of attack to emerge was the rediscovery of the ancient siege tower, which seems to have spread from western France to most of Christendom. The extreme vulnerability of this device to being burnt by resolute defenders, not surprising since the work had to be erected within a few yards of the wall to be effective, has been demonstrated repeatedly in the chronicle accounts quoted. Nonetheless, western rulers in particular seem to have chosen to stay with it. The relative strength of the defence as against the attack is shown by the number of sieges that lasted a very long time, or failed because of that very factor. The measures needed to create an effective blockade of a major walled city can only have been achieved with detailed planning and careful organisation, not elements commonly associated with medieval warriors. The difficulties of feeding a besieging army during a long siege would have been nightmarish, and it is more remarkable that so many blockades were seen through to a successful conclusion than that so many failed. In the east, both the Byzantines and the Muslims remained loyal to the mine, the continued absence of which in the west is remarkable and can only be explained by a shortage or absence of people who knew the skills involved. As the century drew to a close, however, the two still different worlds of east and west would be drawn together in drastic ways and with dramatic consequences by an unprecedented movement, following the preaching of the crusade by Pope Urban II in 1095.

[1] See, for example, Dixon and Marshall, "The great tower in the 12th century", and on the broader social role of the castle itself, Coulson (all eight items in the Bibliography).

6 Franks and Saracens: the early crusades

The First Crusade

THE ARMED PILGRIMAGE that later generations would call the First Crusade unleashed a massive movement in the west, and fascinated both contemporary annalists and later historians. With the exception of a period of persecution by the Egyptian Fatimid Caliph al-Ḥākim in the early eleventh century, Christians had, in fact, been able to visit Jerusalem without hindrance, and Muslim, Jewish and Christian peoples lived side by side under a tolerant Muslim rule, although the situation became more difficult during the periodic Seljuk incursions. In Europe, however, the pope and emperor had been in armed conflict for a generation, and in Francia and Germany, the feuding and local warfare of a growing number of armed knights trying to secure their own land and position had escaped the control of princely rulers and was having catastrophic consequences for the peasantry and the church, often the defenceless target for unlicenced brutality. In one sense, the First Crusade was an extension of the "Peace of God", the attempt led by the church, but also taken up by princes for their own good political reasons, to lessen and control the ferocious and dangerous feuding that characterised the eleventh century.

Pope Urban II's call mobilised immense numbers of people from all social classes. The "People's Crusade" of ordinary folk, who demonstrated their Christian fervour by massacring Jewish communities across Germany and Hungary, were slaughtered by the Seljuk Turks when they crossed the Bosphorus. The more formidable army of the princes arrived in 1097, and was given the assistance of its Byzantine hosts, after the emperor had first secured the agreement of the leaders that anywhere conquered that had once been part of the empire would be returned to Constantinople's rule. It is impossible to know the size of this army, but it is clear that the initial force was, by the standards of the times, very large.[1] An array of princes of the second order had brought significant forces, chiefly from Francia, Burgundy, Provence, Lorraine, Flanders, and Italy. There were no leaders of royal rank. It is probable that many were motivated by religious fervour, recognising the big sacrifices many had to make to come, but for some there may have been calculations of self-interest. Success in war promised a way out of poverty

[1] After a careful analysis, France, *Victory in the East*, 141–2, concluded that the army that began the siege of Nicaea in 1097 may have been around 50,000–60,000 including non-combatants, with a core of 7,000 or more knights. The People's Crusade is estimated at 20,000.

through ransoms and plunder, while for the greater leaders there was the prospect of winning lands and cities of their own, something quite beyond their reasonable expectations back home. In the event of failure, they would at least go to heaven with their sins forgiven.

The key leaders – Bohemond, son of Robert Guiscard, the Norman duke of Apulia, and his nephew Tancred, Count Raymond IV of Toulouse (or of Saint-Gilles), Godfrey of Bouillon, Robert of Normandy – were all experienced in war. Those who came from the south had direct experience of fighting Muslim opponents. Those who came from southern Italy, as many of the Normans present did, would have also fought against their Byzantine hosts. For many of those from more northern parts, the great stone fortresses and walled cities they would encounter would have been a novel sight, but the siege techniques that were developed through the eleventh century stood them in good stead. The great sieges that would dominate the history of the crusade would show that, although there were different emphases, there was little to distinguish the opposing sides in terms of techniques. Instead, what would be learnt would be a sharpening of skills, and lessons about building defences that would return to Europe. Furthermore, the idea that medieval warfare was chaotic and disorganised is again shown to be false. The organisational skills required to transport the crusading army across Asia Minor and see it through the siege of Antioch and successfully on to Jerusalem were remarkable. The victory of the First Crusade was not due to superior technology, nor to religious motivation. Extraordinary organisational ability made it possible, but what contributed more than any other factor was the break-up of the Muslims into local and competing powers. Although the Muslims were all one undifferentiated mass of "Saracens" and "pagans" as far as the Christians were concerned (and the crusaders were all "Franks" to the Muslims), the differences mattered a great deal. When the crusaders crossed the Bosphorus, the Seljuk kingdom was divided into several petty principalities. After crossing Asia Minor, they faced the feuding rulers of Antioch, Aleppo, Hims and Damascus, while the rulers of Egypt were happy to take advantage of their northern neighbours' problems to advance their interests in Palestine and Syria, disputed territory since the arrival of the Seljuks. Had they faced the united forces of the Middle East, there was no way that the crusade could have achieved its objective.

By supplementing the writings of Christian eyewitnesses (Albert of Aachen, Fulcher of Chartres, Raimundus de Agiles, an anonymous writer (probably Norman from southern Italy), Radulphus of Caen, and Robertus of Reims) with the surviving accounts by Muslim contemporaries, (such as the chronicles of Damascus (by Ibn al-Qalānisī) and Aleppo, and compilations by later writers), a balanced and remarkably detailed picture can be drawn. There is also the sharply-observed work of the Byzantine princess Anna Comnena.

Nicaea, 1097

The first objective for the crusaders was the major city of Nicaea (modern İznik), capital of a Seljuk principality since the Turkish victory at Manzikert in 1071. Alexius Comnenus was naturally keen to have the Latin army recapture this important place for him. The city, although important, did not have modern defences. Little had changed since the capture by an Arab army in 725, when the single, but relatively strong, late Roman curtain wall had been undermined. Since then, earthquakes had caused damage in 740 and in 1065, and more damage had been caused during a revolt in 975 (on which see below). It has to be assumed that the earthquake damage had been repaired by 1097, a likely scenario given that Nicaea had become a Seljuk capital. At nine metres in height and between three and four metres thick, the wall was as strong as anything in western Europe, although now very old, but it is not clear how substantial was the ditch in front. However, the large number of projecting towers (more than one hundred) ensured effective flanking to protect the face of the rampart (see Plate 2).[1]

The emperor had nearly succeeded in persuading the Turks to surrender the town to his benevolent rule rather than face the wrath of the crusaders, a course of action recommended in the face of the overwhelming strength of the besiegers. However the Seljuk sultan Kılıç (Kilij) Arslan I, who had been away fighting rivals for Melitene (Malatya), brought up a relieving army. This hardened the resolve of the garrison. However, this army was defeated in bloody battle, and the heads of the dead were propelled over the walls by catapult. Albert of Aachen said that the attackers had been there for seven weeks before the order to build siege weapons was given, others suggested (perhaps more probably) that this was done very quickly. The Christian forces were familiar with a range of equipment: reference has already been made to stone-throwing artillery, and all the sources agree that siege towers were also ordered. Count Raymond set up mangonels to bombard one tower, but "whatever the force with which the shots were directed, it was impossible to break or dislodge a single stone of the ancient edifice, provided with an almost indissoluble cement" (Albert of Aachen). Eventually, by increasing the number of machines and by dint of continuous bombardment, a few stones were dislodged. A major method of attack was the battering ram. All the versions, with the exception of an unclear reference by Anna Comnena, imply either that there was no ditch requiring to be filled up first, or at most a shallow one presenting but little obstacle, a major weakness of the defence.

The attack was pressed on all fronts, as the crusaders had assigned a portion of the wall to each of their commanders. Albert describes a machine made of oak timbers to shelter an approach to the foot of the wall, in which were twenty knights, but all were killed when it collapsed on them. The role of the siege towers at Nicaea seems to have been to provide shelter for miners to sap the foot of the wall, by clearing defenders from the immediate area. The defenders responded with inflammable mixtures, which destroyed some of the attacking machines.

[1] Schneider and Karnapp, *Die Stadtmauer von İznik*, 1–8, 10–11.

Raymond of St Gilles ordered the construction of a circular tower protected with leather hides outside and entwined wickerwork inside. From it, trained sappers with iron tools were to work on breaking down the Gonatas tower, so-named in Greek from its "kneeling" shape when it had been undermined and brought down in the siege of a rebel a century earlier (Anna Comnena). The anonymous describes how after several days of attack, a hole was excavated and propped up with timbers, then burnt so that the masonry collapsed. However, because the breach occurred in the middle of the night "like a crack of thunder" the crusaders were not in position to exploit it, and the Turks were able to repair the damage.

The Byzantines were close by but not taking part in the siege operations. Alexius had until recently been a sworn enemy of the Normans, whom he had every right to regard as treacherous aggressors. He had, however, maintained contact with the Turks in Nicaea. When he observed that the sultan was regularly sending assistance across the lake to the city, he ordered boats built. In order to cover his intentions, he sent a contingent of 2,000 missile troops to assist the crusaders in the next day's planned assault on the walls, and then sent more troops against the city using the boats. The Turks accepted his generous terms of surrender and admitted his forces to Nicaea, and the next the astonished crusaders knew was the appearance of imperial standards on the walls above them.[1]

The main defence of Nicaea had been, probably, the solidity of the walls, requiring a massive attack by sapping. Despite all their numbers and resources, which were sufficient to press the attack on all sides at once, and to build and operate as many machines as there were engineers present to oversee, the crusading army had in fact managed to breach the wall in only one place, and that without further consequence. It was clear that with the defeat of the relieving army the city's fall was inevitable and was merely pre-empted by the emperor's manoeuvre to make sure that it was the Byzantine standard that was raised over it. Anna Comnena suggested that Alexius believed that Nicaea would never fall to crusader assault, but this is contradicted by the speed with which the Byzantine moved to take it himself.

The great army, joined by a Byzantine contingent under Tatikios to watch over imperial interests, set off on what was described as the "short route" across Asia Minor. There were further battles in which the Latin knights defeated and scattered the Seljuk forces arrayed against them, most notably at Dorylaeum (near Eskişehir) on 1 July 1097. The support offered by the local Armenian leaders, who seized control of a number of fortified towns on the road, provided the army with a secure passage and a vital rear base. Tancred diverted to capture Tarsus, the major fortress on the frontiers of Cilicia that had been such a key target of the Byzantines in the tenth century. The Norman had ordered the preparation of an

[1] Comnena, *The Alexiad*, 333–40; *Anonymi Gesta Francorum*, 13–14; Albert of Aachen, *Histoire des faits et gestes*, 1: 83–92; Raimundus de Agiles, *Historia Francorum qui ceperunt Jerusalem*, 595–6; Fulcher of Chartres, *A history of the expedition to Jerusalem*, 82; Robertus of Reims, *Historia hierosolimitana*, 756–9. Detailed analysis of the siege warfare is contained in the invaluable work by Rogers, *Latin siege warfare in the twelfth century*, in which studies of all the sieges to be examined here can be found.

attack to fill the ditches and launch an escalade, but the Turkish defenders fled overnight. The Byzantines were able to re-establish their control of Cilicia. By October, the Christian army had reached the former Byzantine city of Antioch (Antakya).

Antioch, 1097–8

Few of the crusading army would ever have seen a city the size of Antioch before, protected by a wall of immense length marching across up and down hills, over ravines and streams. The vast space included several points of very high ground, with a small citadel at the highest point constructed by the Byzantines after 969, and the more approachable side against the river Orontes (Turkish Asi, Arabic 'Āṣī). The walled area was around three kilometres long and two deep and the wall was studded with a large number of two-storey towers (between 300 and 450 of them). The construction work, dating mainly to a reconstruction on Justinian's orders in the 540s, was of high quality. It seems that the wall was only two metres thick, with the wall-walk carried out on corbels, but the accounts refer to a fore-wall (at least on the flat ground). The six main gates were very strongly built and there were numerous small postern gates.[1] The citadel had 14 small towers, but its main defence was its inaccessibility: it could only be reached by a single narrow path from the side of the city (Plate 3). The garrison has been estimated at around 5,000, which was scarcely adequate to defend such a great circuit of walls, and indeed could only do so because large parts of the circuit on the mountainous side were unapproachable, and the crusader forces had to be deployed against only limited sectors anyway. The population was a mix of all the different ethnic groups and religions of the region, and Turkish warriors. The commander, Yagı Siyan, must have had some reason to fear internal difficulties as well. On top of that, he was also no friend of the neighbouring Muslim leaders, amongst whom a complex feud of changing alliances had taken place over the recent years while the crusade had loomed.

A meeting of the crusader leaders decided that there was little prospect of being able to take Antioch by assault. Tatikios argued for a repeat of the distant blockade used by the Byzantines in 968, and it was resolved to institute a blockade, but to do so from close to the walls. Bishop Adhemar of Le Puy, the papal legate, is said to have stated that Antioch had "strong walls which cannot be destroyed by iron or stones, joined with unknown and indissoluble cement, constructed of enormous stones."[2] Therefore, an attack based on trying to sap or mine a way in was unlikely to succeed. The size and topography of Antioch made it impossible to consider a total shut-in, so instead the various contingents of the army were

[1] Procopius, *Buildings*, 165–9, describes how Justinian reduced the perimeter of the city walls after the destruction caused by an earthquake and capture by the Persians, and diverted the river Orontes so that it ran closer to the walls. See also Bouchier, *A short history of Antioch*, 9–12, reproducing illustrations from earlier visitors when the walls still stood, and Rey, *Étude sur les monuments de l'architecture militaire des Croisés*, 189–92.

[2] Albert of Aachen, *Histoire des faits et gestes*, 1: 143.

assigned different parts of the perimeter on the flatter ground. The relatively small size of the garrison minimised the risk of a sortie inflicting any serious defeat, and the seizure by its Armenian population of the walled town of 'Artah on the road to the east (now Reyhanlı)[1] cut off any chance of relief and made it possible to follow the adopted strategy, combining the benefits of both a close siege and a more distant blockade: the garrison could not be prevented from sallying out in areas where the attackers had no forces, but having done so they would be unable to achieve anything. For further protection, the crusaders dug a ditch between their own camps and the city. There began a long siege. A siege fort (called Malregard) was built by Bohemond's army to the north of the city, and a bridge of boats constructed across the Orontes which enabled the besiegers to approach much nearer the walls and to bottle up the defenders still further. Another siege fort, called La Mahomerie (Mosque tower) would be built right up against the bridge gate on the west, and a third called Tancred's tower was constructed against the southern wall. Stones from old tombs are described being recycled in the construction of the fort. Such a close attack must have placed a great strain on both sides, with raiding and skirmishing taking place regularly. The garrison clearly had artillery, too: they used *petraria* and *fundibala* to hurl back the heads of those unfortunate enough to have been captured in the skirmishing.[2]

Substantial numbers of ships from Genoa, Venice, Pisa and elsewhere kept up a regular traffic of supplies, with Byzantine Cyprus being a key staging post, and the ports along the adjacent coast had fallen into crusader hands at some undetermined point. The securing of these lines of communication was a decisive element in the success of the besiegers, because without them they could never have maintained their presence in front of Antioch for as long as they did. Severe shortages did occur during the winter of 1097/8, with significant losses in people and horses for the besiegers, while the immense circuit of Antioch meant that the Turkish forces, which had occupied all the surrounding countryside without being able to break the siege, were able to maintain a limited contact with the city on the hilltop side. Duqāq of Damascus attempted relief in December 1097, but was beaten back and in February 1098 Raḍwān of Aleppo tried as well, but was defeated by Bohemond.[3] This victory enabled the besiegers finally to complete the blockade of Antioch with the construction of two further forts against the remaining gates of the city on the south side. It was on the night of 2/3 June 1098 that the crusaders got into the city, mounting the walls at night with ladders through the agency of one Firuz, in command of a section of the wall on the south, who had offered to betray Antioch to the crusaders. Once again the city fell not to attack but by betrayal, as the besiegers opened a postern to admit a section of Bohemond's army.[4] The

[1] Radulphus of Caen, *Gesta Tancredi*, 641; Robertus of Reims, *Historia hierosolimitana*, 776, 784.

[2] Fulcher, *A history of the expedition*, 94.

[3] Such is France's interpretation of conflicting evidence, *Victory in the East*, 248–51.

[4] Fulcher, *A history of the expedition*, 98; Albert of Aachen, *Histoire des faits et gestes*, 1: 280–87; Robertus of Reims, *Historia hierosolimitana*, 805; Radulphus, *Gesta Tancredi*, 653–4; Ibn al-Qalānisī, *The Damascus chronicle*, 47; *Anonymi Gesta Francorum*, 45; Kemāl

survivors of the garrison fled to the citadel, and the anonymous chronicler, who appears to have taken part in the break-in, described the crusaders throwing up a wall of stone and lime to cut it off, and constructing a *castellum* and machines.[1]

The capture was timely. Within days, a large army led by Kerboğa, ruler of Mosul, arrived on the scene. He had been entrusted with the relief of Antioch by Baghdad, but had stopped off en route to try to recover Edessa (recently seized by Baldwin),[2] only to find that instead of relieving the city, it was now in Christian hands, and that he would himself have to besiege it. There followed the sudden (and very timely) discovery of the holy lance in Antioch, which inspired the crusader army to a great victory in battle. In the defeat, the Muslim army abandoned its camp including the machines it had brought along. The road to Jerusalem now beckoned at last.

Ma'arrat al-Nu'mān, 1098

But the crusader leaders spent the summer in Antioch disputing control of the city (Bohemond securing it) and their next action. In the autumn, part of the army marched east to try to capture the small town of Ma'arrat al-Nu'mān (Marre), which had resisted previous attempts. The siege began on 27 November. Lying on flat ground and not very strong, having a ditch in front only of part of the wall, the town did not seem a very difficult objective, but was nonetheless to involve the crusaders in several attacks and the preparation of an array of equipment before they took it. Count Raymond of Toulouse's forces were the first to attack, and had apparently concluded that the defences were sufficiently weak to justify an immediate escalade. However, this was carried out with just two ladders, and "these short and fragile", so the attack failed, although it was suggested it might have succeeded with just four additional ladders. The next step was to build siege equipment. Only two sources talk of artillery: Raimundus de Agiles mentions artillery (*petraria*) and Radulphus of Caen writes of the reverberations of the *tormenta*, but all the sources agree on the vital part played by a wooden siege tower, on four wheels, designed to overlook the wall and provide cover while shelters were put in place to enable miners to get to work at the foot. Recognising the danger, the defenders tried to keep their enemies at bay by hurling great stones from their stone throwers, which killed many attackers, but could not ultimately prevent the machine being driven up against the wall. From its top, knights hurled stones down to drive the Muslims away. A further assault took place, but once again without success. Finally, a concerted attack was organised on 11 December, in which it is recorded how one knight got to the top of the wall, only to have his ladder break because of the weight of people following him up, but in the mean time the sappers finally broke through and there was a dreadful massacre of the

el-Dīn, *Extraits de la Chronique d'Alep*, 585–6, all cover various aspects of the long siege and the capture.

[1] *Anonymi Gesta Francorum*, 59.

[2] Matthew of Edessa, *Armenia and the crusades*, 170. It appears that no serious assault against the strong defences was undertaken by Kerboğa's army.

inhabitants by the Provençal troops. Meanwhile, Bohemond, whose forces had arrived after the beginning of the siege, had been attacking from the north side of the town and broke in simultaneously to complete the capture.[1] Some of the full horror of the fighting here is evidenced from accounts of the poorest of the besiegers being driven by starvation to eat the flesh of dead enemies.

Jerusalem, 1099

Eventually, the crusader armies set off for Jerusalem. They left behind many fortified places on the coast, and tried but abandoned attacks on 'Arqā and Tripoli, deciding to take advantage of the harvest season to make progress instead. There was no resistance, although a prepared opponent could have seriously impeded the march. This was because the territory had recently been taken from the Turks by al-Afḍal, the Fatimid vizier and de facto ruler of Egypt, who had taken advantage of the crusade to attack his Muslim rivals. He had assaulted Jerusalem with 40 siege engines and made several breaches in the walls.[2] Al-Afḍal had been negotiating with the crusader leaders since they had arrived in Syria and was unprepared to resist an invasion. For this reason the coastal cities were in no position to impede the march, and the city of Jerusalem itself seems to have been unready to resist a siege. On 7 June, the besiegers arrived with perhaps 15,000 fighting men. What did they face?

The city was described (by Fulcher) as having a width of "four bow shots". The protected space was 86 hectares, and the wall was four kilometres long, and around three metres wide. The strongest point was the citadel called by the crusaders the Tower of David on the west, constructed of solid masonry, "large square blocks sealed with lead", standing atop one part of the podium of the old Herodian fortress, and the large quadrangular tower to its north. The eleventh century had witnessed several changes to the city's defences. The first reconstruction, following an earthquake in 1033, reduced the area enclosed. It seems that al-Afḍal's breaches had been repaired. Later Ottoman rebuilding of the walls makes it difficult to determine what was there in 1099: there was a low (three-metre-high?) outer wall around part of the circuit, quite close to the main wall, erected by the Seljuks from the remains of flattened monasteries, and a ditch — possibly as much as nineteen metres wide and seven deep, unless this was the result of later strengthening by Saladin[3] — running for a kilometre along the northern side, which the crusaders had first to fill this before they could bring their siege equipment to bear. The south and west enjoyed protection from natural ravines.

Additionally, the crusaders were a very long way from the nearest friendly

[1] Lively accounts in *Anonymi Gesta Francorum*, 75–7; Raimundus, *Historia Francorum*, 625–7. Both were present. Ibn al-Qalānisī, *The Damascus chronicle*, 45; Fulcher, *A history of the expedition*, 112–13; Kemāl el-Dīn, *Extraits*, 587.

[2] Ibn al-Athīr, *al-Kāmil fī al-Ta'rīkh*, in *Arab historians of the crusades*, 10.

[3] Based on the recent study by Boas, *Jerusalem in the time of the crusades*, 43–4, 46, 48–9, plans at pp. xvi and 77. For detailed studies, see Prawer, "The Jerusalem the crusaders captured", 1–13; Pringle, "Crusader Jerusalem".

town, and in an environment short of food and water, with the threat of relief by al-Afḍal imminent. It was therefore essential to capture Jerusalem speedily: they tried an initial attack with ladders after just a couple of days, which failed, leading the army's commanders to consult and to decide on the construction of a suitable array of siege equipment: stone throwers, battering rams and siege towers. This introduced another problem: the shortage of timber in the area, which meant risky forays into hostile territory to obtain supplies. Genoese sailors solved this by bringing from their ships at Jaffa (the defences of which had been dismantled by its fleeing garrison) all sorts of ironware and tools to help build the weaponry. This took some four weeks. Two siege towers were placed against the Tower of David. Teams of soldiers were assigned to build shelters or scaling ladders. Women and children were sent out to collect twigs from which protective matting was woven to protect the timbers. Defending stone throwers bombarded the siege towers trying to break them up and several of the accounts wrote of naphtha and pitch being used as incendiaries by the garrison. On the south side, one account told of a machine using an iron hook on chains.[1]

All the while, the artillery duel continued. An unfortunate man from the garrison was captured while trying to escape with a message for the Egyptians, and was propelled back towards the town from a stone thrower: but he weighed too much for the machine, so landed at the foot of the wall instead — apart from its gruesome aspects, an interesting sidelight on the capabilities of the stone thrower. In the end, the crusaders carried out a remarkable manoeuvre on the night of 9–10 July, shifting the point of their attack to a different sector of the wall altogether, which involved moving a tower and a battering ram about a kilometre from where it had been built near the quadrangular tower to a much more weakly defended sector in the north-west of the city, where the ground was also level. Three stone throwers were deployed with the specific aim of driving the defenders from the walls opposite. The garrison tried to hang sacks over the walls to break the impact, whereupon the besiegers responded with burning arrows to set alight the sacks.

Godfrey of Bouillon pressed his siege tower forward at this point, after first using the mighty battering ram to break through the low outer wall. Having done so, it was necessary for the Christians to burn their own machine in order to make the space for the wooden tower behind, while the defenders, ironically, were trying to keep it in one piece by trying to douse the flames! The tower was driven close up to the wall, where, we are told, it was safe from the stone throwers, which seem to have been deployed in the streets behind the wall rather than on the wall-top itself. Albert of Aachen described how they could no longer target the siege tower because of "the houses and towers surrounding them": if they moved further back, they would overshoot, if forward, the stones would bounce back onto their own side. From its top, which must have at least fifteen metres high to overtop the rampart, the crusaders cleared the wall, allowing attackers to storm in by laying timber planks across from the tower, and by bringing up scaling ladders. At the same time, a fierce attack was being directed at the opposite side of the city by

[1] France, "The text of the account of the capture of Jerusalem", 645.

Raymond of Toulouse, thus dividing the forces of the garrison and preventing it from deploying to prevent the break-in. The consequence was another dreadful massacre, with the victorious crusading army securing its goal, literally marching to the holy sepulchre through a sea of blood.[1] The defenders of the Tower of David prudently surrendered to Raymond.

The fall of the city on 15 July 1099, and the subsequent defeat in battle (near Ascalon, now Ashqelon) of the belated Fatimid relief attempt, crowned an astonishing achievement of dedication and organisation, even taking into account the divided state of their opponents. The repeated battles quickly acquainted the Franks with the tactics of their enemies and how to deal with them, and the sieges must also have involved significant numbers of craftsmen in the construction of the weaponry, while many thousands of crusading warriors must have acquired siege-warfare experience in the course of the conquest.

Cities of the coast

The capture of Jerusalem itself settled little. Without control of the coastal cities, it would be impossible to sustain any new regime at the city. The failure of the crusaders to take Ascalon in the aftermath of their victory there left the rulers of Egypt with a fortified foothold pointed right at the heart of the new state, and it would be more than fifty years before it was finally taken. At the same time, the remaining leaders of the crusade were securing for themselves a land to rule, variously based on Edessa, Antioch, Jerusalem or Tripoli. Promises to restore once-imperial territory to the Byzantines were conveniently forgotten: they had, after all, done nothing to recapture it for themselves. The most significant and continuing problem for the new rulers was shortage of manpower. Although the new subject population was not itself necessarily hostile, the surrounding states were always potentially so, and the army of the First Crusade was greatly diminished, by casualties, disease, and contingents returning to their homes. Baldwin (recognised as king of Jerusalem in 1101) had just a few hundred knights, while there were forces of similar size under the command of Tancred, Raymond and Bohemond. There would be a continuous supply of new arrivals throughout the time of the Latin kingdom, but the numbers were rarely large, and those who tried to come by land were often severely weakened by having to fight their way through Turkish Asia Minor — as happened in 1101.

The year 1101 saw an army assembled to attack Arsuf (or Arsur), which had resisted capture by the army en route to Jerusalem in 1099. On that occasion, a Christian force of 3,000 had prepared engines and mangonels over a period of seven weeks, while the garrison had deployed their own "small mangonels". A crusader siege tower had been prepared, as in all the other sieges we have observed, but this one had been burnt down by the defenders. Undaunted by the loss, another machine had been prepared and propelled across the ditch. Once again,

[1] Albert of Aachen, *Histoire des faits et gestes*, I: 320–39; *Anonymi Gesta Francorum*, 86–90; Fulcher, *A history of the expedition*, 116 ff.; Raimundus, *Historia Francorum*, 656–9; Radulphus, *Gesta Tancredi*, 691 ff.; Robertus of Reims, *Historia hierosolimitana*, 863–7.

it had been set alight. The crusader attempt to douse the flames with water had been useless — this was some kind of naphtha or Greek Fire. Eventually, they had abandoned the attack.[1] The outcome this time was different. A new element in the equation was agreement made by the king with the Genoese that in return for the help of their fleet, they would be granted a share of captured cities. The inhabitants negotiated a safe departure with their possessions to Ascalon after just three days of siege (April 1101).[2] Immediately, Baldwin marched on Caesarea Maritima, described as having a strong wall. He ordered the construction of petraria, and a siege tower, using the masts and oars from the ships. The tower was to be made taller than the town's walls so that it could be used to drive the defenders from the wall top. After 15 days, however, and before the tower was ready, although it appears that the crenellations of the wall may have been damaged by bombardment from the stone throwers, the town was stormed with scaling ladders and the Muslim population slaughtered (if male) or enslaved. An immense wealth fell into the hands of the attackers,[3] who had been denied such plunder by the terms of surrender at Arsuf.

In 1103, Baldwin tried to secure Acre, and deployed his siege engines. The defences, which included an outer wall, were very strong, and the garrison resisted bravely, also being assisted on this occasion by sea borne reinforcements. The success of the Muslims in firing the attackers' siege towers may have decided them to abandon the attempt for the time being, but Baldwin tried again in 1104, assisted by a fleet of 70 Genoese ships. After twenty days of attack with unspecified machines, surrender was negotiated. With its large and secure harbour, Acre was a place of great importance to the kingdom, and of great value to the Italians who secured substantial shares of it.[4]

Bohemond had certainly not forgotten his and his father's previous attempts to expand at the expense of the Byzantine empire. Having returned to southern Italy, in 1107 he took an army to resume the attack on Dyrrhachium (Durazzo, now Durrës), which had fallen back into the hands of the empire. Anna Comnena described movable sheds equipped with rams, and others designed to protect sappers. A ram in a shed covered with hides was provided with wheels, but when in place at the foot of the wall, the wheels were removed and the whole secured with poles. However, the wall was strong enough to resist the attack, so the instrument was abandoned and burnt by the defenders.[5] After a year's efforts in vain, peace was at length agreed between Bohemond and Alexius, whereupon most of the Norman army made their way to the Latin kingdom to support its efforts.

The target in 1109 was the city of Tripoli, in the territory claimed by Count Raymond's son and successor Bertrand. Tripoli had powerful defences, with deep ditches and more than one line of walls. It had been under landward blockade since the crusade had arrived, with a siege castle (called Mount Pilgrim) being

1 Albert of Aachen, *Histoire des faits et gestes*, 1: 385–92.
2 Fulcher, *A history of the expedition*, 152.
3 Ibid., 153–4.
4 Ibid., 174, 176; Ibn al-Athīr, *al-Kāmil fī al-Ta'rīkh*, in *Arab historians of the crusades*, 16.
5 Comnena, *The Alexiad*, 400–401; Fulcher, *A history of the expedition*, 192–3.

constructed on the opposite hill, where the immense stone castle still called Qal'at Sanjīl (Saint-Gilles) still stands. Neither machines nor crossbows had made any impact on the defences when the combined forces of the kingdom arrived by land, perhaps not surprising if the slightly earlier account of the construction of the walls of "hewn stone" was the case, and the Genoese by sea, leading the town to negotiate a surrender after several weeks.[1] Next on the king's list was Beirut, of which almost nothing of the original defences now survives. King Baldwin and Count Bertrand of Tripoli pressed an attack in February 1110, again helped by a Genoese squadron. The attack lasted some 75 days. A battery of stone throwers bombarded the walls, "and shook them with their continuous blows" (Albert of Aachen), and siege towers were again prepared and driven up against the walls. The first two were demolished by stones from defenders' catapults but the third succeeded: Fulcher reported that this time, it was used to gain access to the wall, leading to the city's capture.[2]

The Latin kingdom's successes continued with the capture of Sidon in the same year, this time assisted by a contingent of crusaders from Norway, led by their king Sigurd I Jorsalfare in 55 ships. Again the repertoire of siege equipment was prepared, while the Norwegians blockaded from the sea, and in the process kept the Muslim fleet based at Tyre from interfering. Stone throwers were used by both sides. The Franks again turned to their by now regular weapon, the wooden siege tower, in which were placed crossbowmen, whose bolts would clear the enemy from the immediate area. This time, the defenders used a novel method of dealing with the threat. They dug out from under the walls at night with a view to undermining the tower, but the king was warned and saved the tower by the simple expedient of moving it from the danger area. It was also recorded how the tower was protected by branches of vines and dampened ox-hides, and was mounted on wheels. Finally, after some six weeks, with the tower now in a position to bring about a direct attack, in order to save their lives the garrison negotiated surrender, and the king was pleased to take over an unsacked city (December 1110).[3]

Meanwhile, a significant conflict had been unfolding further east as a new leader, Mawdūd ibn Altūntāsh of Mosul, had launched a counteroffensive in alliance with the ruler of Damascus directed at recovering the county and city of Edessa. Mawdūd was apparently not prepared to take the city on directly, but camped in the surrounding country and ravaged it. Count Baldwin at Edessa appealed for help to King Baldwin, then besieging Beirut, but the latter did not send help until the latter place had fallen (May 1110). The Frankish army succeeded in reprovisioning Edessa, a necessary step in light of the Turkish destruction of the

[1] Albert of Aachen, *Histoire des faits et gestes*, 1: 148; Fulcher, *A history of the expedition*, 195; "The First and Second Crusades from an anonymous Syriac chronicle", 73. Tripoli's walls: Nāṣir-i Khusraw, *Diary of a journey through Syria and Palestine*, 7.

[2] Ibn al-Qalānisī, *The Damascus chronicle*, 100; Albert of Aachen, *Histoire des faits et gestes*, 1: 152–4; Fulcher, *A history of the expedition*, 196–7.

[3] Albert of Aachen, *Histoire des faits et gestes*, 1: 166–7; Fulcher, *A history of the expedition*, 199–201; Ibn al-Qalānisī, *The Damascus chronicle*, 107; Matthew of Edessa, *Armenia and the crusades*, 203–5.

food sources of the area. Mawdūd, who had pulled off rather than face battle, later returned to the attack (in 1112), and nearly gained the city when a traitor let down a ladder from a corner tower, but the defenders resisted and the attack was abandoned.[1] In the same year, the Aleppo chronicle records that Tancred laid waste the lands of Aleppo and attacked the fortified town of Athārib, pressing local Muslim labourers into service to construct siege engines, mangonels and "immense siege towers". The noise of the bombardment could be heard over many kilometres. The defenders tried to communicate their plight to Damascus by carrier pigeon, but the bird was killed by a Frankish archer and the town surrendered in December.[2]

In November 1111, the Christians turned their attention to Tyre. The city was recognised to be a major challenge, as it had been throughout its history because of its island location and powerful landward defences.[3] Alexander the Great's epic siege was known of. This time the attack lasted four months. Seventy-five days were spent constructing two siege towers with battering rams, with a trench all around them to protect them against sudden sorties by the defenders. One tower was apparently more than twenty five metres tall, the other more than twenty. They were moved forward on 11 February, and the defenders responded with Greek Fire (perhaps naphtha), which destroyed the smaller tower. The Franks filled in no fewer than three ditches to push the remaining tower into place against one of the bastions of the wall. The defenders had a novel method of defeating this attack. They had previously built a wooden underpinning in front of their tower, which they now set alight. The effect of the blaze was to loosen the facing of the wall which, tumbling down in front, created an obstacle sufficient to prevent the tower being moved close enough to the wall to do any harm. Another version of the siege says (perhaps more convincingly) that the defenders resisted the siege towers by adding wooden towers of their own atop the ramparts opposite the siege towers, thus reversing the advantage of height, and from their new locations were able to set the Frankish towers alight. The attackers also brought battering rams to bear from the base of their towers, which succeeded in dislodging some of the stones. These were resisted with grappling irons, lowered with great skill over the walls to grasp the ends of the rams and pull them up to break them. The rams were said to be more than thirty metres long. Eventually, after a series of setbacks, the siege was abandoned.[4]

Despite this reversal, with relatively small numbers of knights at their disposal, the rulers of Jerusalem, Antioch and Tripoli had been able to capture one after another the cities of the Mediterranean coast. The presence of a fleet had often been a decisive factor: cut off from any prospect of relief or reinforcement, the defenders were far more likely to seek terms of surrender. The ships also offered,

[1] "The First and Second Crusades", 82, 84–5; Fulcher, *A history of the expedition*, 197–8.

[2] Kemāl el-Dīn, *Extraits*, 547–8.

[3] Nāṣir-i Khusraw, *Diary*, 11, noticed the powerful masonry, including the protection of joints against water penetration, and the small extent exposed to landward attack when he visited in 1047.

[4] Ibn al-Qalānisī, *The Damascus chronicle*, 122–6; Fulcher, *A history of the expedition*, 203–4.

on more than one occasion, a source of timber, so rare on land, with which to construct the siege equipment. Although artillery is frequently recorded, it was clearly the wooden siege tower that represented the decisive offensive weapon in these sieges. Whereas the mangonels and petraries could inflict damage on the thinner upper levels of walls, or perhaps on the weaker outer walls where these existed, there is no account suggesting that this was ever decisive. Instead, it was the mobile tower, with or without a ram in its lower level, that would guarantee success if it could be protected from burning. It may well have been that defending garrisons were small in number, because we do not read of attempts to sally out to burn the towers, or it may have been that, as at Tyre, measures were taken to defend the precious weapons from such a fate. Although we have seen accounts of walls being brought down by sapping, usually under the shelter of a siege tower, undermining was not attempted: presumably the skills were not available.

There were other hazards to the fortifications too. In 1114 (on 29 November), a great earthquake destroyed the north gate of Antioch and parts of the walls and demolished several castles in the county, including Athārib.[1] The ruler of Antioch had to hasten to restore the damage to the city's defences, and to the most valuable of the outlying castles, by impressing labour from the population. In 1115, Bursuq, the new Muslim commander at Mosul, attacked Kafarṭāb with "petraries and other engines" in such a way that the wall was demolished and a huge breach opened, allowing entry and the fall of the castle, which was then demolished (September 1115). It was made clear which engines destroyed the wall, from the eyewitness account of Usāma ibn Munqiḍ, the young lord of Shaizar, who wrote of his experiences. Usāma described how miners from Khorasan dug a tunnel from their own positions under the outer wall until reaching the foundations of their target tower, where they enlarged the tunnel, then stuffed the tunnel with dried timber. "As soon as the fire began to have effect, layers of mortar began to fall. Then a crack was made, became wider and wider . . . then the tower fell. But only the outer face fell" so the attackers retired having been much harmed by the defenders' stones. Bursuq then marched on to Zardanā and "painstakingly built siege engines", but was defeated and killed by Roger of Antioch.[2]

Meanwhile, the king of Jerusalem ordered the construction of the castle of Montréal (Shawbak), the first recorded as being built afresh by the crusaders.[3] It was located so that its garrison could control the surrounding countryside, three days' journey from the Red Sea and four from Jerusalem. It was built in "a short time" in 1115–16 by not many men, on a small mountain. The original extent of the work is hard to recover under much later work: it probably comprised a fairly

[1] Kemāl el-Dīn, *Extraits*, 607; Fulcher, *A history of the expedition*, 210; Galterius of Antioch, *The Antiochene wars*, 81: an invaluable account written by a key figure in the government of Antioch.

[2] Galterius, *The Antiochene wars*, 92–6; Fulcher, *A history of the expedition*, 211–14; Usāma ibn Munqiḍ, *An Arab-Syrian gentleman and warrior in the period of the crusades*, 102–3. During the conflict an expedition from Ascalon attacked the Christian town of Joppa by sea and tried to scale the walls but was repelled by the citizens (Fulcher).

[3] Fulcher, *A history of the expedition*, 215.

simple stone rampart enclosing the hilltop. The shortage of water was dealt with by ingeniously contriving a long tunnel through the rock down to cisterns fed by streams. The next year, another hilltop castle was built at Li Vaux Moïse (near Petra), intended like Montréal to control the adjacent trade routes. The rectangular fortress (approximately 100 × 35 metres) sits on a ridge and has been rendered unapproachable from any side except that of the entrance, which itself is carved through solid rock. There are a few towers, provided with arrow slits.[1]

Siege warfare under a new generation

The struggle in the next period was concentrated in two zones: the unfinished business of the Muslim cities of the Mediterranean coast, and the new Christian state of the county of Edessa, which by its location across the river Euphrates represented a threat, and an insult, to the Muslim heartlands and a potential bridgehead for future Latin or Byzantine expansion into territories not controlled since the original Arab conquests. Edessa was additionally, for the Christians, a city of great symbolic importance. Its recapture in 1144 would be a signal sufficient to launch the second crusade.

Much of the fighting took place in the country between Antioch and Edessa. In 1118, Roger of Antioch attacked the Muslim castle at A'zāz, and drove a mine under the walls, propped up with beams which were then set ablaze; the wall tottered and fell and the Franks stormed in, massacring the garrison.[2] The following year, Roger and most of his army were wiped out by the Turks at the Battle of the Field of Blood (Ager Sanguinis). Back in Antioch, the patriarch mobilised everyone who was left including the clergy to secure the city, assigning men to the towers, while a desperate plea for help went to king Baldwin at Jerusalem. The Turkish leader İlgazi (Il-Ghazi), ruler of Aleppo, however, rather than attempt to follow up the victory by recapturing Antioch, attacked the town of Athārib, on the road between Antioch and Aleppo. Clearly, either the defences had been repaired in rapid time since the earthquake of 1114, or else the damage had not been quite so widespread as the chronicler suggested. Galterius of Antioch (Walter the Chancellor) gave an account of the siege with fascinating detail. After an initial attempt to carry the walls by storm had failed, the Turk sent men from several sides at once to excavate underground caves and filling them with dried wood so that they would collapse when kindled. While this was being prepared, "three or four times a day" they would "deprive the towers of their defences and destroy the defenders with savage blows by way of petraries".[3] The town having surrendered, İlgazi moved onto Zardanā, and brought with him the same artillery as had

[1] The two sites described by Kennedy, *Crusader castles*, 22–6. It is also possible that Montréal stood on the site of an early Byzantine fort. For this and their place within the network of castles in Transjordan, see Vannini, Tonghini and Vanni Desideri, "'Medieval' Petra and the crusader–Islamic frontier", 272–4.

[2] "The First and Second Crusades", 85.

[3] Galterius, *The Antiochene wars*, 146; Kemāl el-Dīn, *Extraits*, 620.

proved so effective at Athārib, and began a bombardment by both day and night until the town surrendered. Meanwhile, reinforcements arrived to secure Antioch, and İlgazi's army was defeated in battle in August to end the immediate threat.

There are two very interesting developments in these sieges. First, the skilful use by the Antiochenes of mining at A'zāz demonstrates the adoption of a technique more commonly used by their opponents. Secondly, the effectiveness of the artillery deployed by İlgazi is unusual. It is suggested that not only were his weapons able to drive the defenders from the wall tops, as before, but that they were able to demolish the tops of the walls themselves, implying a degree of accuracy and a size and weight of missile only hinted at in other sieges. Galterius refered to the same weapon being transported to the subsequent unsuccessful siege of Zardanā, suggesting a particularly powerful instrument. There is however nothing to demonstrate that the *type* of weapon used was different from the traditional stone thrower.

In 1123 Nūr al-Dawlah Belek, İlgazi's successor at Aleppo, pulled off a great victory for the Muslims and captured both Joscelin I, count of Edessa (r. 1119–31), and King Baldwin in besieging the distant fortress of Harput (Kharput, now Elazığ). The castle stood on a rock which Belek ordered undermined. When the props placed along the tunnel were fired, the excavation fell in and took with it the tower above "with a loud noise". Rescued by the Armenians, the king was captured again after a further siege in which Belek's great engines "battered the walls without ceasing", making a breach within a few days (the anonymous Syriac chronicle). Other sources however indicate that the breach was definitely created by undermining.[1] The castle, about the defences of which we do not know, was captured on both occasions, but the captivity of such leading figures had little effect on the conflict.

Tyre, 1124

The kingdom of Jerusalem mobilised a substantial force with the intention, finally, of capturing Tyre. The key was the arrival of the Doge of Venice with a fleet to take part in the attack. Although the king was still a prisoner, the patriarch and leading figures of the kingdom were able to raise sufficient funds to "pay knights and hired footmen" (Fulcher) for the expedition. The strength of the city's location has been indicated during the previous, failed, attack in 1111. William, who was to become archbishop of the city, described the defences. On the seaward side, there was a double wall and equally-spaced towers. On the landward, there was a triple wall, the inner defences being very high with massive towers "so close they almost touch". The Latins deployed on the landward approaches, digging a ditch to protect their own encampment, and used timber and ironwork brought by the Venetians to build siege equipment. The centrepieces were once again siege towers of sufficient height to overlook the city, alongside machines "capable of hurling great

[1] "The First and Second Crusades", 93; Matthew of Edessa, *Armenia and the Crusades*, 229–30; Fulcher, *A history of the expedition*, 253.

stones which would shatter walls and towers". The two towers were covered with dampened oak planks, according to one account, so "that naphtha might have no effect". The defenders attacked the towers with their own artillery, hurling "huge stones" against them which temporarily drove back the Christians. One weapon in particular was causing the greatest damage. William recorded that no one in the Latin camp possessed the skill to aim their own weapons, so an Armenian expert was summoned from Antioch. On arrival, he "showed great skill in directing and hurling . . . destroying every target assigned". The garrison attempted to sally out to burn the threatening towers and other equipment at a time when the besiegers were off their guard. Fulcher recorded that they succeeded, William suggested that they did not, and Fulcher was not present. In the end, it was hunger and blockade that led to the surrender of Tyre (July 1124), brought about by the close blockade, to which the Venetian fleet had been essential, rather than any progress in breaking through any of the three lines of walls.[1]

While the siege of Tyre had been in progress, the Muslim leaders fought over control of Manbij, north-east of Aleppo, attacked with artillery by Belek but relieved by Count Joscelin.[2] The next year, Bursuq of Mosul joined with Tuğtekin of Damascus to renew the attack on Zardanā, but finding it too strong went instead against A'zāz. Twelve catapults were deployed, but the key Muslim weapon was once again mining, which brought down two stretches of wall. However, a Christian army drawn from Jerusalem, Antioch and Edessa defeated the Turks in pitched battle and they fled, burning their siege engines.[3]

The next accounts concern the attack on Baalbek in 1132 by Shams al-Malik, in dispute with his brother who held the ancient and well-fortified city in eastern Lebanon. After defeating the inhabitants in battle outside the gates, Shams ordered catapults set up against the gates, whereupon his brother offered submission. The next attack was against the town of Banyas (the ancient Caesarea Philippi), lying in the plains, which was attacked with what were described as "vaulted shields" providing cover for miners to breach the walls to the town, with archers trying to keep defenders away. The Frankish garrison fled to the citadel before surrendering.[4]

Asia Minor in the 1130s

Meanwhile, in the 1130s, the Byzantine emperor John II Comnenus (r. 1118–43) began to lead frequent expeditions into the former imperial lands of western Asia Minor and Armenia. In 1132 he captured from the Turks the town of Kastamon

[1] William of Tyre, *A history of deeds done beyond the sea*, 9–19; Fulcher, *A history of the expedition*, 265–6; "The First and Second Crusades", 95–6. William of Tyre was not present as a witness at these events (being born around 1130).

[2] Matthew of Edessa, *Armenia and the Crusades*, 232; Fulcher, *A history of the expedition*, 262–3; Ibn Jubayr, *The travels*, 269.

[3] Matthew of Edessa, *Armenia and the Crusades*, 234–6; "The First and Second Crusades", 97; Fulcher, *A history of the expedition*, 278–80.

[4] Ibn al-Qalānisī, *The Damascus chronicle*, 216–17.

(now Kastamonu) in Paphlagonia, which surrendered when the scaling ladders were readied for use, apparently under cover of artillery. The Turks of Danishmend recovered the town, bringing the emperor back in 1135. The Byzantines then recovered Kastamon and went on to assault Gangra (Çankırı), lining up batteries of stone throwers. The walls, however, were too strong to suffer damage in this way, so the attackers changed tactics, and moved their artillery onto higher ground in order to shoot over the walls into the town. The greater range entailed using lighter stones, but their effect on the houses and inhabitants was catastrophic, and the Turks surrendered, although they would also recover this town, starving out the garrison.[1] One might think from the encomium penned to welcome the return to his capital of the victorious Byzantine emperor that he had achieved a glorious success. However, Kastamon was actually a Comnenus family estate, and controlled a region once prosperous but now in decline. The ruins suggest that the castle was a regional centre on a ridge above the town, approachable from one side. With dimensions of 155 metres long by 30–50 metres wide, the most vulnerable side had two walls with large rectangular towers, the inner wall being at a higher level and having one semi-circular and one triangular bastion. The stonework may have been of early Byzantine date.[2]

In 1136 John came with a powerful army to do battle with the Armenian prince Leo I, determined to restore imperial control over Cilicia. Adana and Tarsus fell. The castle of Baka (Vahka, now Feke) was taken after a long blockade, and accompanied by tales of heroic single combat. The town of Anarzaba (Anazarbus, now Anavarza) was more important. Situated on a high rock, it was defended by strong walls and the defenders deployed many engines of their own. While the Byzantines' "round stones . . . hit the towers", the defenders hurled heavy stones and hot iron pellets, which easily burnt the reed-covered shelters for the attackers' engines. A sortie completed the destruction. Frustrated but not deterred, the emperor ordered the construction of new machines, and platforms of clay bricks. The Armenians' hot iron proved ineffective against the mud, and the town surrendered after 35 days as the walls were damaged.[3] These accounts provide evidence of the skill of the Byzantine artillerists.

In April 1138 John joined forces with Raymond of Antioch. The large combined Christian army rapidly captured al-Bīrā (Bile, now Birecik) on the east bank of the Euphrates after just seven days, the towers giving way under the hail of stones, and arrived outside Aleppo itself on 20 April but carried on because the country was waterless. Athārib capitulated the next day. They next attacked Kafarṭāb with mangonels, and it offered no resistance. From there they marched against Shaizar.

[1] Choniates, *O city of Byzantium*, 12–13; this source, not written before the turn of the century, nonetheless is full of remarkable detail. Cinnamus, *Deeds of John and Manuel Comnenus*, 19–21; this writer, an imperial secretary writing in the 1180s, was biased towards his employers.

[2] Prodromus, *Historische Gedichte*, 33–67. For an Italian translation, see *Poeti bizantini*, II: 221–2 (ll. 65–75).

[3] Choniates, *O city of Byzantium*, 15–16; Matthew of Edessa, *Armenia and the crusades* (continuation), 241–2.

The castle still stands, crumbling, on a steep ridge overlooking a bridge across the Orontes, the approachable side cut off by a dramatic ditch. There was a large garrison drawn from the whole surrounding area, which first fought outside the walls but was driven back inside. An initial attempt at storm was repulsed, and the next day, according to the recollections of the town's lord, "they set up frightful mangonels which they had brought with them . . . These mangonels could throw a stone to a distance farther than the distance covered by the arrow." The Byzantines forced a breach in the town's wall, "which the mangonels had opened". The battery was formidable: 18 mangonels and other weapons called *la'ba*. The stones would crush a man's head. The defences of the castle itself however held out until, after ten days, the news of an army marching from the east to bring relief led the Byzantines to abandon the siege and burn their formidable battery.[1] There are some inconsistencies in the different accounts, but logic suggests that the attack took place from the side of the town, which is low and flat, and that the town wall was probably weak and may have been of mud brick, enabling the Byzantine army to break through quite easily. William of Tyre, John Cinnamus and Niketas Choniates suggest that Usāma bought off the emperor with lavish presents, which he would not have needed to do if relief was imminent. Whichever was the truth, the castle of Shaizar remained uncaptured, and it would remain an independent Muslim emirate throughout the whole period of the Latin kingdoms in the east.

Syria and Palestine, 1139–42

1139 witnessed a fierce attack on Baalbek by a new leader, 'Imād al-Dīn Zengi of Aleppo. A great battery of artillery was deployed against the town, 14 catapults hurling 1,000 stones a day on the place. This rate of fire (if true) would have been perfectly attainable, but would have required numerous teams of rope-pullers to operate the weapons, working in relays, and a large number of men employed to collect and shape the ammunition. Baalbek surrendered on terms once a breach had been made,[2] the siege having lasted from 28 August until 10 October, with all the sources agreed that it was the bombardment that was critical on this occasion. Zengi had also made his mark in capturing the Latin frontier castle of Mont Ferrand (Bar'īn, near Hims) in 1137, in which King Fulk of Jerusalem (1131–43) had been trapped. William of Tyre records that the "walls shook under the impulse . . . Millstones and huge rocks fell in the midst of the citadel, shattering houses." It was, however, hunger that forced the surrender of the castle after the garrison had eaten their horses and everything else edible.[3]

In 1140, it was the turn of the Christians, who attacked the oft-disputed

[1] Usāma ibn Munqiḍ, *An Arab-Syrian gentleman*, 143–4; Kemāl el-Dīn, *Extraits*, 676–8; William of Tyre, *A history of deeds done beyond the sea*, 95–6; Choniates, *O city of Byzantium*, 17–18; Matthew of Edessa, *Armenia and the crusades* (continuation), 242; Cinnamus, *Deeds of John and Manuel Comnenus*, 24.

[2] Ibn al-Qalānisī, *The Damascus chronicle*, 255; "The First and Second Crusades", 274; Kemāl el-Dīn, *Extraits*, 681.

[3] William of Tyre, *A history of deeds done beyond the sea*, 90.

frontier town of Banyas, arriving outside on 1 May. They set up "the hurling engines called petraries" whence the stones "shook walls and crushed buildings". The besiegers however decided that they would make no progress without their traditional weapon, the siege tower, but were hindered by the complete absence of suitable building material. Now, once again, this operation had an unexpected form: the besiegers included the ruler of Damascus, who sent there for "tall beams . . . which had long ago been set aside for such a purpose." Thus equipped the tower was constructed, the approach flattened and the weapon "attached to the ramparts", whereupon the garrison surrendered on terms (12 June).[1] Meanwhile, the Byzantines were once again engaged further north, and it is recorded that in 1142 John Comnenus secured the submission of fortified islands in Lake Pousgase (Beyşehir Gölü), north of Attaleia (Antalya), constructing floating platforms by lashing boats together and then mounting artillery on them.[2] Once again, the skill of Byzantine engineers is revealed by this improvisation.

The capture, recapture and destruction of Edessa, 1144–6

Edessa had powerful defences and two citadels. After it had survived several attacks by the Persians in the early sixth century, Justinian had further strengthened the walls. In the mid eighth century, the caliph had ordered them razed, only for local nobles to pay for their reconstruction in 814. The upper citadel had been reinforced by the Byzantine general George Maniakes in 1032, and this had held out against the Arabs in 1036, and had resisted the victorious Turks after Manzikert. Recently, in 1086, a wall with towers had been thrown up between the town and the citadel.[3] Maniakes' citadel was a long narrow enclosure 350 metres long but only some 30 wide, with a double wall on the outside face, standing on rocky ground high above the city, with one narrow gate.[4]

Zengi had set his sights on recovering this important county from the Christians during the absence of its lord, Joscelin, and most of his warriors. Artillery was set up all around, and mines were started on the north side, against the wall around the "Gate of Hours". The digging had the effect of crumbling the wall above by destroying the foundations. The mine was then filled in and naphtha was put in "that they might burn like a torch". The wall "swayed and then fell". The defenders, apparently aware of what was going on, had constructed a defence inside the wall, but were undone by the collapse of a longer stretch of the wall than they had expected. When they refused Zengi's offer of terms at this point, the inevitable assault followed, and after a heroic resistance in the breach the town finally fell and was put to the sack. Thousands tried to reach the citadel but the

[1] Ibn al-Qalānisī, *The Damascus chronicle*, 265–6; William of Tyre, *A history of deeds done beyond the sea*, 109–12.

[2] Choniates, *O city of Byzantium*, 22; Cinnamus, *Deeds of John and Manuel Comnenus*, 26.

[3] See Segal, *Edessa, "the blessed city"*, for a detailed historical account.

[4] Detailed plan in Hellenkemper, *Burgen der Kreuzritterzeit in der Grafschaft Edessa*, plate 76, and photographs at plate 3.

gates were shut, and the approach was very narrow, so they died in a horrendous crush, Zengi himself stepping in to halt the carnage. Two days later, with no hope of relief, the citadel also surrendered.[1] In 1146, Joscelin managed to gain re-entry to his former city and perpetrated a massacre of the Muslims therein. The new ruler at Aleppo, Zengi's son Nūr al-Dīn, raced to repair the loss, sending 10,000 horsemen by forced march to Edessa. Joscelin's small force was quite unable to resist such an overwhelming assault. Twenty of his knights took refuge in a tower known as the Water Tower. The Turks "set to work to undermine and in less time than it takes to tell it fell down".[2] The city was subject to a worse sack than it suffered in 1144.

The Second Crusade and Ascalon

The first of the crusader states had fallen, and the significance was not lost on contemporaries. Relatively exposed though it was, Edessa had nonetheless stood in an area with a strong Armenian Christian presence and good links to its north. The blow to the Christians was appreciable. A call for a new crusade was issued to repair the damage, and it was answered by two of the greatest rulers of Europe, Louis VII of France and Emperor Conrad III of Germany. The result was an unmitigated fiasco, for which Latin writers blamed the Byzantines rather than the Muslims. The treatment of the immense hosts as they marched overland might have been mistaken for that of an enemy from the accounts of dreadful deeds, deceit and cheating that survive. The army that finally reached the Latin principalities was sadly diminished.

No Byzantine forces accompanied the crusade for good reason. In 1147, Roger II, king of Sicily and duke of Apulia and Calabria, took advantage of a local dispute to seize the town and citadel of Kerkyra (Corfu), then proceeded to raid the coast of Greece, sacking Thebes and Corinth, nowhere meeting any resistance from the Byzantines. Perhaps it was no surprise that Manuel I Comnenus (r. 1143–80) should have felt little sympathy with the Franks pouring through his empire. In 1148, as the Second Crusade made its fateful progress to Damascus, and Serbs and Hungarians raided in the Balkans, Manuel was trying to recover Kerkyra from the Sicilians. The problem was that the citadel stood on a rock rising sheer from the sea, and was protected by strong walls that had been reinforced by the garrison. Manuel assembled a large army and fleet, and along with his Venetian allies, engaged in a prolonged siege. To start with, their attacks were fruitless. While the great height of the defences ensured that every missile hurled from the walls was effective, their low emplacement made the stone throwers of the besiegers useless against the defenders or the wall tops. The best the Byzantines could do was to lob back the shot hurled on them by the Sicilians' mangonels. The emperor tried

[1] "The First and Second Crusades", 283–4; Ibn al-Qalānisī, *The Damascus chronicle*, 267–8; Kemāl el-Dīn, *Extraits*, 685–6; Matthew of Edessa, *Armenia and the crusades* (continuation), 243–4.

[2] Ibn al-Qalānisī, *The Damascus chronicle*, 275; Matthew of Edessa, *Armenia and the crusades* (continuation), 244–5.

to break the deadlock by devising an immense ladder, built like a tower from ships' planks and masts, to reach from the ships to the foot of the citadel. This was actually achieved, and some brave soldiers scrambled to the top, but the ladder then broke. There followed a dispute with the Venetians that ended in fighting. Manuel, however, was not prepared to lose face over the island, and remained pressing a blockade until hunger drove the defenders to seek terms in July 1149.[1] The use of the ship-borne ladder had not been a success, but perhaps remained in the memory of the Venetians, who would use the same method in 1204 to conquer Constantinople.

Meanwhile, a council of war of the kings of France, Germany and the Latin states resolved on 25 May 1148 to attack Damascus, the largest of the neighbouring Muslim statelets, albeit one whose rulers had sometimes found common cause with the Christians in the past. Alerted by reports of the arrival of large armies, the commander at Damascus, Tuğtekin's mamluk emir Muʿīn al-Dīn Unur organised his defences. Equipment was prepared, wells and watering places were filled in, and weak spots repaired. The city lay in a flat plain, watered and irrigated but without any natural protection. The Latin writers recorded that much of its strength was provided not by the walls (which were of ancient origin) but by the dense network of orchards that surrounded the city, with trees planted close together and each orchard marked off by a wall, offering to attackers a close-knit hedge probably even harder to negotiate than a single wall, and to defenders ample opportunity to impede and harass an attack. Anyway, the vast Christian army attacked on 24 July and fought its way through the obstacles, to find itself confronted by the garrison across the river "using bows and ballistae" to keep them from the water. Obviously desperate for a drink, the German emperor in person led his knights on a foot charge to clear the Muslims from the bank. The besiegers then cut down trees to provide themselves with a wooden stockade to protect their camp. But they were unable to make any progress. Damascus was large, and the emir's appeal to his neighbours was answered by a continuous stream of reinforcements that enabled the Muslims to turn defence into attack, while Nūr al-Dīn and his brother Saif al-Dīn Ghāzī marched west to help too. William of Tyre has a story that the Damascenes bribed some of the Franks to persuade the kings to move their camp to a more exposed position, alleging that the walls there were of mud and would not require machines to break down, but this is rather unlikely. In truth, the crusaders had found themselves trapped in an arid landscape outside a large hostile city with a constantly reinforced army inside. It was not long before they decided to abandon their enterprise and withdraw across the desert, harried all the way. It was an ignoble and utterly fruitless end to a vast expedition.[2]

Some consolation for the Latin cause would come with the capture of Ascalon, the coastal town that they had failed to take a half century before, and that since

[1] Choniates, *O city of Byzantium*, 43–52; Cinnamus, *Deeds of John and Manuel Comnenus*, 45 ff.

[2] William of Tyre, *A history of deeds done beyond the sea*, 186–92; Ibn al-Qalānisī, *The Damascus chronicle*, 283–7; Ibn al-Athīr, *al-Kāmil fī al-Ta'rīkh*, in *Arab historians of the crusades*, 59–62.

then had been a base for naval operations as well as threatening by raids the southern part of the kingdom of Jerusalem. By 1152 it had walls of solid masonry bonded by very hard cement, wide, thick and tall, with a line of outworks, and was provided with wells and cisterns. There was a barbican in front of the east gate. Disarray among the Fatimids now made it a tempting target. An army drawn from the whole kingdom and including the forces of the religious orders, Templars and Hospitallers, arrived with an arsenal of siege weapons and the siege began in January 1153, with the attackers having ships with which to complete a blockade. The ships also provided timber for the construction of a very high siege tower, carefully protected against fire. The same source of timber was used to build more stone throwers and mantlets. Once the protective embankment had been levelled, the tower was moved forward and from its top, missiles were directed into the town. The garrison sortied out to set the tower alight, but the wind blew back the flames which instead of destroying the tower, loosened and crumbled the wall, collapsing it overnight. Forty Templar knights decided to keep the glory for themselves and entered the breach but denied anyone else access. They paid dearly for this stupidity. Having killed them all, the Egyptians repaired the breach with ships' timbers. The siege became a blockade, but the Egyptians failed to send naval forces to try to break it, and in August the garrison offered to surrender on terms, which were granted.[1] It would be the last victory for the Latin kingdom: meanwhile, to the east, Nūr al-Dīn had created a unified Muslim power in Syria.

Fortifications of the crusaders and Armenians

Little significant innovation took place in the design of the fortifications of the region. In the north, the Armenians had been dwelling in the Cilician and Edessa regions since being moved there following the Byzantine reconquest of the late tenth century, and the area they had inherited was dotted with major fortresses of earlier times (Tarsus, Adana, Anarzaba and Edessa) and also the hilltop refuges and watch towers dating from the time that the Taurus mountains had been a frontier zone between the eastern Roman empire and the Arab caliphate. Remains of any medieval period are few, and those from the eleventh century are rare indeed.

Nonetheless, one study has isolated no fewer than 20 distinct features of the castles of Cilicia. Many of these were determined by the geography. The castles consisted of walls following the contours, with well constructed stonework keyed into the rubble cores, and a slight batter at the foot. Being built on rock, no special foundations were required, and indeed in most cases they were immune from the risk of mining, nor were there any ditches because the terrain ruled out siege towers. Rarely were outworks needed, because of the sites. In common with traditional fortifications of the region, towers were round or semicircular, entrance

[1] William of Tyre, *A history of deeds done beyond the sea*, 219–31; "The First and Second Crusades", 301–2.

approaches were covered by walls or towers, and like the Muslim defences, gate passages were usually bent at 90°. Cisterns were provided in every case, and there were no donjons. The slot machicolations in the gatehouses covering the entrance passage are said to have been an Armenian invention, with the slot a maximum of 40 centimetres wide. How many of the Cilician sites date from the new kingdom of the late eleventh century, and how many were there before, it is impossible to know.[1]

A similar picture obtains in the county of Edessa. Here there are donjons at Anarzaba and Sis (now Kozan), and possibly within the citadel of Edessa (Şanlıurfa) itself, although the first two may date from later occupation by the Hospitallers or Teutonic Knights, or the reign of King Leo, while the earlier creation of a donjon by the Frankish rulers of Edessa would have been perfectly understandable. There is no evidence to indicate the origin of many castles, although the style, of narrow hilltop sites protected by a curtain wall, suggests the eleventh century, with noblemen exerting local power the likely builders.[2]

Donjons also feature in the early castles thrown up by the crusaders further south. While their rule was based in the already powerfully fortified cities, their control of the land depended on extending the system of lordship which they had brought with them from the west, and it was no surprise that the new Frankish landlords should build the type of castle with which they were familiar. But donjons were also built inside the existing walled towns and cities. At Byblos (Gibelet, now Jbail) there is a rectangular tower of the early twelfth century inside a rectangular bailey, featuring reused columns from the ancient Roman city (Plate 24). At Tarṭūs (Tortosa), an important port in the north of the county of Tripoli, there is a very large donjon 35 metres square inside a complex of castle and town defences, while here the cathedral of Our Lady was also provided (in the thirteenth century) with defensive corner towers. In the country, it was not often possible to use existing defences as a point of departure, although there is a donjon added to the existing Byzantine defences at Saône (Qal'at Sahyun / Ṣalāḥ el-Dīn). This was part of a rebuilding carried out by the castle's lords, perhaps as early as the 1120s. They were presumably responsible for the immense ditch, thirty metres deep, twenty metres wide and nearly 150 metres long, which cuts off the long spur on which the castle and its associated village stands from the easiest approach route. This work must have involved immense labour. Rising sheer from the ditch sides, the crusaders also constructed extended curtain walls with large square towers to add to the protection of this end, and the two-storey donjon, 24.5 metres square, stood in the middle of the east curtain, over the ditch. The wall heads of the surviving towers show defences at two levels: arches at roof level and then a raised wall walk with crenellations and arrow slits. It is hard to be sure of the dating of each feature here, because there is a mixture of building styles, and while this probably had something to do with the existing knowledge of local masons, it may also suggest reworking at different times before the castle fell in 1188.

[1] All from R W Edwards, *The fortifications of Armenian Cilicia*, 4–19, 25–34.

[2] Hellenkemper, *Burgen der Kreuzritterzeit*, 255–70.

Standing today rather as a symbol and exemplar of crusader castle architecture, the immense pile of Crac des Chevaliers (Qalʿat al-Ḥuṣn) came into the possession of the Knights Hospitaller in 1144. A previous Arab fortification had stood at the site, known from earlier occupancy as the castle of the Kurds (Ḥiṣn al-Akrād). The Knights' original defences comprised a solid curtain wall on the highest point of the ridge, with a few slightly-projecting rectangular towers, now for the most part contained within the massive rebuilding that turned Crac into the strongest and most sophisticated fortress of the Latin states.[1] At this initial stage, however, it would have appeared little different from other Frankish fortresses in the region.

The castle of Saint-Gilles (Qalʿat Sanjil) at Tripoli itself, originally built as we have seen to blockade the city, was constructed around a quite small donjon (17 × 11 metres) with a bailey without flanking towers. Much of the original building is now lost. North of here, there are two further rural castles of the twelfth century that illustrate the normal practice of the time. Chastel Rouge (Castrum Rubrum, Qalʿat Yaḥmūr) is a small two-storied vaulted donjon standing in the middle of a roofed courtyard with two opposite corner towers (perhaps added later by Arab conquerors), and an external stairway to the roof. It was owned by the counts of Tripoli, who passed it later to the Hospitallers. It is now a pen for farm animals! Chastel Blanc (Burj Ṣāfītā) is much grander, a donjon of 31 × 18 metres, of three floors standing (the lowest of which has walls four metres thick) originally inside two walled courtyards on a high hill with spectacular views from the roof. There is a cistern below the tower. The present remains (Plate 25) date from the reconstruction of the tower for the Templars (with an immense chapel occupying the ground floor) after serious damage in an earthquake and the handing over of the castle to the Order in the 1170s.[2]

In the south, several other castles were constructed to offer a shield against raiders from Ascalon during the reign of King Fulk (r. 1131–43). Bethgibelin (Bayt Jibrin, now Beit Guvrin) seems to have been an enclosure 50 metres square, with towers at each corner, inside Byzantine city walls. Blanchegarde (Tell al-Ṣāfī) seems from William of Tyre's description to have been similar. In all cases, ancient sites provided building materials. Skilled masons must have been employed to carry out the construction itself. The most impressive castle in the south was built in the 1140s by a local lord, Payen (or Pagan) the Butler, at Karak (Kerak), who moved his caput there from Montréal, the new site being close to the Dead Sea. Karak stands at the end of a ridge, cut off by a deep ditch on its approachable side,

[1] Kennedy, *Crusader castles*, 145–7. The massive literature on Crac deals mainly with the castle as it stood in the thirteenth century.

[2] Byblos: Kennedy, *Crusader castles*, 64–8; Müller-Wiener, *Castles of the crusaders*, 64; Tartus: Hanna, *The castles and archaeological sites*, 9–19; Kennedy, *Crusader castles*, 132–8; Saône: Saadé, *Le château de Saladin*; Hakim and Jawish, *Saladin und seine Burg*; Kennedy, *Crusader castles*, 84 ff.; Saint-Gilles: Kennedy, *Crusader castles* (brief), 62; Chastel Rouge: Müller-Wiener, *Castles of the crusaders*, 52; Kennedy, *Crusader castles*, 72; Burns, *Monuments of Syria*, 193; Hanna, *The castles and archaeological sites*, 58–60; Safita: Kennedy, *Crusader castles*, 138 ff.; Burns, *Monuments of Syria*, 213; Hanna, *The castles and archaeological sites*, 50–54. For the greatest details and dating, see Deschamps, *La défense du comté de Tripoli et de la principauté d'Antioche*, 249–88.

and constructed of rough masonry, with rectangular flanking towers and plentiful arrow slits. The side facing the town over the ditch was a plain wall protected with more arrow slits, and the gate was a small, narrow entrance. A large reservoir was dug at this most vulnerable side, that would serve as a very effective moat in time of need. One would expect to find a donjon but there is now no sign, much having been rebuilt after its capture.[1] Several of these castles seem to have been constructed with an explicit military purpose of offering protection against Muslim attacks from Ascalon.[2] Others had the function of enabling their occupants to protect people using the road from Jaffa to Jerusalem, the main route for pilgrims: the three castles of Yazur (Casel des Pleins or Casale Balneorum; now Azor), Latrun (Toron des Chevaliers; see below for its Iberian connections), and Yalu (or Yalo, Chastel Arnoul; now in Canada Park) stood along the route and were all entrusted to the control of the Templars at different times. Yazur's remains comprise a rectangular tower in a bailey, 12.6 × 12.8 metres. Yalu is first mentioned in 1106 as a royal foundation, subsequently destroyed then rebuilt by the patriarch of Jerusalem in 1132/3 in the form of a quadrangle 50 × 40 metres, before being entrusted at some unknown time to the Temple.[3]

Muslim counterattack and Christian offensive in the western Mediterranean

In the Spanish peninsula, the victories of the Castilians and Leonese in the eleventh century had established a new frontier zone much further south than before, and had provoked the arrival of a new dynasty of religiously-motivated rulers from Morocco to seek to reverse the losses which followed on from the political fragmentation of al-Andalus. In the first half of the twelfth century, Christian advances were to continue to the point where the Almoravids, having failed, were themselves overthrown by a new religious movement in their Moroccan homelands, the Almohads.

The Almoravids restored a united Muslim state in al-Andalus, but whenever they needed to find sufficient resources to launch a major offensive, they had to bring over large armies from North Africa. At first, the new power halted the Christian advance, and for several years there was conflict across the width of the peninsula, with fortresses changing hands in Portugal, but the key to restoring Muslim fortunes was recapturing Toledo, and a serious attempt was made by the Almoravid ruler 'Alī ibn Yūsuf in 1110/11. A detailed account is given in the *Chronicle of the emperor Alfonso* (Alfonso VII (r. 1126–57), so called because of his pre-eminence among the Spanish kings), a key piece of evidence for this period. 'Alī prepared scaling ladders, and great machines of iron and wood "to overcome

[1] Kennedy, *Crusader castles*, 45–9. Kennedy also supplies the information on the other southern castles listed in this paragraph.

[2] Such at least is the explicit description from an enemy of the building of Gaza: Usāma ibn Munqiḍ, *An Arab-Syrian gentleman*, 34.

[3] Pringle, "Templar castles between Jaffa and Jerusalem".

the towers of the city of Toledo". Destroying the outlying castles at San Servando and Aceca (Azecha, now Villaceca de la Sagrada), 'Alī arrayed the machines "in opportune sites" and attacked with all manner of weapons. The Muslims prepared "furtively" to attack the base of a tower situated "at the head of the bridge" opposite San Servando, and then tried to set their frame alight by projecting inflammables from their ballistas. The defenders, however, managed to dowse the flames. Furious, 'Alī ordered an all-out attack the next day, and from every direction stones, darts, arrows and javelins were hurled from the engines and attacks made with rams and scaling ladders. The defence held, however, and from inside the town other artillery returned the bombardment. On the seventh day, the garrison launched a sortie and succeeded in burning all of the siege engines. The great Muslim army, although it managed to take a number of towns and for example destroyed the walls of Madrid and Talavera (de la Reina), was unable to take the strongest places "which in our language are called *alcazares*". Ibn al-Kardabūs recorded that 'Alī could make no progress because "the Christians took refuge in impregnable fortresses and castles".[1] Alfonso then ordered the walls of Toledo reinforced from river to river.

In 1101, Valencia was recovered from the followers of the Cid, again by long blockade.[2] In 1111, the business of the reconquest was delayed by civil strife and warfare between the Spanish kings themselves. The bishop of Santiago de Compostela had to besiege the "Castrum Daravum", deploying "tormenta . . . ballista ceteraque machinamenta" to bring about surrender. In the opportunity provided, the Muslims retook Santarém.[3] In this territory, exposed to raids, having strong defences seems to have been a regular concern for the ruling bishops. Already in 1108 we find him ordering the rebuilding of the "Castrum Honesti" at Santiago with outworks (*propugnacula*) and a "high tower" so that if the Muslims attacked, they could be resisted with stones. A dozen years later, on the advice of his canons and the princes, the bishop ordered the strengthening of the very same castle with "a very strong tower" in the middle of the castle: a clear suggestion of a kind of donjon that shared its residential and symbolic functions of lordship with a definitely defensive purpose.[4]

The Balearics

Not for the first time, the rulers of the Italian city states now beginning to establish their power and independence, intervened again in the western Mediterranean. In 1113 the Pisans organised a major expedition to try to wrest the Balearic islands from the Muslim emir Mubassir Nasīr al-Dawla. It was a large undertaking that must have stretched Pisa's resources. The poetic account that survives described

[1] *Chronica Adefonsi imperatoris*, 75–9; *Anales toledanos*, 367–8; Ibn al-Kardabūs, *Historia de al-Andalus*, 136–42.

[2] *Chronicon lusitanum*, 418–19; *Anales toledanos*, 386–7.

[3] *Historia compostelana*, 136; *Chronicon lusitanum*, 419.

[4] *Historia compostelana*, 73–4, 306.

the preparation of numerous ships, including those specifically used to carry the materials for the construction of siege towers, bridges, ladders, rams, mantlets and catapults (*balista*).[1] The Pisans set sail on 6 August 1113. Their circuitous route took them to Catalonia, where they made a beneficial deal with Count Raymond Berenguer III, and recruited many additional supporters from the region including the returned crusader William V of Montpellier, and no doubt many others who had fought with the Provençal contingents on the First Crusade. They came ashore in Ibiza in June 1114. The island's capital was strongly located with marshy terrain around it, as well as having a triple wall, each protected by a ditch. While preparing their heavier equipment, the attackers began to probe the defences with assaults and attempted escalades. They had to fill the ditch before bringing rams into action, all the while resisting the bombardment from the defenders, but were unable to bring the machine to bear. A siege tower was also readied although clearly not used. Instead, the walls were severely damaged by a combination of bombardment and sapping, and after tough fighting in the breach the Pisans and their allies finally broke in, leading the Muslim commander to surrender (rather than retire to the inner defences).[2]

August saw the army transfer to Mallorca, landing near the capital, Palma. This place had a citadel, and an old and a newer city, each protected by high towered walls and ditches. The defenders had also had plenty of time to collect supplies and prepare the defences. The attack began as at Ibiza, with skirmishing and probing, while siege towers and armoured mobile shelters (*tecta*) were prepared and launched in a massive assault. The garrison resisted with equal vigour, hurling great stones, erecting their own wooden extensions to negate the height advantage of the Pisan towers, and sallying with inflammable materials ("sulfure flammas") to set light to the mantlets. Although the Christian accounts do not directly suggest it, the assault was clearly defeated with such effect that no further offensive operations were undertaken until early 1115. Then, in February, four siege towers led the renewed assault, attempts to burn them were defeated, and instead the Muslim's own wooden edifice was successfully destroyed by fire, creating the situation in which the Pisans and their allies could storm the outer defences. The Muslim king withdrew behind the next line of walls. In March a further assault led by the siege towers and a bridge thrown across the ditch carried the next line and it was at the beginning of April that the citadel finally fell to sustained attack.[3] The operations in the Balearic islands were remarkable for their sustained level of effort and the immense resources that must have been mobilised to achieve two successful conquests of vigorously defended cities. While the fact that evidence is provided only by two sources both of which were concerned to glorify just one of the participants must lead to some caution in accepting their accounts at face value, the likelihood is that the description of the sieges themselves is fairly reliable. It is striking how the full panoply of methods familiar from the crusades was

1 *Liber maiolichinus de gestis Pisanorum illustribus*, 10–11 (ll. 121–5).

2 Ibid., 43–58 (ll. 975–1380); description of the walls at ll. 1255–64.

3 Ibid., 88–129 (ll. 1550–3460); *Gesta triumphalia per Pisanos facta*, 91–3.

deployed against the islands, with the result here that both cities eventually fell to assault. It is also striking that in each case, although neither Ibiza nor Palma could expect any relief from outside, each managed to hold out for many months against such an array. The triumph of the Christians was itself short-lived, because neither of the conquerors had the resources to hold on to their gains, and the Balearics returned to Muslim rule in 1119.

Zaragoza, 1118

The big event of the early years of the century on the mainland was the capture of Zaragoza, which had been the capital of a separate Muslim principality since the early days of the Arab conquest. There are different versions of the events of the fall of Zaragoza, which was a strongly defended city with Roman walls and a palace of the ruling family, within its citadel (of which little of this date remains), all within a mountainous region with its own heavily fortified outlying towns and castles. However, over the previous two decades, the Aragonese had gradually been hemming the territory in with the capture of the surrounding towns (Monzón, Almenara, Huesca, Barbastro). In 1118, King Alfonso I of Aragon with armies from his own kingdom and Navarre, and with the support of Gaston IV of Béarn from Gascony, and other troops from France, undertook the siege. It was recorded that Gaston was a returning crusader and that he brought with him particular siege warfare skills as a result. The Christian army deployed "lofty wooden towers . . . on wheels" and "thundering machines . . . as well as 20 mangonels". The outcome of the siege was the surrender of Zaragoza in December after an operation of unknown length, and different accounts of whether an Almoravid army attempted relief but was defeated, or arrived too late. All sources agree however that despite the noteworthy siege weaponry deployed, the city fell through hunger.[1] It does not appear that, as in the Balearics, returning crusaders brought any new weaponry with them, as siege towers were of course a western export in the first place. It may be that refinements were made. Certainly, the Compostelans, a few years later, were able to build an array of equipment to attack a castle protected by ditch and rampart and to bring about its downfall by mining.[2]

The 1130s in Iberia

Warfare between the Christian kingdoms saw sieges of towns and castles involving blockades and assaults where powerful defences proved again and again that, where the attacker could not afford to remain for months on end, the advantage

[1] *Historia de la corona de Aragón*, 66–8; Maqqarī, *The history of the Mohammedan dynasties in Spain*, I: 303; *Anales toledanos*, 389. For a modern study, see Lacarra, "La conquista de Zaragoza por Alfonso I", 68–92.

[2] *Historia compostelana*, 444. The siege was in 1126, of the "turris Taberiali" held by 36 knights.

lay with the defence.[1] Meanwhile, the close connections across the medieval world are illustrated by the story of Count Rodrigo González de Lara. Having been relieved of his position as the emperor Alfonso's commander in Toledo (1135), he took his crusading across the Mediterranean and built the castle of Toron (Latrun) near Ascalon, which he then entrusted to the Templars. It seems that this castle was a rectangular enclosure 72 × 55 metres enclosing a tower 14 metres square with walls three to four metres thick, an outer court of uncertain date, and a gate protected by a portcullis. An outer enclosure is probably of later date.[2]

The city of Coria was Alfonso's target in 1138, and the count sent to take the place deployed a full arsenal once again. Wooden siege towers rose above the ramparts, and mantlets protected miners and sappers. Knights, archers and slingers used the siege tower, but once the commanding count had been killed by an arrow, the attack was abandoned. Meanwhile, Alfonso brought an army to capture the Muslim castle at Oreja (near Ontígola). This was recorded as being powerfully fortified and having a large garrison. Alfonso ordered his engineers to construct siege towers and other weapons, but the account suggests that the vital blow was the success of the besiegers in blocking the castle's water supplies. The defenders negotiated surrender when 'Alī informed them he could not challenge the Christian army to break the siege. The emperor then resumed the attack on Coria, and apparently deployed the same weaponry as on the previous, aborted attempt: a wooden tower to dominate the wall, catapults and mining to destroy the towers. This time the garrison were not so fortunate, and as 'Alī was again unable to offer the prospect of relief, they surrendered to save their lives (June 1142).[3]

The 1140s: North Africa, Lisbon and Almería

While the Second Crusade wound its disastrous way to Damascus, other Christian armies attacked Islam in the west at the same time, and Roger II (r. 1130–54), recently recognised as king of Sicily, was expanding his realm, although it would be hard to ascribe any crusading motives to a rapacious monarch as happy to attack the emperor at Constantinople as the emir at Tripoli. In 1143, a Sicilian fleet was sent by King Roger to attack Tripoli. Arriving on 26 June, they created a breach in the wall by attaching hooks, which sounds like they were sapping. On this occasion, the defenders sortied and drove off the "Franks". Three years later, a larger fleet resumed the attack. On this occasion there was dissension in the town, as a result of which a stretch of wall was left unwatched, allowing the invaders to get in by escalade. The solid masonry of the defences was thus of no avail.[4]

[1] *Chronica Adefonsi imperatoris*, 16, 18, 19, 24, 41 (Morón de la Frontera, Castrojeriz, Bayonne).

[2] Ibid., book I. Kennedy, *Crusader castles*, 55 (where the author suggests that the origin of this castle is unknown); Pringle, "Templar castles", 94–102.

[3] *Chronica Adefonsi imperatoris*, 107 ff.; *Anales toledanos*, 390–91.

[4] Bakrī, *Description de l'Afrique septentrionale*, 19, a text completed in 1068. Tripoli is described as having a strong stone wall destroyed by the Franks in Idrīsī, *Description de l'Afrique et de l'Espagne*, 143. Idrīsī was a contemporary.

New occupiers then spent six months strengthening and restoring the walls and digging a deep ditch, which the writer, Ibn al-Athīr, could still see as he compiled this account.[1] The Sicilians aimed to establish a number of bridgeheads on the North African coast through which they controlled a very lucrative trade.

The attack on the major Muslim port and city of Almería, the city refounded with powerful *ṭabīʿa* walls by ʿAbd al-Raḥmān III in the tenth century,[2] involved naval forces supplied by Genoa, and a Genoese chronicle is a major source for the event. The Italian city had become a significant political force as it strengthened its trading relations across the Mediterranean, alongside the other Italian city-states of Venice and Pisa whose aid to the crusading cause was always at a price. In 1118, as Zaragoza fell, a Genoese fleet of 80 galleys attacked Pisa, accompanying "four large ships carrying machines and all the instruments necessary". In 1147, having made a treaty with Alfonso about the division of spoils, the Genoese spent five months preparing the substantial fleet of some 180 vessels including ships carrying "many wooden machines". Travelling via Minorca, which they found too strong to attempt, the force landed near Almería and marched around the city. The defenders made a sortie against the landing site, but were defeated and driven back behind their walls. The Genoese then set about constructing "cats" and other machines including mangonels. They were joined outside the city by the emperor Alfonso with a large Leonese army, which had marched overland, taking and destroying the castle of Andújar (July 1147) on the way, followed by Baños de la Encina and Baeza (August). The count of Barcelona also brought an army. The junction of forces took place in August, and the attack was pressed with engines including a siege tower, against which the Muslims concentrated their attack, hurling stones at it from their mangonels, but in vain. Reaching the wall with the tower and other engines, the besiegers succeeded in demolishing a stretch one hundred metres long. The defenders capitulated.

The following year, the Genoese continued their successful operations with the capture of Tortosa, in alliance with the count of Barcelona. Siege towers were prepared from timber collected from woods a long way from the city, and then moved around Tortosa in July 1148. In a complex and dangerous operation, the ditch had to be filled and the towers brought across. Recognising where the danger lay, the Muslim defenders concentrated their attack on the wooden towers, hurling stones up to "200 pounds in weight", and damaging one of the two towers. The assault was maintained however, and the heavy stones from the attackers' stone throwers damaged the walls and destroyed their palaces and houses. The governor sought a truce, promising to surrender if no relief came within 40 days. The Genoese and Catalans entered the city in December. It is not clear how strong the defences of Almería were at this moment. The immense citadel above the town has suffered later reworking, earthquake damage, but mainly from modern restoration. The strength of the original fortress however was certainly considerable, and it was the

[1] Ibn al-Athīr, *al-Kāmil fī al-Taʾrīkh*, extracts in Ibn Khaldūn, *Histoire des berbères*, II: 579–80.

[2] They were also extended in the early eleventh century as the city grew, with the work being of *ṭabīʿa*: see Torres Balbas, *Ciudades hispanomusulmanas*, 561.

combination of a simultaneous land and sea attack with the inability of the then rulers to bring armed relief that led to surrender.[1] Tortosa had been reputed to be very strong when fortified several centuries before (see Chapter 3), but clearly the armoury available to the attackers in 1147 was too strong for it.

A further attack was taking place in Iberia at the same time, as a crusading fleet from northern Europe intending to travel to the Holy Land was persuaded to attack the Muslims in Portugal instead, and conveniently left behind several detailed accounts of their capture of Lisbon. The king of Portugal (recently recognised as a separate entity) had been driving the Muslim southwards, while the latter had taken every opportunity to recapture lost towns and castles. The area immediately north of the river Tagus features regularly in accounts of the construction of castles and their fall, and recapture. One such new castle was Leiria, founded on a hilltop site in 1135 by the Portuguese ruler Afonso I, according to the chronicles being established both for the protection it offered against Muslim raids, and the base it provided for harming the enemy. Despite leaving a garrison of several hundred knights, Leiria, which was perhaps not fully defensible yet, fell to a Muslim counterattack. Within nine years it had changed hands back and forth at least four times, before Afonso brought in help from Coimbra (itself not long in Christian hands) to guarantee possession.[2]

Lisbon was an important centre, situated in an area of such flourishing natural fruitfulness as to attract the astonished notice of the northern European witnesses. Its capture received a great deal of international recognition across Christendom, with accounts being widely circulated. Significant weaponry was once again deployed by the besiegers. But it would still take many months to crack the defence. The Muslim city lay below the fortress (now called São Jorge), protected by a towered rampart descending steeply from the hilltop to the water. The citadel, in the form of an irregular square, had strong walls and 11 square towers, a traditional Muslim format. Lisbon, which had changed hands twice, had undergone a number of refurbishments. The town itself was so prosperous that it regularly outgrew its walls, something that caused problems for both attack and defence when the net closed in 1147. No less problematic was the fact that there were more than usual mouths to feed as a result of the flight there of many Muslims from the surrounding towns of Santarém, Sintra, Almada and Palmela fleeing the Christian advance. Various attempts have been made to establish the trace of the city wall of the twelfth century, but this has been made very hard by subsequent rebuilding, the Lisbon earthquake of 1755, and disappearance beneath more recent buildings. The citadel has also suffered from a drastic "restoration" in

[1] Caffaro, *Annali genovesi*, I: 16 (1116), 33–5; the "Poem of Almería", at the end of the *Chronica Adefonsi imperatoris*, 165–86; *Anales toledanos*, 390. The poem breaks off in mid-verse and makes no mention of the Genoese role; for both sieges, Caffaro, *Ystoria captionis Almerie et Turtuose*, in *Annali genovesi*, I: 79–88.

[2] *Chronicon lusitanum*, 420, 424, 426; *Annales D. Alfonsi Portugallensium regis*, 151–3; *Chronica Gothorum*, 12–14; on Leiria, see Gil, *The finest castles in Portugal*, 124–5. The castle as originally built probably comprised a keep and stone enceinte around two wards, in which a palace was then constructed.

the twentieth century. The wall had an exterior-facing length of some 1250 metres, joining the citadel at each end, and five or six gates, mostly narrow and straight. There were numerous projecting rectangular bastions, mostly open-backed. The surviving Torre de Alfama now stands eight metres high but is lacking its original top.[1] Much of the strength lay in the steeply-climbing site.

The accounts[2] are full of fascinating detail. A force of relatively low-ranking crusaders, led by Count Arnold III of Aarschot, Christian of Gistel and Saher of Archelle, from Flanders, and Simon of Dover, Andrew of London and Hervey Glanvill met forces from Cologne in Dartmouth harbour, whence the combined force of 164 ships set sail on 19 May. They came from Cologne, Flanders, Boulogne, Brittany, Scotland, Normandy, London, Bristol, Southampton and across East Anglia. On 28 June they entered the estuary of the Tagus river. Initial skirmishing succeeded in driving the Muslim troops back from the suburbs. The Anglo-Normans were on the west side of Lisbon, the Germans and Flemings on the east. Around the beginning of August the building of siege equipment began. These were rams, sows (presumably armoured or protected mantlets) and, critically, a siege tower, one on each side down by the shore. This was presumably the only place where the ground was flat enough to permit such a cumbrous instrument to be deployed. The German-Flemish machine was called a "walking tower", the Anglo-Norman was an immense 95 feet high. The assault was launched but was a disaster. The tower on the west become stuck in mud, and after being subjected to three days' and nights' bombardment by the Muslim mangonels it was finally set alight. The ram fared no better.[3] The besiegers regrouped and began to reconstruct the siege towers. The *De expugnatione Lyxbonensi* describes how two "Balearic mangonels [*fundas*]" were built. "All these men having been divided into groups of 100, on a given signal the first 100 retired and another took their places, so that within the space of ten hours 5,000 stones had been hurled". The same men apparently undertook to use five of these weapons "to shatter the walls and towers", but clearly did not achieve this result, and the siege towers would turn out to be the decisive element. Duodechin says that a Pisan began to build one new machine on 8 September, and it was finished in mid October.

Meanwhile, mining had been begun by the men of Cologne ("subterraneas fossas"). But the defenders were alert to the danger and although we are not told how, managed to thwart this first underground attack, causing considerable losses. There was discouragement among the crusaders, although they would have been heartened by the capture of letters between the garrison and the Muslim governor of Évora, the former explaining their serious plight (growing hunger, nowhere to

[1] The most detailed survey is Vieira da Silva, *A cêrca moura de Lisboa*; see also Velloso da Costa, "Lisboa islámica", 296–301. The 1899 plan is reproduced in *De expugnatione Lyxbonensi*, facing p. 13. On the citadel, see Gaspar and Gomes, "O Castelo de S. Jorge", 397–404.

[2] *De expugnatione Lyxbonensi* was apparently written by an Anglo-Norman; there is also a letter from a German, Duodechin, inserted in the *Annales Sancti Disibodi*, 27–8. The two accounts are very close to each other in their detail.

[3] *De expugnatione Lyxbonensi*, 128–36; Duodechin in *Annales Sancti Disibodi*, 28.

bury the dead) and the latter excusing himself from action on the grounds that he had a truce with King Afonso. Anyway, the Germans resumed mining in a new spot, creating a gallery with five entrances, and they were rewarded with success when the mine was fired and 66 metres (200 feet) of wall collapsed. However, because of the steep ground along which the wall ran, the Muslim soldiers who rushed to hold the breach were able to drive off the Flemish and German attack, and when the English and Normans offered to help, they were firmly told to find their own way in; perhaps the creators of the breach suspected that their allies wanted to take advantage of their efforts to lay hands on the spoils. The other account makes the breach much smaller, only some 20 metres, and Da Silva's survey intriguingly noticed a stretch of this length that had been patched in a different stonework near the point where the rampart reached the walls of the citadel.[1] The crusaders on the other side accepted the challenge, and the siege tower on the west was moved forward the next day, towards the Porto do Ferro. From its top, archers and crossbowmen drove the defenders from a wide stretch of the defences: the towers were open-backed, making this task easier. The attackers themselves faced a major problem: the incoming tide cut off the machine and the several hundred men left to protect it. The garrison launched attack upon attack on the isolated tower, but somehow it survived, with the men inside dowsing the flames while resisting the frequent sorties. Mangonels were moved up to bombard the tower, and the engineer himself was wounded, but despite having to face such attacks through two high tides the siege tower remained undamaged, and on the next occasion it was manoeuvred forward to within four feet of the wall. From here, a bridge could be lowered to enable attackers to charge across onto the rampart. At this moment, realising they were a short time from being stormed and facing a bloodletting, the Muslim commander laid down his arms and offered to surrender. This took place on 21 October.[2]

Some of these Christian victories were short-lived. Almería returned to Muslim control in 1157/8. 'Abd al-Mu'min brought an army from Granada. The Genoese and Leonese gave battle outside but were driven back, and the Muslims entered the town with them, forcing them back into the citadel. This was despite the frantic work of the occupiers to strengthen the defences of town and citadel. A relief attempt launched by Alfonso VII was defeated, the emperor dying at Baeza, which opened its gates to the Muslims. Inside were found "numerous machines of war". Almería's garrison surrendered in exchange for their lives.[3]

[1] Vieira da Silva, *A cêrca moura de Lisboa*, 43.

[2] *De expugnatione Lyxbonensi*, 136–76; Duodechin in *Annales Sancti Disibodi*, 27–8; *Chronicon lusitanum*, 426; *Annales D. Alfonsi*, 157; *Chronica Gothorum*, 15; Gil, *The finest castles in Portugal*, 162–5.

[3] Caffaro, *Ystoria captionis Almerie*, in *Annali genovesi*, I: 51; "Un recueil des lettres officielles almohades", 39–41; Ibn Khaldūn, *Histoire des berbères*, II: 192.

The Almohad (al-Muwahhidūn) revolution

The Muslim leader who recaptured Almería was not an Almoravid, but came from a movement that had just completed the destruction of the Almoravid empire in North Africa and al-Andalus. Beginning like most such movements with a single dedicated individual proclaiming himself the representative of true Islam, and his enemies heretics, the new leader Muḥammad ibn Tūmart gathered a number of Berber tribes around him before challenging the Almoravid power in the Maghreb. In 1129, following defeat in battle, the Almoravids were pursued back into their main city of Marrakech, only recently constructed by 'Alī ibn Yūsuf (r. 1106–43), who had surrounded it with walls. On this occasion, the insurgents were driven back. The Mahdi Ibn Tūmart died and was succeeded by 'Abd al-Mu'min. For the new movement, victory depended on capturing the fortified towns of North Africa. The desert tribesmen must therefore rapidly have acquired the skills needed to capture walled towns, just as the Almoravids had before them.

'Abd al Mu'min proved to be both an inspirational leader and a successful commander. His series of victories including captures of Almoravid towns began in 1132. While the Christian powers were taking advantage of the disarray among the Almoravids, the foundation of their kingdom in Morocco was falling away. Tlemcen fell, despite the strength of walls, with the besiegers apparently constructing their own fort against it, then Oran, which was too strong to assault (its walls being built "with art") but surrendered because of a shortage of water (1144). Fès (with its wall built of *ṭabī'a* of mixed clay, sand and lime on a wooden box frame, rendering it very resistant to siege weapons[1]) and Ceuta acknowledged the new power without resistance. The critical event was the siege in 1146 of Marrakech. The powerful walls resisted the Almohad assault for seven (or nine, or eleven: three sources give three different figures) months, but the increasingly desperate defenders tried a sortie. This was driven back, and in the pandemonium the Almohad forces entered the gates at the same time as their fleeing enemies, and the city fell, the victors consolidating their conquest by massacring their opponents and killing 'Alī ibn Yūsuf. The following year, Ceuta rebelled and had to be forced into submission, the walls being razed as punishment. After this, 'Abd al Mu'min began to extend his power into al-Andalus, and most local commanders in general submitted quietly to their new masters. In 1152, the Almohads extended their empire into Ifriqiya, occupying Béjaïa (Bougie) and Algiers and storming Kalâa des Béni Hammad. This last was a castle, prosperous and sited on a ridge, but its walls were stormed from the side on the plain.[2]

[1] Ibn Faḍl Allāh al-'Umarī, *Masālik el abṣār*, 156–7: "more solid than stone and on which machines of war are without effect and do not even leave any traces"; Torres Balbas, *Ciudades hispanomusulmanas*, 560.

[2] 'Abd al-Wāḥid al-Marrākushī, *Histoire des Almohades*, 178–82; Zarkashī, *Chronique des Almohades & des Hafçides*, 6–9; Ibn Khaldūn, *Histoire des berbères*, II: 174–91; "Notes d'histoire almohade I", 376, 383, describes Marrakech as being welcome to Ibn Tūmart

Having recovered Almería, in 1159 the Almohads returned to Africa to recapture Mahdia, which had once again — as in 1087 — fallen to a sea borne attack in 1158, the invaders coming this time from the kingdom of Sicily. Taking advice from the ousted governor, 'Abd al-Mu'min decided not to try to assault the powerful stone defences that cut off the narrow peninsula on which the town stood, and instead he ordered a blockading wall erected to keep the Franks in while his ships prevented access by sea. The town possessed a citadel and a wet ditch as well, but once the Sicilian fleet had been defeated (September 1159) the final surrender of the town could not be prevented, for lack of food, although it held out for six months until January 1160. The Sicilian writer Hugo Falcandus (possibly a pseudonym) attributes the defeat to the treacherous actions of the chancellor Maio, acting in the name of the new king William I (r. 1154–66), but the other accounts all say the relief fleet was defeated in battle.[1]

The conqueror spent twenty days overseeing the restoration of the defences, and installed a garrison and substantial provisions to prevent a recurrence of the loss. He then proceeded along the coast to recover Roger of Sicily's other North African footholds at Tripoli and Sfax,[2] about which another preserved letter from 'Abd al Mu'min gives great detail. The town was strong, with high ramparts, but much of its strength derived from its location in a rocky terrain barren of water and food, close to the desert. Thus the first task facing the Almohad force was to secure sufficient supplies. As part of this process, grain stores were constructed in the besiegers' camp. Driving off sorties by the garrison, they then set about levelling the buildings outside the town so as to be able to surround it closely, and to begin to fill in the ditch. This done, they resolved to continue the attack with artillery (*manājaniq*), and their construction was "providentially aided" by the presence of considerable quantities of "building timber" brought by Christian sailors previously. This was transported to the siege and the engines constructed. Another request for surrender was rejected, so the machines were deployed against the wall as the ditch was levelled. Soon, they were able to capture an outer wall and demolish one of the towers, whereupon the town surrendered.[3]

Almohad fortresses

Since arriving from their desert bases the Almohads had plenty of experience of some of the most advanced and sophisticated building techniques of the age, and of the means of capturing them. These would be reflected in their own new building. The first base of Ibn Tūmart at Tinmel (Tine Mel) in Morocco was fortified with a rectilinear defence with quadrangular towers every 30–35 metres,

because of its important strategic location, and that the first siege lasted 40 days. Town descriptions from Idrīsī, *Description de l'Afrique et de l'Espagne*, 92, 96, 100, 106.

[1] Falcandus, *The history of the tyrants of Sicily*, 78–9.

[2] Ibn Khaldūn, *Histoire des berbères*, II: 193; Ibn al-Athīr, *al-Kāmil fī al-Ta'rīkh*, extracts ibid., II: 581 ff.; 'Abd al-Wāḥid, *Histoire des Almohades*, 196–7; Zarkashī, *Chronique*, 12–13. Description of Mahdia, Idrīsī, *Description de l'Afrique et de l'Espagne*, 126–7.

[3] "Un recueil des lettres officielles almohades", 45–7.

of which the survivor is eight metres by four. The lower stretches of the wall are of large stone blocks, changing higher up to concrete mixed with stone and with timber frameworks. The mixture of building materials has been explained by the need for haste. By the end of the century, with the North African empire consolidated, concrete was used throughout, and the technique of construction had been perfected, as exemplified in the powerful walls of Rabat. Tinmel had a simple gate, bent to the right. At Tit (al-Fiṭr), on the coast south of El Jadida, a *ribat* was built by the Almohads to offer a defence against seaborne assault. Again, there is a mixture of materials in the walls. The walls to seaward have strong towers with effective flanking. There is a substantial provision of slits and machicolations. Bent entrances feature in the well-constructed gates.[1] The important town of Taza was fortified by Abd al-Mu'min from 1135 with double walls and strong towers behind a ditch where the terrain itself did not suffice.[2]

In al-Andalus, the Almohads would also undertake some serious work on fortifications. At the *alcazaba* of Badajóz, where the town had been refounded as we saw in the late ninth century, their work would be substantial, and again carried out at speed because of the continued attacks on it from Portugal. Key features included the extensive use of the *torre albarrana,* but of a distinctly stronger form with better protection for those using it, paralleled in the early-thirteenth-century Torre d'Oro of Seville; a low outer wall (four metres) backing a narrow five-metre intervening space, totally overlooked by the much taller (nine-metre) inner wall, providing a concentric design; and further sophistication of the bent gateway, with a solid four-sided tower projecting a long way to create a dangerous roadway to the outer gateway, which would be paralleled in work at two of the gates (the Puertas de la Justicia and de las Armas) of the Alhambra of Granada.[3]

These refinements to existing designs of Muslim fortifications represent a fusion of the features being developed across the Almoravid and then the Almohad empires, which in both cases saw their heartland as being in North Africa rather than al-Andalus, but in which the most sophisticated features were to be found north of the Straits of Gibraltar. We do not know who built these works, but it must be that much was learnt during the struggles between the rival powers in Iberia. All the rulers and many of the fighting people would have taken part in siege after siege, perhaps from both the sides of the attack and the defence, and although many of the conflicts did not resolve themselves through close assaults, there were enough examples of attack and defence of the wall to show where weaknesses lay, and methods of making improvements. It was clear that the main threat to defences was represented by the enemy getting close enough to break down, sap or mine the wall, so the best defence was to keep the attacker as far from the wall as possible. Hence the importance of flanking towers and outer defences, both of which were features of Almohad building.

[1] Basset and Terrasse, *Sanctuaires et forteresses almohades,* 13–41 (Tinmel), 343–63 (Tit), 382–91 (Tasrhimout). The building of Tinmel is recorded by Ibn al-Athīr, *al-Kāmil fī al-Ta'rīkh,* extracts in Ibn Khaldūn, *Histoire des berbères,* II: 573.

[2] Mabrouk, "La fortification médiévale de la ville de Taza", 845–48.

[3] Torres Balbas, "La alcazaba almohade de Badajoz", 168–200.

In India

Written sources for the Indian subcontinent are very thin for this period, although the establishment of Muslim states in the north of the country was by now having a clear impact on many aspects of life, including fortress design. One state for which there is helpful chronicle evidence for the first half of the twelfth century is Kashmir, in the north-west of the region, where Kalhaṇa, a member of a family well placed to observe the goings-on, wrote a colourful account, the final part of which has exceptional value because it is contemporaneous.

The mountain kingdom was often at war with its neighbours, and there were several civil conflicts as well, all described in sufficient detail to be able to analyse the warfare of the era. King Harṣa (r. 1089–1101), for example, "threw into the mêlée burning arrows smeared over with vegetable oil", confirming the use of fire arrows by Indian armies. Traditional siege methods were employed by this king where the defences — often hilltop fortresses — were too strong to assault. Prājimathikā in the north, about to fall as its food ran out after a month's blockade, survived because the besieging army was bribed to leave. King Jayasimha (r. 1129–49) marched to the relief of his fortress of Awantīpur in 1129: this place, recorded as having strong stone walls, had successfully resisted assault with stone throwers that had created a breach. In the same year, a rebel occupied the fortress of Banasala (Banihal), which was then besieged by royal troops who bombarded with catapults and arrows, the defence replying by rolling down stones. After a month a breach was made in the lower defences, enabling the attackers to seize the water tank. Encouraged, they attempted a storm, but were repelled, and the place only fell when it was betrayed.

In 1144, a royal army besieged the castle of Śirahśilā (by the temple of Śāradā pīṭha near Shardi on the river Nīlam or Kishanganga). For the first four months, no progress was made because the attackers had failed to completely cut off the place, though they had large forces and had cleared the forest to set up their camp, with local villagers pressed into providing labour services. The terrain however enabled them to set up attacking points from which to bombard the castle "with showers of stones" while the river circling the hill site was also the site from which stones were taken as ammunition. Eventually, the siege was pressed more vigorously with a "large quantity of war engines", and the few defenders, then cut off from the river, and submitted to continuous bombardment and day and night noise from drums, surrendered.[1]

Although covering just one corner of the subcontinent, and limited to a few decades of contemporary evidence, this chronicle suggests that the siege techniques that were assumed to be available in earlier centuries, sometimes only on the basis of the classical examples, were well established at this time, while similarly, the fortress builders, whose earlier forms were noted in chapter two,

[1] All from Kalhaṇa, *Rājataraṅgiṇī*, I: 344 (arrows), 360–61 (Prājimathikā, Dudhagāta), II, 116 (Awantīpur), 132–33 (Banasala), 199–207 (Śirahśilā).

maintained the same fundamental approach as before. The evidence does not allow a determination of what form the artillery took, although by now it would be very surprising if it were not of the lever-arm stone thrower kind. Evidence for the adoption of the crossbow suggests that this was only imported under the Delhi sultan Iltutmish (r. 1211–36), although as indicated before, Indian armies did dispose of larger arrow-shooting weapons.[1]

Some conclusions

At the beginning of the crusades, there was little difference between the technical and military abilities of the western and eastern forces that came into collision. In the wood-poor lands of Syria, as in most of Iberia, building in stone or brick was the natural way to proceed.[2] Guaranteeing water supplies was another essential requirement, leading to the creation of large cisterns inside most fortifications. Byzantine and Muslim defences came from the same mould, and generally comprised strong enceintes, protected by towers and ditches where the terrain allowed, with smaller citadels provided in the cities as the defended base of the governor. The power of the Muslims, and their defensive efforts, were mainly concentrated in cities. The Christians brought large numbers of "private" castles and the donjon, the ultimate in private, feudalised designs. However, since the option of building in wood was not available, and the craftsmen and masons employed would have been brought up with local building skills, it was not a surprise that the donjon apart, the style of crusader fortifications was not very different from their neighbours'. Many of the sites noticed here were, like their European counterparts, quite small, although both Karak and Saône were exceptional in their great size, which may have owed much to their locations, pre-existing defences, and intended functions. A general change would come during the twelfth century as many castles were handed over by their lords to the new orders of religious knights, particularly at this time the Hospitallers and the Templars with their original blend of religious, charitable and military functions. Their domestic arrangements, involving a monastic type of common dwelling, called for different designs of the living space within the castle, while the numbers of men, both knights and lesser brethren and servants, required much larger constructions. We have little evidence of any new construction by Muslim rulers in this first part of the twelfth century, except for the repairs to those cities that found themselves newly in a frontier zone, such as Aleppo, Hims and Damascus, while in al-Andalus there is little to suggest that Almoravid building styles differed from those they inherited. It would be their Almohad conquerors who would bring further refinements.

[1] Bhakari, *Indian warfare*, 99, citing Pant, *Studies in Indian weapons and warfare*.

[2] Only one motte (seven metres high) appears to have been uncovered in the Latin states, south-east of Sidon: see Pringle, "A castle in the sand", 188. This interpretation is now less clear following further excavation.

One difference between west and east would have been the sheer scale of siege warfare in the Middle East. The Muslim armies were already organised in a different system, with slave troops and mercenaries in large numbers supplementing the private forces of the leaders, and with a substantial proportion of mounted archers. Large numbers of specialists, both in building artillery and in digging mines, were also available. Yet for most of the period studied here, the crusader armies relied upon the trusted equipment they had known back home, in particular the siege tower. Requiring immense quantities of rare timber, slow to build, difficult to move and vulnerable to burning, the tower nonetheless remained a standard weapon from Nicaea to Ascalon. While the accounts of its use actually suggest that it was very effective, it should also be noted that, apart from inland sites like Jerusalem where other factors intervened, it was rarely effective on its own. Requiring the filling in of ditches and the breaching of outer walls, its use was inevitably costly in manpower and time. There is nothing to compare to the achievement of the crusader capture of Jerusalem by this method in all the other examples. Siege towers were also deployed with decisive effect in several of the major sieges in the west, as well, and the sheer size of the machines recorded suggests a significant degree of technical mastery of design and engineering. Battering rams were also common to West European experience, and again there are accounts of their use. They had the same drawbacks as the siege towers, but at least offered a way to break through a wall. Few of the accounts suggest that rams were located in the bottom floor of siege towers, although many imply that the tower enabled work to be carried out to break through a wall without specifying exactly how.

The Muslim armies all seem to have had access to skilled miners, and the several detailed accounts show a common method of attack which succeeded time and again. There is only one account of the crusaders using mining in the east up to this time (as opposed to reaching the base of the wall and chipping away at it from ground level), Roger of Antioch's attack at A'zāz in 1118, and that might possibly have been by sapping instead. The men of Cologne used mining effectively at Lisbon, and this is one of the earliest clear references to the use of the weapon by westerners. Yet when the Muslim armies used it, they succeeded, and there was as yet no effective defence unless the fortification stood on terrain that prevented such an attack. The defenders of Edessa in 1144 could only think to abandon the undermined wall and construct a new defence inside, but they got their calculations fatally wrong.

One of the great concerns of the Muslim rulers was the absence of native supplies of raw material essential to their building and their siege weaponry: timber and iron. They went to some lengths to secure these vital supplies, and their main suppliers were the very same trading cities, Venice, Genoa, Pisa, Amalfi, Bari and from Sicily, that also provided essential support to the crusader principalities — in exchange for privileged trading relationships. Alongside a regular trade in cloth, furs and silver was another trade in iron, timber and pitch, three items regarded as essential to their military activities by Muslim governments and therefore monopolised by them. In exchange, the Muslims granted privileges to the Italian cities. Most of the trade had, of course, to go through Egypt once the crusader states were established. A specific treaty survives from 1173 in which Pisa undertook to supply Saladin with iron, timber and pitch.[1] The church embargoed such trade, but the search for profit has never been much bothered by such strictures, even when Christian supplies of timber enabled 'Abd al-Mu'min to build the weapons he needed to attack the Christian garrison of Mahdia.

Timber and iron were, of course, essential to the construction of stone throwers, and the use of artillery on a massive scale is clearly indicated by the middle of the twelfth century. This was the mangonel or petrary as it was called by contemporary sources. Whether the generic names concealed different sizes of machine cannot be proved. Even though we know the Byzantines had long before given different names to weapons of different sizes and capacity, their chroniclers of this time use only general terms. There is no reason to doubt that machines of different sizes were employed by those who had inherited the traditions of building and operating them, but none of the chroniclers provides the evidence to confirm this. However, there are repeated indications that some machines were able to project very large stones causing structural damage, while others fired missiles designed to kill people. The substantial and growing number of statements that stones from catapults were damaging walls might merely indicate that it was the much thinner wall top crenellations that were shattered, a very useful purpose for the attacker in denying cover to the defenders if an escalade was attempted. But accounts of walls themselves being shaken and broken suggests more than this. Beyond confirming that heavier stones were now being propelled, which might seriously harm weaker walls, it is not possible to be categorical. But the description of the relays in use at Lisbon is valuable in confirming how efficient this artillery could be. It remains the case, however, that there are hardly any examples where the use of artillery, on its own, was responsible for the fall of a fortification. It would require a new *type* of weapon to produce that effect. This is not to play down the awful effectiveness of a bombardment on the occupants of the place under siege, which was recorded many times, and would be brought about by a sustained bombardment over the walls by a large battery of weapons, which would only be possible with a very large force of artillery, such as the Muslim armies seem increasingly able to deploy. But the horrible impact of bombardments by crusader and Byzantine artillery is confirmed by Usāma ibn Munqiḍ. The other, easily overlooked, role of artillery was in defence, and many of the sieges examined here tell of stone throwers being used to great effect against siege towers, rams and their operators.

It has been argued that the chief effect of this growing array of weaponry was psychological, as so few places were actually captured directly as a result of their use.[2] There is some evidence to support this view, although many smaller places were taken as a result of siege equipment being used, and places of all sizes were sometimes taken by the walls being breached by mining or battering rams. It is, however, wrong to separate the psychological from the material effect of a bom-

[1] Abulafia, "The role of trade in Muslim–Christian contact", 4–10; Ashtor, *A social and economic history of the Near East*, 114, 196–7, 240.

[2] García Fitz, "Tecnología militar y guerra de asedios".

bardment or an attempt to break in. It is irrelevant whether the wall was brought down to induce surrender, or whether the garrison surrendered before the attack started, either way the capture was down to an estimation of the likely outcome. The garrison of a place that surrendered before an assault was far more likely to be able to negotiate favourable terms, whereas once a breach had been opened, the terms would become harsher, and if the place was carried by storm, the defenders could expect no mercy. By 1150, ever larger armies with ever more powerful siege trains were finding more effective ways to reduce the stone-walled, towered and ditched defences commonplace in the south. Yet sieges could remain very time-consuming because of the strength of the sophisticated and powerful defences often reinforcing an inaccessible location. In the north, the situation was different.

7 The twelfth century in northern and central Europe and Byzantium

Background and overview

International travel and international contacts were commonplace for the ruling élites of medieval Europe, and inherent dangers and risks were accepted hazards. The first years of the twelfth century saw thousands of returning crusaders bringing back what they had learnt from the east, while thousands more went in the opposite direction.

This enhanced exchange of experiences between east and west was accompanied by various important fundamental changes taking place in society, some of which would have long-term consequences for the map and history of Europe. National identities were slowly becoming established, apparent in the comments of the chroniclers. It may not have been clear what was meant by "French", but it was increasingly used to distinguish subjects of the Capetians from the Germans and English. The westerners were all Franks to the Muslims, but English, Germans and Netherlanders certainly understood the differences, defined by more than different vernacular languages. The attribution of offensive "national characteristics" to other peoples had its forerunners in the abusive characterisations of other peoples by twelfth-century monks.[1] It is still premature to talk of evolving "nation-states" across Europe, but the process was underway and was accompanied by a consolidation of royal governments relying on bureaucracies to collect taxes and administer justice, rather than the personal rule of the monarch. This process would be of paramount importance in the history of fortifications, because control of fortifications had almost everywhere and at all times been an attribute of royalty, even where it could not be exercised in practice.

During this century, the unusually highly-centralised government of Norman England was emulated as the French monarchy successfully extended its control across the vast land that theoretically acknowledged it as king. German kings, instead, had to recognise the division of their even larger territory into semi-independent principalities. This did not prevent emperors such as Frederick I Barbarossa bestriding the second half of the century as monarchs of awesome

[1] The emergence of national characteristics took place alongside the reality that, for agricultural economies in which the majority of people rarely travelled beyond the nearest market, "community" could be no more than one's village, and "foreigners" could begin with the next village.

authority and reputation, but it did place a far greater emphasis on the personal qualities of the ruler to mobilise resources over which the crown did not possess direct control. One part of Frederick's dominions in particular confronted the German king with a new problem. In Italy, the cities of the densely urbanised north had begun to secure political independence to match their rapidly growing economic power. To them, the German emperor was a foreign tyrant denying their natural right to self-rule, while to the emperors the cities were upstart and common rebels against the authority of a ruler whose right went back through Charlemagne to the Roman empire itself. No one in 1150 could have guessed that it would be the cities that would win. Elsewhere in the peninsula, the Norman kingdom of Sicily would stand out as one of the most spectacular new arrivals, as the dynasty of adventurers consolidated a unique blend of the skills and cultures of Islam, Byzantium, the Lombards and the Normans into one of the great powers of Europe, surviving the frequent hostility of their neighbours. This polyglot kingdom, like the Byzantine empire, was following a different trend from the process of nation-creating happening elsewhere.

Underlying the political evolution were important economic developments.[1] It is widely accepted that the population of Europe had been growing since the tenth century. This was reflected in the clearance of forests for agriculture, and of virgin lands on which to establish people encouraged to move from old lands to new. Because there was little in the way of technological development to increase agricultural productivity, putting new land under the plough was the only way to feed new mouths. Towns were growing in size and significance, most notably in northern Italy and in the southern Netherlands where trade (for example, in wool) and manufacture were expanding, but also and more unevenly across Europe, to meet the needs of the expanding rural population. The consequences of the greater wealth and resources of Venice, Genoa and Pisa for the political and military struggles in the Mediterranean have already been noted.

Some rulers were quick to make sure that this development was to their advantage. Many took the opportunity to establish mutually beneficial relationships with the towns, an increasing number of which began to be established. Economic expansion also meant a greater surplus of income for governments, if it could be exploited. For the rulers who had the systems and the power to increase their revenues, the possibilities for increasing their own political authority were considerable, and during the century there was an increasing reliance on hiring military service in exchange for wages, alongside the restricted free service owed by landowners, which gave the princes a greater freedom of action. It also permitted the hiring of engineers and experts in the construction and attack of fortresses, and increased the resource disparity between the crown and potential noble rebels. The strenuous efforts of kings everywhere to enhance their status as God-given supreme rulers, backed powerfully by the Church, which itself relied on royal protection, reflected the substantial distance the medieval monarch had

[1] There is an excellent summary of the available evidence for medieval economic development in Pounds, *An economic history of medieval Europe*, 90–122.

travelled from being the chief warrior of a war band. But being the chief warrior, along with law giver, remained a central attribute of the monarch, even when he was unable to put a stop to the private warfare that often raged among the nobility and knighthood.

The Church was the single largest landowner of Christendom, and the new wealth saw an early reflection in the great cathedrals that began to rise wherever the resources permitted. The skills and techniques now available across Europe were of the same order as allowed the magnificent architecture of Islam and Byzantium to flourish. The twelfth century would see the development of the striking new style subsequently termed "Gothic". Such building skills were also available to the secular owners of palaces and castles, providing they had the money to finance the masons, carpenters and workers. It was the availability of resources, not the availability of skills that separated northern and central Europe from the south. There was still a significant disparity in art, literacy and other traditional measures of civilisation.

The kingdom of France

Suger, abbot of Saint-Denis, close confidant of King Louis VI, instigator of the rebuilding of his abbey in gothic style, and author of an invaluable life of his king that provides the only detailed contemporary evidence for much of this important reign, was no stranger to military and civil matters despite his ecclesiastical position. His history is replete with classical references, which causes its own problems of interpretation.[1] Where his evidence can be cross-checked with the Anglo-Norman chronicle of Orderic Vitalis, or the Angevin chronicles, a more confident overall picture can be established.

The Capetian dynasty had the unparalleled luxury among the royal families of Europe of an unbroken succession of long-reigning kings, avoiding disputed successions and long minorities. However, as the century turned, King Philip I of France, who had by then reigned for forty years, was little more than king of Paris and its immediate countryside. His nominal vassal the duke of Normandy was now also ruler of wealthy England, and Henry I, last of the sons of the Conqueror, would prove no less determined to hold onto his Norman inheritance, even though he had to dispute it with his surviving elder brother (and returned crusader) Robert. The descendants of Fulk Nerra of Anjou continued to exercise direct authority over much of western France, treating their nominal superior as effectively an equal. The dukes of Aquitaine held sway in virtually total independence, and indeed as often looked to the Pyrenees for their closest engagement as to the north. Only Burgundy remained unproblematically loyal, but its rulers had the same problem asserting their control of their own barons as their Parisian cousins. Furthermore, King Philip had succeeded in alienating the Church by his scandalous adventures in marriage. Throughout his own heartlands, local barons

[1] Grant, *Abbot Suger of St-Denis*, 29, 95.

had built castles in their lordships as bases from which to exercise their power and to challenge their neighbours in private warfare. It fell to the lot of his son Louis VI, known as "the Fat", to restore the fortunes of the Capetians in their own homeland.

When William Rufus met his untimely death in 1100, his younger brother Henry had to secure control of the lordship of Maine by dealing with a rebellion at Le Mans. The garrison of the strongly defended city eventually gave in to Henry and Fulk of Anjou, but not before boasting that they were unafraid of the besiegers' machines, stating that they could shower them with stones from a great height.[1] The chronicler Orderic Vitalis specified that Henry, in upholding the royal dignity, had his castles in Normandy well guarded, and proceeded to list fifteen of them, matching the list of ducal palaces and county castles already identified.

Castles along the Epte

Sites not mentioned here were those that stood along the little river Epte, dividing the Norman from the French Vexin. Intermittent conflicts over this frontier ensured a continuing process of castle building on both sides. Philip I had rebuilt the walls at Mantes(-la-Jolie) and fortified Chaumont-en-Vexin; William Rufus had taken over Meulan and La Roche-Guyon and had reinforced the Epte with the castles of Gisors, Neaufles-Saint-Martin, Dangu and Châteauneuf (Château-sur-Epte). Gisors became the key fortification, standing in a neck of land projecting into the river and regularly reinforced by its Norman and later Angevin holders, and it is remarkably well preserved. The original motte is massive, standing 20 metres high with a base diameter of 70 metres, in the middle of a large bailey, and was undoubtedly that of William Rufus, and the lower stages of the big donjon on the motte were popularly attributed to him. However, there was not time between the throwing up of the motte and the death of the king for a stone tower to have been safely built, and it is more likely that the lower stages date to no earlier than around 1125. Even then, the octagonal shape would be very unusual for so early a date in the west. The stone wall that rings the edge of the motte top and the outer curtain may also date to this time, along with the first of the walls that surrounded the whole town, although most of the rest of the castle probably dates to Henry II's time.[2]

The common feature of almost all the castles of this frontier was a motte, with a donjon added some time during the first quarter of the twelfth century. That of Château-sur-Epte, attributed to around 1087, is enormous, being some 50 metres across (Plate 11). Generally, attributions of any strategic purpose (in the modern sense) for castle building are wide of the mark, and misunderstand the multiple purposes for which castles were constructed. A number of the fortifications along the Epte, however, seem to have been built, or rebuilt, at around the same time,

1 OV, v: 302.

2 Toussaint, *Gisors*, summarises current thinking; see also Pépin, *Gisors et la vallée de l'Epte*, 7–11, 15–44.

and those on the Norman side were placed by routes across the river. Although entrusted to landholding castellans, their upkeep and further development would in many cases result from royal involvement too. The conclusion is inescapable that this frequently-fought-over territory was deliberately castellated to protect the frontier from incursion or invasion. Suger described the frontier as closely "protected by an impressive line of new castles" created by the "foresight" of the English king and Norman duke.[1]

The troubles of Louis the Fat

Young Louis confronted serious difficulties in the lands he would duly inherit from his father in 1108. The still energetic prince was soon fighting members of his own baronage. In 1102, he had to besiege the castles of Count Matthew I of Beaumont-sur-Oise at Luzarches (which the count had seized from its rightful owner) and Chambly (on the Oise). The first he took by storm, after a struggle, and provided a guard for the donjon before restoring it to its owner. Luzarches appears to have been a motte with a rectangular bailey. At Chambly, the prince deployed engines, but in an extraordinary turn of events a great storm apparently terrified his men and led them to flee. Louis's early efforts at sieges had not been blessed with regular success.[2] The following year he besieged the castle of Montaigu near Laon, ritually described as "very strong". It was again a motte located on the top of a hillock, with a defended enclosure. The castle had come into the hands of a man called Thomas de Marle, whom Suger describes as unspeakably evil, so much so that his neighbours sought to starve him out of his new castle by constructing a wooden palisade around the place, apparently strengthened with wooden towers. De Marle slipped out while the work was incomplete and begged Louis's aid, and the prince brought 700 knights to raise the siege. The assembled local nobility, unwilling to do battle with their prince, withdrew and the blockading works were pulled down. Suger notes that de Marle eventually lost Montaigu through the annulment of his marriage.[3]

In 1107, Louis had to attack the castle of Gournay-sur-Marne near Paris after its lord, Hugh of Crécy, had ambushed and kidnapped some merchants travelling the royal highway. The castle controlled an island in the river, which Louis first captured by organising a flotilla of boats, from which his knights and foot soldiers launched a determined attack up the steep sides of the island, eventually forcing the garrison back behind the castle's ramparts, comprising an oak wall atop the earth bank. Impatient for success, the prince ordered an all-out assault, which involved crossing a stream and two embankments. It was, apparently, unsuccessful, because Suger next told of the besiegers constructing a three-storey siege tower from the top of which his archers and crossbowmen prevented the garrison from

1 Suger, *The deeds of Louis the Fat*, 112.

2 Ibid., 31–2; Suger, *Vie de Louis VI le Gros*, 21–2.

3 Suger, *Vie*, 30; Suger, *Deeds*, 38. For summary descriptions of the castles mentioned in these paragraphs, see Salch, *Dictionnaire des châteaux et des fortifications*; Mesqui, *Châteaux forts et fortifications*.

moving around, driving them into holes and shelters. However, they were apparently familiar with the way to prevent attack from siege towers. The besiegers were prepared to push out a bridge from the top to gain access to the rampart, but the defenders were prepared with long beams to tumble the bridge. In the end, a relief army assembled by neighbouring lords was defeated by Louis, and the castle surrendered.[1] The following year, now king, Louis attacked La Ferté-Baudouin (renamed in 1127, now La Ferté-Alais), in which Hugh of Crécy had continued his mischief-making by seizing the castle and its rightful lord. Louis besieged it in December 1108, blocking it off with entrenchments of a ditch and palisade, then carrying the bailey by assault, the defenders retiring to the donjon. From here, however, there was no escape, and the defenders (who did not include Hugh, who had remained outside the castle) surrendered themselves and their captives.[2]

The importance of Gisors was emphasised in 1109, when Henry I met Louis VI and the French king demanded the demolition or surrender of the castle on account of the damage its garrison could do, and in accord with previous treaty obligations. Henry rejected the demand, and there was fighting in which the strength of the castle was confirmed. Suger makes much play of the glory of his own side and the treachery of the other, but in fact Henry had had plenty of problems of his own in Normandy, much of it fomented by Louis. The Battle of Tinchebray in 1106 had confirmed Henry's seizure of the duchy from his brother Robert. The king reissued the Conqueror's laws including those restricting the right to build castles, following this up with the destruction of "adulterina castella" (Orderic Vitalis).[3] In 1110, Louis had his own family problems with his half-brother Philip, to whom he had granted Montlhéry and Mantes. Philip became rebellious, and the king assailed the castle at Mantes with stone-throwing catapults ("mangunella et fundibalaria"), the first time we have heard of these in Louis' siege warfare. The place soon surrendered.[4] At the same time, Henry was dealing with the frequently disobedient Bellême family on Normandy's borders with France, and had to besiege and capture the castle of Bellême itself. To the south, Fulk of Anjou also had his problems, and in 1110 Ralph I of Beaugency and Hugh I of Amboise were besieging the major castle of Montrichard, built as we saw in the previous century (though it did not yet have its surviving stone donjon), with various machines, such that the defenders surrendered.[5]

[1] Suger, *Vie*, 72–4; Suger, *Deeds*, 55–8.

[2] Suger, *Vie*, 96–7; Suger, *Deeds*, 65–8; confirmed by Clarius, *Chronique de Saint-Pierre-le-Vif*, 146.

[3] For a detailed study of the whole development of castles in Normandy, see Yver, "Les châteaux-forts en Normandie".

[4] Suger, *Vie*, 124; Suger, *Deeds*, 82.

[5] *Gesta Ambasiensium dominorum*, in *Chroniques des comtes d'Anjou*, 111.

Sieges of Le Puiset

In 1111 there begins a story that offers real detail on siege warfare techniques and the nature of the castle at this time, when King Louis become embroiled in a war with Hugh II of Le Puiset, who held the castle of that name south-east of Chartres. The castle itself had a tumultuous history, having been captured by Henry I of France around 1031, when the king had had to deploy machines,[1] and besieged again by King Philip in 1078, but unsuccessfully. The castle originally seems to have comprised a large earthwork enclosure with rampart and ditch (more than 400 metres by more than 200), in the style of a Carolingian-age enclosure containing assorted comital buildings and a church, and more than large enough to offer protection to a population from all around. A motte was added to the north-east of the enclosure at some uncertain date. It has been deduced that the next stage of construction must have taken place around 1100. This was the throwing up right in the middle of the 11-hectare enclosure of another, separate fortified complex consisting of a much smaller (1.3-hectare) enclosure with bank and ditch, and a motte on the edge of it with a wooden tower on top. This was the site that faced Louis when he arrived in 1111 following pleas from Theobald IV (Thibaut le Grand), count of Chartres, and neighbouring lords for help against their violent neighbour.

First sending Suger to prepare a base at the estate belonging to Saint-Denis at Toury, the king himself appeared in May and brought a substantial host against the castle. He ordered a direct assault on the enclosure, and while the royal forces attacked on one side, against a gate, Theobald attacked on the other. Louis had prepared carts loaded with timber greased with fat and lard, with a view to burning down the gate. The defence was vigorous, and the steepness of the bank of the ditch caused great difficulty to the attackers, who were frequently hurled back down before being able to try to break down the wooden palisade. The fire-cart was not used, presumably because there was no way to bring it to bear against the rampart or gate. Just when the attack seemed doomed, a "bald priest", unarmed, made his way up the bank and began to attack the palisade by pulling off the various (unspecified) coverings. Shamed, the troops returned to the attack and began to break the wall down with axes, and Louis's and Theobald's men broke in simultaneously. The great size of the enclosure meant that many of the defenders were cut off and could not make it back to the inner castle, but Hugh did. However, he decided that he could not defend the outer defences of this fortification and went straight to the donjon on the motte. Here, observing the hopelessness of his situation, he surrendered and was imprisoned. Louis ordered Le Puiset burnt down, except for the tower on the motte.[2]

Hugh was released, and linking up with the Normans and his former enemy Theobald, rapidly set about rebuilding the castle in the spring of 1112. Perhaps for reasons of time and resources, the rebuilding was of the inner castle only.

[1] *Les miracles de St Benoît*, 243.

[2] Suger, *Vie*, 135 ff.; Suger, *Deeds*, 87–9; on the castle, see Fournier, "Le château du Puiset", with a plan of the remaining works of the castle.

He also plundered Toury, whose wooden defences were quickly manned by the men of Saint-Denis's lands. Louis had ordered the abbot of St Peter's at Sens to erect a rampart and *propugnaculis* suitable for defence.[1] Suger went ahead with a small party and slipped through his enemies into Toury. He told how he could see the three-storey wooden tower from a distance and knew it had not fallen, because Hugh would have burnt it. Louis arrived, and his foes withdrew to Le Puiset, where they formed up behind the embankment of the destroyed outer castle. A counterattack by a force of knights concealed behind the church and other houses in this large enclosure drove out the king's men after they had fought their way in, forcing the king to reassemble his army right back at Toury. Joined by a contingent of Normans taking their total force to 1,300 knights, Hugh and Theobald continued rebuilding Le Puiset. After the Normans had left, Louis brought his reunited army back to the castle, and fought his way in through a gate, driving his enemies back to the inner castle. He posted a line of warriors to peg them in, and then set about erecting a wooden palisade on the other, abandoned motte. This then played the part of a siege castle, albeit one only "a stone's throw" from the target. The king himself took most of his force to Janville, a mile away, and created another fort with a palisade of stakes and interwoven branches. A surprise attack by Theobald having been defeated, the net began to close around Le Puiset, and the large numbers of the rebel army must have rapidly diminished the food stocks. Sensing the inevitable, Theobald abandoned his ally and made terms, and Le Puiset again fell into royal hands, the walls being levelled and wells filled in. Remarkably, even this was not the end, because Hugh rebelled again in 1118 and again lost his castle and inheritance.[2]

Further wars of King Louis

In the same year 1112, the people of Laon formed a commune to resist the misrule of the bishop and the abbey that occupied the highest point of the hill on which the city stood. The bishop called in peasants from episcopal estates to defend the church towers and his palace, not for the first time indicating that ecclesiastical buildings in stone could well serve a defensive purpose. The measures were in vain, as the townspeople stormed the palace and brutally killed the bishop, an affront that brought the inevitable appearance of the king and suppression of the commune. But the conflict drew the attention of the odious Thomas de Marle, who offered his protection to the town. In 1114, Louis ordered Thomas to surrender his "adulterina castella", then marched on one of them, Crécy-sur-Serre. A vigorous attack carried the "first moat". The next day, the place was stormed, including the "strong tower", and all was burnt. Nearby Nouvion (either Nouvion-l'Abbesse, now Nouvion-et-Catillon, or Nouvion-le-Comte just west) then surrendered promptly. Next, Louis had to direct his attention to Amiens, where a local force was resisting the bishop from the "tower", presumably the donjon. The defend-

[1] Clarius, *Chronique de Saint-Pierre-le-Vif*, 170.

[2] Suger, *Deeds*, 95–104.

ers had strengthened themselves by erecting outer walls. The besiegers deployed "enormous" siege towers held by knights, but the defence had men of experience, one of whom arrayed two stone throwers inside to bombard the towers, operated by "80 women to jettison the large stones". Although many of the women were wounded, the two towers were disabled, the knights fled and the defenders sortied to complete the destruction. Recognising that the fortress was too strong to storm, the king reverted to a blockade which eventually brought surrender, but only some two years later.[1]

In 1122, Louis found himself far from home in the Auvergne, at the behest of the bishop of Clermont, who was in flight from the count. He laid siege to the castle of Montferrand, carried the outlying works and drove the garrison back into the donjon. Then he set up engines, presumably stone throwers, which hurled great stones at the tower and brought about its surrender.[2] Louis's unprecedented mobilisation of the forces of the entire kingdom to ward off an invasion by German emperor Henry v in 1124 was a powerful sign of the growing authority of the monarchy. But problems close at home still required his presence, as in 1128 when Louis had to capture the castle of Livry(-Gargan, just 11 kilometres from Paris) as a result of a quarrel between the seneschal Stephen of Garlande and Amalric III of Montfort. Louis deployed siege machines including stone throwers, but his enemies also had them, and both the king and Count Ralph I of Vermandois were injured by missiles thus propelled. These must therefore have been light weapons. The king anyway destroyed the captured fortress and forced Amalric to abandon his claims.[3]

Meanwhile, King Henry's wars along the Norman–French frontier had resumed. In 1123 the king attacked Pont-Audemer, and ordered his carpenters to build siege towers, through the agency of which the defenders were worn down and brought to surrender.[4] He then placed a garrison of his household troops in the fortress "and charged them to protect the country people" (Orderic Vitalis). The following year, Count Waleran de Beaumont (afterwards earl of Worcester) sought to bring provisions to his castle of Vatteville, under blockade by the king's forces, and "unexpectedly stormed" the siege castle (*munitio*) that was blocking the tower (*arx*). The commander of the garrison was "caught up by an artificial hand with iron hooks" while on the rampart. Vatteville-la-Rue is a motte castle with a substantial defended bailey lying on low ground near the Seine, originally built by Roger of Beaumont between 1059 and 1093.[5] The king himself at this time was besieging Brionne, and constructed two siege castles (this time called *castella*), which eventually induced surrender.[6]

[1] Ibid., 106–8. A more detailed description is given by Guibert of Nogent, *A monk's confession*, 152 (Laon), 168 (Thomas), 188 (Crécy), 189 ff. (Amiens). Thomas de Marle finally met his end wounded in a skirmish in 1130: Suger, *Deeds*, 143.

[2] Suger, *Vie*, 234; Suger, *Deeds*, 136.

[3] Suger, *Vie*, 254; Suger, *Deeds*, 144–5.

[4] OV, VI: 340–44.

[5] OV, VI: 346; Flambard Héricher, "La construction dans la basse vallée de la Seine", 93–5.

[6] OV, VI: 354.

Flanders, 1127

March 1127 saw an event that caught the shocked attention of rulers and churchmen everywhere: the murder of Count Charles the Good of Flanders (r. 1119–27) while he prayed at St Donatian's in Bruges by some of his lowborn but now wealthy subjects. Detailed accounts survive of the events (in which the non-noble origin of the killers was as outrageous as the sacrilegious circumstances) and its consequences.[1] Louis granted the heirless countship to William Clito, son of Robert of Normandy, conveniently a firm enemy of Henry I, after descending on Bruges to punish the conspirators, led by the Erembald family, which held the important posts of provost of Bruges and castellan of the city. The *castrum* of Bruges was centrally involved in the unfolding drama. Located on the edge of the old centre of the city, known as the Burcht, absolutely nothing now remains above ground. However, it seems clear from various studies that the original Carolingian count's castle had developed into an administrative quadrangle enclosed by a wall, with part of the site being clerical, and the other half comital, with a big tower (the *steen,* built in the eleventh century as the "lapidea domus comitatus") alongside a separate count's house and other buildings. Originally an island in the river, this was no longer the case in 1127 with the expansion of the whole complex, but it appears to have had direct access by canal to the sea.[2] However, it is apparent from the account that no ditch or moat impeded direct access to the outer wall from the outside.

The outraged citizens attempted to force their way into the walled complex occupied by the Erembalds and their supporters, by burning down the gates and posterns. But they were driven back by arrows and stones "as large as millstones". Two days later (14 March), help arrived from nearby Ghent, whose people "knew how to demolish defences". Following this expert advice, the citizens began the construction of scaling ladders, "at first a wider ladder with rungs constructed according to the height of the walls; to left and right, green branches woven tightly, formed a kind of wall, and in front a similar 'wall' was woven. On this ladder, another, longer and narrower, was superimposed, lying on its side . . . so that it could be slid over the wall of the castle." The attack took place on 18 March, but the massive ladders were extremely heavy, being made of green timber and having to be moved behind strong screens to ward off the defenders' fire. While the men of Ghent struggled to advance their immense construction under fire, young men of Bruges tried to gain the glory and rushed in with smaller ladders. At the same time, others were using mallets and masons' iron tools, trying to break through the stonework. The defenders were able to drive them all off with stones and arrows, and used spears to force back anyone trying to climb a ladder.[3]

The next day, 19 March, the citizens of Bruges managed to outdo their neigh-

[1] Of which the best is Galbert of Bruges, *The murder of Charles the Good*. Suger, *Deeds*, 138–42, also covers the events.

[2] Dhondt, "De vroege topografie van Brugge", 5–7; Verhulst, "Les origines et l'histoire ancienne de la ville de Bruges", 37–61; Noterdaeme, "De burcht van Brugge", 186–200.

[3] Galbert, *The murder of Charles the Good*, 160–72.

bours, and got into the complex by using lighter and more easily manoeuvred ladders. The garrison was sheltering from intense cold inside the count's house. Quickly forming up in battle order inside, the besiegers sent a contingent to clear away the piles of earth and stone the defenders had thrown up inside the gates, no doubt to reinforce them against battering rams or to provide an inner rampart in case they were burnt, and admitted the besiegers from outside. A ferocious conflict followed between the garrison and the citizens as the latter captured the count's house, accessed like other such towers by stairs to the first floor, and the besieged managed to flee across a connecting bridge to the church of St Donatian, from the towers of which they kept their foes at bay with arrows and stones, and were thus able to prevent anyone using the windows of the count's house to shoot at them. When the church doors were stormed on the next day, the conspirators withdrew to the gallery of the church and used it to defend themselves and hurl missiles down into the space below, and the place quickly became a battleground. Clearly, they had had enough time to lay in ample supplies of food and plentiful weapons. The citizens had to leave the exposed inside of the church.

The Flemish baronage began to assert control, and placed guards to prevent the besieged escaping while they themselves set off to meet their now arriving king, Louis. The gruesome fate of the conspirators who tried to escape from their death-trap need not detain us. Meanwhile, on April 11, the newly appointed castellan Gervaise ordered the great construction of the men of Ghent, now called a "wooden tower", taken apart and its strongest beams reused as battering rams to break down the church walls. Galbert listed the siege equipment now being prepared outside, with the artisans being protected from the besieged's arrows by knights holding shields. On April 14, the new weapon was brought against the wall of the church through the dormitory that stood against one side of it, shielded from the defenders. This device apparently incorporated a ladder as well, able to reach up to the height of the gallery. Further woven shields were placed to protect the operators from falling masonry. Then the iron-headed beam was repeatedly driven against the stonework, finally breaking through after being employed non-stop from noon until dark, and the defenders' efforts to damage the ram by burning the roof above the dormitory were in vain. A great host of citizens and royal knights charged in, and the desperate defenders retreated into the church's tower, fighting on the stairways. Again, the attackers had to back away. On April 18, Louis ordered his men to begin to demolish the church towers by sapping away with iron tools. With the blows echoing from underneath their feet and the structure beginning to shake, the terrified defenders of the tower (now 28 in number) threw themselves on the king's mercy and surrendered.[1]

So a church had served as a fortress resisting the citizens of a populous town and the army of the king of France, and everything they could throw at it, for a month. A key element had been the west work, where it has been discovered through excavation that there was a very large stone tower adjoining a circular

[1] Ibid., 214–48. The captured men were condemned to death and hurled one by one from the top of the Steen.

church similar in design to the Palatine Chapel in Aachen.[1] The tower had a central block about ten metres square, with staircases in slightly smaller turrets attached to each side. Such a design would certainly explain the resistance, and how so few defenders could hold out so long until, finally, they were undermined.

The new count quickly alienated the big towns, and rivals appeared for his job. In May 1128, the supporters of Count William were besieged in the castle of Ghent by the citizens. They sallied out to set fire to neighbouring houses, and while the citizens' attention was diverted to deal with the blaze, "they cut down with axes . . . the mangonels with which the citizens were destroying the stone house and tower" of the castle.[2] At this time the castle comprised a stone donjon raised to two stories shortly before this time, with the new basement surrounded by a motte, an attached tower (subsequently demolished), and two baileys with now unknown defences. Not long after, the rival count, Thierry of Alsace (Diederik or Derrick), gathered his forces, which included stone throwers, to attack the "house" at Tielt of one of William's supporters, but was defeated by the count coming to the rescue. William proceeded to besiege the chief magistrate's house in the village of Oostkamp. This place was protected with a ditch and palisades and backed against a river, and apparently a strong garrison. The count deployed mangonels and petraria, and fighting took place across the water as the attackers strove to break through the wooden fence. The count then turned to a siege tower, while the defenders built one opposite. The attackers tried to fill the ditch with hay, grass and brush, which the defenders then set alight with pieces of pitch and burning wax. The smoke however blew back on its creators. William sustained the attack for six days before giving up. Extraordinarily, a simple house had been rendered capable of resisting the forces of the ruler with a fence and ditch, and a stalwart body of defenders who seemed to know what they were doing.[3]

Louis VII succeeded to the throne in August 1137. Suger composed a life of the new king as well, but this does not survive in complete form. But he does record that there was a revolt in Poitiers in 1138, in which the town's *municipium* was occupied, possibly suggesting a donjon or some other construction on or by the motte raised inside the Roman ramparts, and that the new king had to assemble troops including archers and crossbowmen to bring back to Poitiers, which then submitted.[4]

Meanwhile, an account of events taking place in mid-century on France's northern borders (and within the formal bounds of the Empire) provides further fascinating information on the forms, and limits, of siege warfare at this time. Godfrey of Bouillon had sold his castle to the bishop of Liège to finance his crusade (1096), but the place was seized by Count Reginald I of Bar in 1134, and the

[1] Mertens, "The church of Saint Donatian at Bruges".

[2] Galbert, *The murder of Charles the Good*, 290.

[3] Ibid., 302.

[4] Suger et al., *De glorioso rege Ludovici, Ludovici filio*, 152, 157. There is extensive literature about Poitiers, none of the surviving defences of which are earlier than the construction work undertaken on Henry II's orders in the second half of the twelfth century. See Durand, "Les fortifications de Poitiers", 61–74.

occupants raided the bishop's lands. The latter organised his barons to recover the castle. However, this was a daunting task according to the chronicler (who may have been a canon of Liège), because Bouillon stood on a sheer ridge in a narrow meander of the river Semois, cut off by a ditch, such that no one in the place had cause to fear attack by stone throwers, ballistae or archery. The restored site confirms this judgement of its impregnability, which is also interesting insofar as the clerical author should choose to describe the castle by its defensive capacity. It could only be taken by blockade, and that would have to involve large numbers to be effective. The fighting reflected this reality. The count of Namur assaulted a mill at the foot of the rock that was held by the castle's garrison, using crossbows. The bishop set up a blockade, using the noise of trumpets to intimidate the defenders, while the local population provided arms and victuals to the army. A further attack on the mill took place but was again repelled. The site actually contained two fortifications, the stone castle existing under the later rebuilding and, in advance of it but on the same ridge, a motte castle with a wooden tower called the Ramonette, blocking access to the newer construction. The count of Namur continued to show great boldness in the bishop's cause in attacking this older fortress, exchanging stones and darts shot from *scorpiones*, but made little headway, and his attempts to burn it down (the local people carrying inflammable materials up to the foot) was foiled by changing wind directions. The castle was eventually surrendered through negotiation (September 1141). Inaccessibility rendered Bouillon quite untakeable by direct attack, despite the vulnerability of the wooden defences of the Ramonette, but also meant that once a blockade had been closed around it, and could be maintained through the availability of supplies to the besiegers, there was only going to be one outcome.[1]

England, Wales and Normandy from Henry I to the Angevins

While the Capetian king was engaged alongside the German king in the great task of transporting large armies across Europe to the Middle East, the kingdom of England, so secure and wealthy under Henry I's firm rule, fell into the period once dubbed the "Anarchy" as his successor Stephen of Blois fought to hold it against the late king's daughter Matilda, widow of the German emperor Henry V and now married to Count Geoffrey V of Anjou. It was a time in which siege warfare played a significant role and in which the "adulterine" castle forcefully brought itself to the attention of medieval chroniclers and modern historians.

So effective was Henry I's control that the only examples of conflict in the island were to be found beyond the borders of the kingdom, in Wales and in the marches, where the local lords had a free hand and regularly sought to extend their

[1] *Triumphus sancti Lamberti de castro Bullonico*. Matthys, "Les fortifications du 11e siècle", 225–34. There is an official guide leaflet, *La "Ramonette", premier château de Bouillon*, by I Tellier.

dominion into the most fertile of the lands held by the Welsh princes. In 1105/6, the Welsh chronicles relate that the earthwork castle of Rhyd-y-gors (later Carmarthen) had to be provisioned by Richard fitz Baldwin as the Welsh ravaged all the lands about and kept the Anglo-Normans behind their walls. Two years later, it was recorded that Gerald de Windsor built a new castle in his Pembrokeshire lordship, "and he made a dyke and a palisade . . . and a gate with a lock" (possibly this was Cilgerran Castle). In 1110, Gilbert fitz Richard built castles at Llanbadarn Fawr (probably the high ring-work standing above the river Ystwyth near Aberystwyth) and at Din Geraint (the precursor of Cardigan Castle), "where earl Roger had castles before", seeking to re-establish control of the lands of Cardigan, from which the settlers had been ejected by the Welsh in 1094.[1] They continued the approach of replacing Welsh rule in the existing political units with their own lordship, and secured it by creating earthwork castles as quickly as possible.

The Welsh, however, ensured by their resistance that these castles saw much more service as military sites than their English counterparts. A sustained counterattack by the Welsh took place in 1116, when Gruffudd ap Rhys "invaded and destroyed" the castle of Narberth. This was the ring-work remains known as Sentence Castle, standing some way from the thirteenth-century stone castle at this town. He then marched on and attacked Llandovery, belonging to Richard fitz Pons. This was a motte with a bailey. The Welsh succeeded in overrunning the bailey, the buildings of which they burned, but they were thwarted by the tower on the motte, from which defenders drove them back with arrows. Continuing his march rather than seeking to capture the motte,[2] Gruffudd then assaulted Henry de Beaumont's castle at Swansea, once again storming and burning the bailey but being thwarted by the defenders of the tower on the motte. The assault on the Anglo-Normans continued and the Welsh chronicles list a series of castles taken and burnt, not all of which can be identified. The central role of the Norman castle at Rhyd-y-gors (Carmarthen) in controlling the valley of the Tywi and as the chief castle of west Wales is again confirmed by the report in the *Brut y Tywysogyon* that castle-guard there was assigned to the neighbouring lordships in fortnightly turns.[3]

Examples of Welsh princes actually building castles for themselves are still very rare at this time. The cases of which we know were both in Powys, perhaps the principality most closely exposed to Anglo-Norman influence. Here, the prince, Cadwgan ap Bleddyn, had built the first castle at Welshpool (a motte and bailey) in 1111, and had acted to destroy the motte castle thrown up by a too-independent subject at Cymer in Merioneth in 1116, events recorded in all the Welsh chronicles.[4] Familiarity with siege warfare was, however, another matter. Where the defences consisted of relatively small earthworks and palisades, with small numbers

[1] *ByT(H)*, 49, 55, 73; *ByT(P)*, 26, 28, 34; *ByS*, 101, 105, 117; *Annales Cambriae*, 34.

[2] *ByT(H)*, 87, *ByT(P)*, 41, *ByS*, 127.

[3] *ByT(H)*, 89, *ByT(P)*, 41.

[4] *ByT(H)*, 101; *ByT(P)*, 37; *ByS*, 135–7; *Annales Cambriae*, 36. For an insightful modern review of the whole period of post-1066 castle warfare in Wales, see Kenyon, "Fluctuating frontiers".

of defenders, a direct assault on the bailey had proved to be effective. Dislodging the occupants of the tower on the motte had however proved to be a sterner task. Generally, the Welsh seem to have decided that to clamber the steep slopes while exposed to archery was not worth the risks.

After their forays into the lands around Carmarthen, the Welsh insurgents switched their attention back to Cardigan. The castle of Razo the Steward at Ystrad Peithyll (near Penrhyn-coch) was stormed and destroyed along with its occupants, and the Welsh then advanced on the castle at Aberystwyth, where Razo himself was. The commander sent urgently for help and received reinforcements from the nearby castle at Ystrad Meurig. Meanwhile, the Welsh band viewed the hilltop ring-work from a distance, "thinking to set up engines to shoot at the castle". This is the first reference to the Welsh themselves considering using siege artillery, although the first reference in a Welsh source, as we saw, related to the unsuccessful defence of north Wales castles against Gruffudd ap Cynan in the 1090s. In this instance, nothing became of the thought. Instead, a chaotic skirmish across the bridge leading to the castle with archers and a mounted, armoured knight sent out from the garrison ended with the Welsh fleeing.[1]

Devon and Exeter, 1136

Stephen of Blois's recognition as king of England and duke of Normandy in 1135 earned him only a short period of peace. The first major rebellion was by Baldwin de Redvers in 1136, a baron holding much land in Devon and the Isle of Wight. The king had already brought his host to the south-west to deal with a less eminent rebel, Robert of Bampton, whose lordship was based on the castle at Bampton in north Devon, which was built around a motte. Stephen deployed his forces around the castle, posting archers all around. The garrison surrendered quickly when the king hanged one of the rebels who had tried to slip out, and threatened the remainder with the same fate. While there, the king heard of the rebellion of Baldwin, who was collecting provisions in his major castle of Exeter (which was also a royal possession), to the dismay of the citizens. The most detailed source for the reign, the *Gesta Stephani* probably written by Robert, bishop of Bath, gives an accurate description of the ancient Roman walls of the city with the castle (Rougemont) standing in the corner on a high mound, surrounded by walls and towers. Baldwin's men greeted the king's troops with arrows and javelins hurled from the ramparts, and made sallies from "secret posterns".

As more barons arrived in response to the king's summons, the foot soldiers accompanying Stephen were put to good use. They first drove Baldwin's men back into the castle itself, and captured an outwork "raised on a very high mound to defend the castle". It is possible that this outwork was the earthwork recently excavated and known locally as "Danes Castle" (Plate 9).[2] The alternative explanation, that it was a siege castle thrown up against the defences, is on balance more

[1] *ByT(H)*, 93–5; *ByT(P)*, 42–3.

[2] "Medieval Britain and Ireland in 1993", 203–4, no. 50: "At *Danes Castle, Howell Road*". The site is perfectly round with a surrounding ditch 7–9 metres wide. The rampart was 11

likely. The king built siege castles regularly in his wars, but usually only when he could not stay with his whole army before a castle and had to leave a small force to contain it. None of the accounts mention such a device here, either. But its design and location point to it being part of the besiegers' works, from which Stephen pressed a very direct attack on Exeter, enclosing the rebels both from the outside, and from the side of the town, where he broke down the bridge between town and castle. High palisades were erected to protect against the bombardment from the walls. Numerous engines were prepared for the attack, and miners were called up to begin to undermine the rock.

While the siege went on, the king's men ravaged Baldwin's lands and those of his supporters. Desperate, those holding Baldwin's castle at Plympton (another motte and bailey) surrendered that castle to Stephen, which he proceeded to destroy. Another leading tenant of Baldwin's abandoned his own castle as indefensible, and slipped through Stephen's army into Exeter, his men in their chain-mail being indistinguishable from the royal troops.[1] After three months, the wells of Exeter castles suddenly failed. The garrison held on, drinking only wine, and indeed using wine to extinguish the lighted torches with which the king's men bombarded them. When this too ran out, the king's barons persuaded him to allow the castle's defenders to surrender with their lives.[2] Baldwin himself was ensconced on the Isle of Wight, where he was concentrating on reinforcing his castle of Carisbrooke.

Serious trouble for the king began in 1138, and was set off with the rebellion of Miles de Beauchamp, castellan of Bedford Castle. As the royal army approached, Miles requisitioned everything useful he could lay his hands on for the castle (again a motte and bailey). Stephen surveyed the site, then placed archers at points from which they could shoot over the walls. He ordered engines built to damage the rampart and walls. However, the strength of the castle, with a strong donjon and a high wall, and a vigorous garrison, made a rapid capture unlikely, so he left most of his forces there with orders to starve it out if they could not take it with engines while he left for other pressing business. Once more, without hope of relief, the defenders were compelled to surrender.[3] However, the growing number of exiles was plotting to support the replacement of Stephen with Matilda, and meanwhile King David I of Scotland was taking Stephen's preoccupation as a good opportunity for an invasion of the north of England.

metres wide at the base, but the interior, with a diameter of only 17.5 metres, would have offered little space for a garrison and its supplies.

[1] This tenant was heir to Judhael of Totnes, who had held both Totnes and Barnstaple castles from the Conqueror. Both were of motte-and-bailey design. The thus abandoned castle must have been Barnstaple, because the family had lost Totnes (temporarily) having backed Robert against William Rufus in 1088. *Gesta Stephani*, 38; Rigold, *Totnes Castle*, 6.

[2] *Gesta Stephani*, 30–36; Henry of Huntingdon, *The Chronicle*, 265.

[3] *Gesta Stephani*, 46–50; OV, VI: 510.

The north, 1136–8

This was not the first Scottish incursion: David has shown his intentions already in 1136 when he had crossed the border and seized the castles at Carlisle, Wark on Tweed, Alnwick, Norham and Newcastle, but was thwarted in front of Bamburgh and, later, Durham. Stephen had to race north, and peace was agreed, with the Scottish king retaining Carlisle. In January 1138, Scottish troops tried to take Wark by surprise. The king followed up with his main army, and organised the construction of battering machines to break into the motte-and-bailey castle. The garrison, however, was stalwart, keeping David at bay for three weeks. Stephen now arrived hotfoot from his siege of Bedford, and the Scots withdrew. The English king could not stay in the north, however, and once he had returned to the south to deal with other troubles, David attacked again. In April, he captured the bishop of Durham's castle at Norham. The tiny garrison of only nine knights surrendered, and were criticised for not holding out, "while as yet the wall was in good condition, the tower very strong, and provisions abundant", while the bishop was also blamed for not providing an adequate garrison. David's forces now raided far and wide, but in May the garrison at Wark sallied out to capture a supply train, leading the angry king to try to take the important frontier fortress again, but again in vain. In August, the northern barons under the leadership of the archbishop of York won a great victory over the Scots in the Battle of the Standard. Driven back, they nonetheless continued to attack Wark with "engines and new constructions", which the heroic garrison succeeded in destroying. Finally, in November, and facing starvation, the defenders surrendered the castle, which was demolished.[1]

These accounts are quite revealing of the approach and skills prevalent on the northern marches. The importance of the border castles was apparent. While we are not told very much about the defences of Wark, they were probably of a fairly rudimentary character, but the combination of the strength of the site with the defensibility of the motte and its two baileys, along with the apparent relative inexperience of the Scots in siege warfare, enabled it to hold out for so long. At Norham, it was recorded that the bishop of Durham, Ranulph Flambard, had ordered the castle built only in the 1120s "to curb inroads of the Scots and robbers", and that there had not previously been a garrison there. The castle at Durham itself, which resisted David, had been thrown up by Earl Waltheof in 1072 and strengthened by the bishop in 1128.[2] The castle here was again of the motte-and-bailey design, sitting as now across the top of the rock from the cathedral and

[1] Richard of Hexham, *The acts of King Stephen*, 39–55 (quotation from p. 51); John of Worcester, *The Chronicle*, 52 (Wark).

[2] Simeon of Durham, *History of the Church of Durham*, 716 (continuation); Simeon of Durham, *History of the kings of England*, 558 (Waltheof). Wark had been founded around 1100 by Walter Espec. It was substantially rebuilt in the 1150s. Newcastle was probably a motte castle: the splendid keep was built by Henry II in the 1170s. Alnwick was another motte castle, created by Yves de Vescy around 1100: see Blair, "The early castles of Northumberland", 121–2.

representing the secular aspect of the bishop's power. It is also recorded that this bishop levelled the space between castle and cathedral to reduce the risks of fire. The sources are tantalisingly inexact in the detail of the Scottish siege operations. The baronial castle was a novelty in Scotland, and was introduced along with the Norman settlers being encouraged by the Scottish kings, who themselves had close connections with the Anglo-Norman royal house, throughout the century. Certainly, King David (r. 1124–53) was instrumental in bringing the castle to Scotland. The motte castles scattered across the lowlands are the mark of this process,[1] such as the surviving motte at Annan built by the Norman settler Robert de Brus (Bruce) in this reign. There is specific reference in the sources to Scottish Normans being part of David's invading armies. Perhaps it was from this source that the Scottish army took its first lessons in how to attack motte castles.

Troubles elsewhere, 1138–40

King Stephen was unable to attend to his northern neighbour's aggression, because he was deeply engaged in the other parts of his realm. In May, Geoffrey of Anjou had invaded Normandy, winning the support of Matilda's brother Robert, earl of Gloucester, and securing Caen and Bayeux. Robert held major fortified centres in England, including Canterbury, Leeds (Kent), Bristol and Dover. In the west country, Geoffrey Talbot joined the rebellion and seized Hereford. Other rebels included the possessors of the castles at Wareham, Usk, Dunster, Bourne (Lincolnshire), Overton (now vanished), Ellesmere, and Whittington (all three in Shropshire), and Castle Cary and (East) Harptree (Richmont Castle). Faced with this far-flung challenge to his crown, the king responded energetically, and deployed three armies simultaneously. His queen, Matilda of Boulogne, went to blockade Dover, which surrendered when she summoned ships from Boulogne to close the blockade. Meanwhile Gilbert de Clare secured the surrender of Leeds. The king went in person to the west, recapturing Hereford. While he was there, the castellan of Shrewsbury rebelled. Stephen marched on the town, and succeeded in storming the castle after an attack of one month's duration. The king ordered siege machinery constructed, and was able to fill the ditch with large heaps of wood in order to attack the wall.[2] Bristol, on the other hand, which was Earl Robert's chief stronghold, was too strong for the king, the castle on its mound having powerful "walls and outworks, towers and diverse machines", and there being a large garrison. A column of troops from Bristol attempted to take nearby Bath by surprise, arriving at dawn with scaling ladders. However, they were thwarted by the bishop of Bath's soldiers. Stephen inspected the town and ordered the walls strengthened, and left a strong garrison to constrain the rebels at Bristol. His instructions were for the old Roman walls to be heightened, perhaps reckoning to avoid the risk of another escalade. While in the area, the king then turned suddenly on Cary (a castle site of which almost nothing survives above ground) and

[1] Talbot, "The defences of earth and timber castles", 3–6; Tabraham, *Scotland's castles*, 13–31.

[2] OV, VI: 516–20; John of Worcester, *The Chronicle*, 50–51 (Shrewsbury).

assaulted it with ballistas and stone throwers, forcing it to surrender. Moving then to Harptree, he ordered his men to set fire to the gates while attacking the walls with ladders, again forcing surrender.[1] At this time, the king's cause seemed to be prospering. While he was successfully putting down rebels in the west country, his supporters in Normandy were holding their own against the further invasions of Count Geoffrey, who was forced to abandon his siege of the major ducal castle of Falaise in October, Richard de Lucy's garrison being strong and well provisioned.[2]

Different versions exist of the dramatic attack by the king in 1139 on the immensely wealthy bishop of Salisbury, former chancellor to Henry I, and his nephews, the bishops of Lincoln and Ely. These prelates were said to have been massing arms and equipment within their own castles in order to assist the empress on her return. If that were true, the great castles that these ever-so-worldly bishops had built during recent years would represent a serious threat to the crown. The stone castles at Newark-on-Trent, Sleaford and Devizes were all described as immensely strong as well as being visually impressive, testimony to the power and wealth of their constructors. By seizing the owners, Stephen was able to enforce the surrender of these places and of Malmesbury and Sherborne as well. These castles generally comprised towered stone curtains enclosing substantial inner courtyards. Sherborne had an inner complex of domestic buildings attached to a large donjon, although any defensive purpose would have been largely confined to the outer curtain and ditch. Newark, where the remaining stretch of curtain is from the fourteenth century, retains a gatehouse from its original building, which had only begun around 1129, and stood on the river Trent, controlling the Fosse Way. Devizes was older, probably being a ring-work and bailey of the eleventh century, to which the bishop had added a stone donjon after 1115. To the clerical chroniclers for whom the king's actions seemed devilish, one is tempted to enquire why the princes of the church needed impressive buildings that were also among the most strongly defended in the kingdom.[3]

William de Mohun rebelled in the west. The king raced an army to deal with William's depredations, his ravagers being based in the castle at Dunster. This was on a high rock on the coast, which made one side inaccessible, while the approachable side was covered by a stone wall and towers, and an additional palisade. Stephen left a siege castle close to Dunster, and also based a force at Barnstaple, the effect of these measures being to restrict the rebel's movements and eventually to bring him to heel. Leaving this story to unfold, the king then had to besiege another opponent who had seized Wareham and the nearby castle of Corfe.[4] It was while he was pondering what to do faced with so strong a place, at this time

[1] *Gesta Stephani*, 56, 64, 68. The detail of the events at Bath lends credence to the supposition that its bishop was the author of the chronicle.

[2] OV, VI: 526.

[3] William of Malmesbury, *History*, 24, 25; *Gesta Stephani*, 72–80; Henry of Huntingdon, *The Chronicle*, 271; P White, *Sherborne Old Castle* (where the date of these events is wrongly given); Dixon et al., *Newark Castle studies*, 12, dates the gatehouse to 1130–40; P Marshall, "The twelfth-century castle at Newark", plan at 112.

[4] *Gesta Stephani*, 80–86.

at least a stone-walled inner ward powerfully located on a high rock, that he heard of the imminent arrival in England of Empress Matilda and Robert of Gloucester, whose arrival would turn a situation of frequent but scattered rebellion into one of open civil war. It seems scarcely credible therefore that, his enemies having landed and taken shelter in Arundel Castle, when the king arrived before it, he chivalrously granted his rival an escorted passage to Robert's castle of Bristol once the earl had succeeded in eluding the king's watchers and made his own escape.

The king's next march to the west involved recovering Malmesbury Castle from a local opponent who had seized it, taking by storm the "small castle" at South Cerney (of which nothing remains except a short stretch of ditch), and failing before the more substantial motte castle of Trowbridge (also disappeared), despite building engines, although he left a strong force in Devizes to contain it. Meanwhile the empress's party seized back Hereford, driving the small royal garrison into the castle, which Stephen was unable to relieve.[1] The following year, the real effects of civil war became apparent as local lords were able to choose between two rival power bases, and the chronicles begin to talk of the castles as being not the defenders, but the bases for the ravaging of their neighbourhoods. Key among the defectors were the constable Miles de Gloucester, from his base at Gloucester Castle, and Brian Fitzcount, who held the large earthwork castle at Wallingford that would be from now on a continual target for the king's efforts to break his opponent's power. During the year, Robert and Miles succeeded in extending the land under their control by retaking the castles of Harptree, Sudeley and South Cerney, and capturing and demolishing the siege castle that Stephen had built to contain Wallingford. The *Gesta Stephani* attributes this defeat to the fact that the king had desecrated a church to make into a castle, neither the first nor the last time such an event would occur.[2]

The king's rapid movements included a march all the way to Cornwall to suppress a further rebellion, and then on his return journey he is recorded as demolishing many of the "adulterine castles" springing up in the heart of the war zone in the west country. However, a Flemish follower of Earl Robert's took Devizes Castle by a night attack using leather ladders that reached to the top of the walls, driving the few defenders into (probably) the donjon, which they duly surrendered, having few provisions. His subsequent fate (as he tried to use the castle as a base for his independent operations and to seize by stealth the nearby Marlborough Castle that had been seized by John the Marshal) need not detain us. Meanwhile however, Geoffrey Talbot was still besieging the king's soldiers in Hereford Castle, and enraged the chroniclers by turning the cathedral into a fortress, the graveyard serving as a source of material for a rampart, and catapults (balistae) being placed in the tower. Miles de Gloucester meanwhile pressed an attack with engines on the opposite side. Hereford was by now a powerful motte and bailey with a strong donjon and outer defences,[3] but seems to have fallen.

1 Ibid., 92–6; William of Malmesbury, *History*, 30.
2 *Gesta Stephani*, 92–6; William of Malmesbury, *History*, 32.
3 *Gesta Stephani*, 108–10.

Widespread civil war, early 1140s: Geoffrey de Mandeville

1141 saw dramatic events as first Stephen was captured at the Battle of Lincoln, then Matilda failed to seize her moment, was driven out by the Londoners she had offended, and saw her own army scattered by troops led by William of Ypres[1] and Stephen's queen at the "rout of Winchester", during which events her brother Robert was captured. The two prisoners were exchanged, and civil war continued unabated. At Lincoln, where the king had been besieging the large and strong castle, he had deployed "ballistas and diverse other" machines. Lincoln Castle was one of the Conqueror's foundations, in the corner of the Roman walls of the city, and there is good reason to believe (from the masonry) that it had been provided with a stone wall at an early date. The very unusual two mottes may still have carried timber towers. One chronicler records that Stephen castellated a church, no doubt because it would offer a strongly-built advance work close to the castle. During the events at Winchester, it was noted that Stephen's brother, the bishop, had fortified his dwelling (the defensible palace of Wolvesey) "impregnably, very much like a castle". From this castle, combustibles were hurled into the town and the houses set alight.[2]

It was during this year that the ill-reputed Geoffrey de Mandeville (constable of the Tower of London and holder of various lands including (Saffron) Walden Castle) had advanced his power, securing from the two rival rulers ever better terms and holdings including several castles. Despite the alleged "anarchy", Geoffrey was still very careful to obtain Stephen's permission (later confirmed retrospectively by Matilda) before building himself a new castle at South Mimms, which proved to have a very unusual structure: a motte created around a timber tower, with an entrance tunnelled through the earth itself.[3] The following year saw the empress's party try to strengthen its position, as Robert crossed to Normandy and assisted Geoffrey of Anjou in the securing of ten castles without meeting much resistance. Matilda meanwhile sought to secure her position in the Oxford area with new castles at Woodstock and Radcot, "surrounded by marsh and water", at Cirencester and at Bampton, "on the church tower". The king, marching against these works, found Cirencester abandoned and proceeded to destroy the rampart and stockade, then stormed Radcot and captured Bampton before moving against Oxford itself (lost since his defeat the previous year). Oxford itself was protected by its rivers, and secured with ramparts, while on one side the castle had a large motte and a donjon of "great height". The king's men fought their way into the town, then besieged Matilda in the castle for three months, driving

[1] William of Ypres was an illegitimate descendant of Count Robert of Flanders who had been one of the competitors for the county after the murder of Charles the Good, in which he may have been implicated, and had lost out in the subsequent fighting that we have already considered. He was a loyal and effective mercenary supporter of the king throughout the Anarchy.

[2] *Gesta Stephani*, 110 (Lincoln), 126 (Winchester); William of Malmesbury, *History*, 36 (Lincoln), 42 (Winchester); OV, VI: 540 (Lincoln).

[3] The charter evidence was studied by Round, *Geoffrey de Mandeville* (the charters reproduced at 142, 168); Kent, "Excavations at the motte and bailey castle of South Mimms".

the garrison to starvation (although engines were also deployed). The empress's legendary escape was followed by the fall of the castle, and then by a royal march against Wareham (or perhaps Corfe), where Robert had landed and captured the castle from its royal garrison, "shaken . . . by the engines of the earl". Robert's army defeated Stephen's at Wilton (Wiltshire), and recovered Sherborne Castle.[1]

Geoffrey de Mandeville was deprived of his offices and castles in 1143 and immediately rebelled, basing himself in the fens around Ely, turning Ramsey Abbey into a castle, and ruthlessly ravaging the territory around. Stephen found him impossible to pin down, so he built his own new castles in which to base forces to constrain Geoffrey. These appear to be the unfinished substantial rectangular earthworks at Burwell (Cambs.) and Rampton: unfinished because the rebel died of a wound received when attacking Burwell (September 1144).[2] These sites, then, were siege castles, temporary forts with a strictly military function. Further glimpses of the siege warfare of this time are provided by the chronicles. In 1144, the king made a further attempt to take Lincoln, but gave up after eighty of his men were suffocated when a trench they were digging collapsed. Whether this represented the preliminaries for an attack by mining is not revealed. He had no better fortune when he attacked a fortification being built at Cricklade: although his attack carried the outer rampart, he had not been able to take the inner when Robert of Gloucester brought relief. He had more success at Winchcombe, where the castle is described as being on a high mound surrounded by defences, but with only a small garrison. Here a direct assault covered by archers led to surrender. This castle, of which there is no trace, sounds as if it was a motte.[3]

In the north, where neither side in the civil war exercised any real authority, the local nobility had to order affairs themselves. In August of 1144, William Cumin (former chancellor of King David of Scotland and pretender to the bishopric of Durham) and his retainers occupied a church five miles from Durham and began to convert it into a castle. Three local barons collected their own retainers and marched against him. When they arrived, the church was almost surrounded by a ditch, and the defenders made their stand in the tower. But the attackers broke in through the windows. The collapse of the masonry work of the desecrators ("because of the slightness of the workmanship") no doubt confirmed the role of outraged divine justice here,[4] but the role yet again of churches as potential castles is underlined in this incident, along with the suggestion that apart from a surrounding ditch, the occupants were intending to create a stone fortress.

[1] *Gesta Stephani*, 140–48; William of Malmesbury, *History*, 49–51; OV, VI: 550 (Normandy).

[2] *Gesta Stephani*, 164–6; William of Newburgh, *The history of English affairs*, 67–70. Burwell is some 65 × 40 metres with a ditch six metres deep and three wide. Excavations showed it to be incomplete. Giant's Hill at Rampton is 45 × 40 metres with a partially wet moat 15 metres wide: VCH *Cambs.*, II: 19–20, 37.

[3] *Gesta Stephani*, 170–74; Henry of Huntingdon, *The Chronicle*, 282; William of Newburgh, *The history*, 73. It is not the nearby Sudeley Castle.

[4] Simeon of Durham, *History of the Church of Durham*, 728 (continuation).

Later stages of the civil war

The destruction by Stephen of the new castle begun by Robert at Faringdon in 1145 was seen as something of a turning point, with the empress and her supporters losing the initiative. The castle was constructed with a rampart and a stockade. Stephen marched against it, and ordered the throwing up of his own protective siege works. He then set up siege engines and deployed archers on all sides and bombarded the castle with stones and arrows. Clearly, stone-throwing artillery was in use here. The ferocity of the attack led to the surrender of the substantial garrison, which had suffered great losses.[1] It is hard to reconstruct this important event. Excavations on the site of the folly now standing on the hilltop known as Faringdon Clump found a circular clay rampart surrounding a small inner platform, assumed to be a base for a wooden tower, but no evidence for any outworks, which there must have been because the inner circle was far too small to contain more than a handful of people. A tangle of skeletons was a gruesome hint of fighting, although nothing dateable was found.[2]

In 1147, the hitherto loyal Gilbert fitz Richard de Clare revolted. Stephen seized three of the earl's (unnamed) castles but then came up against Pevensey, where a Norman castle occupied the corner of a late Roman fortress that served as its outer bailey. Stephen was quick to see that this stone fortification, with a resolute garrison, would not fall rapidly, so he resorted to his by now well-established method of leaving a force to contain it. His more common approach of direct attack secured conquest at the unidentified west country "castle of the wood" while the bishop of Winchester brought "Lidelea" to surrender by erecting two siege castles. At Downton (south of Salisbury) the castle built in 1138 as a ring-work and bailey was also forced to surrender to the bishop's forces through starvation created by the presence of a siege castle.[3]

However, in Wales the absence of royal authority left the Welsh princes to compete with the marcher lords without fear of interruption. Castles continued to feature prominently in this relatively small-scale fighting in the southern and western marches. In 1136, the Welsh princes Owain Gwynedd and Cadwaladr, sons of Gruffudd ap Cynan, had invaded Cardigan (Ceredigion), capturing and burning the settler castles across the area. In the following year they continued operations with the capture and destruction of Ystrad Meurig, the promontory castle of Llansteffan and Carmarthen.[4] In 1143, the two brothers were at war, and Owain burnt Cadwaladr's castle at Aberystwyth, presumably the reoccupied Anglo-Norman site, providing another early reference to Welsh adoption of the

[1] *Gesta Stephani*, 182; Henry of Huntingdon, *The Chronicle*, 283.

[2] Leeds, "An adulterine castle on Faringdon Clump", 165–9; Leeds, "Second report", 296.

[3] *Gesta Stephani*, 204–12. Godshill is proposed as Downton's siege castle by Renn, *Norman castles in Britain*, 173. Lidelea is identified as Barley Pound west of Farnham and its siege castle as the earthworks at Powderham and Bentley (Hants.), ibid., 225. Bentley was excavated and found to be a much ploughed-down motte with a ditch seven metres wide and three deep, and a low platform surrounded by a bank, some 75 × 40 metres: Stamper, "Excavations on a mid twelfth century siege castle", 81–7.

[4] *Annales Cambriae*, 40–41; *ByT(H)*, 155–7; *ByT(P)*, 51–2; *ByS*, 145–7.

castle. Meanwhile, the marcher lords returned to the offensive and recaptured lost lands, immediately setting about the reconstruction of the burnt castles, including particularly Carmarthen. But the balance swiftly turned again, with Cadell ap Gruffudd capturing Dinweiler (probably either Alltyferin, Dinefwr or Pencader), Carmarthen (where the garrison were granted their lives) and Llansteffan "in a severe struggle". There was an immediate counter attack by a host of "French and Flemings" who noted the small number of Welshmen inside and raised ladders against the wall. The defenders waited for the escalade to reach the embrasures then pushed them over into the ditch. The major castles of the Tywi valley thus remained in Welsh hands for the moment. Further west, Cadell tried to capture Wizo's (Wiston) Castle, a large motte and bailey. Initially thwarted by the defences, he took advice from Hywel ap Owain Gwynedd and captured it. So the story continued: the chronicles record increasing numbers of castles being built by the princes, and being attacked and defended in their own disputes. Oswestry Castle was rebuilt by Prince Madog ap Maredudd in 1149; Cadell repaired Carmarthen in 1150 and Ystrad Meurig the following year. Dinefwr, which would become a centre of the power of Prince Rhys ap Gruffudd from 1155, was restored again in 1151, although what was built remains unknown under the later stone castle.[1] The Welsh had become proficient in the capture, and defence, of earthwork castles. No detail is given to allow an appreciation of the extent to which they had learnt other approaches than direct assault.

While the Welsh fought back against the invader, Stephen was unable to finish his wars in England. An increasing role in the Angevin cause was played by Count Geoffrey's son, the young prince Henry. Angevin control of the lands on the French side of the channel was becoming more or less complete. In 1143, Rouen itself had surrendered, but the soldiers of the earl of Warenne, who held the place for the king, refused to surrender the tower (donjon) and castle. Geoffrey therefore besieged it alongside the Norman nobles who had come to terms with him. They built siege engines but could not take the castle, because of its siting and also because of the "stability of the building". In the end, running out of food and with no hope of relief, the castle was surrendered. The chronicle records that Geoffrey placed his battering engines (stone throwers) against the side where in the previous year part of the tower had fallen down. With Rouen gone, the few remaining castles in Normandy still holding for Stephen surrendered, except for Arques, which was under blockade. This place surrendered the following year when its commander, a Flemish monk loyal to his oath to the king, was killed by an arrow.[2]

In 1149 Geoffrey began what proved to be an extremely lengthy siege of a rebellious vassal at Montreuil-Bellay near Saumur. The castle, originally dating from the time of Fulk Nerra, was described as having a double wall, outworks and a very tall and strong tower, whence the defenders could hurl stones, arrows and other missiles. The present castle and fortified town date from much later, so

[1] *Annales Cambriae*, 44–5; *ByT(H)*, 127–30; *ByT(P)*, 56–7; *ByS*, 153–5. Flemish settlers had been encouraged to take land in modern Pembrokeshire by the Anglo-Norman kings and their names feature in a number of the sites and castles.

[2] Torigni, *The chronicles of Robert de Monte*, 720–21.

the original site cannot be reconstructed. The terrain made it very difficult to approach, there being a precipitous ravine in the way, so the traditional means of breaking in by battering ram were hard to use, and siege towers could not be brought close enough. The count was no stranger to siege warfare, and was said to have deployed every type of weapon — *petrarites, fundibulariae, mangonellae* — as well as rams (*arietes*) and wooden siege towers. Breaches were made in the walls, only to be repaired by night. In a much-quoted story, Geoffrey was said in his exasperation to have read Vegetius' famous text and applied the science to his own problem. Splendid though the story is, in fact the chronicles go on to say that he created a kind of Greek Fire to be propelled by his engineers from a mangonel. This struck the timber defences of the breach and consumed them in a spreading fire which presumably cleared a sufficient means of access to allow for an assault and lead to the hurried surrender of the castle, which was then levelled. Greek Fire was unknown to Vegetius, but the experience of countless crusaders would have acquainted western warriors with both the deadly but still secret weapon of the Byzantines, which needed to be projected from a device with a nozzle, and the less mysterious naphtha bombs widely used by the Muslim armies.[1]

The returned crusader Louis VII allied himself with Stephen, leading to sieges of frontier castles across Normandy and Anjou. Vernon (Eure) was attacked by the French for a fortnight in 1152 with assaults and attacks by engine, but without success, while the previous year Henry had ravaged the Vexin, taking and burning castles except for the tower at La Ferté (probably La Ferté-Saint-Samson), which stood on a height.[2] Suger reported on another difficulty faced by the French king further east. Not for the first time in this period, a commune arose in the town of Vézelay to oppose, in 1152, the oppressive government of the abbot. The latter fortified the towers of his monastery (which stood on the high ground at one end of the town) and exchanged shots from bows and ballistas (which may be either crossbows or catapults) with the rebellious townspeople until Louis came to his aid. Presumably this was one example of the military use of ecclesiastical buildings that did not attract pious condemnation![3]

For the final years of the Anarchy some of the key sources are lacking in detail. Wallingford on the river Thames featured, with Stephen quite unable to capture it by force. The inner ward, in the corner of an old Wessex *burh*, had massive earthworks and enclosed a space 165 × 95 metres, with wide water defences. The king was forced therefore to establish siege castles around the fortress. One had been built against the bridge across the Thames facing the castle at Crowmarsh, and in 1152 this was strengthened (or perhaps a new work started, more directly threatening to Wallingford). Siege forts on the other (western) side of Wallingford had

[1] *Historia Gaufredi ducis Normannorum et comitis Andegavorum*, in *Chroniques des comtes d'Anjou*, 217–19; *Gesta Ambasiensium dominorum*, ibid., 92; Torigni, *The chronicles of Robert de Monte*, 63–5. Parts of the history are translated by Bradbury, *The medieval siege*, 85–6.

[2] Torigni, *The chronicles of Robert de Monte*, 69, 72; Suger et al., *De glorioso rege Ludovici*, 161–2.

[3] Suger et al., *De glorioso rege Ludovici*, 174.

been built and destroyed then restored or replaced, such that it is now next to impossible to determine which of the several earthwork remains in nearby villages played what role. It nonetheless looks as if Stephen may have been planning a serious attack on this powerful Angevin outpost. Meanwhile, Count Geoffrey had died and his son Henry had succeeded to Anjou, had been accepted as duke of Normandy, and had taken to wife Eleanor, heiress of Aquitaine. He now sought to add England to his inheritance. His army stormed the town of Malmesbury (where no defences now survive), the infantry scaling the walls with ladders while the garrison retreated to the castle. When the king gave up an attempt to relieve it, the last defenders surrendered.

At some point, Stephen had forced the surrender of the castle at Newbury held by John fitz Gilbert the Marshal, an event of interest because the much later *History* of his son William the Marshal tells how the very young boy was nearly the missile for a stone thrower described in the poem as a *perriere*, which the child innocently saw as a swing, before being reprieved by Stephen.[1] The Angevin meanwhile marched into the Midlands, taking the strong motte castles of Warwick by trickery and Tutbury after a long siege, then Stamford. Ipswich rebelled in his support, but was retaken by the king, and Nottingham Castle, on its daunting rock, resisted the Angevin forces. Turning back south, Henry approached Crowmarsh, described as a castle "on a very high mound" and launched a direct attack, which failed.[2] At this moment the gathering of the forces of both sides face to face led to the peace brokered by the leading barons of both sides, by which Stephen retained the kingdom until his death, with Henry as his heir. Thus the Anarchy ended, perhaps appropriately, to avoid a battle in the shadow of a temporary siege castle, the flattening of which was one of the immediate terms of the settlement. With single authority restored, all accounts confirm that many "adulterine" castles built without formal permission during the long civil war were now ordered destroyed by the king or his heir.

Siege warfare and castles during the Anarchy

Historians now reject as a great exaggeration the traditional view of this period, as encapsulated in the famous words of the *Anglo-Saxon chronicle* as the time when "God and his saints slept". The role of castles, as the source of much of the evil, and as having flourished as adulterine in the sense of unlicenced local power bases, has also been re-examined. The factual evidence for the condition of the people is, of course, wholly lacking. What is fairly clear is that in areas where effective authority did not reach, it was very likely that unopposed lawless-

[1] *L'histoire de Guillaume le Maréchal*, 54 (l. 540) (cf. Crosland, *William the Marshal*, 20); Henry of Huntingdon, *The Chronicle*, 289.

[2] For the rest of this paragraph, see *Gesta Stephani*, 226–38; Henry of Huntingdon, *The Chronicle*, 288–92; "Winchcombe annals", 130; William of Newburgh, *The history*, 123–5. For descriptions of the Wallingford fortifications, see Spurrell, "Containing Wallingford Castle"; Slade, "Wallingford Castle".

ness would prevail, and lesser knights and adventurers would take advantage of the situation to advance their own fortunes at the expense of their neighbours and of villages and churches. For parts of the time, a substantial stretch of territory from East Anglia through to the West Country represented a kind of frontier zone between the two parties where such lawlessness could occur. In areas such as Kent, where royal power was mainly unchallenged and the (relatively) effective systems of rule with which the English kingdom had been familiar, and which had been consolidated by the Normans, normal conditions prevailed.[1]

It is in these disputed areas that the chroniclers tell repeatedly of the creation of unauthorised castles, whence their occupants would ravage, loot and plunder. Sufficient of the chroniclers use the term "adulterine" to suggest that contemporaries expected that castles (at least those of the size determined by existing laws) needed to be formally permitted, even if, as was undoubtedly the case far more often than just the example given above of Geoffrey de Mandeville, the permit was retrospective. Castles were an attribute of lordship. Obtaining formal endorsement was a recognition of the lordship, but not a means of controlling the construction of defensible buildings, which was a view of the role and function of the private castle largely invented by nineteenth-century historians. Where royal power was absent, the great barons might seek to make their own arrangements, as in the treaty of about 1151 between the earls of Chester and Leicester regulating the possession of castles and calling for the dismantling of one. In the arrangement, Ranulf of Chester granted (*dedit*) the castle of Mountsorrel to Robert, earl of Leicester, but reserved the right of access to it. It has been suggested convincingly that this unusual treaty (for Britain) mirrored current practices in much of France, where the abeyance of royal power meant that feudal lords *had* to regulate such matters between themselves. It was to be expected that the great feudatories, who would be the very people who would bring the civil war to an enforced end, would prefer stability in which to further their own ambitions.[2]

There are frequent references to the deployment of siege equipment in the numerous sieges considered above, although these were largely at the disposal of the king or the greatest of his enemies. The indication is that only the great lords had access to the skilled engineers required. It is also apparent, from the first attack on Exeter and by inference from the royal inspection of Bath, that Stephen was well versed both in siege warfare and fortress design. Stone-throwing artillery and ballistas make frequent appearances. However, attempts at mining were very rare, and references to wooden siege towers are lacking altogether in Britain. This may be a reflection of the fact that castles were either assaulted rapidly, or left under blockade, there not being time to construct such weapons.

[1] Eales, "Local loyalties in Norman England".

[2] Coulson, "The castles of the Anarchy"; Coulson, "The French matrix of the castle-provisions". Coulson argues that the idea of "unlicenced" castles is a modern invention and that "adulterine" is only used by Orderic Vitalis and Suger, who suffered from "ecclesiastical castrophobia" ("The castles of the Anarchy", 75). In fact, it is also to be found in the *Gesta Stephani* and in William of Newburgh and implied by William of Malmesbury. See also Amt, *The accession of Henry II in England*, 17–19.

The international connections of the nobility and participation in the crusades would have compensated (in terms of experience) for the fact that there had not been such sustained warfare in the country in living memory, and certainly never had fortifications played so considerable a part. But the methods and the conventions that governed medieval fighting were common across Europe, as were the designs of the castles. Time and time again, we have seen the issue of control of a territory (which was an intrinsic element of lordship) revolve around possession of a castle. The control offered by such possession was over the surrounding land and people, the resources of which provided the essential components of power. Where immediate resort to arms was the normal response within the baronage to political challenges, lordship and control were inevitably entwined with being able to defend these attributes physically and materially, so that for a castle to represent lordship it had also to represent physical power, and to be effective in time of war had to have a garrison to enforce and protect it.[1]

While the great majority of the castles named in the sources were already existing in 1135, other fortifications had to be constructed through the exigencies of the struggle. Many were siege forts, with functions of containing or of blockading enemy garrisons. The precise purpose would determine whether they were placed close by (in order to blockade), or at a distance (to protect territory from their garrisons). Most often, they would take the form of ring-works or mottes, easily thrown up. Thus in form they were identical to castles, although their function was much more limited. It was the character of the struggle that required the king, in particular, to resort to such tactics, although it is clear that he would have preferred to be able to take each target at once, in the interests of both asserting his authority and his control, if only time and circumstances had allowed. Another very common measure taken by all sides was the adaptation of church buildings to act as castles or as siege forts, ignoring the outraged condemnation of the clergy. The presence of an existing stone tower or of sturdy buildings was clearly too great a temptation to be resisted.[2]

Another feature of the conflict was that the great tower or donjon continued to feature not just as a dominating residence of the lord and potent symbol of lordship, but also as a practical instrument of defence, particularly in Wales, but also in England and Normandy. Time and again, the accounts record how with the outer defences fallen, the tower functioned as a place of final resort. Generally the outer defences consisted of earth and timber rather than masonry, and where the site of the castle allowed for a level approach it would be easier for an attacking force to break in without resorting to siege engines. Another consideration sometimes highlighted in the sources was that garrisons would usually be few in number, especially in the castles of barons without the resources to hire extra soldiers, making it harder to hold an extended circuit against a superior attacker.

[1] This is to dispute the conclusions of Speight, "Castle warfare in the *Gesta Stephani*". I am very grateful to Sarah Speight for letting me see a copy of her article before publication.

[2] To the examples cited can be added several more in Foliot, *The letters and charters*, 40 (Slaughter, Glos.), 74 (Malmesbury, Wilts.).

In all the circumstances, therefore, it was not surprising to find defenders falling back upon the donjon (whether of wood or stone). Often, where the attacker (as in Wales) did not have the time or the equipment to organise to attack the great tower, the defenders were safe. Elsewhere, they resisted serious assault, with differing degrees of success.

The first years of Henry II

The "Angevin empire" was a happenstance collection of inheritances in the single hands of Henry II, whose lands stretched from the Scottish border to the Pyrenees. But different traditions of rule operated in each part, and only England and Normandy had experienced the centralised control of the Conqueror and his descendants.

The freedom enjoyed by the Anglo-Norman nobility during a time of civil war had evaporated, and in the first year Henry recovered several major castles from recalcitrant lords, as he sought to ensure that royal castles were returned to royal control, or were entrusted to more loyal crown supporters. The overall trend of restoration was clear. At Henry's accession, it has been shown that there were 225 (identified) baronial castles and only 49 royal. During the Angevin period down to 1215 the balance changed to 179 baronial and 93 royal. Those acquired by the crown around 1154 included Wallingford from Brian Fitzcount, and the midland castles of Nottingham, Stafford, Newcastle-under-Lyme, Tickhill, Bolsover and Peveril (at Castleton) from the earls of Chester. The former bishop's castle of Devizes was not returned to episcopal control; the western castles of Ellesmere, Overton and Whittington were withheld from their heirs; Northampton came back in 1155, and in the same year some of those formerly held by the bishop of Winchester were demolished. In 1157, the castles of Stephen's son William were seized: an array of major strongholds including Norwich, Pevensey, Eye, Lancaster, Lewes, Reigate, Castle Acre, Conisbrough and Sandal. The surrender by the Scots of the northern counties brought back the great castles of Northumberland and Cumbria; East Anglian castles were handed over by Hugh Bigod, and the former Mandeville castles at Pleshey and Walden were pulled down.[1] Thus was the reach of royal control returned to massive territories that had fallen under the sway of others during the Anarchy. Control of the castles was certainly a sign of royal presence, and also a physical presence in the countryside enabling effective government to be restored.

In Wales, meanwhile, the Welsh princes had continued to fight both amongst themselves and with the marcher lords, and the new prince Rhys ap Gruffudd took the castle of Llandovery, followed by the destruction of all the settler castles in Ceredigion, then swept away the castles of Dyfed: the restored defences of Carmarthen held out.[2] The new ruler was clearly a force to be reckoned with, and

[1] The evidence from various parts of the country has been analysed by Amt, *The accession of Henry II*, 17–19, 27, 34, 43–4, 70–79; R A Brown, "A list of castles", 90–93.

[2] *ByT(P)*, 61; *ByT(H)*, 139–41; *ByS*, 163; *Annales Cambriae*, 48.

would be recognised as such by Henry. The king was involved in north rather than west Wales in 1157, restoring control in the march in modern Flintshire. Clearing the forests away, from the shelter of which the Welsh had successfully restricted the English advance, he ordered the construction of a castle at Rhuddlan (presumably a reconstruction of the eleventh-century motte castle) and another at Basingwerk, probably another motte.[1] The next year he was across the channel, and, in an illustration of the detailed knowledge of castle building possessed by rulers, "marked out the site and gave directions to his officers as to the plan on which the castle . . . to be rebuilt" at Pontorson. In 1160, he was at war with the French again, advancing on Beauvais, where he destroyed the "exceedingly strong" castle of Gerberoy — with the exception of a "certain strong tower which his troops could not capture". Remaining in Normandy in 1161, he ordered works to strengthen the castles of his borders, with most resources going into Gisors. Henry's work here is considered to have been the heightening of the massive buttressed octagonal donjon on the motte, renewal of the stone wall around the top of the motte, and the great stone curtain wall with rectangular open-backed towers around the whole castle. He may also have been responsible for the creation of a strong barbican in advance of the gate. Henry's command of siege techniques was further demonstrated in 1168, when he was involved in asserting control of Brittany. Several castles fell to him, and it was specifically recorded that Bécherel was of stone, and after besieging it for several days, he brought up siege engines whereupon the garrison gave in.[2]

New styles of castle building in western Europe

All the accounts, as well as archaeological investigation, confirm that earth-and-timber castles continued to form the majority of castles of the first two-thirds of the twelfth century, both of the motte-and-bailey and of the ringwork design, the choice between the two perhaps motivated by the local geology as much as by any other consideration.[3] For those who could afford the considerable expense involved, rectangular great towers of stone were favoured, and this is the great age of their construction across the lands of France and England particularly. But new developments were also taking place, in the shape of circular great towers, similar in appearance to the German *Bergfried*, which had been a feature since the eleventh century, and, in Anglo-Norman territory in particular,

[1] Torigni, *The chronicles of Robert de Monte*, 82. On the location of the disappeared Basingwerk, see King, *Castellarium anglicanum*, 1: 155.

[2] Torigni, *The chronicles of Robert de Monte*, 82, 84 (Pontorson), 91 (Gerberoy, which has disappeared), 107 (Bécherel); Pépin, *Gisors et la vallée de l'Epte*, 28–30, 40.

[3] The case for this is convincingly made to explain the distribution of the two types of earthwork castle in Glamorgan according to the extent of glacial drift, where the very thin soil made motte construction too demanding a task: RCHM Wales, *Early castles in Glamorgan*, 34–6.

the walling around in stone of the top of a motte to create the shell-keep, with its buildings constructed against the inside of the circuit.

It is now impossible to determine the motivation for the changes in style. It was once supposed that the reason for the move away from rectangular donjons was a recognition of the defensive weakness created with the dead ground at the corners, and the vulnerability of the corner to mining. This purely military reasoning is now accepted to be totally inadequate in understanding the role and function of a great tower, with its function as a private residence of its lord and the material symbol of lordly or royal power. It is anyway rendered nonsensical as an explanation by the continued building of rectangular great towers by rulers and lords who well understood siege techniques for many decades after the arrival of non-rectangular designs (as at Gisors), such as Henry II's great towers at Newcastle, Scarborough and Dover, or in France, the massive donjons built by great lords such as Troyes, Provins and Bar-sur-Aube in Champagne, or Paris, Laon, Senlis and Étampes in the Capetian heartlands, or Châteaudun on the Loir (about 1170). The continued development of the count's castle at Ghent followed the rectangular pattern as well. Houdan, built before 1137 by the Montforts, was unusual, a remarkable quasi-rectangular donjon with round full-height corner turrets.

The round tower at New Buckenham has been dated to around 1140, and it certainly looks like a rather experimental design. Henry II himself ordered a tower that was circular inside, although octagonal outside, at Orford (1165), where the disappearance of the apparently massive outer curtain wall removes the evidence for the serious defences of this castle, while his half-brother Hamelin's slightly later tower at Conisbrough is similar. In Britain, some half of the round donjons are in Wales (although most date from the later part of this century or the beginning of the thirteenth). Shell-keeps have been dated to the twelfth century on the mottes at Launceston and Trematon in Cornwall, Totnes in Devon (although the present walls date from the fourteenth century), Carisbrooke on the Isle of Wight, and many others. The shell-keep at Vatteville-la-Rue on the Seine, in land belonging to the Beaumont dynasty, may date from a rebuilding of the previously damaged castle around 1150.[1] It has been suggested that the presence or absence of suitable stone for the crucial cornerstones may have influenced the decision to move away from the rectangle, which was clearly the most efficient way of creating residential or ceremonial space, or it may just have been a question of style.

Certainly, the long building period required to complete a stone great tower, estimated to have risen on average at as little as three metres per year, lends weight to the argument for the ceremonial and symbolic function of the tower. But this is not to say, as we have shown, that it was not also expected to function

[1] Renn, "Castle fortification in England and adjoining countries", 53–8; Renn, "The round keeps of the Brecon region", 129–39; Heliot, "Sur quelques résidences princières"; Heliot, "La grosse tour de Châteaudun", 225–44; Heliot, "Les origines du donjon résidentiel"; Flambard Héricher, "La construction dans la basse vallée de la Seine", 95. Houdan: Mesqui, *Châteaux forts et fortifications*, 196–7, and the studies cited there. For the other sites in England, see the corresponding English Heritage handbooks, and Remfry, *Buckenham Castles*.

as a defensible building as well. Nor was there any reason why a donjon originally erected with one purpose could not be adapted for another, as has been suggested with the rebuilding of the monumental three-storey tower of the counts of Vianden (Luxembourg), now dated dendrochronologically to the second half of the twelfth century.[1]

There has been debate over the appearance of arrow loops in west European fortifications, a clear indication of the defensive purpose of a building.[2] Once it was asserted that they were unknown before the middle of the twelfth century, while others have argued that the first types in the west date from the late 1080s. Loops offered protection to archers and crossbowmen that was not available from crenellations. However, their more sophisticated development, with a wide field of fire, had to wait for the next century. It has been suggested that it was the renewed appearance of the siege tower late in the eleventh century that drew attention to the importance of creating flanking fire, to counter the ability of attackers to command the wall-top, and that arrow loops were installed as a means of discouraging attacks by escalade at parts of the circuit, thus forcing an enemy to attack the places that were strongest.[3] Certainly, increasingly sophisticated techniques of attack have been observed at work in western Europe during the twelfth century. The most advanced methods would have been encountered in the east, or in al-Andalus, by crusaders. But here, as we have seen, defence still largely took place from the top of the wall, and from the flanking towers that had featured there since Roman times. Machicolation had appeared to cover gate entrances. The role of the archer, east and west, had been significant in siege warfare for a long time, to the extent that we have seen specific reference to the assembly of bowmen and crossbowmen in order to conduct sieges, and to their no less important role in defence. By the following century, more obviously carefully planned splays of arrow loops and spaces for archers would become a very common feature of towers on defensive circuits, and it is of course likely that the first stages of such a development had appeared previously. The evidence, however, permits no firm conclusions as to changes in fortress design to improve protection for the archer in the twelfth century.

Central and northern Europe

The continued involvement of the German kings and emperors in the affairs of Italy provides much of the evidence for siege warfare and fortifications in the twelfth century, but at the same time the expansion to the east of the German kingdom, driven by population growth and the need for new territories, brought closer contact with the Slav tribes of the north and the more established

[1] Zimmer, "Le château de Vianden".

[2] Unless, of course, the opening was really for light. See Chap. 9 for more discussion of loops.

[3] Salamagne, "Origines et diffusion des embrasures de tir".

Slav and Hungarian lands of central Europe. The continued development of the Magyar kingdom also brought contact in the Balkans with the Byzantine empire as a result of the restoration of the latter's Danube frontier.

Across the river Elbe a march had been created in the tenth century by Henry I, and Slav *castra* had been captured. The *castrum* at Brandenburg itself had changed hands back and forth. Margrave Albert I came into conflict with his eastern neighbour, the Polish duke who recaptured Brandenburg, while Albert recovered it with the aid of Wichmann, archbishop of Magdeburg, in 1157.[1] Further south, conflict continued between the Bohemian dukes and their neighbours. In 1114 they tried to gain the castrum at Glatz (Kłodzko), and burnt the *palacium*, the outworks and the tower near the walls. The role of castles in the state-building of the dukes was affirmed by grants of castles recorded in 1115, and orders to rebuild strongholds in Moravia made in 1121.[2]

In Germany the dukes and princes had become more accustomed to their autonomy. Henry V found himself unable to bring to heel his chief opponents, Lothar of Süpplingenburg, duke of Saxony, and the Rhineland bishops. In 1113, his loyal vassal, and heir to the Salian lands, Duke Frederick II of Swabia, dealt with a revolt west of the Rhine while the king was engaged against Lothar, "destroying fortresses". Frederick besieged Bar-le-Duc, which was strongly defended against him, but he stormed it. Mainz, which like its near neighbour Cologne was a city with strong walls and towers, successfully resisted the duke, although it is recorded that in his march down the Rhine he took many castles.[3]

Henry was succeeded as king in 1125 by his Saxon opponent as Lothar III. Lothar's intervention in Bohemia was thwarted by the *indagines*, the barriers created to prevent passage through the forests. The following year, finding the defences of the former royal fortress at Nuremberg too strong, he was engaged trying to take Speyer on the Rhine, which finally surrendered in 1129. He was also concerned with his north-eastern marches: he ordered the construction of a new castle at (Bad) Segeburg, assembled the people of the province to build it, then left it strongly garrisoned. The Slavs rightly saw this as a threat, so when Lothar died and was succeeded by Conrad III, of the Staufer (Hohenstaufen) dynasty, they rose and destroyed several of the new German castles. In 1143, Count Adolf II of Schauenburg and Holstein ordered the reconstruction of Segeburg, and he provided it with a wall, suggesting the original defence must have been of timber. The work included a tower on a rock. He further ordered the construction of Lübeck, on the site of a Slav fort which had been a massive earth and timber stronghold serving as a royal residence, town and seaport: captured and destroyed by a rival in 1138, it was deserted when the Germans resumed occupation (1143).[4]

1 *Chronica marchionum Brandenburgensium*, 116–18.

2 Cosmas of Prague, *Chronica Boemorum*, 205, 213–14, 220.

3 *Annales Hildesheimenses*, 62–3; Otto of Freising, *The deeds of Frederick Barbarossa*, 44–7; Ekkehard, *Chronicon universale*, 247.

4 *Annales Hildesheimenses*, 66–7; Helmold of Bosau, *The chronicle of the Slavs*, 160–69; Fehring, "Origins and development of Slavic and German Lübeck", 251–60.

Scandinavia

The castle, and regular siege warfare, made their appearance in Denmark in the first part of the twelfth century. Danish archaeologists have found no evidence of castle building in the Danish kingdom in the post-Viking period. The eleventh century is, apart from a handful of mottes, the dates of which are extremely hard to determine, a complete blank. But Danish kings and the greater nobility were building castles in the 1100s. Early in the century, Knud Lavard "hired builders" to have two castles built on opposite sides of the sound leading to Slesvig (the Slien, Schlei in German), with a chain between to block the passage to raiders from the Baltic-coast Slavs. In 1133, King Niels (r. 1104–34) besieged a castle at Roskilde, which had been established in 1127 and reinforced in 1132 with earthworks and timber. In order to achieve the capture, the monarch had to draw on the expertise of Saxons dwelling in the town to construct engines, including stone throwers, through the agency of which the rampart was brought down. In 1161, Søborg Castle was captured by the king through treachery: this seems to have consisted of a ring-work containing a tower and palace. The oldest known Danish donjon is at Bastrup (between Slangerup and Farum), dated to mid-century, and is said to have been modelled on the old tower at Hamburg. It has four storeys and very thick, plain walls. The absence of any outbuildings leaves questions about its function. Another early donjon stands at Helsingborg, round with walls 4.25 metres thick. The donjon at Søborg was octagonal (where it had been built by Archbishop Eskil of Lund in the 1140s), as was the neighbouring royal castle at Dronningholm. King Valdemar I's donjon on the island of Sprogø, from the 1160s, was rectangular. Other castles of this period were of earth and timber, not unreasonably given the long Danish tradition of timber building and the shortage of building stone in the country, a situation that encouraged the early adoption of brick as the main building material in the second half of the century. When Valdemar captured Søborg, he extended it with new buildings of brick, while the same material was used when the former wooden castles of Vordingborg and Kalundborg were rebuilt in the 1170s.[1]

Norway had been largely isolated from the processes occurring in Europe for some time, even the account of the castle-capturing exploits of the future king Harald Hardrada while serving in the Byzantine Varangian guard in the 1030s suggesting (if they are not legendary) no skill except courage and cunning.[2] But the Christianisation of the country went hand in hand with the establishment of a more stable kingship. Castles were very few. The royal palace at Bergen was built of timber, and was beautifully crafted, but had collapsed by the time Theodorus

[1] *Knytlinga saga*, 120 (Knud); Saxo grammaticus, *Gesta Danorum*, 1: 361 (Roskilde). On castles, see Stiesdahl, "Die Motten in Danemark"; Stiesdahl, "Die ältesten dänischen Donjons", 279–83; Olsen, "Danish medieval castles at war", 223–31; Liebgott, "Brick-making and castle-building", 109–15.

[2] Snorri Sturluson, *King Harald's saga*, 53–4. The saga, p. 106, attributes the founding of Oslo to Harald (about 1050).

wrote his history of Norway's kings later in the twelfth century.[1] There was a motte with a thirty-metre diameter as part of the royal castle built in Oslo around 1050, and the archbishop had a timber castle at Nidaros (Trondheim) by the 1170s. The *Heimskringla*, Snorri Sturluson's sagas of the Norse kings, tells that King Magnus III (r. 1130–35) defended Bergen in 1135 by setting up a stone-slinging machine, and iron chains to block the passage, and had caltrops made. In the same year, King Ratibor of the Wends (properly duke of Pomerania) launched a raid that involved the defenders of Konghelle (near Kungälv) retreating to their castle. The Wends only had stones and arrows, but the townsfolk imprudently used up all their missiles and stones and had to surrender.[2] The major developments in castles in this area had to await the lead given by later monarchs. Significant castle building was the domain of the crown, the greater nobility and the bishops. While as elsewhere there were clearly important elements of status involved, it is also clear that significant castles shared with this role a function as defensible bases of power, and as elements of a defensive system against outside attackers. In consequence of having so little need for the art, and therefore very little experience, ideas of siege warfare seem to have been very limited, and we have cited the only references for the whole period to siege weaponry in use.

Italy

The main players in Italy remained the Italo-Normans, the papacy and the cities of the north, nominally part of the empire. The Beneventan chronicler recorded that in 1113, a baronial castle was "wonderfully" built at Monte Sableta (perhaps Serra near Montemiletto) as a base by the local baron Robert the Slave for attacking the city of Benevento, but was demolished by the constable. A few years later, another Robert of Montefusco captured the local castle of Tufo by attacking it with siege engines. The following year the same castle, now with "wonderfully constructed defences, surrounded by a rampart and ditch" was attacked with "wooden machines" (siege towers) and captured.

Roger II, king of Sicily, seeking to extend his control over the mainland and having subjected Salerno and Amalfi, attacked the castle at Brindisi by land and sea in 1131/2, and then Bari, deploying numerous machines to take the place inside two weeks. Meanwhile, Prince Robert II of Capua and the men of Benevento attacked the castle of Balba (now Barba) using "wooden machines for destroying the castle's tower", which again sounds like a siege tower, perhaps this time one provided with a battering ram. The king's men secured the castle and saved it. In 1133, Roger's cause was marked with success as he captured Matera, Anzi and Trani. The powerful Count Tancred of Conversano held out in the castle of Montepeloso (Irsina), which Roger stormed using the same "wooden machines"

[1] Theodorus monachus, *Historia de antiquitate regum Norwagiensium*, 64–5; Theodorus, *An account of the ancient history of the Norwegian kings*, 50–51.

[2] Snorri Sturluson, *Heimskringla*, 326; Ekroll, "Norwegian medieval castles", 65–6.

and other warlike equipment. Count Ranulf II of Alife, another of the king's opponents, attacked the same castle while Roger was busy devastating the province of Apulia, and he deployed machines for destroying the walls, which terrified the defenders.[1]

Roger's triumph was challenged by one of the rival popes (this struggle coincided with another papal schism), the Byzantine emperor and the German emperor Lothar. In 1136, their allies of Pisa tried to take an Amalfitan castle at Ravello with *gatti* ("cats"), but Roger relieved it.[2] In 1137, Lothar marched against Roger and besieged Bari. The king had strongly fortified the castle in the city, although it is impossible to recover what was there at this time. The emperor's army deployed numerous and diverse machines against Bari, succeeding in taking the town but not initially the castle, which then fell and was destroyed. He had pressed the attack for forty days. Meanwhile, Melfi fell, and the Pisans attacked Salerno by building wooden siege towers, "marvellously tall and terrible", mangonels, petraries and cats. Once these had been brought up against the wall, the citizens surrendered, but Roger's knights continued to hold out in the "great tower on the mountain".[3]

Roger's cause was now assisted by events elsewhere that removed his great enemies; his local enemies were on their own. In 1138 Roger sought to recover Melfi, but found it too strong, so instead attacked the castle of Tocco Caudio, deploying wooden towers to destroy the towers and walls. By 1139 the whole territory had been recovered except for Bari itself. Roger ordered built wooden siege towers to enable the walls to be pulled down, but in fact it appears he did not try very hard to storm the isolated city. Instead, he constructed no fewer than thirty siege castles, as a result of which Bari surrendered through hunger after two months.[4] Roger proceeded to consolidate his newly unified realm, and one of his measures was the construction of a substantial network of castles, to which he appointed castellans serving in an administrative role.[5] The evidence of this period then confirms that the same techniques originally noted as being deployed during the conquest in the eleventh century continued in common use, but the fortifications against which they were used may have been stronger. The use of siege castles against Bari may indicate no more than that Roger wanted to avoid the heavy casualties of an assault, or perhaps wanted to take the city without it being sacked. If there is a real change since the previous century, it is the far greater presence in the record of the

[1] Falco Beneventanus, *Chronicon*, 163, 177, 180, 192, 206, 216, 219, 221; *Chronicon romanorum pontificum et imperatorum ac de rebus in Apulia gestis*, in *Ignoti monachi . . . Chronica*, 18–19 (Bari).

[2] Maragone, *Gli Annales pisani*, 10.

[3] *Chronica monasterii Casinensis*, 568; *Chronica romanorum pontificum*, in *Ignoti monachi Chronica*, 21; Falco, *Chronicon*, 232–9. Falco says that the tower at Salerno was lost through discord among the attackers; see also Maragone, *Gli Annales pisani*, 11.

[4] *Chronica romanorum pontificum*, in *Ignoti monachi Chronica*, 26; Falco, *Chronicon*, 242–4.

[5] *Chronica romanorum pontificum*, in *Ignoti monachi Chronica*, 26–7; H Houben, *Mezzogiorno normanno-svevo*, 164.

individual castle as a target, as opposed to cities and Byzantine forts. In a realm ruled by Normans and their descendants, we should expect nothing less.

Northern Italy was the cockpit for the next well-recorded instances of siege warfare. In 1148/9 the Pisans were at war with the people of Lucca. The Pisans captured a seaside castle held by 300 Lucchese, using 12 mangonels and scaling ladders after an attack lasting 15 days. But when the Lucchese attacked the Pisan garrison of the castle at Vorno, despite using catapults and cats for 24 days, they were unable to make headway and had to abandon the attempt. In 1150 they tried again, this time isolating it with three siege castles. The defenders sold the place to the besiegers.[1]

Lothar was succeeded by the Staufer Conrad III (r. 1137–52), and on his death the crown passed to his nephew Frederick I, to become known as Barbarossa. There are some particularly detailed accounts of the new emperor's campaigns in Lombardy, where he was opposed by shifting alliances of the wealthy and increasingly independent cities. The emperor took care to strengthen loyal places like Novara, which received "a new wall and rampart", while hostile towers were destroyed. In 1155, the emperor's first target was the city of Tortona, allied to Milan, to which Frederick laid siege on 14 February, establishing a tight blockade to prevent entrance or egress. First his army waded across the river Scrivia to storm the weakly protected town, but then in February he had to attack the citadel which had lofty towers and walls, with "machines and petraries". The defenders concentrated on keeping the emperor's forces beyond the ditch. Another source suggests a wide range of throwing weaponry was deployed, including stones hurled apparently from a ballista, "which is commonly called *manga*", which hitting the weaker top of the wall split into three, killing three knights. Whether this is genuine reporting (which is not borne out by any other evidence) or chronicler's fancy is unclear, but for the most part the siege concentrated on shattering the masonry and the houses within with stones from engines, trying to create a breach with a battering ram, and aiming at defenders with arrows and javelins including those fired from ballistas. They also attempted a mine (*talpa*, mole or miner). However, the defenders countermined and the attempt failed. Around Easter there was a truce, used by the Tortonese to construct new machines. In negotiations, according to one anonymous source, Frederick boasted that with his mangonels and machines built to destroy the walls, he could easily take Tortona. In the end, the town capitulated on 18 April, and its walls and towers were razed.[2]

The dramatic arrival of the German emperor in northern Italy seems to have led to some frantic building activity. Caffaro da Caschifellone records how in 1155

[1] Maragone, *Gli Annales pisani*, 12–13.

[2] Godfrey of Viterbo, *Gesta Friderici*, 310; *De ruina civitatis Terdonae*, in Hofmeister, "Eine neue Quelle zur Geschichte Friedrich Barbarossas", 150–54; *Barbarossa in Italy*, 13; Morena, *De rebus Laudensibus*, 594; Rahewin, continuation of Otto of Freising, *Deeds of Frederick Barbarossa*, 132–141; Rahewin, continuation of Otto of Freising, *Gesta Friderici I.*, 318–36. Mierow, the translator, points out that much of Rahewin's text is lifted from the classical accounts of Josephus. The reference to a torsion-powered stone-throwing ballista above may lead to a similar conclusion about parts of Otto's work as well.

the Genoese commune decided to erect a powerful wall and gates on the landward side of their city, and the work was done in 1158. Men and women conveyed the stone and timber to the city wall day and night until the wall was established within eight days, a feat apparently never heard of in Italy. Nor indeed does it seem humanly possible, and the next sentence in fact shows that this wall had then to be completed and raised to a proper height, so perhaps the initial building period refers to the foundations: a remarkable enough achievement given the length of the wall. The chronicler records further how wooden hoardings (bretasches) were provided from ships' timbers. Fragments of Genoa's twelfth-century walls and several gates remain as testimony to this building spree.[1] At Pisa, "for fear of Frederick", the city built barbicans and began a (new?) stone enceinte that encompassed the whole city by the second year of construction, while the defences were strengthened with wooden "towers and castles and bretasches", suggesting the provision of wooden hoarding around the circuit to protect the base of the walls against mining or sapping.[2] As he progressed through the cities of Lombardy and Liguria, the emperor had ordered the destruction of the towers built by the local aristocracy within the cities, as at Asti and Piacenza.[3] These tall, usually narrow, rectangular towers had become a feature of the Italian cityscape, representing both symbols of domination and convenient strongpoints for competing noblemen in the crowded urban centres. Such towers still stand in some numbers in several cities, such as Bologna and most famously San Gimignano.

Frederick's most powerful enemy was Milan. Before attacking it, he secured the castle of Trezzo on the river Adda,[4] comprising a wall and tower covering a stone bridge across the river, which surrendered as the emperor stormed it. Leaving a garrison — Trezzo would be an important imperial strongpoint in the years ahead — Frederick marched on Milan itself, recorded as being in 1158 strongly defended with a wall with one hundred towers, six gates and a wet moat fed by the river, which had been recently reinforced against Frederick's anticipated visit. One source talks of a new ditch and palisade encompassing both the old wall and suburbs.[5] The imperial army laid siege on 6 August, protecting itself according to classical principles with a palisade of stakes, a necessary precaution because the Milanese made frequent sorties. The Milanese had converted a Roman arch into a strongpoint lying a bowshot from the city wall, constructed of large stones on four columns. The emperor decided to target this outpost, but the classical masonry was too strong to be shattered by his machines, so instead he surrounded

[1] Caffaro et al., *Annali genovese*, 51 (and further work recorded in 1159–60, pp. 54, 60); *Barbarossa in Italy*, 64 (ll. 1816 ff.).

[2] Maragone, *Gli Annales pisani*, 16. For a description, see Tolaini, "La costruzione delle mura di Pisa". The wall of 1155–61 was low, not more than about 4–5 metres, with small towers. Building work would continue until 1261.

[3] *Annales placentini guelfi*, 411–12.

[4] Garlandini, *I castelli della Lombardia*, 133–5. Frederick added three square towers to the simple enceinte. The remaining tower is very tall (42 metres). The original entrance was a gate reached by a ladder.

[5] Milan's walls: *Barbarossa in Italy*, 64; Flamma, *Chronicon maius*, 655–7 (written a century later); Barni, "La lotta contra Barbarossa", 27.

it with archers and slingers and began to break down the columns, which forced the defenders to surrender. Frederick then turned the arch against his foes, having ladders built to allow access to the top, then built a stone thrower on top to hurl stones at the adjacent gate and its defenders, which was in turn destroyed by a Milanese stone thrower set up against it. He also carefully inspected the large trace of ramparts for any weaknesses. However, the victor was the cruel summer weather, which persuaded all parties to make peace.[1] It was short-lived. In April 1159, the Milanese attacked Trezzo with artillery and siege towers. The stone throwers smashed the houses and damaged the walls, and the attackers apparently used both escalade and undermining, leading the exhausted garrison to surrender after three days.[2]

Siege of Crema, 1159/60

The emperor was persuaded by his allies, especially Cremona, to devote his attention to a siege of the city of Crema, protected by a ditch and double wall, and apparently well supplied against a siege. For six months, the resources of the empire were to be tied down in front of this one small place, low-lying and marshy. Frederick had decided to take it by assault. The full range of resources available to a twelfth-century monarch was deployed. A siege tower was raised on a mound, covered with iron to protect it, and filled with archers and slingers to drive the defenders away from their walls. It stood more than thirty metres high, and "nothing like it had been seen this side of the sea" (Morena), with five hundred men needed to move it. The Cremaschi attacked the towers with stones from two catapults and caused considerable harm to the timber. Frederick also deployed artillery, both large and not so large (a distinction between the terms mangonel and petrary that is unusual). Sorties were launched to try to destroy the mangonels, leading to fierce fighting. Later, the emperor ordered the construction of a giant cat ("gattum maximum") or tortoise, which would be used, alongside a smaller shelter, to permit the besiegers to fill in the ditch. Massive labour must have been called for, as two thousand cartloads of earth and timber were brought up for this purpose. The defenders started to construct a wooden tower of their own to oppose the cat, along with a battery of five great mangonels (or nine, according to Otto of Freising) and many petraries. The cat resisted a bombardment of torches, missiles, rocks and arrows, and the moat was filled in. Various incendiary devices were hurled down in vain, with the emperor himself taking a hand to extinguish the fires. It was to prevent damage to his own siege equipment that Frederick at this point ordered that prisoners be bound to the front of his own siege tower, but this did not deter the Cremaschi from firing, with the result that many of the unhappy prisoners were killed, until the emperor repented and had them removed. As the cat was driven up to the wall, ready to use its rams to destroy the masonry,

[1] *Annales placentini guelfi*, 412; Rahewin, *Deeds of Frederick Barbarossa*, 212–17; Morena, *De rebus Laudensibus*, 605–6; *Barbarossa in Italy*, 64 (ll. 1816–31).

[2] Rahewin, *Deeds of Frederick Barbarossa*, 272; Morena, *De rebus Laudensibus*, 609; *Barbarossa in Italy*, 101 (ll. 2853–68).

the Cremaschi mined out to attack the cat and the siege tower, leading to more ferocious hand-to-hand fighting, with the imperial troops finally driving the Cremaschi back through their tunnel. Meanwhile, Frederick's missile troops drove the Cremaschi away from their engines.

The attack was being launched on more than one front, and Duke Conrad (Frederick's brother) and the count palatine were carrying out a similar assault against one of Crema's gates, using many cats and shelters to protect them as they filled the ditch. On 21 January 1160, Frederick was ready for the assault, and ordered his troops to throw a bridge from the tower across to the wall. The Cremaschi drove the attackers back, though their mangonels failed in an attempt to smash the bridge. So for the next four days, a continuous bombardment was maintained with stones and arrows, such that Crema could resist no longer and surrendered on 27 January.[1] The inhabitants were permitted to leave.

The accounts of the siege of Crema refer to the involvement of master engineers, which should be no surprise. The great siege tower was built by Cremonese who had experienced siege warfare in the east, while the tower with the moveable bridge was apparently built by a Genoese named Marchesius, who was richly rewarded by the emperor, while a Cremonese named Tinto Muso di Gatta received a county. But it is noteworthy that despite the scale of the resources used in the attack, and the array of equipment, after six months of effort the town did not actually fall to assault but surrendered through exhaustion. However, to conclude that the whole immense effort to attack with engines was superfluous would be wrong, as it is clear that it was, in the end, the fear of an assault actually breaking in that induced the final capitulation.

Continued conflict in northern Italy

In March 1161 the Milanese captured the castle of Caṣtiglione d'Adda by storm, using mangonels, petraries and cats that were driven to the foot of the walls, although the account says the garrison surrendered for lack of water.[2] Frederick had been devastating the lands of the contumacious city, taking the castle of Iseo (defended by a palisade and moat) when his forces followed the defenders back into the fortress after beating them in battle. Meanwhile, the citizens of Bergamo attacked Volpino (now part of Costa Volpino), preparing catapults and scaling ladders, leading to surrender when it was apparent that there would be no relief. Milan itself would fall in March 1162, after a long blockade carried out with siege castles, rather than the close attack that Frederick had attempted in 1158 and had used at Crema. However, the blockade had been close, and a large army assembled, including a contingent of archers sent by the king of Hungary. The full range

[1] Morena, *De rebus Laudensibus*, 611–19; *Barbarossa in Italy*, 106–10 (ll. 2990–3110); Rahewin, *Deeds of Frederick Barbarossa*, 280–305; Maragone, *Gli Annales pisani*, 18; Codagnellus, *Annales placentini*, 6; *Annales placentini guelfi*, 412–13. For a vivid account, see Giesebrecht, *Geschichte der deutschen Kaiserzeit*, v: 204–12.

[2] Morena, *De rebus Laudensibus*, 620.

of siege equipment is recorded as being deployed, although why this should be necessary as no direct assault took place leads one to question the accuracy of the chroniclers. Having surrendered, the angry emperor ordered Milan destroyed, in particular its walls and towers.[1]

Barbarossa's triumph was short-lived, with a new rebellion of the Lombard cities, and the reoccupation of abandoned Milan, where the defences were restored and a new city founded, called Alessandria (in honour of Pope Alexander III), "surrounded with the deepest and strongest ditches" (Otto of St. Blasien).[2] In 1173, Frederick sent an army under Christian I, archbishop of Mainz, to besiege Ancona, a town well provided with walls and outer defences. The emperor negotiated to obtain the aid of a Venetian fleet so as to close the blockade. An attack was launched with artillery and ballistas, doing serious harm to Ancona. Alerted that the town only had a few days' supplies, the chronicle reports that the imperial army launched an assault by land and sea, but the Anconitans drove them back to their machines, which were burnt with incendiaries, and also captured some of the ships. The besiegers, chastened, sat back to let hunger do its work, but Ancona had sent for help to Constantinople and to Ferrara, whence a relief force was despatched that raised the siege.[3] Meanwhile, the war between Pisa and Lucca continued, with the Pisans attacking the castle of Corvaia (north of Pietrasanta) for a month in 1170 with mangonels, only for the Lucchese to relieve it, then trying the castle of Agnano (near San Giuliano Terme) with the same equipment. Three years later they sent eight galleys carrying mangonels to attempt the castle on the island of Pianosa. Their assault failed, but they returned in secret to take and destroy it.[4]

Frederick's plans in Italy now looked in tatters, and the cities he had ordered destroyed were restored and leagued against him. In November 1174 he brought an army to besiege newly-founded Alessandria, deploying stone throwers, rams, wooden siege towers (of "wonderful construction") and all kinds of missiles. The defenders were helped by a contingent from Piacenza, whose chronicler records proudly their role in hurling back an imperial assault and counterattacking so successfully that the siege tower was burnt, with the Genoese crossbowmen who were on top of it perishing. With relief from Milan and Verona expected, Frederick abandoned the siege, burning his remaining equipment in April 1175.[5] The chronicles tell of one further siege conducted by the emperor in Italy, in 1186, when he allied with the Cremonese to destroy a castle called Castel Manfredi, at Castelleone near Crema. Frederick prepared mangonels, petraries, cats and a siege

[1] Rahewin, *Deeds of Frederick Barbarossa*, 335; *Barbarossa in Italy*, 112–15 (ll. 3154–233; Godfrey of Viterbo, *Gesta Friderici*, 317; Otto of St. Blasien, *Chronici continuatio*, 310–11; Barni, "La lotta contra Barbarossa", 60–67.

[2] Barni, "La lotta contra Barbarossa", 93–5, where the names of several of the engineers are recorded; Otto of St. Blasien, *Chronici continuatio*, 315.

[3] Boncompagno of Signa, *Liber de obsidione Ancone*, 9–45. The author was from Ancona but not a contemporary.

[4] Maragone, *Gli Annales pisani*, 48, 58.

[5] *Annales placentini guelfi*, 414; Otto of St. Blasien, *Chronici continuatio*, 315; Codagnellus, *Annales placentini*, 9; Godfrey of Viterbo, *Gesta Friderici*, 328.

tower, but as the chronicler notes, the castle "was not captured by force, but by treaty".[1]

The kingdom of Sicily had fallen into internal disarray with the reign of King William "the bad" (r. 1154–66) and the minority of his young son William II. The town of Butera was besieged in 1156, but the king, "seeing no way of forcing them to surrender", had to negotiate a peace. In 1162 he was back, encircling Butera with a ditch. The same problem led to the same solution as six years before. William then crossed to the Calabrian mainland and encountered opposition at the town of Taverna, on a high rock. The first attack was repelled with stones and barrels covered with sharpened nails rolled down the hill. A second attack with scaling ladders succeeded, to the perplexity of the chronicler. In 1167, the people of Reggio Calabria opened their gates to the Messinans, while the knights retired to the citadel. Threatened with the arrival of galleys from Messina, "bringing the siege engines needed to capture the keep", they quickly surrendered. In Palermo, a crowd of people drove the chancellor's knights into the bell tower. They debated how to take it. Some were in favour of piling up wood and setting it alight to loosen the stones, others thought that siege engines should be brought up, others that mantlets be used to allow undermining. In the end, the chancellor accepted terms.[2] Thus the citizens of Palermo were well acquainted with the variety of siege techniques available, but in the end resorted to the simplest.

Henry the Lion, the Slavs and Danes

There was also military activity in the northern parts of the German kingdom, where Henry the Lion, duke of Saxony, was engaged in establishing and extending his realm. In the 1160s he was engaged in conquering formerly Slav territories, generally taking over their forts as new centres of administration, as at Ilow (near Neuburg) and Mecklenburg. Campaigning with King Valdemar I of Denmark from 1160, he found that his opponent, the Obotrite prince Niklot, abandoned most of his forts, including one at Dobin on Lake Schwerin, which has been found to have been a site of 300 × 50–100 metres, cut off at each end by a rampart. In 1163, Henry attacked Werle (now part of Kassow), where Niklot was killed. Using the abundant supplies of timber around, he ordered construction of suitable siege equipment "such as he had seen at Milan and Crema". This included a ram for breaking the wall, and a siege tower from which his archers could shoot into the fort. Machines featured again in his attack on Demmin in 1164, while Wichmann, archbishop of Magdeburg, used them against the duke's castle at Haldensleben in 1166, although in vain. Periodically, the Slavs revolted, overrunning the new German lordships, but generally the new castles held out. In 1167, Henry captured Oldenburg in Holstein, which under the Slavic name of Starigard had served as capital of the Slav principality of Wagria. This was a powerful lozenge-

1 *Annales placentini guelfi*, 415.

2 Falcandus, *The history of the tyrants of Sicily*, 72, 125–7, 204, 211–13.

shaped earthwork 225 × 100 metres. In 1169, Henry attacked a German noble's castle, Burg Desenberg (near Warburg). Because the place stood on a high rock, he was unable to use siege weapons, but his miners in digging into the hill found the castle's water supply and were able to stop it, producing surrender.

In 1177, Henry had to retake Demmin. A "machine-builder" named Frederick advised he could take the place inside three days. The year before, the Danes had taken the attack to Stettin, a strongly fortified settlement. The besiegers constructed wooden mantlets to get close to the wooden ramparts, from which they tried to fire the defences. Lacking engines, they brought earth up. The defenders responded from on high with arrows and slings, but in the end sought terms of surrender. In 1179, Archbishop Wichmann, having failed to deal with Haldensleben by traditional means, took it by making a break in the walls of the town and causing it to be flooded. In 1180, Duke Henry, who may have learnt some of his siege techniques from Frederick Barbarossa's methods in Italy, had not however learnt how far he could push the emperor and made a major misjudgement in rebelling against him, being deprived of his substantial lands, despite his great status and power. Henry took refuge in Lübeck, only recently rebuilt as a German settlement, but could not face the combined army of Frederick, which set about constructing siege engines, and the fleet of the Danes which arrived to close the blockade, and had to give in.[1]

The close engagement of Bohemia with the empire and the involvement of its rulers with crusading had exposed the Slavs of that land to the techniques known to their neighbours. Prague Castle itself had been besieged by Conrad in 1142 with "diverse engines of war", and Frederick repeated the treatment in 1178. Prague Castle, founded in the tenth century and noticed for its strength by Ibrāhīm ibn Ya'kub, had been rebuilt in the form of a palace and citadel by Vratislav II or I from 1061 on the high ground on the left bank of the river, while the fortress of Vyšehrad was on the other bank, further south. The town had undergone considerable expansion during the twelfth century, with a stone bridge existing across the Vltava by 1158, but the town walls were not erected until the next century.[2]

The Byzantine empire and its foes

At mid-point in the twelfth century, Manuel Comnenus faced conflict with Byzantium's foes on many borders: the Muslims in the east, his former subjects and neighbours in the Balkans, and the rapacious Roger of Sicily.

[1] Gabriel et al., *Starigard/Oldenburg*, I: 213–15; Ettel and Wiechmann, "Die slawische Fürstenburg von Dobin"; Theodorus monachus, *Annales Palidenses*, 92–4; *Annales Pegavienses et Bosovienses*, 261–3; Saxo grammaticus, *Gesta Danorum*, I: 488–9 (Stettin); Arnold of Lübeck, *Chronica Slavorum*, 38, 48, 63–5 (Lübeck); Helmold of Bosau, *Chronica Slavorum*, 185–6 (Werle), 201 (Demmin), 206 (Haldensleben). For a modern biography, see Jordan, *Henry the Lion*, 80, 92, 101–5, 172.

[2] Vincentius of Prague, *Annales*, 660, 664, 666, 679, 689; Humel, "Research in Prague", 268–78.

Wars involved recovering Kerkyra (Corfu) from the Sicilians, taking forts by storm from the Serbs and Hungarians, and taking Hungarian Zeugminion (Zimony or Zemun) by blockade. In 1155, a combined army of Hungarians, Czechs and Saxons attacked Braničevo, on the Danube east of Belgrade, although the arrival of a Byzantine relief forced them to abandon their siege, burning their battering rams. In the same year, Manuel took his war across the Adriatic. The citadel of Bari was bribed into surrender. His general, John Doukas, attacked and took several of the castles of the area, including Montepeloso and Gravina, but was held up at a castle where "it was hopeless to pry a stone out of it although it was frequently smitten by siege engines". Taranto held out, as the besiegers on this occasion were "deficient in siege engines". The following year the target was Brindisi. The attackers decided that it was pointless to bombard the city walls because, as elsewhere, they were too well constructed to be harmed by stone throwers. Instead, they directed their bombardment over the walls to terrify the citizens, a tactic that led to the surrender of the town, but not of the castle. To attack this, a tortoise was built and driven up to the wall, but once again the Byzantines were defeated by the strength of the masonry — the stones were so close together that they "seemed to be a single stone". So they turned to mining, put in the timber props, fired them, and the outer wall of the castle collapsed, forcing the defenders back to their inner citadel. A Norman relief army ended the Byzantine effort.[1] Later rebuilding makes it very hard to recover the twelfth century fortifications at these sites.

In 1165 the Hungarians laid siege to Zeugminion, on the north bank of the Danube just north of Belgrade, and captured it. The emperor responded with a counter-attack, marching via Sofia and across the Danube. The city was on a high hill above the river, unapproachable from one side. The defenders set up their own stone throwers against the besiegers. The Byzantines conducted a regular attack. Faggots, firewood and mud were collected to fill in the ditch. Stone-throwing catapults were arrayed against the curtain wall, hurling stones "weighing a talent", which "weakened the wall's joints". The emperor's cousin Andronikos Kontostephanos took charge of one machine, "and by using the sling, winch and screw press shook the section of wall between two towers violently. Struck by the weighty stone and sapped by miners as well, the wall began to crumble and fall" (Choniates). The Hungarians had built a wooden gallery on the wall (a hoarding), but the skilful Andronikos managed to destroy this as well with stone shot. With the collapse of the targeted section of wall, the Byzantines were able to storm the breach using ladders.[2] The war with Hungary was ended in 1167, and Manuel turned his attention to strengthening the empire's defences. The walls of Constantinople itself had suffered, "ruined by time" (Cinnamus). Furthermore, across Byzantine Asia Minor, walled villages were created, called *neokastra*, with governors sent from the centre, in order to consolidate the lands reconquered from the

1 Cinnamus, *Deeds of John and Manuel Comnenus*, 83, 91, 105, 110–13.

2 Ibid., 180–84; Choniates, *O city of Byzantium*, 74–6.

Turks, to provide defensible habitations for the tillers of the soil, and places of refuge from the Muslim raiders.

In 1169 there was a rare combined operation between the Byzantines and the Latin crusader states, with an attack on Damietta in Egypt led by Andronikos and King Amalric I of Jerusalem, aimed at the recently established power of Saladin. Following initial success on landing, the expedition tried to set up engines against the walls, but were hindered by the arrows of the defenders. The Byzantine leader proposed an escalade, but Amalric waited to construct siege towers: one of seven stories was planned, but for whatever reason was built against the strongest part of the wall. After fifty days, the besiegers were still not ready. The Nile was blocked with a chain, and reinforcements regularly reached the defenders, and with hunger facing the invaders, the Byzantine launched his own assault. In the face of Muslim stone throwers and archers, the ladders were put in place, when the attack was called off by the Latin, who had (perhaps) done a deal with the defenders. The campaign was abandoned, and the siege weapons burnt (13 December).[1]

These detailed accounts from Byzantine writers demonstrate a number of important points about the nature of their siege warfare in the middle of the century. Against properly constructed stone walls, siege artillery was still largely ineffective, but the skilled attacker knew how to combine the different elements: stone throwers, siege towers, means to provide cover for men to fill ditches and attack the base of the wall, battering rams to crack open a breach in less well constructed stonework, and mining. At the end, escalade remained an option. Where the defences were familiar, as they would have been quite often to the Byzantines (because so many were originally their own), weak spots could be identified and targeted. Unlike so many of the sieges conducted elsewhere in Europe at this time, most of those carried out by the eastern empire involved direct assault rather than blockade, and often they succeeded, with a success rate unknown to other Christian realms. Their opponents included peoples from every background, and all of them appear to have been aware of the same siege techniques, and were perhaps only limited in the number of skilled workers and engineers they could call upon.

With their long-established facility for understanding their neighbours' strengths and weaknesses to draw upon, and an apparently adequate supply of skilled military craftsmen, the empire under the Comneni was certainly recovering its strength and power by 1170. In 1175, Manuel ordered the reconstruction of the defences of Dorylaeum (near Eskişehir), destroyed by the Turks. His first step was to have a palisade thrown up all round, but inside the previous line of fortifications, and the next to excavate wells. Turkish attempts to interfere were driven off. The place was ready to receive a garrison within a remarkable forty days. However, in 1176 Manuel launched a further offensive, and took an immense army complete with a great siege train against the Turks of Konya (Iconium). His progress was slowed by the enormous number of pack animals carrying the

[1] Choniates, *O city of Byzantium*, 92–5; Cinnamus, *Deeds*, 209. The Latin view expressed by William of Tyre, *A history of deeds done beyond the sea*, 364–66, differs in detail but not necessarily in conclusion, also hinting at treachery.

siege equipment, and by the large number of non-combatants brought along to accompany the train.[1] At the abandoned fortress of Myriokephalon near Lake Beyşehir the Turks wiped out the Byzantine army, inflicting as grievous a defeat as they had a century before at Manzikert. Within thirty years, the great empire would go down to catastrophic defeat at the hands of the fourth "crusade". The blow struck at Myriokephalon had played a major role in leading to this outcome: the progressive restoration of the empire from the disaster of the loss of Anatolia after 1071 was ended.

Byzantine fortifications under the Comneni

Alexius Comnenus had begun to recapture former imperial territory in Asia Minor and took advantage of the First Crusade, as we have seen, to recover Nicaea, then the coastal provinces of Ionia and Lydia. Thereafter, the fortresses of Adramyttium (Edremit), Attalia (Antalya) and Didyma (Didim) were restored. John Comnenus built the fortress at Lopadium (Uluabat) in Mysia and, in 1138, Achyraus (Balıkesir). Manuel Comnenus ordered major reconstruction in the 1160s in Pergamon (Bergama), Chliara (Kırkağaç) and Adramyttium.[2] The process of fortification was systematic, and the purpose was to create defences to protect the rural population, and bases for imperial forces. Hence the different types of fortress, familiar from previous centuries, of watch towers, *kastra* and large strongholds continued.

Korykos (Kızkalesi) was restored from 1099 with a double wall and many towers. Much reused spoil was employed in the construction. Lopadium's remains consist of a round tower standing beside a gateway passage. At Achyraus, which stands on a steep bluff over a river, five round towers project from a curtain wall encircling the hilltop. Malagina had been a staging post in the tenth century for Byzantine armies: it was refortified by Manuel in the 1140s. It has been identified with the castle at Metabole, on a high hill north-west of Pamukova. Dominating the approach is a large rectangular platform on a massive domed chamber, with openings on three sides, identified as a twelfth century catapult platform. The walls were reinforced with timber beams. It covers the only approach. The upper circuit of walls with their triangular projecting bastions suggests the seventh century (as at Ankara), built atop a Hellenic fortress whose massive masonry blocks still comprise the lower courses of the wall, but the platform, similar to one at Sardis, owes its construction to the Comneni.[3] At Ephesus on the coast, which remained a major commercial centre, the citadel on the hill was strengthened by the empire during the twelfth century.[4] Most of the standing remains of the inner fortress at Nicomedia (İzmit), U-shaped and square towers, are also Comnenian. In the same area there are several other castles whose fragmentary remains have

[1] Choniates, *O city of Byzantium*, 100, 102–3; Cinnamus, *Deeds*, 220–23.

[2] Glykatzi-Ahrweiler, "Les forteresses construites en Asie Mineure", 184–88.

[3] Foss, "The defences of Asia Minor against the Turks", 145–64; Foss, "Byzantine Malagina and the lower Sangarius", 162–71.

[4] Foss, *Ephesus after antiquity*, 118–22.

been dated to this time. Significant among them are Niketiaton (Eskihisar near Gebze), a small hilltop site of rectangular shape (some 150 × 50 metres), with U-shaped towers, and a well-fortified palace-type building in the north-west of the circuit. Gates were simple arched passages, but well protected from neighbouring towers. This castle has an outer, lower walled ward with seven towers, semi-circular and square, of uncertain date. At Xerigordon (Çoban Kale), identified as the castle of Kibotos that featured during the crusader advance on Nicaea, there is an oval fortress some 180 by 120 metres on a hilltop rising above the main road to the west. It had solid semi-circular towers. Wooden beams strengthened the masonry of the walls. Whether it was first built by the Byzantines or the Turks who occupied the area after Manzikert cannot be known, but it appears to date from this period.[1]

The second period of building at the massive fortress of Kütahya (Cotyaeum) in Anatolia has been provisionally dated by its surveyor to the time when the area returned to Byzantine control (around 1120–50). The key feature was the rebuilding, or addition, of no less than 56 of the fortress's 63 towers, no fewer than 42 of them being new. Either of two storeys or solid to the top, constructed of alternating courses of stone and brick, the towers are U-shaped and project so far in front of the curtain wall that they also project beyond the low, crenellated outer wall, creating small enclosed spaces of uncertain function. The corner towers were circular, and even larger. Gates were simple passages but securely placed between pairs of wall towers. Being on a precipitate hill, there is no ditch.[2] Defence was conducted from the wall tops.

There are clearly common features to the building styles surveyed here, even taking into account the great difficulty of reliable dating. Older fortifications were pressed into service to protect the empire's recovered lands, and required strengthening. Walls were higher and stronger than before, towers were present in greater numbers, and offered a powerful flanking facility. Many were solid platforms, having no possible function other than to strengthen the defence or to mount artillery.

General conclusions

By the late twelfth century, knowledge of siege techniques was common across the whole of Christian Europe just as it was across the Muslim world. For the first time, rulers in Scandinavia, and in Scotland and Wales, shared the ability to attack fortifications. Their actual ability to do so would still depend for the time being on having access to skilled engineers, but there was no reason why this should have been a major difficulty, given the growing demand for such skills as a result of the crusades, the wars in Italy, and the continual growth of towns, trades and commercial relations. Indeed, the records have several times

1 Foss, *Survey of medieval castles of Anatolia*, II: *Nicomedia*, 19–22, 39–43, 50–57, 63–7.

2 Foss, *Survey of medieval castles of Anatolia*, I: *Kütahya*, 68–70, 82–3.

shown craftsmen coming forward to offer their skill in designing siege weapons to perplexed potentates. Of the various combatants noticed, only the still pagan and tribal Slav tribes of the Baltic (and the Christian but still tribal clans of Ireland) seem to have been still ignorant of these arts at this time. Of course, there was a difference between having seen a siege engine in operation and being able to construct or operate one, and few of the kings and nobles would have possessed the hands-on skills apparently demonstrated by Andronikos Kontostephanos. But many of them have been recorded as showing that they knew how to design a fortress, and how to strengthen its defences against contemporary weaponry. They would also know how to inspect a fortress to identify weak points, and what kind of attack to prepare. There is no doubt that the ability to capture defended places played an essential role in the warfare and the politics of the twelfth century, and if unlike the Byzantines it was based on experience rather than books, it seems to have been little the worse for that.

Another trend observed over this period has been the continuing increase in the number of times that castles, as opposed to walled cities, have featured as the target of attack and defence. The evidence for the multiple roles of castles is indisputable. Not only were they centres of lordship, they were also designed to defend that lordship if necessary. Of course, not every castle had the same function, and indeed many castles had different functions at different times. The evidence that *some* few castles, especially those answering directly to the monarch, had at least a partial strategic role, such as those on the river Epte in the Vexin, is quite compelling. The evidence that some castles could play a strategic role for a specific time or purpose, and were seen that way, is also strong. In the process of settling in new and hostile territory, where there was every chance that the displaced local population would react violently, castles also clearly had a role as instruments of conquest, defence and refuge, whether it was in the Welsh marches or in Slav Wagria. There is no doubt as to their military function. Others, unhappily called siege castles, were in fact simply forts with functions limited to the military only, and none of the other attributes of the castle. They were of course identical in form to other castles: but they did not have a role as a centre of lordship or of administration, or as a strong residence for a lord, or as a refuge.

While the earth-and-timber castles prevalent in the western countries were more easily captured, destroyed and rebuilt than the stone-built fortresses and city walls that dominated in the more cultured and richer south and east, and were gradually spreading in the west and north beyond the circle of very wealthy rulers who could afford them, the techniques to attack either were not greatly dissimilar. We have also seen accounts of masonry work so strong as to resist all current weaponry, much of it a tribute to ancient building skills. But even a less substantially built contemporary wall, providing it was of adequate strength, would not fall to an artillery bombardment. In the decades ahead, there would be great changes in the political map. Accompanying these changes would be enormous steps forward in the strength and sophistication of fortifications, and of the means of attacking them.

8 Consolidation and centralisation

Overview

By the third quarter of the twelfth century, the same methods of building, defending and attacking fortifications held good from the Anglo–Scottish border to the river Ganges and the China Sea. During the next half century, there would be significant developments that would turn the balance of advantage firmly in favour of the attacker, in response to which the builders of fortifications had to make very substantial changes in the design and construction of their fortresses. By the early thirteenth century, neither the old Roman walls (except for those of Constantinople) nor (in the west) the earthwork castles and palisades that had continued to do service for centuries would suffice any longer.

In the Christian kingdoms, a gradual process of centralisation, and consolidation, of royal power was taking place, well exemplified in France, where Philip Augustus achieved a significant extension of royal authority and confirmed it with the erection of a new generation of royal castles. The crown in the Anglo-Norman kingdom was also consolidating its power. Henry II continued to push out the boundaries of the Angevin domain, and although his sons then presided over the loss of Normandy, royal authority was restored despite civil war. Combined with this centralisation went investment in a new generation of fortresses, far more powerful and far more expensive than any that had gone before. Some of the new stone castles built during this time were so much stronger than their predecessors as to be in a different league from the strongholds that most of the baronage could afford to put up. Powerful castles became part of the symbolism as well as part of the effective authority of the monarch, and kings were employing a regular corps of engineers, carpenters and masons to oversee the building work, and to build the siege equipment that would represent another element of the greater power of the crown. The surviving records that illustrate the establishment and spread of administrative government no longer dependent on the person of the monarch contain the clear evidence of this process.

The Staufer (Hohenstaufen) dynasty continued its embroilment in its Italian and Mediterranean territories. Barbarossa and his son Henry VI would devote most of the last years of the century to crusading and campaigning in the south. After an interval of civil strife, the dynasty, in the person of Frederick II, would regain the throne of the empire along with that of Sicily. He would be the last of the German rulers to exercise such dominance and glittering authority.

In Iberia, the Almohad resurgence ended with catastrophic defeat. However, in the Middle East, the new Ayyubid dynasty under its great founder Saladin[1] would drive the settler kingdom of Jerusalem back to the coast. The Third Crusade came in time to rescue the remnants. But thereafter the Latins did not again attempt to recover Jerusalem by force. In 1204 they found Constantinople an easier target, and while in the process of setting up feeble Frankish states in the lands of the Byzantine empire, contributed massively to its destruction. In 1218 they went for Egypt, which would be the target again of St Louis in 1249/50. In the process they encountered some massive new defensive works created by Saladin and his successors. The Muslim armies of the Middle East were probably also responsible for the development of the trebuchet, a counterweight-operated throwing machine that was powerful enough to reduce many previous fortifications to rubble. This was the technical development that underlay the move towards ever more massive defences, and meant that only the wealthiest could afford them. The new weapon made its way across Europe.

The reconquest of Jerusalem and the Third Crusade

By 1174, Saladin was established as the ruler of Egypt (where the Fatimid dynasty ended in 1171) and Syria. The creation of a single Muslim polity around the Latin principalities was recognised as a serious strategic threat.[2] At first, the new regime was not secure, and the kingdom of Jerusalem made early efforts to overthrow it. A fleet from Sicily, including six ships carrying siege engines, attacked Alexandria itself in 1177. The array included three *manājaniq* hurling enormous stones, but the arrival of Saladin with an army, in combination with a sortie by the garrison, drove off the invaders. Saladin then turned to securing control in Muslim Syria, having to set up his mangonels against Hims before it capitulated, but otherwise the strong cities resisted little. At the same time, he ordered the strengthening of the fortifications of Cairo, appointing an emir to oversee the work, which began with the building of the citadel, preceded by its deep ditch. Damietta, the target of the 1169 attack, was ordered reinforced, and ships linked by chains provided to block the channel of the Nile. A fort was built at Suez to guard the road, and a new citadel was ordered at Tinnīs in the delta. New taxes were raised to pay for the works programme. It looks very much like a coordinated defence system for the Nile delta, along with work to strengthen castles in Sinai, but Saladin also ordered the walls and towers of Damascus rebuilt.[3]

In 1187 Saladin wiped out the Latin army at the Horns of Ḥaṭṭīn. Decisive

[1] Properly, Ṣalāḥ al-Dīn Yūsuf ibn Ayyūb, but it would seem perverse to use this rather than what is effectively a household version of the name in the west.

[2] The word is used advisedly. What emerges very clearly from the detailed histories of the struggles in the Middle East is the existence of concepts of strategy, and the continued focus of crusaders on attacking Egypt was founded in sensible strategic thought.

[3] Maqrīzī, *Histoire d'Égypte*, 8 (1900–1901), 513–15, 519, 525 (Cairo), 539–43. Maqrīzī

though the battle undoubtedly was, it is worth noting that it was fought because Saladin was closely besieging the castle of Tiberias, having taken the town. The count of Tripoli advised the Christian army that the place was well provisioned and could hold out, but the other leaders chose to give battle anyway. The domino-like collapse of many of the great fortifications of the kingdom of Jerusalem and the county of Tripoli followed the destruction of the army on 3–4 July, which had drawn most of the fighting men from the garrisons. The personnel of the states could only be replaced by new arrivals from Europe. Saladin showed sound strategic sense by pressing his advantage to the full. Acre yielded when Saladin arrived outside the walls (10 July). Toron (Tibnîne) was attacked at once, with mangonels, and surrendered. On 29 July, Sidon capitulated. The next day he assaulted Beirut, which resisted him. He launched an outright assault using naphtha grenades, arrows and crossbows – he had not brought his throwing engines with him. Thwarted, he then had a wooden siege shelter moved to the ditch, which was eventually filled and crossed. The ever reliable miners were then thrown in to bring down the rampart, which was achieved and the town taken, all in three days. Byblos (Gibelet, now Jbail) then gave in. Most of the inland castles of Jerusalem and Tripoli surrendered. Ascalon was given up in exchange for the release of the captive king of Jerusalem.

Meanwhile, the Muslim leader had been negotiating with the Christians in Jerusalem, saying "he would not have his siege engines bombard" the holy city. His demands rejected, Saladin had to lay siege. He tried to set up camp between the Tower of David and St Stephen's (or the Lions') Gate, but despite eight days of fighting outside the walls "not once was he able to erect a petrary, mangonel or engine" on that side. Thereupon he moved his engines to the area between St Stephen's and the Mount of Olives, whence he could see into the city. Eleven large stone throwers were arrayed to bombard the walls, protected by archers, who also provided cover for sappers, who managed in two days to bring down a whole stretch of wall, which tumbled outwards into the ditch. As the assault was organised, the Christians negotiated a surrender. The Muslim recapture of Jerusalem was carried out in rather more civilised a fashion than the Christian conquest of 1099.

Saladin left a force to besiege Karak Castle, and assembled an array of siege engines to tackle the formidable obstacle represented by Tyre, where many of the survivors of Ḥaṭṭīn had fled. The attack began on 22 November, but the battery of 14 mangonels made no impact on the massive walls, and the Egyptian fleet was worsted by a defence invigorated by the recently arrived Conrad, marquis of Montferrat. The Muslims built scaling ladders with which they penetrated the outer walls but found them too short for the inner ramparts. When the attackers turned to mining, the marquis led a sally that drove them back and thereby prevented progress with the otherwise often lethal tactic. Finally, in miserable

was a fourteenth-century historian. I am grateful to Denys Pringle for information about the castles in Sinai.

weather, Saladin recognised that Tyre was too strong for him, and burnt his equipment (or dismounted it, according to another source).[1]

Saladin sought to continue the conquest in 1188. He was aware that three of the greatest monarchs of Christendom were arming themselves to launch the Third Crusade: Richard of England, Philip of France and the mighty emperor Frederick Barbarossa. Time might be short to consolidate the Muslim victory. In March he attacked Belvoir, but saw the castle was well walled and well provisioned. He then marched on Ṭarṭūs (Tortosa), recorded as being protected with two strong towers "almost like castles". In fact, the double ramparts of the castle, serving as a citadel inside the town, were built after Saladin's attack, so the second of these towers must refer to a strong point on the enceinte. The town itself fell to escalade, as did this tower, but the donjon proper, although apparently standing on its own, was 35 metres square, secured by a wet ditch and massive and solid masonry (which is still there below first floor level) with a sloping outer face (talus), and defended by "enormous" crossbows (ballistas?), proved impregnable. It is likely that it had been built by the Templars after the temporary occupation of the town by Nūr al-Dīn in 1152, so was of the latest design and strength.[2] At the time, it was directly beside the sea.

Passing on, the Muslim army stormed and captured Lattakia. Here, "our stones demolished the parapets" ('Imād al-Dīn), but it was mining once again that created the breach and led to surrender. Saladin then turned to tackle Saône (Qal'at Sahyun / Ṣalāḥ al-Dīn). This castle was protected by a wide natural ravine and a man-made ditch cut in the rock, "60 cubits deep" (thirty metres). Skilfully deploying his mangonels on the southern side, one of Saladin's sons, al-Zāhir Ghāzī,[3] managed to score direct hits on the rampart of the outer ward, much weaker at this point, and create a breach through which the Muslims stormed in only one hour after crossing the ravine. The defenders withdrew into the donjon then surrendered. It was noted that the operators were highly skilled men from Khorasan, who, as 'Imād al-Dīn chillingly reported, "stoned the walls . . . as one stones for adultery". Nearby castles gave in, but Saladin had to besiege Bourzey (Qal'at Mīrzā). Standing on a steep hill above the Orontes river, the earlier rectangular Byzantine castle on the summit had been allowed to decay, and a much larger irregular castle of two baileys, a rectangular donjon and irregularly-placed towers had been built around the hill top. The workmanship was crude and rough (to judge by the remains) but the approaches were extremely steep on all sides. The day after arriving, Saladin deployed his artillery to bombard the walls, but because of the height and angle they could not reach it, so he ordered his men to attack on the most accessible side. He had sufficient numbers to divide them into

[1] 'Imād al-Dīn Isfahani, *Conquête de la Syrie et de la Palestine par Saladin*, 38–76; Ibn Shaddād, *Saladin*, 120–22; *The Old French continuation of William of Tyre*, 37–70; *De expugnatione Terrae Sanctae per Saladinum*, 220–46. Between these sources from opposing sides there is remarkable agreement on the key facts of these events.

[2] See the description in Chap. 6 above; Burns, *Monuments of Syria*, 229–32; Hanna, *Syria*, 9–17.

[3] Who was apparently only fifteen years old: Lyons and Jackson, *Saladin*, 135.

three divisions, to each attack in turn while the others rested. The defenders, very few in number, had no such luxury, and were eventually overwhelmed by escalade (13 August).[1] The Templar castle of Trapessac (Darbı Sak) near Kırıkhan north of Antioch was brought down by mines and escalade after a fortnight (13 September). At Safad, where the mangonels' great stones were unable to damage the masonry, and the mines were ineffective, but the castle surrendered nonetheless (early December). The final conquest of the campaign was Belvoir (Kaukab al-Hawa) where, despite awful weather, an attack was carried out by bombardment and mining, bringing down part of the wall and leading to the castle's fall on 5 January 1189.[2]

The garrisons of many of the strong places that dotted the Latin states were unable to hold off the formidable siege train that Saladin clearly had at his disposal. A considerable number of places did resist, but were taken by assault after the Muslim army had created a breach, usually by mining, but also under heavy artillery bombardment. Their usually small garrisons could not prevent the break in, nor was there any hope of relief. Only a few places, and those mainly on the coast such as Tyre, were defended successfully. The normally generous terms of surrender offered by Saladin were an added inducement. By the end of 1188, the sultan's reputation in the Muslim world was unmatched, and has continued so to this day. While the grand equestrian statue of the leader in Damascus triumphing over the Christian chiefs at Ḥaṭṭīn may speak as much of twentieth-century politics as of twelfth, its location by the now restored walls of the vast citadel completed by Saladin's successors is indicative of the key role of these fortifications in his overall strategy.

Saladin could not finish off the crusader states completely, although the territory he recovered in less than two years was remarkable. While the Christians controlled the sea, the Muslims were always likely to be halted at the coast. Now the crowned heads of Europe were racing to the rescue. Philip of France would have the least remarkable journey. Frederick Barbarossa met his end crossing a river in Asia Minor, and only fragments of the German army made it through to friendly territory in Cilicia and Antioch. Meanwhile Richard of England's adventurous route across the Mediterranean involved armed conflict with Isaac Comnenus, the independent ruler of officially Byzantine Cyprus, and following the capture of one castle and the surrender of the strongly-sited Buffavento Castle, Isaac surrendered and the Lusignan dynasty was established in the island. Prior to that, Richard had fought a fierce battle at Messina in Sicily. Here the king, described by one chronicler as having no equal in capturing towns and overthrowing castles, stormed the city walls under a protective hail of crossbows and javelins and forced the gates with a battering ram, completing the capture within five hours. He then ordered the construction of a very tall wooden tower of great

[1] Best account in Ibn Shaddād, *Saladin*, 124–34, and 'Imād al-Dīn, *Conquête de la Syrie et de la Palestine*, 130–39.

[2] Ibn Shaddād, *Saladin*, 143–53; 'Imād al-Dīn, *Conquête de la Syrie et de la Palestine*, 135–53.

strength next to the city, which he called "Mategriffon" in reference to the bridling of the Greeks he had just defeated.[1]

Siege and capture of Acre

Meanwhile, the main crusader army was attempting to capture Acre, which had been strengthened and reinforced by Saladin to withstand a long siege. He had also ordered the fortifications of Latakia (Al Lādhiqīyah), Jablah, Byblos and Beirut to be destroyed in order to deny their use to the landward advance expected from Barbarossa.[2] In uncanny ways, Acre resembles the siege of Antioch by the First Crusaders of nearly a century before. With control of the sea, the Christians were able to maintain their presence in a hostile land, and with the aid of a defended camp could prevent the larger Muslim army outside from breaking the siege. At the same time, the strength and determination of the defenders made for a siege of epic duration, in which despite the deployment of the most advanced weaponry, and the fact that the city did not yet have its later immense defences, the attackers were continually thwarted.

During 1190, the crusading army constructed three wooden siege towers, protected with iron plates. However, the defenders managed to burn them all (5 May), the incendiaries being hurled from mangonels that were obviously very accurate. One of the defending mangonels was described as "incomparable" and remarkably effective in hurling millstones, so it must have been a large device. Its missiles required two men to lift them, and on landing sank a foot into the ground. Many of the Frankish stone throwers fell victim to this weapon. Others were destroyed when the garrison sortied, as it did regularly and sometimes with great success. The scene changed when the western monarchs arrived. First came King Philip, preceded by a contingent including siege engines, which were set in place before his arrival. The king, having inspected the situation, ordered iron-plated screens built to protect the archers and crossbowmen so that they could maintain continuous cover. The Pisans attempted an attack on the Tower of Flies from the sea, where it stood, constructing some kind of tower, and mounting ladders, on their ships. From the tower they could overlook the walls. However, "Greek Fire" destroyed the tower (25 September).[3] The German contingent finally arrived, the defenders noting that they had constructed many new engines, including a ram "covered with iron plates and mounted on wheels". An attack with battering rams organised by the archbishop of Beauvais was launched on 15 October. This machine was described by the Muslims as a "terrifying rolling tower" with two rams as thick as pillars. The defence responded with a bombardment from their artillery of "pillars of freestone and marble", with beams and with trees, alongside "urns

[1] Richard of Devizes, *Chronicon*, 24–5; *Itinerarium peregrinorum et gesta regis Ricardi*, 168; Ambroise, *The history of the Holy War*, 36–7.

[2] *Continuation of William of Tyre*, 89.

[3] Interestingly, what the Christian accounts call Greek Fire is described in the Muslim accounts as naphtha, in line with the technology they had long possessed.

of sulphur, tar, pitch and tallow." The operators could not resist the bombardment and fled, the roof was smashed, and the device went up in flames.

In June 1191, Philip set up mangonels to bombard day and night, one of them so large that its stones gradually disintegrated the wall. King Richard, recently arrived from Sicily, also hurled himself energetically into the proceedings, setting up his own artillery. Ammunition in the form of round stones was collected by scouring the seashore. The defenders noted that the "stones were not without effect on the ramparts. Fear was felt lest the wall should give way" (Ibn Shaddād). In the end, the attackers were close enough to mine under the protection of a cat, which the defenders at first managed to thwart with a countermine, but in the end failed: the mine was fired and a stretch of the rampart collapsed. The garrison, however, gave a vigorous account of itself and repelled the ensuing assault on the breach (3 July). But at the same time the continuous bombardment from the artillery had ruined large sections of the rampart around the focal point of the attack, the "Tour Maudite", and the tower itself, and, having received permission from Saladin to do so, the town surrendered on 12 July 1191.[1]

Darum and Jaffa, 1192

Richard showed his military skills over the coming months, as well as his ruthlessness when he executed all the prisoners from Acre when Saladin was tardy in fulfilling his side of the surrender terms. Philip had returned to France. Richard ordered the refortification of castles on the road to Jerusalem such as the former Templar Casal des Plains (Yāzūr, now Azor), but did not attempt to lead his army inland as far as Jerusalem. He recaptured Ascalon, and from there launched an attack on Darum (Deir al-Balah). This involved loading his siege engines on ships. Once ashore, the king and his men "bare the beams and timbers on their shoulders for a league". The site was protected by a wall with four towers, one of which stood out as a citadel with its own ditch. It was a Frankish creation of the late twelfth century, and when it was attacked by Saladin in 1170, not long after it was built, it was described as having neither ditch nor barbican: even so, although mining brought down the wall at that time, the citadel held out.[2] It may have been strengthened, or perhaps completed, subsequently. The crusaders set their artillery against this tower, and also set miners to work. The defenders tried to raise a mangonel on the roof of the great tower but it was smashed by the besiegers'

[1] Ibn Shaddād, *Saladin*, 203–15; 'Imād al-Dīn, *Conquête de la Syrie et de la Palestine*, 255; *Continuation of William of Tyre*, 76, 105–6; Richard of Devizes, *Chronicon*, 51; *Itinerarium peregrinorum et gesta regis Ricardi*, 63, 81–5, 98–222; Ambroise, *The history of the Holy War*, 48–75. Ambroise was an eye-witness serving Richard I.

[2] William of Tyre, *A history of deeds done beyond the sea*, II: 273; Kennedy, *Crusader castles*, 31. Darum has disappeared but, according to William of Tyre, seems to have been one of the castles erected both to protect against, and to threaten, Ascalon. Ascalon itself was hurriedly restored by Richard I's army between January and April 1192: Pringle, "King Richard I and the walls of Ascalon", 133–43.

artillery. Finally, after a three-day assault, the place was carried and the garrison retreated to their citadel, whence they agreed to surrender.[1]

Later in 1192, Saladin launched a sudden attack on Jaffa, bringing with him his stone throwers and his miners. Ibn Shaddād describes how he targeted a point at an angle of the curtain wall north of the east gate which had been damaged in an earlier attack and rebuilt. Jaffa's defenders were aware of the mining and able to intercept it, and Saladin's miners had to flee. The sultan was undaunted, and ordered new mines dug, while at the same time he set up a battery of three mangonels against the point of the wall where the first mines had been dug. This time the wall collapsed. The defenders, of course aware of where the breach would occur, created a barrier behind which they set a blaze to prevent an assault. Again the Muslim mangonels were called into use, but stones for the ammunition had to be conveyed from distant ravines, there being nothing suitable in the vicinity. Further mining continued too, and this mine, undetected, was fired successfully. The town surrendered and the troops retreated to the citadel. The battery of mangonels was now directed here, with the effect that the garrison could get no rest, and this too was on the verge of falling when Richard arrived by sea, just in time to oust Saladin's men and secure the city again. The king ordered the sailors to prepare palisades from their ships' yard-arms in order to repair the breaches.[2] Possession of the coastal fortifications was critical to his strategy. Stonemasons must have enjoyed a busy few years throughout the time of the Third Crusade, and even busier afterwards as the battered Latin states strove to repair the damage they had suffered!

While Saladin had ordered the destruction of the walls of many of the fortresses he had captured, he and his successors (he died in 1193) followed a different course in Egypt. A new ditch was dug at Cairo in 1196, and Damietta further strengthened. In that year it was recorded that the Egyptian sultan al-Malik al-'Azīz 'Uthmān (r. 1193–8) ordered the destruction of the Giza pyramids to provide stones for the continued reinforcement of Damietta. When advised of the immense difficulty of the project, he restricted himself to removing granite from the smallest Pyramid of Menkaure.[3]

The next attack came from the German emperor Henry VI. Al-'Azīz sent troops to destroy the rebuilt walls of Beirut and went on to attack Jaffa again. They chose the point near the sea where the tower later built by the patriarch Gerald of Lausanne stood, and stormed the town through this the weakest point. The German troops went to besiege the castle of Toron, and the chronicler records the emperor's anger that "it was shameful for so great a lord to have his

[1] Ambroise, *The history of the Holy War*, 124–6; 'Imād al-Dīn, *Conquête de la Syrie et de la Palestine*, 378; *Itinerarium peregrinorum et gesta regis Ricardi*, 353–5. The capture took place on 28 May 1192.

[2] Ibn Shaddād, *Saladin*, 360–66; 'Imād al-Dīn, *Conquête de la Syrie et de la Palestine*, 384–5; *Continuation of William of Tyre*, 117.

[3] Maqrīzī, *Histoire d'Égypte*, 9 (1902), 63, 64. The problem was, of course, the immense size of the stone blocks involved, but the plundering of ancient monuments for modern reuse was a familiar approach to building defences in the middle ages.

men besiege a [mere] castle". Shameful or not, the crusaders attacked with mines. The garrison, who held a large number of Christians captive, offered to surrender on terms, but the Germans were too proud to grant them and pressed on with the mine, under the cover of a bombardment from engines. The wall duly collapsed, but the Muslims managed to hold the breach. That night, the news that Henry VI had died reached the army, along with news of an approaching relief force, and the castle and its captives, which could have been in Christian hands before, remained Muslim.[1] Al-ʿAzīz then joined with his brother al-Afḍal of Damascus to flatten the walls of Ascalon (again), but this was their last collaboration.

The new Citadel of Damascus; the new work at Cairo and Bosra

Al-Afḍal was replaced by his uncle, Saladin's brother al-ʿĀdil (r. 1200–1218), who turned to the task of modernising the defences of Damascus and Cairo. The citadel of Damascus had no natural strength, so relied entirely on man-made defences. The work was done by constructing an outer stone wall around the core of the existing curtain wall, so as not to leave the place vulnerable during building, then creating a water-filled ditch around the outside, supplied by the river, with the flow controlled by a fortified mill. The great citadel standing today largely owes its shape to the dozen years or so of building at this time, although it has been subsequently damaged and restored. However, recent clearance and study is clarifying the medieval structures. The chief features are massive walls and rectangular three-storey towers, twelve of which remain, the largest of which (dating to 1209) having dimensions of 21 by 23 metres and a height of twenty metres. Large blocks of masonry were used, and the outer faces are "rusticated" (have an uneven and projecting face — in some places the projection is around 35 cm — see Plate 18). The largest surviving gate (the North Gate, Bāb al-Ḥadīd or "Lion Gate") had no fewer than five bends to negotiate. The north-east corner was cut off by an internal wall to form a palace. All the towers have vaulting to hold up the floors and roofs. Powerful machicolations stand out from the corners of the towers to cover the base. Unlike earlier Muslim defences, there are numerous arrow slits, including at ground level, served by the covered gallery which runs around the entire circuit behind. It is notable that the walls are not bonded into the towers, presumably a protection against the effects of mining: the fall of the wall would not bring down the tower with it. The curtain walls are up to four to five metres thick, and even the parapets on the wall walk are up to a metre thick.[2] There must have been reasons for so massive a reconstruction of an existing defence: the reason would appear to be the technological advance represented by the appearance

[1] *Continuation of William of Tyre*, 143–5; Maqrīzī, *Histoire d'Égypte*, 9 (1902), 95 (Beirut and Jaffa).

[2] Omran and Dabboura, *The citadel of Damascus*, 45–69; King, "The defences of the citadel of Damascus". There has been a great deal of (continuing) study and excavation of Damascus since the 1990s. See in particular the articles by Hanisch in the Bibliography; Berthier and El-Ajji (eds.), *Études et travaux à la citadelle de Damas*; Hartmann-Virnich, "Les portes ayyoubides de la citadelle de Damas", 287–305.

of the wall-breaking trebuchet. The new defences would be able to resist larger missiles, the rusticated surfaces further reducing the effect of a stone ball as well as creating a very impressive appearance. The towers were strong enough to mount the same weapons for defence. While the North Gate faced the country, the East Gate was inside the city and faced the great mosque: this was, in contrast, graced with splendid carving above the entrance, suggesting a more symbolic purpose, and it led through a large hall with columns into a palatial gallery (although the gate itself remained thoroughly defensible, with the traditional bent entrances and arrow loops).

The argument for such a motivation is reinforced when the work done under the same ruler at Cairo a few years earlier is examined, because here the citadel of Saladin had only been completed a few years before. Saladin's fortress-palace, built by his emir Qaraqūsh, stood on high ground and was part of a grand plan to embrace the whole of the old city of al-Fusṭāṭ and the newer city of al-Qāhirah with one immense wall, the citadel representing a separate complex of defences and accommodation for the new ruler, who at the time had good reason to fear for his security. Qaraqūsh's wall stretched for around 2,000 metres, forming the northern enclosure of the present complex. With rusticated masonry facing and numerous half-round projecting towers, a gallery that ran around inside the circuit of walls, giving access to arrow loops, and bent entrances, the effort to construct it was immense, as witnessed by Ibn Jubayr, who commented on the use of Christian captives. Strong though it was, just a few years later it was massively strengthened, although the ambitious town wall was never finished. Eight enormous additional towers, most of them rectangular and varying from 21 to 30 metres square and 15 to 18 metres high, were built over and across Saladin's walls on the same design as those of Damascus.[1] The curtain itself was thickened in places. While it is impossible now to know the thinking behind it, the work at Cairo so soon after the creation of the citadel suggests a purpose over and above the needs of symbolism. The new citadels were meant to be impregnable. Nor was this the only reconstruction: the Ayyubid princes also reinforced the citadel of Aleppo, strengthening the gate and deepening the ditch. At Bosra (Buṣrā al-Shām), a powerful new curtain with closely-spaced new rectangular towers of similar design to those of Damascus, defended by arrow loops and machicolations, with a passage behind all round, was added to the existing Seljuk defences that had been built onto the original Roman theatre.[2] Only a very wealthy and powerful prince could afford such fortresses, which consumed immense resources.

[1] Rabbat, *The citadel of Cairo*, 10–22, 54–82; Creswell, *A brief chronology of the Muhammadan monuments of Egypt*; Creswell, *The Muslim architecture of Egypt*, II: 4–63, for a detailed survey. For recent excavations of the city walls of Cairo, see the website http://www.diplomatie.gouv.fr/fr/actions-france_830/archeologie_1058/les-carnets-archeologie_5064/afrique-arabie_5068/egypte-muraille-ayyoubide_5555/index.html (accessed Nov. 2009).

[2] Yovitchitch, "La citadelle de Bosra", 84–7.

Byzantium under attack, 1184–1201

The Byzantine empire and the kingdom of Sicily both had to face serious internal conflicts. The reigns of the emperors Andronicus I Comnenus and his successors Isaac II and Alexius III Angelus were to represent the end of the achievements of the Comnenian dynasty, and to witness the beginning of the end of the empire itself. In 1184 the emperor Andronicus had to besiege rebels supporting the claimant Isaac Angelus in Nicaea. The chronicler Niketas Choniates commented on the impregnability of the walls, "all of baked brick". The emperor set up batteries of stone throwers, described rhetorically as "wall-destroying siege engines", and also deployed miners. Continuing a tradition of hands-on military expertise, Andronicus would "examine the sling, secure the withy end and the crank handle" of his engines. However, the defenders were not unaware of the risks, and sallied successfully with the aim of burning the engines. Relying next on a son's maternal affection, Andronicus then tried strapping Isaac's mother to the top of a battering ram, whence however she was rescued by the defence. Nicaea surrendered anyway, and the emperor marched on to Prusa (Bursa), camping on the flat southern side, where there were double ramparts. He deployed stone throwers to such good effect that the walls crumbled, and an escalade succeeded, with the defenders panicking. Presumably these actions related to the weaker outer wall.

Meanwhile, the emperor's enemies allied with the Normans of Sicily, and in 1185 a Sicilian army crossed to Dyrrhachium (Durrës), which surrendered without a fight, leaving the road open to Thessaloniki and Constantinople. There survives an eyewitness account from Thessaloniki by the archbishop, Eustathius, who was scathing about the leadership of the city's defence. He records that the "Frankish" foot soldiers arrived on 6 August and their fleet on 15 August. The garrison commander lied to the emperor about the situation, so no relief was sent. The Sicilians prepared "with professional skill". On the west they deployed "novelties", new types of engines (*elepolei*), but these had little effect "because of their cumbrous size". No other clue is given as to what these were. On the east, they relied on "more traditional methods", using stone throwers, "mostly of a small size, to attack the men fighting on the walls", and two larger versions, one of them called "daughter of the earthquake". Soldiers set about filling the ditch with rubble, while others set about overthrowing the outer wall and undermining the inner. An attack on both sides was launched, resisted with difficulty. Eustathius complained of the poor state of the masonry of the walls facing the sea, and at the lack of range of the defenders' catapults, which could not reach the enemy's. When one of the weapons broke, the commander failed to supply the necessary timber to repair it. Water was short because one of the cisterns cracked open when it was used before the reconstructed basin had solidified. On 15 August there was another assault on the eastern rampart. The women brought up stones for the engines and slingers, and the Sicilian archers were driven back too far to be able to reach the wall. The attack continued, with archers providing cover to allow sappers to hollow out spaces below the wall and collapse it. Giant stones crashed in, "sending battlements tumbling" and causing great damage in the city. The bombardment also prevented the defenders from constructing a second barricade behind the breach. While the smaller engines caused much harm because of their accuracy, their "missiles were like the efforts of children" compared to the giant engine, whose stones were "fully as much as a man could lift". Finally, a great breach was made, as the walls "cracked asunder" from mining, the commander fled, and the city fell.[1]

The Sicilians then marched on the capital, where Andronicus ordered the walls repaired and all the houses on the outside demolished. It did him little good, as he was replaced in a coup d'état by his rival Isaac. When his successor Alexius (r. 1195–1203) reigned, the Turks attacked Dadybra (Safranbolu) in Paphlagonia in 1196 and captured it after four months' siege. Although their engines had maintained a relentless bombardment from the hills outside that had demolished most of the houses inside, it was famine that caused the surrender. The next year, the emperor tried to recover the town of Prosek (Demir Kapija) in Macedonia from the Vlachs. His initial attack swept over the newly built outer wall, but his men did not have the mattocks they needed to "pull down the wall" and had no ladders, so had to retire. Meanwhile the nomadic Vlachs had stone throwers inside, with which they caused heavy casualties. Not for the first time, the operator was said to be a former "Roman", whose former employers had failed to pay him. Come the night, the Vlachs sortied and destroyed the emperor's engines, forcing him to abandon a thoroughly ill-executed attack. War continued in the Balkans. In 1201, Constanţa was taken and the walls razed; then the Bulgarian tsar Ivan I, called Kaloyan or Yoanitsa, attacked Varna, which was heavily defended. Kaloyan built a four-sided siege engine of such dimensions that when it was wheeled to the ditch and turned on its side, it reached across the whole ditch and at the same time reached the top of the wall, so that the same engine was used "as a bridge and a scaling ladder". The town fell inside three days, and we are not told who designed this equipment, clearly calling for considerable design and engineering skills.[2]

The kingdom of Sicily under a new dynasty

The Staufer kings of Germany inherited the crown of Sicily in 1189 but had to fight to secure their legacy, with Barbarossa's son Henry spending much time in southern Italy before finally bringing the wealthy kingdom under his control. His first major opponent was an illegitimate scion of the Norman dynasty, Tancred of Lecce. The conflict revolved entirely around the capture of castles and fortified towns, many of the defences having been constructed during the establishment of the Norman kingdom. Henry VI set about the business in 1191, taking the castle of Arce by force, then besieging Naples from May to August, where he constructed engines that hurled great stones against the defences,

[1] Eustathius of Thessaloniki, *The capture of Thessaloniki*, 68–101; Choniates, *O city of Byzantium*, 159 (Nicaea), 164–5.

[2] Choniates, *O city of Byzantium*, 260, 278–9, 292.

but was forced to abandon the attack by disease in his ranks. In 1193 Berthold, one of his lieutenants, led the army, having to construct ballistas and battering rams as well as stone-throwing artillery in his campaign to reduce rebel castles to obedience, while the defenders made use of the same artillery to strike at the attackers' machines. The emperor returned in person with a large imperial army, and Salerno was captured in 1194 by a regular siege, with the "turris maior" (donjon) having to be bombarded and assaulted. In Apulia, Barletta and Bari were captured by one of Henry's counts, while Palermo also required a siege. Castles fell much more rapidly than the cities, many surrendering when their territory was occupied by the emperor's army.[1] But there is no evidence of any change in the usual twelfth-century methods of attack and defence.

The British Isles

The picture was not that different further west, where there are rich sources of contemporary record. In 1173, Henry II faced a rebellion instigated by his son and intended heir, Henry the Young King, aided by William I, king of Scotland, and Louis VII of France. In England, many castles in the northwest, the midlands and East Anglia were held by rebel barons, and young Henry and his brothers invaded Normandy, taking Driencourt Castle (Neufchâtel-en-Bray) after a siege of two weeks. King Louis and the Young King besieged the major ducal town and castle of Verneuil-sur-Avre and built machines to attack it. Hugh de Lacy, lord of Meath, defended it for a month, but the inhabitants agreed to surrender if there was no relief. King Henry arrived, but the town had fallen anyway. Louis, however, had withdrawn, so Henry ordered the walls repaired while taking the battle to his domestic enemies. These fled into the castle of Dol-de-Bretagne, where Henry pursued them, setting up stone throwers and other engines and producing a very rapid surrender. Meanwhile in England, the king's men besieged Leicester, burning most of the town, which surrendered at the cost of the destruction of its walls and gates. The old motte castle seems to have played no part. Robert III de Beaumont, earl of Leicester, one of the chief rebels, returned from Normandy and made a vain attack on Walton Castle (near Felixstowe, now disappeared) before taking Haughley (a large earth-and-timber castle in Suffolk) with its garrison of 30 knights after a four day attack. The rebellion continued in 1174 with further outbreaks and a Scottish invasion of the north. In June, Richard de Lucy blockaded the motte castle of Huntingdon with a siege castle, a rare return to an old method of attack, but one that succeeded here. In the north, Roger de Mowbray revolted and occupied an old castle site at Kinnard's Ferry (now Owston Ferry), but the local men of Lincolnshire crossed in boats to take and destroy it. The earl of Leicester's castles at Groby and Mountsorrel surrendered, as did the castles of Roger Bigod at Bungay and Framlingham. Meanwhile in Aquitaine,

[1] Godfrey of Viterbo, *Gesta Friderici*, 335–6; *Annales Casinenses*, 314–17; Petrus de Ebulo, *Liber ad honorem Augusti*, 30–38, 80; Otto of St. Blasien, *Chronici continuatio*, 320–28; Ryccardus of San Germano, *Chronica*, 325–7.

Duke Richard's men seized Saintes and tried to turn the cathedral into a castle, but were driven out by forces loyal to the crown. King Louis was besieging Rouen, but when Henry recrossed the channel and marched against him, he abandoned the siege, burning his stone throwers and equipment.

Further north, the Scots had decided to attack the border castle of Wark on Tweed, and King William deployed stone throwers, slingers, archers and crossbowmen, a fair number of them Flemish mercenaries. Wark was held by Roger de Stuteville with 20 knights and a number of sergeants. A vigorous assault was repelled, with the commander advising his men to keep their arrows until the enemy was actually attacking. The stone thrower was set up against the gate of the outer ward with a view to destroying it, and the first shot knocked down a knight, whose life was saved by his armour, so this was clearly a small missile. The catapult failed to destroy the gate, and an attempt to set the (timber) defences alight failed when the wind turned, so Wark was left in English hands and William marched on Carlisle. Here, the governor replied to the Scottish summons that he would be disgraced if he surrendered while provisions remained, so William left the strong castle here and instead occupied Appleby(-in-Westmorland), because "there were no people in it." He then attacked Brough Castle, held by only six knights. An assault carried the stone-walled bailey, and this was followed by an attempt to fire the small stone donjon, whereupon the knights surrendered. After again summoning Carlisle in vain, but extracting a promise to surrender if not relieved, William attacked Prudhoe Castle, which was strongly held, and three days of assaults failed to make a breakthrough. The Scottish invasion ended ignominiously when William was ambushed and captured.[1]

Henry II had managed to crush a widespread revolt. Against the crown, no coalition of geographically scattered rebel barons, even when including neighbouring monarchs, could resist the centralised resources available to the king. The small fortifications used by the rebels in the war were far apart and vulnerable and fell rapidly, whereas the northern castles held for the king were able when garrisoned to hold off the king of Scotland, whose siege techniques seem to have been not up to the task. Henry's own professional forces, combined with the forces of his loyal baronage, had proved too strong.

Ireland in the twelfth century

In 1169 a small force of Anglo-Norman knights crossed the Irish Sea to serve an exiled Irish king, and as part of their reward, their subsequent leader Richard de Clare (Strongbow), earl of Pembroke, inherited and created the earldom of Leinster. The conquest of all the best land in Ireland, from Leinster to Ulster, followed

[1] Roger of Howden, *Gesta regis Heinrici secundi*, 42–75; Roger of Wendover, *Flowers of history*, II: 25–31; on the north, Fantosme, *Chronicle of the war between the English and Scots*, 255–61 (an Old French verse account). There is a more recent translation, *Jordan Fantosme's Chronicle*. Moore, "Anglo-Norman garrisons", 234, points out a discrepancy between Fantosme's figure for Wark's garrison and the Pipe Roll, which gives ten knights and forty sergeants. The overall total however may not have been very different between the two sources.

as new waves of adventurers followed, and King Henry II decided to take the new land into his realm. The bands of soldiers, armoured in common European fashion, and trained to fight on horse or foot, backed up by archers, were invariably victorious over the native Irish, but the conquest was secured by transforming the landscape in a wave of castle building not equalled since Anglo-Saxon England had fallen a century before.

The Irish were not entirely unfamiliar with fortifications. The chronicles continue to describe frequent attacks on *dún*, *duingne* and *cathir*, *caistel* or *caislen*, but it is impossible to be certain what they are describing. The best assumption is that they were ring-forts, sometimes of earth and timber, sometimes of stone according to the local conditions, and archaeology has suggested a pre-conquest date for the rock-cut ditch in front of the later castle at Ferns. In 1100, for example, a joint army from the west led by King Muirchertach destroyed all the "towns, forts and churches" of Inishowen and destroyed a "stone house", and besieged the town of Armagh for two weeks, in vain. In 1128, the *caislen* or *dún* of Bun Gaillmhe (Galway) with two other such places was burned by men of Munster, and in 1131, the "castle" of Athlone was burnt by lightning, and the same "castle" was destroyed by the king of Meath in 1135. Two years later, another Muirchertach besieged the "Danes" in Limerick and compelled them to submit, while in 1162, Athlone was again razed by the king of Meath.[1] The towns originally founded by the Vikings during the tenth and eleventh centuries, such as Dublin, Cork, Waterford and Limerick, were all fortified, although not all in stone. Limerick was burned in 1124 and Cork plundered in 1127. Clearly, the Irish were not unaware of how to take fortifications, although there is no evidence to suggest that they possessed any techniques except for direct assault, and for the most part the defences (insofar as they can now be ascertained) were not of great strength or sophistication.

The Anglo-Norman invaders swiftly came up against the fortifications that existed. Gerald of Wales's account, tinted though undoubtedly it is with propaganda, gives vital details from the side of the conquerors. Robert fitz Stephen, Gerald's uncle, came over with a force of 30 knights, 60 other armoured men and 300 archers. He was joined by an Irish force of 500 to attack Wexford, whose population of some 2000 retired *intra muros*. Under cover of the archers, the town ditch was filled and the town stormed. The defenders replied with stones and drove back the first attack, but the second succeeded. Strongbow arrived in 1170, and with a large force (140 knights and 1000 others) attacked Waterford, and was twice repulsed. The earl noticed a small overhanging building on the wall suspended by a beam. So during the attack he had the beam cut, and the wall collapsed, allowing him to enter the town. Dublin did have stone walls,[2] but the population was unable to prevent the city being immediately overrun by a bold assault. The Norse ruler,

[1] *The annals of Clonmacnoise*, 133, 191, 193, 202, 205; *Chronicon Scotorum*, 333; *The annals of Ulster*, 573; "Extracts from the Annals of Innisfallen", 3 (1915/16), 150; 4–5 (1918–19), 107. Dollen, "Die Entwicklung des Burgenbaus in Irland", 50–52; Flanagan, "Irish and Anglo-Norman warfare", 53–62.

[2] Barry, *The archaeology of medieval Ireland*, 34, 121. Dublin's first stone wall, which has been dated to around 1100, may have been 1.5 metres wide and perhaps 3.5 high, making it

Asculf mac Torcaill, fled by sea, and tried twice to recover his city. The first time was by naval attack up the river Liffey, thwarted by the courage and battle tactics of the defenders. The second time, the Norse joined with the Irish princes who had gathered to resist the new force in the land, and summoning ships from the kingdom of the Isles (Mann, Kintyre, the Hebrides and parts of Lochaber) to close the blockade, they attacked the Anglo-Normans in Dublin. For two months they surrounded the town, but did not apparently try an assault. As food ran short, the besieged army sallied forth unexpectedly, and the Irish were routed.

In 1171 King Henry himself visited the new domains being conquered by his barons, and took charge (his youngest son John being eventually made lord of Ireland). Dublin he placed in the charge of Hugh de Lacy. The country was "planted with castles and cities with keeps and strongholds", according to the *Song of Dermot and the earl*. But the Irish did not sit back and watch. Lacy built a castle at Trim "and threw a trench around it, then enclosed it with a stockade", placing within it knights under Hugh Tyrell. Archaeological investigation of the great stone castle at Trim has confirmed the previous existence of a ring-work around the site of the later donjon. The king of Connacht came with a great army, causing Tyrell to flee, and the rampart was thrown down and levelled and the house burnt. The earl himself chased them off and rebuilt the castle. Other castles are recorded as being built, then destroyed by the Irish kings, in Meath and Ulster in the mid 1170s.[1] In 1175, the earl ordered an attack on the city of Limerick, carried out by Raymond Fitzgerald, called le Gros. The *Song* records the town as being surrounded by "a river, a wall and a dyke". Gerald of Wales records the heroic action of the attackers in fording a raging river Shannon in the face of stones and arrows hurled from the wall, then overrunning the ramparts. Clearly, these too represented no serious obstacle to an attacker who, having just fought their way across a wide river, cannot even have had ladders with them. When the Irish king of Thomond, Domnall Ó Briain, tried to recover Limerick, he was able only to blockade it, and was forced to raise his siege by a relief force.[2]

Castles of the conquest in Ireland

Stone castles were built by the conquerors almost from the start, at the main sites of lordship of the great settler barons. Alongside them, and much quicker to be ready for use, were the timber castles. Although it is not always possible to

a modest defence by the standards of the late twelfth century: Thomas, *The walled towns of Ireland*, 1: 151–2.

[1] Giraldus Cambrensis, *Expugnatio hibernica*, 31–5 (Wexford), 64–6 (Waterford), 67–9, 78–85 (Dublin), 138–9 (burnings); G H Orpen, trans., *The song of Dermot and the earl*, 199 (ll. 2713–16), 231–2 (ll. 3175–98), 237 (ll. 3262–7), 235 (ll. 3223–338) (Trim); *The annals of Loch Cé*, 153, 167, 173. For a more recent translation of the *Song*, see *The deeds of the Normans in Ireland*.

[2] Giraldus, *Expugnatio*, 148–54; *The song of Dermot and the earl*, 249–51. The surviving part of the *Song* frustratingly breaks off during this account. *The annals of Loch Cé*, 153, wrongly records an Irish victory at this siege.

distinguish an Anglo-Norman earthwork from an Irish rath, enough work has been done to establish the fact that a large number of mottes and a small and uncertain number of ring-work castles were erected in the newly conquered provinces between 1170 and around 1230.[1] Sometimes, they were constructed to take advantage of existing ring-forts. Because of the immense number of raths, and the great difficulty of dating, it has been sensibly concluded that at this time it is impossible to establish with any accuracy the role of the ring-work. The distinctive mottes (350 of which have been identified, while another 100 sites could also be mottes) are much easier to study, but as elsewhere the disappearance of their timber structures conceals their original aspect. It is clear that speed of construction was one factor deciding the nature of the castle to be built, but it is likely that the overriding determinant was the availability of resources and the problem of cost. McNeill shows that a motte castle at Clones cost just £19, compared to the £500 minimum for a stone tower.[2] They could also be built very readily from locally available materials and with local labour.

There are some peculiarities of mottes in Ireland that challenge assumptions about the role and function of such castles. Why were they built where they were, and not built in some lordships? Why do so few of them have baileys? Distribution maps show a dense concentration of mottes on the borders of the settler lordships, and a higher percentage of these have baileys than do the castles of the hinterland. It has been suggested that this was for the purpose of sheltering parties of soldiers to act against Irish raiders (or of course to make raids themselves). This is of course possible, but the castles are not close enough together to form a genuine military barrier, and it is wise to be cautious about any suggestion for a "strategic" purpose in the building of private castles in these times. A simpler explanation for the concentration would be that the border areas were the most vulnerable to raiders and the new landowners needed the castle for their own security. On the other hand, there are areas, such as around Cork, where early settlement was known but very few castles exist. The rarity of baileys remains a puzzle. The few excavations that have been done suggest that motte tops sometimes held not only a tower, but also and mainly a hall, indicating that the whole household of the owner could inhabit it (although not of course their horses). The discovery of pits to allow archers to shelter behind gaps in the palisade at the mottes at Lismahon and Clough (Co. Down; Plate 10) indicate the role of defensive provisions.[3] Some

[1] The ground-breaking work was Orpen, "Motes and Norman castles in Ireland", which still repays reading. The defensive function of the more modest moated sites that abound in Ireland, it has been proposed, were limited to a small minority of the total, although the definition is strained by the large size and strength of fortification of some of the most powerful, including a number that must have been used by Irish rulers especially in Roscommon and Sligo: see Barry, "The defensive nature of Irish moated sites".

[2] McNeil, *Castles in Ireland*, 57, citing the Pipe Roll of 1211–12.

[3] Waterman, "Excavations at Clough Castle", 106–19; Waterman, "Excavations at Lismahon", 146–8. McNeil, *Castles in Ireland*, 56–74, for a very clear discussion of all the issues. Dollen, "Die Entwicklung des Burgenbaus", 50–52; Barry, *The archaeology of medieval Ireland*, 37–50. Barry identified 45 ring-works, but McNeil has argued the riskiness of such an approach.

other mottes seem to have lacked any structures on top at all. One can only conclude that either they were never occupied, or else that they are not mottes. Still others may well represent manorial centres erected at a later date.[1]

The Irish were not afraid of attacking castles, and their effectiveness depended entirely on their being held by armed men. These castles of the conquest played therefore a crucial role in securing the land once it had been occupied, but unless adequately defended they were vulnerable to destruction. No great knowledge of siege techniques was required to attack them, and indeed we have seen how little such knowledge the Irish possessed at this time. But the existence of a (small) number of mottes in areas that were and remained under Irish control for a long period suggests that some Irish rulers were not slow to learn the ways of their would-be conquerors, just as the Welsh had been doing.

Wales

The Welsh chronicles continue to record year by year warfare between the Anglo-Norman barons and the Welsh princes, and feuds with each other. Castles, their building and destruction, were a central feature of the events. An increasing number of castles were built by the Welsh. In the south-west, the Lord Rhys ap Gruffudd was establishing a dominant position for himself, at the expense of the marcher lords as well as his Welsh neighbours. In 1156, he built a castle at Aberdyfi, on the border with the Welsh of Gwynedd, and in 1158 he captured Llandovery Castle, following up with the destruction of the (earth-and-timber) castles of Ceredigion (Cardigan).The following year he repeated the treatment in Dyfed, but was unable to take the key fortress of Carmarthen. In 1165, he destroyed Cardigan itself, and took Cilgerran, the promontory site where a strong thirteenth century stone castle still stands. The following year, Anglo-Norman and Flemish forces twice besieged the Welsh in Cilgerran, in vain — the first instance of the Welsh successfully holding a castle against the invaders. In 1167, the castle built by the Welsh at Castle Caereinion was "breached" by these warriors. Meanwhile, in the north, the castles at Llandegla-yn-Iâl, Basingwerk, Rhuddlan and Prestatyn had all been destroyed by the Welsh of Owain Gwynedd.[2] In 1171, Rhys was recorded as building on the site of the castle he had captured at Cardigan one of "stone and mortar", another first for a Welsh prince. Another castle was built by the ruler at Rhayader in 1177, while in 1189 he cleared out the Anglo-Norman castles of St Clears, Abercorran (Laugharne), Llansteffan and Carnwyllion (Llanelli), confirming his control of Dyfed.[3] Kidwelly Castle was rebuilt in 1190. In 1185 the Welsh had besieged Neath (another timber castle replaced by the standing stone version) while ravaging Glamorgan, but on the approach of a relief force they fled, burning the engines they had constructed. These technical skills were recorded again when Rhys's son Maelgwn attacked Ystrad Meurig with

[1] For which see O'Conor, "Motte castles in Ireland".

[2] *Annales Cambriae*, 46–50, *ByT(P)*, 58–65; *ByT(H)*, 131–49; *ByS*, 157–71.

[3] *Annales Cambriae*, 54–7; *ByT(P)*, 67–73; *ByT(H)*, 155–71 (containing the reference to Cardigan at 155); *ByS*, 171–87.

catapults (Welsh *blifyeu*), with which they breached the defences and destroyed it. In 1196, Rhys marched through central Wales, destroying castles at Colwyn (near Hundred House), Clun and (New) Radnor before besieging the large motte and bailey of Painscastle with "blifyeu a magnaleu", successfully. The same year, an English force attacked Prince Gwenwynwyn ab Owain's motte castle at Welshpool: "after laboriously laying siege . . . with diverse engines and contrivances, won it by sending sappers to dig under and make passages underground", whereupon it surrendered. The prince had it back before the end of the year. The references to siege equipment begin to increase. In 1198, Gwenwynwyn attacked Painscastle (obviously now recovered by its Braose owners) "without recourse to catapults and slings", while in 1203 he and Maelgwn ap Rhys took Llangadog with the same combination.[1]

Welsh castle building

By the end of the twelfth century, the leading Welsh princes were familiar with the siege techniques of the age, and although there is no record of who built or operated the equipment, they must have had access to people with the necessary skills. Similarly, the process of castle building accelerated. Here again, the major determinant would have been the availability of resources, in terms of money and of access to skilled labour, to create castles in the same form as those of the Anglo-Normans. Another problem was raising the resources to maintain and hold a castle, whether newly built or taken over from the Anglo-Normans. Rhys's law code stated the obligation on everyone to work on the king's castles.[2] It is no surprise that the earlier constructions seem to have been motte-and-baileys. While at peace with Henry II after 1171, the Lord Rhys began building in stone, with one large castle being put up in each of his five constituent lordships, presumably intended as the *caput*. It is not clear what form Rhys's defences at Cardigan took, but at Nevern, a motte and bailey captured by the Welsh in 1191, a small rectangular stone donjon was added, and there is a similar building lying in indistinguishable ruin at Ystrad Meurig.[3] Whether the prince's castle building amounts to a strategy, as has been claimed by a recent biographer, is less certain. Some of the castles were at places of strategic significance, but not in the sense understood by a later age. Rather, it is more likely that Rhys wanted secure and prestigious bases for his rule and administration, and the castles on his frontiers could additionally provide a means of checking invasions, if properly garrisoned and provisioned. Even less of a "strategic" pattern can be made of the castles being built by the princes of Gwynedd or Powys, and it is logical to assume that castles had the same purposes for the Welsh rulers as they did for their neighbours, with the difference that Welsh castles remained the possessions only of the princes. Indeed, the relatively poor

1 *Annales Cambriae*, 58; *ByT(P)*, 74–82; *ByT(H)*, 175–83; *ByS*, 189–97; *Annales de Margan*, 18 (Neath).

2 Cited by Turvey, *The Welsh princes*, 106–8.

3 On which see R O Jones, "Report of the centenary meeting", 142; R O Jones, "Ystradmeurig Castle".

quality of building evident from the remains of many of the sites (especially in Gwynedd) suggest that resources were always hard to come by,[1] perhaps on a par with the castle-building resources of the marcher lords.

Capetians and Angevins

Philip II Augustus succeeded to the French throne in 1180, and by the end of his long reign the Capetians ruled most of the territory of modern France. This consolidation was accompanied by a continued strengthening of the power of the crown. Part of the process involved a dramatic programme of castle building, and substantial reform of the military resources of the monarchy. But the first part of the reign witnessed unrest and warfare.

In the 1180s the seat of struggles was the north, and Gislebertus of Mons has left an informative record. In 1188, Henry IV, count of Namur, who was in conflict with the duke of Louvain (Brabant), besieged the castle of Bouvignes on the river Meuse, which was strong and well defended. The count used mangonels to destroy the walls as far as the donjon, then prepared "other machines, that is petraries", to attack the donjon, which surrendered. In 1189, Baldwin V, count of Mons or Hainaut, besieged the castle of Merlemont (near Philippeville) for six days, causing much harm before it surrendered. He went on to attack the fortified monastery of Floreffe, where initially his mangonels and petraries caused no damage. But the walls were supported by wood (perhaps a frame), and once he had managed to set this alight and bring it down, the place surrendered. In 1191 the castle of Enghien, which was "too strong to take without machines", surrendered when the garrison saw the count construct petraries. Four years later, the fighting on the Meuse involved Baldwin in besieging Huy with a large army. Having overcome the problem of approaching the fortress, using mangonels to kill or wound the defenders, he once again set up the petraries, whereupon surrender followed.[2] Several times therefore in this chronicle, there is a clear statement of two types of stone-throwing artillery, with the term mangonel used to describe a smaller weapon used chiefly against personnel, whereas the petrary was powerful enough to damage defensive structures, and indeed induced surrender without even being used.

Philip Augustus was also involved in the struggles in the Low Countries, his dispute being with Philip of Alsace, count of Flanders. The castle of Boves near Amiens was a focal point of action. Its defences consisted of a stone donjon on a high ridge built on top of a rectangular motte, with two baileys in a row protected by earthen ramparts (and presumably palisades) and a deep ditch. It could only be approached from this side. It had previously belonged to the Thomas de Marle who had caused such problems to Louis VI, and in 1185 its lord backed

[1] Turvey, *The Lord Rhys*, 74–7; Turvey, "The defences of twelfth-century Deheubarth", discusses the evidence; Turvey, *The Welsh princes*, 104–5, 138–9, 148, 172–3. On the castles, see P R Davis, *Castles of the Welsh princes*; Avent, *Cestyll tywysogion Gwynedd*.

[2] Gislebertus of Mons, *La chronique*, 176, 221, 265, 303.

the count of Flanders against the king. Guillaume le Breton provides invaluable detail in his panegyric of Philip Augustus. The king's men first stormed the outer ward, without using crossbows or ballistas, then created a machine (a *cat*) to provide the cover to enable them to fill the ditch. Miners then sapped under the rampart, propping the excavation up with cut logs that were then fired, creating a breach through which the besiegers stormed. This left the donjon, which was not particularly strong, having walls only 1.15 metres thick on the side of the bailey, although it stood some seven metres above the ward. The French army now set up "Turkish mangonels" and petraries with slings and stones carried by two teams of four men, and the wall began to crumble. Philip of Alsace ended proceedings by making peace.[1] Confirming the evidence of the Mons chronicler, Guillaume indicated that the Turkish mangonels were light weapons, and the petraries were the heavy artillery. His description of their operation confirms that these were still the manually-operated stone throwers rather than new weapons. But the crumbling masonry confirmed how effective these weapons could be against thin walls and unmodernised seigneurial castles.

The defences of Normandy represented a different level of challenge to Philip Augustus, especially as he confronted there one of the military geniuses of that or any age, Richard the Lionheart.[2] But the Angevins laboured under a substantial and, ultimately, fatal disadvantage. While Philip continued to increase the financial resources of his kingdom to allow a relentless drive to conquer Normandy, the Anglo-Norman regime could not compete with the level of expenditure, and the immense drain on the treasury (not of course helped by the vast ransom raised to free Richard from captivity in Germany) in the end undermined the defence of the duchy. Under Richard, the Norman baronage and towns had been loyal, helping the king secure the increased revenues needed, chiefly, for work on the castles, with less being spent on paying soldiers. In fact, one study has suggested that at times during the war-dominated 1190s, the Angevin field army may have been only a few hundreds of men, perhaps increased by feudal levies. Hundreds more would have needed to garrison the castles.[3]

In 1187, Philip attacked Châteauroux with miners, petraries, a battering ram aimed at the gate and a siege tower built higher than the walls: despite this array, the attempted escalade failed and the arrival of Richard brought an end to this struggle.[4] In 1189, Richard had joined Philip to capture a number of his father's castles and had surprised Henry II at Le Mans. As Henry fled, Richard and Philip

[1] Guillaume le Breton, *Philippides*, in his *Œuvres*, II: 18–19 (book I, ll. 296–360); there is a very close French translation, Guillaume le Breton, *La Philippide*. Dion, "Lettre à M. de Caumont", 435–42.

[2] Early evidence of Richard's siege warfare skills is provided in Roger of Wendover, *Flowers of history*, II: 49, an account of his capture of Taillebourg from a rebel in 1180, deploying machines and capturing the castle by entering pell-mell with the retreating garrison. The castle was demolished. Roger compiled his account in the 1220s, based on earlier narratives.

[3] See Moss, "The defence of Normandy", 145–57, for a recent re-examination of the factors that determined the outcome in the duchy.

[4] Guillaume le Breton, *Philippides*, in his *Œuvres*, II: 61–2 (book II, ll. 542–87).

besieged the donjon and captured it by mining in just three days, with its garrison of 30 knights and 60 sergeants. A few days later he also took Tours, by escalade, with its garrison of 80 knights and 100 sergeants.[1]

On his return from the Middle East, Philip began his sustained attack on the Angevin provinces. In 1193, having taken the vital frontier castle of Gisors by treachery, Philip had occupied the whole of the Vexin and advanced against Rouen, held by Robert III, earl of Leicester. Unable to press the siege, the French withdrew, burning their 23 petraries. The following year, Philip was able to prosecute his invasion, while Richard had to cope with rebellion by his brother John. In March 1194, Richard stormed the powerful castle at Nottingham. Many men fell to archery as the outer gates were burnt, then the king ordered petraries set up against the inner defences, which surrendered. He then crossed rapidly to Normandy. Philip was attacking the castle at Verneuil-sur-Avre (now lost under later work), where for eight days he erected engines and began to mine, but fled at the approach of the English king, burning his engines. Richard went on to recover the castle of Beaumont-le-Roger, a payment being recorded for the subsequent demolition of the donjon, and also took back Loches by storm. Philip lost more of his engines in the rout by the English at Fréteval. Chastened but not defeated, in 1195 Philip captured Le Vaudreuil by mining, when the "great part of the walls" fell down, and Dieppe was burnt by the French "discharging Greek Fire".[2]

From 1196, Richard began the construction of the new castle at Les Andelys, which he called Château-Gaillard, an edifice created in the incredibly short time of a little over a year at immense expense (some £7,000) to provide a new defence on the Seine to block progress toward Rouen from the Vexin. The castle proper was part of a system of defence that included an outlying fortification at Radepont, on the road to Rouen, stockades on the river and defences around the town itself, which were linked by walls to the castle. There were (modest) fortified outposts at Cléry (a motte) and at Boutavent, which may have been a stone tower on an island in the Seine at Tosny. Radepont seems to have been a rectangular towered enceinte with a bastion built with rusticated masonry.[3] Maintenance of the royal castles in Normandy was assigned in groups to a surrounding castlery, maintaining a type of system known for centuries. In 1198, for example, the honour of Évreux was due to find "12 dozen cords, cords to bind the mangonels, 7 slings for petraries, 8 balistae, 6200 bolts" and so on.[4]

[1] Roger of Howden, *Gesta regis Heinrici secundi*, 67–9; Roger of Wendover, *Flowers of history*, II: 73–5.

[2] Roger of Howden, *Chronica*, 207, 239–49 (Nottingham), 369–72; Gislebertus, *La chronique*, 583; Roger of Wendover, *Flowers of history*, II: 136–7. The classic is Powicke, *The loss of Normandy*. Modern scholars (including J C Holt, J Gillingham and D Power) have now questioned some of Powicke's conclusions.

[3] Mesqui, *Châteaux forts et fortifications*, 457.

[4] Powicke, *The loss of Normandy*, 208, citing *Magni rotuli Scaccarii Normanniae*, II: 462.

The siege of Château-Gaillard and the conquest of Normandy

When Richard died from a crossbow bolt during the siege of a rebellious subject at Châlus in Aquitaine in 1199, a great deal of work had been done on the defences of Normandy since the reign of his father. Little expense had been spared. Henry II had taken control of many of the fortifications, and enforced his right of entry into others. In 1188 a massive sum of £4,270 was supplied from the English and Norman Exchequers for works at Gisors, Neaufles-Saint-Martin, Châteauneuf (Château-sur-Epte), Dangu and others. Richard spent large sums on castles like Le Vaudreuil and Château-Gaillard. While Henry's work had been to strengthen the existing sites with higher and thicker walls and additional towers, the work at Les Andelys was new. Standing on a plateau high above the Seine, the only approach faced three consecutive baileys. Behind a ditch eight metres deep and ten wide, the first was triangular, with a strong round tower at the apex and four others and the gatehouse to one side. A broad ditch separated this area from the second bailey, again protected by round towers. Inside this, a further ditch covered the inner bailey, with a massive wall composed of semicircular projections to ensure a rounded outer face, and complete flanking. Finally, the donjon was of similar construction, and additionally had a great pointed beak pointing towards the attacker. In addition to the complete provision of flanking for the whole complex, there were stone machicolations on the donjon, a rarity for the time. While the spectacular design of the inner curtain wall had probably taken its inspiration from the curtain erected around the Gravensteen castle of Philip of Alsace at Ghent in 1180, the overall design exemplified the highest level of thinking about defences of the age (Plate 14). Richard had the advantage, of course, of being able to start from scratch.[1] The only serious weakness was the permanent stone bridge giving access to the inner bailey, which Philip had removed after his victory. John created a further — and fatal — weak spot when he had a chapel built against the wall in the southern side of the middle ward, with lower storey openings giving onto the ditch (and traditionally described as latrine chutes).

Yet this mighty castle fell to assault, albeit after a long siege, in 1204. Philip's army worked its way methodically through the layers of defence. Boutavent was taken in 1202, while other frontier fortifications were forced to surrender. Radepont was taken after a month's siege, and Les Andelys itself and the works on the river were captured after fierce fighting, in which the French had to built towers on boats to attack from the river and to break through the stockades under fire. The garrison of the castle proper comprised around 200 fighting men under Roger de Lacy, who in February 1204 expelled the "useless mouths" of the non-combatants who had taken refuge there, to preserve his supplies, at which point he still had another year's provisions. Philip eventually took mercy on their suffering as they huddled in the ditch in winter weather without food. The French king oversaw

[1] Powicke, *The loss of Normandy*, 178–208; Dieulafoy, "Le Château-Gaillard et l'architecture militaire au XIIe siècle", remains an excellent and detailed account of the castle. Roger of Wendover, *Flowers of history*, II: 174–213; Moss, "The defence of Normandy", 157; Pitte, "Château-Gaillard", 163–71, for a study of the whole defensive system.

a regular siege, entrenching his own camp on both sides, building shelters to protect his troops from the autumn and winter weather, and deploying mangonels and petraries. While these cleared the defenders from the walls, men got to the ditch and found a rare area whence they could mine safely. An escalade attempt failed because the ladders were too short, but mining brought down a tower and the defenders retreated to the next line. This fell when some enterprising soldiers found the opening from the ditch into the chapel and climbed in.

That brought the attack to the inner bailey, connected to the outer with a permanent stone bridge. Setting up his engines on the bridge, and using this as a shelter for his miners, the king had them drive under one of the gate towers while also bombarding it with stones. Too late the defenders, who were recorded as operating one large petrary and mangonels and crossbows, tried to countermine; the tower rocked and cracked, and the assault carried the ward. Guillaume le Breton in fact gave prime responsibility for the collapse of part of the tower to the stones of the petrary. After a siege of nine months, the garrison of the strongest castle in the west did not seek to hold out in the donjon and surrendered on 6 March 1204.[1] There are two striking differences between these events and the great sieges of the third crusade. First, Lacy's garrison was entirely passive, concentrating only on defending the points attacked. There were no sorties to try to burn the engines or disrupt the mining. Second, there was no prospect of relief, King John failing to mobilise a field force to rescue his patrimony. Philip had so entrenched himself that, as in the great siege of Acre, no relief force could have broken through anyway once his position was established. Strong walls could only ever be a part of a strategy, and even the strongest fortress will fall unless there is relief or diversion. Not surprisingly, the fall of Château-Gaillard was followed by the completion of the conquest of Normandy.

The loss of Normandy stirred John into attempts to recover parts of his continental possessions. In 1206, he voyaged to La Rochelle, then marched rapidly to Montauban and attacked the castle there for two weeks, by which time his "petraries, ballistas and slings" had destroyed much of the wall enabling him to capture it by escalade.[2] But by and large his efforts were without result, and in 1214 the great coalition of princes he had helped assemble, including Emperor Otto IV of Germany, was defeated at the Battle of Bouvines by Philip. In England, the continued and increasingly ruthless attempts by John to extract resources from his kingdom to sustain the wars culminated in the revolt of much of the English baronage, leading to the concession of Magna Carta, followed by attempts to rescind it that culminated in civil war. A number of sieges of castles featured as the major engagements of this war, which turned one way with the arrival in support of the rebels of Philip's son Louis, then gradually turned back in favour of the crown as royal control was restored following the death of John by the political and military skill of the regents, the papal legate and the now very old but still highly effective

[1] Guillaume le Breton, *Philippides*, in his *Œuvres*, II: 176–209 (book VII throughout); Pitte, "Château-Gaillard", 173–6.

[2] Roger of Wendover, *Flowers of history*, II: 218–19.

William Marshal, whom we first encountered as a child during King Stephen's siege of Newbury some seven decades before.

In 1215, the assembled baronial army spent two weeks in a vain attack on the royal castle at Northampton, having "come without petraries nor engines of war". This now disappeared fortress, seemingly a motte and bailey subsequently rebuilt in stone, thus repaid the regular expenditure on it by the Angevins,[1] although the size of the garrison is not known. It is however recorded that the king had sent letters to all his castle governors ordering them to lay in provisions and arms and to strengthen their garrisons. Clearly, in the contest ahead, the king understood how crucial control of the royal castles so assiduously restored by his father was going to be. The barons, however, apparently also had access to siege artillery, since they were recorded as attacking the Tower of London vigorously with such weapons prior to the granting of Magna Carta. However, the Barnwell Chronicle stated that they called on Louis to provide engineers who could construct engines of war, suggesting that their access to people with the necessary skills was more limited than the king's.[2]

Rochester, 1215

The rebel William d'Aubigné (de Albini) moved to Rochester, presumably to block the royal route to London, and found the castle destitute of the means of defence. Collecting provisions from the townspeople, he assembled a force given by Roger of Wendover as 140 knights with their retinues, and by others as anything up to 420. The king, with a professional army of Fleming and Poitevin mercenaries, first cut the city off from the road to London, then attacked the castle, with its large rectangular donjon standing inside a stone-walled bailey, on 11 October. He was left undisturbed to prosecute the siege while the other barons returned to London, the defences of which they strengthened, after an abortive attempt at relief. The king attacked with petraries, but seeing that they had little effect on the masonry, set miners to work. A breach was created in the curtain through which the army stormed, and the defenders withdrew into the donjon. The miners again set to work, and through their efforts brought down the south east corner of the donjon. A letter survives in which John asked to be sent 40 of the "fattest pigs", whose fat was used to fire the timbers in the mine. The defenders continued to resist from behind the cross wall that divided the great tower in two, but their cause was now hopeless and they surrendered on 30 November. It was one of the greatest sieges to date in Britain and left the baronial cause reeling, with few thereafter willing to place their trust in castle walls.[3]

Some interesting conclusions can be drawn from the event. The relative

[1] Roger of Wendover, *Flowers of history*, II: 307; Walter of Coventry, *Memoriale*, II: 219. VCH *Northants.*, III: 32–5, and references in King, *Castellarium anglicanum*, II: 317.

[2] Walter of Coventry, *Memoriale*, II: 226.

[3] Roger of Wendover, *Flowers of history*, II: 327–39; Walter of Coventry, *Memoriale*, II: 226–7; Ralph of Coggeshall, *Chronicon anglicanum*, 175–6. R A Brown, *Rochester Castle*, 12–15.

strength of the crown as against the baronage was again demonstrated, and the knowledge of siege techniques of even someone so little credited with military skill as King John is confirmed. The reliance on mining to effect a breach confirms that stone throwers were still not able in 1215 to break down masonry defences. The garrison of Rochester was numerous, but it does not seem to have attempted an aggressive defence to disrupt the siege. Finally, conclusions have been drawn about the role of the great tower from its crucial part in the events of 1215. Some scholars now deny it any defensive role, drawing attention to detailed studies of the design (for example) of the elaborate entrance, and stress its function as a visible symbol of power and lordship. It is rightly pointed out that any strong building could serve as a refuge in an emergency. That may be so, but King John, deploying the most advanced technology available to him, and overwhelming force, could not break into this tower except by undermining the structure. If defensive strength was not amongst the purposes of the builders, it was certainly the effect. The siege additionally confirmed the significance of the question of resources: capturing Rochester may have cost John 60,000 marks in wages (perhaps to be translated as a 'very large sum').

In 1216, the royal heir Louis arrived from France with additional forces to join the barons, and the royal cause was put on the defensive. While John's able lieutenant Sir Falkes de Breauté captured Bedford Castle, which surrendered after a truce of seven days to await relief that did not come, Louis laid siege to Odiham (Hampshire). This small castle consisted of two wards and an octagonal tower and had only been completed in the previous ten years, but it was held by just three knights and ten soldiers. This tiny force held out until the French set up siege artillery, whereupon they surrendered after a week. It is remarkable that the besiegers were not able simply to overrun the walls. Much bigger targets followed: Louis turned his attention to Dover, while the barons attacked Windsor where they deployed engines. The garrison under Engelard d'Athée (or de Cigogné) had 60 knights and their retainers, and the attackers failed to make any progress.[1] At the same time, the rebel barons had linked up with King Alexander II of Scotland to attack the royal garrison of Carlisle Castle, the town having surrendered on 6 August, and from later Pipe Roll evidence it has been deduced that the Scots, showing some advance in technique since their inept attempts in 1173–4, had tried to undermine the powerful fortress, although still without success.[2]

Dover, 1216

Louis meanwhile was engaged in the second great siege of this period at the immense royal fortress at Dover, defended by Hubert de Burgh with a large garrison of 140 knights and many sergeants. Louis borrowed petraries from his father and launched a ferocious battery, but the defence was vigorous, with numerous

[1] Roger of Wendover, *Flowers of history*, II: 350, 376; *Annales prioratus de Dunstaplia*, 46–7; Walter of Coventry, *Memoriale*, II: 232.

[2] The evidence from the Pipe Rolls and elsewhere gathered by McCarthy, Summerson and Annis, *Carlisle Castle*, 126. The *Chronicle of Melrose* gives no detail.

sorties that succeeded in pushing the besiegers back. The French were occupying the higher ground facing the outer ditch and wall of the castle, from which they eventually succeeded in undermining the barbican in front of the north gate and storming it. This enabled them to bring their artillery to bear against the gate itself, but once again it was the miners who made the difference, bringing down the eastern of the two gate towers despite an attempt by the defenders to countermine. Hubert de Burgh's men were, however, up to the challenge, and succeeded in defeating the ensuing assault, hurriedly patching up the breach thereafter. The dispirited prince ended the active attempt to capture Dover at this point, although the blockade continued until the final defeat of the French and baronial cause.[1] Dover had been another recipient of great royal expenditure, with the enormous great tower protected by inner and outer curtain walls with rectangular towers and a deep ditch. They had resisted everything, but the mining had revealed a weakness, and in the subsequent repairs, the old gateway was blocked up and a new one created on the site where it stands today. The other noteworthy point from Dover was the active defence waged by the garrison, which significantly inconvenienced the besiegers and eventually brought their assault to an end.

The French prince now marched around southern England, and a substantial number of royal castles fell into his hands. Hertford Castle, a place that otherwise makes no mark in history, caused Louis to deploy siege engines before it surrendered. At nearby Berkhamsted, reached the next day, stood a weakly-walled motte and bailey with a shell keep, and an unusual double ditch. Louis set up his engines again and maintained a steady bombardment. The king ordered it to surrender after two weeks, there being no hope of relief or of holding out. There are eight earthwork "spurs" attached to the outer rampart that were tentatively identified with emplacements for the siege artillery at this castle. They are well sited to bombard the walls of the inner bailey and of the shell keep; but they were built with the rampart, are extremely close to the inner rampart (from which, we know, an active defence was conducted by a professional garrison), and were surely more likely to be part of the defence than the attack.[2]

Many other castles were subject to capture or siege during this traumatic time. Louis had occupied the fortresses of Reigate, Guildford and Farnham without difficulty, presumably finding them undefended. The royal hunting castle of Knepp near West Grinstead certainly was. At Winchester, he had besieged both the king's "big castle" and the "small" castle (the bishop's palace of Wolvesey). He attacked with petraries and mangonels for two weeks to take them, although it was later recorded that once again it was miners who did the damage to a large section of the ramparts. At Marlborough, the siege took seven weeks. On his

[1] Roger of Wendover, *Flowers of history*, II: 374 ff.; Walter of Coventry, *Memoriale*, II: 230–32; *Histoire des ducs de Normandie et des rois d'Angleterre*, 170, 178–9. For a modern account, see Coad, *English Heritage book of Dover Castle*, 38–40.

[2] Roger of Wendover, *Flowers of history*, II: 381–2; *L'histoire de Guillaume le Maréchal*, II: 202–3; *Histoire des ducs de Normandie*, 182; Kendall, "The siege of Berkhampstead Castle", 40–46, and personal observation. A modern study, Remfry, *Berkhamsted Castle*, 18, 26–8, arrived at similar conclusions.

march through East Anglia, Colchester, Orford, Pleshey and Norwich fell to him without a fight with the permission of the king. The royalist party was too weak to resist in the south and east of the country, holding onto only the castles of Dover and Windsor. Elsewhere, however, they were in control, and the rebel cause was weakening "for the castles of the king were numerous and strong" (Walter). Louis and the barons marched to drive off the royalist earl of Chester, who was besieging the Leicestershire castle of Mountsorrel with stone throwers, and then attacked the more important castle of Lincoln in 1217. This was held for the king by its hereditary castellan, the stout lady Nicola de la Haie, who resisted a bombardment from artillery. The besiegers fell victim to an attack skilfully directed by the aged William Marshal, losing their army and many of their leaders to the royalist forces, and effectively ending the war in the royal favour. However, in his last weeks in England, it was noted that Louis took possession of a new weapon. The contemporary author of the *Histoire des ducs de Normandie* observed that at Easter, Louis had a "trebucket", and the Dunstable annalist said he deployed a "tribuchetta" and other machines. At Dover, he had it set up on the mound in front of the castle, whence it then did some damage, and then, after the defeat at Lincoln, he had it broken up.[1]

The minority of Henry III: sieges of Castle Bytham and Bedford

Armed conflict continued in the period of the royal minority, sometimes directed at disgruntled supporters of King Henry's late father. These men were hardened war veterans well skilled in their trades, and overcoming them required a substantial effort on behalf of the government. Over the next few years, isolated rebels resisted royal power for short periods, but succumbed quickly to the far greater resources of the monarchy. In 1218, Newark was recovered by force from its occupant, involving a bombardment with petraries for eight days. In 1221, William de Forz, count of Aumale, rebelled. His men captured Fotheringhay Castle from the earl of Chester by escalade, the place having been empty of soldiers. Castle Bytham was taken from William by force, although it was a powerful fortification, after a siege of just six days, in which siege engines were erected that destroyed the walls and buildings.[2]

Three years later, John's former commander Sir Falkes de Breauté fell out with the king, who besieged Falkes's brother William in Bedford Castle, an event well recorded and for which substantial administrative records survive. The annals of the nearby Dunstable Priory (supervised but not written by the prior Richard

[1] *Histoire des ducs de Normandie*, 188, 192, 195; *L'histoire de Guillaume le Maréchal*, II: 208–16; Roger of Wendover, *Flowers of history*, II: 391–4; *Annales prioratus de Dunstaplia*, 49–50; Ralph of Coggeshall, *Chronicon anglicanum*, 181–5; Walter of Coventry, *Memoriale*, II: 234–7.

[2] Walter of Coventry, *Memoriale*, II: 245–9; *Annales prioratus de Dunstaplia*, 64; Roger of Wendover, *Flowers of history*, II: 429. On Castle Bytham, Lincolnshire, which had a round keep and two stone-walled baileys behind a wide ditch, see Mackenzie, *The castles of England*, I: 428–32.

de Morins) are full of detail. The castle had been strong enough to force Stephen to blockade it back in 1138. When Falkes took control of it in 1215, he apparently spent the following years significantly improving the defences such that it no doubt presented a strong stone wall, with a barbican, forming two rectangular courtyards around the round stone tower on the motte. Starting on 20 June 1224, a petrary and two mangonels were erected on the east, while from the west two mangonels bombarded the "old tower" (donjon). There was another stone thrower on each of the other sides. Two siege towers were constructed, from which crossbowmen could shoot over the walls, and a cat was constructed under the shelter of which miners went to work. Hundreds of those working the siege machinery became casualties. First the barbican was captured, then the outer bailey, in which were found many horses and supplies gathered by the defenders. A very bloody assault then took the wall around the old tower, and the donjon itself was finally brought down by mining. The captured garrison of about 80 knights and sergeants was hanged.

The absent Falkes had believed that Bedford could resist for a year, but despite its strength and provisions, and the prowess of the garrison, it fell in eight weeks: a substantial period to hold up so strong an opponent determined to reduce it. The ability of the crown to mobilise massive resources from across the country is demonstrated by the entries in the Pipe and Close Rolls. On the first day of the siege, the sheriff of London was ordered to supply 20 slings and 10 cables for the mangonels and petraries. 15,000 crossbow quarrels were summoned from Corfe Castle three days later, while Northampton was called upon for oxen or horse hides to make slings for the artillery. London was called upon again for 10 balistae. The local sheriff was required to raise stone cutters and quarriers to work the stone for the ammunition, and on 1 July 30 miners were summoned from Hereford. The stone shot preserved from the siege weighed between 22 and 40 pounds (8 and 15 kgs), reinforcing indirectly the probability that this was not for the new trebuchet. Four days after the fall of the castle, the order was given for the petraries, mangonels and belfry "to be disjointed and carried to Northampton". The sum of ten shillings was paid to two injured miners. The quantity of timber, tools, hides and crossbow bolts mobilised was enormous.[1] It might be that the speed at which the supplies were gathered had something to do with the king's planned expedition to cross to France, a journey that was aborted by the events of Bedford. In consequence, the French were able to capture Niort, Saint-Jean-d'Angély and La Rochelle itself, deploying trebuchets alongside the more traditional mangonels and petraries.[2]

[1] *Annales prioratus de Dunstaplia*, 88–9; Ralph of Coggeshall, *Chronicon anglicanum*, 206–7; Walter of Coventry, *Memoriale*, II: 254. See Fowler, "Munitions in 1224", 117–29, for extracts from the Pipe and Close Rolls. See Greenshields, "The siege of Bedford Castle", for a more modern account. Amt, "Besieging Bedford", re-examines the accounts. Only the motte now remains of the castle.

[2] *Annales prioratus de Dunstaplia*, 86. The garrison of La Rochelle may have been 300 strong.

Centralised royal resources, engineers and castle building in the west

Henry II had consolidated control of many castles into his own hands, and the survival of government records from this time allow it to be seen that there was a corps of specialists available to the crown, masons and engineers whose names reappear frequently in records of payment for work done. Such were Ailnoth (from 1157 to 1189), Richard, Maurice (working at Dover and possibly Newcastle) and others. Others were specifically hired to build engines, such as Urri (1193 at Nottingham, 1201 with John, 1210 at Carrickfergus) and Elyas (at Nottingham, 1194). Nicholas de Aundley had charge of making engines at Knepp and Bramber (1212), Nottingham (1214) and Knaresborough (1215). Others were listed as *minatores, fossators, hurdatores*. They were distinguished by receiving a high rate of pay.[1] The records additionally confirm the development of a royal arsenal of siege equipment. After Bedford, the artillery was not abandoned or destroyed, but dismantled for transport to Northampton Castle. Although the trebuchet first seen in England at Dover was not listed as a weapon used in the sieges of 1217–24 (and the absence of any mention in the Rolls suggests that the chronicle descriptions are accurate), the first reference comes in the Liberate Rolls for 1226, when Jordan, who is making the king's "trebuchetum" at Windsor, was paid. He was paid again for the same task four months later, suggesting either a very slow rate of construction or else more than one machine, and again in 1227 and 1229, confirming his role as the regular builder of this engine ("trebuchettario nostro"). At 12*d* a day, it was a well-paid job. Other references show that mangonels and petraries were stored in royal castles, at Rochester, Hereford, Bristol, Carlisle and London. In 1238, the constable of Bristol was ordered to receive the king's trebuchet, to have it dismantled and "placed safely in a house in the castle well-roofed."[2]

Developments in France

A similar development took place in Capetian France. Early records are rare (having been lost in the fighting with King Richard). But the Norman Cartulary contains records from 1210 demonstrating the presence of royal weapons — ballistas, petraries and mangonels — at Gisors and Falaise (Calvados).[3] Accounts for 1202 list the expenses for the king's *munitiones*, including "400 cords for petraries, 8 large *cabuli*, 1 large *cabulus*, 1 Turkish mangonel". Miners were paid 18 pence a day, masons, pioneers and builders 15 pence. The names of the master masons

[1] Brown, Colvin and Taylor, *The history of the king's works*, I: 51–61; Harvey, *English medieval architects*, 58, 91, 205.

[2] *Calendar of the liberate rolls: Henry III*, I (1226–40): 8, 41, 47–8, 81, 121, 320 (Bristol 1238). Renn, "'Master Jordan, who made the king's trébuchet'", has analysed the same evidence.

[3] *Cartulaire normande de Philippe Auguste et al.*, 33–4.

who worked on the king's castles were listed. Guérin seems to have had particular responsibility for the king's engines, being paid 112 livres 12*s* 4*d* for the cords and equipment for the petraries. Quite what a *cabulus* was is not clear: it presumably derives from a word common in southern France and Catalonia where it meant some kind of (probably heavy) artillery. Its translation (as a "cable") perhaps refers to the propulsion of the mangonel.[1]

One of the most startling changes brought about by Philip Augustus was his castle building. No fewer than twenty round great towers were erected on his orders between 1190 and 1204 on the same plan, and to very much the same dimensions: an external diameter of between 13.5 and 18 metres, an internal diameter of between 6.1 and 8.2. Wall thickness varied from 3.8 to 6.95 metres; the height was from 27 to 31 metres. They all had three floors, a ditch and drawbridge. Most of the defence was from the wall head, there being very few slits. Vaulting inside strengthened the structures. The ditches, being six metres deep and stone lined, provided good protection against mining. The total cost of all the new towers, at around 40,000 livres, was a big expense, but common design seems to have been economical, if that is compared to the equivalent 34,000 spent on Château-Gaillard alone. They were located mainly inside cities, where they could act as citadel and seat of authority as well as arsenals. Some were added to existing fortifications, like the round tower added to the rectangular donjons at Falaise or the old castle of Gisors. Others replaced previous castles as at Verneuil-sur-Avre. In his later years, Philip would oversee a further change in design, with the construction of new, quadrangular castles, with the circular donjon either at one corner, or else the centre of the castle. This design offered much more accommodation. Examples were the Louvre, Dourdan and Yèvre-le-Châtel. The Louvre of Paris was a quadrangle with boldly projecting angle towers, a deep ditch with a sluice, and a round donjon 31 metres high, with an external diameter of 15 metres.[2] Many were granted out to barons, but preserving the royal right of hand over on demand.

Castles were not the king's only work, with new town walls being provided at Paris. Here the new wall on the right bank of the Seine was 7.5 metres high and 2,800 metres long, with open-backed towers every 30 to 50 metres. Amiens, Laon (Plate 16), Compiègne, Melun, Corbeil-Essonnes, Évreux, Vernon, Gisors, Arques (-la-Bataille), Caen, Falaise, Lillebonne, Rouen and Verneuil-sur-Avre were also walled. These were many of the same places that were being provided with the round donjons. It would appear that Philip anticipated he might have to

[1] Contamine, "L'armée de Philippe Auguste", 579–83; Audoin, *Essai sur l'armée royale*, 98–102, 191. Audoin follows the nineteenth-century historian G Köhler in wrongly identifying the *cabulus* as a torsion-powered engine. Chevedden, "The hybrid trebuchet", 210–11, traces the derivation of the term.

[2] Erlande-Brandenburg, "L'architecture militaire au temps de Philippe Auguste"; Erlande-Brandenburg, "Organisation du conseil d'architecture militaire"; Baldwin, *The government of Philip Augustus*, 296–302. The list of new round donjons is: Beauvais*, Bourges*, Cappy*, Chinon, Corbeil-Essonnes*, Dourdan, Dun-le-Roy (now Dun-sur-Auron), Falaise, Gisors, Laon, Lillebonne, Montargis*, Montdidier (Somme)*, Montreuil-sur-Mer, Orléans, Péronne, Rouen, Vernon, Verneuil-sur-Avre, Villeneuve-sur-Yonne. A considerable number survive. Those now disappeared are marked with an asterisk.

defend his new territory from its previous owners, and wanted the most modern defences. In a list made in 1206/10, there was an unprecedented collection of 113 royal castles, the same process of expanding the number of royal fortresses that Henry II had carried through across the channel in the previous period.

It has been convincingly argued that Philip's castle design rapidly found many imitators in the principalities between France and the German empire. Examples of castles constructed by the great dukes and counts or their leading vassals in Holland, Guelders, Liège, Brabant, Hainault, Namur, Luxembourg, Chimay and Bar have been surveyed,[1] and many show thirteenth-century developments along the same lines. It is only fair to point out that this work was being done for up to a century after Philip's death in 1223, by which time the quadrangular plan and round tower style was very much established across the whole of western Europe anyway.

Alongside the design based around a circular donjon there also appeared a significant variant, the castle displaying the same defensive features, but without a donjon at all. Boulogne-sur-Mer is the best preserved example, in the possession of King Philip's son Philip Hurepel following the Battle of Bouvines, and built in the 1220s. The polygonal enclosure presents massive curtain walls and six tall projecting towers to the exterior (Plate 17), while the two gates (one into the town and the other towards the country) are each protected by a double-towered gatehouse. A deep ditch completes the obstacles. A complete range of buildings around the walls internally provides the accommodation. In a sense, the castle as a whole served the purpose of the donjon in relation to the town, as it stands at the most vulnerable corner of a rectangular town wall built at the same time, with twenty projecting semi-circular towers and four gates. Antecedents for the design have been sought at Fère-en-Tardenois, a seven-sided enceinte with seven projecting towers built on top of an older motte by Robert I, count of Dreux, after 1206. Certainly the model was quickly followed elsewhere in the French domains, most notably in the 17-towered curtain of the new castle at Angers from 1230.[2] Another possible source for the donjon-less polygonal castle has been identified as the Frankish castle at Chlemoutsi in Greece, although this is in fact quite different in appearance.

Developments in the British Isles

The new style popularised by the Capetians was being adopted in new castle building in Britain and Ireland too. Castles were the single largest target of royal expenditure under the Angevin kings. While some were on the frontiers, most were inland. The majority of John's castle spending, for example, went on Nottingham, Windsor, Winchester, London, Odiham, Kenilworth and Corfe. Whilst new rectangular, circular or polygonal donjons were key features of most of the new sites, other works included the provision of stronger stone defences for the bailey, provided with projecting towers (as at Dover), and gateways made more

[1] Bragard, "Essai sur la diffusion du château 'philippien'", 141–55.

[2] Mesqui, *Châteaux forts et fortifications*, 19, 68–70, 168; Heliot, "Le château de Boulogne-sur-Mer"; Bon, "A propos des châteaux de plan polygonal", 178–9.

defensible with the provision of barbicans and flanking towers. The records show that, under John again, castle ditches were systematically widened and deepened, reflecting perhaps a greater awareness of the danger posed by mining. So whatever the intended function of the donjons, the expenditure on outer defences reflected a continuing concern with defence.

The experiences of sieges in the civil war confirmed the obsolescence of most older castles from a military viewpoint. The repair of the damage at Rochester involved a new outer wall, and at the point at which the besiegers had broken in a powerful new projecting tower was built, liberally provided with arrow loops. In this respect it bears a resemblance to the Avranches Tower at Dover, which has banks of arrow loops and is normally assigned to the time of Henry II. The old gate at Dover was closed off, provided with a beak and a round tower standing ahead, while the terrain opposite, from which the French had made their assault, was provided with an outwork. The new Constable's Gate was a massive construction involving a linked structure of altogether five towers, a drawbridge and portcullises and numerous arrow slits, and a barbican. The curtain wall at Windsor was greatly strengthened, the Wakefield Tower was built at the Tower of London, and a new barbican at Bristol, new circular towers were added to Winchester. The vast expense of modernising and maintaining existing castles continued to be a drain on the royal finances throughout the reign of Henry III.[1]

But the king also oversaw the building of new castles, and that started at Montgomery in October 1223 gave some indication of the latest style, as it was begun on a virgin site a mile from the old motte and bailey (Hen Domen), no longer considered adequate to resist the Welsh, although apparently maintained as an outpost since it overlooked the ford over the river Severn. Work proceeded over the next few years, and the castle was strong enough to hold off two attacks from Llywelyn ab Iorwerth before it was finished. Only the inner ward was of stone at this point, the middle and outer baileys and the adjacent town having to wait many years for their stone walls. Montgomery stands high on a ridge 360 metres long and consists of a series of defended enclosures protected by wide and deep ditches, each 13.7 metres wide and six metres deep. The middle ward was reached via a drawbridge leading to a gateway flanked by two towers with round fronts. A further bridge led to the inner bailey, the whole front of which was comprised of two enormous round-fronted towers. The irregular ward, following the contour of the rock on which it stood, had a third immense D-shaped tower close to the gate enclosing a well some sixty metres deep, and another smaller tower projecting from the other end of the castle. The gate was protected by doors and a portcullis. The chief apartments appear to have been in the gatehouse, accessed by steps to the first floor.[2] There was no donjon.

Only the richest magnates could afford to emulate royal construction. Ranulf de Blundeville, earl of Chester, held the powerful march facing north Wales, but

[1] Brown, Colvin and Taylor, *The history of the king's works*, I: 64–80, 110–14; R A Brown, *Rochester Castle*, 16; Coad, *English Heritage book of Dover Castle*, 34, 41–3.

[2] Lloyd and Knight, *Montgomery Castle*. This is now superseded by Butler and Knight, *Dolforwyn Castle*. The castle was largely destroyed in the 1640s.

the earl's lands and interests were national. Soon after returning from the Fifth Crusade, he ordered the building of a new castle on the rock above Beeston from 1225. The function of the castle must be guessed at, but it incorporated clear defensive features with a powerful twin-towered gatehouse, portcullis, arrow slits and a deep ditch. His new castle at Bolingbroke in Lincolnshire echoes Boulogne in its shape, and had walls four metres thick, so once again this was no mere symbolic facade. The design of the towers was repeated in the towered stone enceinte built by the earl at the old motte-and-bailey castle at Chartley.[1]

Hubert de Burgh, justiciar of England and earl of Kent, was able to display his understanding of the new building styles more directly. The "Three Castles" of Llantilio (White Castle), Skenfrith and Grosmont in north Gwent, which were a few miles from each other and had always been seen together as an important part of the defences of the march in Wales, were granted to Hubert in 1201. His work at White Castle and Grosmont featured strongly-projecting towers, arrow slits and gatehouses, while at Skenfrith the justiciar built a castle on the model of the French. A large free-standing circular donjon stood in the middle of an irregular rectangular stone enceinte with four corner towers behind a moat. Holes have been found in the curtain that carried the beams to hold hoarding along the top of the wall.[2]

William Marshal, like Hubert, climbed rapidly to fame and fortune under the Angevins, and by fortunate marriages due to royal patronage became one of the greatest in the land. His possessions included the earldom of Striguil, based at Chepstow Castle, that of Pembroke, and vast tracts of land in Ireland, where he became earl of Leinster. Recent analysis now dates the twin-towered gatehouse at Chepstow on its crag above the river Wye to his time. The gate was protected with two portcullises, two doors and holes in the roof providing simple machicolations. The inner portcullis was operated from the first floor of the three storey structure. Provision was made for wooden hoarding around the parapets. Whereas the wall facing this main approach was plain, the wall and tower of the barbican were liberally supplied with arrow loops. If it was constructed on William Marshal's orders, it would be unique in the Angevin realm and among the earliest examples of such sophisticated defensive design. Pembroke, rebuilt entirely by the Marshals after 1200, features a gatehouse with formidable rows of defences: an approach under the wing of a high curtain and tower, a barbican (now restored) and a right-hand turn toward the gate, itself shielded by a projecting tower, then portcullises, gates and machicolations and numerous (though crude) arrow slits. The defences of the inner bailey were scarcely less remarkable. Access was by a bent entrance (the "horseshoe gate") through a strong curtain, and immediately behind this wall

[1] Weaver, *Beeston Castle*; M W Thompson, "Old Bolingbroke Castle", 315–17. On the earl, see Alexander, *Ranulf of Chester*, esp. 93–4, 114–15. On the beginning of Beeston, see *Annales Cestrienses*, 54. Scrivener, "Chartley earthworks and castle".

[2] Radford, *White Castle*; Radford, *Grosmont Castle*; Craster, *Skenfrith Castle*. There is now a new guide to all three, Knight, *The Three Castles*. On Hubert, see Ellis, *Hubert de Burgh*.

stands a cylindrical donjon 23 metres high, defended from its roof by an elaborate gallery.[1]

Different styles were used in William's castles in Ireland: Kilkenny (quadrangular), Carlow and Ferns (rectangular donjon), and the Rock of Dunamase, with two levels of plunging arrow loops, showing a great advance on the provision for archery from the loops of Pembroke. This mixture of styles was also apparent in the other stone castles being erected in Ireland. The nearest to the designs of the continent were the quadrangular castles of Dublin, built after 1213, and Limerick. "King John's castle" in Carlingford had a simple but powerful curtain cutting off a rock, with rectangular towers and a twin-towered gatehouse, built by Hugh de Lacy, lord of Meath, after 1180, and recessed arrow loops. Donjons also continued to be built, such as at Trim, which also had a curtain wall from after 1220, with well spaced projecting towers.[2] At Nenagh in Tipperary, there is an early-thirteenth-century castle consisting of a well-protected twin-towered gatehouse with a drawbridge, and a rounded curtain linking two circular towers and a tall round donjon (whose iconic top is due to nineteenth-century additions).

Amongst this range of styles, which may have had much to do with the personal preferences of the builder, there were common features reflected everywhere in the stone castles of the great lords. These included features that had a defensive function, such as flanking towers covering straight stretches of wall, arrow loops, barbicans to protect gateways and increasingly sophisticated defences of the gate passage itself, machicolations and hoardings to protect the foot of the wall. Where resources permitted, the walls of thirteenth-century castles were also considerably thicker, reflecting a concern to resist the increased power of offensive siege techniques.

Powerful though these new structures were, however, they remain small by comparison with the vast constructions of the Middle East, where greater resources and wealth and greater expertise continued to set examples to westerners, many of whom continued to become acquainted with these constructions through the crusades, although, as we shall now see, two of the next three such expeditions never once confronted a Muslim, and none of them came near to any of the "holy places" except those already controlled by other Christians.

[1] Perks, *Chepstow Castle*; Knight, *Chepstow Castle*; King, "Pembroke Castle"; King and Cheshire, "The town walls of Pembroke". See Avent, "The late twelfth century gatehouse at Chepstow Castle", 28–33, for the dendrochronological redating of the timber and its consequences.

[2] Information from a number of sources, guides and accounts. Overall, see Sweetman, *The medieval castles of Ireland*; McNeil, *Castles in Ireland*.

The Fourth "Crusade" and the new Latin empire of the east

In October 1202, the crusading army decided to head for Alexandria, but first laid siege to the Christian city of Zara (Zadar), in response to the demand of the Venetians to recover the place from the pirates who threatened their trade. It was said by Geoffroy de Villehardouin, a major participant in the events to come, that 300 petraries and mangonels were loaded on the fleet. A five-day bombardment was followed by the preparation of a mine under one of the city's towers, and the citizens then surrendered to the doge. According to Gunther of Pairis, it lasted only three days, but the outcome was the same.[1] Then came the arrival of the exiled prince Alexius Angelus (son of Isaac II), with the offer of an immense fortune if the assembled force would restore him to the empire of Byzantium. So in July 1203, the host gathered before the great walls of Constantinople. The attack came from land and sea. The Venetians protected the vessels with ox hides and attached ladders to the masts and attacked the sea walls, and also deployed petraries and mangonels on the ships. On land the strength of the defences were an effective deterrent, but from the water the Italians were invincible and broke into the city since their improvised devices were higher than the walls. The usurper, Alexius III, fled and Alexius IV was admitted to his realm.[2]

The problem began when only some of the promised reward was paid over. The "crusaders", in order to recover their missing money, besieged Constantinople a second time. Over the winter there was a long-stand off in which stone throwers were arrayed by both sides, without effect. The Greeks strengthened their defences, erecting wooden structures to raise the height of the sea wall, and fitting hoarding, to prevent a recurrence of the previous assault. Forty (or alternatively sixty) petraries on the walls were an additional defence. After Lent, the Latins completed their preparations and attacked in April 1204. Their crossbowmen were highly effective in keeping the Greeks behind the walls, but once again the strength of the defences deterred a direct attack on land, while at sea it was impossible to find a firm base for stone throwers. The doge of Venice came up with the solution, tying ships together to create a more stable platform, then building, in effect, siege towers on the vessels. Strong wooden walls were built to protect the ships from stone throwers, and they were covered with hides to protect against Greek Fire. A preliminary attack broke the chain protecting the Golden Horn, enabling the Latins to capture the Greeks' ships and deploy against the weakest section of the defences. From the mobile seaborne siege towers, the attackers crossed to the wall and captured several towers. This allowed them to land, and the Greek soldiers, who do not appear to have been very disciplined or determined, were

[1] Gunther of Pairis, *The capture of Constantinople*, 80; Villehardouin, *The conquest of Constantinople*, 48–9; Clari, *The conquest of Constantinople*, 44.

[2] Villehardouin, *The conquest of Constantinople*, 70–71; Gunther, *The capture of Constantinople*, 93–4; Choniates, *O city of Byzantium*, 298–9.

routed.[1] Constantinople, for the first time, was in the hands of a besieger. While various Byzantine contenders for the empire set up minor principalities on both sides of the Bosphorus (based in Nicaea, Trebizond and Epirus), the Latins themselves took over the capital and made Baldwin IX, count of Flanders, emperor, while the Venetians began to consolidate their advantage in the Aegean.

The new empire was narrowly based on an army of adventurers and an unwilling native population. It was also immediately faced by enemies in all directions, with the revived kingdom of Bulgaria to the north providing a hostile neighbour. In March 1205, the new emperor first had to besiege Adrianople (now Edirne), helped by the Venetians. The town was defended with a double ditch, wooden scaffolds erected to raise the height of the towers, and 14 stone throwers. The Latins stormed the first ditch, then filled in the second with great difficulty. Then they moved up siege towers equipped with drawbridges, but both machines failed. One sank in the mud, the other, having been seriously damaged by stones, was set alight. The army appears to have had very few miners, and after a final attempt to construct a new siege tower the siege was abandoned, the Latins deciding instead to attack elsewhere: but their camp and engines were swept away by a sudden rain storm in September.[2] In all these accounts it is clear that the stone throwers used by the Latins were of the traditional kind and that the new trebuchet had not yet made its appearance here.

The tsar of Bulgaria, Ivan I or Kaloyan (r. 1197–1207), had access to siege technology too. In 1207, he in turn besieged Adrianople, setting up "30 great petraries", the blows of which through a day and night bombardment "brought the walls crashing to the earth". Undaunted, the defenders constructed timber defences behind the breaches, with a ditch behind. Choniates states that the breach was actually created by mining. Whatever the truth, the nomadic Cumans (Polovtsy) accompanying the king lost interest and went home, forcing him to raise the siege.[3]

Conquest of the Morea

Meanwhile, other Latins had decided to seek their fortunes further west, with the younger Geoffroy de Villehardouin, nephew of the historian, leading an expedition of some 500 knights and sergeants that eventually conquered most of the Peloponnese (Morea) between 1204 and 1209, Thessaloniki having submitted to Baldwin. The small size of the army suggests that resistance was weak. However, the one surviving chronicle to give a detailed account tells of several sieges. According to the historian, at that time there were only twelve castles in the Morea. At Patras they set up stone throwers, taking the town at the first assault. The

[1] Villehardouin, *The conquest of Constantinople*, 89–90; Gunther, *The capture of Constantinople*, 104–5; Clari, *The conquest of Constantinople*, 83–96; Choniates, *O city of Byzantium*, 313.

[2] Villehardouin, *The conquest of Constantinople*, 119–20; Clari, *The conquest of Constantinople*, 119; Choniates, *O city of Byzantium*, 341.

[3] Choniates, *O city of Byzantium*, 348–9; Villehardouin, *The conquest of Constantinople*, 152.

Byzantine commander Leo Sgouros held out, holding the formidable ancient and Byzantine fortress of Acrocorinth on its high rock. The men of Champagne attacked the town below, using stone throwers from all sides. Their crossbowmen cleared the walls of defenders, enabling an escalade, and the Byzantines withdrew to the rock-top fortress. Villehardouin himself attacked Koroni, blockading it with ships while the low curtain wall was bombarded with artillery. This place and Kalamata both surrendered on terms. Other small castles of the interior surrendered when the Franks arrived and set up their weaponry. Villehardouin then distributed the captured lands. Further siege operations followed against Nikli (on the abandoned site of Tegea, now Aléa or Pialí; sometimes wrongly identified with the ancient Amyclae, now Amíkles, called Sclavochorion in the Middle Ages) and Lacedaemonia (Sparta), walled towns that surrendered after a siege of a few days, when the artillery was set up. In 1210 Corinth itself fell. A siege castle was constructed to the south and another to the north. From these, crossbowmen prevented the Greeks from entering or leaving. The surviving work (known as the Penteskouphi) shows an irregular stone curtain (the present remains being of fifteenth-century date) attached to a typically Frankish square donjon, almost 1500 metres from the triple-walled entrance to Acrocorinth. Surrender followed as a result of the blockade. Nauplia (Nafplio) and Monemvasia were the next targets, both ancient sites on promontories. Nauplia surrendered after two years of blockade in 1212. Monemvasia only fell in 1248, after a three-year blockade, demonstrating the impregnability of its promontory site and the limits of the siege technology.[1] Although the Morea chronicle refers throughout to trebuchets, it was written much later. It includes trebuchets at the sieges of Constantinople in 1203/4, although as we have seen this was not attested by any of the contemporary writers, who were actually there. We may therefore disregard this reference.

While the French adventurers conquered their way across the Morea, the emperor, Baldwin's successor Henry, fought Italian forces holding the citadel of Thebes (1209). He deployed mangonels and crossbowmen against the walls from across the ditch, then a knight built a cat and with it managed to get across the ditch. A mass assault was then launched but driven back by fierce resistance. Emperor Henry then ordered the building of a belfry (siege tower) and wooden shelters and ladders, tall and "well-pegged", and the defenders sought a truce then surrendered.[2]

[1] *Crusaders as conquerors*, 109, 112, 119–22, 126, 131, 154–7. On the deeply disputed origins of this chronicle, which seems to be of the fourteenth century, see Lock, *The Franks in the Aegean*, 21–4. On Kalamata, see Bon, "La prise de Kalamata". On the Penteskouphi, see Bon, "The medieval fortifications of Acrocorinth", 265–7.

[2] Henri de Valenciennes, *Histoire de l'empereur Henri de Constantinople*, 111–14. On the scant remains of the walls of Thebes, Symeonoglou, *The topography of Thebes*, 161, identifies three towers on the curtain of the ancient citadel that may have been rebuilds. The fortress was certainly rebuilt by the Franks after their capture. The previous reference to work on it was a repair by Justinian! (Procopius, *Buildings*, 241.)

Castles of Frankish Greece

The new conquerors, a small number of warriors in a foreign land, turned to the castle to consolidate their lordship. These fortifications are extremely hard to interpret. Dating closely the scanty remains of continually reused material is virtually impossible and there are few documentary records, and later Venetian and Turkish occupation overlays much of this period. The several castles with donjons can reasonably lay claim to belong to the time of the first Frankish conquest, such as that at Mistras (begun in 1249), the redoubt constructed inside the vast circuit of Acrocorinth, and Kalamata. Generally though, the sites are all determined by their topography, consist of a single wall of insubstantial thickness (rarely over 1.5 metres) without any kind of projection at the base, with towers widely spaced. Gates are simple openings in the wall. The castle of Geraki, some 16 miles east of Sparta, is an example (Plate 21). It was the head of a small barony by 1254. The castle has a thin curtain following the topography, with a simple entrance and a postern. Only one wall, which happens to survive well amongst the general ruins of the site, has substantial defences, and this faces the side with the most level approach along the ridge. The crenellated wall has a strong square tower at each end, and the corbels remain for both hoarding and machicolations.[1]

The exception to the general rule is at Chlemoutsi, recorded as being built by Villehardouin between 1219 and 1223, although even here it has been argued that it is fact a great deal later and that other neighbouring castles are meant by the chronicle. However, the apparent discovery of a mint also mentioned in the written sources seems to confirm that Chlemoutsi is the castle in question. It has a hexagonal inner fortification, standing on higher ground over a bailey. This upper building encloses a courtyard some 60 × 25 metres. The roof of the two-storey buildings around the interior provides a wide platform for defence. The outer entrance was unusual in being protected by a portcullis.[2] Large by the standards of its neighbours, it nevertheless cannot compare with the sophistication of the great fortresses then being erected in the west. The poverty of resources and perhaps of available building skills told against any other result.

THE FIFTH CRUSADE: SIEGE OF DAMIETTA

BEFORE THEY DECIDED to found a new set of Latin principalities on the ruins of the eastern Roman empire, the conquerors of Constantinople had planned to crusade by attacking Egypt, having concluded that a direct attempt to recover Jerusalem was unlikely to succeed, whereas an attempt to weaken the

[1] Bon, *La Morée franque*, 1: 643–4. Also Traquair, "Laconia", 263–9.

[2] Bon, "Forteresses mediévales de la Grèce centrale"; K Andrews, *Castles of the Morea*, 1953, 222–4; Müller-Wiener, *Castles of the crusaders*, 82–3; Traquair, "Medieval fortresses of the north-western Peloponnesus", 268–81; Sotiriou, "Le château-fort de Chlemoutsi". See Hetherington, *Byzantine and medieval Greece*, for brief individual descriptions. Bon, *La Morée franque*, contains detailed descriptions and plans of most of the sites of this date.

Egyptian state might yield better results. In 1217, the Fifth Crusade adopted the same strategy, and a large army led by John of Brienne, king of Jerusalem, Leopold VI, duke of Austria, and the forces of the Templars and Hospitallers gathered at Acre. One participant described how the crusaders decided to reinforce the defences of the reduced kingdom of Jerusalem. Caesarea was strengthened, and the castle of 'Atlit (Château Pèlerin, "Pilgrim's Castle") constructed on an ancient site on a coastal promontory. Oliver of Paderborn describes how two towers thirty metres wide were built of "hewn and fitted stone" so large that carts pulled by two oxen were required to convey each one.[1] The headland was cut off with a ditch and two concentric ramparts defended by rectangular towers. The new "Pilgrim's Castle" emulated the Muslim citadels in strength.

The army sailed to Damietta, its defences carefully maintained by the Egyptians, on 29 May 1218. The river Nile was blocked by three iron chains run across the river and protected by a tower constructed on an island in the river linked by a bridge to Damietta, which prevented the crusaders from making any progress up river. Capture of this tower therefore was critical for the attackers, and its defence was likewise critical to the Egyptians, for once it was taken, Damietta could be completely cut off from relief or resupply. The tower was held by 100 crossbowmen, archers and soldiers and was equipped with artillery. An attack was prepared by the duke of Austria with the Hospitallers, who prepared two ships carrying ladders, while the Teutonic Knights and the Frisians prepared three, carefully provided with defensive bulwarks. The assault on 24 June failed. The Hospitaller ladders were destroyed by the defenders and many knights killed, while the Germans' ships attacking between the tower and the town were driven back by incendiaries, though not before their ballistae had caused heavy losses among the Muslims. One account suggests that one of the ship's masts broke, hurling the armoured knights to their death in the Nile. One of our witnesses, Oliver of Paderborn, recorded that the tower could not be taken by "petraries or trebuchets", because it was too far away to be reached from the land, nor mined, nor starved. The Germans therefore joined two ships together (in a style we have now witnessed used several times) and built a wooden castle across the decks, well covered in animal skins, with a strong ladder inside. This equipment was brought against the tower under a ferocious bombardment on 24 August, and from the ladder the crusaders succeeded in storming the island. The fires started by the incendiaries were dowsed with "sour liquid and gravel", and the defenders' artillery was unable to prevent the assault, so the island itself was secured and the tower could be stormed from close by. Without hope, the garrison surrendered to the duke of Austria.[2]

Damietta itself could now be completely isolated, although the crusaders seem to have taken some considerable time to achieve this. The town contained a large

[1] Oliver of Paderborn, *The capture of Damietta*, 56–8. The description is borne out by the remains: Kennedy, *Crusader castles*, 124–7. For a fascinating reconstruction of how the crusade was organised and the personnel involved, see Powell, *Anatomy of a crusade*.

[2] Oliver, *The capture of Damietta*, 65–8; Codagnellus, *Gesta obsidionis Damiatae*, 463–5; "Extraits de l'Histoire des patriarches d'Alexandrie", 241–3.

population and had been stocked with plentiful provisions. The Egyptians were not idle, either, making a number of attempts to break the siege, but they were inept sailors and their efforts to use the Nile for a counterattack failed badly, while land attacks came up against a stronglyfortified crusader camp. The death of Sultan al-'Ādil and the discovery of a conspiracy against his son and successor al-Kāmil led to the abandonment of an early attempt at relief. The crusaders broke down the bridge across the Nile to enable their ships up river, but the Egyptians established a strong field fortification on the western side and deployed petraries and trebuchets, with which they managed to sink a number of ships.

The winter of 1218/19 was harsh, and over the months ahead numbers of crusaders returned to their homes (including the Austrian duke in May 1219). In February 1219, an assault on Damietta was prepared, while the Egyptians organised their own attack from outside. This came on 31 March, assisted with a bombardment from the town of incendiaries that burnt much of the assembled siege machinery. The pattern was to be repeated, sometimes with sorties by the garrison destroying the engines. The crusader camp however proved too strong for the relief forces, although it is recorded how the sultan called for all the jars and pots in Miṣr (Fustat) and Cairo to be sent, to be filled with sand for filling the ditch protecting the camp. Meanwhile, conditions in the town worsened as provisions ran low. Alerted to this by refugees from Damietta, the crusaders pressed their attacks. The ditch before the outer wall was filled in and machines moved up the rampart. But the defenders sallied and burnt the cat. At the end of July, the Pisans and Genoese attacked with scaling ladders and managed to set them up against a tower, but once again the defence succeeded in destroying the ladders. In August, the ditch was filled to enable a "miraculously constructed" cat to be brought up, but it too was destroyed in a sortie, along with seven other machines. In September, the crusaders raised a wooden tower from which they hurled stones and incendiaries into Damietta. By then, plague had struck with devastating effect on the besieged, and the effort of the sultan to send in 1,000 reinforcements, swimming across the river, was discovered and the relief was wiped out. At the beginning of November, an attempt was made to scale the walls, with ten ladders, and the enfeebled garrison was unable to resist any more, the town then falling into crusader hands.[1]

Oliver of Paderborn described the defences of the captured city as being of two lines of walls, with a low wall protecting the ditch, and 28 towers on the inner rampart. Each tower contained two or three "tortoises", all of them unharmed except for one wrecked by the duke of Austria's trebuchet. He described how two tunnels had been prepared, for mining, but this attempt had been frustrated. Among the spoils within Damietta were found four trebuchets and many petraries and

[1] Oliver, *The capture of Damietta*, 73–87; Codagnellus, *Gesta obsidionis Damiatae*, 475–501; "Extraits de l'Histoire des patriarches", 248–53. Among the measures taken by the crusaders had been the erection of a motte-like siege castle on the right bank of the Nile, of sand enclosed with clay only. See Pringle, "A castle in the sand", but also the update in Chap. 6 above.

mangonels, with very strong ballistas and many "hand ballistas".[1] Thus, a town possessing the most modern defences, and equipped (as were the attackers) with trebuchets, had succeeded in resisting attack until defeated by disease. Unlike in previous sieges we have described, there is no suggestion that bombardment had made any serious impact on the massiveness of this new generation of walls. It was also clear that the trebuchet was now part of the armoury of both sides, so must have been introduced at some time between 1205 and 1217.[2]

The crusade was not over. The Egyptian fortification at Tinnīs, with seven strong towers and a double ditch, was abandoned by its garrison, and the Muslim cause was in disarray. The walls of Jerusalem were pulled down, so that if the Christians won the city back, as was apparently expected, they would not be able to hold it. The Syrian forces withdrew to Palestine, where they tried in vain to take the restored defences of 'Atlit. Al-Kāmil made a generous offer of peace, but the papal legate, scenting total victory, turned it down. In 1221, a reinforced crusader army advanced from Damietta towards Cairo. It proved a bad move. Knowing the river much better than the Christians, the Egyptians were able to cut off the army from Damietta as the Nile rose, and brought about an enforced peace, whereby the city was returned to their control.

The Albigensian Crusade

During the early thirteenth century, the church turned its attention to dealing with the heretical sect of the Cathars in southern France, commonly but misleadingly called Albigensians after a public dispute between them and Catholic spokesmen at Albi. Their dualist beliefs had appealed largely to ordinary folk, but a considerable number of local nobility, and in particular the powerful count of Toulouse, showed themselves unenthusiastic at outside efforts to crush them, rightly fearing an end to their virtual autonomy. The crusaders were northern French, led by Simon de Montfort, earl of Leicester, and they demonstrated their mastery of siege techniques as they progressed, wading as in the other crusades through a sea of the blood of the unbelievers. In August 1209, the city of Carcassonne, ruled by Viscount Raymond-Roger Trencavel, was attacked by Montfort. The city then was protected by its late-Roman walls, and the outer suburbs were less well protected than the inner core of the place, the "cité". The defences standing today (heavily restored in the nineteenth century), including the count's castle, largely date from the later thirteenth century: much less formidable obstacles faced Simon de Montfort. The southern suburb was carried by escalade, but the northern held out. So stone throwers (petraries) were hurriedly prepared, and while their missiles weakened the top of the wall and drove off the defenders, miners using a cart as cover sapped underneath and brought it down. The defend-

[1] Oliver, *The capture of Damietta*, 95–6.

[2] See Chap. 9 below.

ers retreated to the inner defences, where, driven by thirst (having lost their access to the water supply), they surrendered.[1]

At the siege of Minerve (June–July 1210), a town protected by deep ravines with a castle with a hexagonal donjon, mangonels and petraries were set up against either side, including one of particular size, "the lady and the queen of petraries". These engines breached the ramparts and smashed the seigneurial palace inside. Cut off from their supplies, including access to the streams outside that provided their water, the town yielded. Immediately following their success here the crusaders laid siege to Termes (Aude), a high-ground castle approachable from only one side. It had a rectangular rampart with circular flanking towers (although these may date from the post-capture rebuilding) and a donjon at the highest point. Simon had the engines transported by road to Termes. An isolated outlying tower had to be assaulted first: the defenders abandoned it when a mangonel was deployed. The attack on the castle itself proved a slow operation, and when the rampart was finally crumbling, the garrison built a new one of wood and stone inside. On the other hand, the besiegers were now able to bring their artillery much closer, and whereas "previously they had achieved little, now they shot with accuracy".[2] In the spring of the following year they besieged Lavaur (Tarn), and battered the walls with artillery from 15 March to 3 May. In order to complete the encirclement, the crusaders had to bridge the river. Capture was brought about when the crusaders managed to manoeuvre a cat across the ditch, and when the count of Toulouse arrived, reluctantly, in the crusader camp. The defenders, having at first believed that he was coming to their relief, were in despair, and surrendered at discretion.[3] This capture and the subsequent burning of 300 Cathars led to their less frequent use of fortifications as a means of defence. The remarkable mountain-top castles of Lastours, north of Carcassonne, at which there are now four separate but adjacent constructions, held out at the beginning of the crusade, but their lord negotiated an amnesty during 1211.

In the autumn of 1211, the Toulousains besieged Simon's garrison in Castelnaudary. The count's mangonel "caused us little trouble", said Peter, but then he had "an enormous siege engine made ready which was to be used to knock down the walls". The *Chanson de la Croisade albigeoise* called it a trebuchet. However, at first they could not find any stones to hurl that did not break up under the force of the impact. Then they located three suitable missiles. Their first shot destroyed a tower, their second ruined a hall, but the third shattered. Abandoning the siege,

[1] Petrus Sarnensis, *History of the Albigensian Crusade*, 53; Guilhem of Tudela, *La chanson de la Croisade albigeoise*, trans. Martin-Chabot, 1: 66 (there is also a more recent translation by Gougaud). These are two eye-witness accounts. There is now an English translation, *The song of the Cathar wars*. Guillaume of Puylaurens, *Chronique*, is not contemporary, but contains much detail. There is now an English translation by Sibly and Sibly.

[2] Petrus Sarnensis, *History of the Albigensian Crusade*, 83; Guilhem of Tudela, *La chanson de la Croisade albigeoise*, trans. Martin-Chabot, 1: 113–14, 136–7; Guillaume of Puylaurens, *Chronique*, 35.

[3] Petrus Sarnensis, *History of the Albigensian Crusade*, 133; Guilhem of Tudela, *La chanson de la Croisade albigeoise*, 1: 162–6; Guillaume of Puylaurens, *Chronique*, 38–41.

the Toulousains either left, or burnt, their trebuchet, whose first arrival in the war had certainly made an impression, if not the one intended by its owners.[1]

Further sieges were recorded. Montfort attacked the count of Foix at Saint-Marcel-Campes (January–March 1212), where a bombardment of stones from the petraries drove back a sortie. Hautpoul, on a crag, was a donjon the defenders of which fled after four days, following a bombardment (April). Penne-d'Agenais was another hilltop castle previously strengthened by Richard the Lionheart, this time standing on a sheer crag above the village on one side and the river Aveyron on the other, with a lower court, and a donjon in the upper ward. The defenders' stone throwers caused losses among the crusaders, whereas the attackers' damaged the houses but failed to harm the walls. Finally, the besiegers set up a bigger engine that began to damage the ramparts and the place surrendered, the defenders negotiating successfully for their lives because the besieging army was continually weakened as contingents of crusaders left after completing their agreed 40 days' service. At Moissac (August–September 1212), the crusaders took the place using artillery, a cat and a battering ram, the first time that use of this weapon is recorded in this crusade: perhaps the terrain had prevented its use at other sites.[2] Capture was effected when part of the wall collapsed outward into the ditch.

Most of the castles involved were reconstructed according to modern principles following their capture, but it would seem that in the main they were relatively simple, with rectangular donjons within single curtain walls without flanking towers, and relied as much on their location as on the strength of the stonework, as the local lords responsible for the places were not wealthy. The count of Toulouse was in a different league, and his castle at Toulouse itself, the Château Narbonnais, was a more powerful citadel (now disappeared beneath the Palais de Justice).[3]

In 1213, Montfort pressed his attack on Toulouse itself. The large city could not be closely attacked with the forces at his disposal, but he placed a garrison in the castle of Le Pujol (now part of Sainte-Foy-d'Aigrefeuille) to harass the inhabitants and prevent them collecting in the harvest: another use of the siege-castle technique. The count of Toulouse captured it with engines. King Peter II of Aragon (r. 1196–1213) rallied the forces of Languedoc against Simon, but the latter demonstrated his military prowess by pulling off a great victory at the Battle of Muret, in which the king was killed, and as a result the political boundary of Iberia was pushed back to the Pyrenees. The crusaders extended their operations against the Cathars into the Agenais, and in 1214 attacked the castle of Casseneuil, surrounded by a number of streams, but on level ground. Montfort's artillery

[1] Petrus Sarnensis, *History of the Albigensian Crusade*, 133; Guilhem of Tudela, *La chanson de la Croisade albigeoise*, 1: 216–17, 238–9. Guillaume of Puylaurens, *Chronique*, 44–5, gives a different account, attributing the crusader success to a victorious skirmish outside the walls.

[2] Petrus Sarnensis, *History of the Albigensian Crusade*, 145–59; Guilhem of Tudela, *La chanson de la Croisade albigeoise*, 1: 258–76. Mesqui, *Châteaux forts et fortifications*, 285–6.

[3] For summary descriptions of some of the sites, see Salch, *L'atlas des villes et des villages fortifiées*, 123 (Lastours), 130 (Termes), 341 (Minerve); and now Cowper, *Cathar castles*.

bombarded continuously, destroying buildings inside and damaging the walls, but the critical step was the construction of a siege tower, well protected with hides against the defenders' petraries, five floors high. From within, crossbowmen cleared the walls of defenders and enabled the crusaders to fill the stream and eventually to cross it. The remarkable wooden tower was also got across. It then had to be defended against vigorous attempts by the defenders to destroy it. Some of the defenders fled and the rest were massacred as the attackers forced their way in.[1]

Despite the victory of the crusade (with Simon appointed to replace Raymond as count), there was a significant revolt by the people of Toulouse, their former count Raymond, and other southern barons. Raymond's son was welcomed into the town of Beaucaire on the Rhône and besieged Montfort's garrison in the castle, forcing it to surrender despite a relief attempt by Simon. During the attack, petraries bombarded the castle, smashing buildings and breaching the walls, and demolishing battlements and galleries. Meanwhile, Simon was in turn besieging the besiegers in the town, ordering the construction of a cat and a high siege tower, protected by a stone thrower in front of the town gate. A battering ram was deployed but was thwarted by the defenders' catching up the ram in a rope. An effort to sap the wall was defeated by the defenders' hurling down bags of sulphur, the suffocating smell driving off the crusaders. A final attack with the cat was resisted when the southerners set it alight by hurling pots of burning asphalt at it. Although the machine was apparently rescued, an attempt to storm Beaucaire via its now demolished gate was repelled, and the castle was then left to its fate.

The crusaders had previously ordered the demolition of the walls of Toulouse, using battering rams to topple the rampart into its ditch, and when the rebellion broke out in 1217, they retreated into the Château Narbonnais, while the citizens had hurriedly to reconstruct defences around the city against the inevitable counterattack of Simon de Montfort. At the same time they deployed trebuchets and mangonels in an effort to capture this citadel with "huge stones", while it is also gruesomely recorded that captured prisoners were cut in pieces to be hurled back by a trebuchet. The Toulousains had access to experts, and the *Chanson de la Croisade albigeoise* records the names of the master carpenters Garnier and Bernard Parayre as constructors of trebuchets. There was, of course, plenty of manpower available to the defence, but they had to hold extemporised earthwork defences against one of the military geniuses of the age. All winter long they held on. Simon ordered the building of a cat to help the crusaders fill and cross the ditch. Realising the dangers, the Toulousains aimed their stone throwers at it, and trebuchet shots were said to have smashed into the device. But it was a mangonel operated by women that hurled the stone that was responsible for crushing the head of Simon de Montfort on 25 June (see Plate 29).[2] His son Amaury retreated to Carcassonne, a month later, after one last assault had failed, abandoning the citadel at the same time.

[1] Petrus Sarnensis, *History of the Albigensian Crusade*, 234–7.

[2] Ibid., 274–6; Guillaume of Puylaurens, *Chronique*, 56–69; continuation of Guilhem of Tudela, *La chanson de la Croisade albigeoise*, 1: 213–51 (Beaucaire), 303–433 (Toulouse).

It would take further crusades, led by the prince, presently King Louis VIII, who succeeded to the throne in 1223, to crush Raymond of Toulouse and in the process to bring about the effective integration of Languedoc into the Capetian kingdom. In an initial attack in 1219, Louis brought a vast army to besiege Toulouse, where once again the defence prepared its arsenal of stone-throwing weaponry, its petraries and "calabres" (the *cabulus* already mentioned), and once again called on masters Parayre and Garnier for its trebuchets. Meeting strong resistance, the prince withdrew on this occasion. But he returned in 1226, advancing down the Rhône, using the river for transport, besieged Avignon with petraries, and took the city after three months. The walls were then demolished as punishment. The county of Toulouse fell into royal hands, and once again the city of Toulouse's defences were demolished. The final defeat of the Cathars would take rather longer and involve further sieges. In 1243, the mountain-top refuge of Montségur was the scene of the final operations, and the castle, originally no more than a stone wall around the top of the crest (the surviving fortifications are of the sixteenth century), fell to assault, with the last Cathars being burnt.[1] After a fruitless blockade, intrepid crusaders had scaled the height and stormed a watch post at the far end, 600 metres from the castrum. From here, they pressed forward and erected trebuchets to bombard the "barbican", a hypothesis attested by the excavation here of numerous stone shot (and crossbow bolts). When this was taken, the artillery was moved into it, to bombard the fortress proper, around 100 metres away. A major building programme would be undertaken during the century in the immense territories now newly and firmly part of the Capetian realm.

Al-Andalus and North Africa

The Iberian peninsula remained divided between Christian and Muslim rulers, the latter now with their main base in Morocco rather than al-Andalus. In the west of the peninsula, since the conquest of Lisbon in 1147, the new Portuguese kingdom had become established, initially as an offshoot of the royal house of Castile. King Afonso I (r. 1128–85) had a new royal castle built, and rebuilt, at the frontier site of Leiria, in which people were encouraged to settle and a worthy warrior was appointed to command. In the 1190s the settlement was also walled. Other frontier castles such as Alcocer were captured, and in 1168, the town and castle of Évora, which was entrusted to the military Order of Calatrava. Towns were taken by guerilla attacks at night, Beja fell to an assault with wide scaling ladders (December 1167), and was held onto for four months, and the castle of Trujillo also fell (1165), along with Cáceres (1166).[2]

[1] Guillaume of Puylaurens, *Chronique*, 70–71 (1219), 78–9 (1226), 86 (1227), 113–14 (1243); continuation of Guilhem of Tudela, *La chanson de la Croisade albigeoise*, 1: 461–71 (1219); on the interpretation of events of the final siege, see Peyrou and Henry-Claude, *Montségur*.

[2] Ibn Ṣāḥib al-Ṣalāh, *Al-Mann bil-Imāma*, 134–8; *Annales D. Alfonsi Portugallensium regis*, 158; *Chronica Gothorum*, 15–16. Ibn Ṣāḥib al-Ṣalāh was a servant of the ruler Yūsuf I.

Meanwhile, the Almohads did not entirely neglect their Iberian possessions. The "Moorish castle" at Gibraltar dated from around 1160, when engineers, carpenters, workers and soldiers were organised from all over al-Andalus to begin construction on the fortification, complete with palaces and houses inside the high strong walls, that now stands (much altered in later centuries) above the town.[1] Building work was also carried out under the rule of Abū Ya'qūb Yūsuf I (r. 1163–84) at Córdoba, and also at Seville, including on the defences of the citadel. Little now remains of the Almohad work in the palatial Alcázar of Seville, although the outer wall is reminiscent of the walls of the Cairo and Damascus citadels, but a significant stretch of the surviving city walls dates from this period of reconstruction, with strongly projecting rectangular towers. On the south, the lower stories of the Torre del Oro (Plate 22) remain from an extension of the city wall from the alcázar to the river Guadalquivir in about 1220, when it functioned as one end of a chain used to block the river. Such works were partly to enhance the prestige and authority of the Almohad regime in a land prone to localist revolts. But they also served to bring up to date the defences of cities that now, as a result of the Christian advances, were close to the frontier. Less substantial fortresses also date to this era. The fortified village at Planes in Alicante had an early-thirteenth-century hilltop castle, of *tapial* construction. Its original entrance has been recovered under a late medieval rebuild, and consisted of four separate gates passing through two of the nine mural towers, climbing a narrow ramp before turning at right angles to enter the interior.[2] Even less grand than these works, but still having much practical purpose, were new refuges. That at Castelo Belinho near Portimão in the Algarve, was a simple rectangle provided with a single tower to protect the entrance, and a cistern, dated to the late twelfth century.[3] In the Murcia region has been found what appears to be a fortified village with a defensible barn, closely copying a type found in Morocco, dating probably to not later than 1247.[4]

Fighting took place along the whole frontier. Badajoz had to be rescued from Christian attack by an army sent by the Muslim prince in 1169. Further west, the Muslims laid siege to the town of Huete (1172). It had a deep ditch dug out of the hard rock, defended by a tower "of ancient construction". They were on the verge of taking it when sudden rains refilled the place's empty reservoirs, an event naturally (by the Christians) ascribed to divine intervention.

Chronicles provide little detail of operations such as Yūsuf I's unsuccessful siege of Santarém (1182).[5] In 1185, however, the information position outside the peninsula improves: a Muslim leader captured Mallorca, catching the garrison in the open then assaulting the citadel with mangonels and other machines of

[1] Ibn Ṣāḥib al-Ṣalāh, *Al-Mann bil-Imāma*, 21–3; Maqqarī, *The history of the Mohammedan dynasties in Spain*, II: 314.

[2] Menéndez Fueyo, "La puerta del Castillo de Planes", 153–74.

[3] Antuña, *Sevilla y sus monumentos árabes*, 48, 61–8; Collins, *Spain*, 258–60; Oliveira, "Uma fortificação islâmica do termo de Silves", 39–41.

[4] De Meulemeester, Amigues and Matthys, "Un grenier fortifié murcien", 129.

[5] Ibn Ṣāḥib al-Ṣalāh, *Al-Mann bil-Imāma*, 137–8 (castles of the Algarve), 144, 156 (Badajoz, twice), 218 (Huete); Zarkashī, *Chronique des Almohades & des Hafçides*, 17; 'Abd al-Wāḥid al-Marrākushī, *Histoire des Almohades*, 224; *Chronicon lusitanum*, 425–8.

war, to establish an independent kingdom of his own. A couple of years later, the Almohad ruler was confronted by a revolt in Gafsa in Ifrīqīya (Tunisia). A letter from the emir Ya'qūb reporting his victory back to Marrakech survives. He recounts how he decided to build siege engines against the town, having pacified the surrounding countryside and secured adequate provisions for his army. The mangonels (*manājaniq*) set up a 24-hour bombardment of rocks and incendiaries, under cover of which the walls were mined and stretches brought down. The ditch was filled, and a mobile tower (*burj*) was brought forward. The town then surrendered and had its walls razed as punishment, and one source says there was a massacre.[1] The main concern of the Almohad rulers really was in North Africa: Yūsuf's son and successor Ya'qūb al-Mansūr (Moulay Yacoub) turned the fortress of Rabat into his capital city at this time, creating a great circuit of modern walls, the mosques and the gates.[2]

In 1184, the Christian chronicles record the assembly of a great army from all the kingdoms to march from Seville, and the preparation of machines "for undermining walls". Silves was taken and retaken by the Muslims. In 1195, King Alfonso VIII of Castile was defeated at the Battle of Alarcos near Ciudad Real. Defeat meant the Castilians lost control of much of their southern territory, as Calatrava and many castles fell to the Almohads, including Alarcos itself, where survivors of the battle had taken refuge but surrendered and were exchanged for Muslim prisoners. The new order of monastic knights had been established at the castle of Calatrava (la Vieja) around 1164 and had built a fortress around the hilltop above the river Guadiana. The town was protected by a single wall with 42 towers, and a dam was used to divert the river into the ditch. The citadel had a double wall facing the town, and large rectangular towers have been assigned to the Almohad occupation from 1195 to 1212.[3] The victors did not press their attack on Toledo, and in 1198 they marched as far as Madrid, which stood firm against a not very determined attempt at a siege: it sounds as if the Almohad ruler knew that these cities would be too well-provided for and defended and that he did not have the resources or the time to take them. Instead, a truce was arranged.[4]

Las Navas de Tolosa and its consequences

In 1204, the new Almohad ruler, al-Nāṣir, faced a revolt in Ifrīqīya, when a number of places declared loyalty to Yahyā ibn Ghānīya, the ruler of Mallorca. The Almohad army swiftly brought most of the towns to heel, but Mahdia, in its strong location, resisted and had to be besieged. This took four months before surrender

[1] "Un recueil des lettres officielles almohades", 62–3 (letter of 14 January 1188); 'Abd al-Wāḥid, *Histoire des Almohades*, 237; Zarkashī, *Chronique des Almohades*, 19.

[2] 'Abd al-Wāḥid, *Histoire des Almohades*, 230.

[3] Collins, *Spain*, 95–7.

[4] 'Abd al-Wāḥid, *Histoire des Almohades*, 243–70; *Anales toledanos I*, 394; *Crónica latina de los reyes de Castilla*, 14–15 (Alarcos), 19 (Madrid), 20 (Vitoria etc); *Primera Crónica general de España*, II: 634; Ḥimyarī, *Al-Rawḍ al-mi'ṭār*, in *Christians and Moors in Spain*, III: 135–7.

(January 1206), during which time al-Nāṣir organised to increase and repair his armoury of mangonels, rams and scaling ladders.[1] Five years later, he was back in Iberia, besieging the castle of Salvatierra near Calzada de Calatrava, which fell after a blockade of three months. The Christian sources note that siege weapons of "wonderful size" were used at this siege, and much damage was done to the walls and towers. Alfonso authorised the place's surrender because he could not relieve it. The following year of 1212, a combined army of crusading Christians marched into Muslim territory and laid siege to Calatrava, defended vigorously by its garrison with the aid of many instruments. Queen Berengaria, writing to her sister, the wife of Louis VIII, explained that it was hard to take unless it could be undermined. However, to do this would render the fortress indefensible for the monk-knights of Calatrava who owned it, so (according to the queen) this was why the Christians accepted an offer of surrender. Soon after, the Muslim army suffered a catastrophic defeat at Las Navas de Tolosa, a victory that enabled King Alfonso of Castile to recover lost territory as the fleeing Muslim army abandoned castles it knew it would be unable to hold against the victors, such as Alarcos, and major towns such as Baeza and Úbeda also fell to the northerners. The latter was destroyed because the Castilians did not have the resources to repopulate it. The Toledo chronicle records the campaigning of the king, and then of the archbishop of Toledo, to capture (or sometimes fail to take) neighbouring castles using an array of weaponry identified by its Arabic names.[2] One of the victors of Las Navas de Tolosa was the same King Peter of Aragon who, the following year, would switch from being a crusader to dying at the hands of that other contemporary crusader, Simon de Montfort, at the Battle of Muret.

The victory proved decisive in the long term as well as the short. The next few decades would witness, in effect, the end for al-Andalus, as power fell back into the hands of individual local rulers and further advances by the Christian Spanish kingdoms would capture the great cities of Córdoba and Seville. For the years just summarised, the absence of detail in much of the history, and conflicting accounts of the course (although not the outcomes) of several of the siege operations make it impossible to assess properly the siege warfare of the time, or to identify the first appearance of the counterweight trebuchet, with the solitary clue about weapons of exceptional size deployed by the Almohad army at Salvatierra being consistent in terms of date with the other information for the western Mediterranean world, but it is uncorroborated, and the word itself does not appear in the sources yet. The combined role of the castle in frontier areas — as centre of government, refuge, settlement and trading centre, and military defence — is again confirmed by what we have learned about the nascent kingdom of Portugal's southern and

[1] Ibn ʿIdhārī al-Marrākushī, *Al-Bayān al-mugrib*, 233–4; Zarkashī, *Chronique des Almohades*, 22; ʿAbd al-Wāḥid, *Histoire des Almohades*, 277.

[2] *Crónica latina*, 23–36; letter of Queen Berenguela (Berengaria) in *Christians and Moors in Spain*, II: 16–23; *Primera Crónica general*, II: 687–93; *Anales toledanos I*, 397–401. The equipment was "almajanequis, algaradas . . .", along with wooden towers. There were hispanicised names of familiar Arab words of siege equipment, in this case *manjaniq* (mangonel), *algarada* (ballista). (*Diccionario crítico etimológico*, I).

eastern boundaries and the Muslim works set against them. In a swathe of territory that represented the borders between Christian and Muslim Spain, castles were the normal target for any military operation designed to secure territory (as opposed to the policy common to all parties of raiding and devastating the enemy's countryside). Held by proper garrisons, castles built to the most modern standards showed themselves able to resist even major armies for months, but would always fall in the end if relief was not forthcoming. Following defeats that would rule out any prospect of relief, many were abandoned completely, as in both 1195 and 1212. In the same way as in the Middle East, important frontier castles such as Calatrava might be entrusted to the military orders to provide a firm occupation, and one which was additionally paid for by someone other than the crown.

Central Europe

On the death of the emperor Henry vi, the son of Frederick Barbarossa's defeated rival Henry the Lion secured election as Otto iv, and had in the process to fight a civil war with his foes led by Barbarossa's son Philip of Swabia, while the future Frederick ii, Barbarossa's grandson, was still a young child in Apulia. The conflicts also continued relentlessly in northern Italy, with routine expeditions despatched by rival alliances of great cities to devastate their opponents' lands, and from time to time to lay siege to some of the dense network of rural castles, often involving the deployment of mangonels and petraries, and the capture and burning of the towers and refuges. In 1199, a more significant expedition was led by Boniface i, marquis of Montferrat, alongside the forces of Milan, Brescia and Como to take many castles, and later that year a Milanese-led army destroyed the Cremonese "new castle", burning the donjon ('turrem am domigniono'). The Cremonese came to its relief, and set up a fortified camp by the river Adda, and deployed siege towers, petraries and *trabuchis*. It appears that there followed a major artillery bombardment across the river, with the Milanese responding with mangonels and petraries, but later withdrawing. This is certainly the earliest reference to the appearance of the trebuchet in action in Europe, although a diminutive form of the name, *trabuchellus*, has been identified in a diplomatic document of 1189 from Vicenza.[1] Although this chronicle records year by year the military activities of the cities and their annual siege operations with mangonels, petraries, rams, siege towers and bridges for crossing rivers and ditches, it does not record the trebuchet again until 1229.[2] By this time, it has also been recorded elsewhere in Italy. At Genoa, an ingenious builder was set to build a *trabucho* in 1224, in a description that strongly implies that this was still a rare skill, and the following year the Genoese attacked a castle that could not be

[1] *Codice diplomatico Eceliniano*, 97. The text states unequivocally "nec cum mangano, nec Trabuchello, aut cum prederia, vel ballista vel Archu traham". The reference, which is not listed in any of the established glossaries of medieval Latin, was identified by Chevedden, "The invention of the counterweight trebuchet", 101.

[2] *Annales placentini guelfi*, 420–48.

reached by ladder, so they built a trebuchet very quickly, and within 18 days the castle surrendered, its walls battered (literally, scourged) and its cistern destroyed by bombardment from the weapon.[1] Clearly, this artillery was a wall-breaker.

Further north, the first reference to the new weapon occurred in the long civil war between Otto IV and Philip. In 1212, the emperor invaded the hostile duchy of Thuringia, where the landgrave Hermann I was on the opposite side. Otto took and destroyed two fortresses, including that of Langensalza, then besieged Weissensee. He captured the town "as far as the citadel" using the instrument of war "which is commonly usually called a tribuk".[2] The chronicle entry suggests that the term may have been in common parlance for a while. The date ties in with the other information reported before, of an adoption of the new instrument after 1205 and before 1218 in much of western and central Europe.

At the same time, the expansion of German rule towards the east was continuing, partly clearing new lands for settlement and partly by conquering the Slav tribes that dwelt east of the river Elbe and along the shores of the Baltic, peoples whose continued adherence to pagan belief systems gave a convenient justification for campaigns against them. The available material, and skills, inevitably meant that much of the building continued to be in timber, although the settlers would bring with them masons and other craftsmen and, where stone was unusual, brick could be a more solid medium for building, as in Denmark. Timber anyway was a regular material for siege operations of all kinds. In 1190, Count Adolf III of Holstein used vast quantities of it to block the river at Lübeck to bring about its surrender. In 1203, King Valdemar II of Denmark (r. 1202–41) passed through this town on the way to lay siege to Lauenberg/Elbe. The Dane had the resources to set up engines including artillery, but the fortification was strong enough to resist direct assault, eventually surrendering after a long blockade. His predecessor King Knud VI (r. 1182–1202) had besieged Wolgast, but had also failed in a direct attack, and on this occasion it was the Danes who ran out of provisions and had to give up the siege.[3]

Castle building in Germany

Throughout the Staufer period, there was a consolidation taking place of the power of the great regional lords and bishops, of which new castles became strong visible symbols of hereditary ownership as well as centres of power. Royal power, in the crown lands, was also exercised by royal officials and *ministeriales* building their own castles. Growing prosperity and increasing population created the resources and provided the impetus for a significant building boom. As part of their growing autonomy, the greater lords obtained the right to build fortifications on their own allodial lands without obtaining a licence from the king, a freedom to be increased during the royal crisis of the post-1198 civil war, and ratified by Frederick II in 1220 and again in 1231.

[1] Marchisius the scribe, *Annales*, 155; Bartholomaeus the scribe, *Annales*, 158.

[2] *Annales Marbacenses*, 172.

[3] Arnold of Lübeck, *Chronica Slavorum*, 238; *Knytlinga saga*, 177 (Knud).

Increasingly, German castles were now of stone. As before, their shape would often be determined by their location on a spur of land that could be cut off with a ditch and enclosed with a rampart. Unlike elsewhere, corner and wall towers remained rare, flanking towers not becoming common until the thirteenth century, and gateways were usually simple passages. Arrow slits made their appearance at this time too. The function of the great tower began to change, with the *Bergfried* turning, possibly into a simple last-resort refuge, or perhaps a purely symbolic statement of lordship, while the castle's lord found more comfortable accommodation in a great hall. The *Bergfried* would often stand at the most vulnerable point of the enclosure, and could be provided with temporary shooting positions. From the second half of the twelfth century, square or pentagonal towers equipped with a triangular beak-like projection became popular. Especially in the castles of the greater lords, rusticated masonry became very popular. The round-tower designs advanced by the Capetians did not find many imitators in Germany, where instead the crown would, under Frederick II, advance a rectangular-plan castle with massive rectangular towers. Frederick's predecessors had also been great builders and renovators, and Barbarossa's palaces across the empire reflected the great power and glory of the emperor at Nijmegen, Goslar, Ingolstadt, Kaiserslautern, Hagenau (now Haguenau), Gelnhausen and Kaiserswerth (Düsseldorf), among others. All had strong *Bergfriede* and curtain walls as their defensive elements, but most of the effort went into massive and decorative palatial buildings the similarity of which to Cistercian monastery styles has been noted. Babenhausen (Hessen), began around 1200, was an unusual design, being a rectangle with a square *Bergfried* in the middle, the shape permitted by the lowland site. Wimpfen, founded by Barbarossa above the river Neckar, was a complex 215 metres long and up to 88 wide. There were two massive square towers, one at each end, and a powerful curtain wall enclosed a complex of palace, chapel and accommodation.

Town walls

The Staufer founded new towns, and such towns were also provided with town walls. Towns that would be loyal to the emperor were a part of the political strategy of the imperial family. Several cities were ordered by Barbarossa to provide themselves with walls: Aachen (1172), Ulm (1179), Donauwörth. Elsewhere, the construction could be the result of the initiative of the town's lord, or might come from the citizenry itself, and the citizens would be the people who would take responsibility for organising the watch over the new barriers. Other towns starting to equip themselves with stone walls during this period were Halle (1150), Leipzig (1156), Vienna (1172), Brunswick (1175), Eisenach (1180–90), Koblenz (1182) and Erfurt (1170). Archbishop Johann I of Trier (1190–1212) had his city walled and gated beginning around 1200, perhaps as a means of facing down the neighbouring secular lords who were his rivals, while Philip I, archbishop of Cologne, had the walls of his town of Soest significantly extended from 1179, and of Cologne itself from 1180. Great lords were increasingly seeing the foundation of towns as a way of increasing their power and of cashing in on economic prosperity, but walls were expensive and slow to construct: in many cases, the first construction would be of an earthwork and palisade. Sometimes, this would suffice for a very long time indeed.

Often, the provision of stone walls for towns accompanied the freeing of the town's status, another development now taking place across the empire. Lübeck, founded by Henry the Lion, became free when he fell in 1180. Its walls appear from archaeological evidence to have been demolished and replaced by new stone defences early in the thirteenth century. The walls of Mainz, destroyed at the emperor's orders in 1165, were rebuilt by the citizens in 1200. The walls might of course have a variety of functions. Walls were associated with the status of *civitas* and were a kind of proof of city rank. Control of access and egress had always been a motive for enclosing the site. But a status symbol did not need deep ditches and frequent flanking towers, such as were provided at considerable numbers of towns during this period including Cologne and Strasbourg.[1]

The Baltic

Scandinavians and Germans were beginning to settle in the lands of Courland and Livonia, not without resistance from the natives. The technological superiority of the invaders was, for some years, a significant benefit to them as they confronted the relatively primitive earth-and-timber forts of the Livonian and Lithuanian Slav tribes of the region. The accounts given by the conquerors tell of the armoured Christian soldiers filling in the ditches and storming and burning the forts. At Mežotne (Mesothen) in 1220, the Christian forces arrayed siege equipment against the fort of the Semigallian tribe. A wooden siege tower was prepared, as the ditch was filled in under cover of a bombardment. Stones and arrows caused heavy casualties among the defenders. A large stone thrower was erected to hurl great rocks at the rampart, the duke himself taking charge of the weapon. The first shot hit the wooden gallery on the rampart, crushing it and its unfortunate occupants. The second shot dislodged the timbers of the palisade itself, and the fort was captured.[2] It would not be long before the technology was adopted by the local rulers. The crusaders naturally set about securing their rule by building their own forts and castles. It was not long before the king of Lithuania was recorded as deploying four stone throwers of his own against the Livonian monk-knights' castle of Dünaburg (Daugavpils), although he had to abandon the attack after four weeks, and leave his weapons behind.[3]

[1] Meckseper, "Wärmequellen", 83–103; Hotz, *Pfalzen und Burgen der Stauferzeit*, 53–107, and inventory; Haase, "Die mittelalterliche Stadt als Festung"; Planitz, *Die Deutsche Stadt im Mittelalter*, 106 (Mainz), 153–5 (Barbarossa), 229–35 (defences), 241 (ditches), 244–9 (towers, gates); Glaser, "Die Lübecker Stadtmauern", 194–6; Marx, *Die Ringmauern und die Thore der Stadt Trier*, 22–6; Schwartz, "Soest".

[2] Henricus de Lettis, *The Chronicle*, 180–81; *Livländische Reimchronik*, 136 (ll. 5935–53; capture of Sinteles, probably near Goldingen, now Kuldīga), 157–8 (ll. 6830–73; Grobiņa; Merkes, probably Markaišu near Aizpute), 184 (l. 8030; Mežotne).

[3] *Livländisches Reimchronik*, 188 (ll. 8213–27), 189 (l. 8269). The stone throwers here were called *Bliden* (the German for trebuchet).

Russia in the Twelfth Century

The arrival of the Danes, Swedes and Germans on the eastern shore of the Baltic also brought them into contact with the Christian Russians, although the sharing of religion did not lessen conflict when it came to establishing a foothold in these rich lands, which also controlled prosperous trade routes. The character of the countryside of Russia and its available materials, and the fact that the most common enemies were nomads, naturally had an influence on the defences deployed by the Russian principalities, each based in a major city. By and large, foot soldiers were only of importance in northern Russia until the growth of more settled cultures in the south, where the princes of Kiev had set about organising the construction of defences against the raids of their enemies during the twelfth century. Some took the form of small round forts, usually with an internal diameter of less than 100 metres. Several hundred places were called towns although they were in fact little larger than villages, but there were perhaps one hundred defended settlements in Russia with an area greater than one hectare by around 1200. Mostly, defences were weak, consisting of an earthen bank and timber palisade without towers. In some cases the banks were a mere one metre high, although in larger towns they could be up to ten. Castles would arrive with the establishment of feudal relations in Russian society, and *goroda* begin to appear in the chronicles. The princes also built their own private defences. The noble *goroda* often had to double up as refuges for the rural population. Although brick and stone were known, and indeed used in the big cities likes Kiev and Novgorod, building was overwhelmingly done in timber. Banks were preceded by ditches, and the embankment might also be strengthened with wooden beams.[1]

The shortage of contemporary chronicles for the eleventh to thirteenth centuries (and the fact that several of them have never been translated from Old Russian) limits access to the original sources. There is an often-quoted story by Henricus de Lettis (Henry of Livonia), describing how at Holm (Kirchholm, now Salaspils) in 1206, "the Russians made a little machine like that of the Germans, but not knowing the art of throwing rocks, they hurled them backwards and wounded many of their own men."[2] The implication that all Russians were similarly unfamiliar with stone throwers is not necessarily confirmed by this, although a thorough search has failed to provide much convincing evidence of sophisticated artillery in use before the thirteenth century.[3]

Siege techniques do appear to have been basic at a time when they were had advanced quite substantially both across Europe and in the Mediterranean. In 1181, the town of Noviborg (Torzhok) was taken by the forces of Murom and Ryazan only after a blockade of five weeks. In 1214, Prince Mstislav Mstislavich and

[1] Rappoport, "Die altrussischen Burgwalle", 71–86, summarises research and excavations; Kirpichnikov, 'Die russischen Waffen des 9.–13. Jahrhunderts", 46–65.

[2] Henricus de Lettis, *The Chronicle*, 63.

[3] Kirpichnikov, "Metatel'naya artilleriya", 77 (in Russian). I am deeply grateful to Mr Boris Krivchenko for organising the translation of this significant article.

the army of Novgorod fought against the pagan Chuds, which involved capturing their "forest fortresses". These were earth-and-timber forts of the kind frequently met before. In the accounts of sieges of Vladimir (in Volhynia) in 1097 and 1099, long blockades were the only form of attack, or on other occasions direct assaults under cover of archery. In 1116, Mstislav Vladimirovich (afterwards grand prince of Kiev) increased the size of Novgorod by laying the foundations of a new wall, while it was considered noteworthy that a stone fortress was begun at Staraya Ladoga. In 1123, Volhynia was again targeted by an army including Hungarians, Poles, Czechs and Russians, but Vladimir prepared itself by laying in ample provisions and the would-be besiegers withdrew. In 1184, the Russians captured the Muslim engineer who was accompanying a Polovtsian raiding army to build what are described as giant mechanical crossbows (i.e. ballistas) to throw incendiaries. Kirpichnikov suggests a connection between this capture and the events of the following year, recorded in a literary composition known as *The song of Igor's campaign*. Although describing a catastrophic raid on the nomadic Cumans or Polovtsy in 1185, the text (penned soon after the events) contains clues about the knowledge and extent of siege techniques. An obscure phrase ("live bolts") has been interpreted as signifying Greek Fire, and certainly appears to refer to some kind of incendiary device, while another line refers to "hurling weighty missiles over the clouds", suggesting stones hurled from catapults. Six large metal bolts (weighing more than two kilogrammes, and 1.69 metres long) were discovered buried in the tomb of a Russian prince who died in 1164. They would seem to have been built to operate as incendiaries for projecting from a large ballista and their association with royalty suggests both a rarity and a particular status.

Beyond this, the evidence cannot take us. It would certainly be very strange if the Russians, especially those in regular contact with Byzantium, were ignorant of such weapons, it may just be that, as Kirpichnikov suggests, they had little cause to use them at this time. He cites examples from chronicles of the mid twelfth century and from the eighteenth-century history by Tatishchev (again much later but using lost sources) that refer, unspecifically, to "engines" and a wall being breached at Novgorod-Seversky in 1152. The Greek Fire and incendiary devices identified from *The song of Igor's campaign* would also have required throwing machines, to propel them.[1] In 1219, we are told that a prince fortified a church in Halych, in which he was following the example of many western rulers before him. On this occasion, Mstislav Mstislavich was outraged by the defilement and captured it. Ten years later, the Russian princes attacked Kalisz. The siege technique on this occasion consisted of trying to fill in the ditch and trying to set alight the drawbridge and gate, which the Poles just managed to prevent. The Russians

[1] *The chronicle of Novgorod*, 31, 52 (Noviborg, forest fortresses); *The Nikonian chronicle*, 1: 220–24 (Vladimir), 240 (Mstislav), 244 (Vladimir again). This chronicle was compiled in the sixteenth century but includes much earlier material now lost; see Kirpichnikov, "Metatel'naya artilleriya", 13–15, citing the *Hypatian chronicle* and V N Tatishchev, *Istoriya rossiĭskaya* (Moscow, 1769–84), who has the first reference to stone throwers in use as being 1146, without effect. Tatishchev's pioneering *Russian history* has not yet been translated into any western language.

then inspected the ramparts and located a weak point, whereupon the town surrendered. In 1234, Danilo, prince of Galicia, and Vladimir III Rurikovich, grand prince of Kiev, marched on Chernigov (Chernihiv). A catapult had been set up that hurled stones a "bowshot and a half", and each stone was so large that it took four men to lift it.[1] There would be a change, both in fortification design and in the spreading of modern siege technology, during the thirteenth century.

Norway

The survival of the saga of King Sverre (r. 1177–1202) provides the first detailed evidence of castle building and siege warfare in the Norwegian kingdom. Sverre Sigurdsson was a successful challenger for the throne whose rule was long opposed, so his story revolves around details of warfare. In 1180, he was to be found living in the "wooden castle" of the important town of Nidaros (Trondheim), built outside the town by the archbishop. Sverre improved the defence by having a palisade constructed between castle and sea. By the following year, it was also protected by a deep ditch. Come the winter of 1182, he was constructing another castle on the Steinbjorg. In 1188, Sverre's enemies attacked this site, and the defenders hurled many stones at them, but could not prevent the castle being burnt. By 1194, the king was established in Bergen, where he ordered another castle built on the hill, defended by men equipped with crossbows. In 1196, Sverre set up a catapult to shoot at his enemies' (known as the Baglers') fleet, but the latter landed overnight and smashed the engine. Three years later, two catapults featured in Sverre's defence of Nidaros, protected by breastworks of timber "like bulwarks on a ship". In 1201, the king besieged the Baglers in a blockhouse. He surrounded the fort with a ditch and palisade but his assault was driven off by stones hurled by the defenders. He then tried to construct a "great wicker shield on stout pillars" to push underneath the blockhouse, but it could not be manoeuvred and came to nothing: in the end, the fort surrendered through hunger.[2] So clearly, while Sverre used castles as an instrument of his lordship, and he was aware of the potential of siege equipment, including stone throwers, their use in practice was relatively backward by comparison with his contemporaries further south.

[1] *The Galician-Volynian chronicle*, 27, 35–6, 43.

[2] Karl Jónsson, *Sverrisaga*, 57, 75, 92, 135, 148, 160, 195, 221–9. The nature of the stone throwers is unclear. Falk, *Altnordische Waffenkunde*, 193–4, suggested that they were counterweight, but the early date suggests they were more likely to be of the type of mangonel or petrary.

Summary: a period of transformation, and the arrival of the trebuchet

In the course of this chapter every reference that has been found to the arrival of the new stone thrower, the counterweight trebuchet, has been cited. The advantage of the new weapon over the old was substantial. By replacing human muscle power with the mechanical effect of a counterweight, the machine could not only propel much heavier missiles, it could do so with devastating precision, hitting the same spot every time if so desired. The debate over the introduction of the new weapon will be considered in the next chapter. Whenever the original appearance, the arrival of the trebuchet as a specifically identified wall breaker in both east and west took place during the period considered in this chapter. We have suggested that at least in part, the design of the massive new fortifications created by the Ayyubids in Egypt and Syria owed something to the existence of the new weapon, although it cannot be proven. In the thirteenth century, the Chinese would look to the Arab world for experts to build the new weapons, thus reversing the transmission route of the original traction device. If it did appear in northern Italy in 1189, that must have been among its earliest outings, and if it does not reappear in the written record for a while thereafter, that may be because it would take a while for engineers to acquire the new skills. However, by 1220 such arts were available to rulers everywhere. South-European sources offer yet further possible confusion: while the accounts of the Albigensian Crusade we have cited make definite mention of the trebuchet, the actual description of weapons being operated by the defenders of Toulouse talks of the traditional traction machines. Whatever the reason for this, and there are a number of possibilities including that the writer had not seen the weapon and therefore was simply describing something that he understood, the word was used for the first time, indicating that there was something new on the scene.[1]

The powerful new weapon, with its capacity to destroy weaker stone walls altogether, and certainly to demolish the defences of the wall top, thus making direct assaults much more fruitful exercises, was a valuable addition to the armouries of the rulers of the thirteenth century. But even without the trebuchet, the princes of 1200 onwards were in a much stronger position than their ancestors. Greater prosperity, effectively tapped, enabled princes who were effective to secure an advantage over all their subjects that put them beyond reach: they were no longer merely the first among equals. While politically every prince could only rule by the consent of their greater barons, the greater resources of the crown enabled

[1] See Chap. 9 on the trebuchet. For its introduction, see D R Hill, "Trebuchets", 103–6; Finó, "Machines de jet médiévales", 33–8; L White jr, "The crusades and the technological thrust of the West", 102–3 (a wrong interpretation of Marḍī ibn 'Alī al-Ṭarsūsī); Bonaparte, *Études sur le passé et l'avenir de l'artillerie*, II: 26–9, a classic study of serious value; Needham, "China's trebuchets, manned and counterweighted", 111–16; DeVries, *Medieval military technology*, 137–9 (does not seem to have accessed all the evidence); Tout, "The fair of Lincoln", 218–20, collects early chronicle references.

even such politically disastrous princes as John to defeat coalitions of their enemies. Part of this greater power was reflected in a new generation of castles, and part was in control of the skills and equipment to conduct effective warfare. Saladin and his heirs were able to deploy large numbers of siege machines and had the skilled operatives to work them. No single machine was decisive: the artillery usually created the conditions that enabled the miners or sappers to do their work. In the west, great siege trains accompanied Philip Augustus, Richard and John. Château-Gaillard, built according to the latest design, was captured, whereas Dover, of slightly older design but better defended, held out, even seeing off the first appearance in England of the trebuchet. Elsewhere, weak old castles and town walls could still offer resistance if the attacker did not have the means to press a siege, but against the engines and engineers of the great, resistance was futile. These trends would now continue in Europe down through the thirteenth century, while in the east the new Mongol empire would overturn the existing world order.

PLATE 17. Boulogne (France). The polygonal castle erected under Philip Augustus was one of the first in western Europe to dispense with a separate donjon.

PLATE 18. Damascus (Syria). The immense proportions of the citadel rebuilt by Saladin's successor al-'Ādil are shown by this tower. Note too the machicolations.

PLATE 19. Carcassonne (France). Reconstructed wooden hoarding on the Count's Castle gives a good impression of this common defence against close-quarters attack.

PLATE 20. Carcassonne (France). The interior of the reconstructed wooden hoarding on the Count's Castle.

PLATE 21. Geraki Castle (Greece) is a typical small French castle of the early thirteenth century. Here can be seen near the top of the tower the holes to bear the beams for wooden hoarding.

PLATE 22. Seville (Spain). The now free-standing Torre de Oro was once linked by bridge to the city wall, forming a *torre albarrana*.

PLATE 23. Granada (Spain). Next to the Alhambra stand the immense walls and towers of the citadel, the *alcazaba*.

Plate 24. Byblos (also Jbail, Giblet) (Lebanon): an early-twelfth-century crusader castle with massive donjon (with cistern in the basement) inside small curtain and ditch.

PLATE 25. Burj Ṣāfītā (Chastel Blanc) (Syria): a three-storey donjon of the early twelfth century rebuilt after earthquake, 1170s.

Plate 26. A mangonel in operation. The ropes at the front are being pulled down by a team concealed from view, while the man at the back needs to let go of the sling (which he will have held to keep it steady) immediately! The triple beam of the throwing arm suggests a large engine, while the crosspiece has led some to believe this might be a "hybrid" trebuchet (see pp. 380–82). Pierpoint Morgan Library, New York, MS 638, fol. 23^{v}.

PLATE 27. Soldiers sapping the base of a wall under cover of a movable shelter while defenders try to stop them with stones, spears and incendiaries. Although this illustration dates from the early fourteenth century, the technique had been unchanged for centuries. From *Les grandes chroniques de France*. © British Library Board. All Rights Reserved. Licence Number: BOYBRE03. Royal MS 16 G VI, fol. 74r.

Plate 28. Sandstone trebuchet balls at Pevensey Castle (England), weighing 61 kg. They are lighter than the balls of around 100 kg displayed at Carcassonne, Montségur and Weißensee.

PLATE 29. The early-thirteenth-century plaque in the church of St Nazaire in Carcassonne (France) shows, at bottom right, a mangonel, with a team of rope pullers and a man loading the sling with a shot above. It is said to represent the attack on Toulouse in which Simon de Montfort was killed in 1218.

PLATE 30. The reconstructed trebuchet in front of Warwick Castle (England) is an exact copy of the version built earlier in Denmark (see p. 383).

PLATE 31. This reconstructed trebuchet at Castelnaud (Dordogne, France) is simpler in design than that at Warwick. The counterweight container is empty.

PLATE 32. Mangonel reconstruction at Caerphilly Castle (Wales).

9 The developing technology of attack and the response of the defence

Some questions

Many studies have addressed the social and economic developments of the Middle Ages that underpinned political, religious and cultural developments. We have surveyed the evolution of the single most important form of warfare throughout this epoch, the attack or defence of fortified places. Few indeed were battles in the open field throughout the whole period, and fewer still of those that did occur were "decisive" in the modern sense of the word. At the other end of the spectrum, innumerable examples of warfare in the form of raiding featured, with extremely unpleasant effects on the territories directly affected, but rarely with any larger-scale impact, although it often led to the creation of frontier fortresses for garrisons or refuges for local people, with other significant and unintended longer-term consequences. But conquest of provinces or the overthrow of rulers either depended on, or was confirmed by, the capture of the enemy's walled cities and fortresses. Every conquering nomadic force very quickly had to learn siege-warfare skills if it was to consolidate its domination of the open country. If the word "siege" itself, strictly defined, means attacking a place by "sitting" down around it, our study has adopted a looser definition to look at the whole range of ways in which such warfare was conducted. We have surveyed hundreds of such operations according to original sources (and where possible, reference to surveys or archaeological studies of the sites), to seek as accurately as possible what actually happened. Now it is time to look at some of the questions that run through the whole period.

What was the relationship, if any, between technological improvements, the techniques and practices used in the attack and defence of fortified places, and the outcomes of such struggles? Modern historians and archaeologists have moved a long way from the simplistic explanations once offered, where changes in the design of fortifications were ascribed solely to the needs of defence as these changed to keep up with developing technology. For the builders of walled sites, demonstrating lordship and symbolising power, alongside the more practical purposes of serving as court, palace, prison, treasury, high class residence, monastery, garrison, arsenal or police station, signified much. Understanding the historical reality of medieval times has thereby improved greatly.

But we have shown repeatedly that fortified places were frequently called upon to serve their defensive purpose as well. When a lord or ruler ordered a town to be walled or a castle built, they might not know whether those defences would ever meet a military need, but this does not mean that defence was not among their purposes. Enough records have been cited of rulers ordering the urgent repair of town or castle walls in the face of apparent danger to show that such instructions were more than symbolic. Likewise, those who sought to impose their lordship had to do so, time and again, by capturing or recapturing towns and castles through sieges.

In proposing multiple roles for fortified places — sometimes simultaneous, sometimes differing between times — we have identified the capture or defence of such places as the main component of almost all the warfare of the whole period under discussion, with even exceptions such as the nomadic invasions having sooner or later to learn how to capture walled sites. More modern concepts of strategy that have sometimes been applied, and still find support in local guides, such as talk of castles "controlling" regions, lines of defences and so on, are rarely applicable. They reflect another era: the nature of warfare for most of the medieval epoch precluded such an approach. The logistical problems of providing for an army for any length of time were often insuperable, severely restricting the options available to any commander. Difficult or terrible communications were often a real obstacle to moving equipment or supplies.[1]

Such considerations must cause automatic suspicion at any figures offered for the size of early medieval armies (before the survival of quartermasters' records gives us a real clue), quite apart from reasonable doubt that anyone knew the real answer. Quite often the writer was following current convention to suggest a very large number. To sit down outside a defended fortification for long enough to take it by driving it to surrender through hunger was a method commonly applied where direct assault seemed unprofitable. But the supply problem cut both ways: for every such siege that succeeded, another was abandoned by the attacker through shortage of supplies for the besieging force, or even more were not begun in the first place, the commander deciding that the chances of taking the fortress were too slim to justify taking the action. It can truly be said that a defending garrison that had preserved its town or castle by preventing the siege in the first place had more than done its job.

Throughout military history, writers on the subject have correctly identified the vital importance of morale. Time and again, this unquantifiable quality has determined the outcome of conflicts, and all too often in the events analysed here

[1] The use of roads for travel and trade was of course routine. In the lands of the former Roman empire, the ancient Roman roads were the basis of medieval routes, with tracks and even cross-country travel linking places not so connected. For movement by foot or horse this seems to have been satisfactory, with average daily distances of twenty miles (32 km) being common in England, for example. But the problem was movement by wheeled transport, along with the upkeep of bridges. The movement of large numbers of people and animals with their supplies, and especially the transport of heavy equipment (such as siege engines, even in a dismantled state) was inevitably slow and difficult. Movement by water was much easier, and navigable rivers were regularly used for heavy goods. On the subject of roads in Britain, see Stenton, "The road system of medieval England", 3–21; Hindle, *Medieval roads and tracks*, throughout.

it has been simply impossible to recover the role of this factor in determining the outcome. This, therefore, is another significant caveat to be attached to all that follows here. This factor has led defenders to resist hopeless odds and attackers to force their way through fearful obstacles to bring about a capture. One side's technological or numerical superiority is secondary in such equations, although it might itself impact on morale. Similarly, experience, discipline and courage have determined many outcomes without being susceptible to easy accounting. Professional pride may have steadied the resolve of the increasingly frequently paid troops of the later years of the era, but also rendered them less likely to resist to the death if another option was available.

Hence, too, the many examples we have cited of engineers finding employment with rulers who were until recently the enemy against whom they were paid to work. It was these engineers who were responsible for the construction and direction of the siege trains, the evolution of which, along with the changes in fortress design, form the subject of this chapter. This point also suggests that technological development would never long remain in one set of hands. It was one thing to see, or hear about, some devilish new weapon, but to make effective use of the new device would usually mean having to secure the services of someone who knew how to make it, and someone who knew how to operate it, or, sometimes, being able to lay hands on the components.[1] None of the technology met in these pages remained restricted for long to its original owners, however, with the remarkable exception of "Greek Fire", so secret that even its owners may have lost the art. Developments in building fortresses were even less capable of being kept secret, although here it was a more complex question of the resources available for the slow and expensive business of improving existing defences, alongside the other reasons that might influence the decision to change the walls of a city, for example its changing size. In noting the process of the transmission of ideas, knowledge, skills and actual materials, it is always worth remembering that modern concepts of nation-states were quite unknown, that treason was an offence against an individual not a country, and that even religious differences so deep as to justify Christian crusades and Islamic holy wars did not prevent routine communication between inhabitants of the various states. The frequency with which Christian and Muslim powers allied together against a co-religionist gives a truer perspective to this issue. Likewise, engineers and builders from any background would not find it strange to provide (paid) service to a ruler from another. There were therefore few obstacles to the transmission of technical knowledge except the relative slowness of travel and the limited number of people who possessed the crafts.

[1] We have noted the problem faced by Muslim rulers in the Middle East of obtaining locally the raw materials needed to construct military equipment, and the regular trade with Italian cities by which they purchased (as a government monopoly) iron, pitch and timber from the eleventh century, for military purposes. Religious differences did not stand in the way of good profits. Ashtor, *A social and economic history of the Near East*, 196–7; Abulafia, "The role of trade of Muslim–Christian contact", 9.

Problems of evidence

Surviving early medieval historical records are few in quantity and often less than detailed in the information they provide. The gaps in knowledge about even the largest issues are still immense. The only certainty is that, provided the extant text has not been subsequently re-edited, the terms used, and the descriptions, cannot refer to anything later than the times of the writer. But making sense of the information can itself be a serious problem. With later written material, at least the availability of a much wider range of sources from across different geographical areas makes it easier to verify statements by cross-checking. But using medieval chroniclers to ascertain what happened in any given siege is rarely straightforward. Even if the writer was present, or received an eyewitness report, the best that we can be fairly sure of is that the events described were *possible* for that time, unless there is separate verification. In using any written source as the evidence for the capabilities of technology, for example the effectiveness of artillery, the particular use of mobile towers and battering rams, or undermining, there is always the risk of exaggeration, or pure invention. Not a few chroniclers were guilty of applying a standard list of current (and sometimes also classical, to show off their education) weaponry to any action, just as some modern journalists are not entirely immune from making up a good story when the facts of the real one are too much like hard work to unearth. Descriptions of particular equipment, and of their effect, are very rare, and historians are dangerously reliant on these handfuls of texts.

However, until the later Middle Ages and the arrival of iron weapons, there is no prospect of recovering any actual examples of these awe-inspiring achievements of medieval technology. The effects of the modern reconstructions remain the only *actual* means to confirm claims made about (for example) the range and destructive power of artillery. However, rebuilding machines from medieval descriptions is itself a task prone to numerous risks, there being no engineering drawings to work from, and many details of great importance to the performance of the weapon, familiar to the original builders and operators, have quite simply been lost for ever. As the thousand-year period unfolds, the growing number of illustrations of weapons and equipment have proved to be an extremely valuable source for what they looked like, but they were drawn, for example, to illustrate monastic texts — for purposes very different from that of providing later generations with a guide to their actual dimensions and proportions. Of the earlier weapons, not even these sources remain to provide evidence for the existence of different types of equipment.

Of problems of archaeological evidence, the one that most concerns this study is the difficulty of being certain about dating, especially in a fortress that remained in use through several building phases. One does not have to observe many of the arguments between equally respected archaeologists to grasp how fragile many of the conclusions can be, when not supported by other evidence such as dated inscriptions or coins. Even the sequence of construction may not be definitely ascertainable by excavation, for example where in a complex site different parts have been built over, and rebuilt or abandoned, at different points over a long period of time. But even where the sequence is beyond dispute, the actual dates may well not be, and the possible margins of error can be huge. Where the site was always an earthwork with timber buildings long gone, such problems can be even harder to resolve. Furthermore, this is sometimes exacerbated by the dangerous process whereby whole sequences and interpretations are built on a single original excavation study. It is not unknown to find that the original excavator proposed a hypothesis, but for subsequent discussion to harden the suggestion into fact and construct new theories based thereon. The process of interpretation from such evidence is therefore often problematic. To take a case where the commonly-held orthodoxy is not in doubt, many of the many hundreds of mottes surviving in England and Wales are identified as such not even by excavation, but simply because they look the same as mottes elsewhere. While no one (now) suggests that these examples are anything other than castles of the eleventh or twelfth centuries, the evidence for any individual site that supports its attribution can be as thin as that generalised comparison, and the range of variety found in superficially similar sites that have been excavated and analysed should also sound a note of caution. None of this helps when trying to establish what a particular place looked like at the time of a specific event in its history.

Therefore it is dangerous to base any firm conclusions on individual cases. On the other hand, where there are plenty of examples from the same time, albeit spread over wide areas, it can be safe to reach conclusions which are supported from such a broad spectrum of evidence: and that is what this study is based on.

From the end of antiquity to the beginning of the "Dark Ages"

In trying to establish the relationship between the evolution of techniques of attack and defence and the development of technology, the first issue is to establish the degree of continuity with what had gone before. A number of questions have to be answered to grasp the level of technology available both to the successor states to the western Roman empire and the Byzantine armies, given that the world of late antiquity was very different from what had preceded it. What remained of the apparatus of the imperial state and of the people who had been its functionaries? Did they remain in the service of the Gothic, Vandal or Lombard kings? If so, did they have access to the knowledge to build and operate late Roman military equipment? Had such equipment survived anywhere, given the abandonment of many cities and the shifting balance of population and the economy back to the land, the desertion of the cities and *castra* as places of government, commerce, administration — and defence? To what extent was such knowledge now useful? The writings of such practical theoreticians as Vitruvius and Vegetius survived and were copied, but did anyone now have the resources to put their recommendations into practice? The evidence does not permit definite answers to such questions, but it is possible to arrive at general conclusions.

From Britain's northern frontier to the edges of the Sahara desert, across Asia Minor and Mesopotamia, the occupants were surrounded by the stone- and brick-built fortifications of the previous age, even when, as in much of the western half of this vast area, there had been little in the way of new building for a long period, or in the eastern half such work as had been carried out largely consisted of shrinkage and retrenchment. The Persian realm continued uninterrupted, and through it or through central Asia lay the links with China, that other colossal imperial civilisation. The quantity of buildings encouraged a great deal of re-use, while the internal continuities and international contacts left no reason for any ruler to remain ignorant for long of the levels of technology that were available. However, none of the surviving evidence allows certainty as to exactly what the late Roman siege technology had been capable of, in contrast to the detailed descriptions surviving of siege warfare carried out by classical Greece and Rome. In the successor states, in the conflicts between the eastern Roman empire and the "barbarians", the siege craft proposed as a model by Vegetius and its further development by later writers (as in Maurice's *Strategikon*) was used for around a century following the removal of the last western emperor, before the further decay of that only gloomily evidenced age. The warfare between Byzantium and its eastern neighbours maintained the existing forms without interruption. But what was the starting level?

The inheritance: the onager and ballista, and new forms of defence

The power of the torsion-powered artillery brought to perfection in the ancient world is well-known from the surviving texts of such writers as Hero of Alexandria (first century AD), but was already well known to Philo of Byzantium three centuries before, and was included in Vitruvius' book of around 25 BC, one of the texts that contain (almost) complete instructions on how to construct a catapult. The effectiveness of the artillery of warlords such as Dionysius of Syracuse and of Philip of Macedon and his son Alexander the Great was sufficiently recorded to reveal the main facts. Experiments with models or full-scale reconstructions have produced a number of different estimates of the range of these early weapons. The torsion-powered crossbow-style weapon may well have been able to reach a range of more than 300 metres firing a bolt, which would have made it a very dangerous machine to be facing, and its value mounted on a wall to hit an attacker at long distance would be great, although on the other hand no bolt was going to harm a wall, while further problems lay in the way such weapons could not be depressed to fire downwards. As for stone throwers, shot fired from one set of small reconstructions varied in range between 180 and 300 metres (for respectively 0.6 kg of stone and 0.4 kg of lead). However, more doubt was cast on these claimed ranges by another reconstruction, by professionals in their respective skills, of the stone-throwing ballista used against Jerusalem in 70 AD. Using the reported actual dimensions, including a stone ball weighing 26 kg, and following the principles stated by Philo and Vitruvius, the resulting machine attained a range of only 90 yards (83 metres) before its timber frame (using no less than twenty tons — 20,320 kg — of oak and 4 km of twisted rope) cracked irretrievably under the stress of the torsion after just two firings.[1] Such results must raise a large question mark over the claimed range of this extremely sophisticated device.

As we have seen, though, these complex weapons had ceased to exist by the time that Vegetius wrote, and instead the much simpler *onager* was the mainstay of late Roman siege weapons, alongside the *ballista*, which had now ceased to be a stone-throwing crossbow and had become a bolt-shooter, and very probably one not propelled by torsion (Plate 13). The *onager*'s characteristics have been discussed. Although the reconstructions achieved ranges of 200–300 metres (Schramm) or even 500 metres (Frankland-Payne-Gallwey), the impact of shot at extreme range was going to be considerably less than at closer range, and Frankland-Payne-Gallwey proposed around 130 metres (150 yards) as the optimum range to have an effect on a stone wall.[2] Such closeness to the target called for an entirely different approach for a besieger, especially given the downside of the *onager*, its weight making it a real challenge to change its siting once emplaced, along with the related fact that the aim could only be changed by moving the whole machine, and the considerable skills needed to construct and operate even this (only relatively) simpler device. None of the surviving evidence makes it possible to determine just how effective batteries of *onagri* might be. There have been some statements suggesting that their bombardment had created breaches, although these references are a minority, and not often unambiguous. The experimental reconstructions did not propel shot of sufficient size to test this possibility, compared with the very much larger missiles claimed for the largest of both the older catapults and the late-Roman replacement. Too many additional factors would need to be known to reach safe conclusions, including the solidity and density of the missile, and the solidity and width of the target wall, the range and the angle of impact, and the ability of the machine to continue to strike the same point. Whereas the classical writers had been able to indicate what width of wall was needed to be safe against artillery (Philo said 10 cubits, around 4.5 metres[3]), the regular pattern of late-Roman and Byzantine walls might also have been the result of engineers working to a standard design across the empire. However, it is at least plausible that the strengthening of walls that was a normal feature of fortress building from the third century on was a reflection of a new reality: the empire's fortifications were now required to act as defences.

Changes to fortress design in the late fifth and sixth centuries may also have had other causes, but that of having to adapt to new defensive needs can be inferred. The examples of Justinian's age present the clearest case. Taken together,

[1] The experiment was broadcast by the television channel BBC2 in 2002. Cattle sinew not being available, polyester was tested and found to be of similar elasticity. Otherwise, by and large, the machine complied with the description given by the ancient writers.

[2] The various descriptions and results were analysed by Marsden, *Greek and Roman artillery*, 80–93.

[3] Ibid., 97.

walls strengthened and heightened, new and deeper ditches dug at a greater distance from the main wall and the creation of a lower outer wall (the *proteichisma*), along with the provision of powerfully projecting and often solid bastions or towers make a strong case for such measures being in response to the effectiveness of siege weaponry (as well as in the case of outer ramparts and deeper ditches serving to better resist undermining). Solid bastions and thicker walls were essential to bear the weight and absorb the shock of defensive artillery, as well as to resist bombardment themselves. Outer walls shielded the inner from being hit, and along with the ditch made a besieger start from further out. Where the outer wall was also provided with shooting positions, these enabled defensive ballistas to have a horizontal line of fire and thus overcame the weakness of such weapons to in defending against close-range attacks, sapping and other means of creating breaches. The heightened walls had various benefits as well, including in this context offering a better defence against the siege tower or the routine Persian form of attack, a siege mound. This period offers the strongest case for a direct link between fortress design and siege technology – stronger still since, where an opponent did not possess such skills, such as in Byzantine North Africa, the walls were much less substantial. Questions of resources were also relevant. But the evidence of real siege warfare and the remains of fortifications of the period 450–550 make the most unanswerable case. The skills needed to put the experts' recommendations into effect remained, along with the availability of the resources, key additional factors. The eastern half of the Roman empire and its traditional enemy to the east retained the necessary elements for this formula, whereas in the western successor states a rather different course had been followed.

The late Roman army appears to have had considerably less artillery than its forebears, organised now in separate units, so the weaponry that came into the hands of successor states might have been limited in quantity, and with fewer people who knew what to operate it, or how to make the engines that been constructed in a few arms factories, listed in the *Notitia dignitatum*, which did not long survive into the new times. It is no surprise that the evidence from Gaul, Germania and other places suggests a substantial loss of the skills and techniques preserved by the eastern empire. In particular, the skills required to construct torsion-powered weapons are hard to imagine surviving beyond texts preserved by a few literate scholars, whose opportunity to put such knowledge into practice in the service of new rulers would have become limited. Even the *onager* called for both calculation of the proportions of the various components and their accurate transfer into the actual machine. We cannot now reconstruct any of the process of development, but logic suggests that the torsion-powered stone thrower would have been decreasingly used the further north or west one moved from the eastern Mediterranean. Once lost, the skill would not easily have been recovered, although the apparent failure to rediscover this weapon through the mechanisms of trade and other communication as the new states of the Franks (etc.) took shape could have several explanations, not least that it had also fallen out of use in Byzantium and Persia, to be replaced with the Chinese-inspired design of the *petrobolus*, the lever-powered *mangonellus*, the arrival of which, courtesy perhaps of the Avars, we have dated to the late sixth century.[1]

As to the balance between attack and defence, let us meanwhile bear in mind that despite possession of the latest weaponry, the Avars failed to take Constantinople with its defences already two hundred years old in 626, just as the more recently strengthened walls of Rome had successfully resisted the Gothic host in 537–8. These two great cities had been provided with significant elements of the improved defences designed to resist the latest forms of attack, and the resources of the eastern Roman empire had not been spared in making this provision. So here at least, there is evidence that the defensive measures had served their purpose. Whether they would have done so had they faced more sophisticated attackers is an open question: the Byzantine–Persian wars suggest that a sufficiently well resourced and organised attack could, other factors being equal, overcome new defences. But the immense effort called for confirms the value of the double and strengthened walls and the potential impact of defensive artillery. In discussing the impact of new technology it is necessary to relate the abstract to the actual examples.

The age of the mangonel

As a general rule, there needs to be some added value to the new if it is to become, and remain, as it did with the *mangonellus*, the only type of siege artillery to be used for far longer than the weaponry it replaced. We use the term to signify that lever-arm weapon propelled by human force pulling one end down in order to cause the other to fly up, releasing its missile from a sling (Plate 32).[2] There are sufficient written accounts to confirm its rapid spread, along with innumerable manuscript illustrations. In contrast, those who persist in arguing that torsion-powered weapons survived (in the form of the *onager* renamed the *mangonel*, but not recognising the change of design) can produce precious little evidence. We have noted the occasional reference to weapons that could possibly be torsion-powered, and their use is not impossible.[3] But even if they did, it was

[1] Other analyses of the relevant texts have also arrived at the conclusion that torsion weapons had given way to simpler tension-powered weaponry by or during the sixth century AD, and that this conclusion applied equally to the western and eastern Roman world, pointing to the absence of reference to torsion weapons in Maurice's *Strategikon* or in Procopius, and that thereafter the Greek texts use the term *toxobolistra*, which clearly signifies "bow-ballista" for crossbow type artillery, indicating a non-torsion weapon, rather than the traditional terms for torsion-propelled devices. See Chevedden, "Artillery in late antiquity", 132–63.

[2] Previously we have adopted the modern term "traction trebuchet" to distinguish this weapon from the *onager* (etc.), but while this might have the benefit of instant clarity, it has the disadvantage of being a term unknown to contemporaries. In this chapter, therefore, since we are attempting to trace a process, we will stick to the terms used at the time. The lands of Islam used *manjanīq*, from the same source as "mangonel".

[3] Although in the absence of other confirmatory evidence, the possibility that the chronicler was inventing a description based on his classical reading rather than the actual

the lever-arm mangonel that was the standard weapon from around 600 until it was joined by the (counterweight-powered) trebuchet five or six centuries later.

What additional benefit did this weapon bring? Why did Byzantium, Persia and India switch to this import from China?[1] No text survives in which the change was noted, advised or recommended. One obvious benefit of the mangonel over the torsion *onager* was its greater simplicity of construction, the same advantage that the latter had held over the weapons it had itself supplanted. Although the proportions continued to be of great importance to the performance, a number of different lengths of the throwing arm, or of the sling, could still function. The wooden frame on which the arm was mounted had to be of sturdy construction, but it did not have to absorb the shock delivered by the torsion weapon, nor did it require the properly balanced and tightly strung (or "tuned") coils of sinew or rope that provided the propulsion, but were vulnerable to slackening. All the new weapon needed was human force, and the evidence quoted from the Chinese *p'ao* suggest that this power was capable of being adjusted upwards, in accordance with the size of the machine, to deliver shot varying from small anti-personnel to large wall-breaking stones. Put crudely, the larger the number of rope-pullers, the greater the range or the size of the missile, providing the frame, beam and axle themselves were large and strong enough. The nature of the propulsion involving the deployment of teams of pullers in front of the engine had the drawback that it could not be deployed by defenders on most walls or towers, but this could be overcome by placing it directly behind a rampart, with spotters on the wall. For the attackers, provided they were able to set up suitable protection from crossbows (etc.) for the operators, they could vary their target with relative ease. The earliest accounts of use in the west demonstrated an ability for the bigger machines to hurl huge stones, capable of demolishing rampart tops, and the largest engines would also be able to make breaches in weaker or poorly constructed walls, although this was unusual.

From the scattered accounts of seventh-century sieges, it was found that the mangonel could be used for a number of different purposes. These included laying down a heavy bombardment to drive defenders from the ramparts and cover the approach of troops with scaling ladders, or demolishing a gate, or being used to shoot at defenders' own weapons, or hurling stones or incendiaries over the wall to cause damage, dismay and losses among the besieged. It is also likely that well-drilled teams of operators would be quite capable of generating a very considerable rate of release, far greater than that of the torsion weapons, which required a big exertion of energy to rewind them. Such an advantage would be of great value, for example, if providing the cover for an assault.

For the defender, taking advantage of a higher platform, the possibility of disrupting the attackers by hitting their engines, or at least forcing such devices

events might apply in such cases. For a summary of the arguments, see Purton, "The myth of the mangonel".

[1] As the analysis of how the Islamic/Arab armies gained their siege weapons and skills given above in Chapter 2 strongly suggests they did.

as siege towers to be erected at a greater distance from the defences, along with the casualties that might be caused among the besiegers' troops, were often cited as important elements. Deployed alongside ballistas, defensive stone throwers could certainly be a great benefit to the besieged, and on a number of occasions indeed the histories tell that they were better used by the defence to save their fortress than by the attackers attempting to take it. However relatively simple the mangonel was, it required sufficient knowledge and skill in its construction to ensure employment for those offering these skills, in the absence of which it would be hard to achieve (probably) an effective range. Similarly, if the intention was to continue to hit the same point on the target, this would call for maintaining a consistency of pulling power and of missile weight, also calling for experience.

The spread of the mangonel to the northern successor states and their neighbours cannot be dated with any precision, and the early references to such weapons in the hands of Franks, Saxons and others are not before the eighth century. It is a surprise how long such this development seems to have taken. A number of factors might explain this: absence of need, or of requisite resources. The source material, however, was composed mostly in later years and is therefore not safe to rely on. But we can sure that the Carolingian empire had access to the equipment.

So what connection existed between the latest siege equipment and the development of defensive building? Within the boundaries of the old empires, there was not a great deal of new building. The same trends continued with Byzantine fortresses as before, suggesting that in order to cope with siege methods that were themselves of great age, apart from the new artillery, the same techniques that had suited Justinian still applied. The Muslim leaders continued for a long period to borrow the expertise of the lands they had conquered. Meanwhile, in the rest of Europe, there was no particular reason for the slowly developing new states to consider adding to the city walls they had inherited, it being a hard enough effort to keep such defences in repair, and the enemies they were likely to face had a still lower level of technology. Anyway, the new kingdoms had less access to resources both in terms of finance and of technical expertise to set about significant changes to structures for military purposes when the old ones were more than adequate, and where the traditional form of the hill-fort continued to provide a structure that could include defence among its various functions. The availability of manual labour also restricted the scope of new work, even where the duty to provide labour services to the lord was customary, since peasant farmers had other work essential to the survival of the community. In one of the very few examples of new building work, the defences of the Vatican in the face of serious dangers from seaborne Muslim raiders were erected according to the pattern of the existing late-Roman walls of Rome, exploiting the ready availability of bricks and cut stone from the surrounding site, and calling for the mobilisation of the papacy's own labour force over five years. The resulting wall was of considerable size (although not the same standard as the Aurelian defences), but the loss of its top level means no conclusion can be drawn as to whether the wall was designed to accommodate modern weaponry.

For the seventh, eighth and probably most of the ninth centuries, therefore, it would seem that in the Christian kingdoms little new development of building took place, but that gradually there was a spread of the technology of siege warfare across the whole area. The Muslim world had quickly acquired the knowledge it had formerly lacked, and had been able to apply its own high levels of culture, and ability to mobilise substantial resources, to achieve the highest levels then possible. The much straitened Byzantines likewise retained existing levels of knowledge and skill as they clung to centralised government and army, and it is in their wars with the Caliphs that the evidence exists for the widespread deployment of stone-throwing artillery, sapping and mining as the standard tools of sieges on both sides (Plate 27), as well as the use of battering rams, some kind of siege tower and the effective construction and propulsion of incendiaries. Nonetheless, wherever it has been possible to establish what kind of defences were really involved, they seem to have been either the existing late-imperial Roman walls, or, if the fortress was of more recent construction, a stronghold that was smaller, probably weaker and often less well built. Very few indeed were the constructions of this period that were actually more substantial than those of the classical era. Where buildings were not of brick or stone, as in much of central and eastern Europe, it is not to suggest that timber buildings were inherently less capable of being powerful, well built or highly decorative to indicate that the evidence from this time indicates that for the most part, the design and construction of earthwork and palisade hill-forts, refuges, palaces and towns remained relatively primitive. Where ancient traditions continued to combine with well-resourced states, such as in India and China, then here and only here the classical forms of both defence and attack were maintained.

Building on the old: the changes of the tenth century

Initially in the ninth, but mainly in the tenth centuries, the foundations were laid in the west for a considerable improvement in technique, design and construction that would render this world distinctly different from that of the early "Middle Ages". Chiefly responsible was the improvement in the productivity of the medieval economy, allowing for growing populations, and more trade resulting from the availability of surpluses, which also involved a substantial development of towns. Social and political changes inevitably followed. Successful rulers were those who now disposed of more substantial resources, many of which were used in building, serving the purpose both of emphasising and glorifying the ruler, and of providing more secure defences. A "renaissance" in learning also had an impact, in the form of the provision of classical texts to the highest levels of society. In chapter three we identified that Carolingian princes had the benefit of access to the newly copied writings of Vitruvius and Vegetius, while Arab princes seem to have had access to now lost Persian as well as Greek, Latin and Sanskrit military manuals explicitly dealing with siege warfare from the late eighth century.

But with what impact? It is not possible now to know, only to hypothesise. The texts certainly continued to be copied and used in the education of princes for centuries, so there must have been some practical effect. Although the chief artillery weapon in use had been unknown to the classical authors, their descriptions both of the best practice to be used for attack and defence, and of the construction of other types of siege equipment, along with their recommendations for building techniques for fortifications, were relevant. Such knowledge must have influenced the leaders whose function continued include command in war, in the decisions they made and perhaps in discussions they may have had with the engineers and masters they paid to carry out their works. Otherwise, there is a gap in the process in which the besiegers or the besieged simply had, or constructed, sophisticated engines or defensive works. Occasionally, as in the building of "exquisite new machines" by Charles the Bald in 873, the implication (quite beyond proof) is that this was done at the king's orders, even according to the king's design. Experience would have taught anyone for whom warfare was a regular occupation the effect, and the limits, of the equipment. It did not take a genius to realise that a battering ram or a siege tower was most easily dealt with by keeping it away from the wall or gate, so a ditch was a relatively easy but very useful defence. Similarly, a besieger directing a bombardment with artillery at particular points of the defence was signalling where they intended to concentrate their attack, making possible the use of simple defensive measures to lessen the impact, such as hanging shock-absorbent materials over the ramparts, or constructing a second defence behind the one visible to the enemy. No textbook was needed to discover these ways of neutralising an attack. But this is not to suggest that textbooks had not been read. In comparison, the more cultured rulers of India carried out the suggestions of the classical writers on architecture, including defensive structures, city walls, towers and gateways, according to the stated and traditional plan in quite sufficient frequency to indicate that the knowledge and recommendations of the texts were read and then put into effect.

The Byzantine empire not only had access to the ancient texts, but also, like the caliphate, produced contemporary treatises. Several major texts of the tenth century have been analysed (Chapter 4), and they were penned to serve as practical manuals. The study of actual events seems to confirm that this was what they did. Byzantine rulers and commanders were expected to have studied the Greek-language military theory as part of their training, although we cannot use this to confirm how similar Latin texts from the past were used. The image of the Byzantine general including manuscripts of military manuals among his baggage when on campaign may seem alien to traditional ideas about the Middle Ages, but was probably what happened.

In addition, Constantine Porphyrogenitus' list of siege equipment for the expedition to Crete gave an idea of the evolution of artillery by the middle of the tenth century. We have already identified that the mangonel was developed in different forms based on the same original design, in order to achieve varying tasks. The Chinese original had undergone the identical process, with one of the crucial differences being the number of bamboo poles bound together to vary the power and capacity of the throwing arm. This particular means of development not being available further to the west, the ballistic force of the weapon and the size of its missile could be varied by increasing the size of the machine and the number of rope pullers, along with, as the surviving illustrations have confirmed, variation in the design of the frame itself. If the tenth-century Byzantine empire disposed of twelve types of stone thrower, the four each of the *tetrarea*, *labdarea* and *eilaktion* listed by Constantine Porphyrogenitus, there must have been a common set of designs and specifications for their construction. The arsenal at Constantinople was clearly a busy place, and without much doubt supplied the weapons that were deployed on the city's own walls when called for, but the sources do not permit conclusions about whether the artillery and other machines used on campaign were transported there or built on the spot. The latter has always seemed more likely, given the difficulties of transport, but cannot be proven except when the history specifically referred to it. Therefore, an army must have relied on trained craftsmen. The experience of the professional soldier might be relied on to construct shelters and mobile sheds with protected roofs for approaching the walls to be attacked, but knowledge of the proportions and dimensions were essential to building artillery.

As with the Byzantine, so with the Muslim realms; although we lack from this period such a list as Constantine's, whether it be in al-Andalus, Africa, the Middle East or Sind, the evidence is of a variety of designs of weaponry, and of a corps of experts, alongside the indication that commanders were trained both in theory and practice. In the realms of the Carolingians and their successors, the contemporary sources of Flodoard, Richer, Ademar and, later, Rodulfus Glaber indicated often enough that siege weapons of all kinds were constructed at the place itself, and also confirmed that, as the century progressed, a wide range of equipment was available to the kings, bishops and counts, suggesting also the existence of a growing number of engineers and craftsmen. Mining called for particular skills as well, although there is a distinction to be made between the perilous method of creating a breach by setting up a strong protective shelter at the foot of a wall and knocking a hole through it with pick and shovel, and the considerably more difficult and even more dangerous method of digging a tunnel underneath, propped up with timbers which when burnt would bring down everything above. Rarely do the sources give enough information to determine which of these methods was being discussed, so the conclusion that the more sophisticated methods were in regular use in the more developed states, while elsewhere references to bringing down a wall by digging were more likely to refer to sapping it, is mainly hypothesis.

Alongside developments in the quality and quantity of siege technology during the tenth century, there was also an improvement in the quality, scale and defensibility of fortresses, at least those under the control of central government and its immediate deputies. Strengthened walls and towers and stronger gateway designs featured in the new works carried out by the rulers of Byzantium and the now several caliphates, as they moved forward, after a very long period of reliance on the walls of late antiquity. No claim is made of a direct connection between the two developments, as it was likely that rulers would build greater works than those they had inherited, once they were able to do so, but it would be reasonable to propose that where such works were carried out in places that had been, or might be subject to hostile assault, the latest architectural or engineering advice might be to strengthen these works' defensive capacities at the same time. Where new building was carried out further down the social scale, by lords and even knights holding relatively small areas of land and officially owing it to other more powerful lords who had legal rights to confiscate it in certain circumstances, as in the realms of the Franks, the result was the arrival, and extremely rapid spread across the countryside, of the feudal *castellum*. Such a minor building could not have been fortified like the royal or comital fortresses for many reasons: not least because they were prevented from doing so because of their lower status, as Coulson has identified. In addition, they may have had neither the means nor the need. During this period too, the appearance and development of the donjon, frequently built of timber but sometimes, and later on regularly, constructed from stone, was an innovation of the feudal states, its very name suggesting a role in signifying lordship. By its association with the rest of the castle, however, it also acquired a place in its defence against attack, where its bulk and strength inevitably made a contribution. But it was where there was a rapid spread of the fortified residence of the numerous lesser landowning class that there would be a further shift in the shape of warfare.

Further developments in the eleventh and twelfth centuries

One of the significant changes of this period, then, was the great difficulty for rulers of imposing their will by force when faced by challenge or rebellion, as so many western and central European kings were. A small rural castle occupied by its lord and household was not a factor in any military calculation. It was not defensible even if its occupant wished to resist. But provided with a few armed men and food and water, and held with determination, it could now compel even the most powerful ruler to organise a time- and resource-consuming blockade or assault. On its own, it might be unlikely that its occupant would take this stance. But if it stood in a territory thickly dotted with castles held by lords linked in resistance to a ruler, the latter now faced an impossible task. The endless conflicts of Henry IV with the pope and the latter's allies gave examples of the impossibility of ending this struggle through military victory, and while the attack and defence of walled cities continued to play a significant role, the provincial conflicts between rival supporters had concentrated for the first time on defending and taking each others' castles. The decentralisation of political power that accompanied the parcelling out of lordships to increasingly lower social levels had entailed crucial broadening of access to military power as well. Where once the fortresses belonged only to the crown or its immediate servants, by the end of the eleventh century the control of fortifications across Europe lay in a countless number of hands, and attempts to recover the rights of central government usually provoked civil strife and further dilution. As time progressed, the means of taking and defending castles could now be deployed by lesser nobles than the rulers and their counts and bishops who had once monopolised them.

The beginning of the crusades brought together worlds that had once possessed distinct characteristics. It remained the case that the Muslim world generally located the centres of its power in fortified cities, with the effect that to achieve victory over Muslim rulers involved conquering their cities. Techniques and technology were similar on both sides. However, local conditions as well as different forms of military (and social) organisation had their effect. The shortage of timber in the Middle East, it has been noted, not only determined that building would be carried out in stone (or sometimes rammed earth or forms of concrete, as in al-Andalus) rather than in wood: it must have influenced the choice of siege weapons and techniques as well. The wooden siege tower, a familiar sight in densely forested Europe, made its appearance in the Middle East courtesy of the crusades, although the dramatic success of the instrument at Jerusalem and other sieges of the First Crusade would rarely be repeated in later sieges, as the Muslim commanders became familiar with how to defend against it. The effectiveness of a device that enabled the attacker to overlook the wall, permitting archers and crossbowmen to make it a very dangerous place for a defender, was obvious in covering a direct assault, provided only that the great tower could be protected against fire.

On the other side, the leading place of undermining in the Muslim repertoire was learnt by their opponents and the technology taken back to Europe by returning crusaders. The problems associated with both methods have been sketched above, and their usefulness for an attacker depended on the nature of the target site, the presence of necessary materials, and the availability of suitably skilled workers. There were sufficient references to skilled miners coming from particular regions to show that this was a specialism. It may be that generally these were people who made their normal living from the peaceful pursuit of this craft, and were recruited specifically in wartime.

Some crusading siege operations called on all the techniques that could be assembled: the capture of Lisbon by a multinational crusading force in 1147 enabled us to see, from the background of a number of accounts, the effectiveness and limitation of the various methods available, and the particular skills possessed by national contingents with various techniques of attack. It also confirmed how if the defence was sufficiently resolute, ready and well organised, even creating a breach would not necessarily bring the siege to an end. The vivid descriptions of this siege gave a powerful picture of the sheer intensity and horror of the fighting. The effect of a continuous bombardment by a battery of stone-throwing artillery should not be underestimated in both its physical and its psychological impact. Once put in position, stone throwers served by sufficient relays of operators and provided with a continuous supply of ammunition were theoretically able to sustain a non-stop attack,[1] the whistling noise accompanying the flight being the

[1] Little evidence survives from this early period of the measures that must have been taken to ensure a supply of missiles for the artillery. Comments by a number of the military manuals about making one's missiles from substances that would disintegrate on landing,

prelude to the fearful crash of the impact, and in the target area, determined by the range of the weapons, life must have been terrifying and seriously dangerous.

The greater detail provided by surviving accounts of this period may be the only significant difference with the major sieges of previous centuries, and that the reality of the earlier conflicts was not dissimilar, but it might also be that the equipment and the techniques of the operators of the twelfth century had developed such as to be actually significantly more effective than their predecessors. Given the accompanying evidence for seriously stronger defences being constructed at the major sites that were centres of power, such would be a reasonable hypothesis. It would be unusual if the wider, higher walls and the provision of towers that had much greater projection than before, and in some cases were built solid such that they must have been intended to provide a stable platform for the mangonel, set behind wider and deeper ditches, did not represent an effort to become a match for an opponent's attacking power. Predicting and planning for as yet unknown weaponry was and is a skill largely unknown to any age; devising measures to cope with the last attack, on the other hand, and then compromising them to match ones resources and priorities, has been the hallmark of military planning throughout history. Anyway, what is proposed here is that by the twelfth century, the existing techniques of attack and defence had not been changed, but had evolved to a higher level of capability and effectiveness.

The age of the siege fort

Intense siege conflicts remained rare compared with those that consisted of tedious blockades, or were abandoned through want of resources by the attacker. Very strong fortifications, in difficult locations, provided with adequate supplies of food and water and defended by a professional garrison, were commonplace in the middle east, and meant a long and expensive operation with uncertain prospects of success. The problem facing someone with relatively limited resources but faced by a dense network of hostile fortifications was different, and it was tackled by another innovation of this time, the "siege castle", employed by a number of commanders but brought to being the standard means of attack by King Stephen in England. More accurately called a fort, such places had none of the other attributes of a castle but closely resembled a castle in appearance. Sometimes comprising a full-scale motte with a ward protected by wooden palisades and towers, sometimes of simple earthwork platforms protected by a ditch and a timber fence and gatehouse, but containing accommodation for men and horses, and sometimes taking advantage of an existing stone building such as a church, our study has found examples of such works, apparently, dating from the year 1000 to as late as the thirteenth century, and in locations stretching from

in order to avoid their being reused by the enemy, suggests one concern, but other examples (especially from later times) of the employment of stonemasons to carve the missiles indicate that in some cases, the destructive force of the ball would have been its most important characteristic.

England and Normandy to Sicily, where examples have been found dating from 1059 to as late as 1220, when no less powerful a ruler than Frederick II built a fort "to keep the other [Entella] in surveillance, and placed in [it] a garrison . . . which he [rotated] regularly".[1] It is necessary to say "apparently" because usually it is impossible to know if the surviving remains really were a siege fort, the deduction being based on written records combined with their distance from the assumed target fortress, usually at a few hundred metres.

The common feature is Norman, whether of the northern or the southern conquerors (of whom Frederick of Hohenstaufen was the heir in southern Italy and Sicily), and the common thinking behind the tactic was that an immediate conquest was not likely, but that a close blockade was not possible. Small mobile forces of trained soldiers, however, based in a secure fortification of their own (or often several), could both blockade a populous town or fortress by interdicting supplies, and hinder movement outside the walls. The 22 forts built against Taormina (1079) may have been exceptional, but a very high proportion of operations carried out in this way seem to have succeeded, compared with other forms of attack, and the chief quality required of the besieger would have been patience. Traditional pictures of medieval warfare are once again revealed as myths: instead, one must imagine very bored garrisons of warriors keeping up a daily watch on the blockaded fortress but mostly doing very little, except perhaps foraging for their own sustenance. When the *form* of such forts was replicated later, it was often as an element of a more active offensive against a blockaded fortress (as in Bohemia at the time of the Hussites).

The role of fire

Time and again, the wooden machinery available to attackers fell victim to the measures taken by the defence. The best defence against a siege tower or battering ram, or a battery of stone-throwing artillery, was to prevent its construction in the first place, or failing that, to prevent it being brought to bear close to the walls. Defensive artillery has been identified as critical to this, and among the missiles commonly used were incendiaries. It is time to have a look at the role of fire, both in this form and in the famous "Greek Fire" form, in sieges.

In the oil-rich Middle East, the development of weapons taking advantage of the easy availability of inflammable substances was not surprising. Naphtha was the main ingredient of forms of incendiary devices that could be hurled by artillery against attacking weapons, or by attackers at and over the walls of towns and castles. The composition of the Muslim weapon was often described in the treatises, along with the possibility that it could be propelled through kinds of siphons (syringes, fire lances) as well as being hurled inside specially made jars that must somehow have exploded (or more precisely, burst into flames) on impact:

[1] Lévi Provençal, "Une héroïne de la résistance musulmane en Sicile", 288, quoting an Arabic account of 1233.

it is hard to imagine how such a vessel (especially if projected over a wall) could have been ignited any other way.[1] The availability of defensive measures, such as the protective coatings applied to the exterior of wooden siege machines, suggests the limits of their effectiveness as well. Fire could also be used in other ways, as the descriptions of the use of "red-hot" iron balls or of "burning arrows" shows. Such shot, projected from artillery, would set alight any flammable target they hit, such as dry timber-built siege equipment and shelters. To have such an effect they must however have been heated to a considerable temperature, suggesting a high level of skill. The Chinese had possessed incendiaries for weapons since the turn of the millennium, in which non-explosive gunpowder was a key element, and they too propelled it as grenades in iron vessels, from fire lances and also from flame-throwers that used bamboo tubes, and the Mongols provided a rapid route for transfer of skills from the early thirteenth century.

One weapon that earned widespread fame (or infamy) was Greek Fire, although ironically most of the descriptions using the term did not mean the Byzantine secret weapon at all, but one of the forms of incendiary mentioned above. Despite the intrigue that Greek Fire inspired then and now, attempts to rediscover the complete composition of the weapon have all failed to answer all of the questions about it. Since its initial use was as a weapon that could set enemy ships ablaze despite the presence of water, this underlying function must have determined its main ingredients. As the protective measures used to preserve siege towers, rams and mantlets against incendiaries were often based on nothing more complicated than water, it is apparent that real Greek Fire cannot have been their problem. An experimental reconstruction, based as far as possible on medieval accounts, was able to confirm the composition, using a light crude oil, obtained from eastern Anatolia and the Black Sea coast, mixed with powdered resin (to enable it to stick to the target), which when raised to very high temperature (1,000 °C) had a lethal effectiveness. The remaining questions were the most difficult: how such a fearful flaming fuel was to be safely heated up, projected and ignited over a distance sufficient to reach an enemy.[2] Undoubtedly it did, so perhaps the real key to this weapon was in the siphon, rather than the fuel, and it was this secret that was too well preserved by the Byzantines, so that not only did their opponents never use it, but they even forgot it themselves, the surviving illustrations of siphons being too sketchy to allow for use as an engineering drawing. For whatever reason, it apparently was not deployed from the solid base of a fortress wall but rather from the rocking wooden deck of a ship in battle, in a situation where its powers of ignition and adhesion must surely have made it terrifyingly effective against besieging equipment on more than a few occasions, and even the example of Euripus

1 Considered in more detail in my *History of the late medieval siege*.

2 The experiment, led by Byzantine scholar John Haldon and others, was broadcast on Channel 4 television on 28 July 2003. The reconstruction revealed various difficulties of a serious nature in the actual construction, that given the horrible consequences to the operators of an accident involving such ingredients at such fierce temperatures were far more than "technical", and demonstrate the vital importance of getting the equipment exactly right.

(Negroponte, now Calchis; late ninth century) was a seaside town resisting an attack from the sea.

It would appear, then, that while incendiary devices could be used with devastating effect from around the ninth century in the west, and probably earlier in the east, they were rarely decisive. They could be resisted by protective measures, and in wet climates would be less useful. Adding to the limit of its effectiveness was surely the absence of a means of projecting it any great distance, a weakness not to be overcome until the advent of explosive gunpowder during the fourteenth century transformed the potential of both a missile's destructive power and its range.

Bows, arrows and arrow loops

The bow was a standard weapon of war (and hunting) throughout the period, across every country and region, and featured at almost every siege. Alongside, the crossbow also became standard equipment, its greater penetrative power being a strong attraction to commanders. In the attack and defence of walled places, both forms of weapon were effective in their way, either to drive a defender away from an intended point of attack, or to prevent an attacker succeeding in getting too close. In descriptions of the direct assaults that so often were the culminating point of any attack on a fortress, the chroniclers' references to hails of stones, arrows and bolts give a vivid picture of the scene. During this period there were some important developments in the way defences were modified to take that into account.

In a simple wall, the wall walk provided not only the means of rapidly moving around the defences, it was also the platform from which to shoot at, or throw stones at, an enemy outside. Crenellated parapets were standard, enabling the defender to shelter behind the upright (merlon), and shoot through the intervening gaps. If there was a defect, it was that with the increasing power of artillery, the parapet, necessarily much narrower than the wall itself, was vulnerable to being demolished, leaving no shelter for the defender. Likewise, during the period when siege towers were in vogue, an attacking platform higher than the wall transferred the advantage from the defender to the attacker. If a wall could be rendered uninhabitable, there was nothing to prevent sappers from safely digging away at the foot of the wall, or an attack by escalade, provided the wall was not too high for ladders, something that a deep ditch might well ensure. In any case, the arrow loop arrived, permitting the defenders to shoot through the wall rather than have to expose themselves on top of it. Such at least was the traditionally accepted theory, before historians began to propose a more sophisticated understanding of the role of all aspects of the design of medieval strongholds. When visiting medieval ramparts, it is interesting to stand behind a loop, whether in a wall or inside a tower, and look through it. From time to time, one might be struck by the extremely narrow field of vision. If in addition, one takes into account the dimensions of the weapon itself, it might seem strange that the slit opening could ever have usefully functioned as a shooting position. In many cases this will be because, probably, it was never intended to, but instead performed the necessary task of letting in light, particularly important if the opening served a room inside a tower. If designed for a crossbow, the shape of both the loop, and the shooting position behind it, needed to take into account the lateral dimensions of the weapon and the space needed to rewind and load it between shots.

Fortunately, there have been studies of the use of loops for defensive as opposed to domestic purposes. They were known before the Middle Ages, having been a legacy of the ancient world (attributed to Archimedes at the siege of Syracuse), and they seem to have been introduced early in Muslim fortifications, developing into sophisticated shooting positions by the time of the crusades. In the Byzantine realm, the continuation of this method of defence was only to be expected. Even if Procopius' description overrates the contribution of the emperor Justinian, his detailed account of the strengthening of the wall of Daras is eloquent testimony to the best practice of the early sixth century. The original crenellations were filled in with stones, reducing the opening to a narrow window of one hand's width, through which "arrows could be shot at assailants." When raising the wall height, the additional five metres consisted of outer stonework behind which ran a colonnaded gallery, on top of which the new wall-top battlements and walkway were constructed. While the necessity of avoiding crushing the existing walls beneath the weight of the mass of masonry being added on top must have required the approach adopted by Justinian's engineers, the stated intention of creating a shooting gallery was clearly an added bonus.[1] The wall towers were similarly heightened to three storeys, all provided with the means of defence. The brick-built gallery of the inner citadel at Sardis from the mid seventh century suggests that the technique continued in regular use, as does the building of slits into the walls of the inner fortress of Ancyra (Ankara) after the Byzantine recapture in 859.[2] These examples demonstrate the continuity in this part of the world, using the relatively rare examples that have been dated with a reasonable degree of precision.

When loops arrived in western Christendom has been a subject of greater debate, with the early or mid twelfth century having the most supporters, when they began to feature in the design of castle walls and towers, including some (but not all) donjons,[3] at least for the warring monarchs of Anglo-Norman and Plantagenet England and Capetian France.[4] Previously, defence would have carried

[1] Procopius, *Buildings*, 102–3 (book II, chaps. 1, 13–17). This volume contains (at pp. 104–5) plans of the Byzantine fortress at Sergiopolis (Ruṣāfah) clearly illustrating the type of wall, gallery and tower described by Procopius.

[2] Foss, *Byzantine and Turkish Sardis*, 57; Foss, "Late antique and Byzantine Ankara", 79 ff.

[3] Where the absence altogether of arrow loops in the thirteenth century (like many of those of Henry II, and Richard the Lionheart's donjon at Château-Gaillard, in the twelfth century) give further support to the argument that such buildings were not intended for defence but served other roles.

[4] The date of the first appearance of loops in west European fortresses was discussed by Salamagne, "Origines et diffusion des embrasures de tir", and Renn, "Castle fortification in England and adjoining countries". Renn identified loops as shooting positions in the Avranches Tower at Dover (for crossbows), and at ground level in the south curtain of

out from the top of the wall or tower only, although it is quite possible that loops could have been made in wooden palisades, which remained common, without leaving any record.

As thicker walls became common, the difficulty of providing a shooting position grew. The angle of opening was originally around 20°, but this offered a very limited field of vision (and shot), so by the middle of the thirteenth century openings of up to 50° were appearing, although these building requirements necessitated the creation of the vaulted positions. New fortifications designed during the thirteenth century certainly witnessed some such large-scale shooting positions, culminating in the skilfully designed multi-embrasured multiple shooting positions in the state-of-the-art royal castles of Edward I in north Wales, which succeeded ingeniously in addressing the problem of restricted fields.[1] Another method was to create embrasures at different levels, sited so as to cover all the ground below between them. If Beaumaris Castle were complete, it has been estimated, the outer circuit of the concentric defences would have contained almost 300 different shooting positions made up of loops at different levels, enabling defenders to cover any part of the castle's circuit, from both within the towers and from loops constructed within very large merlons.[2] Since it was unusual for a garrison to be large enough to mount an effective protection of every part of a fortress at the same time, the potential of such forms of concentrated defence to maximise the effectiveness of small numbers of professional archers or crossbowmen was a considerable bonus for the defender.

A reconstruction (in 1995) allowed for a better understanding of the reality of the use and limitations of arrow loops. Experiments were carried out in France at Loches, Coudray-Salbart (in Échiré), Castelnaud(-Fayrac, now -la-Chapelle) and Budos Castles. They discovered that, contrary to popular conceptions, an archer standing in an average loop was actually quite vulnerable to being hit from the outside. The chance of an arrow shot from outside going through the loop ranged from only 2 per cent at forty metres, but was 33 per cent at fifteen metres,

Framlingham Castle (around 1180–1215); at Corfe Castle, he noted loops in rebuilt towers (1201–) that left enormous dead zones, but better designed positions in the outer bailey wall towers (1209–). At the Château de Coucy in northern France, the angle towers of the inner ward include a hexagonally-vaulted interior with five loops that between them covered 270°: Renn, "Castle fortification", 56–9. Renn had previously tested loops at White Castle (Plate 15): Jones and Renn, "The military effectiveness of arrow-loops". Salamagne suggested three parameters decided what kind of loop would be designed: the thickness of the wall, available technical knowledge and resources, and whether the post was designed for bow or crossbow: Salamagne, "Origines et diffusion", 68–70. He also argued (p. 73) that the limited number, and location (at a height of 27–8 metres) of loops in the walls of the new donjons ordered by Philip Augustus must have been intended to provide a defence against attempts at escalade, a theory that is open to question in the light of the experiments dealt with below. If an attacker had got so close as to be able to set ladders against the donjon without being prevented by missiles from the roof, what use would a few arrow slits be? In addition, ladders long enough to reach the top of such donjons would have defied most medieval carpentry.

[1] For example, see Taylor, *Caernarfon Castle and town walls*, 38.

[2] Taylor, *Beaumaris Castle*, 41.

at Coudray-Salbart, and this using amateur archers with less powerful weapons than were available to the medieval archer. The danger varied with the width of the slit (which was between five and ten centimetres at Budos), and the height of the opening, both of which were factors of course that helped determine how wide the angle available to the defender was. Reproduced in battle conditions, in circumstances where attackers could sustain a continuous barrage, the likely consequence of this high level of risk when a besieger had advanced close to the wall would be that the defender would stand in the shooting position for the minimum possible time, and this, combined with the limited field of vision in the first place, suggested that these exchanges of arrows must usually have been "blind" rather than aimed at specific targets. Of course, it was possible for the defending archer to stand back behind the opening, but unless he stood close to the slit, there was the risk that his shot would fail to find the gap and would instead rebound inwards. This experiment confirmed (as was also shown in Renn's survey) that designing a slit with a lateral opening as well, giving the whole the shape of a cross, greatly improved the field of vision.[1] The zone that could be covered from a slit depended on other aspects of the design as well, such as how steeply the sill plunged.

Regardless of the design of a single shooting position, the only way to ensure the minimum amount of "dead ground" outside the rampart was to create a considerable number of positions with interlocking fields of vision, something best secured by building projecting towers at frequent intervals themselves provided with shooting embrasures in their sides, but then also risking having much dead ground in front of each tower. To meet this challenge called for considerable skill both in design and construction, and the fact that we find exactly such developments in the major west-European fortifications during the thirteenth century confirms that this development of defensive technique was probably in response to real concerns. Château-Gaillard, built at the order of one of the great masters of siege warfare of that time (Richard I, in 1196), had few loops and narrow shooting positions only, compared with the multiple embrasures at Caernarfon under his brother's grandson's command, showing the change that had taken place in the meantime. Again, the role of the expert in ensuring that the construction of the ramparts and towers met the latest standards of defensive design confirm the importance of this often nameless group of engineers.

New artillery: the trebuchet

Defences had been strengthened across the medieval world by the end of the twelfth century by raising the height and increasing the width of walls, and greater use of strong projecting towers, at least in major fortresses. This posed a challenge to those who had to capture such powerful strongholds. Part of the answer lay in a new weapon whose appearance we have argued dated to the

[1] Durand, "L'expérimentation de tir dans les châteaux"; Renn, "Castle fortification", 56.

late twelfth century in the Muslim states of the Middle East: the trebuchet, where human propelling force required by the mangonel was replaced by the greater and more consistent force of a counterweight. The transfer of existing technology — the resemblance of the trebuchet to the counterweight system used to operate water-raising devices in the region (the *shādūf*) had long been noted — suggested itself rather obviously, so that one is driven to wonder why it took so long. Or did it?

Hybrid machines?

Some now propose that the trebuchet was not in fact a brand-new weapon, but rather evolved from a "hybrid trebuchet", named by the American scholar Paul Chevedden. Drawing on doubts expressed by other historians, and arguing from the evidence of a significant increase in the effectiveness of siege artillery before the arrival of the pure counterweight machine, Chevedden suggested that what evolved was a "half-way" machine, in which a counterweight was added to, rather than replaced, the human propulsion of the missile (Plate 26).[1] As a result, the total weight of the missile that could be propelled increased dramatically, so the descriptions that either state, or imply, much heavier and more destructive missiles, might be seen as evidence for this new machine. Chevedden relates a long list of sieges that indicate this growing impact, ending back in the seventh century.

There are a number of flaws in the argument. Ultimately, the existence of the hybrid weapon depends on argument from the absence of an explanation for apparently increased firepower. Such arguments are particularly hard to refute! But the absence of any illustration of the suggested design from the time when it was supposed to exist, but at a time that first the mangonel, and then the counterweight trebuchet, were illustrated time and again in contemporary manuscripts, is also a telling counterargument from absence.[2] The words used to describe artillery are wildly inconsistent. While Chevedden reasonably suggests that some sources

[1] Chevedden, "The hybrid trebuchet", 179–216.

[2] Some manuscript illustrations of mangonels ("traction trebuchets") show additions to the pulling end of the arm, in some cases where the size of the machine and therefore the number of ropes needed meant it was necessary to provide the end of the arm with a way of spreading the points at which the ropes were attached to the arm across a wider distance, but also in cases where the engine is plainly small with a correspondingly small crew. The variations often appear in the form of a triangular spread of arms outwards from a single point at the end of the throwing arm to the point at which the ropes were attached. Naturally, such a piece of timberwork must have had a crosspiece at the far end to prevent breakage and to keep the arms at the planned distance apart, but these crosspieces are sometimes shown as being of considerable size. It is therefore possible that they represent a small counterweight, but the absence of scale in the illustrations makes it impossible to prove, leaving the simpler explanation to be preferred. See e.g. Lacroix, *The arts in the Middle Ages*, 95 (fig. 62, a thirteenth-century MS); Cockerell, *Old Testament miniatures* (from a Polish MS of around 1250), 64 (fol. 10^v, where the engine has a simple crosspiece with five hooks for suspending ropes), 116 (fol. 23^v, where a small engine has a more substantial crosspiece, but its small size is confirmed by only having three ropes); the machines shown at pp. 164 and 208 are very similar to the previous two examples. See also the illustrations

do use precise terms, particularly in the technical manuals and treatises, there remains the problem of distinguishing these from those that do not, and also that not one clearly suggested the hybrid design, nor illustrated it. The terms used by the non-specialist chroniclers are predictably varied, but one only has to think of the imprecise words used by many modern journalists to describe weapons to the general public to suggest that to use these accounts as sources of a theory of weapon development would be inherently risky. Even where the words conveyed precise meanings, these themselves inevitably changed over a long period. The important thing at the time was that the person who had to know understood what was meant, a prince undergoing their military education, an engineer preparing the weapon, or the superintendent of the arsenal responding to an order to supply it. Such people did not have to have a picture drawn for them in order to understand what was required, unlike a modern reader.

There is an alternative explanation for the increasing power of the artillery. It was simply that machines became larger and stronger and were better designed. It is well known that relatively simple design changes like the length of the sling could make a big difference to a machine's range. Increasing the size of the frame and the beam, and the number of rope-pullers, could likewise increase the size of the missile. The examples studied by Chevedden call also for a brief comment. It cannot be said that Abbo's account of the mangonel at the siege of Paris by the Vikings describes a hybrid trebuchet, despite Chevedden's assertion.[1] Abbo said that the weapon threw "enormous stones" (naturally without any accompanying dimensions) from a device made of timbers of equal length, and that the missiles had the effect of crushing the attackers' rudimentary shelters and cracking the skull of anyone who got in the way. The implication is that these weapons were lethal to individuals, but that this was their most notable characteristic, which does not support a conclusion about great size. Abbo also said that their name (mangonel) was common, which does not support a suggestion of novelty.

At the other end of the chronology, Chevedden argues that the description of a machine used at Damietta in 1218, where an Arabic text described a machine requiring lead boxes weighing 370 kg and a crew of 600, and threw a 185-kg missile, is the best remaining proof of the hybrid machine.[2] By this date, the Muslim armies had the pure counterweight trebuchet for certain, and would not have needed the supposed hybrid version. The figures are also suspect. Counterweights of ten, twenty or even more times the quoted figure were needed to propel much smaller missiles (see below). As for the size of the crew, one asks how so many men could possibly be fitted into the space below the engine, even if divided

from Petrus de Ebulo (d. 1220), also of lighter mangonels, reproduced in Erben, "Beiträge zür Geschichte des Geschützwesens im Mittelalter", 93.

[1] Chevedden, "The hybrid trebuchet", 196–7. Abbo's description is "Conficiunt longis eque lignis germinates / Mangana quae proprio vulgo libitur vocitantur / Saxa quibus jaciunt ingentia, sed jaculando / Allidunt humiles scenas gentis truculente." See the discussion in Chapter 3.

[2] Chevedden, "The hybrid trebuchet", 186, citing an untranslated continuation of the *History of the patriarchs of Alexandria* by Sāwīrus ibn al-Muqaffaʿ.

into teams. The number could be explained by the division into relays of men to manoeuvre and supply the shot for such an engine. As the thirteenth century progressed, the case for a hybrid weapon becomes ever less convincing as the new, more efficient and less demanding weapon became widespread. The only drawing reproduced of a machine that could be the hybrid weapon (although not then called it) is dated to 1462–3, by which time the trebuchet itself was on its way to being a museum piece.[1]

At best, then, the case for a hybrid siege weapon is unproven, and in our view, unlikely. After all, if the benefits of attaching a counterweight had been so obvious for so long, why did it take so long to go the whole way and replace human power altogether when an option whose power was greater was already in common use, albeit in a different form?[2]

Characteristics and effects of the trebuchet

The counterweight trebuchet was without doubt the most powerful siege weapon of the Middle Ages, and continued in use alongside early gunpowder artillery until the size, power and reliability of the latter finally overtook all previous weapons. Long a subject of fascination, the trebuchet, with its elegant design and devastating power, was also the subject of many surviving manuscript illustrations, making its features, rarely for medieval weapons, a matter of considerable certainty. There have been plentiful attempts to reconstruct and test the device, though few have been full-size. A reconstruction in Scotland in 2000 of the "Warwolf" constructed at Edward I's command against Stirling Castle in 1304 provided the opportunity to test two possible versions of the original. The conclusions verified medieval claims for this weapon.[3] The machine threw sandstone balls weighing 100 kg (250 pounds) at a target 200 yards (185 metres) distant, and managed to hit the wall at only the third attempt, crumbling right through a wall constructed to the model of the outer wall at Harlech or Caerphilly Castles, 1.53 metres thick (five feet). The final test shot completely destroyed the battlements and crushed the parapet. The frame of the engine weighed one ton (1,016 kg), and had a throwing arm 29 feet (9 metres) long. It was found that having a swinging rather than a directly attached counterweight was more efficient, and that the range and trajectory were easily modified by varying the length of the sling. If the counterweight box was too heavy, it threatened to crack the arm unless the hole for its axle was of precisely the right size. In the test, therefore, a modest four tons (4,064 kg) of sand was used instead of the denser lead employed in the original. By putting a wheel at

[1] Chevedden, "The artillery of King James I the conqueror", 48–80, fig. 1, reproducing a manuscript drawing by Ibn Aranbughā Zaradkāsh, *al-Anīq fī al-manājanīq*.

[2] Basista, "Hybrid or counterpoise?", was seen too late to affect this discussion. Basista also rejects the "hybrid" hypothesis and instead suggests (based on the physics) that extra power may have derived from adding a counterpoise.

[3] Broadcast on Channel 4 on 1 June 2000. The historians Michael Prestwich and Richard Holmes were among an international team of medieval historians and civil engineers involved in the project along with US and French colleagues.

each corner of the frame, not only was the device protected against the risk of tipping over, but it was found that this addition added even greater momentum and energy to the shot, although such wheels are rare in illustrations.[1] These results were obtained by modern engineers with no previous experience of the weapon, but their conclusions after a few days working from scratch was that the trebuchet was an easily adjustable weapon and it was deadly accurate.

Such conclusions confirmed the findings of a larger reconstruction at Weißensee, where a replica machine with an arm of 14.47 metres shot 55-kg stones up to 320 metres. This was done with a counterweight of six tons, although the builders claim they could have increased this to twenty. Accuracy was again remarkable, shots landing within a ten-metre radius, and when there was no wind, two shots fell just 20 cm apart.[2]

A machine constructed at the Danish Falsters Minder Museum in 1989 was rigorously tested, and operated flawlessly and with great consistency. The device was only operated with a counterweight of 2,000 kg although it could have held twice that, and with this loading it was able to project a 15-kg cement ball for 168 metres, although the designers calculated that with the full counterweight it would reach 300 metres. They reported that other key factors were the length of the sling (which had to be proportionate to the throwing arm, something that could only be discovered by actual experiment), the placement of the shot within it, and the angle of bend of the iron tip from which the sling was suspended. The twelve-ton (12,192-kg) machine was held together with wooden wedges and could therefore be taken apart, as was done for transport in medieval times. The timber used was beech, with hornbeam for the axle. The frame's dimensions were 8.5 × 7 metres, 4.8 metres high. The throwing arm was 6.5 metres long, of which 5.5 metres were behind the pivot (Plate 30). All the shots landed within a six-metre-square zone, confirming its extraordinary accuracy. Unlike the later test, this test also noted that it took only around five or six minutes to reload the trebuchet.[3] This was confirmed by the findings of the slightly smaller machine constructed at Caerphilly Castle, Wales, in 1992.[4] However, it should be noted that this was considerably faster than the rate of only a few shots per hour that had been noted in other reconstructions, chiefly in France.[5]

[1] But see an illustration in Conrad Kyeser's *Bellifortis* (written around 1400), studied in Berthelot, "Histoire de machines de guerre", fig. 34.

[2] Kirchschlager and Stolle, "Das teuflische Werkzeug", 27–43. The engineer was W Friedemann.

[3] Vemming Hansen, "Experimental reconstruction of a medieval trebuchet"; an earlier report of the same weapon in Vemming Hansen, "Reconstructing a medieval trebuchet". An identical weapon has been operating at Warwick Castle since 2005, with the same characteristics (personal communication from P Vemming Hansen).

[4] Humphries, *Engines of war*; see also R Morgan, "Magister tormentorum".

[5] Beffeyte, *Les machines de guerre*, 12. This booklet describes and illustrates mangonels and trebuchets on display at fourteen French locations, and offers a different range of names for counterweight weapons with a precise application to specific types that is hard to justify from the sources. Some of the information about the performance of particular weapons is also unattributed.

If we put together the results reported from different experiments, starting with that carried out on behalf of Louis-Napoléon by Captain Favé and described in the French emperor's still useful history of artillery, the outcome is shown in the facing table. The information is not directly comparable, and some is not verified. But a common picture emerges that the trebuchet could achieve a range of 200 metres, and a lot more with the appropriate balance of large counterweight and lighter shot. At the top end of the weight range, the effective propulsion of stones weighing 100 kg and more for significant distances, combined with the devastating effect of such missiles on masonry demonstrated in the Scottish reconstruction, are powerful evidence for its destructive capability. Surviving stone shot suggest that the lighter weight was more common, spheres of a diameter of between 44 and 50 cm being normal, and weights of around 15 kg. But more substantial missiles survive in large quantity at the castle of Pevensey. They are of two sizes and two different types of stone. Granite balls weigh around 82 kg, and sandstone versions weigh 61 kg (Plate 28). Large numbers of stone trebuchet missiles of a similar range of size, of around 60 and 80 kg, were found in the excavations at the site of the castle of Montségur, dating from the siege of 1243–4,[1] confirming the implication that by the middle of the thirteenth century this was a normal size. Even heavier missiles (such as found at Carcassonne and displayed in the museum in the Count's castle) show that 100-kg shot (or more) were used in the largest machines, too. Four shot found at Weißensee weighed from 95 to 115 kg.[2] Anything much larger must have been unusual. The size of the counterweight needed to propel some of the reported sizes of missile would have been impossible to contain within the machine's proportions, and it would have been impossible to hurl such missiles any distance. On the contrary, as Captain Favé's initial experiments demonstrated, it was possible for the machine to hurl its missile backwards. Hansen extrapolated that to project a 1,000-kg shot would have needed a counterweight of 27 tons (27,432 kg), and that merely to prevent it from falling directly on top of the engine itself. It would be prudent therefore to regard some of the medieval accounts of gigantic weapons and missiles as greatly exaggerated, or perhaps the result of the confusion that can arise from the variety of weights and measures in use.

Various methods were devised, too, for the vital task of hauling down the throwing arm after each shot in order to reload the engine, a task in which the size of the counterweight would be a critical element. Illustrations reveal that one method was to set up wheels (either one or two) at the rear, to operate pulley systems to allow teams of operators to wind the arm down, although this could have

[1] Displayed in the magnificent museum in the village of Montségur. See also Peyrou and Henry-Claude, *Montségur*, 6.

[2] Finó, "Machines de jet médiévales", 36; King, "The trebuchet and other siege-engines", 469, giving the approximately similar diameters (17–20 inches) of stones found at Pevensey, Norham, Harlech, Burg Steinich and Montrichard; Kirchschlager and Stolle, "Das teuflische Werkzeug", 29. The weights of the Pevensey balls were kindly provided by Pam Braddock of English Heritage. King's article mainly reports on numerous models of trebuchets, but later scholars have argued that one cannot safely extrapolate from the performance of smaller models to that of full-size machine. His argument is also clearly wrong in some of its other assertions about the names *biffa* and *tripantum*.

test/calculation	*counterweight*	*range*	*missile size*
Bonaparte[a]	4,500 kg	175 metres	24-cm bomb
Bonaparte[a]	4,500 kg	120 metres	32-cm bomb
D R Hill (calculation)[b]	10,000 kg	225 metres	37 kg
Villard de Honnecourt (Finó)[c]	26,000 kg	185 metres	—
Frankland-Payne-Gallwey[d]	—	320 metres	111 kg
Krizek 1986[e]	6,000 kg	200 metres	100 kg
Krizek 1990[e]	4,000 kg	445 metres	8–12 kg
Hansen 1989	2,000 kg	168 metres	15 kg
Caerphilly 1992	2,000 kg[f]	120–150 metres	12 kg
Weißensee 1997[g]	6,096 kg	320 metres	55 kg
Beffeyte 1998[h]	—	170 metres	125 kg
Scotland 2002	4,064 kg	185 metres	100 kg

[a] Bonaparte, *Études sur le passé et l'avenir de l'artillerie*, II: 38–41.

[b] D R Hill, "Trebuchets", 110–11.

[c] Villard de Honnecourt, *The sketchbook*, 132, plate 61. Original text dated to 1225–50; it describes the dimensions of the counterweight, from which Finó, "Machines de jet médiévales", 33, calculated the weight.

[d] Frankland-Payne-Gallwey, *The crossbow, medieval and modern*, 307.

[e] Krizek, "Trebuchet reconstructions in Czechoslovakia", 19–20.

[f] About two tons of sand: information kindly provided by P Humphries from his unpublished article, "Medieval warfare replicas: a project at Caerphilly Castle".

[g] Kirchschlager and Stolle, "Das teuflische Werkzeug".

[h] Beffeyte, *Les machines de guerre*, 13. He reports that bad weather stopped the experiment but that it had been expected that this machine could have reached a range of 250 metres.

been time-consuming. The speed of operation obtained in some experiments suggests more efficient methods were available, and clearly the balance and proportions of the arm must have been significant in determining the amount of power needed to reset the machine for another shot.

Nonetheless, since the evidence of several of the latest tests was that experienced engineers could construct machines with even greater range and throwing power, able to knock holes in thicker walls than those tested, and since the evidence of other tests has been that they had spectacularly thrown such items as grand pianos for a considerable distance, it appears incontestable that the impact of such a weapon on sieges was likely to be dramatic. With the earlier weapon, that we have called the mangonel, we had concluded that where the machine was particularly large, and operated by large teams of men (or sometimes women), and the defences not very strong, the artillery could breach a wall, but that more often we might explain such descriptions as being the destruction of the thinner parapets on top of a wall, or perhaps that they meant the outer walls, generally thinner. This hypothesis was confirmed by accounts such as that of the siege of

Thessaloniki in 1185, and it is a hypothesis that matches the facts. The trebuchet was indeed capable of knocking a breach in a wall: it has been shown that it could (albeit against a reconstructed outer wall). The many accounts of sieges that relate such an outcome, therefore, can more readily be trusted.

No test has tried to estimate scientifically the impact of a trebuchet shot on its target. Using a formula to calculate the energy in a counterweight of mass (in kg) × gravity × height (in metres), the energy in joules transferred to the missile (and assuming very little loss to friction) can be estimated (roughly, but sufficient for this purpose). The results for machines with a throwing arm of nine metres and counterweights of (1) four tons (4,064 kg) and (2) 12 tons (12,184 kg), are:

(1) 600,000 joules, equivalent to 140 grams of TNT;

(2) 1,830,000 joules, equivalent to 440 grams of TNT.

A modern hand grenade contains approximately 150 grams of TNT, so smaller trebuchets would have limited impact, but larger machines could have dramatic effect on a target, particularly given their accuracy.[1]

There was still a place for other forms of artillery. Manually-operated (traction) mangonels still served, as they had for hundreds of years, to attack personnel and weaker defences, or to project missiles over walls and into the undefended streets and houses behind in a medieval version of bombardments designed to terrorise non-combatants.[2] They were (usually) smaller, lighter and cheaper, too.[3] The quantity of material needed to construct a large trebuchet, as noted in the reconstructions, was indeed substantial, and if church roofs might supply a ready source of lead cheaper than it could be obtained today, the quantity of good-quality timber needed to ensure stability and durability under great strain, with the possibility of having to transport it some distance as an added cost, remained. The precision required of the craftspeople building it also called for the availability of experts who possessed rare and therefore well-rewarded skills.

It is a reasonable conclusion that where the word "trebuchet"[4] was used for the

[1] I am very grateful to Steve Freeman of Leamington Spa for devising the formula and how to calculate the TNT equivalent.

[2] This division of functions was clearly stated in the Norwegian *Speculum regale* of around 1240, discussed in my *History of the Late Medieval Siege*.

[3] Tarver, "The traction trebuchet", rebuilt a small mangonel at Toronto in 1991, on Ṭarsūsī's 1187 model, with a frame 3.6 × 5 metres, the arm 4.9 metres, and the sling one metre. He found best results with a pulling crew of 12, and the best range 137 metres (with a shot weighing 1.9 kg). A remarkable rate of shot of 4–6 per minute was achieved.

[4] Whether there is a clue in the word itself is hard to prove. That *trebuchet* was adopted by the Plantagenets might hint that they might have first met the French version, which itself may have derived from the Occitan *trebuca*. That language may also have provided the original term *bride* that became much more widespread throughout Europe in the middle of the thirteenth century in the Italian form *biffa* and the Germanic form *blide*, all words used to describe counterweight weapons; but it already had other forms such as *trabuquet*, *trabuca*, *trabuche* (in Iberia and Italy), or the German *tribuk/triboke*, and an almost infinite number of different Latin spellings. If of French origin it may derive from *trebucher*, to swing. It is also entirely possible that the Occitan, if it was the first European term, was itself derived from Arabic (as had happened with the derivation of *calubre*, a word describing a rope-powered catapult), from Catalan *algarrada* and Arabic *'arrada(t)*. (Chevedden,

first time, it is to report the presence of a new weapon, albeit not necessarily its first use. The origin of the trebuchet in the Middle East or the eastern Mediterranean has yet to be proved, but seems likely. The generic use of descriptive terms for stone throwers in the Arabic texts has made identifying its actual first use in that region particularly hard. That the trebuchet became identified as the "Frankish mangonel" in later Arabic histories does not necessarily mean that the weapon had originated among the "Franks", although it might refer to the Latin states in the Middle East or to those across the Mediterranean. The 1199 chronicle reference to *trabuchis* at Cremona has been doubted on the simple grounds that it is too early, although the only proof for the doubt would be that it is a later insertion in the text. Nonetheless, sufficient references in other sources suggest its arrival in the Italian peninsula and the German kingdom between 1205 and 1218. To add to the confusion, the use of the word to describe what were definitely mangonels during the Albigensian Crusade of the 1220s might be an early case of a new word quickly becoming a generic term. It is the context and the detail, as well as the name, that indicated the brief use by Prince Louis of France of a trebuchet at the end of his campaign in England in 1217, which he had to abandon with the defeat of his cause. Likewise, at the other end of the western world, the siege of Damietta in 1218 gave good evidence for the use of the trebuchet by both sides. It was in use in Iberia shortly afterwards, by which time it was clear from contemporary chronicles that a number of different local varieties were already known.

Whatever the date and the route by which the new weapon made its first appearance, by the third decade of the thirteenth century knowledge of its existence was commonplace. By then too, the different forms of the weapon had been written down, allowing later students to note the distinctions of design and size, where the form of the counterweight — fixed, hanging, or in one or two containers — was a major differential. The variety of terminology so quickly in use means that assumptions can only be made about any particular description when it was accompanied by further detail to confirm it.

Its greater destructive power and accuracy changed the face of siege warfare. But this was rarely completely straightforward. Undoubtedly, it helped the Mamluk sultans complete a rapid reconquest of the Latin east, although it was usually by a combination of three elements — mining, artillery, and lack of morale among the defenders — that victory was achieved. Nonetheless, the captured fortresses included some of the largest, strongest and most heavily garrisoned defences in the world, and the sort of place that had previously fallen only in exceptional circumstances.[1]

There are no ends or beginnings in history, but every study has to start and end somewhere. We have traced significant continuities between the classical and medieval worlds in Europe and the Middle East. We have shown where and how

"The hybrid trebuchet", 210–11, is very useful on this.) On the origin of *biffa*, see Poerck, "L'artillerie à ressorts médiévale", 35–44.

[1] See my *History of the late medieval siege* for discussion and analysis of the Mamluk conquest of the Latin kingdom.

technologies and technical ideas from China were cross-fertilised with those of the Roman successor states via Central Asia and the Muslim world, in turn acting on Africa and the Indian subcontinent, while Chinese influence was intermittently effective in Indochina, Korea and Japan. We have demonstrated that there were very few revolutionary changes throughout the period, but rather incremental changes and developments both in weaponry, and in the form and strength of fortifications, while making the point that it is necessary to put the latter in their historical context: it was the effective exploitation of the economic growth that created the wealth that enabled rulers to construct or modernise their walled and towered cities and fortresses during this whole era. Evolving technologies reached an apex in the twelfth and thirteenth centuries, without there having been a point at which any single dramatic alteration in siege warfare might be pinpointed.

Time line

Notes:

"etc." indicates a number of sieges in the same country or area.
Not all sieges covered in the chapters are listed in this time line, in some cases because the exact date is uncertain.

Year	*Significant reign/event*	*Western Europe, Scandinavia*	*Central/Eastern Europe, Italy, Baltic*	*Balkans, Middle East, North Africa*	*Rus', Central Asia, China, India*
451	Attila in Gaul	Orléans			
452			Aquileia		
476	Formal end of western Roman empire				
482	Clovis (d. 511) (Franks)				
489			Ravenna		
493	Theodoric the Great (d. 526) (Italy)				
502				Amida (and 503)	
511		Avignon			
512		Vienne		Amida	
527	Justinian (d. 565) (Byzantine emperor)				
531				Martyropolis	
532		Vollore, Chastel-Marlhac			
537	Gothic wars		Rome, Rimini		

Year	*Significant reign/event*	*Western Europe, Scandinavia*	*Central/Eastern Europe, Italy, Baltic*	*Balkans, Middle East, North Africa*	*Rus', Central Asia, China, India*
541		Zaragoza, Tarragona, Marseilles			
546					Yu-Pei
548		Barcelona			Ying-ch'uan
551				Petra	
573				Daras	
581	Sui dynasty (China)				
585		Saint-Bertrand de Comminges			
586		Cabaret, Beaucaire			
587			Arzanene, Maserti		
589		Carcassonne			
590			Pavia		
591			Ocras		
595			Illyrian campaign		
597				Thessaloniki	
600			Padua		
603			Cremona		
604				Daras	
610	Heraclius (d. 641) (Byzantine emperor)				
614				Jerusalem	

617				Thessaloniki	
618	Tang dynasty (China)				
626				Constantinople	
627				Medina	
628				Tblisi	
630				Ta'if	
632	Mohammed dies				
634				Damascus	
637				Ctesiphon	
640				Babylon	
641				Alexandria	
642				Tripoli (Libya)	
648				Lapathos	
653				Chalcedon	
662			Benevento		
672					Bukhara
673		Narbonne, Nîmes			
680				Caesarea	
683				Mecca	
695				Carthage (–698)	Debal (Sind)
701			Bergamo		
706					Baikand
711	Conquest of Spain				
712					Multan

Year	*Significant reign/event*	*Western Europe, Scandinavia*	*Central/Eastern Europe, Italy, Baltic*	*Balkans, Middle East, North Africa*	*Rus', Central Asia, China, India*
716			Cumae	Caspian region	
717				Constantinople (–718)	
718			Narni		
721		Toulouse, Angers			
725				Nicaea	
736		Lyons			
737		Avignon			
741		Laon			
749				Wāsiṭ	
750	Abbasid Caliphate			Melitene, Theodosiopolis (Erzurum)	
751	Pepin the Short (d. 768) (Franks)				
753		Narbonne			
755			Pavia, Rome		
756	ʿAbd al-Raḥmān I (d. 788) (al-Andalus)				
761		Bourbon, Clermont			
762		Thouars, Bourges			
768	Charlemagne (d. 814) (Holy Roman Emperor)				

773			Pavia	
775				Amorium (also 778/9)
776			Eresburg	
781				Nakoleia
803		Barcelona, Tortosa		
809				Serdica
812				Mesembria
814	Louis the Pious (d. 840) (Francia)			
820				Constantinople
821			Naples	
824				Crete
832			Conquest of Sicily (–878)	
837				Amorium
840	Charles the Bald (d. 877) (Francia); Lothair I (d. 855) (Germania)			
842			Bari	
857		Chartres		
866	Alfonso III (d. 910) (Léon)			
867	Basil I (d. 886) (Byzantine emperor)			
870		Dumbarton		

Year	*Significant reign/event*	*Western Europe, Scandinavia*	*Central/Eastern Europe, Italy, Baltic*	*Balkans, Middle East, North Africa*	*Rus', Central Asia, China, India*
871	Alfred the Great (d. 899) (Wessex)		Salerno		
873		Angers		Melitene	
878			Syracuse	Germanikeia	
885		Paris			
887	Arnulf (d. 899) (Germania)				
893	Oleg (d. 924) (Kiev)	Chester			
894			Bergamo		
896			Milan		
903				Byhor	
904				Thessaloniki	
907	"Five Dynasties" (China)				
910	'Ubaid Allāh al-Mahdi (d. 934) (Fatimid Egypt)				
912	'Abd al-Raḥmān III (d. 961) (al-Andalus)				
913	Constantine Porphyrogenitus (d. 959) (Byzantine emperor)	Écija, Évora, etc.			

915		Mérida		
917		Carmona		
921		Towcester		
922			Monterubbio	
924			Pavia	
926				Melitene
927		Bobastro		
932		Toledo, Saint-Quentin, Château-Thierry		
936	Otto I (d. 973) (Holy Roman Emperor)			
937		Zaragoza		
938		Laon		
943		Dublin; Tamworth, Leicester		
946		Senlis		
949		Laon		Theodosiopolis
950		Limerick, Waterford; Amiens		
951		Brienne		
954		Edinburgh; Mainz		
955			Augsburg	
959				Heraklion, Amida
964				Tarsus, Adana, etc.
965				Tarsus

Year	*Significant reign/event*	*Western Europe, Scandinavia*	*Central/Eastern Europe, Italy, Baltic*	*Balkans, Middle East, North Africa*	*Rus', Central Asia, China, India*
968				Antioch, Manzikert, etc.	Kiev
972				Preslav, Silistra	
975				Baalbek, Beirut, etc.	
976	Basil II (d. 1025) (Byzantine emperor)				
978				Martyropolis	
985		Barcelona; Verdun			
986				Kilis, Apamea	
987	Hugh Capet (d. 996) (France); Fulk Nerra (d. 1040) (Anjou)				
988		Laon			Cherson
991		Melun			
997		Santiago de Compostela; Auxerre			
998	Maḥmūd of Ghazni (d. 1030)		Rome		
999		Dublin			
1000		Sens			
1001		Exeter			
1003			Vodena (Edessa), Schweinfurt		
1004			Vidin		

1006		Durham, Valenciennes			Multan
1013		Dublin			
1015		Montbazon			
1017			Nimptsch		
1019	Yaroslav I the Wise (d. 1054) (Kiev)				
1022			Troia		
1026		Montboyau			
1027		Amboise			
1031				Edessa	
1032		Sens			
1034			Murten		
1035	William I (the Conqueror) (d. 1087) (Normandy, then England)				
1036				Tbilisi	
1037			Ravenna		
1038	Tuğrul Bey (d. 1063) (Seljuk sultan)	Langeais			
1041			Hainburg, Pressburg		
1042			Trani		
1047		Brionne		Constantinople	
1048		Domfront			
1052		Alençon	Capua, Pressburg		

Year	*Significant reign/event*	*Western Europe, Scandinavia*	*Central/Eastern Europe, Italy, Baltic*	*Balkans, Middle East, North Africa*	*Rus', Central Asia, China, India*
1053		Arques-la-Bataille			
1054		Ambrières		Manzikert	
1055		Antwerp			
1056	Henry IV (d. 1106) (Holy Roman Emperor)				
1057		Málaga, Guadix, Coimbra, etc.			
1059			Reggio		
1063	Alp Arslan (Seljuk Sultan) (d. 1072)	Mayenne			
1064		Barbastro		Ani	
1066		Amboise			
1067		Exeter			Minsk
1069		York, Ely			
1070				Edessa	
1071	Yūsuf ibn Tāshfīn (d. 1106) (Almoravid)		Palermo, Bari, etc.	Belgrade	Vladimir, Przemyśl, Chelm
1072			Palermo		
1073			Harzburg, Lüneburg, etc.		
1074			Cologne		
1075		Norwich			

1076			Salerno		
1077			Würzburg	Damascus	
1078					Chernigov
1079		Gerberoy	Taormina		
1081	Alexius I Comnenus (d. 1118) (Byzantine emperor)		Rome	Dyrrhachium (Durrës)	
1083				Harran, Kastoria	
1084		Toledo	Augsburg		
1086			Agrigento	Edessa	
1087				Mahdia	
1088		Rochester	Butera		
1089		Aledo			Bukhara, Samarkand
1090		Brionne, Córdoba		Chios	
1091		Courcy, Mont-Saint-Michel	Mantua		
1092			Canossa, Nakło nad Notecią		
1093		Valencia			Torchesk
1094		Argentan, Aberlleiniog			Chernigov
1095	1st Crusade proclaimed	Tynemouth, etc., Ejea, Bamburgh			Yur'yev, Starodub
1096		Pembroke, Huesca			
1097		Almenara, Murviedro		Nicaea, Antioch	
1098				Ma'arrat al-Nu'mān	

Year	*Significant reign/event*	*Western Europe, Scandinavia*	*Central/Eastern Europe, Italy, Baltic*	*Balkans, Middle East, North Africa*	*Rus', Central Asia, China, India*
1099	Kingdom of Jerusalem			Jerusalem	Vladimir
1100		Valencia; Le Mans			
1101		Toledo, Valencia		Arsuf, Caesarea	
1102		Arundel, Bridgnorth; Luzarches, Chambly			
1103		Montaigu		Acre (and 1104)	
1105	Roger II (d. 1154) (Sicily)		Nuremberg		
1106			Cologne		
1107		Gournay	Białogard	Dyrrhachium	
1108	Louis VI the Fat (d. 1137) (France)	La Ferté Baudouin	Czarnków		
1109		Toledo	Nakło nad Notecią, Glogau	Tripoli (Lebanon)	
1110		Mantes, Bellême, Montrichard		Beirut	
1111		Santarém; Le Puiset		Tyre	
1112		Le Puiset, Laon, etc.		Edessa, Athārib	
1113		Bar-le-Duc	Białogard		
1114		Ibiza, Mallorca; Crécy-sur-Serre, Amiens	Glatz		
1115				Kafarṭāb, Zardanā	

1116		Swansea, Aberystwyth, etc.		
1118		Zaragoza	Pisa	A'zāz
1119				Athārib, Zardanā
1122		Montferrand		
1123		Pont-Audemer		Kharput
1124		Vatteville-la-Rue		Tyre, Melitene, A'zāz, etc.
1125		Brionne		
1126			Nuremberg	
1127		Bruges	Speyer	
1128		Ghent		
1129				Marrakech
1132			Bari	Baalbek
1133	'Abd al-Mu'min (d. 1163) (Almohad)	Roskilde	Montepeloso	
1134		Bayonne, Bouillon		
1135	Stephen (d. 1154) (England)	Leiria; Bergen		Kastamonu
1136		Bampton, Exeter		Adana, Tarsus, etc.
1137		Carmarthen, etc.	Bari, Salerno	
1138		Coria, Oreja; Poitiers, Montaigu, Falaise; England/Wales × 11		Bile, Aleppo, Shaizar, etc.
1139		England × 7	Bari	Baalbek

Year	*Significant reign/event*	*Western Europe, Scandinavia*	*Central/Eastern Europe, Italy, Baltic*	*Balkans, Middle East, North Africa*	*Rus', Central Asia, China, India*
1140		Wallingford, Devizes, etc.		Banyas	
1141		Lincoln, Oxford, etc.			
1142			Prague		
1143		Carmarthen, etc.; Rouen		Tripoli (Libya)	
1144		Lincoln, etc.; Arques		Edessa, Oran	
1145		Faringdon			
1146				Edessa, Tripoli (Libya), Marrakech	
1147		Almería; Pevensey, etc.	Ceuta, Corfu		
1148	2nd Crusade	Lisbon, Silves		Damascus	
1149		Montreuil-Bellay			
1151				Zeugminion	
1152	Frederick I Barbarossa (d. 1190) (Holy Roman Emperor)	Vernon, Vezelay; England × 8			
1153				Ascalon	
1154	Henry II (d. 1189) (England)				
1155			Tortona, Bari, Taranto, etc.		
1156			Brindisi, Butera		

1157		Almería		
1158		Cardigan	Milan; Glogau	
1159			Crema	Mahdia
1160		Gerberoy		Tripoli (Libya), Sfax
1161		Søborg		
1162			Butera	
1163			Werle	
1164			Demmin (and 1177)	
1165		Carmarthen		Zeugminion
1166		Cáceres	Haldensleben (and 1179)	
1167		Beja	Reggio; Oldenburg	
1169		Wexford		Alexandria, Damietta, Hims
1170		Waterford, Dublin		
1172		Huete		
1173		Verneuil -sur-Avre; England × 3	Ancona, Alessandria	
1174		England × 5; Rouen		
1175	Saladin (d. 1193) (Egypt, Syria)	Limerick		
1176			Stettin	
1178			Prague	
1180	Philip II Augustus (d. 1223) (France)		Lübeck	

Year	*Significant reign/event*	*Western Europe, Scandinavia*	*Central/Eastern Europe, Italy, Baltic*	*Balkans, Middle East, North Africa*	*Rus', Central Asia, China, India*
1182		Santarém			
1184				Prusa, Nicaea	
1185		Mallorca; Boves; Neath, etc.		Dyrrhachium, Thessaloniki	
1187				Sidon, Beirut, Jerusalem, etc.	
1188		Bouvignes		Palestine/Syria × 7	
1189	Richard I (d. 1199) (England); 3rd Crusade	Le Mans, etc.	Messina	Acre	
1190			Lübeck		
1191		Enghien	Naples		
1192				Darum (Deir al-Balah), Jaffa	
1193		Rouen			
1194		Nottingham; Loches, Verneuil-sur-Avre	Salerno, Bari, Barletta		
1195		Huy, Le Vaudreuil; Alarcos			
1196		Painscastle			
1197				Toron; Prosek	
1198		Madrid, Vitoria			
1199		Nidaros	Cremona		

1201				Constanța, Varna	For the 12th and 13th century in Central Asia and the Far East, see my *History of the late medieval siege*.
1202	4th Crusade	Radepont, etc.	Zadar		
1203		Château-Gaillard		Constantinople	
1205				Adrianople; Mahdia; Franks in Greece	
1206			Kirchholm		
1207		Montauban		Adrianople	
1209	Albigensian Crusade	Carcassonne		Thebes	
1210		Minerve		Corinth	
1211		Termes, etc.; Salvatierra			
1212		Hautpoul, Moissac etc.; Calatrava	Langensalza, Burg Weissensee; Nauplio		
1213		Toulouse			
1214		Casseneuil			
1215		Northampton, Rochester			
1216		England × 11			
1217		Toulouse			
1218	5th Crusade	Newark		Damietta	
1220			Mesothen		
1221		Fotheringhay, Castle Bytham			

Year	*Significant reign/event*	*Western Europe, Scandinavia*	*Central/Eastern Europe, Italy, Baltic*	*Balkans, Middle East, North Africa*	*Rus', Central Asia, China, India*
1223		Toulouse			
1224		Bedford			
1226		Avignon			
1229			Kalisz		
1234					Chernigov

Glossary

Note: Italics within definitions refer to cross-references within the Glossary. Some of the words defined here remained in use for a thousand years, others were coined anew, or alternatively transferred from an older meaning. The lack of certainty attached to interpreting many of these terms may be frustrating, but unless the context makes it clear, there is a risk that the word in use conveyed a different meaning for the writer than it did for others at the same, or different, times. Sometimes, a word has a precise significance for one user, but is a generic term for another.

agger	(Latin). Lit. rampart of which earth is the main component.
alcázar, alcazaba	(Spanish, from Arabic "al-qaṣr", "al-qaṣbah"). Citadel, usually in the form of the ruler's defended structure attached or within a town or city.
algarrada	(Catalan). Catalan version of *ʿarrada.*
almajanequi	(al-Andalus), from Arabic *al-manjanīq*, stone thrower.
aplekta	(Greek). Byzantine fortified camps.
aries, pl. *arietes*	(Latin) Battering ram(s). Sometimes in the generic sense of "siege weapon".
ʿarrada, pl. *ʿarradat*	(Arabic, from Syriac). Arabic catapult, general term but may have meant a large crossbow-type bolt-firing weapon; cf. *ballista.*
arrow slit	A slit cut in a wall to enable an archer to shoot out from behind, introduced in Byzantine fortresses of the seventh century and universally used, with numerous variants. French version "archère".
arx	(Latin). Specifically, the citadel, but also a generic term for castle or fortress, implying a significant building.
aula	(Latin). Hall, sometimes describing a whole castle, or the donjon.
Balearic sling	Trans. of "Balearis funda", artillery term in al-Andalus met in the twelfth and thirteenth centuries. No evidence to confirm what type of weapon is meant, although the name suggests some kind of trebuchet. There is no necessary connection with the Balearic islands as it may derive from Greek "ballo", to throw.
ballista	(Latin). Originally, a stone-throwing torsion engine invented by the Greeks and used by the Romans. In later Roman times, the name was used to describe a large-scale crossbow-type weapon that may (or may not) have been torsion-powered. In medieval times, a large crossbow shooting bolts, etc.
barbican	An outer defence erected in front, to protect a gate.

bashura	(Arabic). Outer defensive wall, often signifying the outer enclosure.
bastion	Any form of tower or projecting defensive work, occasionally, by transference, the whole work. From early modern times, was used much more precisely to signify an artillery platform projecting from the *enceinte*.
belfry	A siege tower, designed to overtop the defensive wall to enable attackers to shoot down, and sometimes provided with a bridge to enable attackers to storm the defences.
bent entrance	Originally Muslim design of gate, requiring the entrant to turn at right angles one or more times before reaching the interior. Became widespread.
berefridum	(Latin). See *belfry*. Numerous variants of the name.
Bergfried	(German). Tall, often narrow tower in many German and central European castles, with a very restricted interior. Possible use as symbolic power statement or a watchtower.
berm	(From French). The space between a ditch and the rampart.
biffa	Type of trebuchet.
Blide, Blyde	(German). Trebuchet.
blifyeu	(Welsh). Catapult (generic term).
brachiolla	(Latin). Early term for artillery, possibly a *ballista*.
bretêche, bretache	(French). See *hoarding* and *machicolation*.
burj	(Arabic). Generic term for a tower (also spread into Urdu).
Burgward	(German). Castle guard duty.
burh	(Anglo-Saxon). Fortified site.
calabre, calibus	(Occitan). Trebuchet.
caltrops	Sharpened metal spikes spread in ditches to wound attackers.
caput	(Latin). Lit. "head", the chief place of a lordship.
castellation (sometimes *in-*)	Lit. the creation of a *castle*, for example by fortifying an existing building, although applied generally to any kind of building or settlement.
castellum	(Latin) Literally, a diminutive of *castrum*, common word for *castle*, sometimes a purely military construction such as a siege castle/fort.
castle	Dwelling of a feudal lord (including ecclesiastical), or the monarch, serving one or many of multiple functions. In medieval usage, almost invariably fortified by implication. Word also used by contemporaries for both permanent and temporary fortifications.
castrum	(Latin). Lit. a fortified place, sometimes used to describe a walled town, sometimes a castle. In medieval use, might apply also to a (sometimes only slightly) fortified village; see also (Greek) *kastron*.
cat	Wooden shelter for besieger (from Latin "gattus").
contravallation	Outward facing rampart protecting a besieging army from attack from outside.
crenellation	Parapet on top of wall, with openings for use by defending archers (etc.). The intervening stonework known as *merlons*.
dabbāba, pl. *dabbābat*	(Arabic). A siege tower (from early eighth century).

donjon	Literally, the "lordship tower" or lord's residential or ceremonial tower (or hall), the most important part of the castle; also had a variety of uses. Rarely used as a citadel (or defence of last resort, as was once commonly held), but sometimes played that part in practice.
drawbridge	Bridge across a ditch or moat that could be raised and lowered by the occupants of the town or castle. Various designs.
dun, dún	(Irish/Scottish). A fortified site.
eilaktion	(Greek). One of the four types of stone thrower in the ninth/tenth-century Byzantine arsenal.
enceinte	(French). The enclosing wall or rampart around town, castle, village, church, etc.
fascine	Bundle of any suitable natural material, used for filling in ditches or to construct barricade.
foneval	(Catalan). Stone thrower.
fortalice	(numerous variants, from Latin). Usually small fortified place.
gorod	(Polish and Russian). Fortress.
grad	(Russian). Fortress.
Greek Fire	Although a term used widely to describe incendiaries, exact meaning applied to Byzantine invention designed to burn even in water. Discussed in detail in Chapters 2 and 9.
Hauptburg	(German). Lit. main castle, applying to the inner section where the lord resided, equivalent to inner ward or bailey.
helepolis, elepolis	(Greek). Usually defined as a "siege tower", a mobile timber tower built high enough to enable attackers on the top to shoot down onto a wall, and sometimes used with a bridge to enable attackers to storm a wall top. It has also been suggested that in some Byzantine texts it means nothing more precise than a "dominant weapon".
ḥiṣn, pl. *ḥuṣūn*	(Arabic). Fortified rural refuge(s).
hoarding	Timber gallery built out from the top of a wall to enable defenders to shoot or drop stones etc. onto attackers at the foot of the wall. French *bretêche*.
Hus	(German). Lit. house, signifying a defensible place.
indagines	(Latin). Earthwork/timber barricades blocking routes (Hungary).
incastellamento	(Italian). The process of establishing defended villages, from the tenth century.
kastron	(Greek). See *castrum*. Definition varied, sometimes meaning a fortified refuge, and later more precisely a castle.
kataphygia	(Greek). Byzantine fortified towns or refuges.
keep	Later word in modern English usage, describing a *donjon*.
kremlin, krom	(Russian). Citadel.
labdarea	(Greek). One of the four types of stone thrower in the tenth-century Byzantine arsenal.
liaisai	(Greek). Vine or willow Byzantine screens to protect soldiers at a siege.
licence to crenellate	Formally, the right to turn a building into a castle by providing it with battlements. In reality, usually signified recognition of

noble status for the owner, whether or not the building had any defensive function.

lupus — (Latin), lit. "wolf", another term for a siege weapon.

ma'aqul — (Arabic). Rural refuge (from eighth century).

machicolation — Murder hole. A permanent form of *hoarding*, comprising stone projections from the front of a tower or wall, with holes for dropping stones etc. onto attackers at the foot of the wall. In French, "meurtrière". Might be continuous along a wall, or a single structure e.g. over a door, in which case it is a "box machicolation".

manjanīq (madjanik), pl. *manājanīq* — (Arabic). Arabic version of the *mangonel*, usually a generic term for stone-throwing artillery.

mangana — (Latin). See *mangonel*.

manganikon — (Greek). See *mangonel*. In the tenth century, apparently referred to a specific weapon.

mangola — (Latin). See *mangonel*. One of many variant spellings.

mangonel — Both a generic term for medieval stone-throwing artillery, and to describe specifically manually (traction-) powered weapons. Sometime wrongly used as a translation for *onager*.

mantlet — Roofed timber protection for a besieger, implying a smaller protection for one or two, but also generic term for siege implements; also a wooden shutter between merlons, protecting defenders of a wall.

miles, pl. *milites* — (Latin). Roman soldier, medieval "knight".

mining — The excavation of the ground below a defence, with a view, usually, to cause the collapse of masonry above when the timber frames holding up the mine were burnt. One defence against this was the "countermine", involving digging out from inside the fortification to intercept the mine before it reached the wall.

ministeriales — (Latin). A specifically German institution of socially unfree knights, bound to service of their lord, but sometimes wealthy and powerful.

moat — Water-filled ditch.

motte (mota) — Artificial mound usually of earth, the higher part of a *motte and bailey* castle; the motte serving as base for the lord's tower (donjon) or for a fort in e.g. a siege castle.

motte and bailey — Very widespread form of *castle*, esp. from the tenth to twelfth centuries. The *motte* was the mound on which the lord's residence (hall or tower) might sit, below it the "bailey" (sometimes more than one) was usually an enclosed space for other buildings of lesser status.

munitio — (Latin). Fortress.

murage, Lat. *muragium* — Grant of a tax to be levied by a town for the purpose of constructing a town wall.

mus, musculus — (Latin). Lit. mouse, another term for a timber shelter for the besieger.

naphtha, naft — (Greek, Arabic). Petroleum, with particular reference to naphtha incendiaries, later transferred to mean gunpowder.

neokastron — (Greek). Byzantine walled village (twelfth century).

onager, pl. *onagri* — (Latin). One-armed torsion-powered late-Roman stone thrower.

p'ao, phao — (Chinese). A manually-powered stone thrower, with many different sizes and specifications covered by this general term.

petraria, petrobolos — (Latin, Greek). French "perrière", also "petrary" and many variants. General term for stone-throwing artillery, sometimes given a more specific meaning as a smaller, traction-powered weapon by comparison with the counterweight trebuchet.

phrouria — (Greek). Garrison fort.

portcullis — (from French). Commonly, a wooden or metal grille dropped from above to close an entrance. Universally used.

propugnacula — (Latin). General term for outer defences of a fortress, variously an outer wall, or bastions.

proteichisma — (Greek). Lower, outer defensive wall, sometimes with a more general significance of an outer defence.

ribat — (Arabic). Fortified religious building, sometimes a dwelling for Islamic volunteer warriors.

rocca — (Italian). Castle.

sapping — See also *mining*. A form of attack on a defence involving trying to remove stones at the base of a wall to bring about the collapse of the masonry above.

scorpio, pl. *scorpiones* — (Latin). Occasional alternative term for *onager*. Lit. "scorpion".

siege castle — Temporary fortification erected to enable its garrison to attack, or cut off, a besieged place. Often had the same form as contemporary *castles* (esp. when these were predominantly earth and timber) but did not share other castle functions. By the fourteenth century, more often called a fort.

steen — (Flemish). Tower or more broadly, castle.

ṭabīʿa — (Arabic). Rammed earth; compacted baked mud around wooden frames to make a very strong concrete for walls and towers.

talpa — (Latin). Lit. "mole", battering ram.

talus — Sloping outer face of wall or tower, keeping attacker at a distance, providing protection against sapping, also allowing defender to bounce stones off it.

tapial — (Spanish). See *ṭabīʿa*.

testudo, pl. *testudines* — (Latin) Lit. "tortoise", usually, by medieval times, a moveable roofed timber shelter to protect troops attacking the base of a wall; sometimes a generic term for a "siege weapon". In Roman times it had meant a military formation of soldiers using their shields to protect their heads and front.

tetrarea — (Greek). One of the four types of stone-thrower in the tenth century Byzantine arsenal.

thouria — (Greek). Garrison fort.

tormentum — (Latin). Originally, an instrument for twisting, etc. By extension, applied to mean torsion-powered artillery. In time, became generic term for artillery.

torre albarrana — (al-Andalus). External tower linked at the level of the wall walk with the adjacent main wall.

tower house	Modern term coined to describe a residential tower, sometimes with serious defences, built usually by lesser nobility. Sometimes had outer buildings and a "bawn", although traces of these have often not survived.
trebuchet	(uncertain derivation). The counterweight-powered stone thrower from the end of the twelfth century. Many variant names.
tripodia	(Latin). Uncertain, possibly a roofed wooden siege instrument.
urbs, pl. *urbes*	(Latin). City, but sometimes conveyed meaning of "fortified".
vár	(Hungarian). Castle.
Vorburg	(German). Outer castle area.
vinea	(Latin). Another word for a timber shelter at a siege.

Sources

There are many reference books containing information on the words used in this study, but no single glossary. The following have been particularly useful:

Alonso, M. *Enciclopedia del idioma: diccionario histórico y moderno de la lengua española (siglos XII al XX); etimológico, tecnológico, regional e hispanoamericano*. 3 vols. Madrid, 1958.

Bradbury, J. *The Routledge companion to medieval warfare*. London, 2004.

The encyclopedia of Islam. New edn. Ed. P J Bearman et al. 12 vols. plus Index. Leiden, 1960–2009.

Mittelhochdeutsches Handwörterbuch: zugleich als Supplement und alphabetischer Index zum mittelhochdeutschen Wörterbuche von Benecke–Müller–Zarncke. Ed. M Lexer. 3 vols. Leipzig, 1872–8.

Mora-Figueroa, L de. *Glosario de arquitectura defensiva medieval*. Cadiz, 1994.

Revised Medieval Latin word-list from British and Irish Sources. Ed. R E Latham. Repr. with supplement. Oxford, 1980; repr. 1994.

Schiller, K, and Lübben, A. *Mittelniederdeutsches Wörterbuch*. 6 vols. Bremen, 1875–81.

Bibliography

Abbreviations

The following abbreviations are used for frequently cited titles in the bibliography. Abbreviations used in the footnotes are appended to the pertinent entries below, and also tabulated on p. xiv above.

AC	*Archaeologia cambrensis*
AM	*Archéologie médiévale*
ANS	*Anglo-Norman studies*
AQ	Ausgewählte Quellen zur deutsche Geschichte des Mittelalters
BAR	*British archaeological reports*
BM	*Bulletin monumental*
CFHB	Corpus fontium historiae byzantinae
CG	*Château Gaillard*
CSHB	Corpus scriptorum historiae byzantinae
EHR	*English historical review*
MGH AA	*Monumenta Germaniae historica: antiquissimi auctores*
MGH SrG	*Monumenta Germaniae historica: scriptores rerum germanicarum*
MGH SrL	*Monumenta Germaniae historica: scriptores rerum langobardorum*
MGH SS	*Monumenta Germaniae historica: scriptores*
ns	new series (neue Folge, etc.)
PG	*Patrologia cursus completus: series graeca*, ed. J P Migne
PL	*Patrologia cursus completus: series latina*, ed. J P Migne
RA	*Revue archéologique*
RHC	Recueil des historiens des Croisades
RIS	Rerum italicarum scriptores
RS	Rolls series
SSCISAM	Settimane di studi del Centro Italiano di Studi sull'Alto Medioevo

Primary sources

'Abd al-Ḥamīd ibn Yaḥyá. *Das Sendschreiben des 'Abdalḥamīd B. Yaḥyā (gest. 132/750) an den Kronprinzen 'Abdallāh B. Marwān II*. Trans. H Schönig. Veröffentlichungen der Orientalischen Kommission, 38. Stuttgart, 1985.

'Abd Allāh ibn Buluggīn. "Deux nouveaux fragments des 'Mémoires' du roi Zīride 'Abd Allāh de Grenade". Ed. É Lévi-Provençal. *Al-Andalus*, 6 (1941), 1–63.

——. *El siglo XI en 1a. persona: las "Memorias" de 'Abd Allāh, último rey Zirí de Granada destronado por los Almorávides (1090)*. Trans. É Lévi-Provençal and E García Gómez. Alianza tres, 57. Madrid, 1980.

'Abd al-Wāḥid al-Marrākushī. *Histoire des Almohades d'Abd el-Wâh'id Merrâkechi.* Trans. E Fagnan. Algiers, 1893.

Abbo of Saint-Germain. *De bello parisiaco.* Ed. G H Pertz. *MGH SrG* i. Hannover, 1871.

——. *Le siège de Paris par les Normands.* Ed. and trans. H Waquet. Paris, 1942.

Abū Ḥāmid al-Ġarnaṭī. *Tuḥfat al-albāb.* Partial German trans. in I Hrbek, "Ein arabischer Bericht über Ungarn (Abū Hāmid al-Andalusī al-Garnatī, 1080–1170)", 207–11. *Acta orientalia Academiae Scientarum Hungaricae,* 5 (1955), 205–30.

Actus pontificum Cenomannis in urbe degentium. Ed. G Busson and A Ledru. Le Mans, 1901.

Adalbert. *Vita Heinrici ii. imperatoris.* Ed. G Waitz. In *MGH SS* iv: 792–814. Hannover, 1841.

——. *Continuatio Reginonis.* Ed. A Bauer and R Rau. AQ 8. Darmstadt, 1971.

Adam of Bremen. *History of the archbishops of Hamburg-Bremen.* Trans. F J Tschan. New York, 1959.

Ademar of Chabannes. *Chronique.* Ed. J Chavanon. Collection de textes pour servir à l'étude et à l'enseignement de l'histoire, 20. Paris, 1897.

'Afīf, Shams Sirāj. *Tārīkh-i Fīroz Shāhī.* In *The history of India, as told by its own historians,* iii: 269–373. Ed. H M Elliot and J Dowson. London, 1871.

Aineias the tactician. *How to survive under siege.* Trans. D Whitehead. Oxford, 1990.

Ajbar machmuâ (Colección de tradiciones): crónica anónima del siglo xi. Ed. and trans. E Lafuente y Alcántara. Colección de obras arábigas de historia y de geografía, i. Madrid, 1867.

Albert of Aachen. *Histoire des faits et gestes dans les regions d'outre-mer depuis l'année 1095 jusqu'à l'année 1120 de Jésus-Christ.* Trans. M Guizot. 2 vols. Collection des mémoires relatifs à l'histoire de France, depuis la fondation de la monarchie française jusqu'au 13e siècle, 20–21. Paris, 1824.

Alpertus of Metz. *De diversitate temporum libri ii.* Ed. G H Pertz. In *MGH SS* iv: 700–723. Hannover, 1841.

Amatus of Monte Cassino. *Storia de' Normanni.* Trans. V de Bartholomaeis. Fonti per la storia d'Italia, 76. Rome, 1935.

Ambroise. *The history of the Holy War.* In *Three Old French chronicles of the crusades: The history of the Holy War; The history of them that took Constantinople; The chronicle of Reims.* Trans. E N Stone. Seattle, 1939.

Ammianus Marcellinus. *Rerum gestarum libri.* In *Ammianus Marcellinus,* i–iii. Ed. and trans. J C Rolfe. Loeb classical library. Cambridge, Mass./London, 1935–9.

Anales toledanos i, ii. In *España sagrada: theatro geographico-historico de la Iglesia de España,* xxiii: 381–409. Ed. E Florez. 2nd edn. Madrid, 1799.

The Anglo-Saxon chronicle. Trans. G N Garmonsway. Everyman's library. Rev. edn. London, 1954. [*ASC*]

Anglo-Saxon poetry. Trans. R K Gordon. Everyman's library. Rev. edn. London, 1954.

*Annála Connacht = The annals of Connacht (*a.d. *1224–1544).* Ed. and trans. A M Freeman. Dublin, 1944.

Annales alammanici. In *MGH SS* i: 22–60. Ed. G H Pertz. Hannover, 1826.

Annales augustani a. 973–1104. In *MGH SS* iii: 123–36. Ed. G H Pertz. Hannover, 1838.

Annales Barenses a. 605–1043. In *MGH SS* v: 51–6. Ed. G H Pertz. Hannover, 1844.

Annales Cambriae. Ed. J Williams ab Ithel. RS 20. London, 1860.

Annales Casinenses. In *MGH SS* xix: 303–20. Ed. G H Pertz. Hannover, 1897.

Annales Cestrienses, or Chronicle of the abbey of S. Werburg, at Chester. Ed. and trans. R C Christie. Record society for the publication of original documents relating to Lancashire and Cheshire, 14. London, 1886.

Annales D. Alfonsi Portugallensium regis. Ed. in M Blöcker-Walter, *Alfons I. von Portugal: Studien zu Geschichte und Sage des Begründers der portugiesischen Unabhängigkeit*. Geist und Werk der Zeiten, 14. Zurich, 1966.

Les annales de l'abbaye Saint-Pierre de Jumièges: chronique universelle des origines au XIIIe siècle. Ed. and trans. J Laporte. Rouen, 1954.

Annales de Burton, A.D. 1004–1263. In *Annales monastici*, I. Ed. H R Luard. RS 36/1. London, 1864.

Annales de Burton. In *The Templars: selected sources*. Trans. M Barber and K Bate. Manchester/New York, 2002.

Annales de Margan, A.D. 1066–1232. In *Annales monastici*, I. Ed. H R Luard. RS 36/1. London, 1864.

Annales de Oseneia, A.D. 1016–1347. In *Annales monastici*, IV. Ed. H R Luard. RS 36/4. London, 1869.

Annales de Theokesberia, A.D. 1066–1263. In *Annales monastici*, I. Ed. H R Luard. RS 36/1. London, 1864.

Annales Hildesheimenses. Ed. G Waitz. *MGH SrG* VIII. Hannover, 1878.

Les annales impériales de l'Annam. Trans. A de Michels. 1 vol. in 3. Paris, 1889–94.

Annales Laureshamenses. In *MGH SS* I: 22–39. Ed. G H Pertz. Hannover, 1826.

Annales Magdeburgenses a. 1–1188; 1453–1460. In *MGH SS* XVI: 105–96. Ed. G H Pertz. Hannover, 1859.

Annales Marbacenses a. 631–1375. Ed. R Wilmans. In *MGH SS* XVII: 142–80. Hannover, 1856.

Annales Mettenses priores; accedunt additamenta Annalium Mettensium posteriorum. Ed. B von Simson. *MGH SrG* X. Hannover, 1905.

Annales Mettenses a. 687–930. In *MGH SS* I: 314–36. Ed. G H Pertz. Hannover, 1826.

Annales Pegavienses et Bosovienses a. 1000–1227. In *MGH SS* XVI: 232–70. Ed. G H Pertz. Hannover, 1859.

Annales placentini guelfi a. 1012–1235. In *MGH SS* XVIII: 411–57. Ed. G H Pertz. Hannover, 1863.

Annales prioratus de Dunstaplia, A.D. 1–1297. In *Annales monastici*, III. Ed. H R Luard. RS 36/3. London, 1866.

Annales prussici; see Franciscani Thorunensis Annales prussici

Annales Quedlinburgenses. In *MGH SS* III: 22–90. Ed. G H Pertz. Hannover, 1839.

Annales regni Francorum, inde ab a. 741 usque ad a. 829, qui dicuntur Annales Laurissenses maiores et Einhardi. Ed. G H Pertz and F Kurze. *MGH SrG* VI. Hannover, 1895.

Annales Sanctae Columbae Senonensis a. 708–1218. In *MGH SS* I: 102–9. Ed. G H Pertz. Hannover, 1826.

Annales Sancti Disibodi a. 891–1200. Ed. G Waitz. In *MGH SS* XVII: 4–30. Hannover, 1861.

Annales scanici. In *Danmarks middelalderlige annaler*. Ed. M C Gertz, M Lorenzen, and E Jørgensen; trans. E Kroman. Copenhagen, 1980.

Annales vedastini a. 877–900. In *MGH SS* I: 516–31. Ed. G H Pertz. Hannover, 1826.

Annales vedastini = Jahrbücher von St. Vaast. Ed. B von Simson, trans. J von Jasmund. In R Rau, *Quellen zur karolingischen Reichsgeschichte*, II: 289–337. AQ 6. Berlin, 1960.

Annales Vindocinenses. In *Recueil d'annales angevines et vendômoises*. Ed. L Halphen. Collection de textes pour servir à l'étude et à l'enseignement de l'histoire, 37. Paris, 1903.

Annales qui dicuntur Xantenses a. 790–873; Supplementum Annalium Xantensium a. 640–789. In *MGH SrG* XII: 1–39. Ed. B von Simson. Hannover, 1909.

The annals of Clonmacnoise: being annals of Ireland from the earliest period to A.D. 1408. Trans. C Mageoghagan (1627), ed. D Murphy. Dublin, 1896; repr. Felinfach, 1993.

The annals of Fulda. Trans. T Reuter. Manchester, 1992.

Annals of Jessulmer. Ed. in J Tod, *Annals and Antiquities of Rajasthan*, II: 216–88. London, 1829–32; repr. 1972.

The annals of Loch Cé: a chronicle of Irish affairs from A.D. *1014 to* A.D. *1590*. Ed. and trans. W M Hennessy. RS 54. London, 1871.

Annals of Mewar. Ed. in J Tod, *Annals and Antiquities of Rajasthan*, I: 211–652. London, 1829–32; repr. 1972.

The annals of St-Bertin. Trans. J L Nelson. Manchester, 1991.

The annals of Ulster (to A.D. *1131)*, part I: *Text and translation*. Ed. and trans. S Mac Airt and G Mac Niocaill. Dublin, 1983.

Anonymi Gesta Francorum et aliorum Hierosolymitanorum. Ed. B A Lees. Oxford, 1924.

"ANONYMUS", P. magister known as. *Gesta Hungarorum*. Ed. E Jakubovich. In *Scriptores rerum hungaricarum tempore ducum regumque stirpis Arpadianae gestarum*, I. Gen. ed. I Szentpétery. Budapest, 1937.

ANONYMUS VALESIANUS II. *Chronica theodoriciana*. In *Ammianus Marcellinus*, III. Ed. and trans. J C Rolfe. Loeb classical library. Cambridge, Mass./London, 1939.

Arab historians of the crusades. Trans. F Gabrieli and E J Costello. Islamic world series. London, 1969.

ARBEO, bishop of Freising. *Vitae sanctorum Haimhrammi et Corbiniani*. Ed. B Krusch. *MGH SrG* XIII. Hannover, 1920.

'ARĪB IBN SA'D al-Qurṭubī. *An account of the establishment of the Fatemite dynasty in Africa: being the annals of that province from the year 290 of the Heg'ra to the year 300, extracted from an ancient Arabic Ms. ascribed to El Mas'ûdi, belonging to the ducal library of Saxe-Gotha*. Ed. and trans. J Nicholson. Tübingen/ Bristol, 1840.

Aristakēs Lastivertc'i's History. Trans. R Bedrosian. New York, 1985.

ARNOLD of Lübeck. *Die Chronik Arnolds von Lübeck*. Trans. J C M Laurent. Die Geschichtschreiber der deutschen Vorzeit: 13. Jahrhundert, 3. Berlin, 1853.

——. *Chronica Slavorum*. Ed. J M Lappenberg and G H Pertz. *MGH SrG* XIV. Hannover, 1868.

ARNULF. *Gesta archiepiscorum Mediolanensium*. In *MGH SS* VIII: 1–31. Ed. L C Bethmann and W Wattenbach. Hannover, 1848.

Arnulfi diplomata. Ed. P Kehr. MGH Diplomata regum Germaniae ex stirpe Karolinorum, III. Berlin, 1940.

ASSER. *Life of King Alfred*. In *Alfred the great: Asser's Life of King Alfred and other contemporary sources*. Trans. S Keynes and M Lapidge. Harmondsworth, 1983.

ASTRONOMUS. *Vita Hludovici imperatoris*. In *MGH SS* II: 604–48. Ed. G H Pertz. Hannover, 1829.

ATTALEIATES, Michael. *Historia*. Ed. I Bekker. CSHB 34. Bonn, 1853.

BAIHAḲĪ, Abu'l Faẓl. *Tārīkhu-s Subuktigīn*. In *The history of India, as told by its own historians*, II: 53–154. Ed. H M Elliot and J Dowson. London, 1867.

BAKRĪ, Abū 'Ubayd 'Abd Allāh. *Description de l'Afrique septentrionale par Abou-Obeïd-el-Bekri*. Trans. W MacGuckin de Slane. Paris, 1859.

BALĀDHURĪ, Aḥmad ibn Yaḥyá. *Futūhu-l buldān*. In *The history of India, as told by its own historians*, I: 113–30. Ed. H M Elliot and J Dowson. London, 1867.

——. *Dal Futuh 'al buldan*. In *Biblioteca arabo-sicula: versione italiana*, I. Ed. and trans. M Amari. Turin/Rome, 1880.

——. *Origins of the Islamic state*. Trans. of *Futūh al-buldān* by P K Hitti. 2 vols. New York, 1916.

BAN Gu. *The history of the former Han dynasty by Pan Ku*. Trans. H H Dubs with Jen T'ai and P'an Lo-chi. 3 vols. Baltimore/London, 1938–55. [vol. II 1944]

Barbarossa in Italy. Trans. of *Carmen de gestis Frederici I imperatoris in Lombardia* by T Carson. New York, 1994.

BAR HEBRAEUS. *The Chronography of Gregory Abû'l Faraj, the son of Aaron, the Hebrew physician, commonly known as Bar Hebraeus: being the first part of his Political history of the world*, I: *English translation*. Trans. E A W Budge. Oxford, 1932.

BARTHOLOMEUS the scribe. *Annales a. 1225–1248*. *MGH SS* XVIII: 156–225. Ed. G H Pertz. Hannover, 1863.

The Bayeux tapestry: the complete tapestry in colour. Ed. D M Wilson. London/New York, 1985.

BENZO, bishop of Alba. *Ad Heinricum IV. imperatorem libri VII*. Ed. K Pertz. In *MGH SS* XI: 591–681. Hannover, 1854.

BERNOLD of Constance. *Chronicon*. In *MGH SS* V: 385–467. Ed. G H Pertz. Hannover, 1844.

BERTHOLD of Reichenau. *Annales a. 1054–1080*. *MGH SS* V: 264–326. Ed. G H Pertz. Hannover, 1844.

BONCOMPAGNO of Signa. *Liber de obsidione Ancone (a. 1173)*. Ed. G C Zimolo. RIS 6/3. Bologna, 1937.

BONIZO, bishop of Sutri. *Liber ad amicum*. Ed. E Dümmler. In *MGH Libelli de lite* I: 568–620. Hannover, 1891.

Brenhinedd y Saesson, or The king of the Saxons: BM Cotton MS Cleopatra B v, and the Black book of Basingwerk, NLW MS 7006. Ed. and trans. T Jones. University of Wales, Board of Celtic Studies, History and law series, 25. Cardiff, 1971. [*ByS*]

BROOKS, E W. "A Syriac chronicle of the year 846". *Zeitschrift der Deutschen morgenländischen Gesellschaft*, 51 (1897), 569–88.

——. "The Arabs in Asia Minor (641–750), from Arabic sources". *The journal of Hellenic studies*, 18 (1898), 182–208. Additions and corrections, ibid., 19 (1899), 31–3.

——. "The campaign of 716–18, from Arabic sources". *The journal of Hellenic studies*, 19 (1899), 19–31.

BRUNO of Magdeburg. *Liber de bello Saxonico*. In *MGH SS* V: 327–84. Ed. G H Pertz. Hannover, 1844.

Brut y Tywysogyon, or The chronicle of the princes: Peniarth Ms. 20 version. Trans. T Jones. University of Wales, Board of Celtic Studies, History and law series, II. Cardiff, 1952. [*ByT*(P)]

Brut y Tywysogyon, or The chronicle of the princes: Red book of Hergest version. Ed. and trans. T Jones. University of Wales, Board of Celtic Studies, History and law series, 16. Cardiff, 1955. [*ByT*(H)]

BRYENNIUS, Nicephorus. *Histoire*. Ed. and trans. P Gautier. CFHB 9. Brussels, 1975.

The Burghal Hidage. Ed. and trans. A R Rumble. In *The defence of Wessex: the Burghal Hidage and Anglo-Saxon fortifications*, 14–35. Ed. Hill, D, and Rumble, A R. Manchester, 1996.

Des byzantiner Anonymus Kriegswissenschaft. In *Griechische Kriegsschriftsteller*, II/2. Ed. and trans. H Köchly and W Rüstow. Leipzig, 1855.

CAFFARO di Rustico da Caschifellone et al. *Annali genovesi di Caffaro e de' suoi continuatori, dal MXCIX al MCCXCII*. Ed. L T Belgrano and C I di Sant'Angelo. 5 vols. Fonti per la storia d'Italia, 11–14*bis*. Genoa and Rome, 1890–1929.

CAGNY, Perceval de. *Chroniques*. Ed. H Moranville. Paris, 1902.

CAHEN, C. "La campagne de Manzikert d'après les sources musulmanes". *Byzantion* 9 (1934), 613–42.

Caithreim Cellachain Caisil = The victorious career of Cellachan of Cashel, or The wars between the Irishmen and the Norsemen in the middle of the 10th century. Ed. and trans. A Bugge. Oslo, 1905.

Calendar of documents relating to Scotland preserved in Her Majesty's Public Record Office, London. Ed. J Bain. 4 vols. Edinburgh, 1881–8.

Calendar of the close rolls preserved in the Public Record Office. 47 vols., organised by reign. London, 1892–1963.

Calendar of the liberate rolls preserved in the Public Record Office: Henry III. 6 vols. London, 1916–64.

Calendar of the patent rolls preserved in the Public Record Office. 73 vols., organised by reign. London, 1891–1986.

CAMENIATA, Joannes. *De excidio Thessalonicae.* In *PG* CIX: 519–654. Paris, 1863.

——. *Die Einnahme Thessalonikes durch die Arabe in Jahre 904.* Trans. G Böhlig. Byzantinische Geschichtsschreiber, 12. Graz, 1975.

Campaign organization and tactics. In *Three Byzantine military treatises.* Ed. and trans. G T Dennis. Washington, D.C., 1985.

Capitularia regum Francorum. Ed. A Boretius. 2 vols. *MGH Leges,* Sectio II. Hannover, 1883.

Carmen de bello Saxonico. Ed. O Holder-Egger. *MGH SrG* XVII. Hannover, 1889.

Carmen in victoriam Pisanorum. Ed. in H E J Cowdrey, "The Mahdia campaign of 1087", 23–9. *EHR* 92 (1977), 1–29.

Carolingian chronicles: royal Frankish annals and Nithard's Histories. Trans. B W Scholz with B Rogers. Ann Arbor, Mich., 1972.

Cartulaire normande de Philippe Auguste, Louis VIII, St. Louis et Philippe le Hardi. Ed. L Delisle. *Mémoires de la Société des antiquaires de la Normandie,* 2nd ser. 6 (1852), 1–346.

CASSIODORUS Senator. *The Variae of Magnus Aurelius Cassiodorus Senator: being documents of the kingdom of the Ostrogoths in Italy,* Trans. S J B Barnish. Translated texts for historians, 12. Liverpool, 1992.

CEDRENUS, Georgius. *Compendium historiarum.* Ed. and trans. W Xylander. *PG* CXXI–CXXII: 368. Paris, 1864.

Chach-nāma. In *The history of India, as told by its own historians,* I: 131–211. Ed. H M Elliot and J Dowson. London, 1867.

CHERBONNEAU, A. "Relation de la prise de Tébessa par l'armée arabe en l'an 45 de l'hégire". *Revue africaine,* 13 (1869), 229–38.

CHONIATES, Niketas. *O city of Byzantium: annals of Niketas Choniatēs.* Trans. H J Magoulias. Byzantine texts in translation. Detroit, 1984.

Christians and Moors in Spain. Ed. and trans. C Smith, C Melville and A Ubaydli. 3 vols. Hispanic classics. Warminster, 1988–92.

Chronica Adefonsi imperatoris. Ed. L Sánchez Belda. Madrid, 1950.

"Chronica de Walliae and other documents from Exeter Cathedral Library MS 3514". Ed. T Jones. *Bulletin of the Board of Celtic Studies,* 12 (1946), 27–44.

Chronica domni Rainaldi archidiaconi Sancti Mauricii Andagavensis. In *Chroniques des églises d'Anjou,* 1–16. Ed. P Marchegay and E Mabille. Paris, 1869.

Chronicae Sancti Albini Andegavensis in unum congestae. In *Chroniques des églises d'Anjou,* 17–61. Ed. P Marchegay and E Mabille. Paris, 1869.

Chronica gallica a. CCCLII *et* DXI. In *MGH AA* IX: 615–66. Ed. T Mommsen. Berlin, 1892.

Chronica imperatorum Heinrico v. dedicata. In *Frutolfs und Ekkehards Chroniken und die anonyme Kaiserchronik*, 211–65. Ed. and trans. F-J Schmale and I Schmale-Ott. AQ 15. Darmstadt, 1972.

Chronica Gothorum. In *Portugaliae monumenta historica: Scriptores*, I: 5–17. Lisbon, 1856.

Chronica marchionum Brandenburgensium. Ed. G Sello. *Forschungen zur brandenburgischen und preußischen Geschichte*, I (1888), 111–80; 3 (1890), 609–11.

Chronica monasterii Casinensis. Ed. H Hoffman. *MGH SS* XXXIV. Hannover, 1980.

Chronica Naierensis (Chronica hispana saeculi XII, pars II). Ed. J A Estévez Sola. Corpus christianorum, continuatio mediaevalis, 71a. Turnhout, 1995.

Chronica Sancti Benedicti Casinensis. In *MGH SrL*, 467–89. Ed. G Waitz. Hannover, 1878.

Chronicle of AD 1234. Latin trans. J.-B. Chabot as *Anonymi auctoris Chronicon ad annum Christi 1234 pertinens*, III. Corpus scriptorum Christianorum orientalium, 109. Louvain, 1937. *See also* DIONYSIUS I of Tel-Maḥrē.

Chronicle of Melrose. In *The Church historians of England*, IV/1: 77–242. Trans. J Stevenson. London, 1856.

The chronicle of Novgorod, 1016–1471. Trans. R Michell and N Forbes. Camden Society, 3rd ser. 25. London, 1914.

Chronicon Albeldense, auctore forte Dulcidio Salmaticensi episcopo, continuatore autem Vigila, monacho Albeldensi. In *PL* CXXIX: 1123–46. Paris, 1853.

Chronicon Bergomense guelpho-ghibellinum ab anno MCCCLXXVIII usque ad annum MCCCCVII. Ed. C Capasso. RIS 16/2. Bologna, 1926.

Chronicon Caesaraugustanorum reliquiae a. CCCCL–DLXVIII. In *MGH AA* XI: 221–4. Ed. T Mommsen. Berlin, 1894.

Chronicon Conimbricense. In *Portugaliae monumenta historica: Scriptores*, I: 1–5. Lisbon, 1856.

Chronicon de Lanercost, M.CC.I.–M.CCC.XLVI. Ed. J Stevenson. Bannatyne Club, 65. Edinburgh, 1839.

Chronicon lusitanum. In *España sagrada: theatro geographico-historico de la Iglesia de España*, XIV. Ed. E Florez. 2nd edn. Madrid, 1786.

Chronicon Moissiacense a saeculo quarto usque ad a. 818 et 840. *MGH SS* I: 280–313. Ed. G Pertz. Hannover, 1826.

Chronicon paschale, 284–628 AD. Trans. M Whitby and M Whitby. Translated texts for historians, 7. Liverpool, 1989.

Chronicon salernitanum a. 747–974. In *MGH SS* III: 467–561. Ed. G Pertz. Hannover, 1839.

Chronicon salernitanum. Ed. U Westerbergh. Studia latina Stockholmiensia, 3. Stockholm, 1956.

Chronicon Sancti Maxentii Pictavensis. In *Chroniques des églises d'Anjou*, 349–433. Ed. P Marchegay and E Mabille. Paris, 1869.

Chronicon Scotorum: a chronicle of Irish affairs, from the earliest times to A.D. 1135; with a supplement, containing the events from 1141 to 1150. Ed. and trans. W M Hennessy. RS 46. London, 1866.

Chronicon Sublacense (aa. 593–1369). Ed. R Morghen. RIS 24/6. Bologna, 1927.

Chronicon Turonense magnum (ab anno J.C. 81 ad annum 1127). In *Recueil des chroniques de Touraine*. Ed. A Salmon. Tours, 1854.

Chronicon vulgo dictum Chronicon Thomae Wykes, A.D. 1066–1289. In *Annales monastici*, ed. H R Luard, IV. RS 36/4. London, 1869.

Die Chronik Alfons III. Ed. J Prelog. Europäische Hochschulschriften, 3rd ser. 134. Frankfurt, 1980.

Chronique byzantine de l'an 811. Ed. in I Dujčev, *Medioevo bizantino-slavo*, II: *Saggi di storia letteraria*. Rome, 1968.

La chronique de Morigny (1095–1152). Ed. L Mirot. Collection de textes pour servir à l'étude et à l'enseignement de l'histoire, 41. Paris, 1909.

La chronique de Nantes (570 environ – 1049). Ed. R Merlet. Collection de textes pour servir à l'étude et à l'enseignement de l'histoire, 19. Paris, 1896.

La chronique de Nestor . . . accompagnée de notes et d'un recueil de pièces inédites touchant les anciennes relations de la Russie avec la France. Trans. L Paris. 2 vols. Paris, 1834–5.

Chronique d'Ernoul et de Bernard le trésorier. Ed. L de Mas Latrie. Paris, 1871.

La chronique hypatienne (selection). In *La Russie ancienne: IXe–XVIIe siècles*, 91–5. Ed. M Laran and J Saussay. Documents pour l'histoire des civilisations. Paris, 1975.

Chroniques des comtes d'Anjou et des seigneurs d'Amboise. Ed. L Halphen and R Poupardin. Collection de textes pour servir à l'étude et à l'enseignement de l'histoire, 48. Paris, 1913.

CINNAMUS, Joannes. *Deeds of John and Manuel Comnenus*. Trans. C M Brand. Records of civilization: sources and studies, 95. New York, 1976.

CLARI, Robert de. *The conquest of Constantinople*. Trans. E H McNeal. New York, 1936; repr. Toronto, 1996.

CLARIUS. *Chronicon Sancti Petri Vivi Senonensis*. In *Spicilegium, sive Collectio veterum aliquot scriptorum qui in Galliae bibliothecis delituerant*, II. Ed. L d'Achery. Paris, 1723.

——. *Chronique de Saint-Pierre-le-Vif de Sens, dite de Clarius*. Ed. and trans. R-H Bautier and M Gilles with A-M Bautier. Sources d'histoire médiévale. Paris, 1979.

COCKERELL, S C. *Old Testament miniatures: a medieval picture book with 283 paintings from the Creation to the story of David*. London, 1969.

CODAGNELLUS, Johannes. *Annales placentini*. Ed. O Holder-Egger. *MGH SrG* XXIII. Hannover, 1901.

——. *Gesta obsidionis Damiatae 1217–1219*. In *MGH SS* XXXI: 463–503. Ed. O Holder-Egger. Hannover, 1903.

Codice diplomatico Eceliniano. Ed. G Verci, *Storia degli Ecelini*, III. Bassano, 1779.

Cogadh Gaedhel re Gallaibh = The war of the Gaedhil with the Gaill, or The invasions of Ireland by the Danes and other Norsemen. Ed. and trans. J H Todd. RS 48. London, 1867.

COMNENA, Anna. *The Alexiad*. Trans. E R A Sewter. Harmondsworth, 1969.

Compilation anonyme sur la défense des places fortes. Trans. E Caillemer. *Mémoires de la Société d'émulation du Doubs*, 4th ser. 6 (1870), 383–439.

Conradi I., Heinrici I. et Ottonis I. diplomata. Ed. T. Sickel. MGH Diplomata regum et imperatorum Germaniae, I. Hannover, 1879.

CONSTANTINE VII PORPHYROGENITUS, emperor. *De ceremoniis aulae byzantinae libri duo*. Ed. and trans. J J Reiske. 3 vols. CSHB. Bonn, 1829.

——. *De administrando imperio*. Ed. G Moravcsik, trans. R J H Jenkins. Budapest, 1949.

——. *Vom Bauernhof auf den Kaiserthron: Leben des Kaisers Basileios I., des Begründers der Makedonischen Dynastie*. Trans. L Breyer. Byzantinischer Geschichtsschreiber, 14. Graz, 1981.

——. *Three treatises on imperial military expeditions*. Ed. and trans. J F Haldon. CFHB 28. Vienna, 1990.

Constitutiones et acta publica imperatorum et regum, II: *Inde ab a.* MCXCVIII. *usque ad a.* MCCLXXII. Ed. L Weiland. *MGH Leges, sectio IV*, II. Hannover, 1896.

Continuatio byzantia arabica a. DCCXLI; *Continuatio hispana a.* DCCLIV. In *MGH AA* XI: 323–69. Ed. T Mommsen. Berlin, 1894.

Conventio inter Guillelmus Aquitanorum comitem et Hugonis Chiliarchum. Ed. J Martindale. *EHR* 84 (1969), 528–48.

Corpus of early Arabic sources for West African history. Trans. J F P Hopkins. Fontes historiae africanae, series arabica, 4. Cambridge, 1981.

Corpus scriptorum muzarabicorum. Ed. J Gil. 2 vols. Madrid, 1973.

COSMAS of Prague. *Chronica Boemorum*. Ed. B Bretholz. *MGH SrG* ns II. Berlin, 1923.

The coucher book of Selby, from the original MS. in the possession of Thomas Brooke, 1: to which is prefixed Historia Selebiensis monasterii. Ed. J T Fowler. Yorkshire Archeological and Topographical Association record series, 10. Durham, 1891.

La cronaca dei conti e dei principi longobardi di Capua dei codice Cassinense 175 e Cavense 4 (815–1000). Ed. N Cilento. *Bullettino del Istituto storico italiano per il Medio Evo e Archivio muratoriano*, 69 (1957), 1–65.

Cronica di Cambridge. In *Biblioteca arabo-sicula: versione italiana*, 1. Ed. and trans. M Amari. Turin/Rome, 1880.

Crónica latina de los reyes de Castilla. Ed. and trans. L C Brea. Cádiz, 1984.

Les cronicques de Normendie (1223–1453): réimprimées pour la première fois d'après l'édition rarissime de Guillaume le Talleur, mai 1487, avec variantes et additions tirées d'autres éditions et de divers manuscrits. Ed. A Hellot. Rouen, 1881.

Crusaders as conquerors: the Chronicle of Morea. Trans. H E Lurier. Records of civilization: sources and studies, 69. New York, 1964.

Crusader Syria in the thirteenth century: the Rothelin continuation of the History of William of Tyre, with part of the Eracles or Acre text. Trans. J Shirley. Crusade texts in translation, 5. Aldershot, 1999.

The deeds of the Normans in Ireland = La geste des Engleis en Yrlande: a new edition of the chronicle formerly known as The song of Dermot and the earl. Ed. and trans. E Mullally. Dublin, 2002.

De expugnatione Lyxbonensi = The conquest of Lisbon. Ed. and trans. C W David. Records of civilization: sources and studies, 24. New York, 1936.

De expugnatione Terrae Sanctae per Saladinum libellus. In *Radulphi de Coggeshall Chronicon anglicanum; De expugnatione Terrae Sanctae libellus; Thomas Agnellus: De morte et sepultura Henrici regis Angliae junioris; Gesta Fulconis filii Warini; excerpta ex Otiis imperialibus Gervasii Tileburiensis*. Ed. J Stevenson. RS 66. London, 1875.

De gestis Herwardi Saxonis. In *Hereward: the siege of the Isle of Ely and involvement of Peterborough and Ely monasteries; together with De gestis Herewardi Saxonis (The exploits of Hereward the Saxon), researched and compiled in the 12th century by monastery historians*. Trans. T A Bevis. March, 1982.

DIONYSIUS I of Tel-Maḥrē. *Chronique de Denys de Tell-Mahré, quatrième partie*. Ed. and trans. J-B Chabot. 2 vols. Bibliothèque de l'École des hautes études, 112. Paris, 1895.

——. *The secular history* (extract from the *Chronicle of AD 1234*). In *The seventh century in the West-Syrian chronicles*, 111–220. Trans. A Palmer et al. Translated texts for historians, 15. Liverpool, 1993.

DŁUGOSZ, Jan. *Annales seu Chronicae incliti regni Poloniac*. 6 vols. Ed. J Dąbrowksı et al. Warsaw, 1964–97.

Documents relatifs au comté de Champagne et de Brie, 1172–1361. Ed. A Longnon. 3 vols. Collection de documents inédits sur l'histoire de France. Paris, 1901–14.

DONIZO. *Vita Mathildis –1115*. Ed. L Bethmann. In *MGH SS* XII: 348–409. Hannover, 1856.

DUDO of Saint-Quentin. *De moribus et actis primorum Normanniae ducum*. Ed. J Lair. Caen, 1865.

——. *History of the Normans*. Trans. E Christiansen. Woodbridge and Rochester, N.Y., 1998.

The earliest Welsh poetry. Ed. and trans. J P Clancy. Transactions. London, 1970.

Early charters of the cathedral church of St. Paul, London. Ed. M Gibbs. Camden Society, 3rd ser. 58. London, 1939.

Early sources of Scottish history, 500 to 1286. Ed. A O Anderson. 2 vols. Edinburgh, 1922.

ECSEDY, I. "Uigurs and Tibetans in Pei-t'ing (790–791 A.D.)". *Acta orientalia Academiae Scientarum Hungaricae*, 17 (1964): 83–104.

EINHARD. *The life of Charlemagne*. In *Einhard and Notker: Two lives of Charlemagne*. Trans. L Thorpe. Harmondsworth, 1969.

EKKEHARD. *Chronicon universale*. Ed. D G Waitz. In *MGH SS* VI: 33–265. Hannover, 1844.

ELIAS BAR SHĪNĀYA. *La chronographie d'Élie bar-Šinaya, métropolitain de Nisibe*. Trans. L J Delaporte. Bibliothèque de l'École des hautes études, 181. Paris, 1910.

Eman e dechreu historia hen Gruffud vab Kenan vab Yago; see Vita Griffini filii Conani r. Venedotiae vel Northwalliae.

ENNODIUS, Magnus Felix. *Opera*. Ed. F Vogel. *MGH AA* VII. Berlin, 1885.

Epistolae karolini aevi, III. Ed. E Dümmler. *MGH Epistolae* V. Berlin, 1899.

Epistolae karolini aevi, IV. Ed. E Perels and E Dümmler. *MGH Epistolae* VI. Berlin, 1925.

Epistolae karolini aevi, V. Ed. E Caspar et al. *MGH Epistolae* VII. Berlin, 1928.

Epistolae merovingici et karolini aevi, I. Ed. E Dümmler, W Gundlach and W Arndt. *MGH Epistolae* III. Berlin, 1892.

Epistolae saeculi XIII *e regestis pontificum romanorum selectae*, III. Ed. C Rodenberg. MGH Epistolae. Berlin, 1894.

ERCHEMPERTUS. *Historia Langobardorum beneventorum*. Ed. G H Waitz. In *MGH SrL*, 231–64. Hannover, 1878.

ERMOLDUS NIGELLUS. *Carmen elegiacum de rebus gestis Ludovici Pii, ab anno 781 usque ad annum 826*. In *PL* CV: 551–640. Paris, 1851.

EUSTATHIUS, archbishop of Thessaloniki. *The capture of Thessaloniki*. Ed. and trans. J R Melville Jones. Byzantina Australiensia, 8. Canberra, 1988.

EUTHYMIUS ZIGABENUS. *Panoplia dogmatica*. In *PG* cxxx: 9–20. Paris, 1865.

EVAGRIUS scholasticus. *Ecclesiastical history*. In *A history of the Church from* A.D. *322 to the death of Theodore of Mopsuestia,* A.D. *427, by Theodoret, bishop of Cyrus, and from* A.D. *431 to* A.D. *594 by Evagrius*. Trans. rev. E Walford. Bohn's ecclesiatical library. London, 1854.

Ex miraculis S. Wigberhti. Ed. G Waitz. In *MGH SS* IV: 224–8. Hannover, 1841.

"Extracts from the Annals of Innisfallen". *Kerry Archaeological Magazine*, 1 (1911/12), 415–20, 480–88; 2 (1912/13), 32–40, 93–7, 173–8; 3 (1915/16), 140–51, 293–300; 4–5 (1917–18), 106–15, 279–87.

"Extraits de l'Histoire des patriarches d'Alexandrie relatifs au siège de Damiette sous le règne d'al-Malik al-Kāmil". Ed. E Blochet. *Revue de l'Orient latin*, 11 (1905/8), 240–60.

Eymund's saga. In *Vikings in Russia: Yngvar's saga and Eymund's saga*. Trans. H Pálsson and P Edwards. Edinburgh, 1989.

FALCANDUS, Hugo. *The history of the tyrants of Sicily by "Hugo Falcandus", 1154–69*. Trans. G A Loud and T Wiedemann. Manchester medieval sources series. Manchester, 1998.

FALCO Beneventanus. *Chronicon*. Ed. and trans. S Gatti. *Cronisti e scrittori sincroni napoletani editi ed inediti = Cronisti e scrittori della dominazione normanna nel regno di Puglia e Sicilia*, I. Ed. G del Re. Naples, 1845.

FANTOSME, Jordan. *Chronicle of the war between the English and the Scots in 1173 and 1174*. In *The Church historians of England*, IV/I: 243–88. Trans. J Stevenson. London, 1856. Repr. in *Contemporary chronicles of the Middle Ages: sources of twelfth-century history*. Felinfach, 1988.

——. *Jordan Fantosme's Chronicle*. Ed. and trans. R C Johnston. Oxford and New York, 1981.

Feudal society in medieval France: documents from the county of Champagne. Trans. T Evergates. Philadelphia, 1993.

"The First and Second Crusades from an anonymous Syriac chronicle". Trans. A J Tritton with notes by H A R Gibb. *Journal of the Royal Asiatic Society*, (1933), 69–101.

FLAMMA, Gualvaneus de la. *Chronicon maius*. In his *Chronicon extravagans et Chronicon maius*. Ed. A Ceruti. Miscellanea di storia italiana, 7. Turin, 1869.

FLODOARD of Reims. *Les Annales*. Ed. P Lauer. Collection des textes pour servir à l'étude et à l'enseignement de l'histoire, 39. Paris, 1905.

FLORENCE of Worcester. *The Chronicle of Florence of Worcester, with the two continuations: comprising annals of English history, from the departure of the Romans to the reign of Edward I*. Trans. T Forester. Bohn's antiquarian library. London, 1854.

FOLIOT, Gilbert. *The letters and charters of Gilbert Foliot, abbot of Gloucester (1139–48), bishop of Hereford (1148–63), and London (1163–87)*. Ed. Z N Brooke with A Morey and C N L Brooke. London, 1967.

FOWLER, G H. "Munitions in 1224". *Publications of the Bedfordshire Historical Record Society*, 5 (1920), 117–29.

Fragmentary annals of Ireland. Ed. and trans. J N Radner. Dublin, 1978.

Fragmentum Chronici Fontanellensis, sive Sancti Wandregisili. In *Recueil des historiens des Gaules et de la France*, VII: 40–43. Ed. M Bouquet. Paris, 1749.

FRANCE, J. "The text of the account of the capture of Jerusalem in the Ripoll MSS, Bibliothèque Nationale (Latin) 5132". *EHR* 103 (1988), 640–57.

Franciscani Thorunensis Annales prussici (941–1410). Ed. E Strehlke. In *Scriptores rerum prussicarum*, III: 13–399. Leipzig, 1866.

FREDEGAR. *Chronicorum liber quartus cum continuationibus = The fourth book of the Chronicle of Fredegar with its continuations*. Ed. and trans. J M Wallace-Hadrill. Nelson's medieval texts. London, 1960.

FRUTOLF. *Chronica*. In *Frutolfs und Ekkehards Chroniken und die anonyme Kaiserchronik*. Ed. and trans. F-J Schmale and I Schmale-Ott. AQ 15. Darmstadt, 1972.

FULCHER of Chartres. *A history of the expedition to Jerusalem, 1095–1127*. Trans. F R Ryan. Knoxville, Tenn., 1969.

GALBERT of Bruges. *The murder of Charles the Good*. Trans. J B Ross. Rev. edn. Harper torchbooks. New York, 1967.

The Galician-Volynian chronicle. Trans. G A Perfecky. Harvard series in Ukrainian studies, 16/2. Munich, 1973.

GALLUS anonymus. *Cronicae et gesta ducum sive principum Polonorum = Kronika czyli Dzieje Książąt i władców polskich*. Ed. K Maleczyński. Monumenta poloniae historica = Pomniki dziejowe Polski, ser. 2, 2. Cracow, 1952.

GALTERIUS, chancellor of Antioch. *Walter the Chancellor's The Antiochene wars*. Trans. T S Asbridge and S Edgington. Crusade texts in translation, 4. Aldershot, 1999.

GARDĪZĪ, 'Abd al-Ḥayy ibn al-Ẓaḥḥāq. "On the Slavs". Trans. in C A Macartney, *The Magyars in the ninth century*, 65–6, 210–12. Cambridge, 1930.

——. "Gardizi on India". Trans. V Minorsky. *Bulletin of the School of Oriental and Asian Studies*, 12 (1948), 625–50.

GENESIUS, Josephus. *Historia de rebus Constantinopolitanis*. In *PG* CIX: 985–1154. Paris, 1863.

——. *Genesios on the reigns of the emperors*. Trans. A Kaldellis. Byzantina Australiensia, 11. Canberra, 1998.

GEORGIUS monachus. *Chronicon breve*. Ed. E von Muralt. *PG* CX. Paris, 1863.

GEORGIUS Pisida. *Bellum Avaricum*. In *PG* XCII: 1259–94. Paris, 1860.

——. *Giorgio di Pisidia: Poemi, 1: Panegirici epici*. Ed. and trans. A Pertusi. Studia patristica et byzantina, 7. Ettal, 1960.

GERHARDUS of Augsburg. *Vita sancti Oudalrici episcopi*. Ed. G Waitz. In *MGH SS* IV: 377–428. Hannover, 1841.

GERMANUS II, patriarch of Constantinople. "Homélie de S Germain sur la déliverance de Constantinople". Ed. V Grumel. *Revue des études byzantines*, 16 (1958), 183–205.

Gesta Ambasiensium dominorum. In *Chroniques des comtes d'Anjou et des seigneurs d'Amboise*. Ed. L Halphen and R Poupardin. Collection de textes pour servir à l'étude et à l'enseignement de l'histoire, 48. Paris, 1913.

Gesta Berengarii imperatoris. In *MGH Poetae latini* IV/1: 354–403. Ed. P de Winterfeld. Berlin, 1899.

Gesta comitum Barcinonensium. Ed. and trans. L Barrau Dihigo and J Massó Torrents. Cróniques catalanes, 2. Barcelona, 1925.

Gesta Herwardi incliti exulis et militis. In *Lestorie des Engles solum la translacion Maistre Geffrei Gaimar*, 1: 339–404. Ed. T D Hardy and C T Martin. Rerum Britannicarum medii aevi scriptores, 19. London, 1888.

Gesta sanctorum patrum Fontanellensis coenobii; Gesta abbatum Fontanellensium. Ed. F Lohier and J Laporte. Rouen, 1936.

Gesta Stephani. Ed. and trans. K R Potter. 2nd edn. Oxford medieval texts. Oxford, 1976.

Gesta triumphalia per Pisanos facta de captione Hierusalem et civitatis Maioricarum et aliarum civitatum et de triumpho habito contra Ianuenses. In *Gli Annales pisani di Bernardo Maragone*, fasc. 2. Ed. M Lupo Gentile. RIS VI/2. Bologna, 1930.

GHEVOND vardapet. *Histoire des guerres et des conquêtes des Arabes en Arménie*. Trans. G V Chahnazarian. Paris, 1856.

GILDAS. *The ruin of Britain, and other works*. Ed. and trans. M Winterbottom. Arthurian period sources, 7. London, 1978.

GIRALDUS Cambrensis. *Expugnatio hibernica = The conquest of Ireland*. Ed. and trans. A B Scott and F X Martin. A new history of Ireland: ancillary publications, 3. Dublin, 1978.

GISLEBERTUS of Mons. *Chronicon Hanoniense*. Ed. W Arndt. In *MGH SS* XXI: 481–601. Hannover, 1869.

——. *La chronique de Gislebert de Mons*. Ed. L Vanderkindere. Commission royale d'histoire: Recueil de textes pour servir à l'étude de l'histoire de Belgique. Brussels, 1904.

GIULIANO of Cividale. *Fragmenta Chronici Forojuliensis*. Ed. L Muratori. RIS 24. Milan, 1738.

GLABER, Rodulfus. *Historiarum libri quinque = The five books of the Histories*. Ed. and trans. J France. In *Rodulfi Glabri Historiarum libri quinque; Vita domni Willelmi abbatis*. Oxford medieval texts. Oxford, 1989.

GODFREY of Viterbo. *Gesta Friderici*. In *MGH SS* XXII: 307–34. Ed. G Waitz. Hannover, 1872.

——. *Gesta Heinrici VI*. In *MGH SS* XXII: 334–8. Ed. G Waitz. Hannover, 1872.

'GOS LO-TSĀ-BA GŽON-NU-DPAL. *The blue annals*. Ed. G N Roerich. 2 vols. Asiatic Society monograph series, 7. Calcutta, 1949–53.

Les grandes chroniques de France: selon que elles sont conservées en l'église de Saint-Denis en France, 1: Les premières sources de l'histoire de France jusqu'a la mort de Dagobert Ier. Ed. P Paris. Paris, 1836.

GREGORY I, pope. *Registrum epistolarum*. Ed. P Ewald and L Hartmann. 2 vols. *MGH Epistolae* I–II. Berlin, 1891–9.

GREGORY of Tours. *The history of the Franks*. Trans. L Thorpe. Harmondsworth, 1974.

GUIBERT of Nogent. *A monk's confession: the memoirs of Guibert of Nogent*. Trans. P J Archambault. University Park, Pa., 1996.

GUILELMUS Apuliensis. *La geste de Robert Guiscard*. Ed. and trans. M Mathieu. Istituto siciliano di studi bizantini e neoellenici: Testi e monumenti, 4. Palermo, 1961.

GUILELMUS Pictaviensis. *Gesta Guillelmi ducis Normannorum et regis Anglorum = Histoire de Guillaume le Conquérant, par Guillaume de Poitiers*. Ed. R Foreville. Les classiques de l'histoire de France au Moyen Âge, 23. Paris, 1952.

GUILHEM of Tudela et al. *La chanson de la Croisade albigeoise*. Ed. and trans. E Martin-Chabot. 3 vols. Paris, 1931–61.

——. *La chanson de la Croisade albigeoise*. Trans. H Gougaud. Paris, 1984.

——. *The song of the Cathar wars: a history of the Albigensian Crusade*. Trans. J Shirley. Crusade texts in translation, 2. Aldershot, 1996.

GUILLAUME of Nangis. *Chronique latine de 1113 à 1300, avec les continuations de cette chronique de 1300 à 1368*. Ed. H Géraud. 2 vols. Publications de la Société de l'histoire de France, 33, 35. Paris, 1843.

GUILLAUME of Puylaurens. *Chronique*. Trans. F P G Guizot. Croisade contre les albigeois, 1. Carcassone, 1970.

——. *The chronicle of William of Puylaurens: the Albigensian Crusade and its aftermath*. Trans. W A Sibly and M D Sibly. Woodbridge, 2003.

GUILLAUME le Breton. *La Philippide, poëme*. Ed. F Guizot. Collection de mémoires relatifs à l'histoire de France, depuis la fondation de la monarchie jusqu'au 13e siècle, 12. Paris, 1825.

——. *Œuvres*. In *Œuvres de Rigord et de Guillaume le Breton, historiens de Philippe-Auguste*. Ed. H F Delaborde. 2 vols. Publications de la Société de l'histoire de France, 210, 224. Paris, 1882–5.

GUNTHER of Pairis. *The capture of Constantinople: the Hystoria constantinopolitana*. Ed. and trans. A J Andrea. Philadelphia, 1997.

GUY, bishop of Amiens. *The "Carmen de Hastingae proelio"*. Ed. and trans. C Morton and H Muntz. Oxford medieval texts. Oxford, 1972.

Hacon's saga. In *Icelandic sagas and other historical documents relating to the settlements and descents of the Northmen on the British isles*, IV: *The saga of Hacon, and a fragment of The saga of Macon*. Trans. G W Dasent. Rerum britannicarum medii aevi scriptores, 88/4. London, 1894.

ḤAMZAH IBN AL-ḤASAN al-Iṣfahānī. *The annals of Hamzah al-Isfahani*. Trans. U M Daudpota. *Journal of the K R Cama Oriental Institute*, 22 (1932), 58–120.

HASAN, S A. "Munejjum Bashi's account of the reign of sultan Malik Shah". *Islamic studies*, 3 (1964), 445–56.

HELGALDUS of Fleury. *Vie de Robert le Pieux = Epitoma vitae regis Rotberti Pii*. Ed. and trans. R-H Bautier and G Labory. Sources d'histoire médiévale, 1. Paris, 1965.

HELMOLD of Bosau. *The chronicle of the Slavs*. Trans. F J Tschan. Records of civilization: sources and studies, 21. New York, 1935.

——. *Chronica Slavorum a. 800–1172*. Ed. I M Lappenberg. In *MGH SS* XXI: 1–99. Hannover, 1868.

HENRICUS de Lettis. *The Chronicle of Henry of Livonia*. Trans. J A Brundage. Documents from medieval Latin. Madison, Wisc., 1961.

HENRI de Valenciennes. *Histoire de l'empereur Henri de Constantinople*. Ed. J Longnon. Documents relatifs à l'histoire des croisades, 2. Paris, 1948.

HENRY of Huntingdon. *The Chronicle of Henry of Huntingdon: comprising the history of England, from the invasion of Julius Caesar to the accession of Henry II*. Trans. T Forester. London, 1853; repr. Felinfach, 1991.

HERMANN of Reichenau. *Herimanni Augiensis Chronicon = Hermann von Reichenau Chronik*. Ed. R Buchner, trans. K Nobbe. In *Quellen des 9. und 11. Jahrhunderts zur Geschichte der hamburgischen Kirche und des Reiches*, 615–707. AQ 11. Darmstadt, 1961.

HERMANN of Wartberg. *Chronicon Livoniae*. Ed. E Strehlke. In *Scriptores rerum prussicarum*, II: 9–178. Leipzig, 1863.

HERO of Byzantium. *Liber de machinis bellicis; necnon Liber de geodaesia*. Trans. F Barozzi. Venice, 1572.

HERO of Byzantium. *Siegecraft: two tenth-century instructional manuals*. Ed. and trans. D F Sullivan. Dumbarton Oaks studies, 36. Washington, D.C., 2000.

ḤIMYARĪ, Muḥammad ibn 'Abd al-Mun'im. *La péninsule ibérique au Moyen-Âge d'après le Kitāb arrawḍ al mi'ṭṭārfī ẖabar al-akṭār d'Ibn 'Abd al Mun'im al-Ḥimyarī*. Ed. and trans. E Lévi-Provençal. Publications de la Fondation de Goeje, 12. Leiden, 1938.

L'histoire de Guillaume le Maréchal, comte de Striguil et de Pembroke, régent d'Angleterre de 1216 à 1219. Ed. and trans. P Meyer. 3 vols. Publications de la Société de l'histoire de France, 255, 268, 304. Paris, 1891–1901.

Histoire de la Géorgie, depuis l'antiquité jusqu'au XIXe siècle. Trans. M Brosset. 7 vols. St Petersburg, 1849–58.

Histoire des ducs de Normandie et des rois d'Angleterre; suivie de La relation du Tournoi de Ham, par Sarrazin, trouvère du XIIIe siècle. Ed. F Michel. Publications de la Société de l'histoire de France, 18. Paris, 1840.

Historia compostelana, sive De rebus gestis D. Didaci Gelmirez. Ed. E Florez. *España sagrada: theatro geographico-historico de la Iglesia de España*, XX. Madrid, 1765.

Historia de la corona de Aragón (la más antigua de que se tiene noticia), conocida generalmente con el nombre de Crónica de San Juan de la Peña. Ed. T Ximénez de Embun. Zaragoza, 1876.

Historia de Leone Bardae Armenii filio. In *Leonis grammatici Chronographia*. Ed. I Bekker, trans. J Goar. CSHB 34. Bonn, 1842.

Historia Roderici. Ed. in R Menendez Pidal, *La España del Cid*, II: 901–67. 5th edn. Madrid, 1956.

Historia Sancti Florentii Salmurensis. In *Chroniques des églises d'Anjou*, 205–328. Ed. P Marchegay and E Mabille. Paris, 1869.

Historia Silense. Ed. F Santos Coco. Textos latinos de la Edad Media española: Crónicas, 2. Madrid, 1921.

"History of Katsina". Trans. H R Palmer. In *Sudanese memoirs: being mainly translations of a number of Arabic manuscripts relating to the central and western Sudan*, III: 77–80. 3 vols. in 1. Lagos, 1928; repr. London, 1967.

A history of Sharvān and Darband in the 10th–11th centuries = Fuṣūl min Ta'rīkh al-Bāb wa-Sharwān. Ed. and trans. V Minorsky. Cambridge, 1958.

Hlotharii Capitulare Papiense, 832. Febr. In *Capitularia regum Francorum*, II: 59–63. Ed. A Boretius. *MGH Leges*, Sectio II. Hannover, 1883.

HOFMEISTER, A. "Eine neue Quelle zur Geschichte Friedrich Barbarossas, *De ruina civitatis Terdonae*: Untersuchungen zum 1. Römerzug Friedrichs I." *Neues Archiv der Gesellschaft für ältere deutsche Geschichtskunde*, 43 (1922), 87–157.

HRABANUS MAURUS; *see* Vegetius, *De procinctu romanae militiae*.

HYDATIUS. *Continuatio chronicorum Hieronymianorum ad a.* CCCCLXVIII. In *MGH AA* XI: 1–36. Ed. T Mommsen. Berlin, 1894.

IBN 'ABD AL-ḤAKAM. *Dhikr fatḥ al-Andalus: Ibn Abd-el-Hakem's History of the conquest of Spain.* Ed. and trans. J H Jones. Göttingen and London, 1858.

——. *The history of the conquest of Egypt, North Africa and Spain known as the Futūḥ Miṣr.* Ed. C C Torrey. Yale oriental series, 3. New Haven, 1922.

——. *La conquête de l'Égypte.* Trans. O Tousson. *Bulletin de la Société archéologique d'Alexandrie*, 20 (1924), 213–38.

——. *Conquête de l'Afrique du Nord et de l'Espagne (Futûh Ifrîqiya wa'l-Andalus).* Ed. and trans. A Gateau. 2nd edn. Bibliothèque arabe-française, 2. Algiers, 1947.

IBN ABĪ ZAR' al-Fāsī, 'Alī ibn 'Abd Allāh. *El Cartás: noticia de los reyes del Magreb e historia de la ciudad de Fez por Aben Abi Zara.* Trans. A Huici. Valencia, 1918.

IBN AL-ATHĪR, 'Izz al-Dīn. *Kāmilu-t Tawārīkh.* In *The history of India, as told by its own historians*, II: 244–51. Ed. H M Elliot and J Dowson. London, 1867.

——. *Dal Kamil 'at tawarila.* In *Biblioteca arabo-sicula: versione italiana*, I. Ed. and trans. M Amari. Turin/Rome, 1880.

——. *Annales du Maghreb & de l'Espagne.* Trans. E Fagnan. Algiers, 1898.

——. *The annals of the Saljuq Turks: selections from al-Kāmil fi 'l-Ta'rīkh of 'Izz al-Dīn Ibn al-Athīr.* Trans. D S Richards. London, 2002.

IBN AL-KARDABŪS, 'Abd al-Malik. *Historia de al-Andalus (Kitāb al-Iktifā').* Trans. F Maillo Salgado. Akal bolsillo, 169. Madrid, 1986.

IBN AL-NADĪM, Muḥammad ibn Isḥāq. *The Fihrist of al-Nadīm: a tenth-century survey of Muslim culture.* Ed. and trans. B Dodge. 2 vols. Records of civilization: sources and studies, 83. New York/London, 1970.

IBN AL-QALĀNISĪ, Ḥamzah ibn Asad. *The Damascus chronicle of the crusades, extracted and translated from the chronicle of Ibn al-Qalānisī.* Trans. H A R Gibb. University of London historical series, 5. London, 1932.

IBN AL-QUṬĪYAH, Muḥammad ibn 'Umar. *The history of the conquest of al-Andalus.* Trans. J M Nichols. PhD dissertation, University of North Carolina, 1975.

IBN BĪBĪ, Nāṣir al-Dīn Ḥusayn Ibn Muḥammad. *Die Seltschukengeschichte des Ibn Bībī.* Trans. H W Duda. Copenhagen, 1959.

IBN FAḌL ALLĀH AL-'UMARĪ, Aḥmad ibn Yahyā. *Masālik el abṣār fi mamālik el amṣār*, I: *L'Afrique, moins l'Égypte.* Trans. M Gaudefroy-Demombynes. Bibliothèque des géographes arabes, 2. Paris, 1927.

IBN ḤAWQAL, Muḥammad. *Configuration de la terre (Kitab Surat al-Ard).* Trans. J H Kramers and G Wiet. 2 vols. Beirut, 1964.

IBN HAYYĀN, Abū Marwān Hayyān ibn Khalaf. *Crónica del califa 'Abdarrahmān III an-Nāsir entre los años 912 y 942 (al-Muqtabis V).* Trans. J Viguera and F Corriente. Textos medievales, 64. Saragossa, 1987.

IBN 'IDHĀRĪ al-Marrākushī, Aḥmad ibn Muḥammad. *Histoire de l'Afrique et de l'Espagne intitulée Al-Bayano'l-mogrib.* Trans. E Fagnan. 2 vols. Algiers, 1901.

——. *Al-Bayān al-mugrib fi ijtiṣār ajbār muluk al-Adalus wa al-Magrib.* Ed. and trans. A Huici Miranda. 2 vols. Colección de crónicas árabes de la Reconquista, 2. Tetuan, 1953–4.

IBN ISFANDIYĀR, Muḥammad ibn al-Ḥasan. *An abridged translation of the History of Tabaristán compiled about* A.H. *613* (A.D. *1216*). Trans. E G Browne. E J W Gibb memorial series, 2. Leiden/London, 1905.

IBN ISḤĀQ, Muḥammad. *The life of Muhammad.* Abridged by 'Abd al-Malik ibn Hishām; trans. A Guillaume. London/Lahore, 1955.

IBN JUBAYR, Muḥammad ibn Aḥmad. *The travels of Ibn Jubayr : being the chronicle of a mediaeval Spanish Moor concerning his journey to the Egypt of Saladin, the holy cities of Arabia, Baghdad the city of the caliphs, the Latin kingdom of Jerusalem, and the Norman kingdom of Sicily*. Trans. R J C Broadhurst. London, 1952.

IBN KHALDŪN, 'Abd al-Raḥmān ibn Muḥammad. "Historia de los árabes de España". Trans. O A Machada. *Cuadernos de historia de España* [Buenos Aires], 4 (1946), 136–46.

——. *Histoire des berbères et des dynasties musulmanes de l'Afrique septentrionale*. Trans. W Macguckin de Slane. 2nd edn. 4 vols. Paris, 1925–34; repr. several times.

IBN MISKAWAYH, Aḥmad ibn Muḥammad. *The concluding portion of The experiences of the nations*. Vols. I–II, IV–V of *The eclipse of the 'Abbasid caliphate: original chronicles of the fourth Islamic century*. Ed. H F Amedroz, trans. D S Margoliouth. 7 vols. Oxford, 1920–21.

IBN ṢĀḤIB AL-ṢALĀH, 'Abd al-Malik ibn Muḥammad. *Al-Mann bil-Imāma*. Trans. A Huici Miranda. Textos medievales, 24. Valencia, 1969.

IBN SHADDĀD, Bahā' al-Dīn Yūsuf ibn Rāfi'. *Saladin, or What befell Sultan Yûsuf (Salâh ed-Dîn) (1137-1193 A.D.)*. Trans. C W Wilson. Library of the Palestine Pilgrims' Text Society, 13. London, 1897.

IBRĀHĪM IBN YA'KUB al-Isrāīlī. *Relatio Ibrāhīm ibn Ja'ḳūb de itinere slavico, quae traditur apud al-Bekrī = Relacja Ibrāhīma ibn Ja'ḳūba z podróży do krajów słowiańskich w przekazie al-Bekrīego*. Ed. and trans. T Kowalski. Monumenta Poloniae historica, 2nd ser. I. Cracow, 1946.

IDRĪSĪ. *Description de l'Afrique et de l'Espagne*. Ed. and trans. R Dozy and M J de Goeje. Leiden, 1866.

Ignoti monachi Cisterciensis S. Mariae de Ferraria Chronica et Ryccardi de Sancto Germano Chronica priora. Ed. A Gaudenzi. Naples, 1888.

'IMĀD AL-DĪN MUḤAMMAD IBN MUḤAMMAD KĀTIB al-Iṣfahānī. *Conquête de la Syrie et de la Palestine par Saladin (al-Fatḥ al-qussî fî l-fatḥ al-qudsî)*. Trans. H Massé. Documents relatifs à l'histoire des Croisades, 10. Paris, 1972.

Inventaire des archives de la ville de Bruges, ser. I: *Treizième siècle*. Ed. L Gilliodts-van Severen. Bruges, 1871–85. [vol. III, 1875, cit.]

ISIDORE of Seville. *History of the kings of the Goths, Vandals, and Suevi*. Trans. G Donini and G B Ford jr. Leiden, 1966.

Islam: from the Prophet Muhammad to the capture of Constantinople, I: *Politics and war*. Ed. and trans. B Lewis. Documentary history of Western civilization. New York, 1974; repr. New York/Oxford, 1987.

Itinerarium peregrinorum et gesta regis Ricardi. Chronicles and memorials of the reign of Richard I, I. Ed. W Stubbs. RS 38. London, 1864.

JAN of Czarnkow. *Chronicon Polonorum*. In *Monumenta Poloniae historica*, II. Ed. A Bielowski. Lvov, 1872.

Japan in the Chinese dynastic histories: later Han through Ming dynasties. Trans. R Tsunoda. South Pasadena, Calif., 1951.

JIMÉNEZ DE RADA, Rodrigo. *Historia Arabum*. In *Roderici Ximenii de Rada Opera praecipua*. PP. Toletanorum quotquot extant opera, 3. Madrid, 1793. Facs. ed. M D Cabanes Pecourt. Textos medievales, 22. Valencia, 1968.

JOANNES, bishop of Ephesus. *Historiae ecclesiasticae pars tertia*. Ed. and trans. E W Brooks. Corpus scriptorum Christianorum orientalium, 105–6. Paris/Louvain, 1936.

JOANNES BICLARENSIS. *Chronica a.* CCCCXLIV–DLXVII. In *MGH AA* XI: 207–20. Ed. T Mommsen. Berlin, 1894.

JOANNES DE GARLANDIA. *The Dictionarius of John de Garlande*. In *A volume of vocabularies: illustrating the condition and manners of our forefathers, as well as the history of the forms of elementary education and of the languages spoken in this island from the tenth century to the fifteenth*. Ed. T Wright. A library of national antiquities, 1. Liverpool, 1857.

JOHN of Worcester. *The Chronicle of John of Worcester, 1118–1140: being the continuation of the "Chronicon ex chronicis" of Florence of Worcester*. Ed. J R H Weaver. Anecdota Oxoniensia: mediaeval and modern series, 13. Oxford, 1908.

JOHN, bishop of Nikiû. *The Chronicle of John, bishop of Nikiu*. Trans. R H Charles. Text and Translation Society publications. London/Oxford, 1916.

JORDANES. *The Gothic history of Jordanes*. Trans. C C Mierow. Princeton, 1915.

——. *De origine actibusque Getarum*. Ed. F Giunta and A Grillone. Fonti per la storia d'Italia, 117. Rome, 1991.

JOSHUA the Stylite. *The Chronicle of Joshua the Stylite, composed in Syriac A.D. 507*. Ed. and trans. W Wright. Cambridge, 1882.

——. *The chronicle of pseudo-Joshua the Stylite*. Trans. F R Trombley and J W Watt. Translated texts for historians, 32. Liverpool, 2000.

JULIAN, bishop of Toledo. *Historia rebellionis Pauli adversus Wambam Gothorum regem*. In *PL* XCVI: 759–98. Paris, 1851.

KADŁUBEK, Wincenty. *Magistri Vincentii episcopi Cracoviensis Chronica Polonorum, sive Originale regum et principum Poloniae*. Ed. A Przezdziecki. Cracow, 1862.

Kaiserchronik. In *Frutolfs und Ekkehards Chroniken und die anonyme Kaiserchronik*. Ed. and trans. F-J Schmale and I Schmale-Ott. AQ 15. Darmstadt, 1972.

Kalhaṇa's Rājataraṅgiṇī: a chronicle of the kings of Kaśmīr. Trans. M A Stein. 2 vols. London, 1900.

"Kano Chronicle". Trans. H R Palmer. In *Sudanese memoirs: being mainly translations of a number of Arabic manuscripts relating to the central and western Sudan*, III: 99–109. 3 vols. in 1. Lagos, 1928; repr. London, 1967.

KARL JÓNSSON. *Sverrissaga: the saga of King Sverri of Norway*. Trans. J Sephton. Northern library, 4. London, 1899; repr. Felinfach, 1994.

KAUṬILYA. *Essentials of Indian statecraft: Kautilya's "Arthasastra" for contemporary readers*. Selected and trans. T N Ramaswamy. London/New York, 1962.

KEKAUMENOS. *Vademecum des byzantinischen Aristokraten: das sogenannte Strategikon des Kekaumenos*. Trans. H G Beck. Byzantinische Geschichtschreiber, 5. Graz, 1956.

KEMĀL EL-DĪN. *Extraits de la Chronique d'Alep*. In *RHC: Historiens orientaux*, III: 571–690. Paris, 1884.

KHUSRAU, amīr. *Tārīkh-i 'Alāī, or Khazāīnu-l futūh*. In *The history of India, as told by its own historians*, III: 67–92. Ed. H M Elliot and J Dowson. London, 1871.

Knytlinga saga: the history of the kings of Denmark. Trans. H Pálsson and P Edwards. Odense, 1986.

LALOU, M. "Revendications des fonctionnaires du Grand Tibet au VIIIe siècle". *Journal asiatique*, 243 (1955), 171–212.

LAMPERT of Hersfeld. *Annalen*. Trans. A Schmidt. AQ 13. Darmstadt, 1962.

LANDULFUS. *Historia Mediolanensis usque ad a. 1085*. Ed. L C Bethmann and W Wattenbach. *MGH SS* VIII: 32–100. Hannover, 1848.

LAVAGNINI, B. "Siracusa occupata dagli Arabi e l'epistola di Teodosio Monaco". *Byzantion* 29/30 (1959–60), 267–79.

"Lebensbeschreibungen von Heerführern und Würdenträgern des Hauses Sui". Ed. A Pfizmaier. *Denkschriften der kaiserlichen Akademie der Wissenschaften: philosophisch-historische Klasse*, 32 (1882), 281–377.

Leges Henrici primi. Ed. and trans. L J Downer. Oxford, 1972.

Leo vi, emperor. *Tactica, sive De re militari*. In *PG* cvii: 657–1116. Paris, 1863.

Leo diaconus. *Historia*. In *PG* cxvii: 655–924. Paris, 1864.

——. *Nikephoros Phokas, "der bleiche Tod der Sarazenen", und Johannes Tzimiskes: die Zeit von 959 bis 976 in der Darstellung des Leon Diakonos*. Trans. F Loretto. Byzantinische Geschichtsschrieber, 10. Graz, 1961.

Leo grammaticus. *Chronographia*. In *PG* cviii: 1038–186. Paris, 1861.

Le Strange, G. *The lands of the eastern Caliphate: Mesopotamia, Persia and Central Asia, from the Moslem Conquest to the time of Timur*. Cambridge, 1905; repr. London, 1966.

Lévi-Provençal, É. "Une héroïne de la résistance musulmane en Sicile au début du xiiie siècle". *Oriente moderno*, 34 (1954), 283–88.

Liber maiolichinus de gestis Pisanorum illustribus: poema della guerra balearica. Ed. C Calisse. Fonti per la storia d'Italia, 29. Rome, 1904.

Liudprand, bishop of Cremona. *The works of Liudprand of Cremona*. Trans. F A Wright. London/New York, 1930; repr. London, 1993. Includes *Antapodosis, Liber de rebus gestis Ottonis, Relatio de legatione Constantinopolitina*.

The lives of the eighth-century popes (Liber pontificalis): the ancient biographies of nine popes from AD *715 to* AD *817*. Trans. R Davis. Translated texts for historians, 13. Liverpool, 1992.

The lives of the ninth-century popes (Liber pontificalis): the ancient biographies of ten popes from AD *817 to* AD *891*. Trans. R Davis. Translated Texts for Historians, 20. Liverpool, 1995.

Livländische Reimchronik. Ed. L Meyer. Paderborn, 1876.

Ludowici germanici, Karlomanni, Ludowici junioris diplomata. Ed. P Kehr. MGH Diplomata regum Germaniae ex stirpe Karolinorum, 1. Berlin, 1934.

Magnae Moraviae fontes historici, 1: Annales et chronicae. Ed. D Bartoňková et al. Prague/Brno, 1966.

Magni rotuli Scaccarii Normanniae sub regibus Angliae. Ed. T Stapleton. 2 vols. London, 1840–44.

Makīn, Jirjis ibn al-ʿAmīd. *Chronique des Ayyoubides (602–658/1205–6 – 1259–60)*. Trans. A-M Eddé and F Micheau. Paris, 1994.

Malaspina, Saba. *Die Chronik*. Ed. W Koller and A Nitschke. *MGH SS* xxxv. Hannover, 1999.

Malaterra, Gaufredus. *Historia sicula*. In *PL* cxlix: 1087–210. Paris, 1853.

——. *De rebus gestis Rogerii Calabriae et Siciliae comitis et Roberti Guiscardi ducis fratris eius*. Ed. E Pontieri. RIS 5/1. Bologna, 1928.

Mānasāra. *Architecture of Mānasāra*. Trans. P K Acharya. 2 vols. London/Allahabad, 1934.

Mappae clavicula: a little key to the world of medieval techniques. Ed. and trans. C S Smith and J G Hawthorne. Transactions of the American Philosophical Society, ns 64/4. Philadelphia, 1974.

Maqqarī, Aḥmad ibn Muḥammad. *The history of the Mohammedan dynasties in Spain, extracted from the Nafhu-t-tíb min ghosni-l-Andalusi-r-rattíb wa táríkh Lisánu-d-Dín Ibni-l-Khattíb by Ahmed ibn Mohammed al-Makkarí, a native of Telemsán*. Trans. P de Gayangos. 2 vols. London, 1840–43.

Maqrīzī, Taqī al-Dīn Aḥmad ibn Alī. *Histoire des sultans mamlouks de l'Égypte, écrite en Arabe par Taki-Eddin-Ahmed-Makrizi*. Trans. É Quatremère. 2 vols. in 4. Oriental Translation Fund, 49. Paris, 1837.

——. *Histoire d'Égypte de Makrizi*. Trans. E Blochet. *Revue de l'Orient latin*, 6 (1898), 435–

89; 8 (1900–1901), 165–212, 501–33; 9 (1902), 6–163, 466–530; 10 (1903–4), 248–371; 11 (1905–8), 192–260.

MARAGONE, Bernardo. *Gli Annales pisani*. Ed. M Lupo Gentile. RIS 6/2. Bologna, 1930.

MARCELLINUS comes. *Chronicon ad a.* DXVIII *continuatum ad a.* DXXXIV. In *MGH AA* XI: 37–108. Ed. T Mommsen. Berlin, 1894.

MARCHISIUS the scribe. *Annales a. 1220–1224*. In *MGH SS* XVIII: 142–56. Ed. G H Pertz. Hannover, 1863.

"Maronite chronicle" (extract). In *The seventh century in the West-Syrian chronicles*, 29–35. Trans. A Palmer et al. Translated texts for historians, 15. Liverpool, 1993.

MATTHEW of Edessa. *Armenia and the crusades, tenth to twelfth centuries: the Chronicle of Matthew of Edessa*. Trans. A E Dostourian. Armenian heritage series. Lanham, Md., 1993.

MAURICE, emperor. *Das Strategikon des Maurikios*. Ed. G T Dennis, trans. E Gamillscheg. CFHB 17. Vienna, 1981.

A mediaeval prince of Wales: the life of Gruffudd ap Cynan. ed. D S Evans. Felinfach, 1990.

"Memorandum inédit sur la défense des places". Ed A Dain. *Revue des études grecques*, 43 (1940), 123–36.

MICHAEL I the Syrian, patriarch of Antioch. *Chronique de Michel le Syrien, patriarche jacobite d'Antioche (1166–1199)*. Ed. and trans. J-B Chabot. 4 vols. Paris, 1899–1924.

MINHĀJ SIRĀJ JŪZJĀNĪ. *Tabaḳāt-i-Nāṣirī : a general history of the Muhammadan dynasties of Asia, including Hindūstān, from* A.H. *194 (810* A.D.*), to* A.H. *658 (1260* A.D.*), and the irruption of the infidel Mughals into Islam, by the Maulānā, Minhāj-ud-Dīn, Abū-'Umar-i-'Usmān*. Trans. H G Raverty. London, 1881.

Les miracles de Saint Benoît, écrits par Adrevald, Aimoin, André, Raoul Tortaire et Hughes de Saint Marie, moines de Fleury. Ed. E de Certain. Publications de la Société de l'histoire de France, 96. Paris, 1858.

Miracula S. Gorgonii. ed. G H Pertz. In *MGH SS* IV: 239–47. Hannover, 1841.

Miracula S. Wigberhti (excerpts). Ed. G Waitz. In *MGH SS* IV: 224–8. Hannover, 1841.

Mohammed der Prophet: seine Leben und seine Lehre, aus handschriftlichen Quellen und dem Koran geschöpft und dargestellt. Ed. G Weil. Stuttgart, 1843.

Monumenta dominationis pontificae, sive Codex Carolinus, juxta autographum Vindobonense: epistolae Leonis III. *Carolo Augusto, diplomata Ludovici, Ottonis, et Henrici, chartula comitissae Mathildae et codex Rudolphinus ineditus*. Ed. G Cenni. 2 vols. Rome, 1760–61.

MORENA, Otto. *De rebus Laudensibus*. Ed. P Jaffe. *MGH SS* XVIII: 587–634. Hannover, 1863.

MOVSES DASXURANC'I. *The history of the Caucasian Albanians*. Trans. C J F Dowsett. London oriental series, 8. London, 1961.

MUQADDASĪ, Muḥammad ibn Aḥmad. *Description of Syria, including Palestine, by Mukaddasi (circ. 985* A.D.*)*. Trans. G Le Strange. Library of the Palestine Pilgrims' Text Society, 3/3. London, 1896.

NARSHAKHĪ, Muḥammad ibn Ja'far. *The history of Bukhara*. Trans. R N Frye. Publications of the Mediaeval Academy of America, 61. Cambridge, Mass., 1954.

NĀṢIR-I KHUSRAW. *Diary of a journey through Syria and Palestine by Nâsir-i-Khusrau, in 1047* A.D. Trans. G Le Strange. Library of the Palestine Pilgrims' Text Society, 4/1. London, 1888.

NICEPHORUS, patriarch of Constantinople. *Short history*. Ed. and trans. C Mango. Dumbarton Oaks texts, 10; CFHB 13. Washington, D.C., 1990.

NICEPHORUS URANUS. "Douze chapitres inédits de la *Tactique* de Nicéphore Ouranos". Ed. J-A de Foucault. *Travaux et mémoires* [Centre de recherche d'histoire et civilisation de Byzance], 5 (1973), 281–312.

NICEPHORUS URANUS. *Taktika*, chaps. 56–65. Ed. and trans. in E McGeer, *Sowing the dragon's teeth: Byzantine warfare in the tenth century*. Dumbarton Oaks studies, 33. Washington, D.C., 1995.

NICOLAUS of Bray. *Ex Nicolai de Braia carmine De gestis Ludovici VIII regis*. Ed. O Holder-Egger. *MGH SS* XXVI. Hannover, 1883.

Nihongi: chronicles of Japan from the earliest times to A.D. 697. Trans. W G Aston. 2 vols. London, 1896.

The Nikonian chronicle. Trans. S A Zenkovsky and B J Zenkovsky. 5 vols. Princeton, 1984–9.

NĪSHĀPŪRĪ, Ẓahīr al-Dīn, and RASHĪD AL-DĪN ṬABĪB. *The history of the Seljuq Turks: from the Jāmi' al-Tawārīkh, an Ilkhanid adaption of the Saljūq-nāma of Ẓahīr al-Dīn Nīshāpūrī*. Trans. K A Luther. Studies in the history of Iran and Turkey. Richmond, Surrey, 2001.

NORBERT, abbot of Iburg. *Vita Bennonis II. episcopi Osnabrugensis = Leben Bischof Benno II. von Osnabrück*. In *Lebensbeschreibungen einiger Bischöfe des 10.–12. Jahrhunderts*. Trans. H Kallfelz. AQ 22. Darmstadt, 1973.

"The Norman *Consuetudines et iusticie* of William the Conqueror". Ed. in C H Haskins, *Norman institutions*, 277–84. New York, 1918; repr. 1960.

Normanniae nova chronica ab anno Christi CCCCLXXIII. ad annum MCCCLXXVIII. e tribus chronicis mss. Sancti Laudi, Sanctae Catharinae, et Majoris Ecclesiae Rotomagensium collecta. Ed. A Chéruel. Mémoires de la Société des antiquaires de Normandie, 2nd ser. 8/3. Caen, 1851.

"Notes d'histoire almohade i: Six fragments inédits d'une chronique anonyme du début des Almohades". Ed. É Lévi-Provençal. In *Mélanges René Basset: études nord-africaines et orientales*, II: 335–93. Publications de l'Institut des hautes-études marocaines, 10. Paris, 1925.

NOTKER BALBULUS. *Charlemagne*. In *Einhard and Notker: Two lives of Charlemagne*. Trans. L Thorpe. Harmondsworth, 1969.

The Old French continuation of William of Tyre, 1184–97. In *The conquest of Jerusalem and the Third Crusade*. Trans. P W Edbury. Crusade texts in translation, 1. Aldershot, 1998.

OLIVER of Paderborn. *The capture of Damietta*. Trans. J J Gavigan. In *Christian society and the crusades, 1198–1229: sources in translation*. Ed. E Peters. Sources of medieval history. Philadelphia, 1971.

On skirmishing. In *Three Byzantine military treatises*. Ed. and trans. G T Dennis. Washington, D.C., 1985.

ÔRBELYAN, Stepanos. *Histoire de la Siounie*. Trans. M F Brosset. 2 vols. St Peterburg, 1864–6.

ORDERICUS VITALIS. *The Ecclesiastical history of Orderic Vitalis*. Ed. and trans. M Chibnall. 6 vols. Oxford medieval texts. Oxford, 1969–80. [OV]

Oriental sources concerning Nubia. Coll. and trans. G Vantini. Heidelberg, 1975.

Orkneyinga saga: the history of the earls of Orkney. Trans. H Pálsson and P Edwards. London, 1978.

OTTO I, bishop of Freising, continued by RAHEWIN. *Gesta Friderici I. imperatoris*. Ed. G H Waitz. 2nd edn. *MGH SrG* XLVI. Hannover, 1884.

——. *The deeds of Frederick Barbarossa*. Trans. C C Mierow with R Emery. Records of civilization: sources and studies, 49. New York, 1953.

OTTO of St. Blasien. *Chronici ab Ottone Frisingensi conscripti continuatio*. Ed. R Wilmans. *MGH SS* XX: 302–37. Hannover, 1868.

Ottonis II. diplomata. MGH Diplomata regum et imperatorum Germaniae, II/1. Hannover, 1888.

OUYANG Xiu. *Traité des fonctionnaires et Traité de l'armée, traduite de la Nouvelle histoire des T'ang (chap. XLVI–L)*. Trans. R des Rotours. 2 vols. Bibliothèque de l'Institut des hautes études chinoises, 6. Leiden, 1947.

Palestine under the Moslems: a description of Syria and the Holy Land from A.D. 650 to 1500, translated from the works of the mediaeval Arab geographers. Trans. G Le Strange. Publications of the Palestine Exploration Fund. London, 1890.

Pauli continuatio Casinensis. Ed. L Bethmann and G Waitz. In *MGH SrL*, 198–200. Hannover, 1878.

Pauli continuatio tertia. Ed. L Bethmann and G Waitz. In *MGH SrL*, 203–16. Hannover, 1878.

PAULUS diaconus. *Historia Langobardorum*. Ed. L Bethmann and G Waitz. In *MGH SrL*, 12–192. Hannover, 1878.

——. *History of the Langobards*. Trans. W D Foulke. Philadelphia/New York, 1907.

PELAYO, bishop of Oviedo. *Crónica*. In *Las glorias nacionales: grande historia universal de todos los reinos, provincias, islas, y colonias de la monarquía española, desde los tiempos primitivos hasta el año de 1852*, II. Ed. M Ortíz de la Vega. Madrid, 1853.

——. *Chronicon regum Legionensium*. In *The world of El Cid: chronicles of the Spanish Reconquest*. Trans. S Barton and R Fletcher. Manchester medieval sources. Manchester, 2000.

PETER Damian. *Die Briefe des Petrus Damiani*. Ed. K Reindel. 4 vols. MGH Briefe der deutschen Kaiserszeit, 4. Munich, 1983–93.

PETRUS de Ebulo. *Liber ad honorem Augusti*. Ed. G B Siragusa. 2 vols. Fonti per la storia d'Italia, 39. Rome, 1905–6.

PETRUS Sarnensis. *The history of the Albigensian Crusade: Peter of les Vaux-de-Cernay's Historia Albigensis*. Trans. W A Sibly and M D Sibly. Woodbridge, 1998.

PHILO of Byzantium. "Poliorcétique des Grecs: Traité de fortification, d'attaque et de défense des places par Philon de Byzance". Ed. and trans. A de Rochas d'Aiglun. *Mémoires de la Société d'émulation de Doubs*, 4th ser. 6 (1870), 185–300.

PHOTIUS I, patriarch of Constantinople. *The Homilies*. Trans. C Mango. Dumbarton Oaks studies, 3. Cambridge, Mass., 1958.

Pippini, Carlomanni, Caroli Magni diplomata. Ed. E Mühlbacher. MGH Diplomata Karolinorum, 1. Hannover, 1906.

Les plus anciens recueils des Miracles de Saint Démétrius et le pénétration des Slaves dans les Balkans. Ed. and trans. P Lemerle. 2 vols. Le monde byzantin. Paris, 1979.

Poeti bizantini. Ed. and trans. R Cantarella. 2 vols. Milan, 1948.

Poliorkētika kai poliorkiai diaphorōn poleōn = Poliorcétique des Grecs: traités théoriques, récits historiques. Ed. C Wescher. Paris, 1867.

Primera Crónica general de España, que mandó componer Alfonso el Sabio y se continuaba bajo Sancho IV en 1289. 2nd edn. Ed. R Menéndez Pidal et al. 2 vols. Madrid, 1955.

PROCOPIUS. *Buildings*. Ed. and trans. H B Dewing. Loeb classical library. Cambridge, Mass./London, 1940.

——. *History of the wars*. Ed. and trans. H B Dewing. 5 vols. Loeb classical library. Cambridge, Mass./London, 1919–28.

PRODROMUS, Theodore. *Historische Gedichte*. Ed. W Hörandner. Wiener byzantinische Studien, 11. Vienna, 1974.

PSELLUS, Michael. *Fourteen Byzantine rulers: the "Chronographia"*. Trans. E R A Sewter. Harmondsworth, 1966.

RADULFUS of Caen. *Gesta Tancredi in expeditione Hierosolymitana*. In *RHC: Historiens occidentaux*, III: 587–716. Paris, 1866.

RAHEWIN ; *see* Otto I, bishop of Freising

RAIMUNDUS de Agiles. *Historia Francorum qui ceperunt Jerusalem*. In *PL* CLV: 591–668. Paris, 1854.

RALPH of Coggeshall. *Chronicon anglicanum*. In *Radulphi de Coggeshall Chronicon anglicanum; De expugnatione Terrae Sanctae libellus; Thomas Agnellus: De morte et sepultura Henrici regis Angliae junioris; Gesta Fulconis filii Warini; excerpta ex Otiis imperialibus Gervasii Tileburiensis*. Ed. J Stevenson. RS 66. London, 1875.

RASHĪD AL-DĪN ṬABĪB; *see* NĪSHĀPŪRĪ, Ẓahīr al-Dīn.

RĀZĪ, Aḥmad ibn Muḥammad. "La 'Description de l'Espagne' d'Aḥmad al-Rāzī: essai de reconstitution de l'originale arabe et traduction française". Ed. and trans. É Lévi-Provençal. *Al-Andalus*, 18 (1953), 51–108.

RĀZĪ, 'Īsá ibn Ahmad. *Anales palatinos del califa de Córdoba al-Hakam II (360–364 H. = 971–975 J.C.)*. Trans. E García Gómez. Madrid, 1967.

"Un recueil des lettres officielles almohades". Ed. É Lévi-Provençal. *Hesperis* 28 (1941), 1–80.

Recueil de textes relatifs à l'histoire de l'architecture et à la condition des architectes en France au Moyen Âge. Ed. V Mortet and P Deschamps. 2 vols. Collection de textes pour servir à l'étude et à l'enseignement de l'histoire, 44, 51. Paris, 1911–29.

REGINO of Prüm. *Chronicon a. 1–906*. In *MGH SS* I: 537–612. Ed. G H Pertz. Hannover, 1826.

——. *Chronik*. Ed. and trans. E Dümmler, F Kurze and R Rau. In *Quellen zur karolingischen Reichsgeschichte*, III. AQ 7. Darmstadt, 1960.

Regum Burgundiae e stirpe Rudolfina diplomata et acta. Ed. T Schieffer with H E Mayer. MGH Diplomata. Munich, 1977.

Relation du siège de la ville d'Exeja en Aragon. In *Recueil des historiens des Gaules et de la France*, XII: 384–5. Ed. F Clément and M J J Brial. Paris, 1781.

RICHARD of Devizes. *Chronicon Ricardi Divisiensis de rebus gestis Ricardi primi regis Angliae*. Ed. J Stevenson. London, 1838.

RICHARD of Hexham. *The acts of King Stephen, and the Battle of the Standard*. In *The Church historians of England*, IV/1: 33–58. Trans. J Stevenson. London, 1856. Repr. in *Contemporary chronicles of the Middle Ages: sources of twelfth-century history*. Felinfach, 1988.

RICHER of Saint-Rémy. *Historiarum libri quatuor*. In *PL* CXXXVIII: 9–170. Paris, 1853.

——. *Histoire de France (888–995)*. Ed. and trans. R Latouche. 2 vols. Paris: 1930–37.

RIGORD. *Chronique*. In *Œuvres de Rigord et de Guillaume le Breton, historiens de Philippe-Auguste*, I. Ed. H F Delaborde. Publications de la Société de l'histoire de France, 210. Paris, 1882.

RIMBERT, archbishop of Hamburg and Bremen. *Vita Anskarii*. Ed. W Trillmich. In *Quellen des 9. und 11. Jahrhunderts zur Geschichte der Hamburgischen Kirche und das Reiches*. AQ 11. Berlin, 1961.

ROBERTUS of Reims. *Historia hierosolimitana*. In *RHC: Historiens occidentaux*, III: 717–882. Paris, 1866.

ROGER of Howden. *Gesta regis Henrici secundi Benedicti abbatis: the chronicle of the reigns of Henry II and Richard I, A.D. 1169–1192, known commonly under the name of Benedict of Peterborough*. Ed. W Stubbs. 2 vols. RS 49. London, 1867.

——. *Chronica magistri Rogeri de Houedene*. Ed. W Stubbs. 4 vols. RS 51. London, 1868–71.

ROGER of Wendover. *Flowers of history: comprising the history of England from the descent of the Saxons to A.D. 1235, formerly ascribed to Matthew Paris*. Trans. J A Giles. 2 vols. Bohn's antiquarian library. London, 1849; repr. in 4 vols., Felinfach, 1993–6.

The Roman eastern frontier and the Persian wars, part 2: AD 363–628, a narrative sourcebook. Ed. G Greatrex and S N C Lieu. London and New York, 2002.

Romualdo ii, archbishop of Salerno. *Chronicon (a.m. 130 – a.C. 1178)*. Ed. C A Garufi. RIS 7/1. Città di Castello, 1935.

Rotuli litterarum clausarum in Turri Londinensi asservati. Ed. T D Hardy. 2 vols. Publications of the Record Commissioners, 15. London, 1833–44.

Rotuli Normanniae in Turri Londinensi asservati: Johanne et Henrico quinto Angliae regibus. Ed. T D Hardy. Publications of the Record Commissioners, 19. London, 1835.

Rūdhrāwarī, Abū Shujā' Muḥammad ibn al-Ḥusayn. *Continuation of The experiences of the nations*. Vols. iii, vi of *The eclipse of the 'Abbasid caliphate: original chronicles of the fourth Islamic century*. Ed. H F Amedroz, trans. D S Margoliouth. 7 vols. Oxford, 1920–21.

The Russian primary chronicle: Laurentian text. Ed. and trans. S H Cross and O P Sherbowitz-Wetzor. Publications of the Mediaeval Academy of America, 60. Cambridge, Mass., 1953.

Ryccardus of San Germano. *Chronica a. 1189–1243*. In *MGH SS* xix: 321–84. Ed. G H Pertz. Hannover, 1866.

Saxo grammaticus. *Gesta Danorum*. Ed. J Olrik and H Raeder. 2 vols. Copenhagen, 1931–57.

Scylitzes, Joannes. *Byzanz, wieder ein Weltreich: das Zeitalter der makedonischen Dynastie*. Trans. H Thurn. Byzantinische Geschichtsschreiber, 15. Graz, 1983.

Sebeos, bishop of the Bagratuni kingdom. *Histoire d'Héraclius*. Trans. F Macler. Paris, 1904.

Sermo de relatione beati Vedasti a Bellovaco ad proprium locum. Ed. O Holder-Egger. In *MGH SS* xv(1): 402–4. Hannover, 1887.

Shakundi, Ismā'īl ibn Muḥammad. *Elogio del Islam español (Risāla fī faḍl al-Andalus)*. In *Andalucía contra Berbería: reedición de traducciones de Ben Hayyān, Šaqundī, Ben al-Jaṭīb*. Trans. E García Gómez. Barcelona, 1976.

Sidonius Apollinaris. *Poems and letters*. Ed. and trans. W B Anderson. 2 vols. Loeb classical library. Cambridge, Mass./London, 1936–65.

Sigebert of Gembloux. *Chronica cum continuationibus*. Ed. L C Bethmann. In *MGH SS* vi: 268–474. Hannover, 1848.

Simeon of Durham. *History of the Church of Durham*. In *The Church historians of England*, iii/2: 619–711. With *The continuation of the History of Simeon of Durham*, ibid., 712–30. Trans. J Stevenson. London, 1855.

——. *History of the kings of England*. In *The Church historians of England*, iii/2: 423–617. Trans. J Stevenson. London, 1855.

——. *De obsessione Dunelmi*. In *Symeonis monachi Opera omnia*, i. Ed. T Arnold. RS 75. London, 1882.

Simocatta, Theophylactus. *The History*. Trans. M Whitby and M Whitby. Oxford/New York, 1986.

Snorri Sturluson. *Heimskringla: sagas of the Norse kings*. Trans. S Laing, rev. P Foote. Everyman's library. London, 1961.

——. *King Harald's saga*. Trans. M Magnússon and H Pálsson. Harmondsworth, 1966.

The song of Dermot and the earl: an Old French poem from the Carew manuscript no. 596 in the Archiepiscopal Library at Lambeth Palace. Ed. and trans. G H Orpen. Oxford, 1892; repr. Felinfach, 1994.

The song of Igor's campaign: an epic of the twelfth century. Trans. V Nabokov. New York/London, 1960.

Sophronius, patriarch of Jerusalem. *Vers sur la prise de Jérusalem, en mètre anacréontique, selon l'ordre de l'alphabet*. Ed. and trans. in Alphonse comte Couret, "La prise de Jérusalem par les Perses en 614". *Revue de l'Orient chrétien*, 2 (1897), 123–55.

Step'anos Taronec'i. *Des Stephanos von Taron Armenische Geschichte*. Trans. H Gelzer and A Burckhardt. Leipzig, 1907.

Strategios of Mar Saba. *Moine de Saint-Sabas: Prise de Jérusalem par les Perses*. Trans. of Arab version in Alphonse comte Couret, "La prise de Jérusalem par les Perses en 614". *Revue de l'Orient chrétien*, 2 (1897), 123–55.

——. "Antiochus Strategos' account of the sack of Jerusalem in A.D. 614". Trans. of Georgian version by F C Conybeare. *EHR* 25 (1910), 502–17.

——. *La prise de Jérusalem par les Perses en 614*. 2 vols. Ed. of Georgian version and Latin trans. G Garitte. Corpus scriptorum Christianorum orientalium, 202–3. Louvain, 1960.

Suger. *Vie de Louis VI le Gros*. Ed. and trans. H Waquet. Les classiques de l'histoire de France au Moyen Âge, 11. Paris, 1929.

——. *The deeds of Louis the Fat*. Trans. R Cusimano and J Moorhead. Washington, D.C., 1992

Suger et al. *De gloriose rege Ludovici, Ludovici filio*. In *Vie de Louis le Gros, suivie de l'Histoire du roi Louis VII*, 147–78. Ed. A Molinier. Collection de textes pour servir à l'étude et à l'enseignement de l'histoire, 4. Paris, 1887.

Symeon Metaphrastes. *S. Demetrii martyris acta*. In *PG* CXVI: 1081–[1166/1424]. Paris, 1864.

Ṭabarī, Muḥammad ibn Jarīr. *The History of al-Tabarī*. Trans. C E Bosworth et al. 40 vols. Albany, N.Y., 1985–2007.

——. *The early 'Abbāsī empire*. 2 vols. Trans. J A Williams. Cambridge, 1988–9.

——. *La chronique de Tabari*. Trans. H Zotenberg. 6 vols. in 5. Paris, 1989.

Tamīm ibn Baḥr. Report of a journey to the Uighurs. Ed. and trans. in V Minorsky, "Tamīm ibn Baḥr's journey to the Uyghurs", 278–85. *Bulletin of the School of Oriental and African Studies*, 12/2 (1948), 275–305.

Thegan of Trier. *Vita Hludowici imperatoris*. In *MGH SS* II: 585–603. Ed. G H Pertz. Hannover, 1829.

Theodoricus monachus. *Historia de antiquitate regum Norwagiensium*. In *Monumenta historica Norvegiae: latinske kildeskrifter til Norges historie i middelalderen*, 1–68. Ed. G Storm. Christiania, 1880.

——. *An account of the ancient history of the Norwegian kings*. Trans. D McDougall and I McDougall. Viking Society for Northern Research text series, 11. London, 1998.

Theodorus monachus. *Annales Palidenses ab O.c.–1182 et 1390*. In *MGH SS* XVI: 48–98. Ed. G H Pertz. Hannover, 1859.

Theodorus syncellus. *Traduction et commentaire de l'homélie écrite probablement par Théodore le Syncelle sur le siège de Constantinople en 626*. Trans. F Makk. Opuscula byzantina, 3. Szeged, 1975.

Theodosius diaconus. *De expugnatione Cretae acroases quinque*. In *PG* CXIII: 987–1058. Paris, 1864.

Theodosius monachus. *Epistula de Syracusanae urbis expugnatione*. Ed. L Muratori. RIS 1/2: 255–66. Milan, 1725.

Theophanes confessor. *The Chronicle of Theophanes confessor: Byzantine and Near Eastern history, A.D. 284–813*. Trans. C Mango and R Scott with G Greatrex. Oxford, 1997.

Theophanes continuatus; Ioannes Cameniata; Symeon magister; Georgius monachus. Ed. I Bekker. CSHB 43. Bonn, 1838.

Thietmar, bishop of Merseburg. *Chronicon a. 919–1018*. Ed. V C I M Lappenberg. In *MGH SS* III: 723–871. Hannover, 1839.

Thuróczy, János. *Chronica Hungarorum*, 1: *Textus*. Ed. E Galántai and J Kristó. Bibliotheca scriptorum medii recentisque aevorum, ns 7. Budapest, 1985.

——. *Chronicle of the Hungarians*. Trans. F Mantello. Medievalia hungarica, 2. Bloomington, Ind., 1991.

Torigni, Robert de. *Chronicles of the reigns of Stephen, Henry II, and Richard I, IV: The Chronicle of Robert of Torigni, abbot of the monastery of St. Michael-in-peril-of-the-sea*. Ed. R Howlett. Chronicles and memorials of Great Britain and Ireland during the middle ages, 82. London, 1889.

——. *The chronicles of Robert de Monte*. In *The Church historians of England*, IV/2: 673–813. Trans. J Stevenson. London, 1856. Separately repr. Felinfach, 1991.

Triumphus sancti Lamberti de castro Bullonico. Ed. W Arndt. In *MGH SS* XX: 497–511. Hannover, 1868.

'Ubayd Allāh ibn Abī 'Alī Ṣāliḥ. "Un nouveau récit de la conquête de l'Afrique du Nord par les Arabes". Trans. É Lévi-Provençal. In his "Arabica occidentalia, I". *Arabica* 1 (1954), 17–43.

The Uighur empire according to the T'ang dynastic histories: a study in Sino-Uighur relations, 744–840. Ed. and trans. C Mackerras. 2nd edn. Asian publication series, 2. Canberra, 1972.

Uray, G. "Notes on a Tibetan military document from Tun-Huang". *Acta orientalia Academiae Scientarum Hungaricae*, 12 (1961), 223–30.

Usāma ibn Munqiḍ. *An Arab-Syrian gentleman and warrior in the period of the crusades: memoirs of Usāmah ibn-Munqidh (Kitāb al-I'ti-bār)*. Ed. and trans. P K Hitti. Records of civilization: sources and studies, 10. New York, 1929.

'Utbī, Muḥammad ibn 'Abd al-Jabbār. *The Kitab-i-Yamini: historical memoirs of the Amír Sabaktagín, and the Sultán Mahmúd of Ghazna, early conquerors of Hindustan, and founders of the Ghaznavide dynasty; translated from the Persian version of the contemporary Arabic chronicle of al Utbi*. Trans. J Reynolds. Publications of the Oriental Translation Fund, 72. London, 1858.

Vahram Rabuni. *Vahram's Chronicle of the Armenian kingdom of Cilicia during the time of the crusades*. Trans. C F Neumann. Publications of the Oriental Translation Fund, 20/3. London, 1831.

Vasil'ev, A A. *Byzance et les Arabes*. Ed. and trans. H Grégoire and M Canard. 3 vols. in 4. Corpus bruxellense historiae Byzantinae, 1–3. Brussels, 1935–68.

Vaṣṣāf, 'Abd Allāh. *Tazjiyatu-l amsār wa tajriyatu-l āsār*. In *The history of India, as told by its own historians*, III: 24–54. Ed. H M Elliot and J Dowson. London, 1871.

Vegetius Renatus, Flavius. *De procinctu romanae miliciae*. Abridged and adapted by Hrabanus Maurus. Ed. E Dümmler. *Zeitschrift für deutsches Alterthum*, 15 = ns 3 (1872), 443–51.

——. *Epitome of military science*. Trans. N P Milner. Translated texts for historians, 16. Liverpool, 1993.

Villard de Honnecourt. *The sketchbook of Villard de Honnecourt*. Ed. T Bowie. Bloomington, Ind., 1959.

Villehardouin, Geoffroy de. *The conquest of Constantinople*. In *Joinville & Villehardouin: Chronicles of the crusades*. Trans. M R B Shaw. Harmondsworth, 1963.

Vincentius of Prague. *Annales a. 1140–1167; Continuatio Gerlaci abbatis Milovicensis a. 1167–1198*. Ed. W Wattenbach. *MGH SS* XVII: 654–710. Hannover, 1861.

Vita Barbati episcopi beneventani. In *MGH SrL*, 555–63. Ed. G Waitz. Hannover, 1878.

Vita Griffini filii Conani r. Venedotiae vel Northwalliae = Eman e dechreu historia hen Gruffud vab Kenan vab Yago. Latin trans. N Robinson; ed. R Williams. *AC* 3rd ser. 12 (1866), 30–45, 112–31.

Vita Heinrici IV. imperatoris = Das Leben Kaiser Heinrichs IV. Trans. I Schmale-Ott. In *Quellen zur Geschichte Kaiser Heinrichs IV*. AQ 12. Berlin, 1963.

VITRUVIUS POLLIO. *De architectura = On architecture*. Ed. and trans. F Granger. 2 vols. Loeb classical library. Cambridge, Mass./London, 1931–4; repr. 1983.

WACE. *Le roman de Rou*. Ed. A J Holden. 3 vols. Publications de la Société des anciens textes français. Paris, 1970–73.

WALTER of Coventry. *Memoriale fratris Walteri de Coventria: the historical collections of Walter of Coventry*. Ed. W Stubbs. 2 vols. RS 58. London, 1872–3.

WALTER of Guisborough. *The Chronicle of Walter of Guisborough*. Ed H Rothwell. Camden Society, 3rd ser. 89. London, 1957.

WIDUKIND of Corvey. *Res gestae saxonicae a. c.919–973*. Ed. G Waitz. *MGH SS* III: 408–67. Hannover, 1889.

WILBRAND, Graf von Oldenburg. *Wilbrands von Oldenburg Reise nach Palaestina und Kleinasien*. Ed. and trans. J C M Laurent. Hamburg, 1859.

WILLIAM fitz Stephen. *Description of London*. Trans. H E Butler. In F M Stenton, *Norman London*. Historical Association leaflets, 93–4. London, 1934; repr. New York, 1990.

WILLIAM of Jumièges. *Gesta Normannorum ducum*. Ed. J Marx. Rouen/Paris, 1914.

WILLIAM of Malmesbury. *History of his own times*. In *The Church historians of England*, III/1: 381–422. Trans. J Sharpe and J Stevenson. London, 1854; repr. Felinfach, 1988.

WILLIAM of Newburgh. *The history of English affairs*. Ed. and trans. P G Walsh and M J Kennedy. 2 vols. Warminster, 1988.

WILLIAM of Tyre. *L'estoire de Eracles empereur, et la conqueste de la terre d'outremer*. *RHC: Historiens occidentaux*, I–II. Paris, 1844–9.

——. *A history of deeds done beyond the sea*. Trans. E A Babcock. 2 vols. Records of civilization: sources and studies, 35. New York, 1943.

WINCENTY KADŁUBEK. *Chronica Polonorum, sive Originale regum et principum Poloniae*. Ed. and trans. A Przezdziecki. Cracow, 1862.

"Winchcombe annals, 1049–1181". Ed. R R Darlington. In *A medieval miscellany for Doris Mary Stenton*. Ed. P M Barnes and C F Slade. Publications of the Pipe Roll Society, 76 (ns 36). London, 1962.

WIPO. *Gesta Chuonradi II. imperatoris*. Ed. W Trillmich. In *Quellen des 9. und 11. Jahrhunderts zur Geschichte der Hamburgischen Kirche und das Reiches*. AQ II. Berlin, 1961.

——. *The deeds of Conrad II*. In *Imperial lives and letters of the eleventh century*. Trans. T E Mommsen and K F Morrison. New York, 2000.

YAḤYĀ IBN SAʿĪD al-Anṭākī. *Histoire de Yahya-ibn-Saʿïd d'Antioche, continuateur de Saʿïd-ibn-Bitriq*. Ed. and trans. I Kratchkovsky and A Vasiliev. In *Patrologia orientalis*, XVIII: 701–833; XXIII: 347–520. Paris, 1924, 1932.

ZACHARIAS, bishop of Mytilene. *The Syriac chronicle known as that of Zachariah of Mitylene*. Trans. F J Hamilton and E W Brooks. Byzantine texts. London, 1899.

ZARKASHĪ, Muḥammad ibn Ibrāhīm. *Chronique des Almohades & des Hafçides*. Trans. E Fagnan. Constantine, 1895.

ẒIYĀʾ AL-DĪN BARANĪ. *Tārīkh-i Fīroz Shāhī*. In *The history of India, as told by its own historians*, III: 93–268. Ed. H M Elliot and J Dowson. London, 1871.

ZONARAS, Joannes. *Militärs und Höflinge im Ringen um das Kaisertum: byzantinische Geschichte von 969 bis 1118*. Trans. E Trapp. Byzantinische Geschichtsschreiber, 16. Graz, 1986.

Zwentiboldi et Ludowici Infantis diplomata. Ed. T Schieffer. MGH Diplomata regum Germaniae ex stirpe Karolinorum, IV. Berlin, 1960.

Secondary sources

Aarts, B. "Early castles of the Meuse–Rhine border region and some parallels in Western Europe c1000: a comparative approach". CG 18 (1998), 11–23.

Abadal, R d'. "La domination carolingienne en Catalogne". *Revue historique*, 225 (1961), 319–40.

Abulafia, D. "The role of trade in Muslim–Christian contact during the Middle Ages". In *The Arab influence in Medieval Europe*, 1–10. Ed. D A Agius and R Hitchcock. Middle East culture series, 18. Reading, 1994.

Acién Almansa, M. "Sobre la funcion de los *ḥuṣūn* en el sur de al-Andalus: la fortificación en el Califato". In *Coloquio hispano-italiano de arqueología medieval = Colloquio hispano-italiano di archeologia medievale*, 263–74. Granada, 1992.

Addyman, P V. "Late Saxon settlements in the St Neots area: Saxon settlement and Norman castle at Eaton Socon, Beds." *Proceedings of the Cambridgeshire Antiquarian Society* 58 (1965), 38–69.

Akram, A I. *The sword of Allah, Khalid bin al-Waleed: his life and campaigns*. Karachi, 1970.

Alcock, L. "Early historic fortifications in Scotland". In *Hill-fort studies: essays for A H A Hogg*. Ed. G Guilbert. Leicester, 1981.

——. "The activities of potentates in Celtic Britain, AD 500–800: a positivist approach". In *Power and politics in early medieval Britain and Ireland*. Ed. S T Driscoll and M R Nicke. Edinburgh, 1988.

Alcock, L, and Alcock, E A. "Reconnaissance excavations in early historic fortifications and other sites in Scotland, 1974–84", 1: "Excavations near St Abb's Head, Berwickshire, 1980" (with S M Foster); 2: "Excavations in Dunollie Castle, Oban, Argyll, 1978"; 3: "Excavations at Dundurn, Strathearn, Perthshire, 1976–77" (with S T Driscoll); 4: "Excavations at Alt Clut, Clyde Rock, Strathclyde, 1974–1975"; 5: "A: Excavations and other fieldwork at Forteviot, Perthshire, 1981; B: Excavations at Urquhart Castle, Inverness-shire, 1983; C: Excavations at Dunnottar, Kincardineshire, 1984". *Proceedings of the Society of Antiquaries of Scotland*, 116 (1986), 255–79; 117 (1987), 119–47; 119 (1989), 189–226; 120 (1990), 95–149; 122 (1992), 215–87.

Alexander, J W. "The alleged palatinates of Norman England". *Speculum* 56 (1981), 17–27.

——. *Ranulf of Chester: a relic of the Conquest*. Athens, Ga., 1983.

al-Hassan, A Y, and Hill, D R. *Islamic technology: an illustrated history*. Cambridge, 1986.

al-Khalaf, M. "Die 'abbasidische Stadtmauer von ar-Raqqa/ar-Rāfiqa". *Damaszener Mitteilungen*, 2 (1985), 122–31.

Allain, C, and Meunié, J. "La forteresse almoravide de Zagora". *Hespéris*, 43 (1956), 305–23.

Allsen, T. "The circulation of military technology in the Mongolian empire". In *Warfare in inner Asian history*. Ed. N Di Cosmo. Handbook of oriental studies, sect. 8: Central Asia, 6. Leiden/Boston, 2002.

Amt, E. *The accession of Henry II in England: royal government restored, 1149–1159*. Woodbridge, 1993.

——. "Besieging Bedford: military logistics in 1224". *Journal of Medieval Military History*, 1 (2002), 101–27.

ANDREASSI, G, and RADINA, F. *Archeologia di una città: Bari dalle origini al x secolo*. Prometeo, 2. Bari, 1988.

ANDRESSOHN, J C. *The ancestry and life of Godfrey of Bouillon*. Bloomington, Ind., 1947.

ANDREWS, D. "The archaeology of the medieval castrum in central Italy". In *Archaeology and Italian society: prehistoric, Roman and medieval studies*. Ed. G Barker and R Hodges. *BAR* international series, 102. Oxford, 1981.

ANDREWS, F B. *The medieval builder and his methods*. Oxford, 1925. Repr. from *Birmingham Archaeological Society transactions and proceedings*, 48 (1922).

ANDREWS, K. *Castles of the Morea*. Gennadeion monographs, 4. Princeton, 1953 (rev. edn 2006).

ANDREWS, R, et al. "The late antique and medieval fortifications of Butrint". In *Byzantine Butrint: excavations and surveys, 1994–99*, 126–50. Ed. R Hodges, W Bowden and K Lako. Oxford, 2004.

ANGOLD, M. *The Byzantine empire, 1025–1204: a political history*. London, 1984.

ANON. "Digging in to defend Lewes". *Sussex past and present*, 75 (1995), 7.

ANTUÑA, M M. *Sevilla y sus monumentos árabes*. El Escorial, 1930.

APPLEBY, J T. *England without Richard, 1189–1199*. Ithaca, N.Y./London, 1965.

ARAGUAS, P. "Les châteaux des marches de Catalogne et Ribagorce (950–1100)". *BM* 137 (1979), 205–24.

ARMITAGE, E S. *The early Norman castles of the British Isles*. London, 1912.

ARNOLD, B. *Princes and territories in medieval Germany*. Cambridge, 1991.

——. *Medieval Germany, 500–1300: a political interpretation*. European history in perspective. Basingstoke, 1997.

ARSLAN, E. "L'architettura dal 586 al mille". In *Storia di Milano*, II: 500–621. 17 vols. Milan, 1953–66.

ASHER, C B. "Delhi walled: changing boundaries". In *City walls: the urban enceinte in global perspective*, 247–81. Ed. J D Tracy. Studies in comparative early modern history. Cambridge, 2000.

ASHTOR, E. *A social and economic history of the Near East in the Middle Ages*. Berkeley/London, 1976.

AUBENAS, R. "Les châteaux-forts des xe et xie siècles". *Revue historique du droit français et étranger*, 4th ser. 17 (1938), 548–86.

AUDOUIN, É. *Essai sur l'armée royale au temps de Philippe-Auguste*. Paris, 1913.

AUSSARESSES, F. *L'armée byzantine à la fin du vie siècle, d'après le Strategicon de l'empereur Maurice*. Bibliothèque des Universités du Midi, 14. Paris, 1909.

AUZIAS, L. "Les sièges de Barcelona, de Tortosa et d'Huesca, 801–11". *Annales du Midi*, 48 (1936), 5–25.

AVENT, R. *Cestyll tywysogion Gwynedd = Castles of the princes of Gwynedd*. Cardiff, 1983.

——. *Criccieth Castle*. Cardiff, 1998.

——. "The late twelfth century gatehouse at Chepstow Castle, Monmouthshire, Wales". CG 20 (2002), 27–40.

——. *Dolwyddelan Castle; Dolbadarn Castle; Castell y Bere*. Cardiff, 2004.

BAATZ, D. *Bauten und Katapulte des römisches Heeres*. Mavors: Roman army researches, 11. Stuttgart, 1994.

BACHMANN, C. "Das Offnungsrecht: Ius aperturae". In *Burgen in Mitteleuropa: ein Handbuch*. Ed. H W Böhme. 2 vols. Stuttgart, 1999.

BACHRACH, B S. *Merovingian military organization, 481–751*. Minneapolis, 1972.

——. "Military organization in Aquitaine under the early Carolingians". *Speculum* 49 (1974), 1–33.

——. "Early medieval fortifications in the 'west' of France: a revised technical vocabulary". *Technology and culture*, 16 (1975), 531–69.

——. "Fortifications and military tactics: Fulk Nerra's strongholds circa 1000". *Technology and culture*, 20 (1979), 531–49.

——. "The Angevin strategy of castle building in the reign of Fulk Nerra, 987–1040". *The American historical review*, 88 (1983), 533–60.

——. "The practical use of Vegetius' *De re militari* during the early Middle Ages". *The historian*, 47 (1985), 239–55.

——. *Fulk Nerra, the neo-Roman consul, 987–1040: a political biography of the Angevin count*. Berkeley, 1993.

Bachrach, B S, and Aris, R. "Military technology and garrison organization: some observations on Anglo-Saxon military thinking in light of the Burghal hidage". *Technology and culture*, 31 (1990), 1–17.

Bakay, K. "Archäologische Studien zur Frage der Ungarischen Staatsgründung: Angaben zur Organisierung des fürstlichen Heeres". *Acta archaeologica Academiae Scientiarum Hungaricae*, 19 (1967), 105–73.

——. "Hungary". In *The new Cambridge medieval history*, III: *c.900–c.1024*, 536–52. Ed. T Reuter. Cambridge, 1999.

Bakirtzis, N, and Oreopoulos, P. *An essay on Byzantine fortification: northern Greece, 4th–15th centuries*. Byzantine hours: works and days in Byzantium. Athens, 2001.

Baldwin, J W. *The government of Philip Augustus: foundations of French royal power in the Middle Ages*. Berkeley/Los Angeles/London, 1986.

Balil, A. *Las murallas bajo-imperiales de Barcino*. Anejos de *Archivo español de arqueología*, 2. Madrid, 1961.

Banks, P J, and Zozaya, J. "Excavations in the Caliphal fortress of Gormaz (Soria), 1979–1981: a summary". In *Papers in Iberian archaeology*, 674–704. Ed. T F C Blagg, R F J Jones and S J Keay. 2 vols. *BAR* international series, 193. Oxford, 1984.

Barbetta, G. *Le mura e le fortificazioni di Verona*. Le guide, 95. Verona, 1978.

Barišić, F. "Les Miracles de St. Démétrius comme source historique". Summary of *Čuda Dimitrija Solunskog kao istoriski izvori*. Srpska Akademija Nauka: Posebna izdanja, 2. Belgrade, 1953.

——. "Le siège de Constantinople par les Avares et les Slaves en 626". *Byzantion* 24 (1954), 371–95.

Barlow, F. *William Rufus*. Berkeley/London, 1983.

Barnes, H, and Whittow, M. "The Oxford University / British Institute of Archaeology at Ankara survey of medieval castles of Anatolia (1993). Yılanlı kalesi: preliminary report and new perspectives". *Anatolian studies*, 44 (1994), 187–206.

——. "Medieval castles: Antioch-on-the-Maeander". *Anatolian archaeology*, 4 (1998), 17–18.

Barni, G L. "La lotta contra Barbarossa". In *Storia di Milano*, IV: 1–112. 17 vols. Milan, 1953–66.

Barry, T B. *The archaeology of medieval Ireland*. London/New York, 1987.

——. "The defensive nature of Irish moated sites". In *The medieval castle in Ireland and Wales: essays in honour of Jeremy Knight*, 182–93. Ed. J R Kenyon and K O'Conor. Dublin, 2003.

Barthold, W. *Turkestan down to the Mongol invasion*. 3rd edn. Trans. T Minorsky, ed. C E Bosworth. London, 1968.

Bartlett, R J. "Technique militaire et pouvour politique, 900–1300". *Annales* 41 (1986), 1135–59.

BASISTA, M. "Hybrid or counterpoise? a study of transitional trebuchets". *Journal of medieval military history*, 5 (2007), 33–55.

BASSET, H, and TERRASSE, H. *Sanctuaires et forteresses almohades*. Collection Hespéris, 5. Paris, 1932.

BATES, D. *Normandy before 1066*. London, 1982.

——. "The rise and fall of Normandy, 911–1204". In *England and Normandy in the Middle Ages*, 19–35. Ed. D Bates and A Curry. London, 1994.

BAUDRY, M-P. "La politique de fortification des Plantagenêts en Poitou, 1154–1242". *ANS* 24 (2001), 43–69.

BAZZANA, A. "Le château d'Alcala de Chivert". CG 8 (1977), 21–46.

——. "Typologie des édifices castraux de l'ancien Sharq-al-Andalus". CG 9–10 (1982), 301–28.

——. "Un fortin omeyyade dans le Sharq al-Andalus". *Archéologie islamique*, 1 (1990), 87–108.

——. "Le *hisn* et les *ma'aqil* dans l'organisation du peuplement musulmane d'al-Andalus". CG 15 (1992), 19–31.

——. "Le début du château dans l'Espagne septentrionale". CG 16 (1994), 33–43.

BAZZANA, A, CRESSIER, P, and GUICHARD, P. *Les châteaux ruraux d'al-Andalus: histoire et archéologie des husûn du sud-est de l'Espagne*. Publications de la Casa de Velázquez, 11. Madrid, 1988.

BECKWITH, C I. *The Tibetan empire in Central Asia: a history of the struggle for great power among Tibetans, Turks, Arabs, and Chinese during the early Middle Ages*. Princeton, 1987.

BECKWITH, J. *Early medieval art: Carolingian, Ottonian, Romanesque*. World of art library. New York/London, 1964.

BEEBY, S, BUCKTON, D, and KLANICA, Z. *Great Moravia: the archaeology of ninth-century Czechoslovakia*. Exhibition catalogue, British Museum. London, 1982.

BEELER, J H. "Castles and strategy in Norman and early Angevin England". *Speculum* 31 (1956), 581–601.

——. *Warfare in England, 1066–1189*. Ithaca, N.Y., 1966.

BEFFEYTE, R. *Les machines de guerre au Moyen Âge*. Rennes, 2000.

BERCHER, H, COURTEAUX, A, and MOUTON, J. "Une abbaye latine dans la société musulmane: Monreale au XIIe siècle". *Annales* 34 (1979), 525–47.

BERESFORD, G. "The excavation of the deserted medieval village of Goltho (Lincs).". CG 8 (1977): 47–68.

BERTHELOT, M. "Pour l'histoire des arts mécaniques et de l'artillerie vers la fin du Moyen Âge". *Annales de chimie et de physique*, 6th ser. 24 (1891), 433–521.

——. "Histoire des machines de guerre et des arts mécaniques au Moyen Âge". *Annales de chimie et de physique*, 7th ser. 19 (1900), 289–420.

BERTHIER, S, and EL-AJJI, E (eds.). *Études et travaux à la citadelle de Damas, 2000–2001: un premier bilan*. Supplement to *Bulletin d'études orientales*, 53–4 (2001–2).

BESHIR, B J. "Fatimid military organization". *Der Islam*, 55 (1978), 37–56.

BEST, W, GENSEN, R, and HÖMBERG, P R. "Burgenbau in einer Grenzregion". In *799 — Kunst und Kultur in der Karolingerzeit: Karl der Grosse und Papst Leo III. in Paderborn, III: Beiträge zum Katalog der Ausstellung*, 328–45. Ed. C Stiegemann, M Wernhoff. Mainz, 1999.

BESTEMAN, J C. "Mottes in the Netherlands". CG 12 (1985), 211–23.

BEZACIER, L. *Le Viêt-Nam*. Manuel d'archéologie d'Extrême-Orient, 1. partie: Asie du Sud-Est, 2. Paris, 1972.

BHAKARI, S K. *Indian warfare: an appraisal of strategy and tactics of war in early medieval period*. New Delhi, 1981.

Biernacka-Lubańska, M. *The Roman and early-Byzantine fortifications of lower Moesia and northern Thrace*. Trans. L Tokarczyk. Bibliotheca antiqua, 17. Wrocław, 1982.

Biller, T, and Metz, B. "Anfänge der Adelsburg im Elsaß in ottonischer, salischer und frühstaufischer Zeit". In *Burgen der Salierzeit*, ii: *In den südlichen Landschaften des Reiches*, 245–84. Ed. H W Böhme. Römisch-Germanisches Zentralmuseum Mainz, Forschungsinstitut für Vor- und Frühgeschichte: Monographien, 26. Sigmaringen, 1992.

Binding, G. "Spätkarolingisch-ottonische Pfalzen und Burgen am Niederrhein". CG 5 (1972), 23–35.

Bingham, W. *The founding of the T'ang dynasty: the fall of Sui and the rise of T'ang, a preliminary survey*. American Council of Learned Societies: Studies in Chinese and related civilizations, 4. Baltimore, 1941.

Bisson, T N. *The medieval crown of Aragon: a short history*. Oxford, 1986.

Blair, C H H. "The early castles of Northumberland". *Archeologia Aeliana*, 4th ser. 22 (1944), 116–70.

Blanchet, A. *Les enceintes romaines de la Gaule: étude sur l'origine d'un grand nombre de villes françaises*. Paris, 1907.

Bloom, J. "Walled cities in Islamic north Africa and Egypt with particular reference to the Fatimids (909–1171)". In *City walls: the urban enceinte in global perspective*, 219–46. Ed. J D Tracy. Studies in comparative early modern history. Cambridge, 2000.

Boas, A J. *Jerusalem in the time of the crusades: society, landscape and art in the Holy City under Frankish rule*. London, 2001.

Boba, I. *Moravia's history reconsidered: a reinterpretation of medieval sources*. The Hague, 1971.

Boeckler, A. *Deutsche Buchmalerei der Gotik*. Königstein im Taunus, 1959.

——. *Deutsche Buchmalerei vorgotische Zeit*. Königstein im Taunus, 1959.

Bogdanowski, J. English summary. *Architektura obronna w krajobrazie Polski od Biskupina do Westerplatte*. Warsaw, 1996.

Böhme, H W. "Burgen der Salierzeit in Hessen, in Rheinland-Pfalz und im Saarland". In *Burgen der Salierzeit*, ii: *In den südlichen Landschaften des Reiches*, 7–80. Ed. H W Böhme. Römisch-Germanisches Zentralmuseum Mainz, Forschungsinstitut für Vor- und Frühgeschichte: Monographien, 26. Sigmaringen, 1992.

Bohnsack, D. "Das Fundament eines steinernen Rundturmes des 11. Jahrhunderts in der Hamburger Altstadt". CG 2 (1967), 1–4.

Boinet, A C L. *Le vieux Metz: histoire et description de ses monuments*. Paris, 1923.

Bois, G. *The transformation of the year one thousand: the village of Lournand from antiquity to feudalism*. Trans. J Birrell. Manchester, 1992.

Boisselier, J. *Le Cambodge*. Manuel d'archéologie d'Extrême-Orient, 1. partie: Asie du Sud-Est, 1. Paris, 1966.

Bon, A. "The medieval fortifications of Acrocorinth and vicinity". *Corinth*, 3/2 (1936), 128–281.

——. "Forteresses médiévales de la Grèce centrale". *Bulletin de correspondance hellénique*, 61 (1937), 136–208.

——. "A propos des châteaux du plan polygonal". *RA* 6th ser. 28 (1947), 177–9.

——. "La prise de Kalamata par les Francs en 1205". In *Mélanges d'archéologie et d'histoire offerts à Charles Picard à l'occasion de son 65e anniversaire*, i (= *RA* 6th ser. 29): 78–104. Paris, 1949.

——. *La Morée franque: recherches historiques, topographiques et archéologiques sur la principauté d'Achaïe (1205–1430)*. 2 vols. Bibliothèque des Écoles françaises d'Athènes et de Rome, 213. Paris, 1969.

Bonaparte, L-N. *Études sur le passé et l'avenir de l'artillerie*, i–ii. Paris, 1846–51. (Vols. iii–vi, 1862–71, by I Favé.)

Bonde, S. *Fortress-churches of Languedoc: architecture, religion, and conflict in the high Middle Ages*. Cambridge 1994.

Bonenfant, P. "Les premiers remparts de Bruxelles". *Annales de la Société royale d'archéologie de Bruxelles*, 40 (1936), 7–47.

Bošković, Đ. "Monuments et stations archéologiques en Serbie", ii: "Serbie centrale". Summary of *Arheološki spomenici i nalazišta u Srbiji*, ii: *Centralna Srbija*. Belgrade, 1956.

Bošković, Đ, Stričević, Đ, and Nikolajević-Stojković, I. "L'architecture de la basse antiquité et du Moyen Âge dans les régions centrales des Balkans". In *Actes du xii*e *Congrès international d'études byzantines* [Ohrid 1961], ii: 155–247. Belgrade, 1963.

Bosworth, C E. "Ghaznevid military organisation". *Der Islam*, 36 (1961), 37–77.

——. "Military organisation under the Būyids of Persia and Iraq". *Oriens* 18–19 (1965–6), 143–67.

——. "The city of Tarsus and the Arab–Byzantine frontiers in early and middle 'Abbāsid times". *Oriens* 33 (1992), 268–86.

——. "The Byzantine defence system in Asia Minor and the first Arab incursions". In his *The Arabs, Byzantium and Iran: studies in early Islamic history and culture*. Aldershot, 1996.

Boüard, M de. "Les petites enceintes circulaires d'origine médiévale en Normandie". CG 1 (1964), 21–35.

——. "Quelques données françaises et normandes sur le problème de l'origine des mottes". CG 2 (1967), 19–26.

——. "Le château de Caen". *Congrès archéologique de France*, 132 (1978 for 1974), 9–21.

Bouchier, E S. *A short history of Antioch, 300 b.c. – a.d. 1268*. Oxford, 1921.

Bowlus, C R. *Franks, Moravians and Magyars: the struggle for the middle Danube, 788–907*. Philadelphia, 1995.

——. "The military organization of Carinthia and Pannonia (818–846)". In *Gesellschaftsgeschichte: Festschrift für Karl Bosl zum 80. Geburtstag*, i: 168–78. Ed. F Seibt. Munich, 1988.

Boyer, M N. *Medieval French bridges: a history*. Publications of the Mediaeval Academy of America, 84. Cambridge, Mass., 1976.

Bozzo, C D. "La porta di città nel medioevo come 'testo' semiotico". In *Fortifications, portes de ville, places publiques, dans le monde méditerranéen*. Ed. J Heers. Cultures et civilisations médiévales, 4. Paris, 1985.

Brachmann, H-J. "Research into the early history of the Slav populations in the territory of the German Democratic Republic". *Medieval archaeology*, 27 (1983), 89–106.

——. "Zum Burgenbau salischer Zeit zwischen Harz und Elbe". In *Burgen der Salierzeit*, i: *In den nördlichen Landschaften des Reiches*, 97–148. Ed. H W Böhme. Römisch-Germanisches Zentralmuseum Mainz, Forschungsinstitut für Vor- und Frühgeschichte: Monographien, 25. Sigmaringen, 1991.

Bradbury, J. *The medieval archer*. Woodbridge, 1985.

——. *The medieval siege*. Woodbridge, 1992.

——. *Philip Augustus, king of France, 1180–1223*. The medieval world. London, 1998.

Bradley, J, and Halpin, A. "The topographical development of Scandinavian and Anglo-Norman Cork". In *Cork: history and society*. Ed. P O'Flanagan and C G Buttimer. Dublin, 1993.

Bragard, P. "Essai sur la diffusion du château 'philippien' dans les principautés lotharingiennes au xiiie siècle". *BM* 157 (1999), 141–67.

BRATHER, S. "Karolingerzeitlicher Befestigungsbau in wilzisch-abodritischen Raum: die sogenannten Feldberger Höhenburgen". In *Frühmittelalterlicher Burgenbau in Mittel- und Osteuropa* [Nitra 1996], 115–26. Ed. J Henning and A T Ruttkay. Bonn, 1998.

BRAUNE, M. "Die mittelalterliche Befestigungen der Stadt Tortosa/Ṭarṭūs: Vorbericht der Untersuchungen 1981–1982". *Damaszener Mitteilungen*, 2 (1985), 45–54.

——. "Die Stadtmauer von Damaskus". *Damaszener Mitteilungen*, 11 (1999), 67–85.

BRAUNE, M, and AL-ZAYM, H. "Stadttore und Stadtmauer von Damaskus". In *Damaskus — Aleppo: 5000 Jahre Stadtentwicklung in Syrien*, 180–87. Ed. B Bollmann. Beihefte der Archäologischen Mitteilungen aus Nordwestdeutschland, 28. Mainz, 2000.

BROGIOLO, G P. "Dwellings and settlements in Gothic Italy". In *The Ostrogoths from the migration period to the sixth century: an ethnographic perspective*. Ed. S Barnish and F Marazzi. Studies is historical archaeoethnology, 7. Woodbridge, 2007.

BROOKS, N P. "The administrative background to the Burghal Hidage". In *The defence of Wessex: the Burghal Hidage and Anglo-Saxon fortifications*, 128–50. Ed. Hill, D, and Rumble, A R. Manchester, 1996.

BROWN, P. *Indian architecture: Islamic period*. 5th edn. Bombay, 1968.

——. *Indian architecture: Buddhist and Hindu periods*. 2nd edn. Bombay, 1976.

BROWN, R A. "A list of castles, 1154–1216". *EHR* 74 (1959), 249–80. Cited from repr. in his *Castles, conquest and charters: collected papers*. Woodbridge, 1989.

——. *Rochester Castle, Kent*. 2nd edn. English Heritage handbooks. London, 1969; repr. 1986.

——. *Dover Castle, Kent*. 4th edn. English Heritage handbooks. London, 1988.

——. *Castles, conquest and charters: collected papers*. Woodbridge, 1989.

BROWN, R A, COLVIN, H M, and TAYLOR, A J. *The history of the king's works*, 1: *The Middle Ages*. London, 1963.

BU'LOCK, J D. *Pre-Conquest Chester, 383–1066*. A history of Cheshire, 3. Chester, 1972.

BUCHTHAL, H. *Miniature painting in the Latin kingdom of Jerusalem*. Oxford, 1957; repr. London, 1986.

BUR, M. "The social influence of the motte-and-bailey castle". *Scientific American* 248/5 (May 1983), 132–40.

BURGESS, J. *Memorandum on the antiquities at Dabhoi, Ahmedebad, Thân, Junâgadh, Girnâr, and Dhank*. Archaeological Survey of Western India, 2. Bombay, 1875.

BURGESS, J, and COUSENS, H. *The antiquities of the town of Dabhoi in Gujarât*. Edinburgh, 1888.

BURNOUF, J. "Les mottes castrales en Alsace". *Revue d'Alsace*, 111 (1985), 3–45; 112 (1986), 3–48.

BURNS, R. *Monuments of Syria: an historical guide*. Rev. edn. London/New York, 1999.

BURRIDGE, P. "The castle of Vardounia and defence in the southern Taygetos". In *The archaeology of medieval Greece*, 19–28. Ed. P Lock and G D R Sanders. Oxbow monographs, 59. Oxford, 1996.

BUTLER, A J. *The Arab conquest of Egypt and the last thirty years of Roman dominion*. 2nd edn, ed. P M Fraser. Oxford, 1978.

BUTLER, L, and KNIGHT, J K. *Dolforwyn Castle, Montgomery Castle*. Cardiff, 2004.

BUTLER, R M. "Late Roman town walls in Gaul". *Archaeological journal*, 116 (1959), 25–60.

BÜTTNER, R. "Befestigungsanlagen im Wienerwald um die Jahrtausendwende". *Anzeiger der Österreichische Akademie der Wissenschaften, philosophisch-historische Klasse*, 93 (1956), 320–44.

——. *Burgen und Schlösser an der Donau*. 1st edn. Vienna, 1964.

Büttner, R. "St. Andrä vor dem Hagenthale". In *Handbuch der historischen Stätten Österreich, 1: Donauländer und Burgenland*, 511–12. Ed. K Lechner. Stuttgart, 1970.

Cahen, C. "Les changements techniques militaires dans le Proche Orient médiéval et leur importance historique". In *War, technology and society in the Middle East* [London 1970], 113–24. Ed. V J Parry and M E Yapp. London, 1975.

——. *The formation of Turkey: the Seljukid sultanate of Rūm, eleventh to fourteenth century*. Trans. and ed. P M Holt. A history of the Near East. Harlow, 2001.

Calderini, A. "Milano archeologica". In *Storia di Milano*, 1: 465–696. 17 vols. Milan, 1953–66.

Callmer, J. "The archaeology of Kiev, ca. a.d. 500–1000: a survey". In *Les pays du Nord et Byzance (Scandinavie et Byzance)* [Uppsala 1979], 29–52. Ed. R Zeitler. Acta Universitatis Upsaliensis: Figura, ns 19. Uppsala, 1981.

Calmette, J. "Le siège de Toulouse par les Normands en 864". *Annales du Midi*, 29 (1917), 171–83.

Cara Barrioneuvo, L. *La Almería islamica y su Alcazaba*. Biblioteca de autores y temas almerienses, 19. Almería, 1990.

Carrington, P. *The English Heritage book of Chester*. London, 1994.

Catarino, H. "O castelo de Paderne (Albufeira): resultados de la primeira intervenção arqueológica". *Arqueologia medieval*, 3 (1994), 73–87.

Caudel, M. *Les premières invasions arabes dans l'Afrique du Nord (21–78 h. – 641–697 j.c.)*. Bibliothèque d'archéologie africaine, 3. Paris, 1900.

Cavalieri Manasse, G. "Le mura teodoriciane di Verona". In *Teoderico il grande e i Goti d'Italia* [Milan 1992], ii: 634–44. Atti dei congressi, 13. Spoleto, 1993.

Cazzato, M. *Guida ai castelli pugliesi, 1: La provincia di Lecce*. Le guide verdi, 25. Galatina, 1999.

Charbonneau-Lassay, L. "Les châteaux de Loudon, d'après les fouilles archéologiques de M. Moreau de la Ronde". *Mémoires de la Société des antiquaires de l'Ouest*, 3rd ser. 8 (1915), 1–486.

Chavannes, É. *Documents sur les Tou-Kiue (Turcs) occidentaux*. St Petersburg/Paris, 1903; repr. Taipei, 1969.

Chelhod, J, et al. *L'Arabie du Sud: histoire et civilisation*. 3 vols. Islam d'hier et d'aujourd'hui, 21, 22, 25. Paris, 1984–5.

Cherf, W J. "A Justinian fortification near Thermopylae". Abstract in *7th Annual Byzantine Studies Conference* [Boston 1981]: *Abstracts of Papers*, 7–8.

——. "Carbon 14 and Prokopios' *De aedificiis*". Abstract in *8th Annual Byzantine Studies Conference* [Chicago 1982]: *Abstracts of Papers*, 44–5.

Chevedden, P E. "Artillery in late antiquity: prelude to the Middle Ages". In *The medieval city under siege*, 131–73. Ed. I A Corfis and M Wolfe. Woodbridge, 1995.

——. "The artillery of King James i the conqueror." In *Iberia and the Mediterranean world of the Middle Ages: essays in honor of Robert I. Burns, S.J., ii: Proceedings from "Spain and the western Mediterranean"* [Los Angeles 1992], 47–94. Ed. P E Chevedden, D J Kagay, and P G Padilla. The medieval Mediterranean, 8. Leiden, 1996.

——. "The hybrid trebuchet: the halfway step to the counterweight trebuchet." In *On the social origins of medieval institutions: essays in honor of Joseph F. O'Callaghan*, 179–222. Ed. D J Kagay and T M Vaan. The medieval Mediterranean, 19. Leiden, 1998.

——. "Fortifications and the development of defensive planning during the crusader period". In *The Circle of War in the Middle Ages: essays on medieval military and naval history*, 33–43. Ed. D J Kagay and L J A Villalon. Warfare in history, 6. Woodbridge, 1999.

——. "The invention of the counterweight trebuchet: a study in cultural diffusion". *Dumbarton Oaks papers*, 54 (2000), 71–116.

CHEVEDDEN, P E, et al. "The trebuchet". *Scientific American*, 273/1 (July 1995), 66–71.

CHIBNALL, M. "Orderic Vitalis on castles". In *Studies in medieval history presented to R. Allen Brown*, 43–56. Ed. C Harper-Bill, C J Holdsworth and J L Nelson. Woodbridge, 1989.

CHRISTENSEN, A. *L'Iran sous les Sassanides*. Copenhagen, 1936.

CHRISTIANSEN, E. *The northern crusades: the Baltic and the Catholic frontier, 1100–1525*. New studies in medieval history. London, 1980.

CHRISTIANSEN, T E. "Archaeology and history: the Viking fortress Trelleborg". In *Danish medieval history: new currents* [Copenhagen 1979], 221–2. Ed. N Skyum-Nielsen and N Lund. Copenhagen, 1981.

CHRISTIDES, V. *The conquest of Crete by the Arabs (ca. 824): a turning point in the struggle between Byzantium and Islam*. Athens, 1984.

——. "Once again on Caminiates' 'Capture of Thessalonika'". *Byzantinische Zeitschrift*, 74 (1981), 7–9.

CHRISTIE, N J. "The archaeology of Byzantine Italy: a synthesis of recent research". *Journal of Mediterranean archaeology*, 2 (1989), 249–93.

——. *The Lombards: the ancient Longobards*. The peoples of Europe. Oxford, 1995.

CHRISTIE, N, and GIBSON, S. "The city walls of Ravenna". *Papers of the British School at Rome*, 56 (1988), 156–97.

CILENTO, N. *Italia meridionale longobarda*. 2nd edn. Milan/Naples, 1971.

CLARKE, H, and AMBROSIANI, B. *Towns in the Viking age*. Rev. edn. London, 1995.

Clavis patrum graecorum. Ed. M Geerard. 5 vols. Corpus christianorum, series graeca. Turnhout, 1974–87.

CLEPHAN, R C. "Notes on Roman and Medieval military engines". *Archaeologia Aeliana*, ns 24 (1903), 69–114.

COAD, J. *English Heritage book of Dover Castle and the defences of Dover*. London, 1995.

COLLINS, R. *Early medieval Spain: unity in diversity, 400–1000*. 2nd edn. New studies in medieval history. Basingstoke, 1995.

——. *Spain: an Oxford archaeological guide*. Oxford, 1998.

COLLOQUE DE CAEN (P Bonnassie et al.). "Les fortifications de terre en Europe occidentale du xe au xiie siècles". *AM* 11 (1981), 5–124.

COLOMBO, A. "Le mura di Milano comunale e la pretesa cerchia di Azzone Visconti". *Archivio storico lombardo*, 50 (1923), 277–324.

CONRAD, H. *Geschichte der deutschen Wehrverfassung, 1: Von der germanischen Zeit bis zum Ausgang des Mittelalters*. Munich, 1939.

CONTAMINE, P. "L'armée de Philippe Auguste". In *La France de Philippe Auguste: le temps des mutations* [Paris 1980]. Ed. R H Bautier. Paris, 1982.

CONTI, F. *Castelli del Piemonte*. 3 vols. Milan, 1975–80.

CORSI, P. "Costante II in Italia". *Quaderni medievali*, 3 (1977), 32–72; 5 (1978), 57–108.

CORVISIER, C. "La tour d'Ostrevant à Bouchain: un donjon de type anglo-normand au comté de Hainaut". *BM* 157 (1999), 216–82.

COULSON, C L H. "Fortresses and social responsibility in late Carolingian France". *Zeitschrift für Archäologie des Mittelalters*, 4 (1976), 29–36.

——. "Fortress policy in Capetian tradition and Angevin practice: aspects of the conquest of Normandy by Philip ii". *ANS* 6 (1983), 13–37.

——. "The castles of the Anarchy". In *The Anarchy of King Stephen's reign* [Battle 1992], 67–92. Ed. E King. Oxford, 1994.

Coulson, C L H. "Freedom to crenellate by licence: an historiographical revision". *Nottingham mediaeval studies*, 38 (1994), 86–137.

——. "The French matrix of the castle-provisions of the Chester–Leicester *conventio*". *ANS* 17 (1994), 65–86.

——. "Cultural realities and reappraisals in English castle-study". *Journal of medieval history*, 22 (1996), 171–207.

——. "The sanctioning of fortresses in France: 'feudal anarchy' or seigneurial amity?" *Nottingham mediaeval studies*, 42 (1998), 38–104.

——. "Peaceable power in English castles". *ANS* 23 (2000), 69–95.

Coupland, S. "Carolingian arms and armor in the ninth century". *Viator* 21 (1990), 29–50.

Cousens, H. *The antiquities of Sind, with historical outline*. Archaeological Survey of India Imperial series, 46. Calcutta, 1929.

Cowdrey, H E J. "The Mahdia campaign of 1087". *EHR* 92 (1977), 1–29.

——. *The age of Abbot Desiderius: Montecassino, the papacy, and the Normans in the eleventh and early twelfth centuries*. Oxford, 1983.

Cowper, M. *Cathar castles: fortresses of the Albigensian Crusade, 1209–1300*. Fortress, 55. Oxford, 2005.

Coyne, P. "The fortifications of Byzantine Carthage: new archaeological data". Abstract in *10th Annual Byzantine Studies Conference* [Cincinnati 1984]: *Abstracts of Papers*.

Craster, O E. *Skenfrith Castle, Monmouthshire = Castell Ynysgynwraidd, Sir Fynwy*. London, 1970.

Creighton, O. "Early Leicestershire castles: archaeology and landscape history". *Transactions of the Leicestershire Archaeological and Historical Society*, 71 (1997), 21–36.

——. "'Castles of communities': medieval town defences in England, Wales and Gascony". CG 22 (2006), 75–86.

Creighton, O, and Higham, R. *Medieval town walls: an archaeological and social history of urban defence*. Stroud, 2005.

Cressier, P. "La fortification médiévale islamique au Maroc: éléments de bibliographie". *Archéologie islamique*, 5 (1995), 163–96.

Creswell, K A C. *A brief chronology of the Muhammadan monuments of Egypt to* A.D. *1517*. Cairo, 1919.

——. *Early Muslim architecture, Umayyads, early 'Abbāsids & Ṭūlūnids*, part 2: *Early 'Abbāsids, Umayyads of Cordova, Aghlabids, Ṭūlūnids, and Samānids,* A.D. *751–905*. Oxford, 1940.

——. "Fortification in Islam before A.D. 1250". *Proceedings of the British Academy*, 38 (1952), 89–125.

——. *The Muslim architecture of Egypt*. 2 vols. Oxford, 1952–9.

Crosland, J. *William the Marshal: the last great feudal baron*. London, 1962.

Crow, J. "Alexios Komnenos and Kastamonu: castles and settlements in middle Byzantine Paphlagonia". In *Alexios I Komnenos* [Belfast 1989], I: 11–33. Ed. M Mullett and D Smythe. Belfast Byzantine texts and translations, 4. Belfast, 1996.

Crow, J, and Hill, S. "Amasra, a Byzantine and Genoese fortress on the Black Sea". *Fortress* 5 (1990), 3–13.

——. "The Byzantine fortifications of Amastris in Paphlagonia". *Anatolian studies*, 45 (1995), 251–65.

Curnow, P E, and Thompson, M W. "Excavations at Richard's Castle, Herefordshire". *Journal of the British Archaeological Association*, 3rd ser. 32 (1969), 105–27.

Dagron, G. "Guérilla, places fortes et villages ouverts à la frontière orientale de Byzance vers 950". *Castrum 3: Guerre, fortification et habitat dans le monde méditerranéen au Moyen Âge* [Madrid 1985], 43–8. Ed. A Bazzana. Madrid, 1988.

DAIN, A. "Les stratégistes byzantins". *Travaux et mémoires* [Centre de recherche d'histoire et civilisation de Byzance], 2 (1967), 317–92.

DARK, K. *Britain and the end of the Roman Empire*. Stroud, 2000.

DAVIDSOHN, R. *Forschungen zur älteren Geschichte von Florenz*. 4 vols. Berlin, 1896–1908.

——. *Storia di Firenze: le origini*. 2 vols. Florence, 1907–9.

DAVIES, N. *God's playground: a history of Poland, 1: The origins to 1795*. Oxford, 1981.

DAVIES, R R. "Kings, lords and liberties in the march of Wales, 1066–1272". *Transactions of the Royal Historical Society*, 5th ser. 29 (1979), 41–61.

——. "Henry I and Wales". In *Studies in medieval history presented to R. H. C. Davis*, 133–47. Ed. H Mayr-Harting and R I Moore. London, 1985.

——. *The age of conquest: Wales, 1063–1415*. Oxford, 1987; repr. 1991.

DAVIES, S. *Welsh military institutions, 633–1283*. Studies in Welsh history, 21. Cardiff, 2004.

DAVIS, P R. *Castles of the Welsh princes*. Swansea, 1988.

DAVIS, R H D. "The authorship of the *Gesta Stephani*". *EHR* 77 (1962), 209–32.

DEAN, B. "A crusaders' fortress in Palestine: a report of excavations made by the Museum, 1926". *Bulletin of the Metropolitan Museum of Art, New York*, 22/9 (Sept. 1927), part II.

DEARDEN, B. "Charles the Bald's fortified bridge at Pîtres (Seine): recent archaeological investigations". *ANS* 11 (1989), 107–12.

——. "Charles the Bald's blockage of the Seine at Pont de l'Arche". *Medieval life*, 8 (1997/8), 11–17.

DEARDEN, B, and CLARK, A. "Pont-de-l'Arche or Pîtres? a location and archaeomagnetic dating for Charles the Bald's fortifications on the Seine". *Antiquity* 64 (1990), 567–71.

DEBORD, A. "*Castrum* et *castellum* chez Adémar de Chabannes". *AM* 9 (1979), 97–113.

DECAENS, J. "Les enceintes d'Urville et de Bretteville-sur-Laize: contribution archéologique aux recherches sur l'habitat fortifié". *Annales de Normandie*, 18 (1968), 311–76.

——. "La motte d'Olivet à Grimbosq (Calvados): résidence seigneuriale du xie siècle". *AM* 11 (1981), 167–201.

——. "Les châteaux de la vallée de l'Huisne dans le Perche". *ANS* 17 (1994), 3–20.

——. "Les châteaux du XIe au XIIIe siècle dans l'actuel département de l'Orne: essai d'inventaire et de classement". In *Autour du château médiéval: des fortifications dans l'Orne et ses régions circonvoisines*, 25–35. Ed. B Fajal. Alençon, 1998.

DEKKER, C. "Les châteaux dans le principauté ecclesiastique d'Utrecht, dans leur contexte politique et sociale". CG 13 (1987), 5–20.

DEL ARCO Y GARAY, R. *Algunas indicaciones sobre antiguos castillos, recintos fortificados y casas solariegas del Alto Aragon*. Huesca, 1913.

——. *El castillo real de Loarre*. Madrid, 1917.

DE MARCO, M. *Torri e castelli del Salento: itinerario costiero*. Lecce, 1994.

DE MEULEMEESTER, J. "Mottes castrales du comté de Flandre: état de la question d'après les fouilles récentes". CG 11 (1983), 101–16.

——. "La fortification de terre et son influence sur la développement urbain de quelques villes des Pays-Bas méridionaux". *Revue du Nord*, 74 (1992), 13–28.

——. "Structures défensives et résidences princières: les châteaux à motte du comté de Looz au XIe siècle". CG 15 (1992), 101–11.

——. "Le début du château: la motte castrale dans les Pays-Bas méridionaux". CG 16 (1994), 121–30.

De Meulemeester, J, Amigues, F, and Matthys, A. "Un grenier fortifié murcien hispano-musulman et mudégare". *Boletín de arqueología medieval*, 7 (1993), 129–34.

De Meulemeester, J, and Matthys, A. "Châteaux et guerres: quelques exemples issus des fouilles de châteaux à motte dans l'espace belge". CG 19 (2000), 33–42.

De Meulemeester, J, Matthys, A, and Poisson, J-M. "Structures enmottées: une comparaison d'exemples fouillés récemment en Belgique et en Rhône-Alpes". In *Military studies in medieval Europe* [Bruges 1997], 139–48. Ed. G de Boe and F Verhaeghe. Papers of the "Medieval Europe Brugge 1997" conference, 11. Zellik, 1997.

Dennis, G T. "Byzantine heavy artillery: the helepolis". *Greek, Roman and Byzantine studies*, 39 (1998), 99–115.

Deschamps, P. *Les châteaux des croisés en Terre Sainte*, I: *Le Crac des Chevaliers: étude historique et archéologique*. 2 vols. Bibliothéque archéologique et historique, Haut-Commissariat de la République Française en Syrie et au Liban, 19. Paris, 1934.

Deschamps, P. *Terre Sainte romane*. La nuit des temps, 21. Saint-Léger-Vauban (Yonne), 1964.

——. *Les châteaux des croisés en Terre Sainte*, III: *La défense du comté de Tripoli et de la principauté d'Antioche: étude historique, géographique, toponymique et monumentale*. 2 vols. Bibliothèque archéologique et historique, Institut français d'archéologie de Beyrouth, 90. Paris, 1973.

DeVries, K. *Medieval military technology*. Peterborough, Ont., 1992.

Deyres, M. "Les châteaux de Foulque Nerra". *BM* 132 (1974), 7–28.

Dhondt, J. "De vroege topografie van Brugge". *Handelingen der Maatschappij voor geschiedenis en oudheidkunde te Gent*, ns 11 (1957), 3–30.

Diccionario crítico etimológico castellano y hispanico. Comp. J Corominas. 6 vols. Madrid, 1980–91.

Dieterich, K. *Byzantinische Quellen zur Länder- und Völkerkunde, 5.–15. Jahrhundert*. 2 vols. in 1. Quellen und Forschungen zur Erd- und Kulturkunde, 5. Leipzig, 1912.

Dieulafoy, M. "Le Château-Gaillard et l'architecture militaire au XIIIe siècle". *Mémoires de l'Academie des inscriptions et belles-lettres*, 36 (1898), 325–86.

Dikshitar, V R R. *War in ancient India*. Madras/London, 1944.

Dion, M A de. "Lettre à M. de Caumont sur le château de Boves-lez-Amiens". *BM* 33 = 4th ser. 3 (1867), 435–44.

Ditten, H. "Prominente Slawen und Bulgaren in byzantinische Diensten". In *Studien zum 8. und 9. Jahrhundert in Byzanz* [Schloß Stein 1979]. Ed. H Köpstein and F Winkelmann. Berliner byzantinische Arbeiten, 51. Berlin, 1983.

Dixon, P, and Lott, B. "The courtyard and the tower: contexts and symbols in the development of late medieval great houses". *Journal of the British Archaeological Association*, 146 (1993), 93–101.

Dixon, P, and Marshall, P. "The great tower in the 12th century: the case of Norham Castle". *Archeological journal*, 150 (1993), 410–32.

Dixon, P, et al. *Newark Castle studies: excavations 1992–1993*. Newark, 1994.

Dodd, A, with Munby, J, and Wilkinson, D R P. "Synthesis and discussion". In *Oxford before the University: the late Saxon and Norman archaeology of the Thames crossing, the defences and the town*, 7–63. Ed. A Dodd. Thames Valley landscapes monographs, 17. Oxford, 2003.

Dollen, B von der. "Die Entwicklung des Burgenbaus in Irland seit der anglo-normannischen Eroberung". *Burgen und Schlösser*, 37 (1996), 50–58.

Dollinger-Leonard, Y. "De la cité romaine à la ville médiévale dans la région de la

Moselle et la Haute-Meuse". In *Studien zu den Anfängen des europäisches Städtewesens.* Vorträge und Forschungen, 4. Stuttgart, 1958.

Dormoy, C. "L'expertise dendrochronologique du donjon de Loches (Indre-et-Loire): des données fondamentales pour sa datation". *AM* 27 (1997), 73–89.

Downey, G. "Byzantine architects: their training and methods". *Byzantion* 18 (1946–8), 99–118.

Drachmann, A G. "Remarks on the ancient catapults". *Actes du VIIe Congrès international d'histoire des sciences* [Jerusalem 1953], 279–83. Collection de travaux de l'Académie internationale d'histoire des sciences, 8. Paris, 1953.

Drage, C. "Urban castles". In *Urban archaeology in Britain*, 117–32. Ed. J Schofield and R Leech. CBA research reports, 61. London, 1987.

Drew, K F. "The Carolingian military frontier in Italy". *Traditio* 20 (1964), 437–47.

Ducouk, H, and Lemerle, P. "L'acropole et l'enceinte haute de Philippes". *Bulletin de correspondance hellénique*, 62 (1938), 4–19.

Durand, P. "L'expérimentation de tir dans les châteaux: de nouvelles perspectives pour la castellologie". *BM* 156 (1998), 257–74.

——. "Les fortifications de Poitiers". In *Les fortifications dans les domaines Plantagenêt, XIIe–XIVe siècles* [Poitiers 1994]. Ed. M-P Baudry. Civilisation médiévale, 10. Poitiers, 2000.

Durdík, T. *Kastellburgen des 13. Jahrhunderts in Mitteleuropa.* Prague, 1994.

Dyer, J F. "Bedfordshire earthworks: the castles". *Bedfordshire magazine*, 8 (1962/3), 267–71.

——. "Danish earthworks". *Bedfordshire magazine*, 8 (1962/3), 235–40.

Eales, R. "Local loyalties in Norman England: Kent in Stephen's reign". *ANS* 8 (1985), 88–108.

——. "Castles and politics in England, 1215–1224". In *Thirteenth century England II* [Newcastle upon Tyne 1987], 23–43. Ed. P R Coss and S D Lloyd. Woodbridge, 1988.

——. "Royal power and castles in Norman England". In *The ideals and practice of medieval knighthood III* [Strawberry Hill 1988], 49–78. Ed. C Harper-Bill and R Harvey. Woodbridge, 1990.

Early, R. "Château de Mayenne: les témoins archéologiques de l'évolution d'un centre de pouvoir entre le xe et le XIIe siècles". CG 20 (2002), 247–62.

Earwood, C, and Malachowska, S. "The medieval town of David-gorod, Belarus". *Antiquity* 67 (1993), 534–47.

Ebhardt, B. *Der Wehrbau Europas im Mittelalter: Versuch einer Gesamtdarstellung der europäischen Burgen.* 2 vols. in 3. Berlin/Stollhamm, 1939–58.

Edwards, J G. "The Normans and the Welsh march". *Proceedings of the British Academy*, 42 (1956), 155–77.

Edwards, R W. "The garrison forts of Byzantine Cilicia". Abstract in *8th Annual Byzantine Studies Conference* [Chicago 1982]: *Abstracts of Papers.*

——. Review of C Foss, *Survey of medieval castles of Anatolia, 1: Kütayha. Speculum* 62 (1987), 675–80.

——. *The fortifications of Armenian Cilicia.* Dumbarton Oaks studies, 23. Washington, D.C., 1987.

Ehrensperger-Katz, I. "Les représentations de villes fortifiées dans l'art paléochrétien et leur dérivées byzantines". *Cahiers archéologiques*, 19 (1969), 1–27.

Ekroll, Ø. "Norwegian medieval castles, 1170–1320: building on the edge of Europe". CG 18 (1998), 65–74.

ELAD, A. "The siege of Al-Wasit (132/749)". In *Studies in Islamic history and civilization in honour of Professor David Ayalon*. Ed. M Sharon. Jerusalem/Leiden, 1986.

ELLIS, C. *Hubert de Burgh: a study in constancy*. London, 1952.

ELLIS DAVIDSON, H R. "The secret weapon of Byzantium". *Byzantinische Zeitschrift*, 66 (1973), 61–74.

ELTON, H. *Warfare in Roman Europe, A.D. 350–425*. Oxford classical monographs. Oxford, 1996.

The encyclopaedia of Islam. New edn. Ed. P Bearman et al. 12 vols. plus Index. Leiden, 1960–2009.

ENGEL, P. *The realm of St Stephen: a history of medieval Hungary, 895–1526*. Trans. T Pálosfalvi. International library of historical studies, 19. London, 2001.

ENGLISH, B. "Towns, mottes and ring-works of the Conquest". In *The medieval military revolution: state, society and military change in medieval and early modern Europe*, 45–61. Ed. A Ayton and J L Price. London, 1995.

ERBEN, W. "Beträge zur Geschichte des Geschützwesens im Mittlealter". *Zeitschrift für historische Waffen- und Kostümkunde*, 7 (1916), 85–102, 117–29.

ERDMANN, C. "Burg und Kirche zu Quedlinburg". Repr. in his *Ottonische Studien*. Ed. H Beumann. Darmstadt, 1968.

——. "Die Burgenordnung Heinrichs I." *Deutsches Archiv für Geschichte des Mittelalters*, 6 (1943), 59–101. Cited from repr. in his *Ottonische Studien*. Ed. H Beumann. Darmstadt, 1968.

ERLANDE-BRANDENBURG, A. "L'architecture militaire au temps de Philippe Auguste: une nouvelle conception de la défense". In *La France de Philippe Auguste: le temps des mutations* [Paris 1980], 595–603. Ed. R H Bautier. Paris, 1982.

——. "Organisation du conseil d'architecture militaire et du corps des spécialistes sous Philippe Auguste". In *Artistes, artisans et production artistique au Moyen Âge* [Rennes 1983], II: 221–4. Ed. X Barrel i Altet. Paris, 1987.

ESCO SAMPÉRIZ, C, and SENAC, P. "Un *hisn* de la Marche Supérieure d'al-Andalus: Piracés (Huesca)". *Mélanges de la Casa de Velázquez*, 23 (1987), 125–50.

——. "Une forteresse de la Marche Supérieure d'al-Andalus: le *hisn* de Sen et Men (province de Huesca)". *Annales du Midi*, 100 (1988), 17–33.

ETTEL, P. "Die Eroberung der Schweinfürter Burgen in der historischen und archäologischen Uberlieferung". CG 19 (2000), 59–68.

ETTEL, P, and WIECHMANN, R. "Die slawische Fürstenburg von Dobin, Mecklenburg-Vorpommern in den Feldzügen des 12. Jahrhunderts". CG 19 (2000), 69–73.

ETTINGHAUSEN R, and GRABAR, O. *The art and architecture of Islam, 650–1250*. Pelican history of art. Harmondsworth, 1987.

EVANS, J A S. "The walls of Thessalonika". *Byzantion* 47 (1977), 361–2.

FAGNANI, F. "Il tracciato delle mura romane di Ticinum". *Bolletino della Società pavese di storia patria*, ns 11 (1959), 3–41.

FALK, H. *Altnordische Waffenkunde*. Videnskapsselskapets skrifter, hist.-filos. klasse, 6. Christiania, 1914.

FARRIS, W W. *Heavenly warriors: the evolution of Japan's military, 500–1300*. Harvard East Asian monographs, 157. Cambridge, Mass./London, 1992.

FASTLINGER, M. "Karolingische Pfalzen in Altbayern". *Forschungen zur Geschichte Bayerns*, 12 (1904), 233–69.

FAVRE, E. *Eudes, comte de Paris et roi de France (882–898)*. Bibliothèque de l'École des hautes études, 99. Paris, 1893.

FAWTIER, R. *The Capetian kings of France: monarchy and nation, 987–1328.* Trans. L Butler and R J Adam. London, 1960.

FEDDEN, R. *Crusader castles: a brief study in the military architecture of the crusades.* London, 1950.

FEHRING, G P. "Origins and development of Slavic and German Lübeck". In *From the Baltic to the Black Sea: studies in medieval archaeology* [Southampton 1986]. Ed. D Austin and L Alcock. One world archaeology, 18. London, 1990.

——. *The archaeology of medieval Germany: an introduction.* Studies in archaeology. Trans. R Samson. London, 1991.

FERREIRO, A. "The siege of Barbastro, 1064–65: a reassessment". *Journal of medieval history,* 9 (1983), 129–44.

FIECCONI, A. "Luoghi fortificati e strutture edilizie del fabrianese nei secoli XI–XIII". *Nuova rivista storica,* 59 (1975), 1–54.

FINGERLIN, G, GARBSCH, J, and WERNER, J. "Gli scavi nel castello longobardo di Ibligo-Invillino (Friuli): relazione preliminare delle campagne del 1962, 1963 e 1965". *Aquileia nostra,* 39 (1968), 57–135.

FINÓ, J F. "Le feu et ses usages militaires". *Gladius* 9 (1970), 15–30.

——. "Machines de jet médiévales". *Gladius* 10 (1972), 25–43.

——. *Forteresses de la France médiévale: construction, attaque, défense.* 3rd edn. Paris, 1977.

FITA, F. "Ceuta visigoda y bizantina durante el reinado de Teudis". *Boletín de la Real Academia de la Historia,* 68 (1916), 622–8.

FIXOT, M. "Les fortifications de terre et la naissance de la féodalité dans le Cinglais". CG 3 (1969), 61–6.

FLAMBARD HÉRICHER, A-M. "La construction dans la basse vallée de la Seine: l'exemple du château de Vatteville-la-Rue (Seine-Maritime)". CG 18 (1998), 93–102.

FLANAGAN, M T. "Irish and Anglo-Norman warfare in twelfth-century Ireland". In *A military history of Ireland,* 52–75. Ed. T Bartlett and K Jeffery. Cambridge, 1996.

FLEURY, G. "Recherches sur les fortifications du Maine". *Revue historique et archéologique du Maine,* 24 (1888), 257–76.

——. "Les fortifications du Maine: La tour Orbrindelle et le Mont-Barbet". *Revue historique et archéologique du Maine,* 29 (1891), 137–54, 279–303.

FLORESCU, G. "Fouilles archéologiques de Capidava, 1928–1936". *Dacia* 5–6 (1938), 351–86.

FOLDA, J. *Crusader manuscript illumination at Saint-Jean d'Acre, 1275–1291.* Princeton, 1976.

FORKE, A. "Der Festungskrieg im alten China". *Ostasiatische Zeitschrift,* 8 (1919/20), 103–16.

FOSS, C. *Byzantine and Turkish Sardis.* Archaeological exploration of Sardis: Monographs, 4. Cambridge, Mass., 1976.

——. "Late antique and Byzantine Ankara". *Dumbarton Oaks papers,* 31 (1977), 29–87.

——. *Ephesus after antiquity: a late antique, Byzantine and Turkish city.* Cambridge 1979.

——. "Late Byzantine fortifications in Lydia". *Jahrbuch der österreichischen Byzantinistik,* 28 (1979), 267–320.

——. "The defenses of Asia Minor against the Turks". *Greek Orthodox theological review,* 36 (1982), 145–205. Repr. in his *Cities, fortresses, and villages of Byzantine Asia Minor.* Aldershot, 1996.

——. *Survey of medieval castles of Anatolia,* 1: *Kütahya. BAR* international series, 261; British Institute of Archaeology at Ankara monographs, 7. Oxford, 1985.

FOSS, C. "Byzantine Malagina and the lower Sangarius". *Anatolian studies*, 40 (1990), 161–83. Repr. in his *Cities, fortresses, and villages of Byzantine Asia Minor*. Aldershot, 1996.

——. *Cities, fortresses, and villages of Byzantine Asia Minor*. Aldershot, 1996.

——. *Survey of medieval castles of Anatolia*, II: *Nicomedia*. British Institute of Archaeology at Ankara monographs, 21. Ankara, 1996.

FOSS, C, and WINFIELD, D. *Byzantine fortifications: an introduction*. UNISA studia, 22. Pretoria, 1986.

FOURACRE, P. "Frankish Gaul to 814". In *The new Cambridge medieval history*, II: *c.700–c.900*, 85–109. Ed. R McKitterick. Cambridge, 1995.

FOURNIER, G. "Le château de Puiset au début du XIIe siècle et sa place dans l'évolution de l'architecture militaire". *BM* 122 (1964), 355–73.

——. "Les campagnes de Pépin le Bref en Auvergne et la question des fortifications rurales au VIIIe siècle". *Francia* 2 (1974), 123–35.

FOWLER, G H. "Munitions in 1224". In *Miscellanea*, 117–32. Ed. J E Morris. Publications of the Bedfordshire Historical Record Society, 5. Bedford, 1920.

FRANCE, J. "La guerre dans la France féodale à la fin du IXe et au Xe siècle". *Revue belge d'histoire militaire*, 23 (1979), 177–98.

——. "The military history of the Carolingian period". *Revue belge d'histoire militaire*, 26 (1985), 81–100.

——. *Victory in the East: a military history of the First Crusade*. Cambridge, 1994.

——. "Technology and the success of the First Crusade". In *War and society in the eastern Mediterranean, 7th–15th centuries*, 163–76. Ed. Y Lev. The medieval Mediterranean, 9. Leiden, 1997.

——. *Western warfare in the age of the crusades, 1000–1300*. Ithaca, N.Y./London, 1999.

——. "Fortifications East and West". In *Muslim military architecture in greater Syria: from the coming of Islam to the Ottoman period* [Aleppo 2003]. Ed. H Kennedy. History of warfare, 35. Leiden, 2006.

FRANKE, H. "Siege and defense of towns in medieval China". In *Chinese ways in warfare* [Cambridge, Mass. 1969], 151–201. Ed. F A Kierman and J K Fairbank. Harvard East Asian series, 74. Cambridge, Mass., 1974.

FRANKLAND-PAYNE-GALLWEY, R W. *The crossbow, medieval and modern, military and sporting: its history, construction and management; with a treatise on the balista and and catapult of the ancients*. London, 1903.

FRANKLIN, S, and SHEPARD, J. *The emergence of Rus, 750–1200*. Longman history of Russia. London, 1996.

FRIES, N. *Das Heereswesen der Araber zur Zeit der Omaijaden nach Ṭabarī*. Tübingen, 1921.

FRYE, R N. "Sasanian system of walls for defence". In *Studies in memory of Gaston Wiet*, 7–15. Ed. M Rosen-Ayalon. Jerusalem, 1977.

FÜGEDI, E. *Castle and society in medieval Hungary (1000–1437)*. Trans. J M Bak. Studia historica Academiae Scientiarum Hungaricae, 187. Budapest, 1986.

FULFORD, M G. "Early motte evidence at Pevensey". *Sussex past & present*, 72 (1994), 7.

GABLER, D, SZŐNYI, E, and TOMKA, P. "The settlement history of Győr (Arrabona) in the Roman period and in the Middle Ages". In *Towns in medieval Hungary*, 9–25. Ed. L Gerevich. Budapest, 1990.

GABRIEL, I, et al. *Starigard/Oldenburg, Hauptburg der Slawen in Wagrien*, I: *Stratigraphie und Chronologie: archäologische Ausgrabungen 1973–1982*. Offa-Bücher, 52. Neumünster, 1984.

GABRIELI, F. *Muhammad and the conquests of Islam*. Trans. V Luling and R Linell. World university library. London, 1968.

GARAUD, M. "La construction des châteaux et les destinées de la *vicaria* et du *vicarius* carolingiens en Poitou". *Revue historique de droit français et étranger*, 4th ser. 32 (1953), 54–77.

——. *Les châtelains de Poitou et l'avènement du régime féodal, XIe et XIIe siècles*. Mémoires de la Société des antiquaires de l'Ouest, 4th ser. 8. Poitiers, 1967.

GARCÍA FITZ, F. "Tecnología militar y guerra de asedios: la experienca castellano-leonese, siglos XI al XII". In *Military studies in medieval Europe* [Bruges 1997], 33–41. Ed. G de Boe and F Verhaeghe. Papers of the "Medieval Europe Brugge 1997" conference, II. Zellik, 1997.

GARDELLES, J. *Les châteaux du Moyen Âge dans la France du Sud-Ouest: la Gascogne anglaise de 1216 à 1327*. Bibliothèque de la Société française d'archéologie, 3. Geneva/Paris, 1972.

GARDINER, R. "Crusader Turkey: the fortifications of Edessa". *Fortress* 2 (1989), 23–35.

GARLANDINI, A. *I castelli della Lombardia*. Milan, 1991.

GASPAR, A, and GOMES, A. "O Castelo de S. Jorge: da fortaleza islâmica à alcáçova cristã". In *Mil anos de fortificações na Península Ibérica e no Magreb (500–1500)* [Palmela 2000]. Ed. I C Ferreira Fernandes. Lisbon, 2001.

GAUTHIEZ, B. "Hypothèses sur la fortification de Rouen au onzième siècle: le donjon, la tour de Richard II et l'enceinte de Guillaume". *ANS* 14 (1992), 61–76.

GAYA NUÑO, J A. "Crónica arqueológica de la España musulmana, XIII", [1]: "Gormaz, castillo califal". *Al-Andalus* 8 (1943), 431–50.

GEORGIEV, G I, IVANOV, T, and DZINGOV, G. "Archaeological investigations in Bulgaria in the last few decades". *Bulgarian historical review*, 2 (1975), 85–7.

GEORGIEVA, S. "Études archéologiques des cités du bas Moyen Âge en Bulgarie". In *Actes du premier Congrès international des études balkaniques et sud-est européennes* [Sofia 1966], II: *Archéologie, histoire de l'antiquité, arts*. Ed. D Dimitrov et al. Sofia, 1969.

GEREVICH, L. "The rise of Hungarian towns along the Danube". In *Towns in medieval Hungary*, 26–50. Ed. L Gerevich. Budapest, 1990.

GERKAN, A von. *Die Stadtmauern*. Milet: Ergebnisse der Ausgrabungen und Untersuchungen seit dem Jahre 1899, 2/3. Berlin/Leipzig 1935.

GERŐ, L. *Castles in Hungary*. Trans. G Gulyás. Budapest, 1969.

GIBB, H A R. *The Arab conquests in Central Asia*. James G Forlong fund, 2. London, 1923.

——. "The armies of Saladin". *Cahiers d'histoire égyptienne*, 3rd ser. 4 (1951), 304–20. Cited from repr. in his *Studies on the civilization of Islam*, 74–90. Ed. S J Shaw and W R Polk. London, l962; repr. Princeton, 1982.

GIBSON, S, and WARD-PERKINS, B. "The surviving remains of the Leonine Wall". *Papers of the British School at Rome*, 47 (1979), 30–57; 51 (1983), 222–39.

GIESEBRECHT, W von. *Geschichte der deutschen Kaiserzeit*. 5th edn. 6 vols. in 7. Leipzig, 1880–95.

GIL, J. *The finest castles in Portugal*. With photographs by A Cabrita; trans. G F W Dykes. 2nd edn. Lisbon, 1989.

GILLMOR, C M. "The introduction of the traction trebuchet into the Latin West". *Viator* 12 (1981), 1–8.

——. "War on the rivers: Viking numbers and mobility on the Seine and Loire, 841–886". *Viator* 19 (1988), 79–109.

——. "The logistics of fortified bridge building on the Seine under Charles the Bald". *ANS* 11 (1989), 87–106.

GLASER, M. "Die Lübecker Stadtmauern". In *25 Jahre Archäologie in Lübeck: Erkentnisse von Archäologie und Bauforschung zur Geschichte und Vorgeschichte der Hansestadt*. Ed.

G P Fehring. Lübecker Schriften zur Archäologie und Kulturgeschichte, 17. Bonn, 1988.

Glubb, J B. *The great Arab conquests*. London, 1963.

Glykatzi-Ahrweiler, H. "Les forteresses construites en Asie Mineure face à l'invasion Seldjoucide". In *Akten des xi. internationalen Byzantinistenkongresses* [Munich 1958], 182–9. Ed. F Dolger and H-G Beck. Munich, 1960.

Göckenjan, H. *Hilfsvölker und Grenzwächter in mittelalterlichen Ungarn*. Quellen und Studien zur Geschichte des östlichen Europa, 5. Wiesbaden, 1972.

Gode, P K. "The history of the sling (*gophana*) in India and other countries". Repr. in his *Studies in Indian cultural history*, ii. P. K. Gode studies, 5. Poona, 1960.

Godlewski, W. "The fortifications of Old Dongola; report on the 1990 season". *Archéologie du Nil moyen*, 5 (1991), 103–27.

——. "Old Dongola: the fortifications, 1994". *Polish archaeology in the Mediterranean*, 6 (1995), 93–7.

——. "Old Dongola: Kom A, 1995". *Polish archaeology in the Mediterranean*, 7 (1996), 115–20.

——. "Old Dongola: Kom A, 1996". *Polish archaeology in the Mediterranean*, 8 (1997), 179–87.

——. "Old Dongola: the early fortifications". *Cahiers de recherche de l'Institut de papyrologie et d'égypytologie de Lille*, 17/2 (1997), 175–9.

Goffart, W. *Barbarians and Romans,* a.d. *418–584: the techniques of accommodation*. Princeton, 1980.

Gohlke, W. "Das Geschützwesen des Altertums und des Mittelalters". *Zeitschrift für historische Waffenkunde*, 5 (1909/11), 291–8, 354–62, 379–93, 579–92.

Gojda, M. *The ancient Slavs: settlement and society*. The Rhind lectures, 1989–90. Edinburgh, 1991.

Gómez-Moreno, M. *Catálogo monumental de España: provincia de León*. 2 vols. Madrid, 1925–6; repr. León, 1979.

——. *Catálogo monumental de la provincia de Ávila*. 2 vols. in 3. Ávila, 1983.

Goodall, J. *Pevensey Castle, East Sussex*. London, 1999.

Grabar, A, and Manoussacas, M. *L'illustration du manuscrit de Skylitzès de la Bibliothèque nationale de Madrid*. Bibliothèque de l'Institut hellénique d'études byzantines et post-byzantines de Venise, 10. Venice, 1979.

Graff, D A. *Medieval Chinese warfare, 300–900*. Warfare and history. London, 2002.

Grant, L. *Abbot Suger of St-Denis: church and state in early twelfth-century France*. The medieval world. London, 1998.

Graudonis, J. "Der Bau des Befestigungswerkes der Burg Turaida". In *Castella Maris Baltici ii* [Nyköping 1993]. Ed. M Josephson and M Mogren. Archaeologia Medii Aevi Finlandiae, 2. Nyköping, 1996.

Gravett, C. *Medieval siege warfare*. Elite series, 28. London, 1990.

Green, J A. *The government of England under Henry I*. Cambridge studies in medieval life and thought, 4th ser. 3. Cambridge, 1986.

Greenshields, M F. "The siege of Bedford Castle". *Bedfordshire magazine*, 4 (1954), 183–90.

Gregory, T E. "Kastro and diateichisma as responses to early Byzantine frontier collapse". *Byzantion* 62 (1992), 235–53.

——. *Isthmia: excavations . . . under the auspices of the American School of Classical Studies at Athens*, v: *The hexamilion and the fortress*. Princeton, 1993.

Grekov, B, and Iakoubovski, A. *La Horde d'Or: la domination tatare au xiii*e *et au* xive siècle de la Mer Jaune à la Mer Noire*. Trans. F Thuret. Bibliothèque historique. Paris, 1939.

GRENET, F, and ISAMIDDINOV, M. "La prise de Samarcande par Gengis-Khan". *L'histoire*, no. 149 (Nov. 1991), 8–14.

GRENIER, A. *Quatre villes romaines de Rhénanie: Trèves, Mayence, Bonn, Cologne*. Études d'archéologie rhénane. Paris, 1925.

GRIFFITH, P. *The Viking art of war*. Greenhill military paperbacks. London, 1995.

GRINGMUTH-DALLMER, E. "Deutsche und slawische Burgen in einem Grenzraum beiderseits der Mittelelbe (Altmark und Elbe-Havel-Gebiet)". CG 17 (1996), 111–17.

GROSE, F. *The antiquities of England and Wales*, VIII: *Supplement*. New edn. London, 1787.

GROSLIER, G. *Recherches sur les Cambodgiens d'après les textes et les monuments depuis les premiers siècles de notre ére*. Paris, 1921.

GROSSMANN, G U. "Die Kaiserpfalz (Kaiserburg) zu Nürnberg: Literaturbericht und Forschungsstand". CG 19 (2000), 103–14.

GSELL, S. *Les monuments antiques de l'Algérie*, 2 vols. Paris, 1901.

GUERRA, A, and FABIÃO, C. "Uma fortificação omíada em Mesas do Castelinho (Almodôvar)". *Arquelogia medieval*, 2 (1993), 85–102.

GUTIÉRREZ GONZÁLEZ, J A. "Características de la fortificaciones del reino de León en la alta y plena Edad Media". In *Maŷrīṭ: estudios de arqueología medieval madrileña*. Ed. F Valdés. Biblioteca de arqueología medieval hispánica, 1. Madrid, 1992.

GUTIÉRREZ LLORET, S. "Ciudades y conquista: el fin de las *ciuitates* visigodas y la génesis de las *mudun* islámicas del sureste de al-Andalus". In *Genèse de la ville islamique en al-Andalus et au Maghreb occidental* [Madrid 1994 / Granada 1995], 137–57. Ed. P Cressier and M García-Arenal. Madrid, 1998.

GYŐRFFY, G. "Die Entstehung der ungarischen Burgorganisation". *Acta archaeologica Academiae Scientiarum Hungaricae*, 28 (1976), 323–55.

——. *King Saint Stephen of Hungary*. Trans. P Doherty. East European monographs, 403. Boulder, Colo./New York, 1988.

HAASE, C. "Die mittelalterliche Stadt als Festung: Wehrpolitische-militärische Einflußbedingungen im Werdegang der mittelalterlichen Stadt". *Studium generale*, 16 (1963), 378–401. Repr. in his *Die Stadt des Mitteralters*, 1: *Begriffe, Entstehung und Ausbreitung*. Wege der Forschung, 243–5. 1st edn. Darmstadt, 1969.

HAKIM, H, and JAWISH, H. *Saladin und seine Burg*. Damascus, 1998.

HALDON, J F. "Some aspects of Byzantine military technology from the sixth to the tenth centuries". *Byzantine and Modern Greek studies*, 1 (1975), 11–47.

——. *Byzantium in the seventh century: the transformation of a culture*. Cambridge, 1990.

——. *Warfare, state and society in the Byzantine world, 565–1204*. London, 1999.

——. *The Byzantine wars: battles and campaigns of the Byzantine era*. Stroud, 2001.

HALDON, J F, and BYRNE, M. "A possible solution to the problem of Greek Fire". *Byzantinische Zeitschrift*, 70 (1977), 91–9.

HALDON, J F, and KENNEDY, H. "The Arab–Byzantine frontier in the eighth and ninth centuries: military organization and society in the borderlands". *Recueil des travaux de l'Institut d'études byzantines* [Belgrade], 19 (1980), 79–116.

HALL, A R. "Military technology". In *A history of technology*, II: *The Mediterranean civilizations and the Middle Ages, c.700 B.C. to A.D. 1500*. Ed. C Singer et al. Oxford, 1956.

HALL, B S. New introduction to J R Partington, *A history of Greek Fire and gunpowder*. Repr., Baltimore, 1999.

HALL, J W. "Kyoto as historical background". In *Medieval Japan: essays in institutional history*, 3–38. Ed. J W Hall and J P Mass. New Haven/London, 1974.

HALMER, F. *Burgen und Schlösser zwischen Baden, Gutenstein, Wiener Neustadt*. Niederösterreichs Burgen und Schlösser, 1/2. Vienna, 1968.

HALPHEN, L. *Charlemagne et l'empire carolingien*. Évolution de l'humanité, 33. Paris, 1947.

HANISCH, H. "Der Nordwestturm der Zitadelle von Damaskus". *Damaszener Mitteilungen*, 5 (1991), 183–215.

——. "Der Nordostabschnitt der Zitadelle von Damaskus". *Damaszener Mitteilungen*, 7 (1993), 274.

——. "Die Maschikulis der Zitadelle von Damaskus". *Damaszener Mitteilungen*, 9 (1996), 227–62.

HANNA, Z. *Syria: the castles and archaeological sites in Tartous*. Trans. B Khoutry and R Botrus. Damascus, 1994.

HARPER, R P. "Upper Zohar: a preliminary excavation report". In *The defence of the Roman and Byzantine East* [Sheffield 1986], 1: 329–36. Ed. P Freeman and D Kennedy. *BAR* international series, 297. Oxford, 1986.

——. *Upper Zohar, an early Byzantine fort in Palaestina Tertia: final report of excavations in 1985–1986*. British Academy monographs in archaeology, 9. Oxford, 1995.

HARPHAM, R. "Heron's cheiroballistra (a Roman torsion crossbow)". *Journal of the Society of Archer-Antiquaries*, 40 (1997), 13–18.

HARRISON, D. *The early state and the towns: forms of integration in Lombard Italy, A.D. 568–774*. Lund studies in international history, 29. Lund, 1993.

HARRISON, P. "Church fortifications in La Thiérache". *Fort* 22 (1994), 3–20.

——. "The tower churches of Gower". *Gower* 46 (1995), 15–23.

——. "The fortified village of Ushguli in the Georgian Caucasus". *Fort* 25 (1997), 3–33.

——. *Castles of God: fortified religious buildings of the world*. Woodbridge, 2004.

HARRISON, R M, CHRISTIE, N, LIGHTFOOT, C S, et al. "Amorium excavations 1989[–1993]: the second[–sixth] preliminary report". *Anatolian studies*, 40 (1990), 205–18; 41 (1991), 215–29; 42 (1992), 207–22; 43 (1993), 147–62; 44 (1994), 105–28.

HART, V G, and LEWIS, M J T. "Mechanics of the onager". *Journal of engineering mathematics*, 20 (1986), 345–65.

HARTMANN-VIRNICH, A. "Les portes ayyoubides de la citadelle de Damas: le regard de l'archéologie du bâti". In *La fortification au temps des Croisades* [Parthenay 2002], 287–311. Ed. N Faucherre, J Mesqui and N Prouteau. Archéologie et culture. Rennes, 2004.

HARVEY, J. *English medieval architects: a biographical dictionary down to 1550, including master masons, carpenters, carvers, building contractors and others responsible for design*. With A Oswald. 2nd edn. Gloucester, 1984.

HASAN, S A. "Munejjim Bashi's account of Sultan Malik Shah's reign". *Islamic studies* 3 (1964), 429–69.

HASSALL, J M, and HILL, D. "Pont de l'Arche: Frankish influence on the West Saxon burh" *Archaeological journal*, 127 (1970), 188–95.

——. "Viking warfare: the siege of Paris, 885–6". *Ago* 9 (1971), 17–18; 10 (1971), 16–22.

HATTERSLEY-SMITH, K M. *Byzantine public architecture between the fourth and early eleventh centuries A.D., with special reference to the towns of Byzantine Macedonia*. Makedoniki bibliotheki, 83. Thessalonica, 1996.

HEERINGEN, R M van. "The construction of Frankish circular fortresses in the province of Zeeland (SW Netherlands) in the end of the ninth century". CG 18 (1998), 241–9.

HEERS, J. "Porta Aurea à Gênes: bourg de religieux, bourg d'immigrés". In *Fortifications, portes de ville, places publiques, dans le monde méditerranéen*. Ed. J Heers. Cultures et civilisations médiévales, 4. Paris, 1985.

HEIDINGA, H A. "The Hunneschans at Uddel reconsidered". CG 13 (1987), 53–62.

HEINE, H-W. "Burgen der salischen Zeit in Niedersachsen: ein Überblick". In *Burgen der Salierzeit, 1: In den nördlichen Landschaften des Reiches*, 9–84. Ed. H W Böhme. Römisch-Germanisches Zentralmuseum Mainz, Forschungsinstitut für Vor- und Frühgeschichte: Monographien, 25. Sigmaringen, 1991.

——. "Zu Burgen der Salierzeit in Niedersachsen". CG 15 (1992), 189–204.

——. "Frühmittelalterlicher Burgen in Niedersachsen". In *Frühmittelalterlicher Burgenbau in Mittel- und Osteuropa* [Nitra 1996], 137–49. Ed. J Henning and A T Ruttkay. Bonn, 1998.

——. "Die Harlyburg bei Vienenburg-Wiedelah, Ldkr. Goslar: eine Reichs- und Territorialburg des 13. Jahrhunderts". *Nachrichten aus Niedersachsens Urgeschichte*, 70 (2001), 323–31.

HELIOT, P. *Histoire de Boulogne et du Boulonnais*. Lille, 1937.

——. "Le château de Boulogne-sur-Mer et les châteaux gothiques de plan polygonal". *RA* 6th ser. 27 (1947), 41–59.

——. "Sur quelques résidences princières bâties en France du xe au xiie siècles". *Le Moyen Âge*, 61 (1955), 27–51.

——. "Le Château-Gaillard et les forteresses des xiie et xiiie siècles en Europe occidentale". CG 1 (1963), 56–73.

——. "Les châteaux-forts en France du xe au xiie siècle à lumière de travaux recents". *Journal des savants* (1965), 483–514.

——. "La genèse des châteaux de plan quadrangulaire en France et en Angleterre". *Bulletin de la Société nationale des antiquaires de France* (1965), 238–57.

——. "La grosse tour de Châteaudun et les donjons du xiie siècle". *Bulletin de la Société nationale des antiquaires de France* (1968), 225–46.

——. "Les origines du donjon résidentiel et les donjons-palais romans de France et d'Angleterre". *Cahiers de civilisation médiévale*, 17 (1974), 217–34.

HELLENKEMPER, H. *Burgen der Kreuzritterzeit in der Grafschaft Edessa und im Königreich Kleinarmenia: Studien zur histor. Siedlungsgeographie Südost-Kleinasiens*. Geographica historica, 1. Bonn 1976.

HENG C K. *Cities of aristocrats and bureaucrats: the development of medieval Chinese cityscapes*. Honolulu, 1999.

HENNE, A, and Wauters, A. *Histoire de la ville de Bruxelles: nouvelle édition du texte original de 1845 augmentée de nombreuses reproductions de documents*. Ed. M Martens. 4 vols. Brussels, 1968–9.

HENNING, J. "Ringwallburgen und Reiterkrieger: zum Wandel der Militärstrategie im ostsächsisch-slawischen Raum an der Wende vom 9. zum 10. Jahrhundert". In *Military studies in medieval Europe* [Bruges 1997], 21–31. Ed. G de Boe and F Verhaeghe. Papers of the "Medieval Europe Brugge 1997" conference, 11. Zellik, 1997.

HENSEL, W. "Types des fortifications slaves du haut Moyen-Âge". *Archaeologia polona*, 2 (1959), 71–84.

HERBIN, J-C. "Les châteaux dans la *Geste des Loherains*". In *Le château médiéval et la guerre dans l'Europe du nord-ouest: mutations et adaptations* [Valenciennes 1995], 9–19. Ed. A Salamagne and R Le Jan. Villeneuve d'Ascq, 1998.

HERNÁNDEZ JIMÉNEZ, F. "The Alcazaba of Mérida". In K A C Creswell, *Early Muslim architecture, Umayyads, early 'Abbāsids & Ṭūlūnids*, part 2: *Early 'Abbāsids, Umayyads of Cordova, Aghlabids, Ṭūlūnids, and Samānids, A.D. 751–905*, 197–207. Oxford, 1940.

HERRMANN, J. "Neue Arbeiten über die altrussischen Wall- und Wehranlagen". *Praehistorische Zeitschrift*, 37 (1959), 227–36.

HERRNBRODT, A. "Die frühmittelalterlichen Ringwalle des Rheinlandes". CG 3 (1969), 67–76.

HESLOP, T A. "Orford Castle, nostalgia and sophisticated living". *Architectural history*, 34 (1991), 36–58.

HETHERINGTON, P. *Byzantine and medieval Greece: churches, castles, and art of the mainland and Peloponnese*. London, 1991.

HIEBERT, F T. "The oasis and city of Merv (Turkmenistan)". *Archéologie islamique*, 3 (1992), 111–27.

HIGHAM, R. "Public and private defence in the medieval South West: town, castle and fortress". In *Security and defence in south-west England before 1800*, 27–50. Ed. R Higham. Exeter studies in history, 19. Exeter, 1987.

HIGHAM, R, and BARKER, P. *Timber castles*. London, 1992.

HIGHAM, R, and SAUNDERS, A. "Public and private defence in British medieval towns". *IBI bulletin*, 50 (1998), 117–28.

HILL, D. *A history of engineering in classical and medieval times*. London, 1984.

——. "Gazetteer of sites". In *The defence of Wessex: the Burghal Hidage and Anglo-Saxon fortifications*, 189–231. Ed. Hill, D, and Rumble, A R. Manchester, 1996.

HILL, D R. "Trebuchets". *Viator* 4 (1973), 99–116.

——. "Siegecraft from the sixth to the tenth century". In *De rebus bellicis*, part 1: *Aspects of the De rebus bellicis: papers presented to E. A. Thompson*, 111–17. Ed. M W C Hassall. *BAR* international series, 63. Oxford, 1979.

HILLENBRAND, C. *The crusades: Islamic perspectives*. Edinburgh, 1999.

HINDLE, P. *Medieval roads and tracks*. 3rd edn. Shire archaeology, 26. Princes Risborough, 1998.

HINZ, H. "Rennenburg". In *Kirche und Burg in der Archäologie des Rheinlandes: Austellung im Rheinischen Landesmuseum Bonn, 31. Oktober – 31. Dezember 1962*. Kunst und Altertum am Rhein, 8. Düsseldorf, 1962.

——. "Wehrkirchen und Burgenbau". CG 9–10 (1982), 117–44.

——. "Die schwedischen Kirchenkastale auf Gotland". CG 9–10 (1982), 433–44.

——. "Die ostskandinavischen Wehrkirchen". CG 11 (1983), 165–6.

HIRAI, K. *Feudal architecture of Japan*. Trans. H Sato and J Ciliotta. The Heibonsha survey of Japanese art, 13. New York/Tokyo, 1973.

A history of the county of Cambridge and the Isle of Ely. 10 vols. plus Index. The Victoria history of the counties of England. London, 1938–2002. [VCH *Cambs.*]

A history of the county of Essex. 9 vols. plus Bibliography, Supplements. The Victoria history of the counties of England. London, 1903– . [VCH *Essex*]

A history of the county of Northampton. 6 vols. The Victoria history of the counties of England. London, 1902–2007. [VCH *Northants.*]

HODGES, R, and WHITEHOUSE, D. *Mohammed, Charlemagne & the origins of Europe: archaeology and the Pirenne thesis*. 2nd impr. with addenda. London, 1989.

HODGKIN, T. *Theodoric the Goth: the barbarian champion of civilisation*. Heroes of the nations, 4. New York and London, 1891 (often repr.).

HOFFMEYER, A B. "Military equipment in the Byzantine manuscript of Scylitzes in Biblioteca Nacional in Madrid". *Gladius*, 5 (1966), 8–160.

HOFFSUMMER, P, HOFFSUMMER, A, and WERY, B. "Naissance, transformations et abandon de trois places fortes des environs de Liège: Chèvremont, Franchimont et Logne". CG 13 (1987), 63–80.

HOGG, O F G. *Clubs to cannon: warfare and weapons before the introduction of gunpowder*. London, 1968.

HOLL, I. "The development and topography of Sopron in the Middle Ages". In *Towns in medieval Hungary*, 96–102. Ed. L Gerevich. Budapest, 1990.

HOLLISTER, C W. *The military organization of Norman England*. Oxford, 1965.

——. "The campaign of 1102 against Robert of Bellême". In *Studies in medieval history presented to R. Allen Brown*, 193–202. Ed. C Harper-Bill, C J Holdsworth and J L Nelson. Woodbridge, 1989.

HOLST, N von. *Der Deutsche Ritterorden und seine Bauten: von Jerusalem bis Sevilla, von Thorn bis Narwa*. Berlin, 1981.

HOLT, J C. "Politics and property in early medieval England". *Past & present*, no. 57 (Nov. 1972), 3–52.

HONG, Y. *Weapons in ancient China*. New York, 1992.

HONIGMANN, E. *Die Ostgrenze des byzantinischen Reiches von 363 bis 1071, nach griechischen, arabischen, syrischen und armenischen Quellen*. Vol. III of A A Vasil'ev, *Byzance et les Arabes*. Corpus bruxellense historiae Byzantinae, 3. Brussels, 1935.

HOPE, W H St J. "Gundulf's Tower at Rochester, and the first Norman cathedral church there". *Archaeologia* 49 (1885/6), 323–34.

HOREDT, K. "Die befestigte Ansiedlung von Moresti und ihre frühgeschichtliche Bedeutung". *Dacia* ns 1 (1957), 297–308.

HOTZ, W. *Pfalzen und Burgen der Stauferzeit: Geschichte und Gestalt*. Darmstadt, 1981.

HOUBEN, H. *Mezzogiorno normanno-svevo: monasteri e castelli, ebrei e musulmani*. Nuovo medioevo, 52. Naples, 1996.

HOWARD-JOHNSTON, J D. "The siege of Constantinople in 626". In *Constantinople and its hinterland* [Oxford 1993], 131–42. Ed. C Mango and G Dagron with G Greatrex. Publications of the Society for the Promotion of Byzantine Studies, 3. Aldershot, 1995.

HUDSON, P. *Archeologia urbana e programmazione della ricerca: l'esempio di Pavia*. Biblioteca di *Archeologia medievale*, 1. Florence, 1981.

HUICI MIRANDA, A. *Historia musulmana de Valencia y su región: novedades y rectificaciones*. 3 vols. Valencia, 1969–70.

HULL, R W. *African cities and towns before the European conquest*. New York, 1976.

HUMEL, V. "Research in Prague: an historical and archaeological view of the development of Prague from the ninth to the middle of the fourteenth century". In *From the Baltic to the Black Sea: studies in medieval archaeology* [Southampton 1986]. Ed. D Austin and L Alcock. One world archaeology, 18. London, 1990.

HUMPHRIES, P H. *Engines of war: replica medieval siege weapons at Caerphilly Castle*. 2nd edn. Cardiff, 1996.

HURLEY, M F, and SCULLY, O M B, with McCutcheon, S W J. *Late Viking age and medieval Waterford: excavations 1986–1992*. Waterford, 1997.

HUSAIN, B M C. *Cheshire under the Norman earls, 1066–1237*. A history of Cheshire, 4. Chester, 1973.

IMPEY, E, and LORANS, E. "Le donjon de Langeais (Indre-et-Loire) et son environnement: étude historique et archéologique". *BM* 156 (1998), 9–64.

JACQ-HERGOUALC'H, M. *L'armement et l'organisation de l'armée khmère aux XIIe et XIIIe siècles, d'après les bas-reliefs d'Angkor Vat, du Bàyon et de Bonteay Chmar*. Publications du Musée Guimet: recherches et documents d'art et d'archéologie, 12. Paris, 1979.

JAMES, E. *The Merovingian archaeology of south-west Gaul*. 2 vols. *BAR* supplementary series, 25. Oxford, 1977.

JANDORA, J W. "Developments in Islamic warfare: the early conquests". *Studia islamica*, no. 64 (1986), 101–13.

JANSSEN, W. "Mittelalterlicher Burgenbau am Niederrhein: zum Verhältnis von archäologischem Befund und schriftlicher Bezeugung". *Zeitschrift für Archäologie des Mittelalters*, 3 (1975), 121–8.

JARNUT, J. *Bergamo, 568–1098: Verfassungs-, Sozial- und Wirtschaftsgeschichte einer lombardischen Stadt im Mittelalter*. Beihefte zur *Zeitschrift für Sozial- und Wirtschaftsgeschichte*, 67. Wiesbaden, 1979.

——. "La conquista di Bergamo nell'894: una battaglia decisiva fra l'imperatore Guido e re Arnolfo". *Archivio storico bergamasco*, 1 (1981), 25–34.

JÁRREGA DOMÍNGUEZ, R. "Consideraciones sobre la cronología de las murallas tardorromanas de Barcelona: ¿una fortificación del siglo V?". *Archivo español de arquelogía*, 64 (1991), 326–35.

JÄSCHKE, K-U. *Burgenbau und Landesverteidigung um 900: Überlegungen zu Beispielen aus Deutschland, Frankreich und England*. Vorträge und Forschungen, Sonderband, 16. Sigmaringen, 1975.

JENKINS, R J H. "Cyprus between Byzantium and Islam, A.D. 688–965". In *Studies presented to David Moore Robinson on his seventieth birthday*, 1006–14. Ed. G E Mylonas and D Raymond. 2 vols. (continuously paginated). St Louis, 1951–3.

JERPHANION, G de. *Mélanges d'archéologie anatolienne: monuments préhelléniques, gréco-romains, byzantins et musulmans de Pont, de Cappadoce et de Galatie*. Mélanges de l'Université Saint-Joseph (Beyrouth), 13/1. Beirut, 1928.

JIMÉNEZ GARNICA, A M. "Settlement of the Visigoths in the fifth century". In *The Visigoths from the migration period to the seventh century: an ethnographic perspective* [San Marino 1996], 93–128. Ed. P Heather. Studies in historical archaeoethnology, 4. Woodbridge, 1999.

JIREČEK, C. "Beitrage zur antiken Geographie und Epigraphik von Bulgarien und Rumelien". *Monatsberichte der königlich preussischen Akademie der Wissenschaften zu Berlin* (1881), 434–69.

——. "Archäologische Fragmente aus Bulgarien". *Archäologisch-epigraphische Mitteilungen aus Österreich-Ungarn*, 10 (1886), 43–104, 129–209.

JOHNSON, S. *Late Roman fortifications*. Batsford studies in archaeology. London, 1983.

JONES, M. "The defence of medieval Brittany: a survey of the establishment of fortified towns, castles, and frontiers from the Gallo-Roman period to the end of the Middle Ages". *Archaeological journal*, 138 (1981), 149–204.

JONES, P, and RENN, D. "The military effectiveness of arrow-loops". CG 9–10 (1982), 445–56.

JONES, R O. "Report of the centenary meeting held at Aberystwyth 2nd-6th Sept., 1946; Excursions : Ystradmeurig Castle". *AC* 99 (1947), 129–60.

——. "Ystradmeurig Castle". *Ceredigion*, 1 (1950), 38–42.

JORDAN, K. *Henry the Lion: a biography*. Trans. P S Falla. Oxford, 1986.

KAEGI, W E. *Byzantium and the early Islamic conquests*. Cambridge, 1992.

KAFKA, K. *Wehrkirchen Niederösterreichs: Wehrkirchen, Wehrkirchhöfe, Wehrkirchtürme*. Niederösterreichs Burgen und Schlösser, 5. Vienna 1969.

KAJZER, L, and SALM, J. "Burg und Stadt in mittelalterlichen Polen". In *Castrum bene 6: Burg und Stadt* [Prague 1999]. Ed. T Durdík. Prague, 1999.

——. "Die Burgen und die Struktur des Besiedlung im mittelalterlichen Zentralpolen". In *Castrum bene 7: Burgen und Siedlungsstruktur* [Nitra 2004], 87–104. Ed. A Kuttkay, M Ruttkay, and P Bednár. Nitra, 2004.

KAPELLE, W E. *The Norman Conquest of the North: the region and its transformation, 1000–1135*. London, 1979.

KAUFMANN, J E, and KAUFMANN, H W. *The medieval fortress: castles, forts and walled cities of the Middle Ages*. London, 2001.

KAZHDAN, A P. "Some questions addressed to the scholars who believe in the authenticity of Kaminiates's 'Capture of Thessalonica'". *Byzantinische Zeitschrift*, 71 (1978), 301–14.

KENDALL, J M. "The siege of Berkhampstead Castle in 1216". *Antiquaries journal*, 3 (1923), 37–48.

KENNEDY, H. "From *polis* to *madina*: urban change in late antique and early Islamic Syria". *Past & present*, no. 106 (Feb. 1985), 3–27.

——. *Crusader castles*. Cambridge, 1994.

——. "From antiquity to Islam in the cities of al-Andalus and al-Mashriq". In *Genèse de la ville islamique en al-Andalus et au Maghreb occidental* [Madrid 1994 / Granada 1995]. Ed. P Cressier and M García-Arenal. Madrid, 1998.

——. *The armies of the caliphs: military and society in the early Islamic state*. Warfare and history. London/New York, 2001.

KENT, J P C. "Excavations at the motte and bailey castle of South Mimms, Herts., 1960–67". *Bulletin of the Barnet and District Local History Society*, 15 (1968), whole issue.

KENYON, J R. "Fluctuating frontiers: Normanno–Welsh castle warfare, *c.*1075 to 1240". CG 17 (1996), 119–26.

KENYON, J R, and THOMPSON, M. "The origin of the word 'keep'". *Medieval archaeology*, 38 (1994), 175–6.

KING, D J C. "The defences of the citadel of Damascus: a great Mohammedan fortress of the time of the crusades". *Archaeologia* 94 (1951), 57–96.

——. "The castles of Breconshire". *Brycheiniog* 7 (1961), 71–94.

——. "The defence of Wales 1067–1283: the other side of the hill". AC 126 (1977), 1–16.

——. "Pembroke Castle". AC 127 (1978), 75–121.

——. "The trebuchet and other siege-engines". CG 9–10 (1982), 457–69.

——. *Castellarium anglicanum: an index and bibliography of the castles in England, Wales, and the islands*. 2 vols. Millwood, N.Y./London, 1983.

KING, D J C, and CHESHIRE, M. "The town walls of Pembroke (12th century)". AC 131 (1982), 77–84.

KING, D J C, and SPURGEON, C J. "The mottes in the Vale of Montgomery". AC 114 (1965), 69–86.

KING, G, LENZEN, C J, and ROLLEFSON, G O. "Survey of Byzantine and Islamic sites in Jordan: second preliminary season report, 1981". *Annual of the Department of Antiquities, Hashemite Kingdom of Jordan*, 27 (1983), 385–436.

Kirche und Burg in der Archäologie des Rheinlandes: Austellung im Rheinischen Landesmuseum Bonn, 31. Oktober – 31. Dezember 1962. Kunst und Altertum am Rhein, 8. Düsseldorf, 1962.

KIRCHSCHLAGER, M, and STOLLE, T. "Das teuflische Werkzeug: Entstehung und Geschichte der Weißenseer Steinschleuder". In *... wurfen hin in steine/ gröze und niht kleine ... : Belagerungen und Belagerungsanlagen im Mittelalter*, 27–46. Ed. O Wagener and H Laß. Beihefte zur Mediavistik, 7. Frankfurt, 2006.

KIRPICHNIKOV, A N. "Metatel'naya artilleriya i oboronitel'nye sooruzheniya drevnei Rusi [The throwing artillery of Old Russia]". *Materialy i issledovaniya po arkheologii SSSR* [Materials and investigations in the archaeology of the USSR], 77 (1958), 7–51.

——. "Die russischen Waffen des 9.–13. Jahrhunderts und orientalische und westeuropäische Einflüsse auf ihre Entwicklung". *Gladius* 7 (1968), 45–74.

——. "Steinfestungen Nordrußlands aus der Sicht der neueren bautechnisch-archäologischen Forschungen". *Burgen und Schlösser*, 24 (1983), 66–79.

Kirsten, E. "Die byzantinische Stadt". In *Berichte zum XI. Internationalen Byzantinisten-Kongreß* [Munich 1958], V/3: 1–48. Munich, 1960.

Knight, J K. "The road to Harlech: aspects of some early thirteenth-century Welsh castles". In *Castles in Wales and the Marches: essays in honour of D. J. Cathcart King*, 75–88. Ed. J R Kenyon and R Avent. Cardiff, 1987.

——. *Chepstow Castle and port wall; Runston church; Chepstow bulwarks camp*. Rev. edn. Cardiff, 1991.

——. *The end of antiquity: archaeology, society and religion, AD 235–700*. Stroud/Charleston, S.C., 1999.

——. *The Three Castles: Grosmont Castle, Skenfrith Castle, White Castle, Hen Gwrt medieval moated site*. 2nd edn. Cardiff, 2000.

Knost, S. "L'architecture militaire ayyoubide en Égypte et en Syrie". In *L'Orient de Saladin: l'art des Ayyoubides* [exhibition, Paris 2001–2], 84–7. Ed. S Makariou. Paris, 2001.

Kobyliński, Z. "Early medieval hillforts in Polish lands in the 6th to 8th centuries: problems of origins, function, and spatial organization". In *From the Baltic to the Black Sea: studies in medieval archaeology* [Southampton 1986], 147–56. Ed. D Austin and L Alcock. One world archaeology, 18. London, 1990.

Kolias, T G. *Byzantinische Waffen: ein Beitrag zur byzantinischen Waffenkunde von den Anfängen bis zur lateinischen Eroberung*. Byzantina vindobonensia, 17. Vienna, 1988.

Kostochkin, V V. "Fortress architecture of early Russia". Summary of *Krepostnoe zodchestvo drevnei Rusi*. Trans. V Friedman. Moscow, 1969.

Kozák, K. "Adatok a győri püspökvár történetéhez". *Arrabona* 3 (1961), 55–6; 4 (1962), 67. German summary.

Krahe, F W. *Burgen des deutschen Mittelalters: Grundriß-Lexikon*. Würzburg, 1994.

Krautheimer, R. *Early Christian and Byzantine architecture*. The Pelican history of art. 3rd edn. Harmondsworth, 1981.

——. *Rome: profile of a city, 312–1308*. Princeton, 1980.

Kremer, A von. *The Orient under the caliphs*. Trans. S Khuda Bukhsh. Calcutta, 1920.

Kreutz, B M. *Before the Normans: southern Italy in the ninth and tenth centuries*. Middle Ages series. Philadelphia, 1991.

Krischen, F. *Die Landmauer von Konstantinopel, I: Zeichnerische Wiederherstellung mit begleitendem Text*. Denkmäler antiker Architektur, 6. Berlin, 1938.

Krizek, L. "Trebuchet reconstructions in Czechoslovakia". *Military illustrated*, no. 47 (April 1992), 19–20.

Lacarra, J M. "La conquista de Zaragoza por Alfonso I (18 diciembre 1118)". *Al-Andalus* 12 (1947), 65–96.

Lacroix, P. *The arts in the Middle Ages, and at the period of the Renaissance*. London, 1870.

——. *Military and religious life in the Middle Ages, and at the period of the Renaissance*. London, 1874.

Lafuente, J, and Zozaya, J. "Algunas observaciones sobre el castillo de Trujillo". In *España entre el Mediterráneo y el Atlántico* [Granada 1973], II: 119–27. Granada, 1976.

Lair, J. *Le siège de Chartres par les Normands (911)*. Caen, 1902.

Landels, J G. *Engineering in the ancient world*. Ancient culture and society. London, 1978.

Langen, R. "Die Bedeutung von Befestigungen in der Sachsenkriegen Karls des Großen". *Westfälische Zeitschrift*, 139 (1989), 181–211.

Lapsley, G. "Some castle officers in the twelfth century". *EHR* 33 (1918), 348–59.

LAVAGNINI, B. "Siracusa occupata dagli Arabi e l'epistola di Teodosio monacho". *Byzantion* 29–30 (1959–60), 267–79.

LAWRENCE, A W. "Early medieval fortifications near Rome". *Papers of the British School at Rome*, 32 (1964), 89–122.

——. "The castle of Baghras". In *The Cilician kingdom of Armenia*, 34–84. Ed. T S R Boase. Edinburgh, 1978.

——. "A skeletal history of Byzantine fortification". *Annual of the British school at Athens*, 78 (1983), 171–233.

LEBLING, R. "The pirates of St. Tropez". *Medieval life*, 14 (2000), 26–31.

LECLÈRE, A. *Histoire de Cambodge depuis le 1er siècle de notre ère: d'après les inscriptions lapidaires, les annales chinoises et annamites et les documents européens des six derniers siècles*. Paris, 1914.

LEEDS, E T. "An adulterine castle on Faringdon Clump, Berkshire". *Antiquaries journal*, 16 (1936), 165–78.

——. "An adulterine castle on Faringdon Clump (second report)". *Antiquaries journal*, 17 (1937), 294–8.

LEFEBVRE Saint-Ogan. *Compiègne*. Paris, 1887.

LE MAHO, J. "Fortifications de siège et 'contre-châteaux' en Normandie (XIe–XIIe siècle)". CG 19 (2000), 181–9.

LEMERLE, P. "La chronique improprement dite de Monemvasie: le contexte historique et légendaire". *Recueil des études byzantines*, 21 (1963), 5–49.

LE PATOUREL, J. "The Norman colonization of Britain". In *I Normanni e la loro espansione in Europa nell'alto Medioevo*, 409–38. SSCISAM 16 [1968]. Spoleto, 1969.

——. *The Norman empire*. Oxford, 1976.

LE STRANGE, G. *Baghdad during the Abbasid caliphate: from contemporary Arabic and Persian sources*. Oxford, 1900; repr. London/New York, 1972.

LEV, Y. "The Fatimid army, A.H. 358–427/968–1036 C.E.: military and social aspects". *Asian and African studies*, 14 (1980), 165–92.

LEVILLAIN, L. "Essai sur le Comte Eudes, fils de Harduin et de Guérinbourg, 845–871". *Le Moyen Âge*, 47 (1937), 153–82, 233–71.

LÉVI-PROVENÇAL, É. "Alphonse VI et la prise de Tolède (1085)". *Hespéris* 12 (1931), 33–49. Cited from repr. in his *Islam d'Occident: études d'histoire médiévale*, 111–28. Islam d'hier et d'aujourd'hui, 7. Paris, 1948.

——. "La toma de Valencia por el Cid: según las fuentes musulmanas y el original árabe de la 'Crónica general de España'". *Al-Andalus* 13 (1948), 91–156.

——. *Histoire de l'Espagne musulmane*. Rev. and enl. edn. 3 vols. Paris and Leiden, 1950–53.

LEWCOCK, R, et al. "The urban development of Ṣan'ā'". In *Ṣan'ā', an Arabian Islamic city*, 122–41. Ed. R B Serjeant and R Lewcock. London, 1983.

LEWIS, A R. "La féodalité dans le Toulousain et la France méridionale, 850–1050". *Annales du Midi*, 76 (1964), 243–59.

LEWIS, C P. "The Norman settlement of Herefordshire under William I". *ANS* 7 (1984), 195–213.

LEWIS, J M. *Carreg Cennen Castle*. Cardiff, 1990.

LEYSER, K. "The battle at the Lech, 955: a study in tenth-century warfare". *History* 50 (1965), 1–25. Cited from repr. in his *Medieval Germany and its neighbours, 900–1250*, 43–67. London, 1982.

——. "Henry I and the beginnings of the Saxon empire". *EHR* 83 (1968), 1–32. Cited from repr. in his *Medieval Germany and its neighbours, 900–1250*, 11–42. London, 1982.

Leyser, K. "Ottonian government". *EHR* 96 (1981), 721–53. Cited from repr. in his *Medieval Germany and its neighbours, 900–1250*, 69–101. London, 1982.

——. "Early medieval warfare". In *The battle of Maldon: fiction and fact*, 87–108. Ed. J Cooper. London, 1993. Repr. in *Communications and power in medieval Europe: the Carolingian and Ottonian centuries*, 29–50. Ed. T Reuter. London, 1994.

Lézine, A, *Mahdiya: recherches d'archéologie islamique*. Archéologie méditerranéenne, 1. Paris, 1965.

——. "Sur deux châteaux musulmanes d'Ifriqiya". *Revue des études islamiques*, 39 (1971), 87–93.

Líbal, D. "Burgen und Städte als tschechisches Erbgut". In *Castrum bene 6: Burg und Stadt* [Prague 1999]. Ed. T Durdík. Prague, 1999.

Liddiard, R. *Castles in context: power, symbolism and landscape, 1066 to 1500*. Bollington, 2005.

Liebgott, N-K. "Brick-making and castle-building". CG 18 (1998), 109–18.

Liu M-T. *Die chinesischen Nachrichten zur Geschichte der Ost-Türken (T'u-küe)*. 2 vols. Göttinger asiastische Forschungen, 10. Wiesbaden, 1958.

Llewellyn, P. *Rome in the Dark Ages*. London/New York, 1993.

Lloyd, J D K, and Knight, J K. *Montgomery Castle, Powys = Castell Trefaldwyn*. 2nd edn. London, 1981.

Lloyd, J E. *A history of Wales, from the earliest times to the Edwardian conquest*. 2 vols. 2nd edn. London, 1912.

Lock, P. "The medieval towers of Greece: a problem in chronology and function". In *Latins and Greeks in the eastern Mediterranean after 1204* [Nottingham 1988], 129–45. Ed. B Arbel, B Hamilton and D Jacoby. London, 1989.

——. *The Franks in the Aegean, 1204–1500*. London, 1996.

Logan, P N. "The walled city of Kano". *Journal of the Royal Institute of British Architects*, 36 (1929), 402–6.

Lombard-Jourdan, A. *Aux origines de Paris: la genèse de la rive droite jusqu'en 1223*. Paris, 1985.

López del Álamo, M P, and Rubio Visiers, M J. "Las murallas de Buitrago de Lozoya". In *Maŷrīṭ: estudios de arqueología medieval madrileña*, 33–43. Ed. F Valdés. Biblioteca de arqueología medieval hispánica, 1. Madrid, 1992.

Lot, F. "Le pont de Pitres". *Le Moyen Âge*, 18 (1905), 1–27. Cited from repr. in *Recueil des travaux historiques de Ferdinand Lot*, II. Paris, 1970.

Lot, F, and Halphen, L. *Le règne de Charles le Chauve (840–877)*, 1: *840–851*. Bibliothèque de l'École des hautes études, 175. Paris, 1909.

Louise, G. "La mise en place du réseau castral sur les confins de la Normandie et du Maine". In *Autour du château médiéval* [Alençon 1997], 7–24. Ed. B Fajal. Alençon, 1998.

Lourie, E. "A society organised for war: medieval Spain". *Past & present*, 35 (1966), 54–76.

Lutsch, H. *Verzeichnis der Kunstdenkmäler der Provinz Schlesien*, II: *Die Landkreise des Reg.-Bezirks Breslau*. Breslau, 1889.

Lyons, M C, and Jackson, D E P. *Saladin: the politics of the Holy War*. University of Cambridge oriental publications, 30. Cambridge, 1982.

Mabrouk, S. "La fortification médiévale de la ville de Taza (Maroc)". In *Mil anos de fortificações na Península Ibérica e no Magreb (500–1500)* [Palmela 2000]. Ed. I C Ferreira Fernandes. Lisbon, 2001.

MacAdam, H I. "Some notes on the Umayyad occupation of north-east Jordan". In *The defence of the Roman and Byzantine East* [Sheffield 1986], II: 531–47. Ed. P Freeman and D Kennedy. *BAR* international series, 297. Oxford, 1986.

McAleer, J P. "The so-called Gundulf's Tower at Rochester Cathedral: a reconsideration of its history, date and function". *Antiquaries journal*, 78 (1998), 111–76.

Macartney, C A. "The attack on 'Valandar'". *Byzantinisch-neugriechische Jahrbücher*, 8 (1931), 159–70.

Maccabruni, C. *Pavia: la tradizione dell'antico nella città medievale*. Pavia, 1991.

McCarthy, M R, Summerson, H R T, and Annis, R G. *Carlisle Castle: a survey and documentary history*. English Heritage: Archaeological reports, 18. London, 1990.

McDonald, R A. *The Kingdom of the Isles: Scotland's western seaboard, c.1100–c.1336*. *Scottish historical review* monographs series, 4. East Linton, 1997.

McGeer, E. "Byzantine siege warfare in theory and practice". In *The medieval city under siege*, 123–9. Ed. I A Corfis and M Wolfe. Woodbridge, 1995.

MacIvor, I. *Dumbarton Castle*. Rev edn. Edinburgh, 1993.

Mackenzie, J D. *The castles of England: their story and structure*. 2 vols. London, 1897.

McKitterick, R. *The Frankish kingdoms under the Carolingians, 751–987*. London, 1983.

McLynn, F. *1066: the year of the three battles*. London, 1998; repr. 1999.

McNamee, C. *The wars of the Bruces: Scotland, England and Ireland, 1306–1328*. East Linton, 1997.

McNeil, T E. *Castles in Ireland: feudal power in a Gaelic world*. London, 1997.

Makhdoomee, M A. "Mechanical artillery in medieval India". *Journal of Indian history*, 15 (1936), 189–95.

——. "The art of war in medieval India". *Islamic culture*, 11 (1937), 461–78.

Maldonaldo, B P. "La muralla primitiva árabe de Tudela". *Anuario de estudios medievales*, 16 (1986), 29–42.

——. *Tratado de arquitectura hispanomusulmana*, II: *Ciudades y fortalezas*. Madrid, 1999.

Manzano Martínez, J, Bernal Pascual, F, and Calabuig Jordán, R. "El castillo de Sta. Catalina del Monte (Verdolay-Murcia): un hisn de época musulmana". *Verdolay* 3 (1991), 107–24.

Manzano Moreno, E. *La frontera de al-Andalus en época de los Omeyas*. Biblioteca de historia, 9. Madrid, 1991.

Marsden, E. *Greek and Roman artillery: historical development*. Oxford, 1969.

——. Commentary on Ammianus Marcellinus. In *Greek and Roman artillery: technical treatises*. Ed. E Marsden. Oxford, 1971.

Marshall, C. *Warfare in the Latin East, 1192–1291*. Cambridge studies in medieval life and thought, 4th ser. 18. Cambridge, 1992.

Marshall, G. "The Norman occupation of the lands in the Golden Valley, Ewyas, and Clifford, and their motte and bailey castles". *Transactions of the Woolhope Naturalists' Field Club* (1936–8), 141–58.

Marshall, P. "The twelfth-century castle at Newark". In *Southwell and Nottinghamshire: medieval art, architecture, and industry*, 110–25. Ed. J S Alexander. British Archaeological Association conference transactions, 21. Leeds, 1998.

Marshall-Cornwall, J. "The military geography of the Welsh Marches". *Geographical magazine*, 30/1 (May 1957), 1–12.

Martens, M. "Les survivances domaniales du castrum carolingien de Bruxelles à la fin du Moyen Âge". *Le Moyen Âge*, 69 (1963), 643–54.

MARTIN, J-M. *La Pouille du VIe au XIIe siècle*. Collection de l'École française de Rome, 179. Rome, 1993.

MARTÍNEZ LILLO, S. "Algunos aspectos inéditos en la fortificación musulmana de Talavera de la Reina". In *Actas del II Congreso de arqueología medieval española*, II: 199–205. Madrid, 1987.

——. "Estudio sobre ciertos elementos y estructuras de la arquitectura militar andalusí: la continuidad entre Roma y el Islam". *Boletín de arquelogía medieval*, 5 (1991), 11–37.

MARX, J. *Die Ringmauern und die Thore der Stadt Trier*. Trier, 1876.

MASON, J F A. *William the first and the Sussex Rapes*. 1066 commemoration series, 3. Rev. edn. London, 1972.

MASPERO, G. *Le royaume de Champa*. Paris/Brussels, 1928.

MATARASSO, F. *The English castle*. London, 1995.

MATTHYS, A. "Les fortifications du XIe siècle entre Lesse et Semois". In *Burgen der Salierzeit*, I: *In den nördlichen Landschaften des Reiches*, 225–80. Ed. H W Böhme. Römisch-Germanisches Zentralmuseum Mainz, Forschungsinstitut für Vor- und Frühgeschichte: Monographien, 25. Sigmaringen, 1991.

MAURICI, F. *Castelli medievali in Sicilia dai bizantini ai normanni*. La pietra vissuta, 5. Palermo, 1992.

MAYERSON, P. "The first Muslim attacks on Southern Palestine (A.D. 633–634)". *Transactions and Proceedings of the American Philological Association*, 95 (1964), 155–99.

MECKSEPER, C. "Wärmequellen". In *Burgen in Mitteleuropa: ein Handbuch*, I. Ed. H W Böhme. 2 vols. Stuttgart, 1999.

"Medieval Britain and Ireland in 1993". Compiled by B S Nenk, S Margeson and M Hurley. *Medieval archaeology*, 38 (1994), 184–293.

MEEHAN, B. "The siege of Durham, the battle of Carham and the cession of Lothian". *Scottish historical review*, 55 (1976), 1–19.

MEGAW, A H S. "The arts in Cyprus", B: "Military architecture". In *The art and architecture of the crusader states*, 196–206. Ed. H W Hazard. A history of the crusades, 4. Madison, Wisc., 1977.

MELLEVILLE, M. *Histoire de la ville de Laon et de ses institutions civiles: judiciaires, féodales, militaires, financières et religieuses, monuments, antiquités, mœurs, usages, impôts, finances, commerce, population, etc.* 2 vols. Laon, 1846.

MENÉNDEZ FUEYO, J L. "La puerta del Castillo de Planes (Alicante): una aportación al estudio de las puertas en recodo de fortificaciones de ámbito rural en época almohade". *Boletín de arqueología medieval*, 9 (1995), 153–77.

MENANT, F. "'Fossata cavare, portas erigere': le rôle des fossés dans les fortifications médiévales de la plaine padane". In *Mélanges d'archéologie et d'histoire medievales en l'honneur du Doyen Michel de Boüard*, 277–85. Mémoires et documents de l'École des chartes, 27. Geneva, 1982. (Also *Aevum* 61 (1982), 205–16.)

MERTENS, J. "The church of Saint Donatian at Bruges". In Galbert of Bruges, *The murder of Charles the Good*, 318–20. Trans. J B Ross. Rev. edn. Harper torchbooks. New York, 1967.

MESQUI, J. *Châteaux forts et fortifications en France*. Paris, 1997.

——. "La tour maîtresse du Donjon de Loches". *BM* 156 (1998), 65–125.

——. "Les tours à archères dans les domaines Plantagenêt français, 1160–1205". In *Les fortifications dans les domaines Plantagenêt, XIIe–XIVe siècles* [Poitiers 1994]. Ed. M-P Baudry. Civilisation médiévale, 10. Poitiers, 2000.

MESQUI, J, and FAUCHERRE, N. "L'enceinte médiévale de Césarée". *BM* 164 (2006), 83–94.

METZLER, J, ZIMMER, J, and BAKKER, L. *Ausgrabungen in Echternach*. Luxembourg: Ministère des Affaires culturelles et de la ville d'Echternach, 1981.

MEULET, R. "Guide archéologique du Congrès de Chartres en 1900". *BM* 64 = 7th ser. 4 (1899/1900), 274–7.

MEYER, W, and WIDMER, E. *Das grosse Burgenbuch der Schweiz*. 2nd edn. Zurich, 1978.

MEYER-PLATH, B, and SCHNEIDER, A M. *Die Landmauer von Konstantinopel, II: Aufnahme, Beschreibung und Geschichte*. Denkmäler antiker Architektur, 8. Berlin, 1943.

MICHAILOV, S. "Fouilles des sites du Moyen Âge". *Bulletin d'archéologie sud-est européenne*, 2 (1971), 50–56.

MICHAUD, J F. *Bibliothèque des Croisades*. 4 vols. Paris, 1829.

MILNE, G, with Webber, M, and Pearce, J. "London Wall". In G Milne with N Cohen et al. *Excavations at medieval Cripplegate, London: archaeology after the Blitz, 1946–68*, 9–36. London, 2001.

MINORSKY, V. "New light on the Shaddādids of Ganja". In his *Studies on Caucasian History*. Cambridge oriental series, 6. London, 1953.

MIRKOVIĆ, M. "Sirmium: its history from the I. century A.D. to 582 A.D." *Sirmium: archaeological investigations in Syrmian Pannonia*, I (1971), 5–90.

MONNERET DE VILLARD, U. *La Nubia medioevale: mission archéologique de Nubie, 1929–1934*. 4 vols. Cairo, 1935–57.

MONREAL, L, and RIQUER, M de. *Els castells medievals de Catalunya*, II. Falco, 2. Barcelona, 1958.

MOORE, J S. "Anglo-Norman garrisons". *ANS* 22 (2000), 205–59.

MORGAN, D O. *Medieval Persia, 1040–1797*. A history of the Near East. London, 1988.

MORGAN, R. "Magister tormentorum". *Postern* 7 (1997), 8–11.

——. "The great golden gate of Kyiv". *Postern* 11 (2000), 11–12.

MORILLO, S. *Warfare under the Anglo-Norman kings, 1066–1135*. Woodbridge, 1994.

MOSS, V. "The defence of Normandy, 1193–8". *ANS* 24 (2002), 145–61.

MÜHLE, E. "Die topographisch-städtebauliche Entwicklung Kievs vom Ende des 10. bis zum Ende des 12. Jahrhunderts im Licht der archäologischen Forschungen". *Jahrbücher für Geschichte Osteuropas*, ns 36 (1988), 350–76.

——. *Die städtischen Handelszentren der nordwestlichen Rus': Anfänge und frühe Entwicklung altrussischer Städte (bis gegen Ende des 12. Jahrhunderts)*. Quellen und Studien zur Geschichte des östlichen Europa, 32. Stuttgart, 1991.

MÜLLER-WIENER, W. "Mittelalterliche Befestigungen in südlichen Jonien". *Istanbuler Mitteilungen*, 11 (1961), 5–122.

——. "Die Stadtbesfestigungen von İzmir, Sığacık und Çandarlı: Bemerkungen zur mittelalterlichen Topographie des nördlichen Jonien". *Istanbuler Mitteilungen*, 12 (1962), 59–114.

——. *Castles of the crusaders*. Trans. J M Brownjohn. London, 1966.

MUNBY, J, and WILKINSON, D R P. "The defences: detailed studies of sites on the defences of late Saxon and medieval Oxford". In *Oxford before the University: the late Saxon and Norman archaeology of the Thames crossing, the defences and the town*, 135–86. Ed. A Dodd. Thames Valley landscapes monographs, 17. Oxford, 2003.

MURDOCH, J. *A history of Japan, I: From the origins to the arrival of the Portuguese in 1542 A.D.* London, 1903; repr. 1925.

MUS, P. "Études indiennes et indochinoises, III: Les balistes du Bayon". *Bulletin de l'École française d'Extrême-Orient*, 29 (1929), 331–40.

MUTHERICH, F, and GAEHDE, J E. *Carolingian painting*. London, 1977.

NADOLSKI, A. "Ausgrabungen in Łęczyca". In *Frühe polnische Burgen: Berichte über archäologische Grabungen*. Weimar, 1960.

NAGY, E. "Rapport préliminaire des fouilles d'Esztergom". *Acta archaeologica Academiae Scientiarum Hungaricae*, 23 (1971), 181–94.

NAVARRO SUÁREZ, F J, and MATEO SAURA, M A. "La fortificación islámica de La Luz (Murcia)". *Boletín de arqueología medieval*, 7 (1993), 229–48.

NEEDHAM, J. "China's trebuchets, manned and counterweighted". In *On pre-modern technology and science: a volume of studies in honor of Lynn White, jr.*, 107–45. Ed. B S Hall and D C West. Humana civilitas, 1. Malibu, Calif., 1976.

——. *The shorter Science and civilisation in China*, v. Ed. C A Ronan. Cambridge, 1995.

NEEDHAM, J, and YATES, R D S. *Science and civilisation in China*, v/6: *Military technology: missiles and sieges*. Cambridge, 1994.

NEES, M, and KREINER, F. "Neuss". In *Deutsches Städtebuch*, III/3: *Rheinisches Städtebuch*, 319–24. Ed. E Keyser. Stuttgart, 1956.

NELSON, J L. *Charles the Bald*. The medieval world. London, 1992.

NELSON, L H. *The Normans in South Wales, 1070–1171*. Austin, Tex., 1966.

NICHOLAS, D. *The growth of the medieval city: from late antiquity to the early fourteenth century*. A history of urban society in Europe. London, 1997.

NICOLAÏ, A. "La villa gallo-romaine de Pontius Leontius (Burgus)". *Transactions de la Société archéologique de Bordeaux*, 46 (1929), 1–23.

NICOLLE, D. *The armies of Islam, 7th–11th centuries*. Men-at-arms series, 125. London, 1982.

——. *Armies of the Muslim conquest*. Men-at-arms series, 225. London, 1993.

——. *Yarmuk, 636 AD: the Muslim conquest of Syria*. Osprey military campaign series, 31. London, 1994.

——. *Medieval warfare source book*, II: *Christian Europe and its neighbours*. London, 1996.

——. "The early trebuchet: documentary and archaeological evidence". In *La fortification au temps des Croisades* [Parthenay 2002], 269–78. Ed. N Faucherre, J Mesqui and N Prouteau. Archéologie et culture. Rennes, 2004.

NOLLA, J. "Ampurias en la Antigüedad tardía: una nueva perspectiva". *Archivo español de arquelogía*, 66 (1993), 207–24.

NORTHEDGE, A. "The fortifications of Qal'at 'Amman ('Amman citadel): preliminary report". *Annual of the Department of Antiquities, Hashemite Kingdom of Jordan*, 27 (1983), 437–60.

NORWICH, J J. *The Normans in the South, 1016–1130*. London, 1967.

NOSOV, E N. "Ryurik Gorodishche and the settlements to the north of Lake Ilmen". In *The archaeology of Novgorod, Russia; recent results from the town and its hinterland*, 5–66. Ed. M A Brisbane. Society for Medieval Archaeology monograph series, 13. Lincoln, Neb., 1992.

NOSSOV, K S. *Indian castles, 1206–1526: the rise and fall of the Delhi Sultanate*. Fortress, 51. Oxford, 2006.

——. *Medieval Russian fortresses, AD 862–1480*. Fortress, 61. Oxford, 2007.

NOTERDAEME, J. "De vroegste geschiedenis van Brugge, III: De burcht van Brugge". *Handelingen van het Genootschap voor geschiedenis gesticht onder de benaming Société d'émulation de Bruges*, 112 (1975), 171–204.

NOVÁKI, G. "Die topografischen Eigentümlichkeiten der ungarischen Burgen in 10.–11. Jahrhundert am nordliche Randgebiet der grossen Tiefebene". *Acta archaeologica Academiae Scientiarum Hungaricae*, 28 (1976), 359–69.

NYBERG, H S. "Die Sasanidische Westgrenze und ihre Verteidigung". In *Septentrionalia et orientalia: studia Bernhardo Karlgren a. D. III Non. Oct. anno MCMLIX dedicata*,

316–26. Kungl. Vitterhets historie och antikvitets akademiens handlingar, 91. Stockholm, 1959.

OBOLENSKY, D. *Byzantium and the Slavs*. Crestwood, N.Y., 1994.

O'CONOR, K D. "Motte castles in Ireland: permanent fortresses, residences and manorial centres". CG 20 (2002), 173–82.

OIKONOMIDES, N. "Mesembria in the 9th century: epigraphical evidence". Repr. in his *Byzantium from the ninth century to the Fourth Crusade: studies, texts, monuments*. Aldershot, 1992

O'KEEFFE, T. "The fortifications of Western Ireland, AD 1100–1300, and their interpretation". *Journal of the Galway Archaeological and Historical Society*, 50 (1998), 184–200.

OLIVEIRA, L F. "Uma fortificação islâmica do termo de Silves: o Castelo Belinho". *Arqueologia medieval*, 6 (1999), 39–46.

OLMO ENCISO, L. "Restos defensivos de la ciudad visigoda de Recópolis". In *Homenaje al Prof. Martín Almagro Basch*, IV: 67–74. Madrid, 1983.

OLSEN, R A. "Danish medieval castles at war". CG 9–10 (1982), 223–36.

OMRAN, H, and DABBOURA, G. *The citadel of Damascus: a thirteenth-century fortress of the Ayyubid Sultan al-'Adil*. Damascus, 1997.

OPOLOVNIKOV, A, and OPOLOVNIKOVA, Y. *The wooden architecture of Russia: houses, fortifications, churches*. Trans. J Whitby and R Whitby. London, 1989.

ORPEN, G H. "Motes and Norman castles in Ireland". *EHR* 22 (1907), 228–54, 440–67.

ORTIZ ECHAGÜE, J. *España: castillos y alcázares*. Madrid, 1956; repr. 1964.

OTT, H. "Die Burg Zähringen und ihre Geschichte". In *Die Zähringer*, I: *Eine Tradition und ihre Erforschung*, 5–17. Ed. K Schmid. Sigmaringen, 1986.

OTWAY-RUTHVEN, A J. *A history of medieval Ireland*. 2nd edn. London/New York, 1980.

OWEN, D D R. *William the Lion, 1143–1214: kingship and culture*. East Linton, 1997.

PÄCHT, O, and THOSS, D. *Französische Schule [1]*. Die illuminierte Handschriften und Inkunabeln der Österreichischen Nationalbibliothek, 1. Vienna, 1974.

PAINTER, S. "Castellans of the plain of Poitou in the eleventh and twelfth centuries". *Speculum* 31 (1956), 243–57.

PALLISER, D M. "Town defences in medieval England and Wales". In *The medieval military revolution: state, society and military change in medieval and early modern Europe*, 105–20. Ed. A Ayton and J L Price. London, 1995.

PAN, Jixing. "The origin of rockets in China". In *Gunpowder: the history of an international technology* [Bath 1994]. Ed. B J Buchanan. Bath, 1996.

PANAITE, A, and MAGUREANU, A. "Some observations concerning several late Roman and early Byzantine fortifications from the lower Danube". In *The Roman and late Roman city* [Veliko Tŭrnovo 2000]. Ed. L Ruseva-Slokoska, R Ivanov and V Dinchev. Sofia, 2002.

PANI ERMINI, L. "Città fortificate e fortificazione delle città italiane fra V e VI secolo". *Rivista di studi liguri*, 69–70 (1993–4), 193–206.

PANOUILLÉ, J-P. *Carcassonne in the days of the siege*. Trans. J Abbott. Paris, 1992.

PANT, G N. *Studies in Indian weapons and warfare*. New Delhi, 1970.

PARMENTIER, H. *Inventaire descriptif des monuments čams de l'Annam*. 2 vols in 4. Publications de l'École française d'Extrême-Orient, 11–12. Paris, 1909–18.

——. "Notes d'archéologie indochinoise [1]". *Bulletin de l'École française d'Extrême-Orient*, 23 (1923), 267–300.

——. "Tables géographiques des monuments du Champa et du Cambodge". In G Coedès and H Parmentier, *Listes générales des inscriptions et des monuments du Champa et du Cambodge*. Hanoi, 1923.

PARTINGTON, J R. *A history of Greek Fire and gunpowder*. Cambridge, 1960; repr. Baltimore, 1999.

PASQUIER, F. "Vente du château de Castelviel de Rosanes en 1023, par Bérenger, comte de Barcelone". *Annales du Midi*, 30 (1918), 417–18.

PEARSON, A. *The Roman shore forts: coastal defences of southern Britain*. Stroud, 2002.

PEDDIE, J. *Alfred: warrior king*. Stroud, 1999.

PEERS, C. *Pevensey Castle, Sussex*. English Heritage handbooks. 2nd edn. London, 1953; repr. 1985.

——. *Imperial Chinese armies*, II: *590–1260 AD*. Men-at-arms series, 295. London, 1996.

PENTZ, P. "A medieval workshop for producing 'Greek Fire' grenades". *Antiquity* 62 (1988), 89–93.

PÉPIN, E. *Gisors et la vallée de l'Epte*. New edn. Petites monographies des grands édifices de la France, 79. Paris, 1963.

PERKS, J C. *Chepstow Castle, Monmouthshire*. 2nd edn. London, 1967.

PERROY, É. "Les châteaux du Roannais du XIe et XIIe siècle". *Cahiers de civilisation médiévale*, 9 (1966), 13–23.

PETRIKOVITS, H von. "Fortifications in the north-western Roman Empire from the third to fifth centuries A.D." *The journal of Roman studies*, 61 (1971), 178–218.

PETRIN, S. "Baden". In *Handbuch der historischen Stätten Österreich*, I: *Donauländer und Burgenland*, 207–9. Ed. Lechner, K. Stuttgart, 1970.

PETROVICS, I. "The role of towns in the defence system of medieval Hungary". In *La guerre, la violence et les gens au Moyen Âge* [Amiens 1994], I: *Guerre et violence*, 263–71. Ed. P Contamine and O Guyotjeannin. Paris, 1996.

PEYROU, J, and HENRY-CLAUDE, M. *Montségur: last refuge, last rampart of the Catharist Church*. Trans. B Degans. La mémoire des pierres. Gavaudun, 1998.

PHILIPPSON, A. *Die griechischen Landschaften: eine Landeskunde*. 4 vols. in 8. Frankfurt am Main, 1950–59.

PHLEPS, H. "Die bauerliche Wehrbaukunst der Siebenburger Sachsen". In Zillich, H, *Siebenburgen: ein abendländisches Schicksal*, 106–11. Die blauen Bücher. Königstein in Taunus, 1957.

PICARD, C. "La fondation de Badajoz par 'Abd al-Rahman ibn Yūnus al-Jillikī (fin IXe siècle)". *Revue des études islamiques*, 49 (1981), 215–29.

PIPER, O. *Burgenkunde: Bauwesen und Geschichte der Burgen, zunächst innerhalb der deutschen Sprachgebietes*. New edn. Frankfurt/Munich, 1967.

PITTE, D. "Château-Gaillard, dans la défense de la Normandie orientale (1196–1204)". *ANS* 24 (2002), 163–75.

PLANITZ, H. *Die deutsche Stadt im Mittelalter: von der Römerzeit bis zu den Zunftkämpen*. Graz, 1954.

PLATT, C. *The English medieval town*. London, 1976.

POERCK, G de. "L'artillerie à ressorts médiévale: notes lexicologiques et étymologiques". *Archivum latinitatis Medii Aevi (Bulletin du Cange)*, 18 (1943–4), 35–49.

POHL, W. *Die Awarenkriege Karls des Grossen, 788–803*. Militärhistorische Schriftenreihe, 61. Vienna, 1988.

——. "The barbarian successor states". In *The transformation of the Roman world, AD 400–900*, 33–47. Ed. L Webster and M Brown. London, 1997.

POISSON, J-M. "Villars-les-Dombes". *AM* 19 (1989), 352–3.

POPA, R. "La porte de nord de la forteresse byzantine de Păcuiul lui Soare". *Dacia* II (1967), 271–92.

——. "La porte nord de la forteresse byzantine du Xe siècle de Păcuiul lui Soare et ses relations avec l'architecture militaire byzantine". In *Actes du premier Congrès international des études balkaniques et sud-est européennes* [Sofia 1966], II: *Archéologie, histoire de l'antiquité, arts*, 569–77. Ed. D Dimitrov et al. Sofia, 1969.

——. "A tenth-century Byzantine castle on the Danube". *Fortress* 16 (1993), 16–24.

Popović, V. "A survey of the topographical urban organization of Sirmium in the late empire". *Sirmium: archaeological investigations in Syrmian Pannonia*, 1 (1971), 128–33.

Poppe, A. "The building of the church of St. Sophia in Kiev". *Journal of medieval history*, 7 (1981), 15–66. Repr. in his *The rise of Christian Russia*. London, 1982.

Porcher, J. *French miniatures from illuminated manuscripts*. Trans. J Brown. London, 1960.

Porter, P. *Medieval warfare in manuscripts*. London, 2000.

Poulter, A. "The use and abuse of urbanism in the Danubian provinces during the later Roman Empire". In *The city in late antiquity* [Nottingham 1988], 99–135. Ed. J Rich. Leicester–Nottingham studies in ancient society, 3. London, 1992

——. *Nicopolis ad Istrum, a Roman, late Roman and early Byzantine city: excavations, 1985–1992*. With contributions by T Blagg et al. *Journal of Roman studies* monographs, 8. London, 1995.

Pounds, N J G. *The medieval castle in England and Wales: a social and political history*. Cambridge, 1990.

——. *An economic history of medieval Europe*. 2nd edn. London/New York, 1994.

Powell, J M. *Anatomy of a crusade, 1213–1221*. Middle Ages series. Philadelphia, 1986.

Powers, J F. *A society organized for war: the Iberian municipal militias in the central Middle Ages, 1000–1289*. Berkeley, 1988.

Powicke, M. *The loss of Normandy, 1189–1204: studies in the history of the Angevin empire*. 2nd edn. Manchester, 1961.

Prawer, J. "The Jerusalem the crusaders captured: a contribution to the medieval topography of the city". In *Crusade and settlement: papers . . . presented to R. C. Smail* [Cardiff 1983], 1–16. Ed. P W Edbury. Cardiff, 1985.

Prestwich, J O. "The military household of the Norman Kings". *EHR* 96 (1981), 1–35.

Prestwich, M. *Armies and warfare in the Middle Ages: the English experience*. New Haven/London, 1996.

——. "The garrisoning of English medieval castles". In *The Normans and their adversaries at war: essays in memory of C. Warren Hollister*, 185–200. Ed. R P Abels and B S Bachrach. Warfare in history, 12. Woodbridge, 2001.

Pringle, D. *The defence of Byzantine Africa from Justinian to the Arab conquest: an account of the military history and archaeology of the African provinces in the sixth and seventh centuries*. 2 vols. *BAR* international series, 99. Oxford, 1981.

——. "King Richard i and the walls of Ascalon". *Palestine exploration quarterly*, 116 (1984), 133–47. Repr. in his *Fortification and settlement in crusader Palestine*. Aldershot, 2000.

——. "Crusader Jerusalem". *Bulletin of the Anglo-Israel Archaeological Society*, 10 (1990–91), 105–13. Repr. in his *Fortification and settlement in crusader Palestine*. Aldershot, 2000.

——. "Town defences in the crusader kingdom of Jerusalem". In *The medieval city under siege*, 69–121. Ed. I A Corfis and M Wolfe. Woodbridge, 1995.

——. "A castle in the sand: mottes in the crusader East". CG 18 (1998), 187–91.

——. "Templar castles between Jaffa and Jerusalem". In *Welfare and warfare* [London 1996], 89–109. Ed. H Nicholson. The military orders, 2. Aldershot, 1998.

Pringle, D, et al. "Qal'at Jiddin: a castle of the crusader and Ottoman periods in Galilee". *Levant* 26 (1994), 135–66.

Puig i Cadafalch, J, Falguera, A de, and Goday, J. *L'arquitectura romànica a Catalunya*. 3 vols. Barcelona, 1909–18.

Purser, T S. "Castles of Herefordshire, 1066–1135". *Medieval history*, 4 (1994), 72–81.
Purton, P. "The myth of the mangonel: torsion artillery in the Middle Ages". *Arms and armour*, 3 (2006), 79–90.
Rabbat, N O. *The citadel of Cairo: a new interpretation of royal Mamluk architecture*. Islamic history and civilization, 14. Leiden, 1995.
Radford, C A R. *White Castle = Castell Gwyn, Gwent*. Cardiff, 1962.
——. *Grosmont Castle, Gwent*. London, 1973.
Ranitzsch, K H. *The army of Tang China*. Stockport, 1995.
Rappoport, P A. "Die altrussischen Burgwalle". *Zeitschrift für Archäologie*, 1 (1967), 61–85.
——. "Russian medieval military architecture". *Gladius* 8 (1969), 39–62.
Rauch, C, with Herrmann, F. *Kreis Bingen*. Die Kunstdenkmäler im Volksstaat Hessen: Provinz Rheinhessen. Darmstadt, 1934.
Ravegnani, G. *Castelli e città fortificate nel VI secolo*. Rōmania, 1. Ravenna, 1983.
Reilly, B F. *The contest of Christian and Muslim Spain, 1031–1157*. A history of Spain. Oxford, 1992.
——. *The medieval Spains*. Cambridge medieval textbooks. Cambridge, 1993.
Reinaud, [J T], and Favé, [I]. "Du feu grégeois, des feux de guerre, et des origines de la poudre à canon chez les Arabes, les Persans et les Chinois". *Journal asiatique*, 4th ser. 14 (1849), 257–327.
Remfry, P M. *A guide to castles in Radnorshire*. Monuments in the landscape, 3. Almeley, 1996.
——. *Buckenham Castles, 1066 to 1649*. Worcester, 1997.
——. *The Herefordshire Beacon, 1043 to 1154*. Worcester, 1997.
——. *Berkhamsted Castle, 1066 to 1495*. 2nd edn. Dacorum Heritage Trust publications, 29. Berkhamsted, 1998.
Renn, D F. "The round keeps of the Brecon region". *AC* 110 (1961), 129–43.
——. *Norman castles in Britain*. 2nd edn. London, 1973.
——. "The castles of Rye and Winchelsea". *Archaeological journal*, 136 (1979), 193–202.
——. "Castle fortification in England and adjoining countries from 1150 to 1250". In *Le château médiéval et la guerre dans l'Europe du nord-ouest: mutations et adaptations* [Valenciennes 1995]. Ed. A Salamagne and R Le Jan. Villeneuve d'Ascq, 1998.
——. "'Master Jordan, who made the king's trébuchet'". *Arms & armour*, 1 (2004), 25–32.
Renoux, A. "Recherches historiques et archéologiques sur le château de Fécamp, ancien palais des ducs de Normandie". CG 7 (1975), 183–200.
——. "Le château des ducs de Normandie à Fécamp, Xᵉ–XIIᵉ siècle". *AM* 9 (1979), 5–33.
Reuter, T. "The end of Carolingian military expansion". In *Charlemagne's heir: new perspectives on the reign of Louis the Pious (814–40)*, 391–405. Ed. P Godman and R Collins. Oxford, 1990.
Rey, G. *Étude sur les monuments de l'architecture militaire des Croisés en Syrie et dans l'île de Chypre*. Collection de documents inédits sur l'histoire de France, 1. Paris, 1871.
Reynolds, S. *Fiefs and vassals: the medieval evidence reinterpreted*. Oxford, 1994.
Richard, J. "Châteaux, châtelains et vassaux en Bourgogne aux XIᵉ et XIIᵉ siècles". *Cahiers de civilisation médiévale*, 3 (1960), 433–47.
Rigold, S E. *Eynsford Castle, Kent*. English Heritage handbooks. London, 1964; repr. 1984.
——. *Totnes Castle, Devon*. 2nd edn. English Heritage handbooks. London, 1979; repr. 1987.

RINGEL, J. *Césarée de Palestine: étude historique et archéologique*. Association de Publications près les Universités de Strasbourg, 152. Paris, 1975.

ROBERT, E. "Guerre et fortification dans la *Philippide* de Guillaume le Breton: approches archéologiques". In *Military studies in medieval Europe* [Bruges 1997], 7–19. Ed. G de Boe and F Verhaeghe. Papers of the "Medieval Europe Brugge 1997" conference, 11. Zellik, 1997.

ROBINSON, I S. *Henry IV of Germany, 1056–1106*. Cambridge, 1999.

ROESDAHL, E. "The end of Viking-age fortifications in Denmark, and what followed". CG 12 (1985), 39–47.

——. "The Danish geometrical Viking fortresses and their context". *ANS* 9 (1987), 108–26.

ROGER, J. "L'enceinte basse de Philippes". *Bulletin de correspondance hellénique*, 62 (1938), 20–41.

ROGERS, R. "Aspects of the military history of the Anglo-Norman invasion of Ireland, 1169–1225". *The Irish sword*, 16 (1986), 135–44.

——. *Latin siege warfare in the twelfth century*. Oxford historical monographs. Oxford, 1992.

ROLAND, A. "Secrecy, technology, and war: Greek Fire and the defense of Byzantium, 678–1204". *Technology and culture*, 33 (1992), 655–79.

ROSSER, J. "The Castle of the Forty Columns at Paphos, Cyprus". Abstract in *8th Annual Byzantine Studies Conference* [Chicago 1982]: *Abstracts of Papers*.

ROUCHE, M. "Vinchy: le plus ancien château à motte". In *Mélanges d'archéologie et d'histoire medievales en l'honneur du Doyen Michel de Boüard*, 365–70. Mémoires et documents de l'École des chartes, 27. Geneva, 1982.

ROUND, J H. *Geoffrey de Mandeville: a study of the Anarchy*. London, 1892.

——. *The King's serjeants & officers of state with their coronation services*. London, 1911.

ROVIRA I VIRGILI, A. *Història nacional de Catalunya*, III. Barcelona, 1924.

ROWLANDS, I W. "The making of the March: aspects of the Norman settlement in Dyfed". *ANS* 3 (1981), 142–57.

ROYAL COMMISSION on Ancient and Historical Monuments in Wales. *An inventory of the ancient monuments in Glamorgan*, III: *Medieval secular monuments*, part 1a: *The early castles, from the Norman Conquest to 1217*. London, 1991. [RCHM Wales, *Early castles in Glamorgan*]

ROYAL COMMISSION on Historical Monuments. *An inventory of the historical monuments in Essex*. 4 vols. London, 1916–23. [RCHM *Essex*]

——. *An inventory of the historical monuments in Middlesex*. London, 1937. [RCHM *Middlesex*]

——. *An inventory of the historical monuments in the City of York*, II: *The defences*. London, 1972. [RCHM *York*]

RUBIO VISIERS, J, and LÓPEZ DEL ÁLAMO, O. "Talamanca del Jarama: fortificación y defensa". In *Maŷrīṭ: estudios de arqueología medieval madrileña*. Ed. F Valdés. Biblioteca de arqueología medieval hispánica, 1. Madrid, 1992.

RUNCIMAN, S. *A history of the first Bulgarian empire*. London, 1930.

SAADÉ, G. *Le château de Saladin*. Damascus, nd.

SAILHAN, P. "Typologie des archères et cannonières: les archères des châteaux de Chauvigny". *Bulletin de la Société des antiquaires de l'Ouest et des musées de Poitiers*, 4th ser. 14 (1978), 511–26.

SALAMAGNE, A. "Origines et diffusion des embrasures de tir dans l'architecture militaire de la fin du XIIe siècle: une réévaluation". In *Le château médiéval et la guerre dans l'Europe du nord-ouest: mutations et adaptations* [Valenciennes 1995], 61–75. Ed. A Salamagne and R Le Jan. Villeneuve d'Ascq, 1998.

SALCH, C-L. *L'atlas des villes et des villages fortifiées en France (Moyen Âge).* With J Burnouf. Atlas de la France médiévale, 2. Strasbourg, 1978.

——. *Dictionnaire des châteaux et des fortifications du Moyen Âge en France.* Strasbourg, 1979.

SAMUEL, M. "A tentative chronology for tower houses in west Cork". *Journal of the Cork Historical and Archaeological Society,* 103 (1998), 105–24.

SAND, G. *Clés et défense d'une ville: Bruxelles et son histoire = Sleutel en verdediging van een stad: Brussel en haar geschiedenis.* Brussels, 1984.

SANTORO, L. "Defence of rural production sites in Terra di San Benedetto (Montecassino)". *IBI Bulletin,* 50 (1998), 23–6.

SARKAR, J N. *The art of war in medieval India.* New Delhi, 1984.

SAULCY, F de. "Note sur des projectiles à main, creux et en terre cuite, de fabrication arabe". *Mémoires de la Société nationale des antiquaires de France,* 35 = 4th ser. 5 (1874). 18–34.

SAUNDERS, A D. "Attitudes towards defence at English monastic houses and granges". *IBI Bulletin,* 50 (1998), 51.

SAUNDERS, J J. *A history of medieval Islam.* London/New York, 1965.

SAUVAGET, J. *Les monuments historiques de Damas.* Beirut, 1932.

——. "Châteaux umayyades de Syrie: contribution à l'étude de la colonisation arabe aux Iᵉ et IIᵉ siècles de l'Hégire". *Revue des études islamiques,* 35 (1967), 1–49.

SAUVAIRE, H. "Matériaux pour servir à l'histoire de la numismatique et de la métrologie musulmanes". *Journal asiatique,* 7th ser. 14 (1879), 455–533; 15 (1880), 228–77, 421–78; 18 (1881), 499–516; 19 (1882), 23–77, 97–163, 281–327; 8th ser. 3 (1884), 368–445; 4 (1884), 207–321; 5 (1885), 498–506; 7 (1886), 124–77, 394–468; 8 (1886), 113–65, 272–87, 479–536; 10 (1887), 200–259.

SAVIN, C. "Machines de guerre et ingénieurs du Moyen Âge". In *Artisanat et matériaux: la place des matériaux dans l'histoire des techniques.* Ed. M C Amouretti and G Comet. Cahier d'histoire des techniques, 4. Aix-en-Provence, 1998.

SAWYER, B, and SAWYER, P. *Medieval Scandinavia: from conversion to Reformation, circa 800–1500.* The Nordic series, 17. Minneapolis, 1993.

SAYAR, I M. "The empire of the Salčuqids of Asia Minor". *Journal of Near Eastern studies,* 10 (1951), 268–80.

SCHAFER, E. "The last years of Ch'ang-an". *Oriens Extremus,* 10 (1963) 133–79.

SCHMIEDT, G. "Le fortificazioni altomedievali in Italia viste dall'aereo". In *Ordinamenti militari in Occidente nell'alto Medioevo,* 860–927. SSCISAM 15. Spoleto, 1968.

SCHNEIDER, A M, and KARNAPP, W. *Die Stadtmauer von İznik (Nicaea).* Istanbuler Forschungen, 9. Berlin, 1938.

SCHNEIDER, G. *Erzbischof Fulco von Reims (883–900) und das Frankenreich.* Münchener Beiträge zur Mediävistik und Renaissance-Forschung, 14. Munich, 1973.

SCHOENFELD, E J. "Anglo-Saxon *burhs* and continental *Burgen*; early medieval fortifications in constitutional perspective". *Haskins Society journal,* 6 (1994), 49–66.

SCHOFIELD, J. *The building of London: from the Conquest to the Great Fire.* A Colonnade book. London, 1984.

SCHRADER, C R. "A handlist of extant manuscripts containing the *De re militari* of Flavius Vegetius Renatus". *Scriptorium* 33 (1979), 280–305.

SCHWARTZ, H. "Soest". In *Deutsches Städtebuch,* III/2: *Westfälisches Städtebuch,* 330–40. Ed. E Keyser. Stuttgart, 1954.

SCRIVENER, A. "Chartley earthworks and Castle". *Transactions of the North Staffordshire Field Club*, 39 (1904/5), 143–9.

SEEBACH, C-H. "Der Stand der Werla-Forschung". CG 5 (1972), 165–74.

SEETZEN, U J. *Reisen durch Syrien, Palästina, Phönicien, die Transjordan-Länder, Arabia Petraea und Unter-Aegypten*. Ed. F Kruse et al. 4 vols. Berlin, 1854–9.

SEGAL, J B. *Edessa, "the blessed city"*. Oxford 1970.

SÉNAC, P. "Une fortification musulmane au nord de l'Ebre: le site de La Iglesieta". *Archéologie islamique*, 1 (1990), 123–45.

SERJEANT, R B, and LEWCOCK, R. "The church (al-Qalīs) of Ṣan'ā' and Ghumdān castle". In *Ṣan'ā', an Arabian Islamic city*, 44–8. Ed. R B Serjeant and R Lewcock. London, 1983.

SETTIA, A A. "Castelli e villaggi nelle terre canossiane fra x e XIII secolo". In *Studi Matildici [III]* [Reggio Emilia 1977], 281–307. Modena, 1978.

——. "La struttura materiale del castello nel secoli x e xi: elementi di morfologia castellana nelle fonti scritte dell'Italia settentrionale". *Bollettino storico-bibliografico subalpino*, 77 (1979), 361–430.

——. "Le fortificazioni dei Goti in Italia". In *Teoderico il grande e i Goti d'Italia* [Milan 1992], 1: 101–31. Atti dei congressi, 13. Spoleto, 1993.

——. "Il primo incastellamento nella Bergamesca". In his *Proteggere e dominare: fortificazioni e popolamento nell'Italia medievale*. I libri di Viella, 13. Rome, 1999.

SHAW, M, CHAPMAN, A, and SODEN, I. "Northampton". *Current archaeology*, no. 155 = 13 (1997), 408–15.

SHENTON, C. "The location of the siege camp at Berwick-upon-Tweed, 1333". *Archaeologia Aeliana*, 5th ser. 29 (2001), 253–6.

SHEPARD, J. "Slavs and Bulgars". In *The new Cambridge medieval history*, II: *c.700–c.900*, 228–48. Ed. R McKitterick. Cambridge, 1995.

SHKOLIAR, S A. "L'artillerie de jet à l'époque Sung". In *Études Song*, ser. 1: *Histoire et institutions*, II: 119–42. Ed. F Aubin. Paris, 1971.

SHRADER, C R. "A handlist of extant manuscripts containing the *De re militari* of Flavius Vegetius Renatus". *Scriptorium* 33 (1979), 280–305.

SIDARUS, A. "A propósito das ameias nas fortificações do Sudoeste Peninsular até ao século x". *Arqueologia medieval*, 2 (1993), 211–14.

SIMPSON, W D. *The castle of Bergen and the Bishop's Palace at Kirkwall: a study in early Norse architecture*. Aberdeen University studies, 142. Edinburgh, 1961.

SINCLAIR, T A. *Eastern Turkey: an architectural and archaeological survey*. 4 vols. London, 1987–90.

SIRCAR, J N. "Siegecraft in medieval India". *Itihas: journal of the Andhra Pradesh Archives*, 4 (1976), 92–108.

SLADE, C F. "Wallingford Castle in the reign of Stephen". *Berkshire archaeological journal*, 58 (1960), 33–43.

SMITH, T P. "Why did medieval towns have walls?". *Current archaeology*, no. 94 = 8 (1984), 376–9.

SOEDEL, W, and FOLEY, V. "Ancient catapults". *Scientific American*, 240/3 (March 1979), 120–28.

SOMMA, M C. "Strutture fortificate altomedievali in Italia centrale: l'esempio della Marsica tra VIII e XI seculo". In *Military studies in medieval Europe* [Bruges 1997]. Ed. G de Boe and F Verhaeghe. Papers of the "Medieval Europe Brugge 1997" conference, II. Zellik, 1997.

SOTIRIOU, G. "Le château-fort de Chlemoutsi et son atélier monétaire de tournois de Clarentia". *Journal internationale d'archéologie numismatique*, 19 (1916/17), 273–9.

SOUTHERN, P, and DIXON, K R. *The late Roman army*. The archaeology of the Roman empire. London, 1996.

SPEIGHT, S. "Early medieval castles in Nottinghamshire". *Transactions of the Thoroton Society of Nottinghamshire*, 98 (1994), 58–70.

——. "Churches as castles: the misuse of religious buildings during the reign of Stephen". *Castle Studies Group newsletter*, 13 (1999/2000), 11–12.

——. "Castle warfare in the *Gesta Stephani*". CG 19 (2000), 269–74.

SPUFFORD, P. *Handbook of medieval exchange*. With W Wilkinson and S Tolley. Royal Historical Society guides and handbooks, 13. London, 1986.

SPULER, B. *Iran in früh-islamischer Zeit: Politik, Kultur, Verwaltung und öffentliches Leben zwischen der arabischen und der seldschukischen Eroberung, 633 bis 1055*. Veröffentlichungen der orientalischen Kommission, 2. Wiesbaden, 1952.

SPURGEON, C J. "The medieval town defences of Glamorgan". *Studia celtica*, 35 (2001), 161–212.

SPURRELL, M. "Containing Wallingford Castle, 1146–1153". *Oxoniensia* 60 (1995), 257–70.

STAMPER, P A. "Excavations on a mid twelfth century siege castle at Bentley, Hampshire". *Proceedings of the Hampshire Field Club & Archaeological Society*, 40 (1984), 81–9.

STAŇA, C. "Mährische Burgwalle in 9. Jahrhundert". In *Die Bayern und ihre Nachbarn* [Stift Zwettl 1982], II. Ed. H Friesinger and F Daim. Denkschriften der Österreichischen Akademie der Wissenschaften, 180. Vienna, 1985.

STANČEV, S. "L'architecture militaire et civile de Pliska et de Preslav à la lumière de nouvelles données". In *Actes du XIIe Congrès international d'études byzantines* [Ohrid 1961], III. Belgrade, 1964.

STANČEVA, M. "Sofia au Moyen Âge à la lumière de nouvelles études archéologiques". *Byzantinobulgarica* 5 (1978), 212–16.

STAŇKOVA, J, ŠTURSA, J, and VODĚRA, S. *Pražská architektura: významné stavby jedenácti století*. Prague, 1991.

STEENSTRUP, J. "Les normands et leurs invasions: étude critique". In *Les invasions normandes en France*, 55–141. Ed. J Steenstrup. Le mémorial des siècles, 17. Paris, 1969.

STEHKAMPER, H. "Die Stadt Köln in der Salierzeit". In *Die Salier und das Reich*, III: *Gesellschaftlicher und ideengeschichtlicher Wandel im Reich der Salier*, 75–152. Ed. S Weinfurter et al. 2nd edn. Sigmaringen, 1992.

STEIN, A. *Innermost Asia: detailed report of explorations in Central Asia, Kan-Su, and Eastern Īrān*. 4 vols. Oxford, 1928.

STENTON, F M. "The road system of medieval England". *The economic history review*, 7 (1936/7), 1–21.

——. *The first century of English feudalism, 1066–1166*. The Ford lectures, 1929. 2nd edn. Oxford, 1961.

STIESDAHL, H. "Die Motten in Danemark". CG 2 (1967), 94–9.

——. "Die ältesten dänischen Donjons". CG 8 (1977), 279–86.

——. "Types of public and private fortifications in Denmark". in *Danish medieval history: new currents* [Copenhagen 1979], 207–20. Ed. N Skyum-Nielsen and N Lund. Copenhagen, 1981.

STOKES, A D. "The background and chronology of the Balkan campaigns of Svyatoslav Igorevich". *Slavonic and East European review*, 40 (1961/2), 44–57.

——. "The Balkan campaigns of Svyatoslav Igorevich". *Slavonic and East European review*, 40 (1961/2), 466–96.

STOLBERG, F. *Befestigungsanlagen im und am Harz von der Frühgeschichte bis zur Neuzeit: ein Handbuch*. Forschungen und Quellen zur Geschichte des Harzgebietes, 9. Hildesheim, 1968.

STRIČEVIĆ, Ð. "Les fouilles archéologique [*sic*] d'Heraclea Lyncestis (1936–1938)". *Héraclée*, 1 (1961), 35–42.

STRICKLAND, M. "Military technology and conquest: the anomaly of Anglo-Saxon England". *ANS* 19 (1997), 353–82.

SULLIVAN, D. "Tenth century Byzantine offensive siege warfare: instructional prescriptions and historical practice". In *To empolemo Vyzantio (9os–12os ai.) = Byzantium at war (9th–12th c.)* [Athens 1996], 179–200. Ed. K Tsiknakis. Athens, 1997.

SUPPE, F C. *Military institutions on the Welsh Marches: Shropshire, AD 1066–1300*. Studies in Celtic history, 14. Woodbridge, 1994.

——. "The persistence of castle-guard in the Welsh Marches and Wales: suggestions for a research agenda and methodology". In *The Normans and their adversaries at war: essays in memory of C. Warren Hollister*, 201–25. Ed. R P Abels and B S Bachrach. Warfare in history, 12. Woodbridge, 2001.

SWARZENSKI, H. *Vorgotische Miniaturen: die ersten Jahrhunderte deutscher Malerei*. Die blauen Bücher. Königstein im Taunus, 1927.

SWEETMAN, D. *Irish castles and fortified houses*. Dublin, 1995.

——. "The hall house in Ireland". *Archaeology Ireland*, 12 (1998), 13–16.

——. *The medieval castles of Ireland*. Cork, 1999; Woodbridge, 2000.

ŚWIĘTOSŁAWSKI, W. *Arms and armour of the nomads of the Great Steppe in the times of the Mongol expansion (12th–14th centuries)*. Trans. M Abramowicz. Studies on the history of ancient and medieval art of warfare, 3. Łódź, 1999.

SYMEONOGLOU, S. *The topography of Thebes from the Bronze Age to modern times*. Princeton, 1985.

SZÁDECZKY-KARDOSS, S. "Der awarisch-türkische Einflüss auf die byzantinische Kriegskunst um 600 (Anmerkungen zum *Strategikon* des Maurikios)". In *Turkic–Bulgarian–Hungarian relations: VIth–XIth centuries*, 63–71. Ed. G Káldy-Nagy. Studia turco-hungarica, 5. Budapest, 1981. Repr. in his *Avarica: über die Awarengeschichte und ihre Quellen*, 203–13. Opuscula byzantina, 8. Szeged, 1986.

SZAFRAŃSKI, W. "Swobnica (Wildenbruch) in Westpommern". In *Frühe polnische Burgen: Berichte über archäologische Grabungen*. Weimar, 1960.

——. "Die frühmittelalterliche Hauptburg Płock in Masowien". In *Frühe polnische Burgen: Berichte über archäologische Grabungen*. Weimar, 1960.

TABRAHAM, C. *Scotland's castles*. Historic Scotland. London, 1997.

TAFRALI, O. *Topographie de Thessalonique*. Paris, 1913.

ṬĀHA, 'A D. *The Muslim conquest and settlement of North Africa and Spain*. Exeter Arabic and Islamic series. London/New York, 1989.

TALBOT, E J. "The defences of earth and timber castles". In *Scottish weapons and fortifications, 1100–1800* [Edinburgh 1976, 1978], 1–9. Ed. D H Caldwell. Edinburgh, 1981.

TARAMELLI, A. "Cretan expedition, XXI: Gortyna". *American journal of archaeology*, 6 (1902), 101–65.

TARVER, W T S. "The traction trebuchet: a reconstruction of an early medieval siege engine". *Technology and culture*, 36 (1995), 136–67.

TAUSEND, K. "Hunnische Poliorketik". *Grazer Beitrage*, 12/13 (1985/6), 265–81.

TAYLOR, A J. "Three early castle sites in Sicily: Motta Camastra, Sperlinga, and Petralia Soprana". CG 7 (1975), 209–11.

——. *Beaumaris Castle*. Rev. edn. Cardiff, 1985.

TAYLOR, A J. *Caernarfon Castle and town walls*. Rev. edn. Cardiff, 1989.
——. "Master Bertram, *ingeniator regis*". In *Studies in medieval history presented to R. Allen Brown*, 289–315. Ed. C Harper-Bill, C J Holdsworth and J L Nelson. Woodbridge, 1989.
TAYLOR, P. "The endowment and military obligations of the see of London: a reassessment of three sources". *ANS* 14 (1992), 287–312.
TEODOR, D G. *The east Carpathian area of Romania in the V–XI centuries A.D.* Trans. N Hampartumian. *BAR* international series, 81. Oxford, 1980.
TERRASSE, H. "Crónica arqueológica de la España musulmana, XXXIII", [1]: "La forteresse almoravide d'Amergo". *Al-Andalus* 18 (1953), 389–400.
——. "Les forteresses de l'Espagne musulmane". *Boletín de la Real Academia de la Historia*, 134 (1954), 455–83. (Also pubd separately, Madrid, 1954.)
——. "Crónica arqueológica de la España musulmana, L", [3]: "Un vestige des fortifications oméiyades de Ceuta". *Al-Andalus* 27 (1962), 244–6.
TETTAU, W von. *Beschreibende Darstellung der älteren Bau- und Kunstdenkmäler der Stadt Erfurt und des Erfurter Landkreises*. Beschreibende Darstellung der älteren Bau- und Kunstdenkmäler der Provinz Sachsen und angrenzenden Gebiete, 13. Halle, 1890.
THOMAS, A. *The walled towns of Ireland*. 2 vols. Blackrock, 1992.
THOMPSON, E A. *The Goths in Spain*. Oxford, 1969.
THOMPSON, M W. "Old Bolingbroke Castle". *Archaeological journal*, 131 (1975), 314–17.
——. "The military interpretation of castles". *Archaeological journal*, 151 (1994), 439–45.
——. *Kenilworth Castle*. London, 1996.
TIKHOMIROV, M. *The towns of ancient Rus*. Trans. Y Sdobnikov. Moscow, 1959.
TILLMANN, C. *Lexikon der deutschen Burgen und Schlösser*. 4 vols. Stuttgart, 1957–61.
TODD, M. *The walls of Rome*. Archaeological sites. London, 1978.
TOLAINI, E. "La costruzione delle mura di Pisa negli anni 1155–1161 secondo gli *Annales* del Maragone". *Bollettino storico pisano*, 36–8 (1967–9), 15–36.
TONDELLI, L. "Scavi archeologici a Canossa: le tre mura di cinta". In *Studii gregoriani: per la storia di Gregorio VII e della riforma gregoriana*, IV: 365–71. Ed. G B Borino. Rome, 1952.
TORRES Balbás, L. "Crónica arqueológica de la España musulmana, VIII", [2]: "La alcazaba almohade de Badajoz". *Al-Andalus* 6 (1941), 168–203.
——. *Ciudades hispanomusulmanas*. 2nd edn. Madrid, 1985.
TÓTH, E. "Die karolingische Burg von Sabaria-Szombathely". *Folia archaeologica*, 29 (1978), 151–82.
TOUBERT, P. *Les structures du Latium médiéval: le Latium méridional et la Sabine du IXe siècle à la fin du XIIe siècle*. 2 vols. Bibliothèque des Écoles françaises d'Athènes et de Rome, 221. Rome, 1973.
TOUSSAINT, P. *Gisors: le château*. Gournay-en-Bray, 1993.
TOUT, T F. "The fair of Lincoln and the *Histoire de Guillaume le Maréchal*". *English historical review*, 18 (1903), 240–65. Cited from repr. in his *Collected papers*, II. Manchester, 1934.
TOUZET, P. "L'architecture militaire du VIIe au XIIIe siècle d'après les enluminures de manuscrits". *Bulletin trimestriel de la Société des antiquaires de Picardie* (1977), 17–35.
TOY, S. *The strongholds of India*. Melbourne/London, 1957.
——. *The fortified cities of India*. London, 1965.
TRAHERNE, C. "The conquest of Glamorgan". *AC* 133 (1984), 1–7.
TRAQUAIR, R. "Laconia, I: Mediaeval fortresses". *Annual of the British School at Athens*, 12 (1905/6), 259–76.

——. "Medieval fortresses of the north-western Peloponnesus". *Annual of the British School at Athens*, 13 (1906/7), 268–84.

TUDOR, D. "Sucidava". *Dacia* 5–6 (1938), 387–422; 7–8 (1941), 359–400; 11–12 (1948), 145–208.

TURNBULL, S. *Siege weapons of the Far East, II: AD 960–1644*. New vanguard, 44. Oxford, 2002.

——. *Samurai: the world of the warrior*. Oxford, 2003.

TURNER, H L. *Town defences in England and Wales: an architectural and documentary study, AD 900-1500*. London, 1971.

TURNER, R V, and Heiser, R R. *The reign of Richard Lionheart, ruler of the Angevin empire, 1189–1199*. The medieval world. Harlow, 2000.

TURVEY, R. "The defences of twelfth-century Deheubarth and the castle strategy of the Lord Rhys". *AC* 144 (1995), 103–32.

——. *The Lord Rhys, prince of Deheubarth*. Llandysul, 1997.

——. *The Welsh princes: the native rulers of Wales, 1063–1283*. The medieval world. London, 2002.

TUULSE, A. *Die Burgen in Estland und Lettland*. Tartu, 1942.

UNDERWOOD, R. *Anglo-Saxon weapons and warfare*. Stroud, 2001.

USLAR, R von. *Studien zu frühgeschichtlichen Befestigungen zwischen Nordsee und Alpen*. Beihefte der Bonner Jahrbücher, 11. Graz, 1964.

VACQUER, T. "La découverte d'une partie du Grand Pont de Paris bâti par Charles-le-Chauve". *RA* 12 (1855/6), 502–7.

VALDÉS FERNÁNDEZ, F. "El urbanismo islámico de la Extremadura leonesa: cuatro pautas de desarrollo". In *Genèse de la ville islamique en al-Andalus et au Maghreb occidental* [Madrid 1994 / Granada 1995]. Ed. P Cressier and M García-Arenal. Madrid, 1998.

VALLERY-RADOT, J. "Quelques donjons de Philippe Auguste". *Bulletin de la Société national des antiquaires de France* (1964), 155–9.

VÁMBÉRY, A. *History of Bokhara from the earliest period down to the present: composed for the first time after oriental known and unknown historical manuscripts*. 2nd edn. London, 1873.

VAN BERCHEM, D. "Recherches sur la chronologie des enceintes de Syrie et de Mésopotamie". *Syria* 31 (1954), 254–70.

VAN MILLINGEN, A. *Byzantine Constantinople: the walls of the city and adjoining historical sites*. London, 1899.

VANNINI, G, TONGHINI, C, and VANNI DESIDERI, A. "'Medieval' Petra and the crusader–Islamic frontier: archaeological mission of the University of Florence". CG 20 (2002), 271–6.

VANTINI, G. *Christianity in the Sudan*. Bologna, 1981.

VAN WERVEKE, H, and VERHULST, A E. "Castrum en Oudburg te Gent". *Handelingen der Maatschappij voor geschiedenis en oudheidkunde te Gent*, ns 14 (1960), 3–62.

VASILIEV, A A. *The Russian attack on Constantinople in 860*. Publications of the Mediaeval Academy of America, 46. Cambridge, Mass., 1946.

VELKOV, I; *see* WELKOV, I

VELLOSO DA COSTA, L. "Lisboa islámica". In *Arqueologia medieval española: II congreso* [Madrid 1987], II. Madrid, 1987.

VEMMING HANSEN, P. "Reconstructing a medieval trebuchet". *Military illustrated*, no. 27 (Aug. 1990), 9–16.

——. "Experimental reconstruction of a medieval trebuchet". *Acta archaeologica*, 63 (1992), 189–208.

VENEDIKOV, I, et al. *Nessèbre*. 2 vols. Sofia, 1969–80.

VENEDIKOV, I, TSCHIMBULEVA, S, and LASAROV, M. *Nessebar*. Sofia, 1971.

VERCAUTEREN, F. "Comment s'est-on défendu, au IXe siècle, dans l'empire franc contre les invasions normandes?" *Annales de la Fédération archéologique et historique de Belgique*, 30 (1935), 117–36.

VERHULST, A E. "Les origines et l'histoire ancienne de la ville de Bruges (IXe–XIIe siècle)". *Le Moyen Âge*, 66 (1960), 37–63.

VERNADSKY, G. *A history of Russia*, II: *Kievan Russia*. New Haven, 1948; repr. 1973.

VETTERS, H. "Von der spätantiken zur frühmittelalterlichen Festungsbaukunst". In *Ordinamenti militari in Occidente nell'alto Medioevo*. SSCISAM 15. Spoleto, 1968.

VIEIRA DA SILVA, A. *A cêrca moura de Lisboa: estudo historico-descriptivo*. Lisbon, 1899.

VOLGER, W. *Der Ursprung und der älteste Zustand der Stadt Lüneburg: ein Versuch*. Lüneburg, 1861.

VRYONIS, S, jr. "The evolution of Slavic society and the Slavic invasions in Greece: the first major Slavic attack on Thessaloniki, A.D. 597". *Hesperia* 50 (1981), 378–90.

WACHTER, B. "Burgen im hannoverschen Wendland". *Zeitschrift für Archäologie des Mittelalters*, 25–6 (1997–8), 155–72.

WALES, H G Q. *Ancient south-east Asian warfare*. London, 1952.

WALKER, D. *The Normans in Britain*. Historical Association studies. Oxford, 1995.

WALLACE-HADRILL, J M. *The barbarian West, 400–1000*. 3rd edn. London, 1967.

WALLACKER, B E. "Studies in medieval Chinese siegecraft: the siege of Yü-pi, A.D. 546". *The journal of Asian studies*, 28 (1968/9), 789–802.

——. "Studies in medieval Chinese siegecraft: the siege of Ying-ch'uan, A.D. 548–549". *The journal of Asian studies*, 30 (1970/71), 611–22.

WAND, N. "Die Büraburg bei Fritzlar: eine fränkische Reichsburg mit Bischofssitz in Hessen". In *Frühmittelalterlicher Burgenbau in Mittel- und Osteuropa* [Nitra 1996]. Ed. J Henning and A T Ruttkay. Bonn, 1998.

WARD-PERKINS, B. *From classical antiquity to the Middle Ages: urban public building in northern and central Italy, AD 300–850*. Oxford historical monographs. Oxford, 1984.

——. "Quarries and stoneworking in the early Middle Ages: the heritage of the ancient world". In *Artigianato e tecnica nella società dell'alto Medioevo*, II: 525–44. SSCISAM 18. Spoleto, 1971.

WATERMAN, D M. "Excavations at Clough Castle, Co. Down". *Ulster journal of archaeology*, 3rd ser. 17 (1954), 103–63.

——. "Excavations at Lismahon, Co. Down". *Medieval archaeology*, 3 (1959), 139–76.

WATSON, B. "The Norman fortress on Ludgate Hill in the City of London, England: recent excavations, 1986–1990". CG 15 (1992), 335–45.

WEAVER, J. *Beeston Castle*. London, 1987.

WEDLAKE, W J. "City walls of Bath, the church of St James, South Gate, and the area to the east of the church of St James". *Proceedings of the Somersetshire Archaeological and Natural History Society*, 110 (1965/6), 85–107.

WEIGEL, S. "Die Stadtbefestigung Neutitscheins". *Zeitschrift des Deutschen Vereines für die Geschichte Mährens und Schlesiens*, 23 (1919), 160–68.

WELKOV, I. "Eine Gotenfestung bei Sadowetz (Nordbulgarien)". *Germania: Anzeiger der römisch-germanischen Kommission des Deutschen Archäologischen Instituts*, 19 (1935), 149–58.

WELSBY, D A. *The medieval kingdoms of Nubia: pagans, Christians and Muslims along the middle Nile*. London, 2002.

Werner, E T C. *Chinese weapons*. Publications of the Royal Asiatic Society, North China Branch, extra vol. Shanghai, 1932.

Wetzstein, J G. *Reisebericht über Hauran und die Trachonen, nebst einem Anhange über die sabäischen Denkmäler in Ostsyrien*. Berlin, 1860.

Whimperley, A, and Murphy, D. *The walls of Chester: a guided tour*. A Dalesman mini-book. Clapham, N. Yorks., 1977.

Whitby, M. "Procopius and the development of Roman defences in Upper Mesopotamia". In *The defence of the Roman and Byzantine East* [Sheffield 1986], II: 717–35. Ed. P Freeman and D Kennedy. *BAR* international series, 297. Oxford, 1986.

——. "Procopius' descriptions of Dara (buildings II.1–3)". In *The defence of the Roman and Byzantine East* [Sheffield 1986], II: 737–83. Ed. P Freeman and D Kennedy. *BAR* international series, 297. Oxford, 1986.

White, L, jr. "Tibet, India, and Malaya as sources of Western medieval technology". *The American Historical Review*, 65 (1959/60), 515–26.

——. "The crusades and the technological thrust of the West". In *War, technology and society in the Middle East* [London 1970], 97–112. Ed. V J Parry and M E Yapp. London, 1975.

White, P. *Sherborne Old Castle*. English Heritage handbooks. 2nd edn. London, 1979; repr. 1986.

White, P. "Castle gateways during the reign of Henry II". *Antiquaries journal*, 76 (1996), 241–7.

Whittow, M. *The making of Orthodox Byzantium, 600–1025*. New studies in medieval history. Berkeley/London, 1996.

Wiemken, R C, and Holum, K G. "The joint expedition to Caesarea Maritima: eighth season, 1979". *Bulletin of the American Schools of Oriental Research*, 244 (1981), 27–52.

Wiet, G. *Baghdad: metropolis of the Abbasid caliphate*. Trans. S Feiler. Norman, Okla., 1971.

Wiggins, K. "Warfare underground". *British archaeology*, 71 (2003), 8–13.

Wightman, W E. "The palatine earldom of William fitz Osbern in Gloucestershire and Worcestershire (1066–1071)". *EHR* 77 (1962), 6–17.

——. *The Lacy family in England and Normandy, 1066–1194*. Oxford, 1966.

Wilkes, J J. *Dalmatia*. History of the provinces of the Roman Empire. Cambridge, Mass./London, 1969.

Williams, G A. "Welsh raiding in the twelfth-century Shropshire/Cheshire March: the case of Owain Cyfeiliog". *Studia Celtica*, 40 (2006), 89–115.

Winnig, A. *Der Deutsche Ritterordern und seine Burgen*. Die blauen Bücher. Königstein im Taunus, 1939; repr. 1956.

Yanin, V L. "The archaeological study of Novgorod: an historical perspective". In *The archaeology of Novgorod, Russia; recent results from the town and its hinterland*, 67–106. Ed. M A Brisbane. Society for Medieval Archaeology monograph series, 13. Lincoln, Neb., 1992.

Yovitchitch, C. "La citadelle de Bosra". In *La fortification au temps des Croisades* [Parthenay 2002]. Ed. N Faucherre, J Mesqui and N Prouteau. Archéologie et culture. Rennes, 2004.

Yver, J. "Les châteaux-forts en Normandie jusqu'au milieu du XIIe siècle". *Bulletin de la Société des antiquaires de Normandie*, 53 (1955/6), 28–115.

Zadora-Rio, E. "L'enceinte fortifiée du Plessis-Grimoult, résidence seigneuriale du XIe siècle". CG 5 (1972), 227–39.

ZADORA-RIO, E. "Les essais de typologie des fortifications de terre médiévales en Europe". *AM* 15 (1985), 191–6.

ZAK, J. "Mittelalterliche kegelförmige Burghügel in Polen". CG 11 (1983), 289–91.

ŻAKI, A. "Topografia wcześniośredniowiecznejo Przemyśla". *Acta archaeologica carpathica*, 1 (1958), 95–6 (French summary).

ZAKY, A-R. "Military literature of the Arabs". *Islamic culture*, 30 (1956), 163–72.

ZDRAVKOVIĆ, I, and JOVANOVIĆ, V. "La forteresse de Zveçan, située au Moyen Âge à la frontière entre Byzance et la Serbie". In *Actes du XIIe Congrès international d'études byzantines* [Ohrid 1961], III: 423–8. Belgrade, 1964.

ZETTEL, H. *Das Bild der Normannen und der Normanneneinfalle in westfrankischen, ostfrankischen und angelsachsischen Quellen des 8. bis 11. Jahrhunderts*. Munich, 1977.

ZEUNE, J. "Salierzeitliche Burgen in Bayern". In *Burgen der Salierzeit*, II: *In den südlichen Landschaften des Reiches*, 177–232. Ed. H W Böhme. Römisch-Germanisches Zentralmuseum Mainz, Forschungsinstitut für Vor- und Frühgeschichte: Monographien, 26. Sigmaringen, 1992.

ZILLICH, H. *Siebenburgen: ein abendländisches Schicksal*. Die blauen Bücher. Königstein in Taunus, 1957.

ZIMMER, J. "Le château de Vianden: un bâtiment représentatif du 12e et 13e siècle". CG 20 (2002), 289–305.

ZIMMERMANN, M. "La prise de Barcelone par Al-Mansur et la naissance de l'historiographie catalane". In *L'historiographie en Occident du Ve au XVe siècle* [Tours 1977], 191–218. Annales de Bretagne et des Pays de l'Ouest, 87. Rennes, 1980.

ZOZAYA, J. "Islamic fortifications in Spain: some aspects". In *Papers in Iberian archaeology*, II: 636–73. Ed. T F C Blagg, R F J Jones and S J Keay. *BAR* international series, 193. Oxford, 1984.

Index

Printed and bound by CPI Group (UK) Ltd, Croydon, CR0 4YY
13/07/2026
02165169-0001

Recent Advances in Corrosion Science

Special Issue Editor

Jacek Ryl

MDPI • Basel • Beijing • Wuhan • Barcelona • Belgrade • Manchester • Tokyo • Cluj • Tianjin

Special Issue Editor
Jacek Ryl
Gdansk University of Technology
Poland

Editorial Office
MDPI
St. Alban-Anlage 66
4052 Basel, Switzerland

This is a reprint of articles from the Special Issue published online in the open access journal *Materials* (ISSN 1996-1944) (available at: https://www.mdpi.com/journal/materials/special_issues/Corrosion_Science).

For citation purposes, cite each article independently as indicated on the article page online and as indicated below:

LastName, A.A.; LastName, B.B.; LastName, C.C. Article Title. *Journal Name* **Year**, *Article Number*, Page Range.

ISBN 978-3-03936-176-2 (Hbk)
ISBN 978-3-03936-177-9 (PDF)

Contents

About the Special Issue Editor

Jacek Ryl is an Associate Professor in the Faculty of Chemistry, Gdansk University of Technology. He has received a Ph.D. (2010) and habilitation (2018) in Chemical Technology at GUT. His principal area of scientific activity lies in applied electrochemistry and surface physic-chemistry, in particular corrosion science, electrochemical sensors, and materials for energy storage and conversion. He is involved in the development of instantaneous impedance techniques with multisine perturbation, dedicated to the assessment of nonstationary processes. Jacek Ryl is coauthor of over 90 peer-reviewed papers (h-index=21). In 2015–2020, he was the principal investigator of programs Sonata NCN and Iuventus Plus. His most notable awards include a scientific scholarship by the Minister of Science and Higher Education of Poland (2017) and the IV Division (Engineering Sciences) of the Polish Academy of Sciences award for scientific achievements (2019). Currently, he serves on the Editorial Board for Molecules, MDPI.

materials

MDPI

Editorial

Special Issue: Recent Advances in Corrosion Science

Jacek Ryl

Department of Electrochemistry, Corrosion and Materials Engineering, Gdansk University of Technology, Narutowicza 11/12, 80-233 Gdansk, Poland; jacek.ryl@pg.edu.pl

Received: 13 April 2020; Accepted: 17 April 2020; Published: 19 April 2020

The International Union of Pure and Applied Chemistry (IUPAC) and European Federation of Corrosion (EFC) define corrosion as an irreversible interfacial reaction of a material with its environment which results in its consumption or dissolution, often resulting in effects detrimental to the usage of the material considered. Corrosion failure is a significant problem in any given type of industry, leading to substantial economic consequences, but also often influencing human health and the environment negatively, among other unmeasurable factors. The industry estimates indicate that the total direct cost of corrosion ranges between 3% and 5% of GDP [1], while the indirect costs (outages, delays, revenue losses, etc.) while much harder to evaluate, are estimated to be equal to this. These numbers point out that investments in corrosion protection are, by all means, economically justified.

The dynamic development of the global industry and growing demand for new material technologies generates constantly increasing problems regarding premature material degradation and the requirement to determine corrosion mechanisms and to develop new protection/evaluation approaches. This Special Issue, "Recent Advances in Corrosion Science", brings together fourteen articles and one review, providing a snapshot of the recent activity and development in this field.

The corrosion properties of ferrous metals remain the most popular subject of investigation, which naturally found coverage in numerous research articles present within this Special Issue. The primary source of this versatility is achieved by a proper selection of alloying additives and metalworking, which guarantee the demanded mechanical and physicochemical properties. On the other hand, the alteration of metal structure leads to the formation of galvanic microcells, often translating into various forms of local corrosion. The search for alloying additives enhancing the corrosion resistance without sacrificing the desired characteristics continues, intending to reduce alloy corrosion rate and bring measurable economic profits. Within this Special Issue, you will find multiple original research papers strictly devoted to this issue for both ferrous [2–4] and non-ferrous metals [5–8]. The influence of novel microscopy tools, which enable the direct observation of local corrosion processes, cannot be overestimated. For this reason, I would like to recommend a very interesting and important review prepared by Chen et al. [9], referring to the advances in electrochemical atomic force microscopy (EC-AFM), an outstanding tool to perform real-time in situ corrosion studies of galvanic microcells.

Affecting the corrosion process by electrochemical protection (cathodic or anodic), barrier properties obtained with the use of paints or coatings as well as environment modification with dedicated corrosion inhibitors, are the three primary ways to reduce the corrosion rate found in both principle and industrial studies regarding anti-corrosion technologies. All of these research areas are represented within this Special Issue. The works of Xu et al. [10], Tang. et al. [11] and Ryl et al. [12] reveal various aspects concerning the search for efficient organic corrosion inhibitors and the tools used to evaluate protection mechanisms. The studies of Parchoviansky et al. [13] and Winiarski et al. [14] provide an insight on the development of anti-corrosion composite coatings, while an interesting report from Kania and Sipa [15] shows the improved corrosion resistance of anodic zinc coatings, obtained using a new thermal diffusion process.

It is important to emphasize that, nowadays, corrosion issues are not solely connected with the degradation of metals. Modern composite or semiconductor electrode materials are constantly developed to be used in numerous branches of applied electrochemistry, such as energy storage and conversion, electrochemical sensors and electrocatalytic processes. Their stable performance under aggressive environmental factors is often questionable. Thus, the final manuscript of this Special Issue presents work in this new field, which was devoted to high-temperature oxidation and the degradation of boron-doped diamond nanostructures [16].

Conflicts of Interest: The authors declare no conflict of interest.

References

1. Koch, G.H.; Brongers, M.P.H.; Thompson, N.G.; Virmani, Y.P.; Payer, J.H. *Corrosion Cost and Preventive Strategies in the United States*; FHWA-RD-01-156, R315-01; United States, Federal Highway Administration: Washington, DC, USA, 2001.
2. Liu, D.; Li, Y.; Xie, X.; Zhao, J. Effect of Pre-Corrosion Pits on Residual Fatigue Life for 42CrMo Steel. *Materials* **2019**, *12*, 2130. [CrossRef] [PubMed]
3. Yang, Q.; Zhou, Y.; Li, Z.; Mao, D. Effect of Hot Deformation Process Parameters on Microstructure and Corrosion Behavior of 35CrMoV Steel. *Materials* **2019**, *12*, 1455. [CrossRef] [PubMed]
4. Seikh, A.H.; Sarkar, A.; Singh, J.K.; Mohammed, S.M.A.K.; Alharthi, N.; Ghosh, M. Corrosion Characteristics of Copper-Added Austempered Gray Cast Iron (AGCI). *Materials* **2019**, *12*, 503. [CrossRef] [PubMed]
5. Ren, J.; Guo, E.; Wang, X.; Kang, H.; Chen, Z.; Wang, T. Influence of Alloyed Ga on the Microstructure and Corrosion Properties of As-Cast Mg–5Sn Alloys. *Materials* **2019**, *12*, 3686. [CrossRef] [PubMed]
6. Ouyang, P.; Mi, G.; Li, P.; He, L.; Cao, J.; Huang, X. Non-Isothermal Oxidation Behaviors and Mechanisms of Ti-Al Intermetallic Compounds. *Materials* **2019**, *12*, 2114. [CrossRef] [PubMed]
7. Palcut, M.; Ďuriška, L.; Černičková, I.; Brunovská, S.; Gerhátová, Ž.; Sahul, M.; Čaplovič, Ľ.; Janovec, J. Relationship between Phase Occurrence, Chemical Composition, and Corrosion Behavior of as-Solidified Al–Pd–Co Alloys. *Materials* **2019**, *12*, 1661. [CrossRef] [PubMed]
8. El-Bagoury, N.; Ahmed, S.I.; Ahmed Abu Ali, O.; El-Hadad, S.; Fallatah, A.M.; Mersal, G.A.M.; Ibrahim, M.M.; Wysocka, J.; Ryl, J.; Boukherroub, R.; et al. The Influence of Microstructure on the Passive Layer Chemistry and Corrosion Resistance for Some Titanium-Based Alloys. *Materials* **2019**, *12*, 1233. [CrossRef] [PubMed]
9. Chen, H.; Qin, Z.; He, M.; Liu, Y.; Wu, Z. Application of Electrochemical Atomic Force Microscopy (EC-AFM) in the Corrosion Study of Metallic Materials. *Materials* **2020**, *13*, 668. [CrossRef] [PubMed]
10. Xu, S.; Zhang, S.; Guo, L.; Feng, L.; Tan, B. Experimental and Theoretical Studies on the Corrosion Inhibition of Carbon Steel by Two Indazole Derivatives in HCl Medium. *Materials* **2019**, *12*, 1339. [CrossRef] [PubMed]
11. Tang, M.; Li, J.; Li, Z.; Fu, L.; Zeng, B.; Lv, J. Mannich Base as Corrosion Inhibitors for N80 Steel in a CO2 Saturated Solution Containing 3 wt % NaCl. *Materials* **2019**, *12*, 449. [CrossRef] [PubMed]
12. Ryl, J.; Brodowski, M.; Kowalski, M.; Lipinska, W.; Niedzialkowski, P.; Wysocka, J. Corrosion Inhibition Mechanism and Efficiency Differentiation of Dihydroxybenzene Isomers Towards Aluminum Alloy 5754 in Alkaline Media. *Materials* **2019**, *12*, 3067. [CrossRef] [PubMed]
13. Parchovianský, M.; Parchovianská, I.; Švančárek, P.; Motz, G.; Galusek, D. PDC Glass/Ceramic Coatings Applied to Differently Pretreated AISI441 Stainless Steel Substrates. *Materials* **2020**, *13*, 629. [CrossRef] [PubMed]
14. Winiarski, J.; Niciejewska, A.; Ryl, J.; Darowicki, K.; Baśladyńska, S.; Winiarska, K.; Szczygieł, B. Ni/cerium Molybdenum Oxide Hydrate Microflakes Composite Coatings Electrodeposited from Choline Chloride: Ethylene Glycol Deep Eutectic Solvent. *Materials* **2020**, *13*, 924. [CrossRef] [PubMed]
15. Kania, H.; Sipa. Microstructure Characterization and Corrosion Resistance of Zinc Coating Obtained on High-Strength Grade 10.9 Bolts Using a New Thermal Diffusion Process. *Materials* **2019**, *12*, 1400.
16. Ryl, J.; Cieslik, M.; Zielinski, A.; Ficek, M.; Dec, B.; Darowicki, K.; Bogdanowicz, R. High-Temperature Oxidation of Heavy Boron-Doped Diamond Electrodes: Microstructural and Electrochemical Performance Modification. *Materials* **2020**, *13*, 964. [CrossRef] [PubMed]

materials

MDPI

Review

Application of Electrochemical Atomic Force Microscopy (EC-AFM) in the Corrosion Study of Metallic Materials

Hanbing Chen [1], Zhenbo Qin [2], Meifeng He [1,*], Yichun Liu [3] and Zhong Wu [2,*]

1 School of Materials Science and Engineering, University of Shanghai for Science and Technology, Shanghai 200093, China; chb19940420@163.com
2 Key Laboratory of Advanced Ceramics and Machining Technology (Ministry of Education), Tianjin University, Tianjin 300072, China; qinzhb@tju.edu.cn
3 School of Materials Science and Engineering, Kunming University of Science and Technology, Kunming 650093, China; spsjtu@163.com
* Correspondence: hmf752@usst.edu.cn (M.H.); zhong.wu@tju.edu.cn (Z.W.)

Received: 27 December 2019; Accepted: 26 January 2020; Published: 3 February 2020

Abstract: Electrochemical atomic force microscopy (EC-AFM), a branch of a scanning probe microscopy (SPM), can image substrate topography with high resolution. Since its inception, it was extended to a wide range of research areas through continuous improvement. The presence of an electrolytic cell and a potentiostat makes it possible to observe the topographical changes of the sample surface in real time. EC-AFM is used in in situ corrosion research because the samples are not required to be electrically conductive. It is widely used in passive film properties, surface dissolution, early-stage corrosion initiation, inhibitor efficiency, and many other branches of corrosion science. This review provides the research progress of EC-AFM and summarizes the extensive applications and investigations using EC-AFM in corrosion science.

Keywords: EC-AFM; corrosion; metallic materials

---

1. Introduction

As is known, all materials have a certain service life and will suffer various forms of direct or indirect damage during use. Although the material will be damaged in a variety of forms, corrosion is the most important and common form, which is a gradual process and cannot be restored. The problem of material corrosion occurs in various fields of the national economy, from daily life to industrial and agricultural production, as well as from advanced science and technology to the development of the national defense industry. Corrosion can lead to huge economic losses and even catastrophic accidents, which not only consume a large amount of resources and energy but also pollute the environment, resulting in a huge loss in the gross domestic product (GDP) [1]. Therefore, it is necessary to clarify the failure mechanism of materials in corrosive environments and take reasonable protective measures to achieve the purpose of preventing or controlling corrosion, to improve the life cycle of materials.

Corrosion is the irreparable damage or deterioration of materials caused by chemical, electrochemical, and physical effects of environmental media. In processes of metallic material corrosion, a chemical or electrochemical multiphase reaction occurs on the surface or interface of the metal, resulting in the conversion of the metal to an oxidized (ionic) state. Corrosion of metallic substance is started via the oxidation of metals.

$$M \rightarrow M^{n+} + ne^{-}, \quad (1)$$

$$M + (x + y)H_2O \rightarrow MO_x(OH)_y + (2x + y)H^{+} + (2x + y)e^{-}. \quad (2)$$

In order to maintain the neutral condition, a counter cathodic reaction of the electrolyte occurs [2].

$$2H_2O + 2e^- \rightarrow H_2 + 2OH^-, \quad (3)$$

$$2H^+ + 2e^- \rightarrow H_2, \quad (4)$$

$$O_2 + 2H_2O + 4e^- \rightarrow 4OH^-, \quad (5)$$

$$O_2 + 4H^+ + 4e^- \rightarrow 2H_2O, \quad (6)$$

$$M^{n+} + ne^- \rightarrow M. \quad (7)$$

The basic process of metallic corrosion in an aqueous solution consists of the anodic dissolution of metals and the cathodic reduction of oxidants. The redox reactions (Equations (1)–(7)) involve the transfer of electrons and ions between the metal and the solution. According to the corrosion kinetics, the anodic oxidation current of the metal degradation is equal to the cathodic reduction current of the oxidant at the corrosion potential. When the metal electrode potential is more positive, the rates of cathodic reactions increase and the rates of anodic reactions decrease accordingly. Conversely, as the metal electrode potential becomes more negative, the effect on the reactions is opposite.

The development of corrosion science is inseparable from the advancement of research methods and instruments. Conventional electrochemical measurements can only obtain macroscopic electrochemical information on the surface of the material. In situ electrochemical scanning probe technology with high spatial resolution facilitated the development of corrosion science, elucidating the microstructure and dynamic properties of materials and interfaces at the molecular/atomic level.

The electrochemical atomic force microscope (EC-AFM) was developed in 1991 on the basis of the atomic force microscope (AFM) [3]. It is well known that AFM is a kind of scanning probe microscope (SPM) and an extension of the scanning tunneling microscope (STM) [4]. The STM was invented by Gerd Binnig and Heinrich Rohrer in the 1980s, and it was able to image individual atoms for the first time. In 1982, Binnig and co-workers indicated that vacuum tunneling with an externally controllable tunnel distance is feasible, even under the conditions of room temperature and non-ultra-high vacuum, which was the first step in the development of scanning tunneling microscopy [5]. The 7×7 reconstruction on Si(111) observed in real space with STM was considered the first scientific success [6]. G. Binnig and H. Rohrer went into detail about STM, and they were rewarded with the Nobel Prize in Physics for their work in the field of STM in 1986 [7,8]. At present, STM is still an important method in surface science. However, due to the operation principle of STM being based on the quantum tunneling effect, STM has a severe restriction that requires the samples to be conductive. STM can only directly observe the surface structure of conducting and semiconducting materials. In order to overcome the shortcomings of STM, the AFM was invented by Binnig, Quate, and Gerber in 1986 [4]. The AFM can image the surfaces of a flat sample through the weak interaction force (atomic force) between the probe (mounted to a cantilever) and the sample, which is measured by monitoring the deflection of the microcantilever. Therefore, the AFM is suitable for both conductive and non-conductive samples, and its application field is more extensive. In addition, the AFM is a high-resolution microscope with atomic-scale resolution, which developed into a powerful micro/nanometer-scale surface analysis technique [9,10]. On the basis of different capabilities, a large number of techniques were developed after the invention of AFM, such as Kelvin probe force microscopy (KPFM), magnetic force microscopy (MFM), scanning electrochemical microscopy/atomic force microscopy (SECM–AFM), electrostatic force microscopy (EFM), and EC-AFM. These technologies are not only imaging tools; they can also accurately and quantitatively measure local physical and chemical phenomena.

EC-AFM extends AFM technology to the field of electrochemistry for in situ studying of the solid–liquid interface. The main difference between EC-AFM and AFM in liquids is that the applied potential of the sample with respect to the reference electrode is controlled by a potentiostat [11]. Topographic changes of the sample are achieved by measuring the force between the tip and the substrate in a three-electrode electrochemical cell composed of the counter electrode, reference electrode, and working electrode (substrate). EC-AFM is applied to in situ electrodeposition [12,13], biological science [14], supercapacitors [15], batteries and electrodes [16], corrosion and protection, and so on. After the invention of EC-AFM, its electrolytic cell, imaging mode, and probe performance were continuously improved, which opens a new prospect for the understanding of the corrosion mechanism [17,18].

In this review, we illustrate the application of EC-AFM to metal corrosion with specific examples, including understanding the properties of passive films, surface dissolution, pitting corrosion, selective corrosion, intergranular corrosion, inhibitor efficiency, coating protection, and other branches of corrosion science. The main purpose of this review is to provide the present situation and trend of corrosion research on various metal materials by EC-AFM and outline future prospects for the technique.

2. Principle and Operation Modes of EC-AFM

The basic working principle of EC-AFM is similar to that of AFM, and it shares its key components. Primarily, a nanoscale sharp tip located at the end of the cantilever of the probe is used to feel the interaction force between the tip and the sample, a piezo scanner is used for controlling the movement of the tip or sample in the x-, y-, and z-directions, a feedback control system and feedback circuit are used to control the deflection of the cantilever, and a computer system is used to display the results and control the operational parameters [19,20]. In addition, EC-AFM requires an electrochemical cell that can accommodate the working electrode (WE), counter electrode (CE), and reference electrode (RE), as well as a potentiostat for normal potentiostat control in a three-electrode system. Generally, the electrochemical cell is made from chemical-resistant polycarbonate, which can be used with a wide variety of liquids. Eight-degree nose assemblies are recommended for imaging in liquid because the smaller angle considers the different angles at which the laser goes in and out of the fluid, compared to operation in air. The electrochemical cell typically contains retaining clips, an O-ring gasket, and a liquid cell plate. When assembled, the sample itself comprises the bottom of the liquid container. Therefore, the sample must be large enough for the O-ring to seat. The WE should be relatively small, and each point on the WE should be geometrically equivalent to the counter electrode CE, which ensures that the current and potential across the electrode are evenly distributed. Others are based on conventional electrochemical testing requirements. Figure 1 shows the diagram of a typical electrochemical AFM cell with a potentiostat. During the scanning, the sample to which the potential is applied is placed in the electrolytic cell as the working electrode and scanned with the tip mounted on the cantilever. The force between the tip and the sample is measured by monitoring the deflection of the cantilever. By plotting the deflection of the cantilever versus its position on the sample, a topographic image of the sample is obtained [21].

The operation mode of EC-AFM is also similar to that of AFM, and the instrument can be operated either in constant height mode or constant force mode. Based on the interaction force between the tip and the sample, it is generally divided into different operation modes in the imaging process, such as contact mode, non-contact mode, intermittent tapping (tapping mode), torsional resonance mode, and peak force tapping mode. The first is the contact mode, in which the tip of the probe is always in contact with the surface of the sample, and the force between the tip and the sample is a repulsive force. Since the silicon nitride cantilever is soft enough to be deflected and has a high resonance frequency to avoid vibration instability, a silicon nitride probe is often used in contact mode. The contact mode has the advantage of high resolution and high scan speed, but it may cause damage to the surface of the sample. Both the transverse shear force and the capillary force on the surface may have adverse effects [22]. The second type is the non-contact mode, in which the cantilever vibrates above the sample surface, and the distance between the tip and the sample is usually several nanometers [23]. The force between the tip and the sample is van der Waals attraction. In the non-contact mode, there is no damage to the surface of the sample and the lateral force is the smallest, but the resolution is low and the scanning speed is slow. In order to avoid being stuck to the water film on the sample surface, it is often used to scan hydrophobic surfaces and soft samples. However, the development and improvement of this device in recent years enabled non-contact mode to detect repulsion and obtain atomic resolution images. Miyata used a non-contact-mode AFM to observe the dissolution of calcite in water at high speed [24]. This may well be applicable to the observation of corrosion processes on metal surfaces. The third type is the intermittent tapping (tapping mode), in which the cantilever oscillates at a frequency close to its resonance frequency and the oscillation amplitude is monitored [25]. The cantilever oscillates near its resonant frequency above the surface of the sample, and the probe contacts the sample surface once at the bottom of the oscillation during each vibration period. The difference with the non-contact mode is that the amplitude (amplitude > 20 nm) of the cantilever is larger than that of the non-contact mode (amplitude < 10 nm). It not only reduces damage to the sample and reduces the effects of transverse force, but it also has a higher lateral resolution on most samples. The disadvantage is that the scanning speed is limited.

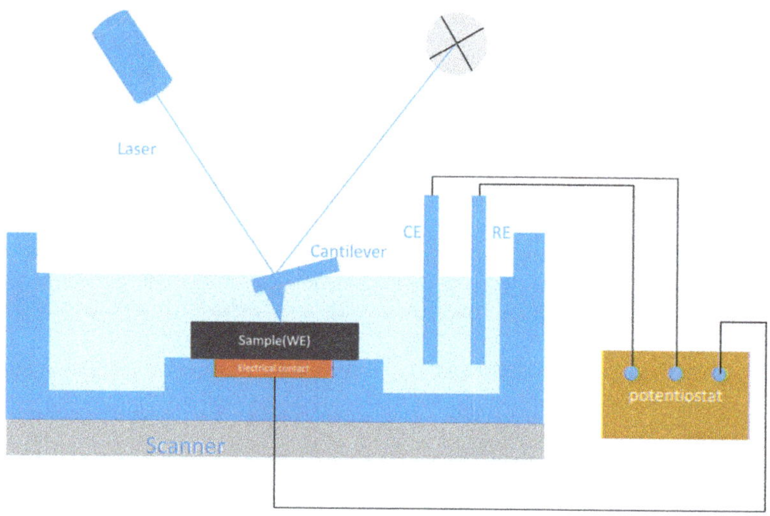

Figure 1. The schematic diagram of electrochemical atomic force microscopy (EC-AFM) electrolytic cell.

In addition, in order to further understand the properties of nanomaterials, a torsional resonance mode (TRmode) AFM was developed, which can measure both vertical and lateral force concurrently [26]. In the torsional resonance mode, the cantilever conducts torsional vibration with the long axis as the center and causes the tip to be in the vibration state. When the probe encounters transverse force on the surface, the system can detect the change in the cantilever torsion vibration and detect the fluctuation in the surface topography of sample [27]. It should be noted that the tapping mode applies a compressive force and the TR mode applies a torsional force; thus, the normal and shear performance is measured in the tapping mode and the TR mode, respectively [28]. Another newly developed operation mode is the peak force tapping mode, in which the tip is oscillated periodically in the vertical direction at a frequency well below cantilever resonance, tapping the sample until the maximum repulsion force reaches the setpoint in each tap [29]. The z piezo is driven with sinusoids rather than a triangular waveform in conventional force–distance (F–D) curves [30]. This mode directly controls the interaction between the tip and the sample, reducing the depth of deformation and the corresponding contact area, minimizing damage to the probe or sample [31,32]. Moreover, the force–distance interactions can be measured directly through peak force tapping mode.

Table 1 summarizes the operation mode of EC-AFM with the necessary parameters added. In the field of corrosion, contact mode and intermittent tapping (tapping mode) are mainly used.

Table 1. Comparison of common operation modes of EC-AFM.

Mode	Probes (Nominal Spring Constants)	Cantilever	Tip–Sample Distance	Force	Scan Rate
Contact mode	Silicon nitride probes (less than tapping and non-contact mode)	The cantilever is deflected: constant force or constant height	<0.5 nm	Repulsive 10^{-9}–10^{-6} N	High scan speeds
Non-contact mode	Silicon probes (20–100 N/m)	The cantilever is oscillated (amplitude < 10 nm)	1–10 nm	Attractive 10^{-12} N	Slower scan speed than tapping and contact mode
Intermittent tapping (tapping mode)	Silicon probes (2–50 N/m)	The cantilever is oscillated (amplitude > 20 nm)	0.5–2 nm (intermittent contact)	Both repulsive and attractive forces	slower scan speed

3. Application of EC-AFM in Metal Corrosion and Corrosion Protection

3.1. The Micro-Area Corrosion

3.1.1. The Corrosion Product Film

EC-AFM can characterize the real-time nanoscale topographic changes of a corrosion product film on a metal surface under different potentials in aqueous environments, which avoids the potential contamination and degradation of the corrosion product film during conventional ex situ characterization and analysis techniques.

A typical example of studying corrosion product films with EC-AFM was presented by Li and co-workers, who studied the surface passivation film and its substructure of carbon steel in a carbonate/bicarbonate solution by EC-AFM in contact mode [33]. They designed a solution container for the EC-AFM, which was large enough to accommodate the three-electrode system and avoid evaporation during testing. The topographic characterization of the sample surface was performed at the passivation potentials of −0.1 $V_{(SCE\text{—saturated calomel electrode})}$, 0.5 $V_{(SCE)}$ and 0.7 $V_{(SCE)}$. The results showed that, at −0.1 $V_{(SCE)}$ and 0.5 $V_{(SCE)}$, as the passivation time increased, the nanoscale features, i.e., the scale-like spots on the surface of the sample, increased in diameter, indicating the growth of the passive film. When the grooves caused by the surface treatment were not observed, a thick passivation film was formed on the surface of the sample. The surface roughness calculated from the topographic profile was small and changed little, indicating that the passive film was relatively uniform. When the potential reached 0.7 $V_{(SCE)}$, the roughness increased and the passive film became non-uniform, as shown in Figure 2.

In addition, based on the morphology, they found that the passive films contained a mixture of Fe_3O_4, Fe_2O_3, and FeOOH when passivated at the active–passive transition potential. The inner layer of the passive film was Fe_3O_4 and the outer layer was Fe_2O_3/FeOOH. When the film-forming potential changed from −0.1 $V_{(SCE)}$ to 0.5 $V_{(SCE)}$, the chemical composition did not change, but the thickness of the inner layer became thicker, leading to an increase in the thickness of the oxide film from about 4.5 nm to 5.8 nm, and the compactness improved, which made the film more protective. On the contrary, when the potential was at 0.7 $V_{(SCE)}$, although the thickness of the film increased, its chemical composition changed greatly and an amorphous structure appeared, which reduced the corrosion resistance of the film.

Padhy et al. [34] investigated the passive film properties of 304L stainless steel in the nitric acid medium in both ex situ and in situ conditions. The ex situ study on the surface morphology of the passive film showed that the surface morphology of 304L stainless steel depended on the stability of the passive film in nitric acid. The morphology features of the passive film in nitric acid medium increased with time and solution concentration. The passive film in 1 M and 4 M nitric acid solution was stable. However, in 8 M and 11.5 M nitric acid, breakdown of the passive film was observed. In situ surface morphology changes of the passive film were monitored by EC-AFM in 0.1 M, 0.5 M, 0.6 M, and 1 M nitric acid. It was found that, at lower concentrations of 0.1 M and 0.5 M nitric acid, the passive film grew in a platelet-like structure. At nitric acid concentrations of 0.6 M and 1 M, platelet-like structures aggregated, homogenized, and began to deplete from the surface, causing selective dissolution. The results of X-ray photoelectron spectroscopy (XPS) analysis showed that the passive film at lower concentration was composed of a hydroxide and oxide layer. However, at high concentrations, the passive film consisted only of an oxide layer. The passivity of 304L stainless steel under the action of low-concentration nitric acid started from the formation of chromium hydroxide, and the surface was in the form of platelet-like structures. As the concentration increased, the hydroxide layer changed to a homogenous oxide layer. As the concentration continued to increase, the protective oxide film was depleted from the structural heterogeneous zone, leading to the opening of the oxidation boundary. The depletion of the oxide layer caused selective dissolution and local corrosion of 304L stainless steel.

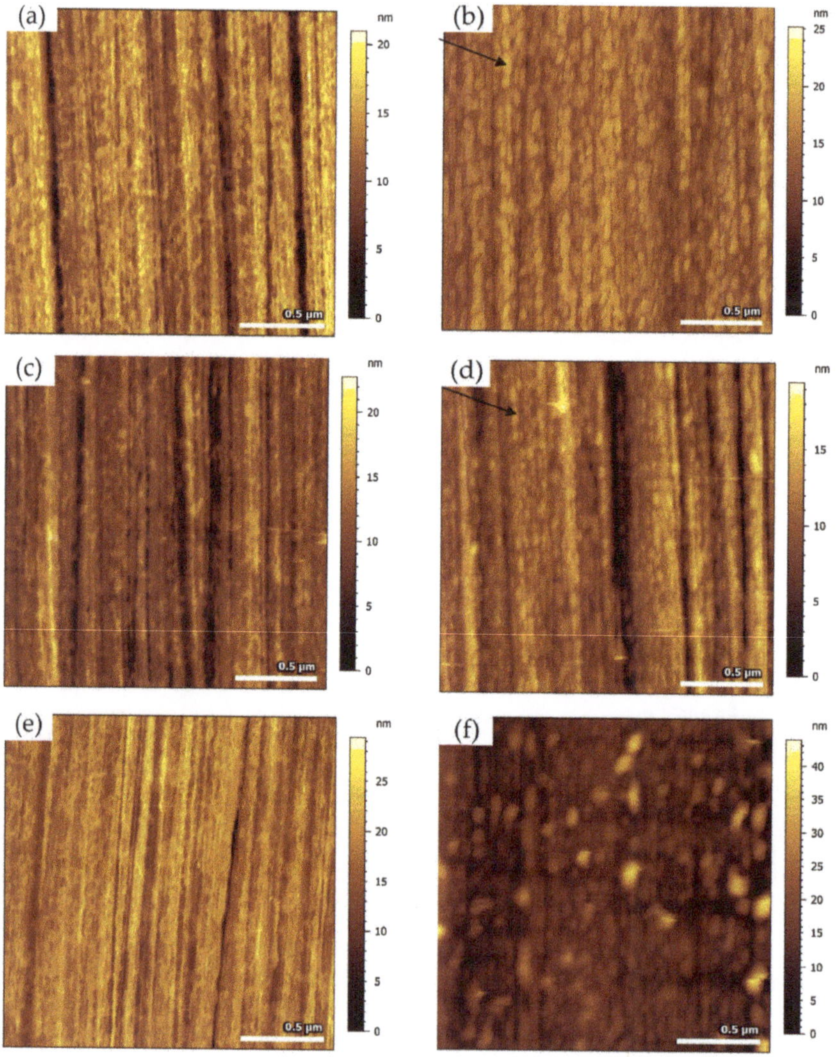

Figure 2. The AFM images of the steel specimen polarized at −0.1 $V_{(SCE—saturated\ calomel\ electrode)}$ for (**a**) 0.5 min and (**b**) 60 min, at 0.5 $V_{(SCE)}$ for (**c**) 0.5 min and (**d**) 60 min, and at 0.7 $V_{(SCE)}$ for (**e**) 0.5 min and (**f**) 60 min. The black arrow points to scale-like spots. Adapted with permission from Reference [33]; copyright 2017 Elsevier Ltd.

Many researchers reported the corrosion process and mechanism of AA 2024 and identified the corrosion products. Kreta et al. [35] explored the changes on the surface during exposure to the sodium chloride environment. In their experiment, the development of the passive film of AA 2024-T3 was analyzed with different analytical tools during exposure to 0.5 M NaCl. In order to track the morphology changes of the protective hydroxide/oxide film, the in situ EC-AFM measurements were carried out on the site without inclusions in a tapping mode. In situ EC-AFM images were taken from samples exposed to different potentials in various time steps. The parameters such as roughness are not only a function of potential, but also a time-dependent function. It can be observed that the surface roughness gradually increased during immersion in sodium chloride solution, and there was a similar increase in surface roughness after the application of external potential, which was mainly due to the formation of an oxyhydroxide layer. In addition, the deposition of the corrosion products led to the formation of a spiky surface, as shown in Figure 3. The profile lines measured over the corroding intermetallic particle are shown in Figure 3e, which allows observing the formation and roughness changes of the corrosion product film. The further increase in potential was most probably responsible for the significant increase in the thickness of the passivation layer. Under a certain potential, the corrosion potential spikes formed but disappeared after the thicker passivating oxyhydroxide layer was formed.

In terms of biomaterials, the inflammatory state of local body fluids in the body may affect the corrosion of implanted metallic biomaterials such as CoCrMo alloys. In the study of Liu and Gilbert, the surface of CoCrMo alloy was imaged by in situ EC-AFM to evaluate the morphology of the oxide film and surface dissolution under different physiologically possible potentials in phosphate buffer saline (PBS) solution and simulated inflammation (SI) solution (PBS solution with 30 mM H_2O_2), respectively [36]. EC-AFM images in SI solution showed different topographic changes compared with PBS solution alone, in which the alloy matrix was corroded and the carbide protruded. In contrast, a less compact oxide layer appeared and gradually covered the outermost surface in the case of PBS solution. The authors speculated that the addition of H_2O_2 changed or removed the passive oxide films and greatly affected anodic dissolution on CoCrMo alloy. In addition, the grain boundaries and carbide boundaries were preferential parts for the dissolution of oxides, which was mainly due to effects of Cr depletion, a heterogeneous structure, and Cr distribution along grain boundaries and carbide boundaries.

In the work of Bearinger and co-workers, the effects of different potentials and hydration on the properties and structure of passive oxide films on the surface of titanium, 6-aluminum, 4-vanadium (Ti-6Al-4V) were investigated by in situ EC-AFM [37]. The AFM imaging of the different surfaces showed that all samples were covered with a protective titanium oxide dome. Due to hydration, the dome area gradually increased, and coalescence occurred with the increase in applied voltage and time. In addition, Bearinger et al. [38] characterized hydration of titanium/titanium oxide surfaces under freely corroding and potentiostatically held conditions using EC-AFM. In contrast to conventional high-vacuum techniques, EC-AFM enables the measurement of morphological surface structures in the in situ hydrated state. The results showed that the titanium surface covered the oxidized dome and grew laterally during hydration. The applied potential altered the growth rate. Under open-circuit potential, growth proceeded approximately six times faster than under −1 V applied voltage. The oxide growth was partly due to the lateral expansion and overgrowth of the dome at the oxide–solution interface. The method was successfully used to study dynamic changes in the surface morphology.

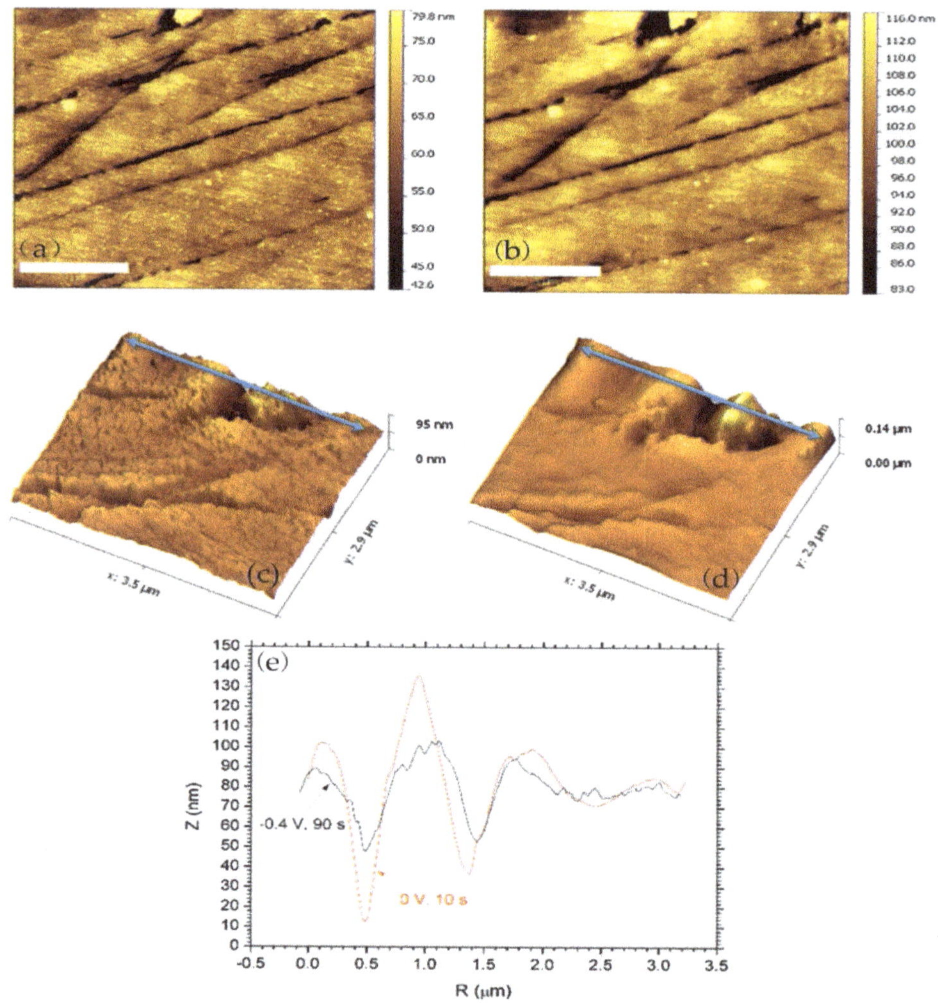

Figure 3. The in situ AFM images of AA 2024: (**a**) 226 min, −0.4 V/Ag/AgCl/KCl_{sat}, 90 s; (**b**) 328 min, 0 V/Ag/AgCl/KCl_{sat}, 10 s. (**c**,**d**) The corroding sites on top of the images (**a**,**b**) as three-dimensional (3D) images. (**e**) The profile lines measured over the corroding intermetallic particle. Reprinted with permission from Reference [35]; copyright 2016 Elsevier Ltd.

Metallic glass has excellent mechanical, magnetic, and chemical properties due to its unique structure. It is noted that the corrosion resistance of metallic glass is not only sensitive to its chemical composition but also inseparable from the structure. The influence of the structure on the corrosion of metallic glass is closely related to the dissolution, passivation, and durability of the passive film. Wang et al. [39] selected $Ni_{50}Nb_{50}$ metallic glass with polymorphic transition during devitrification to clarify the relationship between the amorphous structure and corrosion, especially the stability of the passive film. By comparing amorphous alloys with its crystalline alloys, the breakdown behavior of passive films and characteristics of the surface film were studied in detail. The surface topography of the samples was characterized by in situ EC-AFM in tapping mode. The results confirmed that the surface of both amorphous and crystalline samples could form metastable pits at nanoscale, but the number of nanometer pits on amorphous samples was much smaller than that on crystalline samples.

Moreover, amorphous alloys only had a few deep pits and no shallow pits, as shown in Figure 4a,c, while crystalline alloys had many deep pits and shallow pits, as shown in Figure 4b,d. Although the composition of the passive film on the surface of the two alloys was similar, the amorphous structure could significantly inhibit the formation of pits in corrosive environment. Therefore, amorphous alloys had stronger resistance to pit formation than crystalline alloys.

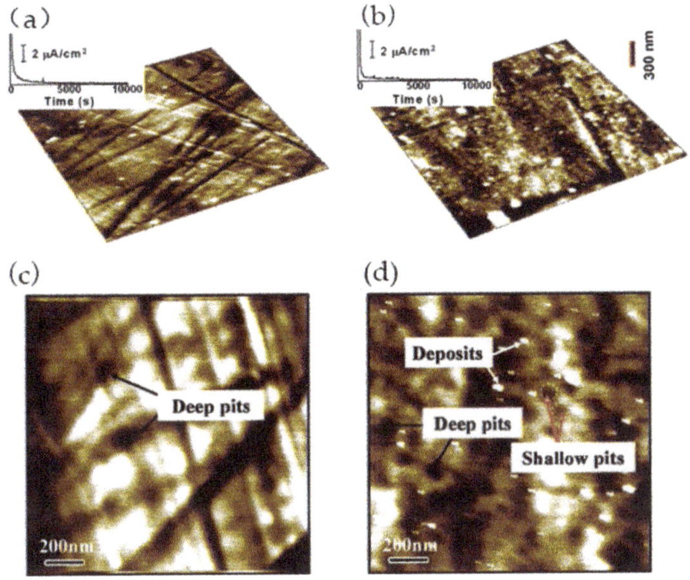

Figure 4. AFM imagines of the surfaces of amorphous and crystallized Ni50Nb50 alloys after polarization at 1.0 V in 1 mol/L HCl solution. (**a**,**b**) Surface morphologies of the amorphous and crystallized samples. Inserts are the corresponding I–t curves during potentiostatic treatments. (**c**,**d**) High-magnification images of selective areas in amorphous and crystallized samples. Adapted with permission from Reference [39]; copyright 2010 Elsevier Ltd.

Phase imaging is an extension of the conventional tapping mode AFM (TM-AFM), through which topographic images and phase images can be obtained simultaneously. The phase images can reflect local nanoscale mechanical properties such as friction, hardness, and viscoelasticity. During the AFM scanning process, different nanometer-scale crystalline phases and their size, shape, and distribution can be identified in the matrix according to the phase change. Zhang et al. [40] studied the formation of the passive film in the Al–Ni–Ce metallic glass by in situ EC-AFM. In situ EC-AFM measurements were performed in open-circuit potential (OCP) conditions and stepped the potential into pitting potential (E_{pit}). They identified α-Al nanophase from the amorphous matrix by TM-AFM. It can be clearly seen from the phase image that the small α-Al islands were randomly distributed on the surface, which had a negative phase shift with respect to the matrix. In the partially magnified phase image, the α-Al island is surrounded by bright ringed patches, as shown in Figure 5.

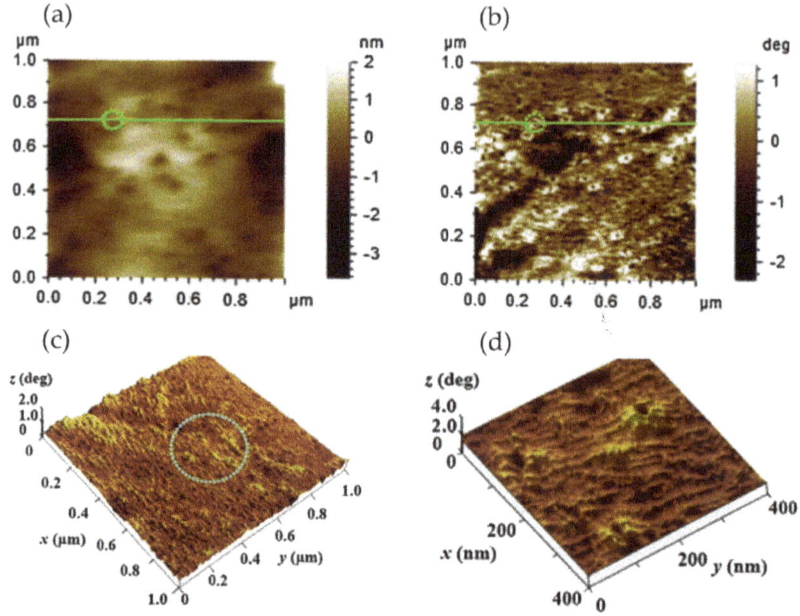

Figure 5. In situ tapping mode (TM) AFM images of the annealed $Al_{88}Ni_{8}Ce_{4}$ amorphous nanocrystalline sample at open-circuit potential (OCP) held for 3 min in 0.01 mol/L NaCl solution. Topography image (**a**) and phase image (**b**). Three-dimensional (3D) images of phase (**c**) and the (**d**) local enlarged phase image of the marked circle in (**c**). Reprinted with permission from Reference [40]; copyright 2014 Elsevier Ltd.

In addition, the nucleation and growth process of the passive film was observed in real time by in-situ EC-AFM, and it was found that the presence of nanocrystalline α-Al precipitates changed the nucleation mechanism of the passive film in the amorphous matrix from instantaneous nucleation to progressive nucleation. Due to the galvanic coupling between the α-Al nanophase (anodic position) and surrounding amorphous matrix (cathodic position), the formation of the corrosion product $Al(OH)_3$ was in the early stage of formation of the passivation film, which was mixed into the passive film, changing the local structure and composition of the passive film. The density of hydroxide in the passive film was lower than that of the oxidation film formed on the amorphous matrix, which reduced the compactness of the passive film and its stability. Therefore, compared with fully metallic glass, metallic glass containing nanometer α-Al precipitation had lower corrosion resistance.

Most previous research focused on steady-state corrosion reactions of steel, such as the study of pipeline corrosion in bicarbonate solutions or carbonate bicarbonate electrolytes. However, some researchers also studied the early stage of steel corrosion. Li et al. discussed the influence of bicarbonate concentration on the topographic evolution and corrosion mechanism [41]. In their work, the early stage of X100 pipeline steel corrosion in bicarbonate solution with different concentrations was characterized by in situ EC-AFM in the contact mode. The relationship between surface roughness and the corrosion process of pipeline steel was established. The pipeline steel was fixed to the bottom of the homemade solution container as the working electrode, which prevented evaporation of the test solution during testing. The authors found that the early stage of the steel corrosion in 0.01 M $NaHCO_3$ solution could be divided into three stages. When the steel was immersed in the solution, the oxides on the surface of the steel began to dissolve, the OCP dropped, and the surface roughness increased rapidly, representing stage I. When the steel was corroded, the OCP further dropped, and the surface roughness increased gradually, representing stage II. In stage III, as the corrosion of steel reached a steady state, the surface roughness and OCP maintained a stable value, and the formation of corrosion products reached a dynamic equilibrium state. However, in solutions with increased bicarbonate concentration, such as 0.1 M and 0.5 M $NaHCO_3$, steel could be passivated. In addition, as the passive capacity of bicarbonate solution increased with the increase in concentration, the morphology of the steel surface in 0.5 M $NaHCO_3$ solution was smoother than that in the 0.1 M solution, consistent with the results of power spectral density (PSD) analysis. Compared with the passive film formed at a low concentration (0.1 M), the passive film formed in a solution with a high concentration (0.5 M) could eliminate the large-scale morphological features.

In addition, AFM–SECM is considered to be a particularly attractive research tool in corrosion science [42]. The clear topographical information obtained by EC-AFM and the electrochemical information provided by SECM can be obtained in a single experiment, which makes up for the deficiencies of the two instruments. Izquierdo et al. [43] modified AFM tips to achieve a bifunctional AFM–SECM tip. Conductive microelectrodes were integrated into an AFM probe to obtain an SCEM signal, through which the changes in surface morphology and current density were achieved to observe the progress of corrosion. Local release of Cu^{2+} ions was monitored by electrochemical reduction and deposition of metal ions on the AFM–SECM probe. The formation and breakdown of passive layers were characterized from the perspective of surface roughness and current density, reflecting the initial growth of the passive layer, subsequent breakdown, and formation of the pit.

In the research on microzone corrosion carried out by our research group, Ding studied the corrosion behavior and the formation of **a** corrosion product film on nickel–aluminum bronze (NAB) in 3.5 wt% NaCl solution. The in situ AFM in contact mode at open-circuit potential was used to observe the formation of the corrosion product film on different phases [44]. Figure 6 shows the in situ AFM topography of the specimen surface.

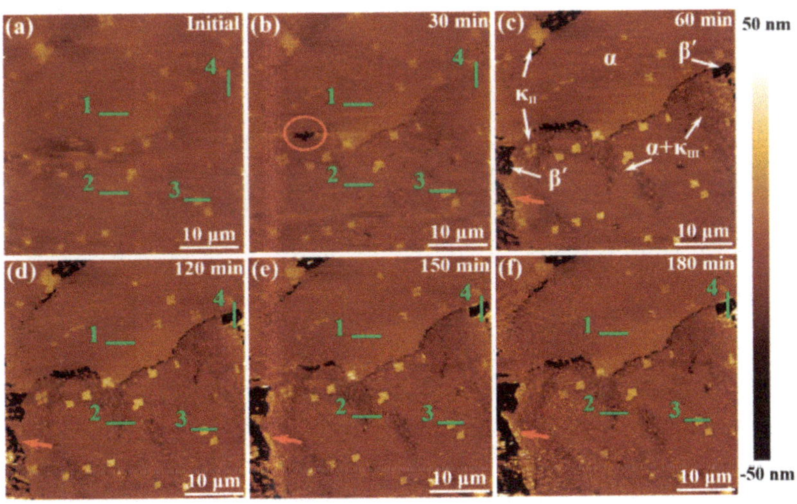

Figure 6. In situ topography images of nickel–aluminum bronze (NAB) specimen surface after exposure to 3.5 wt% NaCl solution at different times: (**a**) initial; (**b**) 30 min; (**c**) 60 min; (**d**) 120 min; (**e**) 150 min; (**f**) 180 min. Site 1 corresponds to the α phase, site 2 corresponds to the $\alpha + \kappa_{III}$ eutectoid structure, and site 3 corresponds to the κ_{II} phase, while site 4 corresponds to the β' phase. Reprinted with permission from Reference [44]; copyright 2019 MDPI.

The α phase exhibited different corrosion behavior at different locations. Serious corrosion occurred at the lamellar α phase within the $\alpha + \kappa_{III}$ eutectoid structure, while the α phase far away from the κ phase was not eroded, mainly due to the formation of many micro galvanic cells in the eutectoid. The κ_{II} and κ_{III} phases exhibited significant corrosion resistance due to the formation of a stable, dense protective film in a short period of time. It is worth noting that the film on the κ_{II} phase was thicker, related to the κ_{II} phase based on Fe_3Al intermetallic compounds and containing more iron. Compared with nickel oxide, iron oxide is fluffier and more unstable; thus, the film on the κ_{II} phase was thicker. Since the metastable martensite structure struggled to form a protective film, the β′ phase suffered the most severe corrosion. When the immersion time reached 150 min, the corrosion product deposition was almost dispersed on the entire surface, and the corrosion depth of each phase was suppressed, as shown in Figure 7. The formation of the corrosion product film tended to be uniform, which prevented the NAB alloy matrix from contacting the corrosive medium and hindered the transport of ions and charges, as well as improved the corrosion resistance of the NAB alloy.

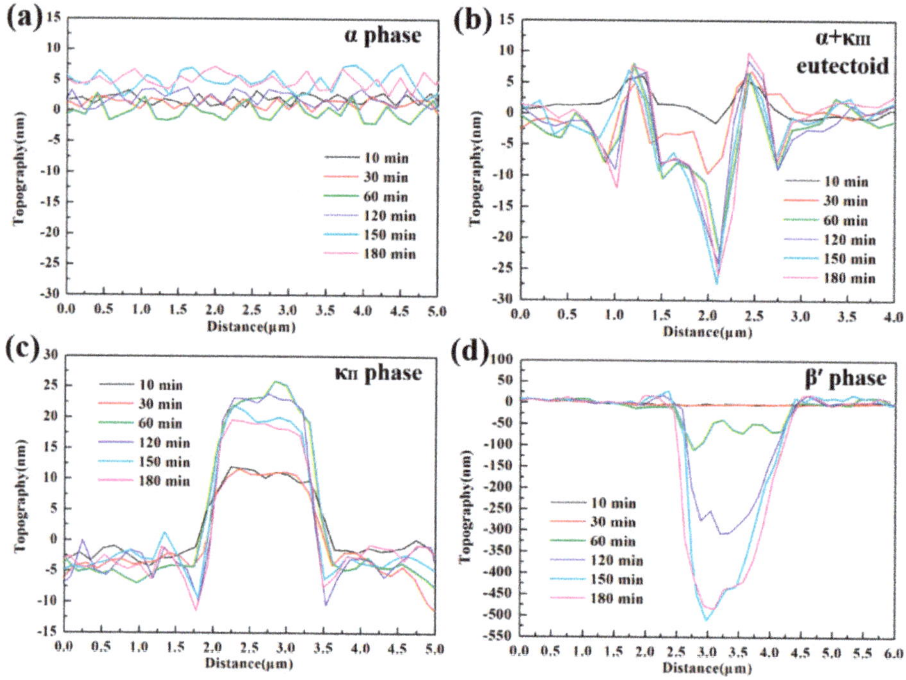

Figure 7. In situ line profiles corresponding to sites 1, 2, 3, and 4 marked in Figure 6, respectively: (**a**) α phase; (**b**) α + κ_{III} eutectoid structure; (**c**) κ_{II} phase; (**d**) β′ phase. Reprinted with permission from Reference [44]; copyright 2019 MDPI.

EC-AFM has great advantages in studying the formation process and microstructure of corrosion product films on a metal surface, especially in characterizing the evolution of early corrosion product films, which is conducive to analyzing the corrosion behavior of metals and promoting the research on the corrosion resistance of metal. The application of EC-AFM in corrosion product films is summarized in Table 2.

Table 2. Summary of the application of EC-AFM in corrosion product films.

	Materials	Solution	Mode	Reference
Corrosion product film	carbon steel	Carbonate/bicarbonate solution	Contact mode	[33]
	304L stainless steel	Nitric acid		[34]
	AA 2024-T3	0.5 M NaCl solution	Tapping mode	[35]
	CoCrMo alloy	Phosphate buffer saline (PBS) and simulated inflammation (SI) solution	Contact mode	[36]
	Ti-6Al-4V	PBS	Contact mode	[37]
	Ni50Nb50 metallic glass	1 mol/L HCl solution	Tapping mode	[39]
	Al–Ni–Ce metallic glass	0.01 mol/L NaCl solution	Tapping mode	[40]
	X100 pipeline steel	0.01 M $NaHCO_3$ solution	Contact mode	[41]
	NAB	3.5 wt% NaCl solution	Contact mode	[44]

3.1.2. Pitting Corrosion

Pitting corrosion was extensively studied in the past few decades, as it has greater destructive and hidden dangers. Generally, the pitting mechanism is related to the properties of impurities on the surface or matrix. Pitting corrosion near inclusions, precipitates, or matrix/impurity interfaces is often caused by the inherent potential difference between the matrix and the inclusions or precipitates. Due to the small anode area, the corrosion rate is very fast, which can lead to sudden accidents. When trying to limit or avoid pitting corrosion, EC-AFM is helpful for improving the understanding of random pitting distribution, including nucleation and growth of pitting.

Pitting corrosion of various stainless steels was the focus of research for many years. As the most used material in the orthopedic and orthodontic bracket, AISI 316 L austenitic stainless steel has good mechanical properties. However, because it is susceptible to localized corrosion in chlorine-containing environments, it is often challenged by corrosive environments in the body. Conradi et al. [45] studied localized corrosion of austenitic stainless steel of the type AISI 316 L and duplex 2205 stainless steel (DSS 2205) using in situ EC-AFM in two different solutions, including simulated physiological solution known as Hank's solution (PS) and artificial saliva (AS). The topography and surface roughness of DSS 2205 and AISI 316 L changed with the increase in chloride ion concentration. EC-AFM topographic images showed that the surface of DSS 2205 had prominent square-like corrosion products in both solutions. Compared with AS, the erosion of PS was manifested by the decrease in the time scale of DSS 2205 surface change due to the increase in chloride ion concentration, as shown in Figure 8. Even though the DSS 2205 steel was exposed to anodic potentials in the region of transpassive oxidation, the samples had high pitting corrosion resistance in both solutions. On the contrary, AISI 316 L steel had stable corrosion resistance to AS. However, when AISI 316 L steel was exposed to PS solution, the sample was easily corroded and showed obvious pitting corrosion with the increase in Cl^- ion concentration, as shown in Figure 9. The surface of AISI 316 L included ellipse-like deposits. This was due to the change in the chemical composition of the matrix material, which formed different growth patterns on the surface of the sample with the growth of chromium, iron, and nickel oxides. Therefore, DSS 2205 had high corrosion resistance compared to AISI 316 L stainless steel. In medical applications, DSS 2205 will be a promising medical material if the nickel hypersensitivity effect can be reduced in patients receiving treatment.

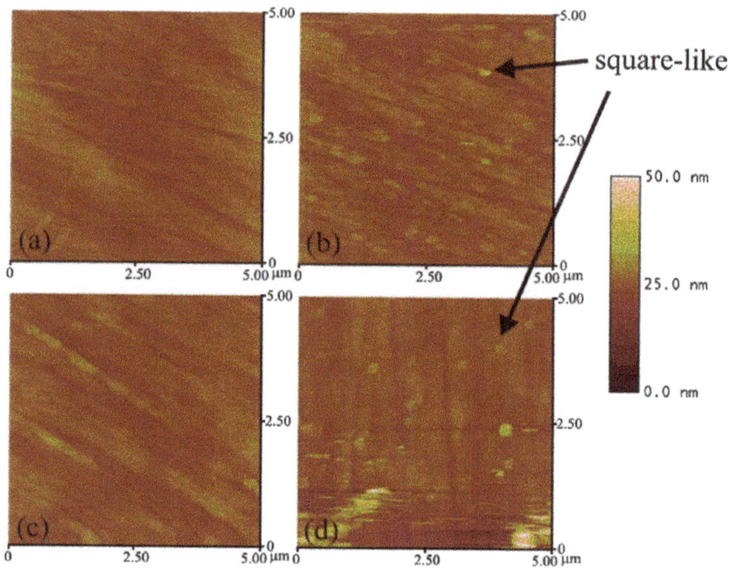

Figure 8. AFM images of DSS 2205 sample: (**a**) in artificial saliva (AS) after being exposed to the potential of 0.8 V for 45 min, (**b**) in AS after additional exposure to the potential of 1 V for 11.6 min, (**c**) in physiological solution (PS) after being exposed to a potential of 0.8 V for 17.2 min, and (**d**) in PS after additional exposure to 1 V for 7.5 min. Adapted with permission from Reference [45]; copyright 2011 Elsevier Ltd.

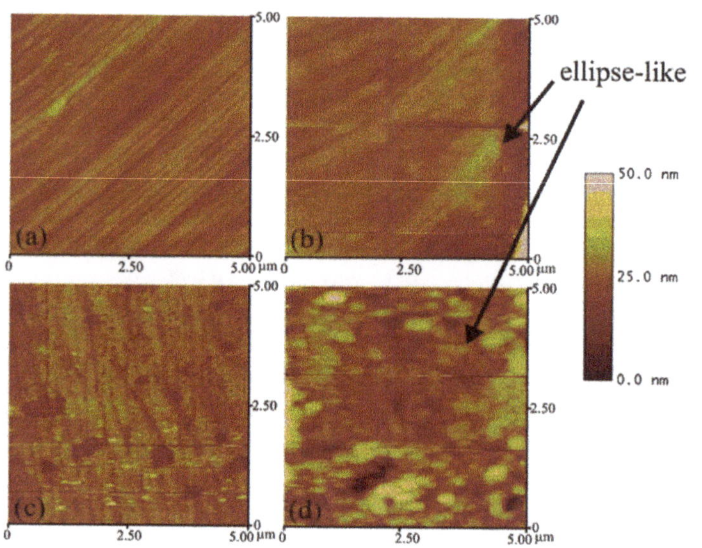

Figure 9. AFM image of AISI 316L sample: (**a**) in AS after exposing the sample to anodic potential of 0.5 V for 10 min, (**b**) in AS after additional exposure of the sample to 0.5 V for 6.6 min, (**c**) in PS immediately after the test cyclic voltammogram in the range of potentials from −0.5 V to 0.8 V, and (**d**) in PS after additional exposure to 0.5 V for 30 s. Adapted with permission from Reference [45]; copyright 2011 Elsevier Ltd.

Martin et al. [46] studied the pitting corrosion of austenitic 304L stainless steel in chloride borate buffer solution using in situ EC-AFM in contact mode. In order to determine whether the pits were randomly distributed at the nanometer scale, the study focused on the location where pits were initiated under controlled potential. In the AFM image, it is clear that the chain of pits (black, right image) was related to the chain of relief islands (white, left image), as shown in Figure 10. However, we can see an intermediate state rather than a typical direct transition, which indicates that the pitting process was a gradual process, as shown in Figure 11. In the final step of surface preparation, the local chemical reactivity of the surface may have led to the formation of a passive film and small oxidized hydroxide aggregate defects. The authors believed that the local chemical defects in the sample preparation process could affect the formation of the passive film, which would eventually cause local differences compared to the rest of the film. These local chemicals or structural defects would reduce the local pitting resistance of the film. In addition, based on the reasonable model that the surface potential under tensile stress would be higher than that under compressive stress, when the strain hardening zone (caused by mechanical polishing) appeared on the surface of the sample, pitting corrosion preferentially initiated in this area.

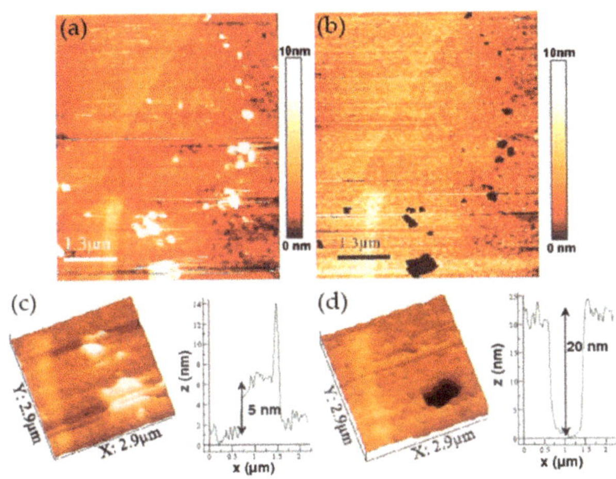

Figure 10. AFM images of the surface at open-circuit potential (**a**) and at pitting potential (**b**). (**c**,**d**) The pit chain (in black, right image) profiles made on images at higher magnification (bottom insets) show that the average relief of the islands was about 5 nm, whereas the average depth of the pits was about 20 nm. Reprinted with permission from Reference [46]; copyright 2008 Elsevier Ltd.

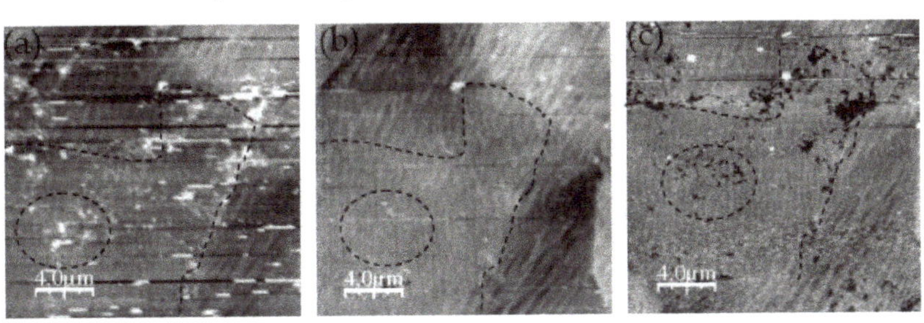

Figure 11. AFM images of the surface, taken at (**a**) open-circuit potential, (**b**) corrosion potential after cathodic scan, and (**c**) pitting potential. Reprinted with permission from Reference [46]; copyright 2008 Elsevier Ltd.

It is generally believed that sulfide inclusions (MnS and mixed oxide/sulfide) in a stainless-steel matrix are most likely to cause pitting corrosion. The products produced by the dissolution of sulfide inclusions form a local corrosive environment, which in turn causes more severe pitting corrosion. Wijesinghe et al. [47] discussed the adverse effect of sulfide inclusion on pitting resistance of stainless steel. The results showed that the sulfide inclusions were clustered on the surface of the stainless steel. As the volume composition of the stainless-steel sulfide increased, the number of inclusions per cluster increased. The pitting corrosion was imaged in real time using in situ EC-AFM, and it was found that pits were formed near the sulfide inclusions, consistent with the three main mechanisms proposed previously, i.e., aggressive local chemical reactions based on inclusion dissolution, stressed oxide, and chromium depletion.

In the work of Zhang and co-workers, the pitting corrosion of the solution- and sensitization-treated austenitic stainless steel SUS304 was studied using in situ AFM in 3.5 wt% sodium chloride solution [48]. This study adopted two observation methods. One was in situ continuous observation, in which the corrosion current was continuously applied to the surface of the sample. Another method was in situ interrupted observation, i.e., the observation was carried out at intervals to minimize the impact of probe scanning on corrosion reactions. In situ observation of the corrosion pit of the solution-treated sample showed that the pit became larger as the corrosion time increased from t = 2.46 to 2.70 ks, but no corrosion products were found, as shown in Figure 12. A series of in situ intermittent observations of the sample after solution treatment revealed two large pits (D and E), but the size of the pits did not change with time. In addition, no corrosion products were found on the two large pits, as shown in Figure 13.

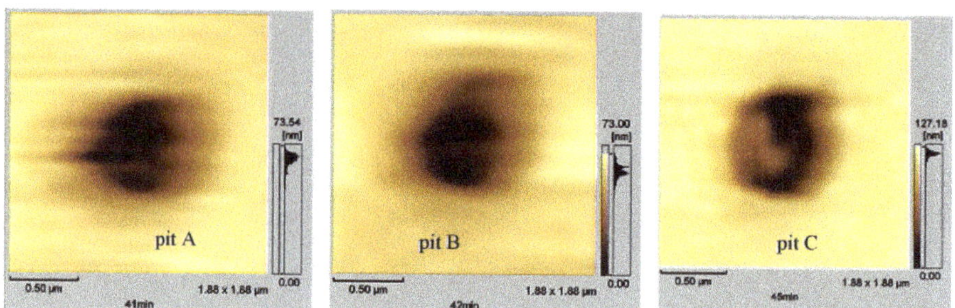

(a) (b) (c)

Figure 12. AFM images of solution-treated SUS304 stainless steel during corrosion in 3.5 wt% sodium chloride solution at 298 K (I = 10 A/m^2): (**a**) t = 2.46 ks; (**b**) t = 2.52 ks; (**c**) t = 2.7 ks. Reprinted with permission from Reference [48]; copyright 2005 Elsevier Ltd.

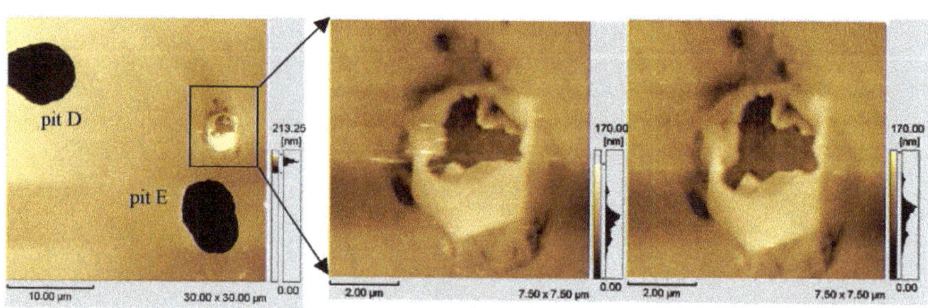

(a) (b) (c)

Figure 13. AFM images of solution-treated SUS304 stainless steel after corrosion in sodium chloride solution at 298 K (I = 5 A/m^2): (**a**) t = 1.5 ks; (**b**) t = 1.5 ks and (**c**) t = 7.2 ks are the magnified images of the framed area of (**a**), showing the presence of the corrosion product. Reprinted with permission from Reference [48]; copyright 2005 Elsevier Ltd.

The authors believed that the corrosion product may have been moved to other locations by the probe or dissolved with the rapid dissolution of the matrix. After the corrosion product was removed, the concentration of local chloride ions and hydrogen ions decreased, and the growth rate of the pits decreased. This means that the corrosion products played an important role in the growth of corrosion pits. When the corrosion product covers the pit, it can lead to an acidic environment and accelerate pitting. On the sensitization-treated sample, the irregular pits were distributed near the grain boundaries. Chromium carbide deposits and pitting occurred in the chromium-depleted area around the carbide, as shown in Figure 14. Cross-sectional profiles of pits along a–b lines in Figure 14a and c–f lines in Figure 14c are shown in Figure 14d. Since no carbide particles were found in the pit according to the cross-sectional profiles, they might have dissolved with pit growth. The dissolution of carbides in pits caused the pit to grow further.

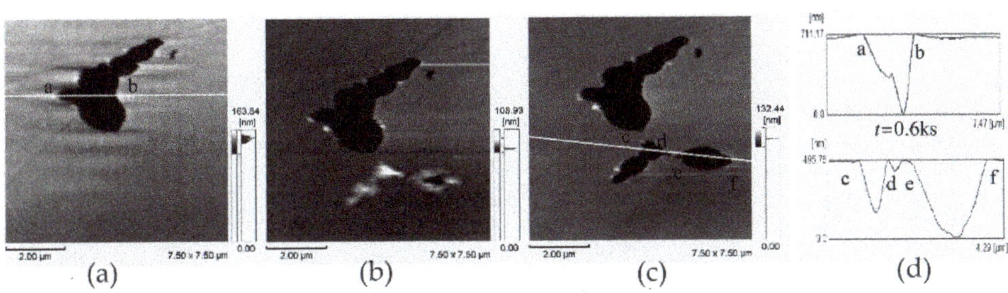

Figure 14. AFM images of sensitization-treated SUS304 stainless steel after corrosion in sodium chloride solution (I = 1 A/m^2): (**a**) t = 0.6 ks; (**b**) t = 0.9 ks; (**c**) t = 1.2 ks. (**d**) Cross-sectional profiles of pits along a–b and c–f lines. Reprinted with permission from Reference [48]; copyright 2005 Elsevier Ltd.

When we study the corrosion of self-passivation metals covered with a semiconductor protective film, there is no doubt that EC-AFM, which does not need the conductivity of the matrix, is more suitable. Qu et al. [49] investigated the corrosion behavior of pure aluminum using in situ AFM in 0.01 mol/L $FeCl_3$ solution. The pitting corrosion process of pure aluminum induced by the potentiodynamic sweep and the repassivation process of active pits were studied. In addition, the effect of mechanical damage on the metal surface caused by AFM tip scratching on the pitting behavior of the sample was emphasized. The topographical images of the sample surface at different immersion times were traced under open-circuit conditions using AFM in contact mode, and the AFM tip scratching process was carried out with a loading force of 800 nN. The results showed that different pitting regions exhibited different pitting activities under the same polarization conditions due to the diversity of physical and electrochemical characterization. The corrosion products contained abundant impure elements such as Fe and Cu. In situ AFM observation of pitting corrosion originated from artificial defects on the aluminum surface showed that physical defects had higher pitting activity and may have been attacked preferentially to pitting corrosion.

In the study of Davoodi and co-workers, the application of the EC-AFM and SECM integrated system for in situ studies of the influence of intermetallic particles on local corrosion of aluminum alloys was introduced [50]. The key to this method is fabricating a dual-mode cantilever/tip that can be used not only as a cantilever for the EC-AFM but also as a microelectrode or nano-electrode tip for the SECM. It can obtain the in situ AFM topographic images and SECM electrochemical current maps simultaneously with micrometer lateral resolution and provide detailed information of localized corrosion related to different kinds of intermetallic particles and the deposition of corrosion products. In their work, when the AA1050 sample was anodically polarized at 300 mV (close to the breakdown potential), the morphology observed using EC-AFM showed many small pits. However, this may have been because many of the active points were close to each other; thus, the local current could not be resolved in the SECM electrochemical current map. During the post-scan, some high-current sites appeared in the lower part of the SECM image, which may have been related to the presence of pits or trenches on the surface, as shown in Figure 15. Preliminary results indicated that the localized dissolution of aluminum alloy may have occurred below the breakdown potential, but only involved a small number of particles.

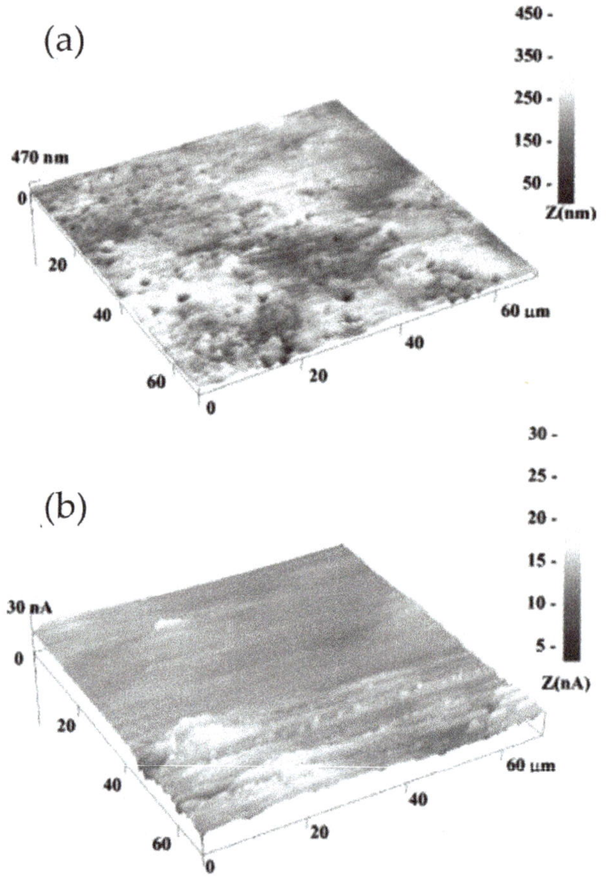

Figure 15. EC-AFM/scanning electrochemical microscopy (SECM) images of AA1050 in 10 mM NaCl+ 5 mM KI at 300 mV anodic polarization, and tip at +750 mV vs. Ag/AgCl: (**a**) topography; (**b**) electrochemical activity map. Adapted with permission from Reference [50]; copyright 2005 Electrochemical Society.

Metallic zinc is used not only for galvanizing steel but also for various applications such as batteries, die casting, and brass metallurgy. It is necessary to study the local corrosion of pure zinc and zinc alloys. Amin et al. [51] studied the passivation breakdown and pitting corrosion of zinc in 0.5 M sodium hydroxide solution containing different concentrations of ClO_3^- or ClO_4^- anions. The passivation breakdown and pitting sensitivity of the $Zn/OH^-/ClO_3^-$ and $Zn/OH^-/ClO_4^-$ interfaces were studied by potentiodynamic anodic polarization measurements, and the topography of pitted Zn surfaces was observed by AFM. The experimental results showed that metastable pitting corrosion was mainly caused by Cl^- ions (generated by reducing ClO_3^- and ClO_4^-). According to the point defect model (PDM), this may have been due to the small size of Cl^- ion, which could occupy anion vacancies and cause pitting corrosion. Therefore, Cl^- was the invasive ion that caused metastable pitting, rather than ClO_3^- and ClO_4^-. However, even in the absence of Cl^-, ClO_3^- and ClO_4^- could induce stable pitting, in which ClO_3^- was more aggressive than ClO_4^-, which could be proven by the roughness calculated according to the AFM topography map. Compared with ClO_4^-, the Ra value was always higher in the presence of ClO_3^-, as shown in Figure 16.

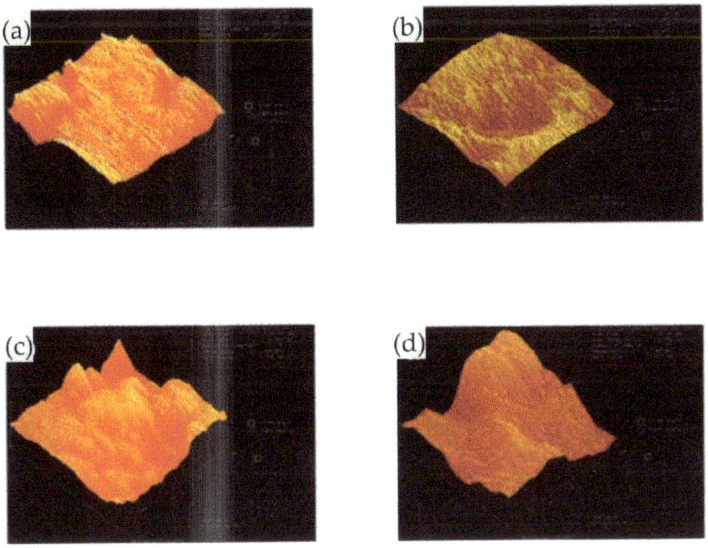

Figure 16. AFM images recorded for Zn in 0.5 M NaOH solution containing 0.05 M (**a**,**b**) ClO_4^- or (**c**,**d**) ClO_3^-. Adapted with permission from Reference [51]; copyright 2013 Springer.

Yong Hwan Kim et al. [52] studied the initial corrosion mechanism of a hot-dip-galvanized surface. The formation of corrosion products, the initiation and growth of pits, and the breakdown of the film were observed using in situ EC-AFM. At the initial stage of passivation, the corrosion product film of $ZnO/Zn(OH)_2$ formed in the dull sector was more unstable than that formed in the bright zone. The authors believed that the relative instability of the passive film in the initial stage was affected by the high-density lath-like structure. The uneven surface structure at the micro-scale, i.e., the lath-like structure, provided a favorable place for the $ZnO/Zn(OH)_2$ formation and pitting.

EC-AFM can provide detailed information of the localized dissolution associated with different kinds of intermetallic particles, as well as the deposition of corrosion products surrounding large particles or covering small pits, including the location of pitting initiation at controlled potentials, and whether pits are randomly distributed at the nanoscale. The application of EC-AFM in pitting corrosion is summarized in Table 3.

3.1.3. Selective Corrosion

Selective corrosion refers to the preferential dissolution of active components in the multicomponent alloy. The relative Volta potential difference is used as the index of selective corrosion caused by microelectronic coupling. Compared with the substrate, the intermetallic phase with positive potential is the cathodic phase. On the contrary, some intermetallics with more negative potential than the matrix show stronger activity during the corrosion process and act as micro-anodes.

Jia et al. [53] investigated the effects of intermetallics on the local corrosion behavior of AZ91 alloy added (La,Ce) mischmetal (MM), including corrosion initiation and propagation. The corrosion morphologies of different intermetallics were imaged using in situ EC-AFM in 0.1 mol/L NaCl solution, and Scanning Kelvin probe force microscopy (SKPFM) was performed to measure the local difference of the relative Volta potential between the intermetallic phases and the α-Mg matrix. In the experiment, the approximate compositions of several intermetallic phases were identified as Al_4(La,Ce), Al_8Mn_4Ce, and decreased b-$Mg_{17}Al_{12}$ phases. SKPFM analysis showed that all intermetallics were noble compared with the α-Mg matrix. The observation of in situ EC-AFM on the initial stages of corrosion indicated that the a-Mg matrix surrounded by b-$Mg_{17}Al_{12}$ or Al_8Mn_4Ce was prone to pitting corrosion. In detail, the b-$Mg_{17}Al_{12}$ phase existed in both AZ91 and AZ91 alloys with (La,Ce) MM. Both elongated and dispersed granular b-$Mg_{17}Al_{12}$ served as effective micro-cathodes, as shown in Figure 17.

Table 3. Summary of the application of EC-AFM in pitting corrosion.

	Materials	Solution	Mode	Reference
Pitting corrosion	AISI 316L and DSS 2205	Physiological solution and artificial saliva	Contact mode	[45]
	304L stainless steel	Chloride borate buffer solution	Contact mode	[46]
	SUS304	3.5 wt% NaCl solution		[48]
	pure aluminum	0.01 mol/L $FeCl_3$ solution	Contact mode	[49]
	AA1050	10 mM NaCl + 5 mM KI	Contact mode	[50]
	Metallic zinc	0.5 M NaOH solution containing different concentrations of ClO_3^- or ClO_4^- anions		[51]
	Hot-dip-galvanized surface	0.01 M NaOH	Contact mode	[52]

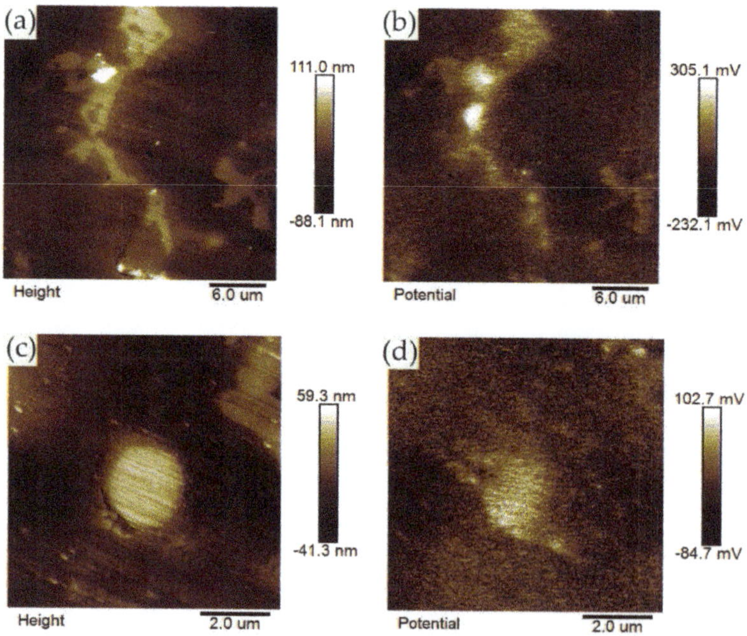

Figure 17. The polished AZ91 alloy (**a**) surface topography and (**b**) relative Volta potential map of elongated b-$Mg_{17}Al_{12}$; (**c**) surface topography and (**d**) relative Volta potential map of granular-like b-$Mg_{17}Al_{12}$. Reprinted with permission from Reference [53]; copyright 2019 Elsevier Ltd.

Al_4(La,Ce) intermetallics, whether featuring an acicular shape or rod shape, were believed to produce significant galvanic coupling. However, the α-Mg matrix near the Al_4(La,Ce) phase did not have obvious corrosion. Moreover, although the Al_8Mn_4Ce phase showed a positive potential difference with respect to the matrix, the corrosion resistance of the alloy was not affected by its galvanic coupling, probably due to its lower amount, as shown in Figure 18.

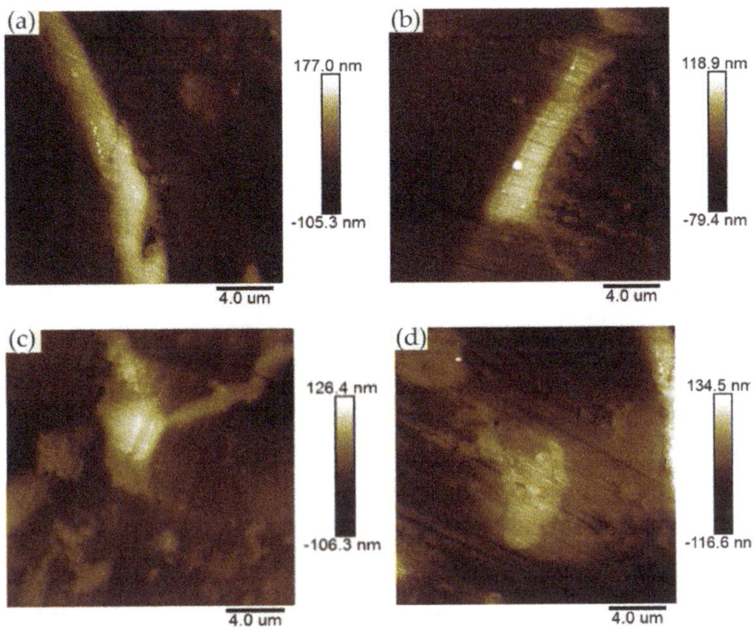

Figure 18. Typical in situ surface images for different intermetallics after immersion in 0.1 mol/L NaCl solution for 60 min: (**a**) acicular-like RE phase, (**b**) rod-like RE phase, (**c**) granular RE phase, and (**d**) dispersed granular b-Mg17Al12 in AZ91 alloy with 1.0% (La,Ce) mischmetal (MM) addition. Reprinted with permission from Reference [52]; copyright 2019 Elsevier Ltd.

It can be seen that the intermetallics with the highest Volta potential differences relative to the matrix did not play the role of effective cathodes, just like the Al_4(La,Ce) phase, which means that the localization of the cathode reaction was not only dependent on the Volta potential differences between the intermetallics and the matrix. The authors believed that the shape of the coupled anode and cathode had an important effect on the current density, i.e., the geometry of the intermetallic compound had an important influence on the micro-galvanic corrosion.

Zhang et al. [54] studied the anodization process of aluminum 6060 alloy under operating conditions and illustrated the effects of intermetallic particles (IMPs) and anodic Al oxide (AAO) film properties. The in situ EC-AFM measurement was performed continuously to monitor the surface topography changes under anodizing potentials in contact mode, revealing the details of localized dissolution and AAO film formation in Al 6060 samples in 0.2 M Na_2SO_4 solution. The extruded Al 6060 alloy mainly contained two types of IMPs: AlFeSi primary particles and Mg_2Si particles. The Volta potential differences obtained by SKPFM showed that, relative to the aluminum matrix, AlFeSi was cathodic, but Mg_2Si was anodic. The in situ EC-AFM measurements showed that AlFeSi particles remained stable, but the local anodic dissolution of Mg_2Si particles occurred during anodization, which was consistent with the SKPFM results. As shown in Figure 19, the large particle in region I was likely an α-AlFeSi particle. The particles in regions II, III, and IV were likely anodic Mg_2Si particles. The protruded large particle in region I remained stable during anodization, whereas the small particles in region II in Figure 19a were dissolved upon the anodization. Moreover, upon the anodization at 1 V, some active dissolution started at certain sites (e.g., III), forming small holes, as shown in Figure 19b. During the anodization at 2, 4, and 8 V (Figure 19c–e), the small hole in area III became deeper, as displayed by the profile lines in Figure 19f, and pronounced localized dissolution occurred in area IV. Two small particles remained stable in the dissolved area (area IV in Figure 19d,e). Furthermore, in the area marked as V, localized dissolution at one site resulted in a small hole, and a deposited particle of a few μm in size formed (Figure 19c), but it disappeared at 4 V, exposing a much deeper pit (Figure 19d). In addition, the growth of AAO films occurred with partial anodic dissolution. The thickness of the anodic barrier film increased linearly with the anode potential, but the growth rate decreased due to local anodic dissolution associated with IMPs in the alloy.

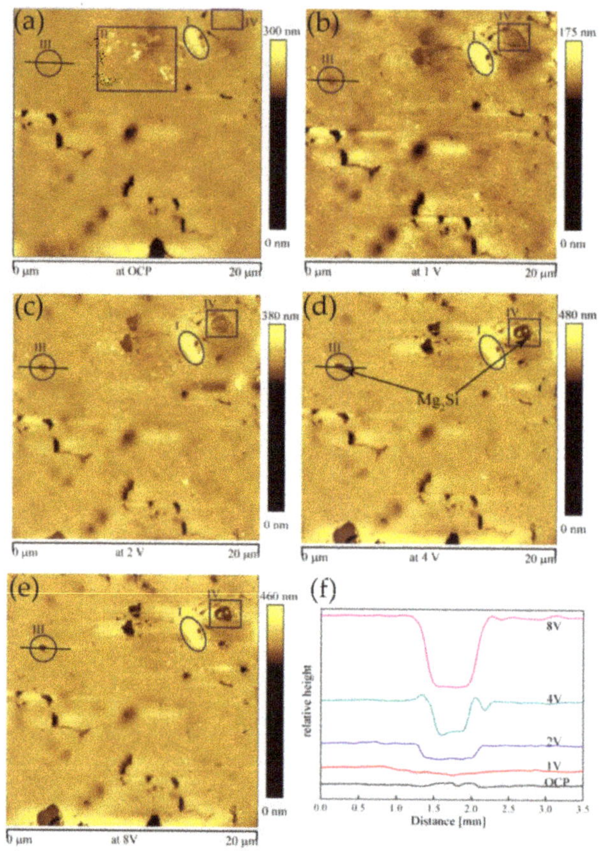

Figure 19. In situ EC-AFM topography images of the Al 6060 alloy in 0.2 M Na_2SO_4 solution at (**a**) OCP, (**b**) 1 V, (**c**) 2 V, (**d**) 4 V, and (**e**) 8 V vs. Ag/AgCl. (**f**) Line profiles across the feature III in (**a**–**d**). Adapted with permission from Reference [54]; copyright 2016 Electrochemical Society.

In some cases, the presence of nitrides can adversely affect the corrosion resistance of the material. Bettini et al. [55] studied the effect of nano-sized quenched-in chromium nitride particles on the corrosion behavior of heat-treated 2205 duplex stainless steel (DSS) in an NaCl solution at room temperature and 50 °C (slightly higher than the critical pitting corrosion temperature). The relative nobility difference between the precipitated nitrides, austenite, and ferrite in the tested materials was evaluated at room temperature by atomic force microscopy-based Kelvin force microscopy (AFM/KFM).

The volt potential mapping at room temperature indicated that the ferritic had a lower relative nobility compared with austenite, and quenched-in Cr_2N particles had a higher relative nobility to the surrounding ferritic matrix. The observation results of EC-AFM in 1 M NaCl solution at room temperature showed that the samples after heat treatment showed a wide range of passivation and a very stable surface up to 1.2 $V_{Ag/AgCl}$, where the selective dissolution of ferrite phase occurred but the quenched-in Cr_2N particles remained stable, as shown in Figure 20. Figure 20d shows the depth line profile from the image obtained after the polarization at 1.2 $V_{Ag/AgCl}$, presenting a depth of ca. 200 nm between the dissolved ferrite phase and the remaining austenite phase. In addition, the very small nitride particles formed during the fast cooling process did not have enough time for the diffusion of elements and were unlikely to form a considerable composition gradient in the surrounding boundary region, which was one of the reasons for its stability. At temperatures above the critical pitting temperature (CPT = 50 °C), rapidly selective dissolution of the austenite phase occurred upon slight anodic polarization, which may have been related to the low content of chromium and molybdenum in the austenite phase. The authors believed that the finely dispersed quenched-in nitrides in the DSS did not cause local corrosion in 1 M NaCl solution. However, the exposure temperature had a great influence on the corrosion resistance of DSS, which changed the selective dissolution behavior of DSS.

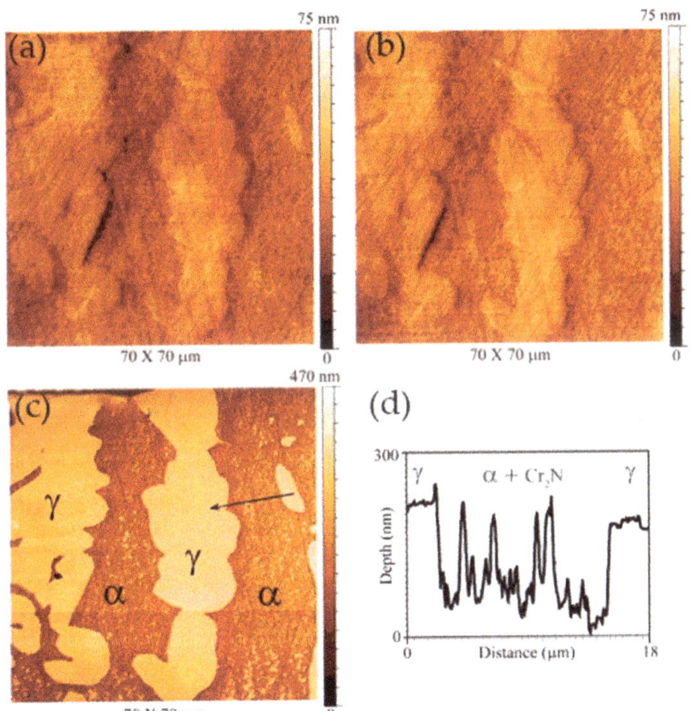

Figure 20. In situ EC-AFM images of the same area of the 2205 HT in 1 M NaCl at room temperature under electrochemical control at different applied potentials: (**a**) OCP, (**b**) 1.1 $V_{Ag/AgCl}$, and (**c**) 1.2 $V_{Ag/AgCl}$. (**d**) Depth line profile from the arrow in image (**c**) showing particles or particle clusters remaining in the dissolved ferrite area. Reprinted with permission from Reference [55]; copyright 2013 Elsevier Ltd.

Yasakau et al. [56] investigated the mechanism of initial steps of localized corrosion at the cutting edges of adhesively bonded Zn (Z) and Zn–Al–Mg (ZM) galvanized steel substrates. The topography of the initially localized corrosion of Z and ZM samples was measured using in situ AFM equipment under anodic polarization in the corrosive solution. The first corrosion pits at the Z cutting edge were mainly formed on the zinc layer, and there was no preferential erosion on the adhesive/zinc or zinc/steel, as shown in Figure 21.

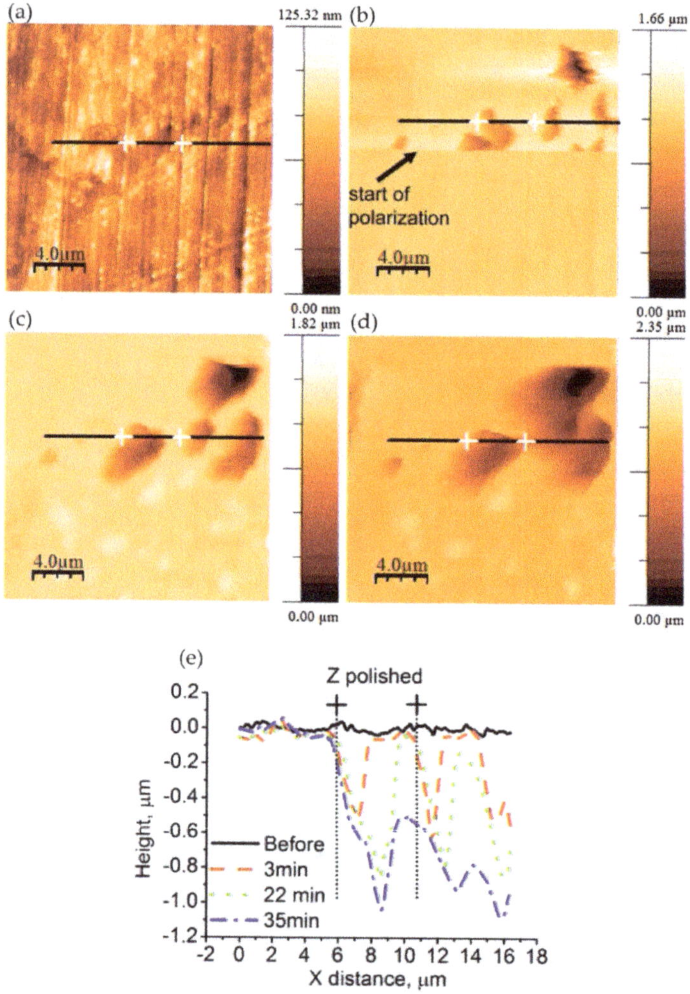

Figure 21. AFM topography maps made on cutting edge of adhesively bonded Zn (Z) substrate before immersion (**a**) and during 24 min (**b**), 1 h 5 min (**c**), and 2 h (**d**) of immersion in 0.001 M NaCl. (**e**) Evolution of topography across the black line profiles drawn in the same zones on AFM maps. Black arrows indicate local pits on the zinc layer. Reprinted with permission from Reference [56]; copyright 2016 Elsevier Ltd.

However, the types of attacks were different at the ZM cut edge, namely, pitting corrosion of the solid solution and selective dissolution at the eutectic zone. For the local corrosion at the adhesive/zinc interface, the first adhesive disbonding area was located near the cutting edge. The second adhesive disbonding zone was located at the buried deep zinc/adhesive interface, where local corrosion occurred in the solid solution phase and eutectic phase of the sample ZM. Under anodic polarization, the corrosion site was located in the eutectic phase at the interface between the eutectic zone and the solid solution zone, similar to the corrosion at the ZM cutting edge, as shown in Figure 22. In addition, scanning vibrating electrode technology (SVET) and electrochemical impedance spectroscopy (EIS) electrochemical tests showed a decrease in corrosion kinetics at the Z and ZM cutting edges, which was due to the blocking effect of the dense film of corrosion products formed on the zinc and steel surfaces.

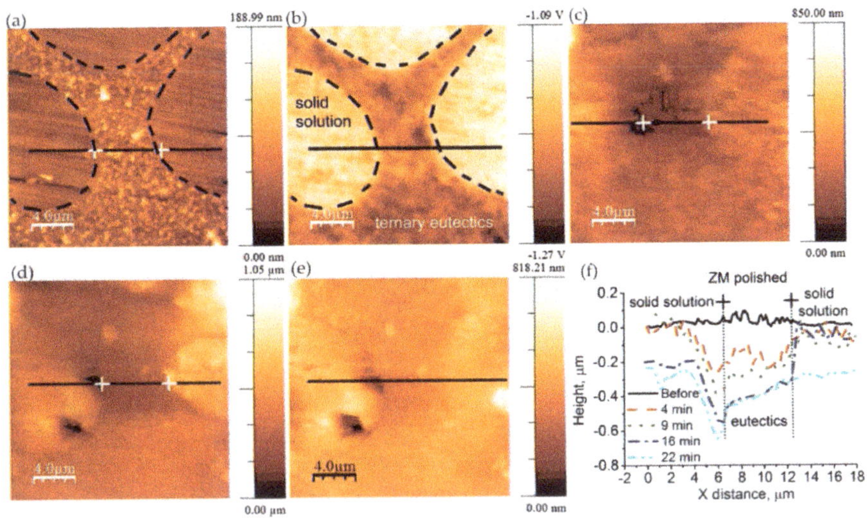

Figure 22. AFM topography (**a**) and SKPFM (**b**) maps of Zn–Al–Mg (ZM) galvanized coating before immersion and during immersion in 0.005 M NaCl for 4 min (**c**), 9 min (**d**), and 22 min (**e**). (**f**) Evolution of topography across the black line profiles. Reprinted with permission from Reference [56]; copyright 2016 Elsevier Ltd.

Depentori et al. [57] studied the corrosion behavior of neodymium-modified titanium alloy $Ti_6Al_4V_2Nd$ in 1.5 wt% NaCl and compared it with a $Ti_6Al_4V_2$ matrix using SKPFM and EC-AFM. Neodymium containing intermetallic compounds was precipitated at β-phase boundaries and inside the grains. The Volta potential maps obtained using SKPFM showed that the Volta potential of intermetallic compounds was lower than that of the Ti matrix. Therefore, intermetallic compounds had strong anodic behavior relative to titanium matrix and were the preferred site for local attack in $Ti_6Al_4V_2Nd$ alloy. When the sample was immersed in 1.5 wt% NaCl solution, EC-AFM observation showed that the volume of the local area on $Ti_6Al_4V_2Nd$ increased significantly, which was due to the formation of hydroxide and oxide. In addition, large amounts of debris were observed. The authors suggested this as a sign that the surface was loosely bound to $Nd(OH)_3$. Cyclic voltammetry showed a clear oxidation reaction without a reduction reaction, and the oxidation peak moved to the right as the exposure time increased, which was a clear sign that the electrochemical reaction was irreversible and that the diffusion barrier formed and increased over time. Corresponding to the results of EC-AFM, the diffusion barrier was the $Nd(OH)_3$ layer formed by the corrosion of a large number of precipitates on the surface, which prevented further oxidation.

Davoodi et al. [58] investigated the difference in corrosion behavior between EN AW-3003 (Rolled 3xxx series Al alloys) and a newly developed Al–Mn–Si–Zr fin alloy. The Volta potential of the two alloys determined using SKPFM showed that the intermetallic particles behaved as cathodes relative to the alloy matrix. Compared to EN AW-3003, the Al–Mn–Si–Zr alloy had fewer particles with larger Volta potential difference with respect to the matrix. In situ AFM measurements showed that ring-like corrosion products were deposited on the EN AW-3003 alloy, while only a few corrosion sites and tunnel-like pits were found on Al–Mn–Si–Zr, as shown in Figure 23.

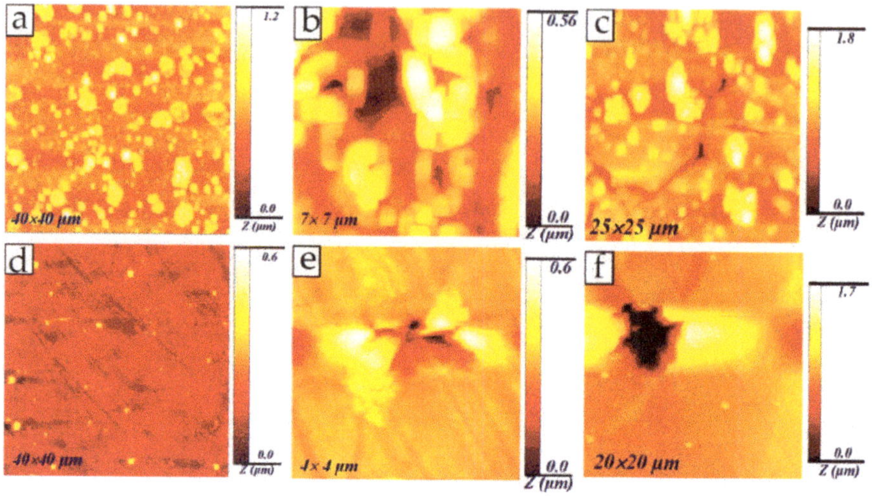

Figure 23. In situ AFM images of (**a**) EN AW-3003 and (**b**) Al–Mn–Si–Zr, after two days; (**e**) EN AW-3003 and (**f**) Al–Mn–Si–Zr, after 3.5 days in Sea Water Acetic Acid Test (SWAAT) solution of pH 4. (**c**,**d**) Magnified images of the framed areas of (**a**,**b**), respectively. Adapted with permission from Reference [58]; copyright 2007 Elsevier Ltd.

Topography and electrochemical currents obtained synchronously by the integrated SECM/AFM systems provided information on the pitting precursor and pitting process. Compared with Al–Mn–Si–Zr alloys, EN AW-3003 alloys had more active sites and extensive localized dissolution, resulting in higher material losses. Interestingly, some of the larger micron-sized intermetallic particles initiated localized dissolution at the boundary region of the particle-matrix, while the fine dispersions were not active. The intermetallic particles in the Al–Mn–Si–Zr alloy were few. Although they could induce selective dissolution and form small tunnel-like holes, because of their small weight loss, they were suitable for fin material in heat exchange applications.

The combination of EC-AFM and scanning Kelvin probe force microscopy (SKPFM) can simultaneously obtain topographical changes and Volta potential maps, which helps to better understand selective corrosion behavior and its mechanism. The application of EC-AFM in selective corrosion is summarized in Table 4.

Table 4. Summary of the application of EC-AFM in selective corrosion.

	Materials	Solution	Mode	Reference
Selective corrosion	AZ91 alloy added (La,Ce) mischmetal	0.1 mol/L NaCl solution	Contact mode	[53]
	Al 6060 alloy	0.2 M Na_2SO_4 solution	Contact mode	[54]
	2205 HT	1 M NaCl solution	Contact mode	[55]
	Cutting edges of adhesively bonded Zn (Z) and Zn–Al–Mg (ZM) galvanized steel substrates	0.001 M NaCl solution	Tapping mode	[56]
	$Ti_6Al_4V_2Nd$	1.5 wt% NaCl solution	Tapping mode	[57]
	EN AW-3003 and Al–Mn–Si–Zr fin alloy	SWAAT solution of pH 4	Contact mode	[58]

3.1.4. Intercrystalline Corrosion

Intercrystalline corrosion is a local corrosion phenomenon that occurs along the grain boundary of a metal material in a corrosive medium and causes the loss of bonding force between grains. When intercrystalline corrosion occurs in the material, there may be no macroscopic change, but the material's strength is almost completely lost, resulting in the sudden destruction of equipment. The main reason for intercrystalline corrosion is the difference in structure and chemical composition between the grains and grain boundaries of the material.

Based on EC-AFM, the contact mode high-speed AFM (HS-AFM) compensates for the shortcomings of AFM with short collection times. The long collection time is a limiting factor for AFM. Contact mode HS-AFM images multiple frames per second, making it orders of magnitude faster than traditional AFM. This enables real-time imaging processes with nano-scale lateral resolution and sub-nanometer-scale height resolution. The increase in speed can not only directly image dynamic nanoscale events, but also macroscopic regions of the sample surface without reducing resolution. It is a valuable imaging tool for in situ observation of nanoscale corrosion initiation events such as metastable pitting, grain boundary dissolution, and short crack formation during stress corrosion cracking.

In the work of Moore and co-workers, local corrosion phenomena such as pitting and intergranular attack (IGA) on thermally sensitized AISI 304 stainless steel were studied using HS-AFM in 1% NaCl [59]. Real-time in situ HS-AFM observations showed that an intergranular pit was formed within 0.5 s during a galvanostatic scan. Intergranular pits were distributed along the grain boundary (GB) in a chain shape, which was caused by the preferential corrosion of GB. Chromium carbide was precipitated along GBs, which resulted in the depletion of the local chromium elements in the area around GBs, greatly reducing the local corrosion resistance of GBs. By using HS-AFM and electrochemical data for the computational model, it was found that the radial diffusion state of the system was reached within 0.01 s, leading to rapid dissolution of materials.

Padhy et al. [60] used EC-AFM to study the surface morphology of austenitic stainless steel in nitric acid medium. In situ EC-AFM results showed that the surface presented a platelet-like structure in low-concentration nitric acid solution (0.1 M, 0.5 M), providing effective protection for the surface. When the concentration of HNO_3 was from 0.1 M to 0.6 M, the roughness decreased, which was related to the thinning of the passive film and marked the beginning of corrosion. From the morphology of 0.6 M HNO_3, breakdown of the passive film and surface dissolution were observed. As the concentration continued to increase to 1 M, the roughness increased due to the intensification of the surface dissolution and selective dissolution of grain boundaries, as shown in Figure 24.

The early stages of localized corrosion are important as mentioned earlier, because they account for most of the lifetime of intergranular corrosion, stress corrosion cracking (SCC), or pitting, and they are more accessible to take remedial action. Williford et al. [61] used EC-AFM to obtain images of the early stages of intergranular corrosion (IGC) in 304L stainless steel. The observation of EC-AFM showed that IGC was present between the carbides, but not completely around the carbide in the early stage. Later, the grain boundaries became wider, the carbides were shortened, and the IGC completely surrounded the carbides. The matrix between carbides began to corrode, which was best explained by the fact that the carbides were cathodic with respect to the matrix; thus, when the matrix dissolved, they were protected by cathodic protection. In addition, the carbide was connected to adjacent grains through the ligaments of the matrix material, which may have meant that the SCC crack front was bridged by the carbide. The SCC crack front could move forward and surround the carbide, creating an area where the carbide bridged the crack.

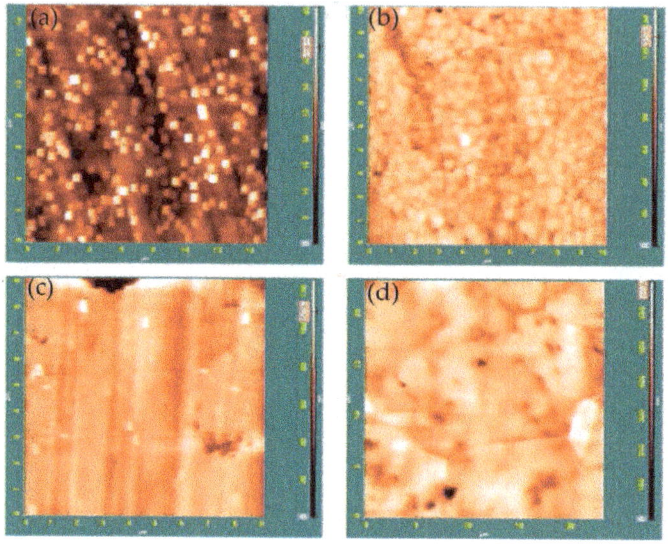

Figure 24. Surface morphology of 304L stainless steel at 1100 mV in (**a**) 0.1 M HNO_3, (**b**) 0.5 M HNO_3, (**c**) 0.6 M HNO_3, and (**d**) 1 M HNO_3. Reprinted with permission from Reference [60]; copyright 2010 Elsevier Ltd.

An AFM with a conductive probe can simultaneously obtain the surface topography and surface potential of the scanned surface. Fu et al. [62] studied the local corrosion of high-chromium cast iron in regions at different distances from interphase boundaries using EC-AFM in the contact mode. According to the measured local potential, a decrease in interface potential between carbide and matrix was observed. Moreover, the corrosion rate of the metal matrix near the primary carbide was significantly higher than that far away from the primary carbide. The morphology map obtained using EC-AFM showed that the area near the primary carbide/matrix interface corroded or dissolved more rapidly than the area away from the interface. Furthermore, the corrosion rate was found to be particularly rapid in areas with sharp edges. The authors suggested that, on the flat side, most of the electrons flowed into the half-space on one side of the matrix. However, at the sharp edge, electrons flowed into a larger space in the matrix, corresponding to a larger total electron flow, which greatly increased the corrosion rate of the sharp edge.

In high-entropy alloys (HEAs), the homogeneous elemental distribution is expected to improve corrosion resistance. In order to develop highly corrosion-resistant HEAs, it is necessary to study the relationship between chemical/microstructure segregation and localized corrosion. In the study of Shi and co-workers, localized corrosion of Al_xCoCrFeNi HEAs was studied using in situ EC-AFM in 3.5 wt% NaCl solution. Surface topography changes at micron/submicron scales were monitored under different anodic potentials [63]. In the experiment, with the increase in aluminum content in HEAs, the microstructure changed from a single FCC (face-centered cubic) solid solution to the FCC phase and (ordered/disordered) BCC (face-centered cubic) phase. The EC-AFM image showed that the uniform single-phase $Al_{0.3}$CoCrFeNi alloy had the best corrosion resistance, and the breakdown of its passive film was in the form of randomly distributed pits, as shown in Figure 25.

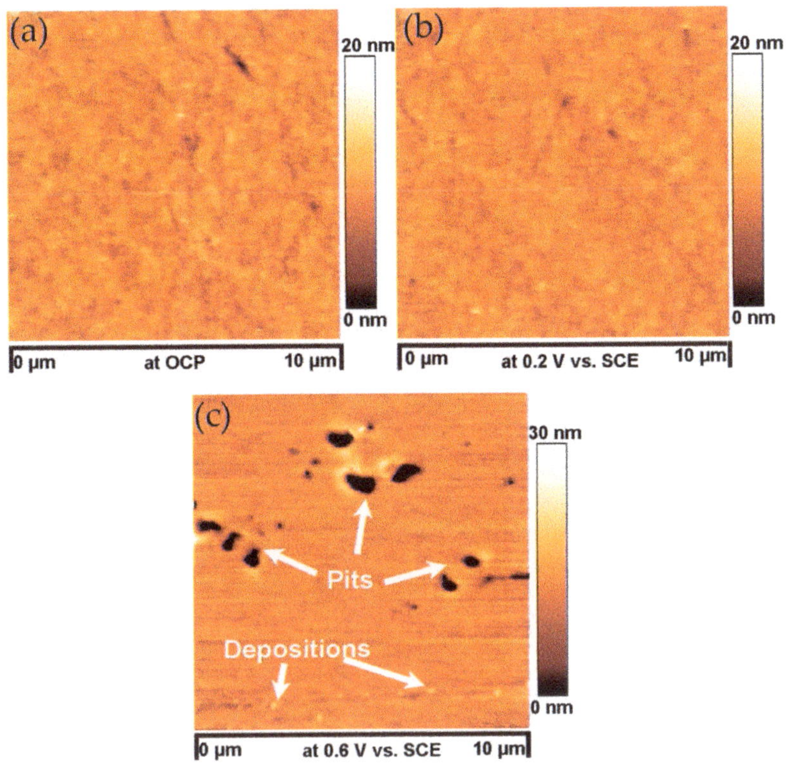

Figure 25. In situ EC-AFM topography images of the $Al_{0.3}CoCrFeNi$ high-entropy alloy (HEA) in a 3.5 wt% NaCl solution after 10 min of exposure at (**a**) OCP, (**b**) 0.2 V, and (**c**) 0.6 V. Reprinted with permission from Reference [63]; copyright 2018 Elsevier Ltd.

With the increase in aluminum content, the BCC phase appeared in the $Al_{0.5}CoCrFeNi$ alloy, resulting in a heterogeneous microstructure. Pits were formed preferentially along the FCC/BCC phase boundary, leading to the initial breakdown of the passive film, which was manifested as the decline in critical pitting potential value in the potentiodynamic polarization curve, as shown in Figure 26.

Pitting corrosion was not observed in the $Al_{0.7}CoCrFeNi$ alloy with the increase in aluminum content and volume fraction of the BCC phase. In contrast, local corrosion along the dendritic/interdendritic boundaries and selective dissolution of the (Al, Ni)-rich ordered BCC phase occurred, as shown in Figure 27. Therefore, the author believed that, as the microstructure changed from single solid solution to multiphase, the breakdown of the passivation film changed from pitting to phase boundary dissolution, which led to a decrease in corrosion resistance.

Bettini et al. [64] studied the effects of carbides on the corrosion/dissolution behavior of biomedical CoCrMo alloys in PBS solution using EC-AFM at different applied potentials. SKPFM results showed that, compared with the matrix, the Volta potential of carbides was higher. In addition, the Volta potential decreased in the boundary region, which may have been related to local depletion of the main alloy elements. This indicated that the carbide boundary had a greater corrosion tendency and was the preferred site for corrosion/dissolution. In situ EC-AFM measurements of the CoCrMo alloys exposed to PBS showed that, at high anodic potential, a dissolution process at carbide boundaries was observed, and an increase in boundary depth was seen in line profiles across these boundaries, as shown in Figure 28. This was consistent with the SKPFM Volta potential mapping, which showed that some of the boundary areas were weak sites for corrosion/dissolution due to lower relative nobility compared to the matrix.

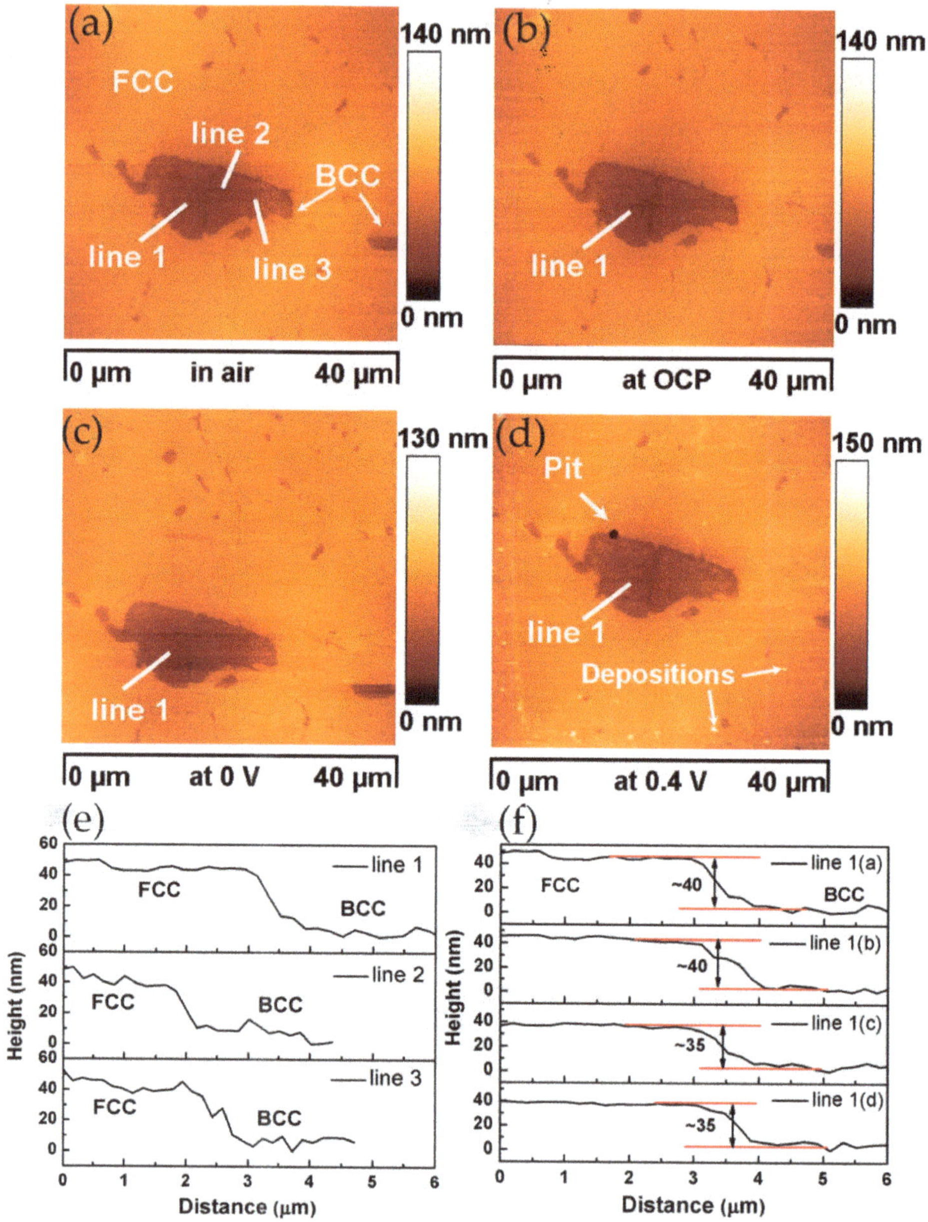

Figure 26. The AFM topography image in air (**a**), and in situ EC-AFM topography images of the $Al_{0.5}CoCrFeNi$ HEA after 10 min of exposure in a 3.5 wt. % NaCl solution at (**b**) OCP, (**c**) 0 V, and (**d**) 0.4 V. (**e**) Three-line profiles across the phase boundaries illustrated in (**a**). (**f**) Vertical profiles along line 1 of the surface in (**a–d**). Reprinted with permission from Reference [63]; copyright 2018 Elsevier Ltd.

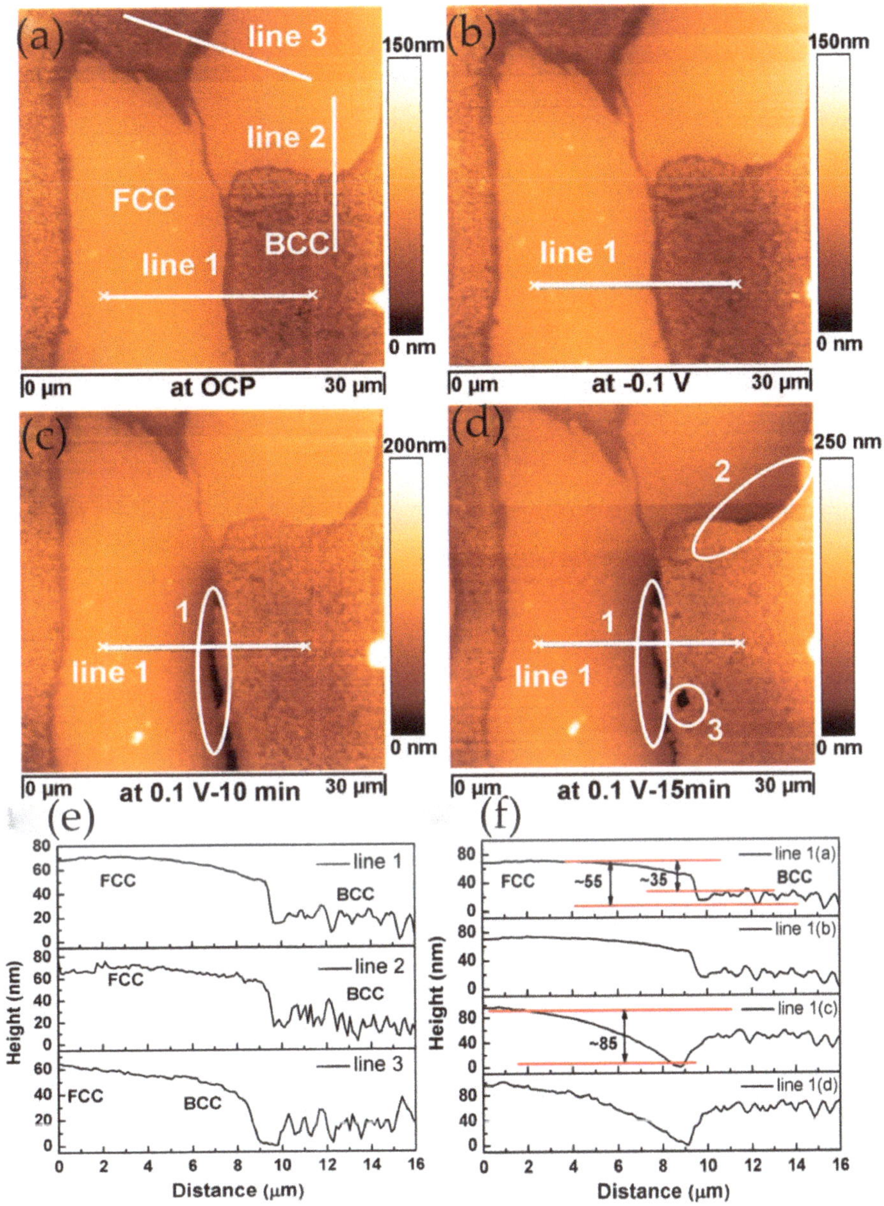

Figure 27. In situ EC-AFM topography images of the $Al_{0.7}$CoCrFeNi HEA after 10 min of exposure in a 3.5 wt% NaCl solution at (**a**) OCP, (**b**) 0.1 V, and (**c**) 0.1 V, and (**d**) after 15 min of exposure at 0.1 V vs. SCE. (**e**) Three-line profiles across the phase boundaries illustrated in (**a**). (**f**) Vertical profiles along line 1 of the surface in (**a**–**d**). Reprinted with permission from Reference [63]; copyright 2018 Elsevier Ltd.

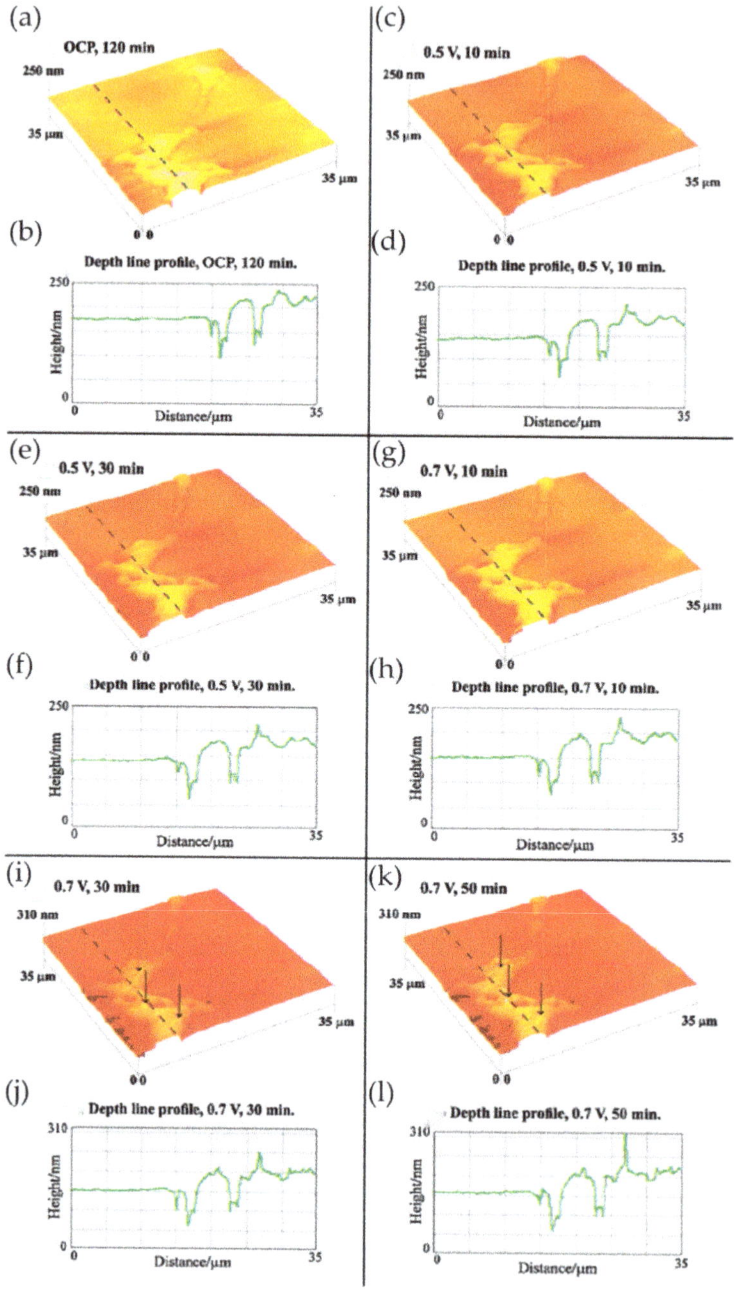

Figure 28. In situ AFM images of the same area of CoCrMo alloy under electrochemical control in PBS solution with pH 7.4, (**a**) after 120 min at OCP, (**c**,**e**) after 10 and 30 min at 0.5 Vsat Ag/AgCl, and (**g**,**i**,**k**) after 10, 30, and 50 min at 0.7 Vsat Ag/AgCl, with marked etching-like dissolution sites on the carbide (where visible). (**b**,**d**,**f**,**h**,**j**,**l**) Depth line profiles at the applied potential and time. Adapted with permission from Reference [64]; copyright 2011 Elsevier Ltd.

Davoodi et al. [65] studied the localized corrosion and preferential dissolution of Al alloys in chlorine solution using an integrated EC-AFM/ SECM system. The integrated EC-AFM/SECM could simultaneously detect topographic changes and electrochemical active sites in the same region and reveals local corrosion processes related to IMPs. The results showed that preferential dissolution occurred in the interfacial region between the alloy matrix and IMPs. The formation of grooves around the larger IMPs indicated that different types of IMPs had different dissolution behaviors. In addition, they found that only a small number of IMPs were involved in the localized dissolution at any given time.

EC-AFM can clearly reveal the formation of trenches and the local dissolution of the grain/phase interface, and it can explore the causes and mechanisms of intercrystalline corrosion in combination with other test tools. Table 5 summarizes the application of EC-AFM in intercrystalline corrosion.

Table 5. Summary of the application of EC-AFM in intercrystalline corrosion.

	Materials	Solution	Mode	Reference
Intercrystalline corrosion	AISI 304 stainless steel	1.0 wt% NaCl solution		[59]
	304L stainless steel	0.6 M and 1 M HNO_3		[60]
	High-chromium cast iron	3.0 wt% NaCl solution	Contact mode	[61]
	$Al_xCoCrFeNi$ HEAs	3.5 wt% NaCl solution	Tapping mode	[63]
	CoCrMo alloys	Phosphate-buffered saline solution	Contact mode	[64]

3.2. *Metal Protection*

3.2.1. Coating Protection

One of the primary protection techniques for metallic materials is the use of a cover layer on the metal surface to avoid direct contact between the metal and the corrosive medium as much as possible. The coating provides an effective barrier to the substrate, slowing the corrosion of the metal. In general, the metal surface coating can be divided into the metal coating and non-metallic coating. The coating not only slows electron transfer between the anode and cathode; it can also act as a barrier to prevent oxygen from penetrating the cathode reaction. Microcrystalline coating, nanocrystalline coating, gradient coating, composite coating, etc. can effectively improve the overall performance of the material, including corrosion resistance.

Li and co-workers studied the electrochemical mechanism and corrosion protection properties of solvent-borne alkyd composite coating containing 1.0 wt% CeO_2 nanoparticles (CeNPs) and 1.0 wt% polyaniline (PANI) for carbon steel in NaCl solution [66]. In their work, the morphology changes of the coatings and redox reactions of PANI at the nanoscale were accurately monitored by linking the volume changes observed using in situ EC-AFM imaging with redox peaks measured using in situ cyclic voltammetry (CV). The results of EC-AFM showed that PANI nanoparticles in the alkyd matrix exhibited contracted morphology in the reduced state, leucoemeraldine base (LB), and expanded morphology in the oxidized state, emeraldine salt (ES). The surface did not change significantly, which indicated that the composite coating was stable in corrosive solutions even under very harsh potential conditions, as shown in Figure 29. OCP and EIS results indicated that the redox reaction of PANI ES/LB forms caused metal passivation, which was an active corrosion protection mechanism.

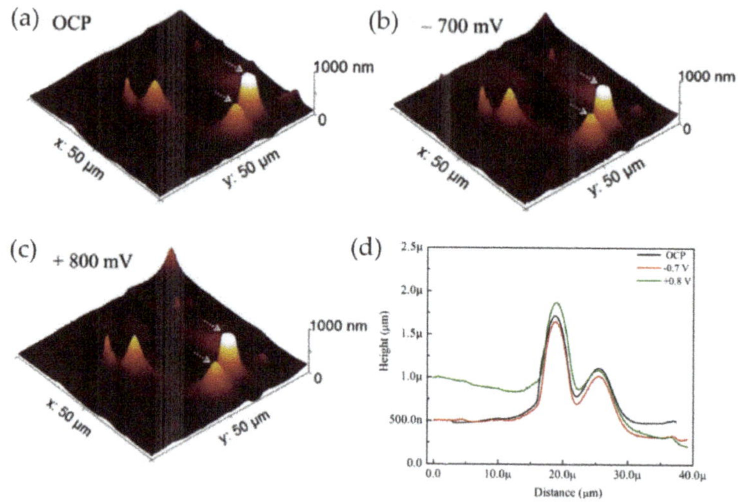

Figure 29. EC-AFM images (3D) of the same scan area of the composite coating in 3.0 wt% NaCl solution obtained after 60 min (**a**) at OCP, (**b**) at −700 mV (vs. Ag/AgCl), and (**c**) at 800 mV (vs. Ag/AgCl). (**d**) Line profiles of CeO_2 nanoparticle (CeNP) aggregates when at the two applied potentials drawn by the crossing line. Reprinted with permission from Reference [66]; copyright 2019 Elsevier Ltd.

The evolution of OCP with exposure time under 3.0 wt% NaCl for composite coating and reference coating showed that the potential value of composite coating was higher than that of reference coating during the whole measurement process, indicating that the corrosion resistance of the composite coating was improved. The increase in OCP value was attributed to the fact that CeNPs significantly improved the barrier effect and slowed down the migration of corrosion ions. Therefore, the synergistic effect of PANI and CeNPs greatly improved the barrier performance and the corrosion resistance of alkyd composite coating.

In a series of studies by Li and his colleagues, they also studied the electrochemical activity of 1.0 wt% *p*-toluene sulfonic acid (PTSA)-doped PANI in solvent-borne alkyd composite coating, as well as its self-healing corrosion protection mechanism on carbon steel in 3.0 wt% NaCl solution [67]. Through the CV (cyclic voltammetry) curves and EC-AFM imaging, it was proven that doped PANI at low percentage had extremely high electrochemical activity, showing contraction at the reduction potential and expansion at the oxidation potential, which provided evidence for a reversible redox reaction between the ES and LB forms. The voltage potential diagram obtained using KFM under air conditions indicated that PTSA-doped PANI had sufficiently high electrochemical activity and stable reoxidation ability to keep the passivation region on the metal surface. The PTSA-doped PANI in alkyd composites caused structural mutations due to energy input, which increased the electrochemical activity and provided doped PANI with a good electrochemical connection to the metal surface. It played an important role in improving the self-healing corrosion protection of composite coatings. In addition, the EIS showed increased resistance of the composite coating, which may have been related to the interaction of PANI particles in the alkyd matrix forming a dense and more resistive network.

Moreover, they studied the corrosion protection properties of a waterborne acrylic composite coating with 1.0 wt% acetic acid-stabilized CeNPs on carbon steel in 3.0 wt% NaCl solution [68]. The results of in situ AFM showed that the CeNPs embedded in the composite coating could greatly reduce the nano-pinholes in the waterborne acrylic coating, as well as significantly improve the stability of the coating, which played an important role in improving the barrier property of the coating. In situ EC-AFM indicated that some CeNPs and aggregates were released from the coating surface during exposure, and then some particles and cerium compounds were precipitated, as shown in Figure 30.

The presence of CeNPs or aggregates acted as nucleation sites to promote precipitation on the coating surface and inside the coating pinholes, thereby preventing the entry of corrosive ions and the corrosion of the metal matrix.

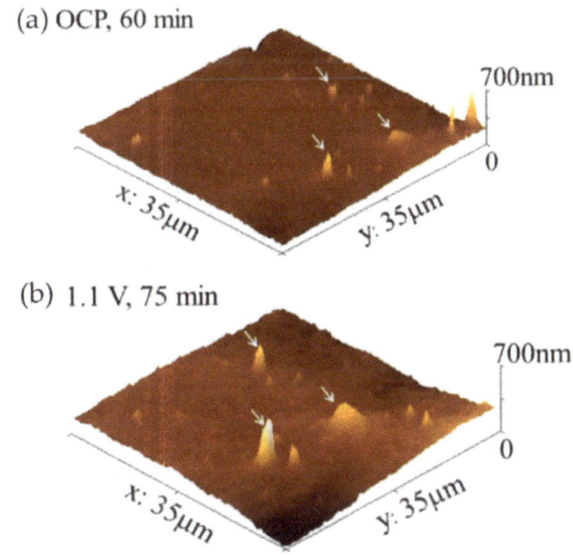

Figure 30. EC-AFM images (3D) of the same area of the composite coating obtained after 14 h of exposure in 3.0 wt% NaCl solution: (**a**) after 60 min at OCP; (**b**) after 75 min at 1.1 V vs. Ag/AgCl. Adapted with permission from Reference [68]; copyright 2015 Electrochemical Society.

Liu et al. [69] studied the pitting behavior of austenitic stainless steel with nanocrystalline (NC) and polycrystalline (PC) microstructures in 3.5 wt% NaCl solution. In situ AFM was used to study the process of passive film formation on the PC alloy and NC alloy under anodic potential. AFM observations showed that the passive film formed rapidly on the PC alloy, and pitting occurred after continuous film formation, which was a slow process of metastable pitting formation and reparation. However, the formation of a passive film on the NC alloy indicated that, due to the accumulation of many small particles on the surface, the oxide particles grew in the original position and eventually became a passive film. The voids and the boundaries of oxide particles may have been inoculation points for metastable pits. Although metastable pits were easy to initiate on the NC coating, the small grain size promoted the diffusion of elements, such that pits could be quickly repaired or healed. Therefore, the pitting mechanism of the NC coating was mainly characterized by rapid metastable pitting initiation and death, and its pitting resistance was higher than that of the PC alloy. In addition, in their other study, the characteristics of both pit initiation and pit growth processes on an austenitic stainless-steel NC coating were monitored using in situ AFM [70]. Pit initiation included the formation of metastable pits and repassivation process. Pitting growth included stable pit growth and material dissolution. The fine grain size promoted the formation and growth of nano-scale oxide particles, which significantly improved the repassivation ability and reduced the probability of stable pit formation. Compared with PC austenitic stainless steel, nanocrystals promoted the formation of metastable pits but reduced the rate of stable nucleation and growth of pits.

In the study of Pan and co-workers, the pitting corrosion of coarse crystal (CC) 304 stainless steel and its NC thin film was studied in 3.5 wt% NaCl solution, especially the influence of nanocrystalline on the pitting process [71]. The whole pit growth process was recorded using in situ AFM and the growth mechanism of stable pits on NC film was understood. The results showed that the initiation site of the pit on the NC film was at the boundary of the oxide particles. As there were lots of boundaries on the surface, metastable pit events on NC films were more likely to occur than those on CC 304 stainless steel, which indicated that nanocrystallization promoted metastable pit processes, as shown in Figure 31.

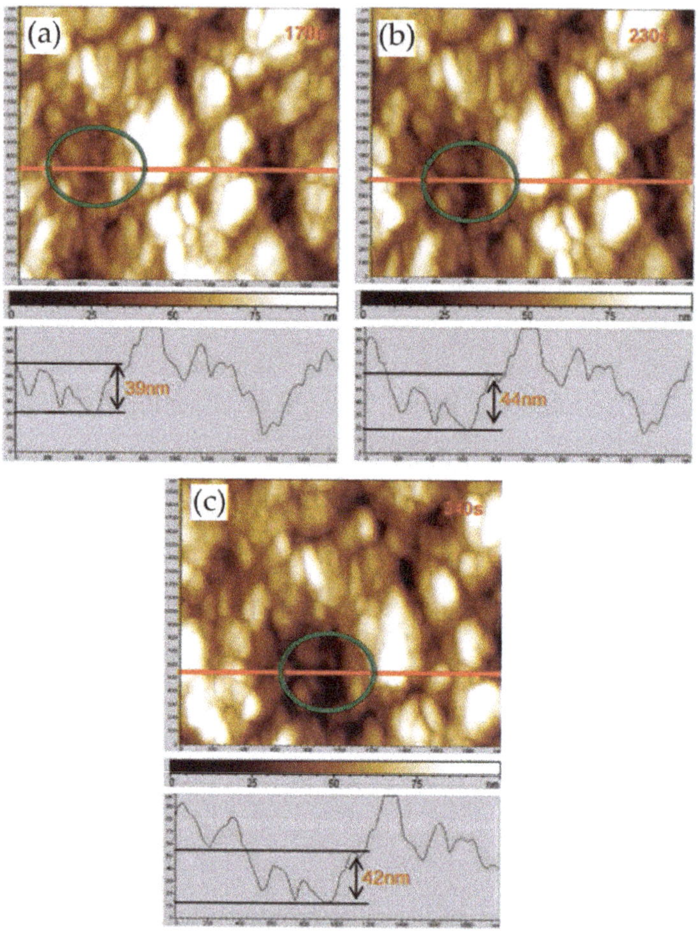

Figure 31. In situ AFM images of nanocrystalline (NC) thin film in the initial pitting stage under anodic polarization in 3.5 wt% NaCl solution: (**a**) 170 s; (**b**) 230 s; (**c**) 340 s. Reprinted with permission from Reference [71]; copyright 2013 Elsevier Ltd.

In addition, the transition from metastable pitting to stable pitting was inhibited due to the excellent repassivation ability of NC films. In the process of NC film deposition, the internal residual stress may have inhibited the formation of lace cover in the process of stable pit growth, and then changed the growth mechanism of the NC film surface stable pit. Therefore, the probability of developing from metastable pitting to stable pitting on NC films was much lower than that on CC 304L stainless steel. Nanocrystallization changed the geometry and growth mechanism of stable pits, slowing down the nucleation and growth process, which improved the pitting corrosion resistance of CC 304L stainless steel. Moreover, they also studied the corrosion behavior of a magnetron-sputtered NC 304L stainless steel coating in 0.05 M H_2SO_4 + 0.2 M NaCl solution, which was compared with conventional rolled CC 304L stainless steel [72]. The nanocrystalline structure reduced the adsorption capacity of Cl^- on the surface and inhibited the incorporation of Cl^- in the passive film. In situ AFM observation showed that the growth rate of the passive film on the NC film was greatly higher than that on the CC 304L stainless steel, as shown in Figure 32. In other words, the nanocrystalline structure improved the growth rate of the passive film and facilitated the healing of the passive film rupture. The composition of the passive film on the NC film determined by XPS had a higher ratio of chromium oxide to iron oxide. The higher content of chromium oxide improved the corrosion resistance of nanocrystalline samples. Since the structure of the passive film was more compact, the ratio of chromium oxide to iron oxide was higher, and the incorporation of Cl^- was less, the corrosion resistance of the NC film was greatly improved.

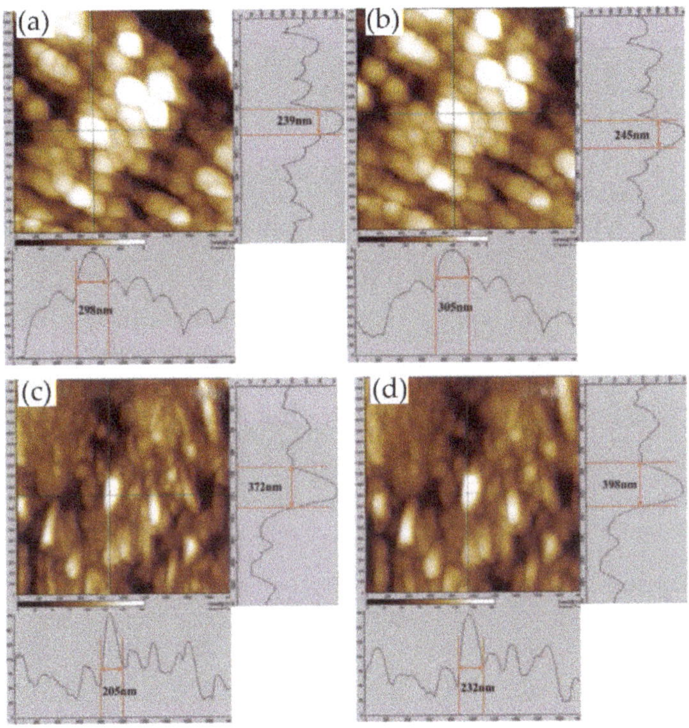

Figure 32. In situ AFM images of coarse crystal (CC) 304L stainless steel (**a**,**b**) and a NC thin film (**c**,**d**) (scale 2 μm × 2 μm) in the growth stage of a passive film under anodic polarization in 0.05 M H_2SO_4 + 0.2 M NaCl solution after passivation of 6 min (**a**,**c**) and 12 min (**b**,**d**). Reprinted with permission from Reference [72]; copyright 2012 Electrochemical Society.

In addition, the passive film growth mechanisms of the NC 304L stainless-steel thin film, deep-rolled bulk nanocrystalline (BN) 304 stainless steel, and CC 304 stainless steel in 0.05 M H_2SO_4 + 0.2 M NaCl solution were studied using electrochemical measurements and in situ AFM [73]. The growth rate of the passive film on the three materials was in the following order: NC thin film > BN304 stainless steel > CC 304L stainless steel. Nanocrystallization changed the nucleation mechanism of passive films from gradual to instantaneous. The passive film on the CC 304L stainless steel and BN 304L stainless steel had a single-layer structure, while the passive film on the NC film had a multi-layer structure.

EC-AFM can not only detect the state of the coating surface through high-resolution imaging, but also produce coating defects by means of probe scraping to obtain direct information on the coating corrosion resistance. The application of EC-AFM in coating protection is summarized in Table 6.

Table 6. Summary of the application of EC-AFM in coating protection.

	Materials	Coating	Solution	Mode	Reference
Coating protection	Carbon steel	Solvent-borne alkyd composite coating containing 1.0 wt% CeO_2 nanoparticles (NPs) and 1.0 wt% polyaniline (PANI)	3.0 wt% NaCl solution	Contact mode	[66]
	Carbon steel	1.0 wt% *p*-toluene sulfonic acid (PTSA)-doped PANI in solvent-borne alkyd composite coating	3.0 wt% NaCl solution	Contact mode	[67]
	Carbon steel	Waterborne acrylic composite coating with 1 wt% acetic acid-stabilized CeO_2 nanoparticles	3.0 wt% NaCl solution	Contact mode	[68]
	Austenitic stainless steel	Nanocrystalline (NC) and polycrystalline (PC) microstructure coatings	3.5 wt% NaCl solution	Tapping mode	[69]
	304L stainless steel	Magnetron-sputtered NC	0.05 M H_2SO_4 + 0.2 M NaCl solution	Tapping mode	[72]

3.2.2. Corrosion Inhibitor Protection

A corrosion inhibitor is a chemical or a mixture of several chemicals that prevents or slows corrosion when present in a corrosive environment (medium) in the proper concentration and form. The addition of corrosion inhibitors can significantly reduce the corrosion rate of metal materials. At the same time, the original physical and mechanical properties of the metal material can be maintained. The advantages of using corrosion inhibitors to protect metal lie in their low dosage, quick effect, low cost, and convenient use. At present, corrosion inhibitors are widely used in machinery, petrochemical, metallurgy, energy and other industries. In some studies, the authors used in situ EC-AFM under real-time operating conditions to detect changes in the corrosion morphology of the sample after the addition of the corrosion inhibitor, and then investigated the properties of the corrosion inhibitor and speculated on the corrosion inhibition mechanism.

The study of non-toxic corrosion inhibitors is important for replacing classical molecules with sulfur, nitrogen, or aromatic functions. Rocca et al. [74] reported the inhibition conditions and mechanisms of linear sodium heptanoate on copper corrosion. In situ EC-AFM under applied potential allowed observing the morphology of the passive film without damaging the layer. The result showed that, when pH = 5.7 and 11, for 0.08 M NaC_7, large non-covering copper heptonate crystals or non-covering copper oxide were formed on the surface, and the corrosion inhibitor efficiency was low in acidic medium. However, when the pH value was 8, for 0.08 M NaC_7, a thin layer of heptanoic was formed, which acted as a barrier and effectively protected the metal matrix. The inhibition of sodium heptanoate was related to the formation of a protective layer consisting mainly of copper heptanoate on copper, and the optimum corrosion inhibition conditions were 0.08 M NaC_7, pH 8 for copper corrosion.

Bertrand et al. [75] used in situ EC-AFM to study the corrosion behavior of a copper surface immersed in various electrolytes under dynamic potential conditions at room temperature. In sodium sulfate and sulfuric acid solutions, dissolution precipitation occurred on the surface, which changed the topographic characteristics of the metallic surface. On the contrary, when copper was immersed in borate or heptanoate solutions, passivation could be clearly observed. Regardless of the oxidation mode (constant potential or open-circuit condition), a thin passive layer was grown on the surface and was stable over time. In the borate medium, the sediments were composed of Cu_2O and CuO oxides, while, in the heptanate electrolyte, a metal soap composed of copper(II) heptanate was detected. In addition, the inhibition mechanism of sodium heptanate was identified, whereby a thin passive layer of copper metal soap was formed on the surface via dissolution precipitation.

The use of polymer corrosion inhibitors attracted attention. On the one hand, they have low cost and good stability. On the other hand, they have multiple adsorption sites that form complexes with metal ions covering the surface and protecting the metal from corrosion. Umoren et al. [76] studied the mechanism of polyacrylic acid (PAA) inhibiting the corrosion of pure cast aluminum in acidic medium and the synergistic effect with iodide ion addition. The in situ AFM morphology of the surface showed that PAA was adsorbed onto the surface of the aluminum and its arrangement was more orderly in the presence of iodide ions, as shown in Figure 33, resulting in higher inhibition efficiency. The authors believed that the alumina film was replaced by the adsorption of KI on aluminum, and then the PAA was adsorbed onto the KI such that the PAA molecules were arranged on the aluminum surface in an orderly manner, which enhanced the inhibition process.

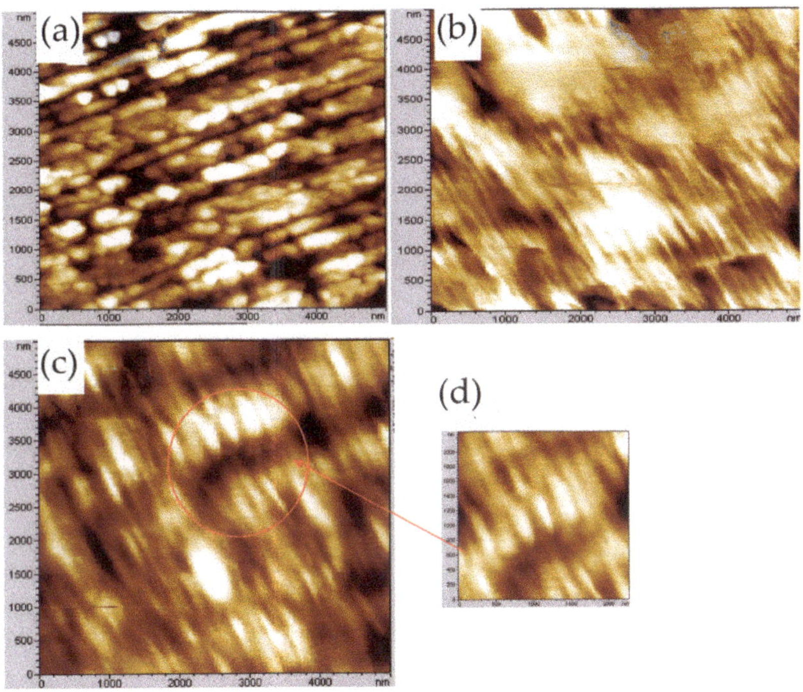

Figure 33. In situ AFM image of Al in 0.5 M H_2SO_4 in the presence of polyacrylic acid (PAA) + KI at different potentials: (**a**) −1.0 V, (**b**) −0.70 V, and (**c**) −0.50 V. (**d**) Magnified images of the framed area of (**c**). Adapted with permission from Reference [76]; copyright 2013 Taylor & Francis.

Zhang et al. [77] investigated the electrochemical corrosion behavior of AISI321 stainless steel in 36% ethylene glycol–water solution. It can be seen from the results of EC-AFM that the passive film became more complete and the number of defects decreased with the increase in polarization time. Moreover, as the passive potential increased, the particle diameter increased and surface defects decreased. When the potential ranged from −0.15 to 0.45 V, an N-type oxide film adhered to the surface of the sample. However, when the potential was between 0.45 and 0.75 V, a P-type oxide film was formed on the surface. The passive film formed at high potential had fewer defects and excellent protective performance. Therefore, AISI321 stainless steel could be passivated in ethylene glycol solution and the passive potential ranged from −0.15 V to 0.75 V. With the increase in passive potential, the protective performance of the passive film was significantly improved.

In the study of Nikhil and co-workers, ethyl-2-cyano-3-(4-(dimethylamino) phenyl) acrylate (ECDPA) and ZnO nanosheet composites were synthesized and used as corrosion inhibitors for copper in 1 M HCl. ECDPA acted as a barrier to acid molecules after adsorbing copper, delaying the corrosion of copper in hydrochloric acid [78]. EC-AFM analysis confirmed that ZnO nanoparticles promoted the adsorption of ECDPA. Adsorption/deposition of a small number of inhibitor molecules was found when ZnO was not added. After the addition of ZnO, the number of inhibitor molecules on the surface was significantly increased, and the size of the inhibitor molecule became larger as the ZnO calcination temperature increased, as shown in Figure 34.

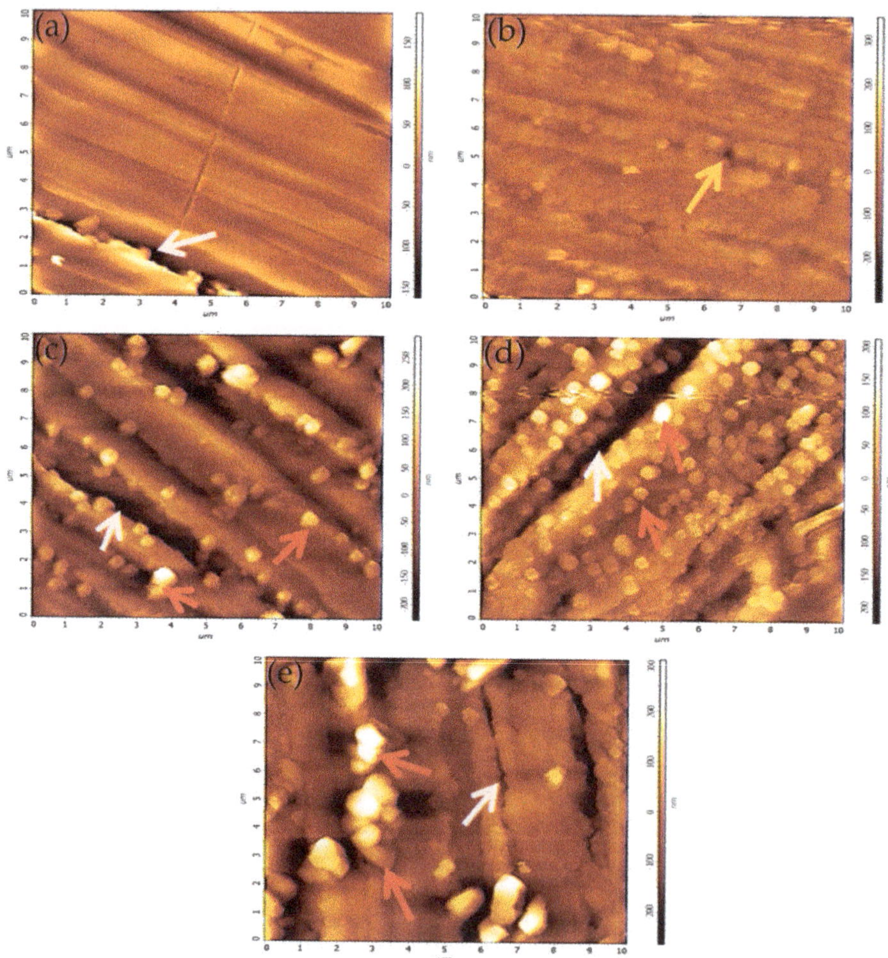

Figure 34. EC-AFM two-dimensional (2D) images of (**a**) polished copper, and copper (**b**) corroded, as well as inhibited by (**c**) ethyl-2-cyano-3-(4-(dimethylamino) phenyl) acrylate (ECDPA), (**d**) EZ3 (ECDPA–ZnO at 300 °C), and (**e**) EZ5 (ECDPA–ZnO at 500 °C) in 10 min of immersion in 1 M HCl at −0.005 V. Reprinted with permission from Reference [78]; copyright 2019 Elsevier Ltd.

The corrosion resistance tests of ECDPA, EZ3 (ECDPA–ZnO at 300 °C), and EZ5 (ECDPA–ZnO at 500 °C) showed that the composite had better protection performance than ECDPA alone. The maximum inhibition efficiency of ECDPA was approximately 75%, while the composite could be further improved to 78% (EZ3) and 81% (EZ5). The improved corrosion inhibition performance of ECDPA–ZnO may have been related to the inclusion of ZnO nanoparticles, which promoted the adsorption of ECDPA over Cu.

Shaban et al. [79] studied the inhibition of dibenzylsulfoxide (DBSO) and *p*-chlorobenzohydroxamic acid (*p*-Cl-BHA) on copper corrosion in 0.5 M NaCl and 0.1 M Na_2SO_4, respectively. In situ AFM was used to observe the corrosion and inhibition process of the electrode surface. When DBSO was not added, the surface became rougher due to the dissolution of copper. In the presence of DBSO, the surface was not subject to severe corrosion and a relatively smooth surface was formed, as shown in Figure 35. Therefore, the authors thought that DBSO inhibited the corrosion of copper in sulfate solutions by converting it to a more stable, less soluble sulfide compound.

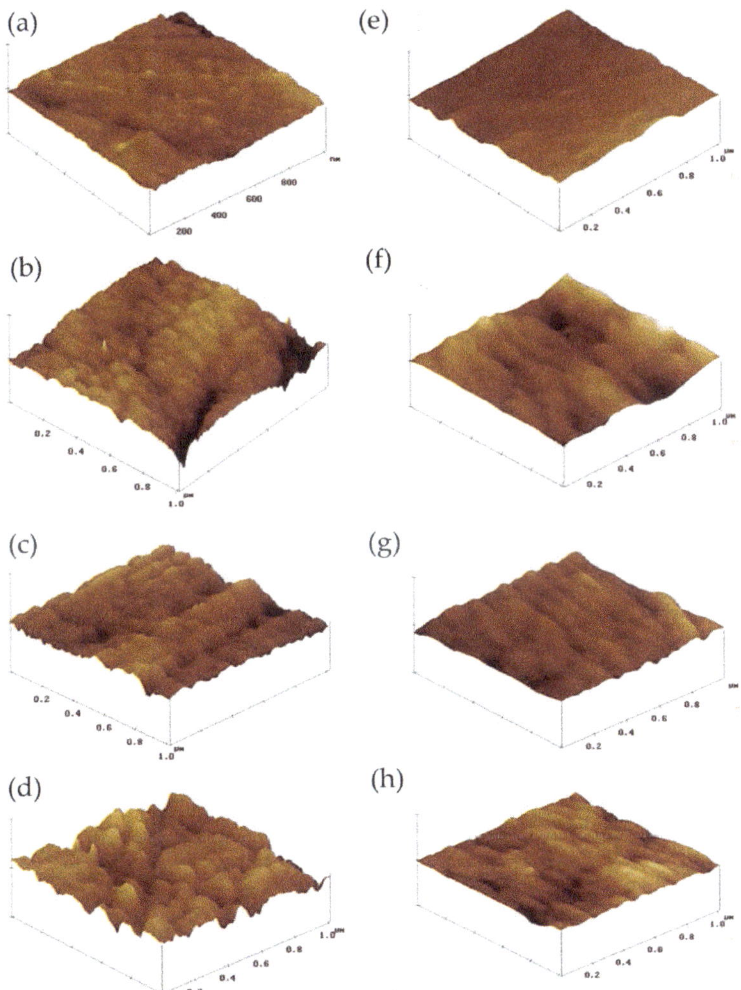

Figure 35. Morphological changes of copper in solution of 0.1 M Na_2SO_4 (**a–d**) and 0.1 M Na_2SO_4 + 0.5 mM dibenzylsulfoxide (DBSO) (**e–h**), at different times: (**a**,**e**) 0 min; (**b**,**f**) 15 min; (**c**,**g**) 30 min; (**d**,**h**) 45 min. Reprinted with permission from Reference [79]; copyright 1998 Springer.

In the absence of *p*-Cl-BHA, the formation and growth of pitting corrosion were detected. The addition of *p*-Cl-BHA significantly impeded the localized corrosion of copper and inhibited the production of corrosion products that occurred when *p*-Cl-BHA was not added, as shown in Figure 36. Thus, the formation of a stable complex by adsorbing a corrosion inhibitor on a corroded surface effectively hindered further dissolution of the metal.

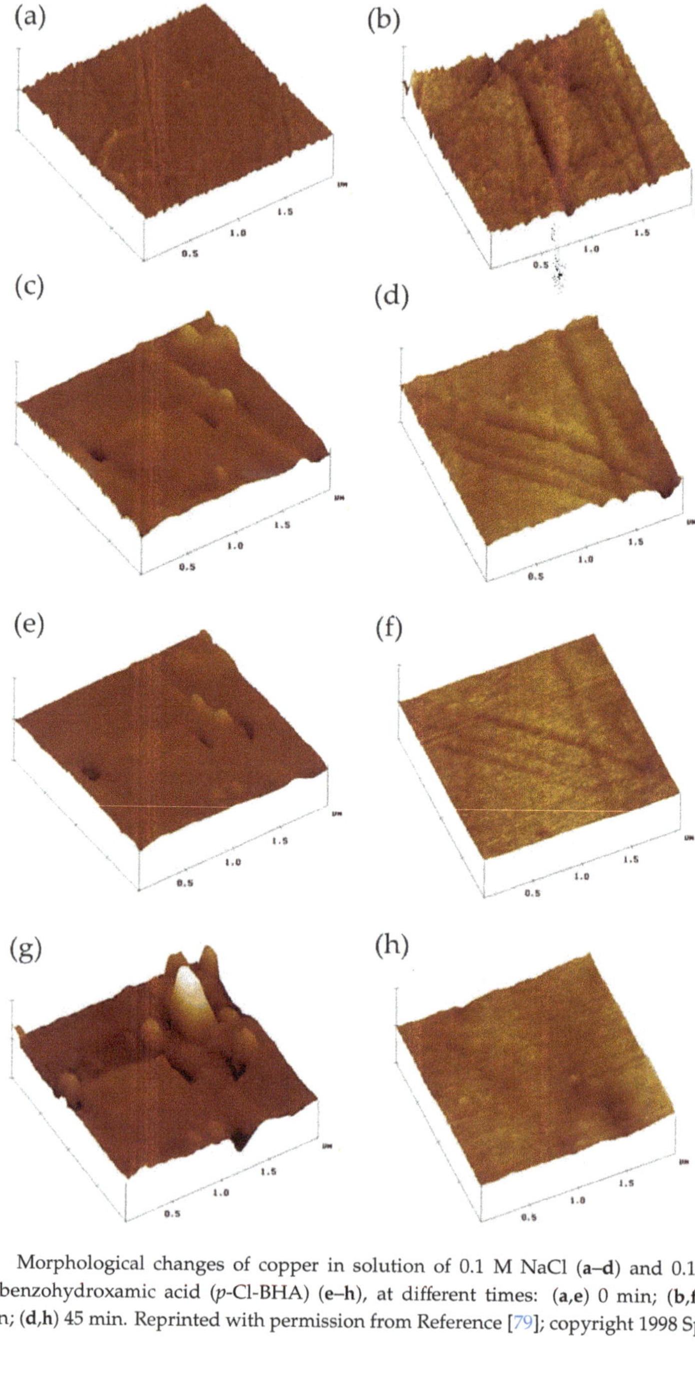

Figure 36. Morphological changes of copper in solution of 0.1 M NaCl (**a–d**) and 0.1 M NaCl + *p*-chlorobenzohydroxamic acid (*p*-Cl-BHA) (**e–h**), at different times: (**a**,**e**) 0 min; (**b**,**f**) 15 min; (**c**,**g**) 30 min; (**d**,**h**) 45 min. Reprinted with permission from Reference [79]; copyright 1998 Springer.

Cruickshank et al. [80] studied the anodic dissolution of polycrystalline copper in acid medium with or without corrosion inhibitor using EC-AFM. In 0.5 M H_2SO_4, the preferential corrosion of some grain surfaces occurred and then dissolved along grain boundaries. The addition of benzotriazole (BTA) formed a protective film that effectively inhibited the dissolution of copper. When the anode potential reached 200 mV, the protective film was stable. However, when the potential was as high as 300 mV, the film underwent local breakdown. In addition, Li et al. [81] investigated the effect of BTA on corrosion inhibition of copper using in situ EC-AFM in 0.01 M $NaHCO_3$. The addition of BTA caused the copper surface to form a BTA film, effectively protecting the copper from erosion. The pitting potential of the copper surface covered by the BTA film could be significantly increased by more than 700 mV. Therefore, the author believed that, in fact, pitting was not a concern in the presence of BAT.

Through the morphology testing of EC-AFM, the adsorption–desorption states of corrosion inhibitors on an electrode surface can be intuitively understood, which is very beneficial for the exploration of electrochemical mechanisms of corrosion inhibitors. The application of EC-AFM in corrosion inhibitor protection is summarized in Table 7.

Table 7. Summary of the application of EC-AFM in corrosion inhibitor protection.

	Materials	Corrosion Inhibitor	Solution	Mode	Reference
Corrosion inhibitor protection	Copper	Linear sodium heptanoate	0.1 M Na_2SO_4 solution	Contact Mode	[74]
	Pure cast aluminum	Polyacrylic acid (PAA) + KI	0.5 M H_2SO_4	Tapping Mode	[76]
	AISI321 stainless steel	36% ethylene glycol–water solution	Ethylene glycol–water solution	Tapping Mode	[77]
	Copper	Ethyl-2-cyano-3-(4-(dimethylamino) phenyl) acrylate (ECDPA) and ZnO nanosheet composites	1 M HCl		[78]
	Copper	Dibenzylsulfoxide (DBSO) and *p*-chlorobenzohydroxamic acid (*p*-Cl-BHA)	0.5 M NaCl and 0.1 M Na_2SO_4	Contact Mode	[79]
	Polycrystalline copper	Benzotriazole (BTA)	0.5 M H_2SO_4	Contact Mode	[80]
	Copper	Benzotriazole (BTA)	0.01 M $NaHCO_3$	Contact Mode	[81]

4. Conclusions and Perspectives

This review introduced the evolution, principles, and operation modes of EC-AFM, as well as illustrated its application in corrosion science through specific examples.

In summary, EC-AFM can not only perform real-time in situ research on micro-area corrosion (passivation) in the field of corrosion electrochemistry; it also has higher resolution, which provides detailed information on corrosion phenomena, such as activation and passivation at submicroscopic scales, especially in the early stages of corrosion. The surface topography changes of the target sample can be quantitatively analyzed with the information of surface height changes with applied voltage or time, helping us better study the corrosion process and mechanism. In recent years, EC-AFM was continuously improved in electrolytic cells, imaging modes, and probes. Moreover, EC-AFM was combined with SKPFM, SVET, and SECM to study the surface state of metals, greatly expanding the scope of application.

However, EC-AFM also has some problems that affect research, which brings about difficulties obtaining reliable surface topography features, such as drift during the scanning process, alignment of the laser, changes in the refractive index of the electrochemical medium as the process of corrosion goes on, finding accurate resonant frequencies, etc. In addition, the EC-AFM scanning rate is limited, and some rapid interface reactions cannot be monitored in real time.

In the future, the development of EC-AFM will be toward multi-functionality, high sensitivity, high speed, and high efficiency. The continuous improvement of EC-AFM will help carrying out more in-depth research on the in situ dynamic corrosion process, as well as promoting the study of corrosion science.

Author Contributions: Conceptualization, Z.W., M.H., and Y.L.; formal analysis, H.C., Z.Q., and Z.W.; writing—original draft, H.C.; writing—review and editing, H.C. and Z.W. All authors have read and agreed to the published version of the manuscript.

Funding: This research was financed by the National Key Research and Development Program of China (2018YFB0703500) and the National Natural Science Foundation of China (No. 51971155, No. 51771120, and No. 51304136).

Conflicts of Interest: The authors declare no conflicts of interest.

References

1. Kamachi Mudali, U.; Padhy, N. Electrochemical scanning probe microscope (EC-SPM) for the in situ corrosion study of materials: An overview with examples. *Corros. Rev.* **2011**, *29*, 1–2. [CrossRef]
2. Pondichery, S. A Study on the Effect of Magnetic Field on the Corrosion Behavior of Materials. Ph.D. Thesis, The University of Texas, Arlington, TX, USA, 2014.
3. Manne, S. Electrochemistry on a gold surface observed with the atomic force microscope. *J. Vac. Sci. Technol. B Microelectron. Nanometer Struct.* **1991**, *9*, 2. [CrossRef]
4. Binnig, G.; Quate, C.F.; Gerber, C. Atomic force microscope. *Phys. Rev. Lett.* **1986**, *56*, 930–933. [CrossRef] [PubMed]
5. Binnig, G.; Rohrer, H.; Gerber, C.; Weibel, E. Tunneling through a controllable vacuum gap. *Appl. Phys. Lett.* **1982**, *40*, 178–180. [CrossRef]
6. Binnig, G.; Rohrer, H.; Gerber, C.; Weibel, E. 7 × 7 reconstruction on Si (111) resolved in real space. *Phys. Rev. Lett.* **1983**, *50*, 120. [CrossRef]
7. Binnig, G.; Rohrer, H.J. Surface imaging by scanning tunneling microscopy. *Ultramicroscopy* **1983**, *11*, 157–160. [CrossRef]
8. Binnig, G.; Rohrer, H.J.S. Scanning tunneling microscopy. *Surf. Sci.* **1983**, *126*, 236–244. [CrossRef]
9. Manne, S.; Hansma, P.K.; Massie, J.; Elings, V.B.; Gewirth, A.A. Atomic-resolution electrochemistry with the atomic force microscope: Copper deposition on gold. *Science* **1991**, *251*, 183–186. [CrossRef]
10. Ohnesorge, F.; Binnig, G. True atomic resolution by atomic force microscopy through repulsive and attractive forces. *Science* **1993**, *260*, 1451–1456. [CrossRef]
11. Utsunomiya, T.; Yokota, Y.; Fukui, K.-I. Electrochemical Atomic Force Microscopy. In *Compendium of Surface and Interface Analysis*; The Surface Science Society of Japan, Ed.; Springer: Singapore, 2018.
12. Reggente, M.; Passeri, D.; Rossi, M.; Tamburri, E.; Terranova, M.L. Electrochemical Atomic Force Microscopy: In Situ Monitoring of Electrochemical Processes. In *Nanoinnovation 2016*; Rossi, M., Dini, L., Passeri, D., Antisari, M.V., Eds.; AIP Publishing: Melville, NY, USA, 2017.
13. Xu, Z.; Qi, D.M.; Jiang, L.; Chen, Y.; Zhang, Z.; Zhang, J.Q. Progress of Research on Underpotential Deposition-II. Research Techniques and Application of Underpotential Deposition. *Acta Phys. Chim. Sin.* **2015**, *31*, 1231–1250.
14. Nault, L.; Taofifenua, C.; Anne, A.; Chovin, A.; Demaille, C.; Besong-Ndika, J.; Cardinale, D.; Carette, N.; Michon, T.; Walter, J. Electrochemical Atomic Force Microscopy Imaging of Redox-Immunomarked Proteins on Native Potyviruses: From Subparticle to Single-Protein Resolution. *ACS Nano* **2015**, *9*, 4911–4924. [CrossRef] [PubMed]
15. Deng, J.; Nellist, M.R.; Stevens, M.B.; Dette, C.; Wang, Y.; Boettcher, S.W. Morphology Dynamics of Single-Layered $Ni(OH)_2$/NiOOH Nanosheets and Subsequent Fe Incorporation Studied by & ITin Situ&IT Electrochemical Atomic Force Microscopy. *Nano Lett.* **2017**, *17*, 6922–6926. [PubMed]
16. Wang, S.W.; Zhang, W.C.; Chen, Y.N.; Dai, Z.W.; Zhao, C.C.; Wang, D.Y.; Shen, C. Operando study of Fe_3O_4 anodes by electrochemical atomic force microscopy. *Appl. Surf. Sci.* **2017**, *426*, 217–223. [CrossRef]
17. Valtiner, M.; Ankah, G.N.; Bashir, A.; Renner, F.U. Atomic force microscope imaging and force measurements at electrified and actively corroding interfaces: Challenges and novel cell design. *Rev. Sci. Instrum.* **2011**, *82*, 023703. [CrossRef]

18. Martin-Olmos, C.; Rasool, H.M.; Weiller, B.H.; Gimzewski, J.K. Graphene MEMS: AFM Probe Performance Improvement. *ACS Nano* **2013**, *7*, 4164–4170. [CrossRef]
19. Zhang, D.X.; Zhang, H.J.; Lin, X.F. In-situ study of corrosion with an atomic force microscope scanning in liquids. In *Nanophotonics, Nanostructure, and Nanometrology*; Zhu, X., Chow, S.Y., Arakawa, Y., Eds.; SPIE: Bellingham, DC, USA, 2005.
20. Zhong, J. From simple to complex: Investigating the effects of lipid composition and phase on the membrane interactions of biomolecules using in situ atomic force microscopy. *Integr. Biol.* **2011**, *3*, 632–644. [CrossRef]
21. Butt, H.J.; Cappella, B.; Kappl, M.J. Force measurements with the atomic force microscope: Technique, interpretation and applications. *Surf. Sci. Rep.* **2005**, *59*, 1–152. [CrossRef]
22. Marti, O.; Drake, B.; Hansma, P.K. Atomic force microscopy of liquid-covered surfaces: Atomic resolution images. *Appl. Phys. Lett.* **1987**, *51*, 484–486. [CrossRef]
23. Martin, Y.; Williams, C.C.; Wickramasinghe, H.K. Atomic force microscope–force mapping and profiling on a sub 100-Å scale. *J. Appl. Phys.* **1987**, *61*, 4723–4729. [CrossRef]
24. Miyata, K.; Tracey, J.; Miyazawa, K.; Haapasilta, V.; Spijker, P.; Kawagoe, Y.; Foster, A.S.; Tsukamoto, K.; Fukuma, T.J. Dissolution processes at step edges of calcite in water investigated by high-speed frequency modulation atomic force microscopy and simulation. *Nano Lett.* **2017**, *17*, 4083–4089. [CrossRef]
25. Alessandrini, A.; Facci, P. AFM: A versatile tool in biophysics. *Meas. Sci. Technol.* **2005**, *16*, R65–R92. [CrossRef]
26. Huang, L.; Su, C. A torsional resonance mode AFM for in-plane tip surface interactions. *Ultramicroscopy* **2004**, *100*, 277–285. [CrossRef] [PubMed]
27. Bhushan, B.; Kasai, T. A surface topography-independent friction measurement technique using torsional resonance mode in an AFM. *Nanotechnology* **2004**, *15*, 923–935. [CrossRef]
28. Kasai, T.; Bhushan, B.; Huang, L.; Su, C. Topography and phase imaging using the torsional resonance mode. *Nanotechnology* **2004**, *15*, 731–742. [CrossRef]
29. Hu, S.; Mininni, L.; Hu, Y.; Erina, N.; Kindt, J.; Su, C. *High-Speed Atomic Force Microscopy and Peak Force Tapping Control*; SPIE: Bellingham, DC, USA, 2012.
30. Alsteens, D.; Dupres, V.; Yunus, S.; Latge, J.P.; Heinisch, J.J.; Dufrene, Y.F. High-resolution imaging of chemical and biological sites on living cells using peak force tapping atomic force microscopy. *Langmuir* **2012**, *28*, 16738–16744. [CrossRef] [PubMed]
31. Liu, G.; Liu, J.; Sun, H.; Zheng, X.; Liu, Y.; Li, X.; Qi, H.; Bai, X.; Jackson, K.A.; Tao, X. In situ imaging of on-surface, solvent-free molecular single-crystal growth. *J. Am. Chem. Soc.* **2015**, *137*, 4972–4975. [CrossRef] [PubMed]
32. Luo, D.; Yang, F.; Wang, X.; Sun, H.; Gao, D.; Li, R.; Yang, J.; Li, Y. Anisotropic etching of graphite flakes with water vapor to produce armchair-edged graphene. *Small* **2014**, *10*, 2809–2814. [CrossRef]
33. Li, Y.; Cheng, Y.F. Passive film growth on carbon steel and its nanoscale features at various passivating potentials. *Appl. Surf. Sci.* **2017**, *396*, 144–153. [CrossRef]
34. Padhy, N.; Paul, R.; Kamachi Mudali, U.; Raj, B. Morphological and compositional analysis of passive film on austenitic stainless steel in nitric acid medium. *Appl. Surf. Sci.* **2011**, *257*, 5088–5097. [CrossRef]
35. Kreta, A.; Rodošek, M.; Slemenik Perše, L.; Orel, B.; Gaberšček, M.; Šurca Vuk, A. In situ electrochemical AFM, ex situ IR reflection–absorption and confocal Raman studies of corrosion processes of AA 2024-T3. *Corros. Sci.* **2016**, *104*, 290–309. [CrossRef]
36. Liu, Y.; Gilbert, J.L. Effect of simulated inflammatory conditions and potential on dissolution and surface oxide of CoCrMo alloy: In situ electrochemical atomic force microscopy study. *Electrochim. Acta* **2018**, *262*, 252–263. [CrossRef]
37. Bearinger, J.P.; Orme, C.A.; Gilbert, J.L. In situ imaging and impedance measurements of titanium surfaces using AFM and SPIS. *Biomaterials* **2003**, *24*, 1837–1852. [CrossRef]
38. Bearinger, J.P.; Orme, C.A.; Gilbert, J.L. Direct observation of hydration of TiO_2 on Ti using electrochemical AFM: Freely corroding versus potentiostatically held conditions. *Surf. Sci.* **2001**, *491*, 370–387. [CrossRef]
39. Wang, Z.M.; Zhang, J.; Chang, X.C.; Hou, W.L.; Wang, J.Q. Structure inhibited pit initiation in a Ni–Nb metallic glass. *Corros. Sci.* **2010**, *52*, 1342–1350. [CrossRef]
40. Zhang, S.D.; Liu, Z.W.; Wang, Z.M.; Wang, J.Q. In situ EC-AFM study of the effect of nanocrystals on the passivation and pit initiation in an Al-based metallic glass. *Corros. Sci.* **2014**, *83*, 111–123. [CrossRef]

41. Li, Y.; Cheng, Y.F. In-situcharacterization of the early stage of pipeline steel corrosion in bicarbonate solutions by electrochemical atomic force microscopy. *Surf. Interface Anal.* **2017**, *49*, 133–139. [CrossRef]
42. Izquierdo, J.; Fernández-Pérez, B.M.; Eifert, A.; Souto, R.M.; Kranz, C. Simultaneous Atomic Force—Scanning Electrochemical Microscopy (Afm-Secm) Imaging of Copper Dissolution. *Electrochim. Acta* **2016**, *201*, 320–332. [CrossRef]
43. Izquierdo, J.; Eifert, A.; Kranz, C.; Souto, R.M. In situ investigation of copper corrosion in acidic chloride solution using atomic force—Scanning electrochemical microscopy. *Electrochim. Acta* **2017**, *247*, 588–599. [CrossRef]
44. Ding, Y.; Zhao, R.; Qin, Z.; Wu, Z.; Wang, L.; Liu, L.; Lu, W. Evolution of the Corrosion Product Film on Nickel-Aluminum Bronze and Its Corrosion Behavior in 3.5 wt % NaCl Solution. *Materials* **2019**, *12*, 209. [CrossRef]
45. Conradi, M.; Schön, P.M.; Kocijan, A.; Jenko, M.; Vancso, G.J. Surface analysis of localized corrosion of austenitic 316L and duplex 2205 stainless steels in simulated body solutions. *Mater. Chem. Phys.* **2011**, *130*, 708–713. [CrossRef]
46. Martin, F.A.; Bataillon, C.; Cousty, J. In situ AFM detection of pit onset location on a 304L stainless steel. *Corros. Sci.* **2008**, *50*, 84–92. [CrossRef]
47. Wijesinghe, T.L.S.L.; Blackwood, D.J. Real time pit initiation studies on stainless steels: The effect of sulphide inclusions. *Corros. Sci.* **2007**, *49*, 1755–1764. [CrossRef]
48. Zhang, Q.; Wang, R.; Kato, M.; Nakasa, K. Observation by atomic force microscope of corrosion product during pitting corrosion on SUS304 stainless steel. *Scr. Mater.* **2005**, *52*, 227–230. [CrossRef]
49. Qu, J.E.; Guo, X.P.; Wang, H.R.; Huang, J.Y. Corrosion behavior of pure aluminum in $FeCl_3$ solution. *Trans. Nonferrous Met. Soc. China* **2006**, *16*, 1460–1466. [CrossRef]
50. Davoodi, A.; Pan, J.; Leygraf, C.; Norgren, S. In situ investigation of localized corrosion of aluminum alloys in chloride solution using integrated EC-AFM/SECM techniques. *Electrochem. Solid-State Lett.* **2005**, *8*, B21–B24. [CrossRef]
51. Amin, M.A.; Abd El-Rehim, S.S.; Aarão Reis, F.D.A.; Cole, I.S. Metastable and stable pitting events at zinc passive layer in alkaline solutions. *Ionics* **2013**, *20*, 127–136. [CrossRef]
52. Kim, Y.H.; Rae, C.Y.; Kim, K.H.; Chung, W.S. Study of initial stage corrosion of hot-dip-coated zinc surface using in situ AFM. *J. Electrochem. Soc.* **2004**, *151*, B319–B324. [CrossRef]
53. Jia, R.; Yu, S.; Li, D.; Zhang, T.; Wang, F.; Zhong, C. Study on the effect of mischmetal (La,Ce) on the micro-galvanic corrosion of AZ91 alloy using multiscale methods. *J. Alloys Compd.* **2019**, *778*, 427–438. [CrossRef]
54. Zhang, F.; Nilsson, J.-O.; Pan, J. In Situ and Operando AFM and EIS Studies of Anodization of Al 6060: Influence of Intermetallic Particles. *J. Electrochem. Soc.* **2016**, *163*, C609–C618. [CrossRef]
55. Bettini, E.; Kivisäkk, U.; Leygraf, C.; Pan, J. Study of corrosion behavior of a 22% Cr duplex stainless steel: Influence of nano-sized chromium nitrides and exposure temperature. *Electrochim. Acta* **2013**, *113*, 280–289. [CrossRef]
56. Yasakau, K.A.; Kallip, S.; Lisenkov, A.; Ferreira, M.G.S.; Zheludkevich, M.L. Initial stages of localized corrosion at cut-edges of adhesively bonded Zn and Zn-Al-Mg galvanized steel. *Electrochim. Acta* **2016**, *211*, 126–141. [CrossRef]
57. Depentori, F.; Forcellini, C.; Andreatta, F.; Marin, E.; Maschio, S.; Brunke, F.; Benfer, S.; Siemers, C.; Fedrizzi, L.; Fürbeth, W. Oxidation of neodymium precipitates in a $Ti_6Al_4V_2Nd$ alloy in sodium chloride solution. *Mater. Corros.* **2016**, *67*, 277–285. [CrossRef]
58. Davoodi, A.; Pan, J.; Leygraf, C.; Norgren, S. Integrated AFM and SECM for in situ studies of localized corrosion of Al alloys. *Electrochim. Acta* **2007**, *52*, 7697–7705. [CrossRef]
59. Moore, S.; Burrows, R.; Picco, L.; Martin, T.L.; Greenwell, S.J.; Scott, T.B.; Payton, O.D. A study of dynamic nanoscale corrosion initiation events using HS-AFM. *Faraday Discuss.* **2018**, *210*, 409–428. [CrossRef] [PubMed]
60. Padhy, N.; Ningshen, S.; Kamachi Mudali, U.; Raj, B. In situ surface investigation of austenitic stainless steel in nitric acid medium using electrochemical atomic force microscopy. *Scr. Mater.* **2010**, *62*, 45–48. [CrossRef]
61. Williford, R.E.; Windisch, C.F., Jr.; Jones, R.H. In situ observations of the early stages of localized corrosion in Type 304 SS using the electrochemical atomic force microscope. *Mater. Sci. Eng. A* **2000**, *288*, 54–60. [CrossRef]

62. Fu, N.; Tang, X.; Li, D.Y.; Parent, L.; Tian, H. In situ investigation of local corrosion at interphase boundary under an electrochemical-atomic force microscope. *J. Solid State Electrochem.* **2014**, *19*, 337–344. [CrossRef]
63. Shi, Y.; Collins, L.; Balke, N.; Liaw, P.K.; Yang, B. In-situ electrochemical-AFM study of localized corrosion of Al CoCrFeNi high-entropy alloys in chloride solution. *Appl. Surf. Sci.* **2018**, *439*, 533–544. [CrossRef]
64. Bettini, E.; Eriksson, T.; Boström, M.; Leygraf, C.; Pan, J. Influence of metal carbides on dissolution behavior of biomedical CoCrMo alloy: SEM, TEM and AFM studies. *Electrochim. Acta* **2011**, *56*, 9413–9419. [CrossRef]
65. Davoodi, A.; Pan, J.; Leygraf, C.; Norgren, S. Probing of local dissolution of Al-alloys in chloride solutions by AFM and SECM. *Appl. Surf. Sci.* **2006**, *252*, 5499–5503. [CrossRef]
66. Li, J.; Ecco, L.; Ahniyaz, A.; Pan, J. Probing electrochemical mechanism of polyaniline and CeO_2 nanoparticles in alkyd coating with in-situ electrochemical-AFM and IRAS. *Prog. Org. Coat.* **2019**, *132*, 399–408. [CrossRef]
67. Li, J.; Huang, H.; Fielden, M.; Pan, J.; Ecco, L.; Schellbach, C.; Delmas, G.; Claesson, P.M. Towards the mechanism of electrochemical activity and self-healing of 1 wt% PTSA doped polyaniline in alkyd composite polymer coating: Combined AFM-based studies. *RSC Adv.* **2016**, *6*, 19111–19127. [CrossRef]
68. Li, J.; Ecco, L.; Ahniyaz, A.; Fedel, M.; Pan, J. In Situ AFM and Electrochemical Study of a Waterborne Acrylic Composite Coating with CeO2Nanoparticles for Corrosion Protection of Carbon Steel. *J. Electrochem. Soc.* **2015**, *162*, C610–C618. [CrossRef]
69. Liu, L.; Li, Y.; Wang, F. Pitting mechanism on an austenite stainless steel nanocrystalline coating investigated by electrochemical noise and in-situ AFM analysis. *Electrochim. Acta* **2008**, *54*, 768–780. [CrossRef]
70. Liu, L.; Li, Y.; Wang, F. Influence of nanocrystallization on pitting corrosion behavior of an austenitic stainless steel by stochastic approach and in situ AFM analysis. *Electrochim. Acta* **2010**, *55*, 2430–2436. [CrossRef]
71. Pan, C.; Liu, L.; Li, Y.; Wang, F. Pitting corrosion of 304ss nanocrystalline thin film. *Corros. Sci.* **2013**, *73*, 32–43. [CrossRef]
72. Pan, C.; Liu, L.; Li, Y.; Zhang, B.; Wang, F. The Electrochemical Corrosion Behavior of Nanocrystalline 304 Stainless Steel Prepared by Magnetron Sputtering. *J. Electrochem. Soc.* **2012**, *159*, C453–C460. [CrossRef]
73. Pan, C.; Liu, L.; Li, Y.; Wang, S.; Wang, F. Passive film growth mechanism of nanocrystalline 304 stainless steel prepared by magnetron sputtering and deep rolling techniques. *Electrochim. Acta* **2011**, *56*, 7740–7748. [CrossRef]
74. Rocca, E.; Bertrand, G.; Rapin, C.; Labrune, J.C. Inhibition of copper aqueous corrosion by non-toxic linear sodium heptanoate: Mechanism and ECAFM study. *J. Electroanal. Chem.* **2001**, *503*, 133–140. [CrossRef]
75. Bertrand, G.; Rocca, E.; Savall, C.; Rapin, C.; Labrune, J.C.; Steinmetz, P.J. In-situ electrochemical atomic force microscopy studies of aqueous corrosion and inhibition of copper. *J. Electroanal. Chem.* **2000**, *489*, 38–45. [CrossRef]
76. Umoren, S.A.; Pan, C.; Li, Y.; Wang, F.H. Elucidation of mechanism of corrosion inhibition by polyacrylic acid and synergistic action with iodide ions byin-situAFM. *J. Adhes. Sci. Technol.* **2013**, *28*, 31–37. [CrossRef]
77. Zhang, X. Corrosion behavior of AISI321 stainless steel in an ethylene glycol-water solution. *Int. J. Electrochem. Sci.* **2019**, *14*, 2683–2692. [CrossRef]
78. Ji, G.; Prakash, R. Composites of Donor-π-Acceptor type configured organic compound and porous ZnO nano sheets as corrosion inhibitors of copper in chloride environment. *J. Mol. Liq.* **2019**, *280*, 160–172.
79. Shaban, A.; Kálmán, E.; Telegdi, J.; Pálinkás, G.; Dóra, G.J. Corrosion and inhibition of copper in different electrolyte solutions. *Appl. Phys. A* **1998**, *66*, S545–S549. [CrossRef]
80. Cruickshank, B.J.; Gewirth, A.A.; Rynders, R.M.; Alkire, R.C. In situ observations of shape evolution during copper dissolution using atomic force microscopy. *J. Electrochem. Soc.* **1992**, *139*, 2829–2832. [CrossRef]
81. Li, J.; Lampner, D.J.C.; Physicochemical, S.A.; Aspects, E. In-situ AFM study of pitting corrosion of Cu thin films. *Colloids Surf. A* **1999**, *154*, 227–237. [CrossRef]

materials

MDPI

Article

Effect of Pre-Corrosion Pits on Residual Fatigue Life for 42CrMo Steel

Dezheng Liu [1,*], Yan Li [1], Xiangdong Xie [2,*] and Jing Zhao [1]

1 Hubei Key Laboratory of Power System Design and Test for Electrical Vehicle, Hubei University of Arts and Science; Xiangyang 441053, China
2 School of Urban Construction, Yangtze University, Jingzhou 201800, China
* Correspondence: liudezheng@hbuas.edu.cn (D.L.); xdxie@yangtzeu.edu.cn (X.X.)

Received: 13 June 2019; Accepted: 1 July 2019; Published: 2 July 2019

Abstract: The effect of pre-corrosion pits on residual fatigue life for the 42CrMo steel (American grade: AISI 4140) is investigated using the accelerated pre-corrosion specimen in the saline environment. Different pre-corroded times are used for the specimens, and fatigue tests with different loads are then carried out on specimens. The pre-corrosion fatigue life is studied, and the fatigue fracture surfaces are examined by a surface profiler and a scanning electron microscope (SEM) to identify the crack nucleation sites and to determine the size and geometry of corrosion pits. Moreover, the stress intensity factor varying with corrosion pits in different size parameters is analyzed based on finite element (FE) software ABAQUS to derive the regression formula of the stress intensity factor. Subsequently, by integrating the regression formula with the Paris formula, the residual fatigue life is predicted and compared with experimental results, and the relationship of the stress intensity factor, pit depth, and residual fatigue life are given under different corrosion degrees. The fatigue life predicted by the coupled formula agrees well with experiment results. It is observed from the SEM images that higher stress amplitude and longer pre-corroded time can significantly decrease the residual fatigue life of the steel. Additionally, the research work has brought about the discovery that the rate of crack extension accelerates when the crack length increases. The research in this paper also demonstrates that the corrosion pit size can be used as a damage index to assess the residual fatigue life.

Keywords: pre-corrosion pits; residual fatigue life; 42CrMo steel; stress intensity factor

1. Introduction

Most engineering materials are subject to corrosion, and corrosion research has received much attention from different perspectives [1–7]. The 42CrMo (American grade: AISI 4140) is a commonly used ultrahigh-strength steel. Due to its high strength, toughness, and hardenability, 42CrMo steel is widely used in quenched and tempered heavy forgings to build components such as pressure vessels, gears, vehicle axles, and deep oil drilling rod subs [8,9]. However, their corrosion resistance is relatively poor and this material is prone to corrosion. Corrosion pits can weaken the strength of the structure and decrease its fatigue life [10,11], since the fatigue cracks are easy to nucleate at corrosion pits and propagate rapidly under dynamic loads [12]. Thus, to further understand corrosion effects on the strength and residual fatigue life of 42CrMo material, the stress state around corrosion pits should be reasonably estimated.

In recent years, fatigue performance of pre-corroded metallic alloys has been studied extensively. For example, recent works [13–18] have reported the description of the stress concentration arising from corrosion pits based on the finite element (FE) method. Sharland [13] developed a mechanistic model of the propagation stage of an established pit or crevice to simulate the evolution of pit geometry and to describe the corrosion process. However, Wang [14] indicated that a detailed description of both the chemical reactions and the ionic transport was not included in it [13]. Through the use of interaction function to simulate the effect of the corrosion potential and corrosion current density on the corrosion process, Xu [15] developed an FE model for the simulation of the mechanoelectrochemical effect of pipeline corrosion. Because the fatigue strength of the metallic alloys decreases differently under the different corrosion depths and stress amplitudes [16], Pidaparti [17] established a new FE model to investigate the stress state around corrosion pits. However, the proposed FE model in [17] cannot be used to depict the growth process of corrosion pits with time. Through the use of the damage tolerance method to predict the corrosion fatigue life for 7075-T6 stainless steel, Huang [18] indicted that the nucleation time of pits and the material constant of crack propagation can affect the residual fatigue life of stainless steel significantly. On the basis of [16–18], Zheng [19] proposed that the decrease of fatigue life depends on the pre-corroded time and the size of pre-corrosion pits. Most recent studies on the fatigue life of steel in corrosion environments only consider the factors such as mass loss rate and corrosion pit depth, and rarely involve the stress intensity factor varying with corrosion pits in different size parameters. The stress concentration caused by corrosion pits can dramatically influence the basic macroscopic mechanical properties of a steel [20]. Despite the extensive research on the corrosion of metallic alloys [21–24], there are few detailed studies of the effect of pre-corrosion pits on residual fatigue life for 42CrMo steel.

In this study, the effect of stress distribution around corrosion pits on the stress intensity factor is firstly investigated. The crack nucleation sites and the geometry of corrosion pits are examined by a surface profiler and a scanning electron microscope (SEM), and then the relationship between the fatigue life and the maximum applied stress is obtained by fatigue tests of different pre-corroded specimens. Based on experiment results, an FE model is conducted to investigate stress distribution around corrosion pits, and regression formula of the stress intensity factor varying with corrosion pits in different size parameters is then derived. Subsequently, by coupling the regression formula with the Paris formula [25], the fatigue performance is predicted by the FE method and then compared with experimental tests. Under the different stress amplitudes and pre-corroded times, the FE predictions are finally validated by comparison with the experiment results.

2. Materials and Methods

2.1. Material and Specimen Preparation

To better understand the stress intensity factor varying with corrosion pits in different size parameters and to accurately depict the residual fatigue life of 42CrMo steel subject to corrosion, it is necessary to quantitatively investigate the effects of pre-corrosion levels on the fatigue behavior of 42CrMo steel. Experimental specimens were fabricated using the 42CrMo steel (American grade: AISI 4140), which was provided by Baowu Steel Company (Wuhan, China). Specimens were manufactured in the form of flat bare sheets with a thickness of 6 mm by wire cutting machine, the sizes of specimens are provided in Figure 1. The chemical composition was measured using an FLS980-stm Edinburgh fluorescence spectrometer (Edinburgh Instruments Ltd., Livingston, UK). A comparison of the measured chemical composition of 42CrMo specimens and the standard of GB/T3077-2015 (the national standard of the People's Republic of China for alloy structural steel) for 42CrMo steel is presented in Table 1. It can be seen from Table 1 that the chemical composition of the specimens used in this study meets the requirement of the national standard.

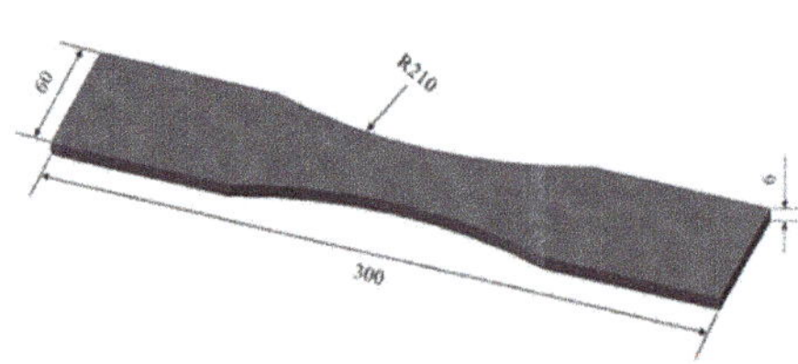

Figure 1. Schematic diagram of specimen sizes (mm).

Table 1. A comparison of the measured chemical composition of 42CrMo specimens and the standard of GB/T3077-2015 for 42CrMo steel (wt.%).

wt.%	C	S	P	Cr	Ni	Mn	Si	Cu	Mo
42CrMo	0.42	0.002	0.022	1.08	0.02	0.69	0.28	0.02	0.18
GB/T3077-2015	0.38–0.45	≤0.035		0.90–1.20	≤0.30	0.50–0.80	0.17–0.37	≤0.25	0.15–0.25

Twelve smooth samples were numbered and divided into four groups. The accelerated neutral salt spray corrosion experiment was established to achieve the pre-corroded steel plates based on the standard of JIS H 8502.5 [26]. A schematic drawing of the sample preparation and the pre-corroded experiment procedure is shown in Figure 2. The sample preparation was carried out in five steps: (1) samples were manufactured in the form of flat bare sheets with the dimensions in Figure 1 by wire cutting machine; (2) the surface of each sample was burnished by the 1000 grit abrasive paper to dispel burrs; (3) samples were rinsed in distilled water; (4) samples were degreased through the use of acetone solvent; (5) all surfaces of samples were rinsed with deionized water and dried by a blower for the pre-corroded experiment.

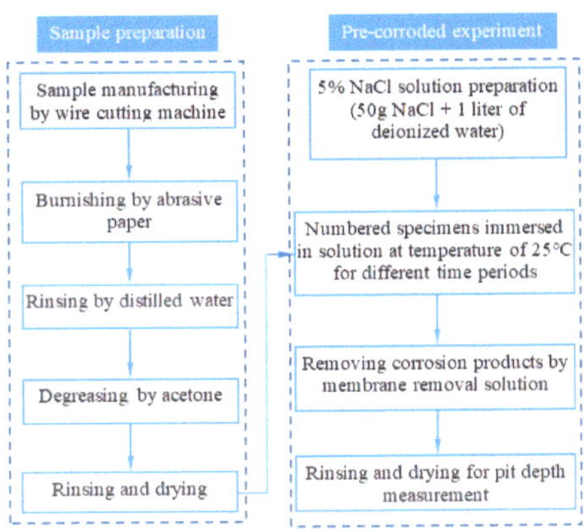

Figure 2. Schematic drawing of the sample preparation and the pre-corroded experiment procedure.

In this study, the pre-corroded experiment procedure can be devised as four main steps: (1) the pure NaCl crystals of 50 g were weighed by an electronic balance and poured into one liter of deionized water and stirred with glass rod at an ambient temperature of (25 ± 1 °C) for 5 min; (2) four groups of numbered samples were alternately immersed in 5% NaCl solution at an ambient temperature of (25 ± 1 °C) for 0 h, 24 h, 48 h, and 96 h, respectively; (3) the corrosion products were removed by membrane removal solution (prepared with 100 mL HCl, 100 mL deionized water and 0.6 g $C_6H_{12}N_4$); (4) all surfaces of samples were rinsed with deionized water and dried by a blower for pit depth measurement.

2.2. Pre-Corrosion Pit Measurements and Fatigue Test

To achieve the morphology of the corrosion surface and to determine the size and geometry of pre-corrosion pits, a non-contact Dektak150 surface profiler (Veeco Instruments Shanghai Co., Ltd., Shanghai, China) and an S-4800 scanning electron microscope (Hitachi, Tokyo, Japan) were used in this study. A region of 30 mm × 20 mm (along the directions of longitude and transverse, respectively) for each specimen was approximately arranged at the area of pre-estimated fatigue fracture. It is noted that the rate of corrosion changes steadily with the increase of pre-corroded time [27], thus the relationship between the corrosion time and the depth of pre-corroded pits can be expressed by an exponential function. Based on reference [28], the coefficient of exponential function can be determined by fitting the measured data, and exponential function expression can be described as:

$$\varphi = 1.235\ t^{0.775} \tag{1}$$

where φ is the depth of a pre-corroded pit (μm) and t is the pre-corroded time (h).

Based on the standard of ASTM E468-90 [29], the fatigue performance of corroded specimens (24 h, 48 h, and 96 h) were evaluated by a hydraulic universal testing machine (Instron-8803, Norwood, MA, USA) under a cyclic load amplitude with a frequency of 10 Hz. The loading operation was force controlled with a proportional error of ±1%. A group of un-corroded specimens were also tested as a control for fatigue studies. The stress ratio was taken as 0.1, and the maximum applied stresses were 100 MPa, 200 MPa, and 300 MPa, respectively. After testing, the fractured specimens were processed by cutting machine and the morphology of the corrosion surface was observed by surface profiler and scanning electron microscope.

The pre-corrosion pit reduced the fatigue life substantially, particularly the interacting pits and sharp pits [30]. In this study, the Paris theory [25] was applied to analyze the relationship between the morphology of fatigue cracks and the stress state around corrosion pits. According to Paris theory [25], the stress intensity factor of cracks can be described as:

$$K_I = \sigma\ \sqrt{\pi a} f(\beta) \tag{2}$$

$$\beta = \frac{a}{b} \tag{3}$$

where K_I is the stress intensity factor of the crack, σ is the external stress under the plane stress state, and a is the crack length and b is the specimen width.

2.3. FE Model

The FE analysis model of the 42CrMo steel in the form of a corroded flat sheet with a thickness of 6 mm was created using the HyperWorks software (HyperWorks 11.0, Altair Corp., Troy, MI, USA). On the basis of the experiment test, a single semi-elliptical pit model was conducted, and the symmetry plane of the flat sheet and refined mesh around the pit are shown in Figure 3.

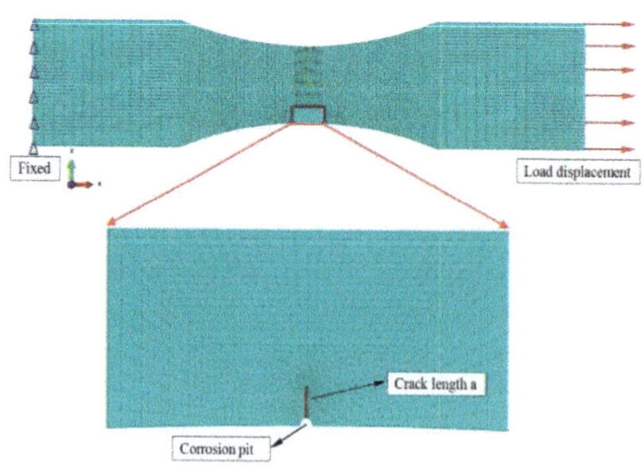

Figure 3. The FE analysis model.

The CPS8 quadrilateral element type was adopted in the FE model and cracks were prefabricated on the surface through the use of an assigned seam function that was provided by HyperWorks. The sweeping method [31] was used to generate the mesh in the integral region and the singularity of the mesh was controlled by the mesh regeneration technique [32]. The use of the above techniques [31,32] with local refinement and appropriate local mesh density can improve the precision of the calculation.

Subsequently, the FE analysis model with a preset corrosion pit was imported into the ABAQUS software (ABAQUS 6.10, Dassault Systemes Simulia Corp., Johnston, RI, USA) to analyze the stress and fatigue performance of the specimens.

3. Results and Discussion

3.1. Corrosion Surface Characterization

When the 42CrMo specimens were immersed in NaCl solution, the Cl element in the solution adhered to the surface of specimens and reacted with the Fe^{2+} in the metal to form a soluble clathrate, which resulted in the anodic dissolution of the metal surface and the formation of corrosion pits. With the increase of pre-corroded time, more metal dissolved in the anode and the size of corrosion pit became larger and larger [33]. The specimens pre-corroded for 0 h, 24 h, 48 h, and 96 h were examined by an S-4800 scanning electron microscope and the morphologies of corrosion surface characterization are shown in Figure 4.

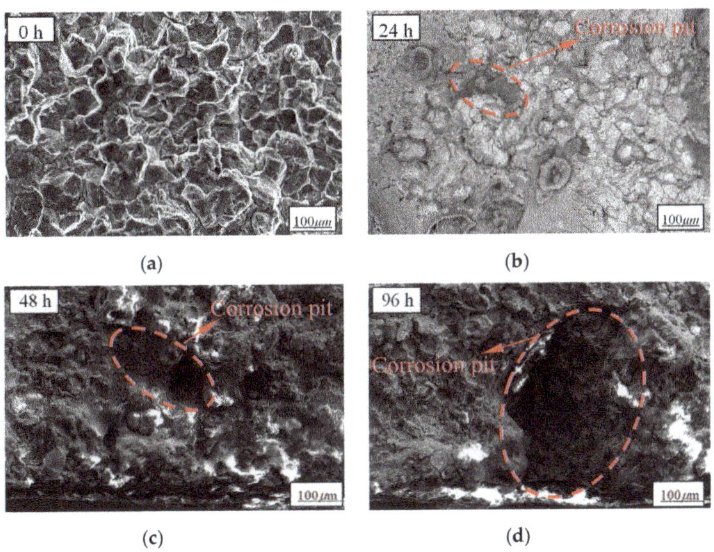

Figure 4. Corrosion surface characterization: (**a**) pre-corroded for 0 h; (**b**) pre-corroded for 24 h; (**c**) pre-corroded for 48 h; and (**d**) pre-corroded for 96 h.

According to Figure 4, corrosion surface characterization can be clearly observed and the obvious corrosion pits after pre-corrosion can be found within the crack source area in each diagram. As shown in Figure 4a, there was no corrosion pit on the surface of specimens without pre-corrosion. With the pre-corrosion treatment, Figure 4b shows that there were local micro-pits and pits distributed densely. With the increase of pre-corroded time, Figure 4c,d shows that corrosion pits became larger since the small pits were interconnected to form larger corrosion pits. A series of electrochemical corrosion processes occurred when the specimen was immersed in the NaCl solution. In this study, the corrosion process can be generally divided into two stages: 1) the initiation of the pits stage and 2) the pit development stage. In the first stage, adsorption and agglomeration of chloride ions took place at certain weak sites and the corrosion pits formed. In the second stage, the pits developed as a result of the anodic dissolution of the metal [34–36]. Thus, the small pits were interconnected to form larger corrosion pits.

To obtain the effect of the stress distribution around corrosion pits on the stress intensity factor and the residual fatigue life of corroded steel, the size of pre-corroded pits should be accurately assessed and the depth of pre-corroded pits can be measured by a surface profiler. The schematic diagram of pre-corrosion pits under the pre-corroded times 24 h, 48 h, and 96 h is shown in Figure 5. From Figure 5, one can see that the depth of the maximum corrosion pit can be regarded as the radius of a semi-circular surface crack. The depth of the maximum corrosion pit was measured by a surface profiler and an SEM.

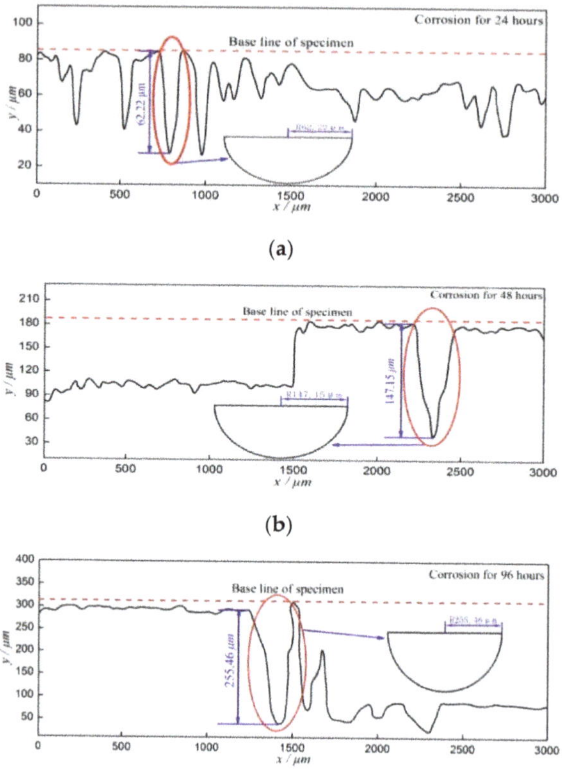

Figure 5. The schematic diagram of pre-corrosion pits under the different pre-corroded times: (**a**) pre-corroded for 24 h; (**b**) pre-corroded for 48 h; and (**c**) pre-corroded for 96 h.

3.2. Fatigue Test Results and Discussion

Pre-corrosion not only reduces residual fatigue life, but also leads to the change of the stress intensity factor. Under the cyclic load with a frequency of 10 Hz, the local stress concentration around the corrosion pits accelerated the process of fatigue damage and resulted in the reduced fatigue life of the specimen. Three constant stress levels (100 MPa, 200 MPa, and 300 MPa) were carried out to pre-corroded specimens for 0 h, 24 h, 48 h, and 96 h. The fatigue testing results of specimens with different exposure times are shown in Table 2.

Table 2 shows that the fatigue life decreased obviously with the increase of pre-corroded time under the same constant stress level. Furthermore, the fatigue life decreased approximately by more than 12% when the pre-corroded time was doubled, and the fatigue life decreased approximately by more than 11% when the applied stress level was doubled, indicating that both pre-corroded time and applied stress level can significantly affect the fatigue life of specimens, which shows a good agreement with the previous works [37,38]. Although the external load is normally invariant in practical engineering structures and residual fatigue life usually depends on corrosion surface condition and the nominal stress level, the raised nominal stress caused by the reduction of the cross-sectional area of components due to corrosion still decreases the fatigue life of steel [39]. After fatigue testing, the fracture surface of the corroded specimens under 100 MPa was observed by SEM, and the results are shown in Figure 6.

Table 2. The fatigue testing results of specimens with different exposure times.

Pre-Corroded Time (h)	Fatigue Life ($\times 10^5$ Cycles)		
	100 MPa	200 MPa	300 MPa
0	11.26	9.95	7.29
24	9.81	7.91	5.47
48	7.52	6.21	4.09
96	5.03	3.91	2.61

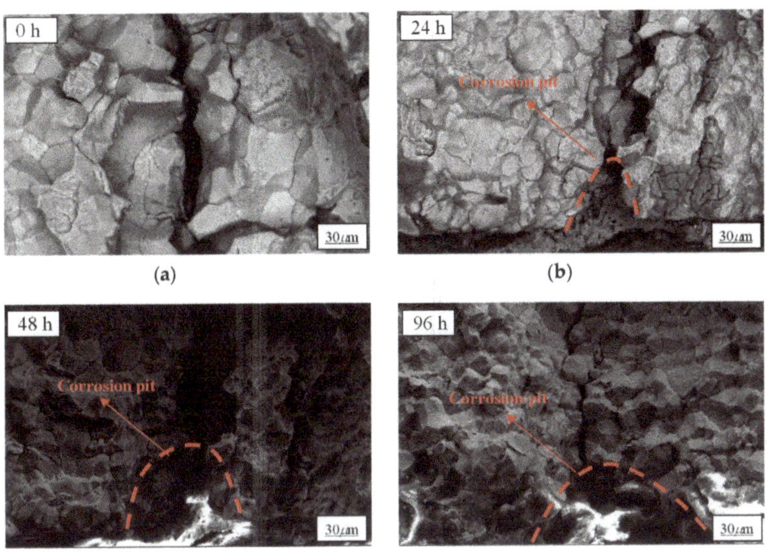

(a) (b) (c) (d)

Figure 6. The fracture surface of the corroded specimens under 100 MPa: (**a**) pre-corroded for 0 h; (**b**) pre-corroded for 24 h; (**c**) pre-corroded for 48 h; and (**d**) pre-corroded for 96 h.

Figure 6a shows the fracture surface of the non-corroded specimen under the constant stress of 100 MPa, and it can be seen that the crack propagated along grain boundaries. Figure 6b–d reveal that the crack originated at the tip of the corrosion pit and propagated along the grain boundaries, and the white highlight area phenomena was caused by the deposition of corrosive elements. Due to the existence of corrosion defects, stress concentration occurred around the sharp intrusion at the pit bottom and accelerated the process of fatigue damage.

3.3. FE Investigations

Based on Equations (1)–(3), ABAQUS was applied to compute the stress intensity factor corresponding to different corrosion pit morphology characteristics, which were depicted in the previous section. The same material properties and loading conditions in Section 2.2 were applied to the FE models, and the computed stress distributions around the crack tip are shown in Figure 7.

The ashy area in Figure 7 represents the yielding zone in specimens. The numbers 1, 2, and 3 in Figure 7 denomination represent the constant stress level of 100 MPa, 200 MPa, and 300 MPa, respectively. From Figure 7, it can be seen that, under the same pre-corroded time, the yielding zone increased with the increase of applied loads through the transverse comparison. Under the same applied loads, it can be seen that the yielding zone increased with the increase of the size of pre-corrosion pits. Moreover, the FEM investigations revealed that the rate of crack extension accelerated when the crack length increased.

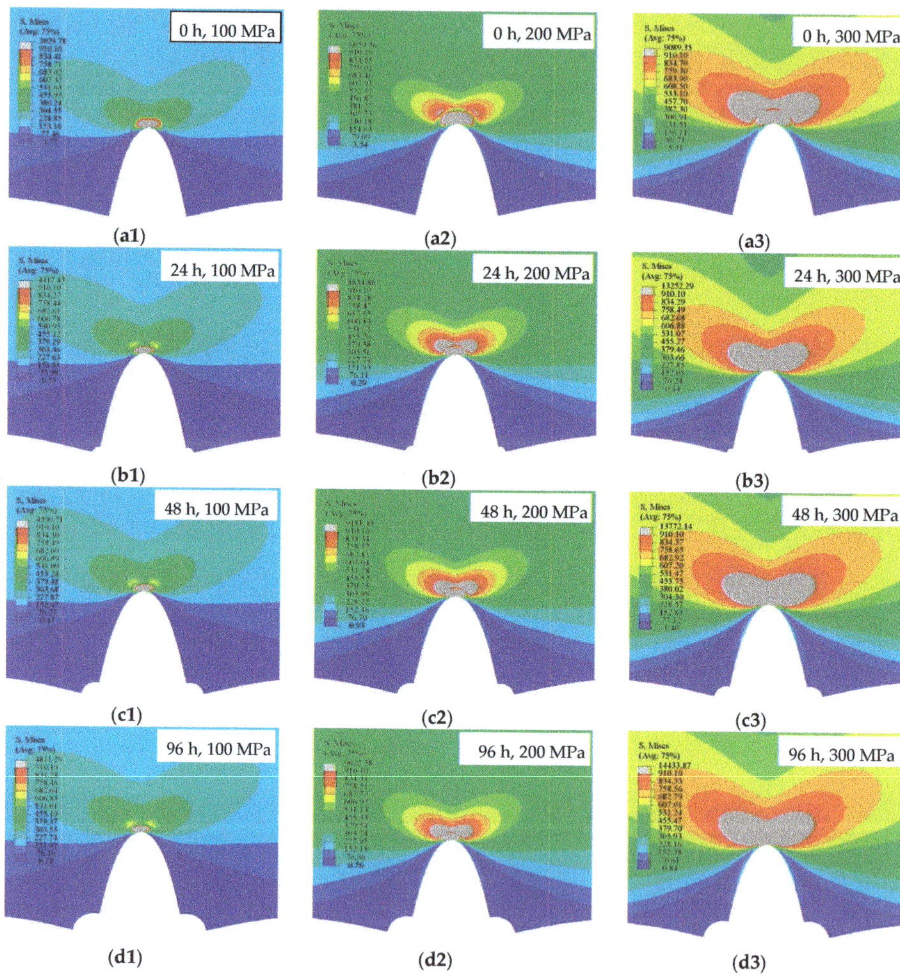

Figure 7. The stress distributions around the crack tip. (**a1**), (**a2**), and (**a3**): non-corroded specimen under the different constant stress levels; (**b1**), (**b2**), and (**b3**): 24 h pre-corroded specimen under the different constant stress levels; (**c1**), (**c2**), and (**c3**): 48 h pre-corroded specimen under the different constant stress levels; and (**d1**), (**d2**), and (**d3**): 96 h pre-corroded specimen under the different constant stress levels.

3.4. Regression Formula of the Stress Intensity Factor

By the adoption of Equations (2) and (3), the stress intensity factor around the crack tip in the process of crack propagation for non-corroded specimens can be calculated and compared with the stress intensity factor that was computed by ABAQUS software to validate the accuracy of the FE simulation results. A comparison of the stress intensity factor by theoretical Paris formula and FE simulation is shown in Figure 8.

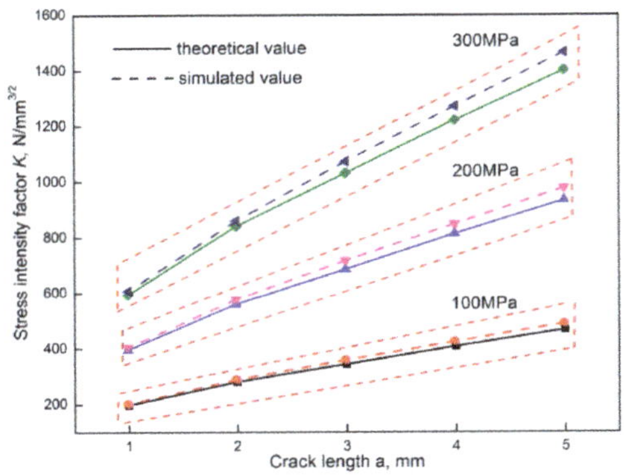

Figure 8. Comparison of the stress intensity factor by theoretical Paris formula and FE simulation.

The maximum percentage error of the calculated stress intensity factor between the theoretical Paris formula and the FEM is 4.7%, which indicates that the theoretical calculation agreed well with the simulated results. Through the integration of the ashy area around the crack tip in Figure 7, the stress intensity factor of the crack tip for pre-corroded specimens can be obtained and the relationship between the stress intensity factor at the crack tip and the depth of the pre-corrosion pits can be established.

Figure 9 shows the relationship between the stress intensity factor at the crack tip and the size of pre-corrosion pits. By using the numerical fitting method, the regression formula to represent the relationship between the stress intensity factor and the depth of the pre-corrosion pits is derived as:

$$\frac{K_{IS}}{K_{I0}} = 0.273e^{3.378\varphi} + 0.731 \tag{4}$$

where K_{IS} is the pre-corroded stress intensity factor, K_{I0} is the non-corroded stress intensity factor, and φ is the depth of corrosion pits. The fatigue life can be estimated according to the Paris formula [25]. The Paris formula is shown as follows:

$$\frac{da}{dN} = C\Delta K^m \tag{5}$$

where da/dN is the fatigue crack propagation, ΔK is the stress intensity factor at the tip of cracks, and C and m are the equation coefficients, respectively. For the 42CrMo steel, the coefficients of C and m are taken as 4.349×10^{-12} and 3.07, respectively [40]. Through the integration of Equation (5), the fatigue life of 42CrMo steel can be estimated as:

$$N = \int \frac{da}{C(\Delta K)^{3.07}} \tag{6}$$

By combining Equations (4) and (6), the relationship between the fatigue life of pre-corroded and non-corroded specimens can be written as:

$$N_C = \left(0.273e^{3.378\varphi} + 0.731\right)^{-3.07} N_0 \tag{7}$$

where N_C is the fatigue life of pre-corroded specimens and N_0 is the fatigue life of non-corroded specimens.

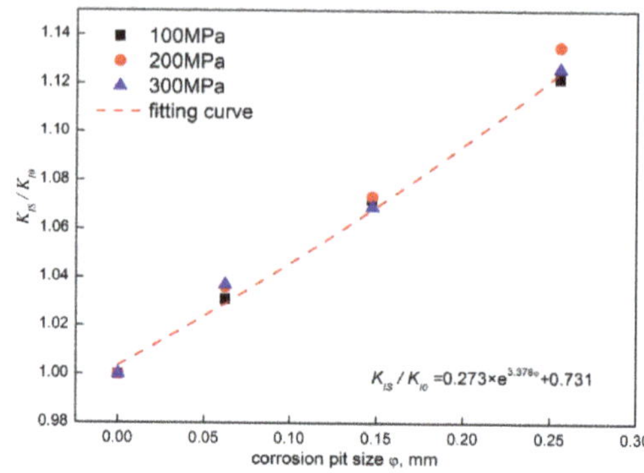

Figure 9. The relationship between the stress intensity factor and the depth of pre-corrosion pits.

Under different stress ranges and corrosion degrees, the residual fatigue life of pre-corrosion specimens is predicted through the use of Equation (7). The comparison of predicted fatigue life and experimental fatigue life is summarized in Table 3.

Table 3. Comparison of predicted fatigue life and experimental fatigue life.

Corrosion Pit Size (μm)	Load (Stress/MPa)	Experimental Values ($\times 10^5$ Cycles)	Predicted Values ($\times 10^5$ Cycles)	Errors (%)
62	100	9.82	10.13	3.15
	200	7.53	7.85	4.08
	300	5.02	5.31	5.46
147	100	8.24	8.73	5.61
	200	6.32	6.75	6.37
	300	4.21	4.53	7.06
255	100	5.62	6.61	8.17
	200	4.43	5.24	8.85
	300	2.95	3.25	9.23

Table 2 shows that the fatigue life predicted by the coupled formula agreed well with experiment results, and the percentage errors between predicted values and experimental values were less than 10%. In the case of low corrosion degrees and low stress levels, the percentage errors were less than 5%. With the increase of pre-corroded time and stress levels, the percentage error was increased to approximately 9.3%. The stress intensity factor at the tip of cracks was calculated on the basis of the linear elastic model and the morphology of the pre-corrosion pits simulated by FEM took pre-set defects in the middle of the specimen, while the practical corrosion pits had the characteristics of the uneven distribution and size of the specimen. Thus, the percentage errors between the predicted values and the experimental values were slightly increased with the increase of pre-corroded time and stress levels.

4. Recommendation for Future Research

In this research, the effect of pre-corrosion pits on the residual fatigue life of the 42CrMo steel was experimentally and numerically studied. It was demonstrated in this paper that the corrosion pit size can be used as a damage index to assess the residual fatigue life. Therefore, the monitoring of corrosion pit size is important not only to know the corrosion status, but also to predict the residual fatigue life of the steel. With the recent development of structural health monitoring technology [41–44], especially the piezoceramic-based active sensing method [45–47], electromechanical impedance (EMI) approach [48–50], and imaging technology [51,52], it is now possible to monitor corrosion pit number and size [10]. Future work should include the real-time monitoring of the onset and growth of the corrosion pit and the prediction of the residual fatigue life of the 42CrMo steel specimens. Future works should also focus on the quantitative analysis of corrosion pits to improve the predicted accuracy of the residual fatigue life for the corroded 42CrMo steel.

5. Conclusions

In summary, this study investigated the effect of pre-corrosion pits on the residual fatigue life of the 42CrMo steel. It was found that higher stress amplitude and longer pre-corroded time significantly decreased the residual fatigue life of the steel. The investigation of the residual fatigue life of pre-corrosion specimens was conducted through the use the regression formula and experimental measures, and the fatigue life predicted by the regression formula agreed well with experiment results. Moreover, it was found that the rate of crack extension accelerated when the crack length increased. It is also demonstrated in this paper that the corrosion pit size can be used as a damage variable to assess the residual fatigue life. The recommendation for future research is to monitor corrosion pit characteristics and to predict the residual fatigue life of the 42CrMo steel.

Author Contributions: Conceptualization, D.L.; methodology, D.L. and J.Z.; software, D.L. and Y.L.; validation, Y.L.; formal analysis, X.X.; writing—original draft preparation, D.L.; writing—review and editing, X.X.; visualization, Y.L.; supervision, X.X.; project administration, D.L.

Funding: This research was funded by Hubei Superior and Distinctive Discipline Group of "Mechatronics and Automobiles" (No. XKQ2019009).

Conflicts of Interest: The authors declare no conflict of interest.

References

1. El-Bagoury, N.; Ahmed, S.I.; Ahmed Abu Ali, O.; El-Hadad, S.; Fallatah, A.M.; Mersal, G.A.M.; Ibrahim, M.M.; Wysocka, J.; Ryl, J.; Boukherroub, R.; et al. The Influence of Microstructure on the Passive Layer Chemistry and Corrosion Resistance for Some Titanium-Based Alloys. *Materials* **2019**, *12*, 1233. [CrossRef]
2. Orlikowski, J.; Ryl, J.; Jarzynka, M.; Krakowiak, S.; Darowicki, K. Instantaneous Impedance Monitoring of Aluminum Alloy 7075 Corrosion in Borate Buffer with Admixed Chloride Ions. *Corrosion* **2015**, *71*, 828–838. [CrossRef]
3. Figueira, R. Electrochemical Sensors for Monitoring the Corrosion Conditions of Reinforced Concrete Structures: A Review. *Appl. Sci.* **2017**, *7*, 1157. [CrossRef]
4. Li, W.; Xu, C.; Ho, S.C.M.; Wang, B.; Song, G. Monitoring concrete deterioration due to reinforcement corrosion by integrating acoustic emission and FBG strain measurements. *Sensors* **2017**, *17*, 657. [CrossRef] [PubMed]
5. Ryl, J.; Arutunow, A.; Tobiszewski, M.T.; Wysocka, J. Aspects of intergranular corrosion of AISI 321 stainless steel in high-carbon-containing environments. *Anti-Corros. Method. Mater.* **2014**, *61*, 328–333. [CrossRef]
6. Huang, Y.; Gang, T.; Chen, L. Interacting effects induced by two neighboring pits considering relative position parameters and pit depth. *Materials* **2017**, *10*, 398. [CrossRef] [PubMed]
7. Zhao, W.; Zhang, T.; Wang, Y.; Qiao, J.; Wang, Z. Corrosion Failure Mechanism of Associated Gas Transmission Pipeline. *Materials* **2018**, *11*, 1935. [CrossRef] [PubMed]
8. Zhu, Z.; Lu, Y.; Xie, Q.; Li, D.; Gao, N. Mechanical properties and dynamic constitutive model of 42CrMo steel. *Mater. Des.* **2017**, *119*, 171–179. [CrossRef]
9. Sola, R.; Poli, G.; Veronesi, P.; Giovanardi, R. Effects of Surface Morphology on the Wear and Corrosion Resistance of Post-Treated Nitrided and Nitrocarburized 42CrMo4 Steel. *Metall. Mater. Trans. A* **2014**, *45*, 2827–2833. [CrossRef]
10. Du, G.; Kong, Q.; Wu, F.; Ruan, J.; Song, G. An experimental feasibility study of pipeline corrosion pit detection using a piezoceramic time reversal mirror. *Smart Mater. Struct.* **2016**, *25*, 037002. [CrossRef]
11. Ryl, J.; Wysocka, J.; Cieslik, M.; Gerengi, H.; Ossowski, T.; Krakowiak, S.; Niedzialkowski, P. Understanding the origin of high corrosion inhibition efficiency of bee products towards aluminium alloys in alkaline environments. *Electrochim. Acta* **2019**, *304*, 263–274. [CrossRef]

12. Fang, B.; Eadie, R.; Chen, W.; Elboujdaini, M. Passivation/immersion method to grow pits in pipeline steel and a study of pit nucleation and growth resulting from the method. *Brit. Corros. J.* **2013**, *44*, 32–42. [CrossRef]
13. Sharland, S.M.; Jackson, C.P.; Diver, A.J. A finite-element model of the propagation of corrosion crevices and pits. *Corros. Sci.* **1989**, *29*, 1149–1166. [CrossRef]
14. Wang, W.; Sun, H.; Sun, L.; Song, Z.; Zang, B. Numerical simulation for crevice corrosion of 304 stainless steel in sodium chloride solution. *Chem. Res. Chin. Univ.* **2010**, *26*, 822–828.
15. Xu, L.; Cheng, Y.; Xu, L. Development of a finite element model for simulation and prediction of mechanoelectrochemical effect of pipeline corrosion. *Corros. Sci.* **2013**, *73*, 150–160. [CrossRef]
16. Ishihara, S.; Saka, S.; Nan, Z.Y.; Goshima, T.; Sunada, S. Prediction of corrosion fatigue lives of aluminium alloy on the basis of corrosion pit growth law. *Fatigue Fract. Eng. Mater.* **2006**, *29*, 472–480. [CrossRef]
17. Pidaparti, R.M.; Koombua, K.; Rao, A.S. Corrosion pit induced stresses prediction from SEM and finite element analysis. *Int. J. Comput. Meth. Eng. Sci. Mech.* **2009**, *10*, 117–123. [CrossRef]
18. Huang, Y.; Ye, X.; Hu, B.; Chen, L. Equivalent crack size model for pre-corrosion fatigue life prediction of aluminum alloy 7075-T6. *Int. J. Fatigue* **2016**, *88*, 217–226. [CrossRef]
19. Zheng, X.; Xie, X.; Li, X. Estimation model for steel wire crack propagation and its application in calculation of pre-corrosion fatigue life. *Chin. Civil. Eng. J.* **2017**, *50*, 101–107.
20. Apostolopoulos, C.A. The influence of corrosion and cross-section diameter on the mechanical properties of b500 c steel. *J. Mater. Eng. Perform* **2009**, *18*, 190–195. [CrossRef]
21. Li, W.; Ho, S.C.M.; Song, G. Corrosion detection of steel reinforced concrete using combined carbon fiber and fiber Bragg grating active thermal probe. *Smart Mater. Struct.* **2016**, *25*, 045017. [CrossRef]
22. Huo, L.; Li, C.; Jiang, T.; Li, H. Feasibility study of steel bar corrosion monitoring using a piezoceramic transducer enabled time reversal method. *Appl. Sci.* **2018**, *8*, 2304. [CrossRef]
23. Zhang, Y.; Tan, G.; Wang, S.; Cheng, Y.; Yang, S.; Sun, X. Flexural characteristics evaluation for reinforced concrete affected by steel corrosion based on an acoustic emission technique. *Appl. Sci.* **2019**, *9*, 1640. [CrossRef]
24. Tang, M.; Li, J.; Li, Z.; Fu, L.; Zeng, B.; Lv, J. Mannich Base as Corrosion Inhibitors for N80 Steel in a CO_2 Saturated Solution Containing 3 wt % NaCl. *Materials* **2019**, *12*, 449. [CrossRef]
25. Paris, P.C.; Gomez, M.P.; Anderson, W.P. A rational analytic theory of fatigue. *Trends Eng.* **1961**, *13*, 9–14.
26. Japanese Industrial Standard. Methods of corrosion resistance test for metallic coatings. In *JIS H 8502*; Japanese Standards Association: Tokyo, Japan, 1999.
27. Venkatraman, M.S.; Cole, I.S.; Emmanuel, B. Model for corrosion of metals covered with thin electrolyte layers: Pseudo-steady state diffusion of oxygen. *Electrochim. Acta* **2011**, *56*, 7171–7179. [CrossRef]
28. Nagarajan, S.; Karthega, M.; Rajendran, N. Pitting corrosion studies of super austenitic stainless steels in natural sea water using dynamic electrochemical impedance spectroscopy. *J. Appl. Electrochem.* **2007**, *37*, 195–201. [CrossRef]
29. ASTM E468–90. *Standard Practice for Presentation of Constant Amplitude Fatigue Tests Results for Metallic Materials*; American Society for Testing and Materials: West Conshohocken, PA, USA, 2004.
30. Xu, S.; Wang, Y. Estimating the effects of corrosion pits on the fatigue life of steel plate based on the 3D profile. *Int. J. Fatigue* **2015**, *72*, 27–41. [CrossRef]
31. Qian, J.; Zhang, Y.; Zhao, H. A fast sweeping method for static convex hamilton–jacobi equations. *J. Sci. Comput.* **2007**, *31*, 237–271. [CrossRef]
32. Min, C.K.; Kim, K.Y.; Lee, K.J. Electrical impedance imaging of phase boundary in two-phase systems with adaptive mesh regeneration technique. *Int. Commun. Heat. Mass.* **2005**, *32*, 954–963.
33. Wang, J.; Wang, J.; Shao, H. The corrosion and electrochemical behavior of pure aluminum in additive-containing alkaline methanol–water mixed solutions. *Mater. Corros.* **2015**, *60*, 269–273. [CrossRef]
34. Li, X.; Fan, C.; Wu, Q.; Dong, L.; Yin, Y.; Wang, C. Effect of solution pH, Cl^- concentration and temperature on electrochemical behavior of ph13-8mo steel in acidic environments. *J. Iron. Steel. Res. Int.* **2017**, *24*, 1238–1247. [CrossRef]
35. Wen, Z.; Lan, W.; Zhao, S.; Cao, X.; Deng, H. Corrosion Behavior of 42CrMo Steel in Chloride Solution. *Surf. Technol.* **2017**, *42*, 216–220.
36. Duddu, R. Numerical modeling of corrosion pit propagation using the combined extended finite element and level set method. *Comput. Mech.* **2014**, *54*, 613–627. [CrossRef]

37. Lin, H.; Zhao, Y.; Ožbolt, J. The bond behavior between concrete and corroded steel bar under repeated loading. *Eng. Struct.* **2017**, *140*, 390–405. [CrossRef]
38. Perez, M.R.; Domínguez, A.G.; Palin, T.; Bathias, C. Very high cycle fatigue analysis of high strength steel with corrosion pitting. *Key Eng. Mat.* **2010**, *449*, 104–113. [CrossRef]
39. Gamboa, E.; Atrens, A. Material influence on the stress corrosion cracking of rock bolts. *Eng. Fail. Anal.* **2005**, *12*, 201–235. [CrossRef]
40. Jang, C. Approximation method for the calculation of stress intensity factors for the semi-elliptical surface flaws on thin-walled cylinder. *J. Mech. Sci. Technol.* **2006**, *20*, 319–328. [CrossRef]
41. Soh, C.K.; Annamdas, V.G.M.; Bhalla, S. Applications of structural health monitoring technology in Asia. Struct. *Struct. Health Monit.* **2015**, *16*, 324–346.
42. Song, G.; Li, W.; Wang, B.; Ho, S.C.M. A review of rock bolt monitoring using smart sensors. *Sensors* **2017**, *17*, 776. [CrossRef]
43. Annamdas, V.G.; Radhika, M.A. Electromechanical impedance of piezoelectric transducers for monitoring metallic and non-metallic structures: a review of wired, wireless and energy-harvesting methods. *J. Intel. Mater. Syst. Str.* **2013**, *24*, 1021–1042. [CrossRef]
44. Song, G.; Wang, C.; Wang, B. Structural health monitoring (SHM) of civil structures. *Appl. Sci.* **2017**, *7*, 789. [CrossRef]
45. Kong, Q.; Robert, R.H.; Silva, P.; Mo, Y.L. Cyclic crack monitoring of a reinforced concrete column under simulated pseudo-dynamic loading using piezoceramic-based smart aggregates. *Appl. Sci.* **2016**, *6*, 341. [CrossRef]
46. Jiang, T.; Zhang, Y.; Wang, L.; Zhang, L.; Song, G. Monitoring fatigue damage of modular bridge expansion joints using piezoceramic transducers. *Sensors* **2018**, *18*, 3973. [CrossRef] [PubMed]
47. Wang, F.; Huo, L.; Song, G. A piezoelectric active sensing method for quantitative monitoring of bolt loosening using energy dissipation caused by tangential damping based on the fractal contact theory. *Smart Mater. Struct.* **2017**, *27*, 015023. [CrossRef]
48. Zhang, J.; Zhang, C.; Xiao, J.; Jiang, J. A PZT-based electromechanical impedance method for monitoring the soil freeze–thaw process. *Sensors* **2019**, *19*, 1107. [CrossRef] [PubMed]
49. Wang, F.; Ho, S.C.M.; Huo, L.; Song, G. A novel fractal contact-electromechanical impedance model for quantitative monitoring of bolted joint looseness. *IEEE Access* **2018**, *6*, 40212–40220. [CrossRef]
50. Annamdas, V.G.M.; Yang, Y. Practical implementation of piezo-impedance sensors in monitoring of excavation support structures. *Struct. Control. Health Monit.* **2012**, *19*, 231–245. [CrossRef]
51. Lu, G.; Li, Y.; Song, G. A delay-and-Boolean-ADD imaging algorithm for damage detection with a small number of piezoceramic transducers. *Smart Mater. Struct.* **2016**, *25*, 095030. [CrossRef]
52. Lu, G.; Li, Y.; Zhou, M.; Feng, Q.; Song, G. Detecting damage size and shape in a plate structure using PZT transducer array. *J. Aerosp. Eng.* **2018**, *31*, 04018075. [CrossRef]

materials

MDPI

Article

Effect of Hot Deformation Process Parameters on Microstructure and Corrosion Behavior of 35CrMoV Steel

Qiumei Yang [1], Yajun Zhou [1,2,*], Zheng Li [1] and Daheng Mao [1]

1 School of Mechanical and Electrical Engineering, Central South University, Changsha 410083, China; YangQMei141@126.com (Q.Y.); lizheng2016@csu.edu.cn (Z.L.); mdh@csu.edu.cn (D.M.)

2 National Key Laboratory of High-Performance Complex Manufacturing, Central South University, Changsha 410083, China

* Correspondence: zhouyjun@csu.edu.cn; Tel.: +86-731-8887-7244

Received: 16 March 2019; Accepted: 1 May 2019; Published: 6 May 2019

Abstract: Hot deformation experiments of as-cast 35CrMoV steel, with strain rates of 0.01 s^{-1} and 10 s^{-1}, deformation temperatures of 850, 950, and 1050 °C, and an extreme deformation reaching 50%, were carried out using a Gleeble-3810 thermal simulator. Electrochemical corrosion experiments were conducted on the deformed specimens. The microstructure was observed by optical microscope (OM), and the corrosion morphology and corrosion products of the specimens were investigated by scanning electron microscopy (SEM), energy disperse spectroscopy (EDS), confocal laser scanning microscopy (CLSM), and X-ray diffraction (XRD) techniques. The results show that the grain size increased gradually with an increase in the deformation temperature at the same strain rate, whereas the corrosion resistance deteriorated. At the same deformation temperature, the grain size becomes smaller as the strain rate increases, which enhances the corrosion resistance. This is mainly attributed to the fine grains, which can form more grain boundaries, increase the grain boundary area, and accelerate the formation of the inner rust layer at the beginning of corrosion. Moreover, fine grains can also refine the rust particles and enhance the bonding strength between the inner rust layer and the matrix. The denseness and stability of the inner rust layer increases as the corrosion process progresses, thereby improving corrosion resistance.

Keywords: thermal deformation parameters; 35CrMoV steel; grain size; electrochemical corrosion

1. Introduction

It is well-known that 35CrMoV steel is a high-quality steel with good hardenability, creep properties, high strength, and a high fatigue limit. It is generally used to manufacture gear, marine power equipment, advanced turbine blowers and impellers, compressor engines, and other important parts which operate under high stress [1]. Under certain conditions, such as high stress and complex marine conditions (especially for complex marine conditions with strong corrosiveness), Cl^- has a strong penetrating power that can cause pitting corrosion of carbon steel and easy penetration of the surface defects. Furthermore, the invasion of Cl^- can force the surface of the passivation film to rupture, which will aggravate the stress corrosion cracking failure of the steel [2–4]. Therefore, in addition to the mechanical properties, its corrosion performance also seriously threatens the service life, safety, and efficiency of the marine components.

During thermal processing, different thermal deformation parameters have significant effects on the microstructure and properties of the material [5,6]. Numerous studies have been done into the evolution of the microstructure during thermal deformation in recent years [7–9]. Kingkam et al. [10] investigated the effects of the deformation temperature and strain rate on the dynamic recrystallization behavior of high-strength low-alloy steel. It was observed that an increase of deformation temperature led to an increase of grain size and the effects of the microstructure on corrosion behavior were briefly discussed. Xiao et al. [11] observed that the smallest dynamic recrystallization grains were obtained at a higher strain rate and lower deformation temperature during hot deformation. Moreover, high temperatures and low strain rates are beneficial for grain growth. Lin et al. and Xu et al. established the constitutive equation and process diagram of 25Cr3Mo3Nb steel. It is reported that the change of grain size is closely related to the strain rate and, when the strain rate is 1 s^{-1}, a uniform distribution and fine grains can be obtained [12]. Additionally, Huang [13] studied the dynamic recrystallization of 35CrMo steel and found that it was affected considerably by the deformation temperature and strain rate and was difficult to recrystallize at a lower temperature and higher strain rate. Lin et al. [14] studied the microstructure of a Ni-Fe-Cr-based superalloy by thermal compression experiments and the experimental results were consistent with the established dynamic recrystallization model. Quan et al. [15] obtained fine-grained, equiaxed-crystal re-crystallized structures of ASS alloys generated by the thermal deformation activation of dynamic recrystallization. These studies explored the evolution of the microstructure during the thermal deformation process, but rarely involved changes in microstructure that affected the corrosion resistance.

Some studies have exhibited that grain refinement can significantly improve the corrosion resistance of the material [16]. For instance, Argade and Alvarez-Lopez [17,18] showed that grain boundaries can serve as corrosion barriers and considered the role they play in delayed corrosion. Balyanov et al. believed that the rapid passivation of ultrafine Ti makes ultrafine crystals, which have a stronger corrosion resistance than coarse-grained industrial pure Ti [19]. Aung [20] discovered that small grains can reduce the corrosion rate of the AZ31B magnesium alloy. Zhao discussed the effect of the microstructure on the corrosion performance of the AZ91 alloy, and pointed out that the presence of a second phase and surface film can act as a corrosion barrier to prevent corrosion of the matrix [21–23]. Schino explored the inter-granular corrosion rate of AISI 304 austenitic stainless-steel and found that it decreases with a decrease of grain size in an H_2SO_4-$FeSO_4$ solution. This is due to the increase of grain boundary per unit-volume and the decrease of Cr consumption caused by carbide precipitation, as the inter-granular corrosion rate goes down with the increase of the grain boundary area [24]. The results of Pradhan et al. showed that a fine-grained microstructure has similar or better corrosion resistance to a coarse-grained structure under the optimum distribution of grain boundary characteristics [25]. In addition, it was suggested that the grain boundary is favorable for passivation, high-density dislocations, and grain boundaries, and that grain refinements are propitious to form passivation film as grain refinement is helpful for the rapid formation of a passivation film on the surface, which can reduce the corrosion rate [26]. It was considered that coarse crystal samples are more susceptible to sensitization and inter-granular corrosion than fine-grained samples because their anode and cathode area are lower [27]. It was reported that the degree of sensitization decreases exponentially with the increase of the grain boundary surface-area [28–30], which means that Cr is diffused from the inside to the grain boundary, thereby reducing the growth of carbide precipitates, delaying the sensitization process, and improving corrosion resistance. The above studies show that the grain size, in both magnesium alloy and stainless-steel, affects the corrosion behavior of the material and that grain refinement can improve the corrosion resistance of the material. However, research on the influence of the microstructure evolution of 35CrMoV steel on the corrosion performance in the process of thermal deformation needs further study. Therefore, by adjusting the thermal processing parameters to study the changes of the microstructure in the thermal deformation process, and then analyzing the influence of grain size on corrosion behavior, more appropriate thermal deformation process parameters are obtained, which can provide technical guidance for obtaining large-size forgings with a fine grain size and good performance.

In this study, as-cast 35CrMoV is taken as the research object and thermal compression experiments are carried out at different strain rates and temperatures. The microstructure of 35CrMoV steel after thermal deformation is characterized by optical microscope (OM) and the grain sizes of specimens with different thermal deformation parameters are obtained by a professional micro-image analysis system. The corrosion behaviors of different grain-size specimens obtained under different thermal deformation process parameters are studied in a 3.5 wt% NaCl solution through a three-electrode system. Furthermore, the corrosion performance of the samples is evaluated by electrochemical impedance spectroscopy (EIS) and TAFEL curves, and the corrosion morphology are observed by the energy disperse spectroscopy (EDS), confocal laser scanning microscopy (CLSM), and Hyperfield 3D microscopic systems. Therefore, we can determine the influence of grain size on corrosion resistance and reveal the influence mechanism of the microstructure, obtained under different parameters, on corrosion resistance.

2. Materials and Methods

2.1. Materials

Our research material, 35CrMoV steel, is a typical low-alloy steel for industrial use and its chemical composition is shown in Table 1. The as-cast 35CrMoV steel specimens of dimensions Φ 400 × 800 mm and hot compression specimens with size Φ 10 × 15 mm were cut at 1/2R (R is the radius of the as-cast, 200 mm) by a wire cutting machine. Then, the thermal compression test was carried out by using the Gleeble-3810 thermal simulator. The graphite sheet was placed between the two ends of the specimens, as well as the contact surface of the pressure head, before the beginning of the experiment, and lubricant was applied onto the two ends of the specimens so that the specimens could be subjected to uniform deformation during heating and compression. Before the beginning of the experiment, the graphite sheet was placed between the two ends of the specimens and the contact surface of the indenter, and lubricant was applied onto the two ends of the specimens to keep the temperature uniformly distributed during heating and compression. The process of thermal compression is shown in Figure 1. First, it was heated to 1200 °C, at a heating rate of 10 °C/s^{-1}, for 5 min and then cooled to the deformation temperature (850 °C, 950 °C, 1150 °C), at a cooling rate of 10 °C/s^{-1}, with heat preservation for 1 min to eliminate the temperature gradient of the specimens. Then, an extreme deformation, reaching 50%, was carried out at a strain rate of 0.01–10 s^{-1}. Finally, the specimens were quenched in water immediately after compression and the same heat treatment was carried out, the specimens were heated to 850 °C for 30 min, then oil-cooled, re-heated to 620 °C for 30 min, and finally, were water-cooled.

Table 1. Composition of GB 35CrMoV steel (wt%).

Chemical Composition	C	Si	Mn	Mo	S	P	Cr	V	Fe
Measured	0.36	0.23	0.29	0.26	0.009	0.035	1.19	0.14	Bal

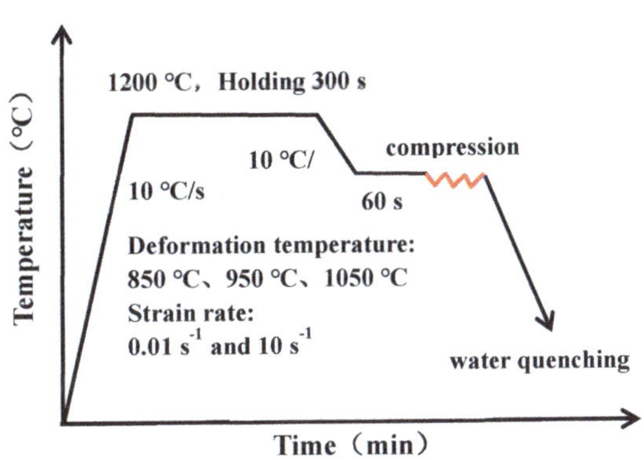

Figure 1. Schematic representation of hot compression test.

2.2. Microscopic Analysis

The microstructure of the specimens was observed by an Olympus DSX500 optical microscope, and the average grain size was calculated by an artificial cut-off method, according to the standard of GB/T6394-2002. Before the metallographic experiment, the specimens, with a diameter of 10×5 mm^2, were taken along the axial direction of the compressed specimens by a wire cutter, and then they were ground, from coarse to fine sandpaper, on a grinding machine and polished with diamond spray until the surface of the specimens was bright. On this basis, the specimens were corroded in a self-made corrosion solution (2.5 g picric acid + 50 mL distilled water + 2 g dodecyl benzenesulfonic acid, sodium salt), heated, and kept at 60–80 °C for 4–8 min. After the corrosion was finished, the specimens were removed from the surface with cotton, washed with distilled water, and dried with hot air, so that the microstructure of the specimens could be observed under an optical microscope.

2.3. Electrochemical Measurements

Three dynamic potential polarization experiments were performed on the electrochemical workstation (Chi660e, CH Hua, Shanghai, China) using a three-electrode corrosion measurement system. Before the experiment, the specimens were ground to 1500# on the grinder and the polishing cloth was mechanically polished with diamond spray polish until the surface was smooth and free from scratches. Finally, it was washed with distilled water, ethanol, and then dried with hot air. The potentiodynamic polarization was measured by a three-electrode system in a 600 mL corrosion medium cell containing 3.5 wt% NaCl with a saturated calomel (Ag/AgCl) electrode as a reference electrode and platinum-plate as an auxiliary electrode. The surface area of the specimens exposed to the solution was 10×5 mm^2. The specimens were immersed in the corrosive medium for 2400 s to obtain a stable open-circuit potential and, on this basis, electrochemical impedance spectroscopy (EIS) was carried out with a frequency range of 0.01–10^5 Hz and an AC excitation signal amplitude of 10 mV, which can obtain the phase angle diagram of Nyquist and Bode. Then, the appropriate equivalent circuit model was selected and the EIS results were found by equivalent-fitting with the ZSimpWin software. Ultimately, the potentiodynamic polarization curve was measured in the range of −1 to 0 V vs. SCE, at a scanning rate of 0.5 mVS^{-1}, and the corrosion potential (E_{corr}), corrosion current density (I_{corr}), anode slope (B_a), and cathodic slope (B_c) were obtained by Tafel fitting.

2.4. Corrosion Surface Analysis

In order to better observe the corrosion morphology of the specimen surface and analyze the elements of the corrosion products, the surface of the specimens was first dried with a compressed air gun and then was observed by scanning electron microscope (SEM) (EvoMA 10C Zeissi Jena, Oberkochen, Germany). Moreover, the chemical elements of the surface corrosion products were analyzed by energy disperse spectroscopy (EDS) and the surface morphology of corrosion specimens is obtained by using Hyperfield 3D microscopic systems (Keyence VHX-5000, Osaka, Japan) and confocal laser scanning microscopy (CLSM) (Zeiss Axio LSM700, Oberkochen, Germany).

3. Results and Discussion

3.1. Microstructure

The microstructure, after hot deformation at different deformation temperatures and strain rates, is shown in Figure 2. The deformation temperatures and strain rates have a significant effect on the microstructure of the 35CrMoV steel. There is a dynamic recrystallization structure in all of the samples and the grain size distribution is not uniform, as displayed in Figure 2. A statistical analysis of the grain size in Figure 2 was carried out and the statistical results are shown in Figure 3. At the strain rate of 0.01 s^{-1}, the grain size increased gradually from 12.5 to 15.4 μm when the deformation temperatures increased from 850 to 1050 °C (increasing by 20%), which is consistent with the law at the strain rates of 10 and 0.01 s^{-1}. The results show that, at the same strain rate, the grain growth with the deformation temperature quickly increases, resulting in coarse grains. This can be attributed to higher deformation temperature making the dislocation motion more intense and more favorable to grain boundary migration, which enhances the diffusion and dislocation slip of vacancy atoms, increasing the nucleation of re-crystallization and the dynamic re-crystallization rate. As a result, small grains are continuously swallowed by large grains and the microstructure is coarsened. When the strain rates are increased from 0.01 to 10 s^{-1} at the deformation temperature of 850 °C, the grain sizes decrease gradually from 12.5 to 10.7 μm, decreasing by 16%. It can be noticed that, as the strain rates are increased from 0.01 to 10 s^{-1} at the deformation temperatures of 950 and 1050 °C, the grain change is consistent with that at 850 °C, which indicates that an increase in strain rate is conducive to grain refinement and acquiring a fine-grained structure at the same deformation temperature. This is mainly due to larger strain rates accelerating the deformation of the samples, and so more strain storage-energy is generated, making the dynamic re-crystallization nucleation rate activate [31]. Thus, there are more phase deformation nuclei and a higher nucleation rate. Moreover, the re-crystallized grain does not have enough time to grow in this case, which reduces the dynamic recovery rate. In addition, the higher strain rate leads to enhanced accumulated strain energy and dislocation density in grains, which makes the dynamic re-crystallization easier to nucleate. It can also be seen from Figure 2 that the number of re-crystallized grains formed along the grain boundary at a strain rate of 10 s^{-1} is higher than that at a strain rate of 0.01 s^{-1}. Furthermore, the dynamic recrystallization rate is lower at lower-deformation temperatures, which can be attributed to the decrease of the grain boundary mobility at low deformation temperatures and high strain rates, as well as a tendency toward incomplete dynamic re-crystallization.

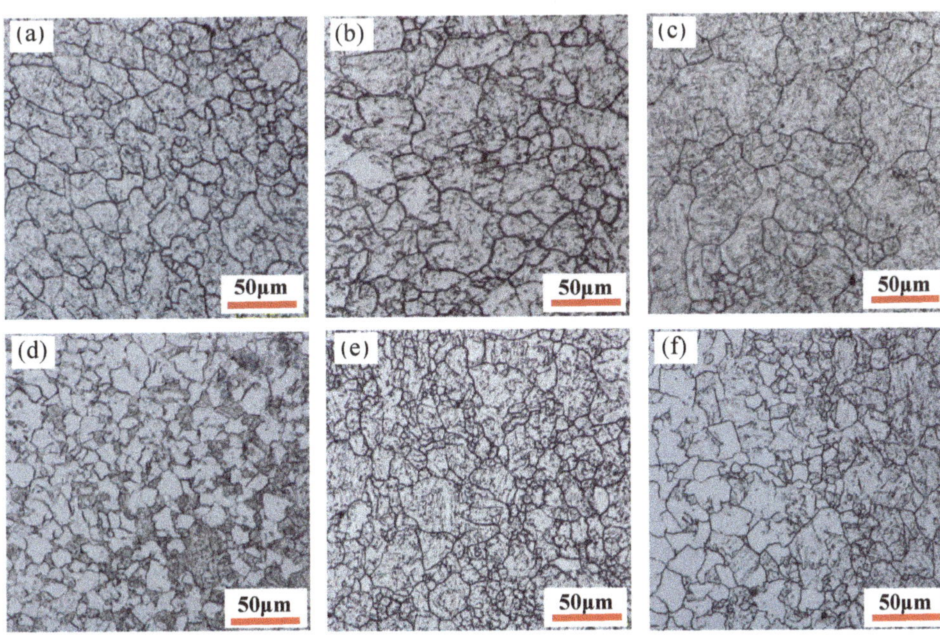

Figure 2. Microstructure of 35CrMoV steel at different deformation temperatures and strain rates: (**a**) 0.01 s^{-1}, 850 °C; (**b**) 0.01 s^{-1}, 950 °C; (**c**) 0.01 s^{-1}, 1050 °C; (**d**) 10 s^{-1}, 850 °C; (**e**) 10 s^{-1}, 950 °C; (**f**) 10 s^{-1}, 1050 °C.

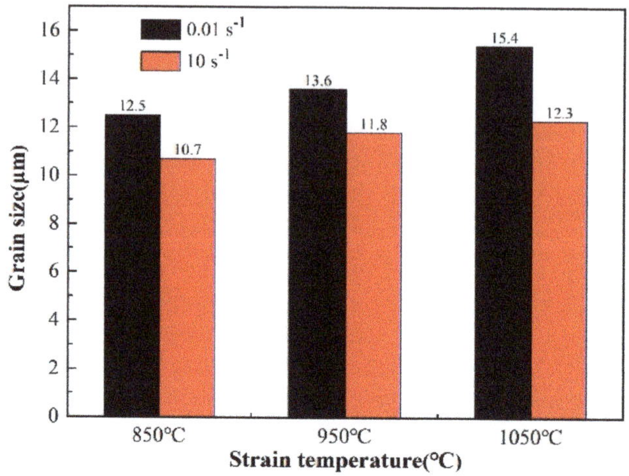

Figure 3. Grain size of 35CrMoV steel at different deformation temperatures and strain rates.

3.2. Electrochemical Analysis

3.2.1. Potentiodynamic Polarization Curve

In this section, the effects of grain sizes, obtained under different process parameters, on the corrosion resistance of 35CrMoV steel are discussed by potentiodynamic polarization experiments. Moreover, the electrochemical impedance spectroscopy and polarization curves are measured under the steady open-circuit potential (OCP). The change curve of an open-circuit potential, measured in 3.5 wt% NaCl solution, with respect to immersion time, is shown in Figure 4. It can be seen from Figure 4 that the change in the OCP of the studied samples follows a similar trend. At the beginning of the experiment, the potential decreases rapidly and all of the samples slowly reach the OCP when the immersion time is 2400 s. A steady OCP is observed at length.

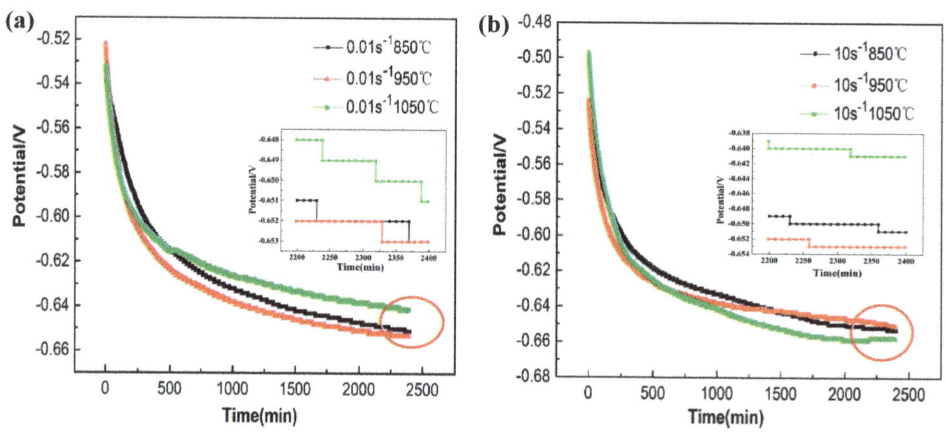

Figure 4. The open-circuit potential (OCP) curves of 35CrMoV steel in 3.5 wt% NaCl solution: (**a**) 0.01 s^{-1}, 850–1050 °C; (**b**) 10 s^{-1}, 850–1050 °C.

Figure 5 shows the potentiodynamic polarization curve of 35CrMoV steel immersed in 3.5 wt% NaCl solution after 40 min. It can be easily seen from Figure 5a,b, that the polarization curves obtained at different strain rates and deformation temperatures are similar in morphology and the

I_{corr} displacement is obvious. There is no passivation and the anode region is an active dissolution of metal, which are the results of the characteristic adsorption of Cl^- onto the surface of the specimen, preventing the formation of the passivation film [1]. In addition, the E_{corr} shift of the polarization curve and obvious changes in the anode region can also be observed, where the anode region represents the dissolution of the matrix at high potential and the cathode region delegates the cathodic hydrogen evolution reaction associated with water reduction [12].

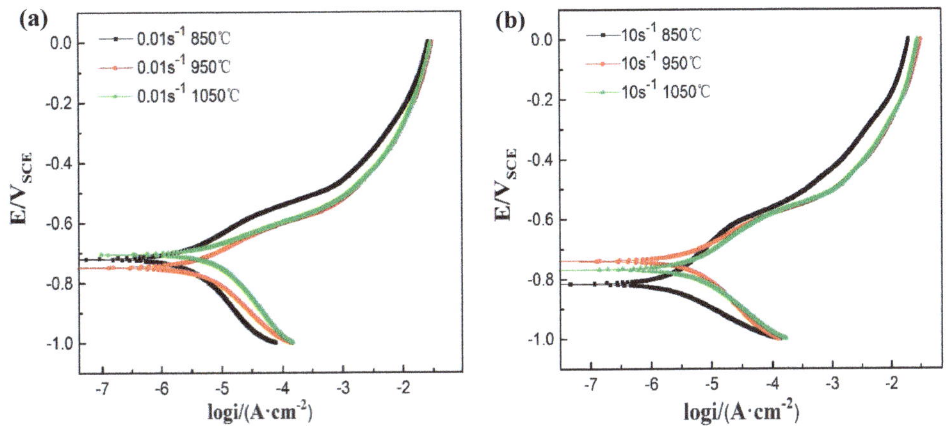

Figure 5. Polarization curves of 35CrMoV steel in 3.5 wt% NaCl solution: (**a**) 0.01 s^{-1}, 850–1050 °C; and (**b**) 10 s^{-1}, 850–1050 °C.

Tafel extrapolation is a rapid and effective method used to discuss the corrosion trend and behavior, and the corrosion rate obtained from it is mainly related to the initial surface corrosion [32]. The values of E_{corr}, I_{corr}, B_a, and B_c obtained from the polarization curves using Tafel extrapolation means (as shown in Figure 5) are listed in Table 2. It can be observed in Figure 5 and Table 2 that I_{corr} increases with an increase of deformation temperature at the same strain rate, and I_{corr} at the deformation temperature of 850 °C is 2.4 μA/cm^2, whereas I_{corr} at the deformation temperature of 1050 °C is 5.4 μA/cm^2. In comparison, the values I_{corr} obtained from 850 and 1050 °C increased by 2.2 times as much. Furthermore, I_{corr} increased from 2.3 to 5.3 μA/cm^2, which is an increase of 2.3 times, at 10 s^{-1}. Therefore, it can be concluded that the higher the deformation temperature of the hot working is, the more easily corrosion occurs. Besides, at the same deformation temperature, I_{corr} decreases with an increase in strain rate and the I_{corr} at 850 °C and 10 s^{-1} is smaller than that at the strain rate of 0.01 s^{-1}. It is well known that the smaller the value of I_{corr} is, the slower the corrosion reaction rate and the higher the corrosion resistance of the material. Accordingly, this shows that the corrosion resistance of 35CrMoV steel can be improved by obtaining smaller grains at high strain rates and low deformation temperatures.

Table 2. Electrochemical test results of 35CrMoV steel.

Sample		E_{corr}/V_{SCE}	$R_p/(\Omega\cdot cm^2)$	$I_{corr}/(\mu A/cm^2)$	$B_a/(mV/dec)$	$B_c/(mV/dec)$
0.01 s^{-1}	850 °C	−0.720	12,290.8	2.441	87.73	57.20
	950 °C	−0.747	6418.7	4.143	105.11	58.40
	1050 °C	−0.704	4083.5	5.393	145.06	52.38
10 s^{-1}	850 °C	−0.815	12,927.8	2.332	55.69	88.50
	950 °C	−0.739	6795.4	4.354	90.24	56.71
	1050 °C	−0.767	6072.5	5.256	69.64	66.59

3.2.2. Electrochemical Impedance Spectroscopy characteristics

Generally, EIS is an effective technique for analyzing and studying the corrosion reaction between the structure of the alloy oxide film and the electrode interface, which can be used to evaluate the corrosion performance of an alloy. Figure 4 shows the curve of the OCP when measured by different specimens. For example, the stable OCP at 850 °C and 0.01 s^{-1} is about 0.6509 V_{SCE} and the EIS is recorded in the frequency range 0.1–100,000 Hz at the steady potential (0.6509 V_{SCE}). The EIS of 35CrMoV steel is represented by the Nyquist plots and phase angle diagram displayed in Figures 6 and 7. It can be seen from Figure 6 that the EIS of the studied 35CrMoVsteel can be greatly influenced by different deformation temperatures and strain rates. This is because the grain sizes of the specimens, acquired with the different thermal deformation parameters, showed in a comparison that the fine grains can make the grain-boundary area increase and the corrosion rate is accelerated at the early stage of corrosion, resulting in the formation of a dense oxide layer and a delaying and deepening of the corrosion. Consequently, the results show that the grain sizes obtained with the different thermal deformation parameters have a significant effect on corrosion resistance.

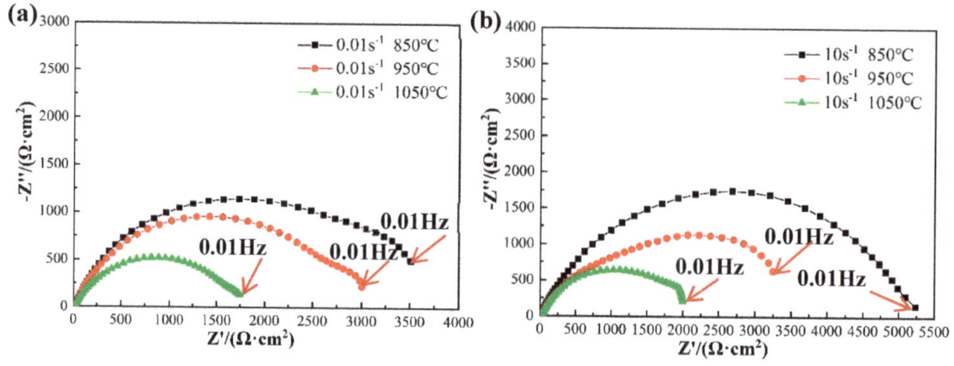

Figure 6. Nyquist plots of the 35CrMoV steel in 3.5 wt% NaCl solution: (**a**) 0.01 s^{-1}, 850–1050 °C; (**b**) 10 s^{-1}, 850–1050 °C.

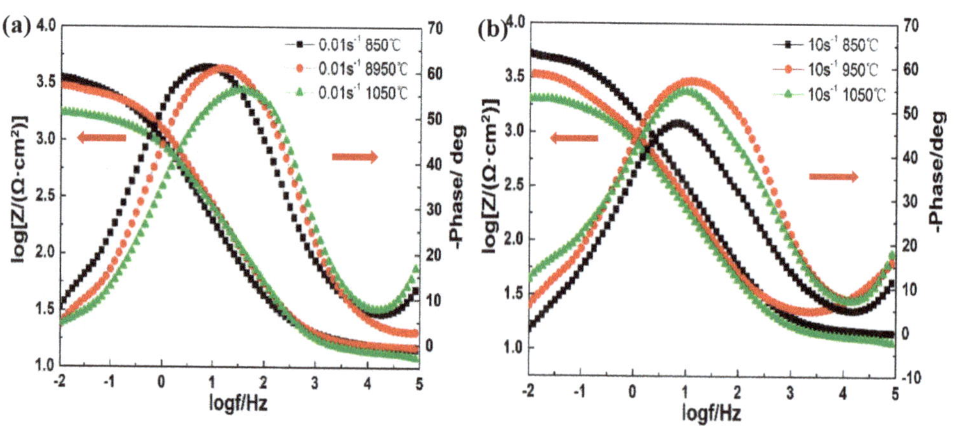

Figure 7. The Bode plots of 35CrMoV steel in 3.5 wt% NaCl solution: (**a**) 0.01 s^{-1}, 850–1050 °C; (**b**) 10 s^{-1}, 850–1050 °C.

The radius of the capacitive reactance arc is positively correlated with its corrosion resistance [21,33–35], the larger the arc radius, the lower the corrosion rate of the specimens in the corresponding solution [36–38]. It can be observed from Figure 6 that all samples have similar semi-circular shapes, where the arc lines are similar and their diameters are different, indicating that the corrosion rate of the specimens is different but the corrosion mechanism is the same [20]. In addition, a larger arc radius can be seen at 850 °C, with strain rates of 0.01 and 10 s^{-1}, which indicates that its corrosion resistance is better. At the same deformation temperature, the arc radius increases gradually with an increase of strain rate, which implies that the corrosion resistance is better (at the same strain rate), the arc radius decreases with an increase of deformation temperature, which indicates that the corrosion resistance is worse. Moreover, the arc-curve radius also reflects the impedance of the electron transfer process onto the electrode surface. It can be concluded that the hindrance can be enhanced and the corrosion rate can be decreased by a larger arc radius. As for metals, it can be noticed that the large resistance to electron transfer means that the gain and loss of electrons does not occur easily, indicating that the metals are difficult to corrode. Therefore, it is clear from Figure 6 that it is most difficult for corrosion to occur at 10 s^{-1} and 850 °C, the result with the largest arc radius, which shows that a higher strain rate and lower deformation temperature can obtain smaller grains and better corrosion resistance.

Figure 7 shows the phase angle plots of the 35CrMoV steel with different strain rates and deformation temperatures. The phase angle plots can be divided into the high-, middle-, and low-frequency regions. From Figure 7 it can be seen that the phase angle in the high-frequency region (100–1000 KHz) is small (close to zero), indicating that the impedance of this frequency range is mainly a solution impedance, and that the resistance behavior is independent of the time constant. The phase angle of the intermediate-frequency region (1–1000 Hz) reaches the maximum value, which is a typical characteristic of solubilization. The maximum value of the phase angle of the samples at 0.01 s^{-1}, 850 °C and 10 s^{-1}, 850 °C moves slightly towards the low-frequency direction. It was found that the capacitance of the double layer is increased [39]. Meanwhile, the presence of a wide angular front in the intermediate- and low-frequency regions implies an interaction between the two relaxation processes, as shown in Figure 7. As per the above analysis, it can be concluded that the two time constants are the reflection of the charge transfer resistance and the corrosion product resistance [40]. In addition, the impedance value, |Z|, of the low-frequency region (0.01–1 Hz) shows the impedance of corrosion reaction, which is one of the parameters used to evaluate the corrosion performance. It indicates that the larger |Z| is, the better the corrosion resistance [41–43]. Additionally, it can be seen from Figure 7 that the impedance value of the specimens decreases with an increase in the deformation temperature, which indicates that the corrosion resistance of the sample surface decreases. Furthermore, at the same deformation temperature, the impedance value increases with the increase of the strain rate. This shows that the corrosion resistance of the sample is stronger. Consequently, the impedance value, |Z|, of the specimen, at 10 s^{-1} and 850 °C, is observed in Figure 7, which is consistent with the results of the polarization curve, indicating the corrosion resistance is at its best and the corrosion product film on the surface is performing better.

In order to further quantitatively analyze the electrochemical corrosion behavior of the 35CrMoV steel in 3.5 wt%NaCl solution, an equivalent electrical circuit (R(Q (R (QR)) is established to fit the EIS spectra, which can well-express the corrosion mechanism of the metal interface or solution. The equivalent circuit consists of two time constants [44], as shown in Figure 8. The EIS, fitted by the ZsimpWin software (Buokamp, MA, USA), and the fitting results are shown in Table 3. The five different elements used in the equivalent circuit have different physical meanings and can be divided into three different categories. The first part, R_s, is the solution resistance of the electrolyte, which is used to characterize the kinematic velocity of the chloride produced on the surface of the specimens in the electrolyte. In the second part, Q_f and R_f are used to imply the capacitance and resistance of the corrosion product respectively, which can reflect the diffusion of ions in the corrosion product layer and the formation of the corrosion product on the substrate surface during electrochemical corrosion. Additionally, R_f is an important parameter used to characterize the protective effect of the oxide layer. It can be found that a larger corrosion product resistance means better protection of the oxide layer and a greater resistance to ion movement. In the third part, Q_{dl} and R_{ct} are used to characterize the double-layer capacitance and charge transfer resistance of the reaction interface respectively, which are related to the electrochemical corrosion reaction between the sample matrix and the electrolytic solution. Furthermore, Q (in the second and third parts) is a constant phase angle element (CPE), which describes the physical quantity when the parameter of interface capacitance, C, deviates due to dispersion effects. In general, the reactions in the second and third parts occur in the middle- and low-frequency regions of EIS, which are related to the corrosion resistance of the specimens. The corrosion resistance of the samples is evaluated by combining the corrosion product resistance R_f with the charge transfer resistance R_{ct} ($R_{corr} = R_f + R_{ct}$) [1,33,35,40]. Generally, the corrosion resistance enhances as R_{corr} increases. The polarization resistances (R_{corr}) of the two groups of all of the samples studied are shown in Figure 9. It can be seen from Figure 9 that R_{corr} decreases with an increase of 35CrMoV steel thermal deformation temperature at the same strain rate, indicating that the corrosion rate is accelerated. This is because an increase in the 35CrMoV hot deformation temperature is conducive to grain growth, resulting in a coarse grain size and lower corrosion resistance. At the same temperature, R_{corr} increases with an increase in the strain rate and, more importantly, the highest R_{corr} appears at 10 s^{-1} and 850 °C, indicating that the fine grains obtained at high strain rates and low deformation temperatures can make the oxide layer denser and improve the corrosion resistance.

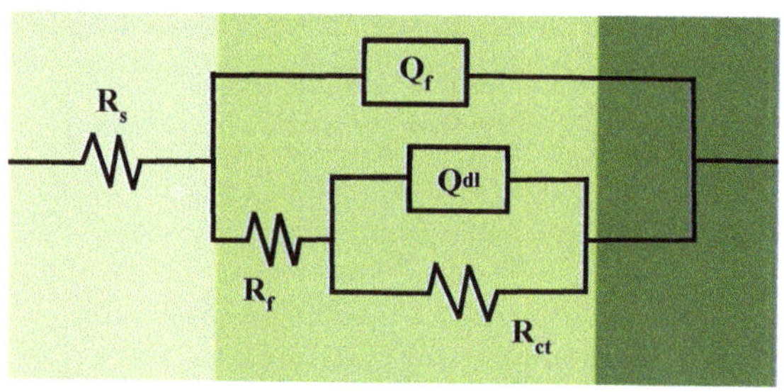

Figure 8. The equivalent electrical circuit of the 35CrMoV steel in 3.5 wt% NaCl solution.

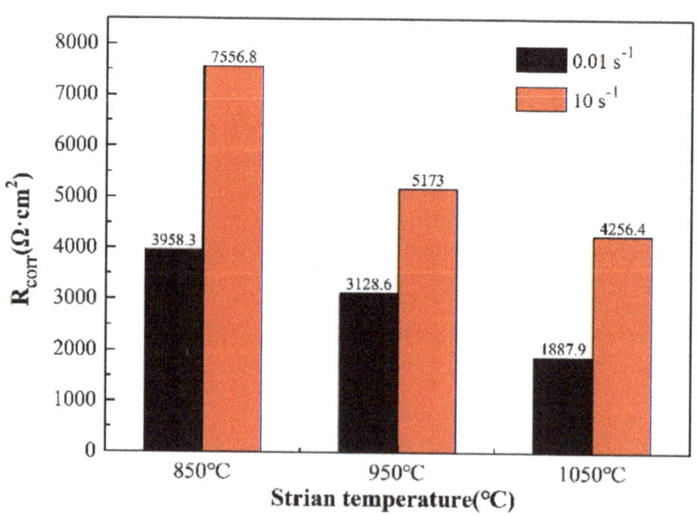

Figure 9. The corrosion rate of the 35CrMoV steel in 3.5 wt% NaCl solution.

Table 3. Spectra fitting results of experimental 35CrMoV steel in 3.5 wt% NaCl solution.

Sample		R_s ($\Omega\cdot cm^2$)	Y_f ($\Omega^{-1}\cdot cm^{-2}\cdot s^n$)	n	R_f ($\Omega\cdot cm^2$)	Y_{dl} ($\Omega^{-1}\cdot cm^{-2}\cdot s^n$)	n	R_{ct} ($\Omega\cdot cm^2$)
$0.01\ s^{-1}$	850 °C	15.66	2.316×10^{-4}	0.7438	3469.00	2.081×10^{-2}	1.000	489.3
	950 °C	15.21	1.544×10^{-4}	0.7619	2768.00	1.585×10^{-2}	0.9732	360.6
	1050 °C	12.84	1.637×10^{-4}	0.7451	961.30	5.139×10^{-4}	0.3708	926.6
$10\ s^{-1}$	850 °C	17.85	3.779×10^{-4}	0.6669	26.83	1.88×10^{-4}	0.8759	7530.0
	950 °C	14.42	1.236×10^{-4}	0.7733	3713.00	3.016×10^{-4}	0.8368	1460.0
	1050 °C	36.56	4.464×10^{-5}	1.0000	189.40	3.157×10^{-4}	0.5485	4067.0

3.3. Corrosion Products and Corrosion Mechanism Analysis

In order to further discuss the effect of the grain sizes, obtained under different deformation conditions, on the corrosion resistance of the specimens, the two- and three-dimensional morphology of the specimens surface is observed, on the basis of the potentiodynamic polarization experiment. The two-dimensional corrosion morphology of four typical specimens with deformation temperatures of 850 °C and 1050 °C, at the strain rates of $0.01\ s^{-1}$ and $10\ s^{-1}$ respectively, immersed in 3.5 wt% NaCl solution for 1 h and placed for 3 days, is depicted in Figure 10. It can be observed from Figure 10 that the surface of the specimens display local corrosion, where the corrosion holes are in the direction of depth. The black and yellowish-brown areas are corrosion areas, and the bright white areas are un-corroded areas. The corrosion pits on the surface are covered with rust spots and corrosion products, as well as corrosion pits of different sizes. At the same strain rate, with an increase in the 35CrMoV hot deformation temperatures, the local corrosion gradually expands and connects to a larger area of corrosion. Furthermore, the corrosion area becomes smaller and the corrosion degree weakens with an increase of the strain rate at the same deformation temperature. The three-dimensional morphology of the a′, b′, c′, and d′ (see Figure 10) regions under the CLSM is shown in Figure 11. The size of the etch pits are measured, along the Y direction of the specimens, with the CLSM. The etch pit depths of (a′), (b′), (c′), and (d′) in Figure 10 are 17.17, 49.55, 13.64, and 30.45 μm, respectively. At the same strain rates ($0.01\ s^{-1}$ and $10\ s^{-1}$), the surface is locally corroded and there are corrosion pits at the deformation temperatures of 850 °C and 1050 °C, but the corrosion pit at the deformation temperature of 1050 °C is large and deep, which indicates that the corrosion resistance is poor at higher deformation temperatures. This is as the dynamic re-crystallization rate of the sample steel increases at high deformation temperatures, and the re-crystallized large grains continue to swallow small grains and grow. Additionally, at the same deformation temperature, the corrosion surface is smoother and the corrosion pit decreases with an increase of strain rate, which is consistent with the characterization in Figure 10. This is because the structure of the double layer at the interface between the matrix iron and the corrosion product film is prone to preferential adsorption of Cl^-, which results in the surface deposition of Cl^- fluids, which then combine with cations on the oxidation film to form soluble chloride and corrosion pits. However, the PH value of the corrosion pit decreases with the continuous hydrolysis of chloride, as well as the dissolution of the anode metal of the corrosion pit, and the external Cl^- is invaded into the corrosion pit through the corrosion product film, which makes the corrosion proceed further. In this cycle, the deepening and expansion of the corrosion pit depth are the results of the corrosion catalysis of Cl^-.

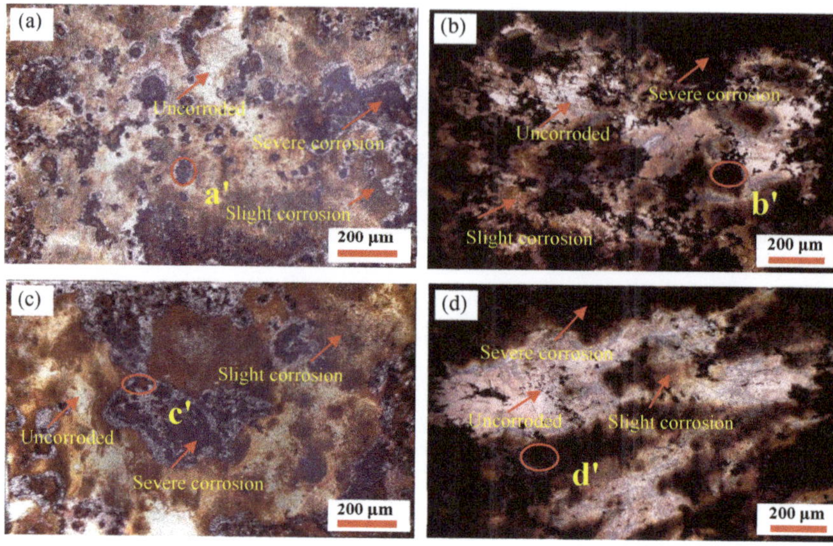

Figure 10. Corrosion morphology of the 35CrMoV steel: (**a**) 0.01 s^{-1}, 850 °C; (**b**) 0.01 s^{-1}, 1050 °C; (**c**) 10 s^{-1}, 850 °C; (**d**) 10 s^{-1}, 1050 °C.

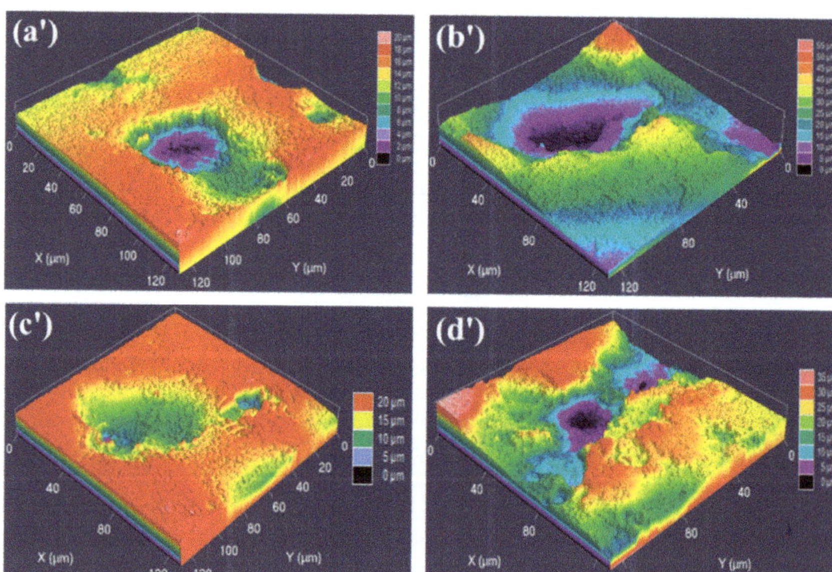

Figure 11. Three-dimensional corrosion morphology of the 35CrMoV steel: (**a′**) 0.01 s^{-1}, 850 °C; (**b′**) 0.01 s^{-1}, 1050 °C; (**c′**) 10 s^{-1}, 850 °C; (**d′**) 10 s^{-1}, 1050 °C.

Figures 12 and 13 show the surface morphology of the specimens after polarization in 3.5 wt% NaCl solution. It can be observed from Figures (a), (b), and (c) in Figures 12 and 13 that there are cracks and platelet crystal-like corrosion product films on the surface of the specimens. At the same strain rate, the surface corrosion product-coverage area increases and the corrosion product accumulates with an increase of the deformation temperature. However, at the same deformation temperature, the accumulation of corrosion products decreases because of an increase in the strain rate, indicating that the corrosion degree is weakened. The cracks on the corroded surface are displayed in Figure 12 and in Figure 13a′–c′, and are due to dehydration after the sample immersion test [45]. The corrosion cracks are formed at the initial stage of corrosion which, with an increase of temperature, become wider, deeper, and looser at the same strain rate, and the corrosion products fall off slightly, indicating that corrosion is deepening. However, at the same deformation temperature, the corrosion cracks on the surface of the specimens become narrower, shallower, denser, and less detached as the strain rate increases, indicating that the corrosion rate is weakened. The platelet-like corrosion products (Figures 12 and 13d′–f′) generated on the surface are due to the deepening of the surface corrosion, resulting in rust being generated and the corrosion products accumulating. In other words, with an increase of deformation temperature, the gap in platelet corrosion products becomes wider and looser at the same strain rate, which is more likely to cause oxidative corrosion on the surface, favorable to the further invasion of the corrosive medium which accelerates the corrosion. Furthermore, an increase of the deformation temperature is beneficial to grain growth, which then affects the corrosion rate at the same strain rate. Additionally, the cracks are narrower and the platelet-corrosion product film is densified and uniform at the same deformation temperature, due to strain rate increases. The deformation process becomes rapid, the dynamic recovery rate decreases, and the re-crystallized grains have not yet fully grown, resulting in smaller grains.

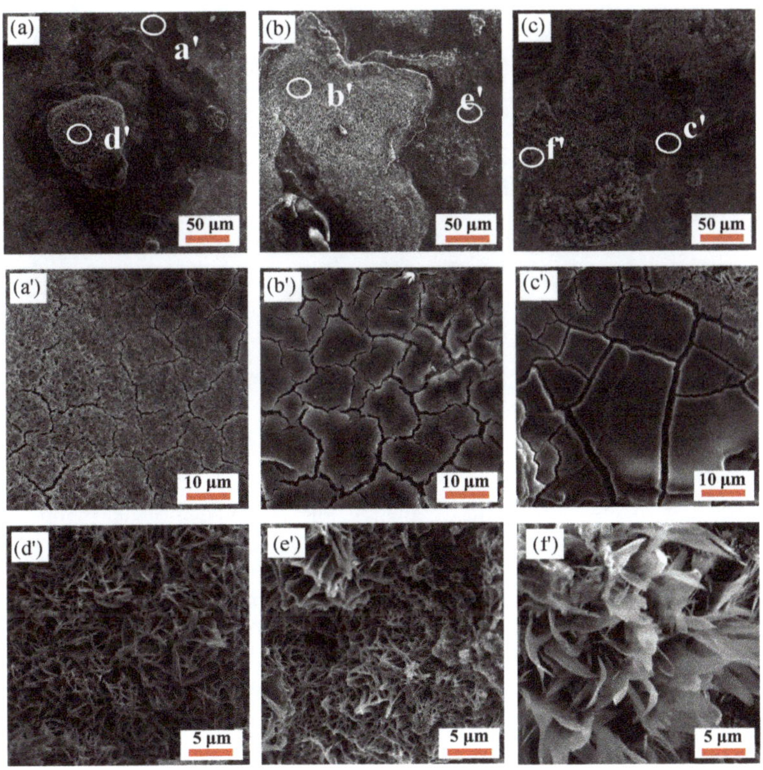

Figure 12. Scanning electron microscopy (SEM) corrosion morphology of the 35CrMoV steel: (**a**,**a′**,**d′**) 0.01 s^{-1}, 850 °C; (**b**,**b′**,**e′**) 0.01 s^{-1}, 950 °C; and (**c**,**c′**,**f′**) 0.01 s^{-1}, 1050 °C.

The results show that the impedance modulus at the EIS (Figure 6) is the appropriate parameter to characterize the protective performance of the corrosion product film [46–48]. From Figure 6a, it can be seen that no scattering fluctuations were observed in all of specimens at 0.01 Hz, which indicates that the corrosion product film has great stability and can effectively protect the matrix from Cl^- in 3.5 wt% NaCl solution, as well as preventing further corrosion [41].

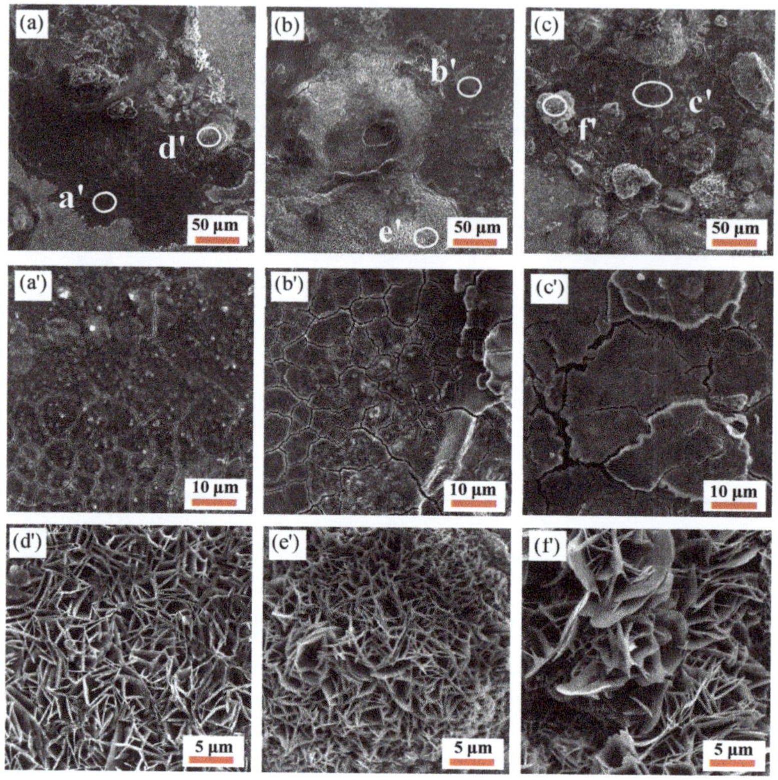

Figure 13. SEM corrosion morphology of the 35CrMoV steel: (**a**,**a′**,**d′**) 10 s^{-1}, 850 °C; (**b**,**b′**,**e′**) 10 s^{-1}, 950 °C; and (**c**,**c′**,**f′**) 10 s^{-1}, 1050 °C.

The surface of the specimens was scanned by EDS and two groups of specimens, with different strain rates at deformation temperature 850 °C, were selected for analysis. Figure 14 shows the corrosion morphology and EDS analysis of the surface of the 35CrMoV steel at deformation temperature 850 °C at 0.01 and 10 s^{-1}. Both of the specimens can be observed to have corrosion cracks (Figure 14(a1,b1)) and platelet-like products (Figure 14(a2,b2)) [20], produced on the surface of the 35CrMoV steel. Additionally, there is an obvious accumulation of massive corrosion products, and the thickness of the corrosion layer is uneven. The corrosion cracks (Figure 14(a1,b1)) of the specimens all contain O, Fe, C, and Cr, as revealed by EDS, and the oxygen content on the surface decreases obviously with the increase of the strain rate. Meanwhile, the surface of the platelet-like product (Figure 14(a2,b2)) is mainly composed of O, Fe, C, and Cr, and the oxygen content in 0.01 s^{-1} is obviously higher than that in 10 s^{-1}, indicating that the corrosion product is mainly oxide. Furthermore, the content of Cr in 10 s^{-1} obviously increased and the presence of Cr elements on the surface of the corrosion product film can effectively prevent the deep erosion of Cl^-, improve the chemical stability of the corrosion product film, and prevent further corrosion.

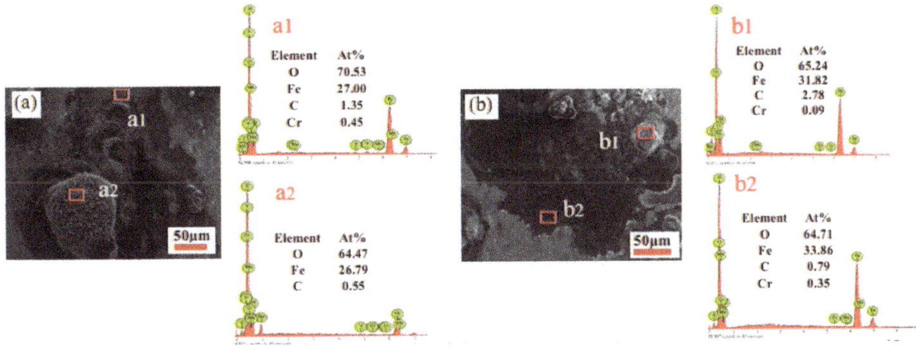

Figure 14. Corrosion morphology and energy disperse spectroscopy (EDS) analysis of the 35CrMoV steel at the same temperature (850 °C) and at different strain rates: (**a**) 0.01 s^{-1}; and (**b**) 10 s^{-1}.

X-ray diffraction analysis was conducted to determine the types of corrosion products (see Figure 15a) and the matrix (see Figure 15b) after corrosion on the surface of the samples. The energy spectrum analysis results of the 35CrMoV steel matrix and corrosion products after corrosion are displayed in Figure 15. It can be seen from Figure 15 that the surface of the matrix after heat treatment mainly contains Fe, and some studies have shown that its structure is tempered sorbite [49,50]. In summary, the corrosion products on the surface after corrosion are mainly FeOOH. Combined with the EDS analysis of Figure 14, it can be concluded that O, Fe, and C are the main components of the corrosion product. Therefore, it can be inferred that the cathode in the corrosion solution is depolarized by oxygen, due to adequate oxygen supply, and the surface is covered by a thin water film.

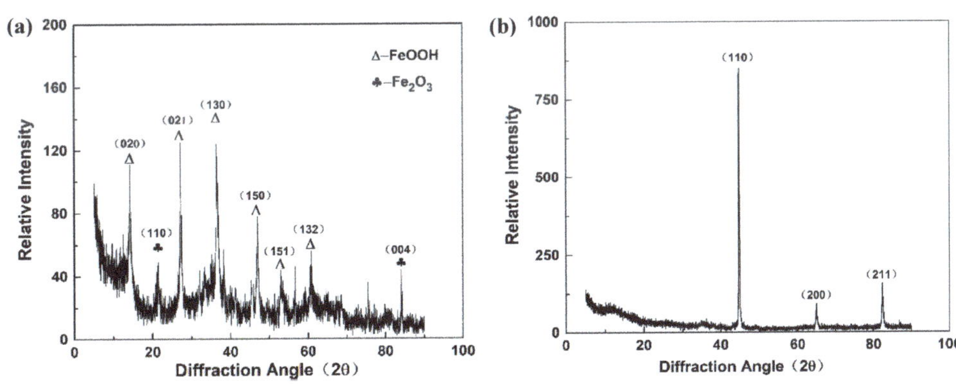

Figure 15. X-ray diffraction (XRD) analysis of the 35CrMoV steel: (**a**) corrosion products; and (**b**) matrix.

In the early stages of corrosion, Cl^- adsorbed on the metal surface play an erosion role and Fe is dissolved from the anode to form Fe^{2+}, combined with Cl^- to form $FeCl_2{\cdot}4H_2O$ [51], and further decomposed to form $Fe(OH)_2$. However, $Fe(OH)_2$ is unstable and decomposed into FeO or oxidized to FeOOH by the O_2 dissolved in the water film (see Equations (1) and (2)) [1,52–54]. In addition, some studies have shown that Fe $(OH)_2$ can continue to be oxidized and dehydrated to form Fe_2O_3 and Fe_3O_4, which are both dense and difficult to decompose. Accordingly, the reduction in corrosion rate is the result of the Fe_2O_3 and Fe_3O_4 hindering the diffusion of oxygen and Cl^-, which exist in the inner rust layer [55]. According to the Evans theory, the corrosion products will partially dissolve due to the strong erosive Cl^- in the solution. Many cracks will occur in the rust layer, which can provide a gap between the corrosion media (such as O_2) and spread, causing the corrosion to continue.

$$Fe^+ + 2OH^- \rightarrow Fe(OH)_2 \tag{1}$$

$$4Fe(OH)_2 + O_2 \rightarrow 4FeOOH + 2H_2O \tag{2}$$

Grain refinement leads to an increase in grain boundary area per unit volume at lower deformation temperatures and higher strain rates. It is well known that the potential at grain boundaries is lower than inside the grain, which is as the crystal defect density at the grain boundary is large. In addition, the existence of a potential difference (PD) constitutes the grain–grain boundary corrosion micro-battery. As a result, the grain boundary (serving as an anode) has priority in corrosion. In the same corrosion environment, there is a certain potential difference between the grain and the grain boundary. The local anodic corrosion current density experienced by the grain boundary is relatively small when corroded, so it will not intrude very deep holes and cracks, and it will give the rust layer a good compactness. The grain size directly affects the corrosion rate at the beginning of corrosion. In contrast to the coarse-grain rate, the fine-grain corrosion rate is faster, which is mainly due to the fact that grain refinement increases the grain boundary area, and the grain boundary is increased in the microstructure. Besides this, the grain boundary is a high-active zone, which may make it vulnerable to pitting corrosion, leading to surface unevenness and increases in the anode surface area, and causing the electrochemical reaction to proceed rapidly [56]. This would result in a larger corrosion area, heavier corrosion, and faster anodic dissolution, which is conducive to the rapid formation of a protective inner rust layer. Thus, a finer-grain of the specimens is more favorable to the flatness of the rust layer/matrix interface in the specimens and a more-uniform dissolved matrix. At the same time, the rust particles are refined and there is a reduction of the cracks and pores in the rust layer, which can cause the matrix and rust layer to combine firmly. Furthermore, the rust layer becomes thicker, denser, and more stable and the alloying elements on the matrix surface play an important role in the development of corrosion [57,58], resulting in an increase in the self-corrosion potential of the sample while enhancing the corrosion resistance and reducing the dissolution rate of the anode. In addition, the current carrying density of the oxide film was reduced and the dissolution of the oxide film slowed down due to grain refinement [59,60], accordingly improving the corrosion resistance [61].

4. Conclusions

The microstructure of 35 CrMoV steel under different hot deformation conditions and the short-term corrosion behavior in 3.5 wt% NaCl solution was discussed, and some important conclusions, as follows, were obtained:

(1) Deformation temperatures and strain rates have an important influence on the microscopic structure. At the same strain rate, the grain size increased with an increase in deformation temperature and decreased with an increase in strain rate at the same strain temperature. Amongst all of the hot deformation parameters studied, the grain size at a strain rate of 10 s^{-1} and a deformation temperature of 850 °C was the smallest (10.7 μm), indicating that a finer-grain structure can be obtained at lower deformation temperatures and higher strain rates.

(2) The difference in grain sizes had a significant effect on the corrosion resistance of 35CrMoV steel. At the same strain rate, the grain size augmented with an increase in temperature and the corrosion resistance decreased with an increase in the grain size, within a certain corrosion time. Furthermore, the corrosion resistance was the worst when the grain size was the largest (15.4 μm). At the same deformation temperature, the higher the strain rate, the smaller the grain size, and therefore, the better the corrosion resistance of the specimens within a certain period of time. The smallest grain size (10.6 μm) was observed at a strain rate of 10 s^{-1} and a deformation temperature of 850 °C, resulting in the lowest corrosion rate, which indicates low-deformation temperature and high-strain rate can

achieve the effect of refining grains and improving corrosion resistance. However, the long-term corrosion trends of these materials need to be studied further.

(3) The differences in the corrosion resistance of 35CrMoV steel were related to the grain boundary change after grain refinement. Grain refinement increased the grain boundary area and the grain boundary, which accelerated the formation of a protective oxide film at the initial stage of corrosion, improved the stability of the oxide film, and prevented the corrosion from proceeding deeply, thereby improving corrosion resistance.

Author Contributions: Conceptualization, Q.Y.; Data curation, Z.L.; Formal analysis, Z.L.; Funding acquisition, Y.Z. and D.M.; Methodology, Q.Y.; Project administration, Y.Z.; Resources, D.M.; Writing—original draft, Q.Y.; Writing—review & editing, Q.Y. and Y.Z.

Funding: This research was funded by the national 973 project of China, grant number No.2014CB046702 and the experimental cost were funded by the national 973 project of China.

Acknowledgments: The authors would like to acknowledge the financial assistance provided by the Major State Basic Research Development Program of China (No. 2014CB046702).

Conflicts of Interest: The authors declare no conflict of interest.

References

1. Jiang, X.; Zhou, Y.; Shi, C.; Mao, D. Effects of Ultrasonic-Aided Quenching on the Corrosion Resistance of GB 35CrMoV Steel in Seawater Environment. *Metals* **2018**, *8*, 104. [CrossRef]
2. Niu, L.; Cheng, Y. Corrosion behavior of X-70 pipe steel in near-neutral pH solution. *Appl. Surf. Sci.* **2007**, *253*, 8626–8631. [CrossRef]
3. Alizadeh, M.; Bordbar, S. The influence of microstructure on the protective properties of the corrosion product layer generated on the welded API X70 steel in chloride solution. *Corros. Sci.* **2013**, *70*, 170–179. [CrossRef]
4. Shimura, T.; Aramaki, K. Prevention of passive film breakdown on iron by coverage with one-dimensional polymer films of a carboxylate ion self-assembled monolayer modified with alkyltriethoxysilanes. *Corros. Sci.* **2004**, *46*, 2563–2581.
5. Zhang, B.; Zhang, B.; Ruan, X.; Zhang, Y. The Hot Deformation Behavior and Dynamic Recrystallization Model Of 35 crmo Steel. *Acta. Metall. Sin.* **2003**, *16*, 183–191.
6. Wang, S.; Huang, Y.; Xiao, Z.; Liu, Y.; Liu, H. A Modified Johnson-Cook Model for Hot Deformation Behavior of 35CrMo Steel. *Metals* **2017**, *7*, 337. [CrossRef]
7. Chen, X.-M.; Lin, Y.; Wen, D.-X.; Zhang, J.-L.; He, M. Dynamic recrystallization behavior of a typical nickel-based superalloy during hot deformation. *Mater. Des.* **2014**, *57*, 568–577. [CrossRef]
8. He, D.-G.; Lin, Y.; Chen, J.; Chen, D.-D.; Huang, J.; Tang, Y.; Chen, M.-S. Microstructural evolution and support vector regression model for an aged Ni-based superalloy during two-stage hot forming with stepped strain rates. *Mater. Des.* **2018**, *154*, 51–62. [CrossRef]
9. Ren, F.; Chen, F.; Chen, J.; Tang, X. Hot deformation behavior and processing maps of AISI 420 martensitic stainless steel. *J. Manuf. Process.* **2018**, *31*, 640–649. [CrossRef]
10. Kingkam, W.; Zhao, C.-Z.; Li, H.; Zhang, H.-X.; Li, Z.-M. Hot Deformation and Corrosion Resistance of High-Strength Low-Alloy Steel. *Acta Met. Sin.* **2018**, *32*, 495–505. [CrossRef]
11. Xiao, Z.-B.; Huang, Y.-C.; Liu, Y. Plastic Deformation Behavior and Processing Maps of 35CrMo Steel. *J. Mater. Eng. Perform.* **2016**, *25*, 1219–1227. [CrossRef]
12. Xu, L.; Chen, L.; Chen, G.; Wang, M. Hot deformation behavior and microstructure analysis of 25Cr3Mo3NiNb steel during hot compression tests. *Vacuum* **2018**, *147*, 8–17. [CrossRef]
13. Huang, Y.; Wang, S.; Xiao, Z.; Liu, H. Critical Condition of Dynamic Recrystallization in 35CrMo Steel. *Metals* **2017**, *7*, 161. [CrossRef]
14. Wen, D.; Lin, Y.; Zhou, Y. A new dynamic recrystallization kinetics model for a Nb containing Ni-Fe-Cr-base superalloy considering influences of initial δphase. *Vacuum* **2017**, *141*, 316–327. [CrossRef]
15. Quan, G.; Mao, A.; Zou, Z.; Luo, G.; Liang, J. Description of Grain Refinement by Dynamic Recrystallization Under Hot Compressions for As-Extruded 3Cr20Ni10W2 Heat-Resistant Alloy. *High Temp. Mater. Process.* **2015**, *34*, 697–713.

16. Song, D.; Ma, A.-B.; Jiang, J.-H.; Lin, P.-H.; Yang, D.-H. Corrosion behavior of ultra-fine grained industrial pure Al fabricated by ECAP. *Trans. Nonferrous Met. Soc.* **2009**, *19*, 1065–1070. [CrossRef]
17. Alvarez-Lopez, M.; Pereda, M.D.; Del Valle, J.; Fernandez-Lorenzo, M.; Garcia-Alonso, M.C.; Ruano, O.; Escudero, M.L.; Ruano, O. Corrosion behaviour of AZ31 magnesium alloy with different grain sizes in simulated biological fluids. *Biomaterials* **2010**, *6*, 1763–1771. [CrossRef]
18. Argade, G.; Panigrahi, S.; Mishra, R.; Mishra, R. Effects of grain size on the corrosion resistance of wrought magnesium alloys containing neodymium. *Corros. Sci.* **2012**, *58*, 145–151. [CrossRef]
19. Balyanov, A.; Kutnyakova, J.; Amirkhanova, N.; Stolyarov, V.; Valiev, R.; Liao, X.; Zhao, Y.; Jiang, Y.; Xu, H.; Lowe, T.; et al. Corrosion resistance of ultra fine-grained Ti. *Scr. Mater.* **2004**, *51*, 225–229. [CrossRef]
20. Aung, N.; Zhou, W. Effect of grain size and twins on corrosion behaviour of AZ31B magnesium alloy. *Corros. Sci.* **2010**, *52*, 589–594. [CrossRef]
21. Zhao, Y.; Wu, G.; Jiang, J.; Wong, H.M.; Yeung, K.W.; Chu, P.K. Improved corrosion resistance and cytocompatibility of magnesium alloy by two-stage cooling in thermal treatment. *Corros. Sci.* **2012**, *59*, 360–365. [CrossRef]
22. Song, G.-L.; Bowles, A.L.; StJohn, D.H. Corrosion resistance of aged die cast magnesium alloy AZ91D. *Mater. Sci. Eng. A* **2004**, *366*, 74–86. [CrossRef]
23. Hsiao, H.; Tsai, W. Effect of heat treatment on anodization and electrochemical behavior of AZ91D m agnesium alloy. *J. Mater. Res.* **2005**, *20*, 2763–2771. [CrossRef]
24. Yu, X.; Chen, S.; Liu, Y.; Ren, F. A study of intergranular corrosion of austenitic stainless steel by electrochemical potentiodynamic reactivation, electron back-scattering diffraction and cellular automaton. *Corros. Sci.* **2010**, *52*, 1939–1947. [CrossRef]
25. Pradhan, S.; Bhuyan, P.; Mandal, S. Individual and synergistic influences of microstructural features on intergranular corrosion behavior in extra-low carbon type 304L austenitic stainless steel. *Corros. Sci.* **2018**, *139*, 319–332. [CrossRef]
26. Zhang, T.; Shao, Y.; Meng, G.; Cui, Z.; Wang, F. Corrosion of hot extrusion AZ91 magnesium alloy: I-relation between the microstructure and corrosion behavior. *Corros. Sci.* **2011**, *53*, 1960–1968. [CrossRef]
27. Qian, J.; Chen, C.; Yu, H.; Liu, F.; Yang, H.; Zhang, Z. The influence and the mechanism of the precipitate/austenite interfacial C-enrichment on the intergranular corrosion sensitivity in 310 S stainless steel. *Corros. Sci.* **2016**, *111*, 352–361. [CrossRef]
28. Singh, R.; Chowdhury, S.G.; Kumar, B.R.; Das, S.K.; De, P.; Chattoraj, I. The importance of grain size relative to grain boundary character on the sensitization of metastable austenitic stainless steel. *Scr. Mater.* **2007**, *57*, 185–188. [CrossRef]
29. Singh, R.; Chowdhury, S.; Chattoraj, I. Modification of Sensitization Resistance of AISI 304L Stainless Steel through Changes in Grain Size and Grain Boundary Character Distributions. *Metal. Mater. Tran. A.* **2008**, *39*, 2504–2512. [CrossRef]
30. Ahmedabadi, P.M.; Kain, V.; Muralidhar, K.V.; Samajdar, I. On the role of residual strain in controlling sensitisation of twin-boundary engineered type 304 stainless steel. *J. Mater.* **2013**, *432*, 243–251. [CrossRef]
31. Lin, Y.; Wu, X.-Y.; Chen, X.-M.; Chen, J.; Wen, D.-X.; Zhang, J.-L.; Li, L.-T. EBSD study of a hot deformed nickel-based superalloy. *J. Alloys Compd.* **2015**, *640*, 101–113. [CrossRef]
32. Zhao, M.-C.; Liu, M.; Song, G.-L.; Atrens, A. Influence of pH and chloride ion concentration on the corrosion of Mg alloy ZE41. *Corros. Sci.* **2008**, *50*, 3168–3178. [CrossRef]
33. Chen, J.; Zhou, Y.; Shi, C.; Mao, D. Microscopic Analysis and Electrochemical Behavior of Fe-Based Coating Produced by Laser Cladding. *Metals* **2017**, *7*, 435. [CrossRef]
34. Zhang, H.; Zou, Y.; Zou, Z.; Wu, D. Microstructure and properties of Fe-based composite coating by laser cladding Fe–Ti–V–Cr–C–CeO_2 powder. *Opt. Laser Technol.* **2015**, *65*, 119–125. [CrossRef]
35. Lin, Y.; Liu, G.; Chen, M.-S.; Zhang, J.-L.; Chen, Z.-G.; Jiang, Y.-Q.; Li, J. Corrosion resistance of a two-stage stress-aged Al–Cu–Mg alloy: Effects of external stress. *J. Alloys Compd.* **2016**, *661*, 221–230. [CrossRef]
36. Lv, J. Effect of grain size on mechanical property and corrosion resistance of the Ni-based alloy 690. *J. Mater. Sci. Technol.* **2018**, *34*, 1685–1691. [CrossRef]
37. Awasthi, S.; Pandey, S.K.; Juyal, A.; Pandey, C.P.; Balani, K. Synergistic effect of carbonaceous reinforcements on microstructural, electrochemical, magnetic and tribological properties of electrophoretically deposited nickel. *J. Alloys Compd.* **2017**, *711*, 424–433. [CrossRef]

38. Jiang, K.; Li, J.; Liu, J. Electrochemical codeposition of graphene platelets and nickel for improved corrosion resistant properties. *RSC Adv.* **2014**, *4*, 36245. [CrossRef]
39. Rai, P.K.; Shekhar, S.; Mondal, K. Development of gradient microstructure in mild steel and grain size dependence of its electrochemical response. *Corros. Sci.* **2018**, *138*, 85–95. [CrossRef]
40. Lin, Y.; Liu, G.; Chen, M.-S.; Huang, Y.-C.; Chen, Z.-G.; Ma, X.; Jiang, Y.-Q.; Li, J. Corrosion resistance of a two-stage stress-aged Al–Cu–Mg alloy: Effects of stress-aging temperature. *J. Alloys Compd.* **2016**, *657*, 855–865. [CrossRef]
41. Cai, C.; Song, R.; Wang, L.; Li, J. Surface corrosion behavior and reaction product film deposition mechanism of Mg-Zn-Zr-Nd alloys during degradation process in Hank's solution. *Surf. Coatings Technol.* **2018**, *342*, 57–68. [CrossRef]
42. Jin, W.; Wu, G.; Feng, H.; Wang, W.; Zhang, X.; Chu, P.K. Improvement of corrosion resistance and biocompatibility of rare-earth WE43 magnesium alloy by neodymium self-ion implantation. *Corros. Sci.* **2015**, *94*, 142–155. [CrossRef]
43. Wu, P.-P.; Xu, F.-J.; Deng, K.-K.; Han, F.-Y.; Zhang, Z.-Z.; Gao, R. Effect of extrusion on corrosion properties of Mg-2Ca-χAl (χ = 0, 2, 3, 5) alloys. *Corros. Sci.* **2017**, *127*, 280–290. [CrossRef]
44. Wu, G.; Feng, K.; Shanaghi, A.; Zhao, Y.; Xu, R.; Yuan, G.; Chu, P.K. Effects of surface alloying on electrochemical corrosion behavior of oxygen-plasma-modified biomedical magnesium alloy. *Surf. Coatings Technol.* **2012**, *206*, 3186–3195. [CrossRef]
45. Xu, R.; Shen, Y.; Zheng, J.; Wen, Q.; Li, Z.; Yang, X.; Chu, P. Effects of one-step hydrothermal treatment on the surface morphology and corrosion resistance of ZK60 magnesium alloy. *Surf. Coat. Technol.* **2016**, *206*, 3186–3195. [CrossRef]
46. Chen, Y.; Wang, X.; Li, J.; Lu, J.; Wang, F. Long-term anticorrosion behaviour of polyaniline on mild Steel. *Corros. Sci.* **2007**, *49*, 3052–3063. [CrossRef]
47. Jin, W.; Wang, G.; Lin, Z.; Feng, H.; Li, W.; Peng, X.; Qasim, A.M.; Chu, P.K. Corrosion resistance and cytocompatibility of tantalum-surface-functionalized biomedical ZK60 Mg alloy. *Corros. Sci.* **2017**, *114*, 45–56. [CrossRef]
48. Park, J.; Lee, G.; Nishikata, A.; Tsuru, T. Anticorrosive behavior of hydroxyapatite as an environmentally friendly pigment. *Corros. Sci.* **2002**, *44*, 1087–1095. [CrossRef]
49. Liu, L.; Li, Y.; Wang, F. Electrochemical Corrosion Behavior of Nanocrystalline Materials—A Review. *J. Mater. Sci. Technol.* **2010**, *26*, 1–14. [CrossRef]
50. Yao, Y.; Zhou, Y. Effects of deep cryogenic treatment on wear resistance 4 and structure of GB 35CrMoV steel. *Metals* **2018**, *8*, 502. [CrossRef]
51. Zhang, H.; Li, X.; Du, C.; Qi, H.; Huang, Y. Raman and IR spectroscopy study of corrosion products on the surface of the hot-dip galvanized steel with alkaline mud adhesion. *J. Raman Spectrosc.* **2009**, *40*, 656–660. [CrossRef]
52. Eliyan, F.F.; Alfantazi, A. Corrosion of the Heat-Affected Zones (HAZs) of API-X100 pipeline steel in dilute bicarbonate solutions at 90 °C—An electrochemical evaluation. *Corros. Sci.* **2013**, *74*, 297–307. [CrossRef]
53. Kimura, M.; Kihira, H.; Ohta, N.; Hashimoto, M.; Senuma, T. Control of Fe $(O,OH)_6$ nano-network structures of rust for high atmospheric-corrosion resistance. *Corros. Sci.* **2005**, *47*, 2499–2509. [CrossRef]
54. Hao, X.; Su, P.; Xiao, K. Effect of different NaCl concentration on corrosion products of weathering steel. *Corros. Prot.* **2009**, *30*, 297–299.
55. Fu, A.; Cheng, Y. Characterization of corrosion of X65 pipeline steel under disbonded coating by scanning Kelvin probe. *Corros. Sci.* **2009**, *51*, 914–920. [CrossRef]
56. Li, Y.; Wang, F.; Liu, G. Grain Size Effect on the Elect ro-chemical Corrosion Behavior of Surface Nanocrystallized Low-Carbon Steel. *Corrosion* **2004**, *60*, 891–896. [CrossRef]
57. Yamashitaa, M.; Miyukia, H.; Matsudaa, Y.; Naganoa, H.; Misawa, T. The long Term Grouth of The Protective Rust Layer Formed on Weathering Steel by Atmospheric Corrosion during a Quarter of a Century. *Corros. Sci.* **1994**, *2*, 283. [CrossRef]
58. Misawa, T.; Asami, K.; Hashimoto, K.; Shimodaira, S. The mechanism of atmospheric rusting and the protective amorphous rust on low alloy steel. *Corros. Sci.* **1974**, *14*, 279–289. [CrossRef]
59. Wang, X. Effect of ultrafine grain boundary on corrosion resistance of metals. *Corros. Pro.* **2015**, *36*, 695–699.

60. Lin, Y.; Jiang, Y.-Q.; Zhang, X.-C.; Deng, J.; Chen, X.-M. Effect of creep-aging processing on corrosion resistance of an Al–Zn–Mg–Cu alloy. *Mater. Des.* **2014**, *61*, 228–238. [CrossRef]
61. Lin, Y.; Jiang, Y.; Xia, Y.; Zhang, X.; Zhou, H.; Deng, J. Effects of creep-aging processing on the corrosion resistance and mechanical properties of a typical Al-Cu-Mg alloy. *Mater. Sci. Eng. A* **2014**, *605*, 192–202. [CrossRef]

materials

MDPI

Article

Corrosion Characteristics of Copper-Added Austempered Gray Cast Iron (AGCI)

Asiful H. Seikh [1,*], Amit Sarkar [2], Jitendra Kumar Singh [3,*], Sohail M. A. Khan Mohammed [4], Nabeel Alharthi [1,5] and Manojit Ghosh [6]

1 Centre of Excellence for Research in Engineering Materials, King Saud University, Riyadh 11421, Saudi Arabia; alharthy@ksu.edu.sa
2 Department of Metallurgical and Materials Engineering, Jadavpur University, Kolkata 700032, India; amitsarkar553@gmail.com
3 Department of Architectural Engineering, Hanyang University, Ansan 15588, Korea
4 Department of Mechanical and Industrial Engineering, Ryerson University, Toronto, ON M5B 2K3, Canada; sohailmazher5@gmail.com
5 Mechanical Engineering Department, College of Engineering, King Saud University, Riyadh 11421, Saudi Arabia
6 Department of Metallurgy and Materials Engineering, Indian Institute of Engineering Science and Technology, Shibpur 711103, India; manojit_ghosh1@rediffmail.com
* Correspondence: aseikh@ksu.edu.sa (A.H.S.); jk200386@hanynag.ac.kr (J.K.S.); Tel.: +96-61-467-0760 (A.H.S.); +82-31-436-8159 (J.K.S.)

Received: 16 January 2019; Accepted: 2 February 2019; Published: 6 February 2019

Abstract: The aim of this investigation was to assess the corrosion behavior of gray cast iron (GCI) alloyed with copper. Alloyed GCI specimens were austempered isothermally at varying temperatures. After austenitizing at 927 °C, the samples were austempered at different temperatures ranging from 260 to 385 °C with an interval of 25 °C for 60 min. As a result, these samples developed an ausferrite matrix with different percentages of austenite. The resulting microstructures were evaluated and characterized by optical microscope (OM), scanning electron microscope (SEM), and X-ray diffraction (XRD). The corrosion characteristics were determined using potentiodynamic polarization tests and electrochemical impedance spectroscopy (EIS) of these samples. These tests were carried out in a medium of 0.5 M H_2SO_4 and 3.5% NaCl solution. It was observed from the potentiodynamic polarization results that with increasing austempering temperature, the corrosion rate decreased. All results of the EIS were in accordance with a constant phase element (CPE) model. It was found that with an increase in austempering temperature, the polarization resistance (R_p) increased. The austenite content was also found to influence the corrosion behavior of the austempered gray cast iron (AGCI).

Keywords: austempered gray cast iron; austempering temperature; microstructure; potentiodynamic polarization; electrochemical impedance spectroscopy

1. Introduction

Gray cast iron (GCI) is a potential engineering material, which has a diverse range of applications including use in sophisticated automotive parts [1]. The wide applications of GCI are possible due to its unique properties such as good thermal conductivity, relatively low melting temperature, high damping capacity, and excellent castability [1,2]. The damage of the GCI components at the exterior parts through electrochemical corrosion has been the predominant restricting mechanism against enhancing its life span [3]. The presence of graphitization is a distinguishing feature of the deterioration properties of GCI [3,4]. Attempts have been made to combat the problem of corrosion with the help of alloying additions with the aim of modifying the microstructure from ferrite to fine pearlite. It is also pertinent to mention that Si plays an important role in controlling the corrosion behavior of GCI: The higher the Si content, the higher the corrosion resistance [5]. Additionally, it is well known that single-phase microstructures like austenite, ferrite, and martensite perform better in corrosive media compared to two-phase mixtures like bainite, pearlite, and tempered martensite [6].

Several researchers [7–10] have tried to assess the effect of heat treatment and composition on the microstructure and corrosion behavior of austempered ductile iron (ADI). Prasanna et al. [7] and Banerjee et al. [8] studied the effect of the austempering treatment on the microstructure and corrosion properties of ductile iron. They found that both mechanical properties and corrosion resistance were enhanced due to the austempering of cast iron. Afolabi et al. [9] observed that the austempering temperature and time influenced the microstructure of the ductile iron, and thus its corrosion behavior was affected by the compositional structures. Hsu and Chen [10] concluded that the enhancement of corrosion resistance in ADI was due to the presence of retained austenite as a result of austempering. Similar studies with GCI are also common where the corrosion resistance and the mechanical properties were improved dramatically by tailoring the heat treatment pattern (tempering, austempering, and quenching [11–14]) and by alloy additions [8]. Further improvement of the mechanical properties, compared to those of conventional GCI, was observed in austempered gray cast iron (AGCI) due to the formation of a matrix of ausferritic structures (ferrite and stabilized austenite) or bainitic ferrite during austempering [12,15]. Thus, the domain of applicability of AGCI is even wider than that of GCI due to its favorable combination of enhanced mechanical properties [11–15] and improved wear characteristics [15–17]. The present literature, however, is lacking in reporting the corrosion behavior of AGCI, although a lot of work can be found on testing the corrosion behavior of ADI [7–10].

The above scientific observations indicated the necessity of the present investigation into the effect of austempering temperatures on the microstructure and corrosion behaviors of copper-alloyed AGCI, in order to establish a correlation between them.

2. Materials and Methods

2.1. Sample Preparation

Samples of GCI were prepared from cupola melts in a production foundry. The molten metal, at a temperature of 1420 °C was inoculated with 0.25 wt. % of FeSi-based inoculants in the cupola. During tapping, 0.5% Cu pieces of electrolyte grade were added to the metal stream for the sake of alloying. The specimens were cast in the form of standard 30-mm Y-shaped blocks in sand molds as shown in Figure 1a. Corrosion test coupons (Figure 1b) of suitable size (Φ 10 mm $\times$ 10 mm) were machined from the as-cast Y blocks. The uniform distribution of fine type-A graphite flakes is promoted by the inoculants during solidification [18]. Cu is soluble in austenite and increases the hardness, strength, corrosion resistance, and transformation time for the austempering process [19,20]. Cu has been accepted as an affordable alloying element for several engineering applications. As a result, the replacement of expensive Ni by Cu may become more prevalent. The final chemical composition (wt. %) of GCI was determined using a spectroscopy spark analyzer as shown in Table 1.

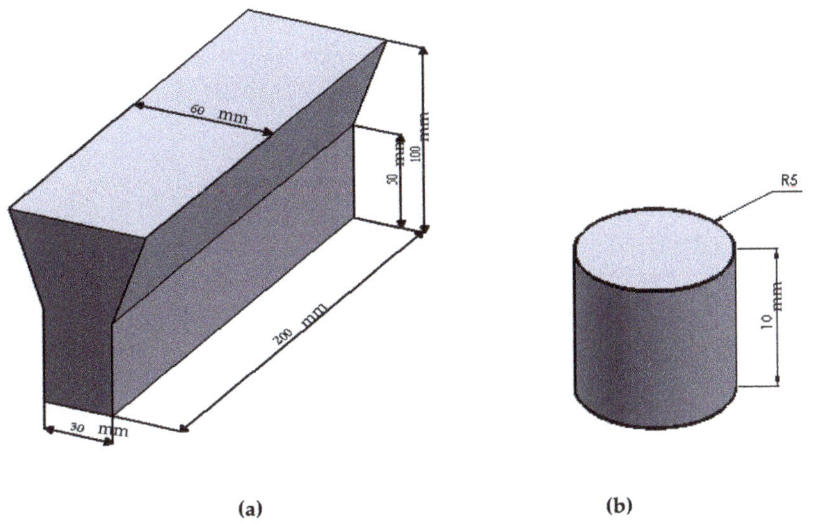

Figure 1. (**a**) Dimension (mm) of the Y-block casting and (**b**) schematic of the corrosion test piece (Φ 10 mm × 10 mm).

Table 1. Chemical composition of GCI (wt. %).

Element	C	Si	Mn	P	S	Cu	Fe
Composition	3.46	2.27	0.53	0.019	0.01	0.50	Bal.

2.2. Heat Treatment of Samples

The samples were initially heated to an austenitizing temperature ($T\gamma$ = 927 °C) and held for 60 min in order to develop a fully austenitic structure (γ). The samples were then rapidly cooled in a molten salt bath comprising 53% KNO_3, 40% $NaNO_2$, and 7% $NaNO_3$ at six different austempering temperatures (T_A), 260, 285, 310, 335, 360, and 385 °C, for 60 min followed by air cooling to complete the phase transformation. Figure 2a schematically represents the entire heat treatment schedule for the austempering process and Figure 2b is the corresponding continuous cooling transformation (CCT) diagram.

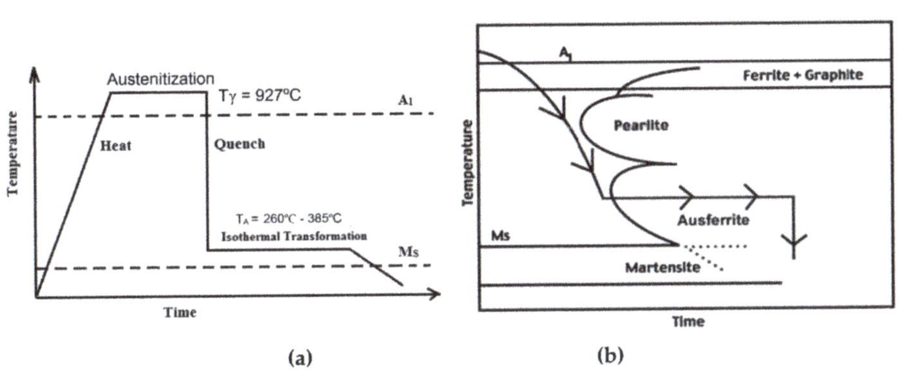

Figure 2. Schematic diagram of (**a**) heat treatment schedule for austempering and (**b**) CCT diagram for the proposed composition.

2.3. Metallography and X-ray Diffraction (XRD)

Samples were prepared for metallographic observation using standard polishing techniques. Moreover, the samples were etched using a 2% nital solution for observation under a scanning electron microscope (SEM, JSM 6360, Jeol techniques, Tokyo, Japan). The volume fractions of austenite were calculated by X-ray diffraction (XRD,) analysis as described by Dasgupta et al. [21]. The XRD data were collected using a Rigaku, Ultima III diffractometer (Japan) with a monochromatic copper Fe-Kα radiation (1.54 Å) at 40 kV and 30 mA. Scanning was done at a rate of 1°/min from 30 to 90° to observe the peaks, which were later analyzed using Jade 7 software (7.1.08). The peak positions were analyzed for the (111), (220), and (311) planes of austenite (FCC) and the (110), (200), and (211) planes of ferrite (BCC). The carbon content in austenite (C_γ) at various austempering temperatures was calculated using the following equation:

$$C_\gamma = \frac{a_\gamma - 3.548}{0.044}, \tag{1}$$

where a_γ is the lattice parameter calculated from the angular position of the austenite peak [22].

2.4. Electrochemical (Corrosion) Test

The heat-treated samples were subjected to electrochemical measurements in a 0.5 M H_2SO_4 and 3.5% NaCl solution at 25 °C (±2 °C). The electrochemical studies were performed in a triplicate set of samples to obtain reproducible results. A cell composed of three electrodes was created, including a graphite one, which acts as a counter electrode; a saturated calomel electrode (SCE), which acts as a reference electrode; and the GCI sample which acts as the working electrode (WE), for potentiodynamic polarization and electrochemical impedance spectroscopy (EIS) measurements. The WE area was fixed at 1 cm^2. Prior to the tests, the samples were ground and polished using SiC papers of 2500 grit size and rinsed in deionized water followed by immersion in the solution for 30 min in order to stabilize the open circuit potential value. The potentiodynamic polarization tests were carried out from −1.0 to +1.0 V at a scan rate of 1 mV/s, whereas, EIS tests were performed over a frequency ranging from 100 kHz to 0.01 Hz.

3. Results and Discussion

3.1. Microstructure and XRD Analysis

Figure 3a,b shows the optical and SEM microstructures of the as-cast gray iron sample, respectively. The matrix of as-cast gray iron is primarily composed of pearlite besides some randomly distributed ferrite. The as-cast specimens were austempered for 60 min and the resulting changes in the microstructure are presented in Figures 4a–f and 5a–f using an optical microscope (OM) and SEM, respectively. The dark, etched needle-like structures represent bainitic ferrite, while the brighter ones represent a mixture of austenite and bainitic ferrite. We can see that the effect of the austempering temperature on the microstructure of austempered irons was significant. It was observed that at lower temperatures (i.e., 260–285 °C), very fine needles of bainitic ferrite and austenite were formed and the volume fraction of ferrite was larger. As the austempering temperature increased, the needles of the bainitic ferrite were coarsened along with an increase in the austenite content. Similar observations in ADI were earlier reported by Patutunda et al. [23] and Yang et al. [24].

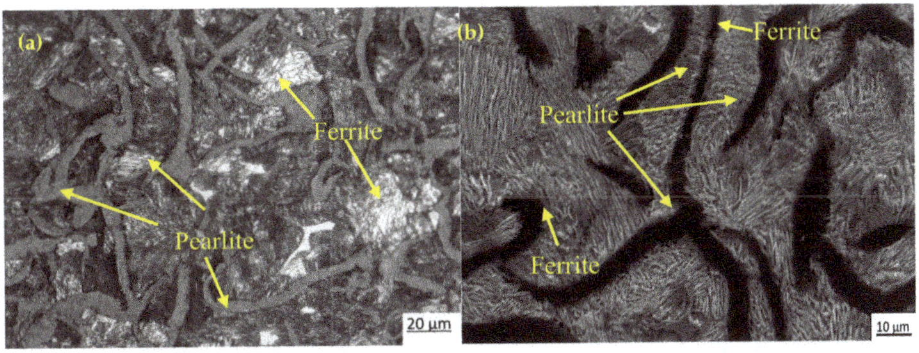

Figure 3. Micrographs of as-cast specimens: (**a**) optical micrograph and (**b**) SEM image.

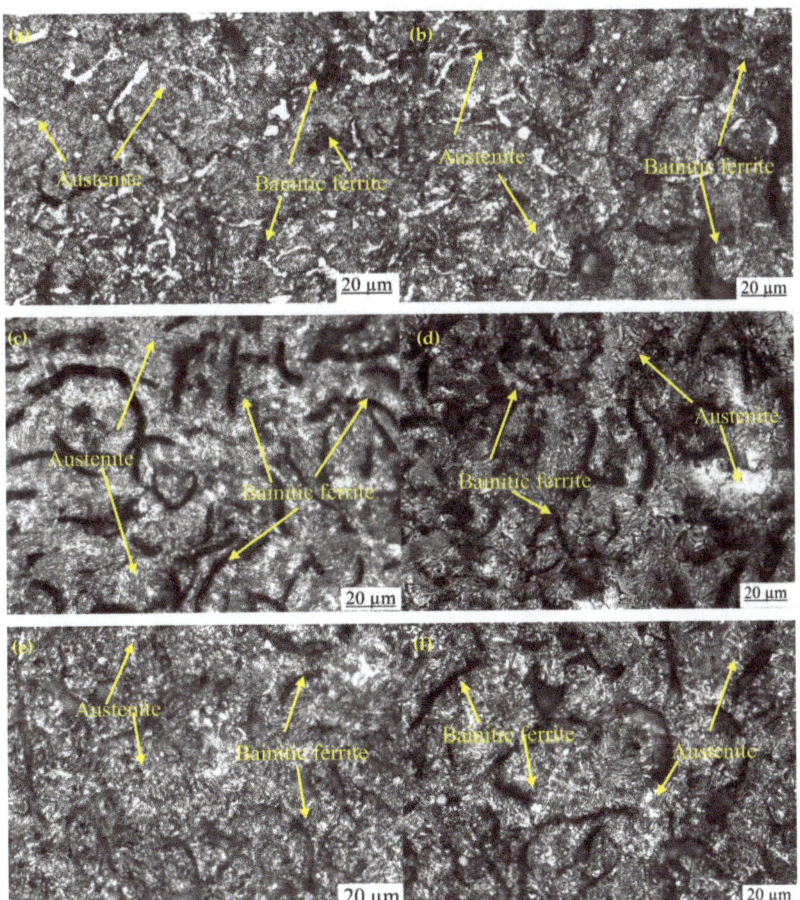

Figure 4. Optical micrographs of samples austempered for 60 min at (**a**) 260 °C, (**b**) 285 °C, (**c**) 310 °C, (**d**) 335 °C, (**e**) 360 °C, and (**f**) 385 °C.

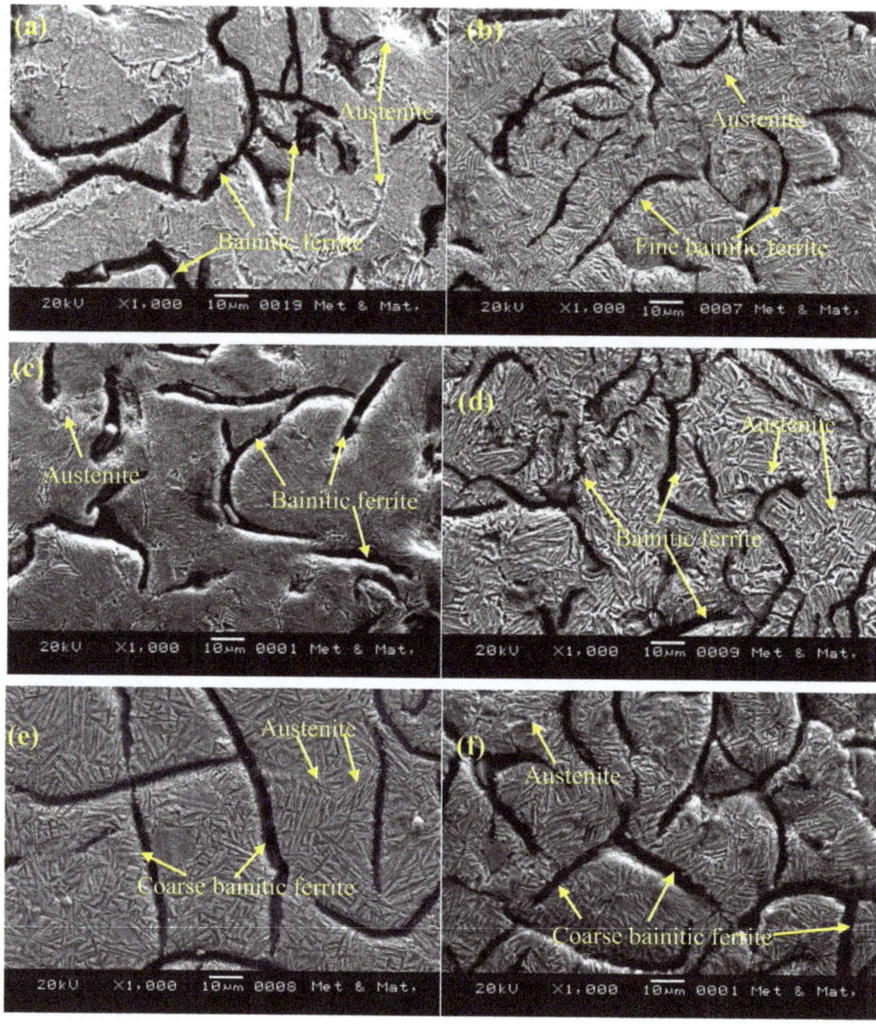

Figure 5. SEM images of samples austempered for 60 min at (**a**) 260 °C, (**b**) 285 °C, (**c**) 310 °C, (**d**) 335 °C, (**e**) 360 °C, and (**f**) 385 °C.

Figure 6 presents the quantitative analysis of the XRD pattern. It is evident from the figure that the austempering temperature has a significant effect on the XRD patterns. It was seen that with changing heat treatment temperature, the amount of austenite was changed. The phases detected include ferrite and austenite. Figure 7 shows the volume fraction of austenite and the carbon content of austenite in the AGCI samples as a function of different austempering temperatures. The volume faction of austenite was calculated by Jade7 software built in the XRD. It may be noted from Figure 7 that the austenite content increases with an increase in the austempering temperature. Greater supercooling at a lower austempering temperature resulted in finer ferrite and austenite as also reported by Patutunda et al. [23]. It is well known that the transformation reaction is more likely to be controlled by the nucleation process rather than growth [23,24]. During the process, it is necessary that the carbon must diffuse into austenite through the ferrite zone. At higher austempering temperatures a quite contrasting mechanism prevails due to lower supercooling which makes the nucleation of ferrite slower. This leads to the stabilization of more austenite in addition to incrementing the rate of diffusion of carbon which leads to the formation of coarse ferrite. Thus, the volume fraction of austenite increases with the increase in austempering temperature.

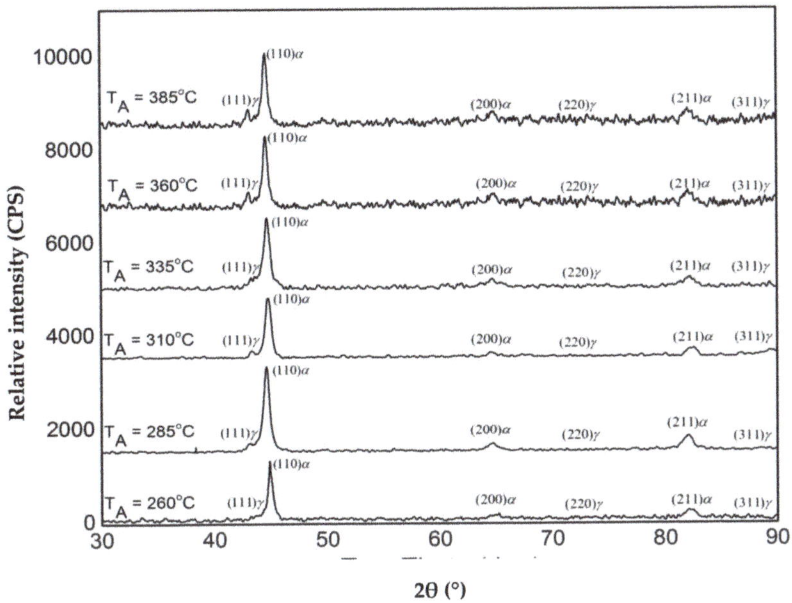

Figure 6. XRD phase analysis of austempered gray cast iron (AGCI) for different austempering temperatures held for 60 min (**a**) 260 °C, (**b**) 285 °C, (**c**) 310 °C, (**d**) 335 °C, (**e**) 360 °C, and (**f**) 385 °C.

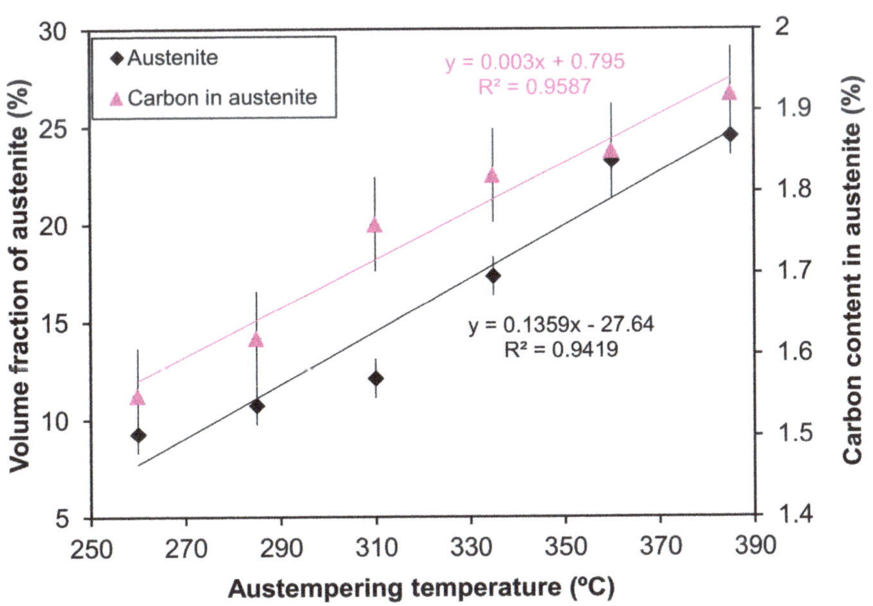

Figure 7. The influence of different austempering temperatures on the volume fraction of austenite and the carbon content of austenite.

3.2. Electrochemical Behavior of As-Cast Gray Iron and AGCI in Different Solutions

The results of potentiodynamic polarization studies of the as-cast gray iron and the AGCI are shown in Figures 8 and 9. The corrosion of iron in neutral 3.5% NaCl solution occurs according to the following equations,

Anodic reaction:

$$Fe \rightarrow Fe^{2+} + 2e, \tag{2}$$

Cathodic reaction:

$$O_2 + 2H_2O + 4e \rightarrow 4OH^-. \tag{3}$$

When iron is in contact with dilute sulfuric acid (0.5 M H_2SO_4), an immediate attack on the metal takes place with the formation of hydrogen gas and ferrous ions, as shown in Equations (4) and (5).

Anodic reaction:

$$Fe \rightarrow Fe^{2+} + 2e^-. \tag{4}$$

Cathodic reaction:

$$2H^+ + 2e^- \rightarrow H_2. \tag{5}$$

The electrochemical parameters are extracted after the extrapolation of the potentiodynamic plots in a Tafel slope. From Figure 7 it is revealed that with increasing austempering temperature, the percentage of austenite increased while the corrosion current density (I_{corr}) decreased and the corrosion potential (E_{corr}) shifted to the cathodic side (Tables 2 and 3). Austenite acts as an anode and ferrite acts as a cathode. The galvanic corrosion is proportional to the cathodic/anodic area. Consequently, with increasing temperature, the austenite percentage increases with a simultaneous decrease in the ferrite percentage. Thus, due to the microstructural homogeneities, distinct localized anodic and cathodic microstructural areas develop, which act as micro-electrochemical cells in the presence of an electrolyte. Thus, the galvanic corrosion decreases in both solutions. From Tables 2 and 3, it can be seen that among the two corrosive mediums, 1 N H_2SO_4 is more corrosive in all cases.

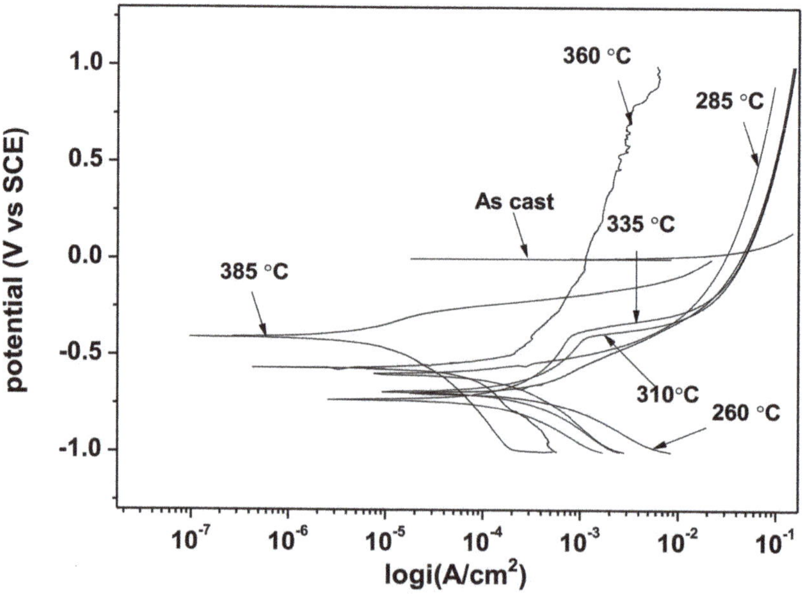

Figure 8. Potentiodynamic polarization curves in 3.5% NaCl solution.

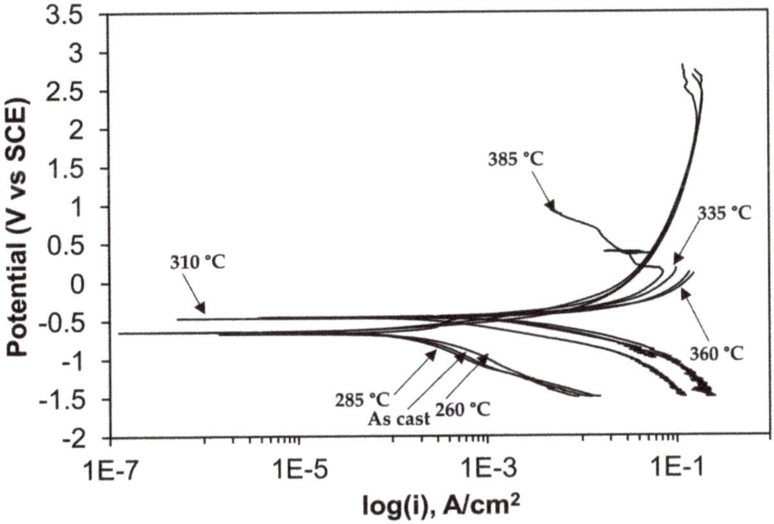

Figure 9. Potentiodynamic polarization curves in 0.5 M H_2SO_4 solution.

Table 2. Potentiodynamic polarization results in 3.5% NaCl.

Sample Condition	I_{corr} (μA/cm^2)	E_{corr} (V)
Austempering Temperature (°C)	-	-
As-cast gray iron	70	–0.80
927 °C–260 °C	62	–0.75
927 °C–285 °C	54	–0.70
927 °C–310 °C	42	–0.62
927 °C–335 °C	30	–0.50
927 °C–360 °C	16	–0.45
927 °C–385 °C	7	–0.40

Table 3. Potentiodynamic polarization results in 1 N H_2SO_4.

Sample Condition	I_{corr} (μA/cm^2)	E_{corr} (V)
Austempering Temperature (°C)	-	-
As-cast gray iron	150	–0.60
927 °C–260 °C	130	–0.55
927 °C–285 °C	110	–0.50
927 °C–310 °C	98	–0.48
927 °C–335 °C	90	–0.42
927 °C–360 °C	82	–0.38
927 °C–385 °C	65	–0.36

Nyquist plots of samples exposed to 3.5% NaCl and 0.5 M H_2SO_4 solutions are shown in the Figures 10 and 11, respectively. All plots show a depressed semicircle pattern in the whole frequency range, indicating that only one time constant exists between the interface of the solid electrode and the solution. Due to the low impedance value at the lower austempering temperature, the Nyquist plots become suppressed. The corresponding plots are shown in the insets of Figures 10 and 11 for exposure to 3.5% NaCl and 0.5 M H_2SO_4 solutions, respectively. All the EIS data match well in a constant phase element (CPE) model. In a CPE model, R_u is the solution resistance, R_p is the polarization resistance, and Y_o is the admittance. The inserted equivalent circuit shown in Figure 12 was used to fit the EIS data, and the fitted polarization resistance (R_p) data are shown in Tables 4 and 5 for 3.5% NaCl and 0.5 M H_2SO_4 solution, respectively. It is known that the diameter of the Nyquist plot represents the R_p. It is also well known that the R_p is inversely proportional to the corrosion rate. With increasing austempering temperature, the diameter of the Nyquist plot increases with a consequent increase in the R_p. It is also seen from Tables 4 and 5 with the error bar (maximum ±7%) that the R_p is higher in 3.5% NaCl solution (Table 4) than it is in 0.5 M H_2SO_4 solution (Table 5) for all cases.

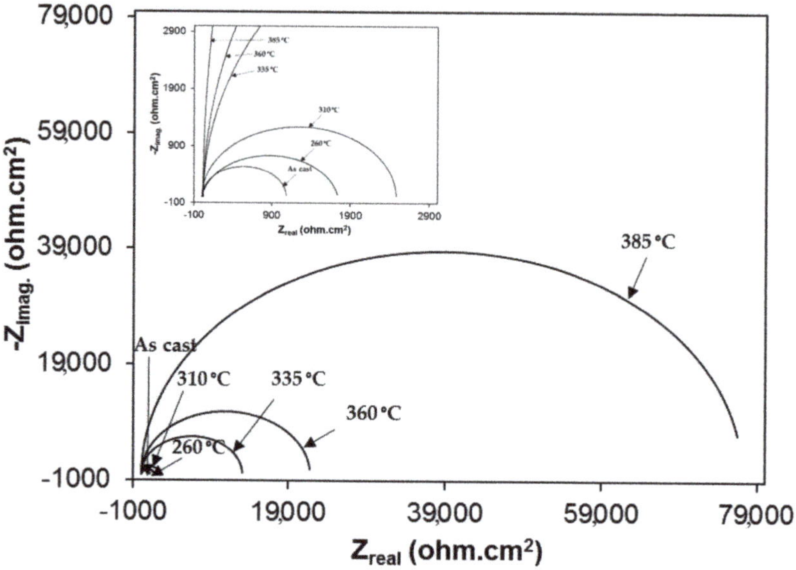

Figure 10. Electrochemical impedance spectroscopy (Nyquist plot) in 3.5% NaCl solution.

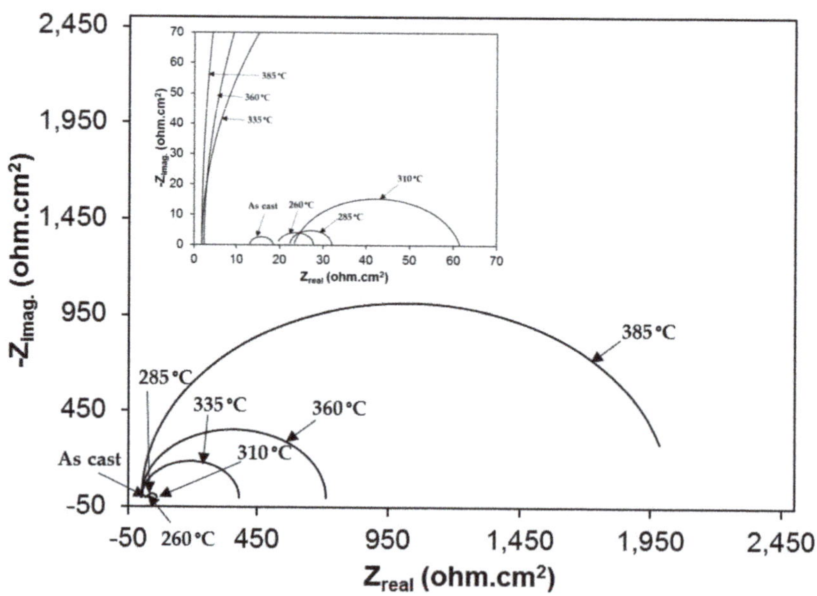

Figure 11. *Cont.*

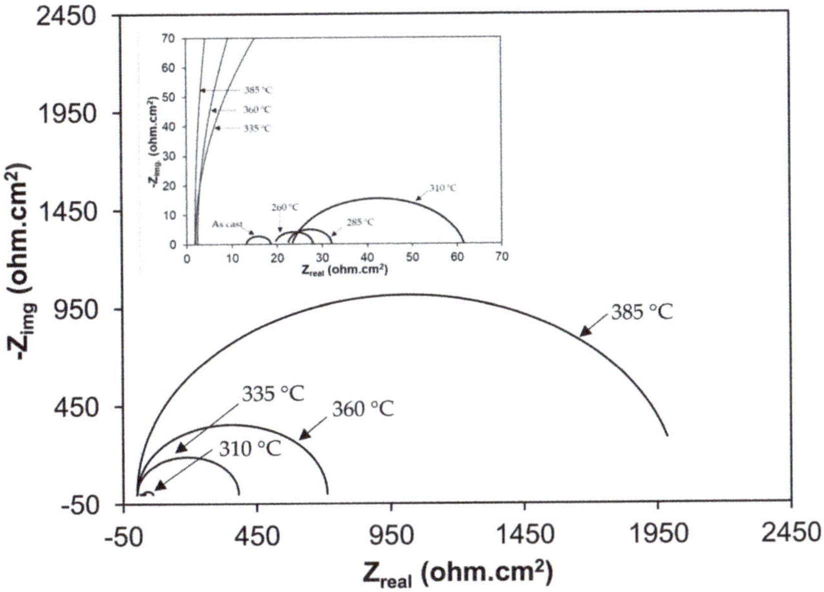

Figure 11. Electrochemical impedance spectroscopy (Nyquist plot) in 1 N H_2SO_4 solution.

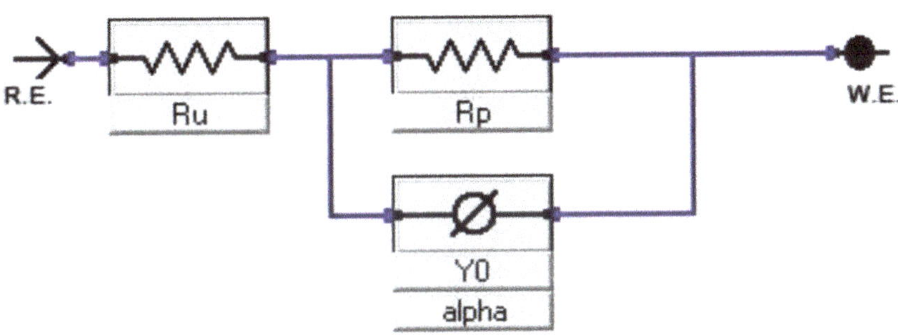

Figure 12. Equivalent circuit of electrochemical impedance spectroscopy (EIS).

Table 4. Electrochemical impedance spectroscopy results in 3.5% NaCl solution.

Sample Condition	R_p (ohm.cm^2)
Austempering Temperature (°C)	-
As-cast gray iron	930 (± 46)
927 °C–260 °C	1850 (± 91)
927 °C–310 °C	2550 (± 102)
927 °C–335 °C	15,000 (± 450)
927 °C–360 °C	20,000 (± 1100)
927 °C–385 °C	80,000 (± 1600)

Table 5. Electrochemical impedance spectroscopy results in 0.5 M H_2SO_4 solution.

Sample Condition	R_p (ohm.cm^2)
Austempering Temperature (°C)	-
As-cast gray iron	20 (± 1)
927 °C–260 °C	28 (± 1)
927 °C–285 °C	32 (± 2)
927 °C–310 °C	62 (± 2)
927 °C–335 °C	450 (± 13)
927 °C–360 °C	700 (± 28)
927 °C–385 °C	2000 (± 85)

3.3. Effect of Austenite Content on Corrosion Behavior

Figures 13 and 14 show the plots of I_{corr} obtained from the potentiodynamic polarization diagram against the volume fraction of the austenite for 3.5% NaCl and 1 N H_2SO_4, respectively. It was observed that with the increasing volume fraction of austenite, the corrosion rate decreased linearly to a sufficient extent in both cases. The linear fit regression value was 0.95 for 3.5% NaCl solution (Figure 13) and 0.84 for 1N H_2SO_4 solution (Figure 14). A regression value close to 1 means the corrosion rate changes linearly with increasing austenite content.

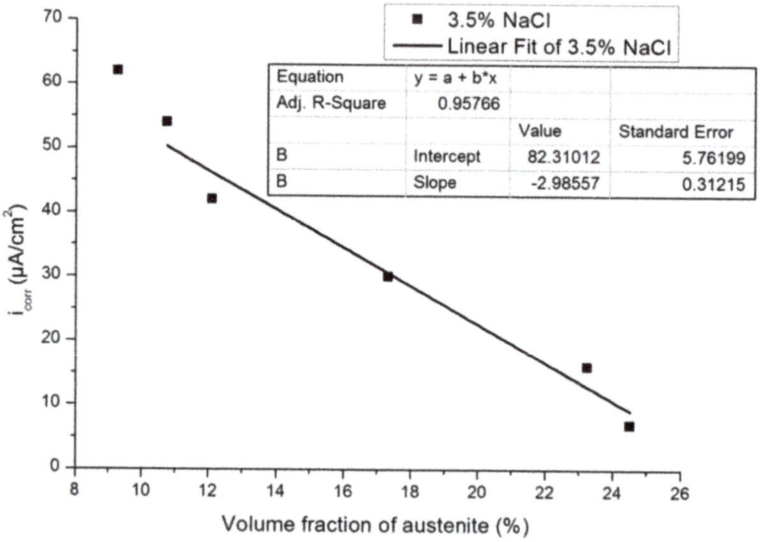

Figure 13. Influence of the volume fraction of the austenite on corrosion rate in 3.5% NaCl.

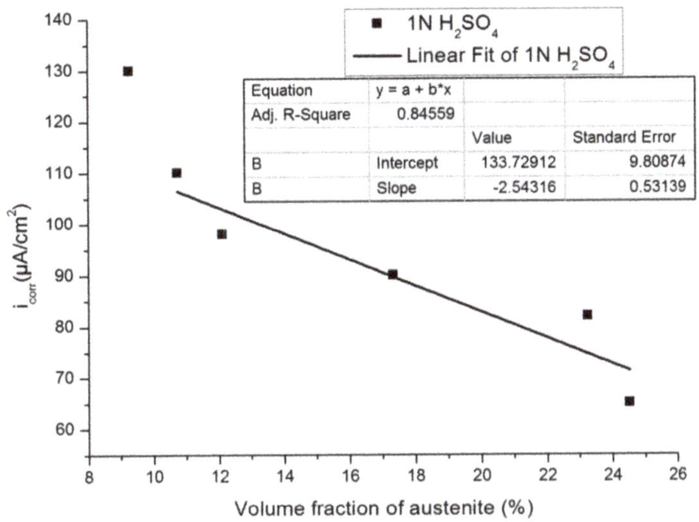

Figure 14. Influence of the volume fraction of the austenite on the corrosion rate in 0.5 M H_2SO_4.

3.4. Microstructure after Corrosion

3.4.1. Optical Images after Corrosion in 3.5% NaCl Solution

Figure 15 shows the OM of as-cast gray iron and AGCI samples dipped in 3.5% NaCl solution. The corrosion products consist of compact structures. It is seen that compactness increases with increasing austempering temperature. It was also observed that in optical images of as-cast gray iron (Figure 15a) with a lower austempering temperature (260 °C and 285 °C), an exfoliation type pattern was present with smaller flake graphite. With increasing austempering temperature (310 °C and above), the exfoliation type pattern disappeared with bigger flake graphite. In addition, intergranular type corrosion was observed in the 3.5% NaCl solution. It is also seen that, in as-cast gray iron at a low austempering temperature, more pitting was seen. However, with increasing austempering temperature the pitting density decreased.

Figure 15. *Cont.*

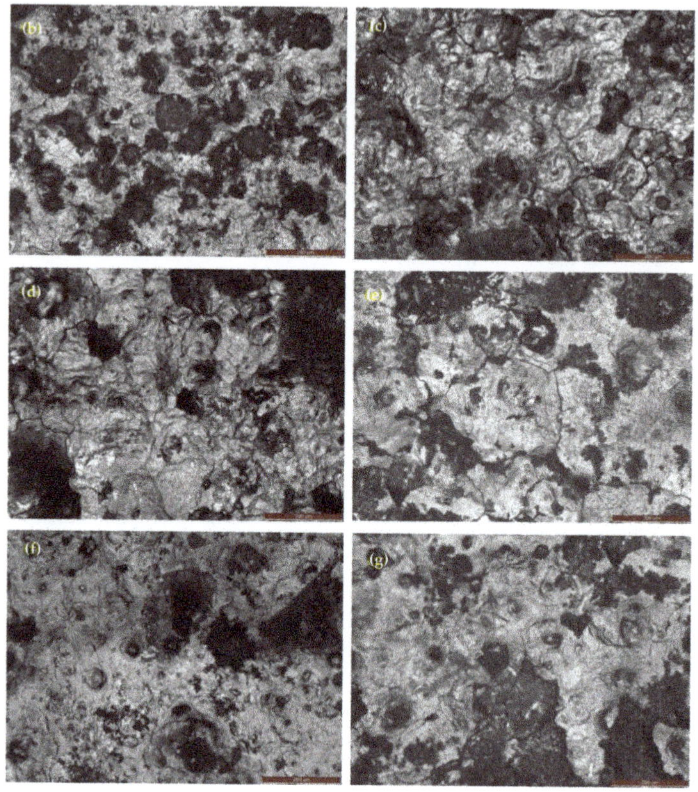

Figure 15. Optical images of iron samples after a potentiodynamic polarization test in 3.5% NaCl solution. (**a**) As-cast gray iron; and AGCI at (**b**) $T\gamma$ = 927 °C, T_A = 260 °C; (**c**) $T\gamma$ = 927 °C, T_A = 285 °C; (**d**) $T\gamma$ = 927 °C, T_A = 310 °C; (**e**) $T\gamma$ = 927 °C, T_A = 335 °C; (**f**) $T\gamma$ = 927 °C, T_A = 360 °C; and (**g**) $T\gamma$ = 927 °C, T_A = 385 °C.

3.4.2. Optical Images after Corrosion in 0.5 M H_2SO_4 Solution

Figure 16 shows the optical images (as-cast gray iron and AGCI) of corrosion products in 0.5 M H_2SO_4 solution. It is seen that pitting formation decreased, with increasing austempering temperature. While with increasing austempering temperature, more metastable pits were formed. So it can be concluded that with increasing austempering temperature, pit formation gradually reduced. At the same temperature, compared to 3.5% NaCl solution, pitting density, radius of pits, and the flake graphite was larger in the case of 0.5 M H_2SO_4 solution. It is also seen that more pits are formed in 1 N H_2SO_4 than in 3.5% NaCl solution.

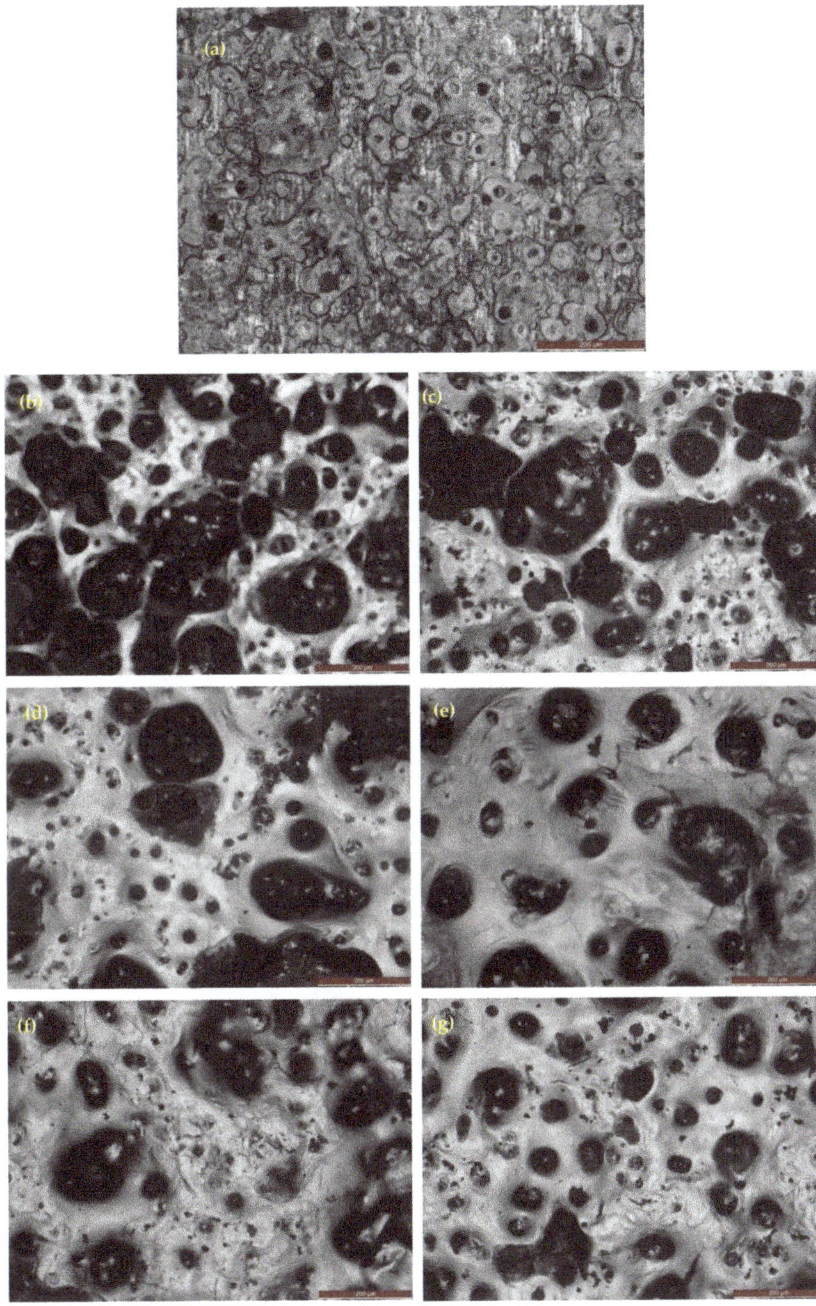

Figure 16. Optical images of iron samples after a potentiodynamic polarization test in 1 N H_2SO_4 solution. (**a**) As-cast gray iron; and AGCI at (**b**) $T\gamma$ = 927 °C, T_A = 260 °C; (**c**) $T\gamma$ = 927 °C, T_A = 285 °C; (**d**) $T\gamma$ = 927 °C, T_A = 310 °C; (**e**) $T\gamma$ = 927 °C, T_A = 335 °C; (**f**) $T\gamma$ = 927 °C, T_A = 360 °C; and (**g**) $T\gamma$ = 927 °C, T_A = 385 °C.

4. Conclusions

The following conclusions can be drawn from the present investigation:

(a) The microstructure of AGCI consists of special bainitic ferrite (α) and high-carbon austenite (Y) which prevents corrosion. Thus, the corrosion-resistance susceptibility of AGCI is higher than that of as-cast gray iron.

(b) At higher austempering temperatures, the volume fraction of austenite increases with a consequent decrease in the corrosion rate.

(c) In the ausferrite matrix, the corrosion rate depends on the austenite content. An increase in the austenite content results in a decrease in the corrosion rate.

(d) Due to an increase in hydrogen generating reactions, 1 N H_2SO_4 is more corrosive than 3.5% NaCl during exposure.

Author Contributions: Data curation: A.H.S. and M.G.; formal analysis: A.S., J.K.S., S.M.A.K.M. and M.G.; funding acquisition: A.H.S. and N.A.; methodology: A.H.S. and A.S.; project administration: A.H.S.; Supervision: A.H.S., S.M.A.K.M., and N.A.; validation: A.H.S., Amit Sarkar, and S.M.A.K.M.; writing—original draft: A.H.S., Amit Sarkar, J.K.S., S.M.A.K.M., N.A., and M.G.; Writing—review and editing: A.H.S., A.S., J.K.S., S.M.A.K.M., N.A. and M.G.

Funding: This research received no external funding.

Acknowledgments: The authors would like to extend their sincere appreciation to the Deanship of Scientific Research at King Saud University for its funding of this research through the Research Group Project No. RG-1439-029.

Conflicts of Interest: The authors declare no conflict of interest.

References

1. Khanna, O.P. *Foundry Technology;* Dhanpath Rai Publishers: New Delhi, India, 2011.
2. *ASM Hand Book, Properties and Selection: Irons, Steels, and High Performance Alloys;* ASM International: Geauga County, OH, USA, 1993.
3. Asadi, Z.S.; Melchers, R.E. Pitting corrosion of older underground cast iron pipes. *Corros. Eng. Sci. Technol.* **2017**, *52*, 459–469. [CrossRef]
4. Sami, A.A.; Saffa, M.H. Ductile and gray cast iron deterioration with time in various NaCl salt concentrations. *J. Eng. Technol.* **2008**, *26*, 1–15.
5. Rana, A.M.; Khan, A.; Amjad, S. Microstructure evaluation in heat treated cast irons. *J. Res. Sci.* **2001**, *12*, 65–71.
6. Muthukumarasamy, S.; Sadiq, B.A.; Sesham, S. High strength ductile irons-as cast bainitic ductile iron andaustempered ductile iron. *Indian Foundry J.* **1992**, *8*, 23–29.
7. Prasanna, N.D.; Muralidhara, M.K.; Mohit, K.A. Mechanical properties and corrosioncharacteristics of IS400/12 grade ductile iron. In Proceedings of the Transactions of 57th International Foundrymen Congress, Kolkata, India, 13–15 February 2009; pp. 89–95.
8. Banerjee, A.; Mitra, P.K.; Chattopadhyay, D.P. Effect of section thickness, 0.5% Cu additionand different austempering temperatures on the microstructure and corrosion properties of ADI. In Proceedings of the Transactions of 61st Indian Foundry Congress, Kolkata, India, 27–29 January 2013; pp. 1–3.
9. Afolabi, A.S. Effect of austempering temperature and time on corrosion behaviour of ductile iron in chloride and acidic media. *Anti-Corros. Methods Mater.* **2011**, *58*, 190–195. [CrossRef]
10. Hsu, C.-H.; Chen, M.-L. Corrosion behavior of nickel alloyed and austempered ductile irons in 3.5% sodium chloride. *Corros. Sci.* **2010**, *58*, 2945–2949. [CrossRef]
11. Kovacs, B.V.; Keough, J.R. Physical properties and application of austempered gray iron. *AFS Trans* **1993**, *93–141*, 283–291.
12. Hsu, C.H.; Shy, Y.H.; Yu, Y.H. Effect of austempering heat treatment on fracture toughness of copper alloyed gray iron. *Mater. Chem. Phys.* **2000**, *63*, 75–81. [CrossRef]
13. Rundman, K.B.; Parolini, J.R.; Moore, D.J. Relationship between tensile properties and matrix microstructure in austempered gray iron. *AFS Trans.* **2005**, *145*, 51–55.

14. Olawale, J.O.; Oluwaseguun, K.M.; Ezemenaka, D.I. Production of austempered gray iron (AGI) using forced air cooling. *Mater. Perform. Charact.* **2014**, *3*, 355–370. [CrossRef]
15. Sarkar, T.; Bose, P.K.; Sutradhar, G. Mechanical and Tribological Characteristics of Copper Alloyed Austempered Gray Cast Iron (AGI). *Mater. Today Proc.* **2018**, *5*, 3664–3673. [CrossRef]
16. Vadiraj, A.; Balachandran, G.; Kamaraj, M. Structure–property correlation in austempered alloyed hypereutectic gray cast irons. *Mater. Sci. Eng. A* **2010**, *527*, 782–788. [CrossRef]
17. Vadiraj, A.; Balachandran, G.; Kamaraj, M. Studies on mechanical and wear properties of alloyed hypereutectic gray cast irons in the as-cast pearlitic and austempered conditions. *Mater. Des.* **2010**, *31*, 951–955. [CrossRef]
18. Harvey, J.N.; Noble, G.A. Inoculation of cast irons—An overview. In Proceedings of the 55th Indian Foundry Congress, Agra, India, 2–4 February 2007; pp. 343–360.
19. Gorny, M.; Tyrala, E.; Lopez, H. Effect of copper and nickel on the transformation kinetics of austempered ductile iron. *J. Mater. Eng. Perform.* **2014**, *23*, 3505–3510. [CrossRef]
20. Bayati, H.; Elliott, R. The concept of an austempered heat treatment processing window. *Int. J. Cast Met. Res.* **1999**, *11*, 413–417. [CrossRef]
21. Dasgupta, R.K.; Mondal, D.K.; Chakrabarti, A.K. Evolution of microstructures during austempering of ductile irons alloyed with manganese and copper. *Metall. Mater. Trans. A* **2013**, *44*, 1376–1387. [CrossRef]
22. Bayati, H.; Elliott, R. Relationship between structure and mechanical properties in high manganese alloyed ductile iron. *Mater. Sci. Technol.* **1995**, *11*, 284–293. [CrossRef]
23. Patutunda, S.K. Development of austempered ductile iron (ADI) with simultaneous high yield strength and fracture toughness by a novel two-step austempering process. *Mater. Sci. Eng. A* **2001**, *315*, 70–80. [CrossRef]
24. Yang, J.; Putatunda, S.K. Improvement in strength and toughness of austempered ductile cast iron by novel two-step austempering process. *Mater. Des.* **2004**, *25*, 219–230. [CrossRef]

materials

MDPI

Article

Non-Isothermal Oxidation Behaviors and Mechanisms of Ti-Al Intermetallic Compounds

Peixuan Ouyang [1,2,3], Guangbao Mi [1,4,*], Peijie Li [2], Liangju He [2], Jingxia Cao [1] and Xu Huang [1]

1 Aviation Key Laboratory of Science and Technology on Advanced Titanium Alloys, AECC Beijing Institute of Aeronautical Materials, Beijing 100095, China
2 National Center of Novel Materials for International Research, Tsinghua University, Beijing 100084, China
3 School of Mechanical and Materials Engineering, North China University of Technology, Beijing 100144, China
4 Beijing Engineering Research Center of Graphene and Application, Beijing 100095, China
* Correspondence: miguangbao@163.com; Tel.: +86-10-62496627

Received: 25 May 2019; Accepted: 27 June 2019; Published: 30 June 2019

Abstract: Non-isothermal oxidation is one of the important issues for the safe application of Ti-Al alloys, so this study aimed to illustrate the non-isothermal oxidation behaviors and the corresponding mechanisms of a TiAl-based alloy in comparison with a Ti_3Al-based alloy. The non-isothermal oxidation behaviors of Ti-46Al-2Cr-5Nb and Ti-24Al-15Nb-1.5Mo alloys in pure oxygen were comparatively investigated with a thermogravimetry-differential scanning calorimetry (TGA/DSC) simultaneous thermal analyzer heating from room temperature to 1450 °C with a heating rate of 40 °C/min. When the temperature rose above 1280 °C, the oxidation rate of the Ti-46Al-2Cr-5Nb alloy sharply increased and exceeded that of the Ti-24Al-15Nb-1.5Mo alloy owing to the occurrence of internal oxidation. When the temperature was higher than 1350 °C, the oxidation rate of the Ti-46Al-2Cr-5Nb alloy decreased obviously due to the generation of an oxygen-barrier β-Al_2TiO_5-rich layer by a chemical reaction between Al_2O_3 and TiO_2 in the oxide scale. Based on Wagner's theory of internal oxidation, the reason for the occurrence of internal oxidation in the Ti-46Al-2Cr-5Nb alloy is the formation of the α phase in the subsurface, while no internal oxidation occurred in the Ti-24Al-15Nb-1.5Mo alloy due to the existence of the β phase in the subsurface with the enrichment of Nb and Mo.

Keywords: titanium aluminides; oxidation; non-isothermal; mechanism; internal oxidation

1. Introduction

TiAl-based alloys have received considerable attention as high-temperature structural materials for aerospace and automotive applications, since they maintain numerous outstanding properties, such as low density (3.9–4.2 g/cm^3), high specific strength, good creep resistance and excellent fireproof performance [1–5]. Nevertheless, the insufficient high-temperature oxidation resistance of the TiAl-based alloys limits their wide application and development [6,7]. Numerous investigations have been carried out to study the high-temperature oxidation behaviors of the TiAl-based alloys, most of which were concerned about the long-term isothermal oxidation at normal service temperature (800–1000 °C) [8–10]. Few studies have been focused on the non-isothermal oxidation behaviors of the TiAl-based alloys, which is also deserving of attention since the alloys are often subject to non-isothermal oxidation in service. For example, aero-engine TiAl components are heated rapidly from room temperature to service temperature during the engine startup period. For another instance, aero-engine TiAl blades are likely to suffer from titanium fire, which consists of ignition and propagation combustion processes under the induction of external energies such as high-energy friction, fracture, melt droplets and so on [11], even though the fireproof performance of TiAl-based alloys is much better than that of conventional titanium alloys. Ignition, which is the precursor process of titanium fire, is essentially a non-isothermal oxidation process [12,13] and the ignition points of titanium alloys are slightly lower than their melting points [11]. Hence, the non-isothermal oxidation of TiAl-based alloys with a rather wide temperature range has a major impact on their safe application in aeroengines.

In addition, the non-isothermal oxidation behaviors of the TiAl-based alloys should be more special than the isothermal oxidation behaviors, since the heating rate is so high that the oxidation may not reach dynamic equilibrium and the temperature range is so wide that phase transition may occur in the alloys. Many studies showed that oxidation temperature and time significantly affect isothermal oxidation behaviors of the TiAl-based alloys [14–17]. Moreover, Vaidya [18] found that during isothermal oxidation, the mass gain of Ti-48Al-2Nb-2Cr alloy exceeds that of Ti-25Al-10Nb-3V-1Mo alloy when the temperature is above 1200 °C, which breaks away from the general understanding that the oxidation resistance of the TiAl-based alloys is better than that of the Ti_3Al-based alloys. This phenomenon is considered to be attributed to the doubly increased oxygen solubility of γ phase in the TiAl-based alloy at 1200 °C [18], while this explanation seems unreasonable since the maximum oxygen solubility of γ phase (~3 at.%) in the TiAl-based alloy is much lower than that of α_2 phase (~13 at.%) in the Ti_3Al-based alloys [19]. Hence, in regards to non-isothermal oxidation, it should be of interest to people what behaviors the TiAl-based alloys would exhibit, whether there exists a similar phenomenon that the oxidation rate of the TiAl-based alloys exceeds that of the Ti_3Al-based alloys at high temperature, as well as what is the reasonable explanation for this phenomenon. On the basis of the above questions, this paper examines the non-isothermal oxidation behaviors of a TiAl-based alloy in comparison with a Ti_3Al-based alloy and illustrates the corresponding oxidation mechanisms. On one hand, this research helps to better understand the ignition mechanisms of TiAl alloys, on the other hand, it might promote the development of new-type TiAl alloys resistant to higher temperature and with less risk to titanium fire by ingredient optimum design.

TGA/DSC simultaneous thermal analysis is a common method to study the non-isothermal oxidation behaviors of metals and their alloys [20–23]. Adopting the TGA/DSC method, G.B. Mi et al. [22] studied the effect of Cr content on the non-isothermal oxidation behaviors of Ti-Cr fire-proof titanium alloys. The results showed that when the Cr content exceeds 10–15 wt.%, the oxidation resistance of Ti-Cr alloys increases with the Cr content due to the precipitation of Cr oxide [22]. Ouyang et al. [23] carried out a research on the non-isothermal oxidation behaviors of a high-temperature near-α titanium alloy (TA29 alloy), meanwhile discussed the effects of lattice transformation and alloying elements on the non-isothermal oxidation behaviors of the TA29 alloy.

In this paper, the non-isothermal oxidation behaviors of a TiAl-based alloy and a Ti_3Al-based alloy in pure oxygen were studied with a TGA/DSC synchronous thermal analyzer heating from room temperature to 1450 °C with a heating rate of 40 °C/min. Combined with microstructural characterization and calculation of oxidation activation energy, the non-isothermal oxidation mechanisms of the TiAl-based alloy were elucidated. Furthermore, the reason for the poorer oxidation resistance of the TiAl-based alloy than that of the Ti_3Al-based alloy at high temperature was revealed based on Wagner's theory of internal oxidation.

2. Experimental

2.1. Specimen Preparation

The nominal compositions of the TiAl-based and Ti_3Al-based alloys studied in this paper are Ti-46Al-2Cr-5Nb (at.%) and Ti-24Al-15Nb-1.5Mo (at.%), respectively. Both the two alloys were prepared by melting, forging, heat treatment and mechanical processing. The microstructure of the as-received Ti-46Al-2Cr-5Nb alloy consists of coarse fully-lamellar $\gamma + \alpha_2$ colonies, as shown in Figure 1a. The microstructure of the as-received Ti-24Al-15Nb-1.5Mo alloy is duplex, being composed of equiaxed primary α_2 and/or O phases, B2 transformed structure consisting of flaky secondary α_2 and O phases and residual B2 phase, as shown in Figure 1b. For non-isothermal oxidation experiments, specimens with dimensions of $3 \times 2 \times 2$ mm^3 were cut from the two kinds of alloy sheets by a computerized numerical control (CNC) dicing saw (SYJ-400, Shenyang kejing instrument company, Shenyang, China. The specimens were ground with sandpapers up to 2000-grit to remove oxide scales and then ultrasonically cleaned with acetone and alcohol.

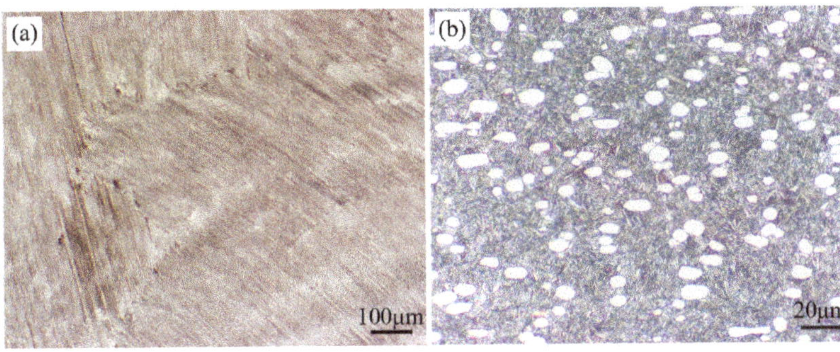

Figure 1. Optical micrographs showing the original microstructure of the (**a**) Ti-46Al-2Cr-5Nb and (**b**) Ti-24Al-15Nb-1.5Mo alloys.

2.2. *Non-Isothermal Oxidation Experiment*

Non-isothermal oxidation experiments were carried out in a TGA/DSC simultaneous thermal analyzer (TGA/DSC 1; Mettler Toledo, Zurich, Switzerland). The specimens of both the Ti-46Al-2Cr-5Nb and Ti-24Al-15Nb-1.5Mo alloys were heated from room temperature to 1450 °C at a heating rate of 40 °C/min under a pure oxygen flow of 50 mL/min. The oxygen flow was added to the system once the heating process was started and the flow was cut off when the heating process was completed. After furnace cooling the oxidized specimens were taken out. For each alloy, the mass-gain, mass-gain rate, and heat flux curves were obtained from three repeated experiments with the relative standard deviations less than 10%. Besides, in order to investigate the non-isothermal oxidation behaviors of the two alloys in detail, the specimens were heated to several interested temperatures below 1450 °C and then were furnace cooled to obtain the corresponding non-isothermal oxidation products.

2.3. *Microstructural Characterization*

Specimens were examined after non-isothermal oxidation using field emission scanning electron microscopy (FE-SEM; Hitachi SU8000, Tokyo, Japan) and X-ray diffraction (XRD; Bruker D8 Advance; Cu Ka, Karlsruhe, Germany) to characterize surface morphologies of the oxide scales and identify oxide phases. For cross-sectional microstructure observation of the oxidized specimens, metallographic specimens were prepared by being placed in the transverse direction, being mounted with cold-setting resin, then being ground with 400–2000 grit SiC sandpapers, being polished with 1.0 μm alumina suspension, being etched with Kroll reagent (92 mL H_2O, 3 mL HF, 5 mL HNO_3) and finally being coated with a thin layer of carbon. The microstructures and elemental distributions on the cross-section of the oxide scales were characterized using an electron probe microanalyzer (EPMA; Shimadzu EPMA-1720H, Kyoto, Japan).

3. Calculation of Oxidation Activation Energy

Similar to traditional titanium alloys, the oxidation mass gains of the Ti-Al alloys also come from oxygen dissolution and growth of oxide scales. The two oxidation behaviors are competitive and one of them governs the mass gain at a certain temperature and time [24–26]. The rate-determining steps of the two oxidation behaviors are respectively the diffusion of O atom in the alloy and the diffusion of O^{2-} in the oxide scale. Thus, it can be assumed that the mass-gain rate of the Ti-Al alloys during non-isothermal oxidation is limited by one-dimensional diffusion of one species A (A is O atom or

O^{2-}), so that the mass-gain rate per unit area of the Ti-Al alloys is proportional to the molar diffusion flux of specie A. Furthermore, based on the Fick's diffusion law and assuming that the diffusion of specie A satisfies Arrhenius kinetics, the relationship among the temperature, the mass gain and the oxidation activation energy of the Ti-Al alloys can be derived, as shown in Equation (1). Details for the derived process can be seen in our previous work [23].

$$-\ln\frac{d(\Delta m)}{dT} - \ln(\Delta m) = -\ln K^{*} + E/RT \tag{1}$$

where Δm is the mass gain per unit area, E is the oxidation activation energy, R is the molar gas constant and K* is constant. Let the left side of Equation (1) be equal to $Y(\Delta m)$, thus $Y(\Delta m)$ is linearly related to the reciprocal of temperature and the product of the corresponding positive slope and R is the oxidation activation energy.

4. Results and Discussion

4.1. Oxidation Mass Gain and Activation Energies

Figure 2 shows the mass-gain and mass-gain rate curves of the Ti-46Al-2Cr-5Nb and Ti-24Al-15Nb-1.5Mo alloys obtained during non-isothermal oxidation. The mass-gain rates of the two alloys were similar when the temperature was lower than 1280 °C. However, the mass-gain rate of the Ti-46Al-2Cr-5Nb alloy increased dramatically when the temperature was higher than 1280 °C, resulting that the mass gain of the Ti-46Al-2Cr-5Nb alloy significantly exceeded that of the Ti-24Al-15Nb-1.5Mo alloy. This phenomenon is not common but is similar to that reported by Vaidya [18]. Nevertheless, the mass-gain rate of the Ti-46Al-2Cr-5Nb alloy turned to decrease obviously when the temperature was above 1350 °C.

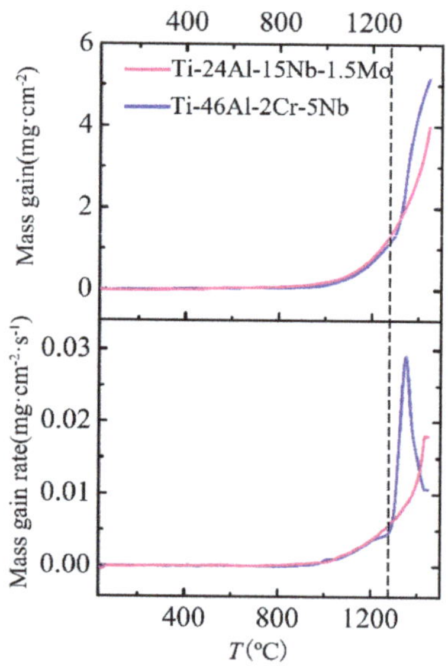

Figure 2. Comparisons of the mass-gain and mass-gain rate curves between the Ti-46Al-2Cr-5Nb and Ti-24Al-15Nb-1.5Mo alloys obtained during non-isothermal oxidation.

According to the evolution of the mass-gain rate, the non-isothermal oxidation process of the Ti-46Al-2Cr-5Nb alloy could be divided into five stages for further study, as shown in Figure 3a. When the temperature was below 870 °C (Stage I), the mass gain of the Ti-46Al-2Cr-5Nb alloy was few and could be neglected. When the temperature rose to 870–980 °C (Stage II), the mass-gain rate increased slowly and the corresponding mass gain was 0.04 mg/cm^2. When the temperature was increased to 980–1280 °C (Stage III), the mass-gain rate increased obviously and the corresponding mass gain was 1.05 mg/cm^2. When the temperature was raised to 1280–1350 °C (Stage IV), the mass-gain rate increased sharply and the corresponding mass gain was 1.68 mg/cm^2. Nevertheless, the mass-gain rate decreased significantly when the temperature was above 1350 °C (Stage V). It is concluded that the non-isothermal oxidation process of the Ti-46Al-2Cr-5Nb alloy consists of five stages, including nearly non-oxidation (Stage I, <870 °C), slow oxidation (Stage II, 870–980 °C), accelerated oxidation (Stage III, 980–1280 °C), severe oxidation (Stage IV, 1280–1350 °C) and decelerated oxidation (Stage V, 1350–1450 °C) stages. However, the non-isothermal oxidation process of the Ti-24Al-15Nb-1.5Mo alloy can be only divided into four stages according to the evolution of the mass-gain rate, as shown in Figure 3b, including nearly non-oxidation (Stage I, <800 °C), slow oxidation (Stage II, 800–1020 °C), accelerated oxidation (Stage III, 1020–1400 °C) and severe oxidation (Stage IV, 1400–1450 °C) stages. Table 1 gives a summary of the temperature ranges corresponding to the different stages of the non-isothermal oxidation process for the Ti-46Al-2Cr-5Nb and Ti-24Al-15Nb-1.5Mo alloys.

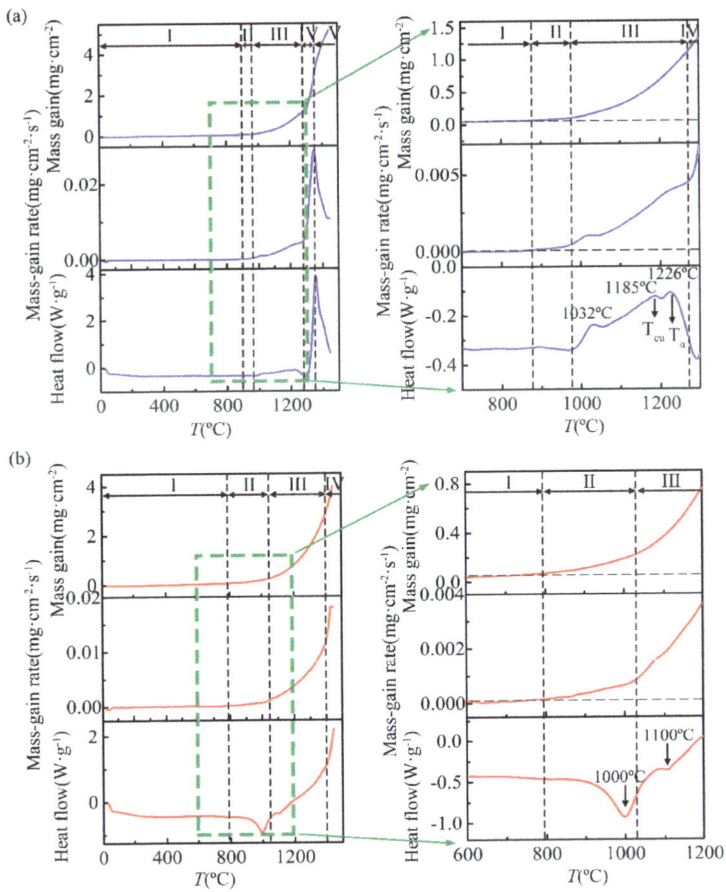

Figure 3. Stage divisions of the non-isothermal oxidation process for the (**a**) Ti-46Al-2Cr-5Nb and (**b**) Ti-24Al-15Nb-1.5Mo alloys according to the evolution of the mass-gain rates.

Table 1. Temperature ranges, oxidation activation energies and matrix phases corresponding to different stages of the non-isothermal oxidation for the Ti-46Al-2Cr-5Nb and Ti-24Al-15Nb-1.5Mo alloys.

Stage	Temperature Range (°C)		Activation Energy (kJ/mol)		Matrix Phases	
	TiAl [1]	Ti_3Al [2]	TiAl [1]	Ti_3Al [2]	TiAl [1]	Ti_3Al [2]
I	<870	<800	-	-	$\gamma + \alpha_2$	α_2 + O + B2
II	870–980	800–1020	217.8	143.2	$\gamma + \alpha_2$	α_2 + O + B2 → α_2 + B2
III	980–1280	1020–1400	249.8	244.8	$\gamma + \alpha_2 \rightarrow \gamma + \alpha \rightarrow \alpha$	α_2 + B2 → B2 → β
IV	1280–1350	1400–1450	985.0	608.6	α	β
V	1350–1450	-	-	-	α	-

[1] TiAl here refers to the Ti-46Al-2Cr-5Nb alloy. [2] Ti_3Al here refers to the Ti-24Al-15Nb-1.5Mo alloy.

Figure 4 presents the $Y(\Delta m)$~$1/T$ curves and the corresponding linear fitting results of the two alloys. The evolutions of the slops of the $Y(\Delta m)$~$1/T$ curves also clearly demonstrate that the non-isothermal oxidation processes of the Ti-46Al-2Cr-5Nb and Ti-24Al-15Nb-1.5Mo alloys are respectively composed of five stages and four stages. It should be noted that since the mass changes of the two alloys at Stage I were mainly induced by the small mass fluctuation of the thermogravimetric analysis equipment, the values of $Y(\Delta m)$ at Stage I fluctuated around a certain value and are not shown in Figure 4. Besides, the $Y(\Delta m)$~$1/T$ curves of both the two alloys at Stages II, III and IV could be positive linearly fitted, so that the oxidation activation energies of the two alloys for the three stages were obtained, as presented in Table 1. It can be seen that the oxidation activation energies of the two alloys at Stage III are similar, while the oxidation activation energies of the two alloys both at Stage II and at Stage IV are quite different, which is related to their oxidation mechanisms and will be discussed in Section 4.3.

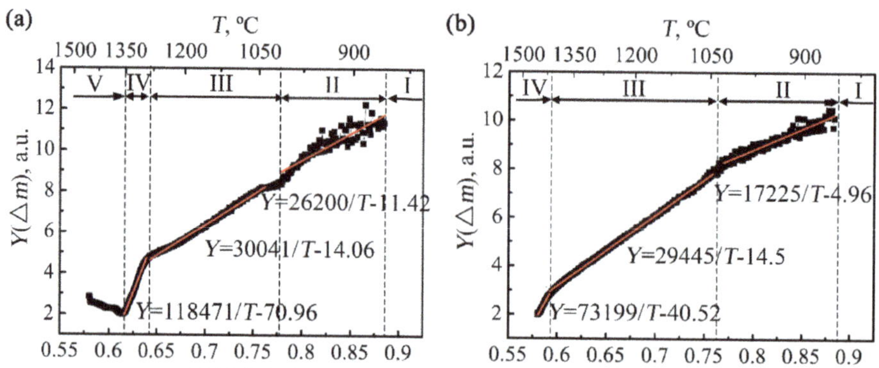

Figure 4. The $Y(\Delta m)$~$1/T$ curves and the corresponding linear fitting results of the (**a**) Ti-46Al-2Cr-5Nb and (**b**) Ti-24Al-15Nb-1.5Mo alloys.

4.2. Matrix Phases

Since the non-isothermal oxidation experiments were carried out in a wide temperature range, the Ti-46Al-2Cr-5Nb and Ti-24Al-15Nb-1.5Mo alloys were likely to undergo phase transitions during non-isothermal oxidation.

Generally, the phase transition types and temperatures of the TiAl-based alloys are affected not only by the aluminum content (45–48 at.%), but also by kinds of other alloying elements (such as Cr, V, Mn, Nb, Ta, W, Si, C, P and B) and their contents (0.1–8 at.%) [27]. According to the Ti-Al phase diagram [28], the heat-flow and thermomechanical analysis (TMA) derivative curves of Ti-(46~47)Al-2Cr-(2~8)Nb alloy [29], the phase transitions of the Ti-46Al-2Cr-5Nb alloy heating from room temperature to 1450 °C are deduced to be $\gamma + \alpha_2 \rightarrow \gamma + \alpha$ and $\gamma + \alpha \rightarrow \alpha$ with the respective temperature range of

1175–1215 °C (T_{eu}) and 1292–1295 °C (T_{α}). The deduced temperature ranges are in accordance with the two exothermic peaks with the respective temperatures of 1185 and 1226 °C in the heat-flow curve of the Ti-46Al-2Cr-5Nb alloy, as shown in Figure 3a. It confirms that the phase transitions of $\gamma + \alpha_2 \rightarrow \gamma + \alpha$ and $\gamma + \alpha \rightarrow \alpha$ occurred in the Ti-46Al-2Cr-5Nb alloy during the non-isothermal oxidation. Besides, it can be seen from Figure 3a that there exists another exothermic peak with a temperature of about 1032 °C in the heat-flow curve of the Ti-46Al-2Cr-5Nb alloy. It is not yet possible to explain the physical meaning of this peak based on the Ti-Al phase diagram. However, it has been reported that there exists a similar exothermic peak in the heat-flow curve of Ti-46Al-1.9Cr-3Nb alloy with typical duplex structure, which is ascribed to equilibrium transformation or uniformity when heating of the non-equilibrium structure generated by thermal mechanical processing [30]. Therefore, the matrix phases of the Ti-46Al-2Cr-5Nb alloy at different stages of the non-isothermal oxidation process could be concluded, as shown in Table 1.

As for the Ti-24Al-15Nb-1.5Mo alloy, the Nb equivalent is about 20 at.% since the beta phase stability of Mo is 3.6 times of that of Nb [31]. Thus, according to the Ti-25Al-Nb phase diagram [31], the phase transitions of the Ti-24Al-15Nb-1.5Mo alloy heating from room temperature to 1450 °C are deduced to be α_2 + O + B2 → α_2 + B2 (1010 °C), α_2 + B2 → B2 (1100 °C) and B2 → β (1220 °C). Since there exist two endothermic peaks with the respective temperatures of near 1000 and 1100 °C in the heat-flow curve of the Ti-24Al-15Nb-1.5Mo alloy during non-isothermal oxidation, as shown in Figure 3b, it demonstrates the occurrence of the former two phase transitions. Besides, no peak corresponding to the last phase transition could be found in the heat-flow curve of the Ti-24Al-15Nb-1.5Mo alloy, which is probably due to the small chemical heat for the disordering from B2 phase to β phase. As a result, the matrix phases of the Ti-24Al-15Nb-1.5Mo alloy at different stages of the non-isothermal oxidation process could be also concluded, as presented in Table 1.

4.3. Non-Isothermal Oxidation Mechanisms

4.3.1. Nearly Non-Oxidation Stage (Stage I)

Similar to the near-α titanium alloy TA29 [23], the oxidation mass gains of the Ti-46Al-2Cr-5Nb and Ti-24Al-15Nb-1.5Mo alloys at Stage I were very small and could be neglected (Figure 3), which is ascribed to the fact that a thin titanium oxide film (200 nm) rapidly forms on the alloys at room temperature and passivates the surface [14].

4.3.2. Slow Oxidation Stage (Stage II)

When the temperature rose to Stage II (870–980 °C for the Ti-46Al-2Cr-5Nb alloy and 800–1020 °C for the Ti-24Al-15Nb-1.5Mo alloy), both the two alloys exhibited slow oxidation behaviors, which is also similar to the near-α titanium alloy TA29 at Stage II with the temperature range of 750–1000 °C [23]. As discussed in our previous work [23], the oxidation mechanism of the TA29 alloy at Stage II is oxygen dissolution with the rate-determining step of oxygen diffusion in the alloy. The oxidation activation energies of the TA29 alloy at Stage II (163.9 kJ/mol) [23] is about 20 kJ/mol higher than that of the Ti-24Al-15Nb-1.5Mo alloy (143.2 kJ/mol). The matrixes of the TA29 and Ti-24Al-15Nb-1.5Mo alloys at Stage II are respectively dominated by the α and α_2 phases. Moreover, it has been reported that the activation energy for oxygen diffusion in the α phase is 10–20 kJ/mol higher than that in the α_2 phase [32], which is equivalent to the difference of the oxidation activation energy between the TA29 and Ti-24Al-15Nb-1.5Mo alloys. Therefore, the oxidation mechanism of the Ti-24Al-15Nb-1.5Mo alloy at Stage II is also the oxygen dissolution in the alloy. It indicates that the thin titanium oxide film (200 nm) passivating the surface at Stage I could not prevent oxygen from diffusing into the alloys at Stage II, which might be due to the increase of the solubility of TiO_2 in the alloys or the cracking of TiO_2 film caused by phase transitions in the oxide film with the temperature increases.

The lower oxidation activation energy of the Ti-24Al-15Nb-1.5Mo alloy in comparison with the TA29 alloy indicates that oxygen atoms diffuse much easier in the former alloy. Since volume diffusion is the main diffusion type in the alloys at such high temperature, the difference of oxygen diffusion ability can be explained by the difference of the lattice structure between the two alloys. The α phase dominating in the TA29 alloy has the close-packed hexagonal structure (hcp, A3), where oxygen atoms tend to occupy the two Ti_6 octahedral interstitial sites [33,34], as shown in Figure 5a. The α_2 phase dominating in the Ti-24Al-15Nb-1.5Mo alloy has the ordered close-packed hexagonal structure (DO_{19}), where oxygen atoms prefer to occupy the Ti_6 octahedral interstitial site instead of the Al_2Ti_4 site [35,36], as shown in Figure 5b. Though the number of the Ti_6 interstitial site in the lattice of the α_2 phase is lower than that in the lattice of the α phase, which leads to the lower oxygen solubility in the α_2 phase than in the α phase, the covalence of Ti-Al bond in the lattice of the α_2 phase makes electrons aggregate between Ti and Al atoms, resulting in the weaker Ti-O bond strength in the lattice of the α_2 phase compared with that in the lattice of the α phase [32]. Therefore, the diffusion resistance of oxygen atoms in the lattice of the α_2 phase is smaller than that in the lattice of the α phase. That is to say, the required activation energy for oxygen diffusion in the α_2 phase is smaller than in the α phase.

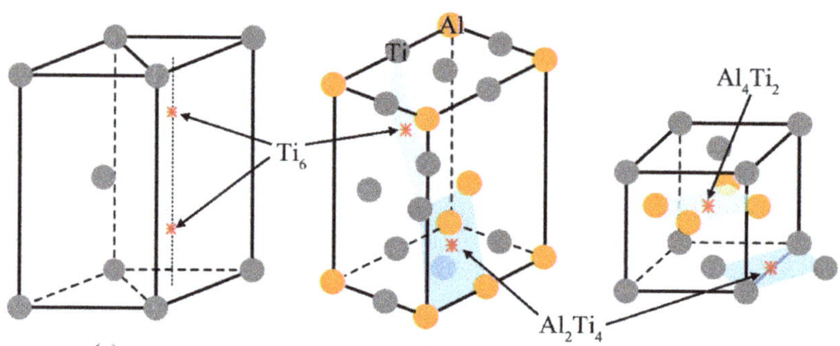

Figure 5. Octahedral interstitial sites for oxygen atoms in the lattices of the (**a**) α, (**b**) α_2 and (**c**) γ phases.

As for the Ti-46Al-2Cr-5Nb alloy, the oxidation activation energy at Stage II (217.8 kJ/mol) is higher than that of the Ti-24Al-15Nb-1.5Mo alloy (143.2 kJ/mol). The γ phase dominating in the Ti-46Al-2Cr-5Nb alloy at Stage II has the ordered face-centered cubic structure (L_{10}), where oxygen atoms can only occupy the interstitial octahedral sites surrounded by both titanium and aluminum atoms (Al_4Ti_2 and Al_2Ti_4), as shown in Figure 5c. Since the covalence of Ti-Al bond makes electrons aggregate between Ti and Al atoms, the Ti-O bond strength in the Al_4Ti_2 and Al_2Ti_4 octahedrons is higher than that in the Ti_6 octahedron. Hence, the oxygen diffusion ability in the lattice of the γ phase is much weaker than that in the lattice of the α_2 phase, leading to the higher oxidation activation energy of the Ti-46Al-2Cr-5Nb alloy than that of the Ti-24Al-15Nb-1.5Mo alloy. The above analysis manifests that the oxidation mechanism of the Ti-46Al-2Cr-5Nb alloy at Stage II is the same with the Ti-24Al-15Nb-1.5Mo alloy, namely, oxygen dissolution in the alloy. Besides, since the oxygen solubility in the γ phase is much lower than that in the α_2 phase, the temperature range corresponding to Stage II for the Ti-46Al-2Cr-5Nb alloy (870–980 °C) is narrower than that for the Ti-24Al-15Nb-1.5Mo alloy (800–1020 °C).

4.3.3. Accelerated Oxidation Stage (Stage III)

When the temperature was raised to Stage III (980–1280 °C for the Ti-46Al-2Cr-5Nb alloy and 1020–1400 °C for the Ti-24Al-15Nb-1.5Mo alloy), the two alloys exhibited accelerated oxidation behaviors and the oxidation activation energies are respectively 249.8 and 244.8 kJ/mol, which is close to the diffusion activation energies of O^{2-} and Ti^{4+} in TiO_2 (234 kJ/mol [37] and 257 kJ/mol [38], respectively). It indicates that the oxidation mechanisms of the two alloys at this stage are mainly the growth of oxide scales dominated by TiO_2. Since the diffusion rates of O^{2-} and Ti^{4+} in TiO_2 are much lower than that of oxygen atom in the alloy, the rate-determining step at this stage is the diffusion of O^{2-} and Ti^{4+} in the oxide scale so that the oxidation activation energy is independent with the matrix phases of the alloys.

Figure 6 demonstrates the surface morphologies and XRD pattern of the oxide scale formed during heating the Ti-46Al-2Cr-5Nb alloy to the end of Stage III. As shown in the XRD pattern (Figure 6d), there are high contents of γ and α_2 phases besides rutile TiO_2 and corundum α-Al_2O_3. It manifests that the oxide is composed of rutile TiO_2 and corundum α-Al_2O_3 (hereafter referred to as TiO_2 and Al_2O_3), and the thickness of the oxide scale is less than the detection depth of X-ray (~20 μm). The oxide scale is a multiple-layer structure (Figure 6a), the outer layer of which consists of coarse TiO_2 crystals (Figure 6b) and is prone to peel off, exposing the inner layer of fine TiO_2 and Al_2O_3 crystals (Figure 6c).

Figure 7 presents the cross-sectional morphology and the corresponding elemental distribution maps of the oxide scale on the Ti-46Al-2Cr-5Nb alloy. The oxide scale is identified to be in the order of the TiO_2 layer/Al_2O_3-rich layer/TiO_2 + Al_2O_3 mixed layer from the outside to the inside. The total thickness of the oxide scale (~13 μm) is indeed less than the detection depth of X-ray, which is consistent with the result of the XRD pattern. An Al-depleted layer that deemed to be α_2 phase [16] is generated in the subsurface of the alloy. Besides, there is a transverse crack between the intermediate and inner layers (Figure 7a). However, the crack is considered to be produced during preparation of the metallographic specimen instead of during oxidation and during cooling, since no such crack is found in the oxide scale formed during heating the alloy to the end of Stage IV and the alloy exhibited much severer oxidation behavior at Stage IV than at Stage III. Thus, the crack is not taken into account when discussing the formation process of the oxide scale at Stage III.

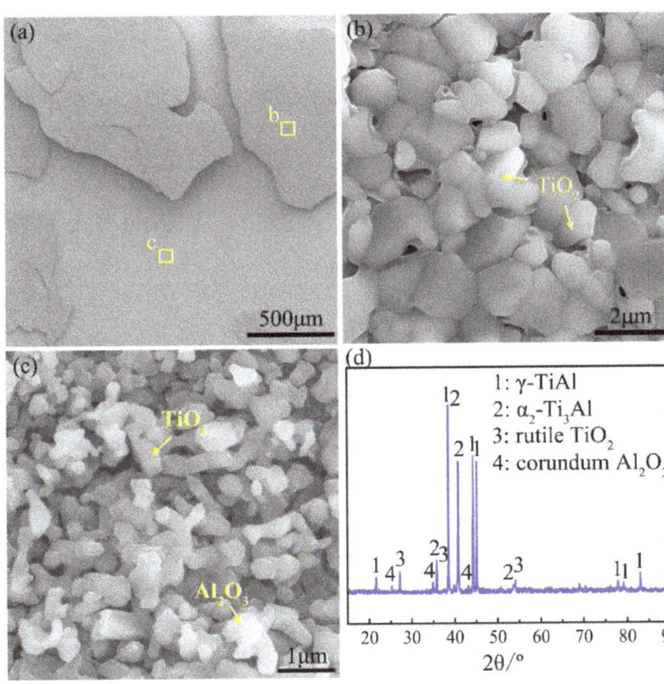

Figure 6. (**a–c**) Surface morphologies and (**d**) X-ray diffraction (XRD) pattern of the oxide formed during heating the Ti-46Al-2Cr-5Nb alloy to the end of Stage III.

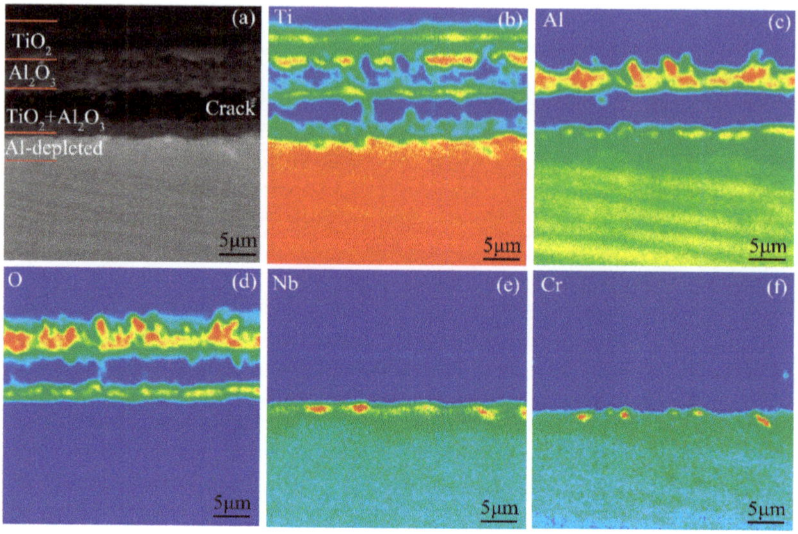

Figure 7. (**a**) Cross-sectional morphology and (**b–f**) the corresponding elemental distribution maps of the oxide scale formed during heating the Ti-46Al-2Cr-5Nb alloy to the end of Stage III.

The oxide scale formed during heating the Ti-46Al-2Cr-5Nb alloy from room temperature to the end of Stage III (Figures 6 and 7) is considered to be mainly generated at Stage III since the Ti-46Al-2Cr-5Nb alloy exhibits nearly non-oxidation behavior at Stage I and the oxidation mechanism at Stage II is oxygen dissolution in the alloy. The three-layer oxide scale structure formed at Stage III (980–1280 °C) is similar to the common oxide scale structure formed during isothermal oxidation at 800–1000 °C [14]. Hence, the growth mechanism of the oxide scale at Stage III could be deduced from the isothermal oxidation mechanism at 800–1000 °C. Firstly, an Al_2O_3 film is rapidly generated on the oxygen-saturated Ti-46Al-2Cr-5Nb alloy [39]. Since the Al_2O_3 film is grown by the inward diffusion of O^{2-} along the Al_2O_3 grain boundaries and the outward diffusion of Al^{3+} along the Al_2O_3 lattice, the Al_2O_3 grain boundaries suffer compressive stress, resulting in curling deformation or even rupture of the Al_2O_3 film [40]. Thus, Ti^{4+} and O^{2-} would diffuse in opposite directions along the cracks of the Al_2O_3 film [40]. As a result, an outer TiO_2 layer and an inner TiO_2 + Al_2O_3 mixed layer are respectively formed on the outside and the inside of the ruptured Al_2O_3 film, as shown in Figure 7. The growth process of the oxide scale at Stage III demonstrates that the growth rate of the oxide scale is mainly controlled by the diffusion of Ti^{4+} and O^{2-} in TiO_2, which is consistent with the result of the oxidation activation energy of the alloy at Stage III. Therefore, it is further confirmed that the oxidation mechanism of the Ti-46Al-2Cr-5Nb alloy at Stage III is the growth of the oxide scale dominated by TiO_2.

As for the Ti-24Al-15Nb-1.5Mo alloy, the oxide scale formed during heating the alloy to the end of Stage III has a three-layer structure, which is in the order of the TiO_2 + Al_2O_3 mixed layer/TiO_2-rich layer/TiO_2(Nb, Mo) + Al_2O_3 mixed layer from the outside to the inside, as shown in Figure 8. Consistent with the oxidation activation energy, the structure of the oxide scale also indicates that the oxidation mechanism of the Ti-24Al-15Nb-1.5Mo alloy is mainly the growth of the oxide scale dominated by TiO_2. Besides, a thin β-Ti layer enriched in Al, Nb and Mo elements was formed in the subsurface of the alloy. The microstructure between the β-Ti layer and the matrix is composed of α-Ti(O) grains and a small amount of β-Ti phase enriched in Al, Nb, and Mo in the grain boundaries. As mentioned in Section 4.3, the matrix is dominated by β phase at the end of Stage III. Thus, it is referred that the microstructure between the subsurface of the alloy and the matrix is stabilized to α phase by the inward-diffusing oxygen, while the microstructure of the subsurface still maintains β phase due to the enrichment of Nb and Mo.

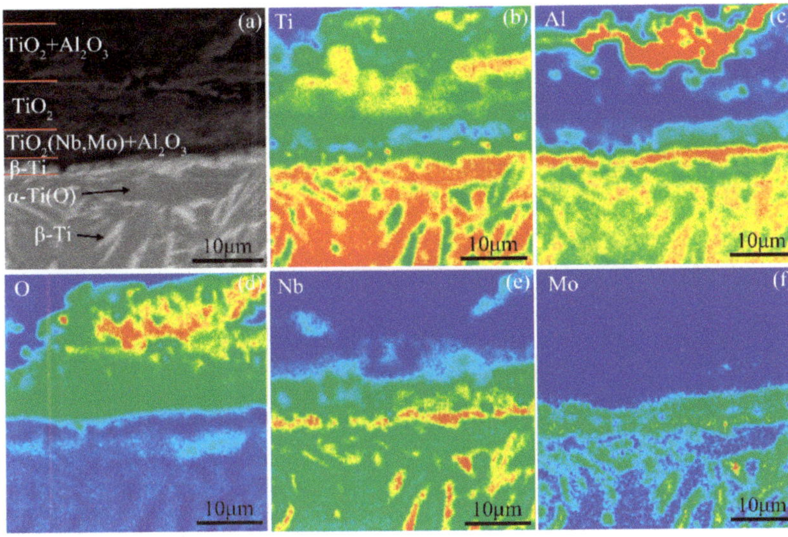

Figure 8. (**a**) Cross-sectional morphology and (**b–f**) the corresponding elemental distribution maps of the oxide scale formed during heating the Ti-24Al-15Nb-1.5Mo alloy to the end of Stage III.

4.3.4. Severe Oxidation Stage (Stage IV)

The two alloys exhibited severe oxidation behaviors when the temperature was raised to Stage IV, while the corresponding temperature range of the Ti-46Al-2Cr-5Nb alloy (1400–1450 °C) is much higher than that of the Ti-24Al-15Nb-1.5Mo alloy (1280–1350 °C). The oxidation activation energies of the Ti-46Al-2Cr-5Nb and Ti-24Al-15Nb-1.5Mo alloys are respectively 985.0 and 608.6 kJ/mol, which is significantly higher than those at Stage III (249.8 and 244.8 kJ/mol). Thus, the oxidation mechanisms at Stage IV should be somewhat different from that at Stage III.

Figure 9 shows the surface morphologies and XRD pattern of the oxide scale formed during heating the Ti-46Al-2Cr-5Nb alloy to the end of Stage IV. As shown in the XRD pattern (Figure 9d), the oxide scale is mainly composed of a large amount of TiO_2 as well as a small amount of Al_2O_3 and β-Al_2TiO_5. Besides, the content of the γ phase is much lower than that detected after heating to the end of Stage III (Figure 6d). It manifests that the thickness of the oxide scale increased significantly at Stage IV compared with that at Stage III, which is consistent with the drastic increase of the oxidation rate at Stage IV (Figure 3). As shown in Figure 9a,b, the outer layer of the oxide scale is still a spalling-prone TiO_2 layer, but it is relatively thin and dense compared with that formed at Stage III (Figure 6a,b). Below the outer layer, a few irregular sintering structures of β-Al_2TiO_5 coexist with relatively regular crystals of TiO_2 and Al_2O_3, as shown in Figure 9c. It can be seen from the TiO_2-Al_2O_3 phase diagram [41] that the reaction of TiO_2 + Al_2O_3 → β-Al_2TiO_5 occurs when the temperature is higher than 1200 °C. Therefore, the irregular sintering structures of β-Al_2TiO_5 was produced by the reaction between TiO_2 and Al_2O_3 in the oxide scale at Stage IV (1280–1350 °C).

Figure 10 presents the cross-sectional morphology and the corresponding elemental distribution maps of the oxide scale. The oxide scale is identified to be in the order of the TiO_2 layer/TiO_2 + Al_2O_3 mixed layer/Al_2O_3-rich layer/TiO_2-rich layer/TiO_2(Nb, Cr) + Al_2O_3 mixed layer from the outside to the inside. In addition, an Al-depleted layer enriched in Nb and Cr elements is generated in the subsurface of the alloy. Moreover, Al_2O_3 oxides are dispersed as islands in the subsurface, indicating the occurrence of internal oxidation at Stage IV. Since internal oxidation deteriorates the oxidation resistance of TiAl-based alloys [14], the severe oxidation behavior of the Ti-46Al-2Cr-5Nb alloy at this stage resulted from the internal oxidation of Al, which is probably the cause for the much higher oxidation energy of the Ti-46Al-2Cr-5Nb alloy at Stage IV than that at Stage III.

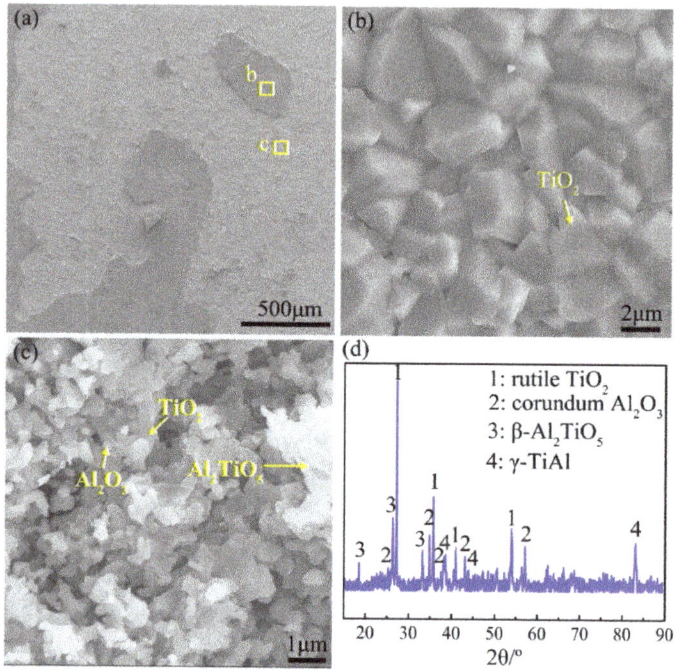

Figure 9. (**a**–**c**) Surface morphologies and (**d**) XRD pattern of the oxide scale formed during heating the Ti-46Al-2Cr-5Nb alloy to the end of Stage IV.

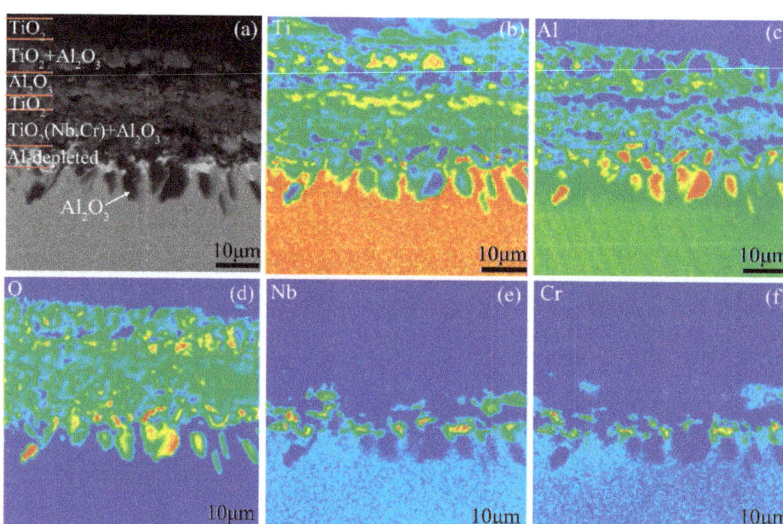

Figure 10. (**a**) Cross-sectional morphology and (**b**–**f**) the corresponding elemental distribution maps of the oxide scale formed during heating the Ti-46Al-2Cr-5Nb alloy to the end of Stage IV.

As for the Ti-24Al-15Nb-1.5Mo alloy, the surface morphologies and XRD pattern of the oxide scale formed during heating the alloy to the end of Stage IV are shown in Figure 11. The XRD result shows that the oxide scale is mainly composed of TiO_2 accompanied with a small amount of β-Al_2TiO_5 (Figure 11d). The outer layer of the oxide scale is crimping (Figure 11a), consisting of coarse rod-like TiO_2 particles and ridged structures composed of fine β-Al_2TiO_5 particles (Figure 11c). The outer layer is prone to exfoliation (Figure 11a), revealing the layer of fine TiO_2 particles (Figure 11b).

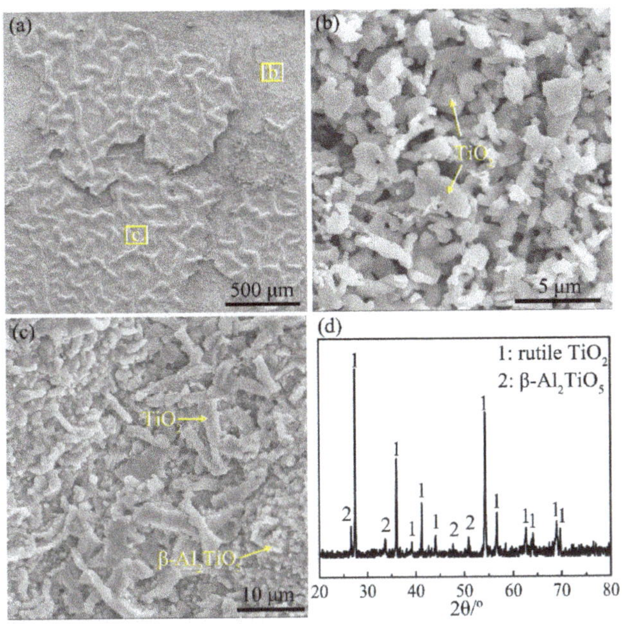

Figure 11. (**a**–**c**) Surface morphologies and (**d**) XRD pattern of the oxide formed during continuously heating the Ti-24Al-15Nb-1.5Mo alloy to the end of Stage IV.

Figure 12 presents the cross-sectional morphology and the corresponding elemental distribution maps of the oxide scale on the Ti-24Al-15Nb-1.5Mo alloy. The structure of the oxide scale is in the order of the TiO_2 + Al_2TiO_5 layer/TiO_2 layer/coarse Al_2O_3 layer/porous TiO_2(Nb, Mo) layer from the outside to the inside. The porous TiO_2(Nb, Mo) inner layer and the coarse Al_2O_3 immediate layer indicate that the Al_2O_3 particles formed in the inner layer dissolved in the surrounding TiO_2(Nb, Mo), then migrated outward and re-precipitated in the immediate layer at Stage IV, the reason for which is that the decreasing oxygen pressure around the inner layer with the thickening of the oxide scale increases the solubility of Al_2O_3 in TiO_2 and reduces the stability of Al_2O_3 [17]. It is reported that the dissolution, migration and re-precipitation of Al_2O_3 destroys the oxygen-blocking ability of the original Al_2O_3 barrier layer, leading to breakaway oxidation [17,42]. However, there is a transverse crack below the oxide scale and whether the crack contributed to the severe oxidation should be discussed. According to the crimping structure and the locally spalling morphology of the oxide scale (Figure 11a), the crack is considered to be induced during cooling after oxidation. The reason is that the oxide scale and the matrix were respectively subjected to compressive and tensile stresses during cooling since the thermal expansion coefficient of the oxide scale is generally smaller than that of the metal, and curling deformation was prone to occur in the oxide scale for stress release. Thus, the severe oxidation behavior of the Ti-24Al-15Nb-1.5Mo alloy at Stage IV is independent of the formation of the transverse crack. Consequently, the severe oxidation at this stage is due to the dissolution, migration and re-precipitation of Al_2O_3, which might lead to the higher oxidation energy at Stage IV than that at Stage III.

In addition, a thin α-Ti(O) layer was formed at the interface between the oxide scale and the alloy, beneath which is a 10-μm-thick layer composed of β-Ti rich in Nb, Mo, Al elements (referred to as *i* in Figure 12a) and several fine α-Ti grains with an orientation nearly perpendicular to the matrix surface (referred to as *ii* in Figure 12a). The microstructure between the β-Ti layer and the matrix is still composed of α-Ti(O) grains (referred to as *iii* in Figure 12a) and a small amount of β-Ti phase in the grain boundaries (referred to as *iv* in Figure 12a). The chemical compositions of these structures are shown in Table 2. The microstructures demonstrate that more oxygen diffused into the subsurface of the alloy at Stage IV in comparison with at Stage III, resulting in the transition from β-Ti to α-Ti(O) in the near subsurface layer.

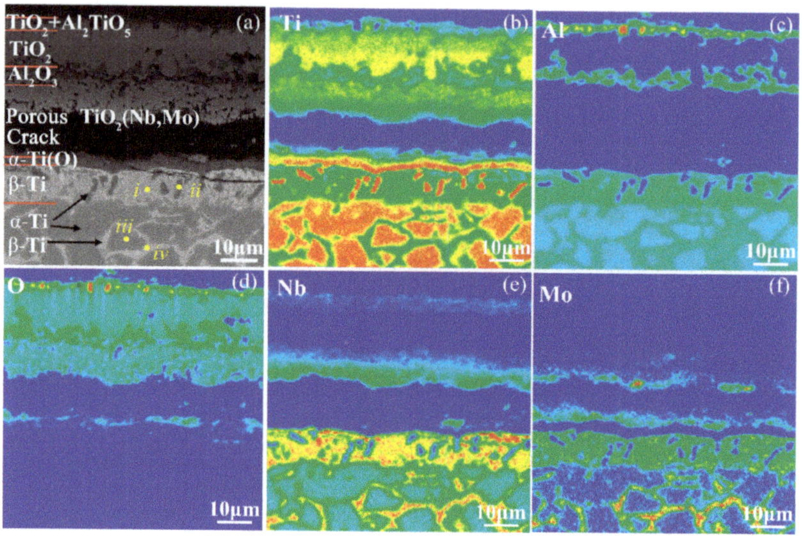

Figure 12. (**a**) Cross-sectional morphology and (**b–f**) the corresponding elemental distribution maps of the oxide scale formed during heating the Ti-24Al-15Nb-1.5Mo alloy to the end of Stage IV.

Table 2. Chemical compositions of the microstructures in the Ti-24Al-15Nb-1.5Mo alloy examined by an electron probe microanalyzer (EPMA) after heating the alloy to the end of Stage IV.

Microstructures in Figure 12a	Compositions (at.%)			
	Ti	Al	Nb	Mo
i	42.99	20.15	34.66	2.2
ii	87.35	4.95	7.7	-
iii	75.62	12.77	11.23	0.37
iv	42.67	19.57	34.77	2.99

4.3.5. Decelerated Oxidation Stage (Stage V)

When the temperature increased to Stage V (1350–1450 °C), the oxidation rate of the Ti-46Al-2Cr-5Nb alloy decreased remarkably. Figure 13 presents the surface morphologies and XRD pattern of the oxide scale formed during heating the Ti-46Al-2Cr-5Nb alloy to the end of Stage V. There is no obvious exfoliation morphology on the surface of oxide scale (Figure 13a). Further, the scale becomes denser (Figure 13b) and the crystals in the surface of the scale become larger (Figure 13c) compared with those formed at Stage IV. As shown in the XRD pattern (Figure 13d), the oxide scale consists of a large quantity of β-Al_2TiO_5 as well as some TiO_2 and Al_2O_3, indicating that the sintering reaction between TiO_2 and Al_2O_3 in the oxide scale aggravated at Stage V.

Figure 13. (**a–c**) Surface morphologies and (**d**) XRD pattern of the oxide scale formed during heating the Ti-46Al-2Cr-5Nb alloy to the end of Stage V.

Figure 14 shows the cross-sectional morphology and the corresponding elemental distribution maps of the oxide scale. The oxide scale from the outside to the inside is identified to be in the order of the TiO_2 layer/Al_2O_3-rich layer/TiO_2-rich layer/Al_2TiO_5 layer/mixed layer of TiO_2 and fine Al_2O_3 flakes. It is interesting that the TiO_2 (referred to as *i* in Figure 14a) and the fine Al_2O_3 flakes (referred to as *ii* in Figure 14a) distribute alternately in the inner layer, the compositions of which are presented in Table 3. White coarse particles (referred to as *iii* in Figure 14a) with an orientation perpendicular to the oxide/substrate interface were generated in the subsurface of the substrate. The detected composition as shown in Table 3 manifests that the white particles might be an alloy or an intermetallic of Nb. Moreover, there are a lot of fine Al_2O_3 flakes distributing around the white coarse particles. Besides, it should be mentioned that there is a transverse crack in the oxide scale, which is not found in the oxide scale formed during heating the alloy to the end of Stage IV. This crack is considered to be produced during cooling after oxidation or during the preparation of metallographic specimens instead of during oxidation, since the oxidation rate at Stage V was significantly reduced in comparison with at Stage IV.

Through comparing the cross-sectional structures of the oxide scales respectively formed by heating the Ti-46Al-2Cr-5Nb alloy to the end of Stage IV and Stage V (Figures 10 and 14), the growth process of the oxide scale at Stage V can be inferred as follows: (1) oxygen from the atmosphere diffused into the alloy and the Al-depleted zones were preferentially oxidized nearby the Al_2O_3 flakes which were generated due to the internal oxidation at Stage IV, thus forming the inner oxide layer structure where TiO_2 and Al_2O_3 flakes distributed alternately; (2) due to the limited solubility of the alloying elements (such as Nb, Cr and Al) in TiO_2, the alloying elements diffused towards the underlying substrate during the growth of the oxide scale, resulting in the formation of massive white coarse particles of Nb alloy or intermetallic in the newly-formed subsurface of the substrate; (3) the continuous inward diffusion of oxygen further induced the occurrence of internal oxidation in the subsurface, leading to the formation of fine Al_2O_3 flakes around the white coarse particles and even the underlying substrate. However, at the same time of the oxide growth, the sintering reaction between TiO_2 and Al_2O_3 in the oxide scale (mainly in the TiO_2 + Al_2O_3 mixed layer) aggravated and an β-Al_2TiO_5-rich layer was formed in the oxide scale (Figure 14a). Since Al_2TiO_5 has higher oxygen resistance than TiO_2 [5,43], the formation of the β-Al_2TiO_5-rich layer could effectively slow down the inward diffusion of oxygen and the oxidation rate decreased. Hence, the decelerated oxidation behavior of the Ti-46Al-2Cr-5Nb alloy at Stage V is due to the generation of an oxygen-barrier β-Al_2TiO_5-rich layer in the oxide scale by the reaction between TiO_2 and Al_2O_3 in large scales.

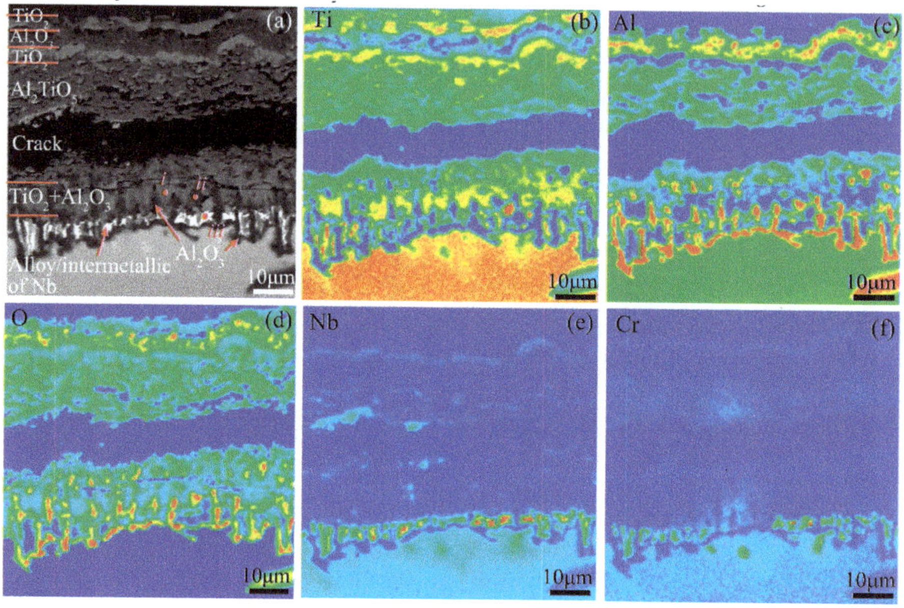

Figure 14. (**a**) Cross-sectional morphology and (**b**–**f**) the corresponding elemental distribution maps of the oxide scale formed during heating the Ti-46Al-2Cr-5Nb alloy to the end of Stage V.

Table 3. Chemical compositions of the microstructures in the inner layer of the oxide and in the subsurface of the Ti-46Al-2Cr-5Nb alloy examined by EPMA after heating the alloy to the end of Stage IV.

Microstructures in Figure 14a	Compositions (at.%)				
	Ti	Al	Nb	Cr	O
i	44.61	3.96	0.24	-	51.19
ii	5.66	26.09	-	-	68.25
iii	29.25	14.25	39.59	16.91	-

4.4. Reasons for the Occurrence of Internal Oxidation in the Ti-46Al-2Cr-5Nb Alloy

In comparison with the Ti-24Al-15Nb-1.5Mo alloy, the Ti-46Al-2Cr-5Nb alloy suffered catastrophic oxidation at the temperature range of 1280–1350 °C due to the occurrence of internal oxidation. In order to improve the non-isothermal oxidation resistance of the Ti-46Al-2Cr-5Nb alloy, it is essential to shed light on the reasons for the occurrence of internal oxidation in the Ti-46Al-2Cr-5Nb alloy instead of in the Ti-24Al-15Nb-1.5Mo alloy. In accordance to the occurrence conditions of internal oxidation [44], it seems that the Ti-24Al-15Nb-1.5Mo alloy mainly dominated by α_2 phase should be more prone to suffer internal oxidation than the Ti-46Al-2Cr-5Nb alloy mainly dominated by γ phase, since both the oxygen solubility and the oxygen diffusion rate in α_2 phase are higher than those in the γ phase and the Al content in the α_2 phase is lower than that in the γ phase. However, this is not the case. Therefore, the actual phases in the Ti-46Al-2Cr-5Nb and Ti-24Al-15Nb-1.5Mo alloys at the temperature range of 1280–1350 °C should be taken into account.

The initial temperature for the occurrence of internal oxidation in the Ti-46Al-2Cr-5Nb alloy (1280 °C) is close to the temperature of $\gamma + \alpha \rightarrow \alpha$ phase transition (about 1226 °C, see Figure 3a). Hence, before the occurrence of internal oxidation (T < 1280 °C), the substrate of the Ti-46Al-2Cr-5Nb alloy was mainly dominated by γ phase and the subsurface of the alloy was dominated by an Al-depleted layer of α_2 phase [16], as illustrated in Figure 15a. However, when the temperature rose above the initial temperature for the occurrence of internal oxidation (T > 1280 °C), the main phases both in the substrate and the subsurface Al-depleted layer of the Ti-46Al-2Cr-5Nb alloy transformed to the α phase, as illustrated in Figure 15b. As for the Ti-24Al-15Nb-1.5Mo alloy at the temperature range for the occurrence of internal oxidation in the Ti-46Al-2Cr-5Nb alloy (1280–1350 °C), the substrate was mainly dominated by β phase and the subsurface was also dominated by β phase due to the enrichment of Mo and Nb elements, even though the microstructure between the subsurface and the substrate was composed of α-Ti(O) grains and a small amount of β-Ti phase, as illustrated in Figure 15c. It is obviously seen from Figure 15 that the phases in the subsurface where internal oxidation occurs are different. Hence, in order to reveal the effect of the phases in the subsurface on the internal oxidation tendency, the four phases involving in the Ti-Al alloys (γ, α_2, α and β) were studied.

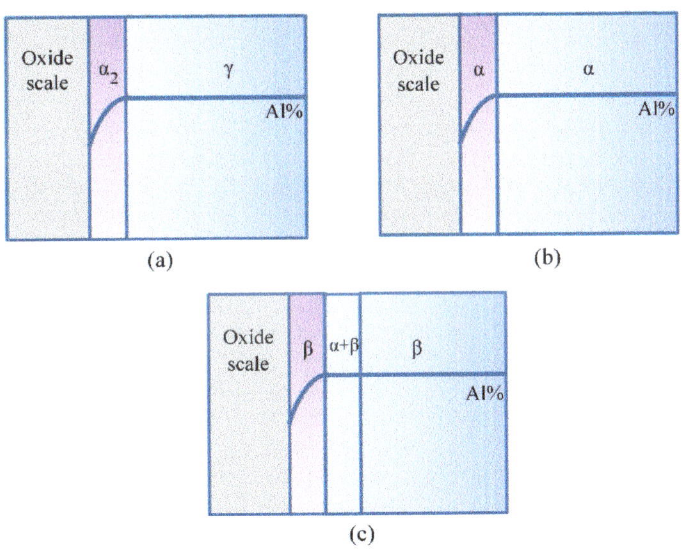

Figure 15. Schematic diagrams of the phase structures for (**a**) the Ti-46Al-2Cr-5Nb alloy when T < 1280 °C, (**b**) the Ti-46Al-2Cr-5Nb alloy when T > 1280 °C and (**c**) the Ti-24Al-15Nb-1.5Mo alloy at the temperature range of 1280–1350 °C.

The tendency of internal oxidation for the different phases in the Ti-46Al-2Cr-5Nb alloy at 1280 °C can be identified based on Wagner's theory [44]. The critical criterion for the transition from internal oxidation to external oxidation is:

$$N_{Al} > N_{crit(Al)} = \left(\frac{\pi g^*}{2}\frac{V_m}{V_{ox}}\frac{D_O}{D_{Al}}N_O\right)^{0.5}, \quad (2)$$

where $N_{crit(Al)}$ is the critical Al content required for the transition from internal oxidation to external oxidation, g^* is a constant factor in the case of an index of 0.3, V_m and V_{ox} are respectively the molar volumes of the Ti-46Al-2Cr-5Nb alloy and Al_2O_3 oxide (in cm^3/mol, V_{ox} = 25.5 cm^3/mol),

D_O and D_{Al} are respectively the diffusion coefficients of O and Al atoms in the lattice of the alloy, N_O and N_{Al} are respectively the oxygen content and Al content in the subsurface of the alloy. The composition in the subsurface of the Ti-46Al-2Cr-5Nb alloy after heating to 1280 °C was detected to be Ti-29.8Al-3.6O-3.0Cr-5.4Nb (in at.%) by EPMA, thus the values for N_O and N_{Al} are respectively 0.036 and 0.298. The values for the other parameters involved in Equation (2) are given in Table 4. It should be noted that the specific value for the oxygen diffusion coefficient in the γ phase has not been reported, but it is readily inferred that the oxygen diffusion coefficient in the γ phase is lower than that in the α_2 phase from their lattice structure differences, as mentioned in Section 4.3.2.

Table 4. Critical Al contents required for the transition from internal oxidation to external oxidation for the different phases in the subsurface of the Ti-46Al-2Cr-5Nb alloy at 1280 °C.

Phases	Parameters				Results
	V_m/(cm^3/mol)	D_O/(m^2/s)	D_{Al} [45]/(m^2/s)	D_O/D_{Al}	$N_{crit(Al)}$
γ	19.7	$<3.66 \times 10^{-15}$	1.97×10^{-14}	<0.186	<0.05
α_2	40.7	3.66×10^{-15} [32]	1.37×10^{-14}	0.267	0.085
α	10.6	3.54×10^{-11} [46]	5.80×10^{-14}	610.3	2.08
β	11.0	3.75×10^{-10} [46]	1.04×10^{-12}	360.6	1.62

Table 4 presents the calculation results of the critical Al contents required for the transition from internal oxidation to external oxidation for the different phases in the subsurface of the Ti-46Al-2Cr-5Nb alloy at 1280 °C. The critical Al contents required for the α_2 and γ phases are respectively 0.085 and less than 0.05, which is much lower than the actual Al content in the subsurface of the substrate (N_{Al} = 0.298). It indicates that if the subsurface of the Ti-46Al-2Cr-5Nb alloy is dominated by γ phase and/or α_2 phase, an external oxide scale would form on the alloy and internal oxidation could not occur. However, the critical Al contents required for the α and β phases are respectively 2.08 and 1.62, which is impractical since the contents are more than 1. This situation is caused by the fact that Wagner's theory of internal oxidation is based on a large number of idealized conditions that often could not be met in the actual system [44]. Nevertheless, Wagner's critical criterion can still reflect the tendency of internal oxidation for the different phases in the subsurface of the Ti-46Al-2Cr-5Nb alloy. The higher the required critical Al content, the more easily internal oxidation will occur. Therefore, the tendency of internal oxidation for the phases is in the order of $\alpha > \beta > \alpha_2 > \gamma$. Thus, it is easy to understand why internal oxidation occurred in the Ti-46Al-2Cr-5Nb alloy when the temperature exceeded 1280 °C, since the main phase in the subsurface of the Ti-46Al-2Cr-5Nb alloy changed from α_2 phase to α phase when the temperature was higher than 1280 °C. In addition, the main phase in the substrate of the Ti-24Al-15Nb-1.5Mo alloy was β phase at the temperature range of 1280–1350 °C due to the enrichment of large amounts of β-stabilizing elements (Nb and Mo), so that no internal oxidation occurred in this alloy. Consequently, it is concluded that the formation of α phase in the subsurface is the basic reason for the occurrence of internal oxidation in the Ti-46Al-2Cr-5Nb alloy. The tendency of internal oxidation in the TiAl-based alloys could be reduced through avoiding the formation of α phase in the subsurface by optimizing alloy ingredients, such as increasing the Al content or adding β-stabilizing elements. However, it should be noted that the optimization of alloy ingredients is complicated and should be further investigated since not only the tendency of internal oxidation, but also other properties of the alloys such as the mechanical properties, should be taken into consideration in practical engineering.

5. Conclusions

1. The non-isothermal oxidation behaviors of the Ti-46Al-2Cr-5Nb and Ti-24Al-15Nb-1.5Mo alloys are similar when the temperature is below 1280 °C, while the Ti-46Al-2Cr-5Nb alloy exhibits poorer oxidation resistance than the Ti-24Al-15Nb-1.5Mo alloy when the temperature exceeds

1280 °C, even though the oxidation rate of the Ti-46Al-2Cr-5Nb alloy decreases significantly when the temperature is above 1350 °C.
2. There are five stages in the non-isothermal oxidation process of the Ti-46Al-2Cr-5Nb alloy, including nearly non-oxidation (<870 °C), slow oxidation (870–980 °C), accelerated oxidation (980–1280 °C), severe oxidation (1280–1350 °C) and decelerated oxidation (1350–1450 °C) stages. The corresponding oxidation mechanisms are as follows: oxygen-barrier effect of the thin titanium oxide film; oxygen dissolution in the alloy; growth of the oxide scale dominated by TiO_2; internal oxidation of Al; formation of an oxygen-barrier Al_2TiO_5-rich layer by reaction between TiO_2 and Al_2O_3 in the oxide scale.
3. There are four stages in the non-isothermal oxidation process of the Ti-24Al-15Nb-1.5Mo alloy, including nearly non-oxidation (<800 °C), slow oxidation (800–1020 °C), accelerated oxidation (1020–1400 °C) and severe oxidation (1400–1450 °C) stages. The oxidation mechanisms for the first three stages are the same with that of the Ti-46Al-2Cr-5Nb alloy, while the oxidation mechanism for the last stage is the dissolution, migration and re-precipitation of Al_2O_3 in the oxide.
4. The tendency of internal oxidation for the different phases in the subsurface of the Ti-46Al-2Cr-5Nb alloy is in the order of $\alpha > \beta > \alpha_2 > \gamma$. The formation of α phase in the subsurface is the basic reason for the occurrence of internal oxidation in the Ti-46Al-2Cr-5Nb alloy. The tendency of internal oxidation in the TiAl-based alloys could be reduced through avoiding the formation of α phase by optimizing alloy ingredients such as increasing Al content or adding β-stabilizing elements.

Author Contributions: Conceptualization, G.M. and P.L.; methodology, P.O.; investigation, P.O.; writing—original draft preparation, P.O.; writing—review and editing, G.M. and P.L.; supervision, X.H. and J.C.; project administration, L.H.; funding acquisition, G.M. and P.L.

Funding: This research was funded by the National Natural Science Foundation of China (Grant No. 51471155) and the Aviation Innovation Foundation of China (Grant No. 2014E62149R).

Conflicts of Interest: The authors declare no conflict of interest.

References

1. Wu, X. Review of alloy and process development of TiAl alloys. *Intermetallics* **2006**, *14*, 1114–1122. [CrossRef]
2. Clemens, H.; Mayer, S. Design, processing, microstructure, properties, and applications of advanced intermetallic TiAl alloys. *Adv. Eng. Mater.* **2013**, *15*, 191–215. [CrossRef]
3. Bewlay, B.; Weimer, M.; Kelly, T.; Suzuki, A.; Subramanian, P. The science, technology, and implementation of TiAl alloys in commercial aircraft engines. *MRS Online Proc. Lib. Arch.* **2013**, *1516*, 49–58. [CrossRef]
4. Cai, J.; Mi, G.; Gao, F.; Huang, H.; Cao, J.; Huang, X.; Cao, C. Research and development of some advanced high temperature titanium alloys for aero-engine. *J. Mater. Eng.* **2016**, *44*, 1–10.
5. Ouyang, P.; Mi, G.; Cao, J.; Huang, X.; He, L.; Li, P. Microstructure characteristics after combustion and fireproof mechanism of TiAl-based alloys. *Mater. Today Commun.* **2018**, *16*, 364–373. [CrossRef]
6. Kim, D.; Seo, D.; Saari, H.; Sawatzky, T.; Kim, Y.W. Isothermal oxidation behavior of powder metallurgy beta gamma TiAl-2Nb-2Mo alloy. *Intermetallics* **2011**, *19*, 1509–1516. [CrossRef]
7. Kim, S.W.; Hong, J.K.; Na, Y.S.; Yeom, J.T.; Kim, S.E. Development of TiAl alloys with excellent mechanical properties and oxidation resistance. *Mater. Design* **2014**, *54*, 814–819. [CrossRef]
8. Legzdina, D.; Robertson, I.; Birnbaum, H. Oxidation behavior of a single phase γ-TiAl alloy in low-pressure oxygen and hydrogen. *Acta Mater.* **2005**, *53*, 601–608. [CrossRef]
9. Lu, W.; Chen, C.; He, L.; Wang, F.; Lin, J.; Chen, G. (S)TEM study of different stages of Ti-45Al-8Nb-0.2W-0.2B-0.02Y alloy oxidation at 900 °C. *Corros. Sci.* **2008**, *50*, 978–988. [CrossRef]
10. Qu, S.; Tang, S.; Feng, A.; Feng, C.; Shen, J.; Chen, D. Microstructural evolution and high-temperature oxidation mechanisms of a titanium aluminide based alloy. *Acta Mater.* **2018**, *148*, 300–310. [CrossRef]
11. Strobridge, T.R.; Moulder, J.C.; Clark, A.F. *Titanium Combustion in Turbine Engines*; Report FAA-RD-79-51, NBSIR 79-1616; US National Bureau of Standards: Washington, DC, USA, 1979.
12. Li, B.; Chen, G.; Zhang, H.; Sheng, C.D. Development of non-isothermal TGA–DSC for kinetics analysis of low temperature coal oxidation prior to ignition. *Fuel* **2014**, *118*, 385–391. [CrossRef]

13. Hosseini, S.G.; Sheikhpour, A.; Keshavarz, M.H.; Tavangar, S. The effect of metal oxide particle size on the thermal behavior and ignition kinetic of Mg-CuO thermite mixture. *Thermochim. Acta* **2016**, *626*, 1–8. [CrossRef]
14. Beye, R.; Verwerft, M.; Hosson, J.D.; Gronsky, R. Oxidation subscale of γ-titanium aluminide. *Acta Mater.* **1996**, *44*, 4225–4231. [CrossRef]
15. Beye, R.; Gronsky, R. Novel phases in the oxidation of γ-titanium aluminum. *Acta Metall. Mater.* **1994**, *42*, 1373–1381. [CrossRef]
16. Rahmel, A.; Schütze, M.; Quadakkers, W. Fundamentals of TiAl oxidation-a critical review. *Mater. Corros.* **1995**, *46*, 271–285. [CrossRef]
17. Becker, S.; Rahmel, A.; Schorr, M.; Schütze, M. Mechanism of isothermal oxidation of the intermetallic TiAl and of TiAl alloys. *Oxid. Met.* **1992**, *38*, 425–464. [CrossRef]
18. Vaidya, R.U.; Park, Y.S.; Zhe, J.; Gray, G.T.; Butt, D.P. High-temperature oxidation of Ti-48Al-2Nb-2Cr and Ti-25Al-10Nb-3V-1Mo. *Oxid. Met.* **1998**, *50*, 215–240. [CrossRef]
19. Das, S. The Al-O-Ti (aluminum-oxygen-titanium) system. *J. Phase Equilib.* **2002**, *23*, 525–536. [CrossRef]
20. Chattopadhyay, K.; Mitra, R.; Ray, K. Nonisothermal and isothermal oxidation behavior of Nb-Si-Mo alloys. *Metall. Mater. Trans. A* **2008**, *39*, 577–592. [CrossRef]
21. Schulz, O.; Eisenreich, N.; Kelzenberg, S.; Schuppler, H.; Neutz, J.; Kondratenko, E. Non-isothermal and isothermal kinetics of high temperature oxidation of micrometer-sized titanium particles in air. *Thermochim. Acta* **2011**, *517*, 98–104. [CrossRef]
22. Mi, G.B.; Huang, X.S.; Li, P.J.; Cao, J.X.; Huang, X.; Cao, C.X. Non-isothermal oxidation and ignition prediction of Ti-Cr alloys. *Trans. Nonferr. Metal Soc.* **2012**, *22*, 2409–2415. [CrossRef]
23. Ouyang, P.; Mi, G.; Li, P.; He, L.; Cao, J.; Huang, X. Non-isothermal oxidation behavior and mechanism of a high temperature near-α titanium alloy. *Materials* **2018**, *11*, 2141. [CrossRef] [PubMed]
24. Kofstad, P. High-temperature oxidation of titanium. *J. Less-Common Metals* **1967**, *12*, 449–464. [CrossRef]
25. Kofstad, P.; Hauffe, K.; Kjollesdal, H. Investigation on the oxidation mechanism of titanium. *Acta Chem. Scand.* **1958**, *12*, 239–266. [CrossRef]
26. Kofstad, P.; Anderson, P.; Krudtaa, O. Oxidation of titanium in the temperature range 800–1200 °C. *J. Less-Common Metals* **1961**, *3*, 89–97. [CrossRef]
27. Overton, J.M. Thermophysical Property and Phase Transformation Determination of Gamma-TiAl Intermetallics. Master's Thesis, Carleton University, Ottawa, ON, Canada, May 2006.
28. Ishikawa, K.; Kazuhiro, R.; Ishida, K. The Ti-Al Binary System. Available online: https://materials.springer.com/msi/phase-diagram/docs/sm_msi_r_10_010909_02_full_LnkDia0 (accessed on 15 May 2019).
29. Terner, M.; Biamino, S.; Ugues, D.; Sabbadini, S.; Fino, P.; Pavese, M.; Badini, C. Phase transitions assessment on γ-TiAl by thermo mechanical analysis. *Intermetallics* **2013**, *37*, 7–10. [CrossRef]
30. Malinov, S.; Novoselova, T.; Sha, W. Experimental and modelling studies of the thermodynamics and kinetics of phase and structural transformations in a gamma TiAl-based alloy. *Mater. Sci. Eng. A* **2004**, *386*, 344–353. [CrossRef]
31. Huang, X. *Advanced Aeronautical Titanium Alloys and Applications*; National Defense Industry Press: Beijing, China, 2012.
32. Koizumi, Y.; Kishimoto, M.; Minamino, Y.; Nakajima, H. Oxygen diffusion in Ti_3Al single crystals. *Philos. Mag.* **2008**, *88*, 2991–3010. [CrossRef]
33. Scotti, L.; Mottura, A. Interstitial diffusion of O, N, and C in α-Ti from first-principles: Analytical model and kinetic Monte Carlo simulations. *J. Chem. Phys.* **2016**, *144*, 084701. [CrossRef] [PubMed]
34. Wu, H.H.; Trinkle, D.R. Direct diffusion through interpenetrating networks: Oxygen in titanium. *Phys. Rev. Lett.* **2011**, *107*, 045504. [CrossRef] [PubMed]
35. Bakulin, A.V.; Latyshev, A.M.; Kulkova, S.E. Absorption and diffusion of oxygen in the Ti_3Al alloy. *J. Exp. Theor. Phys.* **2017**, *125*, 138–147. [CrossRef]
36. Jones, C.Y.; Luecke, W.E.; Copland, E. Neutron diffraction study of oxygen dissolution in α_2-Ti_3Al. *Intermetallics* **2006**, *14*, 54–60. [CrossRef]
37. Unnam, J.; Shenoy, R.; Clark, R. Oxidation of commercial purity titanium. *Oxid. Met.* **1986**, *26*, 231–252. [CrossRef]
38. Venkatu, D.; Poteat, L. Diffusion of titanium of single crystal rutile. *Mater. Sci. Eng.* **1970**, *5*, 258–262. [CrossRef]

39. Maurice, V.; Despert, G.; Zanna, S.; Josso, P.; Bacos, M.P.; Marcus, P. XPS study of the initial stages of oxidation of α_2-Ti_3Al and γ-TiAl intermetallic alloys. *Acta Mater.* **2007**, *55*, 3315–3325. [CrossRef]
40. Taniguchi, S.; Tachikawa, Y.; Shibata, T. Influence of oxygen partial pressure on the oxidation behaviour of TiAl at 1300 K. *Mater. Sci. Eng. A* **1997**, *232*, 47–54. [CrossRef]
41. Goldberg, D. Contribution to study of systems formed by alumina and some oxides of trivalent and tetravalent metals especially titanium oxide. *Rev. Int. Hautes Temp. Refract.* **1968**, *5*, 181.
42. Lang, C.; Schütze, M. TEM investigations of the early stages of TiAl oxidation. *Oxid. Met.* **1996**, *46*, 255–285. [CrossRef]
43. Swain, M.V. *Structure and Properties of Ceramics*; Wiley-VCH: Weinheim, Germany, 1994.
44. Leyens, C.; Peters, M. *Titanium and Titanium Alloys: Fundamentals and Applications*; Wiley: Hoboken, NJ, USA, 2006.
45. Mishin, Y.; Herzig, C. Diffusion in the Ti-Al system. *Acta Mater.* **2000**, *48*, 589–623. [CrossRef]
46. Liu, Z.; Welsch, G. Literature survey on diffusivities of oxygen, aluminum, and vanadium in alpha titanium, beta titanium, and in rutile. *Metall. Trans. A* **1988**, *19*, 1121–1125. [CrossRef]

materials

MDPI

Article

The Influence of Microstructure on the Passive Layer Chemistry and Corrosion Resistance for Some Titanium-Based Alloys

Nader El-Bagoury [1,2], Sameh I. Ahmed [2,3], Ola Ahmed Abu Ali [1], Shimaa El-Hadad [4], Ahmed M. Fallatah [1], G. A. M. Mersal [1,5], Mohamed M. Ibrahim [1,6], Joanna Wysocka [7], Jacek Ryl [7,*], Rabah Boukherroub [8] and Mohammed A. Amin [1,9,*]

1 Department of Chemistry, Faculty of Science, Taif University, P.O. Box 888, Taif 21974, Saudi Arabia; nader_elbagoury@yahoo.com (N.E.-B.); o.abuali@tu.edu.sa (O.A.A.A.); a.fallatah.11@hotmail.com (A.M.F.); gamersal@yahoo.com (G.A.M.M.); ibrahim652001@yahoo.com (M.M.I.)

2 Department of Physics, Faculty of Science, Taif University, Hawiya 888, Saudi Arabia; sameh@sci.asu.edu.eg

3 Department of Physics, Faculty of Science, Ain Shams University, Abbassia 11566, Cairo, Egypt

4 Central Metallurgical Research and Development Institute, P.O. Box 87, Helwan, Cairo, Egypt; shimaamohamad901@yahoo.com

5 Chemistry Department, Faculty of Science, South Valley University, Qena 83523, Egypt

6 Chemistry Department, Faculty of Science, Kafrelsheikh University, Kafr El-Sheikh 33516, Egypt

7 Department of Electrochemistry, Corrosion and Materials Engineering, Chemical Faculty, Gdansk University of Technology, Narutowicza 11/12, 80-233 Gdansk, Poland; joanna.wer.wysocka@gmail.com

8 Univ. Lille, CNRS, Centrale Lille, ISEN, Univ. Valenciennes, UMR 8520-IEMN, F-59000 Lille, France; rabah.boukherroub@univ-lille.fr

9 Department of Chemistry, Faculty of Science, Ain Shams University, Abbassia 11566, Cairo, Egypt

* Correspondence: jacek.ryl@pg.edu.pl (J.R.); maaismail@yahoo.com (M.A.A.); Tel.: +966-54-570-7507 (J.R.)

Received: 5 March 2019; Accepted: 9 April 2019; Published: 15 April 2019

Abstract: The effect of microstructure and chemistry on the kinetics of passive layer growth and passivity breakdown of some Ti-based alloys, namely Ti-6Al-4V, Ti-6Al-7Nb and TC21 alloys, was studied. The rate of pitting corrosion was evaluated using cyclic polarization measurements. Chronoamperometry was applied to assess the passive layer growth kinetics and breakdown. Microstructure influence on the uniform corrosion rate of these alloys was also investigated employing dynamic electrochemical impedance spectroscopy (DEIS). Corrosion studies were performed in 0.9% NaCl solution at 37 °C, and the obtained results were compared with ultrapure Ti (99.99%). The different phases of the microstructure were characterized by X-ray diffraction (XRD) and scanning electron microscopy (SEM). Chemical composition and chemistry of the corroded surfaces were studied using X-ray photoelectron spectroscopy (XPS) analysis. For all studied alloys, the microstructure consisted of α matrix, which was strengthened by β phase. The highest and the lowest values of the β phase's volume fraction were recorded for TC21 and Ti-Al-Nb alloys, respectively. The susceptibility of the investigated alloys toward pitting corrosion was enhanced following the sequence: Ti-6Al-7Nb < Ti-6Al-4V << TC21. Ti-6Al-7Nb alloy recorded the lowest pitting corrosion resistance (R_{pit}) among studied alloys, approaching that of pure Ti. The obvious changes in the microstructure of these alloys, together with XPS findings, were adopted to interpret the pronounced variation in the corrosion behavior of these materials.

Keywords: titanium-based alloys; microstructure; passivity breakdown; pitting corrosion

1. Introduction

Titanium and its alloys are widely used in many industrial applications, because of their highly desirable properties, including very good mechanical properties, excellent corrosion and erosion resistance, and favorable strength to weight ratio [1]. In fact, titanium and its alloys have experienced increased use in the past years as biomaterials, because of their superior biocompatibility, high resistance to localized and generalized corrosion, and their good mechanical properties (fatigue resistance) [2]. Among all titanium and its alloys, the commonly used materials in biomedical area are commercially pure titanium (cpTi) and its ($\alpha + \beta$) Ti6-Al4-V alloy [3–5].

Next to biomedical applications, aerospace sector has dominated titanium use, instead of heavy steel components, in the fabrication of crucial and decisive systems such as airfoils and airframes [6–9]. About 50% of titanium used in the aerospace industry is the ($\alpha + \beta$) Ti-6Al-4V alloy. This alloy possesses a perfect combination of operational and technological properties [10,11]. Titanium alloys have also found widespread applications in a variety of fields such as in chemical and petrochemical sectors due to their excellent corrosion resistance [12]. The outstanding characteristics (such as high specific strength, high fatigue strength, good corrosion resistance, etc.) of the titanium alloys (particularly Ti-6Al-4V) are attributed to a very stable native oxide film (1.5–10 nm) formed on the Ti and Ti-alloy surface upon exposure to atmosphere and/or aqueous environments [13,14]. However, this thin oxide layer can be damaged and thus strongly impacts the bioactivity and other characteristics of the material. To improve the performance of Ti and Ti-alloys for biomedical and aerospace applications, oxidation (anodization) has been applied as a successful approach to improve the material properties [15,16].

The microstructure, formed during various processing methods, is found to greatly influence the mechanical properties of titanium alloys [17]. The microstructure type (bimodal, lamellar and equiaxed) affects the mechanical properties of Ti based alloys [18]. Even though the corrosion of Ti-alloys in different environments has previously beenstudied [19–23], to the best of our knowledge, the literature contains no reports onthe passive layer growth kinetics and breakdown, and subsequent initiation and propagation of pitting corrosion over the surfaces of Ti-6Al-7Nb, Ti-6Al-4V, and TC21 alloys. In this context, the mainobjective of our studywas to assess the effect of microstructure changes of tested alloys on their surface morphology and chemistry using different techniques such as scanning electron microscopy with electron dispersive X-ray spectroscopy (SEM/EDX), X-ray diffraction (XRD), and X-ray Photoelectron Spectroscopy (XPS). The influence of microstructure changes on the anodic behavior and passive layer growth kinetics and breakdown was also investigated. The corrosion resistances were compared using potentiodynamic polarization and impedance spectroscopy tools. All measurements were conducted in 0.9% NaCl solution at 37 °C.

2. Materials and Methods

The working electrodes investigated in this study consisted of three Ti-based alloys, namely Ti-6Al-4V, Ti-6Al-7Nb and TC21; their chemical compositions are presented in Table 1. The as-received titanium alloy samples were prepared by melting in a 500 kg vacuum induction furnace to obtain billets. These billets were then forged and machined into 10 mm diameter bars. The microstructures of these alloys werestudied by Meiji optical microscope (Meiji Techno Co., Ltd., Chikumazawa, Japan) fitted with a digital camera (Meiji Techno Co., Ltd., Chikumazawa, Japan). JEOL JSM5410 and Hitachi S-3400N scanning electron microscopes (LxRay Co., Ltd., Hyogo, Japan) (SEM) were also used for microstructure studies. For this purpose, the specimens were prepared following ASTM E3-11 standard metallographic procedures, and then etched in a mixture of 5 mL HNO_3, 10 mL HF and 85 mL H_2O. The alloys were machined in the form of rods to perform electrochemical measurements. These rods were mounted in a polyester resin offering an active cross-sectional area of ~0.2 cm^2. Prior to conducting any electrochemical analysis, the surface of the working electrode was cleaned and polished using a silicon carbide paper (600-grit) installed on a polishing machine (Minitech 233). The surface was then washed copiouslywith distilled water and rinsed with absolute ethanol (SIGMA-ALDRICH, Steinheim, Germany).

Table 1. Chemical composition of investigated Ti alloys.

Alloy	Chemical Composition, wt %												
	Al	V	Nb	Sn	Zr	Mo	Cr	Si	Fe	C	N	O	Ti
Ti-6Al-4V	5.85	3.94	0.00	0.00	0.00	0.00	0.00	0.00	0.00	0.02	0.03	0.14	
Ti-6Al-7Nb	6.39	0.00	7.78	0.00	0.00	0.00	0.00	0.00	0.00	0.02	0.04	0.12	Bal.
TC21	5.89	0.00	2.41	2.51	1.59	2.27	1.58	0.067	0.05	0.01	0.01	0.13	

Electrochemical measurements were conducted in a standard, double-walled electrochemical cell (Princeton Applied Research, USA) with an inner volume capacity of 200 mL. Temperature of the test solution was maintained constant at the desired value by means of a temperature-controlled water bath (FP40-MA Refrigerated/Heating Circulator) (JULABO GmbH, Seelbach, Germany). The water, after being adjusted to 37 ± 0.1 °C, wasallowed to circulate through external jacket of the cell. The cover of the electrochemical cell had five openings with different sizes. Such openings were designed to be fitted to the working electrode, counter electrode (a long, coiled platinum wire), reference electrode (KCl-saturated calomel electrode (SCE)), a thermometer and a gas inlet/outlet for gas release. The reference electrode was placed in a Luggin capillary, the tip of which was adjusted to be close to the working electrode to minimize iR drop. The cell was connected to a Potentiostat (Autolab PGSTAT30) (Metrohm, Herisau, Switzerland). The test solution was a normal saline (0.9% NaCl). A Millipore Milli-Q water system (Merck Millipore, MA, USA) (18.2 MΩ cm) was used to freshly prepare the saline solution. The salt was of analytical grade and purchased from Sigma-Aldrich (Steinheim, Germany).

Linear sweep voltammetry (LSV), Tafel plots, and Electrochemical Impedance Spectroscopy (EIS) techniques were applied to investigate the uniform corrosion characteristics of the studied alloys. The susceptibility of these alloys to passivity breakdown was evaluated via conducting potentiodynamic polarization measurements. Uniform corrosion measurements were started by stabilizing the working electrode at the rest potential for 2 h, followed by conducting EIS measurements at the respective corrosion potential (E_{corr}) every day for a week of exposure in 0.9% NaCl solution at 37 °C, covering a wide frequency range (100 kHz–10 mHz), with 15 mV perturbation amplitude. Uniform corrosion study was assessed by constructing Tafel plots via sweeping the electrode potential around the Tafel potential ($E = E_{\mathrm{corr}} \pm 250$ mV), applying a sweep rate of 1.0 mV s^{-1}. After that, the electrode was removed from the cell (which was cleaned properly and re-filled with a new fresh test solution), cleaned and polished up to the mirror finish, as described above, and then inserted in the cell for cyclic polarization measurements. Chronoamperometry (CA) technique was also applied using a new set of cleaned and polished electrodes submerged in a cleaned cell filled with a new fresh solution.

Prior to performing cyclic polarization measurements, the working electrode was allowed to stabilize at rest potential for 2 h, then swept linearly, at a sweep rate of 1.0 mV s^{-1}, starting from a cathodic potential of −2.0 V to +8.0 V vs.SCE. The potential sweep was then reversed back with the same sweep rate to reach the start point again, thus forming one complete cycle. To conduct chronoamperometry (current vs. time) measurements, a two-step route was applied. The working electrode was first held at a starting cathodic potential of −2.0 V vs. SCE for 60 s, and then polarized towards the anodic direction at a sweep rate of 1.0 mV s^{-1} untilthe required anodic potential (E_{a}). Finally, the anodic current was measured versus time (5.0 min) by holding the working electrode at E_{a}. To ensure reproducibility, each run was repeated at least three times, where mean values of the various electrochemical parameters and their standard deviations were calculated and reported.

The XRD diffraction patterns were collected for the bulk samples using a SmartLab SE (Rigaku Americas Corporation, Oxford, MS, USA) X-ray diffractometer with Cu Kα (λ = 1.54056 Å) operated at 40 kV and 40 mA. The scanning speed was 0.2°/min and the scanning angle ranged from 20° to 100° in 2θ. Energy dispersive X-ray spectroscopy (EDS) measurements were utilized to determine microstructural composition of investigated alloys as well asevaluate changes in chemical composition as a result of exposure to corrosive media. S-3400N SEM (Hitachi, Tokyo, Japan) was equipped with an UltraDry detector from ThermoFisher Scientific (Waltham, MA, USA). High-resolution X-ray photoelectron spectroscopy (XPS) studies were carried out on an Escalab 250 Xi from Thermofisher Scientific (Waltham, MA, USA), equipped with Al Kα source. Pass energy was 20 eV and the spot size diameter was 650 μm. Charge compensation was controlled through the low-energy electron and low energy Ar^+ ions emission by means of a flood gun (emission current: 150 μA, beam voltage: 2.1 V, filament current: 3.5 A). Avantage software (Thermofisher Scientific, Waltham, MA, USA) was used for deconvolution purposes.

3. Results and Discussion

3.1. Microstructure Investigation

Based on the morphology of α phase, the microstructure of titanium alloys can be classified into equiaxed, lamellar and bi-modal microstructures [24]. The microstructure of Ti-based alloys can be controlled based on their chemical composition, i.e., based on the balance between the α phase stabilizing elements, such as Al, Sn and O, and the forming β phase elements such asV, Mo and Nb [25]. As shown in Figure 1, the microstructure of all studied titanium alloys consisted of bimodal structure of α/β phases. The initial microstructure of Ti-Al-V and Ti-Al-Nb alloys in as-received (forged) state was represented by equiaxed grains of primary αphase (dark), as well as β-transformed structure (light), as shown in Figure 1. The β phase formed in the microstructure of both alloys was globular in shape, but seemed larger in size in Ti-Al-V alloy than in Ti-Al-V alloy. The particle size of β phase in Al-Ti-V alloy was about 0.5–1.5 μm, while its size in Ti-Al-Nb alloy was slightly lower (about 0.25–1 μm), as depicted in Figure 1a,b.

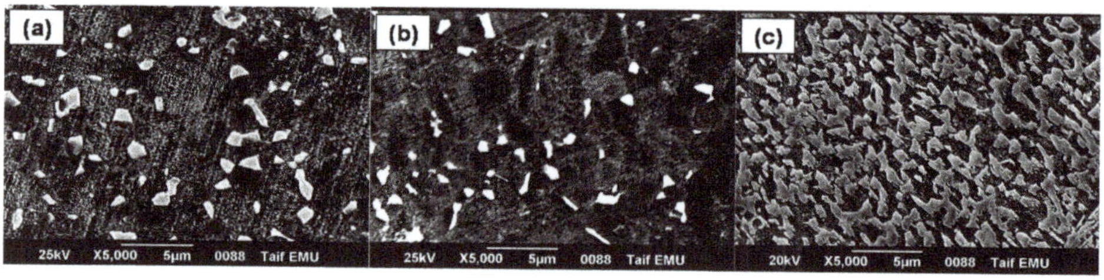

Figure 1. Microstructure of the three investigated Ti alloys: (**a**) Ti-Al-V; (**b**) Ti-Al-Nb; and (**c**) TC21.

Similar to Ti-Al-V and Ti-Al-Nb alloys, the microstructure of TC21 alloy (Figure 1c) contained α and β phases, but displayed different morphologies and volume fractions. The TC21 alloy's β phase consisted of two shapes:an acicular-like structure (Figure 2a) and a blocky shape (Figure 2b). The thickness of the acicular β phase in the TC21 alloy's microstructure ranged from around 0.2 to 0.6 μm, while the size extent of the blocky β phase was about 0.75–1.5 μm. Moreover, the volume fraction of β phase in the microstructure of TC21 alloy was higher than that in the Ti-Al-V and Ti-Al-Nb alloys' microstructure, as depicted in Figure 1.

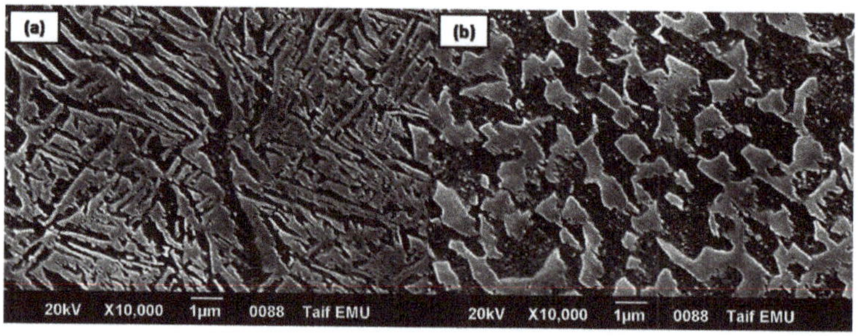

Figure 2. Morphology of β phase in TC21 alloy: (**a**) acicular-like structure; and (**b**) blocky shaped structure.

Table 2 illustrates the volume fraction of α and β phases in the microstructure of the studied titanium alloys. The microstructure of pure Ti has the highest volume fraction of the α phase (~100%) and the lowest volume fraction of β phase (~0.0%). The presence of Al (α-phase stabilizer) and V (β-phase stabilizer) as alloying elements in the chemical composition of Ti-Al-V alloy influenced the volume fraction of α and β phases. The values of the volume fractions of α and β phases (Table 2) in the microstructure of Ti-Al-V alloy were 65% and 35%, respectively. Replacing V with Nb, yielding Ti-Al-Nb alloy, resulted in an obvious enhancement in the volume fraction of α phase (increased to 77%) at the expense of that of the β phase, which decreased to 23%, as shown in Table 2. The volume fraction of both phases in the microstructure of TC21 alloy wasalso altered, probably due to the mutual combination of the alloying elements of that alloy (cf. Table 1). The volume fractions of α and β phases in the microstructure of TC21 alloy recorded almost equal values: 48% for α phase and 52% for β phase (Table 2).

Table 2. Volume fraction of α and β phases in the investigated Ti based alloys.

Alloy	Volume Fraction, %		(α/ β) Ratio
	α Phase	β Phase	
Pure Ti	100	0	–
Ti-6Al-4V	65	35	1.86
Ti-6Al-7Nb	77	23	3.35
TC21	48	52	0.92

To further assess the influence of chemical composition on the microstructure and volume fraction of α and β phases, $[Al]_{eq}$ and $[Mo]_{eq}$ were calculated, where $[Al]_{eq}$ and $[Mo]_{eq}$ represent the alloying elements from α and β phases, respectively [5,26]. Table 3 illustrates the calculated values of $[Al]_{eq}$ and $[Mo]_{eq}$ for the tested Ti-based alloys, following Equations (1) and (2) [5,26].

$$[Al]_{eq.} = [Al] + 0.33[Sn] + 0.17[Zr] + 10[O + C + 2N] \quad (1)$$

$$[Mo]_{eq.} = [Mo] + 0.2[Ta] + 0.28[Nb] + 0.4[W] + 0.67[V] + 1.25[Cr] + 1.25[Ni] + 1.7[Mn] + 1.7[Co] + 2.5[Fe] \quad (2)$$

Table 3. and $[Mo]_{eq}$ for the investigated alloys [26,27].

Alloy	$[Al]_{eq}$	$[Mo]_{eq}$	Ratio
Ti-6Al-4V	8.05	2.64	3.05
Ti-6Al-7Nb	8.59	2.18	3.94
TC21	8.59	5.04	1.71

As shown in Table 3, Ti-6Al-7Nb and TC21 alloys recorded the highest value (8.59) of $[Al]_{eq}$, while the lowest value (8.05) was measured for the Ti-6Al-4V alloy. Additionally, Ti-6Al-7Nb alloy achieved the maximum value of $[Mo]_{eq}$, 3.94, whileTC21 alloy recorded 1.71. Table 3 also depicts the ratio $[Al]_{eq}/[Mo]_{eq}$ for the tested alloys. Ti-6Al-7Nb alloy displayed the maximum ratio, 3.94, while a minimum ratio of 1.71 was determined for the TC21 alloy. The results shown in Table 3 agree well with those in Table 2. The calculated ($[Al]_{eq}/[Mo]_{eq}$) and (α/β) ratios weremaximum in case of Ti-6Al-7Nb alloy, and minimum for the TC21 alloy.

The chemical composition of both phases in all microstructures of the investigated alloys was analyzed using the EDS unit attached to SEM. The averaged chemical composition (wt. %) for each investigated alloy is displayed in Tables 4 and 5. It was evident that the Ti, Al, Sn and Zr elements tended to more segregate to α phase than to β phase [27]. However, V, Nb, Cr and Mo were β forming elements [28], meaning that higher ratios of these elements werefound in β phase rather than in α phase.The detailed EDS linescan/map analyses are discussed in the Supplementary Materials (Figures S1–S3).

Table 4. Chemical composition (wt %) of different phases in Ti-6Al-4V and Ti-6Al-7Nb alloys.

Phase	Ti-6Al-4V Alloy (wt %)			Ti-6Al-7Nb Alloy (wt %)		
	Al	V	Ti	Al	Nb	Ti
α	5.93	3.24	90.83	6.60	7.65	85.75
β	5.57	8.04	86.39	4.88	13.79	81.33

Table 5. Chemical composition (at %) of different phases in TC21 alloy.

Phase	Chemical Composition, at %						
	Al	Cr	Mo	Sn	Zr	Nb	Ti
α	6.38	1.08	1.82	2.45	1.95	1.98	84.34
β	6.21	1.71	2.62	2.24	1.11	2.39	83.72

3.2. X-ray Diffraction Studies

Phase identification was performed by X-ray diffraction (XRD) patterns to define the phases comprising each alloy sample. The diffraction patterns recorded for the studied alloys are compared in Figure 3. The phases were identified by matching the characteristic peaks with the Joint Committee on Powder Diffraction Standards (JCPDS) files [29]. The phases α-Ti (JCPDS#00-044-1294) andβ-Ti (JCPDS#00-044-1288) were common and dominated the composition of the three studied alloys. The Ti-Al-V and TC21 alloys were found to contain solely α-Ti and β-Ti phases, respectively. On the other hand, Ti-Al-Nb alloy contained some Ti and Nb oxides, TiO (JCPDS#00-008-0117) and Nb_6O (JCPDS#00-015-0258).

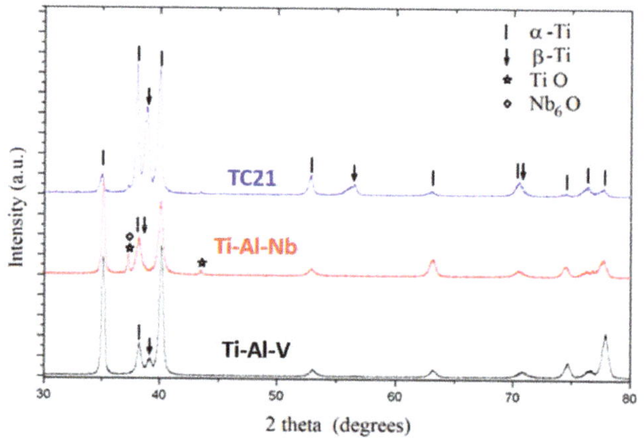

Figure 3. XRD diffraction patterns recorded for the Ti-6Al-4V, Ti-6Al-7Nb and TC21 samples.

An effective procedure for the simultaneous refinement of structural and microstructural parameters based on the integration of Fourier analysis for broadened peaks in the Rietveld method was first proposed by Lutterutti et al. [30] and is implemented in the Maud program [31]. Consequently, weight percent (wt.%), lattice parameters, isotropic crystallite size (D) and r.m.smicrostrain (με) were regarded as fitting parameters in the Rietveld adjustments and refined simultaneously. The structural information for all the refined phases was obtained from the Inorganic Crystal Structure Database (ICSD) [32]. The results obtained for the structural and microstructural analysis are summarized in Table 6 for all alloys. It is worth mentioning that all studied alloys were characterized with considerable degree of preferred orientation, which strongly modified the relative intensities of the Bragg reflections, especially for α-Ti and β-Ti phases. The MAUD program also incorporates correction for preferred orientation [33,34] in the Rietveld adjustments to obtain the best fitting parameters.

Table 6. The structural and microstructural parameters of the three alloys obtained by the Rietveld adjustment of the XRD patterns. Wt. % is the weight percentage of each phase, a and c are the cell parameters of the crystal lattice, D is the average crystallite-size and ε is the lattice microstrain.

	Ti-6Al-4V			
	α-Ti ($P6_3/mmc$)	β-Ti ($Im\bar{3}m$)		
Wt. %	67 (5)	33 (5)		
a (Å)	2.9338 (1)	3.2353 (9)		
c (Å)	4.6780 (3)			
D (nm)	76 (3)	67 (4)		
ε	0.00232 (1)	0.00216 (4)		
	Ti-6Al-7Nb			
	α-Ti ($P6_3/mmc$)	β-Ti ($Im\bar{3}m$)	TiO ($Fm\bar{3}m$)	Nb6O ($P42cm$)
Wt. %	73 (6)	20 (5)	5 (2)	2 (1)
a (Å)	2.9397 (4)	3.2437 (5)	4.1567 (1)	3.3945 (6)
c (Å)	4.6969 (3)			3.249 (3)
D (nm)	67 (2)	53 (2)	79 (12)	100 (2)
ε	0.0028 (1)	0.024 (20)	0.0026 (1)	0.00010 (7)
	TC21			
	α-Ti ($P6_3/mmc$)	β-Ti ($Im\bar{3}m$)		
Wt. %	49 (9)	51 (9)		
a (Å)	2.9407 (7)	3.2521 (14)		
c (Å)	4.6925 (16)			
D (nm)	63 (2)	70 (3)		
ε	0.0021 (2)	0.0035 (3)		

The calculated diffraction patterns from the Rietveld adjustment are plotted with the observed ones for the three alloys in Figure 4. The average R-values obtained for the refinements were about Rwp (%) = 24–27 and Rb (%) = 15–20. The simultaneous refinements of both structural and microstructural parameters produced good matching of the calculated to observed profiles of diffracted intensities. In addition, the incorporation of the preferred orientation models enabled accounting for the variations of the peak intensities of α and β-Ti phases.

In the Rietveld adjustment of the Ti-6Al-4V alloy, the hcp α-Ti (Space group $P6_3/mmc$) together with the bcc β-Ti (Space group $Im\bar{3}m$) dominated the composition of the alloy. In the second alloy, Ti-6Al-7Nb, the formation of some TiO (Space group $Fm\bar{3}m$) and Nb_6O (Space group $P42cm$) was observed and they formed larger crystallites than those formed in the Ti phases. The total weight percent of those oxide phases was less than 10% (Table 6). For the third alloy, TC21, only α- and β-Ti phases were observed in the XRD patterns. No oxide phases were detected due to the slight oxygen content of this alloy. Nevertheless, there weresome mismatches between the wt% values obtained from the Rietveld adjustments and the corresponding wt% values determined with other techniques. This was attributed to the behavior of the preferred orientation of the α-Ti phase observed for the reflection (100), which was relatively stronger for the Ti-6Al-4V and Ti-6Al-7Nb alloys than in the TC21 alloy.

As shown in Table 6, the last two alloys, Ti-6Al-7Nb and TC21, contained relatively higher portions of β-Ti than α-Ti in contrast to the first alloy, Ti-6Al-4V, which hadα-Ti content higher than β-Ti. As known from the literature, Al is an α-stabilizing while V, Nb, Mo and Fe are β-stabilizing elements. Nevertheless, the results indicate that Nb, Mo and Fe hadstronger capabilities to stabilize β-Ti phase than V. These findings corroborate microstructural studies (cf. Section 3.1).

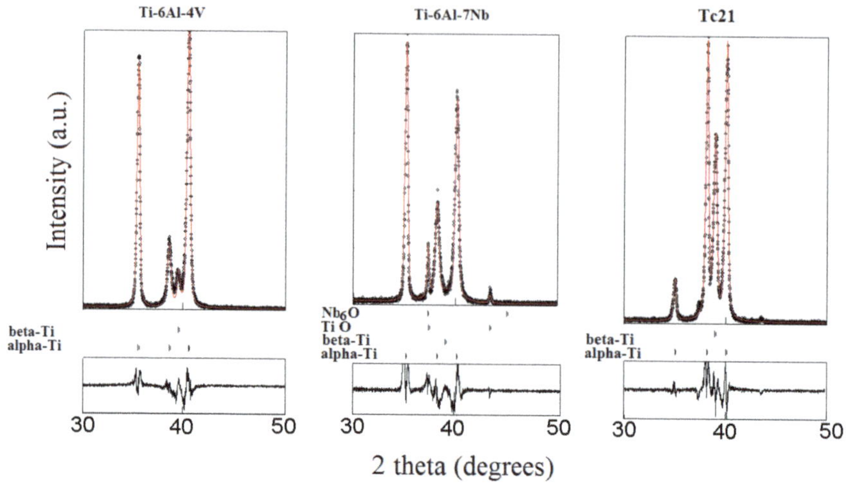

Figure 4. The calculated (red line) and recorded (black dots) diffraction patterns for the three alloys as obtained from the Rietveld adjustments using the MAUD program; the positions of the Bragg reflections of each phase and the difference between the calculated and observed patterns are also presented at the bottom.

3.3. *Electrochemical Measurements*

3.3.1. Uniform Corrosion Studies

Figure 5 illustrates the cathodic and anodic polarization curves for the studied alloys in comparison with pure Ti, after seven days of immersion in 0.9% NaCl solution at 37 °C. As shown in Figure 5, among the studied alloys, TC21 alloy exhibited the lowest cathodic and anodic overpotentials, corresponding to enhanced corrosion rate. On the contrary, Ti-6Al-7Nb alloy displayed the highest overpotentials, close to that of Ti, for both the cathodic and anodic processes, thus referring to its highest corrosion resistance.

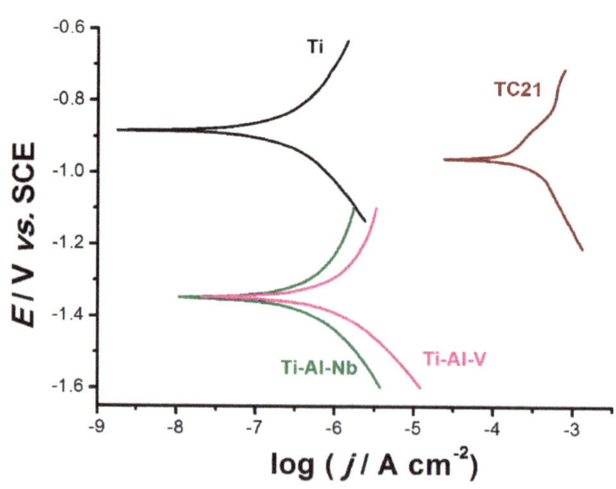

Figure 5. Cathodic and anodic polarization curves recorded for the three tested alloys in comparison with pure Ti, after seven days of exposure in 0.9% NaCl solution at a scan rate of 0.5 mV s^{-1} at 37 °C.

Indeed, Tafel slopes and respective calculated uniform corrosion rates for a metal covered with a semi-conductive passive film raisedstrong doubts, despite the utilization of the Tafel extrapolation method to determine corrosion rates for Ti and some Ti-based alloys [35,36]. In addition, in Figure 5, the polarization curves do not display the expected log/linear Tafel behavior. This was clear in both the anodic and cathodic branches of alloys Ti-6Al-4V and Ti-6Al-7Nb and as well as in the anodic branches of Ti and TC21 alloy, which exhibited some sort of curvature over the complete applied potential range. This in turn made evaluation of Tafel slopes by Tafel extrapolation method, and hence corrosion rates, inaccurate [37–39]. There is, therefore, an uncertainty and source of error in the numerical values of Tafel slopes (β_a and β_c), and possibly in the values of j_{corr}.

The curvature of the anodic branch mightbe attributed to the deposition of the corrosion products and/or passive film formation, as evidenced from XPS studies. With respect to the cathodic branch, since the solution was stationary, diffusion of the electrochemically active species was slow, and concentration polarization couldact to shorten the cathodic linear Tafel region. In the extreme case, linearity mightdisappear altogether, with the cathodic reaction proceeding under combined activation and diffusion control at E_{corr} [37–39]. This counteracted the validity of the Tafel extrapolation method for measuring uniform corrosion rates, which was successfully applied for the charge transfer controlled processes.

EIS measurements were also conducted at the respective E_{corr} during sample's exposure in 0.9% NaCl solution at 37 °C to confirm the polarization data and to assess the kinetics of the uniform corrosion process on alloy surface. The measurements were carried day-by-day allowing for monitoring of uniform corrosion susceptibility [40–42]. Figure 6 displays the impedance plots in Bode projection, recorded for the studied alloys during the second and the last day of exposure. Pure Ti (99.99%) was also included for comparison (see Figure 6a). The impedance spectra recorded on Day 1 (after initial 120 min of conditioning) were highly scattered due to non-stationary conditions at the metal/electrolyte interface, which is a common problem in EIS measurements. This issue became negligible after a few hours of exposure. For this reason, results recorded on Days 2–7 were considered for further analysis. All the impedance plots exhibited a single time constant (capacitive loop). The overall corrosion resistance of each investigated alloy was very high; the impedance modulus |Z| linearly increasedwith frequency decrease, reaching over 10^5 at 0.1 Hz. For each studied alloy, inclination of the phase angle θ was shifted towards lower frequencies on Day 7 of exposure, testifying tothe decreased corrosion process kinetics.

An electric equivalent circuit (EEC) was proposed to analyze the impedance results. When defining the adequate EEC, one must consider whether to include the space charge layer resulting from the semi-conductive nature of titanium oxides. In his studies, Blackwood concluded that thickness of the space charge layer is considerably less than the oxide film itself in the open circuit conditions [43]. On the other hand, its dominant influence was observed under anodic polarization conditions [43–46]. The impedance measurements of titanium oxide films investigated within this manuscript were studied under open circuit conditions, thus the parallel resistance and CPE represented primarily the dielectric properties of the passive layer. Due to absence of additional time constants in the analyzed frequency rage, a simple Randles circuit was proposed with constant phase element (CPE) selected instead of capacitance to account for the dispersion of the time-constant. The parallel resistance represents the sum of charge-transfer limiting effects through the metal/electrolyte interface, dominated by the passive layer resistance R_F [47]. The aforementioned time-constant dispersion originated from thepresence of the charge space layer in the semi-conductive film, thesurface distribution of the time-constant due to the geometric heterogeneity (pits, scratches, and porosity), and diversified surface electric properties due to adsorption processes of passive layer breakdown [48].

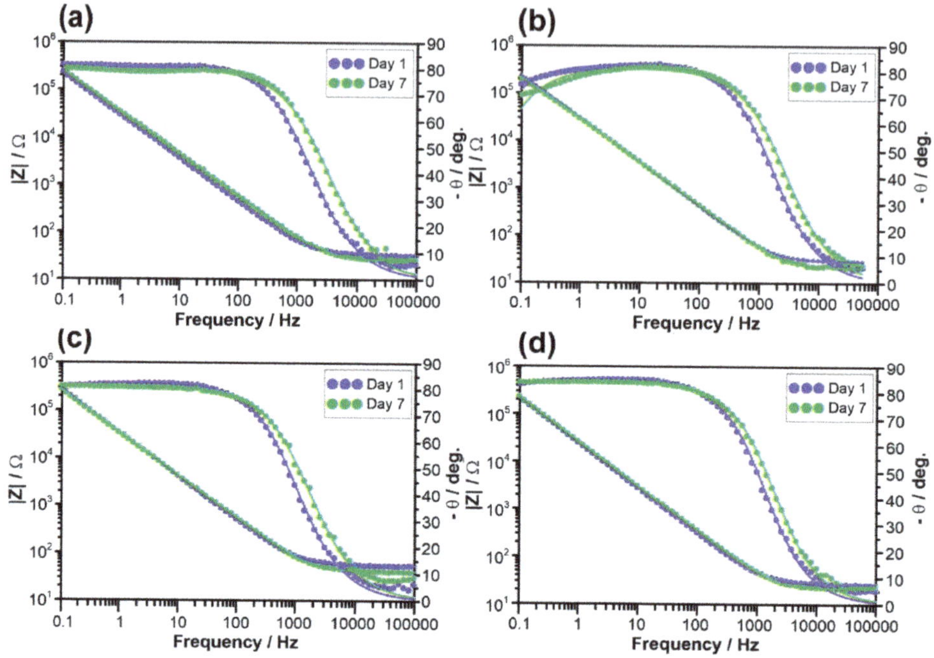

Figure 6. Bode impedance plots on Days 2 and 7 of exposure, recorded for: (**a**) pure Ti; and three tested alloys: (**b**) TC21; (**c**) Ti-6Al-4V; and (**d**) Ti-6Al-7Nb. Studies performed at E_{corr} in 0.9%NaCl solution at 37 °C. Points represent experimental results while the solid line was calculated based on R(QR) EEC.

The CPE impedance $Z_{CPE} = (Q(j\omega)^n)^{-1}$ represents a capacitor with capacitance 1/Q for a homogeneous surface $n \rightarrow 1$. Thus, it is often believed that CPE component n is the heterogeneity factor and its variation can be monitored. CPE describes quasi-capacitive behavior of the passive layer. The effective capacitance C_{eff} can be calculated based on CPE using Hirschorn's model for surface distribution of time constants [49]. The EEC can be schematically written as $R_S(QR_F)$, where R_S is electrolyte resistance. The aforementioned single time-constant EEC covers all the applied frequency range. The fitting quality is represented by solid lines in Figure 6a–d.

Figure 7 depicts the electric parameters obtained on the base of $R_S(QR_F)$ EEC and their changes during the one-week exposure. The higher was the R_F, the lower was the corrosion current density, offering an easy comparison in uniform corrosion resistance of investigated alloys (see Figure 7a). Each investigated alloy was characterized with very high resistance, order of MΩ, owing to a presence of a passive layer tightly covering the metal surface. Nevertheless, for TC21 alloy, $R_F$value wasone order of magnitude lower and slowly but consistently decreased throughout the exposure, revealing its lower corrosion resistance and corroborating DC electrochemical studies.

The analysis of constant phase element (CPE) alloweddrawing conclusions onthe passive layer homogeneity. The effective capacitance C_{eff}, calculated using the surface distribution model, was similar for allinvestigated alloys, falling in a range between 2 and 5 μF. The differentiation may result from differences in passive layer thickness d and to some extent from relative permittivity of alloying additives and their oxides ε_r according to: $C = \varepsilon_0\varepsilon_r d/A$, where ε_0 is the absolute permittivity and A is the electrochemically active surface area. A steady decrease of C_{eff} was attributed to an increase of passive layer thickness, denouncing further passivation of metal in investigated electrolytic conditions. The presence of stable corrosion pits would be visible in theform of rapid increase in C_{eff} [49,50] (likely observed in TC21 alloy on Day 4).

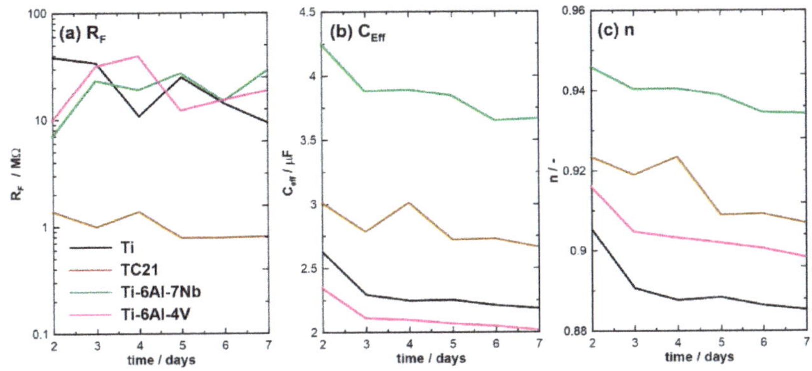

Figure 7. Monitoring of (**a**) passive layer resistance R_F, (**b**) effective capacitance C_{eff} and (**c**) CPE exponent n calculated on the base of R_S (QR_F) EEC for each investigated alloy. The one-week exposure was carried out in 0.9% NaCl solution at 37 °C.

The initial value of CPE exponent n depends on factors such as surface phase distribution and geometric defects remaining after polishing. Its decrease throughout the exposure in corrosive electrolyte reflected the appearance of heterogeneities on analyzed sample surface, which in this case was primarily associated with initial phases of corrosion pits formation (see Figure 8c). This effect was clearly seen in SEM micrographs (see below). Notably, the value of n factor of Ti-6Al-7Nb alloy was both the highest and the least affected by exposure in corrosive media. The aforementioned observation indicated high surface homogeneity, which may be the reason behind outstanding corrosion resistance of this alloy.

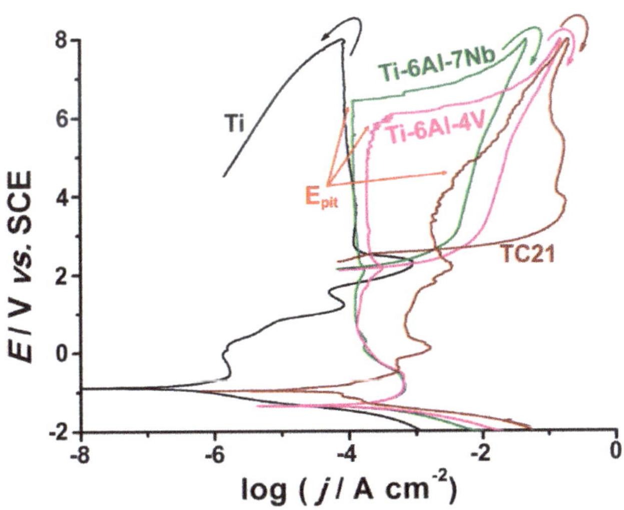

Figure 8. Cyclic polarization curves recorded for the studied alloys in 0.9% NaCl solutions at a scan rate of 1.0 mV s^{-1} at 37 °C.

3.3.2. Cyclic Polarization Measurements

Figure 8 shows typical cyclic polarization curves recorded for the studied alloys between −2.0 V and + 8.0 V (SCE). Measurements were conducted in 0.9% NaCl solution at a scan rate of 5.0 mV s^{-1} at 37 °C.

The polarization curve of TC21 alloy exhibited active dissolution near E_{corr}, followed by an obvious enhancement in the anodic current with the applied potential due to thinning and weakening of the passive layer as a result of the aggressive attack of Cl^- anions. In addition, Ti-6Al-7Nb and Ti-6Al-4V alloys showed active dissolution near E_{corr}, but to a much lower extent than TC21, and, in addition, tended to passivate with a very low current covering a wide range of potential. These findings reflect the weaker passivity of TC21 and its higher tendency to corrode in this solution than Ti-6Al-4V and Ti-6Al-7Nb alloys. On the contrary, as expected, the anodic polarization curve of pure Ti exhibited typical passivity near E_{corr}, referring to its high corrosion resistance.

Passivity of the studied alloys persisted up to reaching pitting potential (E_{pit}). Remarkable changes occurred within the passive region at potentials exceeding E_{pit}. These involved a sudden increase in corrosion current density and formation of a hysteresis loop on the reverse potential scan. These events werea clear sign ofpassivity breakdown, and initiation and propagation of pitting corrosion. Repassivation was only achieved when the reverse scan intersected the forward one within the passive region, below which the working electrode was immune to pitting [51].

A current intermission could be seen on the reverse scan of the three tested alloys. This current discontinuity aws quite clear on the reverse scan of the TC21 alloy, and couldbe observed for Ti-6Al-7Nb and Ti-6Al-4V alloys. We previously reported similar findings during pitting corrosion studies of Zn in nitrite solutions [52]. Recently, Zakeri et al. [53] explored the transition potential and the repassivation potential of AISI type 316 stainless steel in chloride containing media devoid of and containing 0.01 M thiosulfate.

Beyond pit transition potential, the rate of anodic dissolution was diffusion-controlled [51–53]. Such a current transient relationship, when satisfied, referred to an anodic diffusion control process [53]. On reversing the potential scan, the thickness of the salt (pitting corrosion product) film diminished. This decrease in salt film thickness enhanced with back scanning untila certain potential was reached, at which the cation concentration decreased below the saturated concentration. At this stage, salt precipitation was stopped, and the remaining metal salt film dissolved, making the bottom of pits free from salt film. This in turn established an ohmic/activation control (a linear decrease of current density with potential) regime.

Ti-6Al-7Nb alloy's passivity seemed stronger and more stable that of the Ti-6Al-4V alloy. The latter was characterized by a higher j_{pass}, which enhanced with potential untilits E_{pit}, which attained ~50 mV vs. SCE before that of the former. In addition, the pits existing on the surface of Ti-6Al-4V alloy were much more difficult to repassivate than those on the surface of Ti-6Al-7Nb alloy, as the hysteresis loop of the former was much larger than that of the later.

Another important pitting corrosion controlling electrochemical parameter is the pitting corrosion resistance $R_{pit} = |E_{corr} - E_{pit}|$, which defines the resistance against the nucleation of new pits [38]. Referring to Figure 8, it is clear that R_{pit} increased following the order: TC21 << Ti-6Al-4V < Ti-6Al-7Nb. The resistance against growth of the pits also controlled the susceptibility toward pitting corrosion. A specific routine of the software (Autolab frequency response analyzer (FRA) coupled to an Autolab PGSTAT30 potentiostat/galvanostat with FRA2 module) (Metrohm, Herisau, Switzerland.) was used to calculate the areas of the hysteresis loops, related to the charge consumed during the growth of such already formed pits. Again, the hysteresis loop of the TC21 alloy recorded the highest area (charge consumed) among the studied alloys, while the lowest value of the hysteresis loop charge consumed during was measured for Ti-6Al-7Nb alloy. Thus, the resistance against the growth of pre-existing pits was ranked as: Ti-6Al-7Nb > Ti-6Al-4V >> TC21. These findings mean that replacing V by Nb in Ti-6Al-4V alloy promoted alloy's repassivation, thus enhancing its pitting corrosion resistance.

3.3.3. Chronoamperometry Measurements

Chronoamperometry (*j*/*t*) measurements were also carried out to confirm the above results and gain more information about the influence of alloyed V and Nb on the passive layer growth kinetics and breakdown. Figures 9a and 10b depict the *j*/*t* curves measured for the tested alloys at two different

E_a values, far below and close to E_b. Measurements were conducted in 0.9% NaCl solution at 37 °C. The profile of the obtained curves was found to vary according to the chemical composition of the studied alloy and the position of E_a versus E_{pit}. When E_a was located far cathodic to E_{pit}, a j/t profile with two stages was obtained, as shown in Figure 9a. During the first stage, the anodic current (j_a) declined with a rate depending upon chemical composition of the tested alloy, denoting passive layer electroformation and growth [39]. This decay in current then reached a steady-state value (j_{ss}), an almost constant passive current related to j_{pass} (cf. Figure 8), constituting the secondstage of the current. The constancy of j_{ss} originated from a balance between the rates of the passive layer growth (current builds up) and its dissolution (current decays) [54,55].

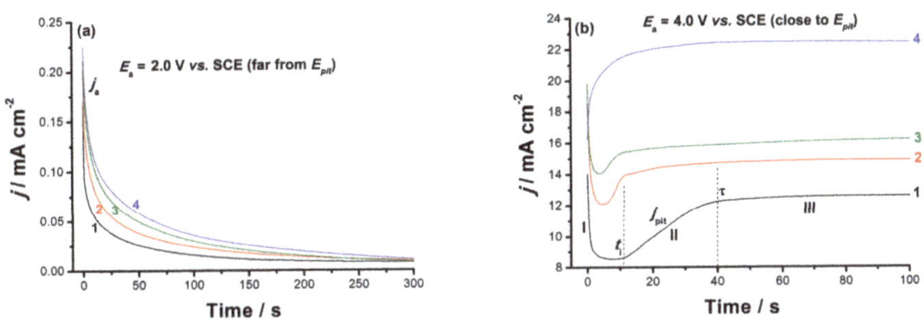

Figure 9. Chronoamperometry (current–time) curves recorded for the studied solder alloys in 0.9% NaCl solution at applied anodic potentials of 2.0 V (**a**) and 4.0 V (**b**) vs. SCE at 37 °C: (1) pure Ti; (2) Ti-6Al-7Nb; (3) Ti-6Al-4V; and (4) TC21.

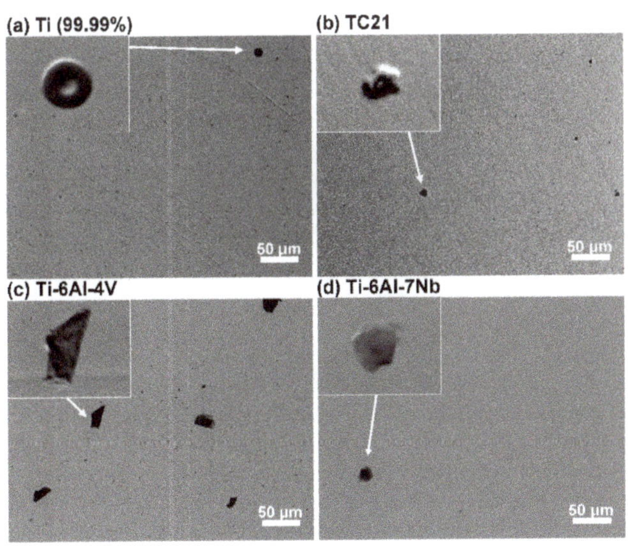

Figure 10. SEM micrographs taken in secondary electron mode for each investigated sample: (**a**) pure Ti as a reference; (**b**) TC21 alloy; (**c**) Ti-6Al-4V; and (**d**) Ti-6Al-7Nb at the end of one-week exposure in 0.9% NaCl at 37 °C. Magnification: × 500. In the inset, back-scatter electron topography mode images of selected surface defects. Magnification: × 2000.

Further, Figure 9a demonstrates that the rate of j_a decay, and consequently the rate of passive layer growth, diminished upon alloying Ti with V and Nb. These results further confirm the influence of the alloying elements V and Nb, with V being more active than Nb, which, when added to Ti, weakened its passivity viadepassivation (destabilizing the passive oxide film through oxide film thinning/dissolution [56]). This in turn madethe passive film more susceptible to pitting.

At an E_a value very close to E_{pit} (Figure 9b), the *j*/*t* curves with three different stages (I–III) were recorded. Similar results were previously obtained in our lab [55,57]. Stage I refersto the passive layer electroformation and growth, as its current fellwith time [54,55,57]. Stage I ended at a certain time (t_i), the incubation time, where Stage I's current reached its minimum value; t_i is defined as the time the adsorbed aggressive Cl^- anions must acquire to locally attack and subsequently remove the passive oxide film [54]. The magnitude of t_i, more specifically its reciprocal value ($1/t_i$), denotes the rate of pit initiation and growth [54,55], and measures the susceptibility of the oxide film to breakdown and initiate pit formation and growth.

Stage II began at t_i and terminated at time τ, and its current aws termed j_{pit} (pit growth current density). j_{pit} increased from the moment just after t_i and continuedto grow until τ, suggesting that the pit formation and growth dominated over passivation during this stage. Ultimately, j_{pit} attaineda steady-state just after the time τ, denoting the onset of Stage III, and remained almost constant until the end of the run. The constancy of Stage III's current was attributed to the hindrance of the current flow (j_{pit}) through the pits sealed off by the pitting corrosion products formed during the events of Stage II, namely pit initiation and growth [55,57]. This hindrance in j_{pit} was balanced by a current increase due to metal dissolution, thus yielding an overall steady-state current.

Close inspection of Figure 9 reveals that j_{pit} increased and t_i shortened, thus referring to accelerated pitting attack, in presence of alloyed V. These results again support the catalytic impact of alloyed V towards pitting corrosion

3.4. Surface Morphology and Composition

After one-week exposure, the investigated samples were reexamined using SEM to evaluate the susceptibility to pitting corrosion. This procedure was carried out after rinsing in ethanol using ultrasonic cleaner. The results of the analysis are exhibited in Figure 10. Defects started to appear on the surface of each analyzed sample throughout the exposure. The micrographs in the inset of Figure 10 were taken using back-scatter electrons (BSE) in topography mode. This allowed bringing out the geometry of aforementioned defects. As can be seen, each analyzed defect formed a bulge above the alloy's surface, testifying to either repassivation once shallow corrosion pits formed or at an early stage of passive layer degradation. Ti-6Al-4V sample was characterized with both the highest amount and the largest defects, reaching 30 μm in diameter. On the other hand, the surface of pure Ti and Ti-6Al-7Nb appeared the most intact. No real corrosion pits were observed on the surface of either investigated alloy at the end of exposure in 0.9% NaCl solution at 37 °C, testifying tothe overall high pitting corrosion resistance.

Nevertheless, the passive layer must have weakened, hence it was possible for corrosion products to adsorb on the metal surface. EDS analysis was carried out on the defects observed on each investigated alloy to qualify their chemical constitution. The exemplary results, obtained for Ti-6Al-7Nb alloy, are summarized in Figure S4 (Supplementary Materials). The chemistry of defects observed for each investigated alloy was similar. The defects were primarily composed of carbon and oxygen, most likely forming metal carbonates typical for early pitting corrosion stages [58]. Small amount of chlorine was also found within defects. Its low amount was distorted by EDS depth of analysis ranging few microns.

The chemistry of the passive layer in each examined case was composed primarily of titanium (IV) oxides, as verified by a strong recorded Ti_{2p} peak doublet, with $Ti_{2p3/2}$ component located each time at 458.6 eV [48,59,60] (Figure 11). Furthermore, there was no sign of titanium oxides at lower oxidation states corroborating the aforementioned result (see Supplementary Materials, Figure S5). Besides the titanium, other alloying additives also tookpart in the passivation process. The strongest signal among the alloying additives was recorded for aluminum oxide Al_2O_3 ($Al_{2p3/2}$ peak at 74.5 eV), ranging between 3.5 and 3.8 at.% for each sample [61,62]. The contribution of VO_2 ($V_{2p3/2}$ at 516.4 eV) in Ti-Al-V and Nb_2O_5 ($Nb_{3d5/2}$ at 207.1 eV) in Ti-Al-Nb alloy did not exceed 0.7 at.% [60,63,64]. The passive film formed on the surface of TC21 alloy was naturally more complex. Besides TiO_2, it wascomposed of Al_2O_3 (3.8 at.%), Nb_2O_5 (0.3 at.%), ZrO_2 (0.4 at.%, $Zr_{3d5/2}$ at 182.4 eV), Cr_2O_3 (0.8 at.%, $Cr_{2p3/2}$ at 576.0 eV), SnO_2 (0.1 at.%, $Sn_{3d3/2}$ at 486.5 eV), MoO_3 and MoO_2 (0.2+0.2 at.%, $Mo_{3d5/2}$ at 232.9 and 229.2 eV, respectively) [65–68].

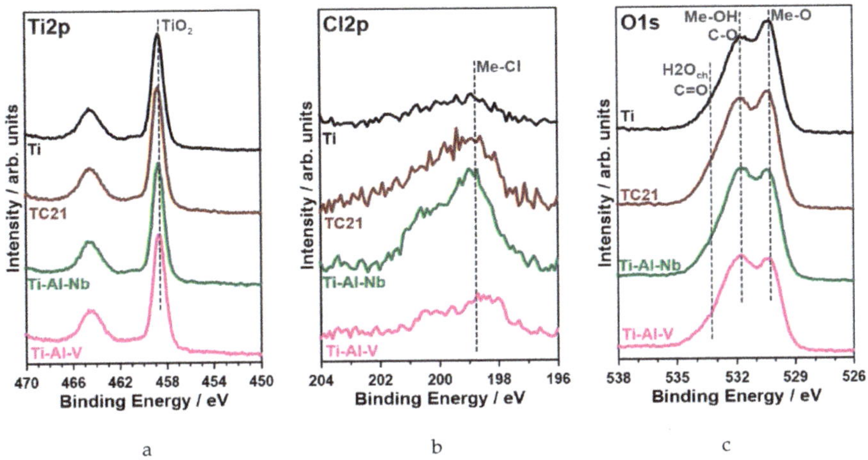

Figure 11. High-resolution XPS spectra recorded in (**a**) Ti_{2p}, (**b**) Cl_{2p} and (**c**) O_{1s} energy range for each investigated alloy after seven days of exposure to 0.9% NaCl solution at 37 °C.

The high-resolution spectra analysis carried out in the Cl_{2p} energy range confirmed the electrochemical and microscopic studies regarding chloride adsorption on the metal surface as a result of seven-day metal exposure to chloride-containing electrolyte. Full chemical analysis is summarized in Table 7. Metal chlorides were found on the surface of each investigated sample, which confirmed metal-chlorine bond formation, shownby a peak doublet: $Cl_{2p3/2}$ at 198.9 eV [48,55,69]. Nevertheless, the amount of adsorbed chlorides Was nearly 2.5 times higher for the TC21 alloy than pure titanium. The chloride concentration obtained for highly resistant Ti-6Al-7Nb alloy was nearly on par withTi sample, and slightly smaller than in the case of Ti-6Al-4V. An interesting conclusion couldbe drawn based on O_{1s} peak analysis for each investigated sample. The spectra were conventionally deconvoluted into three components. Two dominant components located at 530.2 and 531.6 eV were ascribed to Me-O and Me-OH species, respectively. The second component intensity mightbe further influenced by the presence of C-O bonds in carbonates. Its formation may result from prolonged electrolyte exposure as well as adventitious carbon deposition due to air exposure [61,64]. The finding regarding carbonate adsorption on the metal surface wasfurther confirmed by a third O_{1s} component at 532.8 eV, typical for C=O bonds but also chemisorbed water molecules. For clarity purposes, the analysis excluded data recorded for carbon C_{1s}, which was found in large amounts, up to 30 at.%, at binding energies corroborating adventitious carbon and carbonates. Importantly, the highest amount of the adsorbed carbonate species was found on Ti-Al-V sample surface, which is in very good agreement with SEM micrographs presented in Figure 10. The least amount of carbonate species was once more found on the surface of Ti sample.

Table 7. XPS deconvolution results carried out in Ti_{2p}, Cl_{2p} and O_{1s} energy range after seven days of exposure to 0.9% NaCl solution at 37 °C (in at.%).

	Ti_{2p}	Other Additives	O_{1s}			Cl_{2p}
	TiO_2		Me-O	OH/CO	C=O/H_2O	Me-Cl
BE/eV	458.6	*	530.2	531.6	532.8	198.9
Ti	21.3	–	37.5	26.2	13.1	1.9
Ti-6Al-4V	17.4	4.3	28.8	25.7	20.8	3.0
Ti-6Al-7Nb	17.1	4.4	31.1	27.0	18.2	2.2
TC21	17.2	5.8	31.2	25.1	16.1	4.6

4. Conclusions

The effect of microstructure on the uniform and pitting corrosion characteristics of Ti-Al-V, and Ti-Al-Nb alloys were studied. Pure Ti and TC21 alloy were included for comparison. Measurements were conducted in 0.9% NaCl solution at 37 °C employing various electrochemical techniques, and complemented with XRD and SEM/EDS analysis. The obtained results reveal that:

1. The microstructure of the investigated alloys consisted of α matrix, which was strengthened by β phase in all studied alloys except cp-Ti alloy.
2. The volume fraction of β phase in TC21 alloy was higher than in both Ti-Al-V and Ti-Al-Nb alloys. In comparison with other alloys, Ti-Al-Nb alloy had the lowest volume fraction of β phase.
3. Ti-Al-Nb alloy exhibitedthe highest corrosion resistance (lowest corrosion rate) among other alloys.
4. The addition of Nb alloying element at the expense of V in Ti-Al-V alloy decreased the volume fraction of β phase, which improvedthe corrosion resistance of Ti based alloy.
5. The polarization curve of TC21 alloy exhibited active dissolution near E_{corr}, followed by enhanced anodic current with the applied potential, thus revealing weak passivity. On the contrary, Ti-6Al-7Nb and Ti-6Al-4V alloys recorded much lower dissolution rate near E_{corr} followed by a wide potential range of stable passivity.
6. Stable passivity of Ti-6Al-7Nb and Ti-6Al-4V alloys was translated into an obvious anodic (more noble) drift in their pitting potential (E_{pit}) versus that of the TC21 alloy.
7. Chronoamperometry measurements, conducted at a fixed anodic potential, revealed that the rate of passivity breakdown of Ti-6Al-7Nb and Ti-6Al-4V alloys wasmuch lower than that of the TC21 alloy.
8. The XPS results reveal adsorption of chloride and carbonate species on the surface of the investigated alloys with the lowest amount recorded for cp-Ti and Ti-6Al-7Nb alloys, affecting the observed corrosion resistance.
9. Corrosion studies confirmed that the uniform and pitting corrosion rates increased following the sequence: Ti< Ti-6Al-7Nb < Ti-6Al-4V < TC21.

Supplementary Materials: The following are available online at http://www.mdpi.com/1996-1944/12/8/1233/s1. Figure S1—SEM image (a) and EDS spectrum (b) of phase in Ti-Al-V alloy; Figure S2—Line analysis of β phase in TiAlNb alloy; Figure S3—Microstructure of the Ti-Al-Nb alloy (a), and mapping of Ti (b), Al (c), and Nb (d) alloying elements. Figure S4—(a) SEM micrograph with marked areas for EDS analysis, (b) EDS examination at defect and at the surrounding, not corroded area. Figure S5—High-resolution Ti2p XPS spectra with the quality fit (light blue line) for each studied alloy: (a) pure Ti reference, (b) TC21, (c) Ti-6Al-4V, (d) Ti-6Al-7Nb.

Author Contributions: Conceptualization, M.A.A. and R.B.; resources, N.E.-B. and S.E.-H.; investigation, S.I.A. and O.A.A.A, (XRD), A.M.F. and M.M.I. (uniform corrosion), M.A.A. and G.A.M.M. (passive layer breakdown), N.E.-B., J.W. and S.E.-H. (microstructure), J.R. (XPS), and J.R. and J.W. (EIS); writing—original draft preparation, all authors; writing—review and editing, J.R., M.A.A. and R.B.; and project administration, M.A.A.

Funding: This study was funded by the Deanship of Scientific Research, Taif University, Saudi Arabia (Project No. 1-439-6070).

Conflicts of Interest: The authors declare no conflict of interest.

References

1. de Assis, S.L.; Wolynec, S.; Costa, I. Corrosion characterization of titanium alloys by electrochemical techniques. *Electrochim. Acta* **2006**, *51*, 1815–1819. [CrossRef]
2. Geetha, M.; Singh, A.K.; Asokamani, R.; Gogia, A.K. Ti based biomaterials, the ultimate choice for orthopaedic implants—A review. *Prog. Mater. Sci.* **2009**, *54*, 397–425. [CrossRef]
3. Jiang, H. Enhancement of Titanium Alloy Corrosion Resistance via Anodic Oxidation Treatment. *Int. J. Electrochem. Sci.* **2018**, 3888–3896. [CrossRef]
4. Moiseyev, V.N. *Titanium Alloys: Russian Aircraft and Aerospace Applications*; Advances in Metallic Alloys; Taylor & Francis: Boca Raton, FL, USA, 2006; ISBN 978-0-8493-3273-9.
5. Leyens, C.; Peters, M. (Eds.) *Titanium and Titanium Alloys: Fundamentals and Applications*; Wiley-VCH: Weinheim, Germany; John Wiley: Chichester, UK, 2003; ISBN 978-3-527-30534-6.
6. Lütjering, G.; Williams, J.C. *Titanium: With 51 Tables*; Springer: Berlin/Heidelberg, Germany, 2007; ISBN 978-3-540-71397-5.
7. Oberwinkler, B.; Riedler, M.; Eichlseder, W. Importance of local microstructure for damage tolerant light weight design of Ti–6Al–4V forgings. *Int. J. Fatigue* **2010**, *32*, 808–814.
8. Knobbe, H.; Köster, P.; Christ, H.-J.; Fritzen, C.-P.; Riedler, M. Initiation and propagation of short fatigue cracks in forged Ti6Al4V. *Procedia Eng.* **2010**, *2*, 931–940.
9. Fekry, A.M.; El-Sherif, R.M. Electrochemical corrosion behavior of magnesium and titanium alloys in simulated body fluid. *Electrochim. Acta* **2009**, *54*, 7280–7285. [CrossRef]
10. Whittaker, M. Titanium Alloys. *Metals* **2015**, *5*, 1437–1439. [CrossRef]
11. Mountford, J.A., Jr. Titanium—Properties, Advantages and Applications Solving the Corrosion Problems in Marine Service. In Proceedings of the CORROSION, Denver, CO, USA, 7–11 April 2002.
12. Al-Mayouf, A.; Al-Swayih, A.; Al-Mobarak, N.; Al-Jabab, A. Corrosion behavior of a new titanium alloy for dental implant applications in fluoride media. *Mater. Chem. Phys.* **2004**, *86*, 320–329. [CrossRef]
13. García, C.; Ceré, S.; Durán, A. Bioactive coatings deposited on titanium alloys. *J. Non-Cryst. Solids* **2006**, *352*, 3488–3495.
14. Sharma, A.K. Anodizing titanium for space applications. *Thin Solid Film.* **1992**, *208*, 48–54. [CrossRef]
15. Barjaktarević, D.R.; Cvijović-Alagić, I.L.; Dimić, I.D.; Đokić, V.R.; Rakin, M.P. Anodization of Ti-based materials for biomedical applications: A review. *Metall. Mater. Eng.* **2016**, *22*, 129–144. [CrossRef]
16. Qu, Q.; Wang, L.; Chen, Y.; Li, L.; He, Y.; Ding, Z. Corrosion Behavior of Titanium in Artificial Saliva by Lactic Acid. *Materials* **2014**, *7*, 5528–5542. [CrossRef]
17. Hines, J.A.; Lutjering, G. Propagation of microcracks at stress amplitudes below the conventional fatigue limit in Ti-6Al-4V. *Fatigue Fract. Eng. Mater. Struct.* **1999**, *22*, 657–665.
18. Sieniawski, J.; Ziaja, W.; Kubiak, K.; Motyk, M. Microstructure and Mechanical Properties of High Strength Two-Phase Titanium Alloys. In *Titanium Alloys—Advances in Properties Control*; Sieniawski, J., Ed.; InTech: London, UK, 2013; ISBN 978-953-51-1110-8.
19. Gai, X.; Bai, Y.; Li, J.; Li, S.; Hou, W.; Hao, Y.; Zhang, X.; Yang, R.; Misra, R.D.K. Electrochemical behaviour of passive film formed on the surface of Ti-6Al-4V alloys fabricated by electron beam melting. *Corros. Sci.* **2018**, *145*, 80–89.
20. Dadé, M.; Esin, V.A.; Nazé, L.; Sallot, P. Short- and long-term oxidation behaviour of an advanced Ti2AlNb alloy. *Corros. Sci.* **2019**, *148*, 379–387. [CrossRef]
21. Chávez-Díaz, M.; Escudero-Rincón, M.; Arce-Estrada, E.; Cabrera-Sierra, R. Effect of the Heat-Treated Ti6Al4V Alloy on the Fibroblastic Cell Response. *Materials* **2017**, *11*, 21. [CrossRef] [PubMed]
22. Hussein, M.; Kumar, M.; Drew, R.; Al-Aqeeli, N. Electrochemical Corrosion and In Vitro Bioactivity of Nano-Grained Biomedical Ti-20Nb-13Zr Alloy in a Simulated Body Fluid. *Materials* **2017**, *11*, 26. [CrossRef] [PubMed]
23. Zhang, L.; Duan, Y.; Gao, R.; Yang, J.; Wei, K.; Tang, D.; Fu, T. The Effect of Potential on Surface Characteristic and Corrosion Resistance of Anodic Oxide Film Formed on Commercial Pure Titanium at the Potentiodynamic-Aging Mode. *Materials* **2019**, *12*, 370. [CrossRef] [PubMed]
24. Reda, R.; Nofal, A.; Hussein, A.-H. Effect of Single and Duplex Stage Heat Treatment on the Microstructure and Mechanical Properties of Cast Ti–6Al–4V Alloy. *Metallogr. Microstruct. Anal.* **2013**, *2*, 388–393. [CrossRef]

25. El-Bagoury, N.; Ibrahim, K. Microstructure, Phase Transformations and Mechanical Properties of Solution Treated Bi-Modal Titanium Alloy. *Int. J. Eng. Sci. Res. Technol.* **2016**, *5*, 517–525.
26. Zhao, X.; Sun, S.; Wang, L.; Liu, Y.; He, J.; Tu, G. A New Low-Cost β-Type High-Strength Titanium Alloy with Lower Alloying Percentage for Spring Applications. *Mater. Trans.* **2014**, *55*, 1455–1459. [CrossRef]
27. Phukaoluan, A.; Khantachawana, A.; Dechkunakorn, S.; Anuwongnukroh, N.; Santiwong, P.; Kajornchaiyakul, J. Effect of Cu and Co Additions on Corrosion Behavior of NiTi Alloys for Orthodontic Applications. *Adv. Mater. Res.* **2011**, *378–379*, 650–654. [CrossRef]
28. Lee, C.S.; Won, J.W.; Lee, Y.; Yeom, J.-T.; Lee, G.Y. High Temperature Deformation Behavior and Microstructure Evolution of Ti-4Al-4Fe-0.25Si Alloy. *Korean J. Met. Mater.* **2016**, *54*, 338–346. [CrossRef]
29. ICDD. *PDF 2, Database Sets 1-45*; The International Centre for Diffraction Data: Newtown Square, PA, USA, 1995.
30. Lutterotti, L.; Scardi, P. Simultaneous structure and size–strain refinement by the Rietveld method. *J. Appl. Crystallogr.* **1990**, *23*, 246–252. [CrossRef]
31. Lutterotti, L. Total pattern fitting for the combined size–strain–stress–texture determination in thin film diffraction. *Nucl. Instrum. Methods Phys. Res. Sect. B Beam Interact. Mater. At.* **2010**, *268*, 334–340. [CrossRef]
32. Belsky, A.; Hellenbrand!, M.; Karen., V.L.; Luksch, P. New developments in the inorganic crystal structure database (ICSD): Accessibility in support of materials research and design. *Acta Crystallogr.* **2002**, *858*, 364–369. [CrossRef]
33. Dollase, W.A. Correction of intensities for preferred orientation in powder diffractometry: Application of the March model. *J. Appl. Crystallogr.* **1986**, *19*, 267–272. [CrossRef]
34. Will, G.; Bellotto, M.; Parrish, W.; Hart, M. Crystal structures of quartz and magnesium germanate by profile analysis of synchrotron-radiation high-resolution powder data. *J. Appl. Crystallogr.* **1988**, *21*, 182–191. [CrossRef]
35. Cvijović-Alagić, I.; Cvijović, Z.; Mitrović, S.; Panić, V.; Rakin, M. Wear and corrosion behaviour of Ti–13Nb–13Zr and Ti–6Al–4V alloys in simulated physiological solution. *Corros. Sci.* **2011**, *53*, 796–808. [CrossRef]
36. Simsek, I.; Ozyurek, D. Investigation of the wear and corrosion behaviors of Ti5Al2.5Fe and Ti6Al4V alloys produced by mechanical alloying method in simulated body fluid environment. *Mater. Sci. Eng. C* **2019**, *94*, 357–363. [CrossRef] [PubMed]
37. Mansfeld, F. Tafel slopes and corrosion rates obtained in the pre-Tafel region of polarization curves. *Corros. Sci.* **2005**, *47*, 3178–3186. [CrossRef]
38. Flitt, H.J.; Schweinsberg, D.P. A guide to polarisation curve interpretation: Deconstruction of experimental curves typical of the Fe/H2O/H+/O2 corrosion system. *Corros. Sci.* **2005**, *47*, 2125–2156. [CrossRef]
39. Flitt, H.J.; Schweinsberg, D.P. Evaluation of corrosion rate from polarisation curves not exhibiting a Tafel region. *Corros. Sci.* **2005**, *47*, 3034–3052. [CrossRef]
40. Krakowiak, S.; Darowicki, K.; Ślepski, P. Impedance of metastable pitting corrosion. *J. Electroanal. Chem.* **2005**, *575*, 33–38. [CrossRef]
41. Darowicki, K.; Krakowiak, S.; Slepski, P. The time dependence of pit creation impedance spectra. *Electrochem. Commun.* **2004**, *6*, 860–866. [CrossRef]
42. Gerengi, H.; Slepski, P.; Ozgan, E.; Kurtay, M. Investigation of corrosion behavior of 6060 and 6082 aluminum alloys under simulated acid rain conditions: Corrosion behavior of 6060 and 6082 Al alloys under acid rain. *Mater. Corros.* **2015**, *66*, 233–240. [CrossRef]
43. Blackwood, D. Influence of the space-charge region on electrochemical impedance measurements on passive oxide films on titanium. *Electrochim. Acta* **2000**, *46*, 563–569. [CrossRef]
44. Hamadou, L.; Aïnouche, L.; Kadri, A.; Yahia, S.A.A.; Benbrahim, N. Electrochemical impedance spectroscopy study of thermally grown oxides exhibiting constant phase element behaviour. *Electrochim. Acta* **2013**, *113*, 99–108. [CrossRef]
45. Gnedenkov, S.V.; Sinebryukhov, S.L. Electrochemical Impedance Spectroscopy of Oxide Layers on the Titanium Surface. *Russ. J. Electrochem.* **2005**, *41*, 858–865. [CrossRef]
46. Cámara, O.R.; Avalle, L.B.; Oliva, F.Y. Protein adsorption on titanium dioxide: Effects on double layer and semiconductor space charge region studied by EIS. *Electrochim. Acta* **2010**, *55*, 4519–4528. [CrossRef]
47. Jorcin, J.-B.; Orazem, M.E.; Pébère, N.; Tribollet, B. CPE analysis by local electrochemical impedance spectroscopy. *Electrochim. Acta* **2006**, *51*, 1473–1479. [CrossRef]

48. Alqarni, N.D.; Wysocka, J.; El-Bagoury, N.; Ryl, J.; Amin, M.A.; Boukherroub, R. Effect of cobalt addition on the corrosion behavior of near equiatomic NiTi shape memory alloy in normal saline solution: Electrochemical and XPS studies. *RSC Adv.* **2018**, *8*, 19289–19300. [CrossRef]
49. Hirschorn, B.; Orazem, M.E.; Tribollet, B.; Vivier, V.; Frateur, I.; Musiani, M. Determination of effective capacitance and film thickness from constant-phase-element parameters. *Electrochim. Acta* **2010**, *55*, 6218–6227. [CrossRef]
50. Krakowiak, S.; Darowicki, K.; Slepski, P. Impedance investigation of passive 304 stainless steel in the pit pre-initiation state. *Electrochim. Acta* **2005**, *50*, 2699–2704. [CrossRef]
51. Dong, Z.H.; Shi, W.; Guo, X.P. Initiation and repassivation of pitting corrosion of carbon steel in carbonated concrete pore solution. *Corros. Sci.* **2011**, *53*, 1322–1330. [CrossRef]
52. Amin, M.A.; Hassan, H.H.; Abd El Rehim, S.S. On the role of NO2− ions in passivity breakdown of Zn in deaerated neutral sodium nitrite solutions and the effect of some inorganic inhibitors. *Electrochim. Acta* **2008**, *53*, 2600–2609. [CrossRef]
53. Zakeri, M.; Naghizadeh, M.; Nakhaie, D.; Moayed, M.H. Pit Transition Potential and Repassivation Potential of Stainless Steel in Thiosulfate Solution. *J. Electrochem. Soc.* **2016**, *163*, C275–C281. [CrossRef]
54. Amin, M.A.; Abd El-Rehim, S.S.; Aarão Reis, F.D.A.; Cole, I.S. Metastable and stable pitting events at zinc passive layer in alkaline solutions. *Ionics* **2014**, *20*, 127–136. [CrossRef]
55. Amin, M.A.; El-Bagoury, N.; Mahmoud, M.H.H.; Hessien, M.M.; Abd El-Rehim, S.S.; Wysocka, J.; Ryl, J. Catalytic impact of alloyed Al on the corrosion behavior of $Co_{50}Ni_{23}Ga_{26}Al_{1.0}$ magnetic shape memory alloy and catalysis applications for efficient electrochemical H_2 generation. *Rsc Adv.* **2017**, *7*, 3635–3649. [CrossRef]
56. Amin, M.A.; Fadlallah, S.A.; Alosaimi, G.S. Activation of Titanium for Synthesis of Supported and Unsupported Metallic Nanoparticles. *J. Electrochem. Soc.* **2014**, *161*, D672–D680. [CrossRef]
57. Amin, M.A.; Abd El-Rehim, S.S.; El-Sherbini, E.E.F.; Mahmoud, S.R.; Abbas, M.N. Pitting corrosion studies on Al and Al–Zn alloys in SCN−solutions. *Electrochim. Acta* **2009**, *54*, 4288–4296. [CrossRef]
58. Scully, J.R. Localized Corrosion of Sputtered Aluminum and Al-0.5% Cu Alloy Thin Films in Aqueous HF Solution. *J. Electrochem. Soc.* **1990**, *137*, 1365. [CrossRef]
59. Pouilleau, J.; Devilliers, D.; Garrido, F.; Durand-Vidal, S.; Mahé, E. Structure and composition of passive titanium oxide films. *Mater. Sci. Eng. B* **1997**, *47*, 235–243. [CrossRef]
60. Milošev, I.; Kosec, T.; Strehblow, H.-H. XPS and EIS study of the passive film formed on orthopaedic Ti–6Al–7Nb alloy in Hank's physiological solution. *Electrochim. Acta* **2008**, *53*, 3547–3558. [CrossRef]
61. Wysocka, J.; Cieslik, M.; Krakowiak, S.; Ryl, J. Carboxylic acids as efficient corrosion inhibitors of aluminium alloys in alkaline media. *Electrochim. Acta* **2018**, *289*, 175–192. [CrossRef]
62. Amin, M.A.; Ahmed, E.M.; Mostafa, N.Y.; Alotibi, M.M.; Darabdhara, G.; Das, M.R.; Wysocka, J.; Ryl, J.; Abd El-Rehim, S.S. Aluminum Titania Nanoparticle Composites as Nonprecious Catalysts for Efficient Electrochemical Generation of H_2. *ACS Appl. Mater. Interfaces* **2016**, *8*, 23655–23667. [CrossRef]
63. Weibin, Z.; Weidong, W.; Xueming, W.; Xinlu, C.; Dawei, Y.; Changle, S.; Liping, P.; Yuying, W.; Li, B. The investigation of NbO_2 and Nb_2O_5 electronic structure by XPS, UPS and first principles methods: The investigation of NbO_2 and Nb_2O_5 electronic structure. *Surf. Interface Anal.* **2013**, *45*, 1206–1210. [CrossRef]
64. Kharitonov, D.S.; Sommertune, J.; Örnek, C.; Ryl, J.; Kurilo, I.I.; Claesson, P.M.; Pan, J. Corrosion inhibition of aluminium alloy AA6063 T5 by vanadates: Local surface chemical events elucidated by confocal Raman micro-spectroscopy. *Corros. Sci.* **2019**, *148*, 237–250. [CrossRef]
65. Kumar, S.; Kumar, S.; Tiwari, S.; Srivastava, S.; Srivastava, M.; Yadav, B.K.; Kumar, S.; Tran, T.T.; Dewan, A.K.; Mulchandani, A.; et al. Biofunctionalized Nanostructured Zirconia for Biomedical Application: A Smart Approach for Oral Cancer Detection. *Adv. Sci.* **2015**, *2*, 1500048. [CrossRef] [PubMed]
66. Siuzdak, K.; Szkoda, M.; Karczewski, J.; Ryl, J.; Darowicki, K.; Grochowska, K. Fabrication and Significant Photoelectrochemical Activity of Titania Nanotubes Modified with Thin Indium Tin Oxide Film. *Acta Metall. Sin. (Engl. Lett.)* **2017**, *30*, 1210–1220. [CrossRef]
67. Mandrino, D.; Godec, M.; Torkar, M.; Jenko, M. Study of oxide protective layers on stainless steel by AES, EDS and XPS. *Surf. Interface Anal.* **2008**, *40*, 285–289. [CrossRef]

68. Wang, C.; Irfan, I.; Liu, X.; Gao, Y. Role of molybdenum oxide for organic electronics: Surface analytical studies. *J. Vac. Sci. Technol. Bnanotechnol. Microelectron. Mater. Process. Meas. Phenom.* **2014**, *32*, 040801. [CrossRef]
69. Liu, J.; Alfantazi, A.; Asselin, E. Effects of Temperature and Sulfate on the Pitting Corrosion of Titanium in High-Temperature Chloride Solutions. *J. Electrochem. Soc.* **2015**, *162*, C189–C196. [CrossRef]

materials

MDPI

Article

Relationship between Phase Occurrence, Chemical Composition, and Corrosion Behavior of as-Solidified Al–Pd–Co Alloys

Marián Palcut, Libor Ďuriška, Ivona Černičková, Sandra Brunovská, Žaneta Gerhátová, Martin Sahul, Ľubomír Čaplovič and Jozef Janovec *

Faculty of Materials Science and Technology in Trnava, Slovak University of Technology in Bratislava, J. Bottu 25, 917 24 Trnava, Slovakia; marian.palcut@stuba.sk (M.P.); libor.duriska@stuba.sk (L.Ď.); ivona.cernickova@stuba.sk (I.Č.); brunovskas@gmail.com (S.B.); zaneta.gerhatova@stuba.sk (Ž.G.); marian.sahul@stuba.sk (M.S.); lubomir.caplovic@stuba.sk (Ľ.Č.)

* Correspondence: jozef.janovec@stuba.sk; Tel.: +421-918-646-072

Received: 18 April 2019; Accepted: 20 May 2019; Published: 22 May 2019

Abstract: The microstructure, phase constitution, and corrosion performance of as-solidified $Al_{70}Pd_{25}Co_5$ and $Al_{74}Pd_{12}Co_{14}$ alloys (element concentrations in at.%) have been investigated in the present work. The alloys were prepared by arc-melting of Al, Pd, and Co lumps in argon. The $Al_{74}Pd_{12}Co_{14}$ alloy was composed of structurally complex ε_n phase, while the $Al_{70}Pd_{25}Co_5$ alloy was composed of ε_n and δ phases. The corrosion performance was studied by open circuit potential measurements and potentiodynamic polarization in aqueous NaCl solution (3.5 wt.%). Marked open circuit potential oscillations of the $Al_{70}Pd_{25}Co_5$ alloy have been observed, indicating individual breakdown and re-passivation events on the sample surface. A preferential corrosion attack of ε_n was found, while the binary δ phase (Al_3Pd_2) remained free of corrosion. A de-alloying of Al from ε_n and formation of intermittent interpenetrating channel networks occurred in both alloys. The corrosion behavior of ε_n is discussed in terms of its chemical composition and crystal structure. The corrosion activity of ε_n could be further exploited in preparation of porous Pd–Co networks with possible catalytic activity.

Keywords: aluminum alloys; phase characterization; electrochemical corrosion; de-alloying

1. Introduction

Alloys with nominal chemical composition of approximately Al-30 at.% TM (TM stands for one or more transition metals) constitute a specific group of materials called complex metallic alloys (CMAs). These metallic materials contain, besides classical crystalline phases with simple unit cells, structurally complex intermetallic phases (SCIPs) [1]. The SCIPs are composed of giant unit cells and lack translational symmetry. Because of their complex atomic structure, the SCIPs are appealing for thin film applications, coatings, and reinforcement phases in composites [2].

The phase equilibria in the Al–Pd–Co system have been studied by Yurechko et al. [3,4], Černičková et al. [5,6], and Ďuriška et al. [7]. The authors observed six stable ternary phases (W, Y_2, U, V, F, C_2) and a structurally complex ε-family. Selected phases occurring in the Al–Pd–Co system are summarized in Table 1 [3,8]. Their homogeneity ranges at 790 °C are shown in the corresponding isothermal phase diagram section (Figure 1). The cluster-based orthorhombic decagonal quasicrystalline approximant of the ε-family consists of five structures: two binary (ε_6, ε_{28}) and three ternary (ε_{16}, ε_{22}, ε_{34}). Since the ε-family is considered to be a single phase from the thermodynamic point of view, it has been briefly denoted as ε_n. Although two lattice parameters (*a* and *b*) are identical for each structure within the family, the third lattice parameter (*c*) differs for each of the structures since it is associated with the cluster arrangement [3,9]. Contrary to Al–Pd alloys [10], Co was observed to substitute Pd in ε_n in ternary Al–Pd–Co alloys. The Co solubility in ε_n is up to approximately 15 at.% at 790 °C [3]. The Al–Pd–TM ε_n phase is predominantly diamagnetic and has a good electrical conductivity. This phase is brittle and can be easily powdered. Furthermore, it contains Pd, a catalytically active element, which, in combination with a unique crystal structure, provides a variety of different adsorption sites. As such, the Al–Pd–Co SCIPs are interesting for catalytic applications [11].

Table 1. Overview of selected binary and ternary phases occurring in the Al–Pd–Co system and related binaries [3,8].

Phase	Space Group/Symmetry	Lattice Parameter			
		a (nm)	b (nm)	c (nm)	β (°)
		Al–Pd–Co system			
W	$Pmn2_1$	2.36	0.82	2.07	-
Y_2	*Immm*	1.5451	1.2105	0.7590	-
U	C121, C1*m*1 or C12/*m*1	1.9024	2.9000	1.3140	117.26
V	P121, P1*m*1 or P12/*m*1	1.0068	0.3755	0.6512	102.38
F	$P2_1/a\bar{3}$	2.4397	-	-	-
C_2	$Fm\bar{3}$	1.5507	-	-	-
ε_{16}	*Amm*2	2.35	1.68	3.26	-
ε_{22}	orthorhombic	2.35	1.68	4.49	-
ε_{34}	orthorhombic	2.35	1.68	7.01	-
		Al–Pd system			
ε_6	$Pna2_1$	2.35	1.68	1.23	-
ε_{28}	*C*2*mm*	2.35	1.68	5.70	-
Al_3Pd_2 (δ)	$P\bar{3}m1$	0.4227	-	0.5167	-
AlPd (β)	$Pm\bar{3}m$	0.3036	-	-	-
		Al–Co system			
Al_9Co_2	$P2_1/a$	0.85565	0.6290	0.62130	94.76
O-$Al_{13}Co_4$	$Pmn2_1$ or *Pnmn*	0.8158	1.2347	1.4452	-
M-$Al_{13}Co_4$	*C*2/*m*	1.5173	0.81090	1.2349	107.84
Z-Al_3Co	*monoclinic*	3.984	0.8148	3.223	107.97
Al_5Co_2	$P6_3/mmc$	0.76717	-	0.76052	-
AlCo (β)	$Pm\bar{3}m$	0.2854	-	-	-

The corrosion activity of Al-based SCIPs is relatively unknown. It has been found that the electrochemical properties of CMAs differ from those of aluminum metal [12]. The previous studies of Al–Cu–Fe [13,14], Al–Cr–Fe [15], and Al–Cu–Fe–Cr [16] CMAs indicated that the relative amount of alloy phases and their chemical compositions had a major influence on their electrochemical behavior. It was presented that Cr additions significantly improved the corrosion resistance of Al–Cr–Fe and Al–Cu–Fe–Cr alloys [16]. Recent studies of Al–Co CMAs [17–21] have shown that the relative amounts of the alloy's phases and electrical contact between them played an important role in their corrosion performance. The anodic dissolution of different alloy phases was found to take place by a galvanic mechanism. The electrochemical nobility of Al–Co SCIPs was found to increase with increasing Co concentration. The phase crystal structure had only a secondary influence. An exception, however, was found for the structurally complex Z-Al_3Co phase. This phase was found to be more corrosion resistant compared to Al_5Co_2 in chloride-containing environments [19]. The reason for this behavior could stem from the complex crystal structure of Z-Al_3Co, formed by a complex monoclinic unit cell containing large pentagons composed of six small pentagons of monoclinic $Al_{13}Co_4$. The complex structure of this phase is probably stabilized by vacancies. The vacancies may influence the Co diffusivity leading to a protective layer formation on the sample surface.

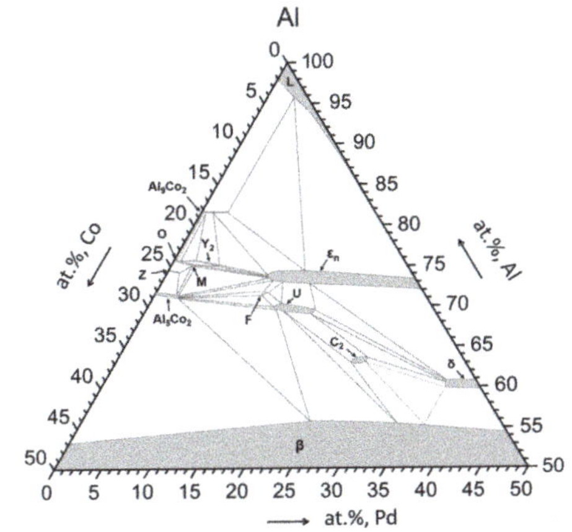

Figure 1. Isothermal section of the Al–Pd–Co phase diagram at 790 °C, redrawn from Reference [3].

The corrosion behavior of Al–Pd alloys in various solutions has been studied in References [22–25]. The results showed a preferential Al dissolution from ε_n (~Al_3Pd). The corrosion attack of the structurally complex ε_n in the Al–Pd alloys led to the formation of a porous, channel-like network [22–25]. This phenomenon is known as electrochemical de-alloying [26], i.e., a corrosion-driven process during which an alloy is decomposed by selective dissolution of the most electrochemically active element (Al). This process results in the formation of nano-porous metal networks composed of noble elements. In the NaCl aqueous solution, the de-alloying of Al–Pd alloys was found to be more pronounced in as-solidified alloys compared to as-annealed samples [24]. The de-alloying of Al–TM alloys has attracted much attention in recent years as a versatile tool for creating nano-porous metal networks with high catalytic activity [25]. Nano-porous ribbons of Pd, Au, Pt, and other precious metals have been fabricated through chemical de-alloying of rapidly solidified Al-based alloys under free corrosion conditions [27].

In the present work, the corrosion performance of $Al_{70}Pd_{25}Co_5$ and $Al_{74}Pd_{12}Co_{14}$ alloys (element concentrations are given in at.%) have been studied by potentiodynamic polarization in 3.5 wt.% NaCl aqueous solution for the first time. The aim of this work is to investigate the effect of both phase occurrence and chemical composition on the alloy's corrosion behavior. Furthermore, the effect of Co concentration on the corrosion behavior of ε_n is studied.

2. Materials and Methods

The alloys with nominal compositions $Al_{70}Pd_{25}Co_5$ and $Al_{74}Pd_{12}Co_{14}$ were prepared by repeated arc-melting of Al, Pd, and Co granules (purity of 99.95%) in argon. After melting, the alloys were rapidly solidified on a water-cooled copper mold, cast in epoxy resin, and metallographically prepared by wet grinding and polishing down to a surface roughness of 1 μm. The as-solidified alloy's phase constitution and microstructure were studied by room temperature X-ray diffraction (XRD) and scanning electron microscopy (SEM), respectively. During XRD experiments, a Panalytical Empyrean PIXCel 3D diffractometer (Malvern Panalytical Ltd., Malvern, UK) with Bragg–Brentano geometry and Co Kα1,2 radiation was used. The measurements were conducted in the 2θ range between 20° and 60°, with the step size 0.0131° and the counting time 98 s per step. For the microstructure observation, a JEOL JSM-7600F scanning electron microscope (JEOL, Akishima, Tokyo, Japan), equipped with an Oxford Instruments X-max 50 spectrometer (Oxford Instruments, Abingdon, UK) and operated by the INCA software (version 5.04), was employed. The microscope was operated at the acceleration voltage of 20 kV. The scanning was performed in regimes of secondary (SEI) and backscattered (BEI) electrons. Furthermore, a scanning transmission electron microscope JEOL JEM ARM200F (JEOL,

Akishima, Tokyo, Japan), operated at 200 kV and equipped with a high-angle annular dark-field detector (HAADF/STEM), was employed to obtain HAADF images. The two-dimensional (2D) projections of crystal structures were calculated in PowderCell software (version 2.4) using the data derived from References [8,28,29].

The corrosion experiments were conducted at room temperature (21 ± 2 °C) in a 500 ml glass vessel filled with an aqueous electrolyte. A three-electrode setup was used. The working electrode consisted of the polished surface of the Al–Pd–Co alloy with an exposed area of about 1 cm^2. A silver–silver chloride electrode immersed in a saturated sodium chloride solution (saturated Ag/AgCl electrode) was used as a reference electrode. The counter electrode was a platinum mesh (2 × 2 cm^2). The corrosion experiments were conducted in an aqueous NaCl solution (concentration 0.6 mol dm^{-3}). The solution was prepared immediately before the experiment by dissolving 35 g of NaCl in 1 liter of de-ionized water (conductivity <20 μS). The electrolyte was not de-aerated before the experiment to simulate real environmental conditions. The progress of the reaction was controlled by a PGU 10 V-1A-IMP-S potentiostat/galvanostat from Jaissle Electronic Ltd. (Waiblingen, Germany).

The surface topography of the corroded samples was analyzed by a Zeiss LSM 700 confocal laser scanning microscope (CLSM, Zeiss, Oberkochen, Germany). The ZEN 2009 software was used for the three-dimensional topographical resolution.

3. Results and Discussion

3.1. Microstructure and Phase Occurrence before Corrosion Testing

The microstructures of the as-solidified $Al_{70}Pd_{25}Co_5$ and $Al_{74}Pd_{12}Co_{14}$ alloys are illustrated in Figure 2. The XRD patterns corresponding to the above alloys are given in Figure 3a,b, respectively. The metal concentrations of microstructure constituents determined by SEM/EDX and their phase assignments are presented in Table 2.

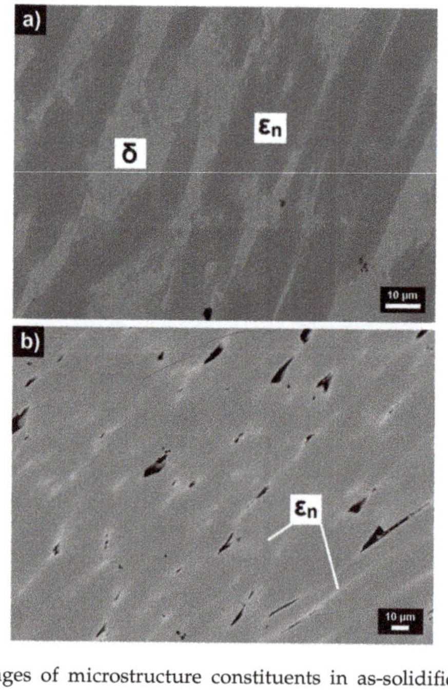

Figure 2. BEI/SEM images of microstructure constituents in as-solidified $Al_{70}Pd_{25}Co_5$ (**a**) and $Al_{74}Pd_{12}Co_{14}$ (**b**) alloys. Black areas in (**b**) correspond to pores. Phases assigned to particular constituents are also marked.

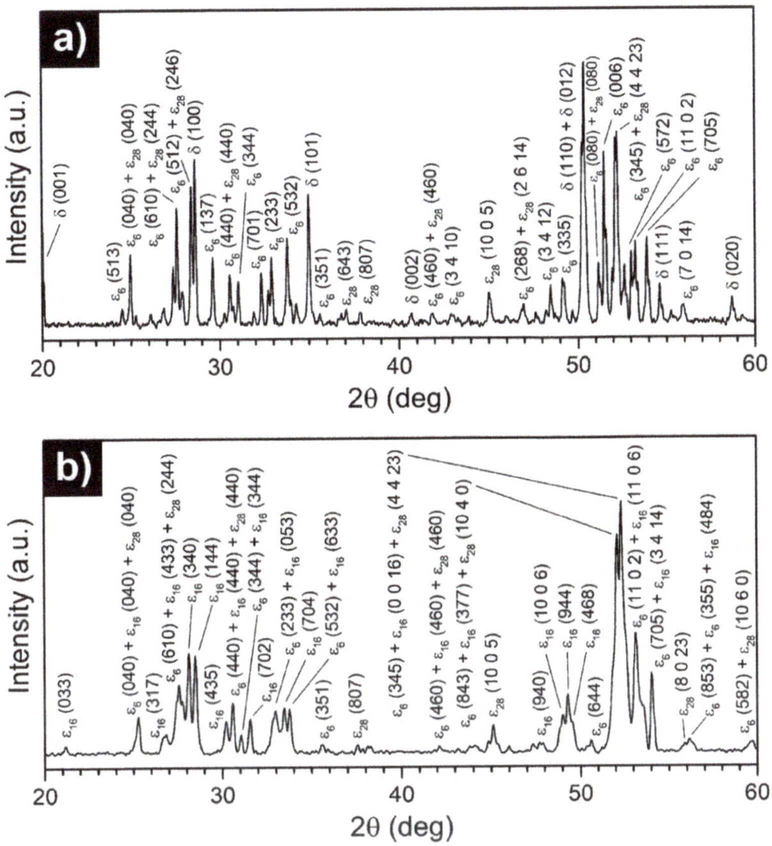

Figure 3. XRD diffraction patterns of as-solidified $Al_{70}Pd_{25}Co_5$ (**a**) and $Al_{74}Pd_{12}Co_{14}$ (**b**) alloys.

Table 2. Metal concentrations and phase assignments of microstructure constituents observed in as-solidified $Al_{70}Pd_{25}Co_5$ and $Al_{74}Pd_{12}Co_{14}$ alloys.

Alloy	Alloy Condition	Identified Phase/Structure	Element Concentration (at.%)			Volume Fraction (%)
			Al	Pd	Co	
$Al_{70}Pd_{25}Co_5$	as-solidified	$\varepsilon_n/\varepsilon_6 + \varepsilon_{28}$	72.5 ± 0.1	18.9 ± 0.3	8.6 ± 0.3	77
		δ	59.8 ± 0.2	39.7 ± 0.4	0.5 ± 0.2	23
$Al_{74}Pd_{12}Co_{14}$	as-solidified	$\varepsilon_n/\varepsilon_6 + \varepsilon_{16} + \varepsilon_{28}$	73.9 ± 1.0	12.0 ± 5.5	14.1 ± 4.6	100

The microstructure of the $Al_{70}Pd_{25}Co_5$ alloy consisted of two different constituents (Figure 2a). The images were acquired in a BEI regime and therefore the bright regions have a higher Pd concentration compared to the dark constituents. The metal concentrations and volume fractions of the bright-grey microstructure constituent (Table 2) indicate that it corresponds to the δ phase (Al_3Pd_2). This assumption was also confirmed by X-ray diffraction (Figure 3a). The visually and chemically homogeneous dark-grey constituent was identified to be a mixture of ε_n structures (Figure 3a). To index diffraction peaks of particular ε_n structures, the data derived from References [8,28,29] were used.

In the XRD pattern of the $Al_{74}Pd_{12}Co_{14}$ alloy (Figure 3b), a combination of ε_6, ε_{16}, and ε_{28} structures was identified. In the related microstructure image, however, a chemically heterogeneous constituent has been observed (Figure 2b). The dark-grey areas had an increased Co concentration, while the bright areas showed a higher Pd concentration compared to the dark-grey areas. The atomic structure of the as-solidified $Al_{74}Pd_{12}Co_{14}$ alloy was observed using HAADF/STEM. Three different structural motives have been recognized in the atomic structure of this alloy (ε_6, ε_{16}, and ε_{28}, Figure 4). For each ε_n structure, specific combinations of phason tiles are characteristic. ε_6 is formed by hexagons only, ε_{16} is represented by the combination of pentagons and nonagons, while ε_{28} comprises all three types of tiles. It has been suggested that transitions between various structures of the ε-family could be associated with a small rearrangement of clusters, resulting in changes in the occurrence and/or configuration of phason tiles. The arrangement of tiles in particular ε_n structures, observed experimentally in this work, was also calculated using the data derived from References [8,28,29]. The 2D projection of the ε_6, ε_{16}, and ε_{28} structures, presented in Figure 5, is in a good agreement with the HAADF/STEM image.

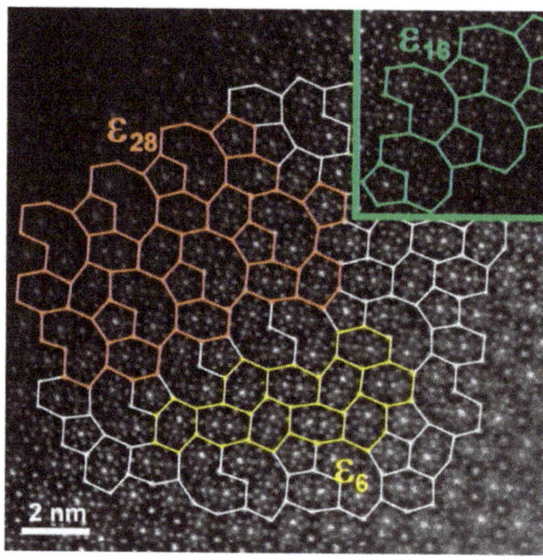

Figure 4. A high-angle annular dark-field (HAADF)/STEM image of the atomic structure of the as-solidified $Al_{74}Pd_{12}Co_{14}$ alloy. Phason tiles, i.e., hexagon, pentagon, and banana-shaped nonagons, are highlighted by solid lines. Yellow, green, and orange structural motifs correspond to ε_6, ε_{16}, and ε_{28}, respectively. For the color interpretation of this figure, the reader is referred to the web version of this article.

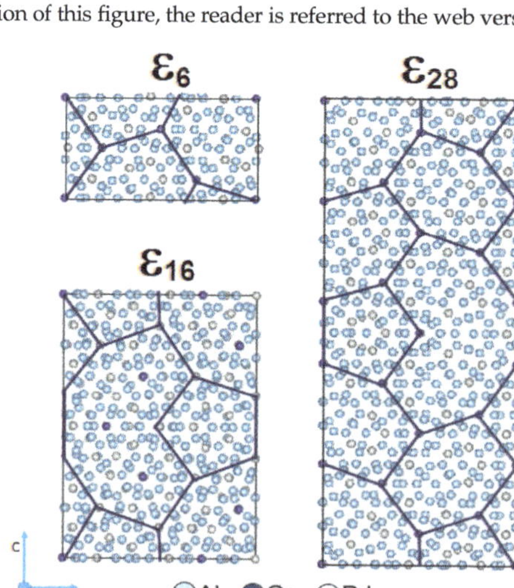

Figure 5. Two-dimensional projection of the crystal structure of ε_6, ε_{16}, and ε_{28}. The phason tiling is denoted by dark-blue lines. For the color interpretation of this figure, the reader is referred to the web version of this article.

Structures of ε_6 and ε_{28} were reported to be binary structural variants of ε_n, while ε_{16} has been described as a ternary ε_n structure [3,10,29,30]. In the latter structure, Co atoms substitute Pd. Therefore, the dark-grey areas (Figure 2b, Table 2), enriched with Co from the Co–Pd balance point of view, could be assigned to the ternary ε_{16} structure in the as-solidified $Al_{74}Pd_{12}Co_{14}$ alloy. Similarly, the bright areas in Figure 2b could be ascribed to the mixture of ε_6 and ε_{28} structures, which lie closer to the Al–Pd binary edge of the Al–Pd–Co ternary system. The bright areas were located preferentially around pores. The pores were formed on the grain boundaries during solidification due to shrinking. Co and Pd concentrations of ε_n changed since de-mixing took place during solidification. The Pd concentration in ε_n increased towards the grain boundary. Thus, the Pd-rich ε_n (ε_6 + ε_{28}) were located preferentially around pores. The Co-rich ε_n (ε_{16}) was located in the center of the grain as this phase structure solidified from the melt. The overall chemical composition of the ε_n phase in the $Al_{74}Pd_{12}Co_{14}$ alloy is presented in Table 2. Due to the presence of ε_{16}, the ε_n phase in the $Al_{74}Pd_{12}Co_{14}$ alloy had a significantly higher Co concentration compared to the $Al_{70}Pd_{25}Co_5$ alloy where the ternary ε_{16} phase has not been identified.

The distributions of particular structures within the ε_n phase were previously studied in the Al–Pd and Al–Pd–Co systems; however, the exact boundaries between structures have not been determined yet. In the Al–Pd system, Yurechko et al. [10] proposed a hypothetical double-phase area (ε_6 + ε_{28}) in between two single-phase areas (ε_6 and ε_{28}). In the partial phase diagram published by Grushko [31], ε_6 and ε_{28} have been positioned in a common "single-phase (ε_6 + ε_{28})" area consisting of two presumably separated subareas adherent to the particular structures. Earlier, the same distribution of ε_6 and ε_{28} was studied by Balanetskyy et al. [28] in the Al–Pd–Fe system at 750 °C. Moreover, the homogeneity ranges of ε_{16} and ε_{22} have been defined. However, the strict boundaries between particular structures have not been described. In the Al–Pd–Co system, Yurechko et al. [3] estimated the boundaries of all the structures within the ε-family. Considering the results obtained using HAADF/STEM in this work and in [29], it can be assumed that the transitions between particular structures are rather open as schematically highlighted in gradient colors (green, red, yellow, and blue) in Figure 6. As follows from this figure, several ε_n structures in the transient area can coexist. This situation can also be seen in the microstructure of the $Al_{74}Pd_{12}Co_{14}$ alloy. The dark-grey areas, corresponding to the ε_{16} structure, fluently transformed to the bright areas represented by the mixture of ε_6 and ε_{28} structures (Figure 2b). The chemical composition of ε_6 is very close to the composition of ε_{28}. Consequently, this bright-grey microstructure constituent in the $Al_{70}Pd_{25}Co_5$ alloy (Figure 2a) can be considered to be homogeneous. Individual ε_6 and ε_{28} structures can be recognized in the HAADF/STEM image only (Figure 4).

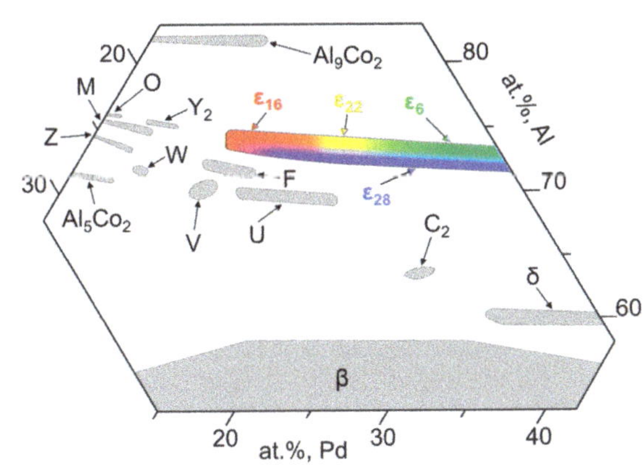

Figure 6. Schematic positions of binary and ternary phases in the isothermal section of a partial Al–Pd–Co diagram, redrawn from Reference [3]. For the color interpretation of this figure, the reader is referred to the web version of this article.

3.2. Corrosion Behavior

Immediately after the sample's immersion in aqueous NaCl, an open circuit potential (OCP) was recorded. The OCPs of the alloys are presented in Figure 7. A distinct behavior has been observed. While the OCP of the $Al_{74}Pd_{12}Co_{14}$ alloy was relatively stable over time, irregular oscillations for the $Al_{70}Pd_{25}Co_5$ alloy have been found. Furthermore, the OCPs of the $Al_{70}Pd_{25}Co_5$ alloy were less negative and a difference of more than 200 mV was found compared to the $Al_{74}Pd_{12}Co_{14}$ alloy.

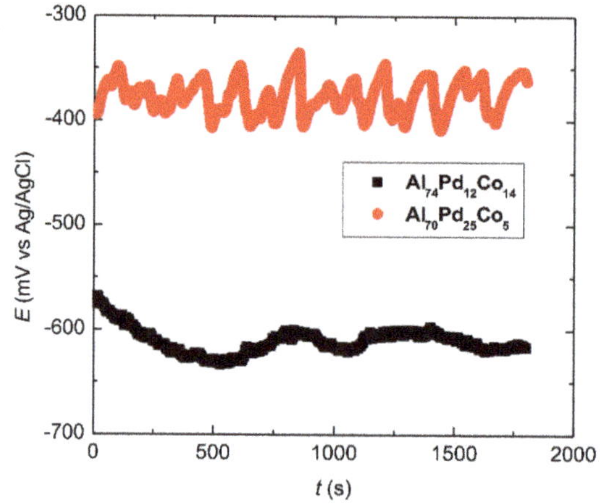

Figure 7. Open circuit potential of the as-solidified $Al_{74}Pd_{12}Co_{14}$ and $Al_{70}Pd_{25}Co_5$ alloys in 0.6 M NaCl. For the color interpretation of this figure, the reader is referred to the web version of this article.

The as-solidified $Al_{74}Pd_{12}Co_{14}$ alloy is a single-phase alloy. The OCP of this alloy therefore corresponds to the electrochemical activity of ε_n. The $Al_{70}Pd_{25}Co_5$ alloy, on the other hand, is a double-phase alloy composed of ε_n and δ (Al_3Pd_2). The less negative OCP of this alloy indicates a higher electrochemical potential of δ. Because of the potential difference between ε_n and δ, local galvanic cells may have been formed on the surface of the $Al_{70}Pd_{25}Co_5$ alloy.

Every physical contact between δ and ε_n corresponds to an elementary galvanic corrosion cell. During galvanic corrosion, there is a net current flow between the cathodic microstructure constituent (δ) and its adjacent matrix (ε_n). The metal ions dissolve into the solution on the anode and electrons released flow to the micro-cathodic area for the reduction process. This causes a redistribution of electrical charge between anodic (ε_n) and cathodic areas (δ), thereby leading to a variation of the OCP. As the OCP is measured at the tip of the Haber–Luggin capillary, it represents the overall contributions of all elementary galvanic cells on the sample surface [32]. These contributions are not correlated. A high number of elementary galvanic cells between ε_n and δ co-exist with each other in the microstructure of the $Al_{70}Pd_{25}Co_5$ alloy (Figure 2a). Their interactions are combined and contribute to the overall corrosion behavior of this alloy.

A further insight into the peculiar corrosion behavior of the Al–Pd–Co alloys was obtained by potentiodynamic polarization. After the OCP measurement, a polarization scanning from −1000 mV to 0 mV (Ag/AgCl) was performed using a sweeping rate of 1 mV s^{-1}. After reaching 0 mV (Ag/AgCl), the polarization direction was reversed and returned back to the initial potential (the direction of the polarization is indicated by open arrows in Figure 8). The resulting cyclic polarization curves are depicted in Figure 8. The forward curves are characterized by the corrosion minimum followed by an increase of the current density at potentials less negative than the corrosion potential. The current density increase was further followed either by stabilization (the $Al_{74}Pd_{12}Co_{14}$ alloy) or even a slight decrease of the current density (the $Al_{70}Pd_{25}Co_5$ alloy). This behavior indicates a passivation of the alloys. The transient behavior was further followed by a sudden current density increase at potentials less negative than −400 mV (Ag/AgCl), indicating a breakdown of the passive film. Upon reverse polarization, a re-passivation of the existing pits occurred. In order to compare equally polarized samples, we reversed the scanning at the fixed potential. The forward curves presented in Figure 8 have been analyzed by Tafel extrapolation [33]. The electrochemical parameters of the alloys (corrosion potential, corrosion current density, and breakdown potential) are listed in Table 3. A re-passivation potential obtained from the reverse curve is also presented. However, caution is required when comparing the individual re-passivation potentials of the alloys. The currents at the vertex were higher for the $Al_{74}Pd_{12}Co_{14}$ alloy and this might have influenced the pit depth and local chemistry [33]. In order to obtain more comparable E_{rp} values, reversing the polarization at a constant current density would be necessary.

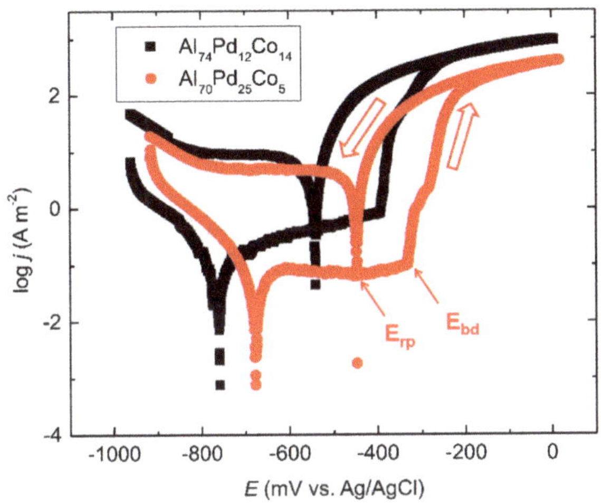

Figure 8. Cyclic potentiodynamic polarization curves of the as-solidified $Al_{74}Pd_{12}Co_{14}$ and $Al_{70}Pd_{25}Co_5$ alloys in 0.6 M NaCl. The polarization direction and positions of breakdown (E_{bd}) and re-passivation potentials (E_{rp}) are indicated by arrows. For the color interpretation of this figure, the reader is referred to the web version of this article.

Table 3. Electrochemical parameters of as-solidified $Al_{70}Pd_{25}Co_5$ and $Al_{74}Pd_{12}Co_{14}$ alloys. Corrosion potentials (E_{corr}) and corrosion current densities (j_{corr}) were obtained by Tafel extrapolation of forward curves (Figure 8).

Alloy	OCP (mV vs. Ag/AgCl)	E_{corr} (mV vs. Ag/AgCl)	j_{corr} (A m^{-2})	E_{bd} (mV vs. Ag/AgCl)	E_{rp} (mV vs. Ag/AgCl)
$Al_{70}Pd_{25}Co_5$	−370 ± 35	−677	0.101	−332	−447
$Al_{74}Pd_{12}Co_{14}$	−607 ± 9	−758	0.176	−393	−540

Based on the above-presented results, a corrosion mechanism of the Al–Pd–Co alloys has been postulated. The corrosion mechanism is depicted in Figure 9. Pitting is a highly localized form of corrosion that happens in the presence of halide anions, such as Cl^- [34]. Initially, a protective alumina scale has been formed on the sample surface, which is indicated by a current density plateau observed upon sample polarization for both alloys. This plateau is observed at potentials of −600 to −300 mV versus Ag/AgCl for $Al_{70}Pd_{25}Co_5$ alloy, i.e., at potentials less negative than is the corrosion potential (Figure 8). In the presence of Cl^-, however, this passive layer has been weakened. Aluminum forms unstable $[AlCl_4]^-$ complexes that dissolve in aqueous solutions. The dissolution of the protective alumina scale in NaCl leaves a naked alloy surface susceptible to further corrosion attack (Figure 9).

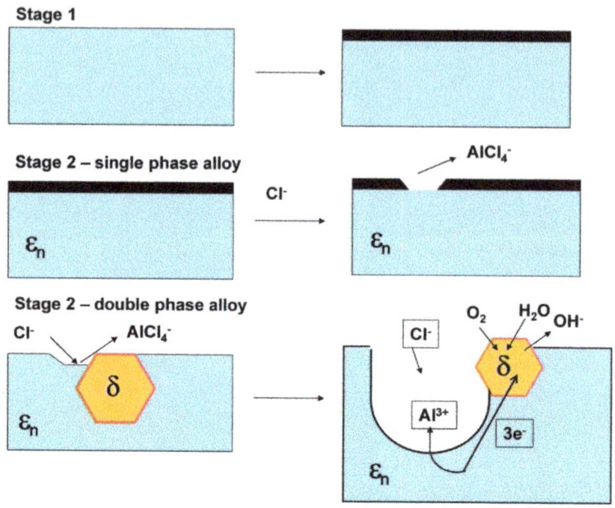

Figure 9. Pitting corrosion mechanism of the Al–Pd–Co alloys. For the color interpretation of this figure, the reader is referred to the web version of this article.

Interactions between co-existing phases in double-phase alloys may play an important role during corrosion [32,35,36]. Once the pitting potential is reached during sample polarization, the compact passivation layer becomes locally disrupted (Figure 9). As a result, Al^{3+} cations are released from the alloy into the solution in the course of the following reaction

$$Al \rightarrow Al^{3+} + 3e^- \quad (1)$$

Reaction (1) leads to positive charge enrichment within the dissolution zone [37]. As a consequence, Cl^- anions of the electrolyte rapidly migrate into the dissolution zone as presented in Figure 9. The released Al^{3+} cations become solvated by water molecules. Consequently, the hydrolysis of $[Al(H_2O)_4]^{3+}$ in aqueous environment takes place in line with the following reaction [34]

$$[Al(H_2O)_4]^{3+} + H_2O \leftrightarrow [Al(H_2O)_3(OH)]^{2+} + H_3O^+ \quad (2)$$

Hydroxo complexes of Al may further react with chloride and water according to the following reactions

$$[Al(H_2O)_3(OH)]^{2+} + Cl^- \rightarrow [Al(H_2O)_2(OH)Cl]^+ + H_2O \quad (3)$$

$$[Al(H_2O)_2(OH)Cl]^+ + H_2O \rightarrow [Al(H_2O)(OH)_2Cl] + H_3O^+ \quad (4)$$

By these reactions, hydrogen cations are released into the pit. Their accumulation yields to a local pH decrease within the dissolution zone, which is known as a self-acidifying effect [34,37]. The presence of H^+ in pits further accelerates the Al dissolution. At the cathode (δ, Figure 9), a reduction of water may take place in accordance with the following reaction

$$O_2 + 2H_2O + 4e^- \rightarrow 4OH^- \quad (5)$$

At the pit walls and possibly in their immediate vicinity, since pH is reduced due to hydrolysis reactions (2) and (4), the most likely prevailing cathodic reactions are

$$O_2 + 4H^+ + 4e^- \rightarrow 2H_2O \quad (6)$$

or

$$2H^+ + 2e^- \rightarrow H_2 \quad (7)$$

The latter reaction takes place in the case of pH having fallen to very low values. As a result, emerging bubbles of H_2 evolve on the alloy surface.

The results presented in Table 3 show that the OCP of the double-phase $Al_{70}Pd_{25}Co_5$ alloy is located between the breakdown and re-passivation potentials. This is also manifested by the OCP oscillations observed between −330 mV (Ag/AgCl) and −400 mV (Ag/AgCl, Figure 7), indicating periodic breakdown and re-passivation events on the sample surface. These observations indicate that this alloy was in a localized corrosion stage already upon the sample's immersion in the electrolyte, contrary to the single-phase $Al_{74}Pd_{12}Co_{14}$ alloy. The passivation stage was found to be more pronounced in the case of the double-phase $Al_{70}Pd_{25}Co_5$ alloy (Figure 8). This could be related to the presence of the noble δ phase in this alloy. For the mono-phasic $Al_{74}Pd_{12}Co_{14}$ alloy, on the other hand, a higher corrosion current density (j_{corr}) has been found, reflecting a higher dissolution rate. Moreover, a more negative corrosion potential for this alloy has been found, indicating a higher corrosion susceptibility. The above-reported differences in the corrosion behavior of the alloys could result from their different microstructures. More information about the specific corrosion attack of different SCIPs has been therefore obtained by investigating the alloys' microstructures after electrode polarization.

The microstructures of the as-polarized alloys are documented in Figure 10. Metal concentrations of the phases after corrosion are summarized in Table 4. For both alloys, a preferential attack of ε_n was found. δ as a nobler phase in the $Al_{70}Pd_{25}Co_5$ alloy has been retained. A de-alloying of Al from ε_n as well as formation of intermittent inter-penetrating channel networks have been observed in both alloys (Figure 10). In the single-phase $Al_{74}Pd_{12}Co_{14}$ alloy, a higher density of intermittent inter-penetrating channels and pits has been found (Figure 10). Moreover, the channels formed a cross-linked network. This behavior is similar to the Al–Pd alloys, where the pits were observed in the interconnection between two channels [24]. In the $Al_{70}Pd_{25}Co_5$ alloy, the pits were observed to be randomly distributed in the channels (Figure 10).

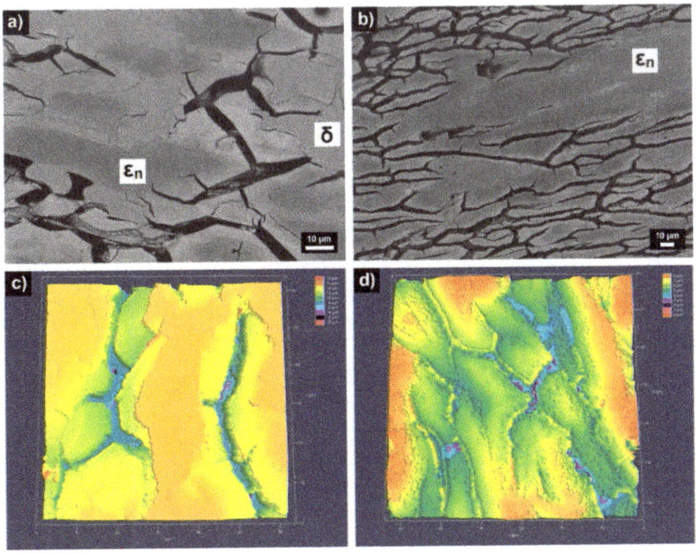

Figure 10. BEI/SEM images (**a**,**b**) and CLSM images (**c**,**d**) of as-polarized $Al_{70}Pd_{25}Co_5$ (**a**,**c**) and $Al_{74}Pd_{12}Co_{14}$ (**b**,**d**) alloys. Phases assigned to particular constituents are also marked. For the color interpretation of this figure, the reader is referred to the web version of this article.

Table 4. Metal concentrations and phase assignments of the as-polarized Al–Pd–Co alloys. Differences in metal concentrations between as-polarized and as-solidified alloys are also presented (compare with data in Table 2).

Alloy	Identified Phase/Structure	Element concentration (at.%)					
		Al	Δ(Al)	Pd	Δ(Pd)	Co	Δ(Co)
$Al_{70}Pd_{25}Co_5$	$\varepsilon_n/\varepsilon_6 + \varepsilon_{28}$	69.0 ± 0.3	−3.5	22.2 ± 0.4	+ 3.3	8.8 ± 0.4	-
	δ	60.0 ± 0.3	-	39.4 ± 0.3	-	0.6 ± 0.2	-
$Al_{74}Pd_{12}Co_{14}$	$\varepsilon_n/\varepsilon_6 + \varepsilon_{16} + \varepsilon_{28}$	71.1 ± 0.9	−2.8	14.3 ± 5.3	+ 2.3	14.6 ± 4.3	-

A dissolution of Al in the $Al_{70}Pd_{25}Co_5$ alloy has been found (Table 4). Simultaneously, the Al concentration in ε_n decreased from 72.5 to 69.0 at.% (Table 4). In the $Al_{74}Pd_{12}Co_{14}$ alloy, a decrease of Al concentration in ε_n from 73.9 to 71.1 at.% has been found. Thus, the level of Al de-alloying was higher in the double phase $Al_{70}Pd_{25}Co_5$ alloy. Moreover, the pits found in this alloy were deeper compared to the $Al_{74}Pd_{12}Co_{14}$ alloy. The formation of cracks observed in the as-polarized alloys could be governed by a combination of de-alloying kinetics and the release rate of internal stresses. As the electrochemical potential is raised in a positive direction, the dissolution rate of the alloy increases (Figure 8). This electrochemical force drives the surface at the de-alloying front further away from the equilibrium [37]. The removal of Al from the alloy phases leads to microcrack initiation. The residual stress accumulated in the alloys during rapid solidification is released during de-alloying.

Corrosion potentials and corrosion current densities of the as-solidified Al–Co [17,18,20,21,38], Al–Pd [24], and Al–Pd–Co alloys are compared in Figure 11. The corrosion potentials of the Al–Co alloys show a significant dependence on the Al atomic fraction. They become more negative with increasing Al concentration. The corrosion potentials of the Al–Pd alloys, on the other hand, are relatively constant with respect to the alloy's overall chemical composition. They are, in fact, more negative than the corrosion potentials of the remaining two groups of alloys. The corrosion currents of the Al–Pd alloys, on the other hand, are higher compared to the Al–Co and Al–Pd–Co alloys (Figure 11b). These observations suggest that the Al–Pd alloys are less corrosion-resistant compared to both the Al–Pd–Co and Al–Co alloys. The corrosion behavior of the Al–Pd–Co alloys is closer to the behavior of the Al–Co alloys. This observation is unexpected, since both alloy groups have a different chemical composition and phase constitution. Moreover, ε_n, the preferentially corroding phase in the Al–Pd–Co alloys, is not present in the Al–Co alloys. Therefore, it can be suggested that Al_3Co SCIPs are nobler compared to binary ε_n (Al_3Pd). This is manifested by the less negative corrosion potentials of the Al–Co alloys compared to the Al–Pd alloys with a similar Al atomic fraction (Figure 11). Furthermore, the Co substitution for Pd significantly improves the corrosion resistance of ε_n. As such, it is not the crystal structure of the phase, but its chemical composition, that plays a major role in the corrosion behavior.

To further probe this hypothesis, we have plotted the corrosion data of other ternary Al–TM systems, found in the literature, together with those of the Al–Pd–Co system. The data survey [13,39–42] is presented in Figure 12. Although the data are scattered due to large variations in alloy chemical compositions, some trends can be identified. The as-solidified Al–Cu–Pd and Al–Cu–Fe alloys have lower corrosion currents compared to the Al–Pd–Co alloys [13,39]. The addition of Pd was found to slightly decrease the corrosion current of the Al–Cu alloys in chloride solution [39]. For the Al_4Cu_9 samples, however, not much effect from Pd has been seen [39]. The corrosion potentials of these alloys are found over a broad range of values. The scatter in E_{corr} values, however, could be caused by variations in their chemical composition. The Al–Cr–Fe alloy is also presented in Figure 12 [40]. This alloy has a more negative corrosion potential due to the absence of noble elements, such as Pd. Furthermore, it has a low corrosion current due to the presence of chromium, which forms a passive layer on the sample surface.

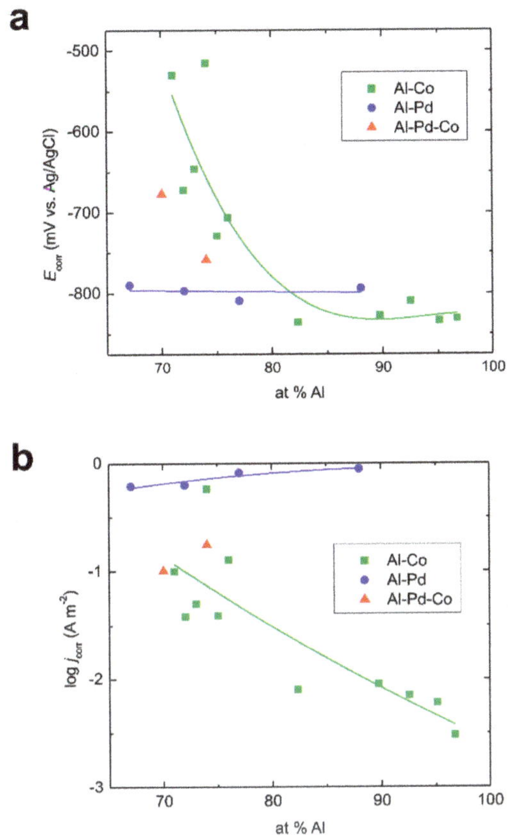

Figure 11. Corrosion potentials (**a**) and corrosion current densities (**b**) of as-solidified Al–Co, Al–Pd, and Al–Co–Pd alloys. Lines are a guide to the eyes only. For the color interpretation of this figure, the reader is referred to the web version of this article.

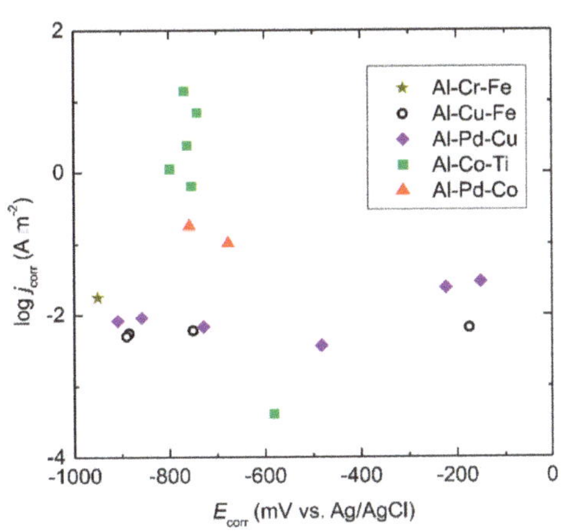

Figure 12. Corrosion current densities versus corrosion potentials of selected ternary Al-based complex metallic alloys (as-cast and as-annealed alloys only). For the color interpretation of this figure, the reader is referred to the web version of this article.

Interesting is the corrosion behavior of as-solidified Al–Co–Ti alloys [41]. These alloys have corrosion potentials comparable to those of the Al–Pd–Co alloys (Figure 12). The concentration of Ti in these alloys was fixed at 2 at.% and the atomic concentration of Co varied between 5 and 30 at.% (Al–xCo–2Ti alloys). As such, the materials design of these alloys had features typical of the Al–Co alloys [41,42]. In general, the corrosion currents of the Al–Co–Ti alloys are higher compared to those of the Al–Pd–Co alloys. An exception was found, however, for the Al–15Co–2Ti alloy since this alloy had a very low corrosion current. This difference is, however, attributable to the fact that the intermetallic particles present in this alloy (Al_9Co_2, $Al_{13}Co_4$, and Al_3Ti) were of different morphologies and volume fractions compared to the remainder of the alloys [41]. These observations indicate that the specific Co concentrations may greatly improve the corrosion performance of the Al–TM alloys. The ε_n phase in the Al–Pd–Co alloys contains a significant amount of Co. The Co additions thus contribute to the corrosion resistance of the Al–Pd–Co alloys and this is especially obvious in the case of the double phase $Al_{70}Pd_{25}Co_5$ alloy.

Al_3Ti and Al_3Fe are noble intermetallic phases with respect to the aluminum matrix [36]. The results presented in this work show that Al_3Co is also relatively noble. These phases are nobler compared to binary ε_n (Al_3Pd). Co substitution for Pd thus significantly improves the corrosion resistance of ε_n. As such, it is not the crystal structure of the phase, but its chemical composition, that plays a major role in the corrosion behavior. The electrochemical behavior of constituent phases may change over time. In a recent study, Zhu et al. studied the evolution of corrosion behavior of intermetallic phases in Al alloys over time [43]. At the early stages, the corrosion attack occurred in the form of de-alloying. However, as the time progressed, the particles became nobler as a result of Al dissolution [43]. This particle ennoblement may accelerate the galvanic dissolution of the surrounding matrix. The corrosion behavior of constituent phases may also change as a result of long-term annealing. The long-term annealing causes element redistribution and reduces stresses accumulated during rapid solidification [19]. A comparative study of as-annealed, near-equilibrium Al–Pd–Co alloys is planned and results will be reported in a future publication.

4. Conclusions

In this work, the corrosion performance of as-solidified $Al_{70}Pd_{25}Co_5$ and $Al_{74}Pd_{12}Co_{14}$ alloys was studied by open circuit potential measurements and potentiodynamic polarization in aqueous NaCl (3.5 wt.%), following an in-depth structural characterization of the alloys. The alloys were prepared by arc-melting of Pd, Al, and Co lumps in argon. Based on the results, the following conclusions can be presented:

1. The $Al_{74}Pd_{12}Co_{14}$ alloy was a single-phase alloy composed of ε_n. In this alloy, a combination of three ε_n structures was identified: ε_6, ε_{16}, and ε_{28}.
2. The $Al_{70}Pd_{25}Co_5$ alloy was a double-phase alloy composed of ε_n and δ (Al_3Pd_2). In this alloy, two ε_n structures were identified: $\varepsilon_6 + \varepsilon_{28}$.
3. Marked open circuit potential oscillations of the $Al_{70}Pd_{25}Co_5$ alloy have been observed, indicating individual breakdown and re-passivation events on the sample surface. A preferential corrosion attack of ε_n was found. Binary δ phase (Al_3Pd_2) was less affected by corrosion.
4. De-alloying of Al from ε_n and formation of intermittent inter-penetrating channel networks occurred in both alloys.
5. The corrosion attack of the $Al_{74}Pd_{12}Co_{14}$ alloy was more significant compared to the $Al_{70}Pd_{25}Co_5$ alloy and resulted in the formation of a de-alloyed and highly porous metallic network. The corrosion susceptibility of ε_n could be further utilized in preparing porous Pd–Co alloys with possible catalytic activity.
6. The Co substitution for Pd significantly improves the corrosion resistance of ε_n. As such, it is probably not the crystal structure of the phase, but its chemical composition, that plays a major role in the corrosion behavior.

7. Specific Co concentrations may greatly improve the corrosion performance of the Al–TM alloys.

Author Contributions: Conceptualization, M.P., L.D., I.C., and J.J.; Funding acquisition, J.J.; Investigation, M.P., L.D., I.C., S.B., Z.G., M.S., and L.C.; Methodology, M.P., L.D., and I.C.; Project administration, J.J.; Supervision, J.J.; Writing—original draft, M.P.; Writing—review & editing, M.P., L.D., and J.J.

Funding: This work was supported by project no. 1/0490/18 of the Grant Agency VEGA of the Slovakian Ministry of Education, Research, Science and Sport, project APVV-15-0049 of the Slovak Research and Development Agency, and project NFP313010T606 (PROGMAT) supported by European Structural Investment Funds.

Acknowledgments: Shinichi Watanabe (JEOL Ltd., Tokyo, Japan) is acknowledged for his assistance with the HAADF/STEM measurements. This paper is dedicated to the memory of our deceased fathers, Ján Palcut and Jozef Gerhát.

Conflicts of Interest: The authors declare no conflict of interest.

References

1. Steurer, W. Twenty years of structure research on quasicrystals. Part I. Pentagonal, octagonal, decagonal and dodecagonal quasicrystals. *Z. Kristallogr.* **2004**, *219*, 391–446. [CrossRef]
2. Dubois, J.M. Properties and applications of quasicrystals and complex metallic alloys. *Chem. Soc. Rev.* **2012**, *41*, 6760–6777. [CrossRef]
3. Yurechko, M.; Grushko, B.; Velikanova, T.; Urban, K. Isothermal sections of the Al-Pd-Co alloy system for 50–100 at.% Al. *J. Alloys Compond* **2002**, *337*, 172–181. [CrossRef]
4. Yurechko, M.; Grushko, B. A study of the Al-Pd-Co alloy system. *Mater. Sci. Eng. A* **2000**, *294–296*, 139–142. [CrossRef]
5. Černičková, I.; Ďuriška, L.; Priputen, P.; Janičkovič, D.; Janovec, J. Isothermal section of the Al-Pd-Co phase diagram at 850 °C delimited by homogeneity ranges of phases epsilon, U, and F. *J. Phase Equilib. Diffus.* **2016**, *37*, 301–307. [CrossRef]
6. Černičková, I.; Čička, R.; Švec, P.; Janičkovič, D.; Priputen, P.; Janovec, J. A study of phase equilibria in the Al-Pd-Co system at 700 °C. In *Aperiodic Crystals (Cairns)*, 1st ed.; Schmid, S., Ed.; Springer: Berlin, Germany, 2013; pp. 133–139.
7. Ďuriška, L.; Černičková, I.; Priputen, P.; Janovec, J. Partial experimental isothermal section of Al-Pd-Co phase diagram for Al-rich corner at 1020 °C. *J. Phase Equilib. Diffus.* **2019**, *40*, 45–52. [CrossRef]
8. Frigan, B.; Santana, A.; Engel, M.; Schopf, D.; Trebin, H.R.; Mihalkovič, M. Low-temperature structure of ξ-Al–Pd–Mn optimized by ab initio methods. *Phys. Rev. B* **2011**, *84*, 184203. [CrossRef]
9. Heggen, M.; Engel, M.; Balanetskyy, S.; Trebin, H.; Feuerbacher, M. Structural variations in ε-type Al-Pd-(Mn,Fe) complex metallic alloy phases. *Philos. Mag.* **2008**, *88*, 507–521. [CrossRef]
10. Yurechko, M.; Fattah, A.; Velikanova, T.; Grushko, B. A contribution to the Al-Pd phase diagram. *J. Alloys Compd.* **2001**, *329*, 173–181. [CrossRef]
11. Armbrüster, M.; Schlögl, R.; Grin, Y. Intermetallic compounds in heterogeneous catalysis-a quickly developing field. *Sci. Technol. Adv. Mater.* **2014**, *15*, 034803. [CrossRef]
12. Massiani, Y.; Yaazza, S.A.; Croussier, J.P.; Dubois, J.M. Electrochemical behaviour of quasicrystalline alloys in corrosive solutions. *J. Non-Cryst. Solids* **1993**, *159*, 92–100. [CrossRef]
13. Huttunen-Saarivirta, E.; Tiainen, T. Corrosion behaviour of Al–Cu–Fe alloys containing a quasicrystalline phase. *Mater. Chem. Phys.* **2004**, *85*, 383–395. [CrossRef]
14. Rüdiger, A.; Köster, U. Corrosion of Al–Cu–Fe quasicrystals and related crystalline phases. *J. Non-Cryst. Solids* **1999**, *250–252*, 898–902. [CrossRef]
15. Beni, A.; Ott, N.; Ura-Binczyk, E.; Rasinski, M.; Bauer, B.; Gille, P.; Ulrich, A.; Schmutz, P. Passivation and localised corrosion susceptibility of new Al–Cr–Fe complex metallic alloys in acidic NaCl electrolytes. *Electrochim. Acta* **2011**, *56*, 10524–10532. [CrossRef]
16. Ura-Binczyk, E.; Homazava, N.; Ulrich, A.; Hauert, R.; Lewandowska, M.; Kurzydlowski, K.J.; Schmutz, P. Passivation of Al-Cr-Fe and Al-Cu-Fe-Cr complex metallic alloys in 1 M H_2SO_4 and 1 M NaOH solutions. *Corros. Sci.* **2011**, *53*, 1825–1837. [CrossRef]
17. Palcut, M.; Priputen, P.; Kusý, M.; Janovec, J. Corrosion behaviour of Al–29at%Co alloy in aqueous NaCl. *Corros. Sci.* **2013**, *75*, 461–466. [CrossRef]

18. Palcut, M.; Priputen, P.; Šalgó, K.; Janovec, J. Phase constitution and corrosion resistance of Al-Co alloys. *Mater. Chem. Phys.* **2015**, *166*, 95–104. [CrossRef]
19. Priputen, P.; Palcut, M.; Babinec, M.; Mišík, J.; Černičková, I.; Janovec, J. Correlation between microstructure and corrosion behavior of near-equilibrium Al-Co alloys in various environments. *J. Mater. Eng. Perform.* **2017**, *26*, 3970–3976. [CrossRef]
20. Lekatou, A.; Sfikas, A.K.; Karantzalis, A.E.; Sioulas, D. Microstructure and corrosion performance of Al-32%Co alloys. *Corros. Sci.* **2012**, *63*, 193–209. [CrossRef]
21. Lekatou, A.; Sfikas, A.K.; Petsa, C.; Karantzalis, A.E. Al-Co alloys prepared by vacuum arc melting: Correlating microstructure evolution and aqueous corrosion behavior with Co content. *Metals* **2016**, *6*, 46. [CrossRef]
22. Zhang, Q.; Zhang, Z. On the electrochemical dealloying of Al-based alloys in a NaCl aqueous solution. *Phys. Chem. Chem. Phys.* **2010**, *12*, 1453–1472. [CrossRef]
23. Palcut, M.; Ďuriška, L.; Špoták, M.; Vrbovský, M.; Gerhátová, Ž.; Černičková, I.; Janovec, J. Electrochemical corrosion of Al-Pd alloys in HCl and NaOH solutions. *J. Min. Metall. Sect. B-Metall.* **2017**, *53*, 333–340. [CrossRef]
24. Ďuriška, L.; Palcut, M.; Špoták, M.; Černičková, I.; Gondek, J.; Priputen, P.; Čička, R.; Janičkovič, D.; Janovec, J. Microstructure, phase occurrence, and corrosion behavior of as-solidified and as-annealed Al-Pd alloys. *J. Mater. Eng. Perform.* **2018**, *27*, 1601–1613. [CrossRef]
25. Zhang, Z.; Wang, Y.; Qi, Z.; Zhang, W.; Qin, J.; Frenzel, J. Generalized fabrication of nanoporous metals (Au, Pd, Pt, Ag, and Cu) through chemical dealloying. *J. Phys. Chem. C* **2009**, *113*, 12629–12636. [CrossRef]
26. Erlebacher, J.; Aziz, M.J.; Karma, A.; Dimitrov, N.; Sieradzki, K. Evolution of nanoporosity in dealloying. *Nature* **2001**, *410*, 450–453. [CrossRef] [PubMed]
27. Wang, X.; Wang, W.; Qi, Z.; Zhao, Ch.; Ji, H.; Zhang, Z. High catalytic activity of ultrafine nanoporous palladium for electro-oxidation of methanol, ethanol, and formic acid. *Electrochem. Commun.* **2009**, *11*, 1896–1899. [CrossRef]
28. Balanetskyy, S.; Grushko, B.; Velikanova, T.Ya.; Urban, K. An investigation of the Al-Pd-Fe phase diagram between 50 and 100 at. % Al: Phase equilibria at 750 °C. *J. Alloys Compd.* **2004**, *374*, 158–164. [CrossRef]
29. Černičková, I.; Švec, P.; Watanabe, S.; Čaplovič, L'.; Mihalkovič, M.; Kolesár, V.; Priputen, P.; Bednarčík, J.; Janičkovič, D.; Janovec, J. Fine structure of phases of ε-family in $Al_{73.8}Pd_{11.9}Co_{14.3}$ alloy. *J. Alloys Compd.* **2014**, *609*, 73–79. [CrossRef]
30. Yubuta, K.; Suzuki, S.; Simura, R.; Sugiyama, K. Structure of ε_{16} phase in Al-Pd-Co system studied by HREM and X-ray diffraction. In *Aperiodic Crystals (Cairns)*, 1st ed.; Schmid, S., Ed.; Springer: Berlin, Germany, 2013; pp. 231–236.
31. Grushko, B. Again regarding the Al-Pd phase diagram. *J. Alloys Compd.* **2013**, *557*, 102–111. [CrossRef]
32. Cheng, Y.L.; Zhang, Z.; Cao, F.H.; Li, J.F.; Zhang, J.Q.; Wang, J.M.; Cao, C.N. Study of the potential electrochemical noise during corrosion process of aluminum alloys 2024, 7075 and pure aluminum. *Mater. Corros.* **2003**, *54*, 601–608. [CrossRef]
33. ASM International, ASM Handbook Committee. *Corrosion: Fundamentals, Testing, and Protection (ASM Handbook Vol. 13A)*, 1st ed.; ASM International: Novelty, OH, USA, 2003.
34. Szklarska-Smialowska, Z. Pitting corrosion of aluminum. *Corros. Sci.* **1999**, *41*, 1743–1767. [CrossRef]
35. Li, J.; Dang, J. A summary of corrosion properties of Al-rich solid solution and secondary phase particles in Al alloys. *Metals* **2017**, *7*, 84. [CrossRef]
36. Birbilis, N.; Buchheit, R.G. Electrochemical Characteristics of Intermetallic Phases in Aluminum Alloys: An experimental survey and discussion. *J. Electrochem. Soc.* **2005**, *152*, B140–B151. [CrossRef]
37. Zhang, Q.; Wang, X.; Qi, Z.; Wang, Y.; Zhang, Z. A benign route to fabricate nanoporous gold through electrochemical dealloying of Al–Au alloys in a neutral solution. *Electrochim. Acta* **2009**, *54*, 6190–6198. [CrossRef]
38. Lekatou, A.G.; Sfikas, A.K.; Karantzalis, A.E. The influence of the fabrication route on the microstructure and surface degradation properties of Al reinforced by Al_9Co_2. *Mater. Chem. Phys.* **2017**, *200*, 33–49. [CrossRef]
39. Lim, A.B.Y.; Neo, W.J.; Yauw, O.; Chylak, B.; Gan, C.L.; Chen, Z. Evaluation of the corrosion performance of Cu–Al intermetallic compounds and the effect of Pd addition. *Microelectron. Reliab.* **2016**, *56*, 155–161. [CrossRef]

40. Li, R.T.; Murugan, V.K.; Dong, Z.L.; Khor, K.A. Comparative study on the corrosion resistance of Al–Cr–Fe alloy containing quasicrystals and pure Al. *J. Mater. Sci. Technol.* **2016**, *32*, 1054–1058. [CrossRef]
41. Debili, M.Y.; Sassane, N.; Boukhris, N. Structure and corrosion behavior of Al-Co-Ti alloy system. *Anti Corros. Methods Mater.* **2017**, *64*, 443–451. [CrossRef]
42. Sassane, N.; Debili, M.Y.; Boukhris, N.E. Structural characterization of a ternary Al-Co-Ti alloy system. *J. Adv. Microsc. Res.* **2018**, *13*, 409–416. [CrossRef]
43. Zhu, Y.; Sun, K.; Frankel, G.S. Intermetallic phases in aluminum alloys and their roles in localized corrosion. *J. Electrochem. Soc.* **2018**, *165*, C807–C820. [CrossRef]

materials

MDPI

Article

Influence of Alloyed Ga on the Microstructure and Corrosion Properties of As-Cast Mg–5Sn Alloys

Jing Ren, Enyu Guo, Xuejian Wang, Huijun Kang, Zongning Chen * and Tongmin Wang *

Key Laboratory of Solidification Control and Digital Preparation Technology (Liaoning Province), School of Materials Science and Engineering, Dalian University of Technology, Dalian 116024, China; 18840830427@163.com (J.R.); eyguo@dlut.edu.cn (E.G.); wangxuejian0618@163.com (X.W.); kanghuijun@dlut.edu.cn (H.K.)

* Correspondence: znchen@dlut.edu.cn (Z.C.); tmwang@dlut.edu.cn (T.W.); Tel.: +86-411-84709500 (Z.C.); +86-411-84706790 (T.W.)

Received: 18 September 2019; Accepted: 5 November 2019; Published: 8 November 2019

Abstract: In this paper, the microstructures and corrosion behaviors of as-cast Mg–5Sn–*x*Ga alloys with varying Ga content (x = 0, 0.5, 1, 2, 3 wt %) were investigated. The results indicated that Ga could not only adequately refine the grain structure of the alloys, but could also improve the corrosion resistance. The microstructures of all alloys exhibited typical dendritic morphology. No Ga-rich secondary phases were detected when 0.5 wt % Ga was added, while only the morphology of Mg_2Sn phase was changed. However, when the addition rate of Ga exceeded 0.5 wt %, an Mg_5Ga_2 intermetallic compound started to form from the interdendritic region. The volume fraction of Mg_5Ga_2 monotonically increased with the increasing Ga addition level. Although Mg_5Ga_2 phase was cathode phase, its pitting sensitivity was weaker than Mg_2Sn. In addition, the standard potential of Ga (−0.55 V) was lower than that of Sn (−0.14 V), which relieved the driving force of the secondary phases for the micro-galvanic corrosion. An optimized composition of 3 wt % Ga was concluded based on the immersion tests and polarization measurements, which recorded the best corrosion resistance.

Keywords: magnesium; immersion test; polarization; microstructure; corrosion resistance

1. Introduction

Magnesium alloys, with their excellent properties of low density and high specific strength, have received extensive attention in automobile, aerospace, electronics, and other industries [1–4]. Among many of the binary alloy systems, the Mg–Sn phase diagram demonstrates that the Mg-rich side has a very shallow α-Mg solvus curve, indicating that it is conducive to maximizing the precipitation of the thermally stable Mg_2Sn particles. This infers that adding Sn potentially helps to enhance the mechanical integrity of the alloy at elevated temperatures [5–9]. On this account, the Mg–Sn-based alloy system is one of the most extensively studied alloys in recent years [6–13].

Aside from the mechanical strength, other factors that limit the application of magnesium alloys are their high chemical activity and poor corrosion resistance. It was documented, however, that alloying with some Sn results in superior corrosion resistance. Ha et al. [14] reported that Sn decreases the cathodic current density and inhibits the cathodic H_2 evolution. Similar effects were also observed in Mg–5Al–1Zn alloy [11] and Mg–7Al–0.2Mn alloy [12]. The reason for inhibition of H_2 evolution by alloying Sn was studied. Ha et al. [13] observed that the main reason for the decrease of H_2 evolution rate was that Sn elements were enriched on the metal surface. They further reported that Sn had a higher potential H_2 evolution than the Mg matrix [15,16]. Moreover, under the condition of a high cathode, Sn could form SnH_4 [14]. That is to say, when the matrix is corroded, the element Sn is spontaneously enriched on the matrix surface, acting as a barrier to further corrosion [13].

The addition of Ga in magnesium alloy has important research value in seawater battery and sacrificial anode [17]. In those applications, Ga enhances the mechanical performance of Mg-based alloys by solid solution and precipitation strengthening [17,18]. It was suggested that Mg–Ga, Mg–In, and Mg–Sn alloys have potential use as biomaterials [15]. Among the aforementioned three binary alloys, Mg–Ga alloys appeared to be the best candidates, considering both the mechanical properties and corrosion behaviors in 0.9 wt % NaCl solution. Marta et al. [19] pointed out that the superior corrosion resistance of Mg–Ga is mainly related to the microgalvanic corrosion between α-Mg and the second phases, and the morphology and distribution of Mg_5Ga_2 phase.

The purpose of this study is to explore the effect of Ga on the microstructure and corrosion resistance of Mg–5Sn alloy. Both potentiodynamic polarization measurements and immersion weightlessness experiments were conducted to evaluate the corrosion property of the experimental Mg–5Sn–*x*Ga alloys. By investigating the microstructures of Mg–5Sn–*x*Ga alloys with varying Ga contents, the immanent relationship between the microstructure and corresponding corrosion behavior is established, and the results are discussed.

2. Materials and Methods

2.1. Specimen Preparation

As-cast Mg–5Sn–*x*Ga (x = 0, 0.5, 1, 2, and 3, all in weight percentage unless otherwise specified) alloys were investigated in this work. The experimental alloys were prepared by melting pure metals in an electric resistance furnace under the protection of an atmosphere containing CO_2 and SF_6 mixture in a ratio of 99:1. Pure Mg (99.99%) ingot was cut into small pieces and put in a magnesia crucible. Once the temperature of the melt reached 973 K, pure Sn (99.99%) and Ga (99.999%) granules were added to the melt. The melt was fully stirred with a graphite impeller for 180 s before it was poured into a cylindrical steel mold preheated at 523 K. The chemical compositions of the alloys were measured by X-ray fluorescence analysis (XRF-1800, Shimadzu, Kyoto, Japan), and the results are summarized in Table 1.

Table 1. Chemical compositions of the experimental Mg–5Sn–*x*Ga (x = 0, 0.5, 1, 2, 3 wt %) alloys analyzed by XRF tests.

Nominal Composition	Element Content (wt %)		
	Sn	Ga	Mg
Mg–5Sn	5.08	-	Bal.
Mg–5Sn–0.5Ga	4.98	0.56	Bal.
Mg–5Sn–1Ga	4.81	1.12	Bal.
Mg–5Sn–2Ga	5.08	2.02	Bal.
Mg–5Sn–3Ga	4.67	2.81	Bal.

2.2. Microstructural Characterization

The samples for microstructural observation were cut 10 mm from the bottom of the ingot. The exposed surfaces were ground by SiC abrasive papers up to 4000 grit and then polished up to 1 μm in a suspension of diamond pastes. The polished surfaces were etched in a mixed solution of picric and acetic acid (10 mL acetic acid, 4.2 g picric acid, 10 mL distilled water, and 70 mL ethanol). The microstructures and corrosion morphologies of the specimens were observed using an optical microscope (OM, Olympus GX51, Olympus Corp., Tokyo, Japan) and a field emission scanning electron microscope (FESEM, Zeiss supra 55, Zeiss Corp., Oberkochen, Germany), equipped with energy-dispersive X-ray spectroscopy (EDS). Phases were identified by an X-ray diffraction instrument (XRD, PANalytical Empyrean, Almelo, The Netherlands) with Cu K_{α} radiation at a scanning speed of 0.142224°/s. An electron microprobe analyzer (EPMA, JXA-8530F PLUS, Tokyo, Japan) equipped with a wavelength dispersive spectrometer (WDS) was used to examine the elemental distribution of selected phases. Grain size was measured using the OM micrographs of the alloys by the intercept method regulated in GB/T 6394-2002:

$$\tau = \frac{L}{M \times N} \tag{1}$$

where τ is the average grain size, L is the measured mesh length, M is the magnification for observation (50× in this work), and N is the number of intercept points on the measured mesh as indicated in Equation (1). The grain sizes were averaged with values measured from six micrographs for each alloy.

2.3. Corrosion Tests

The corrosion behaviors of the as-cast Mg–5Sn–xGa (x = 0, 0.5, 1, 2, and 3 wt %) alloys were investigated using a potentiodynamic polarization test and immersion test. All experiments were carried out at room temperature (298 ± 1 K).

For the electrochemical test, a conventional three-electrode cell was employed. The cell was composed of a working electrode (sample), a reference electrode (Ag/AgCl electrode, saturated KCl with electrode potential of 0.1981 V vs. Standard Hydrogen Electrode), and a Pt plate counter electrode. The test was carried out using a Gamry electrochemical workstation (Reference 600, Gamry Instruments Inc., Warminster, PA, USA). The samples were ground using SiC abrasive papers up to 2000 grit, and the surface of a round specimen with an area of 1.0 cm^2 was exposed to 3.5 wt % NaCl solution. After reaching a steady open circuit potential (OCP), the polarization test was initiated from −0.3 V versus the OCP level of the sample to +0.3 V vs. OCP at a scanning rate of 1 mV/s. Three samples were tested for each alloy to ensure repeatability. The corrosion current density (i_{corr}, mA/cm^2) of the investigated alloys, which was obtained by fitting the cathodic branch of the potentiodynamic polarization curve, was converted to the average corrosion rate (R_i, mm/y) according to the following equation [20,21]:

$$R_i = 22.85 \times i_{corr} \tag{2}$$

The immersion test was performed on three parallel samples for each alloy. Samples with dimensions of 20 mm wide, 25 mm long, and 3 mm thick were mechanically ground using SiC abrasive papers up to 2000 grit. The ground samples were immersed in a 3.5 wt % NaCl solution at 298 K for 72 h. After immersion, the samples were taken out and cleaned with chromate acid (200 g/L CrO_3 + 10 g/L $AgNO_3$) to remove the corrosion products [22,23]. The obtained samples were rinsed with distilled water, cleaned in alcohol, and dried in air. The final specimens were weighed on an analytical balance (ME204E, Mettler Toledo Corp., Greifensee, Switzerland), and the corrosion rate (C_R) for each sample was calculated by Equation (3) [24,25]:

$$C_R = \frac{K \times W}{A \times T \times D} \tag{3}$$

where K is a constant (8.76×10^4), W is the mass loss in g, A is the exposed area of the sample in cm^2, T is the exposure time in hours, and D is the density in g/cm^3.

Another group of samples for observation of corrosion surface morphology, corrosion depth, and products identification were Φ 12 mm in diameter and 3 mm in thickness. These samples were prepared following the same procedure as the samples for microstructural observation mentioned in Section 2.2. Measures were taken to ensure that corrosion occurred only on the transverse surface of the cylindrical samples. After immersion in 3.5 wt % NaCl solution for 24 h, XRD was used to identify the phases in the corrosion product. The surface and cross-section morphologies of the corroded samples were observed in SEM and EPMA, respectively.

3. Results and Discussion

3.1. Microstructure Analysis

The OM micrographs of the as-cast Mg–5Sn–*x*Ga alloys with varying Ga contents are shown in Figure 1. All the alloys present a similar typical dendritic grain structure. The solute elements could have significant effects on the growth behavior of α-Mg grains and the final morphological patterns [26–29]. It was documented that Sn could form a composition undercooling zone of liquid ahead of the solid–liquid interface, leading to dendrite formation [30]. The average grain sizes were measured to be 300.4 ± 18.4, 227.1 ± 19.3, 205.7 ± 16.7, 183.3 ± 12.7, and 155.9 ± 8.6 μm for the Mg–5Sn, Mg–5Sn–0.5Ga, Mg–5Sn–1Ga, Mg–5Sn–2Ga, and Mg–5Sn–3Ga alloys, respectively. The result suggests that the grain size decreased monotonically with increasing Ga content, probably due to the increasing grain growth restriction factor (*GRF*), which can be expressed by Equation (4) for a binary alloy [26]:

$$GRF = m_L C_0 (k_0 - 1) \tag{4}$$

where m_L is the slope of liquidus (assumed to be straight), k_0 the equilibrium distribution coefficient, and C_0 the concentration of a solute. According to the Mg–Ga phase diagram [31], in dilute binary magnesium alloys, the *GRF* value for Ga element is calculated to be 4.04, which is similar to Al and Zn element (GRF_{Al} = 4.32, GRF_{Zn} = 5.31 [26]). Composition undercooling was established by the Ga concentration gradient in the diffusion layer adjacent to the solid–liquid interface, restricting grain growth as a consequence of slow diffusion; and thus, the growth rate is limited and grain size is refined.

As for the case of the Mg–5Sn–0.5Ga alloy, the quantitative results from the WDS analysis of the areas marked by the white rectangles in Figure 1f are given in Table 2. It was found that Mg–5Sn alloy with 0.5 wt % Ga did not cause the formation of any new phases. Most of the Ga and a part of the Sn were either dissolved in the α-Mg matrix or enriched at the interdendritic region (indicated by arrow *a*).

Table 2. The chemical compositions of the areas marked by the white rectangles in Figure 1f using the wavelength dispersive spectrometer (WDS) (in wt %).

Area	Sn	Ga	Mg
a	9.778	1.929	88.293
b	25.115	0.741	74.144
c	1.589	0.167	98.244

Dozens of studies have shown that the microstructural features, such as the grain size and the morphology of second phase, may have impacts on the corrosion resistance of magnesium alloys [32,33]. Grain refinement led to the decreased corrosion rates, possibly because of the enhanced passivity of the oxide film [34]. Because the experimental alloy with 3 wt % Ga had the finest equiaxed grains, it was most likely that the as-cast Mg–5Sn–3Ga alloy would exhibit the greatest corrosion resistance.

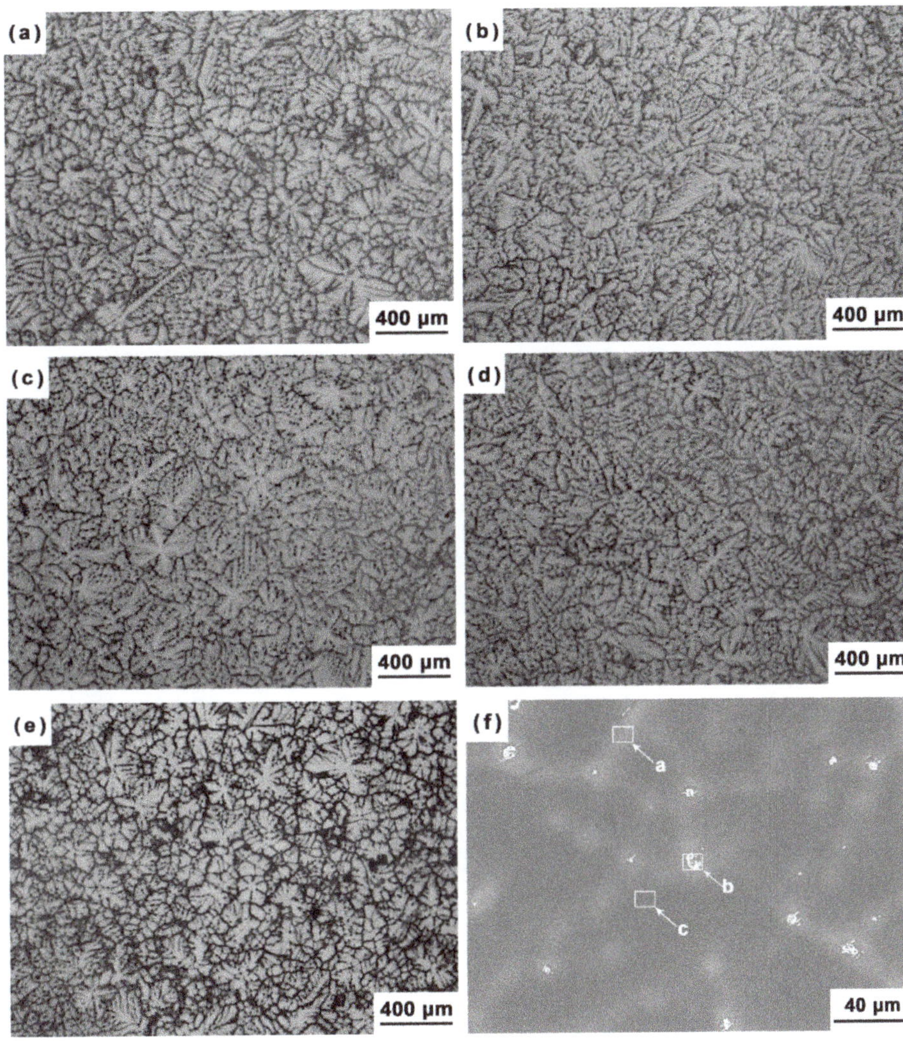

Figure 1. OM micrographs of the Mg–5Sn–*x*Ga alloys: (**a**) Mg–5Sn, (**b**) Mg–5Sn–0.5Ga, (**c**) Mg–5Sn–1Ga, (**d**) Mg–5Sn–2Ga, (**e**) Mg–5Sn–3Ga, and (**f**) backscattered electron (BSE) micrograph of the Mg–5Sn–0.5Ga alloy.

The backscattered electron (BSE) micrographs of the Mg–5Sn–*x*Ga alloys with varying Ga content are shown in Figure 2a–e. The OM and BSE micrographs of the alloys confirm the presence of second phases and element enrichment in the interdendritic regions. The elemental mapping results of the Ga, Mg, and Sn in the Mg–5Sn–3Ga sample using EPMA are shown in Figure 2f–i. Contrasts along the interdendritic contours demonstrate that there are two types of phases with different chemistries. According to the quantitative results measured by WDS in Table 3, the grey eutectic phase indicated by the letter *a* is Mg_5Ga_2 intermetallic compound, and the brighter phase indicated by the letter *b* is Mg_2Sn. The inserts in Figure 2a–e are zoom-in views of the two phases embedded in the interdendritic region. When 0.5 wt % Ga was added, no Ga-rich second phases were detected, according to the composition analysis in Table 2. When the addition of Ga exceeded 0.5 wt %, Ga formed from the interdendritic region in the form of Mg_5Ga_2 phase, while Mg_2Sn did so in the form of divorced eutectic phase.

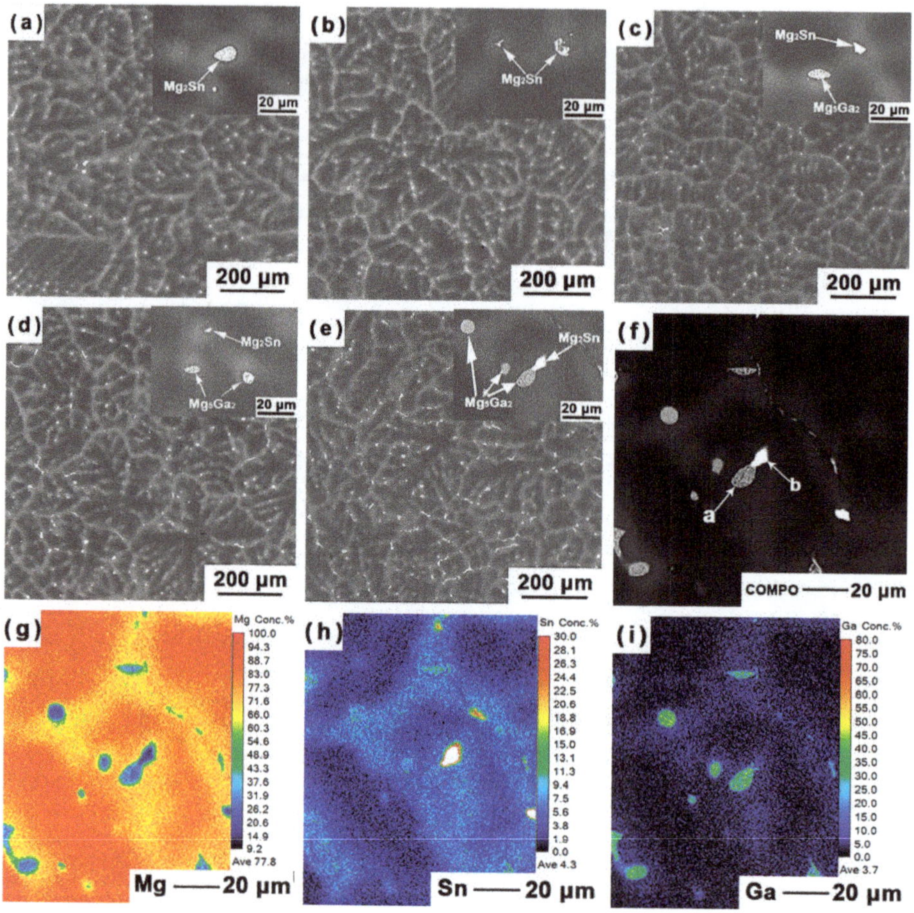

Figure 2. BSE micrographs of the Mg–5Sn–*x*Ga alloys: (**a**) Mg–5Sn, (**b**) Mg–5Sn–0.5Ga, (**c**) Mg–5Sn–1Ga, (**d**) Mg–5Sn–2Ga, (**e**) Mg–5Sn–3Ga. The inserts are zoom-in views showing the second phases. (**f–i**) Elemental mapping of the Mg, Sn, and Ga in the Mg–5Sn–3Ga sample using electron microprobe analyzer (EPMA).

Table 3. The chemical compositions of the phases measured by WDS as indicated by the arrows in Figure 2f (in at %).

Point	Sn	Ga	Mg
a	2.2	19.6	78.2
b	29.3	4.4	66.3

The XRD spectra in Figure 3a further confirmed that the Mg–5Sn alloy consisted of α-Mg matrix and Mg_2Sn. When the Ga content exceeded 1 wt %, a new phase containing Ga was formed. Survey on Joint Committee on Powder Diffraction Standards indicates that the new peaks correspond to the reflections of the Mg_5Ga_2 compound, which normally exists in Mg–Ga binary [15,35] and Mg–Hg–Ga ternary [18] alloys. This is consistent with the quantitative results measured by WDS in Table 3.

The increasing density of the Mg_5Ga_2 peaks is expected with the increase in the Ga content of the alloys. Figure 3b shows the area fraction of the second phases in the investigated alloy calculated by ImageJ 1.47 (US NIH, Bethesda, MD, USA) [36]. It is also clear that the area fraction of Mg_5Ga_2 particles increased with increasing Ga content, while the area fraction of Mg_2Sn particles decreased accordingly.

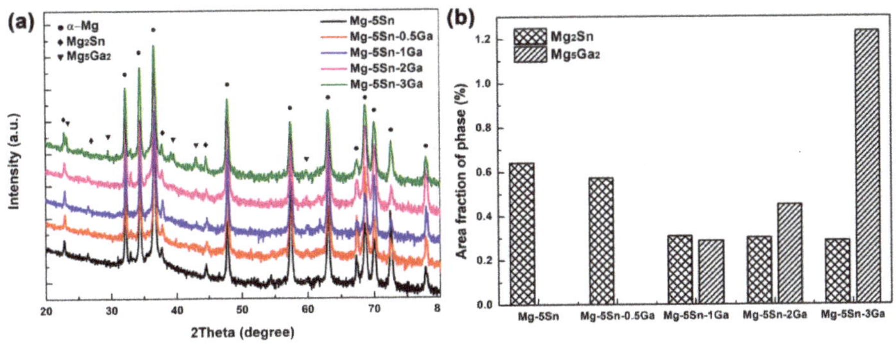

Figure 3. (**a**) X-ray diffraction (XRD) patterns of the Mg–5Sn–xGa alloys; (**b**) the area fraction of the second phases in the investigated alloy.

3.2. Polarization Tests

Figure 4 shows the potentiodynamic polarization curves of the as-cast Mg–5Sn–xGa alloys. The corrosion current density (i_{corr}), corrosion potential (E_{corr}), and the average corrosion rate (R_i) obtained from the potentiodynamic polarization curves are listed in Table 4. The i_{corr} of the as-cast Mg–5Sn–xGa alloys in 3.5 wt % NaCl solution increase in the order of Mg–5Sn–3Ga, Mg–5Sn–2Ga, Mg–5Sn–0.5Ga, Mg–5Sn–1Ga, and Mg–5Sn. The lowest corrosion current density and corrosion potential are 2.79×10^{-2} mA/cm^2 and −1.684 V for Mg–5Sn–3Ga alloy, respectively.

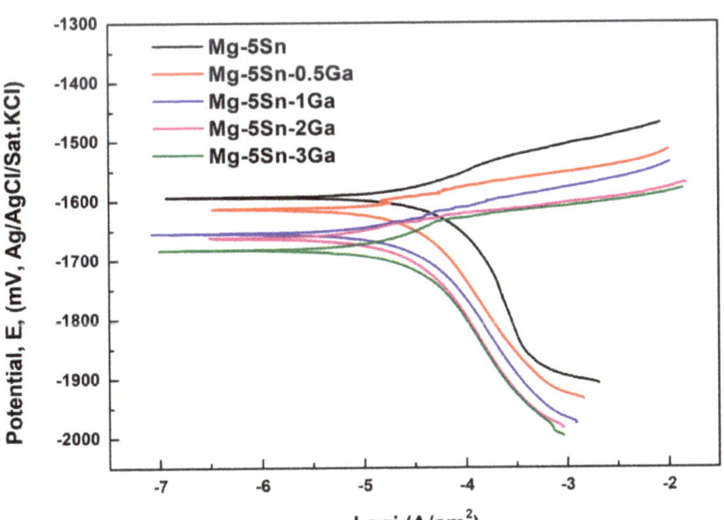

Figure 4. Polarization curves of the Mg–5Sn–xGa alloys in 3.5 wt % NaCl solution.

Table 4. Fitting results of polarization curves of the Mg–5Sn–*x*Ga alloys with varying Ga content in 3.5 wt % NaCl solution.

Alloy	Mg–5Sn	Mg–5Sn–0.5Ga	Mg–5Sn–1Ga	Mg–5Sn–2Ga	Mg–5Sn–3Ga
E_{corr} (V)	−1.596	−1.614	−1.658	−1.665	−1.684
i_{corr} (mA/cm^2)	9.73×10^{-2}	3.28×10^{-2}	3.82×10^{-2}	3.16×10^{-2}	2.79×10^{-2}
R_i (mm/y)	2.223	0.770	0.873	0.722	0.638

It is accepted that the corrosion of Mg-based alloys is generally caused by the cathodic reaction of hydrogen evolution. The anodic reaction of Mg dissolution can be described as [37,38]:

$$\text{Anodic reaction: } Mg \rightarrow Mg^{+} + e^{-} \tag{5}$$

$$Mg^{+} + H_2O \rightarrow Mg^{2+} + OH^{-} + \frac{1}{2}H_2 \tag{6}$$

$$\text{Cathodic reaction: } 2H_2O + 2e^{-} \rightarrow H_2 + 2OH^{-} \tag{7}$$

Ga is reported to have relatively high hydrogen overpotentials [15]. Alloys with Ga are thus anticipated to inhibit the cathodic reaction of Equation (6). The cathodic branches of the polarization curves show that the cathodic current density decreases with increasing Ga content, demonstrating that Ga indeed delays cathodic reaction and thus reduces corrosion rate.

Figure 5a shows the representative anode branches of the polarization curves of the Mg–5Sn–*x*Ga alloys in 0.01 M NaCl solution. Within the whole range of anode polarization, the relation between the passive current density ($i_{passive}$) of the alloys at any potential remains unchanged, so the relation between the current density of the alloys at an eigenvalue potential can be selected to describe the change of the trend of the passive current density of the alloys after adding Ga element. This method is favored by scholars in many studies. Ha et al. [39,40] used this method to study the change trend of anode passive current density at −1.7 V and cathode hydrogen evolution current density at −1.9 V of Mg–5Sn–(1–4 wt %)Zn alloy system. Similarly, Kim et al. [41] studied the polarization curve of Mg–8Sn–1Zn alloy with this method. In order to explore the role that Ga played in the anodic passive layer, Figure 5b compares the passive current density values for the alloys of different Ga contents with the alloy without Ga addition. The $i_{passive}$ values were measured at the anodic potentials of −1.65 V. Evidently, compared with the Mg–5Sn alloy, the $i_{passive}$ values decreased after adding Ga element, among which the Mg–5Sn–0.5Ga alloy exhibited the lowest $i_{passive}$ value. Decreased $i_{passive}$ indicates enhanced passive film stability. As mentioned above, Mg_5Ga_2 phase will be formed when Ga addition exceeds 0.5 wt %, which belongs to the cathode phase and destroys the stability of passive film to some extent. Moreover, the passive film of magnesium alloy is extremely unstable and cannot protect the matrix, so the corrosion rate is mainly controlled by hydrogen evolution reaction of the cathode.

Another factor that also influences the corrosion rate is the microstructural features of an alloy [42–44]. With the increase in Ga element, the area fraction of Mg_2Sn particles monotonically decreases, while the opposite trend is the case for the Mg_5Ga_2, as shown in Figures 2 and 3. Liu et al. [16] reported that both the hydrogen evolution rate and corrosion potential decreased with decreasing volume fraction of the Mg_2Sn particles, which is consistent with the experimental results in this study.

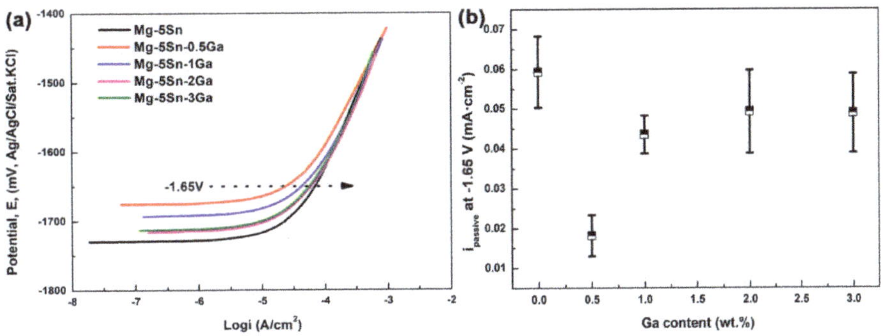

Figure 5. (**a**) The representative polarization curves of the Mg–5Sn–*x*Ga alloys in 0.01 M NaCl solution. (**b**) Passive current density ($i_{passive}$) values measured at −1.65 V based on (**a**). The average and standard deviation were obtained from at least five measurements.

3.3. The Immersion Tests

Figure 6 shows the average corrosion rates of the alloys with continuously increasing Ga content calculated by Equation (2). In essence, the variation trend of the corrosion rate obtained by immersion test is in accord with the potentiodynamic polarization results. The lowest weight loss rate is 1.15 ± 0.1 mm/y for the Mg–5Sn–3Ga alloy. This confirms that the addition of Ga improves the corrosion resistance of Mg–5Sn alloy.

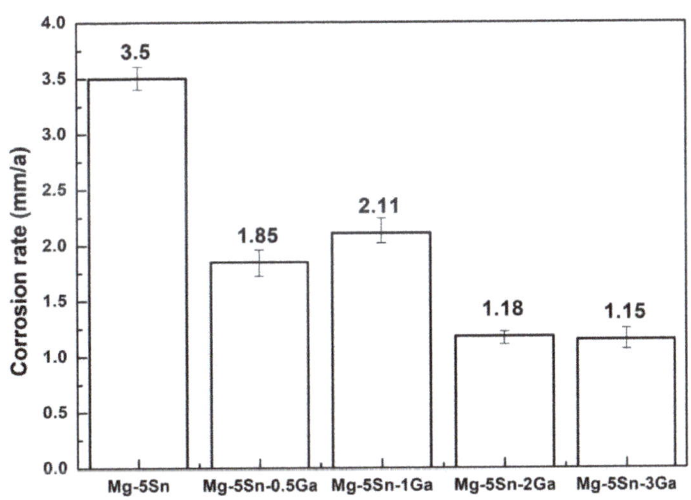

Figure 6. Average corrosion rate of the Mg–5Sn–*x*Ga alloys after immersion in 3.5 wt % NaCl solution for 72 h.

After immersion at 298 K for 6 h, the corrosion initiation sites were observed using SEM and EDS. Figure 7 presents the representative BSE micrographs of three alloys (Mg–5Sn, Mg–5Sn–0.5Ga, and Mg–5Sn–3Ga) and the corresponding EDS spectra of the marked areas. Figure 7a,b reveals that the Mg_2Sn phase was not covered with corrosion films, and cracks were formed in the vicinity of Mg_2Sn phase. Combining with the EDS analysis of points *A* and *B* (Figure 7d,e), it is inferred that corrosion began in the magnesium matrix around the Mg_2Sn phase. Moreover, the adjacent distribution of the two types of phases provides the possibility for analyzing the initial corrosion location. In the Mg–5Sn–3Ga alloy (Figure 7c), the phase at point C is rich in Ga element compared with its neighboring phase (marked by point *D*), suggesting that points *C* and *D* are Mg_5Ga_2 phase and Mg_2Sn phase, respectively. Also shown in Figure 7c is that the surface film on Mg_2Sn phase was broken, while this was not the case for the film on Mg_5Ga_2 phase. Mg_2Sn phase was detached from the matrix after immersion for 6 h. All the aforementioned observations indicate that Mg_2Sn phase, rather than the Mg_5Ga_2 phase, provided the sites to initiate corrosion. Moreover, compared with the larger Mg_2Sn phase, eutectic Mg_5Ga_2 phase is deemed to stay more firmly in the alloy matrix. Therefore, with the increase in Ga content, the increasing area fraction of Mg_5Ga_2 further improved the corrosion resistance of the alloys.

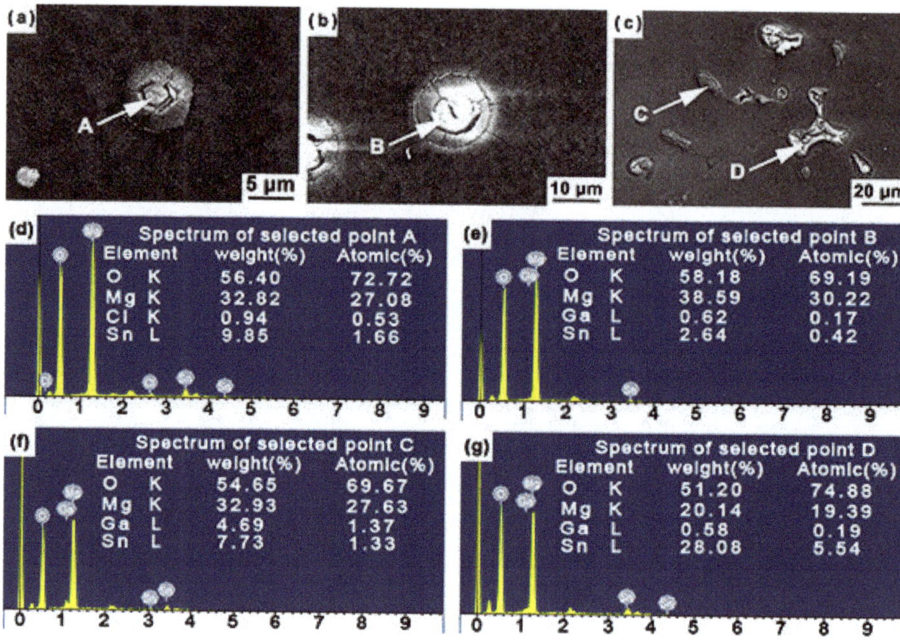

Figure 7. High magnification BSE micrographs showing the alloy surfaces after immersion in 3.5 wt % NaCl solution at 298 K for 6 h: (**a**) Mg–5Sn, (**b**) Mg–5Sn–0.5Ga, (**c**) Mg–5Sn–3Ga; (**d**–**g**) energy-dispersive X-ray spectroscopy (EDS) spectra of the points as indicated by the arrows in the BSE micrographs.

Figure 8 shows the evolution of the surface morphologies of the investigated alloys after immersing for different durations. The surfaces of the Mg–5Sn–*x*Ga alloys are covered by a mixture of dark film and bright corrosion product. A number of deep corrosion pits are observed after immersion (Figure 8a). It is noticeable that the white corrosion products increase with the decrease in Ga content.

It is noted in Figure 8 that the Mg–5Sn–0.5Ga alloy was more seriously corroded than the Mg–5Sn–1Ga alloy, while the immersion tests produced an opposite result. The reason is that in the Mg–5Sn–0.5Ga alloy, there is no Ga-rich second phase, which has a marginal effect on initiating corrosion. In addition, Sn and Ga elements are more passive than Mg (the standard potentials of Mg, Sn, and Ga are −2.36 V, −0.14 V, and −0.55 V, respectively) [45]. The dissolved Ga element was expected to increase the potential of α-Mg, reducing the potential difference between Mg_2Sn and the matrix. When the Ga content exceeded 0.5 wt %, Mg_5Ga_2 was formed and acted as the cathode phase to accelerate the corrosion rate.

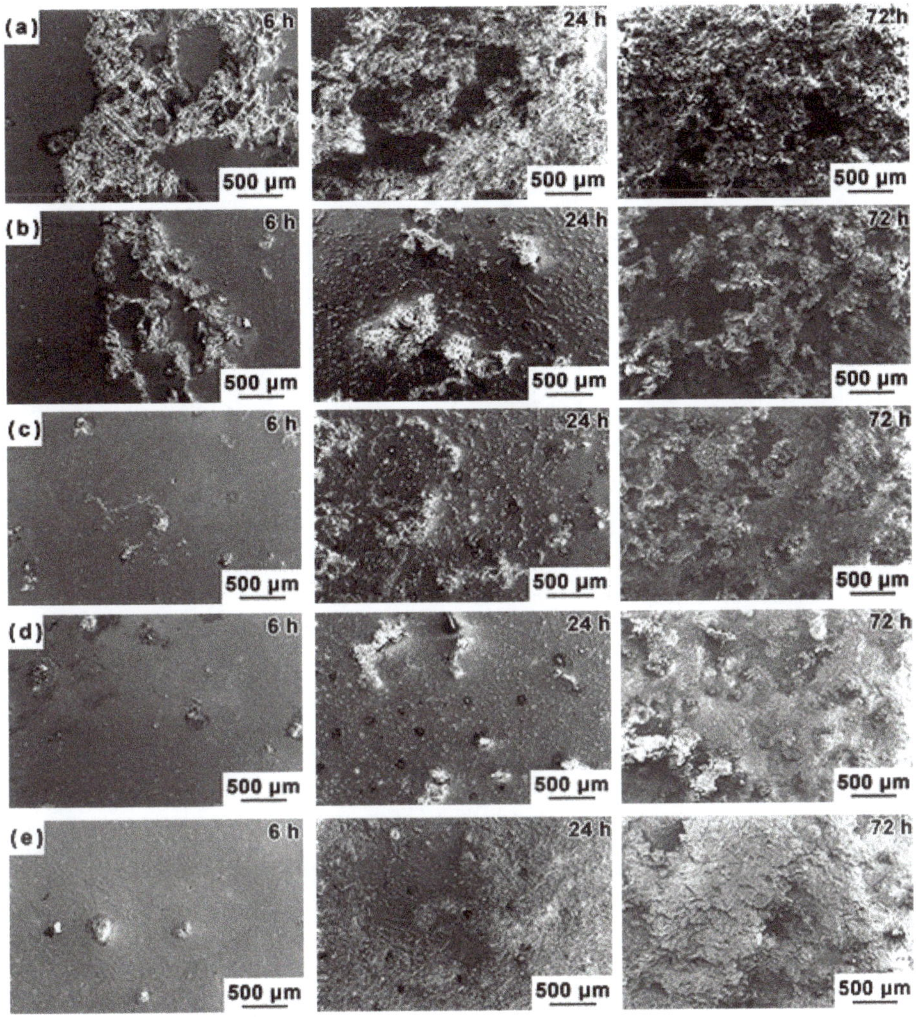

Figure 8. The evolution of the surface morphology of the investigated alloys after soaking for different durations in 3.5 wt % NaCl solution: (**a**) Mg–5Sn, (**b**) Mg–5Sn–0.5Ga, (**c**) Mg–5Sn–1Ga, (**d**) Mg–5Sn–2Ga, and (**e**) Mg–5Sn–3Ga, respectively.

Based on the above analysis, at the initial stage of corrosion, the initiation site of Mg–5Sn–0.5Ga alloy is likely to cause corrosion. However, the corrosion initiation sites of Mg–5Sn–1Ga multiply with immersion time, promoting further corrosion. Therefore, in the case of immersion test, the Mg–5Sn–0.5Ga showed superior corrosion resistance than the Mg–5Sn–1Ga alloy.

Precipitation of $Mg(OH)_2$, as the main corrosion product on the surface, occurs when Mg^{2+} from anodic dissolution meets OH^- from water reduction [37]. The XRD patterns (Figure 9) of the investigated alloys after immersion for 24 h evidence the presence of $Mg(OH)_2$ on the surfaces, and the peak intensities of $Mg(OH)_2$ reflections decrease with increasing Ga content. This indicates that the corrosion resistance of the Mg–5Sn–xGa alloys was improved by the addition of Ga. As the addition amount of Ga was very small, the corrosion products containing Ga were not detected by XRD.

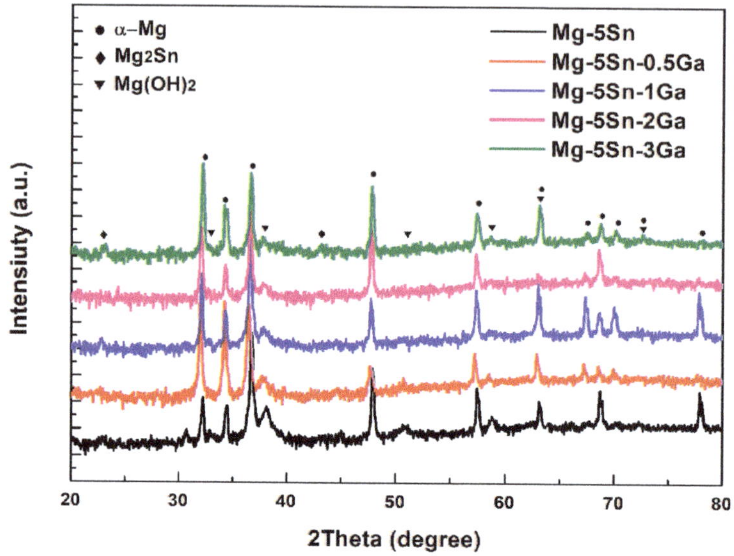

Figure 9. XRD patterns of the Mg–5Sn–*x*Ga alloys after immersion in 3.5 wt % NaCl solution for 24 h.

Figure 10 shows the BSE micrographs of the Mg–5Sn–*x*Ga alloys in cross-section after immersion in 3.5 wt % NaCl solution for 72 h. The samples have good uniformity. After many experiments, it was found that the corrosion depth of all three samples with the same composition showed little discrepancy. As given in Figure 10, corrosion started from the matrix near the second phases and penetrated the matrix along the area of the element-rich region. The thickness of the corrosion layer was the result of measuring the deepest corrosion in a sample. The inserts are an enlarged view of the deepest corrosion of the samples. Again, the thickness of the surface film decreases with increasing Ga content, indicating the positive effect of Ga on influencing the corrosion resistance of the alloys. Figure 10a reveals that the corrosion layer of the Mg–5Sn alloy was 99.1 μm in thickness. The mismatch between the matrix and corrosion product would induce local mechanical stress. Such stress increases with the gradual increasing thickness of the film after immersion [46], and once it reaches a critical value, the surface film will break. The aggressive solution will penetrate the substrate surface through the broken membrane. The relatively complete surface film can protect the substrate to some extent. As a result, the pits grow along a vertical direction.

According to the immersion test, the Mg–5Sn–0.5Ga alloy is more corrosion-resistant than the Mg–5Sn–1Ga alloy. However, this is not consistent with morphologic observations made in the cross-section. The surface film of the Mg–5Sn–0.5Ga alloy (54.7 μm) is thicker than that of the Mg–5Sn–1Ga (38.8 μm) alloy. The calculated mass loss estimates the average corrosion rate in a whole immersion period, while the corrosion section reflects the local corrosion level. It is significant to see that local corrosion occurs around the second phase along the interdendritic region to the interior of the matrix in Mg–5Sn alloy (Figure 10a). The only second phase in the Mg–5Sn–0.5Ga alloy is Mg_2Sn; hence, the two alloys should behave following the same corrosion mode. However, the oxidation state of Ga is higher than Mg in the $Mg(OH)_2$ lattice, which locally induces positive charge in the brucite lattice. Such an increase in positive charge is apt to be balanced by the interaction between Ga element and the harmful Cl^- anions, slowing down the penetration of Cl^- into the hydroxide layer [15].

To sum up, the second phases, Mg_2Sn and Mg_5Ga_2, exhibit more positive corrosion potentials than the α-Mg matrix [42]. They act as the cathode, and α-Mg the anode, resulting in galvanic corrosion in 3.5 wt % NaCl solution. It is worth noting that Mg_5Ga_2 phase has a volta potential difference (VPD) vs. matrix around +75 mV, which will cause marginal local corrosion if the immersion time is not long [19]. Therefore, the presence of Mg_5Ga_2 phase increases the area ratio between cathode and anode, which is not conducive to the corrosion resistance, but the local corrosion sensitivity does not increase. Nevertheless, a problem with Ga addition is the uneconomic cost of metallic gallium. Therefore, higher addition levels of Ga content were not performed in this study.

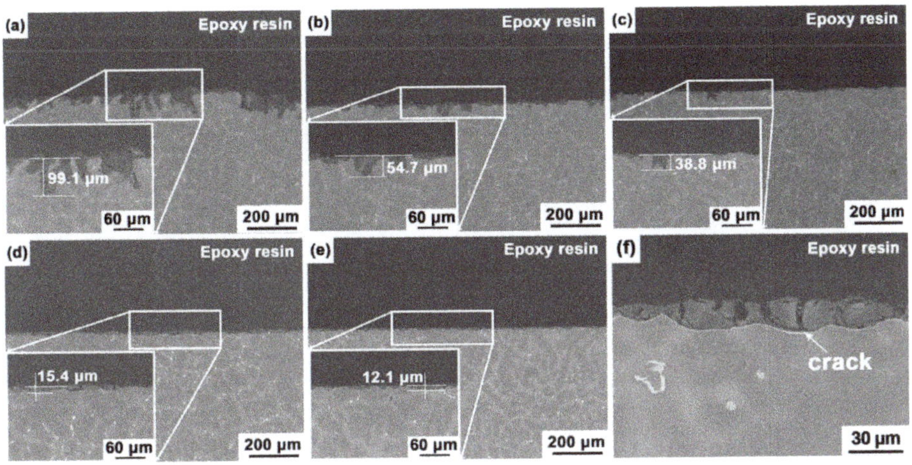

Figure 10. Cross-section morphologies of the Mg–5Sn–*x*Ga alloys after immersion in 3.5 wt % NaCl solution for 72 h: (**a**) Mg–5Sn, (**b**) Mg–5Sn–0.5Ga, (**c**) Mg–5Sn–1Ga, (**d**) Mg–5Sn–2Ga, (**e**) Mg–5Sn–3Ga; the inserts are high-magnification scanning electron microscope (SEM) micrographs of the alloys; (**f**) a typical SEM micrograph showing the cracks on the surface of the Mg–5Sn–3Ga alloy.

4. Conclusions

The main findings of this work are:

1. The microstructures of the Mg–5Sn–*x*Ga alloys similarly present a typical dendritic morphology, regardless of the content of Ga. Ga refines the grain structure adequately. An average grain size of 155.9 ± 8.6 μm was obtained for the Mg–5Sn–3Ga alloy, recording a 48.3% reduction in grain size as compared to the Mg–5Sn alloy;
2. When 0.5 wt % Ga is added to Mg–5Sn alloy, no new phase is formed. When Ga content exceeds 0.5 wt %, a new eutectic phase, identified as Mg_5Ga_2, is found in the interdendritic region. Increasing the Ga content decreases the area fraction of Mg_2Sn phase, while gradually increasing that of Mg_5Ga_2 phase;
3. Immersion test in 3.5 wt % NaCl solution shows that the corrosion of the studied Mg–Sn–Ga alloys is initiated in a pitting mode, which rapidly propagates with intense H_2 evolution on the surface. With the addition of 3 wt % Ga, the overall corrosion rate is decreased significantly. Despite the fact that corrosion is also initiated in a pitting mode, the number of pits is reduced, and the propagation rate is decelerated. The potentiodynamic polarization tests are basically in accordance with the immersion tests.

Author Contributions: Conceptualization, J.R., E.G., X.W., and Z.C.; Formal analysis, J.R.; Funding acquisition, E.G., H.K., Z.C., and T.W.; Investigation, J.R.; Methodology, X.W.; Project administration, X.W.; Resources, E.G., X.W., H.K., and T.W.; Supervision, H.K., Z.C., and T.W.; Writing—original draft, J.R.; Writing—review and editing, J.R., X.W., and Z.C.

Funding: This research was funded by the National Key Research and Development Program of China [grant No. 2016YFB0701203]; the National Natural Science Foundation of China [Grant Nos. 51525401, 51601028, 51774065, 51974058, 51728101, and 51690163]; and the fundamental research funds for the central universities [Nos. DUT18RC(3)042, DUT17RC(3)108].

Conflicts of Interest: The authors declare no conflict of interest. The funders had no role in the design of the study; in the collection, analyses, or interpretation of data; in the writing of the manuscript, or in the decision to publish the results.

References

1. Ghali, E.; Dietzel, W.; Kainer, K. Testing of General and Localized Corrosion of Magnesium Alloys: A Critical Review. *J. Mater. Eng. Perform.* **2004**, *13*, 517–529. [CrossRef]
2. Kang, D.H.; Park, S.S.; Kim, N.J. Development of Creep Resistant Die Cast Mg–Sn–Al–Si Alloy. *Mater. Sci. Eng. A* **2005**, *413–414*, 555–560. [CrossRef]
3. Luo, A.A. Recent Magnesium Alloy Development for Elevated Temperature Applications. *Int. Mater. Rev.* **2013**, *49*, 13–30. [CrossRef]
4. Song, G.; Atrens, A. Understanding Magnesium Corrosion—A Framework for Improved Alloy Performance. *Adv. Eng. Mater.* **2003**, *5*, 837–858. [CrossRef]
5. Mendis, C.L.; Bettles, C.J.; Gibson, M.A.; Hutchinson, C.R. An Enhanced Age Hardening Response in Mg–Sn Based Alloys Containing Zn. *Mater. Sci. Eng. A* **2006**, *435–436*, 163–171. [CrossRef]
6. Cheng, W.; Zhang, Y.; Ma, S.; Arthanari, S.; Cui, Z.; Wang, H.; Wang, L. Tensile Properties and Corrosion Behavior of Extruded Low-Alloyed Mg–1Sn–1Al–1Zn Alloy: The Influence of Microstructural Characteristics. *Materials* **2018**, *11*, 1157. [CrossRef]
7. Sasaki, T.T.; Oh-ishi, K.; Ohkubo, T.; Hono, K. Enhanced Age Hardening Response by the Addition of Zn in Mg–Sn Alloys. *Scr. Mater.* **2006**, *55*, 251–254. [CrossRef]
8. Liu, H.; Chen, Y.; Tang, Y.; Wei, S.; Niu, G. The Microstructure, Tensile Properties, and Creep Behavior of as-Cast Mg–(1–10)%Sn Alloys. *J. Alloys Compd.* **2007**, *440*, 122–126. [CrossRef]
9. Huang, X.; Han, G.; Huang, W. T6 Treatment and its Effects on Corrosion Properties of an Mg–4Sn–4Zn–2Al Alloy. *Materials* **2018**, *11*, 628. [CrossRef] [PubMed]
10. Zhao, Z.; Hua, Z.; Li, D.; Wei, D.; Liu, Y.; Wang, J.; Luo, D.; Wang, H. Effect of Sn Content on the Microstructure, Mechanical Properties and Corrosion Behavior of Biodegradable Mg–X (1, 3 and 5 wt.%) Sn–1Zn–0.5Ca Alloys. *Materials* **2018**, *11*, 2378. [CrossRef] [PubMed]
11. Park, K.C.; Kim, B.H.; Kimura, H.; Park, Y.H.; Park, I.M. Microstructure and Corrosion Properties of Mg–xSn–5Al–1Zn (x = 0, 1, 5 and 9 mass%) Alloys. *Mater. Trans.* **2010**, *51*, 472–476. [CrossRef]
12. Song, G. Effect of Tin Modification on Corrosion of Am70 Magnesium Alloy. *Corros. Sci.* **2009**, *51*, 2063–2070. [CrossRef]
13. Ha, H.; Kang, J.; Yang, J.; Yim, C.D.; You, B.S. Role of Sn in Corrosion and Passive Behavior of Extruded Mg-5 wt % Sn Alloy. *Corros. Sci.* **2016**, *102*, 355–362. [CrossRef]
14. Ha, H.; Kang, J.; Kim, S.G.; Kim, B.; Park, S.S.; Yim, C.D.; You, B.S. Influences of Metallurgical Factors on the Corrosion Behaviour of Extruded Binary Mg-Sn Alloys. *Corros. Sci.* **2014**, *82*, 369–379. [CrossRef]
15. Kubásek, J.; Vojtěch, D.; Lipov, J.; Ruml, T. Structure, Mechanical Properties, Corrosion Behavior and Cytotoxicity of Biodegradable Mg–X (X = Sn, Ga, in) Alloys. *Mater. Sci. Eng. C* **2013**, *33*, 2421–2432. [CrossRef] [PubMed]
16. Liu, X.; Shan, D.; Song, Y.; Chen, R.; Han, E. Influences of the Quantity of Mg2Sn Phase on the Corrosion Behavior of Mg–7Sn Magnesium Alloy. *Electrochim. Acta* **2011**, *56*, 2582–2590. [CrossRef]
17. Feng, Y.; Wang, R.; Peng, C.; Wang, N. Influence of Mg21Ga5Hg3 Compound on Electrochemical Properties of Mg-5%Hg-5%Ga Alloy. *Trans. Nonferr. Met. Soc. China* **2009**, *19*, 154–159. [CrossRef]
18. Feng, Y.; Wang, R.-c.; Peng, C.-q.; Tang, H.-p.; Liu, H.-y. Influence of Mg_5Ga_2 Compound on Microstructures and Electrochemical Properties of Mg-5%Hg-22%Ga Alloy. *Prog. Nat. Sci.* **2011**, *21*, 73–79. [CrossRef]
19. Mohedano, M.; Blawert, C.; Yasakau, K.A.; Arrabal, R.; Matykina, E.; Mingo, B.; Scharnagl, N.; Ferreira, M.G.S.; Zheludkevich, M.L. Characterization and Corrosion Behavior of Binary Mg-Ga Alloys. *Mater. Charact.* **2017**, *128*, 85–99. [CrossRef]
20. Shi, Z.; Liu, M.; Atrens, A. Measurement of the Corrosion Rate of Magnesium Alloys Using Tafel Extrapolation. *Corros. Sci.* **2010**, *52*, 579–588. [CrossRef]
21. Cai, S.; Lei, T.; Li, N.; Feng, F. Effects of Zn On Microstructure, Mechanical Properties and Corrosion Behavior of Mg–Zn Alloys. *Mater. Sci. Eng. C* **2012**, *32*, 2570–2577. [CrossRef]

22. Gu, X.; Zheng, Y.; Zhong, S.; Xi, T.; Wang, J.; Wang, W. Corrosion of, and Cellular Responses to Mg–Zn–Ca Bulk Metallic Glasses. *Biomaterials* **2010**, *31*, 1093–1103. [CrossRef] [PubMed]
23. Ha, H.; Kang, J.; Yang, J.; Yim, C.D.; You, B.S. Limitations in the Use of the Potentiodynamic Polarisation Curves to Investigate the Effect of Zn On the Corrosion Behaviour of as-Extruded Mg–Zn Binary Alloy. *Corros. Sci.* **2013**, *75*, 426–433. [CrossRef]
24. Yang, F.; Kang, H.; Guo, E.; Li, R.; Chen, Z.; Zeng, Y.; Wang, T. The Role of Nickel in Mechanical Performance and Corrosion Behaviour of Nickel-Aluminium Bronze in 3.5 Wt.% Nacl Solution. *Corros. Sci.* **2018**, *139*, 333–345. [CrossRef]
25. Jayalakshmi, S.; Sankaranarayanan, S.; Koh, S.P.X.; Gupta, M. Effect of Ag and Cu Trace Additions on the Microstructural Evolution and Mechanical Properties of Mg–5Sn Alloy. *J. Alloys Compd.* **2013**, *565*, 56–65. [CrossRef]
26. Lee, Y.C.; Dahle, A.K.; StJohn, D.H. The Role of Solute in Grain Refinement of Magnesium. *Metall. Mater. Trans. A* **2000**, *31*, 2895–2906. [CrossRef]
27. Easton, M.; StJohn, D. Grain Refinement of Aluminum Alloys: Part II. Confirmation of, and a Mechanism for, the Solute Paradigm. *Metall. Mater. Trans. A* **1999**, *30*, 1625–1633. [CrossRef]
28. Easton, M.; StJohn, D. Grain Refinement of Aluminum Alloys: Part I. The Nucleant and Solute Paradigms—A Review of the Literature. *Metall. Mater. Trans. A* **1999**, *30*, 1613–1623. [CrossRef]
29. Fu, J.W.; Yang, Y.S. Formation of the Solidified Microstructure in Mg–Sn Binary Alloy. *J. Cryst. Growth* **2011**, *322*, 84–90. [CrossRef]
30. Liu, H.; Zhao, G.; Liu, C.; Zuo, L. Effects of Different Tempers On Precipitation Hardening of 6000 Series Aluminium Alloys. *Trans. Nonferr. Met. Soc. China* **2007**, *17*, 122–127. [CrossRef]
31. Liu, C.; Zhu, X.; Zhou, H. *Magnesium Alloy Phase Atlas*; Central South University Press: Changsha, China, 2006; p. 23.
32. Zhao, C.; Pan, F.; Zhao, S.; Pan, H.; Song, K.; Tang, A. Preparation and Characterization of as-Extruded Mg–Sn Alloys for Orthopedic Applications. *Mater. Des.* **2015**, *70*, 60–67. [CrossRef]
33. Hamu, G.B.; Eliezer, D.; Wagner, L. The Relation Between Severe Plastic Deformation Microstructure and Corrosion Behavior of Az31 Magnesium Alloy. *J. Alloys Compd.* **2009**, *468*, 222–229. [CrossRef]
34. Liao, J.; Hotta, M.; Yamamoto, N. Corrosion Behavior of Fine-Grained Az31B Magnesium Alloy. *Corros. Sci.* **2012**, *61*, 208–214. [CrossRef]
35. Liu, H.; Qi, G.; Ma, Y.; Hao, H.; Jia, F.; Ji, S.; Zhang, H.; Zhang, X. Microstructure and Mechanical Property of Mg–2.0Ga Alloys. *Mater. Sci. Eng. A* **2009**, *526*, 7–10. [CrossRef]
36. Schneider, C.A.; Rasband, W.S.; Eliceiri, K.W. NIH Image to ImageJ: 25 years of image analysis. *Nat. Methods* **2012**, *9*, 671–675. [CrossRef]
37. Zhen, Z.; Xi, T.; Zheng, Y.; Li, L.; Li, L. In Vitro Study on Mg–Sn–Mn Alloy as Biodegradable Metals. *J. Mater. Sci. Technol.* **2014**, *30*, 675–685. [CrossRef]
38. Song, G.; Unocic, K.A. The Anodic Surface Film and Hydrogen Evolution on Mg. *Corros. Sci.* **2015**, *98*, 758–765. [CrossRef]
39. Ha, H.; Kang, J.; Yim, C.D.; Yang, J.; You, B.S. Role of Hydrogen Evolution Rate in Determining the Corrosion Rate of Extruded Mg–5Sn–(1–4 Wt.%) Zn Alloys. *Corros. Sci.* **2014**, *89*, 275–285. [CrossRef]
40. Ha, H.; Kim, H.J.; Baek, S.; Kim, B.; Sohn, S.; Shin, H.; Jeong, H.Y.; Park, S.H.; Yim, C.D.; You, B.S.; et al. Improved Corrosion Resistance of Extruded Mg–8Sn–1Zn–1Al Alloy by Microalloying with Mn. *Scr. Mater.* **2015**, *109*, 38–43. [CrossRef]
41. Kim, H.J.; Kim, B.; Baek, S.; Sohn, S.; Shin, H.; Jeong, H.Y.; Yim, C.D.; You, B.S.; Ha, H.; Park, S.S. Influence of Alloyed Al On the Microstructure and Corrosion Properties of Extruded Mg–8Sn–1Zn Alloys. *Corros. Sci.* **2015**, *95*, 133–142. [CrossRef]
42. Zhao, C.; Pan, F.; Zhao, S.; Pan, H.; Song, K.; Tang, A. Microstructure, Corrosion Behavior and Cytotoxicity of Biodegradable Mg–Sn Implant Alloys Prepared by Sub-Rapid Solidification. *Mater. Sci. Eng. C* **2015**, *54*, 245–251. [CrossRef] [PubMed]
43. Rad, H.R.B.; Idris, M.H.; Kadir, M.R.A.; Farahany, S. Microstructure Analysis and Corrosion Behavior of Biodegradable Mg–Ca Implant Alloys. *Mater. Des.* **2012**, *33*, 88–97. [CrossRef]
44. Yim, C.D.; Yang, J.; Woo, S.K.; Ha, H.; You, B.S. The Effects of Microstructural Factors on the Corrosion Behaviour of Mg–5Sn–X Zn (X = 1, 3 wt%) Extrusions. *Corros. Sci.* **2015**, *90*, 597–605. [CrossRef]

45. Bratsch, S.G. Standard Electrode Potentials and Temperature Coefficients in Water at 298.15 K. *J. Phys. Chem. Ref. Data* **1989**, *18*, 1–21. [CrossRef]
46. Chen, J.; Wang, J.; Han, E.; Ke, W. In Situ Observation of Pit Initiation of Passivated Az91 Magnesium Alloy. *Corros. Sci.* **2009**, *51*, 477–484. [CrossRef]

materials

MDPI

Article

Experimental and Theoretical Studies on the Corrosion Inhibition of Carbon Steel by Two Indazole Derivatives in HCl Medium

Shenying Xu [1,2], Shengtao Zhang [1,*], Lei Guo [3,*], Li Feng [1] and Bochuan Tan [1]

1 School of Chemistry and Chemical Engineering, Chongqing University, Chongqing 400044, China; 20111802011@cqu.edu.cn (S.X.); 20151802007@cqu.edu.cn (L.F.); 20161802006@cqu.edu.cn (B.T.)

2 School of Chemistry and Chemical Engineering, Yibin University, Yibin 644000, China

3 School of Material and Chemical Engineering, Tongren University, Tongren 554300, China

* Correspondence: stzhang@cqu.edu.cn (S.Z.); chygl@gztrc.edu.cn (L.G.)

Received: 2 April 2019; Accepted: 23 April 2019; Published: 24 April 2019

Abstract: In this work, two indazole derivatives, namely 5-aminoindazole (AIA) and 5-nitroindazole (NIA), were investigated as corrosion inhibitors for carbon steel in 1 M HCl solution by experimental and density functional theory (DFT) methods. The electrochemical results indicate that the inhibition ability follows the order of AIA > NIA, which is due to the stronger electron-donating effect of $-NH_2$ of the AIA group than the $-NO_2$ group of NIA. Besides, the frontier orbital theory shows that the AIA exhibits higher reaction activity than NIA, and a more negative adsorption energy for AIA was also obtained, which is consistent with the analysis of the electrochemical measurements. We draw the conclusion that the electron-donating effect makes it easier for AIA to donate electrons to iron atoms to form a stronger protective layer than NIA.

Keywords: corrosion inhibitor; carbon steel; indazole derivatives; electrochemistry; DFT

1. Introduction

Carbon steels are extensively utilized as structural materials in plenty of industrial fields, especially in corrosive environments. In view of this, some protective strategies have been employed to protect steel from corrosion. The addition of corrosion inhibitors has been proved to be an easy and highly effective way to achieve this [1–3]. Organic compounds containing heteroatoms such as N, P, S or O have been used as excellent inhibitors due to their strong electron-donating ability [4–6]. Additionally, due to their high adsorption capacity, indazole derivatives have attracted the attention of several researchers [7,8]. Our previous work revealed that indazole can absorb onto copper surfaces and exhibits a favorable inhibition efficiency for copper corrosion in a 3.0 wt.% NaCl solution [7]. Furthermore, the influence of active adsorption, which centers on the inhibition effectiveness of IA-based inhibitors, was also researched in aggressive solutions [9]. Potentially, it is important to explore the effects of functional groups of indazole-based inhibitors on inhibition effectiveness. Herein, 5-aminoindazole (AIA) and 5-nitroindazole (NIA), as shown in Figure 1, were discovered for the first time to be effective corrosion inhibitors for carbon steel in hydrochloric acid (HCl) solution.

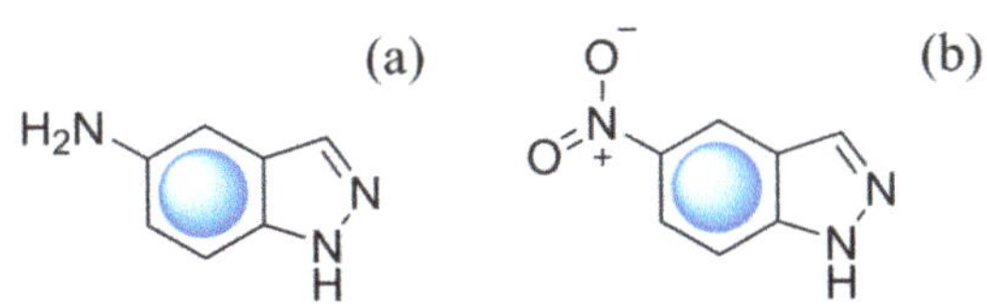

Figure 1. Molecular structures of indazole derivatives, (**a**) 5-aminoindazole; (**b**) 5-nitroindazole.

Specifically, this work focused on the corrosion inhibition of AIA and NIA for Q235 carbon steel in 1 M hydrochloric acid solution by using multiple techniques including electrochemical measurements, scanning electron microscopy (SEM) and DFT calculations. On account of the experimental and theoretical results, the difference of the inhibition mechanisms between AIA and NIA molecules was revealed, which would provide some reference functions to develop more efficient inhibitors for corrosion protection.

2. Experimental Section

2.1. Material and Samples Preparation

The tested inhibitors were AIA and NIA (Aladdin Company, Shanghai, China). The chemical composition of Q235 carbon steel (JIS G3101) was 0.17% C, 0.47% Mn, 0.26% Si, 0.017% S, 0.0048% P and Fe. The specimens for electrochemical measurements were sealed in ethoxyline resin with a 1 cm^2 area exposed as working area. Prior to experiments, all specimens were polished by emery paper from 400 to 1200 grit. Then, samples were cleaned by ethanol and distilled water and dried at room temperature. Analytical reagent grade 37.5% water-diluted HCl was used as the corrosive medium (1 M HCl).

2.2. Electrochemical Measurements

A three-electrode cell was used to do the electrochemical experiments, and all tests were performed in a CHI660B CHI660B electrochemical workstation (Chinstruments, Shanghai, China). Q235 carbon steel, a platinum plate of 1.5 × 1.5 cm^2, and a saturated calomel electrode (SCE) with a Luggin capillary were the working electrode, counter electrode, and reference electrode, respectively. To obtain a steady open circuit potential (OCP), the electrodes were immersed into the corrosive solution for one hour before tests. The potentiodynamic polarization curves were obtained with the potential range from −250 mV to 250 mV vs. OCP at the scan rate of 0.2 mV s^{-1}. Electrochemical impedance spectroscopy (EIS) experiments were performed in a frequency range from 10^{-2} Hz to 10^5 Hz with amplitude of 5 mV at OCP (−0.45 V to −0.49 V). Z-view software was employed to fit the experimental data by an appropriate equivalent circuit.

2.3. Scanning Electron Microscopy

The surface morphologies of Q235 carbon steel specimens were captured with scanning electron microscopy (SEM, Joel-6490LV, Tokyo, Japan) with an accelerating voltage of 20 kV. Before SEM characterization, the samples were immersed for six hours in 1 M HCl with and without AIA or NIA inhibitor.

2.4. Theoretical Simulations

A Gaussian 03W program was employed for quantum chemical calculations based on DFT. The geometry optimized structures of AIA and NIA were obtained through the use of the B3LYP function with a 6-311++G(d, p) basis set. During this process, no imaginary frequency was confirmed and the structures were in their lowest-energy state. Furthermore, DFT was used to calculate the Mulliken charge, dipole moment (μ) and frontier molecular orbitals including the energy of the highest occupied molecular orbital (E_{HOMO}), the energy of the lowest unoccupied molecular orbital (E_{LUMO}) and energy gap ($\Delta E = E_{LUMO} - E_{HOMO}$).

The interactions between AIA or NIA and the Fe (110) surface were modeled in a simulation box (12.4 Å × 9.9 Å × 24.1 Å) with periodic boundary conditions by the $Dmol^3$ program of Material Studio software (BIOVIA, USA). A 4-layer 5 × 5 supercell (the lower two layers were constrained) with 20 Å vacuum slab was used to simulate bulk metal. DFT calculations were treated within the generalized gradient approximation (GGA) function of Perdew–Burke–Ernzerhof (PBE) and the double numerical basis set with polarization functions on hydrogen atoms (DNP). DFT semi-core pseudopots (DSPPs) were used for Fe treatment. The displacement convergence, gradient, and tolerances of energy were 5 ×

10^{-3} Å, 2×10^{-3} Ha·Å^{-1}, and 1×10^{-5} Ha, respectively. The interaction energy ($E_{\text{Fe-inhibitor}}$) between the Fe (110) surface and inhibitor obeyed [10–12]

$$E_{\text{Fe-inhibitor}} = E_{\text{Total}} - E_{\text{Fe}} - E_{\text{inhibitor}} \tag{1}$$

where E_{Total} is the energy of Fe(110) surface and the adsorbed inhibitor molecule, $E_{\text{inhibitor}}$ is the energy of the isolated inhibitor molecule and E_{Fe} is the energy of the steel surface, respectively.

3. Results and Discussion

3.1. Electrochemical Impedance Spectroscopy (EIS) Measurements

Figure 2 shows the electrochemical impedance spectra plots of AIA and NIA, respectively. In Figure 2a,c, it can be seen that the radius of the capacitive resistance arc increased with the growing concentration of inhibitors, which indicates that the protective layer was formed on the steel surface by the adsorption of AIA or NIA and that the corrosion inhibition efficiency of AIA is better than NIA. In addition, the impedance spectrum exhibits a squashed semicircle, which is caused by the formation of a protective layer on the steel surface. Figure 2b,d is Bode plots in the presence and absence of AIA and NIA, respectively. The impedance values and phase angle values increase with the growing concentration of inhibitors. In addition, a time constant can be found in the phase angle, usually due to the relaxation effect of the corrosion inhibitor molecule adsorption [2,13].

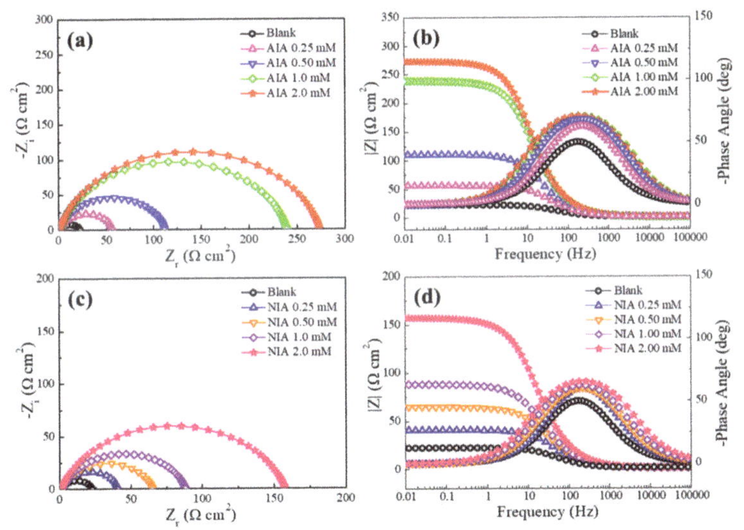

Figure 2. The plots of Nyquist and Bode for carbon steel in 1 M HCl with and without different concentrations of 5-aminoindazole (AIA) and 5-nitroindazole (NIA) at 298 K. AIA: (**a**,**b**); NIA: (**c**,**d**).

The equivalent circuit (Figure 3) is used to fit the impedance spectrum data, and the fitting data are shown in Table 1. In Figure 3, R_s is the solution resistance, R_{ct} is the charge transfer resistance, and CPE is a constant phase element; the impedance of the CPE is expressed as follows [14,15]:

$$Z_{\text{CPE}} = \frac{1}{Y_0 (j\omega)^n} \tag{2}$$

The double-layer capacitance (C_{dl}) can be calculated from CPE parameter values Y_0 and n by the following expression [16]:

$$C_{dl} = \frac{Y_0 \omega^{n-1}}{\sin(\frac{n\pi}{2})} \quad (3)$$

where Y_0 is the CPE constant, n is the phase shift, which can be explained as a degree of surface inhomogeneity, j is the imaginary unit and ω is the angular frequency. The inhibition efficiency η_{EIS} can be expressed by the following equation [17–19]:

$$\eta_{EIS} = (\frac{R_{ct} - R_{ct}^0}{R_{ct}}) \times 100 \quad (4)$$

where R_{ct} and $R_{ct,0}$ are the charge transfer resistances with and without AIA and NIA, respectively. In Table 1, we can see that as the concentration of the corrosion inhibitor increases, the value of the R_{ct} becomes larger, and the value of η_{EIS} also increases simultaneously. When the concentration of the inhibitor was 2 mM, the R_{ct} values of AIA and NIA are 238 and 156.2 Ω cm^2, respectively. According to the Helmholtz model formula, the value of double layer capacitance (C_{dl}) can be expressed as [20,21]

$$C_{dl} = \frac{\varepsilon^0 \varepsilon}{d} S \quad (5)$$

where ε^0 is the dielectric constant of air and ε is the local dielectric constant. S is the surface area of the working electrode, and d is the surface film thickness. Compared with water molecules, the molecular volume of AIA and NIA is significantly larger, and their dielectric constant is smaller than that of water molecules. Therefore, with increasing concentrations of AIA or NIA, the two investigated inhibitors replace the water molecules on the surface of carbon steel continuously, and the value of C_{dl} decreases. Hence, the smaller the C_{dl}, the denser the protective film formed on the surface of the carbon steel by AIA and NIA.

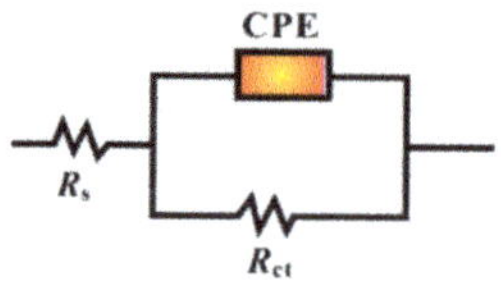

Figure 3. Electrical equivalent circuit used to fit the electrochemical impedance spectroscopy (EIS) experimental data. CPE: constant phase element.

Table 1. Impedance data for Q235 steel in 1 M HCl with various concentrations of NIA and AIA at 298 K.

Inhibitor	C_{inh} (mM)	R_s (Ω cm^2)	R_{ct} (Ω cm^2)	CPE $Y_0 \times 10^{-6}$ (S s^n cm^{-2})	CPE n	C_{dl} (μF cm^{-2})	η_{EIS} (%)
Blank	/	1.43	20.1	504.6	0.85	215.4	/
AIA	0.25	1.26	55.3	214.5	0.85	188.8	62.1
	0.50	1.11	109.9	159.6	0.88	140.4	80.9
	1.00	1.22	238.0	120.3	0.87	104.7	91.2
	2.00	1.36	271.6	113.4	0.87	98.7	92.3
NIA	0.25	1.23	39.6	280.4	0.88	246.8	47.1
	0.50	1.34	63.4	265	0.84	222.6	66.9
	1.00	1.22	86.9	247.4	0.84	207.8	75.9
	2.00	1.14	156.2	185.9	0.83	154.3	86.6

3.2. *Potentiodynamic Polarization Measurements*

Figure 4 shows the electrodynamic polarization curves of carbon steel at 298 K in 1 M HCl solutions with various concentrations of NIA and AIA. The relevant data (in Table 2) are obtained by the extrapolation method. The inhibition efficiency η_P is calculated by the following equation [22–24]:

$$\eta_p = (\frac{i^0_{corr} - i_{corr}}{i^0_{corr}}) \times 100 \quad (6)$$

where $i^0{}_{corr}$ and i_{corr} are the corrosion current density in 1 M HCl solutions without and with the two investigated inhibitors, respectively.

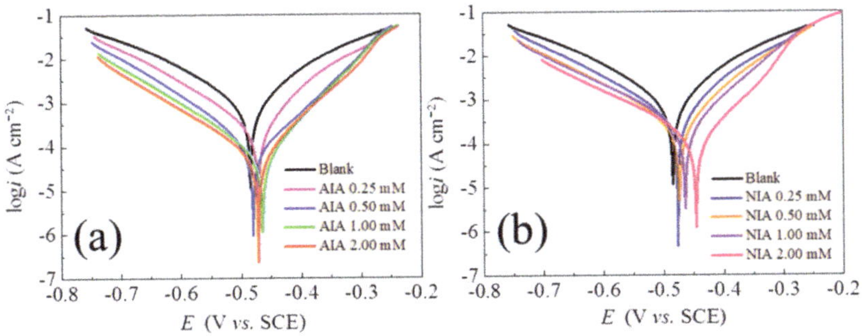

Figure 4. Anodic and cathodic polarization curves for carbon steel in 1 M HCl with various concentrations of (**a**) AIA and (**b**) NIA at 298 K. SCE: saturated calomel electrode.

Table 2. Relevant parameters for Q235 steel in 1 M HCl solution in the absence and presence of different concentrations of AIA and NIA at 298 K from polarization curves.

Inhibitor	C_{inh} (mM)	E_{corr} (V vs. SCE)	β_c (mV dec^{-1})	β_a (mV dec^{-1})	I_{corr} (μA cm^{-2})	η_p (%)
Blank	/	−0.48	119.4	95.3	754.4	/
AIA	0.25	−0.47	112.0	73.9	231.1	69.4
	0.50	−0.48	70.38	78.7	72.2	90.4
	1.00	−0.47	115.3	78.7	59.8	92.1
	2.00	−0.48	81.95	82.7	34.5	95.4
NIA	0.25	−0.48	122.6	82.8	363.7	51.8
	0.50	−0.47	127.7	85.6	248.8	67.0
	1.00	−0.46	132.8	85.8	239.9	68.2
	2.00	−0.45	107.2	74.6	117.4	84.4

It can be seen in Figure 4 that, compared with the values in blank solution, the corrosion potential (E_{corr}) and corrosion current density (I_{corr}) of the polarization curves change obviously with the addition of the two investigated inhibitors. With the addition of AIA and NIA, the I_{corr} decreases and the E_{corr} moves in a positive direction, illustrating that the corrosion reaction is effectively controlled. Clearly, the investigated inhibitors not only reduced the corrosion of the cathode but also reduced the corrosion of the anode. In addition, the shapes of polarization curves are parallel with the increase of concentration for the two investigated inhibitors, indicating that the action mechanism is same under different concentrations of inhibitors. Generally, it is considered as a cathodic inhibitor when the potential change exceeds 85 mV, while it is a mixed inhibitor when the potential change is less than 85 mV [25,26]. From Table 2, the changed values of E_{corr} for the two inhibitors are less than 85 mV, suggesting that AIA and NIA are mixed corrosion inhibitors. The changed values of β_a and β_c also reflect the cathodic and anodic corrosion rates being retarded by the studied inhibitors. Furthermore, the inhibition efficiency (η_p) is also improved with increasing concentrations of the two inhibitors, and the inhibition ability follows the order AIA > NIA, which may be due to the stronger electron-donating effect of $-NH_2$ than the $-NO_2$ [27,28]. The electron-donating effect of the $-NH_2$ makes the electron cloud density of the whole AIA molecule larger than the $-NO_2$ of NIA, which leads to AIA finding it easier to give electrons to iron atoms and having a better protective effect than NIA.

3.3. Morphology Analysis

To obviously display the inhibition differences between AIA and NIA inhibitors, the morphologies of untreated and treated carbon steel were studied. The obtained images are shown in Figure 5. The surface of fresh carbon steel is smooth, while large holes and cracks with a size of 10 μm were observed after immersion in 1 M HCl (Figure 5a,b). The pitting morphology is formed due to the elimination rate of corrosion products slower than the reaction rate between Cl^- and Fe [29,30]. Besides, some small holes and cracks still appeared on the steel surface even after 2 mM NIA was added into the HCl solution (Figure 5d). However, a flat surface was achieved with the addition of 2 mM AIA (Figure 5c); meanwhile the steel surface was covered with the absorbed AIA. Hence, the AIA shows prior inhibition performance than NIA. These results are in good agreement with electrochemical measurements.

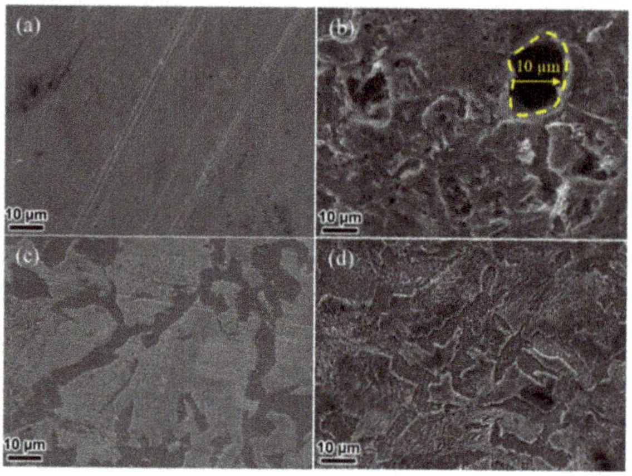

Figure 5. SEM images of (**a**) fresh carbon steel and carbon steel immersed in 1 M HCl solution (**b**) without and with 2 mM (**c**) AIA or (**d**) NIA.

3.4. Computational Study

Computational simulation is an effective way to explain reaction mechanisms. In this work, DFT calculations were used to reveal the adsorption and inhibition performances of AIA and NIA molecules. By comparison, the acidity coefficient (pK_a) of both AIA and NIA displayed higher values than the pH of 1M HCl medium, proving the existence of the protonated molecule. Specifically, there are two pK_a (1.89 and 3.42) for AIA, indicating the existing form of AIA-$2H^+$ in HCl solution. Instead, there is only one pK_a for NIA, indicating the existence of NIA-H^+. The optimized geometry structure and frontier molecular orbitals (the highest occupied molecular orbital (HOMO) and the lowest unoccupied molecular orbital (LUMO)) of AIA-$2H^+$ and NIA-H^+ are shown in Figure 6, and Table 3 shows the energy of HOMO (E_{HOMO}) and LUMO (E_{LUMO}). It is generally known that HOMO is related to the ability of a molecule to donate electrons, and a higher E_{HOMO} value shows a stronger electron-donating ability [31,32]. By contrast, LUMO is associated with the electron-accepting ability of a molecule, and a lower value of E_{LUMO} represents a strong electron-accepting ability [33]. As can be seen from Figure 6c,d, the LUMO of both AIA-$2H^+$ and NIA-H^+ is distributed uniformly around the whole molecule. Conversely, in comparison with AIA-$2H^+$ (Figure 6f), the HOMO of NIA-H^+ was mainly delocalized around the nitro-substituent (Figure 6e), indicating that the electron-donating ability of NIA mainly comes from the nitro-substituent. According to the frontier orbitals, both AIA and NIA tend to interact with the steel surface through the π bond of rings in the way of parallel adsorption configuration.

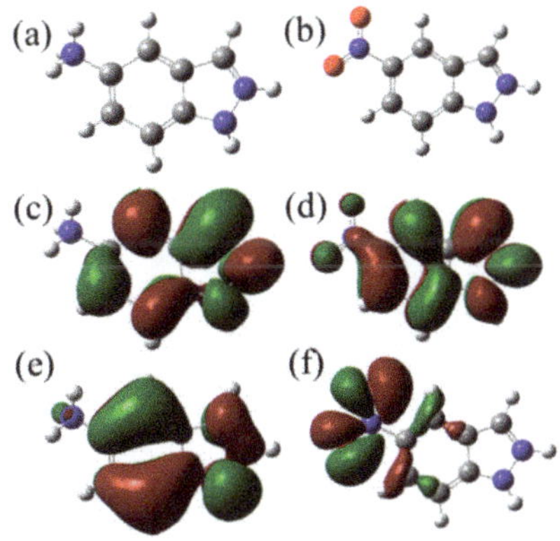

Figure 6. (**a**,**b**) Optimized geometric structures, (**c**,**d**) LUMO orbitals and (**e**,**f**) HOMO orbitals of AIA and NIA inhibitors.

Table 3. Quantum chemical parameters for AIA and NIA by using the B3LYP/6-311 + + G(d,p) method.

Inhibitor	E_{HOMO} (eV)	E_{LUMO} (eV)	ΔE (eV)	μ (Debye)	I (eV)	A (eV)	χ (eV)	γ (eV)	ΔN
AIA	−9.6	−6.2	3.4	5.4	9.6	6.2	7.9	1.7	−0.25
NIA	−11.7	−7.1	4.6	14.3	11.7	7.1	9.4	2.3	−0.53

In addition, the HOMO–LUMO gap (ΔE) is an important parameter to evaluate the stability of inhibitors, and the lower value of ΔE indicates that the inhibitor molecule could more easily adsorb on the metal surface [34,35]. As shown in Table 3, both AIA-2H$^+$ and NIA-H$^+$ have lower values of ΔE (3.4 eV and 4.6 eV, respectively), resulting in their strong ability to accept electrons from the d-orbital of steel as well as the high stability of the [Fe-inhibitor] complexes; namely, the AIA exhibited higher reaction activity than NIA. At the same time, the dipole–dipole (μ) interaction between the inhibitor and metal surface could improve the inhibition efficiency [36]. Herein, the fact that μ_{NIA} is about three times μ_{AIA} indicates that AIA exhibits more appropriate adsorption between the AIA and metal surface than NIA.

The ionization potential ($I = -E_{HOMO}$) and electron affinity ($A = -E_{LUMO}$) could be used to derive the electronegativity (χ) and global hardness (γ). The fraction of the electron transfer (ΔN) between the inhibitor molecules and Fe surface is given by following equation [37–39]:

$$\Delta N = \frac{\chi_{Fe} - \chi_{inh}}{2(\gamma_{Fe} + \gamma_{inh})} \quad (7)$$

where χ_{Fe} and γ_{Fe} are the absolute electronegativity and hardness of the Fe atom; and χ_{inh} and γ_{inh} are the absolute electronegativity and hardness of the inhibitor molecules. A theoretical χ_{Fe} value of bulk Fe is 7 eV/mol, whereas γ_{Fe} is almost zero. χ_{inh} and γ_{inh} are related to I and A [40,41].

$$\chi = \frac{I + A}{2} \quad (8)$$

$$\gamma = \frac{I - A}{2} \quad (9)$$

The direction of electron transfer is manifested by positive or negative ΔN values [42]. From Table 3, both AIA and NIA are electron acceptors. It is noteworthy that the magnitude of ΔN's absolute value is not connected with inhibition efficiency.

AIA-2H$^+$ and NIA-H$^+$ were placed in a simulation box parallel with or perpendicular to the Fe(110) surface. The simulation results showed that both AIA-2H$^+$ and NIA-H$^+$ tended to adsorb in parallel on the Fe(110) surface, as shown in Figure 7. Namely, the indazole and aromatic rings were the adsorption sites, which was in agreement with previous reports. Besides, all the hydrogen atoms upturning after adsorption may be due to the hybridization between Fe and heavy atoms. The AIA molecule is possibly a more efficient inhibitor because of its more negative adsorption energy (−4.65 eV) than NIA (−4.05 eV). This is consistent with the analysis of the electrochemical measurements.

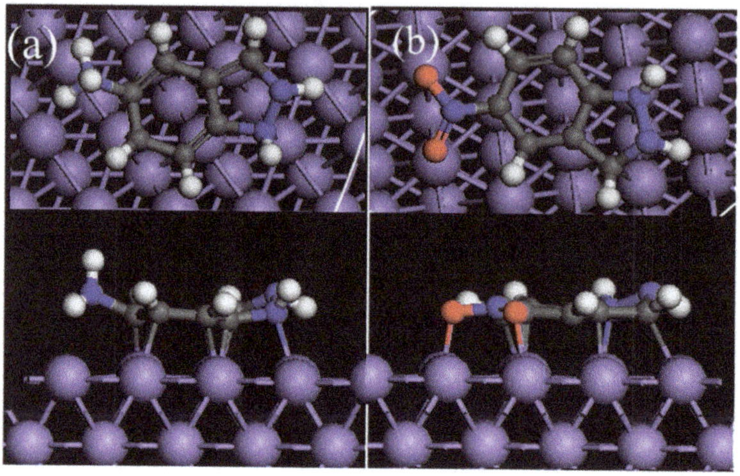

Figure 7. Stable adsorption configurations (side and top view) of (**a**) AIA-2H$^+$ and (**b**) NIA-H$^+$ molecules on the Fe(110) surface.

Figure 8 shows the projected density states of AIA-2H$^+$ and NIA-H$^+$ before and after adsorbing on the Fe(110) surface. By comparing these with the isolate inhibitors, the p orbitals of the adsorbed inhibitors almost disapear, revealing the strong interaction between AIA or NIA and the Fe(110) surface [43]. This is consistant with the inhibition efficencies obtained by experiments.

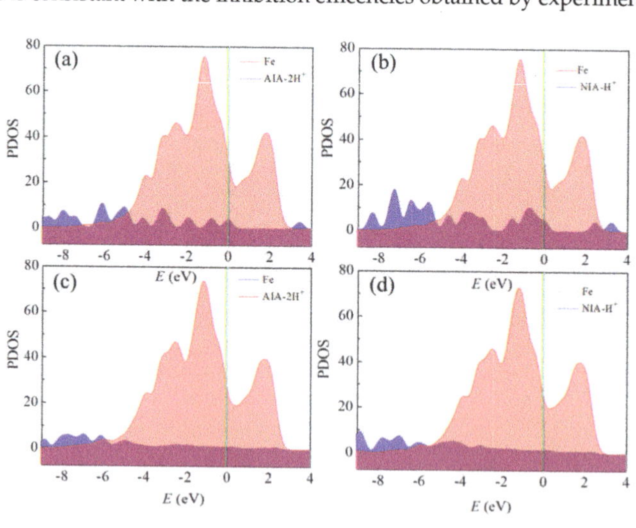

Figure 8. Density states projected of (**a**,**c**) AIA-2H$^+$ and (**b**,**d**) NIA-H$^+$ molecules before and after adsorbing on the Fe(110) surface.

4. Conclusions

In this study, two indazole derivatives, AIA and NIA, were proved to be excellent corrosion inhibitors for carbon steel in 1 M HCl. The inhibition performance was tested by electrochemical methods. Theoretical calculations were also performed to reveal the inhibition mechanism of AIA and NIA. The detailed results are as follows:

(1) The results of electrochemical tests indicated that AIA and NIA are efficient inhibitors for carbon steel in 1M HCl. The inhibition efficiency increased with increasing concentrations of the inhibitors, and the optimal concentration of AIA and NIA is 2 mM. By comparison, the AIA exhibits better inhibition performance than NIA.
(2) The values of the charge transfer resistance increased in the presence of AIA and NIA in EIS tests, indicating that they can protect steel from corrosion by forming a robust protective film. Additionally, the Tafel plots illustrated that both are mixed-type inhibitors.
(3) The results of theoretical calculations explained that the protective effect was due to the electrostatic forces between the AIA-$2H^+$ (or NIA-H^+) and electronegative surface.

Author Contributions: S.Z. and L.G. proposed the concept and were involved in the design of the experiments. S.X. and L.F. performed the experimental work and wrote the main manuscript text. S.X. and B.T. evaluated the inhibition performance using theoretical calculations. All authors were involved in the drafting, revision and approval of the manuscript.

Funding: This research was supported by National Natural Science Foundation of China (21706195, 21878029), Science and Technology Program of Guizhou Province (QKHJC2016-1149), Guizhou Provincial Department of Education Fundation (QJHKYZ2016-105).

Conflicts of Interest: The authors declare that the research was conducted in the absence of any commercial or financial relationships that could be construed as a potential conflict of interest.

References

1. Feng, L.; Zhang, S.; Qiang, Y.; Xu, S.; Tan, B.; Chen, S. The synergistic corrosion inhibition study of different chain lengths ionic liquids as green inhibitors for X70 steel in acidic medium. *Mater. Chem. Phys.* **2018**, *215*, 229–241. [CrossRef]
2. Guo, L.; Zhu, S.; Li, W.; Zhang, S. Electrochemical and Quantum Chemical Assessment of 2-Aminothiazole as Inhibitor for Carton Steel in Sulfuric Acid Solution. *Asian J. Chem.* **2015**, *27*, 2917–2923. [CrossRef]
3. Tan, B.C.; Zhang, S.T.; Liu, H.Y.; Guo, Y.W.; Qiang, Y.J.; Li, W.P.; Guo, L.; Xu, C.L.; Chen, S.J. Corrosion inhibition of X65 steel in sulfuric acid by two food flavorants 2-isobutylthiazole and 1-(1,3-Thiazol-2-yl) ethanone as the green environmental corrosion inhibitors: Combination of experimental and theoretical researches. *J. Colloid Interface Sci.* **2019**, *538*, 519–529. [CrossRef] [PubMed]
4. Tan, J.H.; Guo, L.; Lv, T.M.; Zhang, S.T. Experimental and Computational Evaluation of 3-indolebutyric Acid as a Corrosion Inhibitor for Mild Steel in Sulfuric Acid Solution. *Int. J. Electrochem. Sci.* **2015**, *10*, 823–837.
5. Zhang, S.T.; Tao, Z.H.; Liao, S.G.; Wu, F.J. Substitutional adsorption isotherms and corrosion inhibitive properties of some oxadiazol triazole derivative in acidic solution. *Corros. Sci.* **2010**, *52*, 3126–3132. [CrossRef]
6. Zheng, X.W.; Zhang, S.T.; Li, W.P.; Yin, L.L.; He, J.H.; Wu, J.F. Investigation of 1-butyl-3-methyl-1H-benzimidazolium iodide as inhibitor for mild steel in sulfuric acid solution. *Corros. Sci.* **2014**, *80*, 383–392. [CrossRef]
7. Qiang, Y.J.; Zhang, S.T.; Xu, S.Y.; Yin, L.L. The effect of 5-nitroindazole as an inhibitor for the corrosion of copper in a 3.0% NaCl solution. *RSC Adv.* **2015**, *5*, 63866–63873. [CrossRef]
8. Qiang, Y.; Zhang, S.; Xiang, Q.; Tan, B.; Li, W.; Chen, S.; Guo, L. Halogeno-substituted indazoles against copper corrosion in industrial pickling process: A combined electrochemical, morphological and theoretical approach. *RSC Adv.* **2018**, *8*, 38860–38871. [CrossRef]
9. Qiang, Y.; Zhang, S.; Yan, S.; Zou, X.; Chen, S. Three indazole derivatives as corrosion inhibitors of copper in a neutral chloride solution. *Corros. Sci.* **2017**, *126*, 295–304. [CrossRef]
10. Kokalj, A. Formation and structure of inhibitive molecular film of imidazole on iron surface. *Corros. Sci.* **2013**, *68*, 195–203. [CrossRef]

11. Kokalj, A.; Peljhan, S. Density Functional Theory Study of ATA, BTAH, and BTAOH as Copper Corrosion Inhibitors: Adsorption onto Cu(111) from Gas Phase. *Langmuir* **2010**, *26*, 14582–14593. [CrossRef]
12. Kokalj, A.; Peljhan, S.; Koller, J. The Effect of Surface Geometry of Copper on Dehydrogenation of Benzotriazole. Part II. *J. Phys. Chem. C* **2014**, *118*, 944–954. [CrossRef]
13. Tan, B.; Zhang, S.; Qiang, Y.; Guo, L.; Feng, L.; Liao, C.; Xu, Y.; Chen, S. A combined experimental and theoretical study of the inhibition effect of three disulfide-based flavouring agents for copper corrosion in 0.5 M sulfuric acid. *J. Colloid Interface Sci.* **2018**, *526*, 268–280. [CrossRef]
14. Wang, J. The Inhibition Effect of a Novel Mannich Base on the Corrosion of A3 Mild Steel in 1.0 M Hydrochloric Acid Solution. *Int. J. Electrochem. Sci.* **2016**, *11*, 2621–2637. [CrossRef]
15. Yadav, M.; Gope, L.; Kumari, N.; Yadav, P. Corrosion inhibition performance of pyranopyrazole derivatives for mild steel in HCl solution: Gravimetric, electrochemical and DFT studies. *J. Mol. Liq.* **2016**, *216*, 78–86. [CrossRef]
16. Tao, Z.H.; Zhang, S.T.; Li, W.H.; Hou, B.R. Adsorption and Inhibitory Mechanism of 1H-1,2,4-Triazol-l-yl-methyl-2-(4-chlorophenoxy) Acetate on Corrosion of Mild Steel in Acidic Solution. *Ind. Eng. Chem. Res.* **2011**, *50*, 6082–6088. [CrossRef]
17. Zhang, J. Inhibition Effect and Adsorption Behavior of Two Imidazolium- based Ionic Liquids on X70 Steel in Sulfuric Acid Solution. *Int. J. Electrochem. Sci.* **2018**, *13*, 8645–8656. [CrossRef]
18. Ukpong, I.; Bamgboye, O.; Soriyan, O. Synergistic Inhibition of Mild Steel Corrosion in Seawater and Acidic Medium by Cathodic Protection and Monodora myristica Using Zinc Anode. *Int. J. Corros.* **2018**, *2018*, 1–8. [CrossRef]
19. Kalaiselvi, K.; Chung, I.-M.; Kim, S.-H.; Prabakaran, M. Corrosion resistance of mild steel in sulphuric acid solution by Coreopsis tinctoria extract: electrochemical and surface studies. *Anti Corros. Methods Mater.* **2018**, *65*, 408–416. [CrossRef]
20. Paramasivam, S. Corrosion Inhibition of Mild Steel in Hydrochloric Acid using4- (pyridin-2yl)-N-p-tolylpiperazine-1-carboxamide. *Int. J. Electrochem. Sci.* **2016**, *11*, 3393–3414. [CrossRef]
21. Ma, X.; Jiang, X.; Xia, S.; Shan, M.; Li, X.; Yu, L.; Tang, Q. New corrosion inhibitor acrylamide methyl ether for mild steel in 1 M HCl. *Appl. Surf. Sci.* **2016**, *371*, 248–257. [CrossRef]
22. Tasic, Z.Z.; Mihajlovic, M.B.P.; Radovanovic, M.B.; Antonijevic, M.M. Effect of gelatine and 5-methyl-1H-benzotriazole on corrosion behaviour of copper in sulphuric acid containing Cl^- ions. *J. Adhes. Sci. Technol.* **2017**, *31*, 2592–2610. [CrossRef]
23. Jiang, L.; Ma, H.; Zhang, J.; Lu, Y.; Lu, H.; Meng, X. Electro Polymerization of Polypyrrole Coatings Doped with Different Proton Acids for Corrosion Protection of 304 Stainless Steel. *MATEC Web Conf.* **2017**, *109*, 03007. [CrossRef]
24. El-Sayed Shehata, O. Effect of acetamide derivative and its Mn-complex as corrosion inhibitor for mild steel in sulphuric acid. *Egypt. J.Chem.* **2017**, *60*, 243–259. [CrossRef]
25. Loto, R.T.; Loto, C.A.; Fedotova, T. Electrochemical studies of mild steel corrosion inhibition in sulfuric acid chloride by aniline. *Res. Chem. Int.* **2014**, *40*, 1501–1516. [CrossRef]
26. Tamilarasan, R.; Sreekanth, A. Spectroscopic and DFT investigations on the corrosion inhibition behavior of tris(5-methyl-2-thioxo-1,3,4-thiadiazole) borate on high carbon steel and aluminium in HCl media. *RSC Adv.* **2013**, *3*, 23681–23691. [CrossRef]
27. Tang, Y.-M.; Yang, W.-Z.; Yin, X.-S.; Liu, Y.; Wan, R.; Wang, J.-T. Phenyl-substituted amino thiadiazoles as corrosion inhibitors for copper in 0.5 M H_2SO_4. *Mater. Chem. Phys.* **2009**, *116*, 479–483. [CrossRef]
28. Kovacevic, N.; Kokalj, A. DFT Study of Interaction of Azoles with Cu(111) and Al(111) Surfaces: Role of Azole Nitrogen Atoms and Dipole–Dipole Interactions. *J. Phys. Chem. C* **2011**, *115*, 24189–24197. [CrossRef]
29. Loto, R.T.; Loto, C.A. Pitting corrosion inhibition of type 304 austenitic stainless steel by 2-amino-5-ethyl-1,3,4-thiadiazole in dilute sulphuric acid. *Prot. Met. Phys. Chem. Surfaces* **2015**, *51*, 693–700. [CrossRef]
30. Bensabra, H.; Azzouz, N.; Chopart, J.P. Effect of zinc phosphating treatment on the pitting corrosion resistance of steel reinforcement. *Rev. De Metall. Cah. D Inf. Technol.* **2013**, *110*, 153–163.
31. Kokalj, A. On the HSAB based estimate of charge transfer between adsorbates and metal surfaces. *Chem. Phys.* **2012**, *393*, 1–12. [CrossRef]
32. Kokalj, A.; Peljhan, S.; Finsgar, M.; Milosev, I. What Determines the Inhibition Effectiveness of ATA, BTAH, and BTAOH Corrosion Inhibitors on Copper? *J. Am. Chem. Soc.* **2010**, *132*, 16657–16668.

33. Obot, I.B.; Obi-Egbedi, N. 2,3-Diphenylbenzoquinoxaline: A new corrosion inhibitor for mild steel in sulphuric acid. *Corros. Sci.* **2010**, *52*, 282–285. [CrossRef]
34. Umoren, S.; Obot, I.B.; Gasem, Z.; Odewunmi, N.A. Experimental and Theoretical Studies of Red Apple Fruit Extract as Green Corrosion Inhibitor for Mild Steel in HCl Solution. *J. Dispers. Sci. Technol.* **2015**, *36*, 789–802. [CrossRef]
35. Umoren, S.A.; Obot, I.B.; Madhankumar, A.; Gasem, Z.M. Effect of degree of hydrolysis of polyvinyl alcohol on the corrosion inhibition of steel: Theoretical and experimental studies. *J. Adhes. Sci. Technol.* **2015**, *29*, 271–295. [CrossRef]
36. Ren, X.L.; Xu, S.Y.; Chen, S.Y.; Chen, N.X.; Zhang, S.T. Experimental and theoretical studies of triisopropanolamine as an inhibitor for aluminum alloy in 3% NaCl solution. *RSC Adv.* **2015**, *5*, 101693–101700. [CrossRef]
37. Pearson, R.G. Hard and Soft Acids and Bases. *J. Am. Chem. Soc.* **1963**, *85*, 3533–3539. [CrossRef]
38. Pearson, R.G. Absolute electronegativity and hardness: Application to inorganic chemistry. *Inorg. Chem.* **1988**, *27*, 734–740. [CrossRef]
39. Obot, I.B.; Kaya, S.; Kaya, C.; Tuzun, B. Theoretical evaluation of triazine derivatives as steel corrosion inhibitors: DFT and Monte Carlo simulation approaches. *Res. Chem. Intermed.* **2016**, *42*, 4963–4983. [CrossRef]
40. Khadraoui, A.; Khelifa, A.; Hadjmeliani, M.; Mehdaoui, R.; Hachama, K.; Tidu, A.; Azari, Z.; Obot, I.; Zarrouk, A. Extraction, characterization and anti-corrosion activity of Mentha pulegium oil: Weight loss, electrochemical, thermodynamic and surface studies. *J. Mol. Liq.* **2016**, *216*, 724–731. [CrossRef]
41. Belghiti, M.E.; Karzazi, Y.; Dafali, A.; Obot, I.B.; Ebenso, E.E.; Emran, K.M.; Bahadur, I.; Hammouti, B.; Bentiss, F. Anti-corrosive properties of 4-amino-3,5-bis(disubstituted)-1,2,4-triazole derivatives on mild steel corrosion in 2 M H3PO4 solution: Experimental and theoretical studies. *J. Mol. Liq.* **2016**, *216*, 874–886. [CrossRef]
42. Obot, I.B.; Gasem, Z.M.; Umoren, S.A. Understanding the Mechanism of 2-mercaptobenzimidazole Adsorption on Fe(110), Cu(111) and Al(111) Surfaces: DFT and Molecular Dynamics Simulations Approaches. *Int. J. Electrochem. Sci.* **2014**, *9*, 2367–2378.
43. Kovacevic, N.; Milosev, I.; Kokalj, A. The roles of mercapto, benzene, and methyl groups in the corrosion inhibition of imidazoles on copper: II. Inhibitor-copper bonding. *Corros. Sci.* **2015**, *98*, 457–470.

materials

MDPI

Article

Mannich Base as Corrosion Inhibitors for N80 Steel in a CO_2 Saturated Solution Containing 3 wt % NaCl

Mingjin Tang [1,*], Jianbo Li [1,*], Zhida Li [2], Luoping Fu [1], Bo Zeng [1] and Jie Lv [1]

1 School of Chemistry and Chemical Engineering, Southwest Petroleum University, Chengdu 610500, China; fuluoping@outlook.com (L.F.); zengbo_2014@icloud.com (B.Z.); jlv6@me.com (J.L.)
2 Faculty of Engineering, Computer & Mathematical Sciences, The University of Adelaide, Adelaide, SA 5005, Australia; lzd038@163.com
* Correspondence: tangmingjin@outlook.com (M.T.); ljb0418@163.com (J.L.)

Received: 14 December 2018; Accepted: 30 January 2019; Published: 1 February 2019

Abstract: In this paper, a corrosion inhibitor containing nitrogen atoms and a conjugated π bond was synthesised, and its final product synthesised by the optimal conditions of the orthogonal test results is named multi-mannich base (MBT). The corrosion inhibition effect on the N80 steel sheet of the corrosion inhibitor was evaluated in a CO_2 saturated solution containing 3 wt % NaCl; the corrosion rate was 0.0446 mm/a and the corrosion inhibition rate was 90.4%. Through electrochemical and adsorption theory study, MBT is a mixed corrosion inhibitor that mainly shows cathode suppression capacity. The adsorption of MBT on the surface of the steel sheet follows the Langmuir adsorption isotherm; it can be spontaneously adsorbed on the surface of the N80 steel sheet, which has a good corrosion inhibition effect. The surface of the N80 steel sheet was microscopically characterised by atomic force microscope (AFM). It can be seen from the results that the N80 steel sheet with MBT added is significantly different from the blank control group; the surface of the steel sheet is relatively smooth, indicating that MBT forms an effective protective film on the surface of N80 steel, which inhibits the steel sheet.

Keywords: corrosion inhibitor; electrochemical; AFM; CO_2 corrosion

1. Introduction

To improve oil and gas recovery, CO_2 injection is used in the exploitation of oil, natural gas, coalbed methane and shale gas [1]. However, CO_2 injection is accompanied by CO_2 corrosion [2]. Corrosion will decrease the mining yield, resulting in oil wells failure, shutting down and even cause safety accidents, which will have a serious impact on the development of the oilfield and the economic benefits of the market [3–6]. The most common and effective way to control corrosion is to use chemical additives that reduce corrosion at very small dosages [7,8]. In recent years, various corrosion inhibitors have been successfully applied to the industry, and the corrosion has been controlled obviously and effectively [9,10].

So far, phosphorus and nitrogen-containing compounds have been widely used as corrosion inhibitors, including organic phosphorus, nitrogen-containing compounds and phosphorus-containing organic water-soluble polymers [11]. Phosphorus-containing corrosion inhibitors have excellent corrosion inhibition properties. However, they have some fatal defects; for example, phosphorus compounds are difficult to degrade in water and usually lead to eutrophication. The widespread use of this chemical may cause oxygen deficiency in the water, which leads to the death of aquatic organisms [12]. Therefore, in the last few years, the industry has not only required the effectiveness of chemical compounds but also their safety [13]. Today, chemical emissions are strictly controlled through legislation. Therefore, finding an alternative solution, namely the effective control of corrosion by green corrosion inhibitors, is the most important research direction at the moment. Chemical products identified as "green" are based on criteria: non-toxic and degradable [14,15].

In this paper, salicylaldehyde, diethylenetriamine, formaldehyde and acetone were selected to synthesise corrosion inhibitor MBT containing nitrogen atoms with lone pairs and a conjugated π bond. The structure was characterised by infrared radiation spectroscopy (IR) and nuclear magnetic resonance spectroscopy (^{1}H NMR). The product was evaluated for corrosion inhibition by a weight loss method. Finally, the mechanism of corrosion inhibition was studied by atomic force microscopy (AFM) and electrochemical methods.

2. Experiment

2.1. Main Experimental Materials and Instruments

Salicylaldehyde $C_7H_6O_2$, diethylenetriamine $C_4H_{13}N_3$, ethanol C_2H_6O, acetone C_3H_6O, formaldehyde HCHO: AR, Chengdu Kelong Chemical Reagent Factory, Chengdu, China.

WQF-520 infrared spectrometer: Beijing Ruili Analytical Instrument (Group) Co. Ltd., Beijing, China; Bruker AVANCE III HD 400 nuclear magnetic resonance spectrometer: Bruker Corporation, Billerica, MA, USA; IVICMSTAT electrochemical workstation: Ivium Technologies, Eindhoven, The Netherlands; SPM-9600 atomic force microscope: Shimadzu Corporation, Kyoto, Japan.

2.2. Synthesis of Corrosion Inhibitor

First, in a three-necked flask equipped with a condenser and a thermometer, 20 mL of ethanol was added as a solvent. When the reactant reached a certain temperature, a certain amount of salicylaldehyde and diethylenetriamine were added in a certain ratio. After a period of reaction, the reaction was terminated to give the product intermediate mannich base (MB-1).

Then, a certain amount of acetone and formaldehyde were added in order, being kept at a constant reaction temperature for a certain period, and then the reaction was terminated to obtain the target product mannich base (MB-2). The synthesis of MB-1 and MB-2 is shown in Schemes 1 and 2.

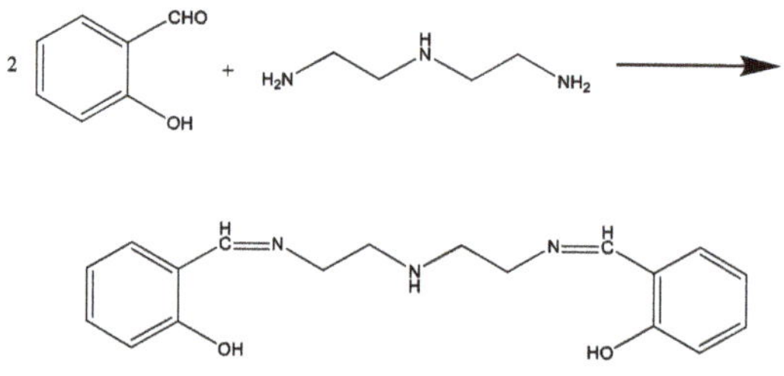

Scheme 1. The synthesis of MB-1.

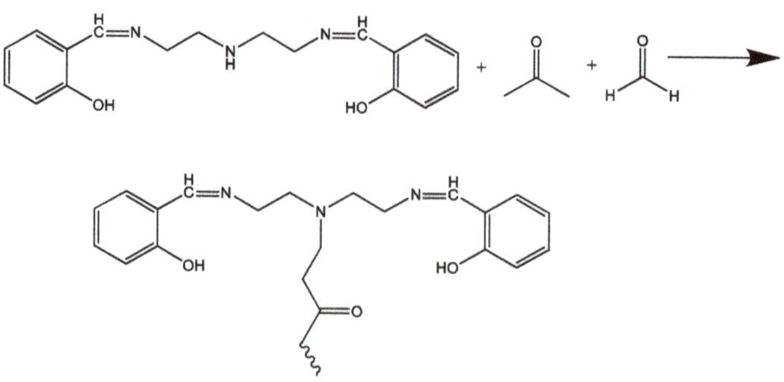

Scheme 2. The synthesis of MB-2.

2.3. Weight Loss Measurements

A hanging N80 steel test piece (50.00 mm × 13.00 mm × 1.50 mm: length, width, height) was ground with sand paper to give a surface with a uniform metal luster [16]. To calculate the superficial area of the test piece, the carbon steel was deoiled with putty powder and then immersed in ethanol for 5 min, dried with an air blower, wrapped in filter paper, maintained in a dryer and weighed to 0.0001 g.

Then the test piece was immersed in a bottle containing 500 mL of CO_2 saturated solution and 3 wt % NaCl saline water with and without inhibitors at 70 °C for 72 h. According to the weight loss of each test piece and Equations (1) and (2), the rate of corrosion of the test piece and efficiency of the corrosion inhibitor at different concentrations could be calculated [4].

$$v = \frac{8.76 \times 10^4 \times \Delta m}{\rho A t}, \tag{1}$$

where ν, Δm, ρ, A and t are the corrosion rate (mm/a), weight loss (g), test sample density (g/cm^3), exposed sample area (cm^2) and immersion time (h), respectively. The inhibition efficiency (η) of the inhibitor was calculated by Equation (2):

$$\eta(\%) = \left(1 - \frac{v}{v^0}\right) \times 100, \tag{2}$$

where ν and ν^0 are the corrosion rates calculated from the weight loss with and without the inhibitor.

2.4. Electrochemical Measurements

The potentiodynamic polarisation curve method was used to measure the Tafel polarisation curve. When the self-corrosion potential (Ec) of the system was stable, the cathode and anode scanning was from 250 mV to 300 mV. The polarisation curves of the N80 steel sheets with different concentrations of corrosion inhibitors were measured [4]. According to the relevant theory of electrochemistry, the corrosion inhibition rate (η) of the corrosion process can be calculated by Equation (3) [17]:

$$\eta = \frac{i^0 - i}{i^0} \times 100, \tag{3}$$

where i^0 and i are the corrosion current densities without and with the inhibitor, respectively.

The electrochemical impedance spectroscopy (EIS, Ivium Technologies, Eindhoven, The Netherlands) experiments were performed by applying a sinusoidal voltage signal of 10 mV; the frequency range was 0.01–10^{-5} Hz; and the alternating current excitation signal amplitude was 10 mV. Impedance diagrams of the N80 steel sheets with different concentrations of corrosion inhibitors were measured. The inhibition efficiency (η) was determined from the EIS data using Equation (4):

$$\eta = \frac{R_t - R_t^0}{R_t} \times 100, \tag{4}$$

where R_t and $R_t{}^0$ are the charge transfer resistance in the absence and presence of inhibitors, respectively.

2.5. AFM Analysis

Atomic force microscopy (AFM, Shimadzu Corporation, Kyoto, Japan) was used for characterizing the surface morphology and measuring surface roughness in a CO_2 saturated solution containing 3 wt % NaCl without and with corrosion inhibitors. The two- and three-dimensional AFM images of steel specimens were taken in the range from 0 to 5 μm at room temperature [18].

3. Results and Discussion

3.1. IR Characterisation Results

The IR spectrum of MB-2 is shown in Figure 1. It is demonstrated that the stretching vibration peak attributed to the carbonyl group in salicylaldehyde does not appear at 1740 cm^{-1} to 1720 cm^{-1}. The stretching vibration peak attributed to the C=N bond at 1644 cm^{-1} indicates that the salicylaldehyde reacts with diethylenetriamine to form a compound containing C=N and the aldehyde group disappears. The absorption peak at 3457 cm^{-1} and 1270 cm^{-1} is attributed to the stretching vibration of the O–H bond and C–O bond of phenol; the absorption peak at 2989 cm^{-1} and 2923 cm^{-1} is attributed to the stretching vibration of methyl (CH_3) and methylene (CH_2); the bending vibration of the N–H bond belonging to the secondary amine did not occur at 1580 cm^{-1} to 1490 cm^{-1}, indicating that the hydrogen on the secondary amine reacted with acetone and formaldehyde, and the secondary amine disappeared into a tertiary amine; the absorption peak at 1718 cm^{-1} is attributed to the stretching vibration of C=O. IR characterization results showed that the synthesized MB-2 was the target product.

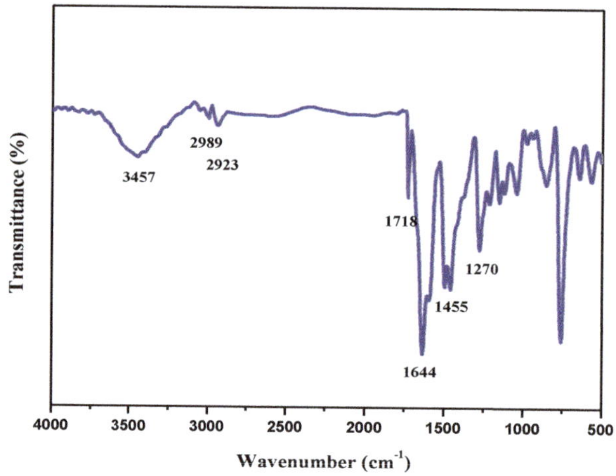

Figure 1. IR spectrum of MB-2.

3.2. Optimization of MB-2 Synthesis Conditions

Through orthogonal experiments, the three-factor four-level $L_{16}(4^3)$ orthogonal table was selected to investigate the influencing factors of MB-2 synthesis, including the ratio of reactants, reaction time and reaction temperature. Considering that both methyl groups of acetone react, the amount of acetone is fixed to adjust the ratio of MB-2 to formaldehyde. According to the product synthesised by the orthogonal experiment, when the amount was 200 mg/L, the corrosion inhibition performance was evaluated in a CO_2 saturated solution containing 3 wt % NaCl.

The experimental factors and levels are shown in Table 1. The results of orthogonal experiments are shown in Table 2. It can be seen from Table 2 that the optimal reaction conditions for the synthesis of MB-2 are: reaction temperature 85 °C; reactant ration(MB-2):n(formaldehyde):n(acetone) = 2:2:1; and reaction time 6 h. The final product synthesised by the optimal conditions of the orthogonal test results is named MBT, and the reaction equation for the synthesis is shown in Scheme 3.

Table 1. Experimental factors and levels.

Level	A n(MB-2):n(HCHO):n(CH_3COCH_3)	B Temperature/°C	C Time (h)
1	1:1:1	70	4
2	1.5:1.5:1	75	5
3	2:2:1	80	6
4	2.5:2.5:1	85	7

Table 2. Orthogonal experimental results.

NO.	A	B	C	Corrosion Rate (mm/a)
1	1	1	1	0.180
2	1	2	2	0.143
3	1	3	3	0.076
4	1	4	4	0.059
5	2	1	2	0.135
6	2	2	1	0.139
7	2	3	4	0.068
8	2	4	3	0.056
9	3	1	3	0.098
10	3	2	4	0.073
11	3	3	1	0.102
12	3	4	2	0.067
13	4	1	4	0.120
14	4	2	3	0.084
15	4	3	2	0.079
16	4	4	1	0.106
I	289.2	259.8	303.2	
II	296.7	302.9	301.0	
III	341.0	338.7	307.8	
IV	308.6	334.2	322.7	
K_1	0.115	0.133	0.132	
K_2	0.100	0.110	0.106	
K_3	0.085	0.081	0.079	
K_4	0.097	0.072	0.080	
R	0.03	0.061	0.053	
Effect order			B > C > A	
Optimal level	A_3	B_4	C_3	

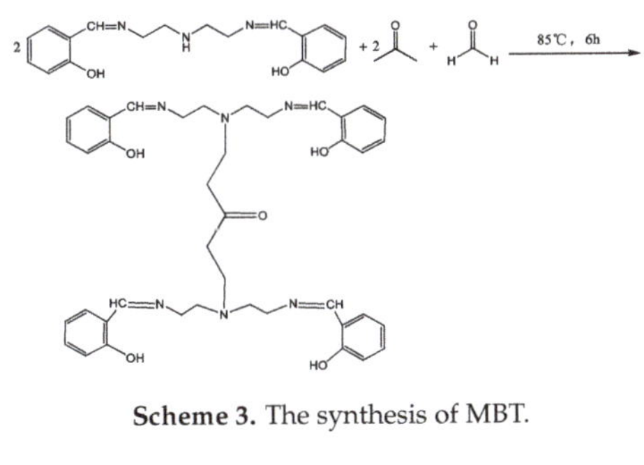

Scheme 3. The synthesis of MBT.

3.3. ^1H NMR Analysis

The ^{1}H NMR spectrum of MBT is shown in Figure 2. It can be seen from Figure 2 that the absorption peak at chemical shift δ = 12.00~11.00 is attributed to –OH; the absorption peak at δ = 8.56~8.45 is attributed to –CH=N; the absorption peak at δ = 7.36~6.79 is attributed to benzene (–CH); δ = 3.68~3.65, 3.02, the absorption peak at 2.63 is attributed to –CH_2 at different positions; and the absorption peak at δ = 2.5 belongs to the solvent peak of Dimethyl sulfoxide (DMSO) and is characterized by IR. Combined with IR characterisation, the results indicate that MBT is the target product of synthesis.

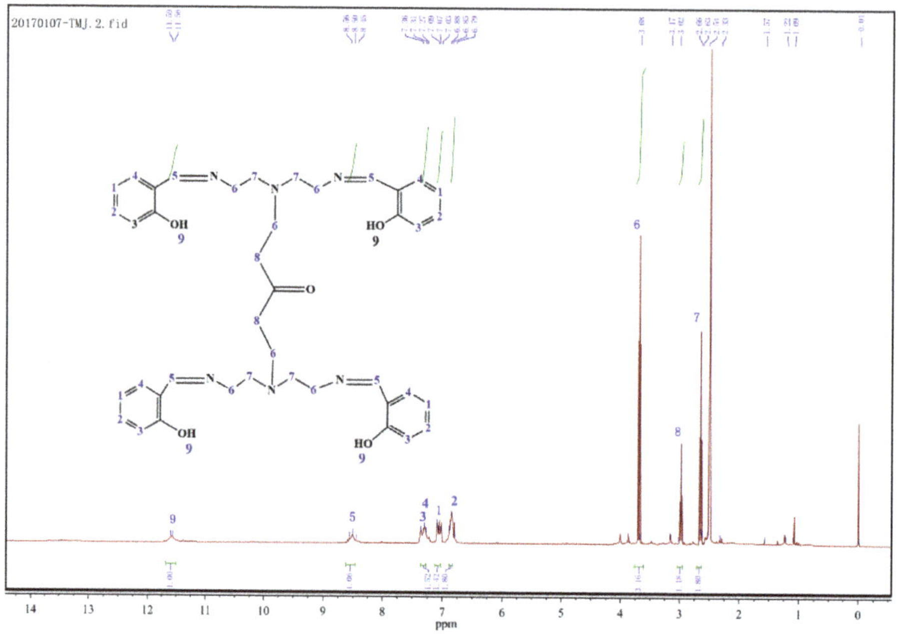

Figure 2. ^1H NMR spectrum of MBT.

3.4. Concentration Effect of MBT on Corrosion Inhibition

The corrosion inhibition performance of the corrosion inhibitor MBT was evaluated by the static weight-loss method. Test conditions: temperature 70 °C, pH = 5.6, ρ(NaCl) = 30 g/L, and constant temperature water bath 72 h. The corrosion inhibition effect of MBT on a N80 steel sheet in a CO_2 saturated solution containing 3 wt % NaCl was investigated under different dosages. The result is shown in Figure 3.

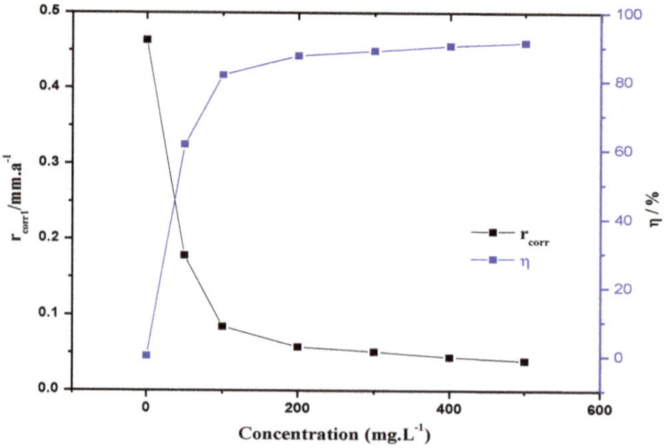

Figure 3. Effect of the dosage of MBT on the corrosion rate and corrosion inhibition rate of the N80 steel sheet.

According to Figure 3, in the case of different dosages of MBT (0, 50, 100, 200, 300, 400, 500 mg/L), when the amount of MBT was 200 mg/L, the N80 steel sheet corrosion rate was 0.0579 mm/a. The corrosion rate of the steel sheet decreases with the increase of the corrosion inhibitor concentration, indicating that MBT can form an effective protective film on the metal surface. When the amount of corrosion inhibitor is 400 mg/L, the corrosion rate is 0.0446 mm/a, and the corrosion inhibition rate can reach 90.4%.

3.5. Temperature Effect of MBT on Corrosion Inhibition

In order to study the corrosion inhibitor for a CO_2 saturated solution containing 3 wt % NaCl, the corrosion inhibition effect at different temperatures (40, 50, 60, 70, 80, 90 °C), the dosage of corrosion inhibitor in this experiment is 400 mg/L, and the evaluation time 72 h.

As shown in Figure 4, the increasing temperature increases the corrosion rate in the absence (r_{cooro}) and appearance of MBT solution (r_{coor1}). This is because as the temperature increases, the adsorption capacity of the corrosion inhibitor decreases, the desorption capacity increases, and the corrosion rate of the steel sheet itself increases as the temperature is higher.

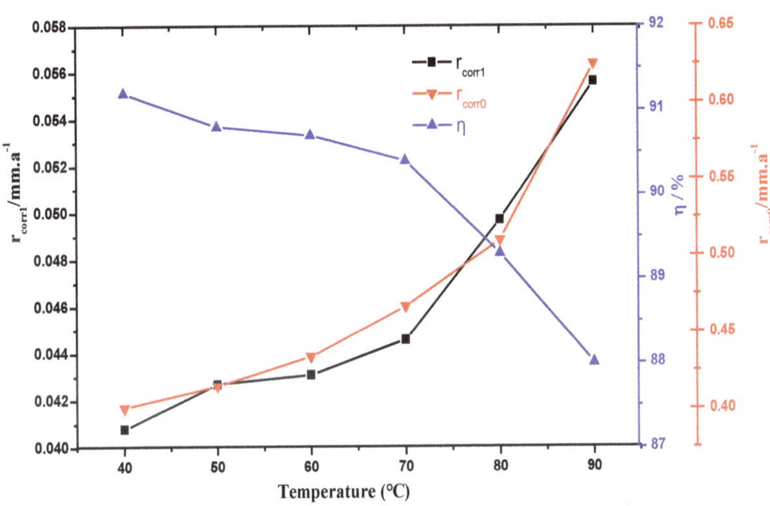

Figure 4. Evaluation of temperature on corrosion rate and corrosion inhibition rate.

3.6. Discussion on Corrosion Inhibition Mechanism

3.6.1. Polarization Curve Data Analysis

The inhibition mechanism of MBT was studied by the Tafel polarization curve method. Different concentrations (0 mg/L, 50 mg/L, 100 mg/L, 200 mg/L, 300 mg/L) of MBT were added, and the corrosive medium was a CO_2 saturated solution containing 3 wt % NaCl at a temperature of 25 °C. The Tafel curve is shown in Figure 5.

It is shown in Figure 5 that when the different concentrations of MBT are added, the polarization curve moves downward as a whole, that is, the corrosion current density decreases. As the concentration of the corrosion inhibitor increases, the corrosion current density decreases. Since the corrosion current density is proportional to the corrosion rate, the concentration of the MBT increase made the corrosion rate drop, which is consistent with the results of the weight loss experiment. It can be seen from Figure 5, that after the addition of MBT, the polarization curves of the cathode and anode move in the direction of low corrosion current density, so MBT reacts to both the cathode and anode.

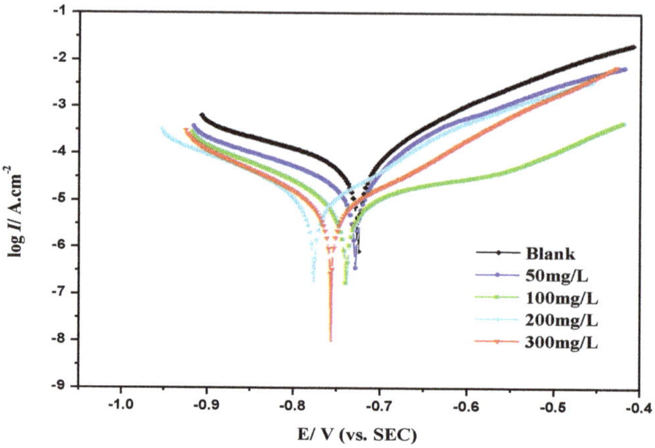

Figure 5. Tafel curve of N80 steel with different concentrations of MBT.

The corrosion potential (Ecorr), corrosion current density (Icorr), Tafel slope (βa and βc) and corrosion inhibition rate (η) calculated by the equation are shown in Table 3. It can be seen in the table that the corrosion inhibitor MBT moves the self-corrosion potential of the N80 steel sheet negatively; it shows that the corrosion inhibitor blocks the cathode process, and the self-corrosion potential moves in the negative direction (fluctuating around 30 mV) [19,20], so MBT is a mixed type corrosion inhibitor mainly for suppressing the cathode.

Table 3. Electrochemical parameters of different MBT concentrations.

Inhibitor	Conc. ($mg \cdot L^{-1}$)	E_{corr} (V)	I_{corr} ($\mu A \cdot cm^{-2}$)	Ba ($mV \cdot dec^{-1}$)	Bc ($mV \cdot dec^{-1}$)	η (%)
MBT	0	−0.725	60.99	10.21	5.11	-
	50	−0.727	26.10	9.86	5.98	57.2
	100	−0.737	12.62	5.59	7.58	79.3
	200	−0.775	10.06	12.06	7.23	83.5
	300	−0.755	7.99	9.74	8.07	86.9

3.6.2. Electrochemical Impedance Spectroscopy Data Analysis

In order to further study the corrosion inhibition mechanism of MBT on steel sheets, the impedance of the system was measured by an AC impedance method. Figure 6 is an impedance spectrum of MBT with different concentrations added.

It can be seen from Figure 6 that the Nyquist diagram of the blank sample and the added corrosion inhibitor is a set of capacitive reactance arcs with a semicircular shape; however, the impedance of the system changes significantly after adding different concentrations of MBT compared with the blank. In addition, the capacitive anti-arc increases with the increase of corrosion inhibitor concentration, and the Bode-modulus value and the phase angle Bode-phase angle increase obviously, indicating that the corrosion inhibitor can be adsorbed on the surface of the N80 steel sheet to increase the transmission resistance to corrosion of the steel sheet, thereby reducing the corrosion rate of the metal. The EIS diagram was fitted using ZSimpWin software (ZSimDemo3.30d), and the equivalent circuit was obtained as R(C(R(QR))), as shown in Figure 7.

In Figure 7, R_S represents the solution resistance; R_{ct} represents the transfer resistance of the N80 steel sheet corrosion reaction charge; R_t and CPE represent the metal interface film resistance and constant phase element (CPE) to replace a double layer capacitance with a more accurate fit [21], the impedance of CPE was calculated following Equation (5); and C_{dl} represents the capacitance. The fitting parameters are shown in Table 4.

$$Z_{CPE} = \frac{1}{Y_0 j w^n}, \tag{5}$$

where Y_0 is CPE constant; j is the imaginary number; w is the angular frequency; and n is the phase shift (which represents the deviation from ideal behavior).

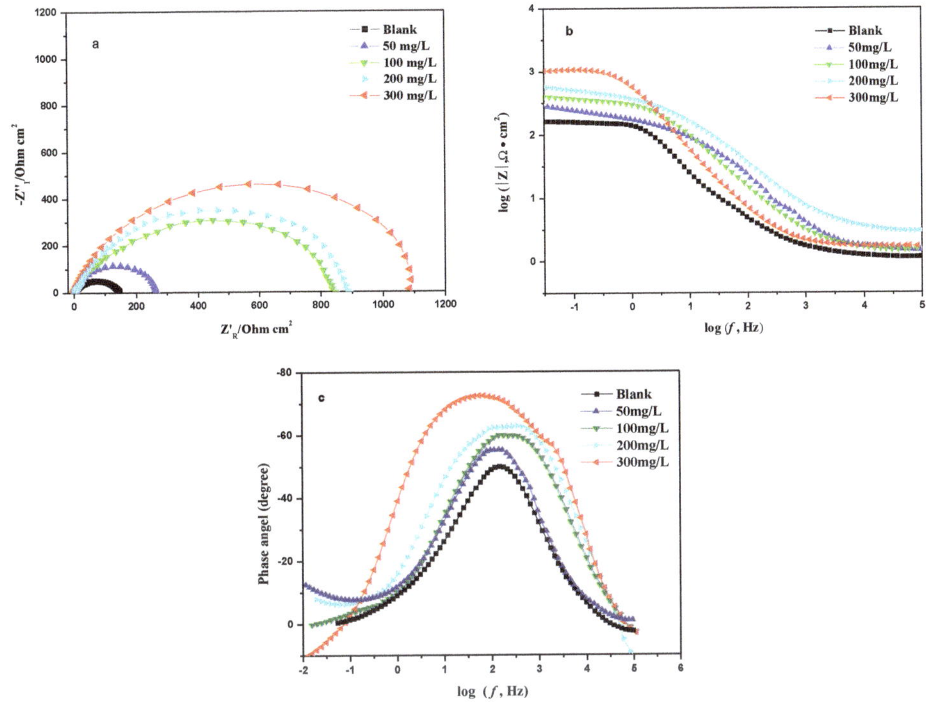

Figure 6. Nyquist (**a**), Bode-modulus (**b**) and Bode-phase angle (**c**) diagrams of N80 steel sheets with different concentrations of MBT.

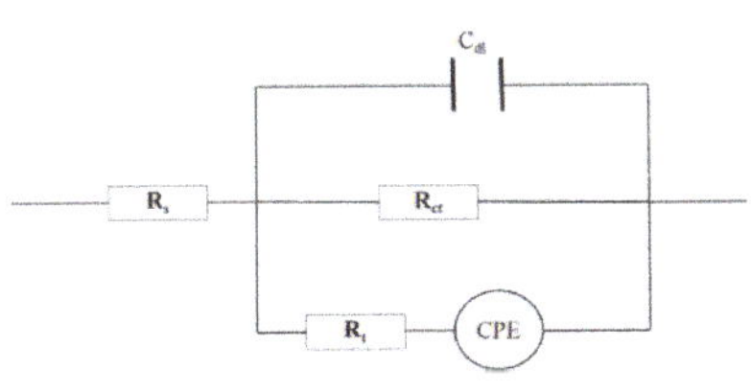

Figure 7. Fitting equivalent circuit diagram.

From the data analysis in Table 4, as the intrusion time of N80 steel increases, the charge transfer resistance R_{ct} and film resistance R_t in the different corrosion inhibitor concentration solutions increase with the concentration of the corrosion inhibitor; the compactness of R_t and MBT adsorption film formation is related to the thickness, which indirectly indicates that the MBT molecule is effectively adsorbed by the N80 steel sheet. The solution resistance R_s does not change much and is significantly smaller than the film resistance R_t and charge transfer resistance R_{ct}. Combined with the analysis of the Nyquist diagram, it is fully demonstrated that the corrosion inhibition effect is mainly absorbed by the corrosion inhibitor molecules on the metal surface, which changes the interface properties of the N80 steel surface and the electric double layer structure, thus achieving a good corrosion inhibition effect [22]. The calculated corrosion inhibition effect is basically consistent with the results obtained by the above weight loss experiment and the polarization curve method.

Table 4. Electrochemical impedance spectroscopy fitting parameters of different concentrations of MBT.

Inhibitor	Conc. (mg/L)	R_s $\Omega{\cdot}cm^2$	C_{dl} $\mu F{\cdot}cm^{-2}$	R_{ct} $\Omega{\cdot}cm^2$	CPE Y_0 $\mu S{\cdot}cm^{-2}{\cdot}s^n$	CPE n	R_t (Ω)	η%
MBT	0	1.579	6.02	136.7	15.33	0.82	13.57	-
	50	4.507	13.96	255	8.83	0.69	23.56	46.4
	100	1.758	7.90	871.7	54.22	0.73	58.71	84.3
	200	1.571	12.13	941.2	53.71	0.67	37.77	85.5
	300	1.867	18.82	1020	10.92	0.61	42.47	86.6

3.7. MBT Adsorption Mode

The MBT molecule is adsorbed on the surface of the metal to form a dense adsorption film, which acts as a corrosion inhibitor, and the stronger the adsorption capacity of the molecule on the metal surface, the better the corrosion inhibition effect [23]. The adsorption of MBT molecules on the metal surface is based on the structure, spatial distribution of the groups in the MBT molecule and the surface morphology of the metal; the adsorption modes are Temkin, Langmuir, Frumkin and Freundkick.

The above four adsorption isotherms were used to fit the data obtained by the polarisation curve method and the impedance method. It was found to be in accordance with the Langmuir adsorption equation, and the Langmuir adsorption curve is shown in Figure 8.

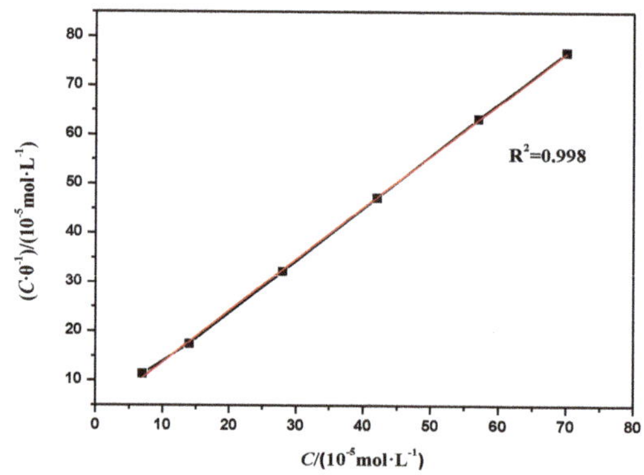

Figure 8. Langmuir isothermal adsorption curve of corrosion inhibitor MBT.

From Figure 8, it can be concluded that K_{ads} = 1.701 × 104 L/mol and R^2 = 0.998, indicating that C/θ has a relatively good linear relationship with concentration C. The results show that at 70 °C, in a CO_2 saturated solution containing 3 wt % NaCl, the adsorption of MBT on the surface of the N80 steel sheet follows the Langmuir isotherm adsorption. The formula for the adsorption free energy ΔG_{ads} of the corrosion inhibitor MBT is as follows:

$$\Delta G_{ads} = -RTln(55.5K_{ads}), \quad (6)$$

Calculate $\Delta G_{ads} = -39.25KJ/mol < 0$ according to the adsorption equilibrium constant K_{ads} and Equation (6), indicating that the corrosion inhibitor MBT is capable of spontaneous chemical adsorption on the N80 steel sheet, which has a good corrosion inhibition effect.

3.8. Corrosion Inhibition Surface Morphology Analysis

In order to study the corrosion inhibition ability of the corrosion inhibitor MBT, the surface of the N80 steel sheet was characterised by AFM. The N80 steel sheet was immersed with different concentrations of corrosion inhibitor MBT in a CO_2 saturated solution containing 3 wt % NaCl, and the water bath was kept at 70 °C for 72 h. The three-dimensional and planar topography of the N80 steel sheet surface is shown in Figure 9.

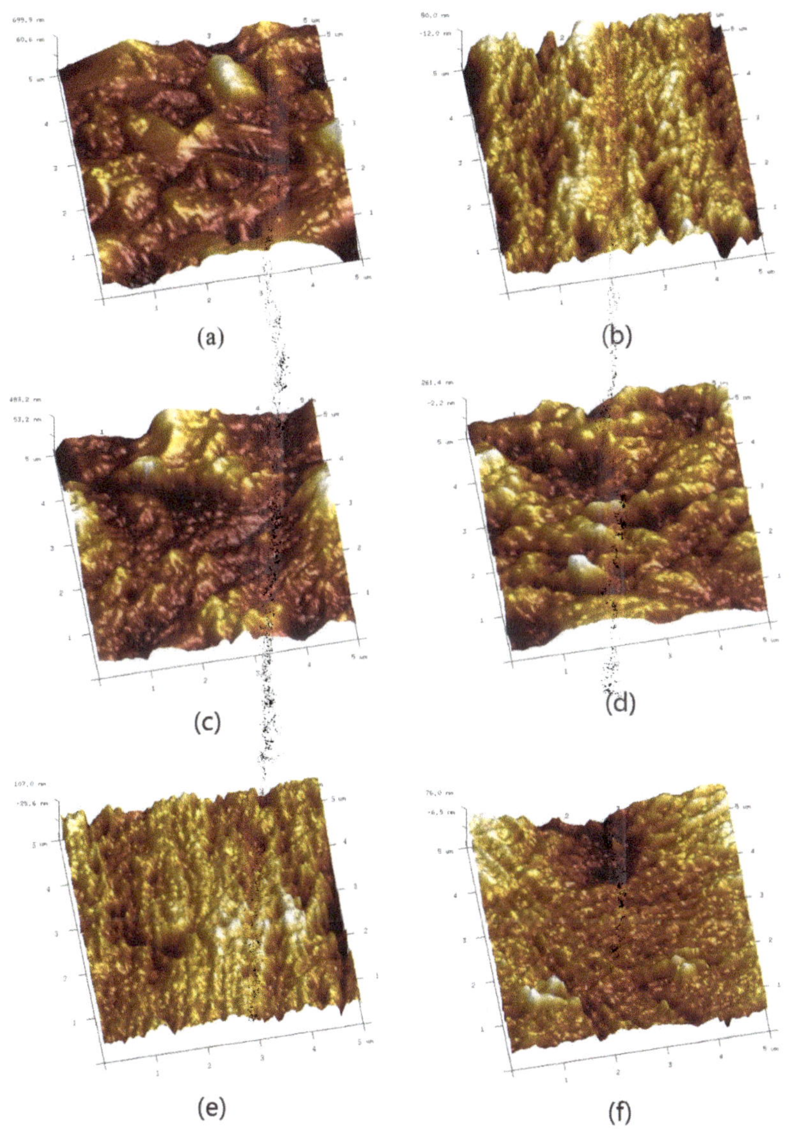

Figure 9. AFM three-dimensional image of N80 steel sheet: (**a**) blank sample; (**b**) initial uncorroded sample; (**c**–**f**) added 50 mg/L, 100 mg /L, 200 mg/L, 300 mg/L MBT.

Figure 9a shows the N80 steel sheet without MBT, it can be seen from the figure that the surface corrosion of the N80 steel sheet is serious and the roughness is high because the metal is strongly damaged in the etching solution; Figure 9b is the polished N80 steel sheet that has not been immersed in the etching solution, it can be seen from the figure that the surface of the steel sheet is relatively smooth and the roughness is very low. Based on Figures 9c–f and 10c–f, the N80 steel sheet added to the corrosion inhibitor MBT is significantly different from that of Figures 9a and 10a. As the concentration of the corrosion inhibitor MBT increases, the corrosion rate of the steel sheet reduces, the surface of the steel sheet is smoother and the roughness is also closer to that of Figure 9b. MBT at a concentration of 300 mg/L is exhibited in Figures 9f and 10f with sandpaper polishing marks. It shows that the surface of MBT carbon steel forms an effective protective film, which prevents the corrosion of the steel sheet.

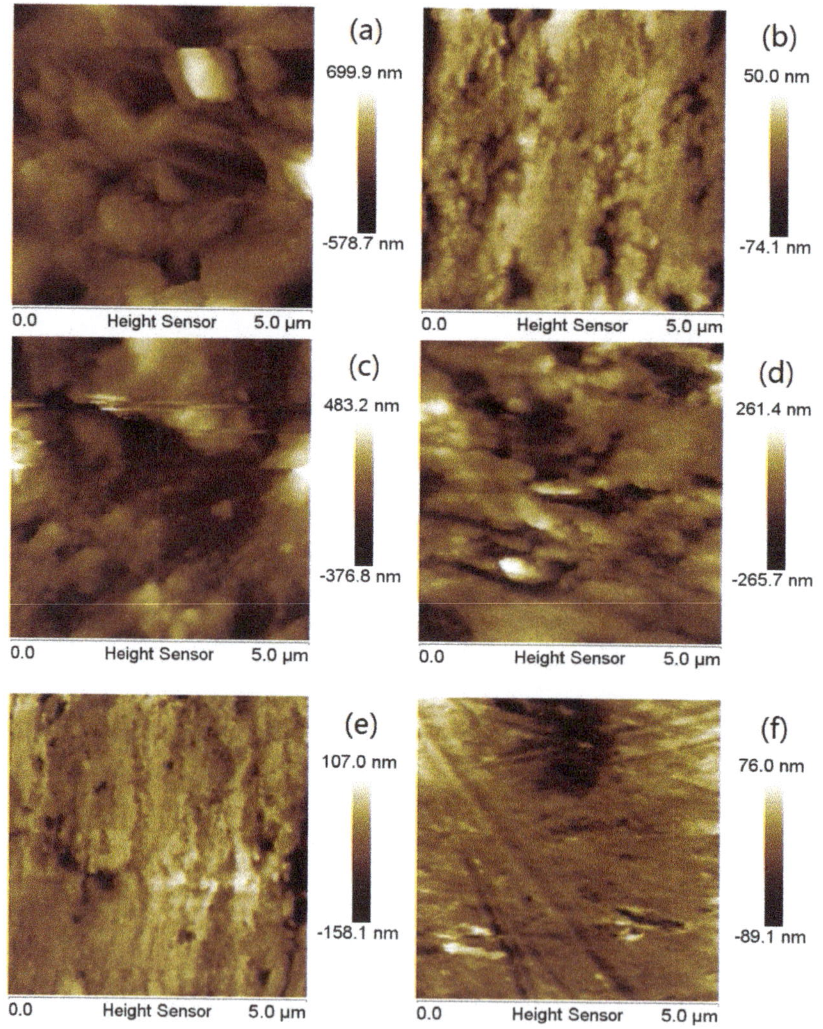

Figure 10. AFM plane topography of N80 steel sheet: (**a**) blank sample; (**b**) initial uncorroded sample; (**c–f**) is added 50 mg/L, 100 mg/L, 200 mg/L, and 300 mg/L MBT.

By analysing the data of the atomic force microscope carried by the analysis software (NanoScope Analysis v1.40), we can obtain the mean square roughness (Rq), the average roughness (Ra) and the maximum height difference (Rmax) of the surface of the N80 steel sheet, which is shown in Table 5. It can be seen from the table, that after the N80 blank is etched, the Rq is 155.0 nm, the Ra is 113.0 nm and the Rmax is 2050.0 nm, which is quite different from the Rmax of 165.0 nm which has not been etched. After adding 300 mg/L corrosion inhibitor MBT, the Rmax of the N80 steel sheet is 192.0 nm, which is very close to the maximum height difference of the unetched N80 steel sheet. Combined with Figures 9 and 10, the surface roughness of the steel sheet can be seen. The appearance of the corrosion inhibitor forms a passivation film on the surface of the N80 steel sheet, which reduces the surface roughness of the steel sheet and effectively prevents the N80 steel sheet from being eroded by corrosive media.

Table 5. AFM parameters after soaking in different concentrations of corrosion inhibitor MBT.

MBT (mg/L) + CM	Ra (nm)	Rq (nm)	Rmax (nm)
0 + 0	13.1	17.3	165.0
0 + CM	113.0	155.0	2050.0
50 + CM	93.0	119.0	873.0
100 + CM	58.1	73.2	597.0
200 + CM	24.4	33.2	369.0
300 + CM	15.9	21.6	192.0

Note: CM (corrosive media) refers to a CO_2 saturated solution containing 3 wt % NaCl.

4. Conclusions

The overall aim of this study was to synthesise and assess the corrosion inhibition effect of MBT. After dosage temperature tests, the corrosion rate was 0.0446 mm/a and the corrosion inhibition rate was 90.4% while the dosage was 400 mg/L in a CO_2 saturated solution of 3 wt % NaCl, in a 70 °C constant temperature water bath for 72 h. As evidenced by polarisation curve analysis, MBT is a mixed type corrosion inhibitor mainly used for suppressing the cathode. The adsorption of MBT on the metal surface contributes to a dense molecular film, changing the interface properties and electric double layer structure, thus achieving a good corrosion inhibition effect. The adsorption follows the Langmuir isotherm adsorption and the calculated $\Delta G_{ads} < 0$ indicates that the chemical adsorption of the corrosion inhibitor is spontaneous on the N80 steel sheet. AFM is applied in microscopic characterisation on the surface. Results demonstrate that the N80 steel sheet added the corrosion inhibitor MBT is different from blank the sample, showing a smooth surface, indicating that MBT contributes to the protective film with a good corrosion inhibition effect on the surface of the steel sheet.

Author Contributions: Conceptualization, J.L. (Jianbo Li) and M.T.; Data curation, M.T. and L.F.; Funding acquisition, J.L. (Jianbo Li); Investigation, J.L. (Jianbo Li), M.T., Z.L., L.F., B.Z. and J.L. (Jie Lv); Writing—original draft, M.T. and Z.L.; Writing—review & editing, M.T., J.L. (Jianbo Li) and Z.L.

Funding: This research was funded by the National Science and Technology Major Project of China, grant number 2016ZX05016-004-008 and The APC was funded by the National Science and Technology Major Project of China.

Acknowledgments: The authors thank the Southwest Petroleum University, Institute, Chengdu, China, for providing the research facilities needed for the above study. This work was supported by the National Science and Technology Major Project of China (No. 2016ZX05016-004-008).

Conflicts of Interest: The authors declare no conflict of interest.

References

1. Liu, Z.; Cheng, Y.; Wang, Y.; Wang, L.; Li, W. Experimental investigation of CO_2 injection into coal seam reservoir at in-situ stress conditions for enhanced coalbed methane recovery. *Fuel* **2019**, *236*, 709–716. [CrossRef]
2. Franco, C.A.; Zabala, R.; Cortés, F.B. Nanotechnology applied to the enhancement of oil and gas productivity and recovery of Colombian fields. *J. Pet. Sci. Eng.* **2017**, *157*, 39–55. [CrossRef]
3. Abelev, E.; Sellberg, J.; Ramanarayanan, T.A.; Bernasek, S.L. Effect of H_2S on Fe corrosion in CO_2-saturated brine. *J. Mater. Sci.* **2009**, *44*, 6167–6181. [CrossRef]
4. Tang, M.; Li, J.; Ye, Z.; Kou, Z.; Fu, L. A Novel Eco-Friendly Scale and Corrosion Inhibitor Modified by β-Cyclodextrin. *Aust. J. Chem.* **2017**, *70*, 933–942. [CrossRef]
5. Liu, H.; Schonberger, K.D.; Peng, C.Y.; Ferguson, J.F.; Desormeaux, E.; Meyerhofer, P.; Luckenbach, H.; Korshin, G.V. Effects of blending of desalinated and conventionally treated surface water on iron corrosion and its release from corroding surfaces and pre-existing scales. *Water Res.* **2013**, *47*, 3817–3826. [CrossRef] [PubMed]
6. Molnár, V.; Fedorko, G.; Krešák, J.; Peterka, P.; Fabianová, J. The influence of corrosion on the life of steel ropes and prediction of their decommissioning. *Eng. Fail. Anal.* **2017**, *74*, 119–132. [CrossRef]
7. Al-Amiery, A.A.; Kassim, F.A.; Kadhum, A.A.; Mohamad, A.B. Synthesis and characterization of a novel eco-friendly corrosion inhibition for mild steel in 1 M hydrochloric acid. *Sci. Rep.* **2016**, *6*, 19890. [CrossRef] [PubMed]
8. Zhang, K.; Xu, B.; Yang, W.; Yin, X.; Liu, Y.; Chen, Y. Halogen-substituted imidazoline derivatives as corrosion inhibitors for mild steel in hydrochloric acid solution. *Corros. Sci.* **2015**, *90*, 284–295. [CrossRef]
9. Pillay, C.; Lin, J. The impact of additional nitrates in mild steel corrosion in a seawater/sediment system. *Corros. Sci.* **2014**, *80*, 416–426. [CrossRef]
10. Hamani, H.; Douadi, T.; Daoud, D.; Al-Noaimi, M.; Rikkouh, R.A.; Chafaa, S. 1-(4-Nitrophenylo-imino)-1-(phenylhydrazono)-propan-2-one as corrosion inhibitor for mild steel in 1 M HCl solution: Weight loss, electrochemical, thermodynamic and quantum chemical studies. *J. Electroanal. Chem.* **2017**, *801*, 425–438. [CrossRef]
11. Vimalanandan, A.; Lv, L.P.; Tran, T.H.; Landfester, K.; Crespy, D.; Rohwerder, M. Redox-responsive self-healing for corrosion protection. *Adv. Mater.* **2013**, *25*, 6980–6984. [CrossRef] [PubMed]
12. Li, J.; Tang, M.; Ye, Z.; Chen, L.; Zhou, Y. Scale formation and control in oil and gas fields: A review. *J. Dispers. Sci. Technol.* **2017**, *38*, 661–670. [CrossRef]
13. Camargo, J.A.; Alonso, Á. Ecological and toxicological effects of inorganic nitrogen pollution in aquatic ecosystems: A global assessment. *Environ. Int.* **2006**, *32*, 831. [CrossRef] [PubMed]
14. Paustovskaya, V.V. Some results of a research in the problem "inhibitors of metal corrosion. Toxicology and industrial hygiene". *Prot. Met.* **2000**, *36*, 89–93. [CrossRef]
15. Nishida, I.; Shimada, Y.; Saito, T.; Okaue, Y.; Yokoyama, T. Effect of aluminum on the deposition of silica scales in cooling water systems. *J. Colloid Interface Sci.* **2009**, *335*, 18. [CrossRef] [PubMed]
16. *SY/T 5273–2000, The Oil Industry Standard of the People's Republic of China*; Petroleum Industry Press: Beijing, China, 2000.
17. Zhang, H.H.; Gao, K.; Yan, L.; Pang, X. Inhibition of the corrosion of X70 and Q235 steel in CO_2-saturated brine by imidazoline-based inhibitor. *J. Electroanal. Chem.* **2017**, *791*, 83–94. [CrossRef]
18. Fu, L.; Yi, F.; Zeng, B.; Hu, C.; Li, J. Study of Synthesis and Corrosion Inhibition of Novel Mannich and Schiff Bases on Carbon Steel in Acid Solution. *Russ. J. Appl. Chem.* **2018**, *91*, 499–509. [CrossRef]
19. Li, X.; Deng, S.; Xie, X. Inhibition effect of tetradecylpyridinium bromide on the corrosion of cold rolled steel in 7.0M H_3PO_4. *Arab. J. Chem.* **2014**, *84*, S3715–S3724.
20. Xiang, Y.G.; Qiao, K.Q.; Cui, Y.S. Corrosion Inhibition of Eggplant Leaf Extract for Carbon Steel in Hydrochloric Acid Medium. *Corros. Prot.* **2016**, *37*, 122–127.
21. Meng, Y.; Ning, W.; Xu, B.; Yang, W.; Zhang, K.; Chen, Y.; Li, L.; Liu, X.; Zheng, J.; Zhang, Y. Inhibition of mild steel corrosion in hydrochloric acid using two novel pyridine Schiff base derivatives: A comparative study of experimental and theoretical results. *RSC Adv.* **2017**, *7*, 43014–43029. [CrossRef]

22. Bentiss, F.; Lebrini, M.; Lagrenée, M. Thermodynamic characterization of metal dissolution and inhibitor adsorption processes in mild steel/2,5-bis(-thienyl)-1,3,4-thiadiazoles/hydrochloric acid system. *Corros. Sci.* **2005**, *47*, 2915–2931. [CrossRef]
23. Feng, L.; Yang, H.; Wang, F. Experimental and theoretical studies for corrosion inhibition of carbon steel by imidazoline derivative in 5% NaCl saturated $Ca(OH)_2$, solution. *Electrochim. Acta* **2011**, *58*, 427–436. [CrossRef]

materials

MDPI

Article

Corrosion Inhibition Mechanism and Efficiency Differentiation of Dihydroxybenzene Isomers Towards Aluminum Alloy 5754 in Alkaline Media

Jacek Ryl [1,*], Mateusz Brodowski [1,2], Marcin Kowalski [1,2], Wiktoria Lipinska [1,3], Pawel Niedzialkowski [4] and Joanna Wysocka [1]

1 Faculty of Chemistry, Gdansk University of Technology, Narutowicza 11/12, 80-233 Gdansk, Poland; mateusz.brodowski96@gmail.com (M.B.); m96kowalski@gmail.com (M.K.); wiktorialipinska154504@gmail.com (W.L.); joanna.wer.wysocka@gmail.com (J.W.)

2 Faculty of Mechanical Engineering, Gdansk University of Technology, Narutowicza 11/12, 80-233 Gdansk, Poland

3 Centre for Plasma and Laser Engineering, The Szewalski Institute of Fluid-Flow Machinery, Polish Academy of Sciences, Fiszera 14, 80-231 Gdansk, Poland

4 Faculty of Chemistry, University of Gdansk, Wita Stwosza 63, 80-308 Gdansk, Poland; pawel.niedzialkowski@ug.edu.pl

* Correspondence: jacek.ryl@pg.edu.pl; Tel.: +48-58-347-1092

Received: 10 August 2019; Accepted: 19 September 2019; Published: 20 September 2019

Abstract: The selection of efficient corrosion inhibitors requires detailed knowledge regarding the interaction mechanism, which depends on the type and amount of functional groups within the inhibitor molecule. The position of functional groups between different isomers is often overlooked, but is no less important, since factors like steric hinderance may significantly affect the adsorption mechanism. In this study, we have presented how different dihydroxybenzene isomers interact with aluminum alloy 5754 surface, reducing its corrosion rate in bicarbonate buffer (pH = 11). We show that the highest inhibition efficiency among tested compounds belongs to catechol at 10 mM concentration, although the differences were moderate. Utilization of novel impedance approach to adsorption isotherm determination made it possible to confirm that while resorcinol chemisorbs on aluminum surface, catechol and quinol follows the ligand exchange model of adsorption. Unlike catechol and quinol, the protection mechanism of resorcinol is bound to interaction with insoluble aluminum corrosion products layer and was only found efficient at concentration of 100 mM (98.7%). The aforementioned studies were confirmed with Scanning Electron Microscopy and X-ray Photoelectron Spectroscopy analyses. There is a significant increase in the corrosion resistance offered by catechol at 10 mM after 24 h exposure in electrolyte: from 63 to 98%, with only negligible changes in inhibitor efficiency observed for resorcinol at the same time. However, in the case of resorcinol a change in electrolyte color was observed. We have revealed that the differentiating factor is the keto-enol tautomerism. The Nuclear Magnetic Resonance (NMR) studies of resorcinol indicate the keto form in structure in presence of NaOH, while the chemical structure of catechol does not change significantly in alkaline environment.

Keywords: aluminum alloy; corrosion inhibitor; alkaline environment; impedance analysis; adsorption; dihydroxybenzene

1. Introduction

Aluminum is the most widely spread metallic element on Earth [1], owing to its unique physico-chemical properties, such as low weight, high thermal and electrical conductivity, high linear expansion coefficient, and good corrosion resistance. It is non-magnetic, non-toxic and may be subjected to repeated recycling [2]. Aluminum and its alloys have been used in almost all industries, in particular in mechanical engineering, the defense industry, aviation, shipbuilding, food and chemical industries, and many others. It is a strategic resource whose consumption is a measure of countries' development and industrialization level. Over the past 50 years, the world production of aluminum has been constantly increasing, with the highest leap occurring in this millennium. Limited access to bauxite ores and their gradual depletion are the main obstacles in the further development of the aluminum industry. Constantly growing demand and utilization of aluminum requires more effective methods for its recycling or protection.

Aluminum and its alloys belong to the group of passivating metals, alongside titanium, chromium and high-alloy steels. Spontaneously formed native oxide layer on the aluminum surface is built primarily of aluminum oxide Al_2O_3. The layer is thermodynamically stable in the pH range between 4 and 9, where aluminum possesses the highest corrosion resistance. On the other hand, the passive layer is not thermodynamically stable in alkaline and acidic media. In the presence of hydroxyl ions, aluminum undergoes oxidation to form $Al(OH)_4^-$, according to the mechanism proposed by Macdonald [3]. One of the most commonly accepted aluminum corrosion mechanisms in alkaline media may be simplified to the form of equations (1–4) [4]. Other, more detailed mechanisms suggests involvement of intermediate products and/or passive layer [5–10].

$$Al \rightarrow Al^{3+} + 3e^- \tag{1}$$

$$Al^{3+} + 3OH^- \rightarrow Al(OH)_3 \tag{2}$$

$$Al(OH)_3 \downarrow + OH^- \rightarrow Al(OH)_4^- \tag{3}$$

The anodic dissolution is accompanied by cathodic process of water electrolysis with hydrogen generation, according to Equation (4):

$$2H_2O + 2e^- \rightarrow H_2 \uparrow + 2OH^- \tag{4}$$

The problem of aluminum corrosion in alkaline media occurs in numerous practical cases, starting from its possible application as an anode material in energy storage devices, through pre-treatment processes prior to anodization or for aesthetic purposes, up to the alkaline character of various cleaning agents used on working elements and constructions [11–14]. One of the most frequently utilized methods for lowering the corrosion rate of aluminum in these environments is the application of corrosion inhibitors.

Various organic inhibitors have been reported to be efficient corrosion inhibitors of aluminum and its alloys. Their inhibition effect depends on the molecule structure, the functional groups being electron donors or acceptors, and the number of such groups per molecule [15,16]. It is worth pointing out that the most effective inhibitors are based on molecules containing heteroatoms, such as oxygen, nitrogen, phosphorus, sulfur and aromatic rings [17,18]. Carboxylic acids in particular have been shown as highly efficient corrosion inhibitors of aluminum and its alloys in aqueous alkaline environments [19–22]. Studies carried out on maleic, malic, succinic, tartaric, citric and tricarballylic acids have revealed changes in corrosion efficiency with the increased amount of –COOH groups and decreased amount of –OH groups within inhibitor molecule [20]. Similar reports involved restriction of aluminum corrosion in alkaline media by polyacrylic acids, where studies shown increase of inhibition efficiency with the increase of molecular weight of inhibitor molecule [23]. Compounds containing nitrogen and/or sulphur have also been proved to be efficient corrosion inhibitors, an example of which may be studies on 3-methyl-4-amino-5-mercapto-1,2,4-triazole (MAMT). Inhibitor molecules adsorption on protected metal surface occurs through amine and thiol functional groups [24].

Lashgari and Malek proved that phenol is a highly efficient corrosion inhibitor of aluminum [25]. Phenols are deprotonated in alkaline environments and transformed into inhibitory active forms of phenoxide and phenolate. Similar conclusions, based on theoretical studies, were later on drawn for p-phenol derivatives, where the author confirmed that the mechanism of inhibition relies on a complicated cycle of protonation/deprotonation of inhibitor molecules in the inner area of the electrical double layer [26]. The process mentioned above leads to local neutralization of corrosive factors and their electrostatic repulsion in the vicinity of an active metal surface. Corrosion inhibition efficiency of p-phenol derivatives depends on several factors, including electron density on oxygen and hydrogen atoms in hydroxyl group, charge transfer, the energy of interaction, molecular activity, electric dipole moment and Gibbs free energy of the dissolution process.

The attention of corrosion scientists is nowadays increasingly focused on application of corrosion inhibitors of natural origin, obtained in accordance with the principles of green chemistry. Green corrosion inhibitors in the form of plant extracts are eco-friendly, non-toxic and biodegradable in neutral environments [27–35]. For example, *Phyllanthus amarus* leaf extract offers nearly 75% efficiency in 2M NaOH solution [36]. The extract contains alkaloids, cyanogenic glycosides, flavonoids, carbohydrates, sugar, proteins, triterpenoids and steroids. Functional groups –OH, $–NH_2$, -SH, present in mentioned above compounds and π-bonds are most likely responsible for inhibition efficiency of *Phyllanthus amarus*. On the other hand, *Piper longum* seed extract, with 94% efficiency at concentration of 0.4 $g{\cdot}L^{-1}$ in 1 M NaOH, contains piperine, piperlongumine, piplartine, piperlonguminine, piperundecalidinine and pipernonaline [37]. The high inhibition efficiency was explained with presence of N-heteroatoms and π-electrons in aromatic rings. The 92% inhibition efficiency of *Gossypium hirsutum* extract in 2 M NaOH most likely originates from presence of O, N or S in amino acids such as: cysteine, lysine, methionine, phenylalanine, arginine, threonine, tyrosine, tryptophan, valine, but also polyphenolic aldehyde and tannins [32]. The authors also observed that higher concentration of active substances in present in leaves rather than seeds of *Gossypium hirsutum*.

Green corrosion inhibitors in the form of extracts from natural products are characterized by a large number of chemical compounds. In such a complex mixture of potential inhibitory compounds it is particularly important to perform phytochemical studies in order to determine the active compounds and their mechanism of interaction, which in many cases appears to be an incredibly difficult task. Therefore, in order to avoid blind-picking during selection of natural extracts containing potentially efficient inhibitor compounds one must get to know in detail the mechanism of interaction of various types of functional groups with protected metal surface as well as how it is modified by other aspects of the molecule structure. A valuable approach towards more effective determination of the active inhibitory compounds may be found in designing the extraction process. Differentiation of the type of solvents or extraction conditions leads to selective extraction of certain groups of compounds. Ryl et al. [38] showed that preparation of bee pollen extracts through extraction in different solvents results in different corrosion inhibition efficiency towards AA5754 in bicarbonate buffer at pH = 11. It has been proved that these differences are caused by varying content of carboxylic acids and phospholipids, which acted as inhibitory active substances in bee pollen extracts.

A certain group of phenol derivatives exhibits very high corrosion inhibition efficiency towards aluminum and its alloys. This group includes catechol, cresol, chlorophenol, resorcinol, nitrophenol and aminophenol [39,40]. Talati and Modi [39] suggested that $–OCH_3$, $–NH_2$ and $–CH_2CHCH_3$ functional groups lower the inhibition efficiency of phenol, while –OH, –Cl, $–NO_2$ increase it. Furthermore, they suggested three different adsorption mechanisms, namely: electrostatic forces, the formation of chelating agents with aluminum ions or covalent bond formation. The authors also observed that the inhibition efficiency decreases with the increase of electrolyte alkalinity. The synergistic interaction of resorcinol with Zn^{2+} ions was further studied by Lakshmi et al. [40], which revealed significant increase in corrosion inhibition efficiency of aluminum. However, all the aforementioned studies were carried out using the gravimetric method, introducing significant uncertainty to the measurements. The formation of the insoluble corrosion products layer on aluminum surface hinders specific determination of weight loss of the analyzed samples.

Not only the type and the number of active functional groups but also their position in the molecule structure may have a significant influence on corrosion inhibition efficiency. The chemical structure of isomer molecules affects modification of their physic-chemical properties such as solubility, while steric hinderance may influence both kinetics and mechanism of the adsorption process. This subject has not been given sufficient attention in the corrosion research; however, several available reports prove the importance of substituents position in the molecule [41–43]. Fouda and Elasmy [41] presented their studies on phenylthiosemicarbazide derivatives as aluminum corrosion inhibitors in 2M NaOH, with efficiency ranging between 75.0% and 98.5%, depending on the derivative. Hassan et al. [42] confirmed that the efficiency of aromatic carboxylic acids depends on the number and position of carboxylic groups and the presence of other substituents in the aromatic ring. The increasing corrosion efficiency was as follows: benzamide < benzaldehyde < acetophenone < benzoic acid < benzophenone (99.99% efficiency).

The electrochemical impedance studies on thiosemicarbazone interaction with aluminum alloys in 1 M HNO_3 revealed an almost 20% higher inhibition efficiency offered by para-substituted compounds in comparison to ortho-substituted ones [44]. There is no general relationship, though. The search for corrosion inhibitors of mild steel in 1 M HCl revealed that ortho-nitroaniline and ortho-methyloaniline show higher corrosion inhibition efficiency in comparison with both meta- and para-substituents, but in the case of phenlylenediamine, meta-substituted functional groups offered the highest efficiency [45]. A similar observation was made on aminophenol-N-salicylidene isomers' action towards zinc in 0.5 M H_2SO_4 [46]. The goal of this work is to evaluate the influence of position of hydroxyl functional groups within dihydroxybenzene molecule on the corrosion inhibition provided by the isomer towards aluminum in alkaline electrolytes. Dihydroxybenzene isomers (catechol, quinol and resorcinol) were utilized, as presence of hydroxyl functional groups is expected to provide reasonable inhibition efficiency in studied electrolytes. In our studies we have implemented newly developed instantaneous impedance tool, i.e., Dynamic Electrochemical Impedance Spectroscopy in galvanostatic mode (g-DEIS), which is capable of accurate and time-efficient determination of the adsorption mechanism differences [19,20,38].

2. Materials and Methods

2.1. Materials

The studied material was aluminum alloy 5754, which had the following alloying additives (in wt.%): Mg 3.6, Fe 0.3, Si 0.3, Cr 0.1, Mn 0.5, Ti 0.1 and Cu 0.1. Cylindrical samples were cut from a rod and subjected to pretreatment procedure in the form of grinding and polishing, carried out on Digiprep 251 (Metkon, Bursa, Turkey) polishing machine. Samples were grinded on a waterproof abrasive papers SiC 600 and 1500, polished with diamond suspensions 6 and 1 μm and mirror-finished on 0.05 μm silica. Following polishing, samples were cleaned and degreased in acetone using ultrasonic cleaner (Polsonic, Warsaw, Poland). Samples were exposed to corrosion studies with 0.5 cm^2 surface area.

The primary electrolyte was the bicarbonate buffer solution. The buffer was prepared using 227 cm^3 0.1 M NaOH and 500 cm^3 0.05 M $NaHCO_3$, diluted with deionized water to 1 dm^3 volume. The obtained buffer had pH = 11 and conductivity of 3.8 $mS \cdot cm^{-1}$. All were analytical purity Sigma Aldrich reagents (Sigma Aldrich, St. Louis, MI, USA). Three dihydroxybenzene isomers were evaluated, namely: resorcinol, catechol and quinol. Their chemical structures are presented on Figure 1. The corrosion inhibition efficiency of the aforementioned compounds was investigated at various inhibitor concentrations c_{inh} = 1, 10 and 100 mM as well as with linearly changing inhibitor concentration during g-DEIS studies related to inhibitor injection into the corrosion cell.

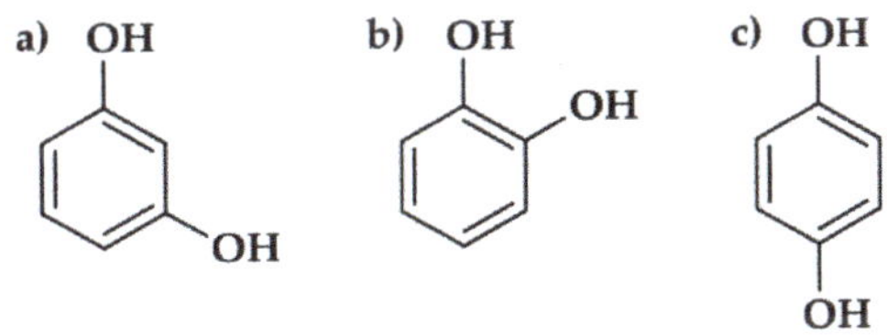

Figure 1. The chemical structure of dihydroxybenzene isomers: (**a**) resorcinol; (**b**) catechol; (**c**) quinol.

2.2. Electrochemical Measurements

The electrochemical impedance studies were performed in a three-electrode setup. Investigated aluminum alloy 5754 was the working electrode (WE), Ag|Ag_2O was the reference electrode (RE) (E° = +0.215 V vs. SHE) while Pt mesh served as the counter electrode (CE). The electrolyte volume in the corrosion cell was 10 mL.

Corrosion inhibition efficiency studies were performed by means of classic Electrochemical Impedance Spectroscopy (EIS) and Dynamic Electrochemical Impedance Spectroscopy in galvanostatic mode (g-DEIS) after the initial conditioning for 15 min. EIS was carried out in potentiostatic mode, at open circuit potential (OCP) conditions. The perturbation signal for EIS measurements was applied in the frequency range between 50 kHz and 40 mHz, with 10 points per decade of frequency and amplitude of 15 mV. Multisinusoidal perturbation signal for g-DEIS studies composed of 29 superimposed elementary signal in the frequency range between 4.5 kHz and 1.0 Hz, with 8 points per decade of frequency. Sampling frequency was 128 kHz. The amplitude of perturbation signal was controlled to assure the amplitude of response signal does not exceed 15 mV. The analysis window for the Short-time Fourier Transformation was 10 s in length. A similar measurement procedure was previously applied in corrosion studies [19,20,38,47–49].

To build the adsorption isotherm, studied corrosion inhibitor was injected from the secondary cell to the corrosion cell through BQ80S microflow peristaltic pump (Lead Fluid, Baoding, China). The flow rate was set as 0.02 mL·min^{-1}. The concentration of studied inhibitor in the secondary cell was set in a way to assure corrosion inhibitor concentration in the corrosion cell (c_{inh}) equal to 10 mM at the end of 6000 s long experiment. The corrosion cell was thermostated at 25 °C. The electrochemical setup used during all corrosion studies is schematically presented on Figure 2.

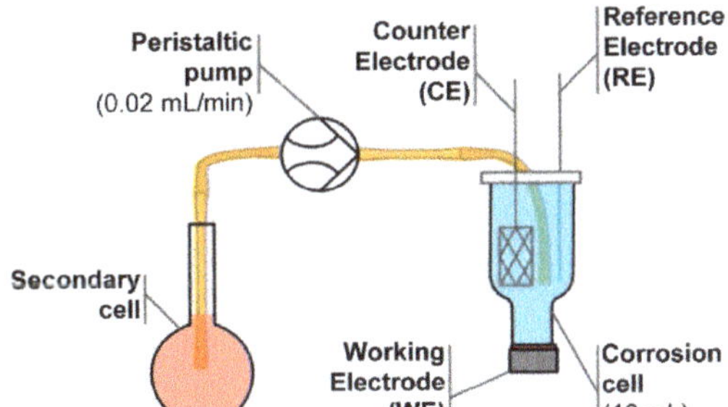

Figure 2. Schematic representation of the setup used during electrochemical studies.

2.3. Equipment

The EIS measurements were carried out using frequency-response-analysis on Gamry Reference 600+ potentiostat (Gamry Instruments, Warminster, PA, USA). The g-DEIS measurement system consisted of Autolab PGSTAT 302N (Metrohm, Herisau, Switzerland) galvanostat connected to PXI-4464 measurement card for AC signal generation and PXI-6124 card for AC/DC signal acquisition, both operating in PXIe-1073 chassis (all from National Instruments, Austin, TX, USA). The microflow peristaltic pump used was BQ80S (Lead Fluid, Baoding, China). The thermostat was Corio CD (Julabo GmbH, Seelbach, Germany)

Microscopy analyses of AA5754 corrosion process were performed on a Scanning Electron Microscope VP-SEM S-3400 N (Hitachi, Chiyoda, Japan), equipped with a tungsten source and operating at 20 kV accelerating voltage. SEM micrographs were done in secondary electron mode.

High-resolution X-Ray Photoelectron Spectroscopy (XPS) analyses were performed on Escalab 250 Xi multispectroscope (ThermoFisher Scientific, Waltham, MA, USA). The spectroscope is equipped in Al Kα X-ray source with a spot diameter of 250 μm. The measurements were carried out at 20 eV pass energy and 0.1 eV energy step size. The charge compensation was provided by means of low-energy electrons and low-energy Ar+ ions emission from the flood gun.

Nuclear Magnetic Resonance (NMR) spectra were recorded on AVANCE III 500 MHz NanoBay spectrometer (Bruker, Billerica, MA, USA). Tetramethylsilane (TMS) was used as an internal standard in all the measurements. The 0.5 mL solution containing 80 mM dihydroxybenzene isomer in D_2O was filled with NaOH. The titrated compounds ratio to NaOH was 1:1, 1:5 and 1:10. ^{1}H-NMR and ^{13}C-NMR spectra were recorded. All the measurements were performed after 24 h in 25 °C and in the same volume of solvent.

3. Results and Discussion

3.1. Dihydroxybenzene Isomers as Corrosion Inhibitors

Electrochemical Impedance Spectroscopy (EIS) was used during the preliminary studies to assess the corrosion resistance of the investigated aluminum alloy immersed in an alkaline environment with the addition of each dihydroxybenzene isomer. The studies were carried out for three different corrosion inhibitor concentrations c_{inh}, namely: 1, 10 and 100 mM. The impedance spectra, presented in the form of Nyquist plots, are plotted in Figure 3. It can be seen by the shape of the impedance semicircles, that studied derivatives offer different level of corrosion protection; however, detailed analysis required fitting of the obtained data with electric equivalent circuit (EEC).

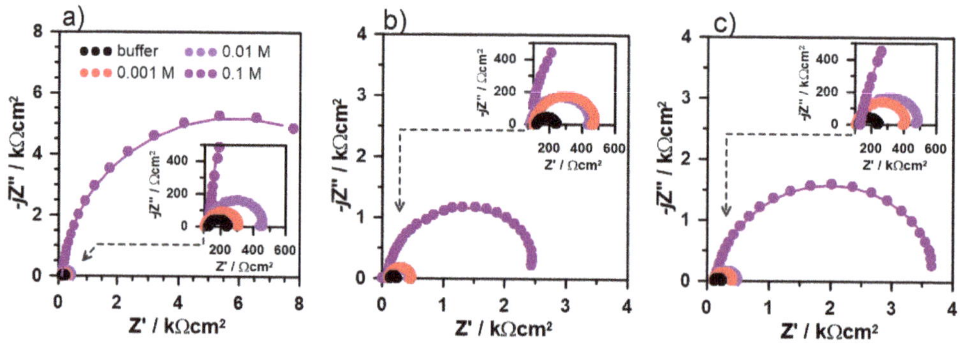

Figure 3. EIS impedance spectra recorded for AA5754 exposed to bicarbonate buffer (pH = 11) with the addition of the studied dihydroxybenzene isomers: (**a**) resorcinol; (**b**) catechol; (**c**) quinol at various concentrations in range 1–100 mM.

There is only one clear time constant present on the obtained impedance spectra, suggesting that the charge transfer through the electrode interface and through the adsorbed inhibitor layer are characterized with similar relaxation times. The authors decided to apply a simple form of Randles electric equivalent circuit (EEC) due to the prominence of one time constant on the impedance spectra in the applied frequency range. The one and only alteration was to replace the capacitance parameter with a constant phase element (CPE) in order to properly consider investigated electrode surface heterogeneity, originating from alloy microstructure and roughness, but even more important presence of local adsorption sites of corrosion inhibitor at its low concentrations. The CPE impedance is given with Equation (5).

$$Z_{CPE} = \left[Q(j\omega)^{\alpha}\right]^{-1} \tag{5}$$

It should be noted that in the boundary case, if $\alpha = 1$, the CPE impedance responds to a capacitor of capacitance Q. Therefore, CPE exponent α is often considered as the homogeneity factor, its decrease represents the increase of surface heterogeneity, while Q reflects the quasi-capacitance of the heterogeneous electrode. Furthermore, the CPE can be used to estimate the effective capacitance of studied electrode C_{eff}, with the use of Hirschorn's approximation for the surface distribution of the capacitance dispersion [50]:

$$C_{eff} = Q^{1/\alpha}\left(\frac{R_e R_{ct}}{R_e + R_{ct}}\right)^{(1-\alpha)/\alpha} \tag{6}$$

where R_e is the resistance of the electrolyte and R_{ct} is charge transfer resistance. The χ^2-distribution of the selected EEC was in the range of 10^{-4}, which is a good result, when taking into consideration EEC simplicity and system non-stationarity.

The shift in the calculated value of charge transfer resistance R_{ct} may be utilized to estimate the inhibition efficiency IE%, using the well-known relationship (7) [19]:

$$IE_{\%} = \left(1 - \frac{R_{ct}^{0}}{R_{ct}}\right) \tag{7}$$

where $R_{ct}{}^{0}$ denounces the measured value of charge transfer resistance in the absence of the inhibitor. The results of impedance data analysis with the R(QR) EEC are summarized in Table 1.

Table 1. Electric parameters of the studied systems obtained from EIS results fitted with R(QR) EEC.

C_{inh}/mM	Buffer	Resorcinol			Catechol			Quinol		
	-	1	10	100	1	10	100	1	10	100
$Q/\mu Ss^{\alpha}\cdot cm^{-2}$	23.9	22	29.4	14.5	24.2	31.1	19	25.5	28	14.5
α/-	0.94	0.94	0.92	0.94	0.91	0.89	0.9	0.93	0.93	0.96
$C_{eff}/\mu F\cdot cm^{-2}$	10.7	10	10.8	8.2	7.5	7.6	6.5	10.7	11.8	9.4
$R_{ct}/k\Omega\cdot cm^{2}$	0.14	0.18	0.3	10.93	0.29	0.38	3.64	0.39	0.35	2.45
$IE_{\%}$/%	-	22.2	53.3	98.7	51.7	63.2	96.2	64.1	60	94.3

Interestingly enough, the addition of resorcinol at the lowest concentration (1 mM) provided the lowest level of corrosion resistance. This feature most likely originates from the altered mechanism of molecules adsorption on the surface of aluminum alloy and was an object of further studies. On the other hand, presence of catechol and quinol offers approx. 60% efficiency already at $c_{inh} = 1$ mM, which does not significantly improve until reaching substantially higher concentrations. This may be observed in particular through effective capacitance C_{eff} changes, which decreased by nearly 30% already at the lowest inhibitor concentrations, compared to reference buffer electrolyte. The most important factor affecting this parameter is the thickness of the adsorbed layer on aluminum surface [20].

Importantly, the highest inhibition efficiency was obtained after the addition of 100 mM of resorcinol. While the difference between IE% at this concentration does not exceed 4%, it should be noted that it corresponds to over 3× and nearly 5× higher R_{ct} values of AA5754 alloy in presence of resorcinol versus catechol and quinol, respectively. The increased efficiency of resorcinol at very high inhibitor concentrations was previously revealed in gravimetric studies [51]. In our opinion, its positive interaction is directly connected to the competitive formation of the corrosion products layer, making an additional barrier to aggressive environment. In the case of catechol and quinol, the barrier properties of the corrosion products layer are less evident, as discussed later.

3.2. The Adsorbed Layer Chemistry

The authors decided to focus on the core of electrochemical and physic-chemical studies on resorcinol and catechol isomers, which is due to the nearly identical response and adsorption mechanism between catechol and quinol. At the same time, there is a significant solubility difference between these two compounds in aqueous electrolytes at 25 °C, hindering possible utilization of quinol as efficient corrosion inhibitor. Quinol is sparingly soluble in water and show tendency for sedimentation over time. Its solubility in water is 5.9 g/100 g in comparison to 43 g/100 g for catechol and 110 g/100 g for resorcinol [52].

The chemistry of the adsorbed dihydroxybenzene layer and corrosion products layer on the surface of studied aluminum alloy 5754 was evaluated with the use of high-resolution XPS analysis. First, samples were pre-exposed to alkaline electrolyte with the addition of studied inhibitor at 10 or 100 mM concentration for a period of 24 h. The XPS spectra in the binding energy (BE) range of C1s, O1s, Al2p and Mg1s peaks were collected and deconvoluted using the fitting procedure described below. The results of the aforementioned deconvolution are summarized in Table 2.

Table 2. Surface chemical composition (in at.%) for AA5754 after 24 h exposure in bicarbonate buffer (pH = 11) with the addition of resorcinol or catechol, based on high-resolution XPS analysis.

Chemical State		BE/eV	Bicarbonate Buffer	Resorcinol		Catechol	
				10 mM	100 mM	10 mM	100 mM
$Al2p_{3/2}$	Al_{ox1}	74.8	1.5	3.2	2.3	3.4	8.0
	Al_{ox2}	76.7	19.8	14.7	15.3	10.5	-
C1s	C-C	284.7	1.4	4.6	5.4	10.4	6.0
	C-OH	286.0	12.1	12.8	16.9	31.4	32.8
	C=O	287.6	2.3	7.6	5.1	3.9	6.1
	COOH	290.3	1.6	2.8	2.4	3.9	8.6
O1s	O^{2-}	531.3	30.8	20.0	22.1	12.6	3.4
	CO/OH	532.6	16.4	15.2	18.5	20.2	29.2
	$C=O/H_2O$	533.7	13.0	18.0	11.0	3.0	5.6
Mg1s	Mg_{ox}	1303.1	1.1	1.1	1.0	0.7	0.3

Figure 4a presents the XPS spectra obtained in Al2p BE range. There is a clear shift in the peak position recorded between resorcinol-exposed and catechol-exposed aluminum samples. First, when exposed to bicarbonate buffer, but also with the addition of resorcinol, the primary peak doublet is located at higher BE values, with $Al2p_{3/2}$ peak at approx. 76.7 eV. This component was labeled Al_{ox2} and ascribed to the non-stoichiometry aluminum corrosion product layer, often observed in studies of this metal in pH range 10–12 [53]. The second component (Al_{ox1}), corresponding to native Al_2O_3 passive film was shifted at −1.9 eV. The presence of native oxide film results from air exposure of samples. The share of non-stoichiometric oxides (Al_{ox2}:Al_{ox1}) is nearly 13:1 when compared to native oxides for sample exposed in bicarbonate buffer, and drops to 5:1 after addition of resorcinol, regardless the concentration. The results suggest presence of corrosion products layer, which contributes the decrease of aluminum corrosion rate in these environments.

On the other hand, the addition of catechol at c_{inh} = 10 mM resulted in Al_{ox2}:Al_{ox1} share of 1:3. The significantly lower contribution from Al_{ox2} may suggest that inhibitory action of catechol efficiently reduces formation of the corrosion products on aluminum alloy 5754 surface. This observation was confirmed for higher concentrations of catechol, where Al2p signal is composed solely of native oxide film.

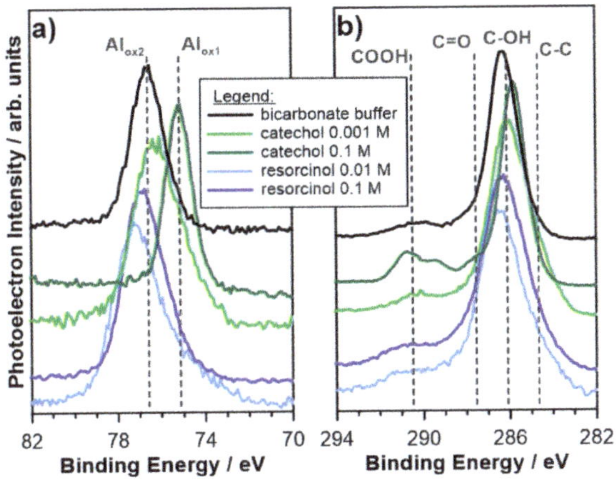

Figure 4. High-resolution XPS spectra recorded in (**a**) Al2p and (**b**) C1s binding energy range for aluminum alloy 5754 after 24 h exposure in bicarbonate buffer (pH = 11) and bicarbonate buffer with the addition of resorcinol or catechol at concentrations 10 and 100 mM.

When compared to bicarbonate buffer exposed sample, the C1s spectra reveal the significant presence of carbon species for both analyzed dihydroxybenzenes and at both concentrations, suggesting that inhibitor molecules indeed take part in formation of the adsorbed layer on aluminum alloy 5754 surface (Figure 4b). The chemistry of this carbon species is strongly altered, however. In the case of catechol, the amount of C-C and C-OH species, based on peaks at 284.7 and 286.0 eV, respectively, is nearly 3× compared to resorcinol at the same concentrations [54–56]. Furthermore, the C1s spectra reveal significantly higher amount of carbonyl and/or carboxyl species for catechol-containing electrolyte (at BE exceeding 287.6 eV). These results corroborate previous findings regarding possible formation of the adsorbed inhibitor layer directly on aluminum surface and negligible participation of the corrosion products layer. On the other hand, in absence of corrosion inhibitor the C1s contribution originates primarily from C-OH and C=O species, testifying the interaction of bicarbonate species with aluminum sample.

The O1s spectra deconvolution is in good accordance with previously discussed results. In the absence of dihydroxybenzene isomers, over 47 at.% of the aluminum surface consists of oxygen atoms, with peak binding energies characteristic either for aluminum metal oxides (531.3 eV) or hydroxides (531.6 eV) [19,20]. Further on, this value slowly decreases for resorcinol-containing electrolyte, still within the range of 40 at.%. However, in the presence of catechol, the amount of O^{2-} species is significantly lower, at 10 and 100 mM. Here, the signal corresponding to peaks located at approx. 532.6 eV is still strong, but with an organic C-O interaction origin, instead. The last deconvoluted O1s component is located at binding energies exceeding 532.7 eV. Its source is adsorbed carboxyl species, but also chemisorbed water molecules [57]. The non-stoichiometric corrosion products layer reveals high hydration level, confirmed with SEM micrographs, therefore higher share of chemisorbed water on surface of bicarbonate buffer-exposed and resorcinol-exposed aluminum alloy 5754 is natural.

Finally, the amount of magnesium, the primary alloying additive, was only slightly altered between various investigated electrolytes. In each case magnesium was found in the form of hydroxides with Mg1s peak BE at 1303.1 eV, its higher contribution for resorcinol testifies the presence of magnesium oxides in the corrosion products layer, having a possible effect on the increased corrosion resistance [58].

The SEM micrographs shown in Figure 5 reveal significant difference in occurrence of the corrosion process for both studied dihydroxybenzene derivatives. The addition of resorcinol does not affect significantly surface topography when compared to reference aluminum alloy sample in buffer electrolyte (Figure 5a). The oxidation is primarily oriented around the alloy microstructure, which can be confirmed through local dissolution of alloy matrix surrounding intermetallic particles [58–60]. These particles appear on the micrographs with bright colors. Based on own studies and the literature survey discussed, the particles are primarily composed of aluminum and alloying additives of Fe, Cr and Mn—each cathodic in nature compared to metal matrix [60,61]. Furthermore, dense network of cracks visible on Figure 5a–c testifies high hydration level of adsorbed layer and corrosion products layer on the electrode surface.

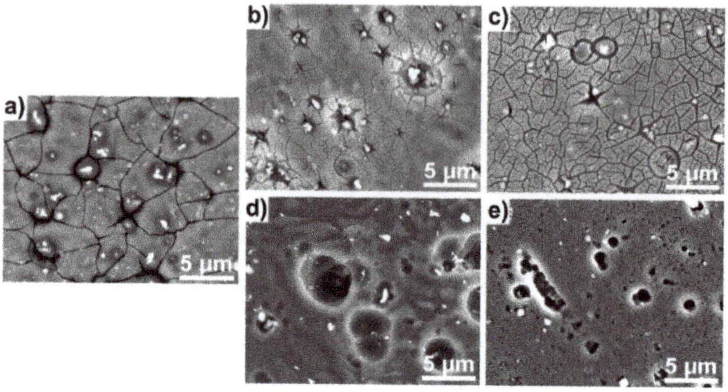

Figure 5. SEM micrographs of aluminum alloy 5754 surface exposed to bicarbonate buffer (pH = 11) for 24 h: (**a**) in absence of corrosion inhibitor; with addition of resorcinol at concentration (**b**) 10 mM and (**c**) 100 mM or with addition of catechol (**d**) 10 mM and (**e**) 100 mM.

The presence of catechol or quinol in the electrolyte consequences in localized corrosion of aluminum, which is most often restricted to anodic Mg-rich phases and areas surrounding cathodic intermetallic particles. The absence of thick corrosion product layer effects in lack of cracks, which otherwise cover metal surface. The localized corrosion is naturally more evident at lower inhibitor concentrations (see Figure 5d), where the spherical shape of caverns should be connected with local areas of hydrogen evolution in coupled cathodic reaction [53].

3.3. *Thermodynamics of the Adsorption Process*

The thermodynamics of dihydroxybenzene isomers adsorption was evaluated based on instantaneous impedance studies, carried out in galvanostatic mode (g-DEIS). The studied inhibitor is injected with linear injection rate (0.02 mL·min^{-1}); thus, the instantaneous value of inhibitor concentration in the corrosion cell is known. Dynamic impedance measurements allow for a precise evaluation of the instantaneous values of electric parameters, where the inhibition efficiency may be estimated from R_{ct} using previously introduced Equation (7). Studies carried out in galvanostatic mode at i_{DC} = 0 ensures constant measurement conditions and lack of an additional polarization component, resulting from corrosion potential changes during inhibitor injection. The approach is characterized with higher accuracy, allowing to obtain larger data set and evaluate the exact inhibitor concentration at which the linear character of the adsorption isotherm is modified due to full electrode surface coverage with inhibitor monolayer. The details of the experimental procedure are presented elsewhere [19,20].

The dynamic impedance spectra presented in the form of the Nyquist plot are shown on Figure 6a,b for resorcinol and catechol, respectively. The shape of the impedance spectra develops during corrosion inhibitor injection, where the increased impedance loop diameter testifies the increase of aluminum corrosion resistance. Fitting impedance spectra with R(QR) EEC allows for determination of dynamic changes of the electric parameters: R_{ct} and CPE within a timeframe of an analytical window length of a Short-Time Fourier Transform function, equal to 10 s in this case. The fitting procedure was carried out using dedicated software based on Nelder-Mead algorithm and build in LabView environment. The resultant χ^2-distribution was typically in 10^{-4} range and did not exceed 2×10^{-3}. Determination of the instantaneous R_{ct} values allowed for calculation of momentary inhibition efficiency IE%, which is also the measure of surface coverage with inhibitor molecules θ ($IE_{\%} = \theta \times 100\%$). According to the principles of the most commonly used Langmuir adsorption isotherm, the adsorption equilibrium constant, K_{ads}, depends on surface coverage θ, which is given with Equation (8):

$$K_{ads}c_{inh} = \left(\frac{\theta}{1-\theta}\right) \tag{8}$$

In the linear range of Equation (8), the adsorption isotherm may serve for calculation of the adsorption Gibbs free energy ΔG, using Equation (9):

$$\Delta G = -RTln(K_{ads} \times 55.5) \tag{9}$$

Importantly, Langmuir isotherm conditions are only fulfilled for concentrations below full coverage with the adsorbent monolayer. One should note, that classical approaches towards adsorption isotherm determination are based on merely few measurement points, where the non-linear behavior resulting from aforementioned situation is difficult to track. On the other hand, the quasi-capacitance parameter obtained during g-DEIS impedance measurements allows estimating the exact concentration required for monolayer formation by inhibitor molecules [20].

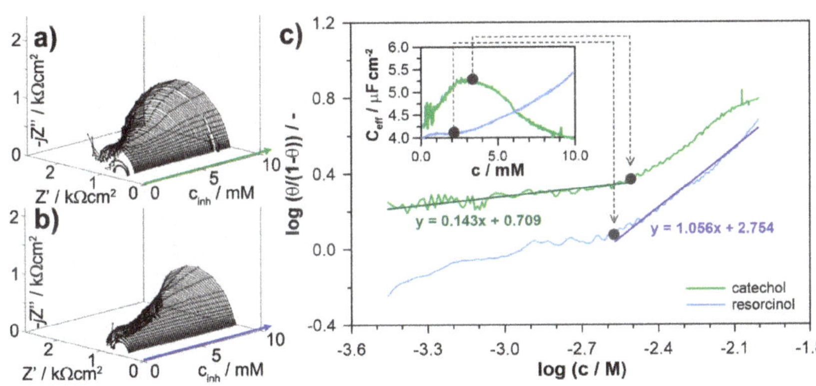

Figure 6. The g-DEIS impedance graphs of aluminum alloy 5754 in bicarbonate buffer (pH = 11), presented in Nyquist projection versus (**a**) catechol and (**b**) resorcinol concentration changes during its injection to corrosion cell. (**c**) Langmuir model of adsorption isotherms drawn based on instantaneous R_{ct} changes, in the inset the instantaneous changes of effective capacitance C_{eff}.

Figure 6c presents the adsorption isotherms obtained with g-DEIS approach and plotted according to the Langmuir model of molecules adsorption. The isotherms were drawn for resorcinol and catechol at concentrations in range 0–10 mM. It can be pointed out that both of these functions are characterized by loss of linear character and both have inflection at concentrations between 2 and 4 mM. However, the effective capacitance C_{eff} measurements plotted in the inset of Figure 6c reveals significant differences in the adsorption mechanism.

There are three primary factors affecting the value of capacitance according to Equation (10), namely, electrochemically active surface area A, permittivity ε, and layer thickness d:

$$C = \frac{\varepsilon_0 \varepsilon A}{d} \tag{10}$$

where ε_0 is vacuum permittivity. Normalization of heterogeneity factor affecting quasi-capacitance Q and estimation of the effective capacitance C_{eff} allows ignoring the effect of electrode heterogeneity. Previous studies have shown that with relatively short measurement duration the key factor influencing instantaneous C_{eff} of the adsorbed layer is its thickness [20,38].

In the case of a catechol-exposed AA5754 electrode, the C_{eff} value increases until reaching its maximum at c_{inh} = 3.5 mM and then decreases. Therefore, it should be assumed that the full coverage of aluminum surface with corrosion inhibitor molecules occurs at this concentration and the following C_{eff} decrease results from the increase in adsorbed layer thickness. The adsorption isotherm still remains linear afterward, but no longer following the aforementioned restriction regarding surface coverage.

The situation is essentially different in the case of resorcinol-exposed AA5754 electrodes, where according to classic EIS and XPS studies the inhibitory action is generally lower. In the whole studied concentration range, up to 10 mM the C_{eff} value effectively increases. The competitive interaction of the corrosion products layer and the resorcinol adsorption layer may have its effect on difficulties in assessing the full coverage of the adsorption layer but also influences layer permittivity. As a result, the adsorption isotherm for resorcinol was following Langmuir model of adsorption at significantly higher inhibitor concentrations than catechol. On the other hand, the estimated inhibitor efficiency at the lowest concentrations is negligible and thus hard to measure. A conclusion should be made that the applicability of the Langmuir adsorption isotherm model for catechol and resorcinol lies in different inhibitor concentration ranges.

Following Equations (8) and (9), the calculated values of Gibbs free energy ΔG of the adsorption process are presented in Table 3. Their negative values in both cases confirm spontaneous adsorption of both studied dihydroxybenzene molecules on aluminum alloy 5754 surface. Nevertheless, the significantly different ΔG values between resorcinol and catechol suggest an altered adsorption mechanism.

Table 3. The applicability range of the Langmuir adsorption model and the obtained thermodynamic parameters for resorcinol and catechol.

Isomer	Langmuir Model c_{inh} Range/mM	Adsorption Equilibrium Constant K_{ads}	Gibbs Free Energy ΔG/kJmol^{-1}
catechol	<3.5	1.39	−10.77
resorcinol	2–10	11.38	−15.98

The more negative Gibbs free energy values are typical in case of chemisorption and formation of chemical bonds between filled π-orbitals in the oxygen atoms and partially unoccupied π-orbitals in the d-block metals. This is the postulated adsorption mechanism of resorcinol. Naturally, the value of this thermodynamic parameter may be further influenced by reported presence of the nonstoichiometric corrosion products layer. On the other hand, our previous studied on carboxylic acids revealed that the less negative free Gibbs energies correspond not only to the electrostatic interaction of the physisorption process but also the ligand exchange model of adsorption, resulting in formation of coordination compounds at the metal interface [19,20]. This is the case of catechol interaction. The lower efficiency of ligand formation by resorcinol and quinol originates from the molecule geometry.

3.4. *The Keto-Enol Tautomerism*

During the long-term exposure tests, the authors observed changes in the color of the studied electrolytes over time, and resorcinol in particular (see inset of Figure 7). These changes were followed by alteration of the electrochemical characteristics over time. On the other hand, for catechol and quinol the long-term inhibition efficiency was significantly higher. The g-DEIS studies were carried out once more to track the exact change in the electrochemical behavior of aluminum alloy 5754 during 24 h exposure. The results of the impedance monitoring are presented in Figure 7.

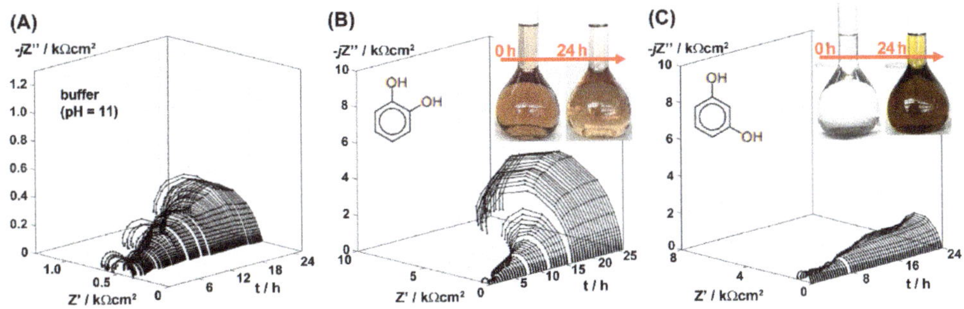

Figure 7. The g-DEIS impedance graphs of aluminum alloy 5754 in Nyquist projection during 24 h exposure in (**A**) bicarbonate buffer (pH = 11); and with the addition of (**B**) 10 mM catechol and (**C**) 10 mM resorcinol.

Analysis of the impedance data makes it possible to draw conclusions regarding the long-term behavior of AA5754 under the studied electrolytic conditions. The slight increase of the semicircle diameter over time, seen on the Nyquist projection for buffer-exposed sample, makes it possible to conclude that the corrosion product layer forming on metal surface provides partial barrier properties and decreases the corrosion rate approximately 2.5×.

When exposed to buffer with addition of 10 mM catechol, the instantaneous charge transfer resistance is slightly higher (typically around 0.4 kΩ) and then gradually increases over time to reach significantly improved inhibition efficiency of ~98% after 24 h exposure. On the other hand, AA5754 exposed to electrolyte containing the same amount of resorcinol shows very small increase of charge transfer resistance over duration of the long-term exposure experiment.

The long-term exposure study allows drawing two important conclusions. First, there must be an additional interaction between studied inhibitor molecules and the electrolyte or the analyzed sample, which further differentiates the electrochemical characteristics of these dihydroxybenzene isomers over time. Second, when performing inhibitor efficiency measurements one has to take into consideration possible changes of investigated system characteristics. This is possible by carrying out fast measurements with techniques that allow non-stationary process analyses (such as g-DEIS). Alternatively, one could employ a sufficiently long conditioning period, which might be different for each studied system. The latter approach, although more accessible, may cause problems in terms of meeting the conditions for many adsorption models.

The authors claim that the mechanism leading to further differentiation of adsorption by catechol and resorcinol on the aluminum alloy surface is the keto-enol tautomerism, which may occur in aqueous alkaline environments. Nuclear magnetic resonance (NMR) studies were performed in order to verify this hypothesis.

^{1}H-NMR measurements were performed in order to determine the presence of possible keto-enol forms in resorcinol in the alkaline conditions or to determine the formation of a salt of those compounds. The studies of proton transfer by ^{1}H-NMR titration present a useful technique to determine the keto-enol equilibria [62,63]. The ^{1}H-NMR spectra of resorcinol dissolved in D_2O with addition of NaOH in molar ration of 1:5 and 1:10 are presented on Figure 8. The ^{1}H-NMR spectra of resorcinol have previously been performed in D_2O [64,65], while the titration of this compound by NaOH has not been investigated.

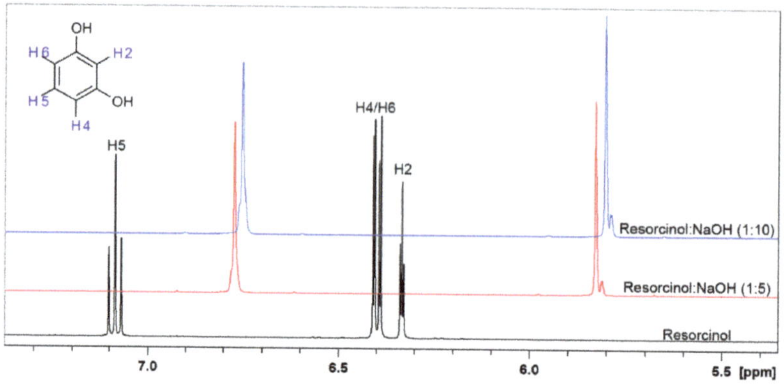

Figure 8. ^{1}H-NMR spectra of resorcinol in D_2O (black line) and in the presence of NaOH in D_2O in the molar ratio 1:5 (red line) and 1:10 (blue line), respectively.

Two triplets are observed on spectra of resorcinol in D_2O, which correspond to H5 and H2, while two doublets correspond to H4 and H6. The spectra of resorcinol after the addition of NaOH in molar ratio 1:5 changed diametrically. The shape, the chemical shifts and the multiplicity are different in comparison to the first one. Two main signals shifted towards negative values are now observed. This phenomenon indicates that the protons present in the structure of resorcinol are changed, further influencing the chemical shifts and the multiplicity. The addition of molar excess of NaOH in (1:10), does not cause any additional changes in ^{1}H-NMR spectrum shape, but the peaks are further shifted.

The determination of the new resorcinol derivative structure was possible after measuring the ^{13}C-NMR spectra, shown on Figure 9. These spectra were performed in D_2O and after addition of NaOH in molar ratio 1:5 and 1:10, similar to previous experiment. It should be noticed that the peaks C4, C5, and C6 present in spectra before and after addition of NaOH do not change their position significantly. On the other hand, peaks C1 and C3, overlapping at 156.8 ppm, change their position to 167.4 ppm with the addition of NaOH, which may indicate that keto form is present in the structure of resorcinol regardless the molar ratio. Additionally, after addition of NaOH the shift of peak C2 is observed from 102.6 ppm to 106.0 ppm, this shift confirms the formation of keto form.

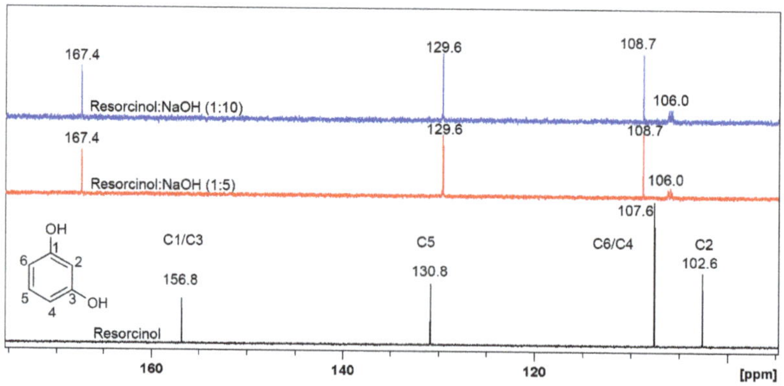

Figure 9. ^{13}C-NMR spectra of resorcinol in D_2O (black line) and in the presence of NaOH in D_2O in molar ratio 1:5 (red line) and 1:10 (blue line), respectively.

In the next step, the ^{1}H-NMR and ^{13}C-NMR spectra were performed for catechol under the same experimental conditions. The ^{1}H-NMR spectra of catechol have previously been performed in aqueous solution, but under acidic pH = 2.4 [66] and in $CDCl_3$ [67]. The shape and chemical shifts of ^{1}H-NMR spectra illustrated on Figure 10 are very similar to those in the literature.

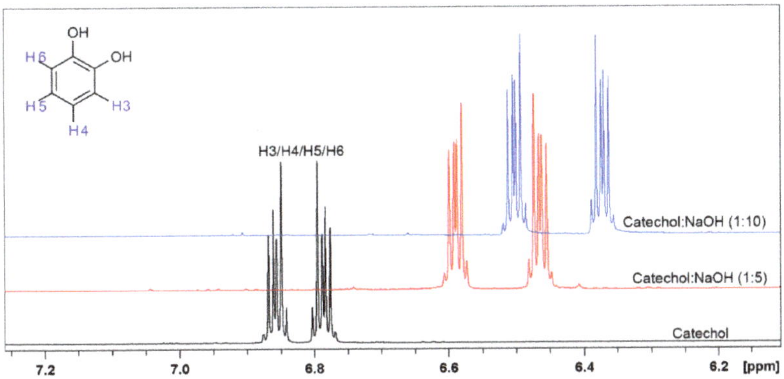

Figure 10. ^{1}H-NMR spectra of catechol in D_2O (black line) and in the presence of NaOH in D_2O in molar ratio 1:5 (red line) and 1:1 (blue line), respectively.

Two multiplets are present regardless of the solution, while the addition of NaOH in ratio 1:5 and 1:10, respectively, do not cause any changes in ^{1}H-NMR spectra shape; however, with increasing molar ratio of NaOH, the spectra are shifting towards more negative values. The above changes clearly indicate that the chemical structure of the catechol does not change significantly, while the shifts may indicate formation of sodium salts of catechol [68].

^{13}C-NMR spectra were performed in order to confirm the formation of sodium salt of catechol in alkaline conditions, as shown on Figure 11. The obtained data confirm previously drawn assumptions. ^{13}C-NMR spectra reveal ^{13}C chemical shifts for catechol in D_2O with the presence of NaOH in molar ratio 1:5 and 1:10. The C1 and C2 atoms connected to the hydroxyl groups in catechol in investigated solutions both give signal at 143.9 ppm in absence of NaOH, but after its addition, the signal is shifted towards 152.6 ppm and 154.9 ppm for ratio 1:5 and 1:10, respectively. The shift of carbon signals directly indicates that catechol in alkaline solutions forms a salt [69]. It is worth noticing that a small change of peak position was also observed for carbon C4/C5 and C3/C6 after additions of NaOH.

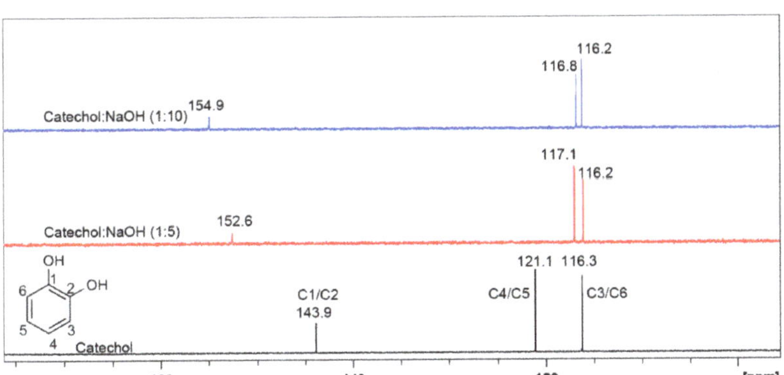

Figure 11. ^{13}C-NMR spectra of catechol in D_2O (black line) and in the presence of NaOH in D_2O in the molar ratio 1:5 (red line) and 1:1 (blue line), respectively.

3.5. Dihydroxybenzene Isomers Adsorption Mechanism

The overall interaction of the studied dihydroxybenzene isomers with aluminum alloy 5754 surface may thus be explained using the scheme presented on Figure 12. Figure 12a illustrates the case of aluminum corrosion in an alkaline environment, according to the two-step mechanism discussed in the introduction section: 1) attack of OH^- ions on Al_2O_3 leading to its dissolution and $Al(OH)_3$ formation, followed shortly after by 2) chemical formation of $Al(OH)_4{}^-$ ions. The corrosion inhibition mechanism for catechol considers formation of ligands with aluminum ions (see Figure 12b), which is hindered in the case of resorcinol. On the other hand, resorcinol depends on the formation of insoluble corrosion products layer, which to a large extent provides a barrier mechanism towards corrosive electrolyte. The molecules chemisorb on the corrosion products layer, which becomes very efficient only at high inhibitor concentrations.

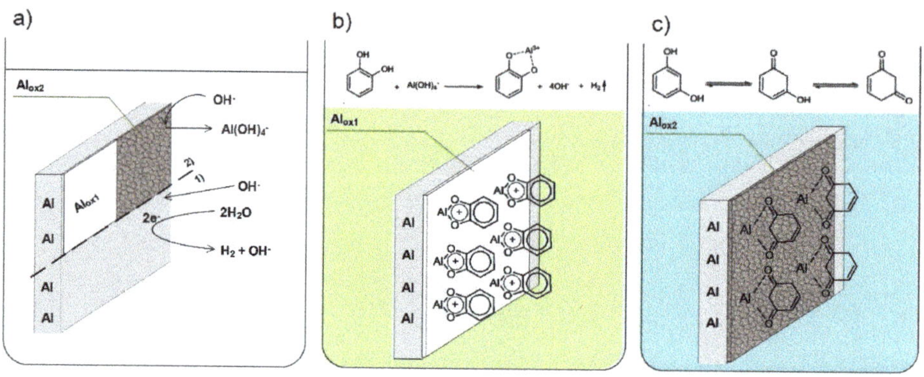

Figure 12. Schematic representation of aluminum corrosion mechanism in aqueous alkaline solutions: (**a**) in absence of corrosion inhibitor and with the addition of (**b**) catechol; (**c**) resorcinol. Phase (1) represents the OH^- attack leading to formation $Al(OH)_3$, phase (2) describes the interaction between OH^- and $Al(OH)_3$, leading to the formation of $Al(OH)_4{}^-$ and a non-stoichiometric insoluble corrosion product layer.

The lower inhibition efficiency of resorcinol at concentrations not exceeding 10 mM is connected with the keto-enol tautomerism mechanism, occurring in aqueous alkaline media and lowering the molecule influence on the corrosion protection (see Figure 12c). Due to local differences in pH in anodic and cathodic zones of electrode/electrolyte interface, the dynamics of the keto-enol tautomerism may be locally altered. The NMR spectra revealed that the process takes place in wide pH range.

4. Conclusions

Understanding the interaction mechanism of inhibitor molecules with the protected metal surface is of key importance in the selection of the most efficient corrosion inhibitors, the most important, in particular, in the case of green inhibitors based on natural extracts. While it is widely known how different functional groups affect the adsorption mechanism, the differences introduced by its location within the inhibitor molecule are often omitted.

In this study, we revealed how the position of hydroxyl groups affects the adsorption mechanism of dihydroxybenzene isomers and offered corrosion resistance toward alumnium alloy 5754 surface in alkaline environment. The utilization of Dynamic Electrochemical Impedance Spectroscopy in galvanostatic mode (g-DEIS) for adsorption isotherm determination made it possible to confirm different forms of dihydroxybenzene interaction. All of the studied inhibitors followed the Langmuir model of adsorption, although we have observed that its applicability lies in different inhibitor concentration ranges.

Resorcinol was found to be characterized by the chemical adsorption mechanism. Its adsorption on aluminum surface is competitive to insoluble corrosion product layer formation, as shown with SEM and XPS studies. This interaction leads to the best inhibitor efficiency at the highest investigated concentration of 100 mM, but is not as efficient at lower concentrations. On the other hand, catechol and quinol follow the ligand exchange model of adsorption. This leads to more efficient adsorption and increases corrosion protection even at lower corrosion concentrations: 1 and 10 mM. The adsorption process dominates insoluble corrosion product layer formation, the presence of which on the analyzed surface was negligible.

The next significant difference lies in the long-term behavior and corrosion protection offered by dihydroxybenzene isomers in alkaline electrolyte. We report that resorcinol molecules undergo keto-enol tautomerism in sodium hydroxide solution, while the aforementioned process was negligible in the case of quinol and catechol. The tautomerism leads to the rebuilding of the inhibitor molecule, electrolyte discoloration, but does not have significant influence on the chemical adsorption mechanism by resorcinol over longer periods of time. It is even possible that the presence of keto-enol tautomerism itself is the reason behind hindered adsorption of resorcinol and offered corrosion resistance. Keto forms were not observed in the structure of catechol and quinol molecules. At the same time, their ability to complex metal ions leads to formation of layers with higher barrier properties and increased corrosion inhibition.

Author Contributions: Conceptualization, J.R.; Methodology, J.R. and J.W.; Investigation, M.B., M.K., W.L. (Electrochemical Studies, SEM), J.R. (XPS) and P.N. (NMR); Resources, J.R.; Writing—Original Draft Preparation, J.R., P.N., J.W. and W.L.; Writing—Review and Editing, J.R. and J.W.; Funding Acquisition, J.R.

Funding: The authors acknowledge the financial support of the Polish Ministry of Science and Higher Education from the budget funds in the period 2016-2019 under Iuventus Plus project IP2015 067574.

Acknowledgments: The authors acknowledge Pawel Slepski and Artur Zielinski from Gdansk University of Technology for development of the software dedicated to effective DEIS data collection and analysis.

Conflicts of Interest: The authors declare no conflict of interest.

References

1. Greenwood, N.N.; Earnshaw, A. *Chemistry of the Elements*, 2nd ed.; Butterworth-Heinemann: Oxford, UK; Boston, MA, USA, 1997; ISBN 978-0-7506-3365-9.
2. Sanders, R.E. Updated by Staff Aluminum and Aluminum Alloys. In *Kirk-Othmer Encyclopedia of Chemical Technology*; John Wiley & Sons, Inc., Ed.; John Wiley & Sons, Inc.: Hoboken, NJ, USA, 2012; ISBN 978-0-471-23896-6. [CrossRef]
3. Macdonald, D.D. Evaluation of Alloy Anodes for Aluminum-Air Batteries. *J. Electrochem. Soc.* **1988**, *135*, 2410. [CrossRef]
4. Chu, D.; Savinell, R.F. Experimental data on aluminum dissolution in KOH electrolytes. *Electrochim. Acta* **1991**, *36*, 1631–1638. [CrossRef]
5. Pyun, S.-I.; Moon, S.-M. Corrosion mechanism of pure aluminium in aqueous alkaline solution. *J. Solid State Electrochem.* **2000**, *4*, 267–272. [CrossRef]
6. Moon, S.-M.; Pyun, S.-I. The corrosion of pure aluminium during cathodic polarization in aqueous solutions. *Corros. Sci.* **1997**, *39*, 399–408. [CrossRef]
7. Adhikari, S.; Hebert, K.R. Factors controlling the time evolution of the corrosion potential of aluminum in alkaline solutions. *Corros. Sci.* **2008**, *50*, 1414–1421. [CrossRef]
8. Adhikari, S.; Lee, J.; Hebert, K.R. Formation of Aluminum Hydride during Alkaline Dissolution of Aluminum. *J. Electrochem. Soc.* **2008**, *155*, C16. [CrossRef]
9. Perrault, G.G. The Role of Hydrides in the Equilibrium of Aluminum in Aqueous Solutions. *J. Electrochem. Soc.* **1979**, *126*, 199. [CrossRef]
10. Brown, O.R.; Whitley, J.S. Electrochemical behaviour of aluminium in aqueous caustic solutions. *Electrochim. Acta* **1987**, *32*, 545–556. [CrossRef]
11. Li, Q.; Bjerrum, N.J. Aluminum as anode for energy storage and conversion: A review. *J. Power Sour.* **2002**, *110*, 1–10. [CrossRef]

12. Liu, Y.; Sun, Q.; Li, W.; Adair, K.R.; Li, J.; Sun, X. A comprehensive review on recent progress in aluminum–air batteries. *Gr. Energy Environ.* **2017**, *2*, 246–277. [CrossRef]
13. Yang, S. Design and analysis of aluminum/air battery system for electric vehicles. *J. Power Sour.* **2002**, *112*, 162–173. [CrossRef]
14. Zhang, X.; Yang, S.H.; Knickle, H. Novel operation and control of an electric vehicle aluminum/air battery system. *J. Power Sour.* **2004**, *128*, 331–342. [CrossRef]
15. Singh, A.; Ansari, K.; Quraishi, M.; Lgaz, H. Effect of Electron Donating Functional Groups on Corrosion Inhibition of J55 Steel in a Sweet Corrosive Environment: Experimental, Density Functional Theory, and Molecular Dynamic Simulation. *Materials* **2018**, *12*, 17. [CrossRef] [PubMed]
16. Al-Suhybani, A.A.; Sultan, Y.H.; Hamid, W.A. Corrosion of aluminium in alkaline solutions. *Mater. Werkst.* **1991**, *22*, 301–307. [CrossRef]
17. Ebenso, E.E.; Isabirye, D.A.; Eddy, N.O. Adsorption and Quantum Chemical Studies on the Inhibition Potentials of Some Thiosemicarbazides for the Corrosion of Mild Steel in Acidic Medium. *IJMS* **2010**, *11*, 2473–2498. [CrossRef] [PubMed]
18. Xhanari, K.; Finšgar, M. Organic corrosion inhibitors for aluminum and its alloys in chloride and alkaline solutions: A review. *Arab. J. Chem.* **2016**, in press. [CrossRef]
19. Wysocka, J.; Krakowiak, S.; Ryl, J. Evaluation of citric acid corrosion inhibition efficiency and passivation kinetics for aluminium alloys in alkaline media by means of dynamic impedance monitoring. *Electrochim. Acta* **2017**, *258*, 1463–1475. [CrossRef]
20. Wysocka, J.; Cieslik, M.; Krakowiak, S.; Ryl, J. Carboxylic acids as efficient corrosion inhibitors of aluminium alloys in alkaline media. *Electrochim. Acta* **2018**, *289*, 175–192. [CrossRef]
21. Brito, P.S.D.; Sequeira, C.A.C. Organic Inhibitors of the Anode Self-Corrosion in Aluminum-Air Batteries. *J. Fuel Cell Sci. Technol.* **2013**, *11*, 011008. [CrossRef]
22. Müller, B. Citric acid as corrosion inhibitor for aluminium pigment. *Corros. Sci.* **2004**, *46*, 159–167. [CrossRef]
23. Amin, M.A.; EI-Rehim, S.S.A.; El-Sherbini, E.E.F.; Hazzazi, O.A.; Abbas, M.N. Polyacrylic acid as a corrosion inhibitor for aluminium in weakly alkaline solutions. Part I: Weight loss, polarization, impedance EFM and EDX studies. *Corros. Sci.* **2009**, *51*, 658–667. [CrossRef]
24. Kumari, P.D.R.; Nayak, J.; Shetty, A.N. 3-Methyl-4-amino-5-mercapto-1, 2, 4-triazole as corrosion inhibitor for 6061 Al alloy in 0.5 M sodium hydroxide solution. *J. Coat. Technol. Res.* **2011**, *8*, 685–695. [CrossRef]
25. Lashgari, M.; Malek, A.M. Fundamental studies of aluminum corrosion in acidic and basic environments: Theoretical predictions and experimental observations. *Electrochim. Acta* **2010**, *55*, 5253–5257. [CrossRef]
26. Lashgari, M. Theoretical challenges in understanding the inhibition mechanism of aluminum corrosion in basic media in the presence of some p-phenol derivatives. *Electrochim. Acta* **2011**, *56*, 3322–3327. [CrossRef]
27. Al-Amiery, A.; Al-Majedy, Y.; Kadhum, A.; Mohamad, A. New Coumarin Derivative as an Eco-Friendly Inhibitor of Corrosion of Mild Steel in Acid Medium. *Molecules* **2014**, *20*, 366–383. [CrossRef] [PubMed]
28. Yang, W.; Wang, Q.; Xu, K.; Yin, Y.; Bao, H.; Li, X.; Niu, L.; Chen, S. Enhanced Corrosion Resistance of Carbon Steel in Hydrochloric Acid Solution by Eriobotrya Japonica Thunb. Leaf Extract: Electrochemical Study. *Materials* **2017**, *10*, 956. [CrossRef]
29. Okeniyi, J.; Loto, C.; Popoola, A. Effects of Phyllanthus muellerianus Leaf-Extract on Steel-Reinforcement Corrosion in 3.5% NaCl-Immersed Concrete. *Metals* **2016**, *6*, 255. [CrossRef]
30. Fares, M.M.; Maayta, A.K.; Al-Qudah, M.M. Pectin as promising green corrosion inhibitor of aluminum in hydrochloric acid solution. *Corros. Sci.* **2012**, *60*, 112–117. [CrossRef]
31. Abdel-Gaber, A.M.; Abd-El-Nabey, B.A.; Sidahmed, I.M.; El-Zayady, A.M.; Saadawy, M. Inhibitive action of some plant extracts on the corrosion of steel in acidic media. *Corros. Sci.* **2006**, *48*, 2765–2779. [CrossRef]
32. Abiola, O.K.; Otaigbe, J.O.E.; Kio, O.J. Gossipium hirsutum L. extracts as green corrosion inhibitor for aluminum in NaOH solution. *Corros. Sci.* **2009**, *51*, 1879–1881. [CrossRef]
33. Abiola, O.K.; Oforka, N.C.; Ebenso, E.E.; Nwinuka, N.M. Eco-friendly corrosion inhibitors: The inhibitive action of Delonix Regia extract for the corrosion of aluminium in acidic media. *Anti-Corros. Methods Mater.* **2007**, *54*, 219–224. [CrossRef]
34. Azzaoui, K.; Mejdoubi, E.; Jodeh, S.; Lamhamdi, A.; Rodriguez-Castellón, E.; Algarra, M.; Zarrouk, A.; Errich, A.; Salghi, R.; Lgaz, H. Eco friendly green inhibitor Gum Arabic (GA) for the corrosion control of mild steel in hydrochloric acid medium. *Corros. Sci.* **2017**, *129*, 70–81. [CrossRef]

35. de Souza, F.S.; Spinelli, A. Caffeic acid as a green corrosion inhibitor for mild steel. *Corros. Sci.* **2009**, *51*, 642–649. [CrossRef]
36. Abiola, O.K.; Otaigbe, J.O.E. The effects of Phyllanthus amarus extract on corrosion and kinetics of corrosion process of aluminum in alkaline solution. *Corros. Sci.* **2009**, *51*, 2790–2793. [CrossRef]
37. Singh, A.; Ahamad, I.; Quraishi, M.A. Piper longum extract as green corrosion inhibitor for aluminium in NaOH solution. *Arab. J. Chem.* **2016**, *9*, S1584–S1589. [CrossRef]
38. Ryl, J.; Wysocka, J.; Cieslik, M.; Gerengi, H.; Ossowski, T.; Krakowiak, S.; Niedzialkowski, P. Understanding the origin of high corrosion inhibition efficiency of bee products towards aluminium alloys in alkaline environments. *Electrochim. Acta* **2019**, *304*, 263–274. [CrossRef]
39. Talati, J.D.; Modi, R.M. O-Substituted Phenols as Corrosion Inhibitors for Aluminium-Copper Alloy in Sodium Hydroxide. *Br. Corros. J.* **1977**, *12*, 180–184. [CrossRef]
40. Lakshmi, D.; Rajendran, S.; Sathiabama, J. Inhibition of Corrosion of Aluminium in Aqueous Solution at pH11 by Resorcinol-Zn2+ System. *Int. J. Nano Corr. Sci. Eng.* **2016**, *3*, 26–42.
41. Fouda, A.S.; Elasmy, A.A. Efficiency of some phenylthiosemicarbazide derivatives in retarding the dissolution of Al in NaOH solution. *Mon. Chem.* **1987**, *118*, 709–716. [CrossRef]
42. Hassan, S.M.; Moussa, M.N.; El-Tagoury, M.M.; Radi, A.A. Aromatic acid derivatives as corrosion inhibitors for aluminium in acidic and alkaline solutions. *Anti-Corros. Methods Mater.* **1990**, *37*, 8–11. [CrossRef]
43. Obi-Egbedi, N.O.; Obot, I.B.; El-Khaiary, M.I. Quantum chemical investigation and statistical analysis of the relationship between corrosion inhibition efficiency and molecular structure of xanthene and its derivatives on mild steel in sulphuric acid. *J. Mol. Struct.* **2011**, *1002*, 86–96. [CrossRef]
44. Khaled, K.F. Electrochemical investigation and modeling of corrosion inhibition of aluminum in molar nitric acid using some sulphur-containing amines. *Corros. Sci.* **2010**, *52*, 2905–2916. [CrossRef]
45. Ramesh Babu, B.; Holze, R. Corrosion and hydrogen permeation inhibition for mild steel in HCl by isomers of organic compounds. *Br. Corros. J.* **2000**, *35*, 204–209. [CrossRef]
46. Talati, J.D.; Desai, M.N.; Shah, N.K. Ortho-, meta-, and para-aminophenol-N-salicylidenes as corrosion inhibitors of zinc in sulfuric acid. *Anti-Corros. Meth Mater.* **2005**, *52*, 108–117. [CrossRef]
47. Ryl, J.; Darowicki, K.; Slepski, P. Evaluation of cavitation erosion–corrosion degradation of mild steel by means of dynamic impedance spectroscopy in galvanostatic mode. *Corros. Sci.* **2011**, *53*, 1873–1879. [CrossRef]
48. Gerengi, H.; Darowicki, K.; Slepski, P.; Bereket, G.; Ryl, J. Investigation effect of benzotriazole on the corrosion of brass-MM55 alloy in artificial seawater by dynamic EIS. *J. Solid State Electrochem.* **2010**, *14*, 897–902. [CrossRef]
49. Gerengi, H. The Use of Dynamic Electrochemical Impedance Spectroscopy in Corrosion Inhibitor Studies. *Prot. Met. Phys. Chem. Surf.* **2018**, *54*, 536–540. [CrossRef]
50. Hirschorn, B.; Orazem, M.E.; Tribollet, B.; Vivier, V.; Frateur, I.; Musiani, M. Determination of effective capacitance and film thickness from constant-phase-element parameters. *Electrochim. Acta* **2010**, *55*, 6218–6227. [CrossRef]
51. Talati, J.D.; Modi, R.M. Dihydroxy-benzenes as Corrosion Inhibitors for Aluminium—Copper Alloy in Sodium Hydroxide. *Br. Corros. J.* **1975**, *10*, 103–106. [CrossRef]
52. Suresh, S.; Srivastava, V.C.; Mishra, I.M. Adsorption of catechol, resorcinol, hydroquinone, and their derivatives: A review. *Int. J. Energy Environ. Eng.* **2012**, *3*, 32. [CrossRef]
53. Wysocka, J.; Krakowiak, S.; Ryl, J.; Darowicki, K. Investigation of the electrochemical behaviour of AA1050 aluminium alloy in aqueous alkaline solutions using Dynamic Electrochemical Impedance Spectroscopy. *J. Electroanal. Chem.* **2016**, *778*, 126–136. [CrossRef]
54. Djordjevic, I.; Choudhury, N.R.; Dutta, N.K.; Kumar, S. Synthesis and characterization of novel citric acid-based polyester elastomers. *Polymer* **2009**, *50*, 1682–1691. [CrossRef]
55. Amin, M.A.; Ahmed, E.M.; Mostafa, N.Y.; Alotibi, M.M.; Darabdhara, G.; Das, M.R.; Wysocka, J.; Ryl, J.; Abd El-Rehim, S.S. Aluminum Titania Nanoparticle Composites as Nonprecious Catalysts for Efficient Electrochemical Generation of H_2. *ACS Appl. Mater. Interfaces* **2016**, *8*, 23655–23667. [CrossRef] [PubMed]
56. Niedziałkowski, P.; Ossowski, T.; Zięba, P.; Cirocka, A.; Rochowski, P.; Pogorzelski, S.J.; Ryl, J.; Sobaszek, M.; Bogdanowicz, R. Poly-l-lysine-modified boron-doped diamond electrodes for the amperometric detection of nucleic acid bases. *J. Electroanal. Chem.* **2015**, *756*, 84–93. [CrossRef]

57. McCafferty, E.; Wightman, J.P. Determination of the concentration of surface hydroxyl groups on metal oxide films by a quantitative XPS method. *Surf. Interface Anal.* **1998**, *26*, 549–564. [CrossRef]
58. Ryl, J.; Wysocka, J.; Jarzynka, M.; Zielinski, A.; Orlikowski, J.; Darowicki, K. Effect of native air-formed oxidation on the corrosion behavior of AA 7075 aluminum alloys. *Corros. Sci.* **2014**, *87*, 150–155. [CrossRef]
59. Wloka, J.; Bürklin, G.; Virtanen, S. Influence of second phase particles on initial electrochemical properties of AA7010-T76. *Electrochim. Acta* **2007**, *53*, 2055–2059. [CrossRef]
60. Yasakau, K.A.; Zheludkevich, M.L.; Lamaka, S.V.; Ferreira, M.G.S. Role of intermetallic phases in localized corrosion of AA5083. *Electrochim. Acta* **2007**, *52*, 7651–7659. [CrossRef]
61. Goswami, R.; Spanos, G.; Pao, P.S.; Holtz, R.L. Precipitation behavior of the ß phase in Al-5083. *Mater. Sci. Eng. A* **2010**, *527*, 1089–1095. [CrossRef]
62. Novak, P.; Skare, D.; Sekusak, S. Substituent, temperature and solvent effects on keto-enol equilibrium in symmetrical pentane-1,3,5-triones. Nuclear magnetic resonance and theoretical studies. *Croat. Chem. Acta* **2000**, *73*, 1153–1170.
63. Billman, J.H.; Sojka, S.A.; Taylor, P.R. Investigations of keto–enol tautomerism by carbon-13 nuclear magnetic resonance spectroscopy. *J. Chem. Soc. Perkin Transection 2* **1972**, 2034–2035. [CrossRef]
64. Gaca, K.Z.; Parkinson, J.A.; Sefcik, J. Kinetics of early stages of resorcinol-formaldehyde polymerization investigated by solution-phase nuclear magnetic resonance spectroscopy. *Polymer* **2017**, *110*, 62–73. [CrossRef]
65. Giles, R.; Kim, I.; Chao, W.E.; Moore, J.; Jung, K.W. Dual Studies on a Hydrogen–Deuterium Exchange of Resorcinol and the Subsequent Kinetic Isotope Effect. *J. Chem. Educ.* **2014**, *91*, 1220–1223. [CrossRef]
66. Lambert, F.; Ellenberger, M.; Merlin, L.; Cohen, Y. NMR study of catechol and some catecholamines. *Org. Magn. Reson.* **1975**, *7*, 266–273. [CrossRef]
67. Luisier, N.; Schenk, K.; Severin, K. A four-component organogel based on orthogonal chemical interactions. *Chem. Commun.* **2014**, *50*, 10233–10236. [CrossRef] [PubMed]
68. Kim, H.; Gao, J.; Burgess, D.J. Evaluation of solvent effects on protonation using NMR spectroscopy: Implication in salt formation. *Int. J. Pharm.* **2009**, *377*, 105–111. [CrossRef] [PubMed]
69. Pretsch, E.; Buhlmann, P.; Badertscher, M. *Structure Determination of Organic Compounds*; Springer: Berlin/Heidelberg, Germany, 2009; ISBN 978-3-540-93809-5.

materials

MDPI

Article

PDC Glass/Ceramic Coatings Applied to Differently Pretreated AISI441 Stainless Steel Substrates

Milan Parchovianský [1,*], Ivana Parchovianská [1], Peter Švančárek [2], Günter Motz [3] and Dušan Galusek [1,2]

1 Centre for Functional and Surface Functionalised Glass, Alexander Dubček University of Trenčín, Študentská 2, 911 50 Trenčín, Slovakia; ivana.parchovianska@tnuni.sk (I.P.); dusan.galusek@tnuni.sk (D.G.)

2 Joint Glass Centre of the Institute of Inorganic Chemistry Slovak Academy of Sciences, Alexander Dubček University of Trenčín and Faculty of Chemical and Food Technology Slovak University of Technology, Študentská 2, 911 50 Trenčín, Slovakia; peter.svancarek@tnuni.sk

3 Ceramic Materials Engineering (CME), University of Bayreuth, D-95440 Bayreuth, Germany; guenter.motz@uni-bayreuth.de

* Correspondence: milan.parchoviansky@tnuni.sk; Tel.: +421-944600882

Received: 16 December 2019; Accepted: 24 January 2020; Published: 31 January 2020

Abstract: In this work, the influence of different cleaning procedures on adhesion of composite coatings containing passive ceramic and commercial glasses was investigated. Two compositions (C2c, D2-PP) of double-layer polymer-derived ceramic (PDC) coating systems, composed from bond coat and a top coat, were developed. In order to obtain adherent coatings, stainless steel substrates were cleaned by four different cleaning procedures. The coatings were then deposited onto the steel substrate via spray coating. Pretreatment by subsequent ultrasonic cleaning in acetone, ethanol and deionised water (procedure U) was found to be the most effective, and the resultant C2c and D2-PP coatings, pyrolysed at 850 °C, indicated strong adhesion without delamination or cracks, propagating at the interface steel/bond coat. In the substrate treated by sandblasting and chemical etching, small cracks in the bond coat were observed under the same pyrolysis conditions. After oxidation tests, all coatings, except for those subjected to the U-treated substrates, showed significant cracking in the bond coat. The D2-PP coatings were denser than C2c, indicating better protection of the substrate.

Keywords: cleaning; bond coat; PDC coatings; fillers

1. Introduction

Because of the increasing costs for metals, there is an effort made to enhance the service life of steel components exposed to aggressive environment, which is commonly used in exhaust gas elements, waste incineration plants or in chemical industry. Refractory stainless steels are highly oxidation and corrosion resistant materials. As metal wear and oxidation/corrosion cause significant economic losses, the development of thermal (TB) and environmental barrier coatings (EBC) is the matter of significant importance.

Due to their extraordinary properties at high temperatures and in chemically aggressive environments, non-oxide and oxide PDC ceramic coatings are suitable for increasing the oxidation and corrosion resistance of metals [1]. Preceramic polymers offer a lot of processing advantages that are not possible with traditional ceramics [2]. For example, organosilicon polymer precursors such as polysiloxanes [3], polycarbosilanes [4], or polysilazanes [5–7], represent a class of hybrid materials which, by suitable heat treatment (pyrolysis in a controlled atmosphere), provide high purity ceramic materials with an adaptable chemical composition and a well-defined structure. These polymers are characterised by an inorganic polymer chain composed of silicon atoms and organic substituents attached to the backbone. Polysilazanes are currently used as precursors for the synthesis of Si_3N_4 and

SiCN ceramics, mainly due to the high ceramic yield after pyrolysis (often >80 wt. %) [8]. Polysilazanes are suitable materials for the preparation of protective coatings due to their excellent oxidation and corrosion resistance, UV stability and high hardness. These polymers have excellent adhesion to a wide variety of different substrates, e.g., metal, composite, graphite and glass. The PDC route provide the application of liquid or diluted polymers by easily scalable methods, such as dip-coating [9,10], spin-coating [11], doctor-blade method [12] or spray-coating [13,14]. The choice of a particular method depends on the future use of the coating, the type and shape of the body to be coated and the deposited layer, the size of the covered area, the thickness of the coating and its desired properties.

The main disadvantage of the organosilicon polymer precursors, however, is shrinkage, often more than 50 vol. %, that occurs during the transformation from polymer to an amorphous ceramic [15]. The undesirable shrinkage of the polymer leads to crack formation and, in extreme cases, complete failure of the coating. To overcome these unwanted problems, the coatings that consist of only liquid polymer have to be loaded with beneficial components called fillers. The fillers are active [16,17] or passive, and include a large variety of materials, including YSZ [18], Si_3N_4 [19], Al_2O_3 [20] and NbC [21] or commercial glasses [22]. The fillers partially or completely compensate the shrinkage, close the pores and increase the coating thickness [23].

The main function of passive fillers is to decrease the bulk fraction of the polymer used, to reduce the amount of gases generated during pyrolysis and, consequently, to alleviate the overall weight loss and shrinkage, and to eliminate the presence of macro-defects by filling the void space in the material without changing its volume. The glass fillers account for densification and sealing of the system, increasing the efficiency of EBC [24]. The service temperature and softening point of the glass filler particles should be matched to increase efficiency of the coating and, in optimum case, heal any defects formed during the coating operation. In our previous work [25], composite PDC coatings with passive fillers and commercial glasses have been developed. Despite using a range of passive fillers, the bulk shrinkage of the polymer precursor has in many cases led to the preparation of porous coatings. Also, the coating often delaminated from the metal substrate, and lost its protective action as EBC.

Another factor ensuring good adhesion of PDC coatings is based on providing an appropriate surface of stainless steel substrate. A number of various pretreatment procedures, such as sandblasting or etching the substrate by different chemical agents, have been described.

In this work, the influence of pretreatment of the AISI441 steel substrate such as sandblasting, etching of the surface or combination of these methods, were investigated in order to choose the most effective type of cleaning and prevent the delamination of the bond coat from the steel substrate. The oxidation tests were performed in order to evaluate the adhesion of the bond coat at higher temperature and longer operating times.

2. Materials and Methods

The preparation of the PDC coating systems consisted of 3 steps: (1) synthesis of passive fillers with compositions in the Al_2O_3-Y_2O_3-ZrO_2 (AYZ) system by sol–gel Pechini method [26], (2) pretreatment of stainless steel by different methods and (3) preparation of double layer coatings consisting of the bond coat and top coat using a combination of a commercial polymer with passive and glass fillers.

2.1. Preparation of the Precursor Powder

A powder in the AYZ system with the composition (in mol. %) 61.49 Al_2O_3, 18.51 Y_2O_3 and 20 ZrO_2 was used as a passive filler was prepared by the modified sol–gel Pechini method [27]. Y_2O_3 (99.9%, Treibacher Industrie AG, Althofen, Austria) was converted into nitrate by dissolving powder oxide in concentrated HNO_3 (65% Centralchem, Bratislava, Slovakia). $Al(NO_3)_3{\cdot}9H_2O$ (p.a., Centralchem, Bratislava, Slovakia) and $ZrOCl_2{\cdot}8H_2O$ (99.9%, Sigma-Aldrich Co. LLC., Darmstadt, Germany) dissolved in deionised water were then added to yttrium nitrate solution. A 1:1 molar ratio solution of $C_6H_8O_7$ (99.8%, Centralchem, Bratislava, Slovakia) and $C_2H_4(OH)_2$ (99%, Centralchem, Bratislava, Slovakia) in deionised water was then added dropwise to the mixture, which was then refluxed under a condenser and heated in an oil bath at a temperature of 85–95 °C for 2 h. Subsequently, the solvent was evaporated under continuous stirring. The product was dried, calcined at 850 °C to a white powder and sieved through a 40 μm sieve. For better usability in relatively thin coatings, the AYZ powder was homogenised and granulated by a freeze-drying process. A flowchart of the process of preparation of AYZ powder is presented in Figure 1.

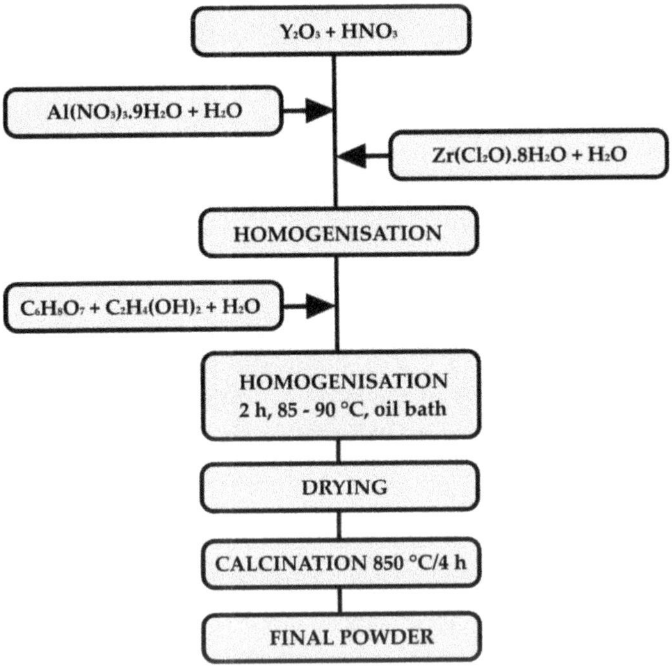

Figure 1. The flowchart of the processing of AYZ powder.

2.2. *Pretreatment of the Stainless Steel*

Ferritic refractory stainless steel grade AISI441, which is commonly used in exhaust gas elements, was used as the metal substrate. Prior to cleaning, the steel sheets were cut into 1 × 1.5 cm^2 plates to make the samples suitable for further characterisation and testing and to prevent damage to the coated samples by further cutting. This was followed by grinding and chamfering the edges and corners of each sample with sandpaper. To produce adhesive coatings without failure, the surface of stainless steel was treated and cleaned to achieve adhesive coatings with sufficient protective capability at temperatures up to 1000 °C. Four different cleaning procedures were applied, i.e., subsequent ultrasonic cleaning in acetone, ethanol and deionised water; sandblasting with glass beads; chemical etching with Kroll's reagent; and a combination of the last two methods. The description of cleaning procedures of the steel is summarised in Table 1.

Table 1. The description of cleaning procedures.

Symbol	Pretreatment of Stainless Steel
U	3-step ultrasonic vibration cleaning in acetone, ethanol and deionised water (10 min each)
S	Sandblasting with glass beads (70–110 μm), ultrasonic cleaning in deionised water
K	Chemical etching—Kroll's reagent (deionised water, HNO_3, HF), 20 s
S+K	Sandblasting with glass beads + etching with Kroll's reagent

2.3. Preparation of the Coatings

A two-layer PDC coating, composed of a bond coat and a ceramic top coat, was applied. The bond coat was prepared from the commercial polymer Durazane 2250 (Merck KGaA, Darmstadt, Germany) by the dip-coating method (dip-coater RDC 15, Relamatic, Glattburg, Switzerland). The pyrolysis of the bond-coat was carried out in air (Nabertherm® N41/H, Nabertherm, Lilienthal, Germany) at a temperature of 450 °C for 1 hour, with heating and cooling rates of 3 K/min. The top coats were prepared from the commercial polymer—Durazane 1800 (Merck KGaA, Darmstadt, Germany), passive fillers and commercial glass. ZrO_2 stabilised with 8 mol. % Y_2O_3 (8YSZ, Inframat® Advanced Materials™, Manchester, CT, USA), AYZ powder prepared by Pechini method and a commercial glass (G018-281, Schott AG, Mainz, Germany) were used as passive fillers. The basic properties of the filler materials are listed in the Table 2.

Table 2. Basic properties of filler materials.

Passive Fillers	d50 (μm)	ρ (g/cm³)	CTE (10^{-6}/K)
8YSZ	0.5	6.1	11.5
AYZ	1–10	4.6	8.6
G018-281	0.5–5	2.7	12.1

Commercially available glass was selected to form a viscous melt at the application temperature of the layers, thereby ensuring the healing of any defects and strengthening of the ceramic top layer. The combination of a liquid commercial polymer Durazane 1800 with glass frits and passive filler particles offers the possibility of designing a large range of compositions. Therefore, the composition of the top layer was designed to match the coefficient of thermal expansion (CTE) of the steel substrate and to reduce the CTE mismatch and increase the compatibility of the metal with the ceramic coating. The CTE of stainless steel was provided by the manufacturer (11.5×10^{-6}/K). The CTE of the prepared coatings were estimated by the rule of mixtures using the CTE of Durazane 1800 (3.0×10^{-6}/K), 8YSZ (11.5×10^{-6}/K), AYZ (8.6×10^{-6}/K) and glass G018-281 (12.1×10^{-6}/K). Two compositions of top coat were tested, in the following text denoted as C2c and D2-PP. The prepared compositions are listed in Table 3.

Table 3. Compositions of the composite top coats after pyrolysis (vol. %).

Sample Name	Durazane 1800	8YSZ	AYZ	G018-281	CTE (10^{-6}/K)
C2c	30	35	-	35	10.1
D2-PP	25	20	20	35	9.6

In the case of the composition C2c, ZrO_2 stabilised with 8 mol. % Y_2O_3 and glass frits were homogenised in a solution of di-n-butyl ether (Acros Organics BVBA, Geel, Belgium) and dispersant (DISPERBYK 2070, BYK-Chemie GmbH, Wesel, Germany). To break up the agglomerates, the suspensions were dispersed in the ultrasound and homogenised for 48 h by stirring with a magnetic stirrer. Subsequently, Durazane 1800 polymer, with 3 wt. % of dicumyl peroxide (DCP) (Sigma-Aldrich Co. LLC., Darmstadt, Germany), was added to the slurry, which was homogenised for an additional 48 hours in a plastic jar with ZrO_2 balls (Ø1 mm). After homogenisation, the suspension was applied to the stainless steel with a bond coat by a spray-coating technique from both sides. The suspension was deposited onto steel substrates by spray coating using a spray coater model 780S-SS (Nordson EFD, East Providence, RI, USA). The nozzle diameter of spray gun was 0.71 mm (0.028"). The final suspension was sprayed under the air pressure of 2.2 bar. The distance between the spray gun and the sample was 10 cm. In the composition D2-PP, the AYZ powder prepared by the modified Pechini sol–gel method was used as an additional passive filler. The coated samples were then pyrolysed in air at 850 °C for 1 hour, at a heating and cooling rate of 3 °C/min. A flowchart of the coating processing is presented in Figure 2.

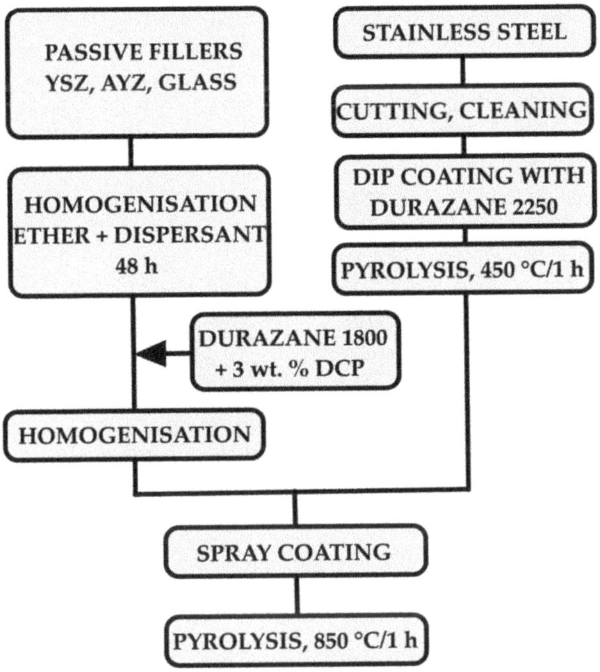

Figure 2. The flowchart of the coatings preparation process.

2.4. Characterisation Methods

X-ray powder diffraction analysis was used to assign the phase composition of the prepared AYZ powder. Diffraction records were measured on an Empyrean DY1098 powder diffractometer (Panalytical, B.V., Almelo, The Netherlands) with a Cu anode and with X-ray wavelength of λ = 1.5405 Å over 2θ angles of 10–80°. Diffraction records were then evaluated using HighScore Plus (v. 3.0.4) using COD2019 (Crystallographic Open Database). Mean Roughness Depth (Rz) was measured using atomic force microscopy (AFM, Brooker Innova, Billerica, MA, USA). Rz was calculated by measuring the vertical distance from the highest peak to the lowest valley within five sampling lengths, then averaging these distances. The surface morphology of pretreated samples was examined by scanning electron microscopy (SEM, JEOL JSM 7600 F, JEOL, Tokyo, Japan). For detailed examination of the coating/metal interface, the cross sections were prepared via mounting in resin followed by grinding and polishing. The inspection of the coatings was then performed using an SEM equipped with an energy-dispersive X-ray spectroscopy (EDXS) detector (Oxford instruments, Abingdon, UK) and was focused on the evaluation of adhesion, homogeneity and possible failures of the coatings.

2.5. Oxidation Tests

The oxidation tests were carried out in a high temperature horizontal electric tube furnace (Clasic 0213T, Clasic, Praha, Czech Republic) in flowing atmosphere of synthetic air (purity 99.5, SIAD Slovakia spol. s.r.o., Bratislava, Slovakia) at a temperature of 950 °C with a heating rate of 5 °C/min and an exposure time of 48 h. The composition of synthetic air is as follows; nitrogen (78%), oxygen (21%), argon (0.9%) and other gases (0.1%).

3. Results and Discussion

3.1. Characterisation of the AYZ Filler

To achieve a homogenous precursor powder to be used as a filler in the prepared coating, the AYZ powder was prepared via a modified Pechini method [27]. The main advantage of this method is that the metallic ions are immobilised in a rigid polymer network, which ensures their homogeneous dispersion on the atomic scale without precipitation or phase segregation. This process allows complete control over the product stoichiometry, even for more complex oxide powders [27]. To facilitate the use of the AYZ powder in the coatings, the powder was refined by milling and freeze drying to avoid agglomeration of precursor powder and achieve the particle size below 10 μm. SEM micrographs of the AYZ powder after freeze drying are shown in Figure 3. From the as-prepared AYZ powder consists of irregular and angular particles resulting from the crushing process, with the sizes ranging from a few to several tens of micrometres.

Figure 3. SEM of AYZ powder.

The XRD pattern of AYZ powder after calcination is shown in Figure 4, confirming the presence of t-ZrO_2, as well as yttrium aluminium garnet ($Y_3Al_5O_{12}$, YAG) as the main crystalline phases formed during calcination. Smaller amount of the mellilite $Y_2Al_2O_6$ phase was also observed after calcination. High background in the diffraction pattern indicates the crystalline phases were embedded in an amorphous (glassy) matrix.

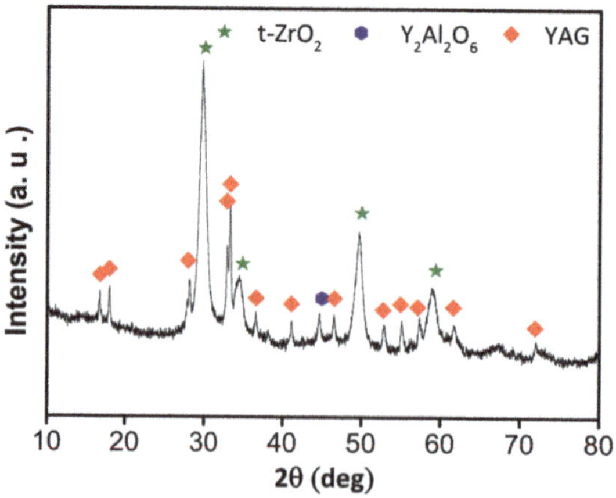

Figure 4. The X-ray powder diffraction (XRD) pattern of AYZ powder after calcination at 850 °C.

3.2. *Treated Steel Surfaces*

The metal surface quality is significant characteristic influencing the adhesion of the protective coating. In the case of double-layer PDC coatings, the weakest point is usually the interface between the bond coat and the metal substrate, due to the presence of impurities or defects on the steel surface, which often lead to delamination of the bond coat during pyrolysis or after corrosion tests. To ensure a high adhesion of the bond coat with the steel, the stainless steel substrate has to be cleaned properly to degrease its surface and remove possible contaminants. The influence of different cleaning procedures on the surface morphology was therefore examined. The Mean Roughness Depth (Rz) was determined using AFM and the surface morphology of pretreated stainless steel samples was examined by SEM. Figure 5 presents the Rz of variously pretreated stainless steel surfaces. The AFM analysis confirmed that different treatment created different sizes of roughness: there is a relation between surface topography and the type of the used cleaning procedure. The Rz was in the range of 0.26 to 1.69 μm. A roughness similar to that observed for untreated surfaces was measured after ultrasonic cleaning, while sandblasting increased the roughness substantially. An increase of Rz was expected to result in stronger bonding at the steel-bond coat interface.

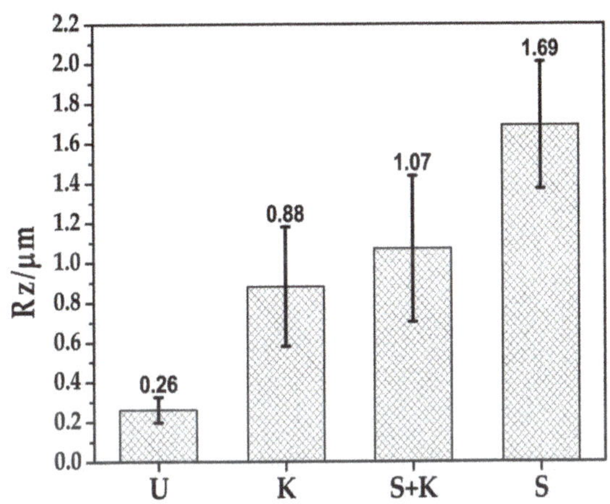

Figure 5. The roughness after pretreatment of stainless steel (U – ultrasonic cleaning, K – chemical etching with Kroll's reagent, S – sandblasting, S+K – sandblasting + chemical etching with Kroll's reagent).

AFM images of the stainless steel cleaned by different methods are shown in Figure 6. SEM-micrographs featuring surfaces of the stainless steel cleaned by different methods are shown in Figure 7. Chemical etching resulted in a surface with irregular topography, with slight roughening of the surface compared to ultrasonically cleaned substrates. More uniform and regular surfaces were obtained by sandblasting with glass beads, or a combination of sandblasting and chemical etching with Kroll's reagent. These treatments resulted in rough surfaces with rounded edges, with the average surface roughness 1.69 μm for sandblasted samples. Chemical etching of sandblasted substrates slightly decreased the average surface roughness, but the sandblasting induced roughening was still significant.

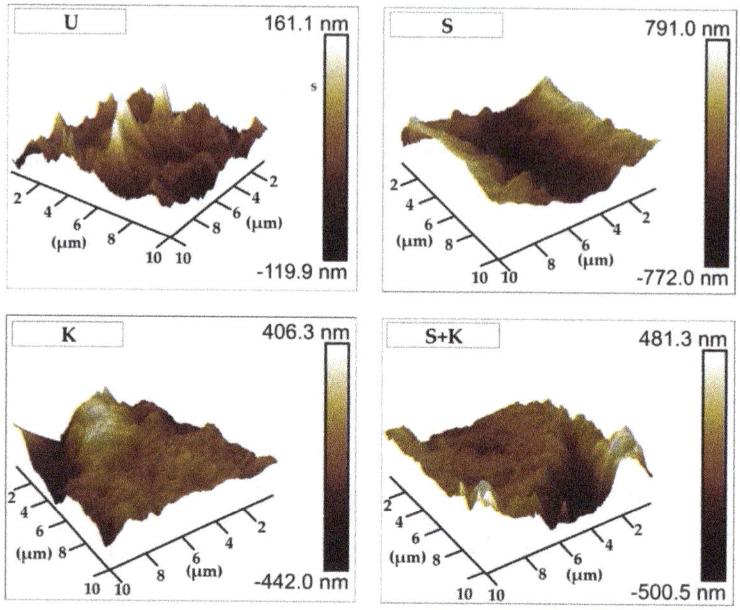

Figure 6. Atomic force microscopy (AFM) images representing roughness of the stainless steel surfaces treated by different cleaning procedures. The applied cleaning procedure is indicated by the abbreviation in the upper left corner of each figure (U – ultrasonic cleaning, K – chemical etching with Kroll's reagent, S – sandblasting, S+K – sandblasting + chemical etching with Kroll's reagent).

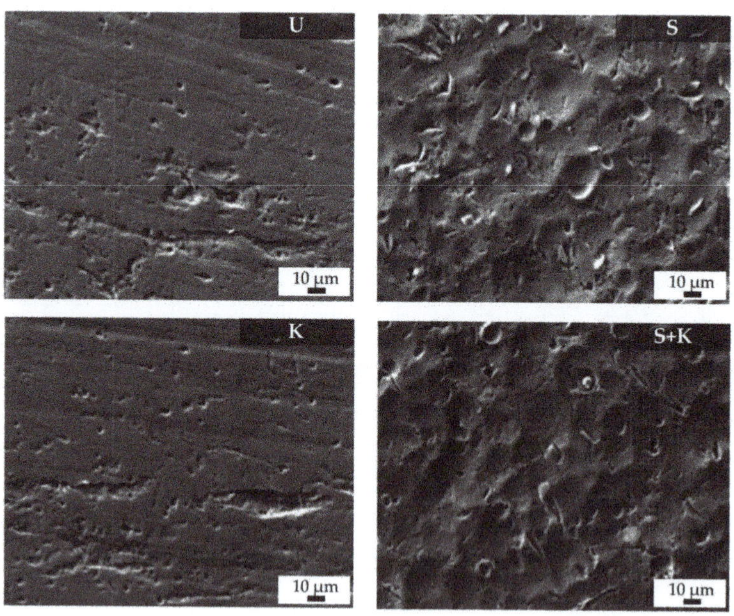

Figure 7. SEM micrographs of the stainless steel surfaces treated by different cleaning procedures. The applied cleaning procedure is indicated by the abbreviation in the upper right corner of each figure (U – ultrasonic cleaning, K – chemical etching with Kroll's reagent, S – sandblasting, S+K – sandblasting + chemical etching with Kroll's reagent).

3.3. Characterisation of the Coatings

The SEM cross-sectional micrographs of compositions C2c and D2-PP were obtained through the metal–ceramic interface to investigate the bonding between the bond coat and variously treated steel substrates (Figure 8 (C2c composition) and Figure 9 (D2-PP composition)). The weakest location in a typical double layer coating is usually the interface between the stainless steel and the bond coat, where the cracks or spallation can occur. The spallation of the bond coat is usually caused by thermal and elastic mismatch between the steel and bond coat, due to the presence of impurities at the steel surface, by changing the chemistry of the steel by chemical etching, or growth of stresses followed by the formation of thermally-grown oxides, mostly due to the weak adhesion of bond coat to steel. Irrespective of the applied surface treatment the bond coat did not delaminate from the steel surface after pyrolysis of the coatings at 850 °C, indicating its good adhesion. An undamaged bond coat approximately 1 μm thick was observed in all cases, which acts as an effective diffusion barrier to oxidation during pyrolysis. Only for the D2-PP coating deposited at the substrate etched by Kroll's reagent the bond coat peeled off from the surface. If the stainless steel was treated by sandblasting or chemical etching or their combination, a few small cracks (marked with white circles) were generated in the bond coat, perpendicular to the substrate surface. The crack formation was attributed to strong adhesion of the bond coat to the metal substrate and, at the same time, by the high volume shrinkage of the polysilazane during heat treatment. No crack formation in the bond coat was observed on samples pretreated via ultrasonic cleaning in acetone, ethanol and deionised water. Cracking could also occur due to uneven and rough sandblasted surface, with sharp edges and peaks acting as stress concentrators in the coatings [24]. The pretreatment by ultrasonic cleaning in acetone, ethanol and deionised water was found to be the most effective process, since no spallation or cracking was observed in the cross section of the bond coat after pyrolysis. Moreover, the ultrasonically cleaned steel surface was uniformly covered by the approximately 1 μm thick bond coat.

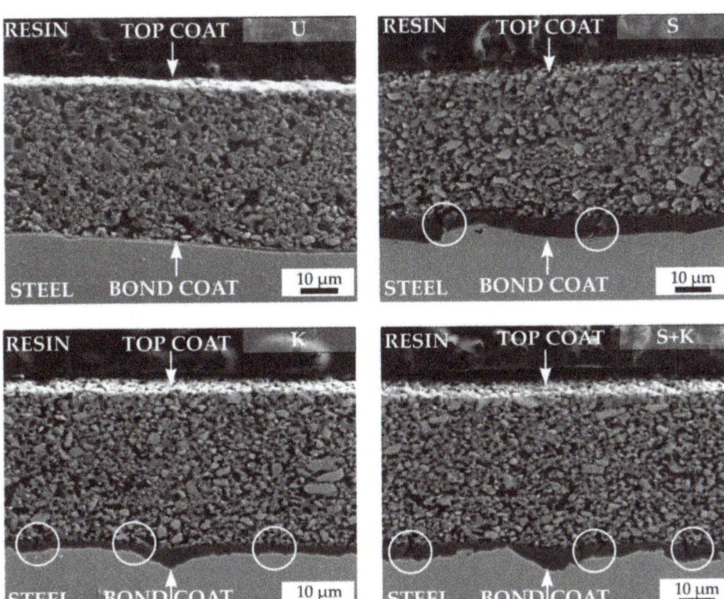

Figure 8. SEM cross sections of pyrolysed C2c coatings. The applied cleaning procedure is indicated by the abbreviation in the upper right corner of each figure (U – ultrasonic cleaning, K – chemical etching with Kroll's reagent, S – sandblasting, S+K – sandblasting + chemical etching with Kroll's reagent).

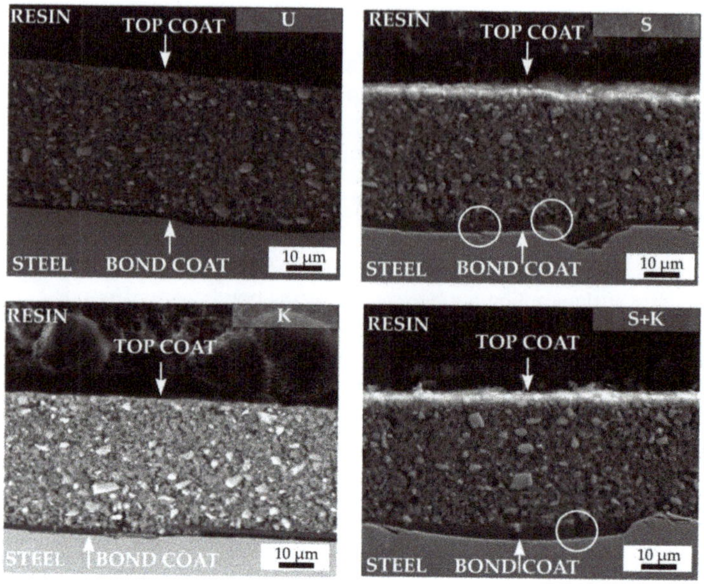

Figure 9. SEM cross sections of pyrolysed D2-PP coatings. The applied cleaning procedure is indicated by the abbreviation in the upper right corner of each figure (U – ultrasonic cleaning, K – chemical etching with Kroll's reagent, S – sandblasting, S+K – sandblasting + chemical etching with Kroll's reagent).

In addition, EDXS mapping was carried out on the cross section of D2-PP-coated steel cleaned by ultrasonic treatment in acetone, ethanol and deionised water to demonstrate the existence of an interface bond metal/base coat and base/top coat. EDXS element maps are shown in Figure 10. The bond coat contains mainly Si and O, since during pyrolysis in air the Durazane 2250 was converted to SiO_2, as confirmed by SEM/EDXS measurement. The presented element maps of Si confirm the formation of the protective bond coat at the steel surface in agreement with the literature [15]. The bond coat enhances the bonding between the steel and the top coat, and it preserves the steel from oxidation/corrosion [15,27]. On steel exposed to ambient environment, a natural oxide layer with adsorbed water is always present. Because of the reactivity of Durazane 2250 with surface-bound –OH groups, steel forms direct metal–O–Si chemical bonds with the base coat, leading to excellent adhesion [28]. The reaction of Durazane 2250 with hydroxyl groups of the substrate surface is described by the following simplified reaction equations [29]:

$$\text{Fe-OH} + {\equiv}\text{Si-NH-Si}{\equiv} \rightarrow {\equiv}\text{Fe-O-Si}{\equiv} + \text{H}_2\text{N-Si}{\equiv} \tag{1}$$

$$\text{Fe-OH} + \text{H}_2\text{N-Si}{\equiv} \rightarrow {\equiv}\text{Fe-O-Si}{\equiv} + \text{NH}_3 \tag{2}$$

and a direct chemical metal–O–Si bonds between the steel and the precursor-derived coatings are formed. Therefore, the adhesive strength of the PDC coatings to the metal surface is very strong. In the case of sandblasting, we assume that it was the surface roughness of the steel that caused the coating delamination, as sandblasting should not affect the concentration of hydroxyl groups present at the steel surface. However, strong adhesion due to formation of the metal–O–Si bonds causes immobility of the coating during pyrolysis, which does not allow the coating to adjust to volume shrinkage of the steel substrate. Moreover, the sharp borders and peaks of the substrate initiate the formation of cracks perpendicular to the substrate surface. The corrosion/oxidation medium is thus able to penetrate through the cracks to the metal surface causing coating delamination. No crack formation in the bond coat was observed in samples pretreated via ultrasonic cleaning. In the case of chemical etching, we assume that this treatment negatively influenced the adhesion of the bond coat because the concentration of hydroxyl groups on steel surface was significantly affected by etching.

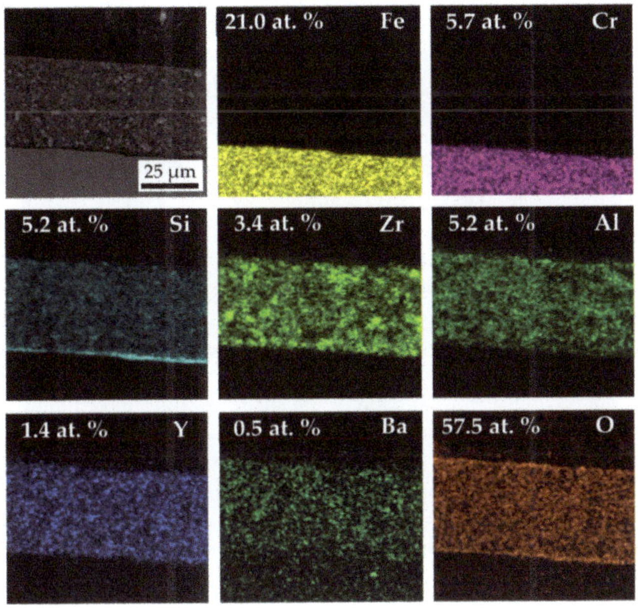

Figure 10. SEM/energy-dispersive X-ray spectroscopy (EDX) of D2-PP coatings documenting distribution of elements in the coating and in substrate.

The main elements of stainless steel Fe and Cr were detected below the protective bond coat. The main component of the bond was Si. The top coat consisted of Zr, Al, Y (from AYZ powder) and Ba (from commercial glass).

The top coat layers of both studied compositions—C2c (Figure 8) and D2-PP (Figure 9)—were almost dense, containing only small closed pores present predominantly at the boundaries between the filler particles and the matrix. The filler particles were well coated with the Durazane 1800 precursor, which builds up the matrix, and acts also as an adhesive between the particles. In the case of the C2c coating, all pores were almost spherical. This indicates that the closed pores could result from bubble formation due to the release of dissolved gases, as well as the expansion of insoluble gases (e.g., oxygen or air) entrapped in the initial pores. The pores could result also from the release of gases such as NH_3, CH_4, and H_2 generated during the polymer to ceramic transformation [15]. Note that pore formation cannot be completely avoided when passive fillers are used in the processing of PDC coatings, and some residual porosity usually remains in the final ceramic top coats. Existing pores provide a transportation path for gaseous products of decomposition escaping the coating. In the case of the D2-PP coating, the addition of the AYZ powder with irregular and angular particles have helped to create a rigid and articulate structure. This structure allowed outgassing of the preceramic polymer pyrolysis products from the system thereby effectively reducing the size and amount of pores. The elimination of larger pores, and thus an increased density, led to a significantly more compact coating in comparison to the C2c composition pyrolysed under the same conditions. Although the top layers were not completely dense, microstructure with residual porosity was beneficial for the thermal stability of coatings, and contributed to the mitigation of residual stresses during the heating and cooling cycles [30].

The mismatch of the CTE between the steel and the coating, together with the Young modulus, are critical factors for the resistance of coatings exposed to high temperatures [31]. By reducing the non-conformance of the CTE and the elastic modulus, the tension in the coating is reduced, resulting in a higher stability of the coatings during thermal loading. Moreover, glasses are suitable material for sealing in PDC coatings. By decreasing both quantities (CTE mismatch and Young modulus), the stresses in the layers can be reduced, leading to better coating thermal stability during the oxidation tests. Coatings with both ceramic and glass fillers should then exhibit better thermal compatibility with the steel substrate than the coatings with active and passive fillers without glass tested in our previous work [26]. As a result, no cracks were observed in top coats, preventing penetration of oxidation agents through the coating and attack of the steel substrate.

Static oxidation tests were performed to assess the efficiency of the selected types of steel substrate cleaning. The aim of the oxidation tests was to determine whether the tested cleaning methods would ensure sufficient adhesion of the bond coat to the metal substrate not only during pyrolysis, but also at long-term operating at high temperature. The tests were performed in the atmosphere of synthetic air at 950 °C at a heating rate of 5 °C/min, with maximum duration of 48 hours. SEM cross-sections of C2c and D2-PP compositions after oxidation tests are shown in Figures 11 and 12, respectively. In all cases, except for ultrasonic cleaning, spallation of the bond coat from the surface of stainless steel was observed. Chemical etching of steel with the Kroll's solution probably caused changes in the microstructure and chemical composition of the steel, resulting in peeling of the coating. For sandblasted samples, or a combination of sandblasting and etching, detachment of the bond coat resulted from high surface roughness: as discussed above, strong adhesion of the coating on a rough surface with sharp edges resulted in formation of cracks perpendicular to the substrate. The cracks serve as a gateway for inward penetration of oxygen, which directly attacks the substrate, creating a layer of oxidation products, and eventually leading the detachment of the bond coat from the substrate. Simple cleaning by ultrasonication in acetone, ethanol and deionised water was found to be the most effective way for achieving a sufficient bonding of the bond coat to the steel substrate after oxidation tests.

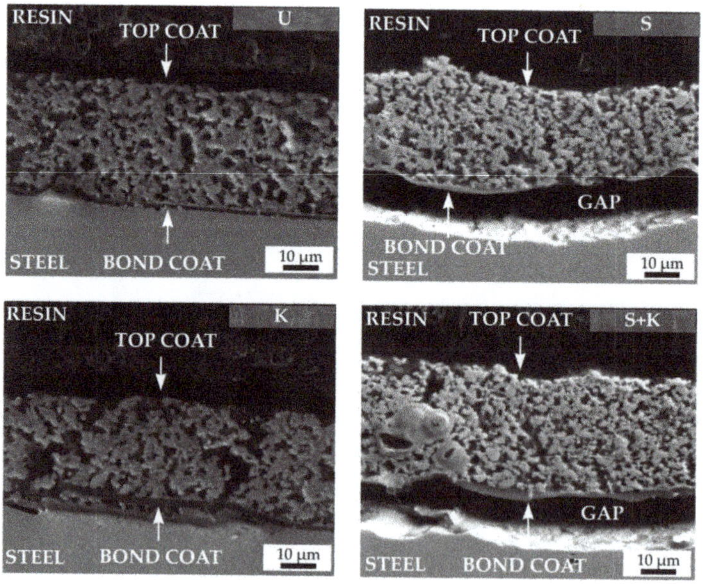

Figure 11. Cross sections of C2c coatings after oxidation tests. The applied cleaning procedure is indicated by the abbreviation in the upper right corner of each figure (U – ultrasonic cleaning, K – chemical etching with Kroll's reagent, S – sandblasting, S+K – sandblasting + chemical etching with Kroll's reagent).

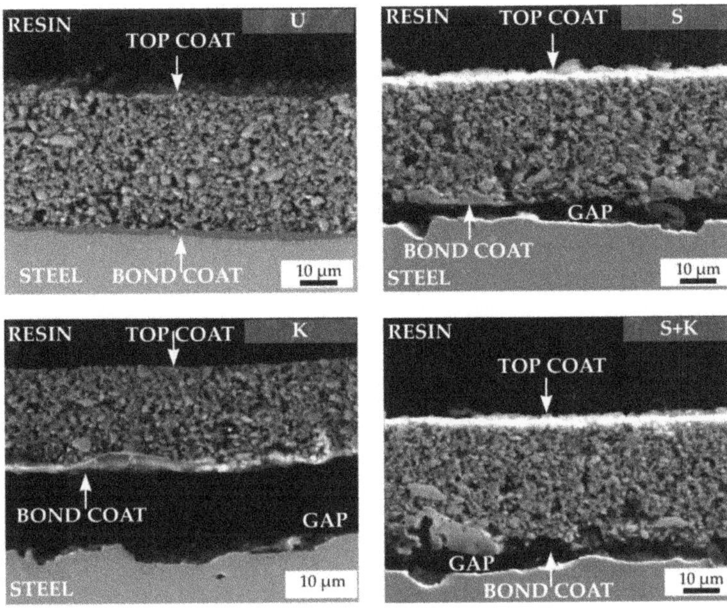

Figure 12. Cross sections of D2-PP coatings after oxidation tests. The applied cleaning procedure is indicated by the abbreviation in the upper right corner of each figure (U – ultrasonic cleaning, K – chemical etching with Kroll's reagent, S – sandblasting, S+K – sandblasting + chemical etching with Kroll's reagent).

Originally, numerous pores were observed across the whole cross-section of the as-prepared PDC glass ceramic coatings. After oxidation tests, a significant increase in the porosity of the layers accompanied by the growth of pores and a decrease of the coating thickness was observed in C2c coating. Decrease of the layer thickness was attributed to differential sintering of individual components (mainly glass fillers) by viscous flow, complemented by inadequate removal of gases entrapped in the ceramic matrix. Due to the softening of the used glass fillers, the cracks in the layers gradually heal, indicating at least partial protection of the metal substrate. After oxidation tests, the D2-PP composition exhibited a compact structure with no increase of the size and amount pores. No cracking or delamination from bond coat was observed.

4. Conclusions

Suitable pretreatment of steel substrate, as well as the using a Durazane 2250 bond coat, are prerequisites for preparation of adherent composite coatings. The most effective cleaning process is a 3-step ultrasonic cleaning in acetone, ethanol and deionised water. Small cracks in the bond coat perpendicular to substrate were observed after pyrolysis in bond coats deposited at substrates treated by sandblasting and chemical etching. After oxidation tests, all coatings, except for those applied to substrates cleaned in an ultrasonic bath, delaminated or showed significant cracking of the bond coat. Combination of PDC with tailored fillers and glass systems enable the processing of dense and crack free coating system on stainless steel. Irregular and angular filler particles favour outgassing of the coating during pyrolysis, reducing the total porosity in the layer, and conferring better protection of the substrate against oxidative environment.

Author Contributions: M.P. performed cleaning procedures, prepared coatings and wrote the manuscript; I.P. prepared the precursor powder, performed cleaning procedures, prepared coatings and performed oxidation tests; P.Š. performed the SEM and EDXS analysis; G.M. and D.G. conceived the study, supervised the project, gave research guidelines, and finalised the manuscript. All authors have read and agreed to the published version of the manuscript.

Funding: Financial support of this work by the APVV 0014-15 grant and the Deutscher Akademischer Austauschdienst (DAAD) grant scheme is gratefully acknowledged. This paper was also created in the frame of the project Centre for Functional and Surface Functionalised Glass (CEGLASS), ITMS code 313011R453, operational program Research and innovation, co-funded from European Regional Development Fund.

Acknowledgments: The authors would like to thank Alexander Horcher and Mateus Lenz-Leite for their help to create the coating solution. The authors would also like to thank Jacob Andrew Peterson for grammar corrections in the manuscript and Branislav Hruška for AFM measurement.

Conflicts of Interest: The authors declare no conflicts of interest.

References

1. Wang, K.; Günthner, M.; Motz, G.; Bordia, R.K. High performance environmental barrier coatings, Part II: Active filler loaded SiOC system for superalloys. *J. Eur. Ceram. Soc.* **2011**, *31*, 3011–3020. [CrossRef]
2. Greil, P. Polymer Derived Engineering Ceramics. *Adv. Eng. Mater.* **2000**, *2*, 339–348. [CrossRef]
3. Torrey, J.D.; Bordia, R.K. Processing of polymer-derived ceramic composite coatings on steel. *J. Am. Ceram. Soc.* **2008**, *91*, 41–45. [CrossRef]
4. Colombo, P.; Paulson, T.E.; Pantano, C.G. Synthesis of Silicon Carbide Thin Films with Polycarbosilane (PCS). *J. Am. Ceram. Soc.* **2005**, *40*, 2333–2340. [CrossRef]
5. Lukacs, A. Polysilazane precursors to advanced ceramics. *Am. Ceram. Soc. Bull.* **2007**, *86*, 9301–9306.
6. Seifert, M.; Travitzky, N.; Krenkel, W.; Motz, G. Multiphase ceramic composites derived by reaction of Nb and SiCN precursor. *J. Eur. Ceram. Soc.* **2014**, *34*, 1913–1921. [CrossRef]
7. Günthner, M.; Pscherer, M.; Kaufmann, C.; Motz, G. High emissivity coatings based on polysilazanes for flexible Cu(In,Ga)Se2 thin-film solar cells. *Sol. Energy Mater. Sol. Cells* **2014**, *123*, 97–103. [CrossRef]
8. Kroke, E.; Li, Y.; Konetschny, C.; Lecomte, E.; Fasel, C.; Riedel, R. Silazane derived ceramics and related materials. *Mater. Sci. Eng. R Rep.* **2000**, *26*, 97–199. [CrossRef]
9. Lenormand, P.; Caravaca, D.; Laberty-Robert, C.; Ansart, F. Thick films of YSZ electrolytes by dip-coating process. *J. Eur. Ceram. Soc.* **2005**, *25*, 2643–2646. [CrossRef]
10. Kappa, M.; Kebianyor, A.; Scheffler, M. A two-component preceramic polymer system for structured coatings on metals. *Thin Solid Film.* **2010**, *519*, 301–305. [CrossRef]
11. Morlier, A.; Cros, S.; Garandet, J.P.; Alberola, N. Thin gas-barrier silica layers from perhydropolysilazane obtained through low temperature curings: A comparative study. *Thin Solid Film.* **2012**, *524*, 62–66. [CrossRef]
12. Barroso, G.S.; Krenkel, W.; Motz, G. Low thermal conductivity coating system for application up to 1000 °C by simple PDC processing with active and passive fillers. *J. Eur. Ceram. Soc.* **2015**, *35*, 3339–3348. [CrossRef]
13. Parchovianský, M.; Barroso, G.; Petríková, I.; Motz, G.; Galusková, D.; Galusek, D. Polymer derived glass ceramic layers for corrosion protection of metals. *Ceram. Trans.* **2016**, *256*, 187–200. [CrossRef]
14. Goerke, O.; Feike, E.; Heine, T.; Trampert, A.; Schubert, H. Ceramic coatings processed by spraying of siloxane precursors (polymer-spraying). *J. Eur. Ceram. Soc.* **2004**, *24*, 2141–2147. [CrossRef]
15. Günthner, M.; Wang, K.; Bordia, R.K.; Motz, G. Conversion behaviour and resulting mechanical properties of polysilazane-based coatings. *J. Eur. Ceram. Soc.* **2012**, *32*, 1883–1892. [CrossRef]
16. Greil, P. Active Filler Controlled Pyrolysis of Preceramic Polymers. *J. Am. Ceram. Soc.* **1995**, *78*, 835–848. [CrossRef]
17. Greil, P. Near Net Shape Manufacturing of Ceramic Components. *Cfi Ceram. Forum Int.* **1998**, *75*, 15–21. [CrossRef]
18. Petríková, I.; Parchovianský, M.; Lenz-Leite, M.; Švančárek, P.; Motz, G.; Galusek, D. Passive Filler loaded Polysilazane-Derived Glass/Ceramic coating system applied to AISI 441 stainless steel, Part 1: Processing and Characterisation. *Int. J. Appl. Ceram. Technol.* **2019**. [CrossRef]
19. Bernardo, E.; Colombo, P.; Hampshire, S. SiAlON-based ceramics from filled preceramic polymers. *J. Am. Ceram. Soc.* **2006**, *89*, 3839–3842. [CrossRef]
20. Labrousse, M.; Nanot, M.; Boch, P.; Chassagneux, E. Ex-polymer SiC coatings with Al2O3 particulates as filler materials. *Ceram. Int.* **1993**, *19*, 259–267. [CrossRef]
21. Scheffler, M.; Dernovsek, O.; Schwarze, D.; Science, M.; Science, M. Polymer / filler derived NbC composite ceramics. *J. Mater. Sci.* **2003**, *8*, 4925–4931. [CrossRef]

22. Schütz, A.; Günthner, M.; Motz, G.; Greißl, O.; Glatzel, U. Characterisation of novel precursor-derived ceramic coatings with glass filler particles on steel substrates. *Surf. Coat. Technol.* **2012**, *207*, 319–327. [CrossRef]
23. Günthner, M.; Kraus, T.; Krenkel, W.; Motz, G.; Dierdorf, A.; Decker, D. Particle-Filled PHPS silazane-based coatings on steel. *Int. J. Appl. Ceram. Technol.* **2009**, *6*, 373–380. [CrossRef]
24. Günthner, M.; Schütz, A.; Glatzel, U.; Wang, K.; Bordia, R.K.; Greißl, O.; Krenkel, W.; Motz, G. High performance environmental barrier coatings, Part I: Passive filler loaded SiCN system for steel. *J. Eur. Ceram. Soc.* **2011**, *31*, 3003–3010. [CrossRef]
25. Parchovianský, M.; Petíková, I.; Barroso, G.S.; Galuskova, D.; Motz, G.; Galusek, D. Corrosion and oxidation behavior of polymer derived ceramic coatings with passive glass fillers on AISI 441. *Ceram. Silik.* **2018**, *62*, 146–157. [CrossRef]
26. Pechini, M.P. Method of Preparing Lead and Alkaline Earth Titanates and Niobates and Coating Method Using the Same to Form a Capacitor. US Patent No. 3330697, 11 July 1967.
27. Günthner, M.; Kraus, T.; Dierdorf, A.; Decker, D.; Krenkel, W.; Motz, G. Advanced coatings on the basis of Si(C)N precursors for protection of steel against oxidation. *J. Eur. Ceram. Soc.* **2009**, *29*, 2061–2068. [CrossRef]
28. Bauer, F.; Decker, U.; Dierdorf, A.; Ernst, H.; Heller, R.; Liebe, H.; Mehnert, R. Preparation of moisture curable polysilazane coatings: Part I. Elucidation of low temperature curing kinetics by FT-IR spectroscopy. *Prog. Org. Coat.* **2005**, *53*, 183–190. [CrossRef]
29. Bahloul, D.; Pereira, M.; Goursat, P.; Yive, N.S.C.K.; Corriu, R.J.; Mixte, R.I. Preparation of silicon carbonitrides from organosilicon polymer. I. Thermal decomposition of the cross-linked polysilazane. *J. Am. Ceram. Soc.* **1993**, *75*, 1156–1162. [CrossRef]
30. Padture, N.P.; Gell, M.; Jordan, E.H. Thermal barrier coatings for gas-turbine engine applications. *Science* **2002**, *296*, 280–284. [CrossRef]
31. Soliman, H.M.; Waheed, A.F. Effect of differential thermal expansion coefficient on stresses generated in coating. *J. Mater. Sci. Technol.* **1999**, *15*, 457–462.

materials

MDPI

Article

Ni/cerium Molybdenum Oxide Hydrate Microflakes Composite Coatings Electrodeposited from Choline Chloride: Ethylene Glycol Deep Eutectic Solvent

Juliusz Winiarski [1,*], Anna Niciejewska [1], Jacek Ryl [2], Kazimierz Darowicki [2], Sylwia Baśladyńska [1], Katarzyna Winiarska [3] and Bogdan Szczygieł [1]

1 Faculty of Chemistry, Department of Advanced Material Technologies, Wrocław University of Science and Technology, 50-370 Wrocław, Poland; anna.niciejewska@onet.pl (A.N.); sylwia.basladynska@pwr.edu.pl (S.B.); bogdan.szczygiel@pwr.edu.pl (B.S.)

2 Faculty of Chemistry, Department of Electrochemistry, Corrosion and Materials Engineering, Gdansk University of Technology, 80-233 Gdańsk, Poland; jacek.ryl@pg.edu.pl (J.R.); kazimierz.darowicki@pg.edu.pl (K.D.)

3 Faculty of Chemistry; Department of Analytical Chemistry and Chemical Metallurgy, Wrocław University of Science and Technology, 50-370 Wrocław, Poland; katarzyna.winiarska@pwr.edu.pl

* Correspondence: juliusz.winiarski@pwr.edu.pl; Tel.: +48-71-320-3193

Received: 30 January 2020; Accepted: 17 February 2020; Published: 19 February 2020

Abstract: Cerium molybdenum oxide hydrate microflakes are codeposited with nickel from a deep eutectic solvent-based bath. During seven days of exposure in 0.05 M NaCl solution, the corrosion resistance of composite coating (Ni/CeMoOxide) is slightly reduced, due to the existence of some microcracks caused by large microflakes. Multielemental analysis of the solution, in which coatings are exposed and the qualitative changes in the surface chemistry (XPS) show selective etching molybdenum from microflakes. The amount of various molybdenum species within the surface of coating nearly completely disappear, due to the corrosion process. Significant amounts of Ce^{3+} compounds are removed, however the corrosion process is less selective towards the cerium, and the overall cerium chemistry remains unchanged. Initially, blank Ni coatings are covered by NiO and $Ni(OH)_2$ in an atomic ratio of 1:2. After exposure, the amount of $Ni(OH)_2$ increases in relation to NiO (ratio 1:3). For the composite coating, the atomic ratios of both forms of nickel vary from 1:0.8 to 1:1.3. Despite achieving lower corrosion resistance of the composite coating, the applied concept of using micro-flakes, whose skeleton is a system of Ce(III) species and active form are molybdate ions, may be interesting for applications in materials with potential self-healing properties.

Keywords: metal coatings; nickel; composite coatings; electrodeposition; XPS; polarization; electrochemical impedance spectroscopy (EIS)

1. Introduction

In recent years, the electrodeposition of metals from ionic liquids (ILs) has become more popular because of the many advantages, such as high solubility metals salts, high conductivity, and wide electrochemical potential windows. Ionic liquids also allow obtaining metal coatings, such as Na and Mg, which is impossible in conventional baths. ILs are non-toxic, non-flammable and can be used in a wide range of temperatures [1–4]. One of the most interesting analogues of ILs are the deep eutectic solvents (DESs). DES is a eutectic mixture of quaternary ammonium salt with hydrogen bond donor species or metal salts. DESs are relatively cheap and easy to prepare [5,6]. DES-based galvanic baths have the advantage that the metals do not passivate during deposition. DESs can be used to electrodeposition a lot of metal coatings: Ni, Cu, Zn, Cr, Sn, Co, Al, Ti, W, Ag, Mg, Se, and Pd [6]. Nickel coatings obtained from DES feature nano-crystalline morphology and lower surface roughness than Ni-coatings from conventional aqueous Watt's baths which produce coatings with a micro-crystalline structure. Furthermore, the microhardness of Ni coatings deposited from DES is 100 HV (Vickers hardness) higher than coatings from aqueous solutions [7]. A used eutectic solvent has a significant impact on the morphology of nickel coatings (size of Ni crystallites) [8]. Moreover, from baths based on eutectic solvents it is possible to obtain Ni coatings on an aluminum substrate. Obtaining bright Ni coatings requires the use of brighteners, which differ from those used in aqueous plating baths [9,10]. The important influence on the coating's deposition from DESs is the water content in a plating bath. By changing the water content, one can control the potential for Ni deposition. Potential Ni impacts the self-limiting growth crystallite and passivation effect [11]. In addition to the above advantages, DESs liquids enable the electrodeposition of nickel alloy coatings: Zn-Ni [12], Ni-Co-Sn [13], Fe-Ni [14], Ni-Co [15], Ni-Mo [16], and nickel composite coatings. Ni-carbon nanotubes composite coatings [17,18] have much better tribological properties than pure Ni [18]. Furthermore, Ni-PTFE (poly(tetrafluoroethylene)) coatings show better wear resistance than nickel coatings and have hydrophobic properties [19]. The addition of SiC to the nickel matrix causes the increase of microhardness [20] and corrosion resistance [21]. Ni-SiO_2 coatings also have better microhardness and wear resistance than Ni coatings [22]. Nickel coatings with a TiO_2 dispersive phase show good electrocatalytic activity [23]. However, co-deposition with micro- or nano-particles of compounds of chemically active elements, e.g. cerium or molybdenum, may affect the protective properties of the composite coatings. Dong et al. prepared cerium molybdate particles with different morphologies, such as the bundle-like, flower-like, and microspheric structures [24]. The nanostructure of Ce-Mo shows excellent adsorption and catalytic properties [24]. Amino acids, for example: glycine, lysine, serine, and glutaminate used at synthesis of cerium molybdate have influence on its morphology [25]. Sajad et al. proved that by using lysine in synthesis one can obtain a pure product (cerium molybdate) which is not contaminated with cerium molybdate oxides [26]. Nanocapsules are also made up of shells from cerium molybdate, while the core was a corrosion inhibitor [27,28] or hollow nanospheres have an antibacterial effect on *Escherichia coli* in the absence of light [29]. Hence their potential application in medicine or the anti-corrosion industry [29]. Patel et al. developed a method of fabrication of cerium zinc molybdate nanopigment with anticorrosive properties [30]. Cerium molybdate nanowires were also added to the coatings obtained by the sol-gel method [31]. Therefore, cerium molybdate is very popular as a corrosion inhibitor in protective coatings for aluminum alloys [32] and magnesium [33,34]. Bhanvase et al. have investigated that 5% content of cerium zinc molybdate nanocontainers, significantly increases the corrosion resistance [35].

The use of cerium molybdenum oxide, proposed in this work, is novel in electrodeposition of nickel composite coatings. Ce and Mo compounds embedded in the coating can modify its protective properties by giving it potential self-healing properties (through the selective release of cerium and molybdenum ions). Moreover, this work uses a deep eutectic solvent based on choline chloride and ethylene glycol as an excellent environment for the co-deposition of microparticles. As a result, deposition is carried out in a stable suspension bath. Microflakes of fairly large size are intentionally used to make it easier to capture the effect of their interaction with a metallic nickel matrix and the corrosive environment. Emphasis is placed on understanding qualitative and quantitative changes in surface chemical composition—X-ray photoelectron spectroscopy (XPS), caused by the presence of microflakes, on the one hand, and the exposure to a corrosive environment, on the other. The corrosion resistance of a newly obtained material is monitored in-situ by electrochemical impedance spectroscopy (EIS) and basic *dc* (direct current) polarization methods. The morphology, structure, and topography of the coatings are investigated by scanning electron microscopy (SEM), X-ray diffraction (XRD), and contact profilometry.

2. Materials and Methods

Cerium molybdenum oxide microflakes were synthesized at low temperature from $Ce(NO_3)_3$ and Na_2MoO_4 precursors through the precipitation method. First, the two precursors were prepared separately by the dissolution of Na_2MoO_4 $2H_2O$ and $Ce(NO_3)_3$ $6H_2O$ in demineralized water in the molar ratio of 2:1, respectively. The solutions were next cooled down to 8 °C overnight in the fridge. Na_2MoO_4 solution was instilled slowly (60 mL min^{-1}) into $Ce(NO_3)_3$ solution under continuous mechanical stirring (200 rpm), which resulted in the formation of a light yellow suspension. After 1 h mixing at 15 °C, the precipitate was filtered under the reduced pressure on a Buchner funnel. The grey-yellowish precipitate was washed several times with demineralized water and then dried in a vacuum dryer.

The plating bath for nickel electrodeposition (100 mL plating volume) was prepared by mixing choline chloride and ethylene glycol in a 1:2 molar ratio and addition of 1 mol dm^{-3} $NiCl_2$ $6H_2O$ at 70 °C. All these reagents were mechanically stirred until a homogeneous green liquid was obtained. Cerium molybdenum oxide microflakes were then added (2 g dm^{-3}) and the bath was homogenized for 30 min (UP50H – Hielscher Compact Lab Homogenizer).

For the electrodeposition, copper disks (1.5 cm diameter, 0.1 cm thick) were used as cathodes. They were polished on 600–1200 grade abrasive paper. Then, the samples were degreased in an ultrasonic cleaner in methanol, acid etched (10 vol.% H_2SO_4), rinsed in demineralized water and again in methanol. Electroplating was carried out in a thermostatic electrolyzer consisting of two anodes (pure nickel) and a cathode placed between them. The process was carried out at a current density 6 mA cm^{-2}, at 70 °C, for 1 h. After this time, the coatings were twice rinsed in demineralized water and methanol. At the end, the coatings were dried and stored in a vacuum desiccator until required.

Surface morphology was analyzed by Quanta 250 (FEI) scanning electron microscope equipped with an Octane Elect Plus SDD microanalyzer (operating at 25 kV, and at 10^{-4} Pa pressure).

The crystal structure and phase composition of cerium molybdenum oxide microflakes and blank and composite nickel coatings were analyzed using powder X-ray diffraction. The XRD diffractograms were recorded on a D 5000 (Siemens) diffractometer with CuKα radiation (λ = 0.15409 nm) at room temperature in the 2θ range: 10–60° for blank and composite Ni coatings, and 5–60° for cerium molybdenum oxide microflakes. Phase identification was carried out by comparing the experimental patterns with the reference patterns collected in the Powder Diffraction Files database (International Centre for Diffraction Data PDF-2 base).

The X-ray Photoelectron Spectroscopy (XPS) measurements were performed using Escalab 250Xi (Thermo Fisher Scientific, United Kingdom). Al Kα monochromatic X-ray source with spot diameter of 250 μm was used. The set up pass energy was 20 eV. Charge compensation was controlled through low-energy Ar^+ ions emission by means of a flood gun, with the final calibration made using Ni2p3 metallic peak component at 852.6 eV. The deconvolution procedure was performed using Avantage software (Thermo Fisher Scientific, Waltham, USA).

The chemical analysis of corrosive solutions was performed by ICP-OES (Inductively Coupled Plasma-Optical Emission Spectrometry) and ICP-MS (Inductively Coupled Plasma-Mass Spectrometry) technique. The Ni concentration was determined using ICP-OES method (Vista MPX, Varian, Australia). Cu, Ce, Mo, and also Ni content was determined using ICP-MS method (XSeries2, Thermo Fisher Scientific, USA). The analyses were performed in the Chemical Laboratory of Multielemental Analysis, Wrocław University of Science and Technology, Poland, accredited by the Polish Centre for Accreditation (AB 696). The laboratory has a measurement procedure of determination of metals in water in its scope of accreditation. The procedure is based on EN-ISO 11885:2009, PN-EN ISO 17294-1:2007, and PN-EN ISO 17294-2:2006.

Corrosion measurements were carried out at 25 °C in 0.05 mol dm^{-3} solution of NaCl in Autolab 400 mL corrosion cell using a Reference 1010E (Gamry, Warminster, USA) potentiostat/galvanostat/ZRA (zero resistance ammeter). The geometric area of the working electrode exposed to the solution was 1 cm^2. A stainless steel rod (geometric area 5 cm^2) and a saturated Ag|AgCl / 3M KCl electrode (Metrohm) mounted in a Luggin capillary were used as the counter and reference electrodes, respectively. Potentiodynamic polarization curves were recorded after 168 hours exposure starting from −0.1 V to +0.5 V vs. open circuit potential (E_{OC}) at a scan rate of 0.166 mV s^{-1}. The polarization resistance (LPR technique) was measured by polarization of the sample starting from −10 mV to +10 mV vs. E_{OC} at a scan rate of 1 mV s^{-1}. Impedance spectra for electrochemical impedance spectroscopy (EIS) were recorded at E_{OC} (potentiostatic mode) with a resolution of 10 pts/dec., in a frequency range from 100 kHz to 0.001 Hz and at a 10 mV signal amplitude. Equivalent circuit modeling, graphing, and analysis of impedance data was performed using ZView® software (Scribner Associates, version 3.5g). The deposition process was studied by cyclic voltammetry (CV) technique using Autolab RDE-2 electrode with a Pt tip (3 mm diameter) in a electrochemical vessel (20–90 mL, Metrohm) with thermostat jacket, Ag and Pt wires as the reference and counter electrodes, in the range of potentials from E_{OC} to −1.5, then 1.5 and ending at E_{OC} with a scan rate of 20 mV s^{-1}. Pt tip instead of Cu was chosen to avoid the influence of any copper ions on the current-potential characteristics.

3. Results and Discussion

3.1. Morphology, Topography, and Phase Structure of Blank and Composite Ni Coatings

After electrodeposition, both blank and composite nickel coatings with a mean thickness of 10 μm were subjected to SEM and XRD analysis. The surface of the Ni coating consisted of spheroidal particles that form randomly arranged larger agglomerates (Figure 1a). The surface of the Ni/CeMoOxide coating, like the Ni coating, consisted of spherical particles, which were noticeably smaller in this case. In addition, there were randomly embedded microflakes of CeMoOxide within the surface. Single cracks were visible, mostly spreading from the places where the microflakes are built-in (Figure 1b). Additional topographic measurements of the surface of both coatings confirmed the increase in roughness of the composite coating, expressed by the R_a parameter equaled to 122.3 ± 17.7 nm (related to 92.6 ± 10.1 nm for blank Ni coating). For comparison, Figure 1c shows the photomicrograph (recorded in a secondary electron - SE mode of SEM) of synthesized cerium molybdenum oxide hydrate powder. The plates are overgrown and form agglomerates with a length exceeding 20 μm, a width below 10 μm, and a thickness of 100–200 nm (Figure 1c).

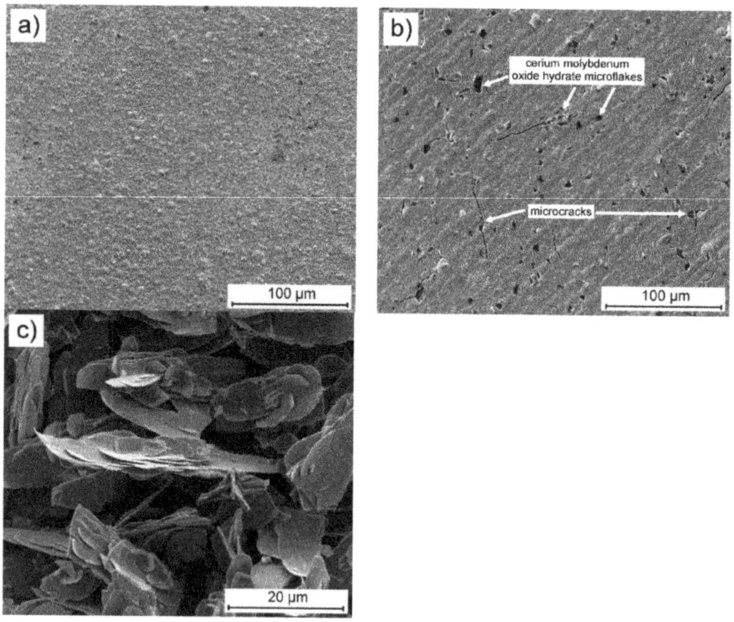

Figure 1. Surface morphology of (**a**) blank and (**b**) composite Ni coatings deposited on the copper substrate at a current density of 6 mA cm^{-2} at 70 °C for 1 h, and (**c**) as-synthesized cerium molybdenum oxide hydrate microflakes for comparison. Pictures were recorded in a SE mode of SEM.

Figure 2 shows the diffractograms for of cerium molybdenum oxide microflakes and blank and composite nickel coating. The thickness of the prepared Ni coatings was sufficient to perform the XRD analysis. The dominant peaks at the diffractograms are characteristic of Ni (111) and (200) crystal planes visible at 2θ 44.6° and 51.9°. The indicated reflections were well indexed to these collected in ICDD JCPDS card No. 00-004-0850 for Ni. Peaks at 2θ 43.3° and 50.4° originated from the Cu substrate (JCPDS card No. 00-004-0836). The other small peaks match well with the cerium molybdenum oxide. The diffractogram recorded for as-prepared microflakes is also shown in Figure 2. All the diffraction peaks correspond to $Ce_2(MoO_4)_3$ $4.5H_2O$ phase (JCPDS card No. 00-031-0333) without the presence of other peaks from any impurities. The strong and sharp diffraction peaks indicate that the prepared cerium molybdenum oxide microflakes were well-crystallized. High-resolution diffractograms were recorded for the blank and composite Ni coatings to estimate the average crystallite size. A copper peak from the substrate overlapped the nickel peak (111), therefore, for the proper determination of FWHM (full width at half maximum) for the Ni (111) peak, a deconvolution using a pseudo-Voigt function was performed. The average crystallite size was calculated using the Scherrer Equation (1):

$$D = \frac{K \cdot \lambda}{B \cdot cos\theta} \tag{1}$$

where D is the mean crystallite size in the direction perpendicular to the (hkl) plane of reflexes in nm, K is a Scherrer constant (0.9) [36], λ = 0.154 nm is the X-ray wavelength used in the measurement, B was calculated from Equation (2):

$$B = \sqrt{(\beta_{FWHM}^2 - \beta_0^2)} \tag{2}$$

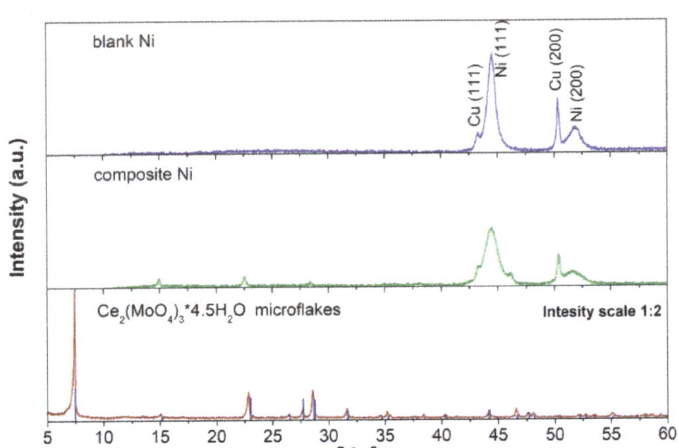

Figure 2. X-ray diffractograms for blank and composite Ni coatings deposited on the copper substrate at a current density of 6 mA cm^{-2} at 70 °C for 1 h. A powder diffraction pattern for as-synthesized cerium molybdenum oxide hydrate is also shown at the bottom of the figure for comparison.

(β_{FWHM} and β_0 is the FWHM of diffraction peak at angle θ and the corrected instrumental broadening - in radian, respectively). After deconvolution, the average crystallite size was estimated at about 10.4 nm and 6.3 nm for blank Ni and composite Ni coatings respectively. The resulting D value indicates the nano-crystalline structure of the electrodeposited nickel. It can be seen that the cerium molybdenum oxide microflakes affects the structure of the coating by co-depositing and reducing the size of the crystallites.

On the basis of the observed changes in the morphology and structure of both types of coating, it is obvious that the cerium and molybdenum compound must have modified the course of cathodic processes. Therefore, cyclic voltammetry curves were recorded under deposition conditions in both (blank Ni and suspension) plating baths. The experiment was repeated twice to avoid accidental results. Representative CV curves are shown in Figure 3. Analyzing the course of the curves, it can be seen that the composite coating deposition can proceed probably with greater efficiency. This can be demonstrated by a much larger anodic peak within the potentials of Ni (deposit) oxidation (Figure 3). It is however not excluded that the larger area of that peak may be associated with, e.g., the oxidation of molybdenum species released at high temperature (70 °C) from the microflakes suspended in the plating bath. However, the cathode curves were not significantly different (Figure 3).

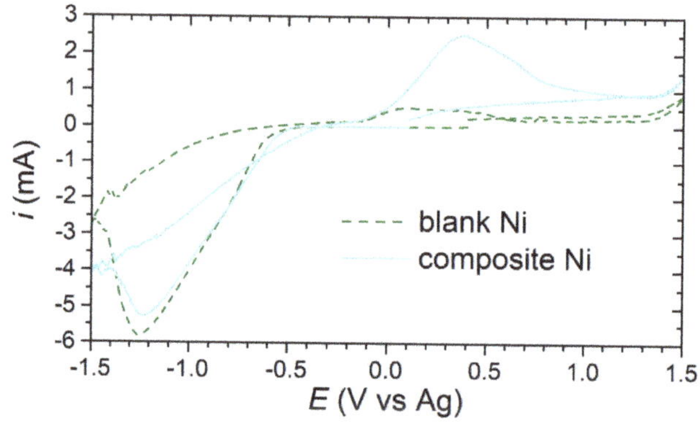

Figure 3. Cyclic voltammetry curves recorded in a blank Ni plating bath and suspension bath containing 2 g dm^{-3} $Ce_2(MoO_4)_3$ $4.5H_2O$ microflakes, at 70 °C on a rotating disk electrode with Pt tip (3 mm diameter), at 600 rpm.

3.2. *Corrosion Resistance*

3.2.1. *dc* Polarization

Polarization resistance (R_p) as a function of exposure time in 0.05 mol dm^{-3} solution of NaCl for blank Ni and composite (Ni/CeMoOxide) coatings is presented in Figure 4. In the case of a blank Ni coating, a clear increase in R_p was observed. During exposure in a NaCl solution, R_p increased up to 536.6 kΩ cm^2 after about 144 hours, then, its value decreased. The R_p-time dependency for the Ni/CeMoOxide coating was different. An increase of R_p to above 100 kΩ cm^2 was visible within the first hours of exposure in a chloride solution. After approximately 72 hours, the R_p value decreased to approximately 77 kΩ cm^2, and then it showed a slight upward trend (Figure 4).

The corrosion potential (E_{corr}) of the Ni/CeMoOxide coating increased quite rapidly in the first hours (Figure 5). After about three days it reached a constant level. The potential-time dependency for the Ni coating was different. An increase in E_{corr} towards more positive values was noticeable throughout the entire time studied. It is possible that such a direction of changes resulted from qualitative changes in the morphology of the layer of nickel oxidation products.

Finally, after 168 hours of exposure—after reaching a certain stability of the potential (at least for the composite coating), polarization curves for Ni and Ni/CeMoOxide coatings were registered. Their course is presented in Figure 6. The anodic branch of the Ni/CeMoOxide coating, compared to the anodic branch of the Ni coating, is characterized by higher corrosion current density values, which suggests more active anodic process on the surface of the former. A section with slowly a growing density of the corrosion current is visible on the anodic side in the course of Ni coating (Figure 6), which may suggest some protective/barrier action of a passive layer.

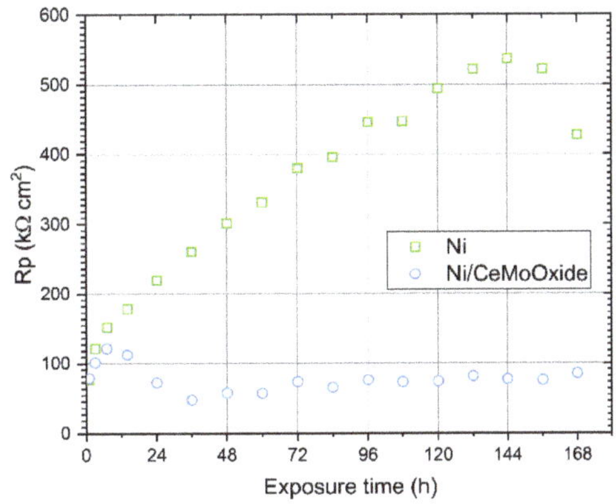

Figure 4. Polarization resistance (R_p) recorded for blank and composite Ni coatings during 168-hour exposure in 0.05 M NaCl solution.

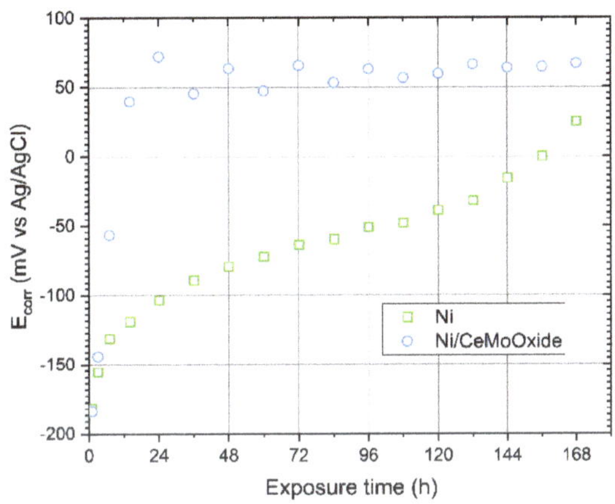

Figure 5. Corrosion potential (E_{corr}) recorded for blank and composite Ni coatings during 168-hour exposure to 0.05 M NaCl solution.

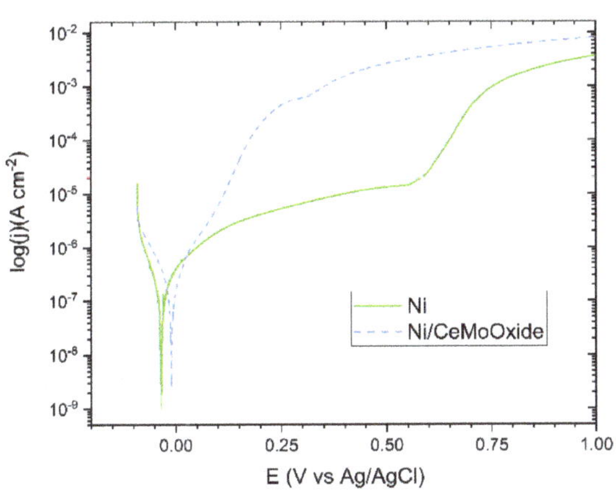

Figure 6. Potentiodynamic polarization curves recorded for blank and composite Ni coatings after 168-hour exposure to 0.05 M NaCl solution.

3.2.2. Electrochemical Impedance Spectroscopy

Impedance spectra of both types of coating were recorded every 24 hours during 7 days of exposure of the samples in 0.05 mol dm^{-3} NaCl solution. They have been presented on a complex Bode plot (Figure 7) to gain better visibility of the changes of the investigated system. The first analysis of the shape of the spectra leads to the conclusion that the corrosion process of Ni coating is characteristic of "corroding coating". The magnitude of impedance |Z| increased during exposure (Figure 7a), which may indicate increasing corrosion resistance, due to the formation of, e.g., a stable passive layer or increasing diffusion constrains (especially after 144 and 168 hours). A similar shape of impedance spectra and similar tendency (increasing) for the impedance modulus have been observed in previous work by Urcezino et al. [37] and J. Winiarski et al. [38].

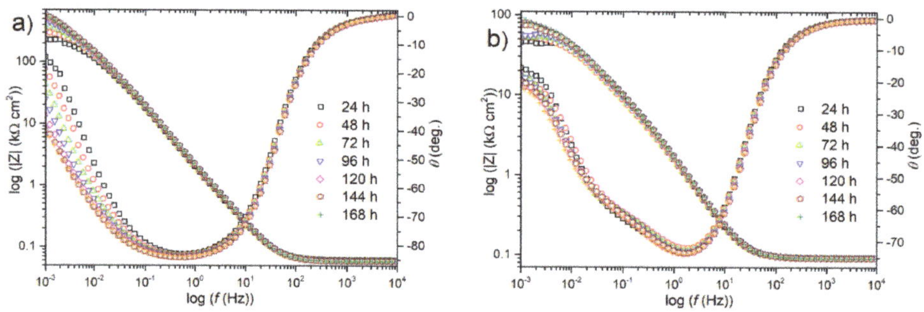

Figure 7. Bode representation of the impedance spectra recorded for (**a**) blank and (**b**) composite Ni coatings for 168 hours exposure in 0.05 M NaCl solution.

Figure 7b presents impedance spectra for composite coating. This material behaved slightly differently, rather like "damaged coating", due to the presence of coating incontinuities. Furthermore, not very clearly separated time constants—caused by capacitive dispersion and overlapping of time constants—were observed (Figure 7b). Therefore, for further discussion, two electric equivalent circuits (EECs) were used for the calculation of the theoretical spectra: a single time constant model for a blank Ni (Figure 8a) and a double time constant model for a composite Ni (Figure 8b).

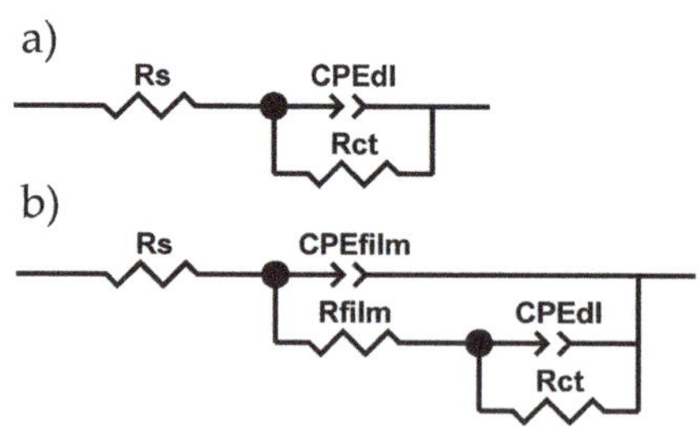

Figure 8. Electric equivalent circuits used for fitting the experimental spectra of (**a**) blank and (**b**) composite Ni coatings.

Both models use a constant phase element (CPE). The impedance of the CPE is defined by Equation (3), where: Y_0 is a time constant parameter (Ω^{-1} cm^{-2} s^{α}), ω is the angular frequency of the AC signal and α is the CPE exponent.

$$Z_{CPE} = Y_0^{-1}(j\omega)^{-\alpha} \quad (3)$$

The other elements used in the EECs presented in Figure 8a,b are: R_s—the resistance of NaCl solution, R_{ct}—the charge transfer resistance associated with nickel oxidation, $Y_{0,dl}$ and α corresponds to a double layer capacitance (C_{dl}). In the circuit for a composite coating (Figure 8b), the properties of a developed surface layer of corrosion product was modeled using R_{film} (the resistance of electrolytic solution in a porous layer) and C_{film} (the dielectric properties of this layer). Both EECs yielded a very good fit to the experimental data (parameter χ_2 in order of 10^{-4}–10^{-5}) and low residual errors (0.2%–8%). However, it should be noted that the fitting was made after manually reducing the frequency range to ca. 10 kHz–0.08 Hz. This interference was intentional to avoid the influence of instability of the measured system, especially in the low frequency range, on the interpretation of the physical meaning of EECs elements and, finally, their values (collected in Tables 1 and 2).

Table 1. Values of electric elements calculated for EEC of blank Ni coating.

Time (h)	R_s/ Ω cm^2	$CPE_{dl, Y0}$/ $Ω^{-1}$ cm^{-2} s^α	$CPE_{dl, α}$	R_{ct}/ kΩ cm^2
24	61	9.0×10^{-5}	0.94	214.8
48	61	8.6×10^{-5}	0.94	299.6
72	61	8.4×10^{-5}	0.93	394.8
96	60	8.3×10^{-5}	0.93	478.8
120	59	8.2×10^{-5}	0.93	552.4
144	59	8.1×10^{-5}	0.94	604.8
168	58	8.2×10^{-5}	0.93	580.8

Table 2. Values of electric elements calculated for EEC of composite Ni coating.

Time (h)	R_s/ Ω cm^2	$CPE_{film, Y0}$/ $Ω^{-1}$ cm^{-2} s^α	$CPE_{film, α}$	R_{film}/ kΩ cm^2	$CPE_{dl, Y0}$/ $Ω^{-1}$ cm^{-2} s^α	$CPE_{dl, α}$	R_{ct}/ kΩ cm^2
24	92	1.0×10^{-4}	0.95	5.6	8.5×10^{-5}	0.66	74.8
48	92	1.0×10^{-4}	0.94	5.9	9.9×10^{-5}	0.69	59.4
72	93	9.9×10^{-5}	0.94	7.0	9.8×10^{-5}	0.68	77.1
96	92	9.8×10^{-5}	0.94	7.9	9.2×10^{-5}	0.67	81.9
120	92	9.7×10^{-5}	0.94	8.6	8.6×10^{-5}	0.67	84.6
144	91	9.6×10^{-5}	0.94	8.2	8.4×10^{-5}	0.65	92.2
168	91	9.5×10^{-5}	0.94	8.8	7.9×10^{-5}	0.64	102.7

For a blank Ni coating, the values of R_{ct} increased with increasing exposure time and finally exceeded 0.5 MΩ cm^2 after 7 days (Table 1). A similar tendency was observed for the R_p determined on the basis of the LPR technique, see Figure 4. This marked increase in R_{ct} was accompanied by only minor changes in CPE parameters (Y_0 and α), whose values indicate the absence of significant changes in (e.g., homogeneity) of the surface at which the corrosion process occurs. In the case of a composite coating (Table 2), it was also noted that the direction of changes of a charge transfer resistance was close to the polarization resistance values obtained from the LPR method (Figure 4). Initially, for the first 48 hours, there was a drop of R_{ct} from 75 to 59 kΩ cm^2, but after 72 hours exposure, the values of R_{ct} increased with increasing time and finally exceeded 100 kΩ cm^2 after 7 days (Table 2). For this coating the contribution of a resistance connected with the presence of a developed surface layer of corrosion products is not to be missed—R_{film} values increased from 5.6 to 8.8 kΩ cm^2 over 7 days exposure (Table 2).

3.3. XPS Surface Analysis of Coatings

In order to better understand the role of microflakes in the corrosion process, both composite and blank Ni coatings were analyzed by X-ray photoelectron spectroscopy (XPS). The high-resolution XPS analyses were performed in the binding energy (BE) range of Ni2p, Cu2p, and O1s photopeaks, with the results presented in Figure 9. Supporting studies were also carried out in the C1s BE range.

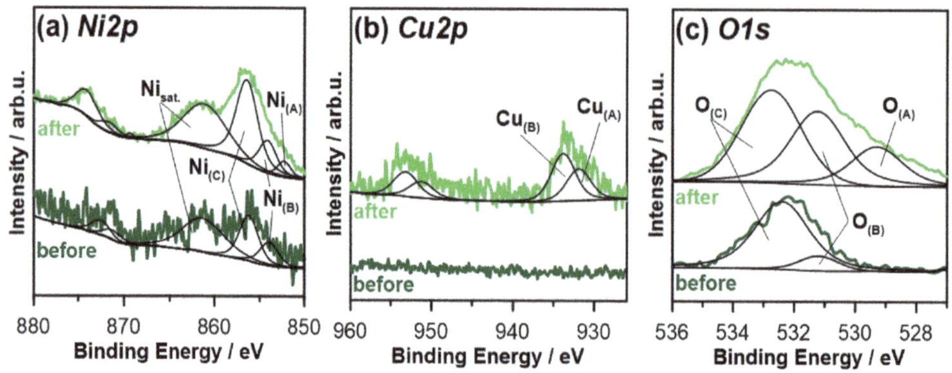

Figure 9. High-resolution XPS spectra recorded for blank Ni coating in (**a**) Ni2, (**b**) Cu2p, and (**c**) O1s peak binding energy range, before and after the exposure to 0.05 M NaCl.

The analysis performed for the blank Ni coating prior to corrosion stability tests reveal a weak signal from nickel in the Ni2p spectra (see Figure 9a), further deconvoluted using two peak doublets $Ni_{(B)}$ and $Ni_{(C)}$. The $Ni2p_{3/2}$ photopeaks of the aforementioned components are located at 854.0 and 856.3 eV, being highly characteristic for nickel(II) oxide NiO and nickel(II) hydroxide $Ni(OH)_2$, respectively [38–41]. This is further confirmed by the location of the Ni satellite peak at 861.3 eV. The small signal intensity originates from substantial surface coverage with adventitious carbon layer, including oxidized carbon species. At the same time, the ion gun etching was not performed, due to significant observable changes in surface chemistry caused by the Ar^+ ions used for sputtering purposes. These findings are further confirmed by the O1s spectra visible at Figure 9c, where the $O_{(B)}$ peak was ascribed to the nickel hydroxide species, while the dominant $O_{(C)}$ component represents the organic carbon-oxygen bonds and carbonates formed during air exposure. The detailed analysis is summarized in Table 3.

Table 3. Spectral deconvolution performed for blank Ni coating.

Element		BE / eV	Blank Ni Before	Blank Ni After
Ni2p	$Ni_{(A)}$	852.4	-	1.2
	$Ni_{(B)}$	854.0	1.9	3.2
	$Ni_{(C)}$	856.3	3.7	9.7
Cu2p	$Cu_{(A)}$	932.5	-	1.4
	$Cu_{(B)}$	933.5	-	2.1
O1s	$O_{(A)}$	529.5	-	15.6
	$O_{(B)}$	530.8	12.2	30.5
	$O_{(C)}$	532.4	82.2	36.7

The one weeklong exposure of the analyzed sample in 0.05 M NaCl solution resulted in partial modification of the observed sample surface chemistry. While the chemical state of Ni and O peaks appears to be unchanged, the photoelectron intensity count is significantly improved, suggesting partial removal of the carbon-based corrosion products layer. Nickel appears primarily in the form of $Ni(OH)_2$ (symbol $Ni_{(C)}$), with the NiO to $Ni(OH)_2$ ratio slightly decreasing from 1:2 to 1:3. Furthermore, the third component $Ni_{(A)}$ emerges as a result of the exposure, its location is characteristic for metallic nickel. Again, the shape of O1s spectra corroborates the results, revealing an increased intensity of hydroxide species $O_{(B)}$ but also nickel oxide species $O_{(A)}$ at approximately 529.5 eV. The amount of adventitious carbon dropped twice.

A characteristic feature of XPS measurements carried out after blank Ni coating exposure to corrosive media is the appearance of the Cu substrate beneath the layer (see Figure 9b). Two peak doublets could be deconvoluted on the basis of the spectra shape: $Cu_{(A)}$ with $Cu2p_{3/2}$ peak at 932.5 eV and $Cu_{(B)}$ at 933.5 eV. The position of the $Cu_{(A)}$ component is characteristic both for metallic copper as well as for Cu_2O [42], on the other hand, the $Cu_{(B)}$ component was ascribed to CuO oxides according to previous literature findings [43–46]. The total copper contribution reached 3.5 at.%.

Similar to blank Ni coating, detailed high-resolution XPS analysis was also carried out for the composite Ni coating, where the analysis was expanded with the Ce3d and Mo3d spectral range. The results of the aforementioned analysis are presented in Figure 10 with the spectral deconvolution summarized in Table 4.

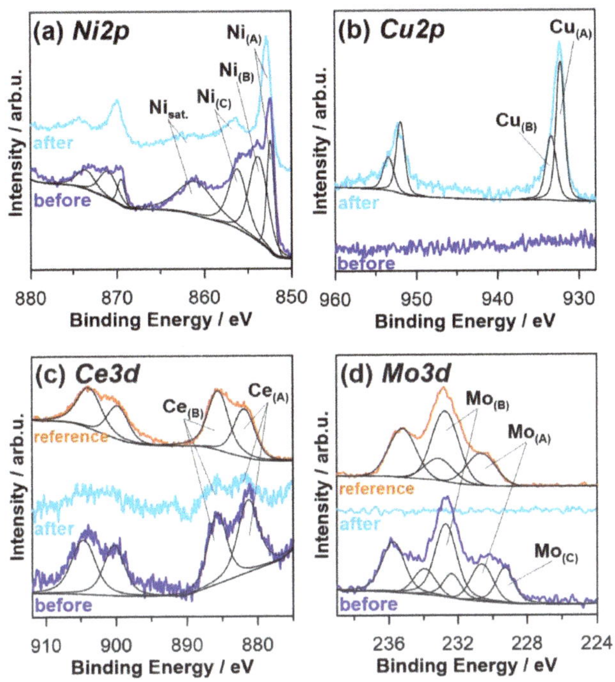

Figure 10. High-resolution XPS spectra recorded for composite Ni coating in (**a**) Ni2p, (**b**) Cu2p, (**c**) Ce3d, and (**d**) Mo3d peak binding energy range, before and after the exposure to 0.05 M NaCl. The chemistry of synthesized cerium molybdenum oxide hydrate powder used for electrodeposition is also shown in (**c**) Ce3d and (**d**) Mo3d spectra ("reference" spectra).

Unlike the blank Ni coating, the nickel chemistry on the surface of the composite coating prior to exposure in corrosive media is well developed, with the total Ni contribution ranging from 38 at.% compared to merely 5 at.% observed in the absence of microflake functionalization. Furthermore, the metallic Ni peak $Ni_{(A)}$ is also well developed, proving significantly higher corrosion resistance of the protective layer under atmospheric conditions. Another difference is the strong contribution of the $Ni_{(B)}$ component, revealing increased NiO to $Ni(OH)_2$ ratio of 1.2:1 (compared to 1:2 for blank Ni prior the exposure).

As mentioned previously, the galvanic Ni coating was modified using a cerium molybdenum oxide hydrate compound. The chemistry resulting from Ce and Mo surface modification is shown in Figure 10c,d, respectively. Comparison with the precursor compound analysis denoted as the reference confirms high spectral similarity, proving functionalization, did not introduce significant chemical changes. The cerium Ce3d spectra reveals high complexity, the deconvolution was performed using two peak doublets, $Ce_{(A)}$ and $Ce_{(B)}$. Both of these were ascribed to the Ce^{3+} component on the basis of the value of peak BE (881.8 and 885.7 eV) and similar intensity, nearly at a 1:1 ratio [47–49]. It is possible, however, that composite coating before exposure contains some Ce^{4+}, which can be inferred from the lowered position of the component at the energy 886 eV.

A similar comparison was made in the Mo3d spectral range for cerium molybdenum oxide hydrate compound and for Ni composite coating functionalized with these microflakes. Both analyses reveal the dominant presence of the component ascribed in Figure 10d as $Mo_{(B)}$. The peak location at 232.7 eV is a very good match with the literature BE value of Mo^{6+}, typically reported in MoO_3 oxides and $(MoO_4)_3{}^{2-}$ molybdates [50–52]. The second component is common for both the analyzed samples, $Mo_{(A)}$, is negatively shifted at approximately −2.1 eV, being characteristic for the Mo^{5+} component. The Mo^{6+}:Mo^{5+} ratio of 2.8:1 remains nearly intact. Finally, the surface analysis of Ni composite coating revealed the appearance of $Mo_{(C)}$ component at 229.5 eV bound with Mo^{4+} oxides. Its absence in cerium molybdenum oxide hydrate microflakes powder may indicate partial reduction of molybdenum during the electrodeposition process [53–55]. According to the literature, the reduction of molybdenum could take place during electrodeposition, because the shape of the voltammetric curve and currents achieved differed from those recorded in a blank Ni electrolyte (Figure 3).

The Ni chemistry on the surface of the composite Ni coating changed as a result of the exposure to the 0.05 M NaCl solution. The second analysis carried out after a one-week period revealed a double diminished Ni contribution associated primarily with the observable removal of the $Ni_{(B)}$ component from the coating surface. It is possible that NiO got, in part, reduced to the metallic Ni in local cathodic areas and, in part, transformed to $Ni(OH)_2$. At the same time, the amount of molybdenum oxides within the protective coating nearly completely disappeared as a result of the corrosion process. The removal of cerium molybdenum oxide hydrate microflakes was further confirmed by significant removal of cerium species, however, it appears that the corrosion process is less selective towards it, since the spectra was still recognizable after the exposure, with nearly intact chemistry. Finally, the secondary effect of the corrosion process is the appearance of the Cu oxides, $Cu_{(A)}$ and $Cu_{(B)}$. Not only the total contribution of Ni, but also Cu, is significantly increased in comparison to blank Ni coating, indicating smaller thickness of the protective layer and/or locally reduced coverage resulting from consumption of cerium/molybdenum species during the corrosion process.

Table 4. Spectral deconvolution performed for composite Ni coating.

Element		BE/eV	NiCeMo Before	NiCeMo After	CeMo Reference
Ni2p	$Ni_{(A)}$	852.5	11.2	5.8	-
	$Ni_{(B)}$	853.7	14.4	3.8	-
	$Ni_{(C)}$	856.3	12.2	5.1	-
Cu2p	$Cu_{(A)}$	932.5	-	5.1	-
	$Cu_{(B)}$	933.3	-	2.9	-
Ce3d	$Ce_{(A/B)}$	881.8/885.7	6.1	2.4	11.6
Mo3d	$Mo_{(A)}$	230.6	1.2	-	5.4
	$Mo_{(B)}$	232.7	3.2	-	15.1
	$Mo_{(C)}$	229.3	1.2	-	-
O1s	$O_{(A)}$	529.5	21.8	0.3	6.0
	$O_{(B)}$	530.8	21.3	11.6	22.5
	$O_{(C)}$	532.4	7.4	63.0	39.4

3.4. Chemical Analysis of Corrosive Solutions

To verify the hypothesis regarding the selectivity of Ce and Mo leaching from microflakes embedded in the composite coating, an additional experiment was planned consisting of a comparative analysis of the composition of NaCl solutions both before and after the corrosion process. For this purpose, blank Ni and Ni composite coatings were immersed for 7 days in 0.05 M NaCl solutions (100 ml volume for each sample). One series of samples was just immersed and left for 7 days, a second one was immersed with an additional continuous stirring of the solution. After 7 days, all solutions were filtered and analyzed by

ICP-OES and ICP-MS. The focus was on the comparative analysis of: Ni, Cu, Ce, and Mo content, with blank 0.05 M NaCl as the reference. The results are summarized in Table 5.

Table 5. Concentrations of Ni, Cu, Ce, and Mo in a corrosive solution.

Element	Units	Blank (Demi Water + 0.05M NaCl)	Blank Ni	Blank Ni_m *	Composite Ni	Composite Ni _m *
Ni	μg/L	<2.0 [2]	2620 ± 370 [1]	4230 ± 590 [1]	2730 ± 380 [1]	6000 ± 840 [1]
Cu	μg/L	2.4 ± 0,4 [2]	2.1 ± 0.3 [2]	2.2 ± 0.3 [2]	29 ± 4 [2]	14 ± 2 [2]
Ce	μg/L	<0.1 [2]	<0.1 [2]	< 0.1 [2]	0.33 ± 0.07 [2]	0.13 ± 0.03 [2]
Mo	μg/L	5.5 ± 0.8 [2]	5.3 ± 0.8 [2]	17 ± 3 [2]	390 ± 55 [2]	374 ± 52 [2]

[1] ICP-OES analysis. [2] ICP-MS analysis. * Solution was mechanically stirred for 7 days exposure of the coatings.

The results presented in Table 5 clearly indicate that Ce did not go into the solution in any significant amounts (similar level before and after corrosion process). At the same time Mo content increased from ca. 5 to 370–390 μg/L. This analysis shows that molybdenum is not only leached from microflakes, but also significant amounts of Mo passed into the corrosive solution. This behavior is quite possible, because the SEM observation of composite coating after 168 hours exposure clearly indicates that some of the microflakes have probably been leached away and those that were visible have been partially dissolved, leaving something like a skeleton/frame (Figure 11a). Furthermore, XPS analysis seems to confirm that molybdenum, when released from micro-flakes, does not form a stable oxide layer on the surface of the coating. Such selective dissolution probably caused voids (well visible in Figure 11a) in the coating, which resulted in a lower corrosion resistance of this composite material.

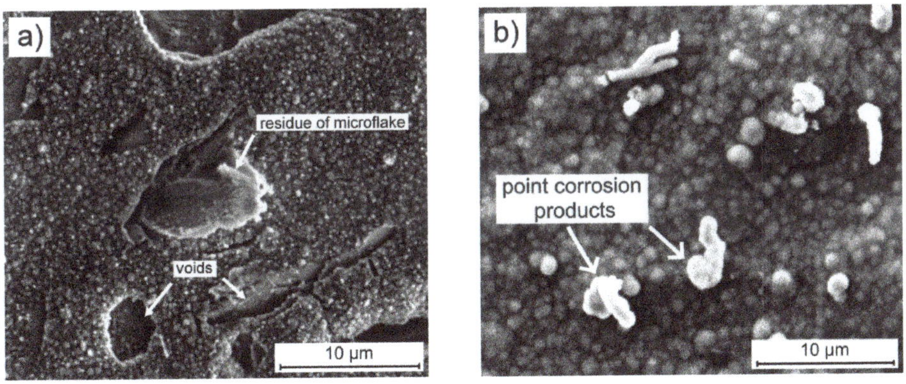

Figure 11. Surface morphology of (**a**) composite and (**b**) blank Ni coatings (deposited on the copper substrate at a current density of 6 mA cm^{-2} at 70 °C for 1 h) after one-week exposure to 0.05 M NaCl solution.

The results related to Ni concentration (Table 5) very likely to confirm that Ni almost completely forms the layer of corrosion products. Only point corrosion products were visible on the surface of blank Ni coating (Figure 11b). Its (nickel) presence in the solution after corrosion at a level similar to that determined in the blank 0.05 M NaCl solution only confirms this assumption. Different behavior was observed for copper (originating from the substrate). In the case of blank Ni coating, Cu was not observed to pass into the NaCl solution, which, in the context of the results of the XPS surface analysis, only confirmed that this element is part of the relatively stable corrosion products on the surface of this coating. Only in the case of composite coating, copper was determined at the level of 14–29 μg/L, which means that this element partially passed in the form of an ion into the solution. This is even more possible because during the dissolution of the microflakes, the copper substrate was gradually exposed to chloride ions.

4. Conclusions

In summary, it can be concluded that:

(1) DES-based plating baths are an excellent environment for the electrodeposition of metal composite coatings.

(2) The addition of cerium molybdenum oxide hydrate microflakes to a plating bath modifies the cathodic process sufficiently to obtain nanocrystalline nickel composite coatings with a smaller (6.3 nm) crystallite size than blank Ni coating (10.4 nm).

(3) The codeposition of large CeMo oxide microflakes caused microcracks in the coating and some deterioration of the protective properties of composite coating.

(4) Charge transfer resistance for both types of coatings increased over 7 days exposure in 0.05 M NaCl solution. However, even after this time, the R_{ct} for a blank Ni exceeded five times (581 kΩ cm^2) that of a composite one (103 kΩ cm^2).

(5) On the basis of spectroscopic studies (XPS, ICP-OES and ICP-MS), it can be stated for composite Ni coating that in the course of corrosion, NiO transforms to $Ni(OH)_2$, while molybdenum oxides nearly completely disappear from the surface of composite coating. The removal of cerium species was less than that of the molybdenum species, which suggests that the corrosion process is more selective towards Mo. The smaller thickness of the protective layer and locally reduced coverage results from the consumption of the Ce/Mo species during the corrosion process.

(6) It would be interesting to develop the described research towards the synthesis of nanoflakes (or nanowires) of mixed cerium and molybdenum compounds. Reducing the particle size to the nano-scale would result in a composite coating with a better dispersed oxide phase. This, in turn, would improve the tightness of the metal coating.

Author Contributions: All authors have read and agree to the published version of the manuscript. Research concept and supervision, J.W.; methodology, J.W. (corrosion, electrochemistry, SEM, powder synthesis), J.R. (XPS), A.N. (electrodeposition, laboratory work, profilometry), K.W. (XRD), S.B. (ICP-OES and ICP-MS); writing—original draft preparation, J.W., A.N., J.R., K.W.; review, K.D., B.S.

Funding: This research was funded by a statutory activity subsidy from the Polish Ministry of Science and Higher Education for the Department of Advanced Material Technologies at Wrocław University of Science and Technology in 2020 year—grant number 8201003902.

Conflicts of Interest: The authors declare no conflict of interest.

References

1. Abbott, A.P.; McKenzie, K.J. Application of ionic liquids to the electrodeposition of metals. *Phys. Chem. Chem. Phys.* **2006**, *8*, 4265–4279. [CrossRef] [PubMed]
2. Abbott, A.; Frisch, G.; Ryder, K. Electroplating Using Ionic Liquids. *Annu. Rev. Mater. Res.* **2013**, *43*, 335–358. [CrossRef]
3. Simka, W.; Puszczyk, D.; Nawrat, G. Electrodeposition of metals from non-aqueous solutions. *Electrochimica Acta* **2009**, *54*, 5307–5319. [CrossRef]
4. Liu, F.; Deng, Y.; Han, X.; Hu, W.; Zhong, C. Electrodeposition of metals and alloys from ionic liquids. *J. Alloy. Compd.* **2016**, *654*, 163–170. [CrossRef]
5. Abbott, A.; Boothby, D.; Capper, G.; Davies, D.; Rasheed, R.K. Deep Eutectic Solvents Formed between Choline Chloride and Carboxylic Acids: Versatile Alternatives to Ionic Liquids. *J. Am. Chem. Soc.* **2004**, *126*, 9142–9147. [CrossRef]
6. Smith, E.L.; Abbott, A.; Ryder, K. Deep Eutectic Solvents (DESs) and Their Applications. *Chem. Rev.* **2014**, *114*, 11060–11082. [CrossRef]
7. Abbott, A.; Ballantyne, A.; Harris, R.; Juma, J.A.; Ryder, K.; Forrest, G. A Comparative Study of Nickel Electrodeposition Using Deep Eutectic Solvents and Aqueous Solutions. *Electrochimica Acta* **2015**, *176*, 718–726. [CrossRef]

8. Wang, S.; Zou, X.; Lu, Y.; Rao, S.; Xie, X.; Pang, Z.; Lu, X.; Xu, Q.; Zhou, Z. Electrodeposition of nano-nickel in deep eutectic solvents for hydrogen evolution reaction in alkaline solution. *Int. J. Hydrogen Energy* **2018**, *43*, 15673–15686. [CrossRef]
9. Abbott, A.; El Ttaib, K.; Ryder, K.; Smith, E.L. Electrodeposition of nickel using eutectic based ionic liquids. *Trans. IMF* **2008**, *86*, 234–240. [CrossRef]
10. Abbott, A.; Ballantyne, A.; Harris, R.; Juma, J.A.; Ryder, K. Bright metal coatings from sustainable electrolytes: The effect of molecular additives on electrodeposition of nickel from a deep eutectic solvent. *Phys. Chem. Chem. Phys.* **2017**, *19*, 3219–3231. [CrossRef]
11. Cherigui, E.A.M.; Sentosun, K.; Mamme, M.H.; Lukaczynska, M.; Terryn, H.; Bals, S.; Ustarroz, J. On the Control and Effect of Water Content during the Electrodeposition of Ni Nanostructures from Deep Eutectic Solvents. *J. Phys. Chem. C* **2018**, *122*, 23129–23142. [CrossRef]
12. Fashu, S.; Gu, C.; Zhang, J.; Huang, M.-L.; Wang, X.-L.; Gu, C. Effect of EDTA and NH4Cl additives on electrodeposition of Zn–Ni films from choline chloride-based ionic liquid. *Trans. Nonferrous Met. Soc. China* **2015**, *25*, 2054–2064. [CrossRef]
13. Vijayakumar, J.; Mohan, S.; Kumar, S.A.; Suseendiran, S.; Pavithra, S. Electrodeposition of Ni–Co–Sn alloy from choline chloride-based deep eutectic solvent and characterization as cathode for hydrogen evolution in alkaline solution. *Int. J. Hydrogen Energy* **2013**, *38*, 10208–10214. [CrossRef]
14. Yanai, T.; Shiraishi, K.; Watanabe, Y.; Nakano, M.; Ohgai, T.; Suzuki, K.; Fukunaga, H. Electroplated Fe–Ni Films Prepared From Deep Eutectic Solvents. *IEEE Trans. Magn.* **2014**, *50*, 1–4. [CrossRef]
15. You, Y.; Gu, C.; Wang, X.; Tu, J. Electrodeposition of Ni–Co alloys from a deep eutectic solvent. *Surf. Coatings Technol.* **2012**, *206*, 3632–3638. [CrossRef]
16. Golgovici, F.; Pumnea, A.; Petica, A.; Manea, A.C.; Brincoveanu, O.; Enachescu, M.; Anicai, L. Ni–Mo alloy nanostructures as cathodic materials for hydrogen evolution reaction during seawater electrolysis. *Chem. Pap.* **2018**, *72*, 1889–1903. [CrossRef]
17. Martis, P.; Dilimon, V.; Delhalle, J.; Mekhalif, Z. Electro-generated nickel/carbon nanotube composites in ionic liquid. *Electrochimica Acta* **2010**, *55*, 5407–5410. [CrossRef]
18. Liu, D.; Sun, J.; Gui, Z.; Song, K.; Luo, L.; Wu, Y. Super-low friction nickel based carbon nanotube composite coating electro-deposited from eutectic solvents. *Diam. Relat. Mater.* **2017**, *74*, 229–232. [CrossRef]
19. You, Y.H.; Gu, C.D.; Wang, X.L.; Tu, J.P. Electrochemical preparation and characterization of Ni-PTFE composite coatings from a non-aqueous solution without additives. *Int. Electrochem. Sci.* **2012**, *7*, 12440–12455.
20. Li, R.; Chu, Q.; Liang, J. Electrodeposition and characterization of Ni–SiC composite coatings from deep eutectic solvent. *RSC Adv.* **2015**, *5*, 44933–44942. [CrossRef]
21. Li, R.; Hou, Y.; Liang, J. Electro-codeposition of Ni-SiO2 nanocomposite coatings from deep eutectic solvent with improved corrosion resistance. *Appl. Surf. Sci.* **2016**, *367*, 449–458. [CrossRef]
22. Li, R.; Hou, Y.; Liu, B.; Wang, D.; Liang, J. Electrodeposition of homogenous Ni/SiO2 nanocomposite coatings from deep eutectic solvent with in-situ synthesized SiO2 nanoparticles. *Electrochimica Acta* **2016**, *222*, 1272–1280. [CrossRef]
23. Protsenko, V.; Bogdanov, D.; Korniy, S.; Kityk, A.; Baskevich, A.; Danilov, F. Application of a deep eutectic solvent to prepare nanocrystalline Ni and Ni/TiO2 coatings as electrocatalysts for the hydrogen evolution reaction. *Int. J. Hydrogen Energy* **2019**, *44*, 24604–24616. [CrossRef]
24. Dong, M.; Lin, Q.; Sun, H.; Chen, D.; Zhang, T.; Wu, Q.; Li, S. Synthesis of Cerium Molybdate Hierarchical Architectures and Their Novel Photocatalytic and Adsorption Performances. *Cryst. Growth Des.* **2011**, *11*, 5002–5009. [CrossRef]
25. Xu, M.-K.; Ouyang, Z.-H.; Shen, Z. Topological evolution of cerium(III) molybdate microflake assemblies induced by amino acids. *Chin. Chem. Lett.* **2016**, *27*, 673–677. [CrossRef]
26. Ayni, S.; Sabet, M.; Salavati-Niasari, M.; Hamadanian, M. Synthesis and characterization of cerium molybdate nanostructures via a simple solvothermal method and investigation of their photocatalytic activity. *J. Mater. Sci. Mater. Electron.* **2016**, *27*, 7342–7352. [CrossRef]
27. Kartsonakis, I.; Kordas, G. Synthesis and Characterization of Cerium Molybdate Nanocontainers and Their Inhibitor Complexes. *J. Am. Ceram. Soc.* **2010**, *93*, 65–73. [CrossRef]
28. Kartsonakis, I.; Balaskas, A.C.; Kordas, G.C. Influence of cerium molybdate containers on the corrosion performance of epoxy coated aluminium alloys 2024-T3. *Corros. Sci.* **2011**, *53*, 3771–3779. [CrossRef]

29. Kartsonakis, I.; Kontogiani, P.; Pappas, G.; Kordas, G. Photocatalytic action of cerium molybdate and iron-titanium oxide hollow nanospheres on Escherichia coli. *J. Nanoparticle Res.* **2013**, *15*, 1759. [CrossRef]
30. Patel, M.; Bhanvase, B.A.; Sonawane, S. Production of cerium zinc molybdate nano pigment by innovative ultrasound assisted approach. *Ultrason. Sonochemistry* **2013**, *20*, 906–913. [CrossRef]
31. Yasakau, K.; Kallip, S.; Zheludkevich, M.; Ferreira, M. Active corrosion protection of AA2024 by sol–gel coatings with cerium molybdate nanowires. *Electrochimica Acta* **2013**, *112*, 236–246. [CrossRef]
32. Yasakau, K.; Tedim, J.; Zheludkevich, M.; Drumm, R.; Shem, M.; Wittmar, M.; Veith, M.; Ferreira, M. Cerium molybdate nanowires for active corrosion protection of aluminium alloys. *Corros. Sci.* **2012**, *58*, 41–51. [CrossRef]
33. Lehr, I.; Saidman, S. Corrosion protection of AZ91D magnesium alloy by a cerium-molybdenum coating-The effect of citric acid as an additive. *J. Magnes. Alloy.* **2018**, *6*, 356–365. [CrossRef]
34. Mu, S.; Du, J.; Jiang, H.; Li, W. Composition analysis and corrosion performance of a Mo–Ce conversion coating on AZ91 magnesium alloy. *Surf. Coatings Technol.* **2014**, *254*, 364–370. [CrossRef]
35. Bhanvase, B.A.; Patel, M.; Sonawane, S. Kinetic properties of layer-by-layer assembled cerium zinc molybdate nanocontainers during corrosion inhibition. *Corros. Sci.* **2014**, *88*, 170–177. [CrossRef]
36. Langford, J.I.; Wilson, A.J.C. Scherrer after sixty years: A survey and some new results in the determination of crystallite size. *J. Appl. Crystallogr.* **1978**, *11*, 102–113. [CrossRef]
37. Urcezino, A.S.C.; Dos Santos, L.P.M.; Casciano, P.N.S.; Correia, A.N.; De Lima-Neto, P. Electrodeposition study of Ni coatings on copper from choline chloride-based deep eutectic solvents. *J. Braz. Chem. Soc.* **2017**, *28*, 1193–1203. [CrossRef]
38. Winiarski, J.; Cieślikowska, B.; Tylus, W.; Kunicki, P.; Szczygieł, B. Corrosion of nanocrystalline nickel coatings electrodeposited from choline chloride:ethylene glycol deep eutectic solvent exposed in 0.05 M NaCl solution. *Appl. Surf. Sci.* **2019**, *470*, 331–339. [CrossRef]
39. Biesinger, M.C.; Payne, B.P.; Lau, L.W.M.; Gerson, A.; Smart, R.S.C. X-ray photoelectron spectroscopic chemical state quantification of mixed nickel metal, oxide and hydroxide systems. *Surf. Interface Anal.* **2009**, *41*, 324–332. [CrossRef]
40. Tang, Y.; Pattengale, B.; Ludwig, J.; Atifi, A.; Zinovev, A.V.; Dong, B.; Kong, Q.; Zuo, X.; Zhang, X.; Huang, J. Direct Observation of Photoinduced Charge Separation in Ruthenium Complex/Ni(OH)2 Nanoparticle Hybrid. *Sci. Rep.* **2015**, *5*, 18505. [CrossRef]
41. Mansour, A.N. Characterization of NiO by XPS. *Surf. Sci. Spectra* **1994**, *3*, 231–238. [CrossRef]
42. Espinos, J.P.; Morales, J.; Barranco, A.; Caballero, A.; Holgado, J.P.; González-Elipe, A.R.; González-Elipe, A.R. Interface Effects for Cu, CuO, and Cu_2O Deposited on SiO_2 and ZrO_2. XPS Determination of the Valence State of Copper in Cu/SiO2and Cu/ZrO2Catalysts. *J. Phys. Chem. B* **2002**, *106*, 6921–6929. [CrossRef]
43. Shaikh, J.; Pawar, R.C.; Devan, R.; Ma, Y.; Salvi, P.; Kolekar, S.; Patil, P. Synthesis and characterization of Ru doped CuO thin films for supercapacitor based on Bronsted acidic ionic liquid. *Electrochimica Acta* **2011**, *56*, 2127–2134. [CrossRef]
44. Poulston, S.; Parlett, P.M.; Stone, P.; Bowker, M. Surface oxidation and reduction of CuO and Cu_2O studied using XPS and XAES. *Surf. Interface Anal.* **1996**, *24*, 811–820. [CrossRef]
45. Morales, J.; Espinos, J.P.; Caballero, A.; Gonzalez-Elipe, A.R.; Mejías, J.A.; González-Elipe, A.R. XPS Study of Interface and Ligand Effects in Supported Cu2O and CuO Nanometric Particles. *J. Phys. Chem. B* **2005**, *109*, 7758–7765. [CrossRef]
46. González-Elipe, A.R.; Holgado, J.P.; Alvarez, R.; Munuera, G. Use of factor analysis and XPS to study defective nickel oxide. *J. Phys. Chem.* **1992**, *96*, 3080–3086. [CrossRef]
47. Holgado, J.P.; Alvarez, R.; Munuera, G. Study of CeO2 XPS spectra by factor analysis: Reduction of CeO_2. *Appl. Surf. Sci.* **2000**, *161*, 301–315. [CrossRef]
48. Al-Doghachi, F.A.J.; Rashid, U.; Taufiq-Yap, Y.H. Investigation of Ce(iii) promoter effects on the tri-metallic Pt, Pd, Ni/MgO catalyst in dry-reforming of methane. *RSC Adv.* **2016**, *6*, 10372–10384. [CrossRef]
49. Beche, E.; Charvin, P.; Perarnau, D.; Abanades, S.; Flamant, G. Ce 3d XPS investigation of cerium oxides and mixed cerium oxide (CexTiyOz). *Surf. Interface Anal.* **2008**, *40*, 264–267. [CrossRef]
50. Li, Z.; Gao, L.; Zheng, S. SEM, XPS, and FTIR studies of MoO_3 dispersion on mesoporous silicate MCM-41 by calcination. *Mater. Lett.* **2003**, *57*, 4605–4610. [CrossRef]
51. Reddy, B.M.; Chowdhury, B.; Reddy, E.P.; Fernández, A. An XPS study of dispersion and chemical state of MoO3 on Al2O3-TiO2 binary oxide support. *Appl. Catal. A: Gen.* **2001**, *213*, 279–288. [CrossRef]

52. Lee, Y.J.; Nichols, W.T.; Kim, D.-G.; Kim, Y.D. Chemical vapour transport synthesis and optical characterization of MoO3thin films. *J. Phys. D: Appl. Phys.* **2009**, *42*, 115419. [CrossRef]
53. Li, H.; Ye, H.; Xu, Z.; Wang, C.; Yin, J.; Zhu, H. Freestanding MoO_2/Mo_2 C imbedded carbon fibers for Li-ion batteries. *Phys. Chem. Chem. Phys.* **2017**, *19*, 2908–2914. [CrossRef] [PubMed]
54. Kim, H.; Cook, J.; Tolbert, S.H.; Dunn, B. The Development of Pseudocapacitive Properties in Nanosized-MoO2. *J. Electrochem. Soc.* **2015**, *162*, A5083–A5090. [CrossRef]
55. Marin-Flores, O.; Scudiero, L.; Ha, S. X-ray diffraction and photoelectron spectroscopy studies of MoO_2 as catalyst for the partial oxidation of isooctane. *Surf. Sci.* **2009**, *603*, 2327–2332. [CrossRef]

materials

MDPI

Article

Microstructure Characterization and Corrosion Resistance of Zinc Coating Obtained on High-Strength Grade 10.9 Bolts Using a New Thermal Diffusion Process

Henryk Kania [1,*] and Jacek Sipa [2,*]

1 Silesian University of Technology, Institute of Materials Engineering, Krasińskiego 8, 40-019 Katowice, Poland

2 Remix S.A., Poznańska 36, 66-200 Świebodzin, Poland

* Correspondence: henryk.kania@polsl.pl (H.K.); j.sipa@remixsa.pl (J.S.); Tel.: +48-68-4755400 (J.S.)

Received: 15 March 2019; Accepted: 25 April 2019; Published: 29 April 2019

Abstract: The article presents the results of research on the application of innovative thermal diffusion zinc coating technology with the recirculation of the reactive atmosphere to high-strength grade 10.9 bolts. The innovation of this method consists in the introduction of reactive atmosphere recirculation and the application of coating powder mix which contains zinc powder and activator. Recirculation of the reactive atmosphere ensures its uniform composition, while the presence of an activator intensifies the process of saturating steel surface with zinc, which boosts the efficiency of active agents. Coatings were created at 440 °C and a heat soaking time of 30–240 min. Coating structure (SEM) was exposed, chemical composition in microsites (EDS) was defined, and coating phase structure (XRD) was identified. The kinetics of coating growth were defined. It was found that the increment of coating thickness was controlled by square root of soaking time. Coatings obtained using innovative thermal diffusion zinc coating technology had a two-layer structure. At the substrate, a compact layer of phase Γ_1 ($Fe_{11}Zn_{40}$) was created, which was covered with a layer of phase δ_1 ($FeZn_{10}$). The new method of thermal diffusion zinc coating will alow for the creation of coatings of very good corrosion resistance while maintaining strength properties of bolts defined as strength class 10.9.

Keywords: thermal diffusion coatings; grade 10.9 bolts; corrosion resistance

---

1. Introduction

Due to their versatility and reliability, bolted joints are among most frequent methods of joining steel structures. They demonstrate a number of advantages, which include easy and quick assembly at minimum cost of maintenance and control but also good strength at variable loads [1]. However, due to specific geometry of such joints, they are the most critical area of construction, which in numerous instances necessitates the application of high-strength bolts made of special, additionally heat-treated steel grades.

The durability of steel structures is determined by the impact of the environment whose aggressiveness speeds up corrosive wear. Therefore, adequate anti-corrosion protection of steel structures is an important factor and is decisive for safe exploitation. Steel structures are most often protected against corrosion by hot dip metallization. Zinc coatings are also used to protect bolts designed for joining structural elements [2]. However, the requirement of a proper thread match between the bolt and nut brings about the limitation of coating thickness [3]. Despite being stronger than construction steel, bolts become the weakest element of a structure in a corrosive environment.

Currently, zinc coatings on bolts are obtained by various methods. Hot dip galvanizing (HDG) is most frequently used to protect bolt surfaces [2]. Bolts are galvanized by a batch hot dip method in baskets with centrifugation [4]. Coating consists of a transient layer of intermetallic phase Fe-Zn, which is covered with an outer layer of zinc extracted from the bath [5,6]. The depth of the external layer is difficult to control, while removing an excess of zinc is not always successful at a conventional hot dip galvanizing temperature of 450 °C. Coatings obtained usually have a thickness of 45 to 65 μm. This makes it difficult to obtain a proper fit between bolts and nuts, while decreased coating thickness lowers the strength of the joint compared to other elements of the structure [7]. Much better results are achieved by high-temperature galvanizing. Galvanizing at a higher temperature allows for the effective removal of excess zinc from the surface of bolts. Coating obtained at a high temperature consists exclusively of intermetallic phase Fe-Zn. The thickness of intermetallic phase layers is more effectively controlled with process parameters. The corrosion resistance of high temperature coatings is approximately two times greater than that of conventional zinc coatings [8]. Removal of excess zinc allows one to maintain the thread fit, while increased corrosion resistance yields bolt strength comparable to other structural elements. High temperature galvanizing of bolts is effected at a temperature of ca. 560 °C [9]. In the case of high-strength quenched and tempered bolts, galvanizing at such a temperature poses a risk of compromising the high-strength properties. Furthermore, the pickling procedure in hydrochloric acid, normally applied in the procedure of surface finishing, creates a risk of hydrogen embrittlement due to hydriding in the case of high-strength bolts.

Another method of protecting bolts against corrosion is electro-galvanizing, which ensures obtaining a uniformly thick coating on a threaded surface. Due to small coating thickness (usually up to 15 μm), the corrosion resistance of the bolts, which is significant at the beginning of exploitation, decreases fairly quickly, which considerably shortens the life of the entire structure. The application of galvanized coatings does not require heating of treated elements during the process, which minimizes the risk of high-strength bolts losing their strength after quenching and tempering. However, this method leads to extensive contamination of the natural environment as well as possible hydrogen embrittlement in high-strength steel [10].

Much better results are achieved when zinc coatings on bolts are obtained by sherardizing. The sherardizing process is run in closed rotary retorts in which bolts are placed together with zinc powder and zinc oxide [11]. The retort is soaked at a temperature of 380–450 °C, which allows for the creation of a layer of intermetallic phase Fe-Zn on the bolts [12–14]. The protective layer created by sherardizing, similar to the one obtained through high-temperature galvanizing, consists exclusively of intermetallic phase Fe-Zn, which permits the control of case depth and ensures high corrosion resistance. Meanwhile, due to a much lower processing temperature, the high-strength bolts retain their strength properties [13]. It is usually enough to have the surfaces cleaned mechanically before sherardizing in order to prevent hydrogen embrittlement. The major drawback of sheradizing is its relatively long processing time. In order to obtain coatings of the required depth of 15–30 μm, 6 to 12 h will be needed, to be further extended if the process temperature is lowered.

2. Innovative Thermal Diffusion Zinc Coating Technology with Reactive Atmosphere Recirculation

The technology of thermal diffusion zinc coating with recirculation of reactive atmosphere is a new technology in zinc coating. The process is run in a closed retort, which is loaded with products to be coated together with powder mixture. The rotating motion of the retort ensures continuous contact of powder mix with the surfaces of processed products. The powder mix is composed of zinc powder, ZnO as a filler, and NH_4Cl as an activator. In the course of heating the powder mix in the retort, gaseous reaction products are created that move towards the surfaces to be coated. On steel substrate adsorption of zinc atoms created from the gaseous mixture as a result of reduction, dissociation, or exchange occurs.

Forming of thermal diffusion coating results from the chemical reaction of vapor deposition [15–17]. Previous studies have shown that NH_4Cl decomposes according to the following reaction [16]:

$$NH_4Cl \rightarrow NH_3 + HCl, \quad (1)$$

Zn powder reacts with HCl vapors to form $ZnCl_2$ by the reaction:

$$Zn + 2HCl \rightarrow ZnCl_2 + H_2, \quad (2)$$

$ZnCl_2$ is reduced by hydrogen to Zn as a result of heating process [16]. During heating, active zinc atoms can also be formed according to Equation [18], as follows:

$$ZnCl_2 \rightarrow Zn + 2Cl, \quad (3)$$

The concentration of active Zn atoms in the powder mixture is higher than on the steel surface. The concentration gradient causes internal diffusion of zinc, which in a gaseous form deposits on the steel surface [18].

Active zinc atoms easily diffuse into the steel substrate along existing vacancies, leading to distortion of the crystallographic structure of the steel substrate. The iron atoms on the surface of the substrate can be activated by the following reactions [18]:

$$2HCl + Fe \rightarrow FeCl_2 + H_2, \quad (4)$$

$$FeCl_2 \rightarrow Fe + 2Cl, \quad (5)$$

Crystal lattice distortion accelerates the diffusion of active Fe atoms outside [18]. During heating, there is a two-way diffusion between Zn and Fe, causing the formation of Fe-Zn intermetallic phases [17].

The process, besides rotary movement of the retort, is aided by an innovative solution that consists of forced circulation of reactive atmosphere along the retort cylinder [19]. This ensures a better and more uniform contact of product surface with powder mix, both on external and internal surfaces as well as on surfaces of intricate shape. Introduction of atmosphere recirculation along the retort cylinder permits uniformity of reactive atmosphere within the entire volume of the retort, which leads to uniform temperature in the working space, lower consumption of powder mix, and more effective use thereof in comparisation to typical sherardizing [19]. Due to recirculation and better use of active agents, it is possible to create the layers of intermetallic phase Fe-Zn [20] on low-carbon steel at a much shorter time compared to conventional sherardizing technology. At the same time, the coatings obtained through this method demonstrate higher corrosion resistance than the hot dip coatings [21]. The new technology constitutes an alternative to traditional process of sherardizing, in which an activator is not used and mixture motion in the atmosphere is determined solely by the rotating movement of the retort.

3. Materials and Methods

Coatings for research were created on a prototype plant equipped with a furnace with a rotary working chamber of capacity 100 kg and forced recirculation of reactive atmosphere. The workload consisted of M10 bolts and threaded bars of strength grade 10.9 made of quenched and tempered 1.7225 steel (42CrMo4). The powder mix was composed of zinc powder with an addition of 15% of zinc oxide ZnO as a filler and 3% of NH_4Cl as an activator. All that was soaked at the temperature of 440 °C and times of 30, 60, 120, and 240 min. Before filling the working chamber of the furnace, the ingredients of powder mix were dried at the temperature of 120 °C for 12 h in order to reduce moisture content to approx. 1%. The activator was added to the mixture right before processing due to its high hygroscopicity. Before processing, bolt surfaces were cleaned mechanically using sandblasting process to prevent the risk of hydrogen embrittlement.

In order to establish the structure of the thus-produced coatings, metallographic tests were performed with a light microscope. The microstructure and chemical composition in micro-areas of coatings were analyzed with a Hitachi S-3400 N (Tokyo, Japan) scanning microscope equipped with a X-ray energy dispersion spectroscope.

Phase analysis by X-ray diffraction was performed with a JEOL JDX-7S X-ray (Tokyo, Japan) diffractometer using a copper anode lamp ($\lambda CuK\alpha$ = 1.54178 Å) powered with 20 mA current at 40 kV, and with a graphite monochromator. Recording was performed with a stepwise approach of 0.05° step and counting time of 3 s in the range of 10 to 90° 2θ. Phases were identified with the help of the ICDD (Newtown Square, PA, USA) PDF-4+ database. Diffractometric tests were performed on skew-ground sample surfaces so that phase structure could be displayed in the entire cross-section of the coating.

Testing of resistance to the impact of neutral salt spray was performed in accordance with the standard EN ISO 9227 in a salt chamber CORROTHERM Model 610 by Erichsen.(Hemer, Germany) Testing was conducted in a 5% spray of sodium chloride in distilled water at the temperature of 35 °C. In order to determine mass changes, gravimetric analyses were performed in the course of testing following 48, 96, 164, 240, 480, 720, and 1000 h of the exposure of samples in the chamber. Corrosion tests were carried out on grade 10.9 screws with thermal diffusion zinc coating obtained at 4408C and heating time of 12 min.

Tensile strength testing was performed on a tensile strength testing machine Inspekt Table 100 by Hegewald und Peschke MPT GmbH (Nossen, Germany) with maximal load 100 kN.

4. Results and Discussion

4.1. Surface and Cross-Sections Appearance

External appearance of zinc coatings obtained on a grade 10.9 bolt with the new technology of thermal diffusion zinc coating with reaction of reactive atmosphere is presented in Figure 1. The coatings obtained do not show any discontinuity, while their surface looks matt grey. No traces of powder mix were found on the thread surface after zinc coating (Figure 1a). Additionally, in a cross-section the coating shows no discontinuity and densely covers the bolt surface both on the head (Figure 1b) and on the tip of thread (Figure 1c) as well as in the groove (Figure 1d). The grey and matt appearance of the coating and its cross-sectional morphology are a proof of intermetallic Fe-Zn phases in its structure. The appearance of cracks located mainly in the upper part of the coating can be observed in the coating structure, which may indicate the brittleness of the coating. The occurrence of transverse cracks is characteristic of the intermetallic phases of the Fe-Zn phases.

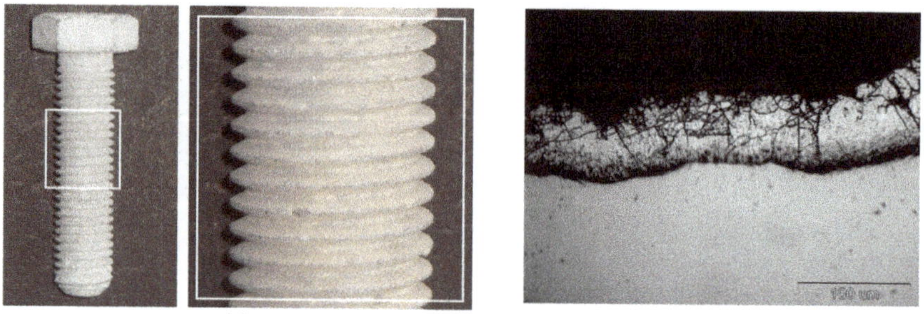

(a) (b)

Figure 1. *Cont.*

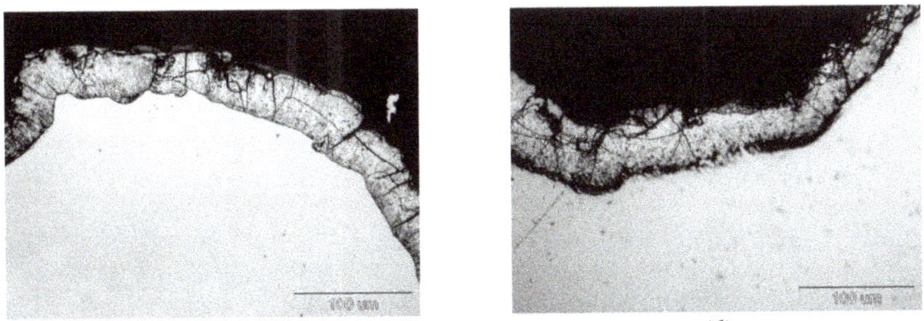

(c) (d)

Figure 1. Appearance of surface (**a**) and cross-section of coating on bolt head (**b**), thread tip (**c**), and thread groove (**d**) of grade 10.9 bolt.

4.2. *Growth Kinetics*

Growth kinetics of coatings obtained on high strength grade 10.9 bolts at the temperature of 440 °C and soaking time from 30 to 240 min is presented in Figure 2. Coating thickness increases with the increase of soaking time. However, the increment of coating thickness becomes slower and slower. After 120 min, a coating of average thickness of 50.61 µm was obtained. During further soaking, the increment was over 2 x slower with thickness of 72.62 µm obtained after the time of 240 min (Figure 2a).

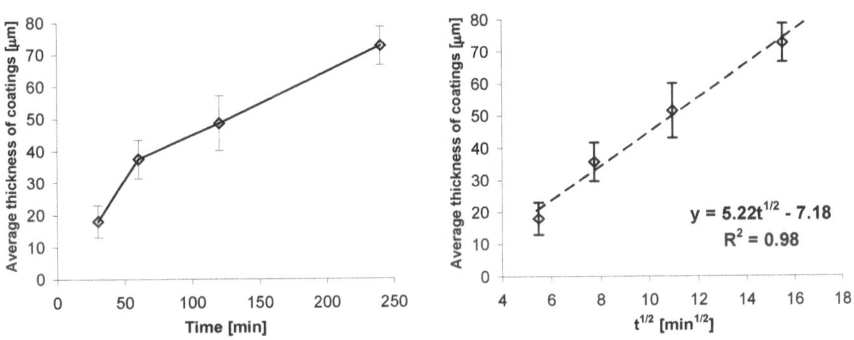

(a) (b)

Figure 2. Growth kinetics (**a**) and plot of coating thickness as a function of $t^{1/2}$ (**b**).

The impact of temperature soaking time on coating thickness is shown in more detail as the dependence of thickness in the square root function of soaking time $t^{1/2}$. For the experimentally defined dependence of coating thickness from soaking time, there is a nearly linear correlation of coating thickness against square root of soaking time $t^{1/2}$. The determined trend function and high value of correlation factor $R^2 = 0.98$ confirm the linear nature of coating thickness increment. Thus, the growth kinetics for a coating may be described by the following equation:

$$y\ (\text{thickness}) = 5.22 \times t^{1/2} - 7.18, \tag{4}$$

The presence of a factor equaling $t^{1/2}$ in the equation describing growth kinetics means that the coating increase is a process controlled by diffusion. Since those coatings are created as a result of zinc diffusion into the substrate, it may be concluded that the kinetics of that phenomenon are controlled by square root of time, which appears to be in compliance with literature data [22]. The equation also shows that the growth of the coating begins after the time that is necessary to reach the temperature allowing the diffusion of ingredients.

4.3. Microstructure (SEM) and Microanalysis (EDS)

The microstructure of a coating obtained at a temperature of 440 °C and time of 120 min is presented in Figure 3. Percentage contents of analysed elements are shown in Table 1. The coating is composed of two layers: outer layer defined as site A and a layer adjoining the substrate marked as site B. In the outer layer (Figure 3b), the presence of 7.2 wt % Fe and 92.8 wt % Zn was found in outer zone (point 1, Table 1) and 11.4 wt % Fe and 88.6 wt % Zn in the zone bordering on the underlying layer (point 2, Table 1). Chemical composition of outer layer is comprised within intermetallic phase δ_1 ($FeZn_{10}$) [23]. Change of concentration of components on the cross-section of that phase is typical of phase δ_1 and may be an evidence of its varied morphology: phases δ_{1p} of palisade structure in the upper zone and phase δ_{1k} of compact structure in the substrate zone [5]. Chemical composition of the underlying layer (Figure 3c) is more homogenous. In the outer layer, there is 19.7 wt % Fe and 80.3 wt % Zn (point 3, Table 1), whereas the substrate layer contains 21.2 wt % Fe and 78.8 wt % Zn (point 1, Table 1). Chemical composition in the underlying layer corresponds to stability range for phase Γ_1 ($Fe_{11}Zn_{40}$) [6].

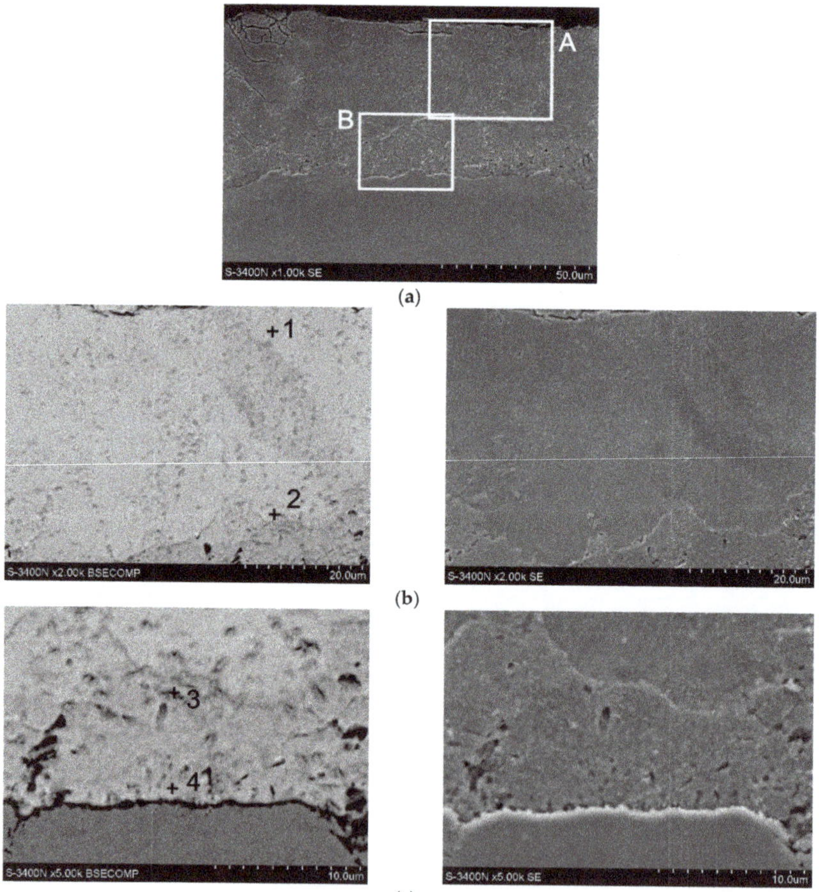

Figure 3. Microstructure (SEM) of coating obtained on high-strength grade 10.9 bolts via thermal diffusion method with reactive atmosphere recirculation: cross-section of the coating with selected areas (**a**), microstructure of the coating in area A (**b**), microstructure of the coating in area B (**c**).

Table 1. Chemical composition at selected micro-areas of coating obtained on high-strength grade 10.9 bolts (measurement points acc. to Figure 3).

Measurement Point	Content of Elements			
	Fe-K		Zn-K	
	wt %	at. %	wt %	at. %
point 1	7.2%	8.3%	92.8%	91.7%
point 2	11.4%	13.1%	88.6%	86.9%
point 3	19.7%	22.3%	80.3%	77.7%
point 4	21.2%	24.0%	78.8%	76.0%

4.4. X-ray Phase Analysis (XRD)

XRD analysis performed on the surface of a skew-ground coating (Figure 4) demonstrates phase δ_1 ($FeZn_{10}$) and phase Γ_1 ($Fe_{11}Zn_{40}$). Considering chemical composition in the coating microsites (Table 1), it may be claimed that the outer layer of the coating is phase δ_1, whereas the layer adjacent to the substrate is phase Γ_1. The research did not confirm any presence of other phase Fe-Zn, although these are stabile in the conditions the coating is created. According to phase equilibrium system, phases Γ (Fe_3Zn_{10}) and ζ ($FeZn_{13}$) are also stabile [24]. Configuration of all stabile phases (Γ, Γ_1, and δ_1 i ζ) occurs in the diffusion layer of coatings obtained by HDG method at the temperature of 450 °C [6]. However, the mechanism of producing a hot dip galvanizing coating is more complex than that, and the transition layer is created as a result of simultaneous processes of diffusive growth, solutioning in liquid zinc, and recrystallization [24].

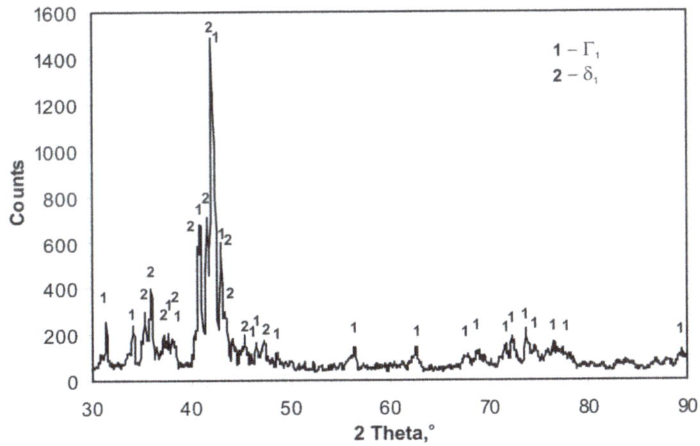

Figure 4. Diffractogram of the surface of skew-ground coating obtained on high-strength grade 10.9 bolts through thermal diffusion method with reactive atmosphere recirculation.

For technologies using powder mixtures as zinc carrier, the available literature does not allow an unambiguous determination of coating phase structure. In conventional sherardizing, a coating is created that is composed of a compact layer of phase δ_1 (marked as $FeZn_7$) and a non-homogenous outer layer of phase ζ ($FeZn_{13}$) [14]. However, Konstantinov [25] claims that the coating is built of phases Γ and δ_1. Liu [26] obtained a similar structure, making his coatings with a pack cementation method in a powder mix with an addition of activator on 1.7225 (42CrMo4) steel. On the other hand, Wortelen [27] claims that the coating obtained through sherardizing in powder mix with an activator is composed of phases Γ, Γ_1, δ_1, and ζ. It must be pointed out that characteristic spectra for phases Γ and ζ coincide largely with the spectra of phases Γ_1 and δ_1; therefore, XRD examination does not allow for unambiguous exclusion of the presence of phases Γ and ζ. However, no other structural components were determined in the microstructure of the tested coating, whose chemical composition might correspond to the range of homogeneity of phases Γ and ζ. This was also confirmed in the research conducted by Chaliampalias [15,28], who obtained a coating composed of phases Γ_1 and δ_1 in the pack cementation process.

4.5. Corrosion Resistance

The surface appearance of grade 10.9 bolts with thermal diffusion after corrosion testing in neutral salt spray is presented in Figure 5. During exposure in a salt chamber, the thermal diffusion coating becomes covered with products of white rust. After 1000 h of exposure in the salt chamber, local rusty discolouring becomes visible on the surface (Figure 5a). White rust products accumulate in thread grooves and expose the thread tips, where the intensity of corrosion is clearly higher. Upon completion of corrosion test, no distinctive permeation of coating into substrate was visible, which was confirmed by structural research on a cross-section of coating (Figure 5b). After testing, a non-uniform depletion of coating thickness was manifested on the cross-section; however, the coating did not yet lose its continuity. The presence of rusty discolourations on the surface of zinc coatings is characteristic of corrosion of intermetallic phase Fe-Zn. Such a phenomenon is observed during exposure of HDG coatings to an environment that contains chlorides [8].

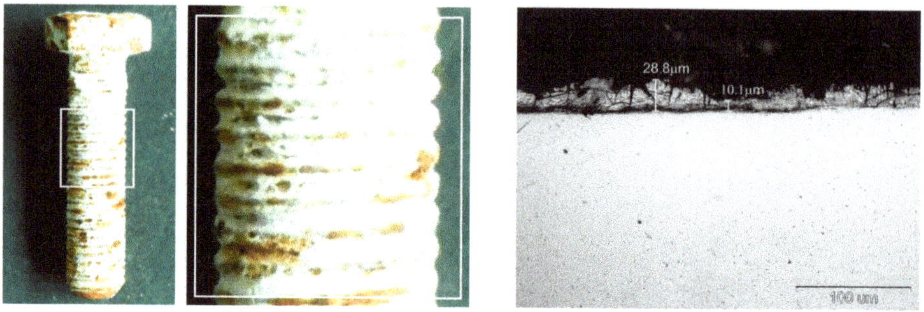

(a) (b)

Figure 5. The appearance (**a**) and structure (**b**) of the thermal diffusion zinc coating after corrosion tests in a salt chamber.

Dependence of mass changes in tested coatings from the time of exposure in the salt chamber is presented in Figure 6a. It may be concluded that during exposure to neutral salt spray, the tested coatings are characterized by increment of mass, which is an evidence of accumulation of corrosion products on their surfaces. Changes of bolt mass, and at the same time corrosive wear of thermal diffusion coating, follow a linear relationship. The trend function, determined in the course of experimental research, and the high correlation factor $R^2 = 0.97$ (Figure 6a) constitute proper linear match with data and confirm the linear character of the wear of thermal diffusion coating.

The linear character of coating wear is also evidenced by coating loss in its cross-section. Average coating thickness changes after 480, 720, and 1000 h of exposure in a salt chamber are presented in Figure 6b. Assuming average initial thickness of 51.52 ± 8.48 μm, they decreased to 18.03 ± 5.53 μm after a 1000 h corrosion test. The trend function, determined through measurements of average thickness of coating and good match of correlation factor $R^2 = 0.96$, confirm the linear character of corrosive wear of coating in its cross-section. Extrapolation of trend line allows an estimation that the total loss of thermal diffusion coating on a bolt's surface and at the same loss of protective properties will happen after approx. 1540 h of exposure in the salt chamber. In comparison to conventional hot dip galvanizing zinc coatings, the durability of the thermal diffusion coating tested in a corrosion test in a salt chamber [22] results in an approx. 2–3-fold increase in the durability time to penetration of the substrate surface (steel). While carrying out the corrosion test in a salt-spray chamber (acc. EN ISO 9227) traditional hot dip galvanizing coating of comparable thickness (average thickness 50.8 ± 6.42 μm), distinctive permeation of coating into substrate was visible after 480 h of exposure in neutral salt-spray environment [21]. For conventional hot dip zinc coating, much more severe corrosion losses can be observed in the initial stage of corrosion process when the outer layer of zinc is subject to corrosion. When this layer is worn out, the corrosion process proceeds more slowly in the Fe-Zn intermetallic layers [8]. Thermal diffusion zinc coating does not show increased corrosion intensity in the initial stage of corrosion process due to the lack of an outer zinc layer. The corrosion process proceeds only in the Fe-Zn intermetallic layers, whose thickness is higher in comparison with conventional hot dip coatings. The average thickness loss of the coating was used to estimate coating durability. Locally, in areas of lower thickness, the coating may be broken through to the base material. Due to the very good protection properties of the coating and the significant share of areas with a higher coating thickness, the protection of the steel in the initial stages of coating penetration will still be preserved.

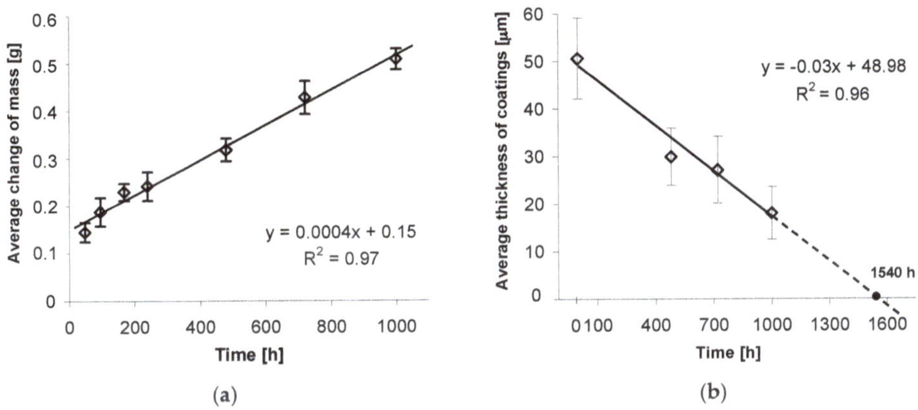

Figure 6. Average change of mass in tested bolts grade 10.9 (**a**) and average change of thermal diffusion coating thickness (**b**) during exposure in salt spray chamber.

4.6. Tensile Test

A diagram of a static tension test on threaded bar grade 10.9 without coating and with thermal diffusion coating is presented in Figure 7. The values of tensile strength limit Rm and conventional yield limit Rp0.2 for three tests on each bar are presented in Table 2. The tensile strength of grade 10.9 threaded bar decreased slightly after creation of thermal diffusion coating. The average tensile strength limit was Rm = 1024 MPa for the uncoated bar and Rm = 1007 MPa for the thermal diffusion coated one, respectively. The process of creating a coating by the new method of thermal diffusion zinc coating did not compromise the tensile strength below the required limit. However, the ratio of conventional yield limit to tensile strength had a similar value of 0.94–0.95 both for the uncoated bar and the thermal diffusion coated one. Tensile strength of the threaded bar coated with the new method of thermal diffusion zinc coating with recirculation of reactive atmosphere fulfilled the tensile strength requirement for grade 10.9.

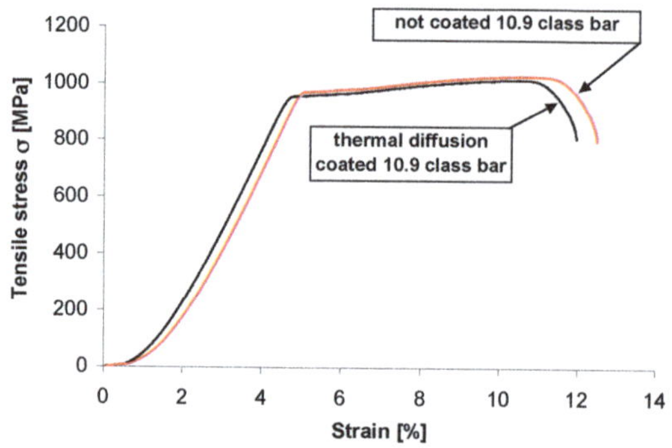

Figure 7. Tensile stress-displacement relationship curve of coated and uncoated 10.9 grade threaded bar.

Table 2. Results of tensile strength test of coated and uncoated 10.9 grade threaded bar.

Type of Threaded Bar	Sample Number	R_m [MPa]	$R_{p0,2}$ [MPa]	$R_{p0,2}/R_m$ [MPa]
not coated 10.9 grade bar	test 1	1031.73	980.87	0.95
	test 2	1016.93	955.24	0.94
	test 3	1023.36	968.32	0.95
thermal diffusion coated 10.9 grade bar	test 1	1004.01	946.00	0.94
	test 2	1001.11	936.56	0.94
	test 3	1018.41	962.87	0.95

In these tests, threaded bars M10 in strength grade 10.9 made of 42CrMo4 steel (1.7225) after quenching and tempering process were used. The total length of the sample was 220 mm, gauge length 160 mm. Special high nuts were screwed on the ends of the bar, and then the sample was placed in the holders of the testing machine and stretched until it was broken.

5. Conclusions

The innovative technology of thermal diffusion zinc coating with recirculation of reactive atmosphere provides effective anti-corrosion protection for high-strength grade 10.9 bolts while retaining their strength properties. The application of reactive atmosphere recirculation ensured homogenization of its composition and a better contact of product surface with powder mix, which permitted one to obtain continuous coatings of a compact structure in a shorter time compared to the conventional sherardizing method.

The research conducted allows for the following conclusions to be made:

1. The new method of thermal diffusion zinc coating with recirculation of reactive atmosphere permits one to obtain coatings 50–72 μm thick in soaking time of 120–240 min at a temperature of 440 °C upon grade 10.9 bolts;
2. The coatings obtained on grade 10.9 bolts feature a two-layer structure. Directly on the substrate, a compact layer of intermetallic phase Γ_1 ($Fe_{11}Zn_{40}$) is created, which is covered with a layer of intermetallic phase δ_1 ($FeZn_{10}$);
3. The coatings obtained in this way feature very good corrosion resistance. It is estimated that corrosion strength of 50 μm thermal diffusion coatings on grade 10.9 bolts in neutral salt spray amounts to approx. 1500 h;
4. Thermal diffusion zinc coating parameters permit retaining the strength properties for bolts defined by strength grade 10.9.

Author Contributions: Conceptualization, H.K., J.S., Experimental research, J.S., Methodology, H.K., Research of structure and corrosion resistance, H.K., Tensile test, J.S., Data curation, H.K., Visualization, H.K. Analysis of results H.K., Project administration, J.S.; Funding acquisition, J.S.

Funding: The research on thermal diffusion zinc coating was conducted within the R&D project "Innovative Technology of Thermal Diffusion Zinc Coating of Key Structural Elements with Application of Reactive Atmosphere Recirculation" (project no. POIG.01.04.00-08-383/13) co-financed by European Union within Operating Programme Innovative Economy from the means of European Regional Development Fund.

Conflicts of Interest: The authors declare no conflict of interest. The funders had no role in the design of the study; in the collection, analyses, or interpretation of data; in the writing of the manuscript; or in the decision to publish the results.

References

1. Lokaj, A.; Klajmonova, K. Comparison of behaviour of laterally loaded round and squared timber bolted joints. *Frattura Integr. Strutt.* **2017**, *11*, 56–61. [CrossRef]
2. Votava, J. Anticorrosion protection of strength bolts. *Acta Univ. Agric. Silvic. Mendel. Brun.* **2012**, *1*, 181–188. [CrossRef]
3. Bickford, J.H. *Introduction to the Design and Behavior of Bolted Joints*; CRC Press Taylor & Francis Group LLC: Boca Raton, FL, USA, 2008.
4. Hot Dip Galvanized Fasteners. Advantages of and Design Considerations for Hot-Dip Galvanized Fasteners. Copyright © by American Galvanizers Association. 2009. Available online: https://galvanizeit.org/uploads/publications/Galvanized_Steel_Fasteners.pdf (accessed on 23 January 2019).
5. Kania, H.; Liberski, P. Synergistic influence of Al, Ni, Bi and Sn addition to a zinc bath upon growth kinetics and the structure of coatings. *IOP Conf. Ser. Mater. Sci. Eng.* **2012**, *35*, 012004. [CrossRef]
6. Wołczyński, W.; Guzik, E.; Janczak-Rusch, J.; Kopyciński, D.; Golczewski, J.; Hyuck, M.L.; Kloch, J. Morphological characteristics of multi-layer/substrate systems. *Mater. Charact.* **2006**, *56*, 274–280. [CrossRef]
7. Townsend, H.E.; Gorman, C.D.; Fischer, R.J. Atmospheric corrosion performance of hot-dip galvanized bolts for fastening weathering steel guiderail. *Mater. Perform.* **1999**, *38*, 66–70.
8. Liberski, P.; Podolski, P.; Kania, H.; Gierek, A.; Mendala, J. Corrosion resistance of zinc coatings obtained in high-temperature baths. *Mater. Sci.* **2003**, *39*, 652–657. [CrossRef]
9. *High Temperature Galvanizing*, 1st ed.; International Lead Zinc Research Organization, Inc.: New York, NY, USA, 1980; pp. 35–36.
10. Kwiatkowski, L. Cynk i ochrona przed korozją. *Ochr. Koroz.* **2004**, *10*, 256–257.
11. Porter, F. *Zinc Handbook: Properties Processing and Use in Design*; Marcel Dekker: New York, NY, USA, 1991.
12. Chatterjee, B. Sherardizing. *Met. Finish.* **2004**, *102*, 40–46. [CrossRef]
13. Smith, C.A. Sherardizing: Part 2. In *Anti-Corros. Method*; M. MCB UP Ltd.: Bingley, UK, 1980; Volume 27, pp. 6–7. [CrossRef]
14. Jiang, J.H.; Ma, A.B.; Fan, X.D.; Gong, M.Z.; Zhang, L.Y. Sherardizing and Characteristic of Zinc Protective Coating on High-Strength Steel Bridge Cable Wires. *Adv. Mat. Res.* **2010**, *97*, 1368–1372.
15. Chaliampalias, D.; Pistofidis, N.; Vourlias, G. Effect of temperature and zinc concentration on zinc coatings deposited with pack cementation. *Surf. Eng.* **2008**, *24*, 259–263. [CrossRef]
16. Pistofidis, N.; Vourlias, G.; Chaliampalias, D.; Chrysafis, K.; Stergioudis, G.; Polychroniadis, E.K. On the mechanism of formation of zinc pack coatings. *J. Alloys Compd.* **2006**, *407*, 221–225. [CrossRef]
17. Shen, T.-H.; Tsai, C.-Y.; Lin, C.-S. Growth behavior and properties of Zn–Al pack cementation coatings on carbon steels. *Surf. Coat. Tech.* **2016**, *306*, 455–461. [CrossRef]
18. Yu, S.; Liu, L. Zn-Fe and Zn-Fe-Y Cementation Coatings for Enhancing Corrosion Resistance of Steel. *Int. J. Electrochem. Sci.* **2017**, *12*, 4782–4794.
19. Kania, H.; Sipa, J. Thermal diffusion zinc coating technology with reactive atmosphere recirculation. Part 1: General description of technology and structure of coatings. *Ochr. Koroz.* **2018**, *11*, 338–345.
20. Kania, H.; Sipa, J. Zinc coating deposition with new thermal diffusion process on low carbon steel substrates. *Hutn. Wiad. Hutn.* **2019**, *1*, 2–8.
21. Kania, H.; Sipa, J.; Skupińska, A. Thermal diffusion zinc coating technology with reactive atmosphere recirculation. Part 2: Corrosion resistance of coatings. *Ochr. Koroz.* **2018**, *12*, 375–382.
22. Marder, A.R. The metallurgy of zinc-coated steel. *Prog. Mater. Sci.* **2000**, *45*, 191–271. [CrossRef]

23. Kubaschewski, O. *Iron Binary Phase Diagrams*; Springre-Verlag: Berlin, Germany, 1982.
24. Tatarek, A.; Saternus, M. Badanie zjawisk rozpuszczania dyfuzyjnego stali reaktywnych w kąpieli cynkowej z dodatkiem bizmutu. *Ochr. Koroz.* **2018**, *7*, 186–190.
25. Konstantinov, V.M.; Buloichyk, I.A. Some aspects of sherardizing implementation during anticorrosion protection of heat-treated metal parts. *IOP Conf. Ser. Mater. Sci. Eng.* **2015**, *71*, 012063. [CrossRef]
26. Liu, L.; Yu, S. A comparative study on Zn and Zn-Y coatings on 42CrMo steel by pack cementation process. *Int. J. Electrochem. Sci.* **2017**, *12*, 9575–9587. [CrossRef]
27. Wortelen, D.; Frieling, R.; Bracht, H.; Graf, W.; Natrup, F. Impact of zinc halide addition on the growth of zinc-rich layers generated by sherardizing. *Surf. Coat. Tech.* **2015**, *263*, 66–77. [CrossRef]
28. Chaliampalias, D.; Papazoglou, M.; Tsipas, S.; Pavlidou, E.; Skolianos, S.; Stergioudis, G.; Vourlias, G. The effect of Al and Cr additions on pack cementation zinc coatings. *Appl. Surf. Sci.* **2010**, *256*, 3618–3623. [CrossRef]

materials
MDPI

Article

High-Temperature Oxidation of Heavy Boron-Doped Diamond Electrodes: Microstructural and Electrochemical Performance Modification

Jacek Ryl [1,*], Mateusz Cieslik [1], Artur Zielinski [1], Mateusz Ficek [2], Bartlomiej Dec [2], Kazimierz Darowicki [1] and Robert Bogdanowicz [2]

1 Department of Electrochemistry, Corrosion and Materials Engineering, Faculty of Chemistry, Gdansk University of Technology, Narutowicza 11/12, 80-233 Gdansk, Poland; matciesl2@student.pg.edu.pl (M.C.); artzieli@pg.edu.pl (A.Z.); kazdarow@pg.edu.pl (K.D.)
2 Department of Metrology and Optoelectronics, Faculty of Electronics, Telecommunication and Informatics, Gdansk University of Technology, Narutowicza 11/12, 80-233 Gdansk, Poland; matficek@pg.edu.pl (M.F.); bartlomiej.dec@pg.edu.pl (B.D.); robbogda@pg.edu.pl (R.B.)
* Correspondence: jacek.ryl@pg.edu.pl; Tel.: +48-58-347-1092

Received: 28 January 2020; Accepted: 19 February 2020; Published: 21 February 2020

Abstract: In this work, we reveal in detail the effects of high-temperature treatment in air at 600 °C on the microstructure as well as the physico-chemical and electrochemical properties of boron-doped diamond (BDD) electrodes. The thermal treatment of freshly grown BDD electrodes was applied, resulting in permanent structural modifications of surface depending on the exposure time. High temperature affects material corrosion, inducing crystal defects. The oxidized BDD surfaces were studied by means of cyclic voltammetry (CV) and scanning electrochemical microscopy (SECM), revealing a significant decrease in the electrode activity and local heterogeneity of areas owing to various standard rate constants. This effect was correlated with a resultant increase of surface resistance heterogeneity by scanning spreading resistance microscopy (SSRM). The X-ray photoelectron spectroscopy (XPS) confirmed the rate and heterogeneity of the oxidation process, revealing hydroxyl species to be dominant on the electrode surface. Morphological tests using scanning electron microscopy (SEM) and atomic force microscopy (AFM) revealed that prolonged durations of high-temperature treatment lead not only to surface oxidation but also to irreversible structural defects in the form of etch pits. Our results show that even subsequent electrode rehydrogenation in plasma is not sufficient to reverse this surface oxidation in terms of electrochemical and physico-chemical properties, and the nature of high-temperature corrosion of BDD electrodes should be considered irreversible.

Keywords: boron-doped diamond; high-temperature treatment; surface oxidation; microstructure defects; electrochemical activity

1. Introduction

Boron-doped diamond (BDD) surfaces are widely studied due to their unique electrochemical and physico-chemical properties [1]. While the most often reported applications are for sensing [2], energy storage [3], or water treatment [4], BDD does play an important role in high-temperature environments [5]. The electrically conductive nanocrystalline boron-doped diamond (BDD) layers can be applied as a protective coating of Si photoelectrodes in sun-driven photoelectrochemical cells in aqueous electrolyte solutions [6]. Furthermore, wastewater treatment with advanced oxidation processes at BDD usually strongly depends on the potentials and current densities. It has been shown that a larger current promotes stability for urea removal at high potentials [7], which could also induce the thermal issues and electrode surface defects [8]. Moreover, the relatively high thermal conductivity coefficient (~700 W/mK) of BDD allows for its use as a heat spreader, replacing the commonly used metal spreaders such as copper, copper/refractory, or copper laminate in high power RF/microwave devices resulting in higher isolation of the ground plane at below 1.5 GHz [9,10].

BDD thin films are found to be particularly attractive as electrodes for electrolysis and electroanalytical applications due to their outstanding properties, which are significantly different from those of other conventional electrodes (e.g., glassy carbon or platinum electrodes [11]). In addition to the innate properties of diamond, such as high thermal conductivity, high hardness, and chemical inertness, the attractive features of conductive BDD include a wide electrochemical potential window in aqueous and non-aqueous media, very low capacitance, and extreme electrochemical stability [12]. BDDs are said to possess higher oxidation resistance at elevated temperatures in comparison to pure diamond [13], which is due to the formation of a B_2O_3 surface layer. Wang and Swain [14] reported a Pt/diamond composite electrode that exhibited superb morphological and microstructural stability during vigorous electrolysis in acidic media at a temperature of around 170 °C and current density around 0.1 A/cm^2. In addition, BDD layers can operate at high temperatures in electronics, an example being diamond-based Schottky barrier diodes for high-power devices [15,16].

High-temperature treatment is also one of the reported routes to modify and oxidize the surface termination of BDD electrodes [17–19]. Other approaches include oxidation under electrochemical polarization [20,21], chemical agents [22,23], ozone [24], plasma and/or UV treatment [25,26], but also natural aging in air [27–29]. The surface termination type significantly differentiates the electric and physico-chemical properties of boron-doped diamond electrodes. Hydrogen termination (HT-BDD) leads to hydrophobic and non-polar behavior, as well as high surface electrical conductivity and low electric transfer resistance [22,25]. The presence of hydroxyl or carbonyl surface bonds and oxidized BDD termination (OT-BDD) results in hydrophilicity and much lower surface electrical conductivity [30,31]. On the other hand, the electrochemical potential window for OT-BDD electrodes is reported to be wider than in the case of hydrogen-terminated ones [31,32]. Importantly, Zielinski et al. [29] reported that oxidation homogeneity of polycrystalline electrodes as well as oxidation rate in general highly depend on the above-mentioned modification route and confirmed that various treatment procedures produce different surface terminating species (C–OH, C–O-C, C=O, C–OOH, etc.), which may have a significant influence on tailoring desired functionalization procedures and BDD electrode characteristics.

Among the multitude of reported oxidation routes, high-temperature oxidation is characterized by a few unique features [29]. First of all, it appears to be the most heterogeneous at low durations, which was claimed to be related to the diverged propensity of differently oriented BDD crystal planes towards surface oxidation. Furthermore, according to studies where X-ray photoelectron spectroscopy (XPS) was performed, high-temperature oxidation tends to produce a significantly larger amount of surface hydroxyl species than any other reported approach. Their high polarity in comparison to other surface species results in very small reported contact angles during drop shape analysis tests. Next to electrochemical anodization, high-temperature treatment is one of the most often used pretreatment methods to clean the impurities on as-prepared BDD electrodes after the chemical vapor deposition (CVD) process [19,33].

High-temperature oxidation may also lead to changes in surface morphology as well as the altered sp^2/sp^3-carbon ratio [34,35]. The surface structure of BDD electrodes exposed to elevated temperatures in an oxygen-containing atmosphere was presented by Jiang et al. [17], who noticed that during such a treatment the amorphous carbon phase converts to the diamond phase and that the diamond grain increases with the decrease of grain boundaries. In addition, the short high-temperature treatments (30 min) led to the arrangement of grain boundaries on the electrode surface, which had a positive effect on their conductivity. An interesting consequence of the high-temperature oxidation is the enhancement of the peak ratio between the diamond peak and the graphitic peak, showing a decreasing amount of sp^2-carbon after the process, resulting in an increased carrier concentration [36,37]. However, prolonged high-temperature oxidation led to disordered grain boundaries and significantly deteriorated conductivity of the BDD electrode. High-temperature treatment may also lead to a diffusion of silicon substrate through the columnar structure of the diamond film and corrosion [33].

In light of the above presented discussion, the aim of this work was to present the effect of high-temperature treatment in air at 600 °C on the utility properties of boron-doped diamond electrodes, in particular for electrochemical applications. To the best of our knowledge, there are no dedicated studies on BDD electroactivity and stability of the properties when subjected to prolonged high-temperature exposure, the homogeneity and the rate of high-temperature corrosion of BDD, and its reversibility.

2. Materials and Methods

BDD films were synthesized in a microwave plasma-assisted chemical vapor deposition system (SEKI Technotron AX5400S, Tokyo, Japan). The substrates were seeded by sonication in nanodiamond suspension for 30 min following the standard procedure [38]. The chamber stage was maintained at 700 °C during the deposition process and the growth time was 6 h. The boron level expressed as [B]/[C] ratio in the gas phase was 10,000 ppm (boron dopant concentrations 2×10^{21} atoms cm^{-3}) [39]. A more detailed description of the thin film synthesis can be found elsewhere [40,41]. Then, the electrode surface was cleaned and hydrogenated. First, metallic impurities were dissolved in hot aqua regia (HNO_3:HCl/1:3), followed by the removal of organic impurities by hot "piranha" solution (H_2O_2:H_2SO_4/1:3) at 90 °C. Microwave hydrogen plasma treatment was performed using 1000 W microwave power and 300 sccm of hydrogen gas flow for 10 min.

The high-temperature treatment was carried out in an MRT-20 furnace (Czylok, Jastrzebie-Zdroj, Poland). The BDD electrodes were treated at 600 °C for 3, 10, 30, or 90 min in air. Afterward, the samples were removed from the furnace and cooled in air. A similar procedure was carried out in other studies [29,30,33]. After the electrochemical and physico-chemical examination, the hydrogenation procedure described earlier was carried out for high-temperature-treated BDD samples in order to determine surface oxidation reversibility.

Electrochemical measurements were carried out in a three-electrode cell. The working electrode was Si/BDD, with Ag/AgCl used as a reference electrode and platinum mesh as a counter electrode. All reagents were analytical purity (Sigma-Aldrich, Saint Louis, MO, USA). The exposed BDD sample area was 0.25 cm^2. Cyclic voltammetry (CV) studies were carried out in the polarization range between −0.9 V and 1.1 V vs. Ag/AgCl, at different scan rates between 5 and 800 mV/s. The electrolyte used was 0.5M Na_2SO_4 with 2.5 mM $K_3[Fe(CN)_6]$ and 2.5 mM $K_4[Fe(CN)_6]$. Scanning electrochemical microscopy (SECM) studies were performed using a three-step motor system (Sensolytics, Bochum, Germany) with 1 μm resolution in each direction, coupled to an Autolab 302 N potentiostat equipped with bipotentiostat module (Metrohm, Herisau, Switzerland). The commercially available ultramicroelectrodes (UMEs) made of platinum wire sealed in glass were used, with a platinum disc diameter of 5 μm. Each probe was polished and rinsed with acetone prior to the study. The electrolytic solution was composed of 2.5 mM $K_4[Fe(CN)_6]$ and 0.5M Na_2SO_4, purged with argon prior to BDD sample examination. The potential applied to the UMEs and the BDD sample was +0.4 and 0.0 vs. Ag/AgCl, respectively. In such a setup, the oxidation of redox species occurs at the UME tip. The SECM maps were recorded with a scanning speed of 5 μm/s and 1 μm steps in the x-y plane. A similar experimental approach was previously applied by authors [40].

Topographic and electrical microscopic measurements were made using an NTegra Prima device by NT-MDT (Moscow, Russia) in contact mode. CTD-NCHR-10 probes from Nanosensors (Neuchatel, Switzerland) were used with the following catalog parameters: L × W × T lever dimensions: 125 × 29 × 4 μm. Lever spring constant was equal to 71 N/m. The contact force determined from the approach curve was 8 μN. The tip radius of the curvature was in the 100–200 nm range according to manufacturer data. Conductivity measurements were made in the scanning spreading resistance mode using a constant voltage of 20 mV. Scanning electron microscopy (SEM) micrographs were taken with an S-3400

N microscope (Hitachi, Tokyo, Japan), with 20 kV accelerating voltage and operating in secondary electron mode.

High-resolution X-ray photoelectron spectroscopy (XPS) analyses were carried out in C1s binding energy range using an Escalab 250Xi multispectroscope (ThermoFisher Scientific, Waltham, MA, USA). The spectroscope was equipped with a monochromatic Al Kα energy source. The applied X-ray spot diameter was 650 μm and the pass energy through the hemisphere analyzer was 10 eV. Prior to operation, the spectroscope was calibrated on Cu and Au single crystals. Charge compensation was controlled through low-energy electron and Ar+ ion flow by means of a flood gun. Spectral deconvolution was performed with Avantage software (v5.973, ThermoFisher Scientific, Waltham, MA, USA) provided by the spectroscope manufacturer.

3. Results and Discussion

The heavy boron-doped diamond electrodes were subjected to electrochemical and physico-chemical examination in order to evaluate the effect of oxidation under high temperature on the charge transfer kinetics. Afterward, the oxidation and corrosion reversibility was evaluated and determined.

3.1. Electron Transfer through High-Temperature-Treated BDD Interface

Figure 1 shows the results of the cyclic voltammetry studies, which were carried out on BDD electrodes before and after high-temperature oxidation in air at 600 °C. The redox couple used within this study, $[Fe(CN6)]^{3-/4-}$, is characterized by an inner-sphere electron transfer (ISET) mechanism, which is said to be more dependent on the electrode homogeneity and applied modification procedures, thus offering a more sensitive approach to track subtle changes of charge transfer kinetics [42]. Figure 1a reveals $[Fe(CN6)]^{3-/4-}$ oxidation/reduction kinetics for each sample, observed at a 50 mV/s scan rate. High-temperature oxidation leads to hindered charge transfer kinetics, demonstrated by the decrease in peak current i_A and i_C when comparing to the as-prepared BDDs, a feature observed even at the shortest treatment duration. The peak current decreased nearly by a factor of 2 after merely 3 min of sample exposure to high temperature, and by a factor of 4 after 10 min. Detailed analysis is presented in Table 1.

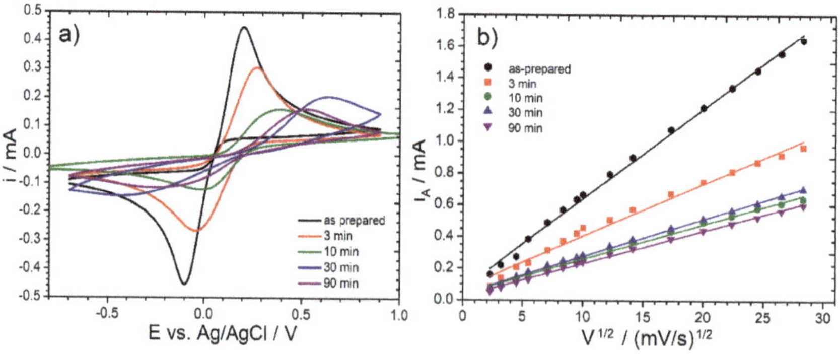

Figure 1. Cyclic voltammetry (CV) studies for boron-doped diamond (BDD) electrodes after high-temperature oxidation at 600 °C. (**a**) CV scans recorded at 50 mV/s for different oxidation durations; (**b**) CV anodic peak current vs. scan rate square root function for each studied electrode. Electrolytic solution: 0.5 M Na_2SO_4 with 2.5 mM $K_3[Fe(CN)_6]$ and 2.5 mM $K_4[Fe(CN)_6]$.

Table 1. Electrochemical properties (anodic peak current i_A, anodic-to-cathodic peak current ratio i_A/i_C, peak separation ΔE) based on CV studies (at 50 mV/s scan rate) of BDD samples after various high-temperature treatment durations.

Parameter	Untreated	High-Temperature Treatment			
		3 min	10 min	30 min	90 min
i_A/mA	0.44	0.29	0.16	0.20	0.16
i_A/i_C/-	0.97	1.14	1.33	1.45	1.43
ΔE/V	0.35	0.31	0.39	1.04	0.72
k_0/cm/s	3.58×10^{-3}	1.96×10^{-3}	1.94×10^{-3}	1.78×10^{-3}	1.91×10^{-3}

It should also be noted that the modification of BDD surface termination type translates into a significant peak separation ΔE increase, the parameter which is inseparably connected with reversibility of the corrosion process. In the case of a single electron transfer reaction, the fully reversible processes should have ΔE equal to 59 mV [43]. The as-prepared BDD sample is characterized by ΔE value of approximately 350 mV. The value is quite high taking into consideration other reported heavy boron-doped electrodes, which is due to the lack of any other electrode pretreatment procedures [41,44]. Nevertheless, the ΔE increase resulting from the applied high-temperature treatment should be explained by the move further away from the diffusion-controlled mechanism, due to the slowing down of the charge transfer kinetics. Prolonged high-temperature treatment is also characterized by the increased anodic and cathodic peak asymmetry, indicating a change in process kinetics, a feature typical for irreversible processes.

The discussed irreversibility of the studied redox process, as well as the decrease in electrode kinetics, are a testimony for surface modification from HT- to OT-BDD as a result of high-temperature treatment. The above-mentioned behavior results in BDD electrode corrosion, causing variability in electrode behavior and worsening its efficiency in electroanalytical studies.

The CV anodic peak vs. the scan rate square root function in the wide scan rate range is illustrated in Figure 1b. These plots show a strong linear trend, with local deviations explained by the heterogeneous nature of the polycrystalline electrodes [45]. To determine the standard reaction rate constant k^0, a numerically determined current function is used, depending on peak separation. The standard rate constant was calculated with the approach proposed by Velasco [46] to estimate completely irreversible processes. The following equation was used:

$$k^0 = 2.415 \exp\left(-0.02\frac{F}{RT}\right) D^{\frac{1}{2}} \left(E_p - E_{\frac{p}{2}}\right)^{-\frac{1}{2}} v^{\frac{1}{2}} \tag{1}$$

where E_p and $E_{p/2}$ are the potentials of the CV peak and half-peak, respectively, and v refers to the rate of change of potential; the diffusion coefficient was assumed as $D = 6.67 \times 10^{-6}$ cm^2/s [47].

Figure 2 reveals the effect of high-temperature treatment on the local distribution of charge transfer kinetics during $[Fe(CN)_6]^{4-}$ oxidation, assessed with scanning electrochemical microscopy (SECM). The relative current values strongly depend on the distance between the SECM tip and the electrode and vary from sample to sample. The current value is less important than its local changes due to electrode charge transfer heterogeneity. A normalization procedure was applied for the Z-axis of these graphs in order to show local discrepancies in the oxidation currents. The procedure based on a negative shift of the tip current displayed on each map to the position where the lowest recorded value equaled zero. A similar procedure was successfully applied in previous studies on heterogeneous charge transfer through the BDD electrode [40].

The obtained SECM micrographs clearly illustrate that the surface distribution of the oxidation currents is significantly increasing already after a short 10 min high-temperature treatment (Figure 2b) as compared to the as-prepared BDD electrode (Figure 2a). The heterogeneous electron transfer results from an altered propensity towards surface oxidation by BDD grain of various crystallographic orientations. A similar observation was previously reported for various methods of BDD surface treatment, in particular through electrochemical anodic polarization [40]. Importantly, increasing the high-temperature treatment length does not significantly increase surface homogeneity. Large tip current discrepancies are still observed after 90 min of oxidation (Figure 2c). At the same time, the as-prepared BDD electrode reveals positive feedback on the approach curve, suggesting low charge transfer resistance at the electrode interface [48]. The longer the high-temperature treatment, the higher was the negative feedback observed, whereas the sample oxidized for 10 min was characterized with large discrepancies of the approach curve shapes, depending on tip landing location. These results corroborate the decrease of the electron transfer kinetics with high-temperature treatment duration and the heterogeneous nature of the oxidation process.

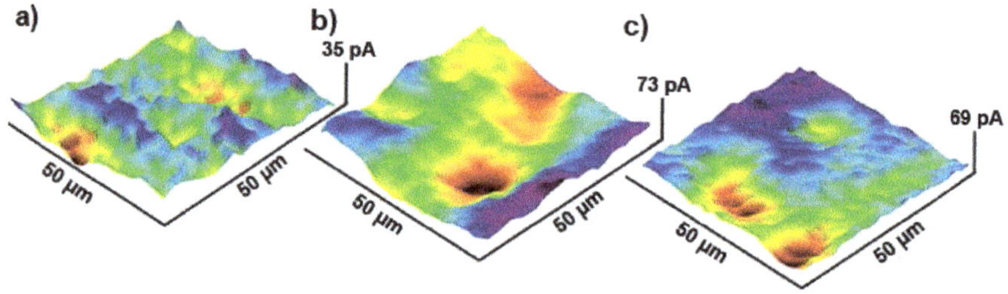

Figure 2. Typical scanning electrochemical microscopy (SECM) maps revealing charge transfer heterogeneity at heavy-doped BDD electrode interface after (**a**) 10 min, (**b**) 30 min, and (**c**) 90 min of high-temperature oxidation at 600 °C in air. Electrolyte: 0.5 M Na_2SO_4 + 2.5 mM $K_4[Fe(CN)_6]$.

3.2. *High-Temperature Oxidation Influence on BDD Physico-Chemical Properties*

The effect of high-temperature oxidation and corrosion on the microstructure of boron-doped diamond electrodes is shown on the SEM micrographs in Figure 3. The first 10 min of exposure to 600 °C do not lead to significant changes in grain structure; however, after this period the material undergoes gradual degradation.

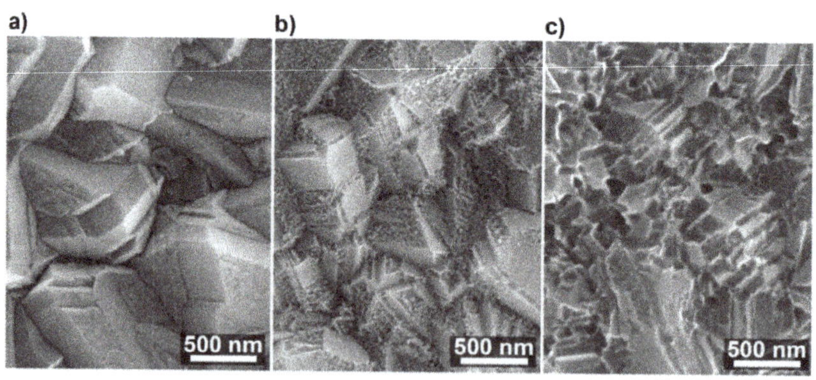

Figure 3. Typical scanning electron microscopy (SEM) micrographs of heavy-doped BDD surface after (**a**) 10 min, (**b**) 30 min, and (**c**) 90 min of high-temperature oxidation at 600 °C in air.

After 30 min of high-temperature oxidation, small and shallow etch pits start to appear throughout the diamond surface (Figure 3b), yet the grain structure is still recognizable. It is said that diamond can react with O_2 and water vapor contained within the atmosphere, to create surface etch pits [37]. Treatment in the oxygen-containing environment leads to high BDD surface etching and corrosion in the relatively short periods and a large surface area with respect to the geometric electrode area [49].

Furthermore, it is evident that certain crystallographic planes on the diamond surface are more prone to surface etching. Ohashi et al. [50] reported steam-activated high-temperature treatment in 600–900 °C to be an efficient nanotexturing tool for diamond surfaces. Notably, it was observed that, in the case of heavy-doped BDD films, the (111) facets are more prone to etching [19]. This observation lies in compliance with the fact that (111) facets have the dominant presence in the texture of studied BDD electrodes [51,52]. Other studies revealed that the propensity towards the modification of BDD surface termination is also dependent on crystallographic orientation [40,51].

Extended exposure at elevated temperature leads to advanced electrode decay, where the grain structure of the polycrystalline electrode almost completely fades away after 90 min of high-temperature oxidation (Figure 3c). Overall, the combined effect of surface area enhancement as well as previously reported [36,37] increased charge carrier concentration within the diamond structure may possibly result in an improved electrochemical response by BDD electrodes.

The detailed topography studies were carried out using atomic force microscopy (AFM) in contact mode. Next, scanning spreading resistance microscopy (SSRM), an AFM technique derivative, allows obtaining maps of local distribution of surface resistance, originating from changes in BDD termination type due to high-temperature treatment. The topographic images (Figure 4) were acquired simultaneously with surface conductivity maps (Figure 5).

The sequence of three-dimensional images of the surface shown in Figure 4 allows observing the evolution of surface morphology consistent with that obtained by scanning electron microscopy (refer to Figure 3). The first clearly represented pyramidal crystallites change into fractured, irregularly shaped structures with high-temperature treatment duration, clearly visible in the red-marked areas. It should be noted that this form of surface development affects the modification of the conductivity distribution through the geometric factor of the contact surface between the surface and various parts of the surface of the AFM pyramid tip [29]. Furthermore, possible surface development increases might contribute to the increase of the electrochemically active surface area.

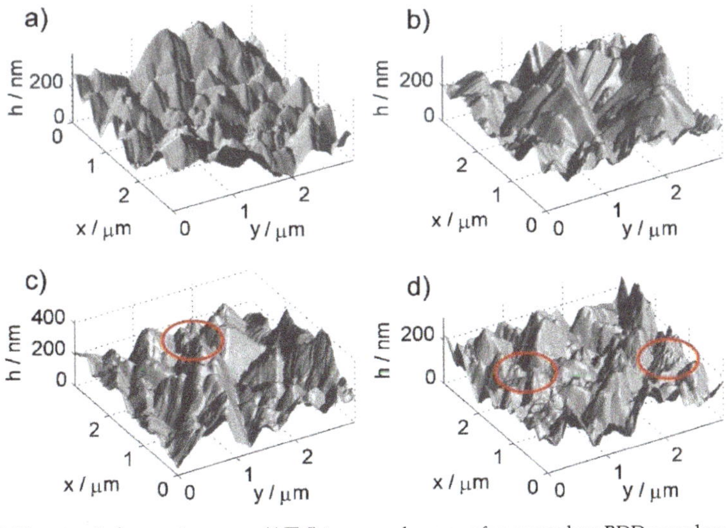

Figure 4. The atomic force microscopy (AFM) topography maps for exemplary BDD samples subjected to high-temperature treatment for different durations: (**a**) reference, untreated sample; (**b**) 10 min; (**c**) 30 min; (**d**) 60 min.

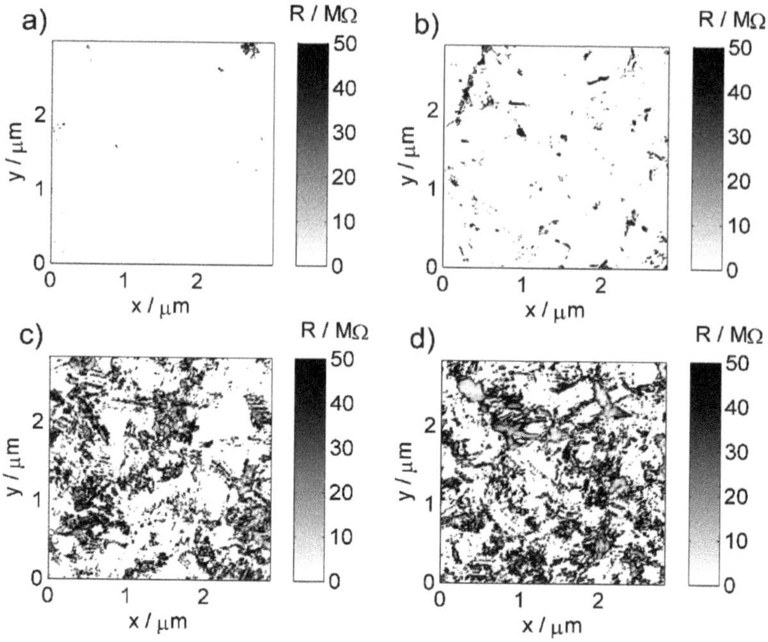

Figure 5. Typical spreading resistance (SSRM) maps for BDD samples subjected to different durations of high-temperature treatment in air at 600 °C: (**a**) reference untreated sample; (**b**) 10 min; (**c**) 30 min; (**d**) 90 min.

For the statistical description of the sample topography, the average roughness parameter was used, defined as:

$$S_a = \frac{1}{MN}\sum_{k=0}^{M-1}\sum_{l=0}^{N-1}\left|z(x_k, y_l) - \mu\right| \tag{2}$$

where $M = N = 256$ are the length and width of the analyzed image in pixels and μ is the average height, defined as:

$$\mu = \frac{1}{MN}\sum_{k=0}^{M-1}\sum_{l=0}^{N-1} z(x_k, y_l) \tag{3}$$

The above statistical parameters were determined for areas of 3 × 3 μm, the scan was repeated 4 times for each sample in different places, and the average roughness contained in Table 2 is the average value from each set of topographic scans.

Table 2. Statistical parameters: average roughness S_a and mean surface resistance R of studied BDD samples, based on AFM and scanning spreading resistance microscopy (SSRM) analyses.

Parameter	Untreated	High-Temperature Treatment			
		3 min	10 min	30 min	90 min
Mean Sa/nm	55.1374	55.860	51.4811	54.4982	36.3451
Mean Rs/Ω	0.1505	0.7993	1.6417	13.0889	16.1148

The value defined in Equation (2) as S_a determines the local height deviations from the average value for a given area. Maxima can be associated with the presence of high crystallites; however, their degradation, which progresses over the course of the heat treatment process, causes their degradation and additionally the formation of fragments of relatively small sizes, contributing to the reduction of maxima.

Figure 5 reveals the results of surface conductance analysis in accordance with the previously presented assumptions. The SSRM technique makes it possible to create a map of the quantity defined as the spreading resistance [53], characterizing the local changes in nanocontact conductivity between the probe tip and the sample surface. A simplified, commonly used formula [54]:

$$R_s = \frac{1}{4\sigma a} \tag{4}$$

where a is the radius of curvature of the probe tip and σ is the specific conductivity of the sample material, indicates the need for a trade-off between spatial resolution and probe durability, related to the thickness of the conductive layer and the radius of curvature a. The probes used in this report have a significantly increased radius of curvature for typical topographic ones (10 nm); however, they provide sufficient resolution to visualize changes in conductivity structure.

There is a visible decrease in the surface resistance average value, as well as a significant change in the mutual relationship between areas with relatively high and low electrical conductivity. For samples subjected to longer exposure, a bimodal character of conductivity can be seen with a gradual increase in the share of low conductivity areas. The observable changes present a good representation of the transition from HT- to OT-BDD under the oxidation agent [29]. Furthermore, it may be observed that high-temperature treatment leads to significant heterogeneity in surface oxidation, where the areas of altered spreading resistance are corresponding to different grain areas of the polycrystalline electrode. This observation may suggest that a varied crystallographic orientation is affecting the oxidation propensity, thus translating to locally variable electrochemical activity, as demonstrated by SECM studies.

It should be noted that the recorded heterogeneity in conductive regions can only be treated as a rough estimate of the variable electrochemical activity due to the difference in the conditions on the nanocontact in the atmosphere of air and in an electrolytic environment. Nevertheless, changes in the spreading resistance should to some extent correlate with changes in the charge transfer resistance on the surface of BDD and additionally reflect the trend of its changes together with the variability of heat treatment conditions. In this sense, the above discussed heterogeneity may correspond to the Compton model of partially blocked electrodes [55,56].

The high-resolution XPS spectra of BDD electrodes after different durations of high-temperature treatment are presented in Figure 6. These data were recorded in C1s peak binding energy range. The spectra were then deconvoluted according to the fitting model presented below and the results of the analysis summarized in Table 3.

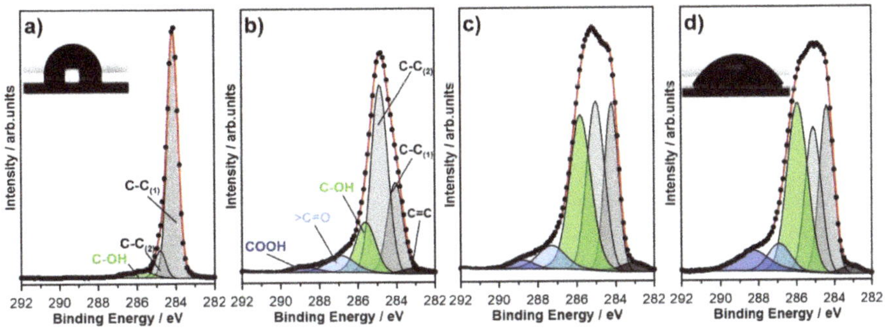

Figure 6. High-resolution X-ray photoelectron spectroscopy (XPS) spectra recorded in C1s binding energy range with applied spectral deconvolution. Spectra recorded for BDD samples: (**a**) as-prepared and after high-temperature treatment in air at 600 °C: (**b**) 3 min; (**c**) 10 min; (**d**) 30 min.

Table 3. Chemical composition (in %) of various carbon chemical states on the surface of untreated BDD electrodes and after high-temperature treatment in air at 600 °C, based on high-resolution XPS analysis.

Chemical State		BE/eV	Untreated	High-Temperature Treatment			
				3 min	10 min	30 min	90 min
C1s	C-C sp^2	**283.2**	2.1	2.4	2.9	1.8	1.5
	C-$C_{(1)}$	284.2	83.5	23.1	23.7	24.4	25.3
	C-$C_{(2)}$	284.9	8.7	49.2	30.3	23.9	20.8
	C-OH	285.6	4.8	15.8	31.6	35.6	36.8
	>C=O	287.0	0.6	7.4	6.6	6.4	6.5
	COOH	288.7	0.3	2.1	4.9	7.9	9.1

The C1s spectra recorded for the as-prepared BDD electrode, not subjected to high-temperature oxidation, reveals relatively simple surface chemistry. The registered spectra may be deconvoluted using three separate components. The primary peak, denoted as C-$C_{(1)}$ and located at approximately 284.2 eV, lies in the energy range characteristic for sp^3-carbon CH species on the hydrogen-terminated diamond surface and sp^3-carbon within the BDD bulk. The exact location of this peak depends on crystallographic texture rather than boron dopant concentration [57,58]. Thus, the peak position is in good agreement with previous studies on heavy boron-doped diamond substrates [20,29,59]. The second notable component, C-$C_{(2)}$, is usually attributed to non-hydrogenated carbon atoms on the BDD surface but also adsorbed polyhydride carbon (CH_x) species [20,60]. The position of this component is typically shifted by +0.7 eV vs. C-$C_{(1)}$, which is also in this case.

The oxidation of BDD surface termination, occurring as a result of high-temperature treatment in air, results in the substitution of hydrogenated terminal bonds with oxygen-containing species: hydroxyl C–OH (peak at 285.6 eV), but also carbonyl >C=O (at 287.0 eV) and carboxyl COOH (at 288.7 eV) groups [41,57,61]. The XPS analyses confirmed that hydroxyl species are the dominant ones on the surface of high-temperature-treated BDD electrode, unlike other types of surface oxidation treatments (electrochemical, oxygen plasma, ozone treatment, etc.) [29]. The HT- to OT-BDD transition is bound to the increase in surface hydrophobicity, as illustrated by the drop shape analyses in the inset of Figure 6a,d.

The total share of oxidized BDD surface area, OT-BDD, counted as a sum of the above-mentioned components is naturally increasing with the duration of sample exposure to the high temperature. The share of OT-BDD surface for the untreated samples was 5.7% and gradually increased up to 43.1% after 10 min of exposure. Prolonged treatment leads to further surface oxidation, but the total OT-BDD share tends to the plateau with only 52.4% oxidized surface after 90 min of treatment. This effect is partially bound to the fact that the XPS analysis is partially obtained from ~5 nm volume beneath the electrode surface. However, similar studies on BDD electrodes, but oxidized by electrochemical or oxygen plasma treatment, have led to significant diminishing of C-$C_{(1)}$ peak, down to 12%–14% [29]. Retaining around 25% of HT-BDD surface even after 90 min of high temperature is well represented and imaged on the SSRM micrographs (see Figure 5), which reveals a significant heterogeneity of oxidized areas on the BDD electrode surface. Following the electric BDD spatial heterogeneity is the electrochemical behavior, determining the conditions of the charge transfer mechanism by a partially blocked electrode.

It was also observed that prolonged high-temperature treatment durations have led to the disappearance of the spectral component, located at approximately 283.2 eV, and originating from surface sp^2-carbon, which is well explained by a previously mentioned theory that such an exposure results in disordered grain boundaries and significantly deteriorated BDD conductivity.

The above discussed effect of high-temperature treatment in air at 600 °C on BDD surface chemistry is schematically presented in Figure 7. The initially hydrogen-terminated surface of the polycrystalline BDD electrode undergoes surface oxidation over very short durations, even after a few minutes of exposure to the oxidation agent. This effect is heterogeneous in nature and hypothetically limited to specific crystallographic planes, based on previous studies [29,51,62]. The authors did not investigate the exact planes of high-temperature interaction within this study. The prolonged treatment leads to the appearance of etch pits on the diamond surface. This effect is said to increase the electroactive surface area; however, it was not confirmed since the OT-BDD surface is characterized by lower values of the standard reaction rate constant. Most importantly, the appearance of the etch pits is heterogeneous and depends on the crystallographic orientation, which is similar to surface oxygen termination. According to the literature survey, the etch pits are primarily initiated at (111) facets [19], which, on the other hand, are the least prone to surface oxidation [63].

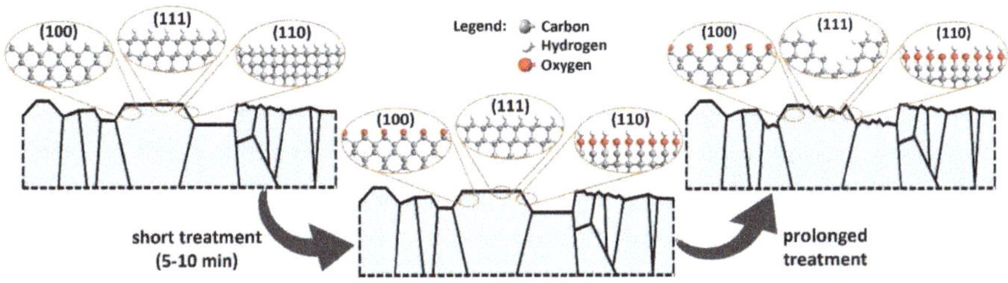

Figure 7. Schematic visualization of the oxidation process under high-temperature treatment in air at 600 °C.

3.3. Reversibility of the High-Temperature BDD Surface Oxidation

A few available routes for surface rehydrogenation have been reported in the literature, where hydrogen plasma treatment is claimed to be the most efficient [64]. Despite growing awareness and interest in this topic, some studies reveal issues with the reversibility of oxidized BDD electrodes [31]. Various surface physico-chemical properties (such as surface chemistry or contact angle) are often reported to be on par with those observed for BDD electrodes prior to their oxidation; however, electrochemical and electric parameters seem to be a subject of irreversible change. The above-mentioned characterization is most often discussed in the case of electrochemically oxidized BDDs. We decided to evaluate the reversibility of the oxidation process under study since both the mechanism of oxidation and the resultant surface chemistry are claimed to be different in the case of high-temperature treatment than any other oxidation route.

Figure 8a reveals the CV scans at 50 mV/s for high-temperature oxidized (at various durations) and then plasma rehydrogenated BDDs, compared to the as-prepared BDD electrode. Our studies show that neither of the investigated electrodes is characterized by improved peak current values. On the other hand, a significant improvement is visible when compared to oxidized BDD electrode kinetics. It appears that very short high-temperature treatments might lead to a certain improvement in electrochemical characteristics, that is, the peak separation ΔE is significantly improved and closer to the theoretical value of diffusion-controlled electrode processes, with high peak symmetry and peak currents equal to nearly 90% of its original value. This feature might be explained by the removal of sp_2-carbon species adsorbed on the BDD electrode surface after the CVD process, demonstrating a successful electrode cleaning. The longer the high-temperature treatment duration, the lower the value of both anodic and cathodic peak currents and the more irreversible the character of the oxidation process. This effect is naturally connected with the degradation of the BDD grain structure. While the deterioration of the electrode kinetics is progressive, an interesting feature may be observed for the sample after 90 min oxidation and rehydrogenation, which is characterized by the most narrowly observed ΔE value. Following the previously defined partially blocked electrode mechanism, it may be concluded that deep etching of BDD grains (refer to Figure 3c) leads to homogenization of the electrode surface and unification of the diffusion fields at the electrode interface, a conclusion supported by SSRM studies. The detailed analysis is summarized in Table 4.

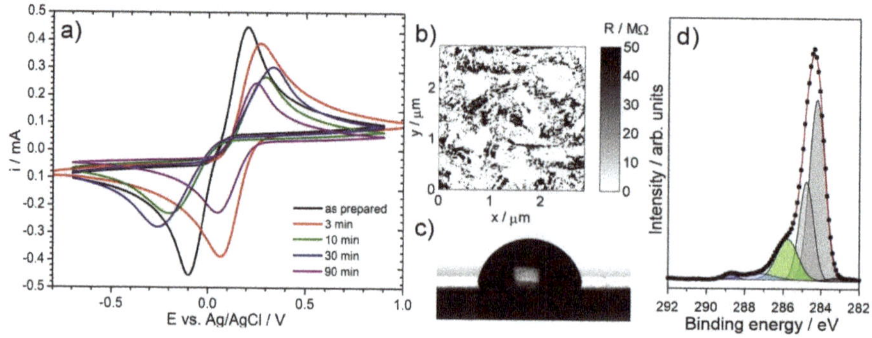

Figure 8. The effect of BDD surface rehydrogenation in plasma: (**a**) CV studies after various durations of oxidation. A scan rate of 50 mV/s. Electrolytic solution: 0.5 M Na_2SO_4 with 2.5 mM $K_3[Fe(CN)_6]$ and 2.5 mM $K_4[Fe(CN)_6]$; (**b–d**) exemplary results for electrodes after 30 min oxidation followed by rehydrogenation: (**b**) SSRM map, (**c**) contact angle analysis, and (**d**) high-resolution XPS C1s spectrum.

Table 4. Electrochemical properties (anodic peak current i_A, anodic-to-cathodic peak current ratio i_A/i_C, peak separation ΔE) based on CV studies (at 50 mV/s scan rate) of BDD samples after various high-temperature treatment durations, followed by sample rehydrogenation in plasma.

Parameter	Untreated	High-Temperature Treated and Rehydrogenated			
		3 min	10 min	30 min	90 min
i_A/mA	0.44	0.39	0.26	0.29	0.24
i_A/i_C/-	0.97	1.00	1.19	0.88	1.06
ΔE/V	0.35	0.22	0.50	0.55	0.20

A similar observation is revealed by SSRM maps, shown in Figure 8b for BDD electrodes after 30 min of high-temperature treatment and rehydrogenation. While the hydrogen plasma is increasing the surface conductance, the Rs is far off from its original value (10.416 MΩ) and the local distribution of electric properties on the BDD surface remains heterogeneous. This is an important observation, confirming that not only the electrochemical but also the electric parameter characteristics have been modified. However, the rehydrogenation process undoubtedly does have a positive effect on BDD surface chemistry, demonstrated by the XPS analysis (Figure 8d). The hydrogenation process led to a significant reduction in the OT-BDD share (from 49.9% to 20.6%) and, in particular, in the number of surface hydroxyl groups (from 35.6% to 15.7%), which decreased more than twice for prolonged oxidized samples. The reappearance of the hydrogen-terminated surface resulted in the increase of surface hydrophobicity, observed with the drop shape analysis (Figure 8c). However, the process is not fully reversible, thus leading to a conclusion regarding the corrosive nature of the high-temperature treatment, in particular at prolonged oxidation durations.

4. Conclusions

In summary, we performed a detailed study of the influence of high temperature on the oxidation processes for heavy boron-doped diamond electrodes. Electrode surface was investigated by both electrochemical and physico-chemical techniques to reveal its charge transfer kinetics and oxidation behavior.

High-temperature oxidation leads to:

- ✓ hindered charge transfer kinetics, demonstrated by the decrease in peak current on CV scans when comparing to the as-prepared BDDs, a feature observed even at the shortest treatment duration;
- ✓ the modification of BDD microstructure and the appearance of small and shallow etch pits throughout the diamond surface, damaging diamond structure at prolonged treatments;

- ✓ the disappearance of the spectral component, located at approximately 283.2 eV, and originating from surface sp^2-carbon in XPS;
- ✓ the bimodal character of BDD conductivity observed by SRRM with a gradual increase in the share of low conductivity areas, dependent on the BDD crystallographic structure.

The change in the electrode's surface electric properties by local surface oxidation is reflected in a more heterogeneous distribution of the diffusion field, observed by SECM. The longer the high-temperature exposure duration, the lower the value of both anodic and cathodic peak currents and the more irreversible is the character of the oxidation process, which is mainly attributed to the corrosion of BDD grain structure and the deterioration of electrochemical activity.

Author Contributions: Conceptualization, J.R. and A.Z.; Methodology, J.R. and A.Z.; Investigation, J.R. (XPS, SEM, electrochemical studies), M.C. (electrochemical studies), M.F. and B.D. (BDD synthesis and structural studies), and A.Z. (AFM and SSRM); supervision, J.R.; writing—original draft preparation, all authors; writing—review and editing, J.R., R.B., and A.Z.; funding acquisition: J.R., K.D., and R.B. All authors have read and agreed to the published version of the manuscript.

Funding: This research was funded by the National Science Centre, grant SONATA number 2015/17/D/ST5/02571, and by the National Centre for Science and Development, grant Techmatstrateg number 347324. The DS funds from the Faculty of Chemistry and the Faculty of Electronics, Telecommunication, and Informatics are also acknowledged.

Acknowledgments: The authors acknowledge Lukasz Burczyk for performing the SECM measurements.

Conflicts of Interest: The authors declare no conflict of interest.

References

1. Sussmann, E.S. *CVD Diamond for Electronic Devices and Sensors*; John Wiley & Sons: New York, NY, USA, 2009; ISBN 978-0-470-74039-2.
2. Muzyka, K.; Sun, J.; Fereja, T.H.; Lan, Y.; Zhang, W.; Xu, G. Boron-doped diamond: Current progress and challenges in view of electroanalytical applications. *Anal. Methods* **2019**, *11*, 397–414. [CrossRef]
3. Yu, S.; Yang, N.; Jiang, X.; Zhang, W.; Liu, S. Conductive Diamond for Electrochemical Energy Applications. In *Nanocarbon Electrochemistry*; Yang, N., Zhao, G., Foord, J., Eds.; John Wiley & Sons: New York, NY, USA, 2020; pp. 171–199. ISBN 978-1-119-46823-3.
4. Nidheesh, P.V.; Divyapriya, G.; Oturan, N.; Trellu, C.; Oturan, M.A. Environmental Applications of Boron-Doped Diamond Electrodes: 1. Applications in Water and Wastewater Treatment. *ChemElectroChem* **2019**, *6*, 2124–2142. [CrossRef]
5. Yang, N. (Ed.) *Novel Aspects of Diamond: From Growth to Applications*; Springer International Publishing: Cham, Switzerland, 2019; Volume 121, ISBN 978-3-030-12468-7.
6. Ashcheulov, P.; Taylor, A.; Mortet, V.; Poruba, A.; Le Formal, F.; Krýsová, H.; Klementová, M.; Hubík, P.; Kopeček, J.; Lorinčík, J.; et al. Nanocrystalline Boron-Doped Diamond as a Corrosion-Resistant Anode for Water Oxidation via Si Photoelectrodes. *ACS Appl. Mater. Interfaces* **2018**, *10*, 29552–29564. [CrossRef]
7. Schranck, A.; Doudrick, K. Effect of reactor configuration on the kinetics and nitrogen byproduct selectivity of urea electrolysis using a boron doped diamond electrode. *Water Res.* **2020**, *168*, 115130. [CrossRef] [PubMed]
8. Mei, R.; Zhu, C.; Wei, Q.; Ma, L.; Li, W.; Zhou, B.; Deng, Z.; Tong, Z.; Ouyang, G.; Jiang, C. The Dependence of Oxidation Parameters and Dyes' Molecular Structures on Microstructure of Boron-Doped Diamond in Electrochemical Oxidation Process of Dye Wastewater. *J. Electrochem. Soc.* **2018**, *165*, H324–H332. [CrossRef]
9. Zhang, Y.; Teo, K.H.; Palacios, T. Beyond Thermal Management: Incorporating p-Diamond Back-Barriers and Cap Layers Into AlGaN/GaN HEMTs. *IEEE Trans. Electron. Devices* **2016**, *63*, 2340–2345. [CrossRef]
10. Williams, G.; Calvo, J.A.; Faili, F.; Dodson, J.; Obeloer, T.; Twitchen, D.J. Thermal Conductivity of Electrically Conductive Highly Boron Doped Diamond and its Applications at High Frequencies. In Proceedings of the 2018 17th IEEE Intersociety Conference on Thermal and Thermomechanical Phenomena in Electronic Systems (ITherm), San Diego, CA, USA, 29 May–1 June 2018; pp. 235–239.
11. Panizza, M.; Cerisola, G. Application of diamond electrodes to electrochemical processes. *Electrochim. Acta* **2005**, *51*, 191–199. [CrossRef]

12. Luong, J.H.T.; Male, K.B.; Glennon, J.D. Boron-doped diamond electrode: Synthesis, characterization, functionalization and analytical applications. *Analyst* **2009**, *134*, 1965. [CrossRef]
13. Herrmann, M.; Matthey, B.; Gestrich, T. Boron-doped diamond with improved oxidation resistance. *Diam. Relat. Mater.* **2019**, *92*, 47–52. [CrossRef]
14. Wang, J.; Swain, G.M. Dimensionally Stable Pt/Diamond Composite Electrodes in Concentrated H_3PO_4 at High Temperature. *Electrochem. Solid State Lett.* **2002**, *5*, E4. [CrossRef]
15. Umezawa, H.; Nagase, M.; Kato, Y.; Shikata, S. High temperature application of diamond power device. *Diam. Relat. Mater.* **2012**, *24*, 201–205. [CrossRef]
16. Ueda, K.; Kawamoto, K.; Soumiya, T.; Asano, H. High-temperature characteristics of Ag and Ni/diamond Schottky diodes. *Diam. Relat. Mater.* **2013**, *38*, 41–44. [CrossRef]
17. Jiang, M.; Yu, H.; Li, X.; Lu, S.; Hu, X. Thermal oxidation induced high electrochemical activity of boron-doped nanocrystalline diamond electrodes. *Electrochim. Acta* **2017**, *258*, 61–70. [CrossRef]
18. Pehrsson, P.E.; Mercer, T.W.; Chaney, J.A. Thermal oxidation of the hydrogenated diamond (100) surface. *Surf. Sci.* **2002**, *497*, 13–28. [CrossRef]
19. Zhang, J.; Nakai, T.; Uno, M.; Nishiki, Y.; Sugimoto, W. Effect of the boron content on the steam activation of boron-doped diamond electrodes. *Carbon* **2013**, *65*, 206–213. [CrossRef]
20. Ryl, J.; Burczyk, L.; Bogdanowicz, R.; Sobaszek, M.; Darowicki, K. Study on surface termination of boron-doped diamond electrodes under anodic polarization in H_2SO_4 by means of dynamic impedance technique. *Carbon* **2016**, *96*, 1093–1105. [CrossRef]
21. Martínez-Huitle, C.A.; Ferro, S.; Reyna, S.; Cerro-López, M.; De Battisti, A.; Quiroz, M.A. Electrochemical oxidation of oxalic acid in the presence of halides at boron doped diamond electrode. *J. Braz. Chem. Soc.* **2008**, *19*, 150–156. [CrossRef]
22. Hayashi, K.; Yamanaka, S.; Watanabe, H.; Sekiguchi, T.; Okushi, H.; Kajimura, K. Investigation of the effect of hydrogen on electrical and optical properties in chemical vapor deposited on homoepitaxial diamond films. *J. Appl. Phys.* **1997**, *81*, 744–753. [CrossRef]
23. Grot, S.A.; Gildenblat, G.S.; Hatfield, C.W.; Wronski, C.R.; Badzian, A.R.; Badzian, T.; Messier, R. The effect of surface treatment on the electrical properties of metal contacts to boron-doped homoepitaxial diamond film. *IEEE Electron. Dev. Lett.* **1990**, *11*, 100–102. [CrossRef]
24. Švorc, Ľ.; Rievaj, M.; Bustin, D. Green electrochemical sensor for environmental monitoring of pesticides: Determination of atrazine in river waters using a boron-doped diamond electrode. *Sens. Actuators B Chem.* **2013**, *181*, 294–300. [CrossRef]
25. Yagi, I.; Notsu, H.; Kondo, T.; Tryk, D.A.; Fujishima, A. Electrochemical selectivity for redox systems at oxygen-terminated diamond electrodes. *J. Electroanal. Chem.* **1999**, *473*, 173–178. [CrossRef]
26. Boukherroub, R.; Wallart, X.; Szunerits, S.; Marcus, B.; Bouvier, P.; Mermoux, M. Photochemical oxidation of hydrogenated boron-doped diamond surfaces. *Electrochem. Commun.* **2005**, *7*, 937–940. [CrossRef]
27. Geisler, M.; Hugel, T. Aging of Hydrogenated and Oxidized Diamond. *Adv. Mater.* **2010**, *22*, 398–402. [CrossRef]
28. Vanhove, E.; de Sanoit, J.; Arnault, J.C.; Saada, S.; Mer, C.; Mailley, P.; Bergonzo, P.; Nesladek, M. Stability of H-terminated BDD electrodes: An insight into the influence of the surface preparation. *Phys. Status Solidi* **2007**, *204*, 2931–2939. [CrossRef]
29. Zielinski, A.; Cieslik, M.; Sobaszek, M.; Bogdanowicz, R.; Darowicki, K.; Ryl, J. Multifrequency nanoscale impedance microscopy (m-NIM): A novel approach towards detection of selective and subtle modifications on the surface of polycrystalline boron-doped diamond electrodes. *Ultramicroscopy* **2019**, *199*, 34–45. [CrossRef]
30. Ghodbane, S.; Haensel, T.; Coffinier, Y.; Szunerits, S.; Steinmüller-Nethl, D.; Boukherroub, R.; Ahmed, S.I.-U.; Schaefer, J.A. HREELS Investigation of the Surfaces of Nanocrystalline Diamond Films Oxidized by Different Processes. *Langmuir* **2010**, *26*, 18798–18805. [CrossRef]
31. Girard, H.; Simon, N.; Ballutaud, D.; Herlem, M.; Etcheberry, A. Effect of anodic and cathodic treatments on the charge transfer of boron doped diamond electrodes. *Diam. Relat. Mater.* **2007**, *16*, 316–325. [CrossRef]
32. Ryl, J.; Bogdanowicz, R.; Slepski, P.; Sobaszek, M.; Darowicki, K. Dynamic Electrochemical Impedance Spectroscopy (DEIS) as a Tool for Analyzing Surface Oxidation Processes on Boron-Doped Diamond Electrodes. *J. Electrochem. Soc.* **2014**, *161*, H359–H364. [CrossRef]
33. Ferro, S.; Dal Colle, M.; De Battisti, A. Chemical surface characterization of electrochemically and thermally oxidized boron-doped diamond film electrodes. *Carbon* **2005**, *43*, 1191–1203. [CrossRef]

34. Shpilman, Z.; Gouzman, I.; Minton, T.K.; Shen, L.; Stacey, A.; Orwa, J.; Prawer, S.; Cowie, B.C.C.; Hoffman, A. A near edge X-ray absorption fine structure study of oxidized single crystal and polycrystalline diamond surfaces. *Diam. Relat. Mater.* **2014**, *45*, 20–27. [CrossRef]
35. Zolotukhin, A.; Kopylov, P.G.; Ismagilov, R.R.; Obraztsov, A.N. Thermal oxidation of CVD diamond. *Diam. Relat. Mater.* **2010**, *19*, 1007–1011. [CrossRef]
36. Show, Y.; Witek, M.A.; Sonthalia, P.; Swain, G.M. Characterization and Electrochemical Responsiveness of Boron-Doped Nanocrystalline Diamond Thin-Film Electrodes. *Chem. Mater.* **2003**, *15*, 879–888. [CrossRef]
37. Szunerits, S.; Boukherroub, R.; Downard, A.; Zhu, J.-J. (Eds.) Nanocarbon chemistry and interfaces. In *Nanocarbons for Electroanalysis*, 1st ed.; John Wiley & Sons: Hoboken, NJ, USA, 2017; ISBN 978-1-119-24395-3.
38. Ficek, M.; Bogdanowicz, R.; Ryl, J. Nanocrystalline CVD Diamond Coatings on Fused Silica Optical Fibres: Optical Properties Study. *Acta Phys. Pol. A* **2015**, *127*, 868–873. [CrossRef]
39. Zieliński, A.; Bogdanowicz, R.; Ryl, J.; Burczyk, L.; Darowicki, K. Local impedance imaging of boron-doped polycrystalline diamond thin films. *Appl. Phys. Lett.* **2014**, *105*, 131908. [CrossRef]
40. Ryl, J.; Burczyk, L.; Zielinski, A.; Ficek, M.; Franczak, A.; Bogdanowicz, R.; Darowicki, K. Heterogeneous oxidation of highly boron-doped diamond electrodes and its influence on the surface distribution of electrochemical activity. *Electrochim. Acta* **2019**, *297*, 1018–1027. [CrossRef]
41. Bogdanowicz, R.; Sawczak, M.; Niedzialkowski, P.; Zieba, P.; Finke, B.; Ryl, J.; Karczewski, J.; Ossowski, T. Novel Functionalization of Boron-Doped Diamond by Microwave Pulsed-Plasma Polymerized Allylamine Film. *J. Phys. Chem. C* **2014**, *118*, 8014–8025. [CrossRef]
42. Richard, W. Kinetic Study of Redox Probes on Glassy Carbon Electrode Functionalized by 4-nitrobenzene Diazonium. *Int. J. Electrochem. Sci.* **2019**, 453–469. [CrossRef]
43. Bard, A.J.; Faulkner, L.R. *Electrochemical Methods: Fundamentals and Applications*, 2nd ed.; Wiley: New York, NY, USA, 2001; ISBN 978-0-471-04372-0.
44. Bogdanowicz, R.; Sawczak, M.; Niedzialkowski, P.; Zieba, P.; Finke, B.; Ryl, J.; Ossowski, T. Direct amination of boron-doped diamond by plasma polymerized allylamine film: Direct amination of boron-doped diamond. *Phys. Status Solidi* **2014**, *211*, 2319–2327. [CrossRef]
45. Actis, P.; Denoyelle, A.; Boukherroub, R.; Szunerits, S. Influence of the surface termination on the electrochemical properties of boron-doped diamond (BDD) interfaces. *Electrochem. Commun.* **2008**, *10*, 402–406. [CrossRef]
46. Velasco, J.G. Determination of standard rate constants for electrochemical irreversible processes from linear sweep voltammograms. *Electroanalysis* **1997**, *9*, 880–882. [CrossRef]
47. Konopka, S.J.; McDuffie, B. Diffusion coefficients of ferri- and ferrocyanide ions in aqueous media, using twin-electrode thin-layer electrochemistry. *Anal. Chem.* **1970**, *42*, 1741–1746. [CrossRef]
48. Holt, K.B.; Bard, A.J.; Show, Y.; Swain, G.M. Scanning Electrochemical Microscopy and Conductive Probe Atomic Force Microscopy Studies of Hydrogen-Terminated Boron-Doped Diamond Electrodes with Different Doping Levels. *J. Phys. Chem. B* **2004**, *108*, 15117–15127. [CrossRef]
49. Kondo, T.; Kodama, Y.; Ikezoe, S.; Yajima, K.; Aikawa, T.; Yuasa, M. Porous boron-doped diamond electrodes fabricated via two-step thermal treatment. *Carbon* **2014**, *77*, 783–789. [CrossRef]
50. Ohashi, T.; Zhang, J.; Takasu, Y.; Sugimoto, W. Steam activation of boron doped diamond electrodes. *Electrochim. Acta* **2011**, *56*, 5599–5604. [CrossRef]
51. Ryl, J.; Zielinski, A.; Bogdanowicz, R.; Darowicki, K. Heterogeneous distribution of surface electrochemical activity in polycrystalline highly boron-doped diamond electrodes under deep anodic polarization. *Electrochem. Commun.* **2017**, *83*, 41–45. [CrossRef]
52. Siuzdak, K.; Bogdanowicz, R.; Sawczak, M.; Sobaszek, M. Enhanced capacitance of composite TiO_2 nanotube/boron-doped diamond electrodes studied by impedance spectroscopy. *Nanoscale* **2015**, *7*, 551–558. [CrossRef]
53. Denhoff, M.W. An accurate calculation of spreading resistance. *J. Phys. D Appl. Phys.* **2006**, *39*, 1761–1765. [CrossRef]
54. Dickens, L.E. Spreading Resistance as a Function of Frequency. *IEEE Trans. Microw. Theory Technol.* **1967**, *15*, 101–109. [CrossRef]
55. Chevallier, F.G.; Fietkau, N.; del Campo, J.; Mas, R.; Muñoz, F.X.; Jiang, L.; Jones, T.G.J.; Compton, R.G. Experimental cyclic voltammetry at partially blocked electrodes: Elevated cylindrical blocks. *J. Electroanal. Chem.* **2006**, *596*, 25–32. [CrossRef]
56. Davies, T.J.; Banks, C.E.; Compton, R.G. Voltammetry at spatially heterogeneous electrodes. *J. Solid State Electrochem.* **2005**, *9*, 797–808. [CrossRef]

57. Ivandini, T.A.; Watanabe, T.; Matsui, T.; Ootani, Y.; Iizuka, S.; Toyoshima, R.; Kodama, H.; Kondoh, H.; Tateyama, Y.; Einaga, Y. Influence of Surface Orientation on Electrochemical Properties of Boron-Doped Diamond. *J. Phys. Chem. C* **2019**, *123*, 5336–5344. [CrossRef]
58. Ghodbane, S.; Ballutaud, D.; Deneuville, A.; Baron, C. Influence of boron concentration on the XPS spectra of the (100) surface of homoepitaxial boron-doped diamond films. *Phys. State Solid* **2006**, *203*, 3147–3151. [CrossRef]
59. Ballutaud, D.; Simon, N.; Girard, H.; Rzepka, E.; Bouchet-Fabre, B. Photoelectron spectroscopy of hydrogen at the polycrystalline diamond surface. *Diam. Relat. Mater.* **2006**, *15*, 716–719. [CrossRef]
60. Girard, H.A.; Simon, N.; Ballutaud, D.; Etcheberry, A. Correlation between flat-band potential position and oxygenated termination nature on boron-doped diamond electrodes. *Comptes Rendus Chim.* **2008**, *11*, 1010–1015. [CrossRef]
61. Catalan, F.C.I.; Hayazawa, N.; Yokota, Y.; Wong, R.A.; Watanabe, T.; Einaga, Y.; Kim, Y. Facet-Dependent Temporal and Spatial Changes in Boron-Doped Diamond Film Electrodes due to Anodic Corrosion. *J. Phys. Chem. C* **2017**, *121*, 26742–26750. [CrossRef]
62. Pleskov, Y.V.; Evstefeeva, Y.E.; Krotova, M.D.; Varnin, V.P.; Teremetskaya, I.G. Synthetic semiconductor diamond electrodes: Electrochemical behaviour of homoepitaxial boron-doped films orientated as (111), (110), and (100) faces. *J. Electroanal. Chem.* **2006**, *595*, 168–174. [CrossRef]
63. Wolfer, M.; Biener, J.; El-dasher, B.S.; Biener, M.M.; Hamza, A.V.; Kriele, A.; Wild, C. Crystallographic anisotropy of growth and etch rates of CVD diamond. *Diam. Relat. Mater.* **2009**, *18*, 713–717. [CrossRef]
64. zevedo, A.F.; Braga, N.A.; Souza, F.A.; Matsushima, J.T.; Baldan, M.R.; Ferreira, N.G. The effect of surface treatment on oxidation of oxalic acid at nanocrystalline diamond films. *Diam. Relat. Mater.* **2010**, *19*, 462–465. [CrossRef]

www.ingramcontent.com/pod-product-compliance
Lightning Source LLC
Chambersburg PA
CBHW061814210926
51850CB00006B/699
9783039361762

MDPI
St. Alban-Anlage 66
4052 Basel
Switzerland
Tel. +41 61 683 77 34
Fax +41 61 302 89 18
www.mdpi.com

Materials Editorial Office
E-mail: materials@mdpi.com
www.mdpi.com/journal/materials